GEAR HANDBOOK

OTHER McGRAW-HILL HANDBOOKS OF INTEREST

HEYEL · The Foreman's Handbook
HUSKEY AND KORN · Computer Handbook
JURAN · Quality Control Handbook
KALLEN · Handbook of Instrumentation and Controls
KING AND BRATER · Handbook of Hydraulics
KNOWLTON · Standard Handbook for Electrical Engineers
KOELLE · Handbook of Astronautical Engineering
KORN AND KORN · Mathematical Handbook for Scientists and Engineers
LASSER · Business Management Handbook
LEGRAND · The New American Machinists' Handbook
MACHOL · System Engineering Handbook
MAGILL, HOLDEN, AND ACKLEY · Air Pollution Handbook
MANAS · National Plumbing Code Handbook
MANTELL · Engineering Materials Handbook
MARKS AND BAUMEISTER · Mechanical Engineers' Handbook
MAYNARD · Industrial Engineering Handbook
MAYNARD · Top Management Handbook
MORROW · Maintenance Engineering Handbook
PERRY · Chemical Engineers' Handbook
PERRY · Engineering Manual
ROSSNAGEL · Handbook of Rigging
ROTHBART · Mechanical Design and Systems Handbook
SHAND · Glass Engineering Handbook
STANIAR · Plant Engineering Handbook
STREETER · Handbook of Fluid Dynamics
TOULOUKIAN · Retrieval Guide to Thermophysical Properties Research Literature
TRUXAL · Control Engineers' Handbook

GEAR HANDBOOK

The Design, Manufacture, and Application of Gears

DARLE W. DUDLEY, Editor

Manager, Advance Gear Engineering
General Electric Company
Lynn, Massachusetts

FIRST EDITION

McGRAW-HILL BOOK COMPANY

New York Toronto London Sydney

GEAR HANDBOOK

Library of Congress Catalog Card Number: 61-7304

17902-6

10 11 12 13 14 15 16 17 18 19 VBVB 898765432

CONTRIBUTORS

MERIWETHER L. BAXTER, Jr., Chief Research Engineer, Gleason Works, Rochester, N.Y. (*Author of Chapter 1, Basic Theory of Gear-tooth Action and Generation*)

ALBERT S. BEAM, Staff Engineer, Vinco Corporation, Detroit, Mich. (*Author of Chapter 23, Gear-inspection Devices*)

SOL BINDER, Chief, Transmission Stress, Vertol Division, Boeing Airplane Company, Morton, Pa. (*Co-author of Chapter 2, Functional Types of Gears*)

F. E. BIRTCH, Manager, Cone-Drive Division, Michigan Tool Company, Detroit, Mich. (*Contribution to Chapter 23, Gear-inspection Devices*)

FRED BOHLE, Manager of Spiroid Division of Illinois Tool Works, Chicago, Ill. (*Contribution to Chapter 5, Gear-tooth Proportions*)

BEN F. BREGI, Vice President—Manufacturing and Research, National Broach and Machine Company, Detroit, Mich. (*Author of Chapter 18, Gear Finishing by Shaving, Honing, and Lapping*)

E. G. CHARTIER, President, Tekniks Engineering Company, Elmhurst, Ill. (*Author of Chapter 17, Gears Produced by Dies*)

WELLS COLEMAN, Chief Gear Analyst, Gleason Works, Rochester, N.Y. (*Contributions to Chapter 5, Gear-tooth Proportions, and Chapter 13, Load Rating of Gears*)

WARNER T. COWELL, Chief Engineer, Gear Engineering Test Center, Gleason Works, Rochester, N.Y. (*Contribution to Chapter 12, Gear Reactions and Mountings*)

S. L. CRAWSHAW, Vice President of Philadelphia Gear Corporation. (*Contribution to Chapter 3, Gear Arrangements*)

PAUL M. DEAN, Jr., Gear Technology Engineer, General Engineering Laboratory, General Electric Company, Schenectady, N.Y. (*Author of Chapter 5, Gear-tooth Proportions, and Chapter 6, Center Distance*)

DARLE W. DUDLEY, Manager, Advance Gear Engineering, General Electric Company, Lynn, Mass. (*Author of Chapter 3, Gear Arrangements. Contribution to Chapter 5, Gear-tooth Proportions. Author of Chapter 10, Gear Materials. Contribution to Chapter 23, Gear-inspection Devices*)

THOMAS W. EDWARDS, Senior Engineer, Engineering Department, Vinco Corporation, Detroit, Mich. (*Contribution to Chapter 23, Gear-inspection Devices*)

M. L. FALLON, Senior Application Design Analyst, Gear Engineering Test Center, Gleason Works, Rochester, N.Y. (*Contribution to Chapter 12, Gear Reactions and Mountings*)

WAYNE H. GLOVER, Vice President and General Manager, Indiana Gear Works, Inc., a Division of the Buehler Corporation, Indianapolis, Ind. (*Author of Chapter 19, Grinding Spur and Helical Gears*)

HARRY J. HART, Sales Engineer, Gleason Works, Rochester, N.Y. (*Co-author of Chapter 20, Bevel- and Hypoid-gear Manufacturing*)

MARTIN A. HARTMAN, Research Specialist, Rocketdyne Division of North American Aviation, Canoga Park, Calif. (*Author of Chapter 12, Gear Reactions and Mountings*)

RICHARD S. HILDRETH, Chief Engineer, Michigan Tool Company, Detroit, Mich. (*Author of Chapter 22, Gear-cutting Tools*)

CLARENCE JOHNSON, Superintendent, Utility Manufacturing Company, North Wilbraham, Mass. (*Contribution to Chapter 17, Gears Produced by Dies*)

STUART J. JOHNSON, Chief Engineer, Small Tools Division, Barber-Colman Company, Rockford, Ill. (*Author of Chapter 16, Gear Cutting*)

T. W. KHIRALLA, Staff Engineer, Curtiss-Wright Corporation, Propeller Division, Caldwell, N.J. (*Author of Chapter 21, Worm and Wormgear Cutting and Grinding Processes*)

CHARLES B. KING, Associate Chief Research Engineer, Gleason Works, Rochester, N.Y. (*Contribution to Chapter 8, Calculation of Gear-tooth Profiles. Co-author of Chapter 20, Bevel- and Hypoid-gear Manufacturing*)

LEO KINGSTON, Chief, Transmission Design, Vertol Division, Boeing Airplane Company, Morton, Pa. (*Co-author of Chapter 2, Functional Types of Gears*)

EDWIN R. KITCHEN, Gear and Cutter Engineer, Farrel-Birmingham Co., Inc., Rochester, N.Y. (*Contribution to Chapter 16, Gear Cutting*)

THEODORE F. KLEM, Gear Engineer, Gleason Works, Rochester, N.Y., (*Contribution to Chapter 23, Gear-inspection Devices*)

HAROLD KRON, Chief Engineer, Philadelphia Gear Corporation. (*Contribution to Chapter 3, Gear Arrangements*)

ROY E. LANDSTROM, Chief Tool Designer, Michigan Tool Company, Detroit, Mich. (*Contribution to Chapter 22, Gear-cutting Tools*)

A. J. LEMANSKI, Supervisor of Gear Engineering, Lycoming-Avco, Stratford, Conn. (*Contribution to Chapter 17, Gears Produced by Dies*)

GEORGE L. McCAIN, Gear Consultant, Birmingham, Mich. (*Author of Chapter 11, Gear Drawings*)

GEORGE W. MICHALEC, Head of Special Products Department, GPL Division—General Precision, Inc., Pleasantville, N.Y. (*Author of Chapter 9, Gear Tolerances*)

WOODROW NELSON, Chief Engineer, The Spiroid Division of Illinois Tool Works, Chicago, Ill. *(Contribution to Chapter 5, Gear-tooth Proportions)*

CARL PARKER, Assistant Chief Tool Engineer, The Fellows Gear Shaper Company, Springfield, Vt. *(Contribution to Chapter 5, Gear-tooth Proportions)*

ROBERT T. PARSONS, President, The Van Keuren Company, Boston, Mass. *(Author of Chapter 24, Numerical Data Tables)*

HARRY PEDERSEN, Chief Design Analyst, Gleason Works, Rochester, N.Y. *(Co-author of Chapter 20, Bevel- and Hypoid-gear Manufacturing)*

FOLKE RICHARDZ, Consulting Specialist, Gearing Products, Westinghouse Electric Corporation, Buffalo, N.Y. *(Author of Chapter 15, Lubrication of Gears)*

GERALD L. SCOTT, Staff Engineer, American Gear Manufacturers Association, Washington, D.C. *(Author of Chapter 4, Elements of Gears and Basic Formulas. Contribution to Chapter 5, Gear-tooth Proportions)*

JOHN C. SEARS, Executive Director of American Gear Manufacturers Association, Washington, D.C. *(Author of Foreword)*

GEORGE SHIMMIN, Sales Manager, Cone Drive Division, Michigan Tool Company, Detroit, Mich. *(Contribution to Chapter 5, Gear-tooth Proportions)*

EUGENE E. SHIPLEY, Project Engineer—Advance Development, General Electric Company, Lynn, Mass. *(Author of Chapter 14, Loaded Gears in Action)*

GILMORE M. SPEAR, Research Engineer, Gleason Works, Rochester, N.Y. *(Co-author of Chapter 20, Bevel- and Hypoid-gear Manufacturing)*

M. F. SPOTTS, Professor of Mechanical Engineering, Northwestern University, Evanston, Ill. *(Author of Chapter 7, Special Calculations for Spur and Helical Gears)*

ANTONI TABAK, Gear Metallurgist, General Electric Company, Lynn, Mass. *(Contribution to Chapter 10, Gear Materials)*

CLARENCE UNDERWOOD, Gear Consultant, Raytheon Company, Wayland, Mass. *(Contribution to Chapter 10, Gear Materials)*

EDWARD C. VARNUM, Head, Operations Research, Barber-Colman Company, Rockford, Ill. *(Author of Chapter 8, Calculation of Gear-tooth Profiles)*

GARY L. VESEY, Mechanical Engineer, Spiroid Division of Illinois Tool Works, Chicago, Ill. *(Contribution to Chapter 23, Gear-inspection Devices)*

E. J. WELLAUER, Director of Research and Development, The Falk Corporation, Milwaukee, Wis. *(Author of Chapter 13, Load Rating of Gears)*

HAROLD WINZELER, Vice President and General Manager, Winzeler Manufacturing Company, Chicago, Ill. *(Contribution to Chapter 17, Gears Produced by Dies)*

FOREWORD

The design and manufacture of a gear that is no better than the requirement for its successful application is an art rather than a science.

Volumes have been written and said in attempts to attain good gears. Many of them have been rather unsuccessful interpretations of what their authors opined previous authors meant or failed to say.

Thus, much of the information comes under what I call "gear folklore" —just about as obsolete as the cogwheel that sired the quality of gearing we need in modern applications.

The authors of this "Gear Handbook" are seasoned and experienced gear men. They have spent most of their careers in the study of gear phenomena, in tempering theory in the furnace of practicality and economic necessity. Several of them have received the highest honor the American Gear Manufacturers Association has to bestow on contributors to the gear art, its "Edward P. Connell" Award.

These dedicated authors set out to give the gear designer and manufacturer a true and accurate recording of the present state of the art.

They have eminently succeeded in so doing. Therefore, I personally and on behalf of AGMA enthusiastically commend this "Gear Handbook" to all who have a genuine desire and economic need to acquaint themselves with the gear art as it is now being practiced by forward-looking, down-to-earth gear designers and manufacturers.

J. C. SEARS, *Executive Director*
American Gear Manufacturers Association

PREFACE

Although a rather surprising amount of material has already been written on the subject of gears, it has been hard for the average person to quickly find the detailed gear information that he needs. A wealth of gear information has already been published in trade standards, magazine articles, manufacturers' catalogues, private company reports, and gear books. The key information to solve the design of a gear, plan tooling for a gear job, analyze a gear failure, or pick an appropriate manufacturing process may be hidden almost anywhere in current available technical literature.

The "Gear Handbook" was conceived as a means of solving this problem. In approximately one thousand pages of moderately fine print all aspects of the gear art are covered. The questions that the gear designer, the gear builder, or the gear user may have are all covered. Condensed information is given on the many types of gears that are now in use. Specific information is given on the great diversity of manufacturing processes that may be used. Considering that the field of gearing goes all the way from the little mass-produced gears for toys, clocks, instruments, and the like, up to the massive main propulsion units used for major ships or airplanes, it can be appreciated that the manufacturing techniques will vary widely.

The people who design and build gears comprise only a small fraction of the people engaged in manufacturing work. However, the people who use gears include almost everyone. It has been said quite truthfully that you can't tell the time of day without gears, you can't wash your clothes without gears, you can't drive to work without gears, and you can't make anything in your shop without gears. The user of gears is often very interested in things like the performance of gear equipment, the life expectancy of gearing, or the relative cost of different ways of making gears. Subject matter for the "Gear Handbook" has been selected with the aim of covering the whole wide field of interest in the gear art.

Chapters 1 through 4 introduce the reader to the basic theory of gearing, define the types of gear teeth, explain the many possible gear arrangements, and define the nomenclature and basic geometry of gears.

All of this is necessary background material to acquaint a person with the language and definitions peculiar to the gear trade.

Chapters 5 through 15 are the prime engineering-design chapters. These cover the elements of design, such as tooth proportions, involute calculations, choice of materials, stress calculations, load rating, tolerancing of gear dimensions, efficiency calculations, and design of lubrication systems.

Chapter 14 presents material on how to test gears and how to analyze the several possible kinds of gear failure.

Chapters 16 through 21 cover the machine tools and methods of gear manufacture. Gear cutting, gear grinding, die forming of gears, gear shaving, gear lapping, and gear honing are all covered in these chapters.

Chapter 22 covers tools to make gears, and Chapter 23 covers inspection tools. Chapter 24 presents many tables of data useful in the gear trade.

Over three dozen people were direct contributors to the project. These people were selected because of their knowledge and competence in the various aspects of gear work. The technical knowledge that this group of people was able to put together over a four-year working period represents a depth of technical knowledge in this field that is far beyond that of any one individual. From my own standpoint it has been a most rewarding experience to be associated with this project and learn from experts in all phases of gear work. In many cases, information collected for the "Gear Handbook" represents the first time this information has ever been put in public print.

I am very grateful to the many contributors for their unselfish effort in getting the required information and laboring through all the trials of putting it into handbook style. In addition, I feel particularly grateful to certain key individuals who were most helpful in providing technical guidance for the project and in making the resources of their organizations available to ferret out key items of information. My special thanks go to the officers and Headquarters staff of the American Gear Manufacturers Association, the "engineering body for the gear industry," for their assistance in checking many parts of this work. I also wish to give special thanks to Wells Coleman of the Gleason Works for his assistance in the bevel-gear field; to Mr. Gerald Brophy of International Nickel Co. for his advice on gear material developments; to Mr. Fred E. Birtch, Cone-Drive Division, Michigan Tool Co., for his assistance on double-enveloping wormgears; to Mr. S. L. Eastman, Cleveland Worm & Gear Division, Eaton Manufacturing Co., for his help on cylindrical wormgears, and to Mr. F. Bohle and Mr. W. Nelson, Spiroid Division of Illinois Tool Works, for their assistance on the Spiroid family of gears. Mr. Laurence Collins and others in the General Electric Company were very helpful in encouraging me to carry out this project and in making available to me extensive test and field data on spur and helical gears.

In spite of the fact that most of the gear technical material was received from contributors and others associated with this project, I must accept the responsibility for having made the final decisions that determined the technical content of this work. I believe this "Gear Handbook" will represent an important step forward in the development of knowledge in the gear field. It makes this knowledge available for the widest possible benefit of the great host of people who buy, design, build, or use geared equipment.

DARLE W. DUDLEY

Lynnfield, Mass.

CONTENTS

GEAR HANDBOOK

Chapter 1

BASIC THEORY OF GEAR-TOOTH ACTION AND GENERATION

By

MERIWETHER L. BAXTER, Jr., *Chief Research Engineer, Gleason Works, Rochester, N.Y.*

This chapter deals with the principles* of gear-tooth action and generation in a broad way; details applicable to specific gearing problems will be found in subsequent chapters. The reader is presumed to have an elementary knowledge of gearing terminology; refer to Chap. 4 for standard gear nomenclature.

PRINCIPLES COMMON TO ALL GEARS

Most of the ideas underlying gear theory are applicable to any type of gear. This chapter expresses these ideas in general terms; subsequent chapters will indicate their application for particular gear types.

1-1. Axial Arrangement and Relative Motion. The information required to define the motion requirements for a pair of rotating gears is

1. The relative disposition of the axes
2. The relative rotational directions
3. The velocity ratio

This article will deal with the various possible combinations of the above.

General Arrangement of Axes. The most general disposition of two gear axes occurs when they do not intersect and are neither parallel nor perpendicular.

Figure 1-1 shows two such axes, X_1 and X_2. The shortest distance between these axes, the *center distance* or *offset,* is measured along the mutual perpendicular to the axes called the *line of centers.*

Figure 1-2 is a view along the line of centers, showing the angular disposition of the axes, with vectors $\bar{\omega}_1$ and $\bar{\omega}_2$ indicating the directions and magnitudes of the angular velocities. The *relative angular velocity* $\bar{\omega}_r$ is the vector difference between the given velocity vectors $\bar{\omega}_1$ and $\bar{\omega}_2$. The *shaft angle* Σ is the angle between the axes which contains the relative angular velocity vector and lies between 0 and 180°.

* See references at the end of the chapter for important earlier technical literature on this subject.

The *kinematic pitch angle* is the angle between the positive direction of the velocity vector for a member and the positive direction of the relative angular velocity vector for that member with respect to its mate. In Fig. 1-2 the kinematic pitch angles are designated γ_1 and γ_2. The sum of the kinematic pitch angles is the shaft angle.

$$\tan \gamma_1 = \frac{\omega_2 \sin \Sigma}{\omega_1 + \omega_2 \cos \Sigma} \tag{1-1}$$

$$\tan \gamma_2 = \frac{\omega_1 \sin \Sigma}{\omega_2 + \omega_1 \cos \Sigma} \tag{1-2}$$

$$\gamma_1 + \gamma_2 = \Sigma \tag{1-3}$$

Note that, when $\Sigma > 90°$, one of the kinematic pitch angles may be greater than 90°.

Special Axial Arrangements. The simpler forms of gearing may be considered as special cases of the general situation.

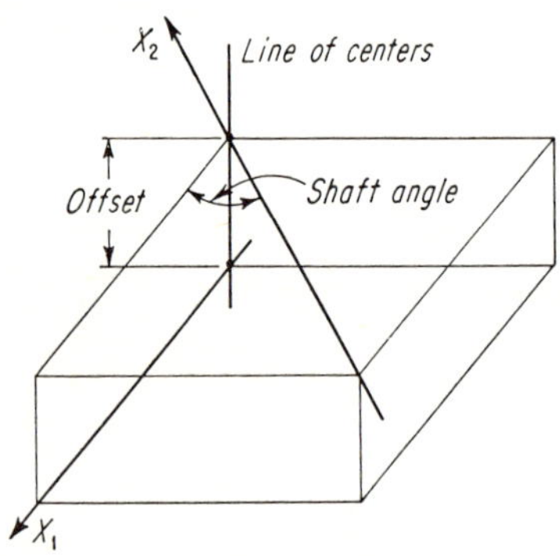

FIG. 1-1. General arrangement of gear axes.

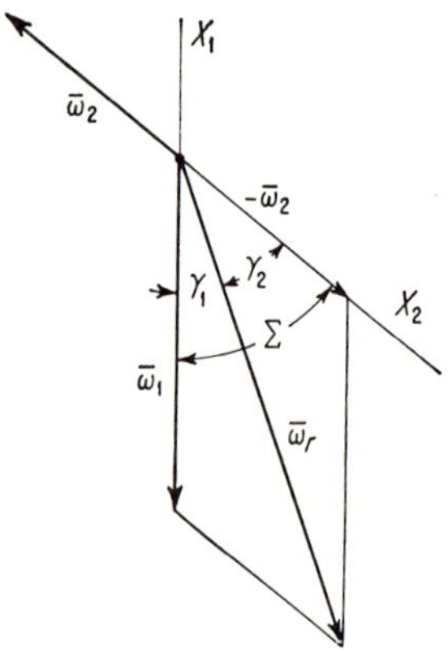

FIG. 1-2. Angular-velocity relationships.

Spur gears have parallel axes; the shaft angle and the kinematic pitch angles are zero.

Bevel gears have intersecting axes; the offset is zero and the kinematic pitch angles become the actual pitch angles. When the shaft angle is 90° the pitch-angle formulas simplify to

$$\gamma_1 + \gamma_2 = 90° \tag{1-4}$$

$$\tan \gamma_1 = \frac{\omega_2}{\omega_1} = \cot \gamma_2 \tag{1-5}$$

1-2. Conjugate Action of Gear-tooth Surfaces. Two surfaces are *conjugate* if each generates or envelops the other under a specified relative motion. Thus, if a tooth surface of one gear is of a rigid material, and the mating tooth of the second gear is of a plastically deformable material, the specified relative motion will cause a surface to be produced on the tooth of the second gear which is conjugate to the tooth surface of the first.

Conjugate surfaces contact at any instant along a *line of contact* which is the locus of all *points of contact* at that instant.

At any point of contact the surfaces have a common *contact normal*, and the relative normal velocity is zero. This is the fundamental requirement for gear-tooth contact and permits us to compute points of contact. The relative tangential velocity is the *sliding velocity*.

Figure 1-3 shows one of a pair of conjugate tooth surfaces and the line of contact existing at one instant. Also shown are the contact normal and the sliding velocity vector at one point in the line of contact.

As conjugate action progresses the line of contact describes a surface in space called the *surface of action.*

The two mating tooth surfaces and the surface of action are interrelated by the condition of conjugacy. *If any one of the three is defined the other two can be determined from the specified relative motion.*

Curvature of Conjugate Surfaces. In a plane section which contains the common contact normal at a specific point of contact we may study the curvatures of the conjugate surfaces. In such a normal section taken tangent to the line of contact the curvatures of the two surfaces are equal. In all other normal sections the teeth must be relatively convex so as to permit the rolling and sliding action characteristic of gear teeth; we say there is *relative curvature* between the surfaces.

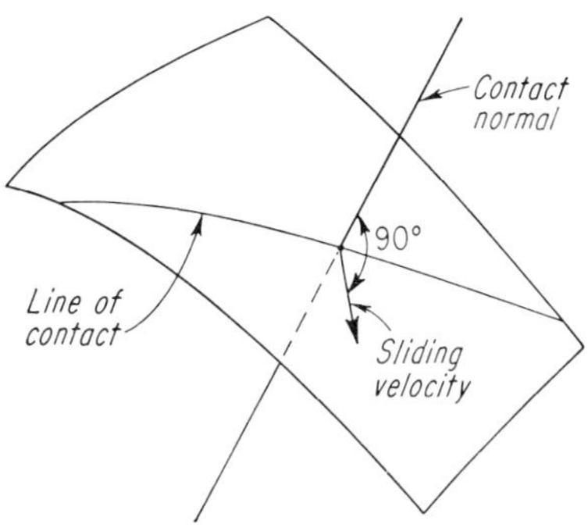

FIG. 1-3. Conjugate surface contact.

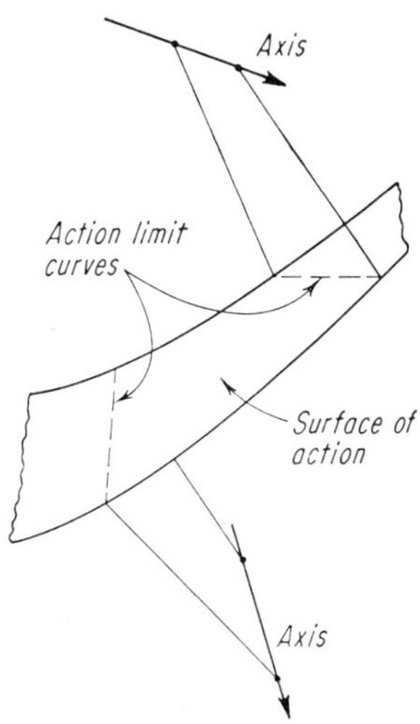

FIG. 1-4. Surface of action.

A singular condition exists along the curve in the surface of action which is the projection of the axis of a gear. When conjugate action reaches a point in this curve, there occurs a cusp in the tooth profile and its radius of curvature is zero. It is not generally possible for real tooth action to progress past this curve, and it may therefore be called the *action-limit curve* (see Fig. 1-4). A similar condition exists in gear generation; an attempt to generate past the action-limit curve results in mutilation of the previously generated surface called *undercut.*

Limits of Tooth Action. The boundaries of the tooth surfaces are established by the shape of the blank in which the teeth are formed. The volume common to the blanks of a pair of gears is the *region of mesh.* Only those portions of the surfaces of action lying within the region of mesh are available for motion transmission.

Furthermore the action-limit curves place an additional limit on tooth action when they lie within the region of mesh. These restrictions therefore establish the boundaries of the *effective surfaces of action.*

In simple forms of gears the teeth are symmetrical and the action is symmetrical. In the more complex forms separate consideration must be given to each side. The two surfaces of action intersect in a curve; this curve should be centrally and symmetrically located within the region of mesh in order to obtain maximum utilization of the available space for gear-tooth action. This consideration is of primary importance in the design of gears for nonintersecting axes.

Total Contact Ratio. The extent of the effective surface of action determines the amount of rotation of a gear during which one tooth surface remains in contact. The ratio of this angle of rotation to the angular pitch is the *total contact ratio.*

The total contact ratio must be at least unity for continuous conjugate action; in practice it is usually greater than 1.2.

GEARS WITH COPLANAR AXES

This section deals with gears having intersecting or parallel axes; for such gears there exists an *axial plane* containing both axes, and we can develop certain concepts not applicable to gears with offset axes.

1-3. Pitch Surfaces and the Instantaneous Axis. See Fig. 1-5. Gears with intersecting axes are called *bevel gears* and can be represented kinematically by *pitch cones* which roll without slipping at the specified velocity ratio and which contact along

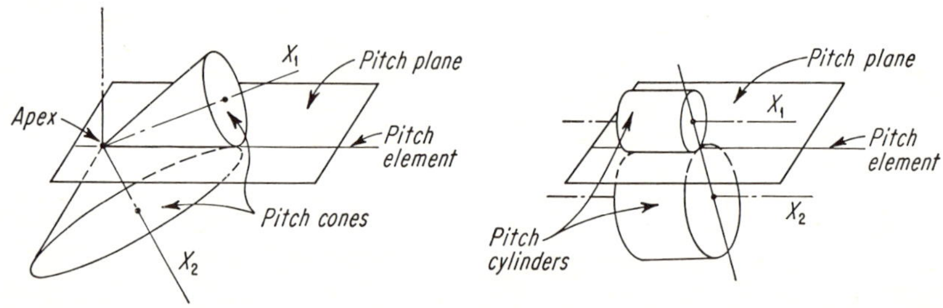

(a) (b)

FIG. 1-5. Pitch surfaces. (a) Intersecting axes. (b) Parallel axes.

a common element called the *pitch element.* The pitch element is therefore the *instantaneous axis of relative motion* of either gear with respect to the other. The two axes and the instantaneous axis all lie in the axial plane and intersect at the *apex.*

Gears with parallel axes are called *spur gears* or *helical gears* and may be considered a special case of bevel gears in which the pitch cones become *pitch cylinders* and the apex is at infinity.

A *pitch angle* is the angle between an axis and the pitch element; the formulas for pitch angles are the same as those given in Art. 1-1 for the kinematic pitch angles of gears with offset axes.

Properties of Pitch Surfaces. The pitch cones of bevel gears and the pitch cylinders of spur and helical gears have the following important properties which make them valuable concepts in analysis and design:

1. Pitch surfaces roll together with no sliding.
2. The pitch element is the instantaneous axis of relative motion.
3. The common tooth surface normal at any point of contact intersects the pitch element.
4. The pitch element is the intersection of the surfaces of action.

It should be noted that there are no pitch surfaces for gears with nonintersecting axes which fulfill any of the first three conditions; the fourth condition can be fulfilled only by choosing specific tooth shapes for the purpose. Pitch cylinders and pitch cones are sometimes referred to in connection with worm and hypoid gearing as a matter of convenience, but they must not be considered to be true pitch surfaces.

Uses of Pitch Surfaces. Since the surfaces of action intersect in the pitch element it is natural to use the pitch surfaces as a basis for designing gear blanks. Thus conical gear blanks are usually used for bevel gears and cylindrical blanks for spur and helical gears. When this is not done practical disadvantages become apparent in the form of pointed teeth or undercut.

The gear teeth are positioned around the pitch surface; the intersections of the tooth surfaces in the pitch surface are the *pitch curves;* the spacing of the pitch curves on the pitch surface is the *pitch.* Pitch may be measured angularly or linearly, and in various directions for various purposes. As the pitch surfaces roll on each other the pitch curves likewise roll together without sliding.

The *pitch plane* is mutually tangent to the pitch surfaces of a pair; it contains the pitch element and is perpendicular to the axial plane. Tooth directions such as pressure angle and spiral angle or helix angle are usually specified with respect to the pitch plane.

1-4. Basic Members. When the pitch angle of a bevel gear is 90° it is a *crown gear,* and its pitch surface is a plane which rotates about the apex while rolling with the pitch cone of its mate. Similarly, when the pitch radius of a spur or helical gear becomes infinite it is a *rack,* and its pitch surface is a plane which has translational motion while rolling with the pitch cylinder of its mate.

A rack or a crown gear is the *basic member* for the family of gears conjugate to it. Tooth dimensions are customarily specified for the basic member; it is also the basis of most generating processes. In certain cases the basic member may be other than a rack or crown gear, but the principles involved are the same.

Two basic members are *complementary* if they can be fitted together face to face with completely coincident pitch and tooth surfaces. Thus we speak of *complementary crown gears* and *complementary racks* (see Fig. 1-6*a*). Any member of the family of gears conjugate to one of a pair of complementary basic members is conjugate to any member of the family conjugate to the other basic member. Two members of the same family, however, are not necessarily conjugate to each other.

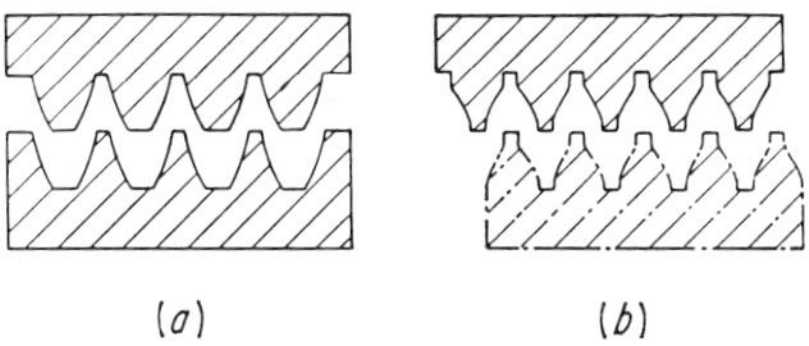

FIG. 1-6. Complementary racks. (*a*) Complementary racks. (*b*) Basic rack for an interchangeable system (self-complementary).

In a set of *interchangeable gears* any member of the set is conjugate to any other. The requirement for interchangeability within a family of gears is that the basic member be complementary to itself (see Fig. 1-6*b*).

Most methods of generating gear-tooth surfaces depend on the idea of a basic member that is simulated by the cutting tool and its motions.

GEAR-TOOTH ACTION IN A PLANE

Since the detailed study of three-dimensional gear-tooth surfaces is difficult, it is natural where possible to reduce the problem to a two-dimensional one. This section describes the action of tooth profiles in a plane.

1-5. The Transverse Plane. A *transverse plane* is perpendicular to the pitch element and therefore mutually perpendicular to the axial plane and the pitch plane. It is natural to study spur- and helical-gear action in a transverse plane since any point on the tooth surface remains in that plane during rotation; such an analysis is therefore exact for the particular section studied.

In a bevel gear a particular point on the tooth does not remain in the same transverse plane during rotation, and hence bevel-gear analysis in a transverse plane is approximate; however, the method is widely used and is sufficiently accurate for many purposes. See Art. 1-12 for a discussion of exact bevel-gear analysis in a sphere.

Figure 1-7 shows the relation between the axial planes of spur and bevel gears and the transverse plane. Table 1-1 lists the notation and formulas discussed in this section.

Figure 1-8 is the transverse section of a pair of gears in mesh. The *turning centers* are O_1 and O_2; they determine the *line of centers* and the *center distance* C. The *pitch circles* are centered at O_1 and O_2 and contact at the *pitch point* P on the line of centers. The *velocity ratio* is the ratio of the tooth numbers and the ratio of the pitch radii.

The *circular pitch* is the arc length on the pitch circles between successive like tooth profiles, and is the same for both gears and the basic rack. The *angular pitch* is the angle subtended by the circular pitch. The *diametral pitch* is equal to π divided by the circular pitch, and for any member is the ratio of its number of teeth to its pitch diameter.

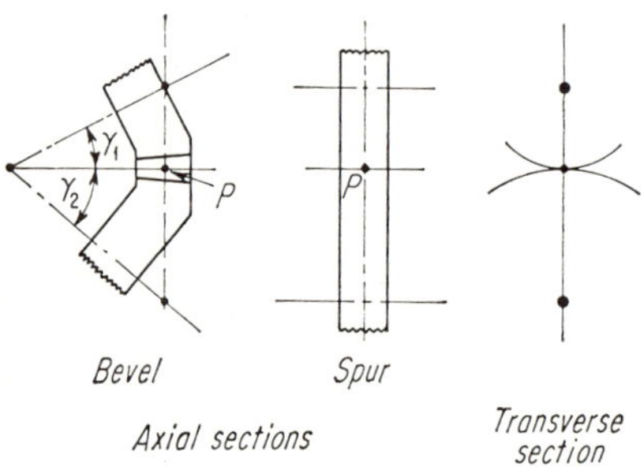

FIG. 1-7. Axial and transverse planes.

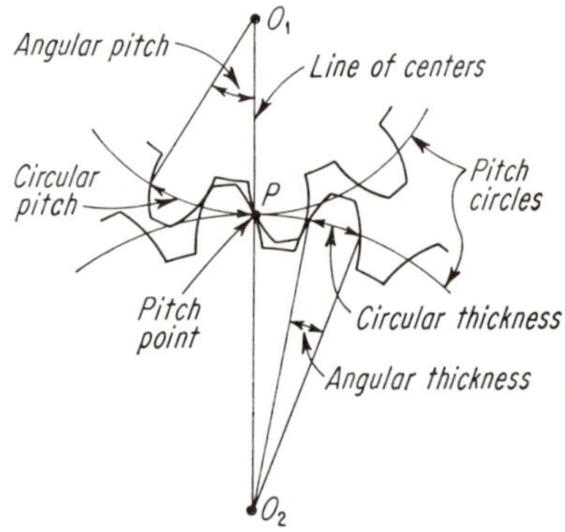

FIG. 1-8. Gear elements in the transverse plane.

The *circular thickness* of the teeth is measured along the pitch circle; the sum of the circular thicknesses of a pair equals the circular pitch minus the *backlash.* The *angular thickness* is the angle subtended by the circular thickness.

1-6. Plane Tooth Action. When the pitch circles are established the tooth-profile curve of the basic rack or of either member may be chosen arbitrarily, and the other profiles may be derived from it by application of the *fundamental law of gear-tooth action: The common normal to the profiles at a point of contact contains the pitch point.*

Figure 1-9 illustrates tooth action in a plane; it will be clear that this is directly analogous to the three-dimensional tooth action described in Art. 1-2.

The *line of action* is the locus of the points of contact as the profiles go through mesh. The point on the line of action closest to a turning center is a *limit of action;* the radius of curvature of the profile associated with that turning center becomes zero at this point and conjugate tooth action cannot in general proceed beyond it. The limits of action should lie outside of the *region of mesh* enclosed by the outside circles of the mating gears, shown shaded in Fig. 1-9.

The portion of the line of action available for proper tooth action, taking into account the region of mesh and the limits of action, is the *effective line of action,* of length Z.

The angle of rotation of a gear during which contact proceeds from one end of the effective line of action to the other is the *angle of contact;* the ratio of the angle of contact to the angular pitch is the *transverse contact ratio,* which must be greater than unity for continuous motion transmission in the particular transverse section under study.

Profile Curvatures. Figure 1-10 shows a pair of tooth profiles in contact at a general point M a distance s from pitch point P. The contact normal at M must pass through P and is inclined at the (instantaneous) pressure angle ϕ.

K_1 and K_2 on the contact normal are the centers of curvature of the profiles at the contact point M and must be related geometrically as shown in the figure for uniform

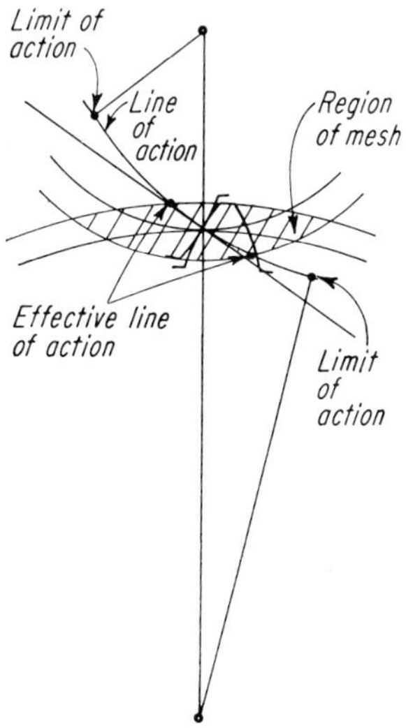

FIG. 1-9. Plane tooth action.

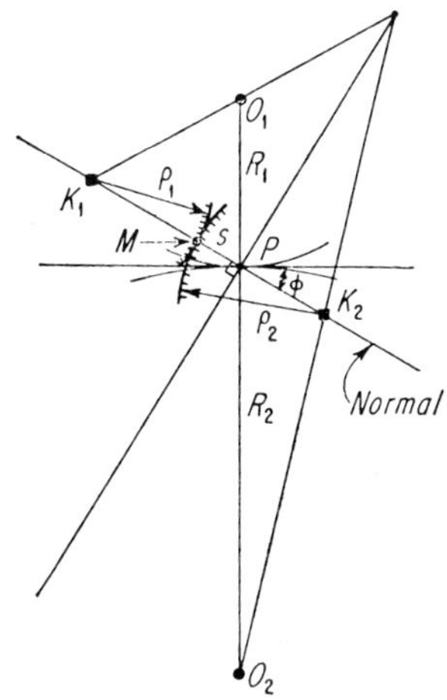

FIG. 1-10. Tooth-profile curvatures.

motion. The algebraic statement of this relationship is the *Euler-Savary equation,* which in our symbols is

$$\frac{1}{\rho_1 + s} + \frac{1}{\rho_2 - s} = \frac{1}{\sin \phi}\left(\frac{1}{R_1} + \frac{1}{R_2}\right) \tag{1-6}$$

where ρ_1 and ρ_2 are the radii of curvature of the profiles at M.

For a pair of profiles to continue to transmit uniform motion this relation must hold at all phases of the mesh.

The *relative curvature* between the profiles is

$$\frac{1}{\rho_0} = \frac{1}{\rho_1} + \frac{1}{\rho_2} \tag{1-7}$$

For contact at the pitch point the relative curvature is given by

$$\frac{1}{\rho_0} = \frac{1}{\sin \phi}\left(\frac{1}{R_1} + \frac{1}{R_2}\right) \tag{1-8}$$

The *sliding velocity* at a point of contact M is the distance from the pitch point s multiplied by the relative rotational velocity. Sliding velocity and relative curvature are important in the study of gear-tooth wear and failure.

1-7. Involute Tooth Profiles. The simplest and most-used tooth form has a basic rack with straight tooth profiles (see Fig. 1-11). The inclination of the rack profiles is the *pressure angle.* The line of action is straight and normal to the rack profiles. The circles about the turning centers and tangent to the line of action are the *base* circles. The tooth-profile curve on each member is the *involute* of its base circle.

Table 1-1. Notation and Formulas, Plane Tooth Action

Notation	Formula
Basic Elements	
C Center distance	$C = R_1 + R_2$
N Number of teeth	
R Pitch radius m_G Ratio	$m_G{}^* = \dfrac{N_2}{N_1} = \dfrac{R_2}{R_1}$
Pitch and Thickness	
p Circular pitch	$p = \dfrac{2\pi R}{N}$
P Diametral pitch	$P = \dfrac{\pi}{p} = \dfrac{N}{2R}$
β_p Angular pitch	$\beta_p = \dfrac{360°}{N}$
t Circular thickness	$t_1 + t_2 + b = p$
τ Angular thickness	$\tau = \dfrac{180°}{\pi}\dfrac{t}{R}$
b Backlash	
p_b Base pitch (involute gears) ϕ_t Pressure angle (transverse)	$p_b = p \cos \phi_t$
Contact Ratio	
β_c Angle of contact	$m_p = \dfrac{\beta_c}{\beta_p}$
m_p Transverse contact ratio	$m_p = \dfrac{Z}{p_b}$ (involute gears)
Z Length of action (involute gears)	
Relative Curvature	
ρ Profile radius of curvature	$\dfrac{1}{\rho_1 + s} + \dfrac{1}{\rho_2 - s} = \dfrac{1}{\sin \phi}\left(\dfrac{1}{R_1} + \dfrac{1}{R_2}\right)$
ρ_0 Radius of relative curvature	$\dfrac{1}{\rho_0} = \dfrac{1}{\rho_1} + \dfrac{1}{\rho_2}$
s Distance of contact point from pitch point	$\rho_0 = \dfrac{1}{C}\left[R_1R_2 \sin \phi - s(R_2 - R_1) - \dfrac{s^2}{\sin \phi}\right]$ (involute gears)
Sliding Velocity	
ω Angular velocity (radians) V_s Sliding velocity	$V_s = s(\omega_1 + \omega_2)$

* Normal gear practice is to take m_G as *greater* than unity.

The *base pitch* is the distance along the line of action between successive like rack profiles and successive like involute profiles. The limits of action are the points of tangency K of the line of action with the base circles; they are also the centers of curvature of the tooth profiles throughout the mesh.

Involute Geometry. The involute of a circle may be defined as the curve traced by a point on a straight line which rolls without slipping on the circle. The circle is called the *base circle* of the involute. A single involute curve has two branches of opposite hand meeting at a cusp on the base circle, where the radius of curvature is zero. All involutes of the same base circle are congruent and parallel; involutes of different base circles are geometrically similar.

Figure 1-12 shows the elements of involute geometry. The generating line G was originally in position G_0, tangent to the base circle at P_0. It then rolled about the base circle through the *roll angle* ϵ to position G_1, when it is tangent to the base circle at K. The point P_0 in the generating line has moved to P, describing the involute curve I. Roll in the opposite direction would have generated the other branch I'.

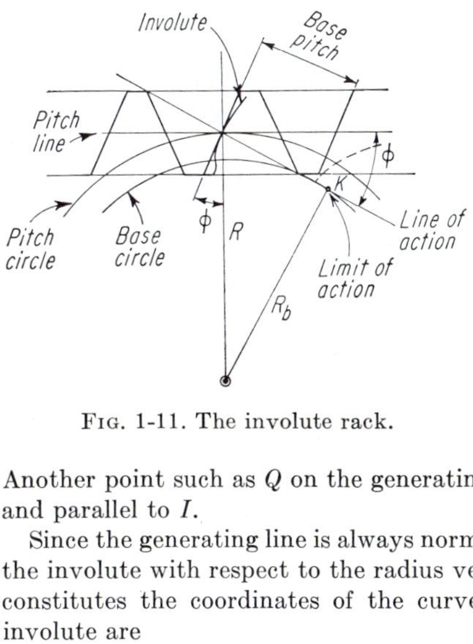

FIG. 1-11. The involute rack.

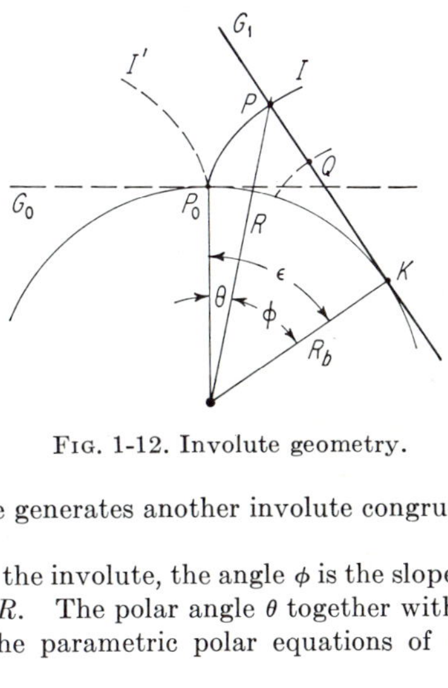

FIG. 1-12. Involute geometry.

Another point such as Q on the generating line generates another involute congruent and parallel to I.

Since the generating line is always normal to the involute, the angle ϕ is the slope of the involute with respect to the radius vector R. The polar angle θ together with R constitutes the coordinates of the curve. The parametric polar equations of the involute are

$$R = R_b \sec \phi \tag{1-9}$$

$$\widehat{\theta} = \tan \phi - \widehat{\phi} \tag{1-10}$$

Note that the roll angle ϵ in radians is equal to $\tan \phi$.

The quantity $(\tan \phi - \widehat{\phi})$ is known as the *involute function* of ϕ and is tabulated in Table 24-14 for use in involute-profile calculations. See also Art. 7-8.

A straight line having a motion comprising (1) a rotation and (2) a translation proportional to the rotation generates an involute of a circle. This statement has a more direct application to involute gearing than the conventional discussion just given.

SPUR AND HELICAL GEARS

Earlier sections of this chapter have dealt with tooth contact in general, pitch surfaces, basic members, and the action of tooth profiles. This section relates more specifically to helical gears, of which spur gears may be considered a special case.

1-8. The Helical Rack. The pitch surfaces of helical gears are cylinders contacting along the pitch element. The *axial plane* contains both axes and the pitch element. A *transverse plane* is perpendicular to the axes, and hence to the axial plane. The *pitch plane* is mutually tangent to the pitch cylinders, contains the pitch element, and is perpendicular to the axial plane and all transverse planes. Figure 1-13 shows these relationships.

A basic helical rack has the pitch plane for its pitch surface and takes for its axial and transverse directions those of the mating gear. It has straight teeth inclined at the helix angle to the axial direction; the intersections of the rack tooth surfaces in the pitch plane are straight *pitch lines* which mesh with the pitch helices of a mating gear. Figure 1-14 illustrates this situation.

A *normal plane* is perpendicular to the pitch plane and to the pitch lines of the rack.

A *tangent plane* is perpendicular to the normal planes, and tangent to a rack tooth surface along its pitch line.

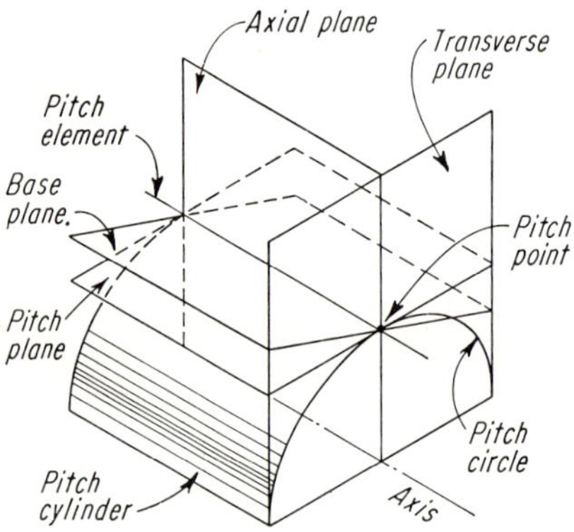

Fig. 1-13. Reference planes—spur gears.

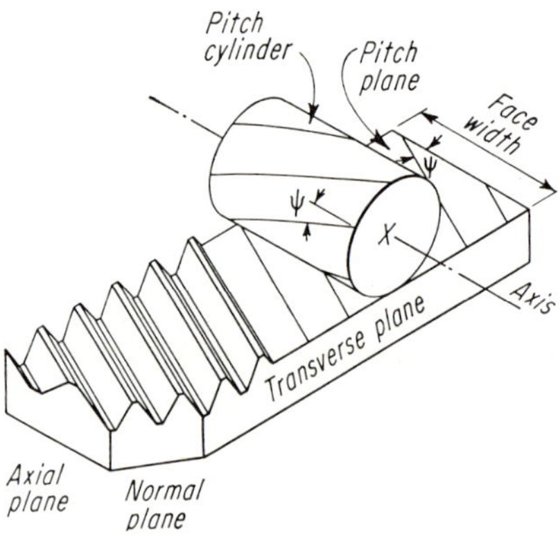

Fig. 1-14. The helical rack.

A *base plane* is perpendicular to the transverse planes and the tangent planes.

The spacing between the tangent planes to like sides of two successive teeth is the *pitch* of the rack. Pitch may be measured in a number of directions, the most common of which are listed here with the customary symbols.

Circular pitch p: in the transverse direction in the pitch plane
Normal pitch p_n: in the normal direction in the pitch plane
Axial pitch p_x: in the axial direction in the pitch plane
Normal base pitch p_N: in the normal direction in the base plane
Base pitch p_b: in the transverse direction in the base plane

The *pressure angle* of a rack is the inclination of the tangent plane with respect to the pitch plane and may be measured in the transverse, normal, or axial planes:

Transverse pressure angle ϕ_t
Normal pressure angle ϕ_n
Axial pressure angle ϕ_x

The intersection of a tangent plane in the pitch plane is inclined at the helix angle ψ with the axial direction. Similarly, the angle made by the intersection of a tangent plane in the base plane with the axial direction is the *base helix angle* ψ_b.

The relationships among the various pitches, pressure angles, and helix angles are given in Table 1-2.

Table 1-2. Pitches and Angles for Helical Gears

	Pitch
Circular	$p = p_b \sec \phi_t = p_n \sec \psi = p_x \tan \psi$
Axial	$p_x = p \cot \psi = p_b \cot \psi_b = p_n \csc \psi = p_N \csc \psi_b$
Normal	$p_n = p \cos \psi = p_x \sin \psi = p_N \sec \phi_n$
Base	$p_b = p \cos \phi_t = p_x \tan \psi_b = p_N \sec \psi_b$
Normal base	$p_N = p_n \cos \phi_n = p_b \cos \psi_b = p_x \sin \psi_b$
	Pressure Angle
Transverse	$\sin \phi_t = \sin \phi_n \sec \psi_b = \tan \phi_x \tan \psi_b$
	$\cos \phi_t = p_b/p = \tan \psi_b \cot \psi$
	$\tan \phi_t = \tan \phi_n \sec \psi = \tan \phi_x \tan \psi$
Normal	$\sin \phi_n = \sin \phi_t \cos \psi_b$
	$\cos \phi_n = p_N/p_n = \sin \psi_b \csc \psi$
	$\tan \phi_n = \tan \phi_t \cos \psi = \tan \phi_x \sin \psi$
Axial	$\tan \phi_x = \tan \phi_t \cot \psi = \tan \phi_n \csc \psi = \sin \phi_t \cot \psi_b$
	Helix Angle
Pitch	$\sin \psi = p_n/p_x = \sin \psi_b \sec \phi_n = \tan \phi_n \cot \phi_x$
	$\cos \psi = p_n/p = \tan \phi_n \cot \phi_t$
	$\tan \psi = p/p_x = \tan \psi_b \sec \phi_t = \tan \phi_t \cot \phi_x$
Base	$\sin \psi_b = p_N/p_x = \sin \psi \cos \phi_n$
	$\cos \psi_b = p_N/p_b = \sin \phi_n \csc \phi_t$
	$\tan \psi_b = p_b/p_x = \tan \psi \cos \phi_t = \sin \phi_t \cot \phi_x$

1-9. Helical Gears. The tooth surfaces of helical gears are helicoids conjugate to the basic rack. Tooth profiles in any transverse plane are identical but shifted in phase through angles proportional to the axial displacement. The intersections of the tooth surfaces in a cylinder coaxial with the gear are helices. The *lead* of all such helical tooth elements is the same, but the helix angle varies with the diameter of the helix as follows:

L = lead of helicoid
R = radius of helix
ψ = helix angle at radius R

$$\tan \psi = \frac{2\pi R}{L} \tag{1-11}$$

The helix angle is usually specified at the pitch radius. The various pitches, pressure angles, and helix angles are the same for a gear as for its basic rack.

In analyzing the action of a pair of gears it is convenient to imagine rack tooth surfaces meshing simultaneously with both members as they mesh with each other, the surfaces of action being the same for either gear meshing with the rack as for the two gears. At any point of contact, therefore, the two gear surfaces and the rack surface are tangent and have a common normal which intersects the pitch element.

When contact occurs at a point on the pitch element the tangent plane of the rack is tangent to the tooth surfaces, and the direction of the instantaneous line of contact in the tangent plane is given by the projection of the pitch element to the tangent plane, as shown in Fig. 1-15. The dashed line lies in the tangent plane, is tangent to the line of contact at P, and is the intersection of a base plane through the pitch element with the tangent plane. The values of the angle w are given by the following equations:

$$\sin w = \sin \psi \sin \phi_t = \tan \psi_b \tan \phi_n \tag{1-12}$$
$$\cos w = \cos \psi \sec \psi_b = \cos \phi_t \sec \phi_n \tag{1-13}$$
$$\tan w = \sin \psi_b \tan \phi_t = \tan \psi \sin \phi_n \tag{1-14}$$

1-10. Involute Helical Gears. Involute helical gears have a basic rack with plane tooth surfaces, and hence straight profiles in the various reference planes. The tangent plane of the rack is the rack tooth surface itself.

The gears have involute profiles in a transverse plane, and their surfaces are *involute helicoids.* Lines of contact at any position of mesh are straight and inclined at the angle w. The surfaces of action are base planes of the rack containing the pitch element, and tangent to the *base cylinder.* The involute helical tooth surfaces meet the base cylinder in the *base helix,* and any line of contact extended is tangent to the base helix on the base cylinder.

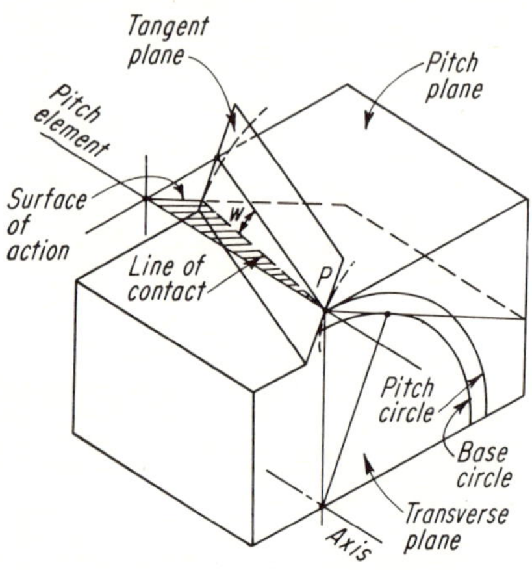

FIG. 1-15. Helical tooth contact.

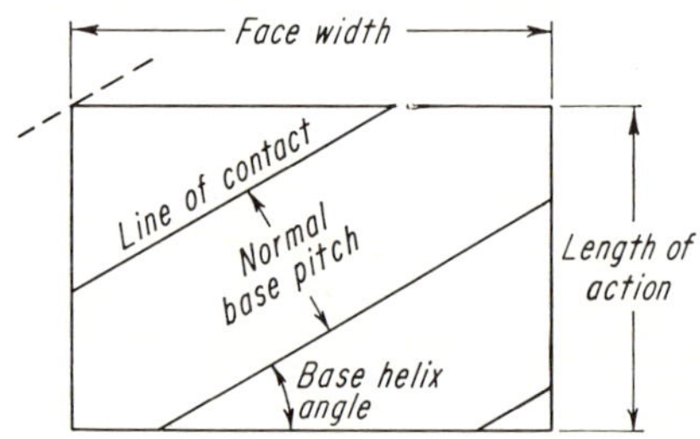

FIG. 1-16. Surface of action—helical gears.

The normal curvature of a tooth surface in the direction of the line of contact is zero, since in this direction the surface contains a straight-line element. In a direction normal to the line of contact the radius of curvature is

$$\rho_w = R \sin \phi_n \tag{1-15}$$

where ϕ_n is the normal pressure angle at radius R. Since ϕ_n is a function of R the following formula may be more useful:

$$\rho_w = \sec \psi_b \sqrt{R^2 - R_b{}^2} \tag{1-16}$$

where R is the only variable for a particular gear.

The *face advance* Q is the distance on a pitch circle through which a tooth moves while contact travels along the pitch element a distance equal to the face width F.

$$Q = F \tan \psi \tag{1-17}$$

The *face contact ratio* is

$$m_F = \frac{Q}{p} \tag{1-18}$$

The total contact ratio is the sum of the transverse contact ratio and the face contact ratio, and is a measure of the average number of teeth in contact.

$$m_t = m_p + m_F \tag{1-19}$$

The actual teeth in contact and the length of the line of contact on each are best seen in a view of the plane of action (Fig. 1-16). The effective surface of action is a rectangle whose width is the face width F and whose length is the length of action in the transverse plane Z. Lines of contact cross this rectangle at the base helix angle ψ_b at a spacing equal to the normal base pitch p_N.

As mesh progresses the lines of contact move across the rectangle, and the number of teeth and the total length of contact lines at any instant can be computed. Such a study is frequently used to estimate the division of load among the teeth for stress determination.

BEVEL GEARS

The general principles of gearing have been described in earlier sections of this chapter; this section extends these principles to bevel gears specifically.

1-11. Reference Planes. Figure 1-17 illustrates the principal reference planes: the *axial plane* contains the intersecting axes; the *pitch plane* is perpendicular to the

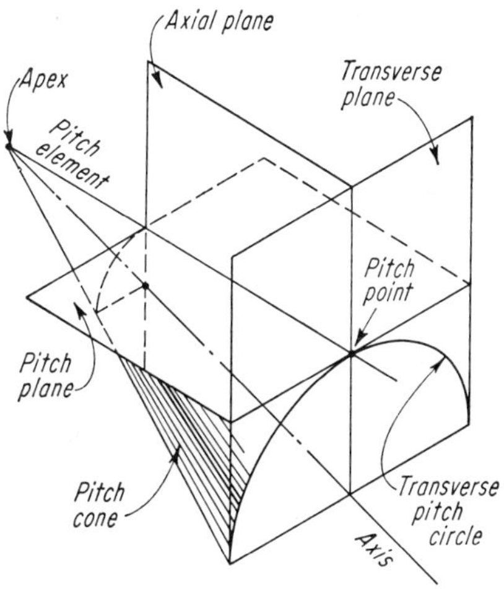

FIG. 1-17. Reference planes—bevel gears.

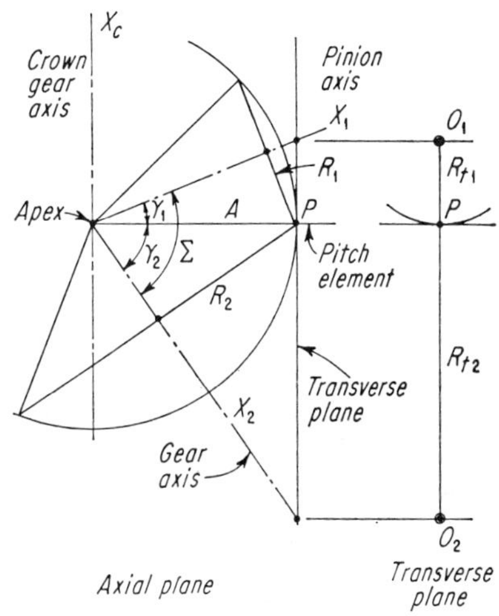

FIG. 1-18. Axial and transverse planes—bevel gears.

axial plane and tangent to the pitch cones; the pitch plane and axial plane intersect in the common *pitch element;* the *transverse plane* is perpendicular to the axial plane and the pitch plane and the pitch element. Any number of transverse planes may be considered, at various distances from the apex.

Other reference planes which will be used are the normal plane and the tangent plane.

The Axial Plane. The left-hand side of Fig. 1-18 is a view of the axial plane. The axes X_1 and X_2 intersect at the *apex* and are disposed at a *shaft angle* Σ. The *pitch angles* are γ_1 and γ_2, which add up to the shaft angle; the ratio of the sines of the pitch angles is the velocity ratio.

The crown-gear axis X_c lies in the axial plane and is perpendicular to the pitch element.

At a specified *cone distance* A along the pitch element from the apex is a pitch point P; the distances of the pitch point from the axes are the *pitch radii* R_1 and R_2. The pitch radii are proportional to the sines of the pitch angles, and their ratio is the velocity ratio.

Formulas relating these quantities and the tooth numbers N_1 and N_2 are listed in Table 1-3.

Table 1-3. Bevel-gear Relations

N	Number of teeth	$m_G = \dfrac{N_2}{N_1} = \dfrac{\sin \gamma_2}{\sin \gamma_1}$
Σ	Shaft angle	$\gamma_1 + \gamma_2 = \Sigma$
γ	Pitch angle	$\tan \gamma_2 = \dfrac{\sin \Sigma}{m_G + \cos \Sigma}$
m_G	Velocity ratio	$\tan \gamma_1 = \dfrac{m_G \sin \Sigma}{1 + m_G \cos \Sigma}$
A	Cone distance	$N_c = \dfrac{N_1}{\sin \gamma_1} = \dfrac{N_2}{\sin \gamma_2}$
R	Pitch radius	$R = A \sin \gamma$
N_c	Teeth in crown gear	$\left.\begin{array}{l} \tan \gamma_1 = \dfrac{N_1}{N_2} \\ \tan \gamma_2 = \dfrac{N_2}{N_1} \\ N_c = \sqrt{N_1^2 + N_2^2} \\ A = \sqrt{R_1^2 + R_2^2} \end{array}\right\} \Sigma = 90°$

The Transverse Plane. The right-hand side of Fig. 1-18 shows a view of a transverse plane containing the pitch point P. O_1 and O_2 are the traces of the axes X_1 and X_2 in the transverse plane.

If we take O_1 and O_2 as the centers of pitch circles tangent at P we may study the action of the gears as a plane problem by the methods given in Arts. 1-6 through 1-7. The equivalent pitch radii are

$$R_t = \frac{R}{\cos \gamma} = A \tan \gamma \tag{1-20}$$

and the equivalent tooth numbers are

$$N_t = \frac{N}{\cos \gamma} \tag{1-21}$$

Study of bevel-gear action and generation in the transverse plane is approximate; nevertheless it gives satisfactory results when the tooth numbers are not extremely small, and the method is widely used. It should be noted that the transverse pressure angle must be used in such work, and not the normal pressure angle which is usually specified for bevel gears.

1-12. Tooth Action in the Sphere. Exact analysis of bevel-gear profiles can be carried out in the surface of a sphere centered at the apex, and whose radius is the cone distance. Spherical trigonometry is used throughout. This method is used for very small tooth numbers, such as are found in automotive differentials, and for certain theoretical investigations.

Figure 1-19 is a view of the sphere looking along the pitch element toward the apex. Lines on the sphere are understood to be arcs of great circles, and their lengths are therefore designated as angles. O_1 and O_2 are the traces of the axes in the sphere and P is the trace of the pitch element. K_1 and K_2 are the centers of curvature for the tooth profiles; the construction illustrated relates these centers for (instantaneously) uniform motion transmission and is analogous to the construction of Fig. 1-10 given for plane tooth action. The corresponding formula, analogous to the Euler-Savary equation, is

$$(\cot \widehat{K_1P} + \cot \widehat{K_2P}) = \csc \phi \, (\cot \gamma_1 + \cot \gamma_2) \tag{1-22}$$

Figure 1-20 is a similar spherical diagram in which a straight-sided crown-gear tooth of pressure angle ϕ has been rotated an angle θ about the crown-gear axis. The point of contact is located by dropping a perpendicular from P to the tooth profile, and its position is specified by the quantities σ and ϕ' in the figure. The following

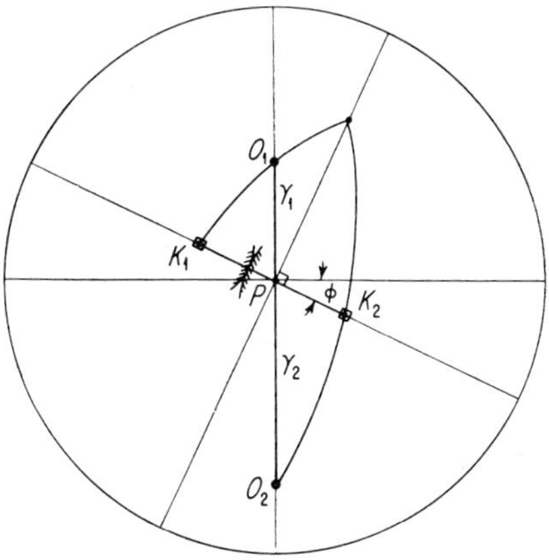

FIG. 1-19. Bevel-gear tooth-profile curvatures.

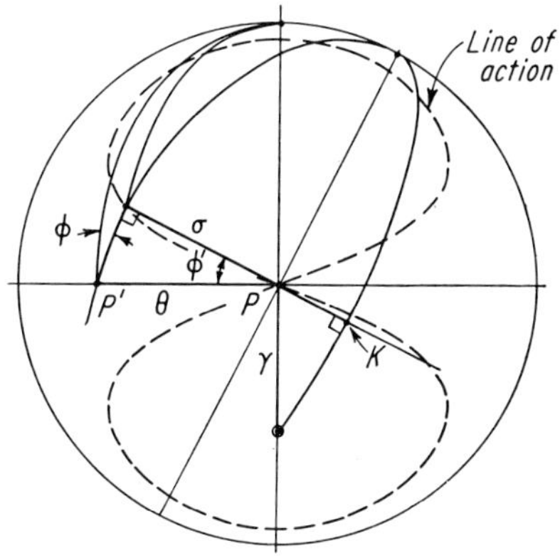

FIG. 1-20. Bevel-gear line of action.

formulas locate such points of contact and may be considered the parametric equation of the line of action in the sphere:

$$\tan \phi' = \tan \phi \sec \theta \tag{1-23}$$

$$\sin \sigma = \cos \phi \sin \theta \tag{1-24}$$

The line of action is quite straight in the vicinity of the pitch point but then forms a "figure eight" on the sphere as shown in the figure. For this reason bevel gears based on a crown gear with straight profiles are sometimes said to have *octoid* teeth.

A bevel gear of pitch angle γ conjugate to this crown gear will have its center of curvature at K where

$$\tan \widehat{PK} = \tan \gamma \sin \phi' \tag{1-25}$$

1-13. Mesh in the Pitch Plane. The intersections of the tooth surfaces in the pitch cone are the *pitch curves*. The pitch curves of a mating pair roll together without sliding as the pitch cones roll on each other. The pitch curves of the crown gear are used to describe the general lengthwise shape of the teeth. Figure 1-21 shows a bevel pitch cone rolling with its crown gear for straight bevel and for spiral bevel teeth.

Straight Bevel Gears. The pitch curves of a straight bevel crown gear are straight lines, intersecting the apex. The spiral angle at any cone distance is zero; the normal pitch and the transverse pitch are equal and vary directly with the cone distance.

Spiral Bevel Gears. The pitch curves of a spiral bevel crown gear are inclined to the pitch element at the *spiral angle* and are usually circular. Figure 1-22 shows the geometry of such a crown gear.

The spiral angle ψ is usually specified at a pitch point P located at the middle of the face at a mean cone distance A. CP is the normal to the pitch curve at P, C is the center of curvature, and r_c is the radius of curvature. The spiral angle is different at different cone distances; at a general cone distance A' the spiral angle ψ' is given by the following formula:

$$\sin \psi' = \frac{A}{A'}\left(\sin \psi + \frac{A'^2 - A^2}{2Ar_c}\right) \tag{1-26}$$

The transverse pitch is proportional to the cone distance, but the normal pitch is not, because of the variation of spiral angle. The normal pressure angle, however, is constant, while the transverse pressure angle varies with cone distance.

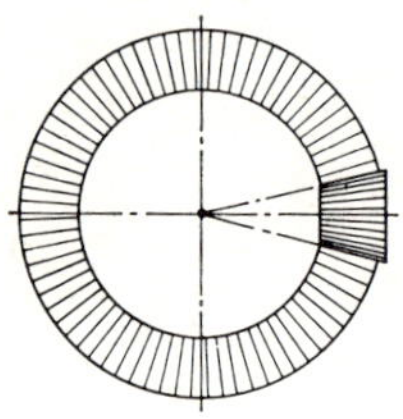

FIG. 1-21. Rolling pitch curves—bevel gears.

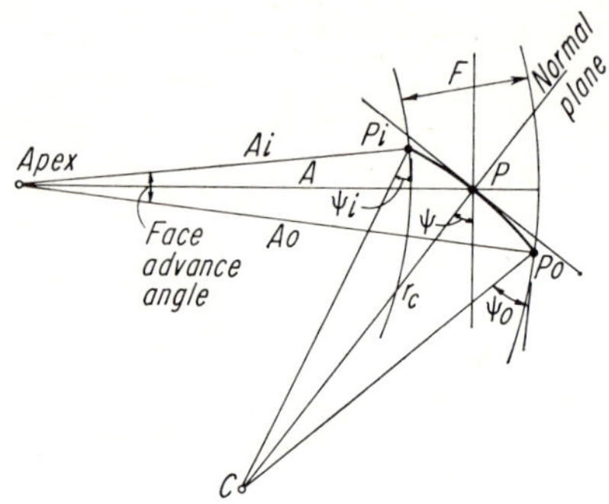

FIG. 1-22. Pitch plane geometry of spiral bevel crown gear.

The *face advance angle* is the angle at the center of the crown gear subtended by the ends of the pitch curve. The *face contact ratio* is the ratio of the face advance angle to the angular pitch of the crown gear.

1-14. The Normal Plane. The normal plane is perpendicular to the pitch plane and to a pitch curve at some specified point, such as P in Fig. 1-22. The *tooth normal* lies in the normal plane and is inclined at the normal pressure angle ϕ to the pitch plane. PC is the trace of the normal plane in the pitch plane.

Spiral-bevel gear-tooth action is frequently studied in the normal plane, in much the same way as in the transverse plane. The equivalent tooth numbers are

$$N_n = N \sec \gamma \sec^3 \psi \tag{1-27}$$

and the equivalent pitch radii are

$$R_n = A \tan \gamma \sec^2 \psi \tag{1-28}$$

Such studies are of course approximate. For details see Art. 8-9.

1-15. Pitch and Angle Relationships. The plane tangent to the tooth surface at a specified point (frequently a pitch point, such as P in Fig. 1-22) is a *tangent plane.* It is of course perpendicular to the tooth normal.

The *base plane* is perpendicular to the tangent plane and contains the tooth normal and the pitch element; it is tangent to the true surface of action and is often used as an approximation of it.

The relationships between the reference planes: pitch, axial, transverse, normal, tangent, base, are exactly analogous for spiral bevel and helical gears; hence the formulas in Table 1-2 relating pitches and angles in various planes hold also for spiral bevel gears, when the term "spiral angle" is substituted for "helix angle." Likewise formulas (1-11), (1-12), and (1-13) give the direction of the line of contact for spiral bevel gears. Since, however, the spiral angle and pitch for spiral bevels change with different cone distances, the values obtained from the helical formulas will be correct only for the specified point in the pitch element.

GEARS WITH NONINTERSECTING AXES

The principles of tooth contact for this most general case of gearing have been discussed in Art. 1-2. The material in this section will deal more specifically with the geometry of hypoid and worm gearing.

1-16. Basic Hypoid Geometry. If we are given two axes X_1 and X_2 as in Fig. 1-23 we can determine their common perpendicular C_1C_2. C_1 and C_2 are the *crossing points* and distance $\overline{C_1C_2}$ is the *offset E*.

We may within reason choose the location of the region of mesh at will; this choice will determine the kind of gearing and certain tooth characteristics. For example, if the region of mesh is on the line of centers C_1C_2 we shall have worm gears, and the location along the line of centers will determine (for a given velocity ratio) the lead angle of the worm. Let us assume a general location of a point P such as that shown in the figure which is centrally located in the region of mesh. Point P projects to the axes at B_1 and B_2, and the perpendicular distances are R_1 and R_2, the radii of the gears.

The location of P may be defined in a number of ways; let us locate it by specifying the radius R_1, the distance $\overline{C_1B_1} = Z_1$, and the angle ι which B_1P makes with a plane perpendicular to the line of centers.

The line D_1D_2 contains P and intersects both axes and is called the *pitch vertical.* The pitch vertical is normal to any surfaces of revolution about the axes which are tangent at P. The plane O_1O_2P is perpendicular to the pitch vertical at P and intersects the axes at O_1 and O_2. It is therefore possible to describe two cones, one about each axis, which are tangent at P, whose apices are at O_1 and O_2, and which are tangent to plane O_1O_2P along the lines O_1P and O_2P.

Let us now define plane O_1O_2P as the *pitch plane* and the cones as the *pitch cones,* bearing in mind that they are not real pitch surfaces in the sense described in Art. 1-3. Distances O_1P and O_2P are the *cone distances* and denoted by A_1 and A_2; the angle O_1PO_2 in the pitch plane is ι', and the pitch angles are Γ_1 and Γ_2.

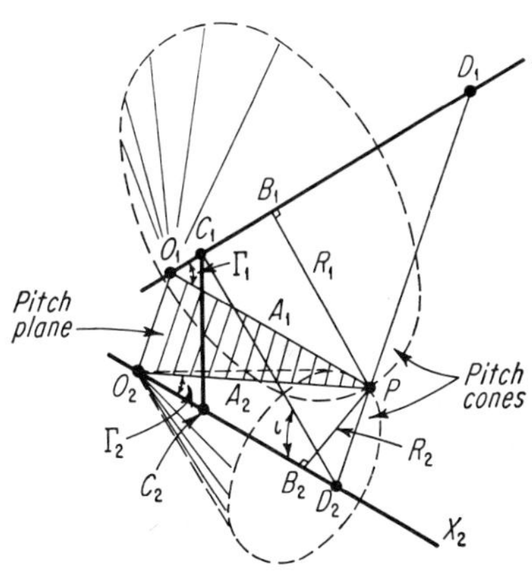

FIG. 1-23. Hypoid geometry.

When P has been specified by the quantities R_1, Z_1, and ι, and the shaft angle is 90°, the other dimensions may be determined from the following formulas:

$$\cot \Gamma_1 = \frac{E \csc \iota - R_1}{Z_1} \tag{1-29}$$

$$\sin \Gamma_2 = \cos \Gamma_1 \cos \iota \tag{1-30}$$

$$R_2 = Z_1 \frac{\cos \Gamma_2}{\sin \Gamma_1} \tag{1-31}$$

$$A_1 = R_1 \csc \Gamma_1 \tag{1-32}$$

$$A_2 = R_2 \csc \Gamma_2 \tag{1-33}$$

$$\sin \iota' = \sin \iota \sec \Gamma_2 \tag{1-34}$$

Figure 1-24 is a view of the pitch plane, looking along the pitch vertical from D_2 toward D_1. Axis X_1 meets the plane at O_1 and projects into line O_1P; axis X_2 meets the plane at O_2 and projects into O_2P. Let us now consider what kind of tooth surfaces we can construct on the pitch cones at P.

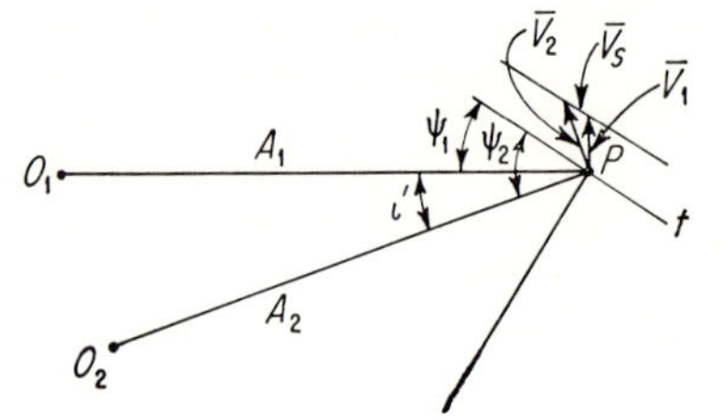

FIG. 1-24. Hypoid pitch plane.

$\overline{V_1}$ and $\overline{V_2}$ are the velocity vectors at P for the two gears; they lie in the pitch plane, and their magnitudes are

$$V_1 = R_1\omega_1 \qquad V_2 = R_2\omega_2$$

where ω_1 and ω_2 are the rotational velocities. The relative velocity vector is $\overline{V_s}$, the *sliding velocity*, and also lies in the pitch plane, making an angle ψ_1 with line O_1P. Any plane containing the pitch tangent t may therefore be taken as a tangent plane for mating tooth surfaces at P, since the relative normal velocity will be zero. The spiral angles (relative to the pitch plane) must therefore be ψ_1 and ψ_2, but the pressure angle may be freely chosen. The formulas for ψ_1, ψ_2, and V_s, the sliding velocity, are

$$\tan \psi_1 = \cot \iota' - \frac{\omega_1}{\omega_2}\frac{R_1}{R_2} \csc \iota' \tag{1-35}$$

$$\psi_2 = \psi_1 + \iota' \tag{1-36}$$

$$V_s = R_1\omega_1(\tan \psi_2 - \tan \psi_1) \tag{1-37}$$

1-17. Hypoid Surfaces of Action. The surfaces of action depend on the shape of the tooth surfaces, and the geometry described above gives no assurance that they will provide satisfactory contact except at the point P. It is desirable that the intersection curve of the surfaces of action lie in, or at least tangent to, the pitch plane and that the surfaces of action for the two sides of the teeth be approximately equally inclined to the pitch plane.

These results can be obtained by choosing the correct curvature of the tooth surfaces in the pitch plane and by using unequal pressure angles for the two tooth sides. The amount of pressure-angle unbalance is ϕ_0, where

$$-\tan \phi_0 = \frac{A_1 \sin \psi_1 - A_2 \sin \psi_2}{A_1 \tan \Gamma_1 + A_2 \tan \Gamma_2} \tag{1-38}$$

Use of this amount of pressure-angle unbalance on the teeth relative to the pitch plane will result in approximately equal contact conditions and equal freedom from undercut on the two sides.

The optimum tooth curvature in the pitch plane is $1/r$, where

$$\frac{1}{r} = \frac{\cos \phi_0}{\tan \psi_2 - \tan \psi_1}\left[\frac{1}{A_2 \cos \psi_2} - \frac{1}{A_1 \cos \psi_1} - \tan \phi_0 \left(\frac{\tan \psi_1}{A_1 \tan \Gamma_1} + \frac{\tan \psi_2}{A_2 \tan \Gamma_2}\right)\right] \tag{1-39}$$

Departure from this value will result in teeth which tend to become pointed at one end of the face and undercut at the other.

By choosing different locations of point P in the region of mesh, it is possible to obtain solutions giving any desired spiral angle and lengthwise tooth curvature; however, this cannot be calculated directly, and iterative methods are used.

1-18. Wormgears. When the region of mesh and point P in the general hypoid geometry lie on the line of centers, we have the special case of wormgearing. The pitch vertical is the line of centers, and the pitch cones become pitch cylinders.

When the shaft angle, center distance, and velocity ratio are known, the spiral angle or lead angle is dependent on the size of the pitch cylinders. This is analogous to the hypoid case where the spiral angle is determined by the location of point P.

For a helicoidal worm and 90° shaft angle the analysis of wormgear action can be reduced to a plane problem and dealt with by the methods of Art. 1-6. The planes of study are perpendicular to the gear axis, and the worm sections in each plane must be determined and treated as a rack. This procedure is very laborious but is correct and will yield the true wormgear profiles and surfaces of action. Where rapid computation facilities are available it is more satisfactory to treat wormgear contact as a space problem by applying basic gearing kinematic principles as outlined in Art. 1-2.

1-19. Crossed-axis Helical Gears. Helical involute gears operating on nonparallel axes transmit uniform motion, but there is theoretically point contact between the tooth surfaces at any instant. Hence they are not conjugate in the sense used in this chapter; the conjugate gear to one of them would be its mating worm gear.

The surface generated on a uniformly rotating gear by a plane having constant translational motion is an involute helicoid whose characteristics can be readily determined by a study in its transverse plane. The surface of action is a plane tangent to the base cylinder and perpendicular to the generating plane. The same plane, simultaneously generating an involute helicoid about a second axis not parallel to the first, will have a different surface of action; both helicoids will contact the generating plane along straight lines of contact, but these lines will not coincide. The intersection of the lines of contact in the tangent plane is a *point of contact* between the two helicoids; the intersection of the two planes of action is the *path of action* for the two mating involute helicoids.

If the common tangent-generating plane is considered as the surface of a rack tooth, we find that in general the two gears will roll with different pitch planes on that rack; and the path of action does not in general intersect the line of centers.

The practical problem most often encountered in work with crossed-axis helical gears is to design them for a specified shaft angle, center distance, and velocity ratio, with integral tooth numbers, and using standard tools. Procedures for such work are given in Art. 6-1.

GEAR-TOOTH GENERATION

The generation of gear-tooth surfaces for conjugate action, or some desired approximation thereto, is a complex subject and will be described in detail in subsequent chapters in connection with gear-producing machinery. We shall describe here some of the principles underlying these methods.

1-20. Gear Generation. In all gear-generating processes the tool simulates one or more teeth of an imaginary *generating gear* by virtue of its shape and motions. The generating gear may be the basic rack or crown gear of a pair or family of gears, or it may be the mating gear itself, or in general it may be any gear known to be conjugate to the gear to be produced.

The most important requirement for the generating gear is that it can be readily simulated by a practical tool with practical mechanical motions.

The tool and its motions must simulate not only the tooth surfaces of the generating gear but also the relative rolling motion of the generating gear with the workpiece. The latter function is usually called the *generating roll.*

When the surface of the tool and the surface of a tooth of the generating gear coincide, the only motion that affects the tooth surface produced is the generating roll. It is clear that the lines of contact between the work and the tool are the same as between the work and the generating gear, and the surface generated will be continuous. This type of generation is found in grinding, lapping, and burnishing processes where individual cutting edges are not required.

More usually a line in the tool contacts the surface of the generating gear at any instant, and a *traverse motion* is required to move the line across the tooth surface of the generating gear. The line of contact may be a cutting edge, or it may be the line of contact of a surface, such as a cone or helicoid, with the generating-gear surface. At any instant the tool contacts the generating-gear tooth in a line; the generating gear contacts the work in a different line; and contact between the tool and the work exists (theoretically) at a point.

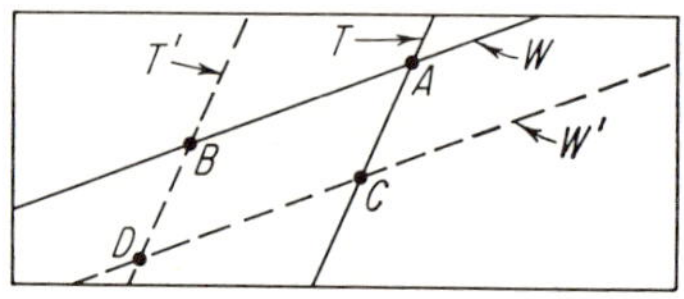

FIG. 1-25. Tooth-surface generation.

Figure 1-25 represents the tooth surface of the imaginary generating gear. Line T is the instantaneous line of contact with the tool; its position is controlled by the traverse motion. Line W is the instantaneous line of contact between the generating gear and the work, and its position is controlled by the generating roll. Point A is the point of contact between tool and work.

If the traverse motion proceeds to T' while the generating roll is stopped, contact will proceed from A to B, and a line AB in the work will be finished. If then the generating roll is advanced so that the line of contact is W', and the traverse is repeated, line CD in the work will be finished. The tooth surface produced will then be made up of a series of facets or *flats* whose envelope is the theoretical tooth surface. The spacing of the flats is determined in this case by the size of the increments of generating roll, and the direction of the flats by the theoretical contact between generating gear and work.

If, on the other hand, the traverse motion is held constant while the generating roll proceeds, the flat AC will be produced; then when the traverse motion is incremented to T', a second flat BD will be produced. Here the direction of the flat is that of the tool contact line, and the spacing is determined by the traverse increments.

When both traverse and generation proceed simultaneously, the direction of flats will be intermediate, such as AD; and to produce other flats to complete the surface it is only necessary to change the phase relationship between traverse and generation.

Various generating processes may be characterized by the following:

1. Kind of generating gear
2. Kind of tool, and its contact with the generating gear
3. Kind of traverse motion
4. Relative rate of traverse and generation
5. Method of producing phase shift

1-21. Fillet Generation. A tooth surface of a generating gear, as simulated by a real tool, has a finite boundary curve which determines the depth of the tooth being generated. This curve, like all others in the generating surface, describes a surface in space tangent to the generated tooth surface and produces the *fillet* which connects the root surface of a gear tooth to its intended generated surface.

Figure 1-26 is a plane diagram illustrating fillet generation. G is the profile of the tooth of the generating gear and P is the profile of the generated tooth. As stated in Art. 1-2, such a profile will have a cusp C when generation reaches a limit of action, after which a second branch of the profile P' will be enveloped theoretically.

The curves labeled F are fillet curves produced by various points in the generating line G. They are tangent to P or P', and their radius of curvature at the point of tangency varies considerably; in fact, F_2 has a zero radius of curvature. The height of fillet on a gear tooth is frequently investigated to avoid *interference* with the mating tooth, and the curvature of a fillet is important in strength calculations.

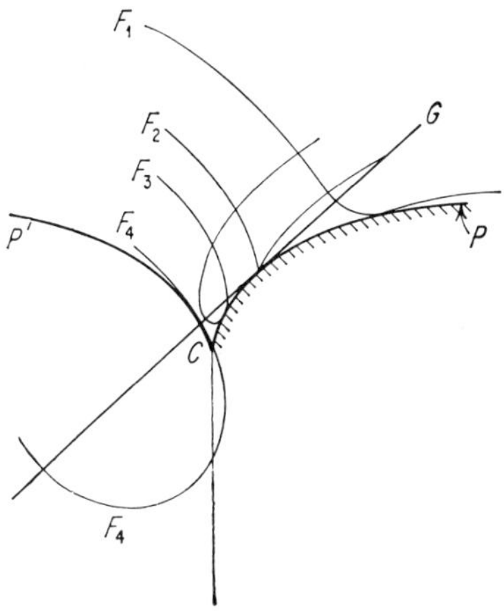

FIG. 1-26. Fillet generation.

Fillet curve F_4 is exceptional in that it is described by a point on G which would theoretically generate a point on the second branch P', and hence is beyond the limit of action. F_4 is tangent to P', as would be expected, but it is not tangent to the real desired profile P; it intersects P at a point above the cusp C at a considerable angle. Such a profile is then said to be *undercut*. Undercut occurs to some degree whenever the tooth surface of the generating gear extends beyond the limits of action *for the generating process*.

Note that the same gear will have different fillet and undercut characteristics when made by different processes.

REFERENCES

1. Willis, Robert: "Principles of Mechanism," John Wiley & Sons, Inc., New York, 1870.
2. MacCord, C. W.: "Kinematics," John Wiley & Sons, Inc., New York, 1883.
3. Grant, George B.: "A Treatise on Gear Wheels," Philadelphia Gear Works, 1899.
4. Buckingham, Earle: "Spur Gears," McGraw-Hill Book Company, Inc., New York, 1928.
5. Candee, Allan H.: Gear Geometry, *Am. Machinist*, July 4, 11, 1929. Reprinted with revisions by the American Gear Manufacturers Association, *AGMA Pub.* 115.01, 1959.
6. Vogel, Werner: "Involutometry and Trigonometry," Michigan Tool Co., 1945.
7. Wildhaber, Ernest: Relationships of Bevel Gears, *Am. Machinist*, Aug. 30, Sept. 27, Oct. 11, Oct. 25, 1945.
8. Wildhaber, Ernest: Basic Relationships of Hypoid Gears, *Am. Machinist*, 1946.
9. Merritt, H. E.: "Gears," Sir Isaac Pitman & Sons, Ltd., London, 1942.
10. Candee, Allan H. "Introduction to the Kinematic Geometry of Gear Teeth," Chilton Co., New York 1961.

Chapter 2

FUNCTIONAL TYPES OF GEARS

By

SOL BINDER, *Chief, Transmission Stress, Vertol Division, Boeing Airplane Company, Morton, Pa.*

LEO KINGSTON, *Chief, Transmission Design, Vertol Division, Boeing Airplane Company, Morton, Pa.*

Many kinds of gear teeth are in general use. The first problem in designing a gearset is to pick the right kind of gears. Although many people may think a certain kind of gear is the best kind of gear, there is really no *best kind.* For each application, though, there is often one kind of gear that is better than all other kinds. This chapter will discuss the problems involved in selecting the right kind of gear. The designer of a gear transmission should consider all the possible gear types and then pick the type that best meets his design objectives.

2-1. Introduction to Gear Types. One factor influencing the choice of the right gear is the geometric arrangement of the apparatus which needs a gear drive. Gears may transmit motion from the shafts oriented in practically any relationship. Where the shafts are parallel, the pitch surfaces of the gears are cylindrical. If the shafts are intersecting, the gears have conical pitch surfaces. Crossed shafts are geared together by gears of various shapes, ranging from the hyperboloid (hypoid) gear to the wormgear drive in the single- (cylindrical) or double-enveloping (globoid) form

Table 2-1. Gear Types

Parallel axes	Intersecting axes	Nonintersecting nonparallel axes
1. Spur gears *a.* External *b.* Internal 2. Helical gears *a.* External *b.* Internal 3. Herringbone or double-helical gears *a.* External *b.* Internal	1. Bevel gears *a.* Straight bevel gear *b.* Zerol bevel gear *c.* Spiral bevel gear 2. Face gear 3. Beveloid	1. Crossed helical 2. Single-enveloping worm 3. Double-enveloping worm 4. Hypoid 5. Spiroid 6. Planoid 7. Beveloid 8. Face gear 9. Helicon

Table 2-2. General Comparison of Gear Types

Type of gear	Approx. range of efficiencies	Max.* widths generally used†	Type of load imposed on support bearings	Nominal range of reduction ratio‡	Nominal max. pitch line velocity, fpm		Member of pair	Methods of manufacture	Methods of refining
					Aircraft and high precision	Commercial			
Parallel Axes									
External spur gears	97–99.5	$F = d$	Radial	1:1–5:1	20,000	4,000	Both	Hob, shape, mill, broach, stamp, sinter	Grind, shave, lap (crossed axis only), hone
Internal spur gears	97–99.5	Dictated by mating gear	Radial	1:5–7:1	20,000	4,000		Shape, mill, broach, stamp, sinter	Grind, shave, lap (crossed axis), hone
External helical gears	97–99.5	$F = d$	Radial and thrust	1:1–10:1	40,000	4,000	Both	Hob, shape, mill	Grind, shave, lap, hone
Internal helical gears	97–99.5	Dictated by mating gear	Radial and thrust	1:5:1–10:1	20,000	4,000		Shape, mill	Shave, grind, lap, hone
Internal herringbone or double helical	97–99.5	Dictated by mating gear	Radial	2:1–20:1	20,000	4,000		Shape	Lap, shave, hone
External herringbone or double-helical gear	97–99.5	$F = 2d$	Radial	1:1–20:1	40,000	4,000	Both	Hob, shape, mill	Grind, shave, lap, hone
Intersecting Axes									
Straight bevel gear	97–99.5	⅓ cone distance	Radial and thrust	1:1–8:1	10,000	1,000	Both	Generating, forming	Grind
Zerol bevel gear	97–99.5	28% of cone distance	Radial and thrust	1:1–8:1	10,000	1,000	Both Gear	Generating Forming	Grind
Spiral bevel gear	97–99.5	⅓ cone distance	Radial and thrust	1:1–8:1	25,000	4,000	Both Gear	Generating Forming	Grind, lap
Face gear	95–99.5	From $0.2d$ at a low ratio to d at a high ratio	Radial and thrust	3:1–8:1	5,000	4,000	Pinion Gear	Same as external spur Shape	Lap, grind, shave, hone Lap
Beveloid	95–99.5	$\frac{5}{p_n}$	Radial and thrust	1:1–8:1	5,000	4,000	Both	Hob, generating	Lap, grind

Nonintersecting Nonparallel Axes

Crossed helical	50–95	$F_1 = 4p_n \sin \psi_1$ $F_2 = 4p_n \sin \psi_2$	Radial and thrust	1:1–100:1	10,000	4,000	Both	Hob, shape, mill	Grind, lap, shave
Cylindrical worm	50–90	$F_w = 5p_n \cos \lambda$ $F_g = 0.67d$	Radial and thrust	3:1–100:1	10,000	5,000	Pinion Gear	Mill, hob Hob	Grind Lap
Double-enveloping worm	50–98	$F_w = 0.9D$ $F_g = 0.9d$	Radial and thrust	3:1–100:1	10,000	4,000	Worm Wheel	Shape, hob Hob, mill	Lap, grind
Hypoid	90–98	$F_g = \frac{1}{3}$ cone distance	Radial and thrust	1:1–10:1	10,000	4,000	Both Gear	Generating Forming	Grind, lap
High-reduction hypoid	50–90	$F_g = 0.15D$	Radial and thrust	10:1–50:1	10,000	4,000	Pinion Gear	Generating Forming	Grind, lap Grind, lap
Spiroid	50–97	$F_p = 0.24D$ $F_g = 0.14D$	Radial and thrust	9:1–100:1	10,000	6,000	Pinion Gear	Mill, hob Hob, mold	Grind, chase
Planoid	90–98	$F = \frac{1}{3}$ cone distance	Radial and thrust	1.5:1–10:1	10,000	4,000	Pinion Gear	Hob Mill, broach	Lap, grind Grind
Helicon	50–98	$F_p = 0.21D$ $F_G = 0.12D$	Radial and thrust	3:1–100:1	10,000	6,000	Pinion Gear	Mill, hob Hob, mold	Grind, chase
Face gear	95–99.5	From 0.2D at a low ratio to D at a high ratio	Radial and thrust	3:1–8:1	10,000	4,000	Pinion Gear	Same as external spur Shape	Same as external spur Lap
Beveloid	50–95	$\frac{5}{p_n}$	Radial and thrust	1:1–100:1	10,000	4,000	Both	Generation, hob	Grind, lap

* Face width given is only a "nominal" maximum. Consult other parts of the Gear Handbook for detail limitations on face width.

† d = pinion pitch diameter
D = gear pitch diameter
λ = lead angle
ψ = helix angle
p_n = normal circular pitch

‡ The gear types showing an upper ratio of 100:1 can be built to 300:1 or higher. The ratio of 100:1 is shown as a normal maximum limit.

The meshing action on gears with crossed shapes involves more sliding on the tooth flanks and therefore greater friction losses than on drives with parallel or intersecting axes. For this reason, high-power installations requiring high efficiency are usually designed for cylindrical or bevel gears. Chapter 13 covers the subject of horsepower ratings for different kinds of gears. Several tables are given which show the capacity of these gears over a range of ratios and sizes. Table 2-1 is a summary of applicable gear types.

Additional factors affecting the choice of the proper gear are equipment available for manufacture, engineering experience available for the design, and limitations of space and weight. Also involved are specific requirements such as vibration resistance and noise level. A general comparison of the various gear types is given in Table 2-2. All these gears have been developed to fill a need. The choice of the gear type to use can be made only by the designer after he weighs all the above requirements. The following example illustrates some factors influencing the choice of the proper gear.

Example. *Conditions:*

1. Two shafts at an angle of 90°. Shafts need not intersect.
2. Input—1,000 hp at 3,000 rpm.
3. Gear ratio = 3:1.
4. Losses must not exceed 0.5 per cent.
5. Weight and space are at a premium.
6. Very little development time is available.
7. Work to be subcontracted; therefore, no consideration to be given to available equipment.

To meet the above requirements the choice may well be a hardened and ground spiral bevel gearset. If any of the seven conditions are changed another choice may readily result. The particular choice made was arrived at for the following reasons:

1. To keep losses at less than 0.5 per cent requires a minimum of sliding on the tooth flanks. Therefore, gears mounted on intersecting shafts are the choice.
2. Hardened and ground gears are dictated by the requirements of minimum space and weight.
3. Spiral bevels are chosen because of their high load and speed capacity.
4. The choice of spiral bevels is further enhanced by the short development time required for this type of gear to ensure proper operation.

PARALLEL AXIS GEARS

2-2. External Spur Gears. The most prevalent and best-known type of gear is the external spur. It is used to provide a constant speed and torque relation between parallel shafts. An infinite number of curves can be used for gear-teeth profiles which will produce conjugate action. The most predominant tooth-profile form used is in the shape of an involute. Small changes in center distance do not affect tooth action.

A spur gear and a spur rack are shown in Fig. 2-1. A spur rack is a spur gear having an infinitely large pitch diameter. Terminology used with these gears is also shown.

The most common pressure angles used for spur gears are 14½, 20, and 25°. The lower pressure angle has the advantage of smoother and quieter tooth action because of the larger profile contact ratio. In addition, lower loads are imposed on the support bearings because of a decreased radial component. The problem of undercutting associated with small numbers of pinion teeth is more severe with the lower pressure angle. The larger pressure angles have the advantages of better load-carrying capacity.

2-3. External Helical Gears. Helical gears are used to transmit power or motion between parallel shafts. A helical gear may be considered as composed of an infinite number of narrow staggered spur gears, the result being that each tooth is slanting across the face as a cylindrical helix. Figure 2-2 shows the terminology of a helical gear and a helical rack. Meshing helical gears must have the same helix angle but opposite hand of helix. (A right-hand pinion meshes with a left-hand gear and a left-hand pinion meshes with a right-hand gear.) As a result of the helix angle, both thrust and radial loads are imposed on the helical-gear support bearings.

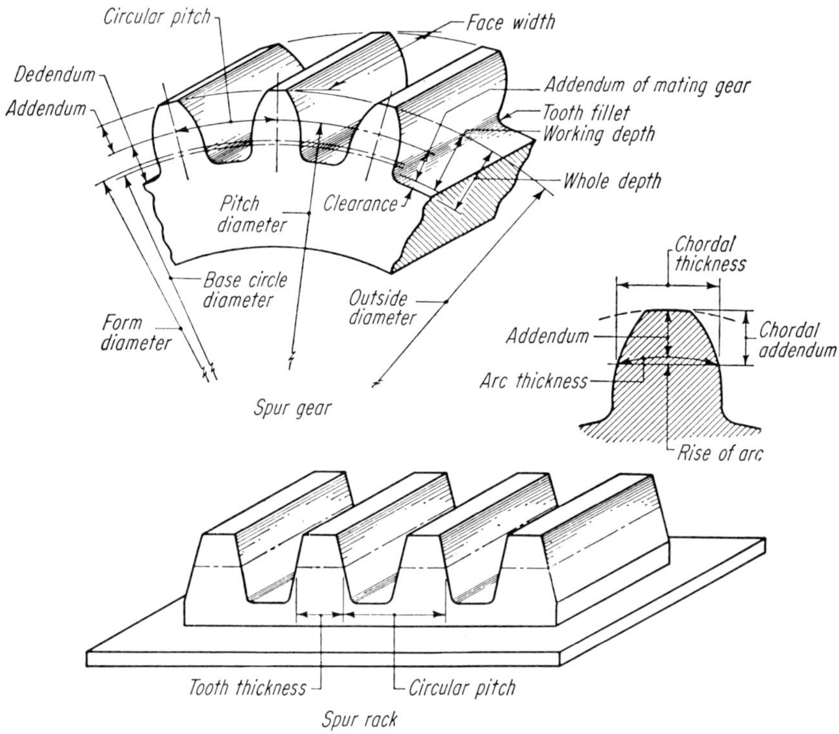

FIG. 2-1. Spur-gear and rack terminology.

For smooth operation, one end of a tooth should be advanced a distance greater than the circular pitch over the other end. A recommended overlap is 2 but 1.1 is a reasonable minimum. With face overlap, engagement takes place progressively with two or more teeth always in contact. As a result, helical gears operate much more smoothly and quietly than spur gears.

The involute is the tooth-profile form most commonly used for helical-gear teeth. Small changes in center distance do not affect tooth action. The helix angles commonly used range from 15 to 35°.

2-4. External Herringbone or Double-helical Gear. Herringbone gears are double-helical gears. They are a combination of right-hand and left-hand helices. Axial thrust load on the support bearings, which is a disadvantage of single-helical gears, is eliminated by equal and opposite thrust reactions in the gear itself. One member of a herringbone gearset should be capable of axial float in order to prevent excessive tooth loads due to mismatch of the two halves of the gear. A double-helix

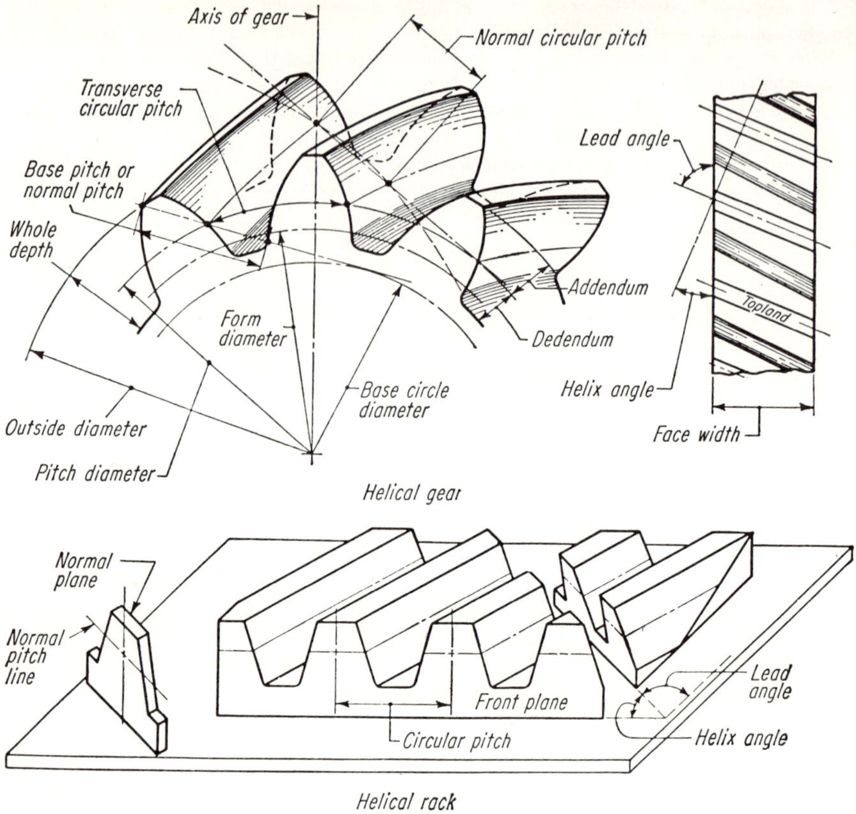

FIG. 2-2. Helical-gear and rack terminology.

gear suffers only one-half the misalignment error of a single helix or spur of the same face mounted with the same error.

All discussion of single-helical gears applies to herringbone gears, except that the helix angle is usually larger for double helicals, since there is no axial thrust.

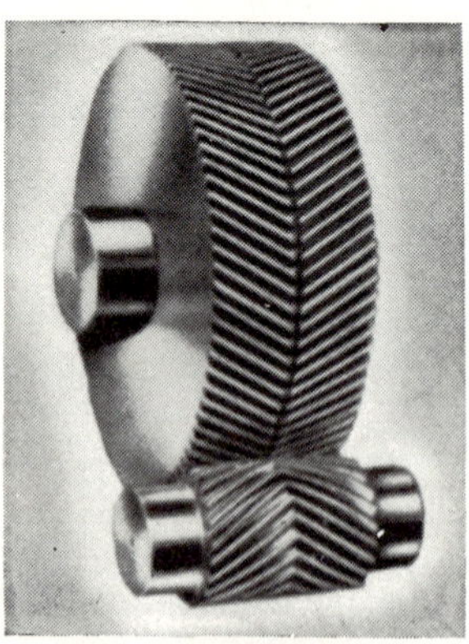

FIG. 2-3. Double-helical or "herringbone" gear.

2-5. Internal Gears. Internal gears are the opposite of external gears in that the teeth point toward the center rather than pointing outward. For the same number of teeth between pinion and gear, the internal gearset will have a greater length of action than an external gearset. The relative sliding of the teeth at the start and end of tooth action is less in the case of the internal gear. Tooth strength of an internal gear is greater than that of an equivalent external gear. The internal gear has several other advantages. It operates at a closer center distance with its mating pinion than external gears of the same size. This often permits a more compact design. The internal gear eliminates the use of an idler gear where it is necessary to have two parallel shafts rotate in the same direction. The internal gear forms its own guard over the meshing gear teeth. This is highly desirable for some types of machinery to prevent accidents.

Internal gears cannot be used where the number of teeth in the pinion is nearly equal to those in the gear. For full-depth 20°-pressure-angle teeth, the internal gear must be at least one and one-half the external gear; gears with smaller ratios require considerable modification of the tooth shape to prevent interference as the teeth enter and leave the mating gear. Internal gears also have the disadvantage that fewer machine tools can produce them.

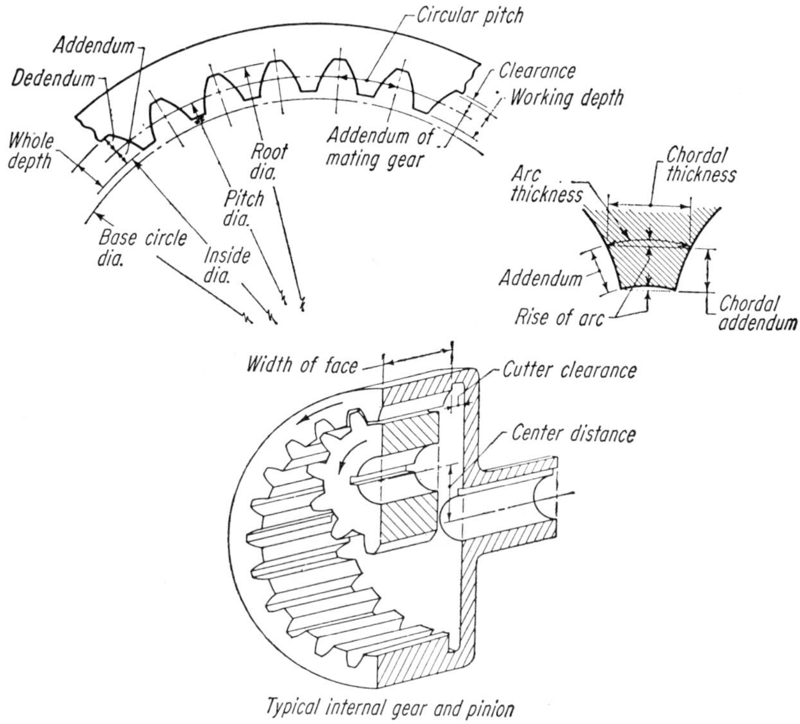

FIG. 2-4. Internal-gear terminology.

An internal helical gear has the same hand and angle of helix as its mating pinion. Figure 2-4 shows the terminology of an internal spur gear.

INTERSECTING AXIS GEARS

2-6. Straight Bevel Gears. Bevel gears are the most common gear used for the transmittal of power between intersecting shafts. The simplest form of the bevel gear is the straight bevel gear. Extensions of the straight teeth intersect at the axis of the gear. The tooth profile in a section normal to the tooth closely approximates that of an involute spur gear having a number of teeth equal to the actual number of teeth in the bevel gear divided by the cosine of the pitch angle (equivalent number of teeth). The teeth are tapered in both thickness and height. The outer part of the tooth or heel is larger than the inner part called the toe. The diametral pitch of the bevel gear is conventionally measured at the heel of the tooth.

Straight bevel gears which have a localized tooth bearing are called Coniflex®* gears. They have the advantage of permitting larger tolerances in the mounting of the gears in assembly without concentrating the load on the end of the teeth.

Straight bevel gears impose thrust and radial loads on their support bearings. The most common pressure angles used for straight bevel gears are 14½ and 20°. Bevel-gear terminology is given in Fig. 2-5.

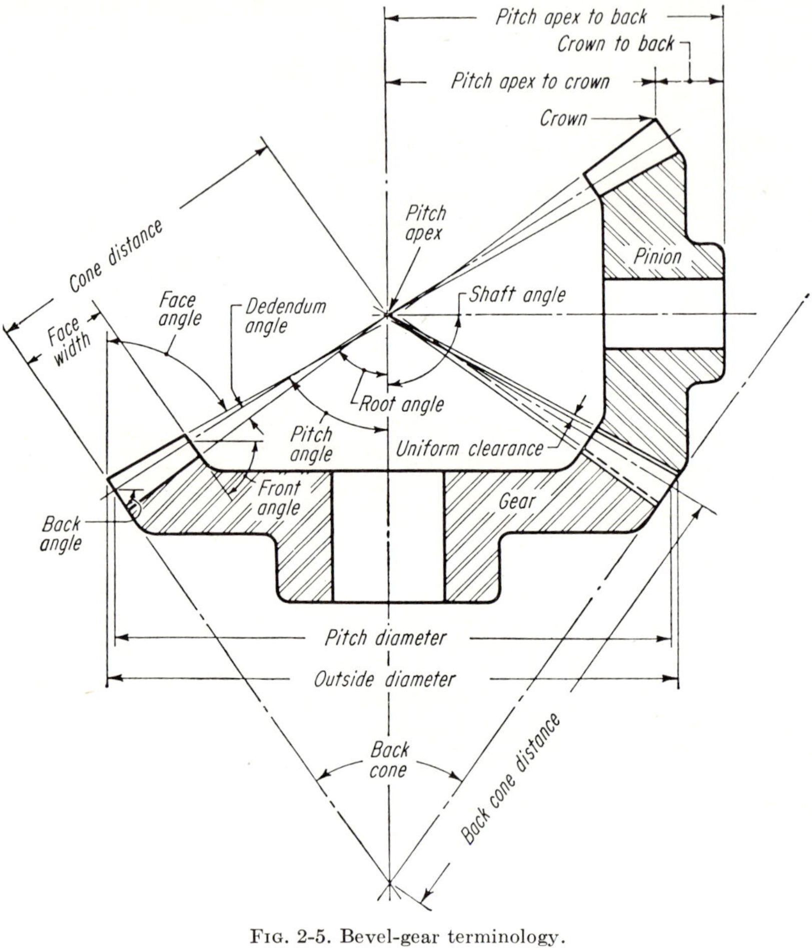

Fig. 2-5. Bevel-gear terminology.

2-7. Zerol®† Bevel Gears. Zerol gears are a special form of spiral bevel gears with curved teeth and having a zero degree mean spiral angle. They can be used in the same type of drives as straight bevel gears. Zerol gears impose the same loads on their support bearings as straight bevels and may be used in the same mountings.

Zerol gears are manufactured with the same type of cutters and on the same machines as spiral bevel gears. This is economically important as it eliminates the need for more than one type of bevel-gear-cutting equipment.

* Coniflex is a registered trade-mark of the Gleason Works, Rochester, N.Y.
† Zerol is a registered trade-mark of the Gleason Works, Rochester, N.Y.

Localized tooth bearing of Zerol gears is accomplished by length mismatch (face-curvature mismatch) and profile modification. The most common pressure angle used for Zerol bevel gears is 20°.

FIG. 2-6. Comparison of Zerol, spiral, and straight bevel gear.

2-8. Spiral Bevel Gears. The relationship between spiral and straight bevel gears is comparable with that between helical and spur gears.

The teeth of spiral bevel gears are curved and oblique, and as a result they have a considerable amount of overlap. This ensures more than one tooth in contact at all time and results in gradual engagement and continuous pitch-line contact.

Spiral bevels because of their continuous contact have larger load-carrying capacity and run more smoothly and quietly than straight bevel gears of the same size.

Spiral bevel gears impose much more thrust load on their support bearings than Zerol bevel gears. A rolling-element thrust bearing is usually required with spiral bevels, although plain bearings are quite often used with straight or Zerol bevel gears. The more common pressure angles for spiral bevel gears are 16 and 20°. The most common spiral angle in use is 35°.

2-9. Face Gears. Face gears, as the name implies, have teeth cut on the end face of a gear. The mating pinion is either a spur or a helical gear. When the axis of the pinion and the gear intersect, the face gear is commonly termed "on center." If the pinion is offset, the face gear is called "off center." The pinion and face gear are usually mounted with a 90° shaft angle. Functionally, this type of gear is akin to bevel gears.

The inclination of the flanks of the face-gear teeth vary, being less at the inner end and greater at the outer end. The sides of the teeth are substantially straight. The shapes of spur face-gear teeth are shown in Fig. 2-7.

The active face width of an "on-center" face gear is limited by trimming of the flank of the tooth at the inner end and by the tooth top land becoming pointed at the outer end.

The pressure angles and pitches used for face gears are similar to spur- or helical-gear practice.

2-10. Beveloid®* Gearing. The Beveloid gear is a completely generalized form of involute gear. It is an involute gear with tapered tooth thickness, root, and outside diameter. The tooth form of the Beveloid gear is shown in Fig. 2-8. Beveloid gears are useful primarily for precision-instrument drives where the combination of high precision and limited load-carrying ability fits the application. The following advantages may be realized with Beveloid gears:

* Registered trade-mark of Vinco Corp., Detroit, Mich.

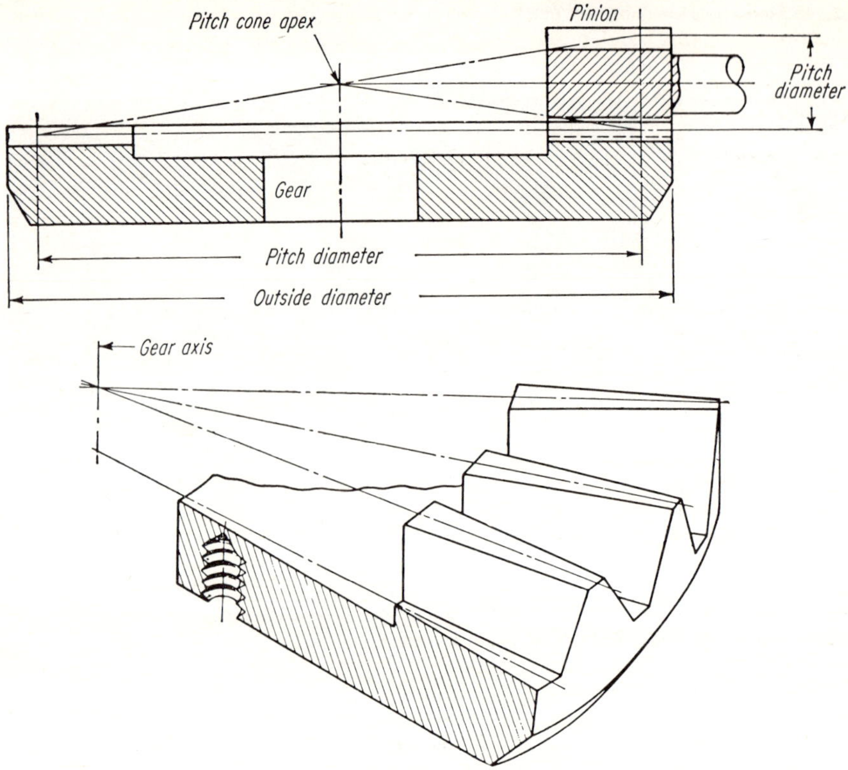

FIG. 2-7. Face-gear terminology.

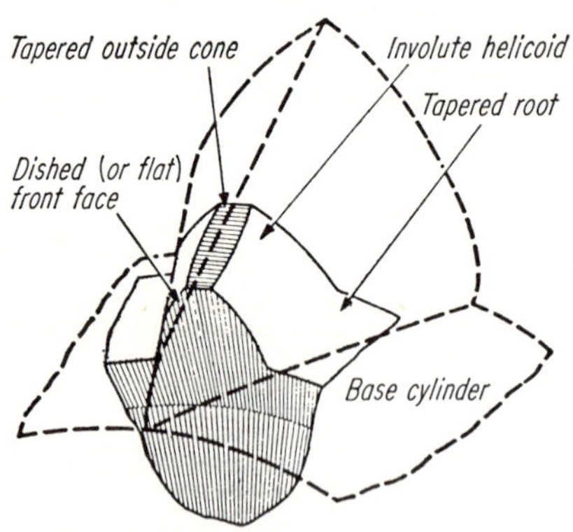

FIG. 2-8. Actual Beveloid tooth (back cone or face invisible).

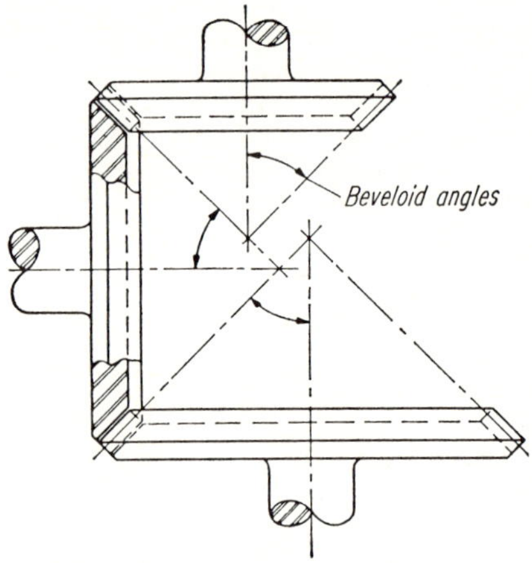

FIG. 2-9. Beveloid combination with intersecting axes. Design not limited by common vertex.

1. It lends itself to high accuracy in manufacturing.
2. The Beveloid operates with all gears derived from a rack, such as spurs, helicals, and worms.
3. Mating Beveloids need not have a common vertex. Refer to Fig. 2-9.
4. Beveloid gears are insensitive to mounting errors.

The following limitations apply to Beveloid gears:

1. Intersecting-axis or crossed-axis Beveloid gears have point contact, and their load-carrying capacity is therefore limited.
2. For small number of teeth and large cone angle, serious undercutting occurs.

NONINTERSECTING-NONPARALLEL-AXIS GEARS

In this family of gears relative sliding takes place along the profile, as in the case for parallel- and intersecting-axis gears, and across the face. The efficiency of these gears depends largely on the percentage of the sliding across the gear face. (Note the large spread in efficiency of this type of gear in Table 2-2.) The higher friction losses of these gears, coupled with the limited contact area under load, limits the use of these gears to the lower-powered applications (for reference, see Chap. 13).

2-11. Crossed-axis Helical Gears. Crossed-axis helical gears are the simplest form of gearing for nonintersecting shafts with conjugate action. They can be thought of as nonenveloping wormgears.

The action of crossed-helical gears consists primarily of screwing or wedging with a resultant high degree of sliding on the tooth flanks. Point contact between mating teeth limits the load-carrying capacity of this type of gearing.

Slight changes in shaft angle and center distance do not affect the conjugant action. Mounting is therefore greatly simplified. Crossed-helical gears can be manufactured by any machine producing helical gears. There is no difference between individual gears of helical and crossed-helical gearsets.

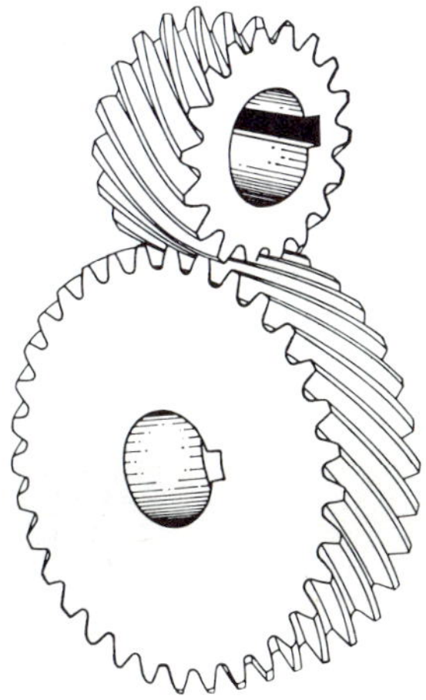

FIG. 2-10. Crossed-helical gear drive.

The mating gears in a helical gearset must have opposite hand of helix. In the crossed-helical gearset, the same or opposite hand of helix can be meshed.

2-12. Wormgears. Wormgears are used to transmit power between nonintersecting and nonparallel shafts. Wormgears are characterized by one member having a screw thread. A single-enveloping wormgear set has a cylindrical worm with a throated gear wrapped around the worm. The double-enveloping wormgear set has both members throated and wrapped around each other. The hand of helix is the same for both mating members.

The worm thread may have a variety of forms but the wormgear must be generated to produce conjugate action. Wormgears, especially double-enveloping wormgears, require very accurate mounting.

The static capacity of wormgears is high. Their running capacity is curtailed by the high generation of heat produced by sliding across the face of the teeth.

Wormgears are sometimes used where irreversibility in a mechanism is required. Irreversibility is possible when the lead angle is equal to or less than the static angle of friction. The lead angle for self-locking should be kept below 5°.

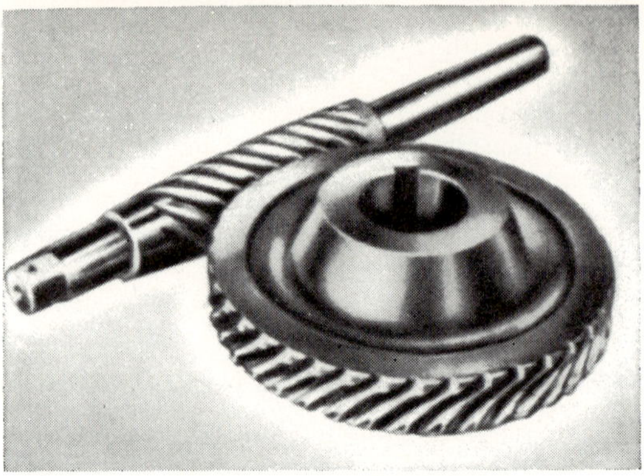

FIG. 2-11. Single-enveloping wormgear drive.

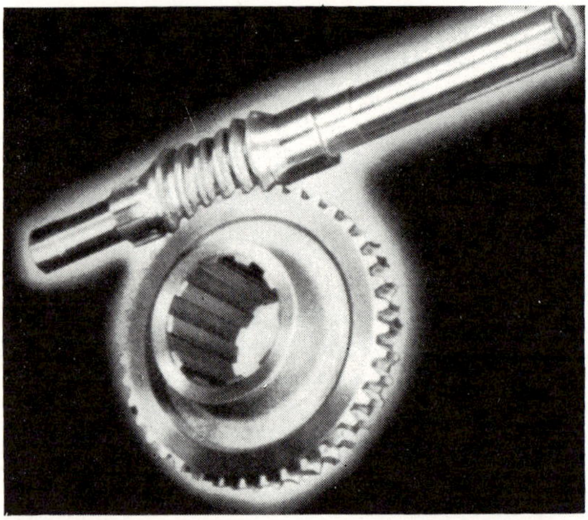

FIG. 2-12. Double-enveloping wormgear drive.

2-13. Hypoid Gears. In appearance hypoid gears resemble spiral bevel gears except that the pinion axis is offset above or below the gear axis. The pitch surfaces of hypoid gears are hyperboloids of revolution. The teeth in mesh have line contact. A hypoid gear has conjugate action. The normal pitch of both members of a hypoid gearset must be the same. Similar to crossed-axis helical gears, the number of teeth in a gear and pinion are not directly proportional to the pitch diameter. This makes it possible to make sturdy pinions without unduly enlarging the driven gear.

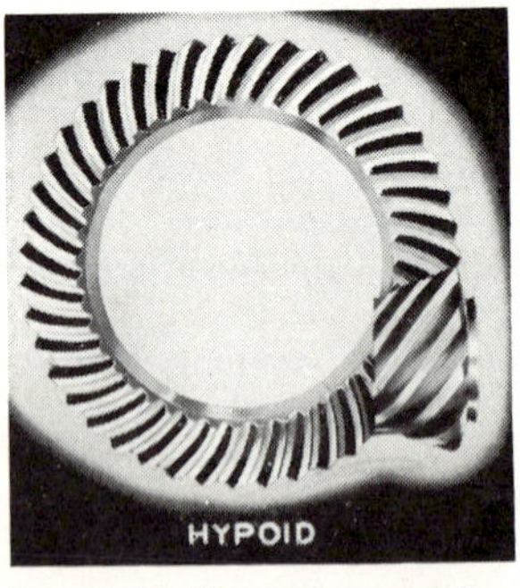

FIG. 2-13. Hypoid-gear drive.

Sufficient offset of the hypoid permits straddle mounting of the pinion and the gear. In operation, hypoid gears are smoother and quieter than spiral bevel gears. However, as in all cases of nonintersecting gearsets, sliding takes place across the face of the teeth, with a resultant loss of efficiency. (See Chap. 14 for typical efficiency curves of all kinds of gears.)

2-14. Planoid Gears®.* Planoid gears are used to connect nonintersecting and nonparallel shafting. Their application is similar to hypoid gearing. Pinion offset is limited to approximately one-sixth to one-third of the gear diameter. Planoids are usually used for ratios in the 1.5:1 to 10:1 range.

The most valuable aspect of this gear type is that, for large-volume applications, production costs may be low. The pinion is generated on a hobber. The gear is produced in a milling machine.

FIG. 2-14. Planoid-gear drive.

A Planoid gearset has line contact. The tooth flanks of the gear are plane surfaces (Fig. 2-14).

2-15. Spiroid Gears®.* Spiroid gears are used to connect nonintersecting and non-parallel shafts.

The performance of spiroid gears is closely related to the wormgear. In Spiroid gearing the worm is conical in shape and the mating member is a face-type gear. Spiroid gears have axis offsets intermediate to those of the hypoid and conventional wormgears (see Fig. 2-15). As with wormgears, a high degree of sliding takes place across the gear face.

The following considerations make Spiroids a highly desirable addition to the family of gears:

1. A large number of teeth are in simultaneous contact (see Fig. 2-15). This leads to compact design even at high ratios.
2. Very high ratios can be accomplished in a single step.
3. Gears can be manufactured by standard hobbing machines.

2-16. Helicon®* Gears. A Helicon gear is a Spiroid gear without taper. The lack of taper permits this gear type to be used at ratios less than 10:1. It can also be used over the range of ratios normally handled by Spiroids.

* Registered trade-marks of Illinois Tool Works, Chicago, Ill.

In the range of ratios where either a Spiroid or a Helicon might be used, the Spiroid has higher load-carrying capacity for a given size. The Helicon pinion is somewhat simpler to make since its threads are not cut on a tapered blank.

FIG. 2-15. Spiroid-gear drive.

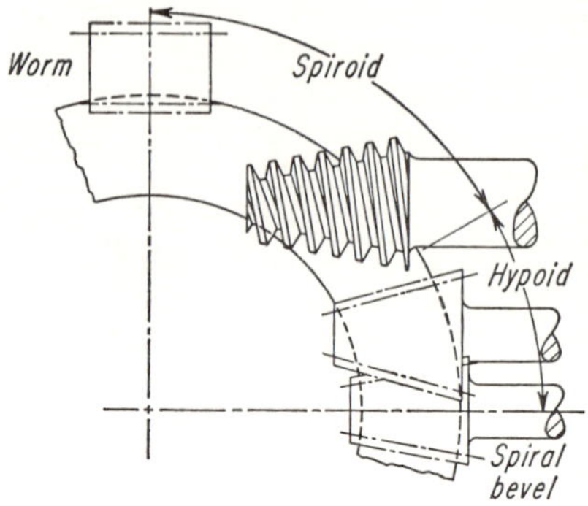

FIG. 2-16. Comparison of worm-, Spiroid-, hypoid-, and bevel-gear drive.

FIG. 2-17. Helicon-gear drive.

In general terms, the geometry of the Helicon is similar to that of helical offset face gears. Detail design and manufacturing practices, though, are somewhat different for these two types.

REFERENCE

1. Bryant, Richard C., and Darle W. Dudley: Which Right-angle Gear System?, *Product Engineering*, Nov. 7, 1960.

Chapter 3

GEAR ARRANGEMENTS*

By

DARLE W. DUDLEY, *Manager, Advance Gear Engineering, General Electric Company, Lynn, Mass.*

Gears may be arranged in a great variety of ways. In the more complex pieces of machinery, the problem of arranging and fitting the gear drives into available space is one of the major gear-design problems.

This chapter makes an over-all survey of the kinds of gear arrangements in common use. The clever designer will appreciate that there are an infinite number of arrangement possibilities for the more complex machinery drives. The arrangements shown in this chapter should be considered as examples of gear arrangements rather than as a complete listing of the possibilities of the field.

Chapter 3 can be best understood if the reader is already familar with the *theory* of gearing given in Chap. 1 and the *kinds* of gear teeth given in Chap. 2.

3-1. Possibilities in Gear Arrangements. In general terms a *gear drive* implies one toothed member that engages another toothed member and transmits motion to the second toothed member. It is usually intended† that the second toothed member should rotate at a uniform angular velocity when the first member is driven at a smooth angular velocity.

The second toothed member may be externally toothed or internally toothed, or it may be a section of a *rack* (a rack is a gear of infinite pitch diameter). A spur or helical pinion can be used to mesh with any of these three gear types. Many of the other gear types cannot mesh with the gear member if it is anything other than externally toothed. In some cases it is the geometry of the gear part that is limiting. In most cases, though, the gear part might be made in theory, but the machinery normally available in the gear trade does not have the capability of making the part.

Some types of gear teeth are *interchangeable* and some are not (see Art. 1-4). In an interchangeable-gear-tooth system, toothed gear wheels *a*, *b*, *c*, and *d* might be made. These might be meshed in a chain with *a* meshing with *b*, *b* meshing with *c*, and *c* meshing with *d*. If these gears were interchangeable, they could be taken apart

* A considerable amount of material for this chapter was furnished by S. L. Crawshaw, Vice President of Philadelphia Gear Corp., and Harold Kron, Chief Engineer, Philadelphia Gear Corp.

† In gearing for watches a nonuniformity in motion is often tolerated to cut down the friction when a gear drives a pinion of small tooth numbers. In computing machines elliptical or other noncircular gears are sometimes used to create a motion corresponding to some mathematical formula.

and assembled and run properly with *a* meshing with *c* or *b* meshing with *d*. Generally speaking *involute* profile gears have the property of interchangeability and noninvolute gears do not. Spur and helical gears are normally made with involute teeth, and when so made they have interchangeability. An exception to this is the single-enveloping wormgear where interchangeability does not exist even when the worm member is made to an involute profile.

In some gear types it is geometrically rather impractical to make the pinion member with as few teeth as one. A spur pinion with only one tooth would not be capable of continuous gear action. On the other hand, a worm with one thread is very practical and is often used. Some gear types can be made with the pinion member having as few as five teeth. Most gear types can have pinions with over 16 teeth. A single-enveloping worm, however, is somewhat unhandy to manufacture at 16 threads, and there is seldom any good design reason to use this many threads on a worm.

Table 3-1 shows in tabular form the meshing possibilities of all the major types of gear teeth. This table is based on normal trade practice and should not be considered as absolutely firm. For instance, an *internal* bevel gear is theoretically possible and has been used in a few special cases. However, normal bevel-gear-manufacturing machines will not make such a gear.

Table 3-1. Gear-meshing Possibilities

Type of gear teeth	Pinion and gear	Pinion and rack	Pinion and internal gear	Interchange-ability *a-b-c-d* or *a-c, b-d*	One-tooth pinion	Pinion of 5 teeth	Pinion of 16 or more teeth
Spur	Yes	Yes	Yes	Yes	No	No*	Yes
Helical	Yes	Yes	Yes	No	No*	No*	Yes
Straight bevel	Yes	No*	No	No*	No	No*	Yes
Zerol®† bevel	Yes	No	No	No	No	No*	Yes
Spiral bevel	Yes	No	No	No	No*	No*	Yes
Hypoid	Yes	No	No	No	Yes	Yes	Yes
Face gear	Yes	No	No	No	No	No*	Yes
Crossed-helical	Yes	Yes	No	Yes	Yes	Yes	Yes
Single-enveloping worm	Yes	No*	No*	No	Yes	Yes	No*
Double-enveloping worm	Yes	No	No	No	Yes	Yes	No*
Beveloid®†	Yes	Yes	No	Yes	No	No*	Yes
Spiroid®†	Yes	No	No	No	Yes	Yes	No*
Planoid®†	Yes	No	No	No	No*	Yes	Yes
Helicon®†	Yes	No	No	No	Yes	Yes	Yes

* Items with asterisk indicate that it is mechanically possible to have this design but that the design is not normally used in the gear trade.

† Trade-marks: Zerol, registered trade-mark of Gleason Works, Rochester, N.Y.; Beveloid, registered trade-mark of Vinco Corp., Detroit, Mich.; Spiroid, Planoid, Helicon, registered trade-marks of Illinois Tool Works, Chicago, Ill.

A worm and an internal gear might be construed as simply a nut and bolt. Nuts and bolts are, of course, very practical. However, they are not normally considered as *gear* parts.

3-2. How the Arrangement May Drastically Affect Ratio, Power, Efficiency, and Gearbox Volume. The arrangement of the gear drive can play a very important part on how easy or hard it is to achieve things like high ratio, high power capacity, high efficiency, or a very compact unit.

In ordinary spur- or helical-gear practice, it is reasonable to handle ratios from 1:1 to about 8:1 in a single reduction. Ratios of even 10:1 are quite good. Suppose a ratio of 120:1 was needed. This would require a triple-reduction unit of spur or helical gears. It would also take a triple-reduction bevel-gear unit. A triple-reduction unit has three sets of pinions and gears, or a total of six toothed parts. With a wormgear set it is possible to handle this much ratio in just a single reduction (only two toothed parts).

For each toothed part in a gearset, a shaft and generally two bearings are required. Where large ratios are needed, the choice of gear arrangement is often determined by finding the arrangement with the fewest number of parts that will do the job adequately.

Table 3-2 gives some general information about how different ratios can be achieved and the number of gear parts required.

Table 3-2. How to Obtain Ratios

Kind of arrangement	Min. No. of toothed parts	Ratio range		
		5:1	50:1	100:1
Single reduction:				
Spur	2	Yes	No	No
Helical	2	Yes	No	No
Bevel	2	Yes	No	No
Hypoid	2	Yes	Yes	Yes
Face	2	Yes	No	No
Worm	2	Yes	Yes	Yes
Spiroid	2	No	Yes	Yes
Planoid	2	Yes	No	No
Simple planetary	3	Yes	No	No
Fixed differential	5	No	Yes	Yes
Planocentric	2	No	Yes	Yes
Harmonic Drive	2	No	Yes	Yes

The power-transmitting capacity of different gear arrangements is quite variable. For instance, there has never been a wormgear set that would handle 5,000 hp as a continuous duty rating. On the other hand, many helical gearsets are in regular service carrying more than 10,000 hp. The upper limit of power for gear arrangement is about as hard to define precisely as it is to define the altitude at which the earth's atmosphere ceases. The upper limit of a gear's capacity is determined by such things as

1. How large a gear can be made by available gear-manufacturing equipment?
2. What is the upper limit on tooth-surface loading and pitch-line velocity for the best material available?
3. Are the job requirements of ratio, input speed, life required, and lubrication favorable for a maximum horsepower?

Table 3-3 shows some nominal maximum horsepower capacities for different gear arrangements. These should be used only as a general guide. In all cases there are a few gearsets in use (or that could be put into use) at considerably higher horsepower ratings. Conversely, the job requirements of many applications make it completely impractical to achieve the ratings shown for the specific application. No horsepower limits are shown for the fixed differential, the planocentric, or the Harmonic Drive. These kinds of units have been used mostly at low horsepowers (less than 50 hp), but they can probably be extended up into fairly high horsepower.

Table 3-3. General Survey of Power and Efficiency

Kind of arrangement	Nominal max. hp	Typical efficiency, %		
		5:1 ratio	50:1 ratio	100:1 ratio
Single reduction:				
Spur	3,000	98		
Helical	30,000	98		
Straight bevel	500	98		
Zerol bevel	1,000	98		
Spiral bevel	5,000	98		
Hypoid	1,000	95	80	60
Crossed-helical	100	95	80	60
Cylindrical worm	750	95	80	60
Double-enveloping worm	1,000	95	80	60
Spiroid	500	95	80	60
Planoid	1,000	95		
Helicon	100	95	80	60
Double reduction:				
Spur	3,000	97	96	94
Helical	30,000	97	96	94
Spiral bevel	5,000	97	96	
Simple planetary	10,000	97		
Fixed differential	......	...	80	60
Planocentric	......	...	90	85
Harmonic Drive	......	...	90	85

For specific data on load rating for a given gear design, Chap. 13 should be consulted.

The efficiency that can be achieved in a gearset is quite variable. The kind of oil, the amount of pitch-line velocity, and the kind of bearings enter quite directly into the efficiency. In a spur gearset, the best designs made under favorable conditions will achieve as much as 99 per cent efficiency. Some spur sets under poor conditions may not even achieve 90 per cent efficiency.

Table 3-3 shows the nominal order-of-magnitude efficiencies for different gear arrangements. These represent general trade experience with good designs but are not the best possible efficiency that can be achieved with an optimum design and a favorable application.

Table 3-3 shows that the efficiency drops off rather rapidly as the ratio increases. Generally this is caused by the fact that the sliding velocity increases rather rapidly as ratio increases. A worm, for instance, at high ratio has a very large amount of sliding.

The fixed-differential gear is capable of high ratio, but it too has a drop-off in efficiency. This is caused by the fact that a large amount of tooth meshing occurs at high torque (see Art. 3-11).

Suppose a 100:1 ratio is needed. So far as arrangement goes the double-reduction approach using either spur, helical, or bevel gears is the easiest way to achieve high efficiency. With a favorable application and a good design an efficiency of 96 per cent can be achieved with helical gears. It is quite impossible to obtain this kind of efficiency with any of the single-reduction gears that can reach a 100:1 ratio.

Chapter 14 covers the general subject of gear efficiency and shows curves for many gear types. This chapter should be consulted for detail information on gear efficiency.

The volume or weight of a gearbox is also considerably affected by the arrangement. The weight of a gearbox to do a given job may vary as much as 50:1! For instance, a single-reduction helical-gear unit made of low-hardness material in industrial practice might weigh 10,000 lb. An aircraft-style planetary made of fully hardened gears in aircraft practice might handle the same horsepower, for the same design life, and with the same degree of reliability as the industrial gear and weigh only 200 lb.

In very general terms, a 50:1 spread in gear weight can be broken down as follows:

Weight ratio for low tooth loads on low-hardness material vs. high loads on high-hardness material	7:1
Weight ratio for heavy construction using thick walls and solid shafts vs. thin walls and hollow shafts	3:1
Weight ratio of best arrangement to save weight vs. normal arrangement	2½:1

The opportunity to save weight in gear designs depends somewhat on the ratio required. For instance, a worm gearset is not particularly lightweight at a low ratio of 5:1. At a high ratio of 50:1 the wormgear becomes rather light because it can still do the job with just two toothed parts, whereas a spur or helical set would take at least twice as many parts to do the job.

Table 3-4. Gearbox Relative Size and Weight

Kind of arrangement	Ratio range			
	5:1	20:1	50:1	100:1
Single reduction:				
Spur, helical, bevel	Small			
Worm	....	Small	Small	Small
Hypoid	Small	Small	Small	Small
Spiroid	....	Small	Small	Small
Planoid	Small			
Double reduction:				
Single power path, helical gears	....	Med. size		
Multiple power path, helical gears	....	Small	Very small	
Epicyclic gears:				
Simple planetary	Very small			
Compound planetary	....	Very small		
Double-reduction planetary	....	Very small	Very small	
Fixed differential	....	Small	Very small	Very small

Table 3-4 shows a general picture of what arrangements might be small for different gear ratios. In using this table, it should be kept in mind that intensity of tooth loading is often more important than arrangement in reducing weight. For instance, a casehardened gearset used in a nominal arrangement might give a lighter unit than a set of low-hardness cut gears used in the most weight-saving arrangement.

In using Table 3-4 it should also be kept in mind that the amount of horsepower must be favorable for the design. Some designs cannot handle high horsepower. These designs are small in weight only in comparison with other designs when the horsepower required is well within their capability.

Generally speaking, the volume of a gearbox is fairly proportional to the weight—provided that the type of construction is the same. Aircraft-type gearboxes with magnesium casings will have more volume for their weight than industrial boxes with cast-iron casings.

3-3. Parallel-axis Arrangements, Simple Mesh. Perhaps the most common gear arrangement is that of a pinion and gear meshing on parallel axes. The teeth for this kind of arrangement may be either spur, helical, or herringbone.*

A single pinion and gear make a *single*-reduction unit, or an S unit in popular terminology. If two such units are in series, it is a *double*-reduction unit or a D unit. A *double*-reduction unit may have all the gears mounted in a common casing, or the unit may be made up of two somewhat separate casings which are attached to each other. In a like manner it is possible to have a *triple*-reduction unit or a *quadruple*-reduction unit.

The ratio for such units is figured quite simply

$$m_G = m_{G_1} m_{G_2} m_{G_3} \text{ (etc.)} \tag{3-1}$$

where

$$m_{G_1} = \frac{N_{G_1}}{N_{P_1}}$$

N_{G_1} = number of gear teeth, first reduction

N_{P_1} = number of pinion teeth, first reduction

Quite a variety of arrangements are possible with parallel axes and simple meshing of external teeth. Besides multiple reductions, there may be multiple power inputs or multiple power outputs. There may also be multiple power paths through the unit. The latter type is analogous to parallel circuits in electrical wiring.

An example of the multiple power path is the *double-reduction twin* unit, or DT gear. In this case a first-reduction pinion meshes with two first-reduction gears. These first-reduction gears in turn are connected to two second-reduction pinions. The second-reduction pinions drive a common second-reduction gear.

The reduction ratio of a multiple-power-path gear unit is calculated by following through any one power path and using Eq. (3-1).

Figure 3-1 shows schematic arrangements of many of the more popular arrangements used in turbine-drive applications. These same arrangements are used frequently in lower-speed motor drives; but the lower-speed units frequently have fewer bearings and flexible couplings than shown in Fig. 3-1. The number of toothed gear elements would be the same in either case.

Figure 3-2 shows a typical high-speed single-reduction unit. Units like this are often used with steam-turbine drives. They may be used with electric motors, diesel engines, or gasoline engines. Many sets are also used with gas turbines.

High-speed gears used for power in industrial plants, public-utility plants, or for drives on large ships have horsepower ratings going from less than 1,000 to over 50,000 hp.

Figure 3-3 shows a DT-type high-speed gear rated at 4,000 hp. Note the close connection of the turbine and the gear unit.

Figure 3-4 shows a typical MD-type marine gear unit with a 17,500-hp rating. In this case two turbines drive a single propeller.

Articulated Gear Unit. In a double-reduction design the first-reduction gear may be solidly mounted on the second-reduction pinion shaft with only two bearings supporting the assembly. Or the pinion and gear may be each mounted on a pair of bearings and the two connected by a flexible quill shaft. The latter arrangement is called an "articulated design." The articulated design is quite popular for the higher-speed higher-horsepower gear units. It has the advantage of providing flexibility between a first-reduction gear and a second reduction which will reduce dynamic† tooth loading, and the fact that both the gear and pinion have their own separate

* Chapter 2 defines the *kinds* of gear teeth. Chapter 4 defines and explains all kinds of gear nomenclature.

† See Chap. 14 for information on dynamic tooth loading.

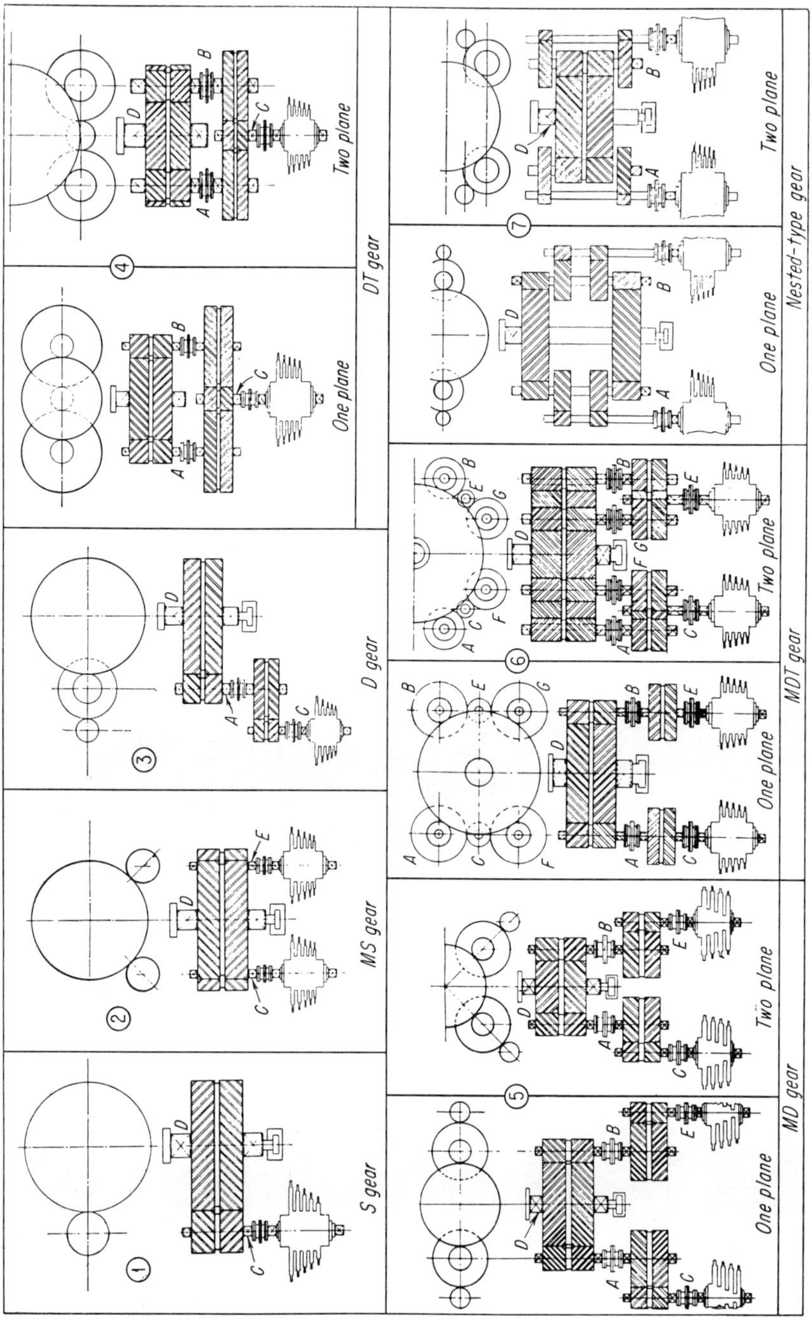

FIG. 3-1. Some schematic arrangements for parallel-axis gears.

FIG. 3-2. High-speed single-reduction unit with oil pump and force-feed system. (*Courtesy of General Electric Co., Lynn, Mass.*)

FIG. 3-3. Partly disassembled DT-type gear unit for ship propulsion rated at 4,000 hp. (*Courtesy of General Electric Co.*)

pair of bearings makes it possible to provide a better control on the contact across the face width at each mesh.

Figure 3-5 shows a schematic arrangement for an articulated MDT unit. Further details of an articulated design are shown in Fig. 3-6. Figure 3-7 shows an MDT unit of articulated design that was used as a main drive unit on a super aircraft carrier.

FIG. 3-4. Marine MD-type gear unit with covers off. Rating 17,500 hp. (*Courtesy of General Electric Co.*)

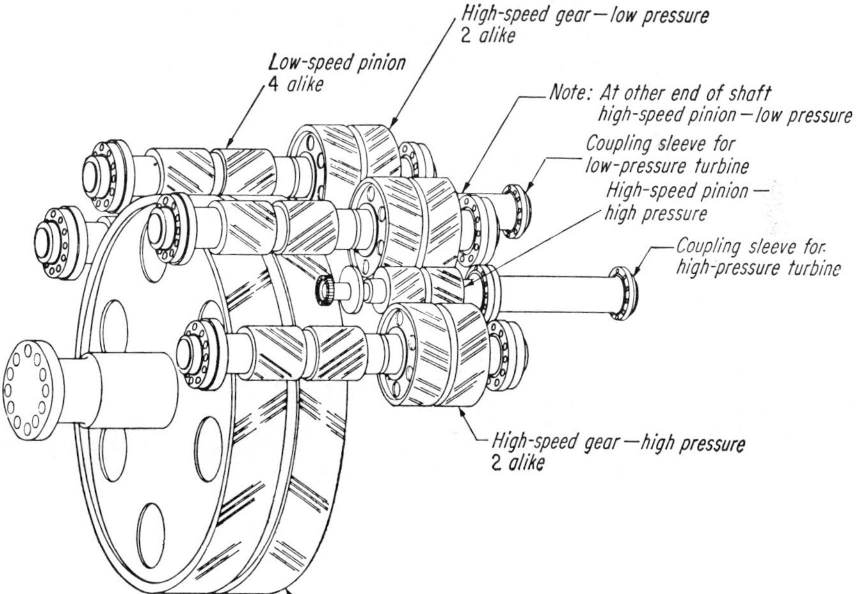

FIG. 3-5. Schematic arrangement of articulated MDT unit.

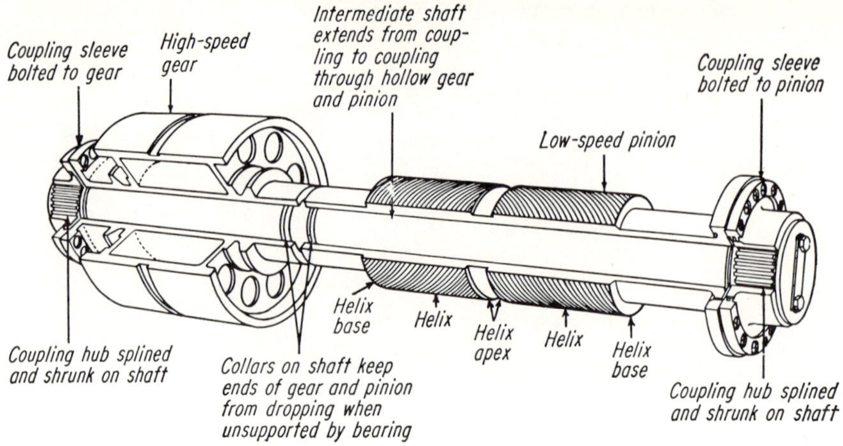

FIG. 3-6. Detail of articulated design of high-speed gear and low-speed pinion. Note intermediate shaft passing through hollow pinion and gear and the fact that both pinion and gear have their own separate bearings.

FIG. 3-7. Side view of MDT unit of type used on super aircraft carrier *Saratoga*. Note articulated design. Rating in excess of 50,000 hp. (*Courtesy of U.S. Navy and General Electric Co.*)

Nested Design. It is possible to make a unit somewhat more compact and use fewer bearings by having one reduction straddle another reduction. This style of gear is called the "nested design." Figure 3-1 in part 7 shows the schematics of a nested design. Figure 3-8 shows a typical double-reduction nested unit.

The nested unit has been widely used for low-speed industrial gearing, and it has been used for high-speed high-horsepower marine units.

FIG. 3-8. Double-reduction nested gear unit. (*Courtesy of Philadelphia Gear Corp., Philadelphia, Pa.*)

FIG. 3-9. Compact arrangement of a quadruple-reduction gear unit. (*Courtesy of Philadelphia Gear Corp.*)

Figure 3-9 shows a very compact quadruple-reduction unit. This style of unit is quite popular in general gear work. This design is not nested.

Parallel-axis Gears Combined with Right-angle Gears. In many cases it works out well to take a basic parallel-axis gear unit and add a first or last reduction of right-angle gearing. This will permit the drive to turn a right-angle corner, and it may have merit from the standpoint of making the drive more compact or using fewer bearings and other parts.

Figure 3-10 shows a triple-reduction unit with a bevel first reduction. Figure 3-11 shows a double-reduction unit with a bevel second reduction. Both these units demonstrate how a right-angle drive can be added to parallel-axis gearing and a compact drive achieved.

Figure 3-12 shows a double-reduction unit with a cylindrical-worm second reduction. A unit like this can handle over-all ratios of 50:1 to 200:1 with ease. The worm set operates at the slower speed where it has its best torque capacity for its

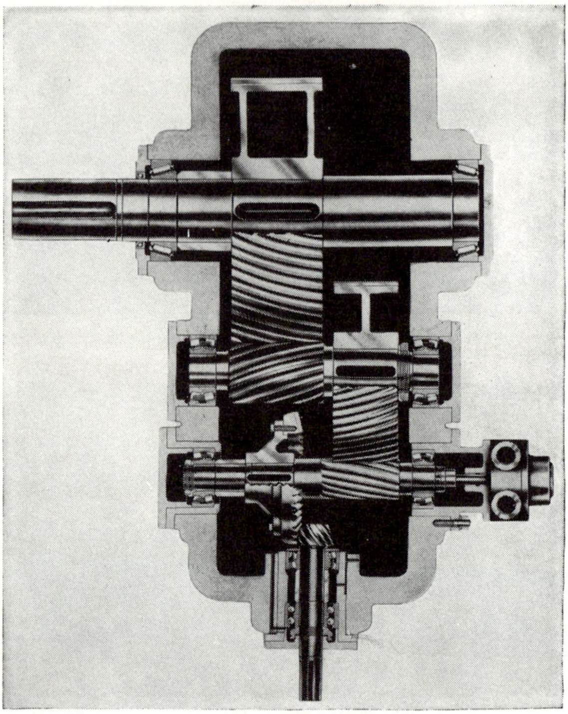

FIG. 3-10. Triple-reduction unit using a bevel first reduction. (*Courtesy of Philadelphia Gear Corp.*)

size. A unit of this type is quite popular in industrial work where the output speed is quite low and the horsepower handled is less than 500.

Design Problems with Parallel-axis Gears. There are several problems in parallel-axis gear design that are more or less peculiar to this arrangement. In picking an arrangement the designer should be alert to the design problems and satisfy himself that he can design the arrangement he has picked so as to obtain a practical design to manufacture and a reliable design in the field.

Some of the problems that may show up are how to

1. Obtain uniform tooth loading across the face width.
2. Compensate for excessive wind-up of long skinny pinions or long shafts.
3. Obtain equal torque through each power path in a multiple-power-path design.
4. Choose numbers of teeth, gear locations, and gear-to-shaft connections so that multiple-power-path gearing will assemble.

FIG. 3-11. Double-reduction unit with a bevel second reduction. (*Courtesy of Philadelphia Gear Corp.*)

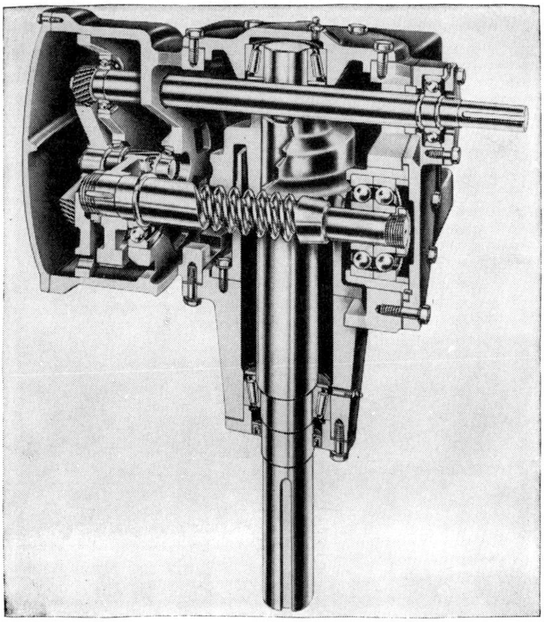

FIG. 3-12. Double-reduction unit with a worm second reduction. Typical of high-ratio units used to drive conveyors. (*Courtesy of Philadelphia Gear Corp.*)

The length-to-diameter ratio of helical and spur gears can be fairly large. In most other gear types the pinion will not have face width more than about one-third the pinion pitch diameter. In spur gearing the L/D ratio may be as great as 1 while in double-helical gearing it may approach 2. Needless to say, high accuracy is needed in boring the casings and machining the teeth when a relatively large face width is used. Low-hardness gearing of about 200 BHN will exhibit some tendency to wear in and improve contact. As much as 0.001″ may be worn off the high ends of the teeth to equalize the contact without too serious a result. Medium- to high-hardness gears have but slight tendency to wear in. If the contact across the face width is not good at full load, the teeth may break before any equalizing wear occurs. (See Chap. 13 for data on how to handle misalignment when load-rating gears.)

Large pinions with an L/D ratio of 1 or more will twist elastically under load an appreciable amount. It is necessary to provide helix modification to compensate for this on large gears where the material is fairly hard and the horsepower is high. (Helix modification was necessary on the gears shown in Figs. 3-4 and 3-7.)

Nested pinions tend to suffer from bowing and twisting of the long shaft between the two pinion halves. Helix compensation is often needed.

In multiple-power-path units, each power path must be able to have the teeth engage simultaneously when a slight torque is applied. In addition errors in tooth spacing and runout must be slight enough in the gears so that the flexibility of the connecting shafts can keep a uniform load on each branch. If the connections are too stiff or the teeth not precise enough, the load will shift rapidly from one power path to another as the gear revolves.

In multiple-power-path gearing, the gears will not assemble if there is a random orientation between the teeth of a gear and a pinion connected by the same shaft. Either the teeth and coupling means must be cut so that the orientation is exactly alike on each like assembly or adjustment must be provided so the proper tooth "timing" is achieved at assembly. Generally the teeth are cut at a random orientation. Then at assembly the parts are rotated until the pinion and gear by chance have their teeth in almost the right orientation. Then the final adjustment is made by some mechanical means like axial shimming (single-helix design), elongating coupling boltholes, redoweling, or using a stepped key. Incidentally, a shrewd choice on tooth numbers can make the timing problem work out much more easily.

3-4. Parallel-axis Epicyclic Gearing. A large family of gear arrangements goes under the general name of "epicyclic" gearing. Generally speaking the epicyclic train has a central "sun" gear, several "planets" meshing with the sun and spaced uniformly around the sun and an "annulus" or ring gear meshing with the planets. The sun and planets are externally toothed while the ring is internally toothed.

The name epicyclic comes from the fact that points on the planets trace out epicycloidal curves in space.

The name "planetary" gear is sometimes used interchangeably with epicyclic to denote the whole family of gears. In a more strict sense the *planetary* is one of the types of epicyclic gears.

There are a large number of possible epicyclic arrangements. These may be divided into three general groups:

1. Single epicyclic trains
2. Compound epicyclic trains
3. Coupled epicyclic trains

Single Epicyclic Trains. The simple epicyclic train consists only of the elements of sun, planet, and ring gear. This arrangement is often called a "simple" epicyclic gear or a "simple" planetary gear to distinguish it from the more complicated arrangements.

Figure 3-13 shows schematic arrangements of the three simple epicyclic gears. These are the *planetary*, the *star*, and the solar *arrangements*. The figure shows only three planets. The number of planets will vary quite considerably in actual practice. With some ratios it has been possible to squeeze in as many as twenty planets. As few as one is sometimes used for light load applications.

The derivation of ratios for planetary units is somewhat complicated. Several references[1,2,3,*] go into the derivation of ratios in considerable detail. Table 3-5 shows the over-all ratio for each of the three single epicyclic trains. If one of these units were driven backward, the ratio would be less than one and would simply be the reciprocal of the ratio as a speed-reducing set.

A range of ratios is shown for each kind of epicyclic gear. Note that only the solar will handle very low ratios. The ratio range of 1.7:1 to 2:1 is a range that is rather difficult to handle with simple epicyclic gearing. The star will work in this range, but it turns out to be rather awkward.

Epicyclic gears will not assemble unless the tooth numbers are properly chosen. Assuming the planets are equally spaced around the sun—which is the usual case—the equations given at the bottom of Table 3-5 must be met. The sum of the ring and sun teeth must be equally divisible by the number of planets. If there were three planets, the ring teeth plus sun teeth would have to divide evenly by three.

The tooth load at each planet is balanced between the sun-planet mesh and the planet-ring mesh. This tangential driving load may be calculated by

$$W_t = \frac{126{,}050\ P_H}{n_s d_s \text{ No. planets}} \tag{3-2}$$

where n_s = rpm of sun
d_s = pitch diameter of sun
P_H = horsepower transmitted

It should be appreciated that Eq. (3-2) determines the theoretical load only when there is perfect sharing of load between the planets. There is quite a problem in load sharing in epicyclic gears plus the possibility of appreciable dynamic overload due to tooth errors. More will be said on this subject later in this article.

Table 3-5. Simple-epicyclic-gear Data

Kind of arrangement	Fixed member	Input member	Output member	Over-all ratio m_G	Range of ratios normally used
Planetary.	Ring	Sun	Cage	$\frac{N_R}{N_S} + 1$	3:1–12:1
Star.	Cage	Sun	Ring	$\frac{N_R}{N_S}$	2:1–11:1
Solar.	Sun	Ring	Cage	$\frac{N_S}{N_R} + 1$	1.2:1–1.7:1

N_S = number of sun teeth
N_P = number of planet teeth
N_R = number of ring (annulus) teeth

To assemble,

$$N_R = N_S + 2N_P \tag{3-3}$$

$$\frac{N_R + N_S}{\text{No. planets}} \text{ must equal an integer} \tag{3-4}$$

* Superscript numbers refer to references at the end of the chapter.

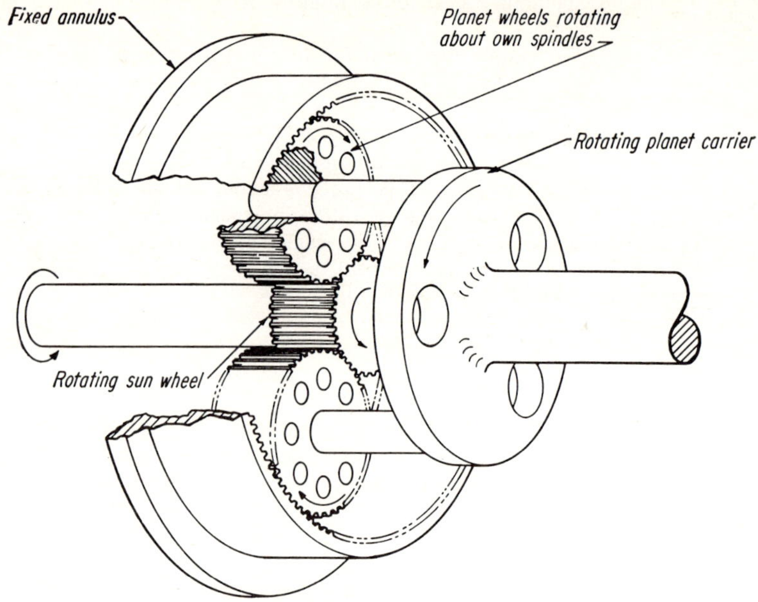

(a)

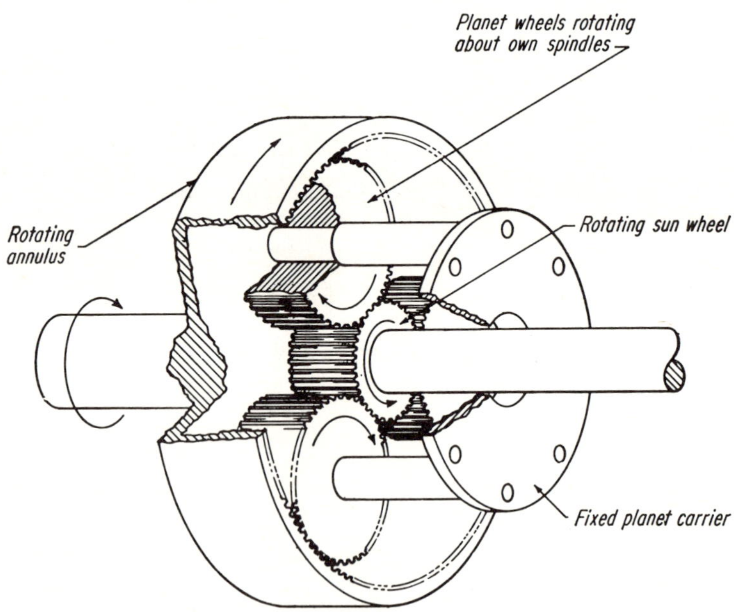

(b)

FIG. 3-13. Simple epicyclic gears. (a) Planetary

Compound Epicyclic Trains. The "compound" epicyclic train has two planet members attached together on a common shaft. Figure 3-15 shows the schematic arrangement of a compound star gear. Note the large planets (or stars) which mesh only with the sun and the small planets which mesh only with the ring gear.

Table 3-6. Compound-epicyclic-gear Data

Kind of arrangement	Fixed member	Input member	Output member	Over-all ratio m_G	Range of ratios normally used
Compound planetary....	Ring	Sun	Cage	$\frac{N_R N_{P_1}}{N_S N_{P_2}} + 1$	6:1–25:1
Compound star.........	Cage	Sun	Ring	$\frac{N_{P_1} N_R}{N_S N_{P_2}}$	5:1–24:1
Compound solar........	Sun	Ring	Cage	$\frac{N_S N_{P_2}}{N_R N_{P_1}} + 1$	1.05:1–2.2:1

N_S = number of sun teeth
N_{P_1} = number of first-reduction planet teeth
N_{P_2} = number of second-reduction planet teeth
N_R = number of ring teeth

To assemble,

$$D_R = D_S + D_{P_1} + D_{P_2} \tag{3-5}$$

where D_R = ring pitch diameter
D_P = planet pitch diameter

$$\frac{N_{P_1} N_R + N_{P_2} N_S}{\text{No. planets}} \text{ must equal an integer} \tag{3-6}$$

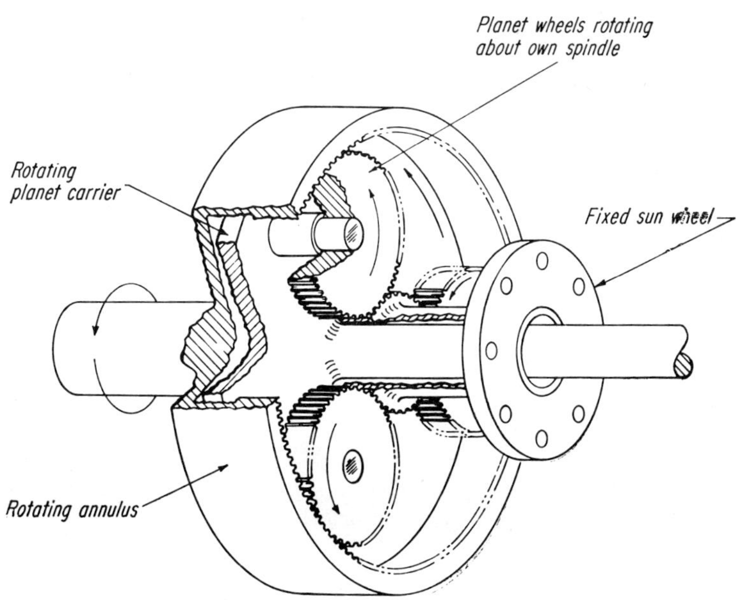

(c) Solar gear

(c)

gear. (b) Star gear. (c) Solar gear.

Table 3-6 shows the basic data for compound epicyclic gears. Note that the compound planetary and the compound star will handle much higher ratios than the simple epicyclic types. The compound solar is still limited to a narrow ratio range.

Table 3-6 shows the rules for assembly of the compound epicyclic. Equal spacing of planet shaft assemblies about the sun is assumed.

FIG. 3-14. Planetary gear as part of a geared motor. (*Courtesy of General Electric Co., Paterson, N. J.*)

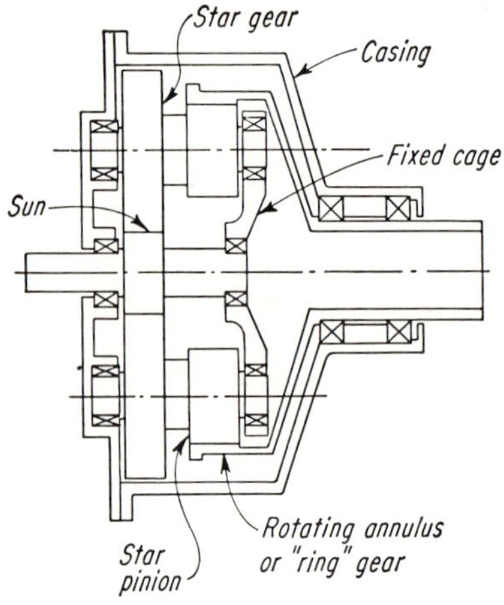

FIG. 3-15. Schematic arrangement of a compound star gear.

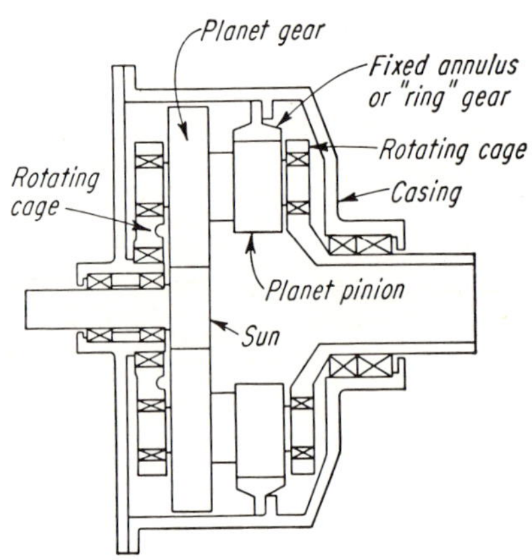

FIG. 3-16. Compound planetary.

Equation (3-2) may be used to get the theoretical driving load at the first-reduction planets. The load at the second-reduction planets is

$$W_{t_2} = \frac{W_{t_1} D_{P_1}}{D_{P_2}} \tag{3-7}$$

In the case of the solar unit Eq. (3-2) will not work since the sun is not rotating. By changing the equation to use the speed of the ring and the diameter of the ring, the driving load can still be obtained. For the compound solar, the load determined from the ring should be considered as W_{t_2}. Equation (3-7) can be worked backward to solve for W_{t_1}, the load at the sun-planet mesh.

Coupled Epicyclic Trains. Coupled epicyclic trains consist of two or more single epicyclic trains arranged so that two members in one train are common to the adjacent train. Figure 3-17 shows some schematic arrangements of coupled epicyclic trains. There are many more possible combinations.

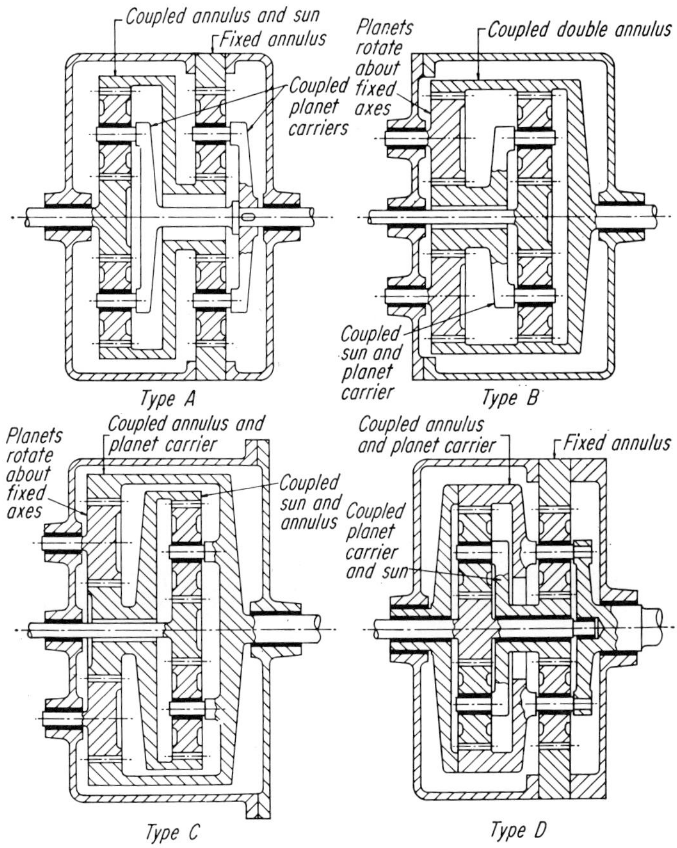

FIG. 3-17. Some schematic arrangements of coupled epicyclic gears.

The calculation of ratios, tooth loads, and efficiencies becomes quite complicated for the coupled epicyclic train. Since these gears are not widely used, detail information will not be given here. The reader is referred to refs. 2 and 4.

Application of Epicyclic Gear Trains. Epicyclic gears have been used for many centuries in the gear trade. James Watt in 1781 patented a sun and planet gear which he used with one of his early engines. There have been many patents on epicyclic gears. Even in the recent period since the Second World War, several refinements in planetary-gear design details have been patented.

The epicyclic gear is widely used in the aircraft industry. Almost all propeller-driven commercial aircraft use an epicyclic gear of some type to drive the propeller. The piston engines do not need too much ratio, and this permits them to use only a single-reduction planetary. The higher-speed gas-turbine engines have used double-reduction gearsets with the last reduction being epicyclic, or they have used compound epicyclic gears.

There has been a wide use of epicyclic gearing in geared motors and other industrial applications where the horsepower was fairly low (usually under 100 hp).

The epicyclic gear requires great precision and very clever use of flexibility to make all the planets work evenly. The best of modern aircraft planetary units probably divide up the load so uniformly that no planet ever sees more than about 125 per cent of the theoretical load. Many early aircraft planetary units (1935–1945 era) appeared to have a nonuniform split of loading that permitted as much as 200 per cent load to be applied to the most heavily loaded planet.

Epicyclic gears have not been widely used in the high-horsepower high-speed industrial or marine field. There is a trend to more use of these units in this field. If modern aircraft-quality units can be provided at industrial-gear prices, then there may be a considerable increase in the use of epicyclic gears to handle thousands of horsepower in land and marine applications.

Epicyclic gears are widely used in tank transmissions, truck transmissions, and automatic-shift transmissions for passenger cars. In these cases the basic epicyclic gear is generally combined with other gears, clutches, brakes, and pumps to make an over-all transmission package. Article 3-9 gives some examples of such transmissions. The horsepower handled by these transmissions is generally under 500.

Problems of Epicyclic Gears. Several of the problems in designing and using epicyclic gears have already been mentioned. A summary of the problems is as follows:

1. Dividing load between planets
2. High bearing loads on planet pins
3. Balance and vibration of rotating cage
4. More complicated assembly
5. Hazard of jamming the whole drive if a piece of metal gets loose in the mesh

3-5. Right-angle-gear Arrangements. Right-angle[11] gears are used in a wide variety of meshing arrangements. Table 3-7 shows a general survey of the field.

In theory all the right-angle gear types could be used on either right angles or angles other than 90°. In practice, though, because of limitations in manufacturing

Table 3-7. Angular-gear Meshing Possibilities

Type of gear teeth	Pinion and gear		Arrangements		
	Right angle	Skew angle	Planetary	Differential	Reversing
Straight bevel	Yes	Yes	Yes	Yes	Yes
Zero bevel	Yes	Yes	Yes	Yes	Yes
Spiral bevel	Yes	Yes	Yes	Yes	Yes
Hypoid	Yes	...	No	Yes	Yes
Face gear	Yes	No	...	Yes	Yes
Crossed-helical	Yes	Yes	No	...	No
Cylindrical worm	Yes	...	No	Yes	No
Double-enveloping worm	Yes	...	No	Yes	No
Beveloid	Yes	Yes	Yes	Yes	Yes
Spiroid	Yes	...	No	Yes	No
Planoid	Yes	...	No	...	Yes
Helicon	Yes	...	No	Yes	No

equipment and other considerations, some of the gears are not ordinarily used on skew angles. Note the dash lines in the skew-angle column of Table 3-7.

Some types can be used in a planetary-style arrangement. When this is done the ring member becomes externally toothed. The data given in Table 3-5 can be used to calculate the ratio of a planetary made with angular gears.

Some angular gears are used in differentials. Even the worm gear may be used as the gear drive to rotate a differential package.

Some right-angle gears lend themselves to reversing drives. Two pinions may be shifted axially to make a quick reverse in a drive.

Figure 3-18 shows a typical bevel-gear speed reducer. Typical applications of a wormgear drive and a Spiroid-gear drive are shown in Figs. 3-19 and 3-20.

One of the more important skew-angle drives is the V drive for boats. This permits the engine to be horizontal and the propeller shaft to pass through the hull of the ship at an angle. Figure 3-21 shows a V-drive unit.

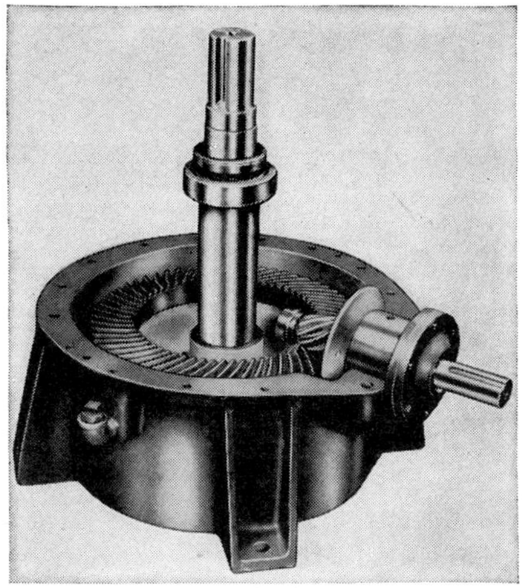

FIG. 3-18. Bevel-gear speed reducer. (*Courtesy of Philadelphia Gear Corp.*)

Figure 3-22 shows the elements of a bevel reverse unit. The bevel planets are attached to a fixed cage. Since the bevel *sun* and the bevel *ring* are really the same size parts, this *star* type of arrangement has a 1:1 ratio. It is used only to reverse direction of rotation.

FIG. 3-19. Motor with wormgear speed reducer. (*Courtesy of Philadelphia Gear Corp.*)

Figure 3-23 shows the parts that could be used to make a 2:1 ratio bevel planetary. In this case the cage would carry the planets and drive the output shaft through one of the hollow bevel gears. One bevel gear would function as an input sun and one would function as a fixed ring.

Bevel planetaries can be made to handle a range of ratios. To do this the input sun is made smaller than the bevel functioning as the ring. The axes of the planets are not at right angles to the input and output shafts in this case.

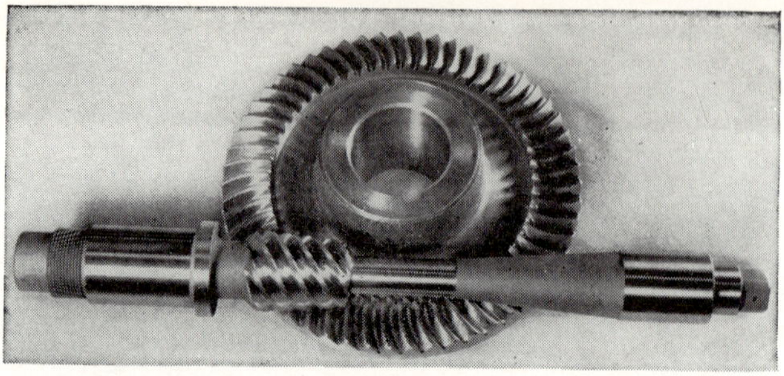

Fig. 3-20. Spiroid-gear drive for a band saw. (*Courtesy of Spiroid Division of Illinois Tool Works, Chicago, Ill.*)

Fig. 3-21. V-drive unit. (*Courtesy of Gleason Works, Rochester, N.Y.*)

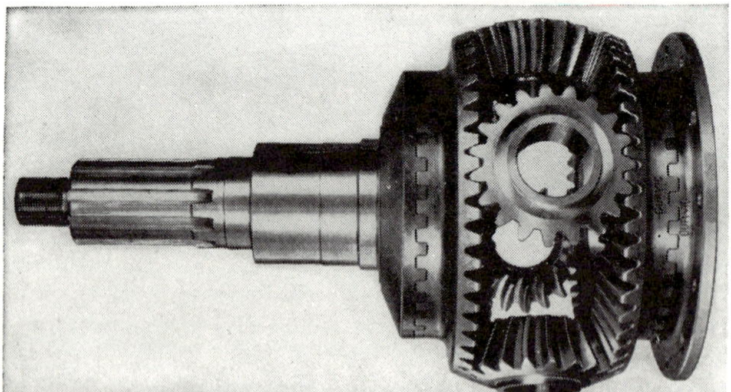

Fig. 3-22. Bevel reverse unit. (*Courtesy of Gleason Works, Rochester, N.Y.*)

FIG. 3-23. Parts for 2:1 ratio bevel planetary. (*Courtesy of Gleason Works, Rochester, N.Y.*)

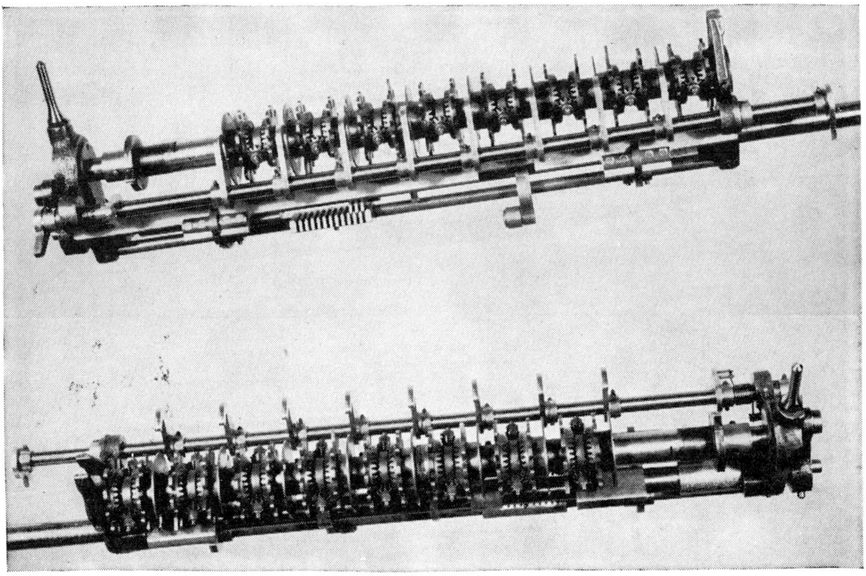

FIG. 3-24. Bevel-gear parts used in a cash-register machine. (*Courtesy of Gleason Works and National Cash Register.*)

Figure 3-24 shows an array of bevel gears used in a cash-register machine. Note how this compact arrangement provides a multiplicity of input shafts and a ready means for reversing any input.

3-6. Differential Gearing. In a general sense a "differential" gear is an arrangement where the normal ratio of the unit can be changed by driving into the unit with a second drive. Figure 3-25 shows a simple bevel-gear differential of the type often used in instruments. Input rotations may be applied to the gears at either end of the bevel planetary set. The output speed is taken off the "spider" or cage shaft. Either input speed may be stopped and the device will still transmit a ratio between the other input and the output. By varying the speed of the second input it is possible to get an infinite series of ratios.

Figure 3-26 shows the elements of a spur-gear differential. Again there are two inputs and one output, or it is possible to have one input and two outputs.

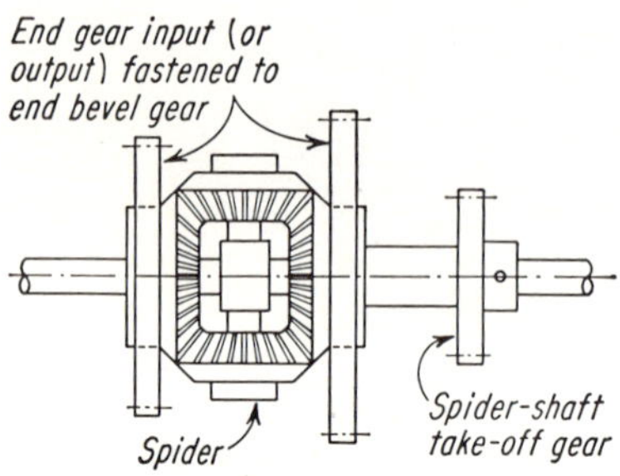

FIG. 3-25. Simple bevel-gear differential.

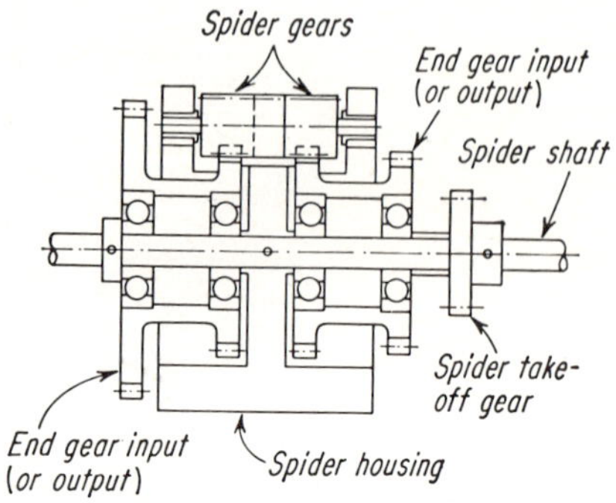

FIG. 3-26. Spur-gear differential.

Figure 3-27 shows a ground spiral bevel differential. One of the most familiar uses of a differential is in the rear end of an automobile. When an automobile goes around a corner, one wheel must make more turns than the other. A differential is needed with one input and two outputs. Figure 3-28 shows a photograph of a typical hypoid differential for an automobile. The large gears are hypoid and the small gears are bevel. The hypoid type of car rear end allows the drive shaft to be lower

FIG. 3-27. Ground spiral bevel differential. (*Courtesy of Gleason Works, Rochester, N.Y.*)

FIG. 3-28. Hypoid differential used in rear end of automobile. (*Courtesy of Gleason Works, Rochester, N.Y.*)

than would be the case with an all-bevel differential. A low drive shaft permits the automobile to be built lower.

Fixed Differential. The differentials just discussed with two inputs and one output or two outputs and one input are used to vary ratio. There is another kind of differential that has a fixed ratio and develops an unusually large ratio. This kind is called a differential because the output speed is the *difference* between the speeds of two parts of the drive that are running at almost the same speed.

There are dozens of arrangements of the (free) differential discussed earlier. Likewise there are dozens of fixed-differential arrangements. Figure 3-29 shows four typical arrangements of fixed differentials.

Table 3-8 shows some general data for the four fixed differentials shown in Fig. 3-29. Note that there is a subtractive item in the expression for over-all ratio for each of

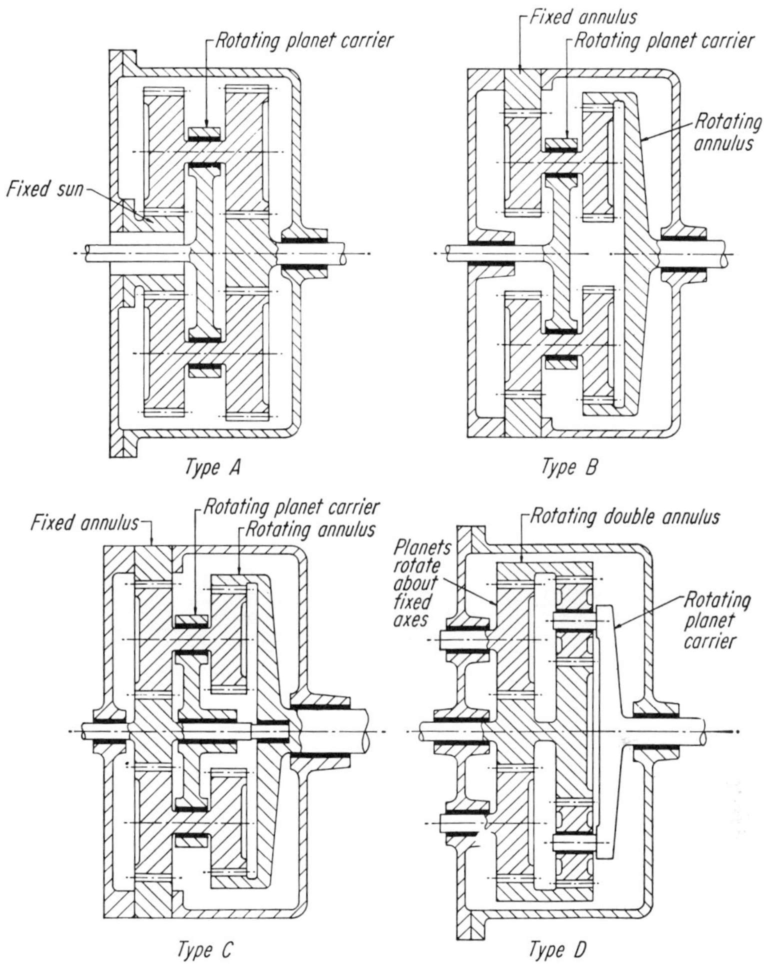

FIG. 3-29. Fixed-differential gear arrangements.

these types. If the pitch was made infinitely fine, it would be possible to get the subtractive term to almost reach zero. This would tend to make the over-all ratio infinitely large. Practical considerations on number of teeth, though, tend to limit the ratio that can be obtained. The ratio range shown represents normal design practice rather than any theoretical limit on the ratio.

The fixed-differential gears tend to be low in efficiency. The differential has a large amount of tooth meshing going on at very high tooth loads. Losses are much higher than for regular gearing. A fixed differential of 100:1 ratio might have an efficiency as low as 25 per cent. With good design a fairly good efficiency can be obtained.

Table 3-8. Fixed-differential-gear Data

Kind of differential*	Fixed member	Input member	Output member	Arrangement	Over-all ratio	Range of ratios normally used
A. All external teeth	No. 1 sun	Cage	No. 2 sun	No. 1 sun/No. 1 planet No. 2 planet/No. 2 sun	$\frac{N_{S_2}N_{P_1}}{N_{S_2}N_{P_1} - N_{S_1}N_{P_2}}$	10:1–50:1
B. Internal gear	No. 1 ring	Cage	No. 2 ring	No. 1 planet/No. 1 ring No. 2 planet/No. 2 ring	$\frac{N_{R_2}N_{P_1}}{N_{R_2}N_{P_1} - N_{R_1}N_{P_2}}$	15:1–100:1
C. Planetary	No. 1 ring	Sun	No. 2 ring	No. 1 sun/No. 1 planet No. 1 planet/No. 1 ring No. 2 planet/No. 2 ring	$\left(\frac{N_{R_1}}{N_{S_1}} + 1\right) \frac{N_{R_2}N_{P_1}}{N_{R_2}N_{P_1} - N_{R_1}N_{P_2}}$	20:1–500:1
D. Star-planetary	No. 1 cage	No. 1 sun	No. 2 ring	No. 1 sun/No. 1 planet No. 1 planet/No. 1 ring No. 2 sun/No. 2 planet No. 2 planet/No. 2 ring No. 1 sun attached to No. 2 sun No. 1 ring attached to No. 2 ring	$\frac{1}{\frac{1}{\left(\frac{N_{R_2}}{N_{S_2}} + 1\right)} - \frac{1}{\frac{N_{R_1}}{N_{S_1}}\left(\frac{N_{S_2}}{N_{R_2}} + 1\right)}}$	15:1–100:1

Follow rules of assembly for compound epicyclic for applicable parts of each differential (see Table 3-6).
* See Fig. 3-29 for schematic arrangement of each kind.

The chief value of the differential is that it permits a large ratio to be obtained in a very compact unit. References 7 and 8 give detail information on how to calculate the efficiency of fixed differentials.

3-7. Contra-rotating Concentric-shaft Drives. There is sometimes a need to have concentric output shafts rotating in opposite directions. Some torpedoes have concentric propellers rotating in opposite directions. A few airplanes have had concentric propellers rotating in opposite directions. Helicopter rotors have also been used in a contra-rotating concentric-shaft arrangement.

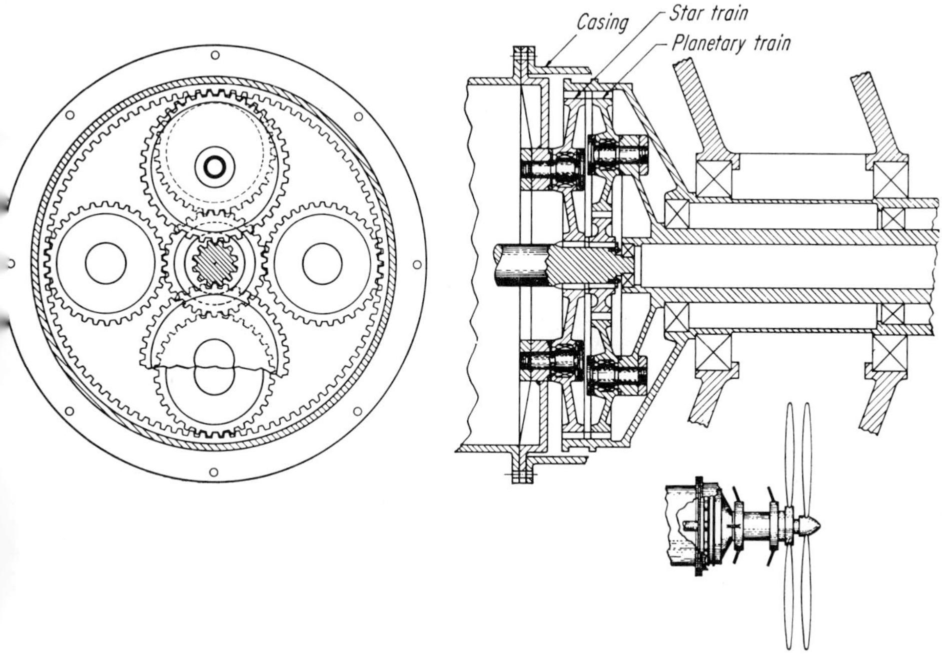

FIG. 3-30. Contra-rotating gear arrangement.

Several kinds of gear drives will accomplish this result. Three general approaches to the problem are

1. Use two final drive pinions on the same shaft. Drive an external gear with one and drive an internal gear with the other. Adjust pinion and gear sizes so that each set has the same ratio.
2. Use a bevel reverse gear after the last stage of the regular reduction gearing. Make the bevel gears hollow and connect the input bevel to the inner output shaft. Connect the output bevel to the outer shaft.
3. Use a combination star- and planetary-type arrangement. Connect one output shaft to the cage. Connect the other to the ring gear. Attach the star ring and the planetary ring together and permit this assembly to be rotatable.

The arrangement shown in Fig. 3-30 illustrates the star-planetary arrangement. In this arrangement most of the power goes through the planetary side of the drive. Only enough to balance the output torques goes through the star side of the drive.

3-8. Speed-changing Transmissions by Shifting. Gear arrangements that change speed by shifting have been widely used in automobiles, trucks, tanks, and many other applications. The gear-shift transmission may have a variety of parallel-axis gears, right-angle-axis gears, and clutch elements.

The normal mode of operation for these transmissions is to declutch the drive, shift a gear or coupling axially, and reengage the clutch.

The general way these transmissions are made can best be understood by looking at cross-section views or cutaway views of typical designs.

FIG. 3-31. Industrial gear-shift transmission. (*Courtesy of Philadelphia Gear Works.*)

FIG. 3-32. Simple reverse gear shift. (*Courtesy of Gleason Works, Rochester, N.Y.*)

Figure 3-31 shows an industrial-type gear-shift transmission. Note the slidable coupling members and the herringbone gear teeth.

Figure 3-32 shows a simple reversing unit using bevel gears. Note the slidable dog clutch.

Figure 3-33 shows a five-speed transmission of the type used in automobiles and trucks. Note the compact arrangement. Note how the upper right-hand gear is slidable on its shaft.

Figure 3-34 shows a 12-speed transmission of the type used in a heavy-duty truck. This is a very compact arrangement with a minimum of parts for a 12-speed transmission.

Figure 3-35 shows an auxiliary transmission. When a vehicle needs more speeds than can be provided by the normal transmission, an auxiliary transmission can be installed in series with the main transmission to give several more speeds.

3-9. Speed-changing Transmissions by Clutches and Brakes. Many transmissions are made so that a gear shift can occur without disengaging the drive (or declutching). These transmissions use things like fluid couplings, brakes, or differentials to change speed while torque is being transmitted. The so-called "automatic" transmission used on most passenger automobiles is of this type. Some tanks and other military vehicles use the automatic transmission.

FIG. 3-33. Five-speed transmission. (*Courtesy of Dana Corporation, Toledo, Ohio.*)

Figure 3-36 shows a fully automatic transmission for on-highway truck use. Note the fluid-coupling parts. Figure 3-37 shows a transmission for off-highway vehicles like a front-end loader. Note the fluid coupling and the several sets of brake or clutch disks.

Figure 3-38 shows a cross-drive transmission for a tracked vehicle like a tank. This transmission has two speeds forward and one reverse. The transmission has the capability of controlling the speed ratio to each drive shaft. A tank is steered by driving one track faster than the other.

The one transmission shown in Fig. 3-38 has almost all the gear arrangements discussed in this chapter in one compact package. There are planetary-type gears, bevel gears, helical gears, skew-axis wormgears, differentials, simple spur gears, and multiple-power paths all in this one transmission. There are 4,007 parts in the unit, the weight is approximately 3,000 lb, and the unit will handle up to 850 gross hp.

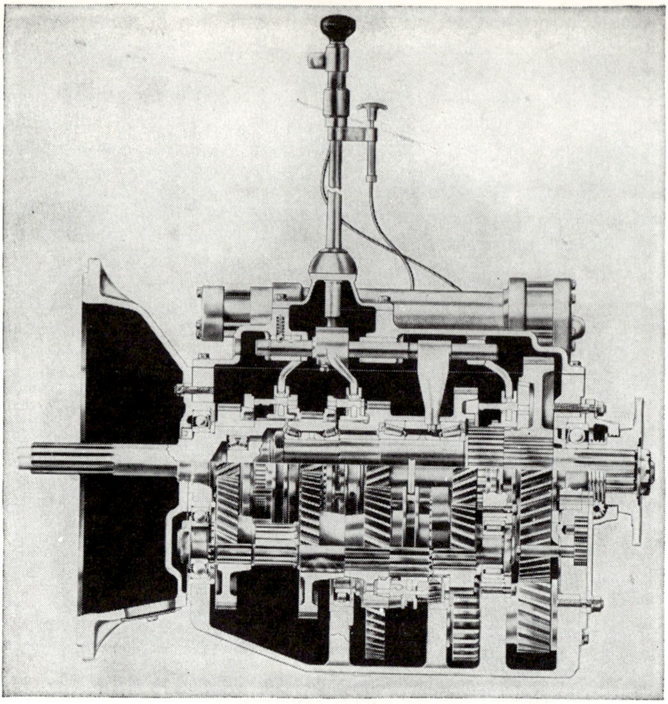

FIG. 3-34. Twelve-speed transmission. (*Courtesy of Dana Corporation, Toledo, Ohio.*)

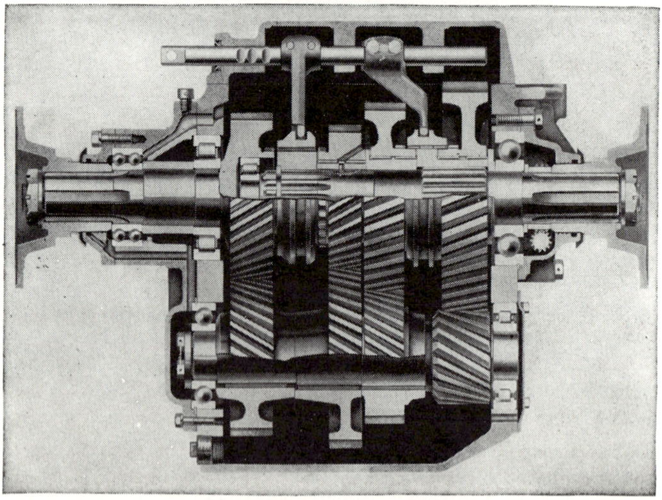

FIG. 3-35. Auxiliary transmission. (*Courtesy of Dana Corporation, Toledo, Ohio.*)

FIG. 3-36. Automatic truck transmission. (*Courtesy of Allison Division, General Motors Corp., Indianapolis, Ind.*)

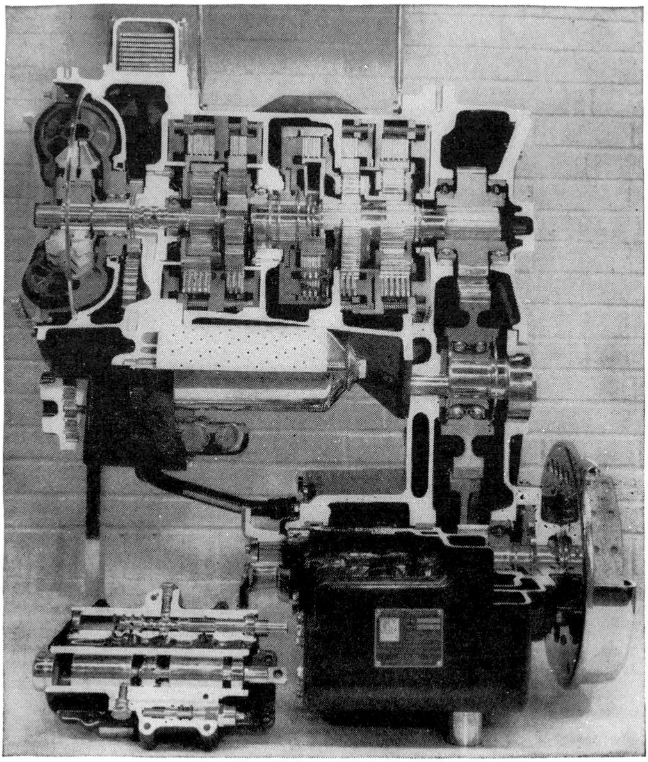

FIG. 3-37. Transmission for material-handling vehicle. (*Courtesy of Allison Division, General Motors Corp.*)

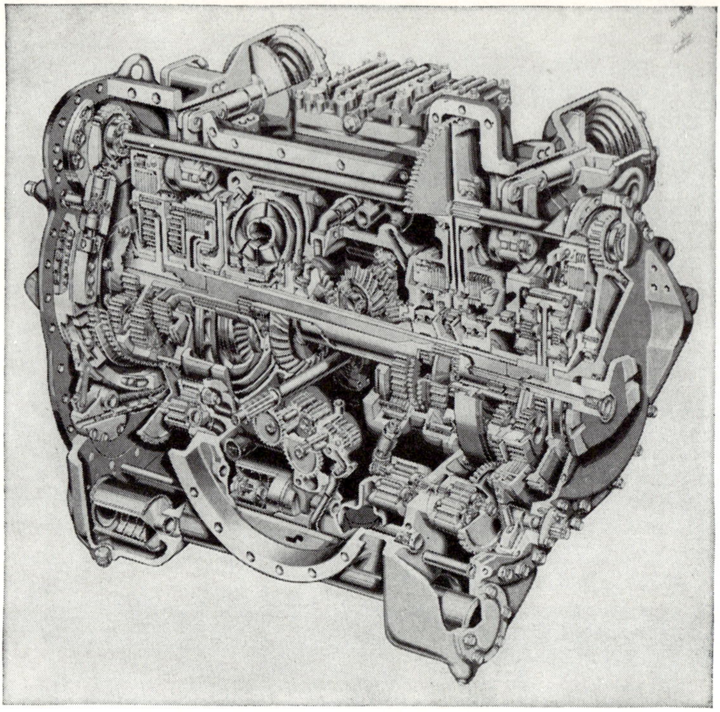

FIG. 3-38. Allison CD-850 cross-drive transmission for military track-laying vehicles. (*Courtesy of Allison Division, General Motors Corp.*)

FIG. 3-39. Marine drive unit rated at 825 hp, 6,840 input speed, 3.22:1 forward speed, and 2.84:1 reverse speed. (*Courtesy of Jered Industries, Inc., Detroit, Mich.*)

FIG. 3-40. Disassembled parts for unit shown in Fig. 3-39. (*Courtesy of Jered Industries, Inc.*)

FIG. 3-41. Partially disassembled view of marine drive unit rated at 2,500 hp. (*Courtesy of Jered Industries, Inc.*)

Figure 3-39 shows a marine unit with a forward and reverse speed. This unit weighs about 520 lb and handles up to 825 hp. Input speed is 6,840 rpm. It has a 3.22:1 forward ratio and a 2.84:1 reverse ratio. The gearing and brakes in the unit make it work as either a planetary unit or a star unit. A planetary unit has its output turning in the same direction as the input. A star unit has an output that turns in the reverse direction from the input. By switching from the star to the planetary, or vice versa, it is possible to change direction of rotation.

Figure 3-41 shows a partially disassembled view of a 2,500-hp marine reverse unit. This unit handles an input speed of 1,000 rpm with a 1:1 forward ratio and a 1.2:1 reverse ratio. The unit weighs a little over 4,000 lb. Figure 3-42 shows the sun, planet, and cage parts.

FIG. 3-42. Sun, planets, and cage structure for unit shown in Fig. 3-41. (*Courtesy of Jered Industries, Inc.*)

The unit shown in Fig. 3-41 has some novel features. In forward drive the cage is held to the sun by a clutch. The whole gear assembly rotates as a unit and there is no tooth meshing. In reverse the first clutch is opened and a second clutch holds the ring gear to the casing. The unit then functions as a planetary gear, but the fact that there are two planets meshing with each other makes the output shaft turn opposite from the input and the ratio is much lower than in a normal planetary.

3-10. High-ratio Arrangements. Besides the fixed-differential gears discussed in Art. 3-6, there are several other gear arrangements that have high-ratio capabilities. These are used under varying trade names. Theoretically they do not belong in any of the gear classifications already discussed in this chapter.

Planocentric Gear. An example of one of the high-ratio types is the kind of gearset called the "planocentric." This unit can have as few gear parts as a spur pinion and an internal gear and yet develop as much as 100:1 ratio! A ratio this great can be readily obtained with a single-thread worm and gear, but it is surprising that this much can be achieved with spur gearing or helical gearing.

Figure 3-43 shows an over-all view of a unit of this type. Figure 3-44 shows the details of the meshing elements.

The pinion is mounted on an eccentric. A bearing separates the pinion from the eccentric and its drive shaft. The pinion is thus under no constraint to rotate at the speed of the input shaft.

The pinion wobbles instead of truly rotating. It does develop a rotation, superimposed on the wobble. Strangely, the rotation of the pinion is at output speed

FIG. 3-43. Planocentric gear, 64:1 ratio, oblique low-speed end. (*Courtesy of General Electric Co., Lynn, Mass.*)

FIG. 3-44. Planocentric gear, 64:1 ratio. High-speed end showing 64-tooth pinion mating with 65-tooth gear. Pinion drives cage pins. (*Courtesy of General Electric Co., Lynn, Mass.*)

instead of input speed! The pinion transmits its rotation to the output shaft by a pin coupling (or other coupling means). The internal gear is fixed to the casing.

The equation for ratio is

$$m_G = \frac{N_P}{N_G - N_P} \tag{3-8}$$

This type of drive has been used for high-torque high-reduction applications in aircraft, marine, and appliance fields. Under favorable conditions the efficiency may be as high as 90 per cent or more.

3-11. Harmonic Drive Arrangements. Harmonic Drive is the name given to a new family of machine systems which use the controlled elastic deflection of one or more parts for the transmission, conversion, or change of mechanical motion. The basic mechanism of Harmonic Drive has the broad multipurpose capabilities of the simple lever and has proved itself adaptable to such diverse forms of mechanical systems as rotary-to-rotary motion transmissions, rotary-to-linear motion converters, linear-to-linear transmissions, and rotary pumps and valves.

Until now, nearly all mechanical systems have been based upon the well-known laws of "rigid-body mechanics," and considerable effort has been devoted to reducing the spring flexure of machine parts to an absolute minimum. Rotating elements have been assumed to remain rigid and to rotate circularly about fixed axes. The following study of Harmonic Drive describes a radical departure from traditional mechanics and requires entry into a new and challenging realm of "nonrigid mechanics," or "elastokinesis"—the new field of elastic-body dynamics.

A close look at the functioning of Harmonic Drive elements is provided in Fig. 3-45. Here the wave generator is elliptoidal in shape and surrounded by a ball bearing.

The Flexspline has 130 external teeth and the circular spline 132 internal teeth, resulting in a 2-tooth difference equal to the number of lobes of the wave generator.

Harmonic Drive is the subject matter of U.S. Patent 2,906,143 and of other United States and foreign patents and patent applications. All proprietary rights are held by the United Shoe Machinery Corporation.

As the wave generator is turned, the Flexspline is progressively deflected to follow the rotating elliptoidal shape. Flexspline and circular spline are held in engagement

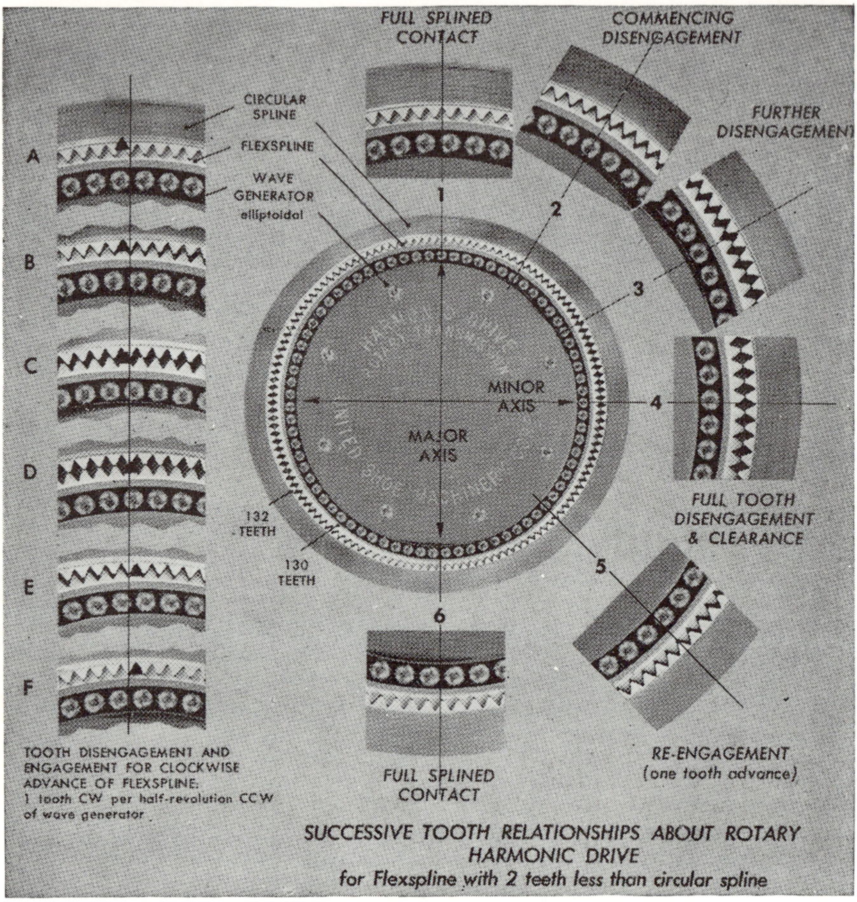

FIG. 3-45. Spline-motion relationships of rotary Harmonic Drive.

at the major axis of the wave generator and are fully disengaged and clearing at the minor axis. At the major axis the teeth are in full spline contact and are rotationally stationary. At the minor axis where the teeth are disengaged, they are in angular motion because of deflection of the Flexspline. Enlarged sections (1 through 6) show successive conditions of spline-tooth engagement and disengagement. At the left in Fig. 3-45, the spline relationship at any one region of the teeth is shown as the wave generator is rotated. It can be seen here how a particular tooth gradually disengages, advances, and reengages, as the wave generator is turned a half revolution. For a full rotation of the wave generator, the Flexspline counterrotates through an angle equivalent to 2 of its 130 teeth, resulting in a reduction ratio of 65:1 magnitude.

Special Properties of the Harmonic Drive. From the foregoing discussion of the basic rotary Harmonic Drive structure, the basis for several unique and inherent properties can be seen:

1. Tooth-motion relationship between Flexspline and circular spline makes unusually high speed reduction or speed increase possible in a single stage.
2. Under load, many spline teeth are in simultaneous engagement, resulting in high torque capacity.
3. Spline teeth come into contact with an almost pure radial motion and have essentially zero sliding velocity, even at high input speeds. Tooth friction losses and tooth wear are thus very low.
4. Because of low friction losses, high mechanical efficiencies can be obtained, which are particularly outstanding at high ratios.
5. Spline teeth in contact and under load are practically stationary. Dynamic loading, under normal operating conditions, is very low.
6. Regions of tooth enga ement and application of load torque are usually diametrically opposed and result in a force couple that is symmetrical and balanced.
7. Harmonic Drive elements are concentric and tend to be self-aligning.
8. Tubular construction of splined elements results in high torsional rigidity and thus less windup under load.
9. Backlash is normally low but can be completely eliminated between Flexspline and circular spline by spring loading the two splines into mesh.
10. With Flexspline formed as an integral section of a flexible cylindrical wall, positive transmission of rotary mechanical motion through the wall can be achieved.

Harmonic Drive can also transmit rotary-to-linear motion into a hermetically sealed chamber. The wall can be part of a welded steel enclosure which is capable of maintaining complete isolation of two environments. For applications requiring the absolute containment of pressure, vacuum, liquids, or contaminants, Harmonic Drive can provide positive and reliable mechanical control and power transmission without the danger and inconvenience of pack glands or other partial seals.

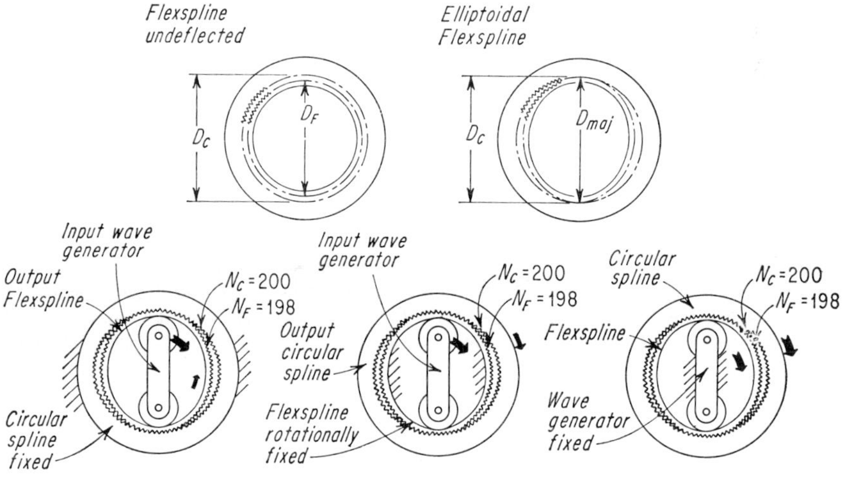

FIG. 3-46. Design relationships and ratios of single-stage rotary Harmonic Drive.

Wave-generator Configurations. The role of the wave generator is to produce controlled deflection of the Flexspline and to make possible the continuous rotation of the resulting elliptoidal shape. In the preceding figures the wave generators are *internal* and deflect the Flexspline outward into engagement with the circular spline. Figure 3-47 shows that wave generators can also be *external* to deflect the Flexspline inward into engagement with the circular spline. In this instance, spline engagement is at diametrically opposite regions on the *minor* axis of the elliptoid.

The most common form of wave generator is the elliptoidal cam, several configurations of which are shown in Fig. 3-47. The wave generators may form the internal

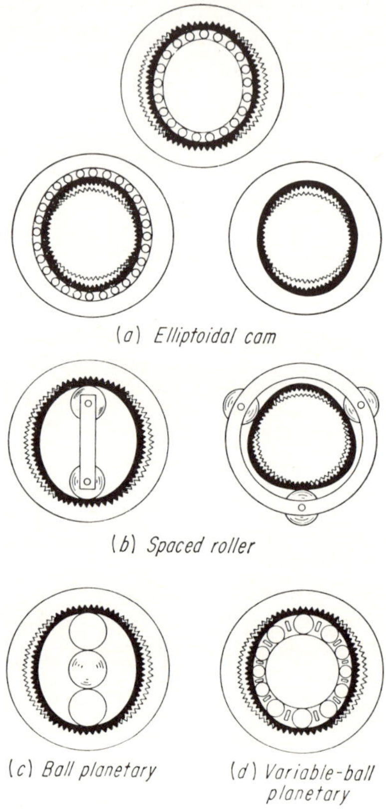

Fig. 3-47. Wave-generator configurations.

or external element of the Harmonic Drive stage and may have plain (or sleeve) bearing contact with the Flexspline or utilize a form of antifriction device. Antifriction bearings may be ball or roller and usually are the retainer type. There are other shapes as well that can be used.

Coupling to Flexspline. Motion of the wave generator or the circular spline can be joined rigidly or rotationally to external equipment by standard coupling means, but the nonrigid Flexspline requires less conventional means of position and motion coupling. Figure 3-45 shows how the Flexspline and circular spline are firmly splined together at the major axis of the elliptoid but move angularly relative to one another at all other points, with maximum relative motion occurring at the minor axis. For most applications, motion of the nonrigid Flexspline is transmitted through a type of coupling which integrates these variations in angular velocity and produces a uniform angular motion.

1. *Internal coupling.* From a performance point of view, the preferred Flexspline couplings are integral, as shown in panel 1 of Fig. 3-48. The Flexspline coupling is in the form of a cup with spline teeth in the cylindrical wall near the open end. Controlled deflection is imposed in the cup at the open end, while the bottom of the cup remains essentially circular. The length of the tubular wall acts as a deflection attenuator and provides a form of integrating coupling between

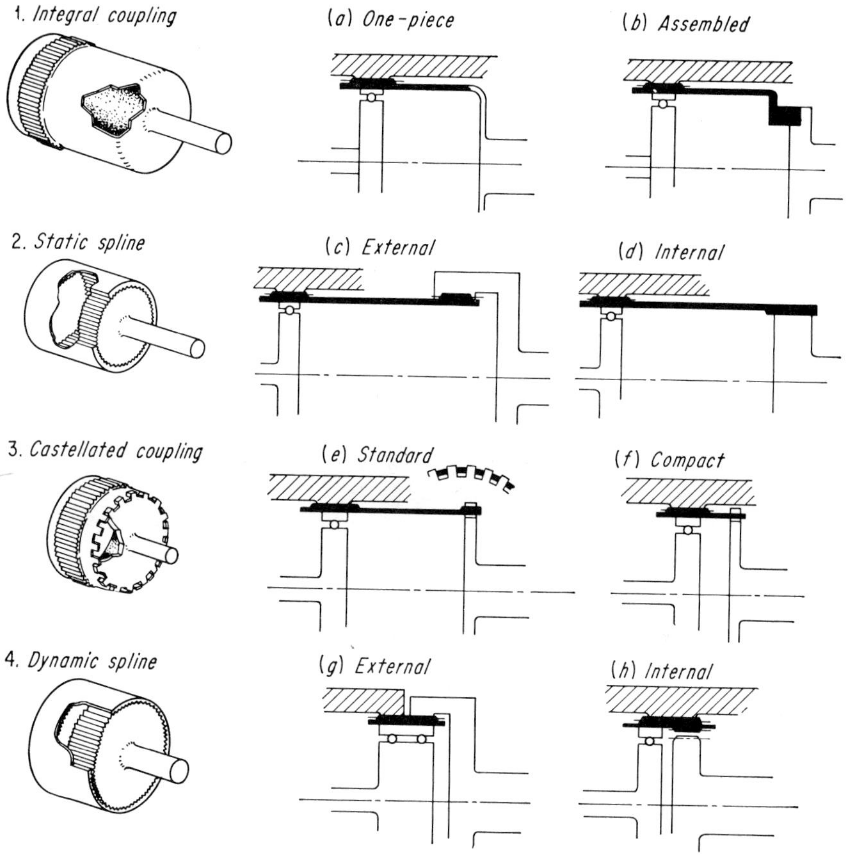

FIG. 3-48. Methods of mechanical coupling to Flexspline.

the spline engagement at the open end and a shaft joined to the cup bottom. The shaft and cup may be formed as one piece or assembled. Integral couplings have high torsional rigidity, low backlash, long life, high torque capacity, and no power loss within the coupling. Harmonic Drive efficiencies, with integral coupling, can be over 90 per cent. The assembled coupling can contain a drive motor, overload clutches, r-f pumping, etc.

2. *Static spline.* The next most efficient coupling is the static spline shown in panel 2 of Fig. 3-48. Here the flexing Harmonic Drive spline is shown at the left end of an open tube, with a standard circular spline being engaged at the right end. Coupling spline teeth can be cut either externally or internally

into the tube wall. The static spline coupling permits slight axial motion of the tube without allowing angular or radial freedom. Operating efficiency of the static spline is slightly less than that of an integral coupling. Fabrication cost is lower, but the static spline requires lubrication.

3. *Castellated coupling.* Another type of coupling is the castellated coupling shown in panel 3. The efficiency, positioning accuracy, and torque capacity of this coupling are governed by the number of lugs and the tube length between spline teeth and lugs. The shorter couplings, however, have about one-fourth the load capacity of an integral coupling since torque is not equally distributed among the lugs. Castellated couplings have lower positioning accuracy and efficiency but are the least expensive and most compact Flexspline couplings for a single-stage drive.

Composite Single-stage Harmonic Drive. With the principal operating elements of Harmonic Drive having been defined and illustrated schematically, it will be helpful to see how these elements function together to form a single Harmonic Drive stage. Figure 3-49 shows a simplified form of Harmonic Drive unit having an

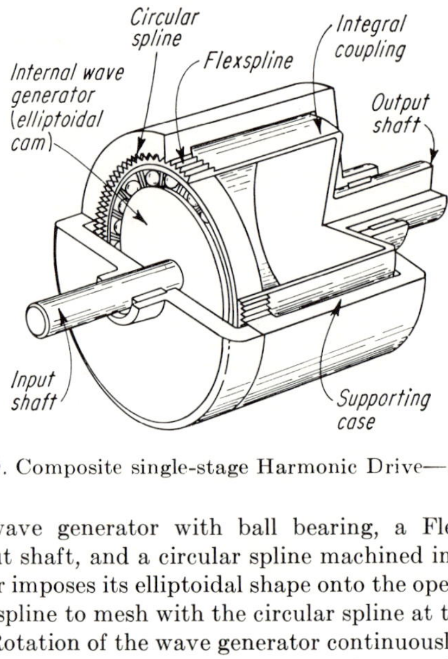

FIG. 3-49. Composite single-stage Harmonic Drive—simplified.

internal elliptoidal wave generator with ball bearing, a Flexspline with integral coupling to the output shaft, and a circular spline machined in the supporting case.

The wave generator imposes its elliptoidal shape onto the open end of the Flexspline cup, causing the Flexspline to mesh with the circular spline at two diametrical regions on the major axis. Rotation of the wave generator continuously advances the regions of spline engagement. The Flexspline rotates relative to the circular spline and transmits its motion through the integral coupling to the output shaft. While the Harmonic Drive in Fig. 3-49 is shown as a speed reducer, it can be used equally well as a speed increaser by interchanging the roles of the input and output shafts.

Harmonic Drive Performance Capabilities. Plotted Harmonic Drive performance curves are based on operational tests conducted with experimental Harmonic Drive units. One piece of test apparatus used is the special stand shown in Fig. 3-50. This was developed by United Shoe Machinery Corporation for the prolonged testing of rotary-to-rotary drives under accurately controlled load conditions. Here independent measurements are made simultaneously on two Harmonic Drive units, which can be tested interchangeably as speed reducer and speed increaser. Units have been driven for hundreds of millions of cycles, and their performance measured to better than 1 per cent accuracy.

Additional comparative data have been taken from Harmonic Drive systems developed for special military and industrial applications.

A band of values is shown in most plots, and this ranges from conditions of standard design and manufacture to conditions of custom design and more exacting methods of manufacture. These broad curves are drawn about "design center" values which

FIG. 3-50. Back-to-back endurance and efficiency test of Harmonic Drive units. (*Courtesy of United Shoe Machinery Corp., Boston, Mass.*)

are conservatively achievable today with Harmonic Drive. The design of any particular drive unit must, as usual, represent a balancing of these various parameters.

1. *Volume vs. output torque.* External volume of a separate 100:1 Harmonic Drive is plotted in Fig. 3-51, graph 1, against output torque. Harmonic Drive is presently capable of producing output torques of 25 lb-in. per cu in. for standard design, and up to 50 lb-in. per cu in. for compact design.
2. *Weight vs. output torque.* The weight of a separate 100:1 Harmonic Drive unit is plotted in graph 2 of Fig. 3-51 against output torque. Harmonic Drive torque output can range from 150 lb-in. per lb for standard design up to 500 lb-in. per lb for special lightweight design. Reducer weight varies with the type of construction, the materials used, and the manner in which the reducer is mechanically coupled with related elements.
3. *Reduction ratio vs. configuration.* In graph 3, the reduction ratios which can be achieved with four different Harmonic Drive configurations are shown. These are the general ranges achievable with steel Flexsplines in units of moderate physical size. In addition to the ratios shown, near-unity ratios and both positive and negative rotations can be achieved with Harmonic Drive. Except at the ultra-high ratios, the plotted ratios apply to speed increase as well as speed reduction. Only a few of the possible Harmonic Drive configurations are indicated here, and these types can be combined.
4. *Efficiency vs. ratio.* The range of efficiencies achievable with Harmonic Drive are shown in graph 4 as a function of a ratio. These figures apply to steel Flexsplines and include the losses of both input and output shaft bearings and couplings. The graph assumes an integral Flexspline. Where Harmonic Drive

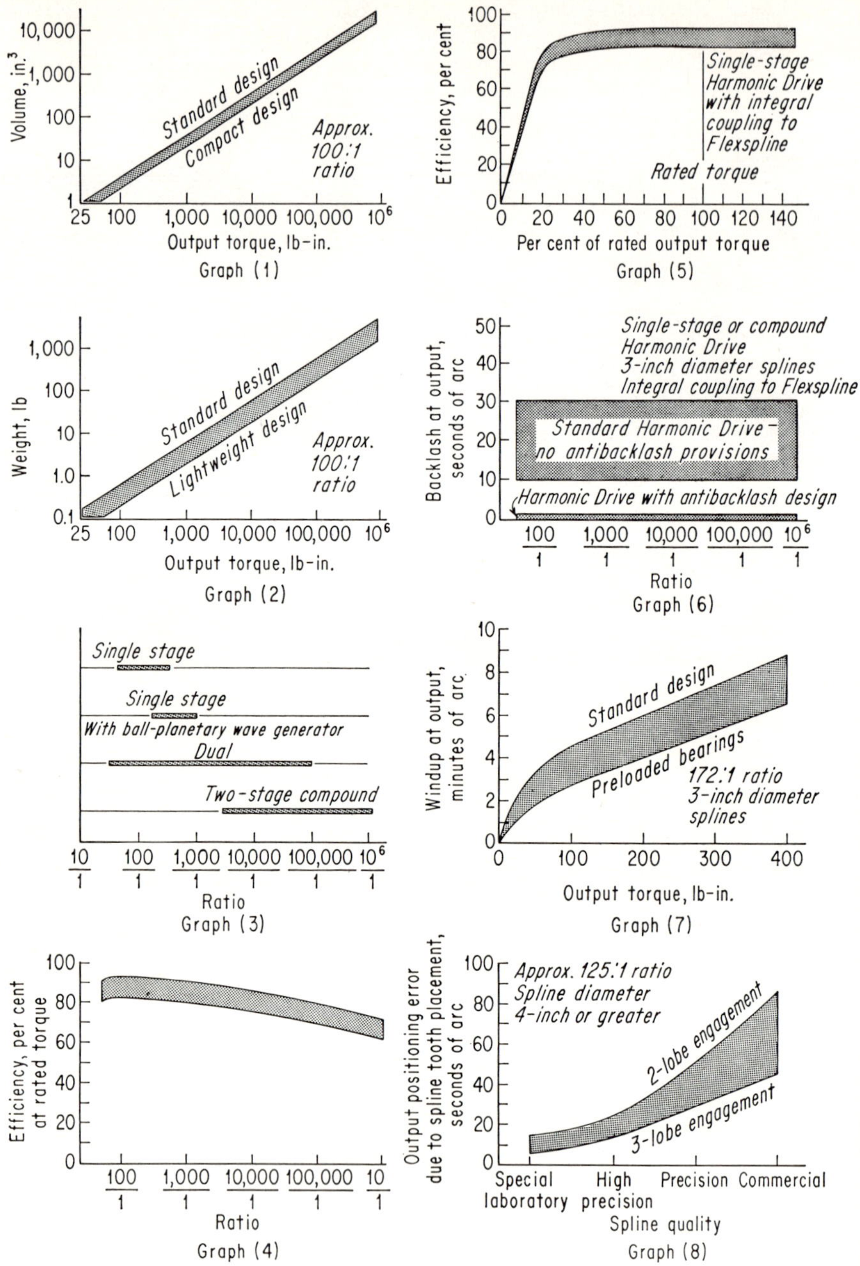

FIG 3-51. Normal capabilities of Harmonic Drive units.

has been used as both speed increaser and speed reducer, it has been found that the operating efficiencies are approximately equivalent.

5. *Efficiency vs. output torque.* Graph 5 shows that Harmonic Drive efficiency rises rapidly with increasing output torque and remains flat over a wide range. This plot pertains to a single-stage Harmonic Drive with integral coupling to the Flexspline.
6. *Backlash vs. ratio.* Backlash at the output of a Harmonic Drive stage is plotted in graph 6 against ratio. This graph is for single-stage or compound drive with approximately 3″-diameter splines and integral coupling to the Flexspline. The upper broad band relates to standard Harmonic Drive design with no special antibacklash provisions taken; the lower line shows substantially zero backlash achievable with minor design modification. An unusual characteristic of Harmonic Drive backlash is that it is essentially constant with increase in ratio; this results from the extremely high ratios possible per stage. Backlash of standard Harmonic Drives can be reduced by increasing the diameter of the splines.
7. *Windup vs. output torque.* Harmonic Drive has inherently low windup for a drive of given physical size and output torque capability. Windup characteristics of Harmonic Drive are illustrated by measurements taken on a specific unit with 172:1 reduction ratio, 380-lb in. rated output torque, and approximately 3″-diameter splines. With the input clamped, graph 7 shows the windup angle developed at the output as a function of the torque applied at the output. The plot shows the windup range between standard design and design using a preloaded output bearing. Windup can be decreased by increasing spline diameter.
8. *Positioning accuracy.* Graph 8 shows the positioning accuracy possible with Harmonic Drive, in terms of the general class of splines being used. While tooth-placement errors in standard gearing appear across the gear radius, errors in two-lobe Harmonic Drive appear across a full diameter and are both statistically and geometrically reduced. Further reduction in error of a three-lobe Harmonic Drive is shown in graph 8. This plot is for spline diameters 4″ and greater. No allowance has been made in these curves for the fact that multiple tooth engagement, unique in Harmonic Drive, tends to average and reduce errors caused by the shape or misplacement of individual spline teeth.

Harmonic Drive Applications. Industrial and military applications include speed reducers, combined motor reducers, linear and rotary actuators, servomechanisms, lifts and jacks, elevators and escalators, power take-offs, rotary-to-translatory motion converters, fine-coarse (two-speed) controllers, torque generators, instrument-panel controls, radio and radar antenna drives, and special turret-traversing mechanisms. Because of its ability to achieve positive mechanical drive through a sealed pressure wall, Harmonic Drive is uniquely suited for hermetically sealed power transmissions, sealed and pressurized valve actuators, mixer and agitator drives, control-rod drives for nuclear reactors, and special marine and medical uses.

REFERENCES

1. Benford, Robert L.: Planetary Gears, *Machine Design*, Aug. 21, 1958, pp. 129–135.
2. Wilkinson, William H.: Four Ways to Calculate Planetary Gear Trains, *Machine Design*, Jan. 7, 1960, pp. 155–159.
3. Sicklesteel, D. T.: Method for Design and Calculation of Planetary Gear Sets, *Design News*, June 15, 1952, pp. 45–49.
4. Poppinga, R.: The Efficiency of Planetary Gear Trains, *Engrs. Digest*, December, 1950, pp. 421–425.
5. Allen, H. N. G.: Epicyclic Gears, *Trans. Inst. Marine Engrs.*, Vol. 64, pp. 79–122, 1952.

6. Michalec, G. W.: Gear Differentials, *Machine Design*, October, December, 1955.
7. Tuplin, W. A.: Compound Epicyclic Gear Trains, *Machine Design*, Apr. 4, 1957, pp. 100–104.
8. Poppinga, R.: The Efficiency of Planetary Gear Trains, *Engrs. Digest*, December, 1950, pp. 421–425.
9. Eisbrenner, Ray D.: Brake on Output Shaft Permits Rapid Forward-reverse Shifting, *Design News*, Mar. 28, 1960.
10. (Editors Report): Planetary-type Reverse Gear Transmits 2500 H.P., *Design News*, Feb. 19, 1959.
11. Bryant, Richard C., and Darle W. Dudley: Which Right-angle Gear System?, *Product Engineering*, Nov. 7, 1960.

Chapter 4

ELEMENTS OF GEARS AND BASIC FORMULAS

By

GERALD L. SCOTT, *Staff Engineer, American Gear Manufacturers Association, Washington, D.C.*

This chapter covers gear nomenclature, standard letter symbols, and abbreviations. Basic formulas and relationships are given for all kinds of gears. The fundamental data given in this chapter will prove very useful in design proportioning of gears (covered in Chaps. 5 and 6) or in load rating of gears (covered in Chap. 13).

Published standards of the American Gear Manufacturers Assoc. (AGMA) are used wherever possible. They constitute the basis and language bridge for today's technical treatises on gearing. Nomenclature for special gear types, not standardized as yet by AGMA, are based on catalogues, special literature, and technical papers.

NOMENCLATURE AND ILLUSTRATIONS

AGMA standard[2,*] "Gear Nomenclature," AGMA 112.03 (ASA B6.10-1954), is the recognized authority on gear nomenclature. The terms, definitions, and illustrations contained in it are considered a complete coverage of the subject and are herein presented. A system is provided for this literature which makes it easily applicable to investigations and discussions of gear problems. Terms are grouped to obtain related continuity. Many of the definitions have been written in a way that makes them depend on one another. In several instances, terms and definitions are grouped in sequence. This method of arrangement leads to a more comprehensive understanding of the concepts and geometrical relations.

AGMA standard[1] 111.03 (ASA B6.5-1954), "Letter Symbols for Gear Engineering," is the authority on general principles of letter-symbols standardization for gearing. Capital and lower-case letters of the English and Greek alphabets are allocated to angles, distances, ratios, rates, factors, etc. The system of symbols and subscripts presented is generally consistent with other American Standard letter symbols and follows the same general principles.

AGMA standard[3] 116.01, "Abbreviations for Gearing," lists the most widely used abbreviations for gearing. Abbreviations are shortened forms of words often used on drawings, in texts, and in tables. However, they are not suitable for use in equations. A typical abbreviation is DP, which stands for diametral pitch, and it will be found on most gear drawings.

* Superscript numbers refer to references at the end of the chapter.

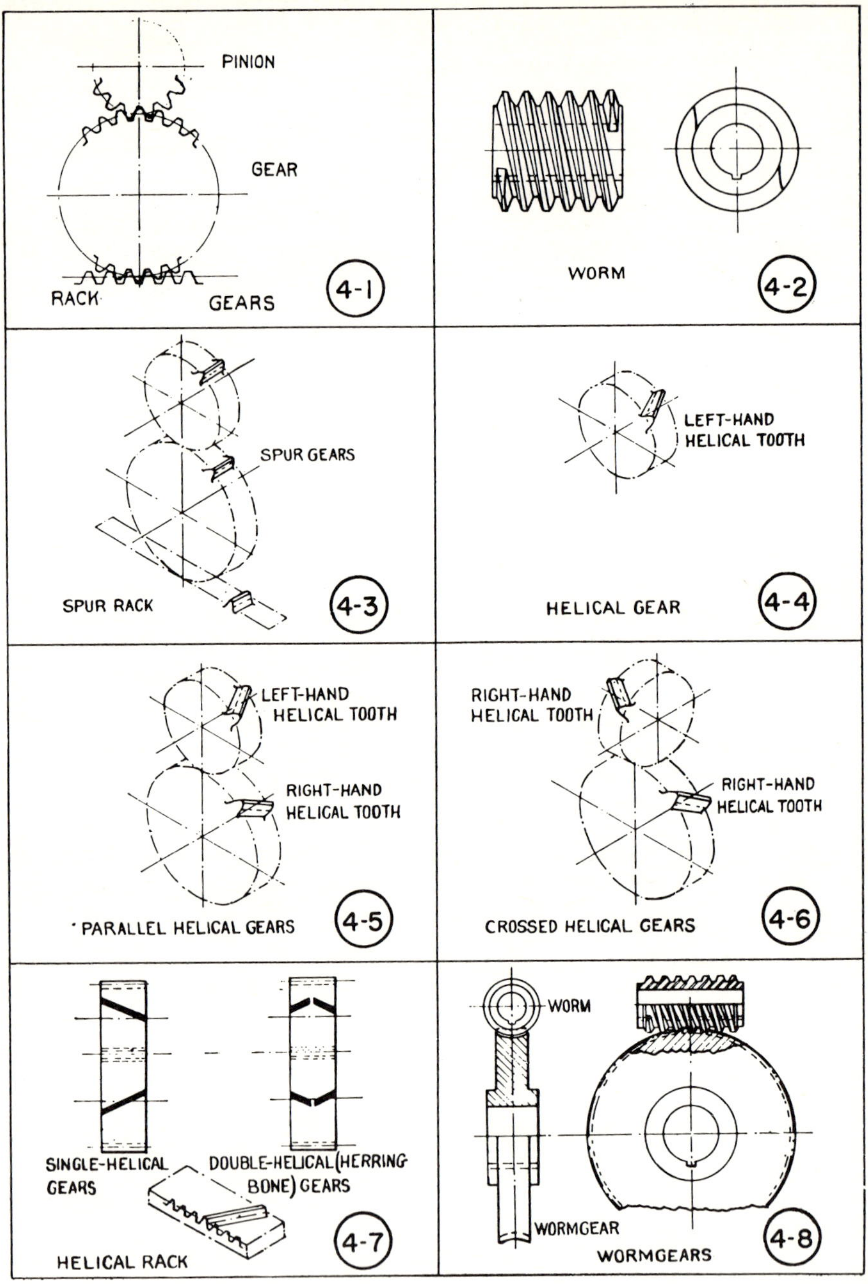

FIGS. 4-1 to 4-8. Nomenclature figures.

4-1. General Designations. *Gears* are machine elements that transmit motion by means of successively engaging teeth (see Fig. 4-1).

A *gear* is any machine part with gear teeth. Of two gears that run together, the one with the larger number of teeth is called the gear (see Fig. 4-1).

A *pinion* is a gear with a small number of teeth. Of two gears that run together, the one with the smaller number of teeth is called the pinion (see Fig. 4-1).

A *rack* is a gear with teeth spaced along a straight line, and suitable for straight-line motion. A basic rack is one that is adopted as the basis of a system of interchangeable gears. Standard gear-tooth proportions are often illustrated on an outline of the basic rack. A generating rack is a rack outline used to indicate tooth details and dimensions for the design of a required generating tool, such as a hob or gear-shaper cutter (see Fig. 4-1).

A *worm* is a gear with one or more teeth in the form of screw threads. Unless otherwise specified, it is cylindrical in form and has helical threads (see Fig. 4-2).

4-2. Kinds of Gears. *Spur gears* are cylindrical in form and operate on parallel axes. The teeth are straight and parallel to the axis (see Fig. 4-3).

A *spur rack* has straight teeth that are at right angles to the direction of motion (see Fig. 4-3).

A *helical gear* is cylindrical in form and has helical teeth (see Fig. 4-4).

Parallel-helical gears operate on parallel axes and the helices are of opposite hand (see Fig. 4-5).

Crossed-helical gears operate on crossed axes and may have teeth of the same or of opposite hand. The term crossed-helical gears has superseded the old term "spiral gears" (see Fig. 4-6).

Single-helical gears have teeth of only one hand on each gear (see Fig. 4-7).

Double-helical gears each have both right-hand and left-hand helical teeth and operate on parallel axes. These are also known as *herringbone* gears (see Fig. 4-7).

A *helical rack* has straight teeth that are oblique to the direction of motion (see Fig. 4-7).

Wormgearing includes worms and their mating gears. The axes are usually at right angles (see Fig. 4-8).

A *wormgear* is the mate to a worm. A wormgear that is completely conjugate to its worm has line contact and usually is cut by a counterpart of the worm. Some forms of hourglass worms and gears are called double-enveloping. A spur gear or helical gear used with a cylindrical worm has only point contact (see Fig. 4-8).

An *enveloping worm* has one or more threads and increases in diameter from its middle portion toward both ends, conforming to the curvature of the gear (see Fig. 4-9).

Bevel gears are conical in form and operate on intersecting axes which are usually at right angles (see Fig. 4-10).

Miter gears are a pair of bevel gears with equal numbers of teeth and with axes at right angles (see Fig. 4-11).

Angular bevel gears are bevel gears in which the axes are not at right angles (see Fig. 4-11).

A *crown gear* is a bevel gear having a plane pitch surface. The crown gear corresponds in bevel gears to the rack in spur gears (see Fig. 4-12).

Straight bevel gears have straight tooth elements which, if extended, would pass through the point of intersection of their axes (see Fig. 4-13).

Spiral bevel gears have teeth that are curved and oblique (see Fig. 4-14).

Zerol® bevel gears* have teeth that are curved but in the same general direction as straight teeth. They are spiral bevel gears of zero spiral angle (see Fig. 4-14).

* Registered trade-mark of the Gleason Works, Rochester, N.Y.

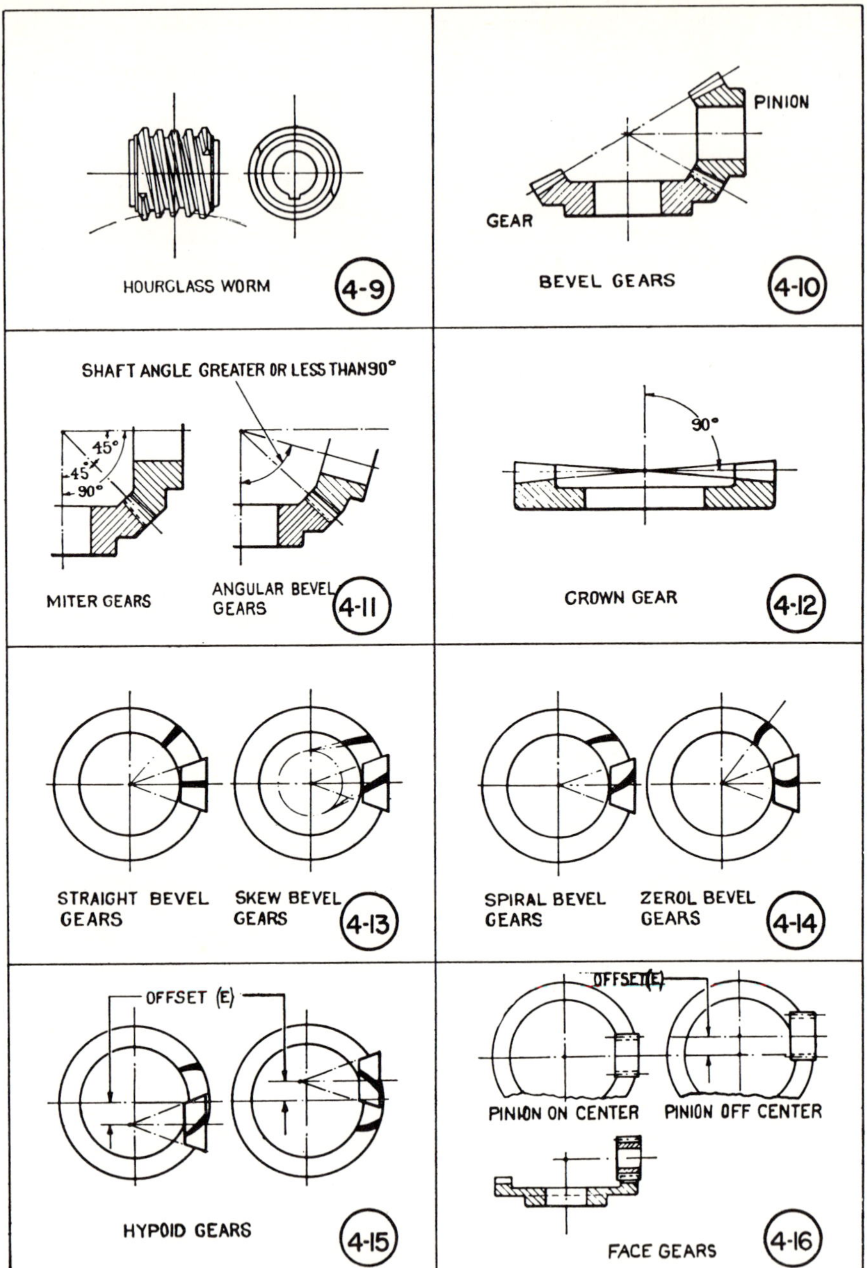

Figs. 4-9 to 4-16. Nomenclature figures.

Skew bevel gears are those for which the corresponding crown gear has teeth that are straight and oblique (see Fig. 4-13).

Hypoid gears are similar in general form to bevel gears but operate on nonintersecting axes (see Fig. 4-15).

Practically all hypoid gears have spiral teeth that are curved and oblique. The axes may be at right angles or otherwise. The tooth surfaces of a hypoid gear and pinion are both cut or generated by the same or similar tools.

Face gears consist of a spur or helical pinion in combination with a conjugate gear of disk form, the axes being usually at right angles, either intersecting or nonintersecting (see Fig. 4-16).

An *external gear* is one with the teeth formed on the outer surface of a cylinder or cone (see Fig. 4-17).

An *internal gear* is one with the teeth formed on the inner surface of a cylinder or cone (see Fig. 4-17).

A *right-hand helical gear or worm* is one in which the teeth twist clockwise as they recede from an observer looking along the axis (see Fig. 4-18).

A *left-hand helical gear or worm* is one in which the teeth twist counterclockwise as they recede from an observer looking along the axis (see Fig. 4-18).

The designations are the same as in the long-established practice for screw threads, both external and internal.

Two external helical gears operating on parallel axes must be of opposite hand. An internal helical gear and its pinion must be of the same hand.

A *right-hand spiral bevel gear* is one in which the outer half of a tooth is inclined in the clockwise direction from the axial plane through the mid-point of the tooth as viewed by an observer looking at the face of the gear (see Figs. 4-19 and 4-20).

A *left-hand spiral bevel gear* is one in which the outer half of a tooth is inclined in the counterclockwise direction from the axial plane through the mid-point of the tooth as viewed by an observer looking at the face of the gear (see Figs. 4-19 and 4-20).

A spiral bevel gear and pinion are always of opposite hand, including the case when the gear is internal.

The designations right-hand and left-hand are applied similarly to spiral bevel gears, Zerol bevel gears, skew bevel gears, hypoid gears, and oblique-tooth face gears.

The direction of curvature in right-hand and left-hand Zerol bevel gears is the same as for spiral bevel gears of similar form, in which the center of curvature is on the same side of the tooth as the cone apex.

In hypoid-gear design, the pinion and gear are practically always of opposite hand, and the spiral angle of the pinion is larger than that of the gear. The hypoid pinion is then larger in diameter than the corresponding bevel pinion.

4-3. Pitch Surfaces. For gears having the ratio of angular velocities constant, and axes either parallel or intersecting—and therefore not including crossed-helical gears, wormgearing, hypoid gears, or offset face gears.

Pitch surfaces are the imaginary planes, cylinders, or cones that roll together without slipping. For a constant-velocity ratio, the pitch cylinders and pitch cones are circular (see Fig. 4-21).

A *pitch plane* is the imaginary surface in a rack or in a crown gear that rolls without slipping with a pitch cylinder or pitch cone of another gear (see Fig. 4-23).

A *pitch cylinder* is the imaginary cylinder in a gear that rolls without slipping on a pitch cylinder or pitch plane of another gear (see Fig. 4-21).

A *pitch cone* is the imaginary cone in a bevel gear that rolls without slipping on a pitch cone of another gear (see Fig. 4-22).

4-4. Boundary Surfaces. The *outside cylinder* is the surface that coincides with the tops of the teeth of an external cylindrical gear (see Fig. 4-24).

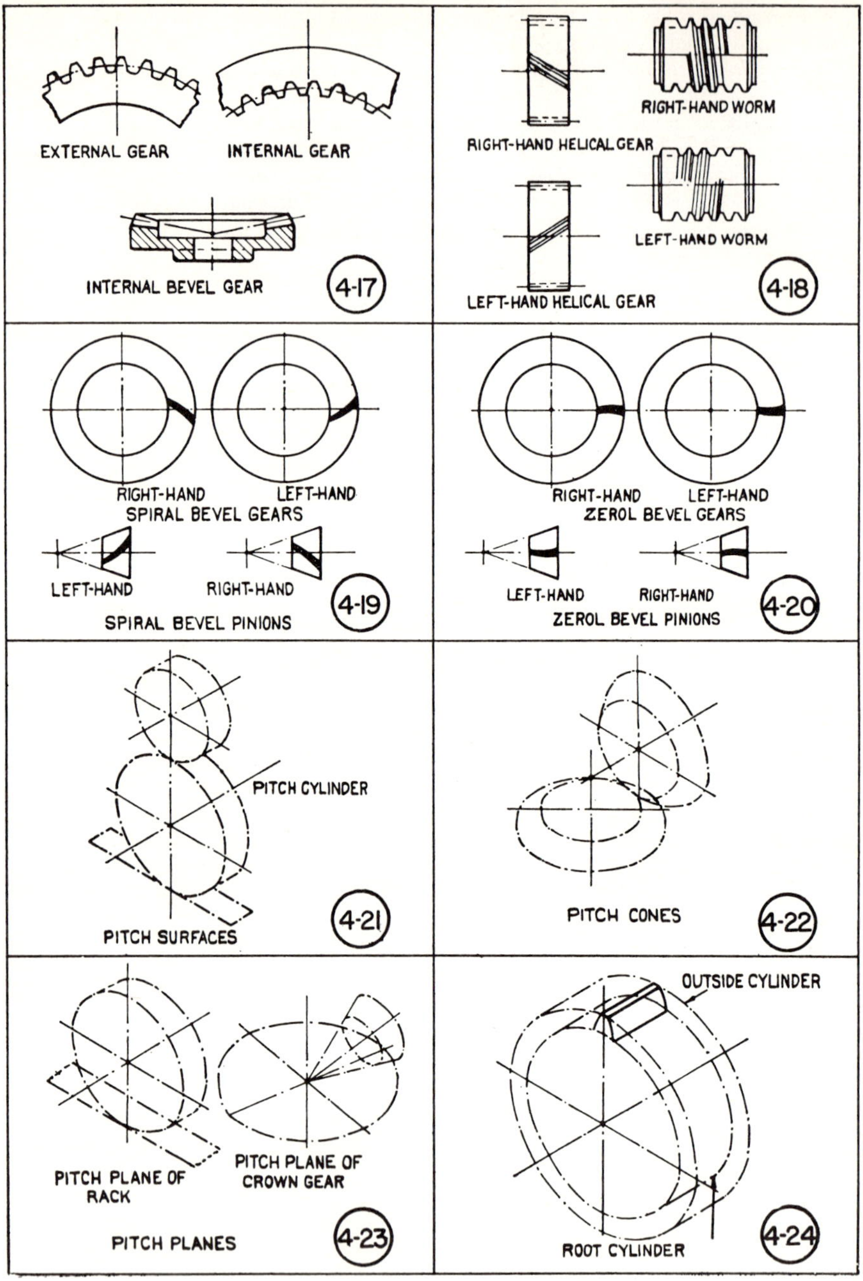

Figs. 4-17 to 4-24. Nomenclature figures.

The *inside cylinder* is the surface that coincides with the tops of the teeth of an internal cylindrical gear (see Fig. 4-52).

The *root cylinder* is the imaginary cylinder tangent to the bottoms of the tooth spaces in a cylindrical gear (see Fig. 4-24).

The *face cone* of a bevel gear contains the tops of the teeth (see Fig. 4-25).

The *root cone* of a bevel gear is an imaginary cone tangent to the bottoms of the tooth spaces (see Fig. 4-25).

The *back cone* of a bevel gear is at the outer ends of the teeth, with its elements perpendicular to those of the pitch cone. The surface of the gear blank at the outer ends of the teeth is customarily formed to such a back cone (see Fig. 4-25).

The *front cone* of a bevel gear is at the inner ends of the teeth, with its elements perpendicular to those of the pitch cone. The surface of the gear blank at the inner ends of the teeth is customarily formed to such a front cone but sometimes may be a plane on a pinion or a cylinder in a nearly flat gear (see Fig. 4-25).

The *crown circle* in a bevel gear is the circle of intersection of the back cone and face cone (see Fig. 4-25).

4-5. Principal Planes. The *axial plane* of a pair of gears is the plane that contains the two axes. In a single gear, an axial plane may be any plane containing the axis and a given point (see Fig. 4-26).

The *pitch plane* of a pair of gears is the plane perpendicular to the axial plane and tangent to the pitch surfaces. A pitch plane in an individual gear may be any plane tangent to its pitch surface. The pitch plane of a rack or crown gear is the pitch surface (see Fig. 4-26).

A *plane of rotation* is any plane perpendicular to a gear axis (see Fig. 4-27).

A *transverse plane* is perpendicular to the axial plane and to the pitch plane. In gears with parallel axes, the transverse plane and plane of rotation coincide (see Figs. 4-26 and 4-27).

A *tangent plane* is tangent to the tooth surfaces at a point or line of contact (see Figs. 4-28 and 4-38).

A *normal plane* is in general normal to a tooth surface at a pitch point, and perpendicular to the pitch plane. In a helical rack, a normal plane is normal to all the teeth it intersects. In a helical gear, however, a plane can be normal to only one tooth at a point lying in the plane surface. At such a point, the normal plane contains the line normal to the tooth surface and is normal to the pitch surface (see Fig. 4-28).

Important positions of a normal plane in tooth measurement and tool design in helical teeth and worm threads are

1. The plane normal to the pitch helix at the side of the tooth
2. The plane normal to the pitch helix at the center of the tooth
3. The plane normal to the pitch helix at the center of the space between two teeth

In a spiral bevel gear, one of the positions of a normal plane is at a mean point and the plane is normal to the tooth trace.

The *principal reference planes* are a pitch plane, axial plane, and transverse plane, all intersecting at a point and mutually perpendicular (see Fig. 4-26).

The *central plane* of a wormgear is perpendicular to the gear axis and contains the common perpendicular of the gear and worm axes. In the usual case with axes at right angles, it contains the worm axis (see Fig. 4-29).

4-6. Principal Directions. These are directions in the pitch plane and correspond to the principal cross sections of a tooth.

The *axial direction* is the direction of the intersection of the axial plane (see Fig. 4-30).

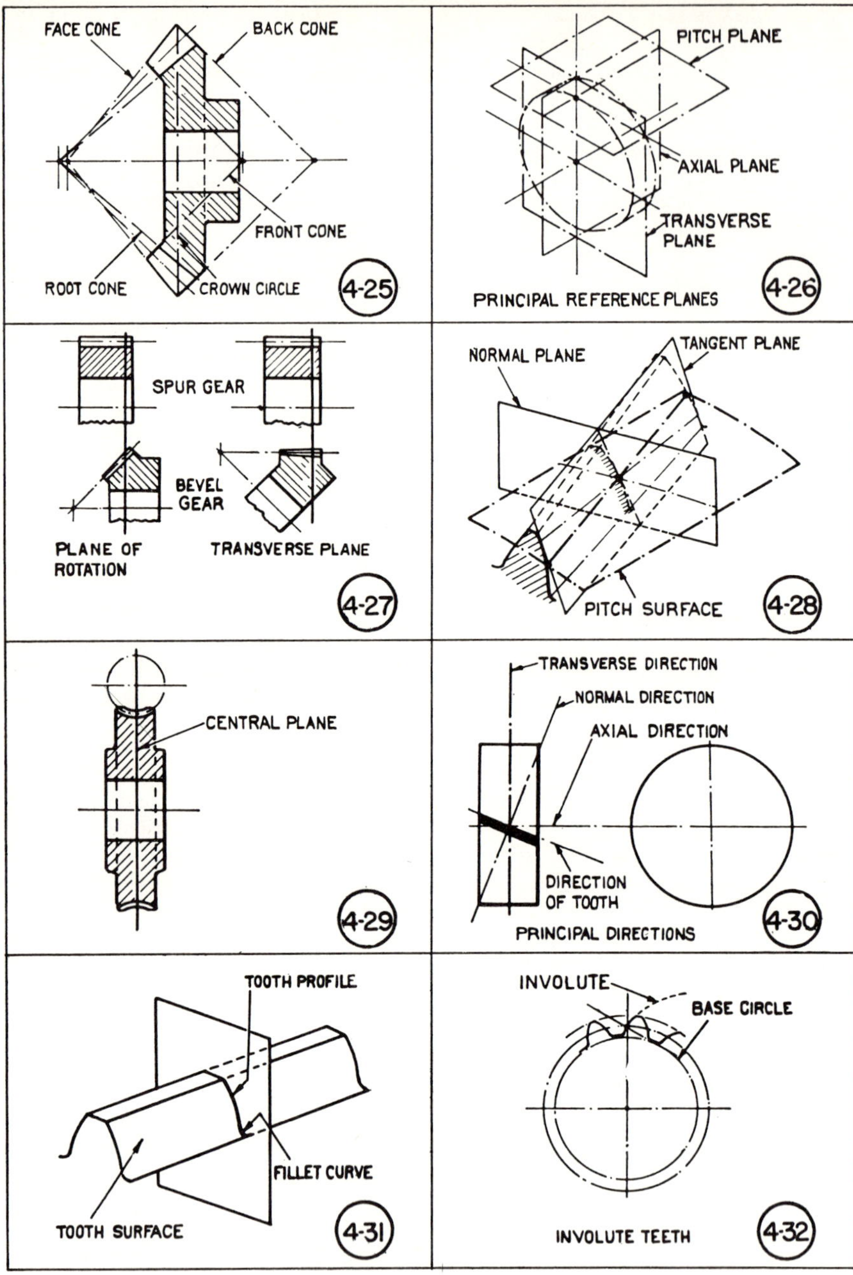

FIGS. 4-25 to 4-32. Nomenclature figures.

The *transverse direction* is the direction of the intersection of the transverse plane (see Fig. 4-30).

The *normal direction* is the direction of the intersection of the normal plane (see Fig. 4-30).

4-7. Elements of Gear Teeth. The *tooth surface* forms the side of a gear tooth (see Fig. 4-31).

A *tooth profile* is one side of a tooth in a cross section. Usually a profile is the curve of intersection of a tooth surface and a plane or surface normal to the pitch surface, such as the transverse, normal, or axial plane (see Fig. 4-31).

The *fillet curve* is the concave portion of the tooth profile where it joins the bottom of the tooth space (see Fig. 4-31).

Involute teeth of spur gears, helical gears, and worms are those in which the profile in a transverse plane (exclusive of the fillet curve) is the involute of a circle (see Fig. 4-32).

The *base circle* is the circle from which involute tooth profiles are derived (see Figs. 4-32 and 4-37).

The *base cylinder* corresponds to the base circle and is the cylinder from which involute tooth surfaces, either straight or helical, are derived (see Fig. 4-33).

A *pitch circle* is the curve of intersection of a pitch surface of revolution and a plane of rotation. According to theory, it is the imaginary circle that rolls without slipping with a pitch circle of a mating gear (see Fig. 4-34).

The *pitch line* corresponds in the cross section of a rack to the pitch circle in the cross section of a gear (see Fig. 4-34).

A *gear center* is the center of the pitch circle (see Fig. 4-34).

The *line of centers* connects the centers of the pitch circles of two engaging gears; it is also the common perpendicular of the axes in crossed-helical gears and wormgears. When one of the gears is a rack, the line of centers is perpendicular to its pitch line (see Fig. 4-34).

The *pitch point* is the point of tangency of two pitch circles (or of a pitch circle and pitch line) and is on the line of centers. The pitch point of a tooth profile is at its intersection with the pitch circle (see Fig. 4-34).

The *addendum circle* coincides with the tops of the teeth in a cross section (see Fig. 4-35).

The *root circle* is tangent to the bottoms of the tooth spaces in a cross section (see Fig. 4-35).

A *point of contact* is any point at which two tooth profiles touch each other (see Fig. 4-36).

The *path of contact* is the path of the point of contact of two engaging tooth profiles (see Figs. 4-36 and 4-39).

The *line of action* is the path of contact in involute gears. It is the straight line passing through the pitch point and tangent to the base circles (see Fig. 4-37).

The *line of contact* is the line or curve along which two tooth surfaces are tangent to each other (see Figs. 4-38 and 4-39).

The *surface of action* is the imaginary surface made up of all positions of the lines of contact of a given tooth surface (see Fig. 4-39).

The *plane of action* is the surface of action in involute parallel-axis gears with either straight or helical teeth. It is tangent to the base cylinders (see Fig. 4-39).

A *tooth trace* is a general term for the curve of intersection of a tooth surface and its pitch surface. The tooth trace of a straight tooth is a straight line (see Fig. 4-40).

The *pitch helix* is the curve of intersection of a tooth surface and its pitch cylinder in a helical gear (see Fig. 4-41).

The *base helix* of a helical involute gear or worm lies on its base cylinder (see **Fig.** 4-41).

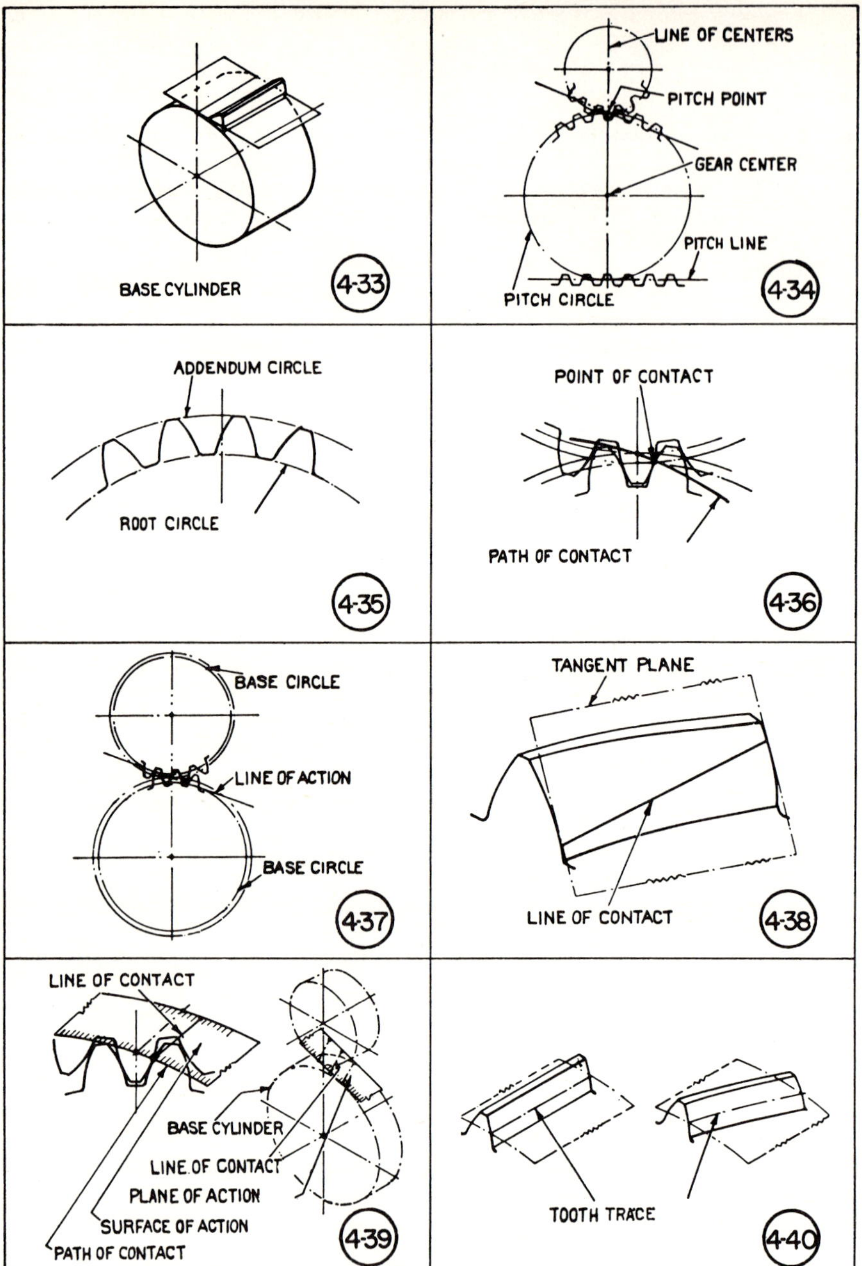

FIGS. 4-33 to 4-40. Nomenclature figures.

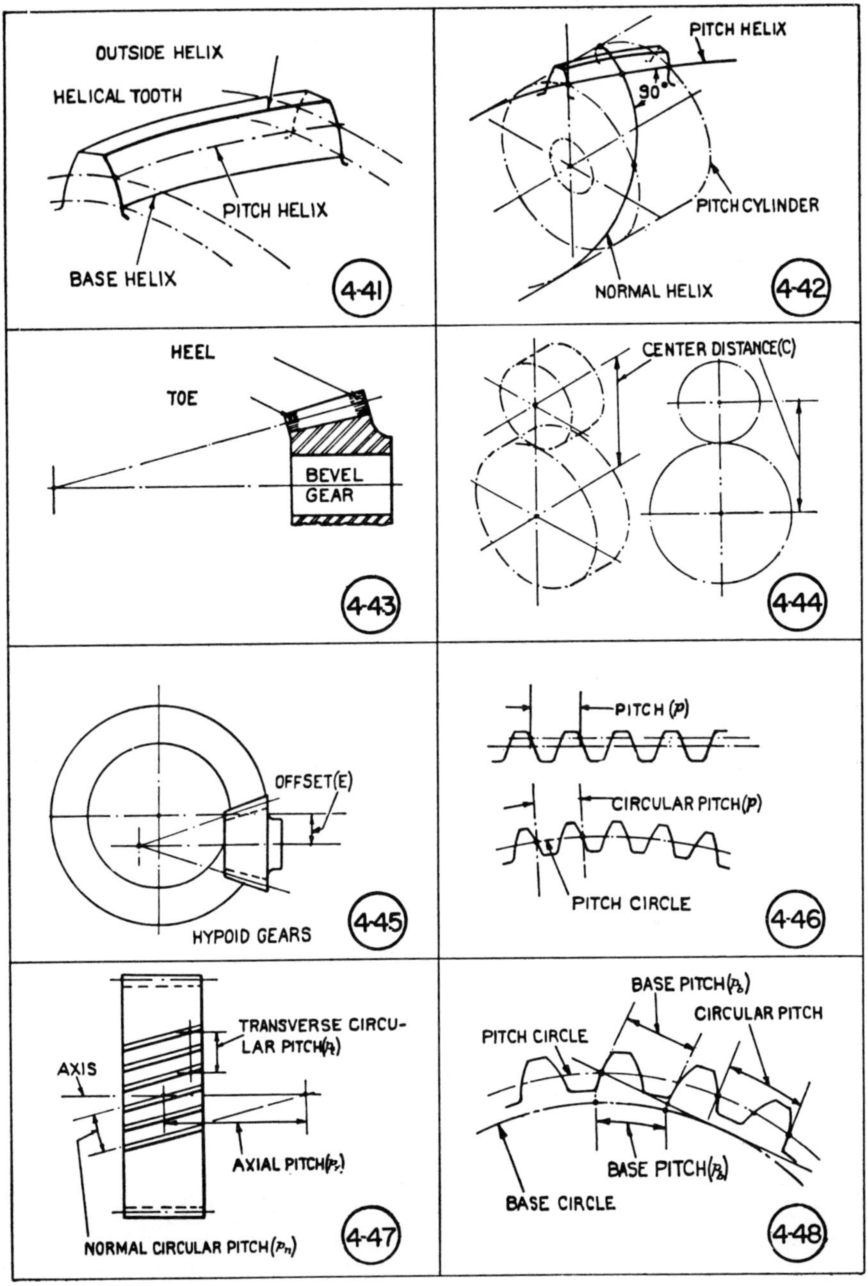

FIGS. 4-41 to 4-48. Nomenclature figures.

A *normal helix* is a helix on the pitch cylinder, normal to the pitch helix (see Fig. 4-42).

The *outside helix* of a helical gear or worm is the intersection of a tooth surface and the outside cylinder (see Fig. 4-41).

The *heel* of a tooth on a bevel gear or pinion is the portion of the tooth surface near its outer end (see Fig. 4-43).

The *toe* of a tooth on a bevel gear or pinion is the portion of the tooth surface near its inner end (see Fig. 4-43).

4-8. Linear and Circular Dimensions. Letter symbols for dimensions conform to AGMA standard 111.03 (ASA B6.5-1954) "Letter Symbols for Gear Engineering" (see Art. 4-12).

(C) *Center distance* is the distance between the parallel axes of spur gears and parallel-helical gears, or the crossed axes of crossed-helical gears and wormgears. Also, it is the distance between the centers of the pitch circles (see Fig. 4-44).

(E) *Offset* is the perpendicular distance between the axes of hypoid gears, and offset face gears (see Figs. 4-15 and 4-45).

(p) *Pitch* is the distance between similar equally spaced tooth surfaces along a given line or curve. The use of the single word pitch without qualification may be confusing, and for this reason specific designations are preferred, like circular pitch, axial pitch, diametral pitch (see Fig. 4-46).

(p) *Circular pitch* is the distance along the pitch circle or pitch line between corresponding profiles of adjacent teeth (see Fig. 4-46).

(p_t) *Transverse circular pitch* is the circular pitch in the transverse plane (see Fig. 4-47).

(p_n) *Normal circular pitch* is the circular pitch in the normal plane, and also the length of the arc along the normal helix between helical teeth or threads (see Fig. 4-47).

(p_x) *Axial pitch* is the circular pitch in the axial plane and in the pitch surface between corresponding sides of adjacent teeth, in helical gears and worms (see Fig. 4-47).

The term axial pitch is preferred to the term linear pitch. The axial pitch of a helical worm and the circular pitch of its wormgear are the same.

(p_b) *Base pitch* in an involute gear is the pitch on the base circle or along the line of action. Corresponding sides of involute gear teeth are parallel curves, and the base pitch is the constant and fundamental distance between them along a common normal in a plane of rotation (see Figs. 4-48 and 4-49).

(p_N) *Normal base pitch* in an involute helical gear is the base pitch in the normal plane. It is the normal distance between parallel helical involute surfaces on the line of action in the normal plane, or is the length of arc on the normal base helix. It is a constant distance in any helical involute gear (see Fig. 4-49).

(p_X) *Axial base pitch* is the base pitch of helical involute tooth surfaces in an axial plane (see Fig. 4-49).

(a) *Addendum* is the height by which a tooth projects beyond the pitch circle or pitch line; also, the radial distance between the pitch circle and the addendum circle (see Fig. 4-50).

(b) *Dedendum* is the depth of a tooth space below the pitch circle or pitch line; also, the radial distance between the pitch circle and root circle (see Fig. 4-50).

(c) *Clearance* is the amount by which the dedendum in a given gear exceeds the addendum of its mating gear (see Fig. 4-50).

(h_k) *Working depth* is the depth of engagement of two gears, that is, the sum of their addendums (see Fig. 4-50).

(h_t) *Whole depth* is the total depth of a tooth space, equal to addendum plus dedendum, also equal to working depth plus clearance (see Fig. 4-50).

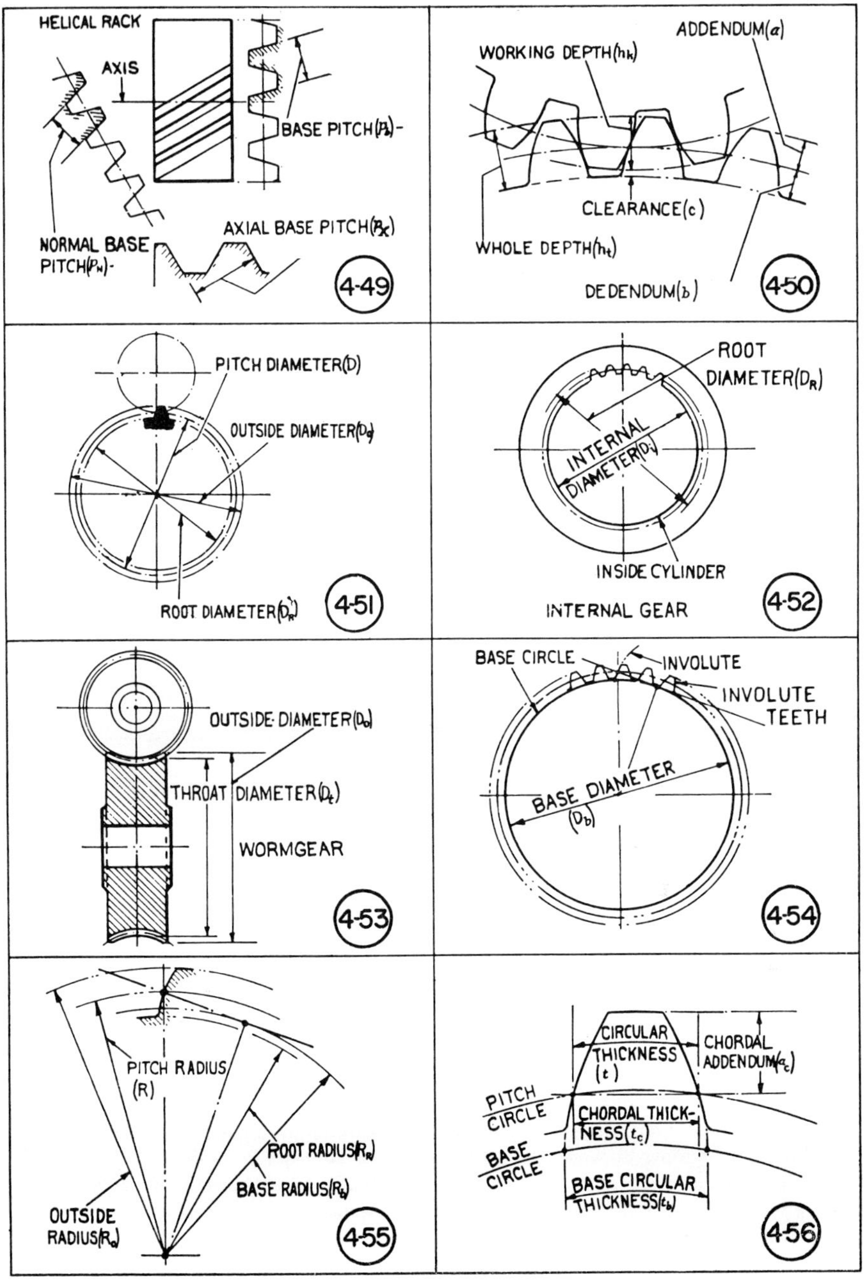

FIGS. 4-49 to 4-56. Nomenclature figures.

(D,d) *Pitch diameter* is the diameter of the pitch circle (see Fig. 4-51). In parallel-shaft gears, the pitch diameters can be determined from the center distance and the numbers of teeth by proportionality. A standard pitch diameter is one calculated according to the standard pitch of the gear-cutting tool.

Operating pitch diameters are the pitch diameters determined from the base circles and the center distance at which gears operate. *Testing pitch diameter* is a value that may be used in the testing or inspection of a gear, and which should be defined in the testing procedure. In a bevel gear, the pitch diameter is understood to be at the outer ends of the teeth unless otherwise specified.

(D_o,d_o) *Outside diameter* is the diameter of the addendum (outside) circle. In a bevel gear it is the diameter of the crown circle. In a throated wormgear, it is the maximum diameter of the blank (see Figs. 4-51 and 4-53). The term applies to external gears; for internal gears, see internal diameter.

(D_R,d_R) *Root diameter* is the diameter of the root circle (see Figs. 4-51 and 4-52).

(D_i) *Internal diameter* is the distance of the addendum circle of an internal gear (see Fig. 4-52).

(D_t) *Throat diameter* is the diameter of the addendum circle at the middle of a wormgear or concave worm (see Fig. 4-53).

(D_b,d_b) *Base diameter* is the diameter of the base cylinder of an involute gear (see Fig. 4-54).

(R,r) *Pitch radius* is the radius of the pitch circle (see pitch diameter and Fig. 4-55).

(R_o,r_o) *Outside radius* is the radius of the addendum circle of an external gear (see Fig. 4-55).

(R_R,r_R) *Root radius* is the radius of the root circle (see Fig. 4-55).

(R_b,r_b) *Base radius* is the radius of the base circle of involute profiles (see Fig. 4-55).

(t_G,t_p) *Circular thickness* is the length of arc between the two sides of a gear tooth, on the pitch circle unless otherwise specified (see Fig. 4-56).

(t_t) *Transverse circular thickness* is the circular thickness in the transverse plane (see Fig. 4-57).

(t_n) *Normal circular thickness* is the circular thickness in the normal plane. In a helical gear it may be considered as the length of arc along a normal helix (see Figs. 4-57 and 4-58).

(t_x) *Axial thickness* in helical gears and worms is the tooth thickness in the axial cross section at the pitch line (see Fig. 4-57).

(t_b) *Base circular thickness* in involute teeth is the length of arc on the base circle between the two involute curves forming the profiles of a tooth (see Fig. 4-56).

(t_c) *Chordal thickness* is the length of the chord subtending a circular-thickness arc (see Fig. 4-56).

(t_{nc}) *Normal chordal thickness* is the chordal thickness in the plane normal to the pitch helix or the tooth curve at the center of the tooth (see Fig. 4-58).

(a_c) *Chordal addendum* is the height from the top of the tooth to the chord subtending the circular-thickness arc (see Fig. 4-56).

(a_{nc}) *Normal chordal addendum* is the chordal addendum in the plane normal to the pitch helix or the tooth curve at the center of the tooth (see Fig. 4-58).

(B) *Backlash* is the amount by which the width of a tooth exceeds the thickness of the engaging tooth on the pitch circles (see Fig. 4-59).

As actually indicated by measuring devices, backlash may be determined variously in the transverse, normal, or axial planes, and either in the direction of the pitch circles or on the line of action. Such measurements should be corrected to corresponding values on transverse pitch circles for general comparisons.

(F) *Face width* is the length of the teeth in an axial plane (see Fig. 4-60).

(F_e) *Effective face width* is the portion that may actually come into contact with mating teeth, as occasionally one member of a pair of gears may have a greater face width than the other (see Fig. 4-60).

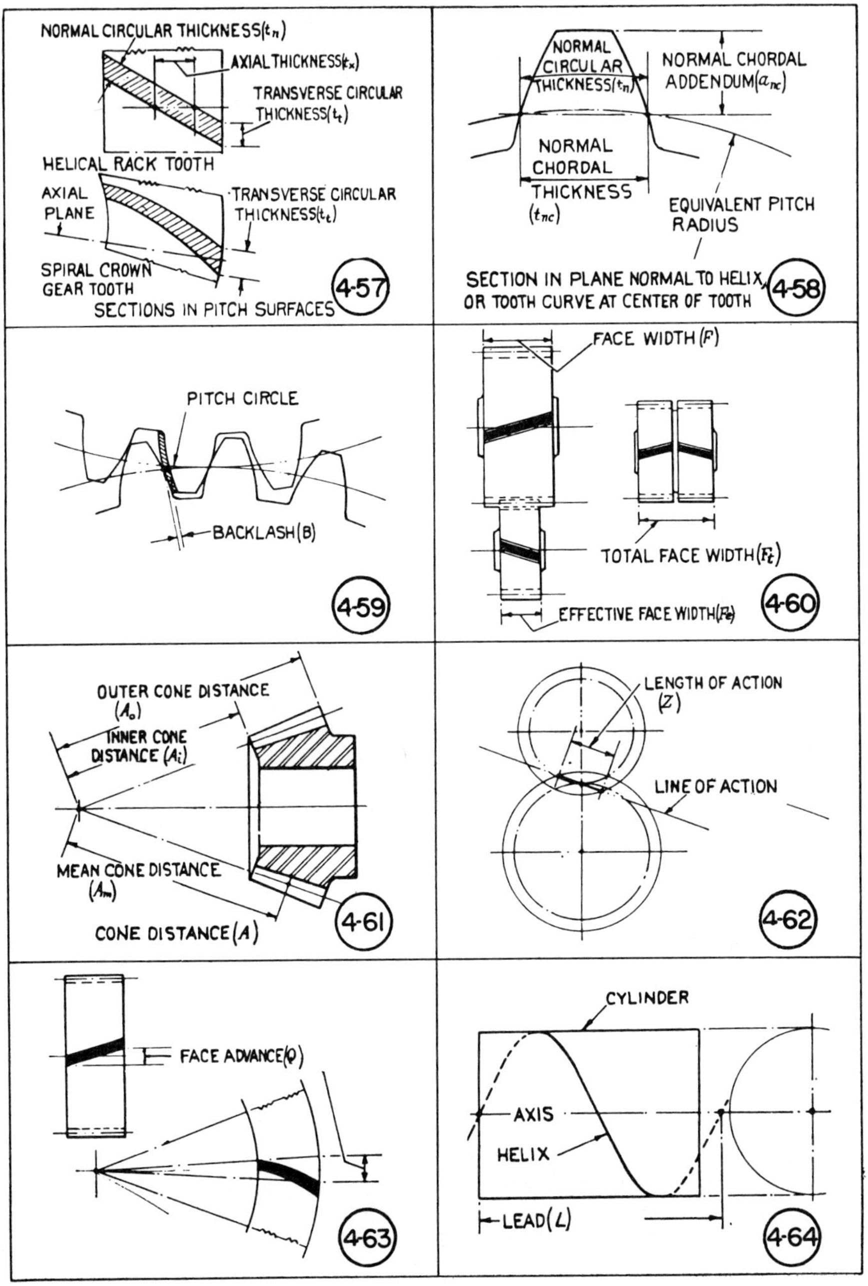

FIGS. 4-57 to 4-64. Nomenclature figures.

(F_t) *Total face width* is the actual dimension of a gear blank that exceeds the effective face width, or as in double-helical gears where the total face width includes any distance separating right-hand and left-hand helices (see Fig. 4-60).

(A) *Cone distance* in a bevel gear is the general term for the distance along an element of the pitch cone from the apex to any given position in the teeth (see Fig. 4-61).

(A_o) *Outer cone distance* in bevel gears is the distance from the apex of the pitch cone to the outer ends of the teeth. When not otherwise specified, the short term cone distance is assumed to be outer cone distance. It is customary to give the diametral pitch, tooth thickness, addendum, and dedendum in bevel gears at the outer cone distance (see Fig. 4-61).

(A_m) *Mean cone distance* in bevel gears is the distance from the apex of the pitch cone to the middle of the face width (see Fig. 4-61).

(A_i) *Inner cone distance* in bevel gears is the distance from the apex of the pitch cone to the inner ends of the teeth, where dimensions for allowable widths of tools and cutters are determined (see Fig. 4-61).

(Z) *Length of action* is the distance on an involute line of action through which the point of contact moves during the action of the tooth profile (see Fig. 4-62).

(Q) *Face advance* is the distance on a pitch circle through which a helical or spiral tooth moves from the position at which contact begins at one end of the tooth curve to the position when contact ceases at the other end (see Fig. 4-63).

(L) *Lead* is the axial advance of a helix for one complete turn, as in the threads of cylindrical worms and teeth of helical gears (see Fig. 4-64).

4-9. Angular Dimensions. (ϕ) *Pressure angle* is the angle between a tooth profile and the line normal to a pitch surface, usually at the pitch point of the profile. This definition is applicable to every type of gear. The term pressure originally meant an angle between the line of pressure and the pitch circle. In involute teeth, pressure angle is often described as the angle between the line of action and the line tangent to the pitch circles (see Fig. 4-65).

Standard pressure angles are established in connection with standard gear-tooth proportions and standard gear-cutting tools. Involute gears operate correctly after change of center distance; and gears designed for changed center distance can be correctly generated by standard cutting tools. According to gear theory, a change of center distance is accompanied by changes in circular pitch, diametral pitch, pitch diameters, pressure angle, and tooth thicknesses. The same involute gear may be used under conditions that change its pitch diameter and pressure angle. Unless there are good reasons for doing otherwise, however, it is convenient and practical to consider that the pitch diameter and pressure angle in an involute gear correspond to the pitch and pressure angle of the cutter or hob used to generate its teeth.

The *operating pressure angle* is determined by the base circles and the center distance at which two gears operate. Various other pressure angles may be used in gear calculations. In helical teeth, the pressure angle may be specified in the transverse plane, normal, or axial plane. For a spur or a straight bevel gear, only one direction of cross section needs to be considered, and the general term pressure angle is customarily used instead of the specific term transverse pressure angle.

Pressure angle is also used in the description and specification of the tools used to cut and generate gear teeth.

(ϕ_t) *Transverse pressure angle* is the pressure angle in the transverse plane (see Fig. 4-66).

(ϕ_n) *Normal pressure angle* is the pressure angle in the normal plane of a helical or spiral tooth. In spiral bevel gears, unless otherwise specified, pressure angle means normal pressure angle at the mean cone distance (see Fig. 4-66).

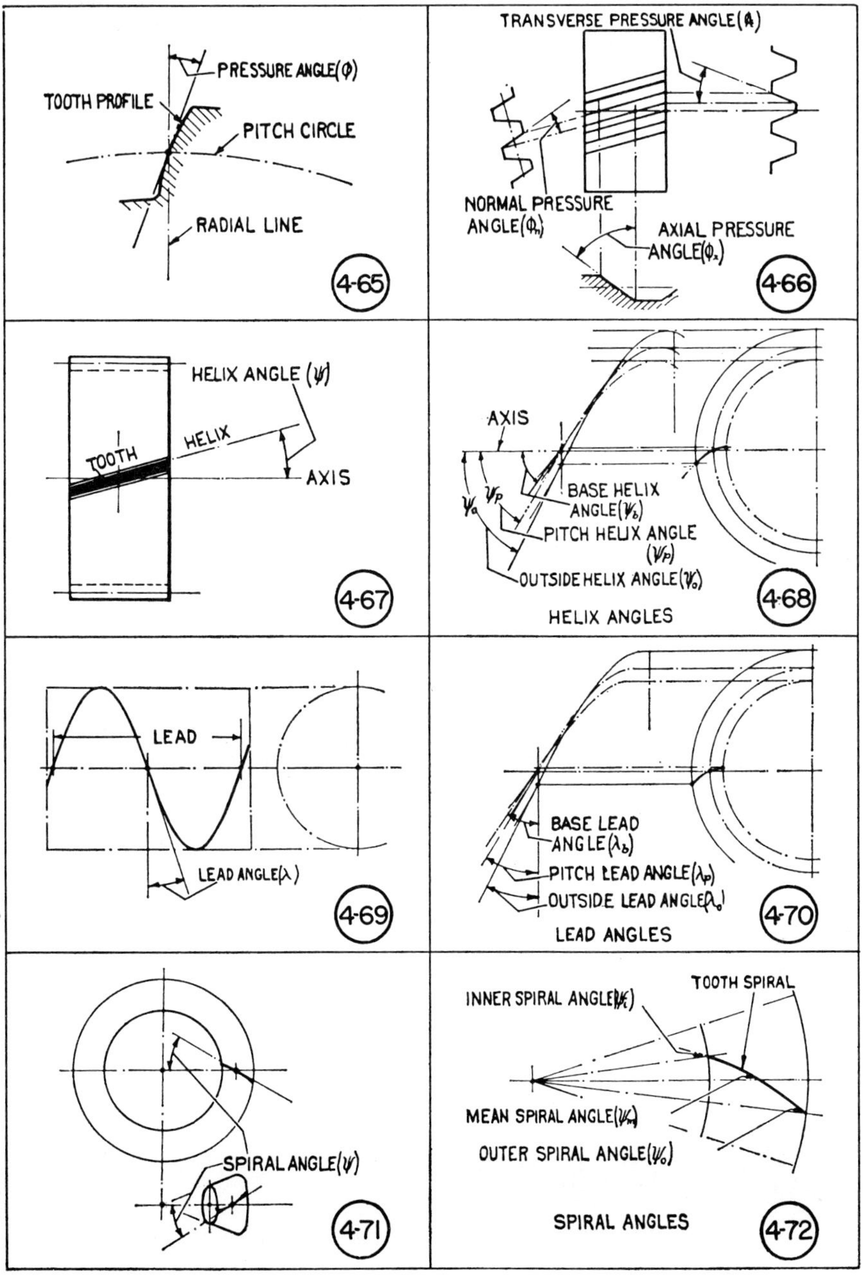

FIGS. 4-65 to 4-72. Nomenclature figures.

(ϕ_x) *Axial pressure angle* is the pressure angle in the axial plane of a helical or spiral tooth (see Fig. 4-66).

(ψ) *Helix angle* is the angle between any helix and an element of its cylinder. In helical gears and worms, it is at the pitch diameter unless otherwise specified (see Fig. 4-67).

(ψ_p) *Pitch helix angle* is the helix angle on the pitch cylinder (see Fig. 4-68).

(ψ_o) *Outside helix angle* is the helix angle on the outside cylinder (see Fig. 4-68).

(ψ_b) *Base helix angle* is the helix angle on the base cylinder of involute helical teeth or threads (see Fig. 4-68).

(λ) *Lead angle* is the angle between any helix and a plane of rotation. It is the complement of the helix angle and is used for convenience in worms and hobs. It is understood to be at the pitch diameter unless otherwise specified (see Fig. 4-69).

In screw-thread practice the term helix angle was formerly used instead of the term lead angle.

(λ_p) *Pitch lead angle* is the lead angle on the pitch cylinder (see Fig. 4-70).

(λ_o) *Outside lead angle* is the lead angle on the outside cylinder (see Fig. 4-70).

(λ_b) *Base lead angle* is the lead angle on the base cylinder (see Fig. 4-70).

(ψ) *Spiral angle,* in a spiral bevel gear, is the angle between the tooth trace and an element of the pitch cone, and corresponds to the helix angle in helical teeth. Unless otherwise specified, the spiral angle is understood to be at the mean cone distance (see Fig. 4-71).

(ψ_m) *Mean spiral angle* is the specific designation for the spiral angle at the mean cone distance in a bevel gear (see Fig. 4-72).

(ψ_o) *Outer spiral angle* is the spiral angle of a bevel gear at the outer cone distance, where the tooth thickness is customarily measured (see Fig. 4-72).

(ψ_i) *Inner spiral angle* is the spiral angle of a bevel gear at the inner cone distance, where the allowable point width of cutting tools usually must be determined (see Fig. 4-72).

(Σ) *Shaft angle* is the angle between the axes of two nonparallel gear shafts. In a pair of crossed-helical gears, the shaft angle lies between the oppositely rotating portions of two shafts. This applies also in the case of worm gearing. In bevel gears, the shaft angle is the sum of the two pitch angles. In hypoid gears, the shaft angle is given when starting a design, and it does not have a fixed relation to the pitch angles and spiral angles (see Fig. 4-73).

(Γ,γ) *Pitch angle,* in bevel gears, is the angle between an element of a pitch cone and its axis. In external and internal bevel gears, the pitch angles are, respectively, less than and greater than 90° (see Figs. 4-74 and 4-75).

(Γ_o,γ_o) *Face angle,* in a bevel gear, is the angle between an element of the face cone and its axis (see Fig. 4-74).

(Γ_R,γ_R) *Root angle,* in a bevel gear, is the angle between an element of the root cone and its axis (see Fig. 4-74).

(α) *Addendum angle,* in a bevel gear, is the angle between elements of the face cone and pitch cone (see Fig. 4-75).

(δ) *Dedendum angle,* in a bevel gear, is the angle between elements of the root cone and pitch cone (see Fig. 4-75).

4-10. Numbers and Ratios. (N) *Number of teeth or threads* is the number of teeth contained in the whole circumference of the pitch circle.

(m_G) *Gear ratio* is the ratio of the larger to the smaller number of teeth in a pair of gears.

(P) *Diametral pitch* is the ratio of the number of teeth to the pitch diameter in inches. Its value is in terms of inches^{-1}. Diametral pitch equals 3.1416 divided by circular pitch ($P = 3.1416/p$).

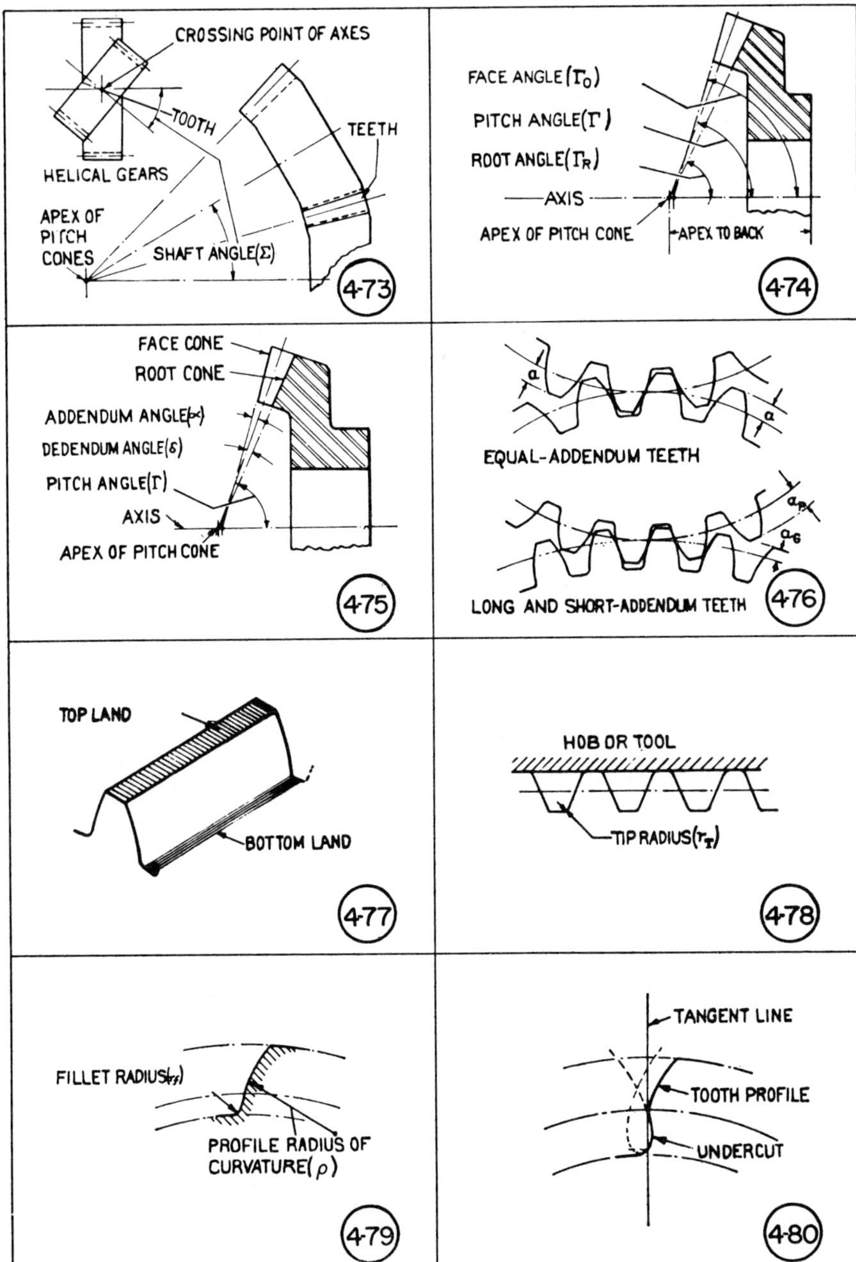

FIGS. 4-73 to 4-80. Nomenclature figures.

(P_t) *Transverse diametral pitch* is the diametral pitch in the transverse plane.

(P_n) *Normal diametral pitch* is the value of diametral pitch in a normal plane of a helical gear or worm. It is calculated by the formula $P_n = P \cos \psi$.

(m) *Module* (inches) is the ratio of the pitch diameter in inches to the number of teeth. It is the reciprocal of the diametral pitch.

(m) *Module* (millimeters) is the ratio of the pitch diameter in millimeters to the number of teeth.

4-11. Miscellaneous Terms and Dimensions. *Full-depth teeth* are those in which the working depth equals 2.000 divided by normal diametral pitch.

Stub teeth are those in which the working depth is less than 2.000 divided by normal diametral pitch.

Equal-addendum teeth are those in which two engaging gears have the same addendums (see Fig. 4-76).

Long-and-short-addendum teeth are those in which the addendums of two engaging gears are unequal (see Fig. 4-76).

Bottom land is the surface at the bottom of a tooth space adjoining the fillet (see Fig. 4-77).

Top land is the surface of the top of a tooth (see Fig. 4-77).

(r_T) *Tip radius* is the radius of the circular arc used to join a side-cutting edge and an end-cutting edge in gear-cutting tools (see Fig. 4-78).

(r_f) *Fillet radius* is the radius of a circular arc approximating the fillet curve. In generated teeth, the fillet curve has a varying radius of curvature (ρ) (see Fig. 4-79).

(ρ) *Profile radius of curvature* is the radius of curvature of a tooth profile, usually at the pitch point or a point of contact (see Fig. 4-79).

Undercut is a condition in generated gear teeth when any part of the fillet curve lies inside of a line drawn tangent to the working profile at its lowest point. Undercut may be deliberately introduced to facilitate finishing operations (see Fig. 4-80).

Equivalent pitch radius is the radius of the pitch circle in a cross section of gear teeth in any plane other than a plane of rotation. It is properly the radius of curvature of the pitch surface in the given cross section. Examples of such sections are the transverse section of bevel gear teeth and the normal section of helical teeth (see Figs. 4-81 and 4-82).

(N_e) *Equivalent number of teeth* is the number of teeth contained in the whole circumference of a pitch circle corresponding to an equivalent pitch radius.

(Q_t) *Arc of action* is the arc of the pitch circle through which a tooth profile moves from the beginning to the end of contact with a mating profile (see Fig. 4-83).

(Q_a) *Arc of approach* is the arc of the pitch circle through which a tooth profile moves from its beginning of contact until the point of contact arrives at the pitch point (see Fig. 4-83).

(Q_r) *Arc of recess* is the arc of the pitch circle through which a tooth profile moves from contact at the pitch point until contact ends (see Fig. 4-83).

(m_c) *Contact ratio* is the ratio of the arc of action to the circular pitch, and sometimes is thought of as an average number of teeth in contact. For involute gears, the contact ratio is obtained most directly as the ratio of the length of action to the base pitch.

(m_t) *Transverse contact ratio* is the contact ratio in the transverse plane.

(m_n) *Normal contact ratio* is the contact ratio in the normal section.

(m_x) *Axial contact ratio* is the ratio of face width to the axial pitch in helical teeth.

(m_F) *Face contact ratio* is the ratio of the face advance to the circular pitch, usually having the same value as axial contact ratio.

(m_t) *Total contact ratio* is the sum of the transverse contact ratio and axial contact ratio, which may be thought of as the average total number of teeth in contact in parallel-helical gears or spiral bevel gears.

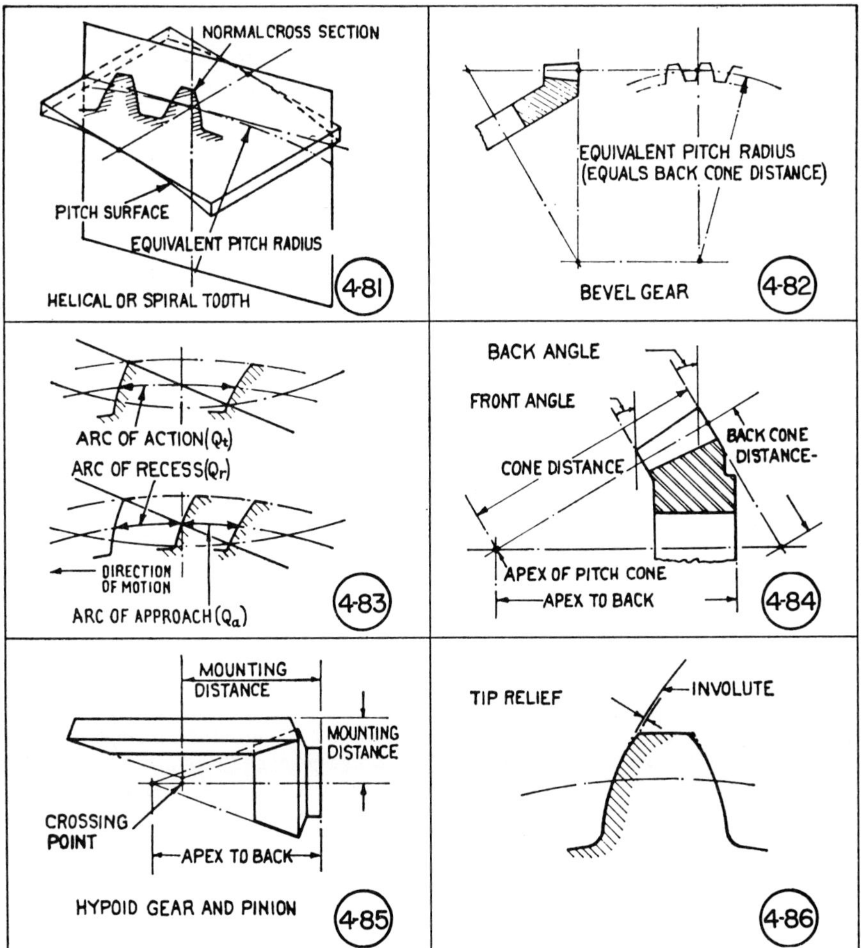

Figs. 4-81 to 4-86. Nomenclature figures.

Back cone distance in a bevel gear is the distance along an element of the back cone from its apex to the pitch circle, and in the direction normal to the pitch cone (see Fig. 4-84).

Back angle, in a bevel gear, is the angle between an element of the back cone and a plane of rotation, and usually is equal to the pitch angle (see Fig. 4-84).

Front angle, in a bevel gear, is the angle between an element of the front cone and a plane of rotation, and usually equals the pitch angle (see Fig. 4-84).

Apex to back, on a bevel gear or hypoid gear, is the distance in the direction of the axis from the apex of the pitch cone to a locating surface at back of blank (see Figs. 4-84 and 4-85).

Mounting distance, for assembling bevel gears or hypoid gears, is the distance from the crossing point of the axes to a locating surface of a gear, which may be at either back or front (see Fig. 4-85).

Tip relief is an arbitrary modification of a tooth profile whereby a small amount of material is removed near the tip of the gear tooth (see Fig. 4-86).

Crossing point is the point of intersection of bevel-gear axes; also the apparent point of intersection of the axes in hypoid gears, crossed-helical gears, wormgears, and offset face gears, when projected to a plane parallel to both axes (see Fig. 4-85).

Form diameter is the diameter of the circle outside of which the tooth profile must conform to the specified involute curve.

Crowned teeth are those modified in a lengthwise direction to produce localized contact.

STANDARD LETTER SYMBOLS

Standard letter symbols for gear engineering were developed to establish a uniform practice in mathematical notation for equations and formulas dealing with toothed gearing. Such equations and corresponding calculations may be in connection with design, application, manufacture, inspection, new methods, and new problems.

Standard letter symbols are not intended or suggested for use on working drawings, manufacturing records, or correspondence, where abbreviations would serve better. The principal use of standard symbols is in mathematical work, published articles, papers, and books, to make understanding easier and confusion less likely. Tables 4-1 and 4-2 show the standard symbols used in gear work.

4-12. Manuscripts. Authors preparing manuscripts dealing with calculations involving gears should adopt the symbols in this section for usual gear terms, and for other terms and symbols in general mathematics, science, and mechanics should refer to other lists of American Standard letter symbols, such as that for the "Mechanics of Solid Bodies" (ASA Z10.3-1948). The terms for which symbols are used should be clearly defined and a table of notation should be given.

4-13. Symbols. Symbols must be distinguished from abbreviations, which are shortened forms of words often used on drawings and in tables. They are not, however, suitable for mathematical work. For example, the symbol for circular pitch is p, whereas the abbreviation is CP.

The allocation of upper-case and lower-case English alphabets and Greek letters for symbols in gearing is shown in Table 4-1. The plan for subscripts is given in Art. 4-14. It is important to distinguish accurately between upper-case and lower-case letters.

The system of symbols and subscripts here presented is generally consistent with American Standard letter symbols and follows the same general principles. It is not restricted to standardized gear terms but instead provides for extension to new problems and practices. A large number of terms and symbols illustrating the application of the system are given in Table 4-2.

Definitions of standard terms in gearing are given in Arts. 4-1 through 4-11.

Table 4-1. Allocation of Letters for Gear-engineering Symbols

A	cone distance	a	addendum; acceleration
B	backlash	b	dedendum
C	center distance	c	clearance
D	diameter	d	diameter
E	offset (of axes)	e	
F	face width	f	
G		g	
H		h	depth; height
I		i	
J		j	
K	factor (general)	k	
L	length; lead	l	lead
M	bending moment	m	ratio; module (inches or millimeters)
N	number of teeth	n	revolutions per unit time
O		o	
P	diametral pitch; power	p	pitch, circular
Q	face advance; arc	q	
R	radius	r	radius
S		s	distance; stress
T	torque	t	tooth thickness
U		u	
V		v	velocity, linear
W	load (total)	w	load per unit length
X	strength factor	x	
Y	tooth-form factor (diametral pitch)	y	tooth-form factor (circular pitch)
Z	length of action	z	

Greek Letters Employed

		A. General Angles in Gears
α	(alpha)	addendum angle
δ	(delta)	dedendum angle
ψ	(psi)	helix angle, spiral angle
θ	(theta)	involute angle
λ	(lambda)	lead angle
Γ, γ	(gamma)	pitch angles
ϕ	(phi)	pressure angle
ϵ	(epsilon)	roll angle
Σ	(sigma)	shaft angle
		B. Miscellaneous Terms (Common to Other Fields)
α	(alpha)	angular acceleration
ρ	(rho)	radius of curvature
θ	(theta)	angular displacement (angle in polar coordinates)
ω	(omega)	angular velocity

4-14. Subscripts. A subscript appearing immediately after and somewhat below the general symbol is used to indicate a value applying to a particular gear or tool, or a value taken at a particular position or in a particular direction. In the case of gears with oblique teeth, it is usually necessary to be specific and to refer to the transverse pitch p_t and the transverse pressure angle ϕ_t in order not to leave any doubt as to whether values are being given for the transverse plane or normal plane.

In simple determinations involving only one gear and one pinion, a convenient way to eliminate subscripts is to use upper-case symbols for values applying to the gear and lower-case letter symbols for values applying to the pinion. This may be allowable even with letters listed, as long as confusion is avoided by properly explaining the notation used. Examples: D, d for diameters; N, n for numbers of teeth; T, t for tooth thicknesses; and Γ, γ for pitch angles.

In allocating subscripts, it should be noted that a value applying to a part such as a gear, pinion, tool, cutter, etc., is in general indicated by a capital-letter subscript, which has the effect of making such designations more conspicuous, thus:

$$a_P = \text{addendum of pinion}$$
$$L_W = \text{lead of worm}$$
$$\phi_T = \text{pressure angle of tool}$$

In general, capital-letter subscripts stand for nouns.

Lower-case letters are used to represent adjectives, which indicate kind, position, direction, etc., thus:

$$p_b = \text{base pitch}$$
$$p_n = \text{normal pitch}$$
$$\phi_x = \text{axial pressure angle}$$

As has been done in mathematics, numerals can be used for subscripts whenever desired, as d_1, d_2, d_3, etc.; but letter subscripts selected from significant words are easier to remember.

4-15. Conflicts. In a few instances the same letter appears as the symbol for two different terms, and of course there are numerous cases of conflict between symbols in different fields. This is not a cause of confusion, unless the two terms or meanings occur in the same equation or problem. Then one way to remove doubt is to apply subscripts. In such a case an author should be careful to list the terms and symbols which he uses.

4-16. Typography. In accordance with the usual practice in published text, letter symbols, letter superscripts, and letter subscripts, whether upper or lower case, should be printed in *italic type*. An exception is Greek capital letters, which are *always vertical.* In manuscripts, italics are indicated by underlining each symbol to be italicized.

Numbers appearing as coefficients, subscripts, superscripts, or exponents should be printed in vertical Arabic numerals. Letter abbreviations, not recommended for use in formulas, should always be vertical. Trigonometric functions should be printed in lower-case vertical type. Standard mathematical notation should be followed. To avoid confusion in reading symbols, and particularly superscripts and subscripts, it is important to select type faces that have italic forms and that make a clear distinction between upper and lower case. For this reason type faces with serifs are not recommended.

ABBREVIATIONS FOR GEARING

Abbreviations are shortened forms of words often used on drawings, in tables, and in texts. However, abbreviations are not suitable for use in equations. The use of abbreviations is approved only when there can be no possible misunderstanding as to their meanings.

Table 4-2. Special Symbols for Gear Engineering

List of Symbols

(For convenience in use, this list includes also the General Symbols)

Term	Symbol
Acceleration, linear	a
Acceleration, angular	α
Addendum	a
Addendum of gear	a_G
Addendum of pinion	a_P
Chordal addendum (for tooth calipers)	a_c
Normal chordal addendum (for tooth calipers)	a_{nc}
Addendum angle (in bevel gears)	α
Addendum angle of gear	α_G
Addendum angle of pinion	α_P
Arc of action	Q_t
Arc of approach	Q_a
Arc of recess	Q_r
Backlash (on pitch circle)	B
Center distance	C
Clearance	c
Cone distance (in bevel gears)	A
Outer cone distance	A_o
Mean cone distance	A_m
Inner cone distance	A_i
Constant (factor)	K
Dedendum	b
Dedendum of gear	b_G
Dedendum of pinion	b_P
Dedendum angle (in bevel gears)	δ
Dedendum angle of gear	δ_G
Dedendum angle of pinion	δ_P
Depth; height	h
Working depth	h_k
Whole depth (total depth)	h_t
Diameters	D, d
Pitch diameter of gear	D, D_G
Pitch diameter of pinion	d, D_P
Internal diameter	D_i
Outside diameter of gear	D_o
Outside diameter of pinion	d_o
Base diameter of gear	D_b
Base diameter of pinion	d_b
Base diameter of cutter	D_{bC}
Equivalent diameter in normal plane	D_n
Root diameter of gear	D_R
Root diameter of pinion	d_R
Throat diameter (of worm gear)	D_t
Distance, linear or along an arc	s
Distance, angular (angle in polar coordinates)	θ
Face advance (in a helical gear, $Q_F = F \tan \psi$)	Q_F
Face width	F
Effective (or active) face width	F_e
Total face width	F_t
Face angle of gear (in bevel gears)	Γ_o
Face angle of pinion (in bevel gears)	γ_e
Factor	K
Helix angle	ψ
Outside helix angle	ψ_o
Base helix angle	ψ_b
Pitch helix angle	ψ_p
Lead (of worm or helical gear)	L, l
Lead angle $\left(\lambda = 90° - \psi, \tan \lambda = \dfrac{l}{\pi D}\right)$	λ

Table 4-2. Special Symbols for Gear Engineering *(Continued)*

Pitch lead angle	λ_p
Outside lead angle	λ_o
Base lead angle	λ_b
Lengths	L, l
Length of line of contact	l
Total length of lines of contact	L
Length of line of action	Z
Approach portion of line of action	Z_a
Recess portion of line of action	Z_r
Load (total)	W
Transverse, or tangential, component of load	W_t
Load component normal to helix (in pitch plane)	W_n
Load normal to surface	W_N
Radial component of load	W_r
Load per unit length	w
Module ($m = D/N$), inches or millimeters	m
Moment, bending	M
Number (of teeth)	N
Number of teeth in gear	N_G
Number of teeth in pinion	N_P
Number of teeth in cutter	N_C
Number of teeth in crown gear	N_c
Equivalent number of teeth	N_e
Number of threads in worm	N_W
Offset (of axes, in hypoid gears and face gears)	E
Pitch, circular	p
Base pitch (normal to involute)	p_b
Transverse pitch	p, p_t
Normal pitch (normal to helix)	p_n
Axial pitch (preferred to linear pitch)	p_x
Normal base pitch (normal to surface)	p_N
Axial pitch of worm	p_{xW}
Axial base pitch	p_X
Pitch, diametral	P, P_d
Normal diametral pitch	P_n, P_{nd}
Pitch angle of gear (in bevel gears)	Γ
Pitch angle of pinion (in bevel gears)	γ
Power	P
Pressure angle	ϕ
Transverse pressure angle	ϕ, ϕ_t
Normal pressure angle	ϕ, ϕ_n
Axial pressure angle	ϕ_x
Pressure angle of cutter	ϕ_C
Axial pressure angle of worm	ϕ_{xW}
Radii	R, r
Pitch radius of gear	R, R_G
Pitch radius of pinion	r, R_P
Outside radius of gear	R_o
Outside radius of pinion	r_o
Pitch radius of cutter	R_C
Base radius of gear	R_b
Base radius of pinion	r_b
Root radius of gear	R_R
Root radius of pinion	r_R
Edge radius of tool	r_T
Fillet radius (when constant)	r_f
Radius of curvature	ρ
Profile radius of curvature in gear	ρ_G
Profile radius of curvature in pinion	ρ_P
Radius of curvature in normal plane	ρ_n
Relative radius of curvature ($1/\rho_0 = 1/\rho_1 + 1/\rho_2$)	ρ_0
Radius of curvature of fillet	ρ_f

Table 4-2. Special Symbols for Gear Engineering *(Continued)*

Gear ratio ($m_G = N_G/N_P$)	m_G
Angular velocity ratio	m_w
Transverse contact ratio	m
Normal contact ratio	m_n
Axial contact ratio	m_x
Contact ratio (of profiles) ($m_p = Z/p_b$)	m_p
Face contact ratio (in axial plane, $m_F = F/p_x$)	m_F
Total contact ratio ($m_t = m_p + m_F$)	m_t
Revolutions per unit of time	n
Roll angle	ϵ
Root angle of gear (in bevel gears)	Γ_R
Root angle of pinion (in bevel gears)	γ_R
Shaft angle	Σ
Spiral angle (in bevel gears)	ψ
Spiral angle of gear	ψ_G
Spiral angle of pinion	ψ_P
Spiral angle at outer cone distance	ψ_o
Spiral angle at mean cone distance	ψ_m
Spiral angle at inner cone distance	ψ_i
Strength factor ($X = t^2/4h$)	X
Stress	s
Stress concentration factor	K
Tooth thickness	t
Circular thickness of gear	t_G
Circular thickness of pinion	t_P
Normal circular thickness	t_n
Tooth thickness on base circle	t_b
Chordal thickness	t_c
Normal chordal thickness	t_{nc}
Transverse circular thickness	t_t
Axial thickness	t_x
Tooth-form factor for circular pitch	y
Tooth-form factor for diametral pitch	Y
Tooth-form factor including stress concentration	Y_K
Torque	T
Velocity, linear	v
Velocity in transverse direction tangential to pitch circle	v_t
Velocity in normal direction (in pitch plane)	v_n
Velocity in axial direction	v_x
Velocity normal to surface	v_N
Velocity, angular	ω
Angular velocity of gear	ω_G
Angular velocity of pinion	ω_P

4-17. Abbreviations. The abbreviation for circular pitch is CP, whereas the letter symbol for circular pitch is the single lower-case italic letter *p*. Abbreviations are always printed in vertical type. Letter symbols are always printed in italics.

Abbreviations commonly used in the field of gearing for texts, tables, and drawings are shown below.

Recommended Abbreviations

AP	axial pitch
APA	axial pressure angle
BD	base diameter
BHA	base helix angle
BP	base pitch
CD	center distance
CP	circular pitch
DP	diametral pitch
FL	flute lead
FW	face width
HA	helix angle
ID	internal diameter; inside diameter
L	lead
LA	lead angle
LH	left-hand
NCP	normal circular pitch
NDP	normal diametral pitch
NPA	normal pressure angle
OD	outside diameter
OHA	outside helix angle
PA	pressure angle
PD	pitch diameter
RH	right-hand
T	teeth
THD	thread
WD	whole depth

GEAR FORMULAS

Most general facts, rules, and principles of gearing are expressed by means of formulas in algebraic symbols. Standardized nomenclature and letter symbols presented in Arts. 4-1 through 4-17 have made it possible to have many formulas for the popular gear types which are easily recognizable without referring to a list of letter symbols.

It is the intention in this section to present those basic gear formulas and relationships which do not require the matter of judgment on the part of the user. Gear elements requiring judgment on proportions, size, etc., are presented in Chaps. 5 and 6.

The basic formula section consisting mainly of tables is broken down into general tables and gear-type tables. The general tables consist of formulas which are applicable to nearly all gear types. The gear-type tables present formulas applicable to only those gear types.

The tables have been set up in a form which it is believed will make them most easily and readily usable for the general gear designer. Tables, wherever possible, have been set up to present formulas expressed in both diametral and circular pitch.

Article 4-19 presents basic formulas and data for several new or special gear types. These special gear types are normally limited in application. However, there are applications where these special gear types are the only solution to certain application problems.

4-18. Basic Gear Formulas. Tables 4-3, 4-4, and 4-5 contain formulas common to most gear types. Table 4-3 contains formulas based upon circular pitch, whereas Table 4-4 contains these same formulas based upon diametral pitch. Table 4-5 contains those formulas and relationships dealing with angles, contact ratios, and conversions.

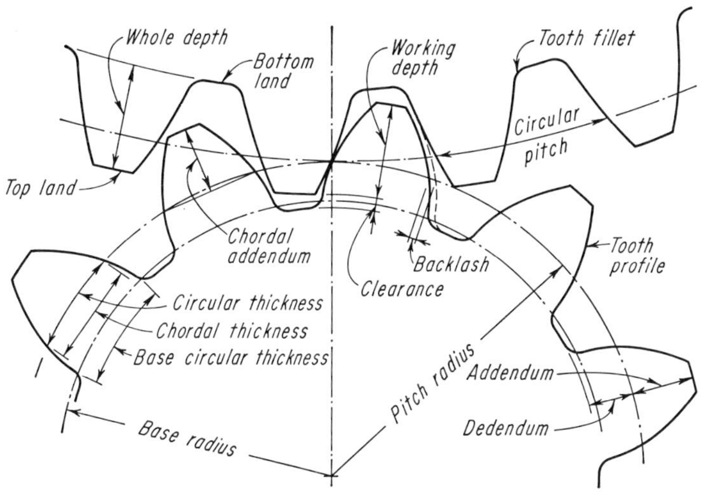

FIG. 4-87. Spur-, helical-, and bevel-gear nomenclature.

Tables 4-6 through 4-13 contain formulas for spur, helical, bevel, and worm-gear families of gearing. The tables sometimes list more than one formula or expression for a gear element. This was done to make the tables adaptable to more than one approach to gear problems.

Formulas for hypoid gears have not been included because of their number and complexity. Hypoid-gear data can be obtained by referring to the manufacturers' technical literature.

Table 4-3. Formulas for Gearing—Circular Pitch

To find	Having	Formula
Circular pitch	Diametral pitch	$p = \frac{3.1416^*}{P}$
Circular pitch	Pitch diameter and number of teeth	$p = \frac{3.1416D}{N}$
Pitch diameter	Number of teeth and circular pitch	$D = 0.3183Np$
Outside diameter	Pitch diameter and addendum	$D_o = D + 2a$
Base diameter	Pitch diameter and transverse pressure angle	$D_b = D \cos \phi$
Number of teeth	Pitch diameter and circular pitch	$N = \frac{3.1416D}{p}$
Approximate thickness of tooth	Circular pitch	$t_t = \frac{p}{2}$

* Exact value of pi is 3.14159265.

Table 4-4. Formulas for Gearing—Diametral Pitch

To find	Having	Formula
Diametral pitch	Circular pitch	$P = \frac{3.1416^*}{p}$
Diametral pitch	Pitch diameter and number of teeth	$P = \frac{N}{D}$
Pitch diameter	Number of teeth and diametral pitch	$D = \frac{N}{P}$
Approximate pitch diameter	Number of teeth and outside diameter	$D = \frac{D_o N}{N + 2}$
Outside diameter	Number of teeth and diametral pitch	$D_o = \frac{N + 2}{P}$
Base diameter	Pitch diameter and transverse pressure angle	$D_b = D \cos \phi$
Number of teeth	Pitch diameter and diametral pitch	$N = DP$
Approximate thickness of tooth	Diametral pitch	$t_t = \frac{1.5708}{P}$

* Exact value of pi is 3.14159265.

Table 4-5. Factors and Relationships

To find	Having	Formula
Millimeters per inch		25.4 = mm per in.
Module (mm)	Pitch diameter (mm) and number of teeth	$\text{Module (mm)} = \frac{D}{N}$
Module (mm)	Diametral pitch	$\text{Module (mm)} = \frac{25.4}{P}$
Module (mm)	Circular pitch	$\text{Module (mm)} = 8.085p$
Diametral pitch	Number of teeth and pitch diameter (in.)	$P = \frac{N}{D}$
Diametral pitch	Module (mm)	$P = \frac{25.4}{\text{module}}$
Diametral pitch	Circular pitch	$P = \frac{3.1416}{p}$
Circular pitch	Diametral pitch	$p = \frac{3.1416}{P}$
Circular pitch	Number of teeth and pitch diameter	$p = \frac{3.1416D}{N}$
Circular pitch (in.)	Module (mm)	$p = \frac{\text{module (mm)}}{8.085}$
Normal pressure angle	Transverse pressure angle and helix angle	$\tan \phi_n = \tan \phi_t \cos \psi$
Transverse pressure angle	Normal pressure angle and helix angle	$\tan \phi_t = \frac{\tan \phi_n}{\cos \psi}$
Normal diametral pitch	Transverse diametral pitch and helix angle	$P_n = \frac{P_t}{\cos \psi}$
Transverse diametral pitch	Normal diametral pitch and helix angle	$P_t = P_n \cos \psi$
Normal circular pitch	Transverse circular pitch and helix angle	$p_n = p_t \cos \psi$
Transverse circular pitch	Normal circular pitch and helix angle	$p_t = \frac{p_n}{\cos \psi}$
Approximate normal tooth thickness	Normal diametral pitch	$t_n = \frac{1.571}{P_n}$
Transverse tooth thickness	Normal tooth thickness and helix angle	$t_t = \frac{t_n}{\cos \psi}$
Lead angle (worm or helical gear)	Lead and pitch diameter	$\tan \lambda = \frac{L}{3.1416D}$
Lead angle	Helix angle	$\lambda = 90° - \psi$
Face advance (helical gear)	Face width and helix angle	$Q_F = F \tan \psi$
Gear ratio	Number of gear teeth and pinion teeth	$m_G = \frac{N_G}{N_P}$
Profile contact ratio	Length of line of action and base circular pitch	$m_p = \frac{Z}{p_b}$
Axial pitch	Circular pitch and helix angle	$p_x = \frac{p}{\tan \psi}$
Face contact ratio (axial plane)	Face width and axial circular pitch	$m_F = \frac{F}{p_x}$
Total contact ratio	Profile contact ratio and face contact ratio	$m_t = m_p + m_F$
Center distance	Number of gear and pinion teeth, and diametral pitch	$C = \frac{N_G + N_P}{2P}$
Center distance	Number of gear and pinion teeth, and circular pitch	$C = \frac{(N_G + N_P)p}{6.2832}$
Center distance	Gear and pinion pitch diameters	$C = \frac{D_G + D_P}{2}$

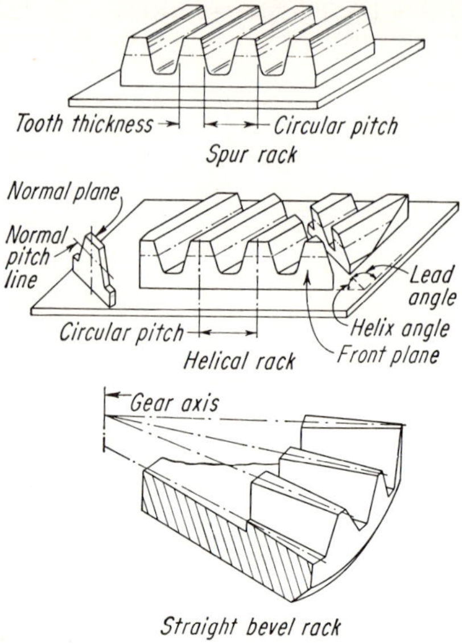

FIG. 4-88. Spur, helical, and bevel racks.

Table 4-6. Spur-gear Formulas

To find	Having	Formula
Diametral pitch	Number of teeth and pitch diameter	$P = \frac{N}{D}$
Diametral pitch	Circular pitch	$P = \frac{3.1416^*}{p}$
Pitch diameter	Number of teeth and diametral pitch	$D = \frac{N}{P}$
Outside diameter	Pitch diameter and addendum	$D_o = D + 2a$
Root diameter	Outside diameter and whole depth	$D_R = D_o - 2h_t$
Root diameter	Pitch diameter and dedendum	$D_R = D - 2b$
Number of teeth	Pitch diameter and diametral pitch	$N = D \times P$
Base-circle diameter	Pitch diameter and pressure angle	$D_b = D \times \cos \phi$
Circular pitch	Pitch diameter and number of teeth	$p = \frac{3.1416D}{N}$
Circular pitch	Diametral pitch	$p = \frac{3.1416}{P}$
Center distance	Number of gear teeth and number of pinion teeth and diametral pitch	$C = \frac{N_G + N_P}{2P} = \frac{D_G + D_P}{2}$
Approximate thickness of tooth	Diametral pitch	$t_t = \frac{1.5708}{P}$ (see design formulas, Chap. 5)

* Exact value of pi is 3.14159265.

Table 4-6a. Internal-spur-gear Formulas

To find	Having	Formula
Pitch diameter:		
Of gear	Number of teeth in gear and diametral pitch	$D = \frac{N_G}{P}$
Of pinion	Number of teeth in pinion and diametral pitch	$d = \frac{N_P}{P}$
Internal diameter (gear)	Pitch diameter and addendum	$D_i = D - 2a$
Center distance	Number of teeth in gear and pinion; diametral pitch	$C = \frac{N_G - N_P}{2P}$
	Pitch diameter of gear and pinion	$C = \frac{D - d}{2}$

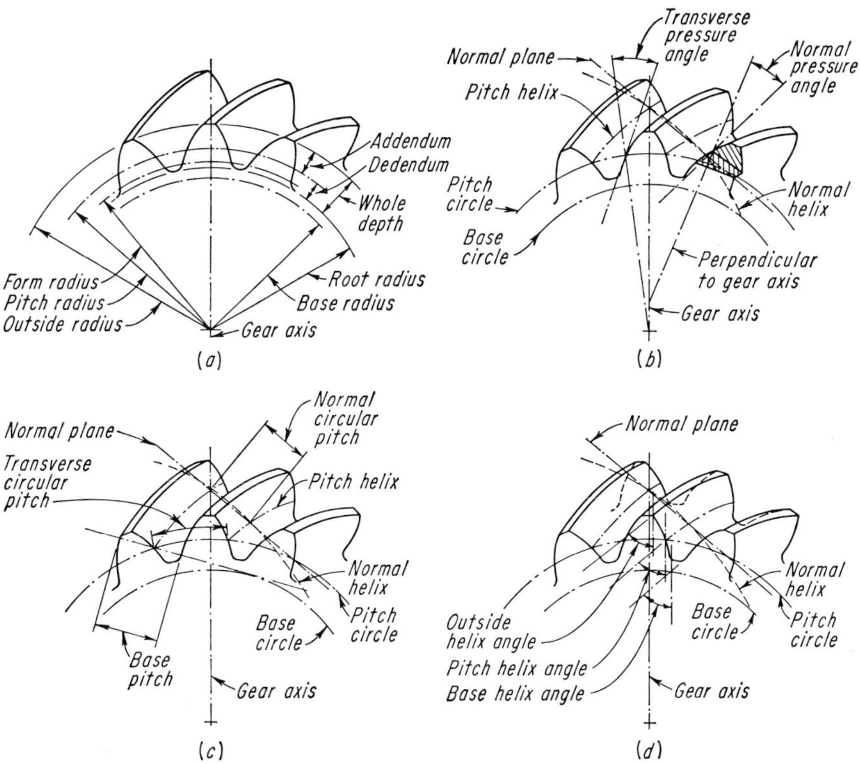

FIG. 4-89. Gear-nomenclature composite. (*a*) Gear radii. (*b*) Gear pressure angles. (*c*) Gear pitches. (*d*) Gear helix angles.

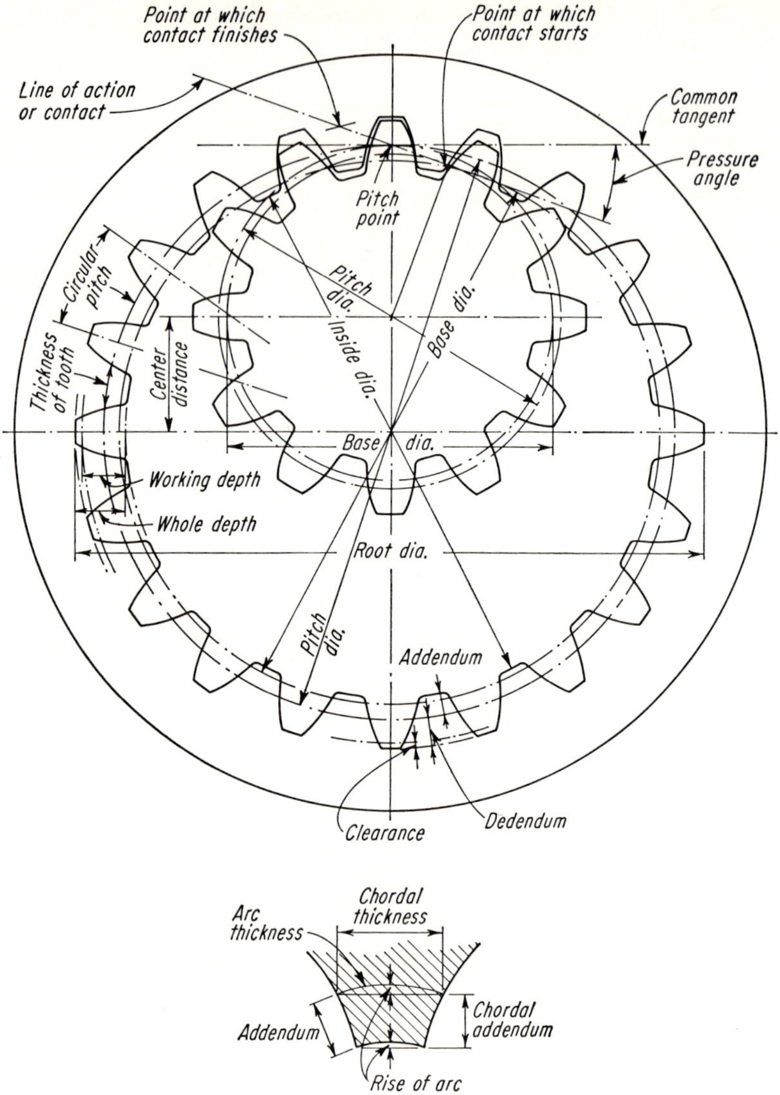

FIG. 4-90. Internal gear and pinion nomenclature.

Table 4-6b. 20° Involute Fine-pitch System for Spur and Helical Gears*

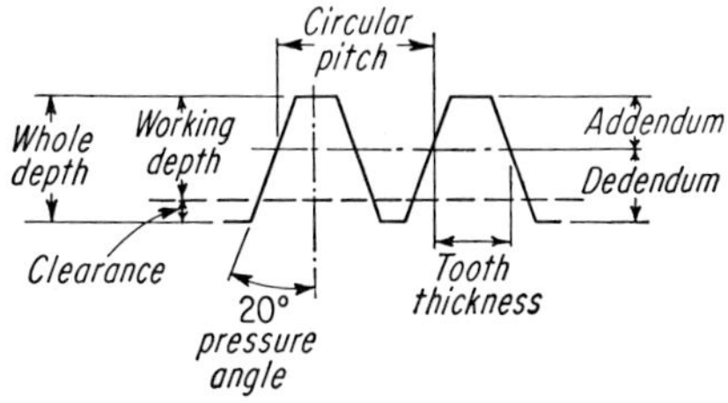

Basic rack

To find	Having	Formula
Addendum..............	Diametral pitch	$a = \dfrac{1.000}{P}$
Addendum..............	Circular pitch	$a = 0.3183p$
Dedendum..............	Diametral pitch	$b = \dfrac{1.200}{P} + 0.002$
Dedendum..............	Circular pitch	$b = 0.3820p + 0.002$
Working depth..........	Diametral pitch	$h_k = \dfrac{2.000}{P}$
Working depth..........	Circular pitch	$h_k = 0.6366p$
Whole depth............	Diametral pitch	$h_t = \dfrac{2.200}{P} + 0.002$
Whole depth............	Circular pitch	$h_t = 0.7003p + 0.002$
Clearance...............	Diametral pitch	$c = \dfrac{0.200}{P} + 0.002$
Clearance...............	Circular pitch	$c = 0.0637p + 0.002$

* Based upon AGMA standard 207.04, "20 Deg Involute Fine-Pitch System for Spur and Helical Gears."

Table 4-7. Helical-gear Formulas

To find	Having	Formula
Normal diametral pitch	Number of teeth, pitch diameter, and helix angle	$P_n = \frac{N}{D \cos \psi}$
Normal diametral pitch	Normal circular pitch	$P_n = \frac{3.1416^*}{p_n}$
Normal diametral pitch	Transverse diametral pitch and helix angle	$P_n = \frac{P}{\cos \psi}$
Normal circular pitch	Normal diametral pitch	$p_n = \frac{3.1416}{P_n}$
Normal circular pitch	Transverse circular pitch	$p_n = p \cos \psi$
Pitch diameter	Number of teeth, normal diametral pitch, and helix angle	$D = \frac{N}{P_n \cos \psi}$
Center distance	Pinion and gear pitch diameter	$C = \frac{D_P + D_G}{2}$
Outside diameter	Pitch diameter and addendum	$D_o = D + 2a$
Approximate normal tooth thickness	Normal diametral pitch	$t_n = \frac{1.571}{P_n}$ (see design formulas, Chap. 5)
Transverse tooth thickness	Normal tooth thickness and helix angle	$t_t = \frac{t_n}{\cos \psi}$
Normal pressure angle	Transverse pressure angle and helix angle	$\tan \phi_n = \tan \phi \cos \psi$

* Exact value of pi is 3.14159265.

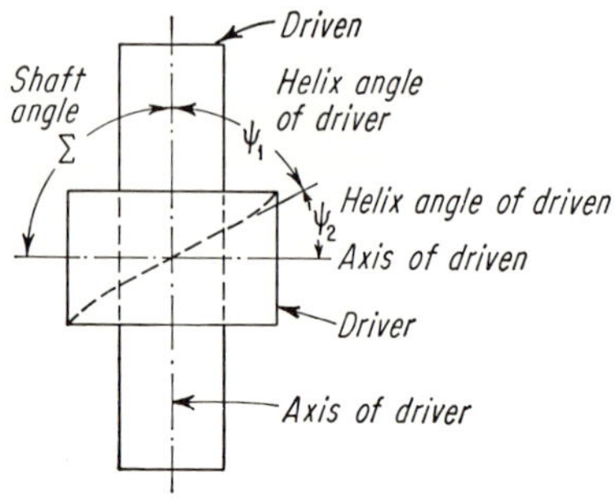

FIG. 4-91. Shaft angle and helix angles of crossed-helical gears.

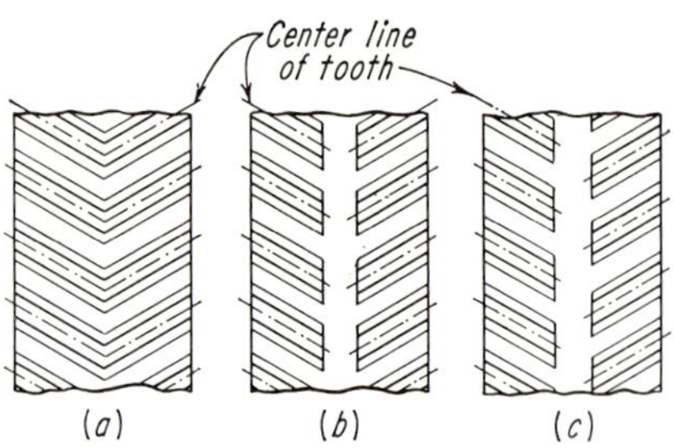

FIG. 4-92. Herringbone and double-helical gear-tooth nomenclature. (*a*) Herringbone—continuous tooth. (*b*) Double-helical (herringbone) with gap. (*c*) Double-helical-staggered tooth.

Table 4-8. Crossed-helical Gear (Nonparallel-shaft, Helicals)*

To find	Having	Formula
Shaft angle	Helix angle of pinion and helix angle of gear	$\Sigma = \psi_P + \psi_G$
Pitch diameter of pinion	Number of teeth in pinion, normal diametral pitch, and helix angle of pinion	$D_P = \dfrac{N_P}{P_n \times \cos \psi_P}$
Pitch diameter of gear	Number of teeth in gear, normal diametral pitch and helix angle of gear	$D_G = \dfrac{N_G}{P_n \times \cos \psi_G}$
Center distance	Pitch diameter of pinion, and pitch diameter of gear	$C = \dfrac{D_P + D_G}{2}$
Speed ratio	Number of teeth in gear and number of teeth in pinion	$m_G = \dfrac{N_G}{N_P}$

* Formulas in Table 4-7 apply to crossed-helical gears except as noted in Table 4-8.

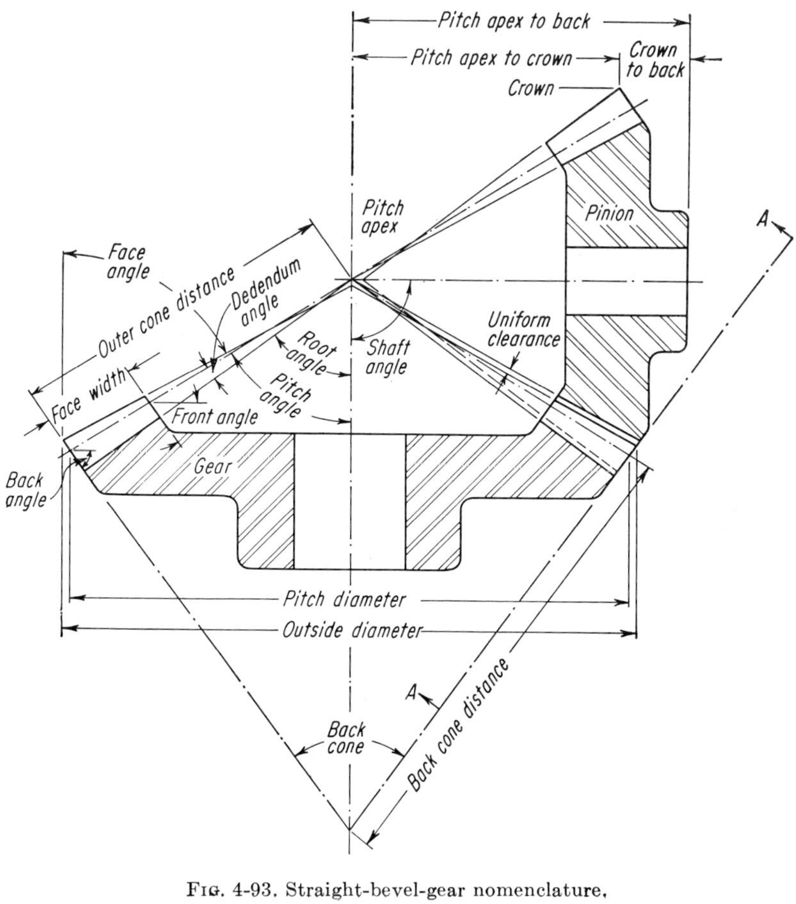

FIG. 4-93. Straight-bevel-gear nomenclature.

Table 4-9. Straight-bevel-gear Formulas
(20° pressure angle, 90° shaft angle)

To find	Having	Formula
Pitch diameter of pinion......	Number of pinion teeth and diametral pitch	$d = \frac{n}{P_d}$
Pitch diameter of gear........	Number of gear teeth and diametral pitch	$D = \frac{N}{P_d}$
Pitch angle of pinion.........	Number of pinion teeth and number of gear teeth	$\gamma = \tan^{-1} \frac{n}{N}$
Pitch angle of gear...........	Pitch angle of pinion	$\Gamma = 90° - \gamma$
Outer cone distance of pinion and gear	Gear pitch diameter and pitch angle of gear	$A_o = \frac{D}{2 \sin \Gamma}$
Circular pitch of pinion and gear	Diametral pitch	$p = \frac{3.1416^*}{P_d}$
Dedendum angle of pinion....	Dedendum of pinion and outer cone distance	$\delta_P = \tan^{-1} \frac{b_{oP}}{A_o}$
Dedendum angle of gear......	Dedendum of gear and outer cone distance	$\delta_G = \tan^{-1} \frac{b_{oG}}{A_o}$
Face angle of pinion blank....	Pinion pitch angle and dedendum angle of gear	$\gamma_o = \gamma + \delta_G$
Face angle of gear blank......	Gear pitch angle and dedendum angle of gear	$\Gamma_o = \Gamma + \delta_P$
Root angle of pinion.........	Pitch angle of pinion and dedendum angle of pinion	$\gamma_R = \gamma - \delta_P$
Root angle of gear...........	Pitch angle of gear and dedendum angle of gear	$\Gamma_R = \Gamma - \delta_G$
Outside diameter of pinion....	Pinion pitch diameter, pinion addendum and pitch angle of pinion	$d_o = d + 2a_{oP} \cos \gamma$
Outside diameter of gear......	Pitch diameter of gear, addendum, and pitch angle of gear	$D_o = D + 2a_{oG} \cos \Gamma$
Pitch apex to crown of pinion..	Pitch diameter of gear, addendum, and pitch angle of pinion	$x_o = \frac{D}{2} - a_{oP} \sin \gamma$
Pitch apex to crown of gear...	Pitch diameter of pinion, addendum, and pitch angle of gear	$X_o = \frac{d}{2} - a_{oG} \sin \Gamma$
Circular thickness of pinion...	Circular pitch and gear circular tooth thickness	$t = p - T$
Chordal thickness of pinion...	Circular tooth thickness, pitch diameter of pinion, and backlash	$t_c = t - \frac{t^3}{6d^2} - \frac{B}{2}$
Chordal thickness of gear.....	Gear circular tooth thickness, pitch diameter of gear, and backlash	$T_c = T - \frac{T^3}{6D^2} - \frac{B}{2}$
Chordal addendum of pinion..	Addendum angle, circular tooth thickness, pitch diameter, and pitch angle of pinion	$a_{cP} = a_{oP} + \frac{t^2 \cos \gamma}{4d}$
Chordal addendum of gear....	Addendum angle, circular tooth thickness, pitch angle, and pitch diameter of gear	$a_{cG} = a_{oG} + \frac{T^2 \cos \Gamma}{4D}$
Tooth angle of pinion........	Outer cone distance, tooth thickness, dedendum of pinion, and pressure angle	$\frac{3{,}438}{A_o}\left(\frac{t}{2} + b_{oP} \tan \phi\right)$ minutes

Table 4-9. Straight-bevel-gear Formulas (*Continued*)

To find	Having	Formula
Tooth angle of gear..........	Outer cone distance, circular tooth thickness, dedendum of gear, and pressure angle	$\frac{3,438}{A_o}\left(\frac{T}{2} + b_{oG} \tan \phi\right)$ minutes

NOTE: $\tan^{-1}$ means "the angle whose tangent is."
* Exact value of pi is 3.14159265.

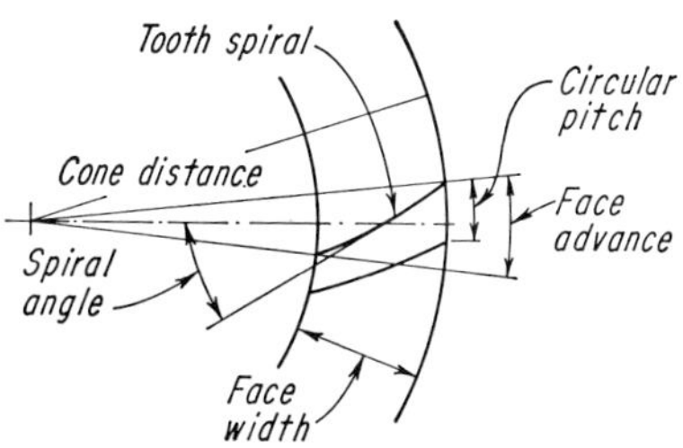

FIG. 4-94. Spiral-bevel-gear spiral angle.

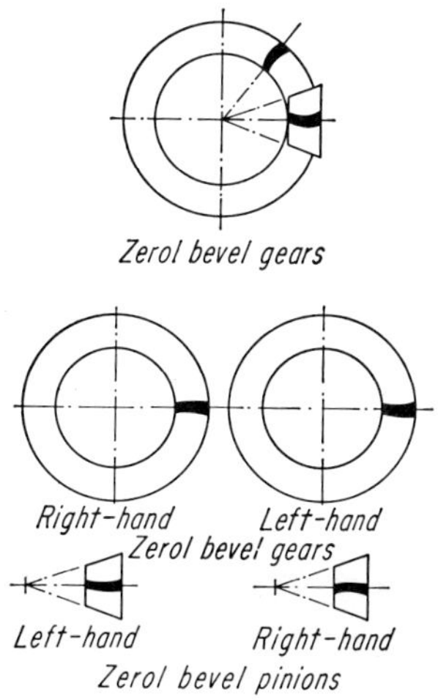

FIG. 4-95. Zerol bevel gear and pinion.

Table 4-10. Spiral-bevel-gear Formulas
(20° pressure angle, 90° shaft angle)

To find	Having	Formula
Pitch diameter of pinion	Number of pinion teeth and diametral pitch	$d = \frac{n}{P_d}$
Pitch diameter of gear	Number of gear teeth and diametral pitch	$D = \frac{N}{P_d}$
Pitch angle of pinion	Number of pinion teeth and number of gear teeth	$\gamma = \tan^{-1} \frac{n}{N}$
Pitch angle of gear	Pitch angle of pinion	$\Gamma = 90° - \gamma$
Outer cone distance of pinion and gear	Pitch diameter of gear and pitch angle of gear	$A_o = \frac{D}{2 \sin \Gamma}$
Circular pitch of pinion and gear	Diametral pitch	$p = \frac{3.1416^*}{P_d}$
Dedendum angle of pinion	Dedendum of pinion and outer cone distance	$\delta_P = \tan^{-1} \frac{b_{oP}}{A_o}$
Dedendum angle of gear	Dedendum of gear and outer cone distance	$\delta_G = \tan^{-1} \frac{b_{oG}}{A_o}$
Face angle of pinion blank	Pitch angle of pinion and dedendum angle of gear	$\gamma_o = \gamma + \delta_G$
Face angle of gear blank	Pitch angle of gear and dedendum angle of gear	$\Gamma_o = \Gamma + \delta_P$
Root angle of pinion	Pitch angle of pinion and dedendum angle of pinion	$\gamma_R = \gamma - \delta_P$
Root angle of gear	Pitch angle of gear and dedendum angle of gear	$\Gamma_R = \Gamma - \delta_G$
Outside diameter of pinion	Pitch diameter, addendum, and pitch angle of pinion	$d_o = d + 2a_{oP} \cos \gamma$
Outside diameter of gear	Pitch diameter, pitch angle, and addendum of gear	$D_o = D + 2a_{oG} \cos \Gamma$
Pitch apex to crown of pinion	Pitch diameter of gear, pitch angle, and addendum of pinion	$x_o = \frac{D}{2} - a_{oP} \sin \gamma$
Pitch apex to crown of gear	Pitch diameter of pinion, pitch angle, and addendum of gear	$X_o = \frac{d}{2} - a_{oG} \sin \Gamma$
Circular thickness of pinion	Circular pitch of pinion and circular thickness of gear	$t = p - T$

* Exact value of pi is 3.14159265.

Table 4-11. Zerol-bevel-gear Formulas
(20° pressure angle, 90° shaft angle)

To find	Having	Formula
Pitch diameter of pinion............	Number of pinion teeth and diametral pitch	$d = \frac{n}{P_d}$
Pitch diameter of gear..............	Number of gear teeth and diametral pitch	$D = \frac{N}{P_d}$
Pitch angle of pinion...............	Number of pinion teeth and number of gear teeth	$\gamma = \tan^{-1} \frac{n}{N}$
Pitch angle of gear.................	Pitch angle of pinion	$\Gamma = 90° - \gamma$
Outer cone distance of pinion and gear	Pitch diameter of gear and pitch angle of gear	$A_o = \frac{D}{2 \sin \Gamma}$
Circular pitch of pinion and gear....	Diametral pitch	$p = \frac{3.1416*}{P_d}$
Face angle of pinion blank..........	Pitch angle of pinion and dedendum angle of gear	$\gamma_o = \gamma + \delta_G$
Face angle of gear blank............	Pitch angle of gear and dedendum angle of pinion	$\Gamma_o = \Gamma + \delta_P$
Root angle of pinion...............	Pitch angle of pinion and dedendum angle of pinion	$\gamma_R = \gamma - \delta_P$
Root angle of gear.................	Pitch angle of gear and dedendum angle of gear	$\Gamma_o = \Gamma - \delta_G$
Outside diameter of pinion..........	Pitch diameter, pitch angle, and addendum of pinion	$d_o = d + 2a_P \cos \gamma$
Outside diameter of gear...........	Pitch diameter, pitch angle, and addendum of gear	$D_o = D + 2a_G \cos \Gamma$
Pitch apex to crown of pinion.......	Pitch diameter of gear, pitch angle, and addendum of pinion	$x_o = \frac{D}{2} - a_P \sin \gamma$
Pitch apex to crown of gear.........	Pitch diameter of pinion, pitch angle, and addendum of gear	$X_o = \frac{d}{2} - a_G \sin \Gamma$
Circular thickness of pinion.........	Circular pitch of pinion and circular thickness of gear	$t = p - T$
Circular thickness of gear...........	Circular pitch, pressure angle, and addendum of pinion and gear	$T = \frac{p}{2} - (a_P - a_G) \tan \phi$

NOTE: $\tan^{-1}$ means "the angle whose tangent is."
* Exact value of pi is 3.14159265.

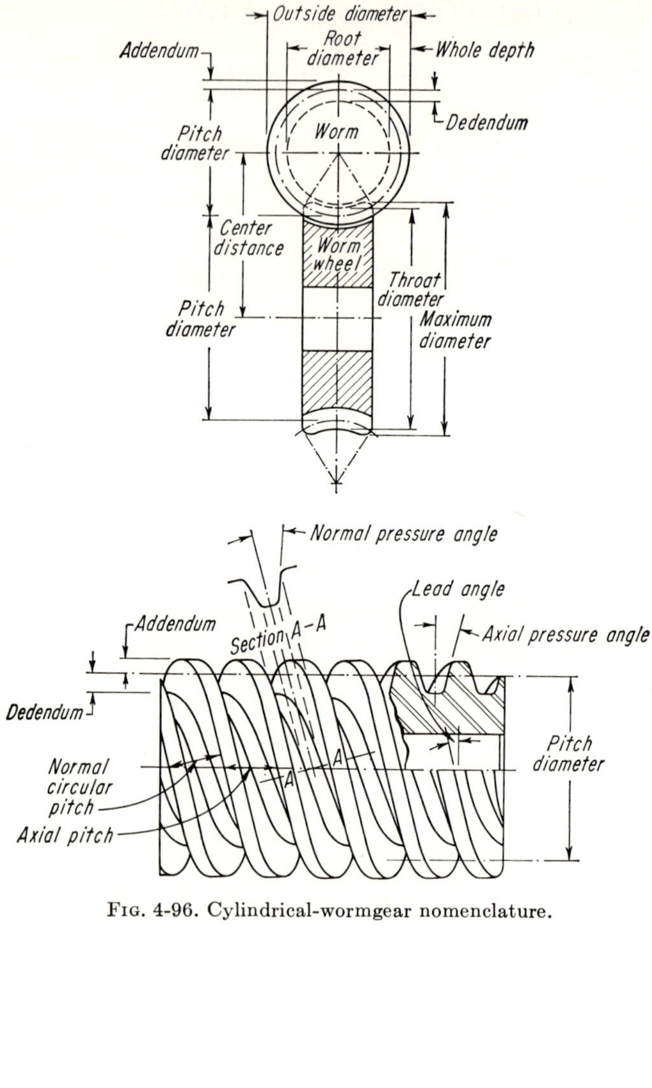

FIG. 4-96. Cylindrical-wormgear nomenclature.

Table 4-12. Wormgearing Formulas

To find	Having	Formula
Worm:		
Lead......................	Number of threads in worm and axial pitch	$L = N_W \times p_x$
Pitch diameter............	Center distance	$d = \frac{C^{0.875}}{1.7}$ to $\frac{C^{0.875}}{3}$
Root diameter.............	Outside diameter of pinion and whole depth of tooth	$d_R = d_o - 2h_t$
Outside diameter..........	Pitch diameter and addendum	$d_o = d + 2a$
Minimum face............	Throat diameter, pitch diameter, and addendum	$f = 2\sqrt{\left(\frac{D_t}{2}\right)^2 - \left(\frac{D}{2} - a\right)^2}$
Lead angle................	Lead and pitch diameter	$\tan \lambda = \frac{L}{3.1416 \times d}$
Normal pitch.............	Axial pitch and lead angle	$p_n = p_x \times \cos \lambda$
Gear:		
Nominal pitch diameter....	Number of teeth in gear and axial pitch	$D = \frac{N_G \times p_x}{3.1416}$
Throat diameter...........	Nominal pitch diameter and addendum	$D_t = D + 2a$
Effective face.............	Pitch diameter and working depth	$F_e = \sqrt{(d + h_k)^2 - d^2}$
Center distance...........	Pitch diameter of gear and pitch diameter of worm	$C = \frac{D + d}{2}$

* Exact value of pi is 3.14159265.

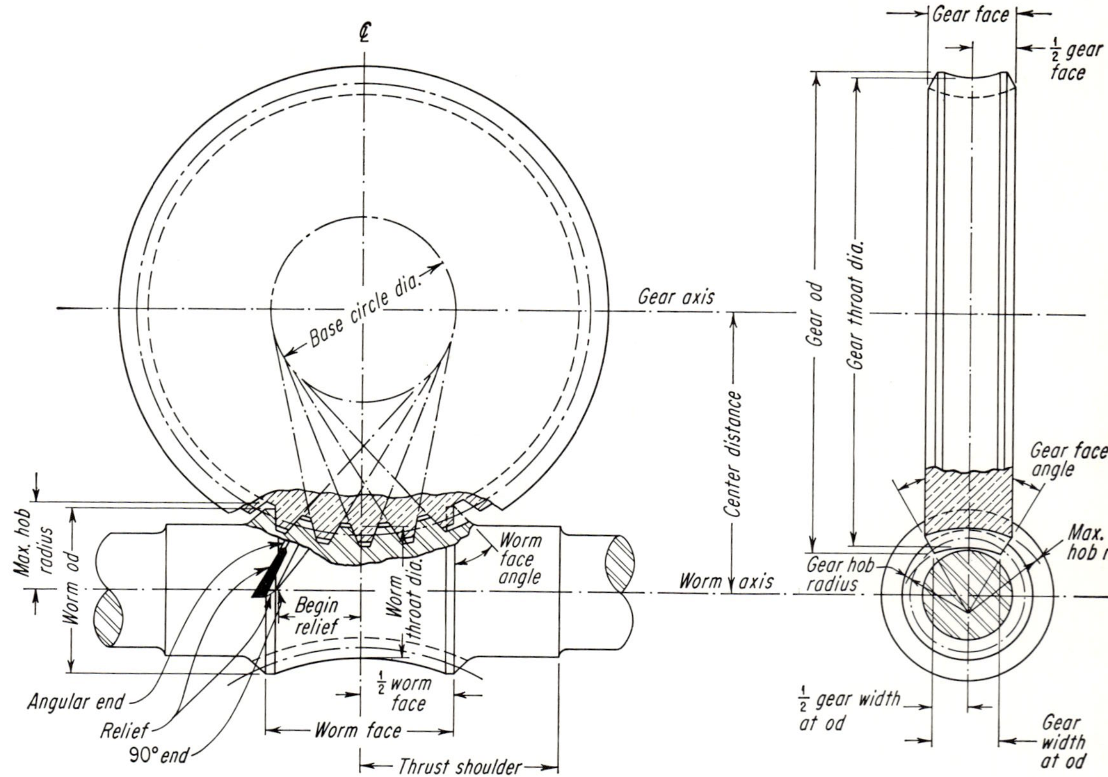

FIG. 4-97. Double-enveloping wormgear layout and nomenclature.

Table 4-13. Double-enveloping Wormgear Formulas

To find	Having	Formula
Worm root diameter*	Pitch diameter of pinion and dedendum of gear; center distance	$d_R = d - 2b_G$ $d_R = \dfrac{C^{0.875}}{3}$ (approx.)
Worm pitch diameter*	Center distance	$d = \dfrac{C^{0.875}}{2.2}$ (approx.)
Gear pitch diameter	Center distance and worm pitch diameter	$D = 2C - d$
Axial circular pitch	Gear pitch diameter and number of teeth in gear	$p_x = \dfrac{\pi D}{N_G}$
Normal circular pitch	Axial circular pitch and pitch cone angle of pinion	$p_n = p_x \cos \lambda$
Whole depth of tooth*	Normal circular pitch	$h_t = \dfrac{p_n}{2}$
Working depth of tooth*	Whole depth of tooth	$h_k = 0.9h_t$
Dedendum*	Working depth of tooth	$b_G = 0.611h_k$
Normal pressure angle*	....	$\phi_n = 20°$
Axial pressure angle	Normal pressure angle and lead angle, deg	$\phi_x = \tan^{-1} \dfrac{\tan \phi_n}{\cos \lambda_c}$
Lead angle at center of worm	Pitch diameter, gear ratio, and worm pitch diameter	$\lambda_c = \tan^{-1} \dfrac{D}{m_G d}$
Lead angle average*	Pitch diameter, gear ratio, and worm pitch diameter	$\lambda = \tan^{-1} \dfrac{0.87D}{m_G\, d}$

* Proportions given in these formulas represent recommendations for Cone-Drive®† gears by the Michigan Tool Co. of Detroit.

† Cone-Drive is a registered trade-mark of the Michigan Tool Co., Detroit, Mich.

The sections on nomenclature and letter symbols should be consulted whenever there is any doubt regarding a particular formula or expression.

4-19. Special Gear Types. There are a few gear types which at this time are considered special types. These are Spiroid®, Helicon®, Planoid®, Beveloid®,* and face gears. *Tapered* gears are a special variety of a face gear. In certain instances very few design data in the form of formulas exist, the actual design being based upon a layout. However, wherever possible, existing data have been listed or tabulated. All data are from manufacturers' catalogues and papers. In many instances new expressions and terms are presented which have not been standardized upon and thus do not appear in the sections on standard nomenclature and letter symbols. However, the new letter symbols can be found listed under Table 4-2.

* Spiroid, Helicon, and Planoid are registered trade-marks of the Spiroid Div., Illinois Tool Works, Chicago, Ill. Beveloid is a registered trade-mark of Vinco Corp., Detroit, Mich.

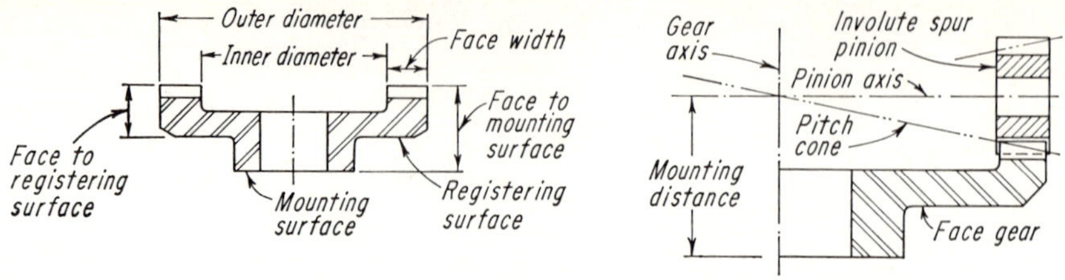

FIG. 4-98. On-center face-gear nomenclature.

Table 4-14. Face Gears—On-center 90° Shaft Angle*

To find	Having	Formula
Gear ratio	Number of pinion and gear teeth	$m_G = \dfrac{N_G}{N_P}$ (from 1.5 to 12.5)
Inside diameter	Diametral pitch, number of gear and pinion teeth, pressure angle	$D_{iG} = \dfrac{1}{P}\sqrt{8N_P + N_G \cos^2 \phi}$
Outside diameter (pointed teeth)	Diametral pitch, number of gear teeth, pressure angle, and tangency pressure angle	$D_{oG} = \dfrac{N_G \cos \phi \sec \phi_o}{P}$ (see Table 4-14*a*)
Max. face width	Face-gear inside and outside diameters	$F_G = \dfrac{D_o - D_i}{2}$
Face to pinion axis	Pinion outside diameter and working depth of tooth	Face to pinion axis $= \dfrac{d_o}{2} - h_k$
Face to mounting surface	Mounting distance and face to pinion axis	Face to mounting surface = mounting distance − face to pinion axis
Min. numbers of teeth	....	See Table 4-15

* Based upon AGMA standard 203.01, "Fine-pitch On-center Face Gears for 20-degree Involute Spur Pinions."

Table 4-14a. Tangency Pressure Angle at Face-gear Outside Diameter

N_P	ϕ_o	sec ϕ_o	N_P	ϕ_o	sec ϕ_o	N_P	ϕ_o	sec ϕ_o
12	40.765	1.32031	25	34.596	1.21481	38	32.837	1.19016
13	39.669	1.29913	26	34.429	1.21237	39	32.730	1.18874
14	38.692	1.28120	27	34.268	1.21005	40	32.627	1.18736
15	37.815	1.26583	28	34.113	1.20783	41	32.525	1.18602
16	37.022	1.25249	29	33.965	1.20572	42	32.427	1.18473
17	36.300	1.24081	30	33.821	1.20369	43	32.332	1.18352
18	36.012	1.23626	31	33.683	1.20176	44	32.240	1.18228
19	35.778	1.23261	32	33.550	1.19990	45	32.150	1.18111
20	35.557	1.22919	33	33.421	1.19812	46	32.061	1.17996
21	35.346	1.22598	34	33.297	1.19640	47	31.976	1.17887
22	35.145	1.22295	35	33.176	1.19475	48	31.893	1.17781
23	34.954	1.22009	36	33.059	1.19317	49	31.810	1.17675
24	34.771	1.21738	37	32.946	1.19164	50	31.734	1.17578

Table 4-15. Minimum Numbers of Teeth in Pinion and Face Gear

Diametral pitch range	Min. numbers of teeth		Diametral pitch range	Min. numbers of teeth	
	Pinion	Gear		Pinion	Gear
20–48	12	18	73–76	19	29
49–52	13	20	77–80	20	30
53–56	14	21	81–84	21	32
57–60	15	23	85–88	22	33
61–64	16	24	89–92	23	35
65–68	17	26	93–96	24	36
69–72	18	27	97–100	25	38

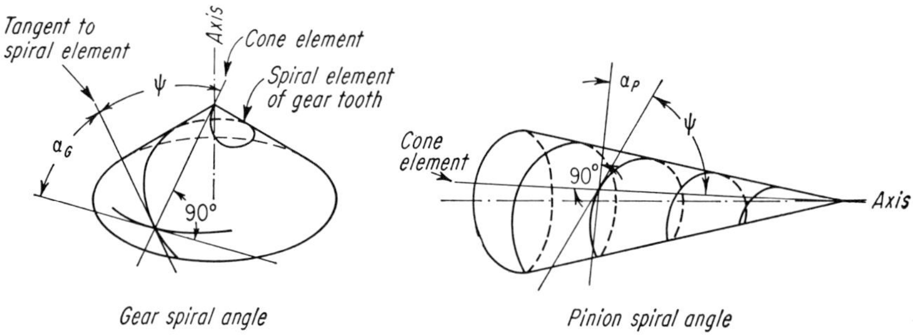

FIG. 4-99. Spiroid pinion and gear spiral angles.

Table 4-16. Spiroid Gears

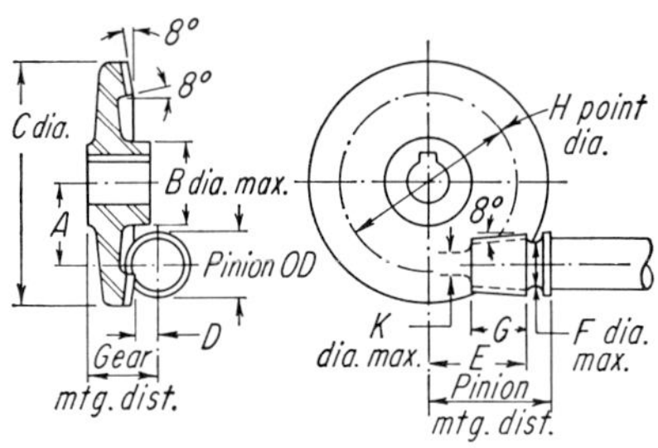

A	*B*	*C*	*D*	*E*	*F*	*G*	*H*	*K*
0.500	0.625	1.500	0.129	0.596	0.2969	0.365	1.101	0.2031
0.750	1.000	2.250	0.176	0.894	0.4219	0.548	1.651	0.2812
1.000	1.375	3.000	0.248	1.192	0.5938	0.731	2.202	0.3750
1.250	1.625	3.750	0.295	1.490	0.7031	0.914	2.752	0.4531
1.500	2.000	4.500	0.338	1.788	0.8281	1.096	3.303	0.5312
1.875	2.625	5.625	0.402	2.236	1.0156	1.370	4.129	0.6406
2.250	3.187	6.750	0.461	2.683	1.1562	1.644	4.954	0.7500
2.750	4.000	8.250	0.536	3.279	1.3594	2.010	6.055	0.8750
3.250	4.750	9.750	0.608	3.875	1.5625	2.375	7.156	1.0156
3.750	5.625	11.250	0.677	4.471	1.7656	2.741	8.257	1.1406
4.375	6.750	13.125	0.757	5.216	2.0000	3.197	9.633	1.2969
5.125	8.000	15.375	0.851	6.111	2.2812	3.745	11.285	1.4844

Table 4-16a. Spiroid-gear Formulas

To find	Having	Formula
Ratio.................	Number of teeth in gear, number of threads in pinion	$m_G = \frac{N_G}{N_P}$
Pinion spiral angle......	Theoretical lead, pinion OD cone angle, pinion pitch radius	$\tan \alpha_P = \frac{L \sec \tau}{2\pi r}$
Gear spiral angle.......	Ratio, pinion pitch radius, gear pitch radius, pinion spiral angle	$\sin \alpha_G = m_G \left(\frac{r}{R_G}\right) \sin \alpha_P$
Pinion pitch point......	Length of pinion primary pitch cone	= ⅓ axial length primary pitch cone from small end

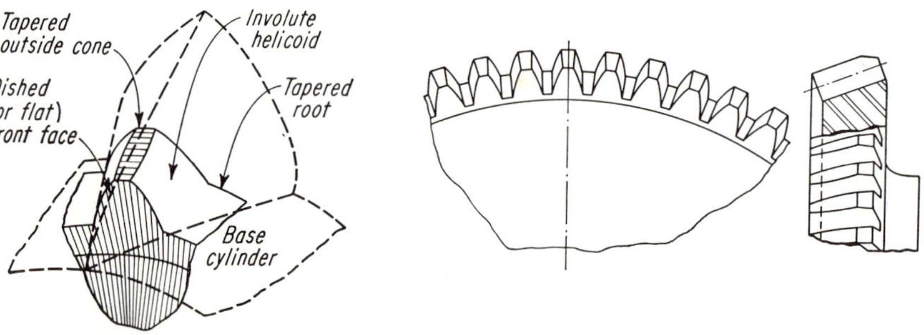

FIG. 4-100. Beveloid tooth and gear. Beveloid gears can be engaged with spur gears, helical gears, worms, racks, and other Beveloid gears. They can be used with intersecting, parallel, or skew shafts, and they are insensitive to mounting errors. Beveloid gears are considered a special gear type at this time. Consult the manufacturer, Vinco Corp., Detroit, Mich., for further information.

REFERENCES

Standards of the American Gear Manufacturers Association and the American Standards Association

1. "Letter Symbols for Gear Engineering," AGMA standard 111.03, ASA standard B6.5-1954.
2. "Gear Nomenclature, Terms, Definitions, and Illustrations," AGMA standard 112.03, ASA standard B6.10-1954.
3. "Abbreviations for Gearing," AGMA standard 116.01.
4. "Tooth Proportions for Coarse-pitch Involute Spur Gears," AGMA standard 201.02 (includes 201.02A and 201.02B).
5. "Zerol Bevel Gear System," AGMA standard 202.02.
6. "Fine-pitch On-center Face Gears for 20-degree Involute Spur Pinions," AGMA standard 203.01.
7. "20-degree Involute Fine-pitch System for Spur and Helical Gears," AGMA standard 207.04, ASA standard B6.7-1956.
8. "System for Straight Bevel Gears," AGMA standard 208.01, ASA standard B6.7-1956.
9. "System for Spiral Bevel Gears," AGMA standard 209.02.
10. "Design of Coarse-pitch Cylindrical Wormgearing," AGMA standard 341.01.
11. "Design of General Industrial Double-enveloping Wormgears," AGMA standard 342.01.
12. "Design of Fine-pitch Wormgearing," AGMA standard 374.03, ASA standard B6.9-1956.

Other References

13. Beam, A. S.: Beveloid Gearing, *Machine Design*, December, 1954.
14. Gleason Works: "Spiral Bevel Gear System."
15. Gleason Works: "20° Straight Bevel Gear System."
16. Gleason Works: "Zerol Bevel Gear System."
17. Illinois Tool Works: "Spiroid Gears."
18. Michigan Tool Company: "Cone Drive Gears."

Chapter 5

GEAR-TOOTH PROPORTIONS*

By

PAUL M. DEAN, Jr., *Gear Technology Engineer, General Engineering Laboratory, General Electric Company, Schenectady, N.Y.*

This chapter deals with the shape and proportions of gear teeth. As a result of the large number of variables that govern the over-all shape of a gear tooth, an almost infinite number of tooth proportions are possible. Fortunately, because of the great amount of attention that has been given to the study of the best shapes for gear teeth they have been reduced to practice in the form of several "standard tooth forms" or "standard systems." The tooth forms covered by these standards are those which have in practice proved to be the most satisfactory for the vast majority of gear applications from the standpoint of performance as well as cost.

This chapter shows not only the standard systems for teeth of various kinds of gears, but in addition, it shows certain modifications to the standard profiles that may be desirable if the gears are to achieve their greatest potential. The modifications discussed fall within the scope of the tooth-form standards. In order to show the reasons for some of these modifications, the kinematics of gear-tooth action is discussed in some detail. Several tooth-form standards that are now out of date have been included for those applications where it is necessary to match a replacement gear to a previous design.

BASIC REQUIREMENTS OF GEAR TEETH

Gear teeth are used to provide a nonslip drive when power or motion is to be transmitted from one shaft to another. When the additional requirement that the motion or power shall be transmitted smoothly is imposed, the shape and size of the teeth become critical.

In order to select the best tooth form for any given gear application, it is necessary to establish an accurate definition of the function that the gear is to perform. In general, a gear train may be called upon to serve one of the following functions. Each requires a special tooth design, markedly different from the others.

* The author wishes to acknowledge the invaluable assistance rendered by Wells Coleman, of the Gleason Works, Rochester, N.Y.; Darle Dudley, of the General Electric Co., Lynn, Mass.; Carl Parker, of the Fellows Gear Shaper Co., Springfield, Vt.; Fred Bohle and Woodrow Nelson, of the Illinois Tool Works, Chicago, Ill.; George Shimmin, of the Michigan Tool Company, Detroit, Mich.; and Gerald Scott, of the American Gear Manufacturers Association, Washington, D.C.

1. Transmission of power from one shaft to another.
   a. The application may be speed-decreasing (torque-increasing). Power gears in an automobile, for example.
   b. The application may be speed-increasing (torque-decreasing). Power gears in a hand-driven cream separator or in a hand-driven grindstone, for example.
   c. Multidirectional drives, those frequently reversed, in which lost motion in the teeth would cause high-impact loads on reversal (such as in rolling-mill drives).
2. Transmission of angular motion from one shaft to another.
   a. Applications where elimination of lost motion is of prime importance as in servo trains.
   b. Applications where absolute angular position is of prime importance as in radar-dish position or gunfire-control equipment.
3. Gearing which must perform two or more of the above functions.

5-1. Definitions of Gear-tooth Elements. In order to explain the special features that gear teeth must have to best perform each of the above functions, it is necessary to define the principal surfaces and boundaries of gear teeth. For more general definitions of gear terms, the reader is referred to Chap. 4.

For simplicity's sake, the following definitions are shown primarily in their relationship to spur and helical gears.

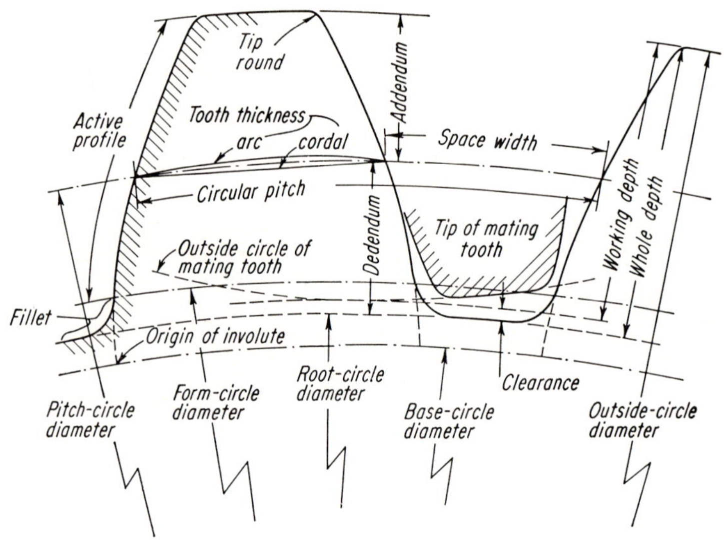

FIG. 5-1. Gear-tooth nomenclature.

Outside diameter (dash line in Fig. 5-2) is the maximum diameter of the gear blank for spur, helical, worms, or wormgears. All tooth elements lie inside this circle. A tolerance on this diameter should always be negative.

Modification diameter (dash line in Fig. 5-2) is the diameter at which any tip modification is to begin. It is a reference dimension and may be given in terms of degrees of roll.

Pitch diameter (dash line in Figs. 5-1 and 5-2) is the theoretical diameter established by dividing the number of teeth in the gear by the diametral pitch of the cutter to be used to produce the gear. This diameter can have no tolerance.

Limit diameter (dash line in Fig. 5-2) is the lowest portion of a tooth that can actually come in contact with the teeth of a mating gear. It is a calculated value and is not to be confused with form diameter. It is the boundary between the active profile and the fillet area of the tooth.

Form diameter (dash line in Fig. 5-2) is a specified diameter on the gear above which the transverse profile is to be in accordance with drawing specifications on profile. It is an inspection dimension and should be placed at a somewhat smaller radius than the limit diameter to allow for shop tolerances.

Undercut diameter (dash line in Fig. 5-2) is the diameter at which the trochoid producing undercut in a gear tooth intersects the involute profile.

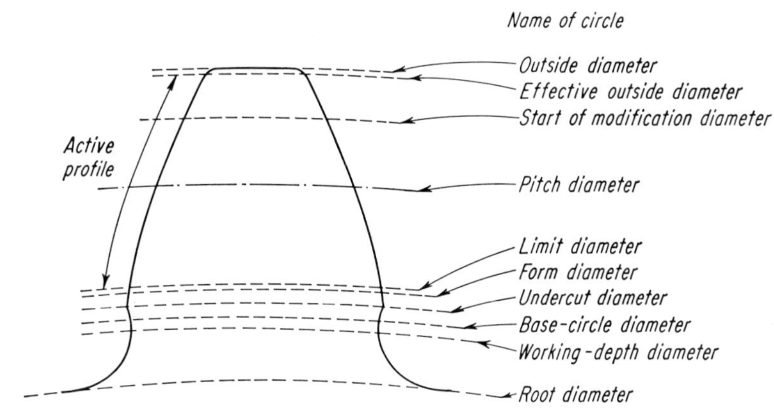

FIG. 5-2. Nomenclature of gear circles.

Base-circle diameter (dash line in Figs. 5-1 and 5-2) is the diameter established by multiplying the pitch diameter (see above) by the cosine of the pressure angle of the cutter to be used to cut the gear. It is a basic dimension of a gear.

Root diameter (dash line in Figs. 5-1 and 5-2) is the diameter of the circle that establishes the root lands of the teeth. All tooth elements should lie outside this circle. The tolerance should be negative.

Active profile (shaded area in Fig. 5-3*a*) is a *surface* and is that portion of the surface of the gear tooth which at some phase of the meshing cycle contacts the active profile of the mating gear tooth. It extends from the limit diameter (see Fig. 5-2) near the root of the tooth to the tip round (see Fig. 5-1) at the tip of the tooth and, unless the mating gear is narrower, extends from one side of the gear or edge round (see Fig. 5-3*k*) at one end of the tooth to the other side of the gear or edge round.

Top land (shaded area in Fig. 5-3*b*) is a *surface* bounded by the sides of the gear (see Fig. 5-3*d*) and active profiles, or if the tooth has been given end and tip rounds (see Fig. 5-3*h* and *i*), the top land is bounded by these curved surfaces. The top land forms the outside diameter of the gear.

Fillet. The fillet of a tooth (shaded area in Fig. 5-3*c*) is a surface which is bounded by the form diameter (see Fig. 5-2) and the root land (if present) (see Fig. 5-3*e*) and the ends of the teeth. In "full-fillet teeth," the fillet of one tooth is considered to extend from the center line of the space to the form diameter.

Sides of gear. The sides of gear (shaded area in Fig. 5-3*d*) are *surfaces* and are the ends of the teeth in spur and in helical gears.

Root land (also known as bottom land). The root land (shaded area in Fig. 5-3*e*) is a *surface* bounded by fillets (see Fig. 5-3*c*) of the adjacent teeth and sides of the gear blank.

Transverse profile (heavy line in Fig. 5-3*f*) is the shape of the gear tooth as seen in a plane perpendicular to the axis of rotation of the gear.

Axial profile (heavy line in Fig. 5-3*g*) is the shape of the gear tooth as seen in a plane tangent to the pitch cylinder at the surface of the tooth. In the case of helical gears, it is the shape of a tooth as seen on a pitch cylinder and may be developed to be shown as a plane.

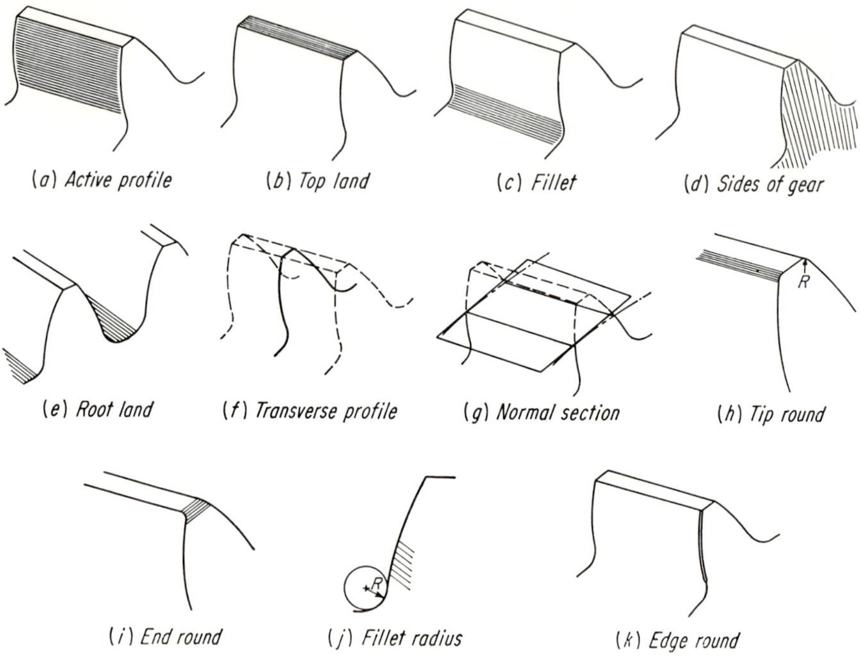

FIG. 5-3. Nomenclature of gear-tooth details (views *a* to *k*).

Tip round (shaded area in Fig. 5-3*h*) is a surface which separates the active profile and the top land. It is sometimes applied to gear teeth either to remove burrs or to lessen the chance of chipping, particularly in the case of hardened teeth. It may also be added as a very mild and crude form of profile modification.

End round (shaded area in Fig. 5-3*i*) is a surface which separates the sides and the top land of the tooth. It is sometimes applied to gear teeth to reduce the chance of chipping, particularly in the case of hardened teeth.

Edge round (shaded area in Fig. 5-3*k*) is the surface which separates the active profiles of the teeth from the sides of the gear. These edges are of importance in the cases of helical gears, spiral bevel gears, and worms, since they become very sharp on the leading edge.

Fillet radius. The fillet radius (dimension in Fig. 5-3*j*) is the minimum radius that a gear tooth may have.

5-2. Basic Considerations for Gear-tooth Design. Gear teeth are a series of cam surfaces that contact similar surfaces on a mating gear in an orderly fashion. In order to drive in a given direction and to transmit power or motion smoothly and with a minimum loss of energy, the contacting cam surfaces on the mating gears must have the following properties:

1. The height and the lengthwise shape of the active profiles of the teeth (cam surfaces) must be such that, before one pair of teeth goes out of contact during mesh, a second pair will have picked up its share of the load. This is called "continuity of action."
2. The shape of the contacting surfaces of the teeth (active profiles) must be such that the angular velocity of the driving member of the pair is smoothly imparted to the driven member in the proper ratio. The most widely used profile shape for active profiles of spur and helical gears that meets these requirements is the involute curve. There are many other specialized curves, each with specific advantages in certain applications. This subject is developed further under "conjugate action."
3. The spacing between the successive teeth must be such that a second pair of tooth-contacting surfaces (active profiles) are in the proper positions to receive the load before the first leave mesh.

Continuity of action and conjugate action are achieved by proper selection of the gear-tooth proportions. Manufacturing tolerances on the gears govern the spacing accuracies of the teeth. Thus, to achieve a satisfactory design, it is necessary to specify correct tooth proportions, and in addition, the tolerances on the tooth elements must be properly specified.

As a general rule, gearing designed in accordance with the standard systems will not have problems of continuity of action or conjugate action. In those cases where it is necessary to depart from the tooth proportions given in the standard systems, the designer should check both the continuity of action and the conjugate action of the resulting gear design.

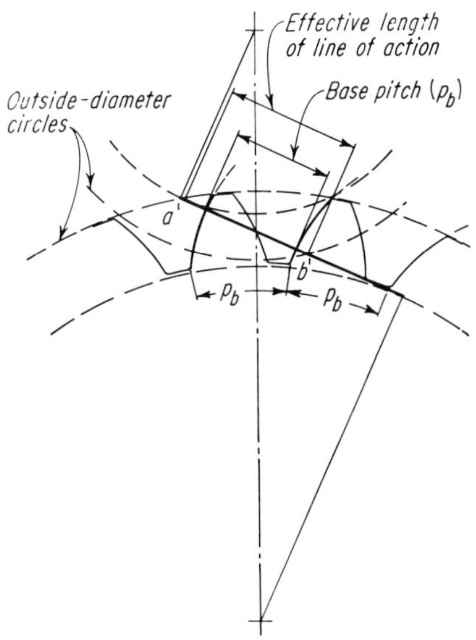

FIG. 5-4. Zone of tooth action.

Continuity of Action. As discussed in Chap. 1, all gear-tooth contact must take place along the "line of action." The shape of this line of action is controlled by the shape of the active profile of the gear teeth, and the length of lines of action is controlled by the outside diameters of the gears (see Fig. 5-4). In order to provide a smooth continuous flow of power, at least one pair of teeth must be in contact at all times. This means that during a part of the meshing cycle two pairs of teeth will be sharing the load. The second pair of teeth will have to be designed such that they will pick up their share of the load and be prepared to assume the full load before the first pair of teeth go out of action.

Control of the continuity of action is achieved in *spur* gears by varying

1. The slope of the line of action (in the case of involute gears, the operating pressure angle).
2. The outside diameters of the pinion and of the gear.
3. The shape of the active profile.
4. The relative sizes of the limit-diameter and undercut-diameter circles (limit circle must be larger).

Control of the continuity of action in *internal*-type spur and helical gears is achieved by varying

1. The slope of the line of action (in the case of involute gears, the operating pressure angle).
2. The inside diameter of the internal gear and the outside diameter of the pinion.
3. The shape of the active profile.
4. The relative sizes of the limit-diameter and undercut-diameter circles. The internal gear will not be undercut, but the external member may. Thus the limit diameter on the pinion must be larger than the undercut diameter.

Control of the continuity of action in *helical* gears is achieved by varying

1. The slope of the line of action (in the case of involute gears, the operating pressure angle).
2. The outside diameters of the pinion and of the gear.
3. The shape of the active profile.
4. The relative sizes of the limit-diameter and the undercut-diameter circles (limit circles must be larger). The above elements (1 through 4) control the continuity of action in the transverse plane.
5. The lead of the tooth.
6. The length of the tooth. These two elements (5 and 6) control the continuity of action in the axial plane. In order to assure continuity of action, the portion of the line of action bounded by the outside-diameter circles (*a-b*) must be somewhat longer than the base pitch (see Fig. 5-4).

The *base pitch* p_b is defined as follows:

$$p_b = p \cos \phi = \frac{\pi \cos \phi}{P_d} \tag{5-1}$$

where p = circular pitch of gear
ϕ = pressure angle of gear
P_d = diametral pitch

Thus either the outside-diameter circles, the operating pressure angle, or the base pitch must be adjusted so that (*a-b*) exceeds p_b by from 20 to 40 per cent.

The most general way of checking continuity of action is by calculating the contact ratio.

A numerical index of the existence and degree of continuity of action is obtained by dividing the length of the line of action by the base pitch of the teeth (see Fig. 5-4). This is called contact ratio m_p.

The AGMA recommends that the contact ratio for spur gears not be less than 1.2.

$$m_p = \frac{L_a}{p_b} = 1.2 \qquad \text{min.} \tag{5-2}$$

A spur-gear mesh has only a transverse contact ratio m_p, whereas a helical-gear mesh has a transverse contact ratio m_p, an axial contact ratio m_f (face contact ratio), and a total contact ratio m_t.

Equations for *contact ratio* are

Spur gears and helical gears

$$m_p = \frac{[(d_o'/2)^2 - (d_b/2)^2]^{1/2} + [(D_o'/2)^2 - (D_b/2)^2]^{1/2} - C' \sin \phi'}{p_b} \tag{5-3}$$

Internal gears, spur and helical

$$m_p = \frac{[(d_o'/2)^2 - (d_b/2)^2]^{1/2} - [(D_{iI}/2)^2 - (D_{bi}/2)^2]^{1/2} + C' \sin \phi'}{p_b} \tag{5-4}$$

Helical gears, axial contact ratio

$$m_f = \frac{F \tan \psi}{p} \tag{5-5}$$

Helical gears, total contact ratio

$$m_t = m_p + m_f \tag{5-6}$$

where d_o' = outside diameter (effective) of pinion
D_o' = effective outside diameter of gear (diameter to intersection of tip round and active profile); see discussion below
D_{iI} = inside diameter (effective), internal gear
d_b = base diameter of pinion
D_b = base diameter of gear
D_{bi} = base diameter of internal gear
C' = operating center distance of pair
ϕ' = operating pressure angle
p = circular pitch (in plane of rotation)
p_b = base pitch
F = length of tooth, axial, the face width of the gear
ψ = helix angle of helical gears

NOTES: In order to achieve correct answers on contact ratio, the following points should be observed.

1. The effective outside diameter d_o' or D_o' is actually the diameter to the beginning of the tip round, usually $D_o' = D_o - 2$(edge round specification) (max.). This value rather than the drawing outside diameter should be used since in many cases manufacturing practices for removing burrs produce a large radius, particularly in fine-pitch gears. Thus a considerable per cent of the addendum may not be effective. If the teeth are given a very heavy profile modification, consideration should be given to performing the calculation under (1) full load, assuming contact to the tip of the active profile, and (2) light load, assuming contact near the start of modification. In this case d_o' or D_o' is selected to have a value close to the diameter at the start of modification. This will give an index to the smoothness of operation at these conditions.
2. This equation also assumes that the form diameter is a larger value than the undercut diameter. If not, use the value of undercut diameter.
3. On occasion, the outside diameter of one or both members is so large relative to the center distance and operating pressure angle that the tip extends below the base circle where it is tangent to the line of action. Since no involute action can take place below the base circle, the value $C \sin \phi'$ (the total length of the line of action) should be substituted in the equation in place of the value of $[(d_o'/2)^2 - (d_b'/2)^2]^{1/2}$ or $[(D_o'/2)^2 - (D_b'/2)^2]^{1/2}$ in the case that one or both become larger than the value of $C \sin \phi'$.

Conjugate Action.* Gear teeth are a series of cam surfaces that act on similar surfaces of the mating gear to impart a driving motion. Many different shapes of surfaces can be used on the teeth to produce uniform transmission of motion. Curves

* See Chap. 1 for a general discussion of gear theory.

which act on each other with a resulting smooth driving action and with a constant driving ratio are called "conjugate curves." The fundamental requirements governing the shapes that any pair of these curves must have is summarized in the "basic law of gearing," which states, *Normals to the profiles of mating teeth must, at all points of contact, pass through a fixed point located on the line of centers.*

In the case of spur- and helical-type gears, the curves used almost exclusively are those of the involute family. In this type of curve, the fixed point mentioned in the basic law is the pitch point. Since all contact takes place along the line of action, and since the line of action is normal to both the driving and driven involutes at all possible points of contact, and lastly, since the line of action passes through the pitch point, it can be seen that the involute satisfies all requirements of the basic law of gearing. Many years ago, the cycloidal family of curves was in common use; however, clockwork gears are about the only application of this type of tooth today. The involute curve has supplanted the cycloid because of its greater ease of design and because it is far less sensitive to manufacturing and mounting errors. In addition, the tools used to generate the cycloid are more difficult to make with the same degree of accuracy as those for the involute.

The use of the involute has become so universal that, with few exceptions, it is hardly necessary to specify on spur- and helical-gear drawings that the involute form is desired. It is only when a designer wishes to use a noninvolute tooth for a special application that it becomes apparent how much work must be done to design a satisfactory profile properly. The results of standardization and the deep understanding of competent manufacturers of gears are appreciated most by those who require quality designs of a nonstandard nature.

The methods of determining conjugate profiles for various gear applications are beyond the scope of this chapter. The designer should be aware, however, that the basic law of gearing can be utilized to determine conjugate tooth profiles other than those shown in this chapter. Several of the unique properties of the involute are utilized by the data in this chapter to provide conjugate gears, made, for the most part, with standard tooling. Each gear will have special properties which will make it best suited for its application. It will, however, not be the type of gear that would result from an indiscriminate application of cookbook gear formulas often found.

Bevel-gear-tooth forms covered by the standards noted in this chapter are *not* involute. In bevel gearing, it is practical to employ a wide variety of conjugate curves, since the problems of interchangeability and of tool costs are minor. This is due to the very great flexibility provided by the generating process used in bevel-gear generators. The tools are usually single-edged, hence relatively easy to make, and are fundamental in nature. One of the most common forms for bevel-gear teeth is called the "octoid" after the shape of its line of action.

Worm gearing, like bevel gearing, is noninvolute. The tooth form of worm gearing is usually based on the shape of the worm; that is, the teeth of the wormgear are made conjugate to the worm. In general, worms can be chased, as on a lathe, or cut by such processes as milling or hobbing, or can be ground. Each process, however, produces a different shape of worm thread and will generally require a different shape of wormgear tooth in order to run properly.

In the case of face gearing, the pinion member is a spur or helical gear of involute form, but the gear tooth is a special profile conjugate only to the specific pinion. Thus pinions having a larger or smaller number of teeth from the number for which the face gear was designed will not run properly with the face gear.

Pitch Diameter. Although pitch circles are not the fundamental circles on gears, they are traditionally the starting point on most tooth design. The pitch circle is related to the base circle, which is the fundamental circle, by the relationships given below. Some authorities list as many as nine distinct definitions of pitch circles. The following definitions cover the pitch circles considered in this chapter.

Standard Pitch Diameter. The diameter of the circle on a gear determined by dividing the number of teeth in the gear by the diametral pitch.

The diametral pitch is that of the basic rack defining the pitch and pressure angle of the gear.

The AGMA gives the following equation for standard pitch diameter in the fine-pitch standards:

$$D = \frac{N}{P_d} \tag{5-7}$$

where D = diameter of standard pitch circle
P_d = diametral pitch of basic rack
N = number of teeth in gear

Operating Pitch Diameter. The diameter of the circle on a gear which is proportional to the gear ratio and the actual center distance at which the gear pair will operate.

A gear does not have an operating pitch diameter until it is meshed with a mating gear. The equations for operating pitch diameters are

External spur and helical gearing

$$d' = \frac{2C'}{m_g + 1} \quad \text{(pinion member)} \tag{5-8}$$

$$D' = \frac{2C'm_g}{m_g + 1} \quad \text{(gear member)} \tag{5-9}$$

Internal spur and helical gearing

$$d' = \frac{2C'}{m_g - 1} \quad \text{(pinion member)} \tag{5-10}$$

$$D' = \frac{2C'm_g}{m_g - 1} \quad \text{(gear member)} \tag{5-11}$$

where d' = operating pitch diameter of pinion
D' = operating pitch diameter of gear
C' = operating center distance of mesh
m_g = gear ratio = $\dfrac{\text{number of teeth in gear}}{\text{number of teeth in pinion}}$

The operating and the standard pitch circles will be the same for gears operated on center distances which are *exactly* standard. The distinction to be made usually involves tolerances on the center distance. Most practical gear designs involve center-distance tolerances which are the accumulated effects of machining tolerances on the center bores and tolerances in bearings (clearances, runout of outer races, etc). Thus gears all operate with maximum and minimum operating pitch diameters. The most important application occurs on gears operated on nonstandard center distances.

In Fig. 5-5, a pair of gears designed to operate on enlarged center distances are shown. Both the standard and operating pitch circles are shown. It will be noted that the pitch circles are related to the base circle as follows:

$$D = \frac{D_b}{\cos \phi} \quad \text{(standard pitch diameter)} \tag{5-12}$$

$$D' = \frac{D_b}{\cos \phi'} \quad \text{(operating pitch diameter)} \tag{5-13}$$

It will also be noted that the standard pitch circles do not contact each other by the amount that the center distance has been increased from standard.

In worm gearing, it is convenient to use pitch circles. In this case, however, it is common practice to make the pitch circle of the gear go through the teeth at a diameter at which the tooth thickness is equal to the space width. In the case of the worm, the pitch circle also defines a cylinder at which the width of the threads and spaces are equal. It is also good practice to modify the worm teeth slightly to achieve the required backlash. When this is done, the space widths are greater than the thread thicknesses, as measured on the "standard" pitch cylinder, by the amount of backlash introduced. The pitch cylinder also defines the diameter at which the lead angle, as well as the pressure angle, is to be measured.

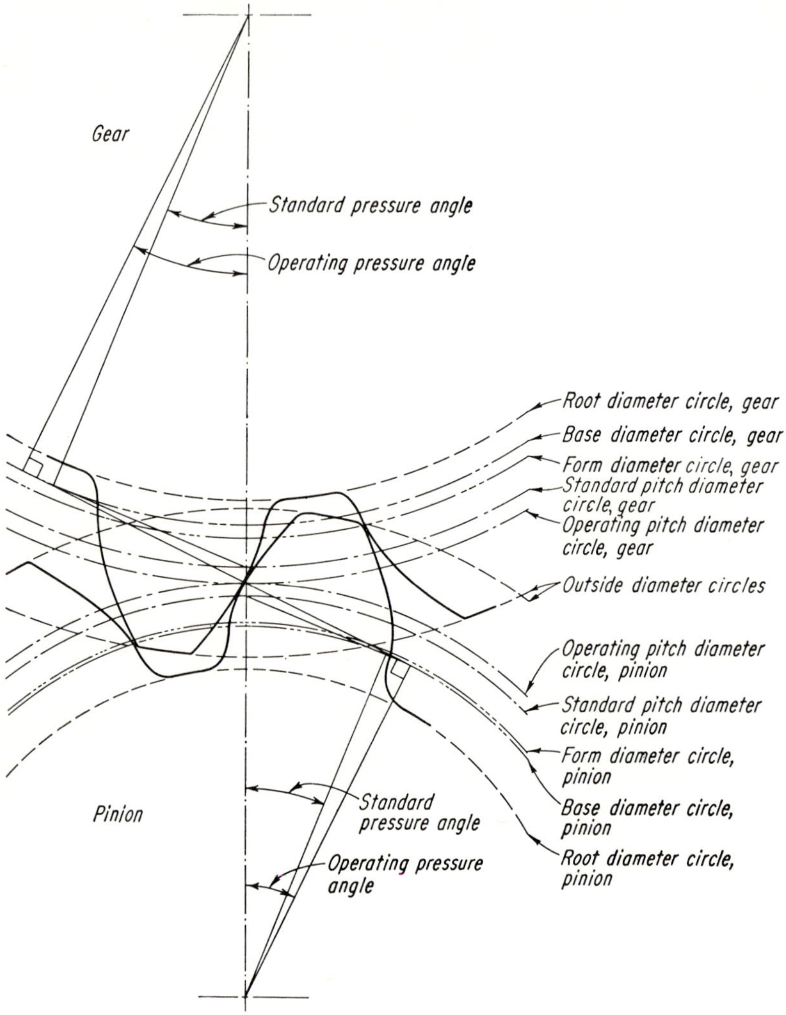

Fig. 5-5. Nomenclature for gear cut to standard pressure angle and operated at an increased center distance.

Zones in Which Involute Gear Teeth Can Exist. Although most gear designs utilize "standard" or "equal-addendum" tooth proportions, it is not always necessary or even desirable to use these proportions. One of the outstanding features of the involute tooth profile is the opportunity it affords for the use of different amounts of addendum and tooth thicknesses on gears of any given pitch and numbers of teeth. These variations can be produced with standard gear-tooth-generating tools. It is not necessary to buy different cutting or checking tooling for each new value of tooth thickness or addendum if the proper tooth thickness–addendum relationship is maintained in the design.

As discussed in other chapters of this handbook, the limits of true involute action are established by the length of the line of action. It has been shown that the end limits defining the maximum usable portion of the line of action are fixed by the points at which this line becomes tangent to the base circles. These limits, a and b, are shown in Fig. 5-6a. The largest pinion or gear that will have correct gear-tooth action is defined by circles which pass through points a and b (see Fig. 5-6b). Therefore, any gear teeth that lie fully within the crosshatched area of Fig. 5-6b will have correct involute action on any portions of the teeth that are not undercut. Undercut limitations are discussed more fully under Undercut.* One should not infer that the largest usable outside diameters will be used simultaneously on both members on a given gear design.

The equations for maximum usable outside diameter are

$$D_{oG} = 2\left[(C' \sin \phi')^2 + \left(\frac{D_G}{2} \cos \phi\right)^2\right]^{1/2} \quad \text{max.} \tag{5-14}$$

$$D_{oP} = 2\left[(C' \sin \phi')^2 + \left(\frac{D_P}{2} \cos \phi\right)^2\right]^{1/2} \quad \text{max.} \tag{5-15}$$

where C' = operating center distance
ϕ = cutting or standard pressure angle
ϕ' = operating pressure angle
D_G = pitch diameter of gear
D_P = pitch diameter of pinion

The operating pitch circles for the gear ratio (2:1) and center distance chosen for this illustration are shown in place in Fig. 5-6c. If this center distance is "standard" for the number of teeth and pitch, the operating pitch circles shown would also be the "standard" pitch circles.

In order to transmit uniform angular motion, a series of equally spaced involute curves are arranged to act on each other. These are shown in place in Fig. 5-6d. On both the pinion and the gear member, the involute curves originate at the base circles and theoretically can go on forever. The more practical lengths are suggested by dash lines. The spacing of the involutes of both members measured on the base circles must be equal, and the interval chosen is called the base pitch. The base pitch of the gear or pinion times the number of teeth in the member must exactly equal the circumference of the base circles.

If similar involute curves of opposite hand are drawn for both base circles, the familiar gear teeth are achieved. It is customary to measure the distance from one involute curve to the next along the standard pitch circle. This distance is called the "circular pitch." It is also customary to make "standard" tooth proportions with tooth thicknesses equal to one-half the circular pitch. The standard addendum used for the gearing systems shown in this chapter is equal to the circular pitch divided by π.

Figure 5-6e shows "standard" teeth developed on the base circles of Fig. 5-6a through d. It will be noted that the addendum of the 12-tooth pinion is equal to that of the gear and that the outside diameters thus established do not reach out to the maximum values established by the line-of-action limits.

In modern gear-cutting practice, the tooth thickness of gears as measured on the "standard" pitch circle is established by the depth to which the generating-type cutter (usually hob or shaper cutter) is fed, relative to the "standard" pitch diameter. In order to obtain a correct whole depth of tooth, the outside diameters of the gear blanks are made larger or smaller than "standard" by twice the amount that the cutter will be "fed in" or "held out" relative to the normal or standard amount that would be used for standard gears. If the cutter is to be held out a distance Δc, the outside diameter of the blank is made $2\Delta c$ larger than standard and the resultant gear is said to be "long-addendum." Figure 5-6f shows an example of a gear in which

* See page 5–15.

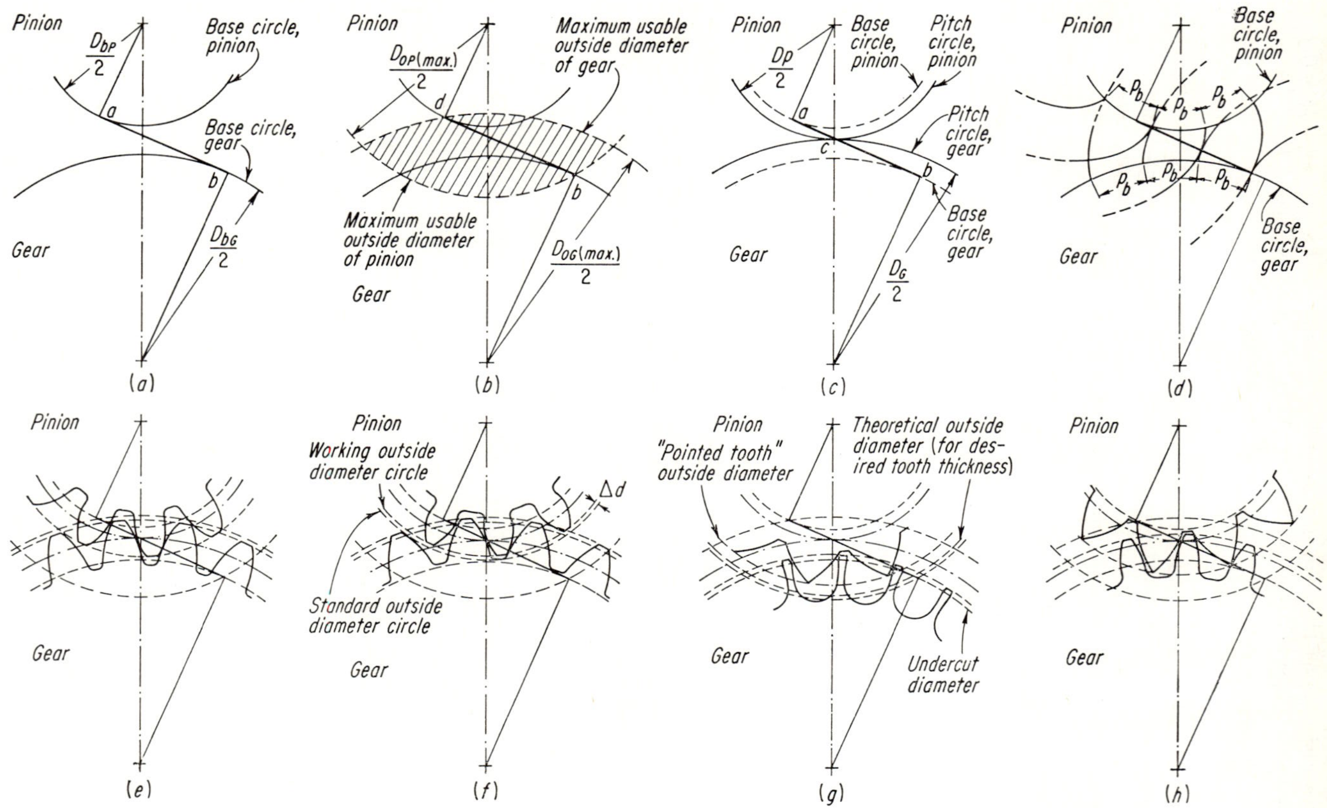

FIG. 5-6. Study of involute-tooth development. (*a*) Maximum usable length of line of action. (*b*) Maximum zone in which conjugate action can take place. (*c*) Operating pitch circles. (*d*) Development of involute profiles. (*e*) Standard addendum tooth proportions. (*f*) Short (pinion) addendum tooth proportions. (*g*) Long (pinion) addendum (abnormal) tooth proportions. (*h*) Long (pinion) addendum (normal) tooth proportions.

the cutter was held out sufficiently so that the outside diameter of the gear was equal to the diameter of the maximum usable outside-diameter circle (see also Fig. 5-6*b*). This is called a "long"-addendum gear. The cutter was fed into the pinion an equal amount, making it "short"-addendum. Reasons for making pinions short-addendum are discussed under Speed-increasing Drives, and the effects of the undercutting produced are discussed under Undercut.

In order to avoid undercut, or to achieve a more equal balance in tooth strength, it is customary to make the pinion addendum "long" and that of the gear "short." If an attempt is made to design a pinion with the maximum usable outside diameter as established by the line of action (see Fig. 5-6*b*) difficulties may arise. In the example shown having 12 and 24 teeth, it is not possible to generate a pinion having such an outside diameter.

In Fig. 5-6*g* is shown the tooth form resulting from such an attempt. The pinion blank was turned to the maximum usable outside diameter, and the cutter was held out an equivalent distance Δc. As a result of the generating action, the sides of the teeth are involute curves "crossed over" at a diameter smaller than the desired outside diameter. Thus the teeth are pointed at the outside diameter and also the whole depth is less than anticipated. This effect is less serious in gears having larger numbers of teeth.

The gear tooth that results from this extreme modification is badly undercut. This effect would not have been so great if the gear had more teeth.

The amount of long and short addendum that may be applied to each member of a gear mesh is limited by the three following considerations.

1. The length of the usable portion of the line of action will form a maximum limit on the outside diameter of a gear. Diameters in excess of this will not provide additional tooth-contact area, since there is no involute portion on the mating gear that can contact this area.
2. The diameter at which the teeth become pointed limits the actual or effective outside diameter.
3. Undercutting may limit the short-addendum gear. The undercut diameter should be always less than the form diameter.

The above three considerations show the extreme limits that bound gear-tooth modifications. The designer should not infer that it is necessary to approach these limits in any given gear design. In the treatment of gear-tooth modifications, some reasons for making given tooth modifications are considered. In each case, however, the designer must be sure that the amount actually used does not exceed the limits discussed here.

Pointed Teeth. The previous sections have shown how gear teeth generated by a specific basic rack can have different tooth thicknesses, and that the outside diameters of such gears are altered from standard as a function of the change in tooth thickness. In practice, these tooth profiles are achieved by "feeding in" or "holding out" the gear-generating tools. It is customary to feed a specific cutter to a definite depth into the gear blank but to achieve long-addendum designs, for instance, by starting with a blank which has a greater than standard outside diameter. The cutter, when working at its full cutting depth, will still be held out from the standard N/P_d pitch circle.

Teeth generated thicker than standard will have tips of a width less than standard, since the cutter must be "held out." For any given number of teeth, the tooth thickness can be increased such that the tip will become pointed at the outside-diameter circle. In Fig. 5-7 are shown four gears all having the same number of teeth, diametral pitch, and pressure angle. In Fig. 5-7*a*, the cutter shown by a rack has been fed to the standard depth. The outside diameter of this gear is standard, $(N + 2)P_d$.

Note the tooth thickness at the tip. In Fig. 5-7*b*, the outside diameter was somewhat enlarged and the cutter fed in the standard whole-depth distance starting from the enlarged outside diameter. This results in the thicker tooth (which would operate correctly with a standard gear on an enlarged center distance) and a somewhat thinner tip. In Fig. 5-7*c*, the outside diameter has been established at the maximum value

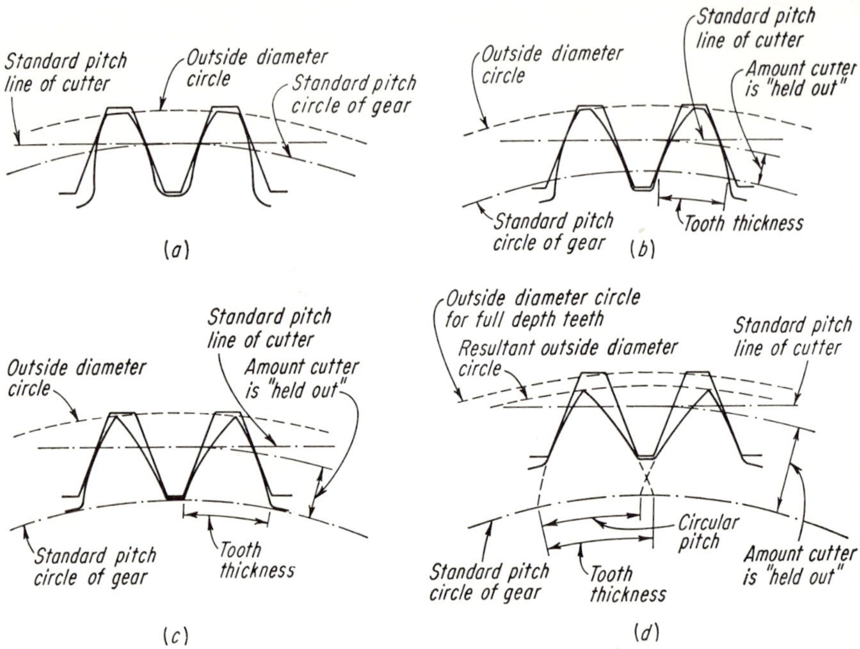

FIG. 5-7. Cutting long-addendum teeth. (*a*) Standard tooth. (*b*) Long-addendum tooth. (*c*) Long-addendum (pointed) tooth. (*d*) Long-addendum (over-enlarged) tooth.

at which it is still possible to achieve a standard whole-depth tooth; the tooth thus generated is pointed. Figure 5-7*d* shows what happens if the maximum outside diameter and tooth thickness are exceeded. The resulting tooth does not have the correct whole depth because the involute curves cross over below the expected outside diameter. This tooth is similar to the one shown on the pinion member in Fig. 5-6*g*.

The amount that the outside diameter of a gear is to be modified is usually a function of the tooth thickness desired. Equation (5-18) gives the relationship usually employed.

The maximum amount that the tooth thickness of a gear can be increased over standard to just achieve a pointed tooth can be calculated from the simultaneous* solution of Eqs. (5-16) and (5-17).

$$\Delta D_o \text{ max.} = N\left(\frac{\cos\phi}{\cos\phi''} - 1\right) - 2 \tag{5-16}$$

$$\Delta D_o \text{ max.} = N\,\frac{(\text{inv}\,\phi'' - \text{inv}\,\phi) - (\pi/2)}{\tan\phi} \tag{5-17}$$

* Equations (5-16) and (5-17) are solved by assuming a series of values for ϕ''. Curves are plotted for ΔD_o max. vs. ϕ''. The point where the curves cross indicate a *simultaneous* solution for the two equations. For instance, a 20-tooth pinion of 1 pitch having a 20° pressure angle and cut with a standard cutter has a crossing point for ϕ'' of 39.75° and a ΔD_o max. of 2.44 in. It should be kept in mind that these equations solve problems for 1 pitch only (divide by pitch to adjust answer for other pitches) and make no allowance for backlash.

The equivalent amount that the tooth thickness must be increased above standard to achieve this increase in outside diameter is given by Eq. (5-18).

$$\Delta T = d(\text{inv}\ \phi'' - \text{inv}\ \phi) - T \qquad (5\text{-}18)$$

where D_o max. = maximum outside diameter at which a tooth having full working depth will come to a point

$= \dfrac{N+2}{P} + \Delta D_o$

T = tooth thickness at standard pitch diameter

ϕ = standard pressure angle of hob or rack

ϕ'' = pressure angle at tip of tooth

Undercut. An undercut tooth is one in which a portion of the profile in the active zone has been removed by secondary cutting action. Under certain conditions, the path swept out by the tip of a generating-type cutter will intersect the involute active profile at a diameter greater than the limit diameter. Such a tooth is said to be undercut since no contact with the mating gear can take place between the undercut and the limit circles.

The amount of undercut will depend upon the type of tool used to cut the gears; in general, a hob will undercut a gear to a greater degree than will a circular shaper-type cutter. Form-type cutters do not normally produce undercut. In calculating the undercut diameter, it is customary to use an equation based on the type of tooling which will produce the greatest undercut, usually a hob. This is done in order to give the shop the greatest freedom of choice in selecting tools. If the most adverse choice will produce a satisfactory tooth profile then all other types of cutter that might be chosen by the shop should prove to be satisfactory, except in those cases where undercut is intended to provide clearance for the tips of the mating teeth.

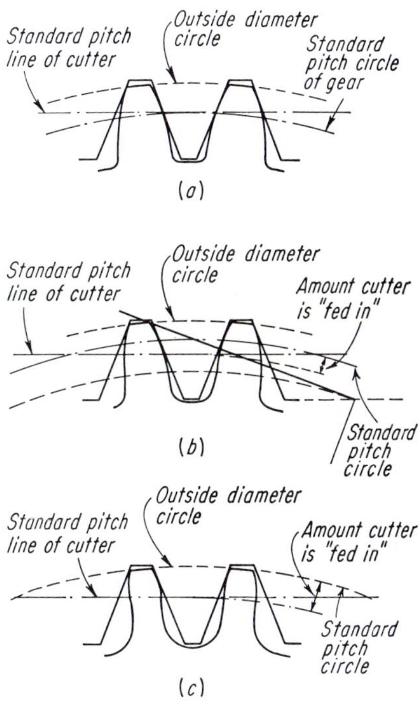

FIG. 5-8. Cutting short-addendum teeth. (*a*) Standard tooth. (*b*) Short-addendum tooth. (*c*) Short-addendum (abnormal) tooth.

Equations for estimating the amount of undercut that will occur in a gear are given in Chap. 4. In general, spur gears of 20° pressure angle having 18 or more teeth, and made to "standard"- or "long"-addendum tooth proportions, will not be undercut. In each of the "standard systems" for different types of gears, the minimum number of teeth recommended for each pressure angle is usually based on undercut.

Helical gears can usually be made with fewer teeth than can spur gears without getting into problems of undercut.

Figure 5-8 shows three gears having the same numbers of teeth, each produced by generating-type cutters. Figure 5-8*a* is the tooth form of a gear having a standard addendum. Figure 5-8*b* shows the shape of the tooth that results when the tip of the rack-type cutter is operating at a depth exactly passing through the point of intersection of the line of action and the base circle. Figure 5-8*c* shows the shape of a tooth which results when the cutter works at a depth somewhat below the intersection of the line of action and the base circle.

Undercutting is a product of generating-type cutters. Gears cut by forming-type cutters do not usually have undercut. In certain cases, undercut may ensure tip clearance in mating gears. Racks operating with gears having small numbers of teeth may show tip interference unless the pinions were generated with special shaper-type cutters or hobs.

5-3. Long- and Short-addendum-gear Design. The previous article shows the geometric limitations on the amount that gears can be made "long"- or "short"-addendum. This section indicates cases in which long and short addendum should be considered.

Modification of the addendum of the pinion, and in most cases the gear member, is recommended for gears serving the following applications:

1. Meshes in which either or both members have small numbers of teeth.
2. Meshes operating on nonstandard center distances because of limitations on ratio or center distances.
3. Meshes of speed-increasing drives.
4. Meshes designed to carry maximum power for the given weight allowance. (This type of gearing is usually designed to achieve the best balance in strength, wear, specific sliding, pitting, or scoring.)
5. Meshes in which an absolute minimum of energy loss through friction is to be achieved.

Addendum Modification for Gears Having Small Numbers of Teeth. Undercutting* is one of the most serious problems occurring in gearing having small numbers of teeth. The amount that gears with small numbers of teeth should be enlarged (made long-addendum) to avoid undercut, has been standardized by the AGMA. The values of modification are based on the use of a hob or rack-type cutter and as a result are more than adequate for gears cut with circular shaper-type cutters. The values of addendum modification recommended for each number of teeth are shown in Table 5-9.

When two gears, each containing a small number of teeth, must be operated together, it may be necessary to make both members "long"-addendum to avoid undercut. Such teeth will have a tooth thickness which is larger than standard and which will necessitate the use of a greater than standard center distance for the pair.

Modifications to avoid undercut fall into three categories.

1. If both members of a pair have less than the critical number of teeth as shown in Table 5-9, both members will have to be made long-addendum, and the operating center distance will have to be increased a proportional amount. Equation (6-5) gives the amount of increase in center distance required.
2. If the pinion member has less than the critical number of teeth, and the mating gear has considerably more, the usual practice is to decrease the addendum of the gear by the amount proportional to the amount that the pinion is increased. This results in a pair of gears free from undercut that will operate on a standard center distance.
3. If the pinion member has less than the critical number of teeth, and the gear just slightly more, a combination of practices 1 and 2 above may be employed. The pinion addendum is adjusted to avoid undercut. The gear addendum

* See Chap. 7, Art. 7-6, for further information on undercutting.

is reduced to the point where undercut would just begin. If this reduction is not enough to compensate for the pinion-addendum increase the remainder of the adjustment required is achieved by an increase in center distance.

An alternate is to increase the pinion addendum by the required amount and increase the center distance by an equivalent amount to make it possible to use a standard gear.

Gears Operating on Nonstandard Center Distances.* It is sometimes required to design a pair of gears to operate on a center distance that is not standard for that total number of teeth and diametral pitch.

In most cases the use of a special nonstandard cutter can be avoided by the use of what is in effect a "long"- or a "short"-addendum design.

Equations (6-5), (6-8), and (6-9), give the total tooth thickness that a pair of meshing gears must have to operate on a given center distance. As was shown under Undercut, the tooth thickness of each member can be reduced from standard by an amount sufficient to produce undercut that will just reach the form-diameter circle without impairing the smooth-running characteristics of the gears. Under Pointed Teeth it was shown that the thickness could be increased to the point where the teeth just became pointed. Therefore, if the sum of the required tooth thickness (to meet the desired center distance) lies between the limits set on tooth thickness by either undercut or pointed teeth, a workable gear design can be obtained. This design may be less than optimum from the standpoint of strength and wear or pitting considerations, however.

This method offers the designer the opportunity to design special gears operating on special, nonstandard pressure angles (22½°, for instance) using perfectly standard gear-cutting tools of 20° pressure angle. In this case the operating center distance is increased to the amount required to achieve the desired pressure angle. The tooth thickness of the gears is increased by the equivalent amount, and the desired result is achieved. This technique is resorted to, to achieve increased load capacity or resistance to pitting, without the need of special tools.

Speed-increasing Drives. Most gear trains are speed-reducing (torque-increasing), and most data on gear-tooth proportions are based on the requirements of this type of gear application. The kinematics of speed-increasing drives is somewhat different, and as a result, special tooth proportions should be considered for this type of gear application. As in the case of conventional drives, the problems to be discussed here are most serious in meshes involving small numbers of teeth.

The first of these problems involves the tendency of the tip edge of the pinion tooth to gouge into the flank of the driving gear tooth. This gouging can come about as a result of spacing errors in the teeth of either member which allow the flank of the gear tooth to arrive at the theoretical contact point on the line of action before the pinion does. The pinion tooth has to deflect to get into the right position or it will gouge off a sliver of gear-tooth side. If the gears are highly loaded, the unloaded pinion tooth entering the mesh will be out of position (lagging) since it is not deflected. The gear tooth is in effect slightly ahead of where it should be. The result is the same as if the gear tooth had an angular-position error. The bearing and lubricating problems at the beginning point of contact are particularly bad. The edge of the pinion tooth tends to act as a scraper and remove any lubricating film that may be present, for some distance along the flank of the tooth.

One possible solution to this problem is to give the tip of the pinion tooth a moderate amount of tip relief. This provides a sort of sled-runner condition which is easier to lubricate and which helps the pinion tooth find the proper position relative to the gear tooth with less impact.

* See Chap. 6 for further information on this subject.

A better solution, which may also be combined with the tip modification, is to modify the tooth thicknesses and addendums so as to get as much of the gear-tooth contact zone in the arc of recess as possible.

The action of gear teeth in the arcs of approach and recess may be likened to a boy pushing or dragging a stick down the street. A gear tooth driving a pinion in the arc of approach is like the case where the boy pushes the stick along ahead of him. It tends to gouge into the ground. The gear-tooth action in the arc of recess is like the case where the boy drags the stick along behind him. It has no tendency to gouge in; it rides up over bumps and is easier to pull. The relative gear efficiencies in each case are discussed under the topic Low-friction Gearing later in this article.

Power Drives (*Optimum Design*). As shown in other chapters, gears fail in one or more of the following ways: actual breaking of the teeth, pitting, scoring, or by wear. In the case of drives made with gears of standard tooth proportions and similar metallurgy, the weakest member is the pinion, and if tooth breakage does occur, it is generally in the pinion. This is a result of the weaker shape of the pinion tooth, as well as the larger number of fatigue cycles that it accumulates. This problem can be relieved to a considerable degree by making the pinion somewhat longer addendum and in so doing increasing the thickness of the teeth and also improving their shapes. If a standard center distance is to be maintained, the gear addendum is reduced a proportional amount. If the proper values are chosen, the pinion tooth strength will be increased and the gear tooth strength somewhat reduced, which will result in almost equal gear and pinion tooth strengths. This will result in an over-all increase in the strength of the gear pair.

Several authorities have suggested addendum modifications which will balance scoring, specific sliding, and tooth strength. Unfortunately, each balance results in different tooth proportions so that the designer has to use proportions which will balance only one feature or else use proportions which are a compromise. Table 5-10 gives values which are such a compromise.

Experimental data seem to indicate that a pair which is corrected to the degree that seems to be indicated by tooth layouts or by calculation for balanced tooth strength will usually result in an overcorrection to the pinion member. The gear is not so strong as form factors seem to indicate. Notch sensitivity in higher hardness ranges seems to be a problem, especially if the gear is to experience a great number of cycles of loading.

In small numbers of teeth, the correction required to avoid undercut on gears operated on standard center distances is excessive, in many cases, in respect to equal tooth strength. An overcorrection in pinion tooth thickness can lead to an excessive tendency to score. As a result, the values of addendum recommended in Table 5-10 represent a compromise among balanced strength, sliding, and scoring.

Gears with teeth finer than about 20 diametral pitch (generally) cannot score, since the tooth is not strong enough to support a scoring load; therefore, the values for addendum increase in fine-pitch gears are somewhat larger than the values for coarse-pitch power gearing.

Low-friction Gearing. In cases where a speed-increasing gear train is to transmit power or motion with the least possible loss of energy, the selection of the tooth proportions is of considerable importance. The sliding should be kept as low as possible, and as much of the tooth action should be put into the arc of recess as possible.

Figure 5-9 shows two involute curves (tooth profiles) in contact at two different points along the line of action. The direction in which the driven pinion tooth slides along the driving gear tooth is shown by the arrows. This example is a speed-increasing drive which is the most sensitive to friction between the teeth. At the pitch point (where the line of centers crosses the line of action) there is no sliding, and it is at this point that the direction of relative sliding of one tooth on the other changes.

The forces shown in Fig. 5-9 are those acting on the driven pinion. The subscripts a are the values considered in the arc of approach and r are those considered in the arc of recess. The normal driving force W_N is the force that occurs at the pitch point and, if there were no friction at the point of gear-tooth contact, would be the force at all other points of contact along the line of action. Since there is friction,

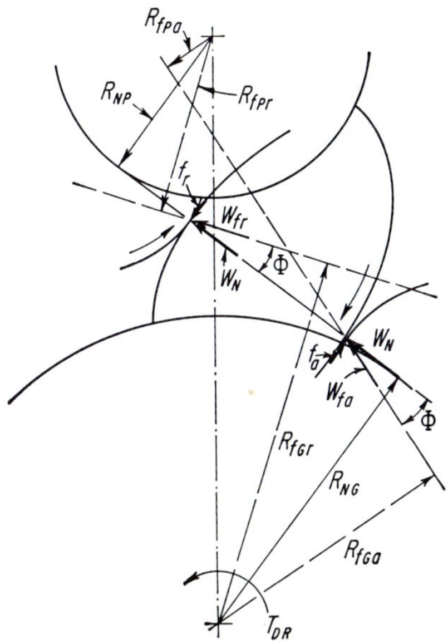

FIG. 5-9. Effect of friction on tooth reactions.

the friction vectors f_a and f_r oppose the sliding of the gear teeth in the arcs of approach and recess. Note the change in direction due to the change in direction of sliding. The angle of friction is Φ and is assumed to be the same in both cases. The torque exerted by the shaft driving the driving gear T_{DR} manifests itself in arc of approach as

$$T_{DRa} = W_N \times R_{NG} \qquad \text{if no friction}$$
$$T_{DRa} = W_{fa} \times R_{fGa} \qquad \text{if friction is assumed}$$

and in the arc of recess

$$T_{DRr} = W_N \times R_{NG} \qquad \text{if no friction}$$
$$T_{DRr} = W_{fr} \times R_{fGr} \qquad \text{if friction is assumed}$$

The resisting moments are, in the arc of approach,

$$T_{DNa} = W_N \times R_{NP} \qquad \text{if no friction}$$
$$T_{DNa} = W_{fa} \times R_{fPa} \qquad \text{if friction is assumed}$$

and the corresponding moments are in the arc of recess

$$T_{DNr} = W_N \times R_{NP} \qquad \text{if no friction}$$
$$T_{DNr} = W_{fr} \times R_{fPr} \qquad \text{if friction is assumed}$$

Note that, in all cases above, single tooth contact is assumed.

Efficiency is output divided by input and in this case is the torque that would appear on the driven shaft when friction losses are considered, compared with the torque that would result if no losses occurred.

Equation (5-19) shows the efficiency of the mesh (single tooth contact) for the contacts occurring in the arc of approach, and Eq. (5-20) shows the efficiency in the arc of recess.

$$E\text{, approach} = \frac{R_{fPa}}{R_{fGa}} \frac{R_{NG}}{R_{NP}} \tag{5-19}$$

$$E\text{, recess} = \frac{R_{fPr}}{R_{fGr}} \frac{R_{NG}}{R_{NP}} \tag{5-20}$$

In the case of speed-increasing drives, the increased efficiency can have considerable significance; cases have occurred in which, for very high ratios and poor lubrication, the speed increaser actually became self-locking.

5-4. Special Design Considerations. Several special considerations should be kept in mind during the evolution of a gear design.

Interchangeability. Two types of interchangeability are related to gearing. The first, and most generally recognized, is "part" interchangeability. This means that, if a part made to a specific drawing is damaged, a similar part made to the same drawing can be put in its place and can be expected to perform exactly the same quality of service. In order to achieve this kind of interchangeability, all parts must be held to carefully selected tolerances during manufacture, and the geometry of the parts must be clearly defined. The second kind is engineering interchangeability. This type provides the means whereby a large variety of different sizes or number of teeth can be produced with a very limited number of standard tools. Parts so made may or may not be designed to have "part" interchangeability. The several "systems" of gear-tooth proportions are based on engineering interchangeability. By careful design and skill in the application of each "system," it is possible to design gears that will perform in almost any application and that can be made with standard tools. Only in the most exceptional cases will special tools be required. The advantages of gearing designed with engineering interchangeability are the lower tool costs and the ease with which replacement or alternate gears can be designed.

Gears procured through catalogue sources are good examples of gears having both "part" and "engineering" interchangeability.

Tooth Thickness. The thickness of gear teeth determines the center distance at which they will operate, the backlash that they will have, and as discussed in previous sections, their basic shape. One of the important calculations made during the design of gear teeth is establishment of tooth thickness.

It is essential to specify the distance from the gear axis at which the desired tooth thickness is to exist. The usual convention is to use the distance that is established by the theoretical pitch circle N/P. Thus, if no other distance is shown, the specified tooth thickness is assumed to lie on the "standard" pitch circle. In certain cases, the designer may wish to specify a thickness, such as the chordal tooth thickness, at a diameter other than the standard pitch diameter. This special diameter should be clearly defined on the gear drawings.

The actual calculation of tooth thickness is usually accomplished by the following procedure.

1. The theoretical tooth thickness is established.
   a. If the gears are of conventional design and are to be operated on standard center distances, the tooth thickness used is one-half the circular pitch.
   b. If special center distances are to be accommodated, Eqs. (5-21) and (5-22) may be used.

$$\cos \phi_2 = \frac{C}{C_2} \cos \phi \tag{5-21}$$

where ϕ_2 = operating pressure angle
C = standard center distance
ϕ = standard pressure angle
C_2 = operating center distance

$$T_P + T_G = \pi + (\text{inv } \phi_2 - \text{inv } \phi)(N_P + N_G) \tag{5-22}$$

where T_P = tooth thickness of pinion member (at 1 diametral pitch)
T_G = tooth thickness of gear member (at 1 diametral pitch)
N_P = number of teeth in pinion
N_G = number of teeth in gear

c. If special tooth thicknesses for tooth-strength considerations are required, the total thickness (sum of pinion and gear) must satisfy Eq. (5-22). If standard centers are used the sum of the modified thickness must equal 1 circular pitch.

2. After the theoretical tooth thickness is established the allowance for backlash is made.
   a. Backlash allowance may be shared equally by pinion and gear. In this case the theoretical tooth thickness of each member is reduced by one-half the backlash allowance.
   b. In case of pinions with small numbers of teeth which have been enlarged to avoid problems of undercut, it is customary to take all the backlash allowance on the gear member. This avoids the absurdity of increasing the tooth thickness to avoid undercut and then thinning the teeth to introduce backlash.

   In case *a*

$$T(\text{actual}) = T(\text{theoretical}) - \frac{B}{2} \tag{5-23}$$

   where B = backlash allowance
   In case *b*

$$T(\text{actual})(\text{pinion}) = T(\text{theoretical}) \tag{5-24}$$
$$T(\text{actual})(\text{gear}) = T(\text{theoretical}) - B \tag{5-25}$$

3. The teeth are lastly given an allowance for machining tolerance. This tolerance gives the machine operator a size or processing tolerance. Usually this is a unilateral tolerance. Chapter 9 deals more fully with the establishment of this value.

   Addendum. The standard value of addendum is $1/P_d$. This is the distance from "standard" pitch-diameter circle to the outside-diameter circle.

   It is often desirable to calculate gear-tooth proportions on the basis of addendum modifications, as against tooth-thickness modifications. In such cases addendum is written

$$a = \frac{1}{P_d} \pm \Delta a \tag{5-26}$$

   where Δa is an addendum increment, added to, or subtracted from, the standard addendum. Values for this increment are given in the fine-pitch standards of the AGMA, for instance.

Addendum should be related to tooth thickness, since special tools will be needed to cut gears if the addendum–tooth thickness relationship is allowed to get too far from proper values. Equation (5-27) shows the amount that the tooth thickness Δt will change for equivalent changes in the depth to which a hob is fed, ΔC. It is normal practice to feed a gear-cutting tool slightly deeper than standard in order to cut backlash into a gear. When more than a small amount of backlash is to be introduced or the tooth thickness is made nonstandard for other reasons, the addendum should be adjusted to permit the use of standard tooling.

$$\Delta t = 2\Delta C \tan \phi = 2\Delta a \tan \phi \tag{5-27}$$

for 14½° pressure angle

$$\Delta t = 0.5172\ \Delta C \tag{5-28}$$

for 20° pressure angle

$$\Delta t = 0.7279\ \Delta C \tag{5-29}$$

NOTE: These are approximations. Equation (6-5) should be used when more accurate results are needed.

In the case of gearing modified to operate on a nonstandard center distance, the gear-addendum modification may be based on the amount that the pinion has been modified and the amount that the center distance has been modified

$$a_G = \frac{2}{P_d} - a_p + \Delta C \tag{5-30}$$

where $\Delta C = C' - C$
C = standard center distance
C' = working center distance

Tooth-profile Modifications. Errors of manufacture, deflections of mountings under load, and deflections of the teeth under load all combine to prevent the attainment of true involute contact in gear meshes. As a result, the teeth do not perform as they are assumed to by theory. Premature contact at the tips or excessive contact pressures at the ends of the teeth give rise to noise and/or gear failures. In order to reduce these causes of excessive tooth loads, profile modification is a usual practice.

Transverse or Profile Modifications. It is customary to consider the shape of the individual tooth profile. Actually, an operating gear presents to the mating gear tooth profiles whose shapes are distorted by errors in the profile shape (involute) and in tooth spacing, as well as by deflections due to contact loads acting at various places along the tooth. Ideally, teeth under load should appear to the mating gear to have perfect involute profiles and to have perfect spacing. This would require a tooth profile of such a shape when under no-load conditions as to deflect into a conjugate profile when subjected to a load. Since loads vary this is not practical. Usual practice is to give tip- or root-profile modifications to otherwise correct involute profiles that will be distorted by contact loads.

Tip modification usually takes the form of a slight thinning of the tip of the tooth, starting at a point about halfway up the addendum. The amount of this modification is based on the probable accumulated effects of the following.

Allowances for Errors of Gear Manufacture. The correct involute profile may not be achieved as a result of manufacturing tolerances in the cutter used to cut the gear teeth or in the machine guiding the cutter. Errors in spacing may also be introduced by the cutter or machine. The result is that the tips of the teeth will attempt to contact the mating gear too soon or too late. Tip modification produces a sort of "sled-runner" shape to help guide the teeth into full contact with the least impact.

Allowances for Deflection under Load. Although the teeth of a gear may have a correct profile under static conditions, the loads imposed may deflect the teeth in engagement to such a degree that the teeth which are just entering mesh will not be in relative positions which are correct for smooth engagement. These tooth deflections cause two errors. The actual tooth profiles are not truly conjugate under load and therefore do not transmit uniform angular motion, and the spacing or relative angular placement of the driving teeth relative to the driven teeth is such that smooth tooth engagement cannot take place. High dynamic loads are the result.

Several methods may be used to compensate for these effects. The most usual is to provide tip relief as discussed above. The flanks of the teeth may also be relieved. Tip and flank relief assume that the normal tooth would tend to contact "too soon." If the tips of the driven gear are made slightly thin, the tooth will be able to get into a position on the line of contact before contact between that tooth and its driving mate actually occurs. The sled-runner effect of the tip modification will allow the tooth to assume full contact load gradually.

In the case of spur gearing the amount of tip relief should be based on the sum of the allowance for probable tooth-to-tooth spacing errors and for assumed deflection of the teeth already in mesh. Equations (5-31) and (5-32) give good first approximations of the amount of tip relief required. The values obtained by these equations should be modified by experience for the best over-all performance.

Modification at first point of contact

$$\text{Modification} = \frac{\text{driving load (lb)} \times 3.5 \times 10^{-7}}{\text{face width (in.)}} \tag{5-31}$$

Remove stock from tip of driven gear.

Modification at last point of contact

$$\text{Modification} = \frac{\text{driving load (lb)} \times 2.0 \times 10^{-7}}{\text{face width (in.)}} \tag{5-32}$$

Remove stock from tip of driving gear.

In general, gear teeth which carry a load in excess of 2,000 lb per in. of face width for more than 1,000,000 cycles should have modification. Those under 1,000 lb per in. of face do not generally require modification.

Bevel-gear teeth are often modified to accommodate mounting misalignments, tooth errors, and deflections due to load. The geometry of bevel-gear teeth allows profile modifications both along the length of the teeth and from root to tip. In the case of spiral and Zerol®* bevel gears cut with face mill-type cutters, the contacting faces of the gear and pinion teeth can be easily made to a slightly different radius of curvature on each member. This is equivalent to crowning of spur gears and is called mismatch.

The tips of the teeth may be given relief, and in some cases the root of the flank is relieved (undercut). This is most commonly done on spiral bevel gears and hypoid gears that are to be lapped. It is done by cutters having a special protuberance called Toprem®* cutters.

Mismatch is calculated into the machine settings and is therefore beyond the scope of this chapter.

Axial Modifications. In general, it is expected that a gear tooth will carry its driving load across the full face width (if a spur gear). Because of deflections in shafts, bearings, or mountings, when under load, or because of errors in the manufacturing of these parts or in the gears, the teeth may not be quite parallel, and end loading may result. Since a heavy load on the end of a gear tooth will often cause it to break off, attempts to avoid loading by relieving the ends of the teeth (crowning) are often made. A gear tooth which has been crowned is slightly thicker at the center section of the tooth than at the ends when measured on the pitch cylinder.

In effect, crowning allows a rocking-chair-like action between the teeth when the shafts deflect into increasingly nonparallel positions. Heavy concentrations of load at the ends of the teeth are avoided. The fact that the whole face cannot act when the shafts are parallel requires that the load imposed upon a set of crowned gears be

* Zerol and Toprem are trade-marks registered by the Gleason Works, Rochester, N.Y.

less than could be carried on a similar noncrowned pair if parallel teeth could be maintained.

In general, the ends of crowned gears are made from 0.0005 to 0.002″ thinner at the ends as compared with the middle, when measured on the circular-arc tooth thickness.

Spur and helical gearing is crowned by means of special attachments on the gear-tooth-finishing machines.

Bevel gearing is crowned, in effect, by the shape of the cutting tools and the way in which they are driven in relation to the teeth. Coniflex®* teeth are straight bevel teeth cut on special generating machines which produce lengthwise crowning. Spiral and Zerol bevel gears and hypoid gears are given the effect of crowning by the use of a different radius of lengthwise curvature on the convex and the concave sides of the teeth.

Throated worm gears can be given the effect of crowning by the use of slightly oversize hobs.

Face gears are given the effect of crowning by the selection of a cutter having one or more teeth than the mating gear.

Root Fillets. The shape and minimum radius of curvature that the root fillet of a tooth will have depends upon the type and design of the cutting tool used to produce the gear tooth. The shape of the root fillet, as well as its radius, and the smoothness with which it blends into the root land and active profile of a gear tooth can have a profound effect on the fatigue strength of the finished gear. The radius of the fillet at the critical cross section of the tooth is controlled on drawings by specifying the minimum acceptable fillet radius. The point of minimum radius occurs almost adjacent to the root land in the case of gears cut by hobs. An equation giving a reasonable evaluation of the minimum radius for teeth cut by hobs is

$$r_f = 0.7\left[r_T + \frac{(h_t - a - r_T)^2}{(d/2\cos^2\psi) + h_t - (a + r_T)}\right] \tag{5-33}$$

where r_f = minimum calculated fillet radius produced by hobbing or generating grinding
r_T = edge radius of generating rack, hob, or grinding wheel
a = addendum of gear
h_t = whole depth of gear
d = pitch diameter of gear
ψ = helix angle (use 0° for spur gears)

The edge radius specified for the generating tool will, in general, depend upon the service the gear is to perform or upon special manufacturing considerations. Table 5-16 shows suggested values of edge radius for various gear applications.

Other types of manufacturing tools can be designed to produce the minimum fillet radius as obtained from Eq. (5-33). The shape of the fillet will be somewhat different, however.

A common method of checking the minimum radius of the coarser-pitch gears is to lay a pin in the fillet zone and note that the contact is along a single line.

The constant 0.7 is to allow a reasonable working tolerance to the manufacturer of the tools. Edge breakdown of hobs and cutters will tend to increase the radius produced. This is particularly true of hobs and cutters for very-fine-pitch gears.

Effective Outside Diameter. It is customary to consider the outside diameter of a gear as the outer boundary of the active profile of the tooth. In several cases this approximation is not good enough.

1. In very-fine-pitch gears that have been burr-brushed, the tip round may be quite large in proportion to the size of the teeth even though it is only a few thou-

* Coniflex is a trade-mark registered by the Gleason Works, Rochester, N.Y.

sandths in radius by actual measurement. Since no part of this radius can properly contact the mating tooth, the outside diameter is, from the standpoint of conjugate action, limited to the diameter where the tip radius starts. "Effective outside diameter" should be used instead of "outside diameter" in calculations of contact ratio.

2. In some gear meshes in which the pinion member contains a small number of teeth, the tips of gear teeth may be found to be extending into the pinion spaces to a depth greater than that bounded by the line of action. In Fig. 5-6*b* the dimension $D_{oP}/2$ is the maximum effective diameter of the pinion. The actual outside diameter of the pinion may exceed this value, however. In calculating contact ratio, the effective diameter as limited by the pinion base circle and center distance should be used.

Width of Tip of Tooth. The tooth thickness at the tip of the tooth is a convenient index of the quality of a gear design. For most power-gearing applications, the thickness of the gear tooth should not be more than 1½ to 2 times that of the pinion tooth at the tip.

If the tooth thickness (arc), at the pitch diameter (standard) is known, the following equation will give its thickness at the tip:

$$t_o = D_o\left(\frac{t}{D} + \text{inv } \phi - \text{inv } \phi_o\right) \tag{5-34}$$

$$\cos \phi_o = \frac{D \cos \phi}{D_o} \tag{5-35}$$

where t_o = thickness at outside diameter D_o
D_o = outside diameter of gear, or diameter where tooth thickness is wanted
t = arc tooth thickness at D, or at known diameter
D = standard pitch diameter, or diameter where t is known
ϕ = standard pressure angle, or pressure angle where t is known
ϕ_o = pressure angle at outside diameter or at diameter where tooth thickness is wanted

Pointed-tooth Diameter. An independent method of checking the quality of a long- and short-addendum gear design is to calculate the diameter at which the teeth would come to a point. If this value is smaller than the value of outside diameter chosen by other means, the design should be recalculated. Equations (5-36) and (5-37) provide one method of calculating pointed-tooth diameters.

$$\text{inv } \phi_{o(P)} = \frac{t}{D} + \text{inv } \phi \tag{5-36}$$

$$D_{o(P)} = \frac{D \cos \phi}{\cos \phi_{o(P)}} \tag{5-37}$$

where $\phi_{o(P)}$ = pressure angle at pointed-tooth diameter
$D_{o(P)}$ = pointed-tooth diameter
See Eq. (5-35) for other symbols.

Backlash. In general, backlash is the lost motion between mating gear teeth. It may be measured along the line of action or on the pitch cylinders of the gears (transverse backlash) and, in the case of helical gears, normal to the teeth.

In a set of meshing gears, the backlash that exists is the result of the actual center distance at which the gears operate, and the thickness of the teeth. Changes in temperature, which may cause differential expansion of the gears and mountings, can produce appreciable changes in backlash.

When establishing the backlash that a set of gears will require, the following should be considered:

1. The minimum and maximum center distance. These values are the result of tolerance build-ups in the distance between the bores supporting the bearings on which the gears are mounted as well as the basic or design values. Antifriction bearings, for example, have runout eccentricity between the outside diameter and the outer ball path and runout between the bore and the inner ball path. As the shaft and the inner race of the bearing rotate, and the outer race creeps in the housing, the center distance will vary by the amount of the eccentricities of these bearing elements.
2. The thickness of the teeth, as measured at a fixed distance from the center on which the gear rotates, will vary because of gear runout. Also the workman cutting the gear is given a tooth-thickness tolerance to work to, which introduces tooth-thickness variations from one gear to the next.

The minimum backlash will occur when all the tolerances react all at the same time to give the shortest center distance and the thickest teeth with the high points of gear runout. The maximum backlash will occur when all the tolerances move in the opposite direction.

Design backlash is incorporated into the mesh to ensure that contact will not occur on the nondriving sides of the gear teeth. Although backlash may be introduced by increasing the center distance, it is usually introduced by thinning the teeth. The minimum value should be at least sufficient to accommodate a lubricating film on the teeth.

Sometimes a statistical approach is used, since there are a sufficient number of tolerances involved; thus the design backlash introduced may not be numerically so large as the possible adverse build-up of tolerances. This approach is particularly handy in instrument gearing, since the maximum backlash allowable usually has to be held to a minimum.

By definition, backlash cannot exist in a single gear! Backlash is a function of the actual center distance on which the gears are operated, and the actual thicknesses of the teeth of each gear.

It is customary to use generally recognized values of center-distance tolerance and gear-tooth tolerances for power gearing. If this is done, and the calculated tooth thicknesses are reduced by the amounts of design backlash as shown in Table 5-11, satisfactory gears should result. In the case of instrument gearing, either the values in Table 5-11 may be used or the tooth thickness actually required and the resulting maximum backlash may be calculated.

The backlash that is measured in gears under actual operation will in all probability be considerable larger than the values given in Table 5-11, since these values are for design and include a correction for normal machining tolerances.

Limit and Form Diameters. These dimensions, shown in Fig. 5-2, are the distances from the axis of spur, helical, and internal gears that the active profiles of the teeth begin. Form diameter is the lowest point (spur and helical gearing) that the mating tooth can contact the active profile (see Fig. 5-5).

STANDARD SYSTEMS OF GEAR-TOOTH PROPORTIONS

A "standard" or " system" of gear-tooth proportions provides a means of achieving engineering interchangeability for gears of all numbers of teeth of a given pitch and pressure angle. Because of the large variety of tooth proportions that are possible, it has been found desirable to standardize on a limited number of tooth systems. These systems specify the various relationships among tooth thickness, addendum, working depth, and pressure angle.

The American Gear Manufacturers Association and the American Standards

Association have provided standard systems for various types of gear teeth. When undertaking the design of new equipment, the designer should be sure that he is working with the latest standards. The earlier systems are shown in the following sections to aid the designer confronted with replacement-gear problems. The designer should also be aware of the limitations of each of the systems. These are clearly stated in the discussion of each system. When a design falls outside these limitations, the designer will have to employ tooth modifications required by the design. Since these standards are not intended to be absolutely rigid, the designer is at liberty to modify the tooth proportions within limits to achieve an optimum design. Previously, information was provided to indicate the more common tooth modifications and the ends achieved by these modifications. As much of the standard system as is applicable should be used. When departing from the proportions given in the standard, consideration should be given to the reasons giving rise to these departures and also the consequences. In some cases, departures will require special tooling, whereas others, if properly handled, will not. The following sections provide the basic data covered by most of the systems in present-day use. For additional information about these systems, reference should be made to the original standards listed. In each of the following systems, the tooth proportions are shown in terms of the basic rack of that system. In each case, all gears designed to the basic tooth proportions will have engineering interchangeability.

5-5. Standard System for Spur Gears. The following data are based on the information contained in AGMA standards 201.02 and 201.02A, "Tooth Proportions for Coarse-pitch Involute Spur Gears" and AGMA standard 207.04, "20-degree Involute Fine-pitch System for Spur and Helical Gears."

Limitations in Use of Standard Tables. Caution should be exercised in using the data contained in Table 5-1. The items shown apply only to gears that meet the following requirements:

1. Standard-addendum gears must exceed the minimum numbers of teeth shown in Table 5-3.
2. Long- and short-addendum designs as derived from Tables 5-1, 5-5, and 5-8 are to be used for *speed-reducing drives only.*

The tooth proportions that result from the data shown in this section or from the application of the original standards will be suitable for most speed-reducing (torque-increasing) applications. All gears, including pinions with small numbers of teeth, designed in accordance with the procedure shown will be free from undercut. In order to avoid the problems of undercut in pinions having fewer than the "minimum standard" number of teeth, each system shows proportions for long- and short-addendum teeth. Gears designed with long- and short-addendum teeth cannot be operated interchangeably on standard centers. In general, such gears should be designed to operate only as pairs. These gears can be cut with the same generating-type cutters and checked with the same equipment as standard-addendum gears. The proportions of long and short addendum shown in Table 5-9 are based on avoiding undercut. Such teeth will not have the optimum balance of strength and wear. Slightly different proportions can be used to achieve equal sliding balanced strength or a reduction in the tendency to score. Table 5-10 shows tooth proportions that have a good balance of strength and a minimum tendency to score.

Standard Tooth Forms That Have Become Obsolete. Because industry design standards are continuously reviewed by the sponsoring organizations to ensure that the standards embody the most modern technology, there now exists a group of obsolete tooth-form standards. On occasion, a designer is confronted with a situation in which a replacement gear must be made to mesh in a gear train conforming to one of these earlier standards.

Table 5-15 shows basic data for some of the obsolete standards. The use of this data for new designs is not recommended.

Brown and Sharpe System. This system, see Table 5-15, was developed by the Brown and Sharpe Company to replace the cycloidal-tooth system. It was therefore given similar tooth proportions. It was intended to be cut by form-milling cutters. The departure from the true involute curve in this system is made to avoid the problems of undercut in pinions having small numbers of teeth. Backlash is achieved by feeding the cutter deeper than standard.

The American Gear Manufacturers Association adopted the principal features of this system in 1932 as the "composite system." Present-day form-milling cutters are normally made to this system.

AGMA 14½° Composite System. This system, based on the Brown and Sharpe system, was standardized in 1932 as an AGMA standard. It is interchangeable with the B and S system. It has been made obsolete by AGMA standard 201.02. Replacement parts may be designed from the data in Table 5-15, which are extracted from AGMA 201.02A.

Fellows 20° Stub-tooth System. In order to achieve a stronger tooth form for special drives, the Fellows Gear Shaper Company developed a stub-tooth system in 1898. This system avoided the problem of tooth interference by the combined means of higher pressure angle and smaller values of addendum and dedendum. The pitch was specified as a combination of two standard diametral pitches, thus 10/12 (read ten-twelve). The circular pitch, pitch diameter, and tooth thickness are based on the first number, i.e., 10, and are the same as for a standard 10-diametral-pitch gear. The addendum, dedendum, and clearance, however, are based on the second number (12) and are the same as for a 12-diametral-pitch gear.

AGMA 14½ Full-depth System. In this system, the tooth proportions are identical with those of the 14½ composite system. The sides of the rack teeth are straight lines and therefore produce involute gear-tooth profiles in the generating process. The minimum number of teeth to obtain full tooth action is 31 unless the teeth are modified.

Cycloidal Tooth Profiles. The cycloidal tooth profile is no longer used for any types of gears except clockwork and certain types of timer gears. A combination of involute and cycloidal tooth profile is found in the now obsolete "composite system." *Clockwork gearing* is based on the cycloid but has been greatly modified for practical reasons.

The cycloidal tooth is derived from the trace of a point on a circle (called the describing circle) rolling without slippage on the pitch circle of the proposed gear. The addendum portions of the tooth are the trace of a point on the describing circle rolling on the outside of the pitch circle. This trace is an epicycloid. The dedendum portion of the gear tooth is formed by a trace of a point on the describing circle rolling on the inside of the pitch circle. This trace is a hypocycloid.

Interchangeable systems of gears must have describing circles which are identical in diameter and teeth which have the same circular pitch. The faces and flanks of the teeth must be generated by describing circles of the same size. The tooth proportions for cycloidal teeth were made to the same proportions as were adopted by the B and S system for gears.

The cycloidal system does not have a standard pressure angle. The operating pressure angle varies from zero at the pitch line to a maximum at the tips of the teeth.

In order to achieve correct meshing, the gears must be operated on centers that will maintain the theoretical pitch circles in exact contact. This was one of the major disadvantages of the cycloidal tooth form.

If the diameter of the describing circles were made equal to the radius of the pitch circle of the smallest pinion to be used in the system, the flanks of the teeth would be radial. This member, in effect, establishes the describing circle diameters of each system. In general, the systems for industrial gears were based on pinions of 12 or 15 teeth.

Clockwork and Timer Tooth Profiles. The tooth form of most clockwork and timer gearing involving some type of escapement differs from other types of gearing because of the peculiar requirements imposed by the operating conditions. Two requirements of this type of gearing are outstanding.

1. The gearing must be of the highest possible efficiency;
2. The gearing is speed-increasing, i.e., large gears driving small pinions, with high ratios (between 6:1 and 12:1) due to the need of minimizing the number of gear wheels and economizing on space.

As is discussed in Art. 5-3 under Low-friction Gearing, the most efficient mode of tooth engagement is found in the arc of recess. In the case of speed-increasing involute-profile drives, this means short-addendum pinions. This requirement may be in direct conflict with the need to make the pinion long-addendum to avoid undercut. Clockwork gearing also does not have the requirement that it transmit smooth angular motion. A typical watch train will come to a complete stop five times a second. Furthermore, the motion and force are in one direction, minimizing the need for accurate control of backlash.

The modified cycloidal tooth form is the most commonly used form for the going train of clocks, timers, and watches. Experience has shown that the various modifications to this basic tooth have little effect on the performance because of the scale effect of tolerances required.

There are no American standard tooth forms for going trains.

Specific Spur-gear Calculation Procedure. The following directions give a step-by-step procedure for determining spur-gear proportions:

1. The application requirements should give the requirements of ratio, input speed, and kind of duty to be performed. The duty may be one of power transmission or of motion transmission.
2. Based on application requirements, decide on what pressure angle to use and plan to use a standard system (if possible) (see Table 5-1).
3. Pick approximate number of pinion teeth.
   a. Consult Table 5-2 for general information.
   b. Check Tables 5-3 and 5-4 for undercut limitations.
   c. If power gearing, check Table 5-5 for data on balancing strength and wear capacity.
4. Having approximate number of pinion teeth, determine appropriate center distance and face width. If power gearing, estimate size from data in Arts. 13-3 and 13-4.
5. Based on number of pinion teeth and center distance, determine approximate pitch. Check Table 5-6 for standard pitches, and if possible use a standard pitch. Readjust pinion tooth numbers, center distance, and ratio to agree with pitch chosen. See Chap. 4 for basic relations of pitch, center distance, and ratio. Also use Eqs. (5-8), (5-9), (5-10), and (5-11) if special "operating" center distance is used.
6. Determine whole depth of pinion and gear. Use Tables 5-7 and 5-1. Divide tabular value for 1 diametral pitch by actual diametral pitch to get design whole depth.

7. Determine addendum of pinion and gear. Consult Tables 5-8, 5-9, 5-10, and 5-1. Follow these rules:
   a. Use enough long addendum to avoid undercut (Table 5-9).
   b. If a critical power job—speed-decreasing—balance addendum for strength (Table 5-10).
   c. If no undercut or power problems, use standard addendum (Table 5-1).
8. Determine operating circular pitch. If standard pitch is used, consult Table 5-12. If special pitch is used, calculate from Chap. 4 formulas. If enlarged center distance is used, determine *operating* circular pitch from *operating* pitch diameters [Eqs. (5-8), (5-9), (5-10), and (5-11)].
9. Determine design tooth thickness. Decide first on how much to thin teeth for backlash. Table 5-11 gives recommended amounts for *power gearing* having in mind normal accuracy of center distance and normal operating-temperature variations between gear wheels and casings. If special designs with essentially "no backlash" or unusual materials and accuracy are involved, consult Chap. 6 for special calculations. If the pinion and gear have equal addendum, make the theoretical tooth thicknesses equal and do this by dividing the circular pitch by 2. If a *long*-addendum pinion is used with a *short*-addendum gear, adjust tooth thicknesses by the following:

$$\Delta t = \Delta a 2 \tan \phi_n$$

   If the pinion only is enlarged and the center distance is enlarged to accommodate a standard gear, enlarge the pinion tooth thickness only (see Table 5-13).

   After the theoretical tooth thicknesses are obtained, subtract one-half the amount the teeth are to be thinned for backlash from the pinion and gear theoretical tooth thicknesses. This is the *maximum design* tooth thickness. Obtain the *minimum design* tooth thickness by subtracting a reasonable tolerance for machining from the maximum design tooth thickness.
10. Recheck load capacity using design proportions just obtained. Use Arts. 13-5, 13-6, and 13-9. If not within allowable limits, change design.
11. If the gear design is for critical power gears, additional items will need calculation:
    a. Root fillet radius [Eq. (5-33)]
    b. Form diameter (Art. 7-4)
    c. Modification of profile [Eqs. (5-31), (5-32)]
    d. Diameter over pins (Art. 7-9)
12. Certain general dimensions must be calculated and toleranced.
    a. Outside diameter, $D + 2(a \pm \Delta a)$
    b. Root diameter, $D_o - 2h_t$
    c. Face width
    d. Chordal addendum and chordal thickness (Art. 7-12)
13. It may be necessary to specify
    a. Tip round (at outside diameter) (Table 5-14)
    b. Edge round (Table 5-14)
    c. Roll angle (Art. 7-4)
    d. Base radius [Eq. (7-9)]
14. In some unusual designs, it may be necessary to check the following to assure a sound design:
    Diameter at which teeth become pointed [Eq. (5-37)]
    Width of top land [Eq. (5-34)]
    Effective contact ratio [Eqs. (5-3) and (5-4), Art. 7-3]
    Undercut diameter (Art. 7-6)

Explanation and Discussion of Items in Table 5-1. Table 5-1 shows the "basic" or standard tooth proportions as given in the AGMA or ASA standards listed at the end of the table.

Table 5-1. Basic Tooth Proportions of Spur Gears, AGMA and ASA Standard Systems

(Tooth proportions for fully interchangeable gears operated on standard center distances)

Symbol	Item	Coarse pitch (coarser than 20P), full depth		Fine pitch (20 and finer), full depth	Explanation No.
ϕ	Pressure angle	20°	25°	20°	1
a	Addendum (basic*)	$\frac{1.000}{P}$	$\frac{1.000}{P}$	$\frac{1.000}{P}$	2
b	Dedendum (min.) (basic†)	$\frac{1.250}{P}$	$\frac{1.250}{P}$	$\frac{1.200}{P} + 0.002''$	3
h_k	Working depth	$\frac{2.000}{P}$	$\frac{2.000}{P}$	$\frac{2.000}{P}$	4
h_t	Whole depth (min.) (basic†)	$\frac{2.25}{P}$	$\frac{2.25}{P}$	$\frac{2.200}{P} + 0.002''$	5
t	Circ. tooth thk. (basic*)	$\frac{\pi}{2P}$	$\frac{\pi}{2P}$	$\frac{1.5708}{P}$	6
r_f	Fillet radius (in basic rack)	$\frac{0.300}{P}$	$\frac{0.300}{P}$	Not standardized	7
c	Clearance (min.) (basic†)	$\frac{0.250}{P}$	$\frac{0.250}{P}$	$\frac{0.200}{P} + 0.002''$	8
c	Clear., shaved or grd. teeth	$\frac{0.350}{P}$	$\frac{0.350}{P}$	$\frac{0.3500}{P} + 0.002''$	9
N'_P	Min. numbers of teeth:*				
	Pinion	18	12	18	10
	Pair	36	24	....	11
t_o	Min. width of top land	$\frac{0.25}{P}$	$\frac{0.25}{P}$	Not standardized	12
	Reference standards:				
	AGMA	201.02	201.02	207.04	13
	ASA			B6.19	14

* These values are basic, for equal-addendum gearing. When the gearing is made long- and short-addendum these values will be altered.

† These values are minimum. Shaved or ground teeth should be given proportions suitable to these processes.

1. *Pressure angle.* The pressure angle and numbers of teeth in a given pair of gears are related in that low pressure angles (14½°) should be avoided for low numbers of teeth (see Table 5-3 and Table 5-4).
2. *Addendum.* For general applications, use equal-addendum teeth when the numbers of teeth in the pinion and in the pair exceed the values shown in items 10 and 11.

   Use long addendums for pinions having numbers of teeth, pressure angles, and diametral pitches as shown in Table 5-9.

   The amount of decrease of addendum in the mating gear is to be limited to values that will not produce undercut.

   The values of long addendum for pinions are limited to speed-decreasing drives. Data on speed-increasing drives are given in Art. 5-3 but are beyond the scope of AGMA standards.

(Continued on page 5-38)

3. *Dedendum.* The values shown in the table are the minimum standard values. Shaved or ground teeth should be given a greater dedendum (whole depth and clearance values); see items 5 and 9. A constant value, 0.002″, is added to the dedendum of fine-pitch gears which allows space for the accumulation of foreign matter at the bottoms of the spaces. This provision is particularly important in the case of very fine diametral pitches.

Table 5-2. Numbers of Pinion Teeth, Spur Gearing

No. of teeth	*Remarks*
7–9	Smallest number of teeth recognized by AGMA 207.04
	a. Requires long addendum to avoid undercut on all pressure angles
	b. If 20°, outside diameter should be reduced in proportion to tooth thickness to avoid pointed teeth
	c. Should be made with 25° pressure angle if feasible
	d. May result in poor contact ratio in very fine diametral pitches because of accumulation of tolerances
	e. See Table 5-4 for minimum number of teeth in mating gear
	f. Subject to high specific sliding and usually have poor wear characteristics
10	Smallest practical number with 20° pressure angle.
	a. Requires long addendum to avoid undercut if 20° pressure angle or less
	b. Contact ratio may be critical in very fine pitches
	c. See Table 5-4 for minimum number of teeth in mating gear
12	Smallest practical number for power gearing of pitches coarser than 16 diametral pitch.
	a. Requires long addendum to avoid undercut if 20° pressure angle or less
	b. Smallest number of teeth that can be made "standard" if 25° pressure angle
	c. About the minimum number of teeth for any good fractional-horsepower gear design where long life is important
	d. See Table 5-4 for minimum number of teeth in mating gear
15	Used where strength is more important than wear
	a. Requires long addendum to avoid undercut if 20° pressure angle or less
	b. See Table 5-4 for minimum number of teeth in mating gear
19	Can be made standard addendum if 20° pressure angle or greater
25	Allows good balance between strength and wear for hard steels. Contact (contact diameter) is well away from critical base-circle region
35	If made of hard steels, strength may be more critical than wear. If made of medium-hard (Rockwell C 30) steels strength and wear about equal
50	Excellent wear resistance. Favored for high-speed gearing because of quietness. Critical on strength on all but low-hardness pinions

4. *Working depth.* The working depth customarily determines the type of tooth; that is, a tooth with a $1.60/P$ working depth is a "stub tooth" whereas a tooth with a $2.0/P$ working depth is called a "full-depth" tooth.
5. *Whole depth.* The value shown is the standard minimum whole depth. It will increase in proportion to the increase in backlash cut into the teeth (unless the outside diameter is also correspondingly adjusted). It will also increase slightly if long- and short-addendum teeth are generated with pinion-shaped cutters. See also items 3, 8, and 9 for increases due to manufacturing processing requirements.
6. *Circular tooth thickness, basic.* This is the basic circular tooth thickness on the standard pitch circle. These values will be slightly altered if backlash is introduced into the gears to allow them to mesh at a standard center distance. These values will be drastically altered in the case of long- and short-addendum designs. Table 5-13 gives values of tooth thickness coresponding to standard values of long and short addendum for small numbers of teeth.

Table 5-3. Minimum Number of Pinion Teeth vs. Pressure Angle and Helix Angle Having No Undercut

Helix angle, deg	Min. No. of teeth to avoid undercut			
	Normal pressure angle, ϕ_n			
	14½	20	22½	25
0 (spur gears)	32	17	14	12
5	32	17	14	12
10	31	17	14	12
15	29	16	13	11
20	27	15	12	10
23	25	14	11	10
25	24	13	11	9
30	21	12	10	8
35	18	10	8	7
40	15	8	7	6
45	12	7	5	5

NOTE: Addendum $1/P_d$; whole depth $2.25/P_d$.

Table 5-4. Numbers of Teeth in Pinion and Gear vs. Pressure Angle and Center Distance

No. of teeth in pinion	No. of teeth in gear and pressure angle			
	14½ coarse pitch*	20 coarse pitch†	20 fine pitch†	25 coarse pitch†
7			42‡	
8			39‡	
9			36‡	
10		25	33	15
11		24	30	14
12	52	23	27	12
13	51	22	25	
14	50	21	23	
15	49	20	21	
16	48	19	19	
17	47	18	18	
18	46			
19	45			
20	44			
21	43			
22	42			
23	41			
24	40			
25	39			
26	38			
27	37			
28	36			
29	35			
30	34			
31	33			

* Gears of this pressure angle not recommended for new designs.

† Gears with these numbers of teeth can be made standard addendum and operated on standard center distances.

‡ Not recommended; but if essential, use these values.

NOTES: Pinions having fewer than 10 teeth are not recommended in any AGMA standard. The fine-pitch standard AGMA 207.04 gives data for pinions from seven through nine teeth but strongly discourages their use. Gears having fewer teeth than shown for any given pinion-gear combination will require an enlarged center distance for proper operation.

Table 5-5. General Recommendations on Numbers of Pinion Teeth for Spur and Helical Power Gearing

The maximum number of teeth in the range is based on providing a suitable balance between strength and wear capacity. The minimum number of teeth is intended to require either no enlargement to avoid undercutting or no more enlargement than necessary.

Range of No. of pinion teeth	Ratio m_G	Diametral pitch P_d	Hardness
19–60 19–50 19–45	1–1.9 2–3.9 4–8	1–19.9	200–240 BHN
19–45 19–38 19–35	1–1.9 2–3.9 4–8	1–19.9	Rockwell C 33–38
19–30 17–26 15–24	1–1.9 2–3.9 4–8	1–19.9	Rockwell C 58–63

NOTE: Small numbers of teeth in the pinion require special tooth proportions to avoid undercut.

Gears containing prime numbers of teeth of 101 and over may be difficult to cut on the gear-manufacturing equipment available. In general, prime numbers over 101 and all numbers over 200 should be checked with the shop to be sure the teeth can be made.

Table 5-6. Recommended Diametral Pitches

Coarse pitch			Fine pitch		
2	4	12	20	48	120
2.25	6	16	24	64	150
2.5	8		32	80	200
3	10		40	96	

These diametral pitches are suggested as a means of reducing the great amount of gear-cutting tooling that would have to be inventoried if all possible diametral pitches were to be specified.

Table 5-7. Equations and References for Whole-depth Calculations

Pitch range	Equation	Ref.	Application	Remarks
Coarse pitch up to 19.99	$\frac{2.25}{P_d}$	Table 5-1	General application	AGMA standard
	$\frac{2.35}{P_d}$	....	Gears to be given finishing operations such as shave or grind	AGMA standard
	$\frac{2.40}{P_d}$	....	Max. strength for full-radius (tip) hobs	
Fine* pitch 20 P_d and finer	$\frac{2.20}{P_d} + 0.002$	Table 5-1	Standard whole depth for general applications	AGMA standard
	$\frac{2.35}{P_d} + 0.002$	....	Whole depth for preshave cutters for shaving or grinding	AGMA standard

* Does not apply to pinions containing nine teeth or less.

Table 5-8. Equations and References for Addendum Calculations, Spur Gears

Type of tooth design	Operating condition	Pinion	Gear
Standard or equal addendum	Numbers of teeth in gear and pinion greater than minimum numbers shown in Tables 5-4 and 5-3	(*a*) $\frac{1.00}{P_d}$; see Table 5-1 (*b*) Value from Table 5-12	(*a*) $\frac{1.00}{P_d}$; see Table 5-1 (*b*) Value from Table 5-12
Long and short addendum	Numbers of teeth in pinion *less* than minimum numbers shown in Table 5-3 and numbers of teeth in gear *more* than minimums shown in Table 5-4	(*c*) $\frac{\text{Table 5-9}}{P_d}$ or (*d*) $\frac{\text{Table 5-10}}{P_d}$ if not less than (*c*)	(*c*) $\frac{\text{Table 5-9}}{P_d}$ or (*d*) $\frac{\text{Table 5-10}}{P_d}$
	Numbers of teeth in pinion *less* than minimum numbers shown in Table 5-4 and numbers of teeth in gear *less* than minimum numbers of teeth shown in Table 5-4	*Increase center distance* sufficiently to avoid undercut, in pinion *and* gear (*e*) $\frac{\text{Table 5-9}}{P_d}$	(*e*) See Art. 5-3
	Gearing designed for balanced strength	(*f*) $\frac{\text{Table 5-10}}{P_d}$ NOTE: If value is *less* than given by Table 5-9 check for undercut	(*f*) $\frac{\text{Table 5-10}}{P_d}$ NOTE: If value is *less* than given by Table 5-9 check for undercut
	Designed for nonstandard center distance	(*g*) Calculate tooth thickness to meet required center distance from Eq. (6-1). Then calculate required addendum from Eq. (5-27)	

Table 5-9. Values of Addendum

No. of teeth in pinion	Coarse-pitch teeth (1 through 19 P_D) per AGMA 201.02 20°			Coarse-pitch teeth (1 through 19 P_D) per AGMA 201.02 25°			Fine-pitch teeth (20 P_D and finer) per AGMA 207.04		
	a_P pinion	a_G gear	Recommended min. No. of teeth	a_P pinion	a_G gear	Recommended min. No. of teeth	a_P pinion	a_G gear	Recommended min. No. of teeth
7	.....	.....	...	.....	.....	...	1.4143*	0.4094	42
8	.....	.....	...	.....	.....	...	1.4189*	0.4679	39
9	.....	.....	...	.....	.....	...	1.4190*	0.5264	36
10	1.468	0.532	25	1.184	0.816	15	1.4151	0.5849	33
11	1.409	0.591	24	1.095	0.905	14	1.3566	0.6434	30
12	1.351	0.649	23	1.000	1.000	12	1.2982	0.7019	27
13	1.292	0.708	22	1.000	1.000	12	1.2397	0.7604	25
14	1.234	0.766	21	1.000	1.000	12	1.1812	0.8189	23
15	1.175	0.825	20	1.000	1.000	12	1.1227	0.8774	21
16	1.117	0.883	19	1.000	1.000	12	1.0642	0.9358	19
17	1.058	0.942	18	1.000	1.000	12	1.0057	0.9943	18
18	1.000	1.000	...	1.000	1.000	12	1.000	1.0000	

NOTE: The values in this table are for gears of 1 diametral pitch. For other sizes divide by the required diametral pitch.

The values of addendum shown are the minimum increase necessary to avoid undercut. Additional addendum can be provided for special applications to balance strength. See Table 5-10.

* These values are less than the proportional amount that the tooth thickness is increased (see Table 5-13) in order to provide a reasonable top land.

Table 5-10. Approximate Values of Addendum for Balanced Strength

M_G (gear ratio) N_G/N_P		Number of pinion teeth, N_P					
		17		25		35	
From	To	Pinion a_p	Gear a_G	Pinion a_p	Gear a_G	Pinion a_p	Gear a_G
1.00	1.08	1.058	0.942	1.012	0.988	1.010	0.990
1.08	1.16	1.058	0.942	1.024	0.976	1.020	0.980
1.16	1.26	1.058	0.942	1.038	0.962	1.032	0.968
1.26	1.37	1.061	0.939	1.053	0.947	1.044	0.956
1.37	1.48	1.080	0.920	1.067	0.933	1.056	0.944
1.48	1.60	1.098	0.902	1.080	0.920	1.065	0.935
1.60	1.75	1.116	0.884	1.093	0.907	1.076	0.924
1.75	1.90	1.130	0.870	1.105	0.895	1.085	0.915
1.90	2.05	1.145	0.855	1.117	0.883	1.094	0.906
2.05	2.22	1.158	0.842	1.127	0.873	1.102	0.898
2.22	2.42	1.172	0.828	1.137	0.863	1.110	0.890
2.42	2.62	1.186	0.814	1.147	0.853	1.118	0.882
2.62	2.84	1.198	0.802	1.156	0.844	1.124	0.876
2.84	3.09	1.210	0.790	1.163	0.837	1.130	0.870
3.09	3.35	1.221	0.779	1.170	0.830	1.135	0.865
3.35	3.65	1.232	0.768	1.177	0.823	1.140	0.860
3.65	3.96	1.241	0.759	1.183	0.817	1.145	0.855
3.96	4.29	1.250	0.750	1.189	0.811	1.150	0.850
4.29	4.64	1.260	0.740	1.193	0.807	1.153	0.847
4.64	5.00	1.267	0.733	1.198	0.802	1.158	0.842
5.00	5.40	1.274	0.726	1.201	0.799	1.160	0.840
5.40	5.88	1.280	0.720	1.203	0.797	1.162	0.838
5.88	6.42	1.288	0.712	1.204	0.796	1.163	0.837
6.42	7.00	1.292	0.708	1.205	0.795	1.164	0.836

NOTE: Do not select values from this table for the pinion member that are smaller than those given in Table 5-9. Straight line interpolation may be used for intermediate numbers of teeth.

Table 5-11. Recommended Backlash Allowance for Power Gearing

Normal diam. pitch P_{nd}	Center distance, in.						
	0–5″	5–10″	10–20″	20–30″	30–50″	50–80″	80–120″
½	.....	.....	.....	.....	0.045	0.060	0.080
1	.....	.....	.....	0.035	0.040	0.050	0.060
2	.....	.....	0.025	0.030	0.035	0.045	0.055
3	.....	0.018	0.022	0.027	0.037	0.042	
4	.....	0.016	0.020	0.025	0.030	0.040	
6	0.008	0.010	0.015	0.020	0.025		
8	0.006	0.008	0.012	0.017			
10	0.005	0.007	0.010				
12	0.004	0.006					
16	0.004	0.005					
20	0.004						
32	0.003						
64	0.002						

Table 5-12. Standard Tooth Parts

Diam. pitch P_d (1/in.)	Circ. pitch p, in.	Module $1/P_d$ addendum a, in.	Whole depth: Shave or grind h_t, in.	Whole depth: Cut teeth h_t, in.*	Tooth thickness $(p/2)$ t, in.	Double depth: Cut teeth $2h_t$, in.	Double depth: Ground $2h_t$, in.
1	3.1415927	1.0000000	2.35000	2.25000	1.5707963	4.5000	4.7000
1¼	2.5132741	0.8000000	1.88000	1.80000	1.2566371	3.6000	3.7600
1½	2.0943951	0.6666667	1.56667	1.50000	1.0471980	3.0000	3.1333
1¾	1.7951958	0.57142857	1.34286	1.28571	0.8975979	2.5714	2.6857
2	1.5707963	0.5000000	1.17500	1.12500	0.7853982	2.2500	2.3500
2½	1.2566371	0.4000000	0.94000	0.90000	0.6283185	1.8000	1.8800
3	1.0471976	0.3333333	0.78333	0.75000	0.5235988	1.5000	1.5667
3½	0.8975979	0.2857143	0.67143	0.64286	0.4487990	1.2857	1.3429
4	0.7853982	0.2500000	0.58750	0.56250	0.3926991	1.1250	1.1750
5	0.6283185	0.2000000	0.47000	0.45000	0.3141593	0.9000	0.9400
6	0.5235988	0.1666667	0.39167	0.37500	0.2617994	0.7500	0.7833
8	0.3926991	0.1250000	0.29375	0.28125	0.1963495	0.5625	0.5875
10	0.3141593	0.1000000	0.23500	0.22500	0.1570796	0.4500	0.4700
12	0.2617994	0.0833333	0.19583	0.18750	0.1308997	0.3750	0.3917
16	0.1963495	0.0625000	0.14688	0.14063	0.0981748	0.2813	0.2938
20	0.1570796	0.0500000	0.11950	0.11200	0.0785398	0.2240	0.2390
24	0.1308997	0.0416667	0.09992	0.09367	0.0654499	0.1873	0.1998
32	0.0981748	0.0312500	0.07544	0.07075	0.0490873	0.1415	0.1509
40	0.0785398	0.0250000	0.06075	0.05700	0.0392699	0.1140	0.1215
48	0.0654498	0.0208333	0.05096	0.04780	0.0327249	0.0956	0.1019
64	0.0490874	0.0156250	0.03872	0.03638	0.0245437	0.0728	0.0774
80	0.0392699	0.0125000	0.03138	0.02950	0.0196350	0.0590	0.0628
96	0.0327249	0.0104167	0.02648	0.02492	0.0163625	0.0498	0.0530
120	0.0261799	0.0083333	0.02158	0.02033	0.0130900	0.0407	0.0432
150	0.0209440	0.0066667	0.01767	0.01667	0.0104720	0.0333	0.0353

* h_t for fine pitch (20 P_d and finer) is $2.2/P_d + 0.002''$.

7. *Fillet radius, basic rack.* The fillet radius shown is that in the basic rack. The tooth-form standard (AGMA 201.02) directs that the edge radius on hobs and rack-type shaper cutters should be equal to the fillet radius in the basic rack. It also directs that pinion-shaped cutters should be designed using the basic rack as a guide so that the gear teeth generated by these cutters will have a fillet radius approximating those produced by hobs and rack-type shaper cutters. It can be calculated approximately by the data shown in Eq. (5-33).

   In the case of 25° pressure angle teeth, the fillet radius shown must be reduced for teeth having a clearance in excess of $0.250/P$. This is discussed in Art. 5-4.

   In the case of fine-pitch teeth, the fillet radius usually will be larger than the clearance customarily given as $c = 0.157/P$ because of edge breakdown in the cutter. The effects of tool wear on fine-pitch gears are proportionally larger than the effects produced by coarse-pitch tools.
8. *Clearance.* The value shown is the minimum standard. Greater clearance is usually required for teeth which are finished by grinding or shaving. In general the value shown in item 9 will be suitable for these processes. See also items 3 and 5.

Table 5-13. Values of Tooth Thickness for Pinions t_P and Gears t_G with Small Numbers of Teeth (Long and Short Addendums)

No. of teeth in pinion	Coarse-pitch system				Fine-pitch system	
	20° pressure angle		25° pressure angle		20° pressure angle	
	Pinion	Gear	Pinion	Gear	Pinion	Gear
7					2.0007	1.1409
8					1.9581	1.1835
9					1.9155	1.2261
10	1.912	1.230	1.742	1.399	1.8730	1.2686
11	1.868	1.273	1.659	1.482	1.8304	1.3112
12	1.826	1.315	1.5708	1.5708	1.7878	1.3538
13	1.783	1.358	1.5708	1.5708	1.7452	1.3964
14	1.741	1.400	1.5708	1.5708	1.7027	1.4389
15	1.698	1.443	1.5708	1.5708	1.6601	1.4815
16	1.656	1.486	1.5708	1.5708	1.6175	1.5241
17	1.613	1.529	1.5708	1.5708	1.5749	1.5667
18	1.5708	1.5708	1.5708	1.5708	1.5708	1.5708
19	1.5708	1.5708	1.5708	1.5708	1.5708	1.5708

NOTE: These tooth thicknesses go with Table 5-9 addendums.
The above values are for 1 diametral pitch.
These basic tooth thicknesses do not include an allowance for backlash.

Table 5-14. Values for Tip Round and Edge Round and End Round

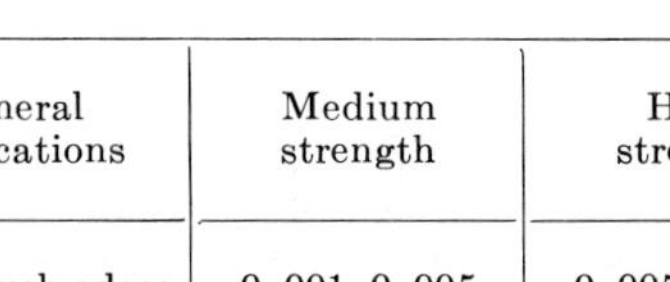

Diam. pitch	Edge round			Tip round and end round
	General applications	Medium strength	High strength	
20 and finer	Burr-brush edges	0.001–0.005	0.005–0.010	0.001–0.005
16	Burr-brush edges	0.003–0.015	0.010–0.025	0.003–0.010
12	Burr-brush edges	0.005–0.020	0.012–0.030	0.005–0.015
10	Burr-brush edges	0.010–0.025	0.015–0.035	0.005–0.015
8	Burr-brush edges	0.010–0.025	0.020–0.045	0.010–0.030
5	Burr-brush edges	0.010–0.025	0.025–0.060	0.010–0.030
3	Burr-brush edges	0.015–0.035	0.040–0.090	0.010–0.050
2	Burr-brush edges	0.015–0.035	0.060–0.125	0.010–0.050

9. *Clearance for shaved or ground teeth.* This is the recommended clearance for teeth to be finished by shaving or grinding.

   In the case of 25° pressure angle teeth, the fillet radius shown must be reduced for teeth having a clearance in excess of $0.250/P$.

10. *Minimum number of teeth in pinion.* These are the lowest numbers of teeth that can be generated in pinions having standard addendums and tooth thicknesses that will not be undercut. Pinions having fewer teeth should be made long-addendum in accordance with Table 5-9.

Table 5-15. Tooth Proportions of Spur Gears, Obsolete Systems‡
(Tooth proportions for fully interchangeable gears operated on standard center distances)

Symbol	Item	Brown and Sharpe system	AGMA 14½° compos-ite system	Fellows 20° stub system	AGMA full-depth composite system
ϕ	Pressure angle, degrees	14½	14½	20	14½
a	Addendum (basic*)	$\frac{1.00}{P}$	$\frac{1.00}{P}$	$\frac{0.800}{P}$	$\frac{1.00}{P}$
b	Dedendum (min.) (basic†)	$\frac{1.157}{P}$	$\frac{1.157}{P}$	$\frac{1.000}{P}$	$\frac{1.157}{P}$
h_k	Working depth	$\frac{2.00}{P}$	$\frac{2.000}{P}$	$\frac{1.60}{P}$	$\frac{2.00}{P}$
h_t	Whole depth (min.) (basic†)	$\frac{2.157}{P}$	$\frac{2.157}{P}$	$\frac{1.80}{P}$	$\frac{2.157}{P}$
t	Circular tooth thickness (basic*)	$\frac{\pi}{2P}$	$\frac{\pi}{2P}$	$\frac{\pi}{2P}$	$\frac{\pi}{2P}$
r_f	Fillet radius (in basic rack)	$\frac{0.157}{P}$	$\frac{0.157}{P}$	Not standardized	1.33 × clearance
c	Clearance (min.) (basic†)	$\frac{0.157}{P}$	$\frac{0.157}{P}$	$\frac{0.20}{P}$	$\frac{0.157}{P}$
c	Clearance, shaved or ground teeth				
N_P'	Min. numbers of teeth*				
	Pinion	32		14	32
	Pair	64			64
	Reference standards:				
	AGMA		201.02A		201.02A

* These values are basic for equal-addendum gearing. When the gearing is made long- and short-addendum these values will be altered.

† These values are minimum. Shaved or ground teeth should be given proportions suitable to these processes.

‡ See Table 5-1.

11. *Minimum numbers of teeth in pair.* This is the smallest number of teeth in pinion and gear that can be meshed on a standard center distance without one member's being undercut. For pairs with fewer teeth the members will have to be meshed on a nonstandard (enlarged) center distance.
12. *Minimum width of top land.* This is the approximate minimum width of top land allowable in standard long-addendum pinions. Increases in addendum which cause the tips of the teeth to have less than this value should be generally avoided.
13. *Reference standard.* The source of the data in each column. AGMA standards published by the American Gear Manufacturers Association.

14. *Reference standard.* The source of the data in each column. ASA standards published by the American Society of Mechanical Engineers.

5-6. System for Helical Gears. The following data are based on the information contained in AGMA standard 207.04, "20-degree Involute Fine-pitch System, for Spur and Helical Gears." In addition, tooth proportions are shown that are used by several of the larger gear manufacturers for the design of helical gears. These tooth proportions are shown since there are no AGMA standards for coarse-pitch helical gears.

In general, helical tooth proportions are based on either getting the most out of a helical-gear design or on using existing tooling. The tooth proportions referred to above have been found to yield very good gears. In some cases, for less critical applications, tools on hand may be used. Very often hobs for spur gears are available, and gears for general service can be made with these tools. Since such hobs have no taper, they are not well suited to cut helix angles much over 30°.

In helical-gear calculations, care should be taken to avoid confusion as to which plane the various tooth proportions are measured in. In some cases, the transverse plane is used. This plane is perpendicular to the axis of the helical-gear blank. In some cases, it is desirable to work in the normal plane.

If a spur-gear hob is used to cut helical teeth, the relationship between the transverse and normal pressure angles and the transverse and normal pitches as well as the base pitches should be established.

$$P_D = \frac{\cos \psi \times \pi}{p_n} \tag{5-38}$$

$$\tan \phi = \frac{\tan \phi_n}{\cos \psi} \tag{5-39}$$

$$p_N = p_n \cos \phi_n \tag{5-40}$$

where P_D = diametral pitch in transverse plane
ψ = helix angle
p_n = circular pitch in normal plane
ϕ_n = pressure angle in normal plane
p_N = normal base pitch (normal to surface)

When gears mesh together or with hobs, racks, shaper, or shaving cutters, their normal base pitches must be equal.

When a gear shaper is to be used to cut helical teeth, special guides are installed in the machine. These impart the twist into the cutter spindle which produces the proper helix angle and lead. The cutter used has the same lead ground into the teeth as is produced by the guides. The cutter and guides will in turn produce a given lead angle on the gear being cut. Practical considerations limit the lead angle of the guides to roughly 35°. In order to control costs, helical gears that are to be shaped should be designed either to a standard series of values of leads or to values of lead available in existing guides.

Equation (5-41) shows the relationship of lead, number of teeth, diametral pitch, and helix angle in a given cutter, guide, or helical gear.

$$\tan \psi = \frac{N\pi}{LP_d} \tag{5-41}$$

where ψ = helix angle
N = number of teeth, cutter, or gear
P_d = diametral pitch
L = lead of cutter

Selection of Tooth Form. The tooth forms shown in Table 5-16 are used for ranges of applications as follows:

Type of gear	*Suggested tooth form*
Single helix, low allowable thrust reaction	Type 1
Single helix, moderate allowable thrust reaction	Types 2 and 3
Double helix, general purpose	Types 4 and 5
Double helix, minimum noise	Type 6
Double helix, high load	Types 20 and 21

The above types each require a separate set of tooling. They have the advantage of getting the most out of good helical-gear design.

When an existing spur-gear hob is to be used to produce the gear, use types 18 and 19. The data in Table 5-17 will be found helpful in making calculations for gears having types 18 and 19 tooth forms. Type 19 is similar to type 18 except that its proportions are based on the use of hobs designed to cut fine-pitch gears (20 diametral pitch or finer).

Selection of Helix Angle. Single-helical gears usually are given lower helix angles than double-helical gears in order to limit thrust loads. Typical helix angles are 15 and 23°. Low helix angles do not provide so many axial crossovers as can be achieved on high-helix-angle gears of a given face width. Double-helix gears have helix angles that typically range from 30 through 45°. Although higher helix angles provide smoother operation, the tooth strength is lower.

In order to get the most quiet gears and at the same time achieve good tooth strength, special cutting tools for each helix angle should be provided. Table 5-16 shows typical tooth proportions. Tooth types 1 through 5 require special cutters, and the helix angle shown should be used. These teeth are stubbed more and more as the helix angle is higher. Tooth types 6 through 12 can be made with any helix angle desired. Types 13 and 14 can also be made with almost any helix angle. These tooth proportions are based on the use of standard spur-gear hobs. Since such hobs are not usually tapered, they do not do so good a job in cutting as do the specially designed helical hobs, and as a result, they should be limited in use to helix angles less than 25°.

Face Width. In the design of helical gears, the face width is usually based on that needed to achieve the required load-carrying capacity. In addition, the face width and lead are interrelated in that it is necessary to obtain at least two axial pitches of face width ($F = 2p_x$) to get reasonable benefit from the helical action, and four or more if high speeds, noise, or critical designs are confronted.

Long- and short-addendum designs are not so common among helical gears as among spur gears. This is because a much lower number of teeth can be cut in helical pinions without undercut. In low-hardness helical gearing, pitting, which is little affected by changes in addendum, is usually the limiting feature. In high-hardness designs, the same proportions as those used for spur-gear addendums should be employed.

Specific Calculation Procedure for Helical Gears. The procedure for helical gears is very similar to that given in Art. 5-5 for spur gears.

1. Determine application requirements: ratio, power, speed, etc.
2. Decide on basic tooth proportions and helix angle (see Table 5-16).
3. Pick appropriate number of pinion teeth (see Tables 5-2, 5-3, and 5-5).
4. Determine approximate center distance and face width. If power gearing, estimate from Arts. 13-3 and 13-4.

Table 5-16. Basic Tooth Proportions for Helical Gears

Tooth form No.	Pitch of teeth: Diametral, Normal	Pitch of teeth: Diametral, Transverse	Pitch of teeth: Circular, Normal	Pitch of teeth: Circular, Transverse	Pitch of teeth: Axial	Pressure angle: Normal	Pressure angle: Transverse	Depth of teeth: Working	Depth of teeth: Whole	Helix angle	Addendum	Tooth thickness (arc)	Edge radius of generating rack
	P_{nd}	P_d	p_n	p	p_x	ϕ_n	ϕ	h_k	h_t	ψ	a	t	r_T
1	1.03528	1	2.89185	3.1416	11.72456	19°22′12″	20°	2.0000	2.3500	15°	1.0000	1.5710	0.350
2	1.0836	1	2.8919	3.1416	7.40113	18°31′22″	20°	2.0000	2.3500	23°	1.0000	1.4460	0.350
3	1.0836	1	2.89185	3.1416	7.40113	18°31′22″	20°	1.8400	2.2000	23°	0.9200	1.4460	0.350
4	1.15470	1	2.72070	3.1416	5.44140	17°29′43″	20°	1.7400	2.0500	30°	0.8700	1.3604	0.300
5	1.22077	1	2.57340	3.1416	4.48666	16°36′ 6″	20°	1.6400	1.9500	35°	0.8200	1.2867	0.300
6	1.41421	1	2.22144	3.1416	3.14159	14°25′58″	20°	1.4200	1.7000	45°	0.7100	1.1107	0.2500
7	1	...	3.1416	......	....	23°14′ 7″	....	2.0000	2.2500	...	1.0000	......	0.240
8	1	...	3.1416	......	....	23°14′ 7″	....	2.0000	2.3300	...	1.0000	......	0.270
9	1	...	3.1416	......	....	21°58′50″	....	2.0000	2.2700	...	1.0000	......	0.310
10	1	...	3.1416	......	....	21°58′50″	....	2.0000	2.3500	...	1.0000	......	0.300
11	1	...	3.1416	......	....	21°58′50″	....	2.0000	2.4000	...	1.0000	......	0.290
12	1	...	3.1416	......	....	21°58′50″	....	2.0000	2.4500	...	1.0000	......	0.275
13	1	...	3.1416	......	....	18°31′35″	....	1.6300	2.0180	...	0.8150	......	0.204
14	1	...	3.1416	......	....	24°39′57″	....	1.8280	2.1170	...	0.9140	......	0.158
15	1	...	3.1416	......	....	24°51′32″	....	......	2.1570	...	0.9260	......	0.220
16	1	...	3.1416	......	....	24°54′59″	....	......	2.1550	...	0.9230	......	0.264
17	1	...	3.1416	......	....	15° 0′ 0″	....	......	2.6150	...	1.1580	......	0.410
18	1	...	3.1416	......	....	....	20°	2.000	2.3500	...	1.0000	......	0.350
19	1	...	3.1416	......	....	....	20°	2.000	$2.200 + 0.002P_d$	...	1.0000		
20	1.15470	1	2.72070	3.1416	5.44140	22°30′	25°33′41″	1.7400	2.0500	30°	0.8700	......	0.250
21	1.10338	1	2.84725	3.1416	6.73717	25°	27°13′35″	1.5000	1.7500	25°	0.7500	......	0.200

5. Determine pitch of teeth.
6. Determine whole depth.
7. Determine addendum of pinion and gear. Use equal addendum for pinion and gear except in special cases where addendum must be increased to avoid undercut or special consideration must be given to increasing the pinion strength.
8. Determine operating circular pitch, operating helix angle, and operating normal pitch if special center distance was used (see Chap. 6).
9. Determine design tooth thickness. If long and short addendum, adjust theoretical tooth thicknesses accordingly. Thin teeth for backlash. Consult Table 5-11 for power-gearing backlash allowance. See Chap. 6 if special "no-backlash" design.
10. Recheck load capacity. Use Arts. 13-5, 13-6, and 13-9. Check number of axial crossovers [Eq. (5-5)]. Check ability of thrust bearings to handle thrust reactions (Chap. 12). If results of these checks are unsatisfactory, change proportions.
11. If gear design is for critical power gears, additional items may need calculation.
    a. Root fillet radius [Eq. (5-33)].
    b. Form diameter Art. 7-4.
    c. Modification of profile [Eqs. (5-31), (5-32)].
    d. Diameter over balls (Art. 7-9).
12. General dimensions to be determined and toleranced.
    a. Outside diameter.
    b. Root diameter.
    c. Face width.
    d. Chordal addendum and chordal tooth thickness (Art. 7-12).

Table 5-17. Tooth Proportions for Helical Gears

Helix angle, deg	Diam. pitch P_d	Circ. pitch P_t	Axial pitch p_x	Pressure angle*	Working depth h_k	Whole depth h_t†
0	1.000000	3.14159	∞	20°00′00″	2.000	2.250
5	0.996195	3.15359	36.04560	20° 4′13.1″	2.000	2.250
8	0.990268	3.17247	22.57327	20°10′50.6″	2.000	2.250
10	0.984808	3.19006	18.09171	20°17′ 0.7″	2.000	2.250
12	0.978148	3.21178	15.11019	20°24′37.1″	2.000	2.250
15	0.965926	3.25242	12.13817	20°38′48.8″	2.000	2.250
18	0.951057	3.30326	10.16640	20°56′30.7″	2.000	2.250
20	0.939693	3.34321	9.18540	21°10′22.0″	2.000	2.250
21	0.933580	3.36510	8.76638	21°17′56.4″	2.000	2.250
22	0.927184	3.38832	8.38636	21°25′57.7″	2.000	2.250
23	0.920505	3.41290	8.04029	21°34′26.3″	2.000	2.250
24	0.913545	3.43890	7.72389	21°43′22.9″	2.000	2.250
25	0.906308	3.46636	7.43364	21°52′58.7″	2.000	2.250
26	0.898794	3.49534	7.16651	22° 2′44.2″	2.000	2.250
27	0.891007	3.52589	6.91994	22°13′10.6″	2.000	2.250
28	0.882948	3.55807	6.69175	22°24′ 9.0″	2.000	2.250
29	0.874620	3.59195	6.48004	22°35′40.0″	2.000	2.250
30	0.866025	3.62760	6.28318	22°47′45.1″	2.000	2.250

* Pressure angle based on 20° normal pressure angle.

† The values shown for whole depth are for coarse-pitch gears. If the gears are to be shaved or ground, use $h_t = 2.35$.

For fine-pitch gears, use $2.2/P_{nd} + 0.002$ for general-purpose gearing and $2.35/P_{nd} + 0.002$ for gear to be shaved or ground.

13. It may be necessary to specify
    a. Tip round or chamfer (Table 5-14).
    b. Edge round or end bevel (Table 5-14).
    c. Roll angle (Art. 7-4).
    d. Base radius [Eq. (7-9)].
14. Generally helical gears are not designed for such critical applications as to be in danger of pointed teeth or undercut.

Table 5-18. Equations and References for Addendum Calculations, Helical Gears

Type of tooth design	Operating condition	Pinion	Gear
Standard or equal addendum	Numbers of teeth in gear and pinion greater than minimum numbers shown in Table 5-3	a = value from Table 5-16 $a = \dfrac{\text{types 1–6}}{P_d}, \dfrac{\text{types 7–19}}{P_{nd}}, \dfrac{\text{types 20–21}}{P_d}$	If standard center distance and values of a_P as shown at (*a*) left, $a_G = h_k - a_P$
		Value from special tools to be used	If nonstandard center distance, see item below
Long and short addendum	Numbers of teeth in pinion *less* than minimum numbers shown in Table 5-3 and numbers of teeth in gear more than minimums shown in Table 5-4	Increase addendum sufficiently to avoid undercut $a_P = a + K_n'/P_n$ See Fig. 5-10 for K_n'	Decrease addendum by amount pinion addendum is increased. If undercut, increase center distance. See item below
	Numbers of teeth in pinion *less* than minimum numbers shown in Table 5-3 and numbers of teeth in gear *less* than minimum numbers of teeth shown in Table 5-4	Increase addendum sufficiently to avoid undercut $a_P = a + K_n'/P_n$	Increase addendum sufficiently to avoid undercut $a_G = a + K_n'/P_n$
		Increase center distance by amount sufficient to accommodate increased addendums of both pinion and gear	
	Designed for nonstandard center distance	Calculate tooth thickness to meet required center distance from Eq. (6-1); then calculate required addendum from Eq. 5-18	

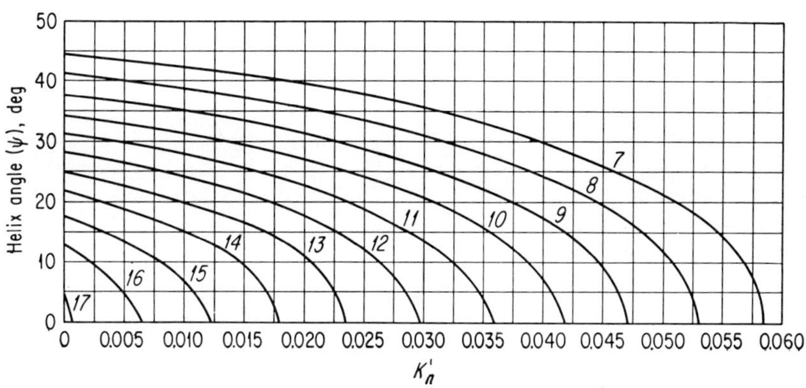

FIG. 5-10. Addendum increment K_n' for 1 diametral pitch.

5-7. System for Internal Gears. At the present time, there is no AGMA standard covering internal gears. The following material is based upon information covered in "Practical Gear Design" by D. W. Dudley.

In general, both spur internal meshes and helical internal meshes may be calculated by the same methods as external meshes. However, in internal gear meshes, there are several problems unique to this type of gearing.

The first of these is tip interference. In this type of interference, the pinion member cannot be assembled radially with the gear. Only axial assembly is possible, and it should be provided for in the design. If a shaper cutter having a number of teeth equal to or greater than the pinion is used to cut the internal gear, it will cut its way into mesh but in so doing will remove some material from the flanks of a few of the teeth that should have been left in place in order to have good tooth operation. This cutting action is also known as trimming. Such teeth will have poor contact and will tend to be noisy. If the proper shaper cutter or a broach is used, this problem will not occur.

The second problem is sometimes known as "fouling." In this case, the internal gear teeth interfere with the flanks of the external toothed pinion if there is too small a difference in numbers of teeth between the pinion and gear members.

Both these problems can be avoided in most gear designs by reducing the addendum of the internal gear (increasing its inside diameter). Table 5-19 shows a group of tooth proportions that will avoid these problems. A rather complicated graphical layout or involved calculations may be required to determine the exact proportions to avoid these problems. Since such calculations are beyond the scope of this chapter, Table 5-19 is offered as a guide.

Special Considerations. In general, the tip-interference problem can be eliminated by providing more than a 17-tooth difference between gear and pinion.

The internal gear teeth have an addendum which extends toward the inside from the pitch circle. The critical dimension of addendum height is maintained by holding inside diameter.

In general, the addendums of internal gears are made considerably shorter than those for equivalent external gears to avoid interference. The effect of the gear wrapping around the pinion tends to increase contact ratio, and also the chances for pinion-fillet interference.

Internal gearsets may be designed so that the pinion can be introduced into mesh at assembly by a radial movement. For any given number of teeth in the pinion, there must be a number of teeth in the gear that is somewhat greater than required if axial assembly were to be allowed.

If a minimum difference in numbers of pinion and gear teeth is desired, the designer must design the gear casings and bearings so that the pinion can be introduced into mesh with the gear in an axial direction.

Table 5-19 gives the minimum numbers of teeth that a gear may have for any given pinion to achieve either radial or axial assembly.

The "hand" of helical internal gears is determined by the direction in which the teeth move, right or left, as the teeth recede from an observer looking along the gear axis. Whereas two external helical gears must be of opposite hand to mesh on parallel axes, an internal helical gear must be of the same hand as its mating pinion.

Specific Calculation Procedure for Internal Gears. Generally speaking, internal gears may be designed by the same procedures as outlined in Art. 5-5 for spur gears or Art. 5-6 for helical gears. However, there are some special considerations.

1. *Number of teeth.* The number of teeth in the pinion is based on the ratio required and also on tooth strength and wear considerations. In addition there must be a sufficiently large difference between the numbers of teeth in the pinion and gear

Table 5-19. Internal Gearing

Numbers of teeth and addendum ratios (for 1 diametral pitch)

Pinion member		Internal gear member					
No. of teeth	Addendum	Min. No. of teeth in gear		Gear addendum			
		Axial assembly	Radial assembly	m_G min. up to 2	m_G 2–4	m_G 4–8	m_G and higher
N_P	a_P	N_G	N_G	a_G	a_G	a_G	a_G
12	1.350	19	26	0.472	0.510	0.582	0.616
	1.510			0.390	0.412	0.451	0.471
13	1.290	20	27	0.507	0.556	0.635	0.673
	1.470			0.419	0.445	0.488	0.509
14	1.230	21	28	0.543	0.601	0.688	0.729
	1.430			0.447	0.479	0.525	0.548
15	1.180	22	30	0.574	0.642	0.733	0.777
	1.400			0.470	0.506	0.554	0.577
16	1.120	23	32	0.608	0.688	0.786	0.834
	1.380			0.487	0.526	0.574	0.597
17	1.060	24	33	0.642	0.734	0.839	0.890
	1.360			0.505	0.546	0.594	0.617
18	1.000	25	34	0.676	0.779	0.892	0.947
	1.350			0.516	0.558	0.605	0.628
19	1.000	27	35	0.702	0.792	0.898	0.950
	1.330			0.539	0.578	0.625	0.648
20	1.000	28	36	0.713	0.802	0.903	0.952
	1.320			0.550	0.590	0.636	0.658
22	1.000	30	39	0.733	0.821	0.912	0.957
	1.290			0.577	0.621	0.666	0.688
24	1.000	32	41	0.750	0.836	0.920	0.960
	1.270			0.599	0.644	0.687	0.709
26	1.000	34	43	0.766	0.849	0.926	0.963
	1.250			0.620	0.666	0.709	0.729
30	1.000	38	47	0.792	0.870	0.936	0.968
	1.220			0.654	0.702	0.741	0.761
40	1.000	48	57	0.836	0.903	0.952	0.976
	1.170			0.718	0.764	0.797	0.814

to avoid problems of tip interference. Table 5-19 gives the minimum numbers of teeth in gear and pinion that can be specified and still be able to achieve either axial or radial assembly.

2. *Number of teeth, gear.* The minimum number of teeth in the gear member that can be used without getting into problems of assembly techniques or tip-interference problems is shown in Table 5-19. The manufacturing organization which will produce the gears should be checked to determine the availability of suitable equipment when the number of teeth in the gear exceeds 200 or when prime numbers over 100 are used. If the internal gear is to be broached, consideration should be given to minimizing the number of teeth in the gear so as to keep broach costs down.
3. *Helix angle.* Since shaper guides are usually required for the cutting of the internal gear, first choice in helix angles should be based on existing guides. Because of the kinematics of the gear shaper, helix angles above about 30° should be avoided. The helix angle of the pinion member is of the same hand as that of the gear.
4. *Diametral pitch.* Since special tools are usually required to produce the internal gear, thought should be given to standardizing on ranges of numbers of teeth and diametral pitches for internal gearing. Table 5-6 shows suggested diametral pitches. Internal gearing can be made long- and short-addendum so that the need of nonstandard pitches to meet special center distances is virtually nonexistent.
5. *Normal circular pitch.* In helical internal gears the normal circular pitch may be specified on the internal gear cutter in the case of existing cutters.

5-8. Standard Systems for Bevel Gears. The following data are based on information contained in AGMA standards 208.02 "Straight Bevel Gears," 209.02 "Spiral Bevel Gears," and 202.02 "Zerol Bevel Gears" and also on data contained in handbooks published by the Gleason Works, "20° Straight Bevel Gear System, 1956"; "Spiral Bevel Gear System, 1959"; and "Zerol Bevel Gear System, 1960."

Discussion of the Standard Systems. 20° *Straight-bevel-gear System.* The tooth form of the gears in this system is based on a symmetrical rack. In order to avoid undercut and to achieve approximately equal strength, a different value of addendum is employed for each ratio. If these gears are cut on modern bevel-gear generators they will have a localized tooth bearing. This resulting Coniflex tooth form is discussed in Art. 5-4. The selection of addendum ratios and outside diameter is limited to a 1:1.5 ratio in width of top lands of pinion and gear. The face cone of the gear and pinion blanks is made parallel to the root-cone elements to provide parallel clearance. This permits the use of larger edge radii on the generating tools with the attendant greater fatigue strength.

Spiral-bevel-gear System. Tooth thicknesses are proportioned so that the stresses in the gear and pinion will be approximately equal with the left-hand pinion driving clockwise or a right-hand pinion driving counterclockwise. The values shown apply for gears operated below their endurance limits. For gears operated above their endurance limits special proportions will be required. Special proportions will also be required for reversible drives on which optimum load capacity is desired. The method of establishing these balances is beyond the scope of this chapter but may be found in the Gleason Works publication "Strength of Bevel and Hypoid Gears."

The tooth proportions are based on the 35° spiral angle. A smaller angle may allow undercut and a reduction in contact ratio.

Zerol-bevel-gear System. The considerations of tooth proportions to avoid undercut and loss of contact ratio as well as to achieve optimum balance of strength are similar in this system to those of the straight and spiral systems.

This system is based on the duplex cutting method in which the root-cone elements do not pass through the pitch-cone apex. The face cone of the mating member is made parallel to the root cone to produce a uniform clearance.

Special Tooth Forms. The teeth of bevel gears can be given special forms to reduce manufacturing costs or for engineering reasons.

Because of the flexibility of the machining system used to cut most bevel gears many forms are possible. In general, it is customary to produce a desired profile on the teeth of the gear member and then to generate a conjugate profile on the teeth of the pinion.

Although it is theoretically possible to use bevel-gear teeth with an involute profile, this is rarely done. The most common curve for bevel gearing is the "octoid," so called because of the shape of its line of action. This curve is much easier to generate than the spherical involute. It closely approximates the involute.

The Formate®* tooth form is often used for gearing of high ratios because of its manufacturing economies. The teeth of the gear member are cut without generation, thus saving time, and the extra generation required to produce a conjugate pair is taken on the pinion. Since there are fewer pinion teeth less time is spent in generation than if both pinion and gear teeth were generated.

Another tooth form suitable for high-speed manufacture is the Revacycle®* straight bevel gear. This special tooth form is generated by a large disk cutter, one space being completed with each cutter revolution.

Both these forms are beyond the scope of this chapter.

Limitations in Bevel-gear Systems. *20° Straight-bevel-gear System.* The data contained in Table 5-20 apply only to gears that meet the following requirements:

1. The standard pressure angle is 20°; in certain cases, depending upon numbers of teeth, other pressure angles may be used (see Table 5-21).
2. In all cases, full-depth teeth are used. Stub teeth are avoided because of the reduction in contact ratio, which may increase noise, and the reduction in wear resistance.
3. Long- and short-addendum teeth are used throughout the system (except for 1:1 ratios) to avoid undercut and to increase the strength of the pinion.
4. The face width is limited to between one-fourth and one-third cone distance. The use of greater face widths results in an excessively small tooth size at the inner ends of the teeth.

Spiral-bevel-gear System. This system is more limited in its application than the straight-tooth system. The data in this system do *not* apply to the following:

1. Automotive rear-axle drives
2. Formate pairs
3. Gears and pinions of 12 diametral pitch and finer, which are usually cut with one of the duplex spread-blade methods
4. Gear cut spread-blade and pinion cut single-side, with a spiral angle of less than 20°
5. Ratios with fewer teeth than those listed in Table 5-21
6. Large spiral bevel gears cut on the planning-type generators where the spiral angles should not exceed 30°

Designs that fall in these restrictions should be referred to the Gleason Works for assistance.

Zerol-bevel-gear System. The data contained in Table 5-20 apply only to Zerol bevel gears that meet the following requirements:

* Registered trade-marks of the Gleason Works, Rochester, N.Y.

1. The standard pressure angle (basic) is 20°. Where needed to avoid undercut, 22½ and 25° pressure angles are standard (see Table 5-21).
2. The face width is limited to 25 per cent of the cone distance since the small-end tooth depth decreases even more rapidly as the face width increases because of the duplex taper. In duplex Zerols, 1″ face is the maximum value in any case.

Table 5-20. Tooth Proportions of Standard Bevel Gears

Item	Type of tooth		
	Straight teeth	Spiral teeth	Zerol teeth
Pressure angle, deg	20 standard[a]	20 standard[d]	20 basic, 22½, or 25, where needed
Working depth	$2.000/P_d$	$1.700/P_d$	$2.00/P_d$
Clearance	$0.188/P_d + 0.002''$	$0.188/P_d$	$0.188/P_d + 0.002''$
Face width	$F \leqq A_o/3$ or $F \leqq 10/P_d$ do not exceed smaller value[b]	$F \leqq 0.3A_o$ or $F \leqq 10/P_d$ whichever is smaller[b]	$F \leqq 0.25A_o$ or $F \leqq 10/P_d$ whichever is smaller. $F < 1.00$ on duplex Zerols[b,g]
Spiral angle	....	35°[e]	0°
Min. No. of teeth in system	13;[c] see Table 5-21	12	13
Whole depth	$\frac{2.188}{P_d} + 0.002''$	$\frac{1.888}{P_d}$[f]	$\frac{2.188}{P_d} + 0.002''$
Diametral-pitch range	....	12 and coarser	3 and finer
AGMA ref	208.02	209.02	202.02

[a] 20° is the standard pressure angle for straight-tooth bevel gears. Table 5-21 shows ratios that may be cut with a 14½° pressure angle.

[b] If the face width exceeds one-third the outer cone distance, the tooth is in danger of breakage in the event that tooth contact shifts to the small end of the tooth.

[c] This is the minimum number of teeth in the basic system (see Table 5-21 for equivalent minimum number of teeth in the gear member).

[d] 20° is the standard pressure angle for spiral bevel gears. Table 5-21 shows ratios that may be cut with 14½ and 16° pressure angle teeth.

[e] 35° is the standard spiral angle. If smaller spiral angles are used, undercut may occur and the contact ratio may be less.

[f] For gears of 10 diametral pitch and coarser the teeth are often rough-cut 0.005 deeper to avoid having the finishing blades cut on their ends.

[g] On duplex Zerol bevel gears 1″ is the maximum face width in all cases.

Specific-bevel-gear Calculation Procedure. A detail calculation procedure is given in Table 5-22 to calculate all standard types of bevel gears. Table 5-20 shows the basic proportions of different bevel-gear types, and Table 5-21 shows the minimum numbers of teeth. To start a calculation like that given in Table 5-22, it is necessary to pick the type of bevel gear, the numbers of teeth, and the shaft angle. It is also necessary to pick an approximate size of gearset. This can be done with the help of Art. 13-3.

Column 4 gives item references. Instructions to go with these item references are:

1. *Number of teeth, gear.* AGMA and ASA standards give only limited information as to the selection of numbers of teeth. Table 5-5 will be found helpful in

selecting the proper numbers of teeth for most power-gearing applications. Table 5-21 gives the minimum numbers of teeth in the gear for each number of pinion teeth and pressure angle. In general, the more teeth that are in the pinion, the more quietly it will run and the greater will be its resistance to wear. The equipment used to produce bevel gearing imposes upper limits on numbers of teeth. In general, if the gear is to contain over 120 teeth if an even number, or above 97 if a prime number, the manufacturing organization that is to produce the gears should be checked for capability.

2. *Number of teeth, pinion.* Table 5-21 gives the minimum numbers of teeth in the pinion for each number of teeth in the gear and for each pressure angle. The minimum number of pinion teeth is based on considerations of undercut.

Table 5-21. Pressure Angle and Ratio, Minimum Number of Teeth in Gear and Pinion That Can Be Used with Any Given Pressure Angle

Pressure angle, deg	Type of bevel gear					
	Straight tooth		Spiral tooth		Zerol tooth	
	Pinion	Gear	Pinion	Gear	Pinion	Gear
20 (standard)	16	16	17	17	17	17
	15	17	16	18	16	20
	14	20	15	19	15	25
	13	30	14	20		
			13	22		
			12	26		
14½	29	29	28	28		
	28	29	27	29		
	27	31	26	30		
	26	35	25	32		
	25	40	24	33	Not used	
	24	57	23	36		
			22	40		
			21	42		
			20	50		
			19	70		
16			24	24		
			23	25		
			22	26		
			21	27		
	Not used		20	29	Not used	
			19	31		
			18	36		
			17	45		
			16	59		
22½	13	13	16	16	14	14
			to		13	15
			16	19		
			15	15		
			to			
			15	24		
			14	14		
			13	15		
25	12	12	13	13	13	13
			to			
			13	14		
			12	12		

In bevel gearing, the ratio of widths of top lands has been limited to values up to 1.5:1. In order to maintain this ratio in the case of pinions with small numbers of teeth, a reduction in tooth thickness is made. Since this reduction is also accompanied by an increase in pressure angle for small numbers of teeth (see Table 5-21) the gears thus modified have a very low design contact ratio. A limit of 12 teeth is thus the minimum practical number for satisfactory operation. Smaller numbers are beyond the scope of this chapter.

3. *Diametral pitch.* In bevel gearing, it is not so essential to adhere to integral values of diametral pitch as it is in most other types of gearing. The tooling used to cut bevel gears is made up of separate tool bits, each working on but one side of a tooth at a time. Thus the principal limitations are that the tool bit have sufficient length to accommodate the whole depth of the tooth and that the point width not be so large as to interfere at the small end of the space width. The designer can therefore use any diametral pitch desired without incurring excessive tool costs.
4. *Shaft angle.* The included angle between the shafts. It is fixed by the mounting design. In many of the calculations, a shaft angle of other than 90° requires special calculations.
5. *Type of tooth.* The calculating procedure given in Table 5-22 covers straight, spiral, and Zerol tooth bevel gears. Tooth forms such as the nongenerated straight-sided teeth used on Formate gears are beyond the scope of this chapter.
6. *Ratio.* The number of gear teeth divided by the number of teeth in the pinion. This value is used in several of the subsequent calculations.
7. *Pitch diameter.* These equations give "standard" pitch diameter, a theoretical dimension. It is customary to consider that this circle lies at the large ends of the teeth. This diameter is a theoretical value and is not given a tolerance.
8. *Pitch angle.* This is the angle between the axis of the gear and an element of the pitch cone. Pitch angles larger than 90° indicate that the gear is an internal bevel gear. Internal bevel gears cannot be produced on most standard bevel-gear generators. The sum of the pitch angles of the pinion and the gear equal the shaft angle.
9. *Outer cone distance.* This is the linear distance from the pitch-cone apex to the back cone measured along the pitch cone.
10. *Face width.* The face width of bevel gears is limited by considerations of both tooth strength and tooling. The greater the face width, the smaller is the pitch of the small end of the tooth. Since deflection of the shafts, mountings, or bearings will occur and will allow the tooth contact to shift to the small ends of the teeth, the minimum size of the teeth at the small end must be sufficient to withstand the loads imposed. In addition, the space width at the roots of the teeth at the small end (toe) and the large end (heel) must be such that the cutter width required to clear the small end is not so narrow as to leave a "flange" at the large end of tooth space.
11. *Equivalent* 90° *ratio.* This engineering value is used to determine the gear addendum for gears operating on shafts at other than a 90° shaft angle.
12. *Working depth.* This is an engineering value.
13. *Addendum.* Values of addendum given by the equations achieve freedom from undercut and balanced wear in the pinion and gear teeth. The values given are at the outer cone distance (large end) of the teeth. Addendum is not given a tolerance. Table 5-23 is included for those cases where it is desired to match a set of gears originally designed to an earlier standard.
14. *Whole depth.* The calculated value of whole depth is the depth to which the finishing tools are intended to work. Gears of 10 diametral pitch and coarser are

roughed 0.005″ deeper than the calculated value of whole depth so that finishing tools will not cut on their ends.

15. *Dedendum.* The calculated value of dedendum is an engineering value used in subsequent calculations. The actual dedendum found on a completed gear is 0.002″ deeper in the case of a straight bevel gear.
16. *Pressure angle.* The standard pressure angle is 20°. Values of 14½, 16, 22½, and 25° may be used for special applications.
17. *Number of teeth in crown gear.* This value is used in the calculation of "change in dedendum angle," for Zerol gears.
18. *Change in dedendum angle.* This is an engineering value used in the calculation of Zerol bevel gears.
19. *Dedendum angle.* An engineering value. In the case of Zerol bevel gears, it consists of two parts, the normal dedendum angle not including the duplex taper, plus an increment to provide duplex taper.
20. *Face angle.* The face angle and the position of the face-cone apex are chosen such as to provide uniform clearance between the top lands and the root lands of the mating teeth. This is done to provide greater tooth strength. It accomplishes this by allowing the use of a larger edge radius on the cutter while avoiding the possibility of fillet interference at the small ends of the teeth. The larger radius reduces the stress concentration at the bottom of the gear teeth.
21. *Outside diameter.* The outside diameter of a bevel-gear blank is the theoretical diameter defined by the intersection of the face and back cones of the blank. In practice, the actual outside diameter will be less because of a rounding off of the sharp edge. Outside diameter is usually a reference dimension. Table 5-14 gives suggested values of edge round.
22. *Pitch apex to crown.* Sometimes known as "cone center to crown," this is a reference or engineering dimension.
23. *Crown to back.* This is a reference dimension in most cases, since edge rounding removes the edge from which measurements can be made. X_{BG} is the mounting distance for the gear, and X_{BP} is the mounting distance for the pinion.
24. *Circular pitch.* An engineering dimension.
25. *Backlash.* The values of backlash shown in Table 5-24 are actually a "backlash allowance." It is intended that the calculated thickness of the teeth be reduced by the values taken from the table. The gears when assembled should have the backlash shown in Table 5-24 at the tight point of mesh. The runout and other errors in the gear members and in the bearings, shafts, and mountings will control the maximum backlash found in the assembly.
26. *Circular tooth thickness.* The engineering tooth thickness required in the gear. The calculation differs from many found in gear literature in that it includes the allowance for engineering backlash. It does not include any manufacturing tolerances. See Fig. 5.13 for factor K for straight bevel gears and Fig. 5-14 for K for spiral-tooth bevel gears.
27. *Chordal tooth thickness.* Straight-bevel-gear teeth can be given tooth-thickness control by means of tooth vernier calipers. This calculation supplies the drawing data for this specification.
28. *Chordal addendum.* This is a setting for tooth vernier calipers when tooth-thickness control is desired. It is the distance from the top land at the large end of the tooth that the given tooth thickness must be maintained.
29. *Root angle.* This is the included angle between the axis of the gear and an element of the root cone. It is a drawing dimension.
30. *Back-angle distance.* Values of bevel-gear-blank dimensions are shown in Fig. 5-15.

Table 5-22. Calculation of Bevel-gear-tooth Dimensions for Standard Bevel-gear-tooth Design

Item	Type of bevel gear	Member	Item ref.*	Symbol	Equation	Example
No. of teeth	All	Gear	1	N_G	....	78
		Pinion	2	N_P	....	31
Diametral pitch	All	Both	3	P_d	....	12
Shaft angle	All	Both	4	Σ	....	90°
Type of tooth	All	Both	5	....	....	Zerol tooth form
Ratio	All	Both	6	m_G	N_G/N_P	$78/31 = 2.516129$
Pitch diameter	All	Gear	7	D_G	N_G/P_d	$78/12 = 6.5000$
		Pinion		D_P	N_P/P_d	$31/12 = 2.5833$
Pitch angle †	All	Pinion	8	γ	When $\Sigma = 90°$, $\gamma = \tan^{-1}\left(\dfrac{N_P}{N_G}\right)$	$\tan^{-1} = 31/78 = 0.39744$
					When $\Sigma < 90°$, $\gamma = \tan^{-1}\dfrac{\sin\Sigma}{m_G + \cos\Sigma}$	$\gamma = 21.67°$
					When $\Sigma > 90°$, $\gamma = \tan^{-1}\left[\dfrac{\sin(180° - \Sigma)}{m_G - \cos(180° - \Sigma)}\right]$	
		Gear		Γ	$\Sigma - \gamma$ If Γ is larger than 90° see notes Check: $\dfrac{\sin\gamma}{\sin\Gamma} = \dfrac{N_P}{N_G}$	$\Gamma = 90° - 21.67$ $= 68.33$

Outer cone distance	All	Both	9	A_o	$\frac{1}{2}(D_P^2 + D_G^2)^{1/2}$, when $\Sigma = 90°$ $\frac{D_G}{2 \sin \Gamma}$, when $\Sigma \neq 90°$	$6.50/2 \times 0.92932$ $= 3.4972$
Face width	Straight spiral	Both	10	F	$\left.\begin{matrix}\frac{A_o}{3}\\ \frac{10}{P_d}\end{matrix}\right\}$ max.; use whichever value is smaller ($0.3A_o$ recommended)	
	Zerol				$\left.\begin{matrix}= 0.275A_o\\ \leqq \frac{10}{P_d}\end{matrix}\right\}$ max.; use whichever value is smaller ($0.25A_o$ recommended)	$= 0.25 \times 3.4973$ $F = 0.8743$ $10/12 = 0.8333$ Use $F = 0.8333$
Equivalent 90° ratio	All	Both	11	m_{90}	When $\Sigma = 90°$ $m_{90} = m_G$ When $\Sigma \neq 90°$ calc. m_{90} $m_{90} = \sqrt{m_G \frac{\cos \gamma}{\cos \Gamma}}$	$\Sigma = 90°$ $m_{90} = m_G = 2.516$ (from item 6 above)
Working depth	Straight Zerol	Both	12	h_k	$2.0/P_d$	$h_k = 2.0/12 = 0.167$
	Spiral				$1.70/P_d$	
Addendum (at heel of tooth)	Straight and Zerol	Gear	13	a_{oG}	$\frac{0.54}{P_d} + \frac{0.460}{P_d m_{90}^2}$	$\frac{0.54}{12} + \frac{0.460}{12 \times (2.516)^2}$ $= 0.051$
	Spiral				$\frac{0.46}{P_d} + \frac{0.390}{P_d m_G^2}$	
	All	Pinion		a_{oP}	$h_k - a_{oG}$	$0.167 - 0.051 = 0.116$

* See instructions, pages 5-51 to 5-53.
† $\tan^{-1}$ means "the angle whose tangent is."

Table 5-22. Calculation of Bevel-gear-tooth Dimensions for Standard Bevel-gear-tooth Design *(Continued)*

Item	Type of bevel gear	Member	Item ref.*	Symbol	Equation	Example
Whole depth (calc.)	Straight Zerol	Both	14	h_t	$\frac{2.188}{P_d} + 0.002$ see Note *a*	$\frac{2.188}{12} + 0.002$ $h_t = 0.184$
	Spiral				$\frac{1.888}{P_d}$ see Note *a* Note *a*: For diametral pitches of 10 and coarser, h_t will frequently be made 0.005″ deeper on *straight*, *Zerol* and *spiral* teeth to avoid having finishing blades cut on their ends	
Dedendum (at heel of tooth)	Straight	Gear	15	b_{oG}	$h_t - a_{oG}$ (calc.)	
		Pinion		b_{oP}	$h_t - a_{oP}$ (calc.)	
	Spiral Zerol	Gear		b_{oG}	$h_t - a_{oG}$ (calc.)	$0.184 - 0.051 = 0.133$
		Pinion		b_{oP}	$h_t - a_{oP}$ (calc.)	$0.184 - 0.116 = 0.068$
Pressure angle	All	Both	16	ϕ	When $\Sigma = 90°$ see Table 5-21	20°
	Straight	Both			When $\Sigma \neq 90°$ see Fig. 5-11	
	Spiral	Both			When $\Sigma \neq 90°$ see Fig. 5-12	
	Zerol	Both			When $\Sigma \neq 90°$ $\sin \phi' = \sqrt{\frac{1.15 b_{oP}}{A_o \tan \gamma}}$ use the nearest larger pressure angle 20, 22½, or 25°	
No. of teeth in crown gear	Zerol	Both	17	N_c	$2P_d A_o$	$2 \times 12 \times 3.4972 = 83.933$

Change in dedendum angle	Zerol	Both	18	$\Delta\delta$	When $\Sigma = 90°$ $\frac{K'}{N_c} - \frac{300\sqrt{D_p \sin \Gamma}}{N_c F} - \frac{14P_d}{N_c}$ where $K' = 6{,}668$ for $\phi = 20°$ $= 4{,}868$ for $\phi = 22\frac{1}{2}°$ $= 3{,}412$ for $\phi = 25°$ When $\Sigma \neq 90°$ $\frac{K'}{N_c} - \frac{300}{F}\sqrt{\frac{1}{N_c P_d(\tan \gamma + \tan \Gamma)}} - \frac{14P_d}{N_c}$	$= \frac{6{,}668}{83.933}$ $- \frac{300\sqrt{2.5833 \times 0.92932}}{83.933 \times 0.8333}$ $- \frac{14 \times 12}{83.933}$ $= 70.796$ min. $= 1.18°$
Dedendum angle	Straight spiral	Gear	19	δ_G	$\tan^{-1} \frac{b_{oG}}{A_o}$	
		Pinion		δ_P	$\tan^{-1} \frac{b_{oP}}{A_o}$	
	Zerol	Gear		δ_G	$\tan^{-1} \frac{b_{oG}}{A_o} + \Delta\delta$	$\tan^{-1} \frac{(0.133)}{(3.4972)} + 1.18 = 3.36°$
		Pinion		δ_P	$\tan^{-1} \frac{b_{oP}}{A_o} + \Delta\delta$	$\tan^{-1} \frac{(0.068)}{(3.4972)} + 1.18 = 2.29°$
Face angle	All	Gear	20	Γ_o	$\Gamma + \delta_P$	$68.33 + 2.29 = 70.62°$
		Pinion		γ_o	$\gamma + \delta_G$	$21.67 + 3.36 = 25.03°$
Outside diameter	All	Gear	21	D_{oG}	$D_G + 2a_{og} \cos \Gamma$	$6.50 + 2(0.051) \times 0.3693 = 6.538$
		Pinion		D_{oP}	$D_P + 2a_{oP} \cos \gamma$	$2.5833 + 2(0.116) \times 0.9293 = 2.799$
Pitch cone apex to crown	All	Gear	22	X_{oG}	$\Sigma = 90°$, $D_P/2 - a_{oG} \sin \Gamma$ $\Sigma \neq 90°$, $A_o \cos \Gamma - a_{oG} \sin \Gamma$	$\frac{2.5833}{2} - 0.051 \times 0.92932 = 1.244$
		Pinion		X_{oP}	$\Sigma = 90°$, $D_G/2 - a_{oP} \sin \gamma$ $\Sigma = 90°$, $A_o \cos \gamma - a_{oP} \sin \gamma$	$6.50/2 - 0.116 \times 0.36926 = 3.207$

* See instructions, pages 5-51 to 5-53.

Table 5-22. Calculation of Bevel-gear-tooth Dimensions for Standard Bevel-gear-tooth Design (*Continued*)

Item	Type of bevel gear	Member	Item ref.*	Symbol	Equation	Example
Crown to back	All	Gear	23	O_{BG}	$X_{BG} - X_{oG}$	3.750 − 1.244 = 2.506
		Pinion		O_{BP}	$X_{BP} - X_{oP}$	8.000 − 3.207 = 4.793
Circular pitch	All	Both	24	p	π/P_d	3.14159265/12 = 0.2618
Backlash (design)	All		25	B	See Table 5-24	0.002
Circular tooth thickness†	Straight	Gear	26	t_G	$p/2 - (a_{oP} - a_{oG}) \tan \phi - \dfrac{K \text{ (from Fig. 5-13)}}{P_d}$	
		Pinion		t_P	$p - t_G$	
	Spiral	Gear		t_G	$p/2 - 1.22\,(a_{oP} - a_{oG}) \tan \phi - \dfrac{K \text{ (Fig. 5-14)}}{P_d}$	
		Pinion		t_P	$p - (t_G + B)$	
	Zerol	Gear		t_G	$p/2 - (a_{oP} - a_{oG}) \tan \phi - \dfrac{K \text{ (Fig. 5.13)}}{P_d}$	$0.2618/2 - (0.116 - 0.051) \times 0.36397 - \dfrac{0}{12} = 0.1072$
		Pinion		t_P	$p - t_G$	0.2618 − 0.1072 = 0.1546
Chordal tooth thickness	Straight	Gear	27	t_{cG}	$t_G - \dfrac{t_G^3}{6D_G^2} - \dfrac{B}{2}$	
		Pinion		t_{cP}	$t_P - \dfrac{t^3_P}{6D_P^2} - \dfrac{B}{2}$	

Chordal addendum	Straight	Gear	28	a_{cG}	$a_{oG} + \dfrac{t_G^2 \cos \Gamma}{4D_G}$	
		Pinion		a_{cP}	$a_{oP} + \dfrac{t_P^2 \cos \gamma}{4D_P}$	
Root angle	Straight spiral Zerol	Gear	29	Γ_R	$\Gamma - \delta_G$	68.33 − 3.36 = 64.97°
		Pinion		γ_R	$\gamma - \delta_P$	21.67 − 2.29 = 19.38°
Back-angle distance	All	Both	30		Back-angle distance = A_o	3.4972
Face-angle distance	All	Gear			$O_{BG} \sin \Gamma_o + \frac{1}{2} D_{oG} \cos \Gamma_o$	2.506 × 0.94334 + ½ × 6.538 × 0.33123 = 3.449
		Pinion			$O_{BP} \sin \gamma + \frac{1}{2} D_{oP} \cos \gamma$	4.793 × 0.42309 + ½ × 2.799 × 0.90609 = 3.296

* See instructions, pages 5-51 to 5-53.
† Thicknesses shown have not been reduced for backlash.

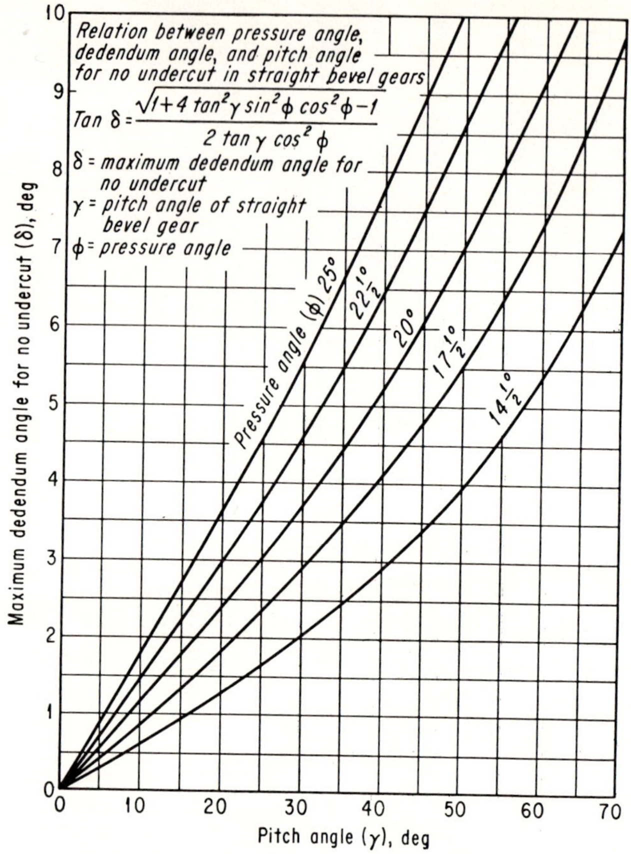

FIG. 5-11. Relation between the dedendum angle and the pitch angle at which undercut begins to occur in generating straight bevel gears using sharp-cornered tools.

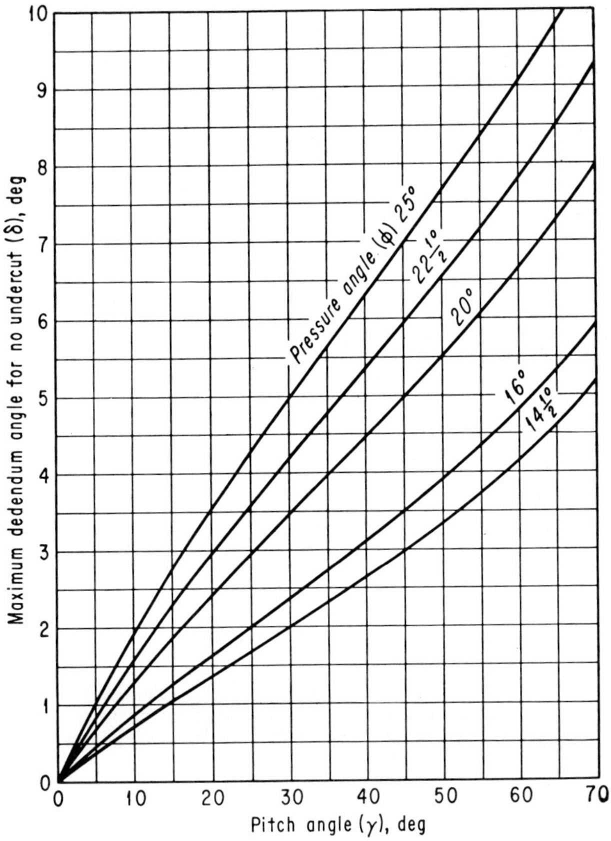

FIG. 5-12. Relation between the dedendum angle and the pitch angle at which undercut begins to occur in generating spiral bevel gears at 35° spiral angle using sharp-cornered tools.

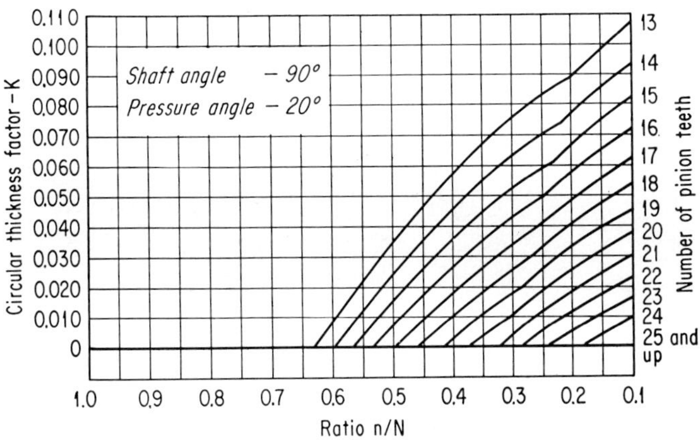

FIG. 5-13. Circular-thickness factor for straight and Zerol bevel gears.

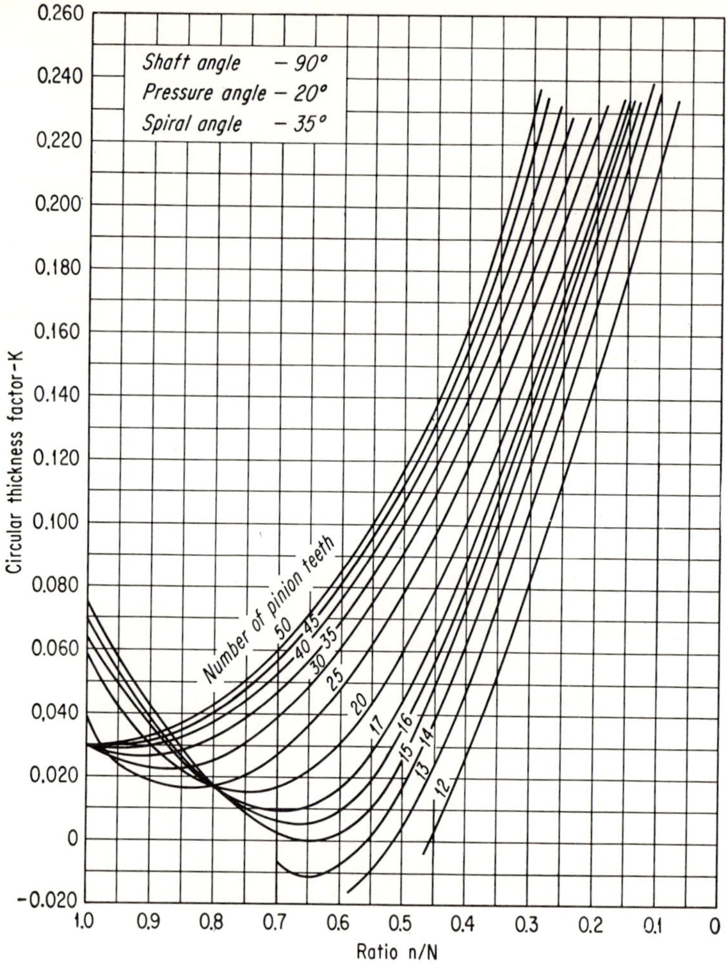

FIG. 5-14. Circular-thickness factor for spiral bevel gears.

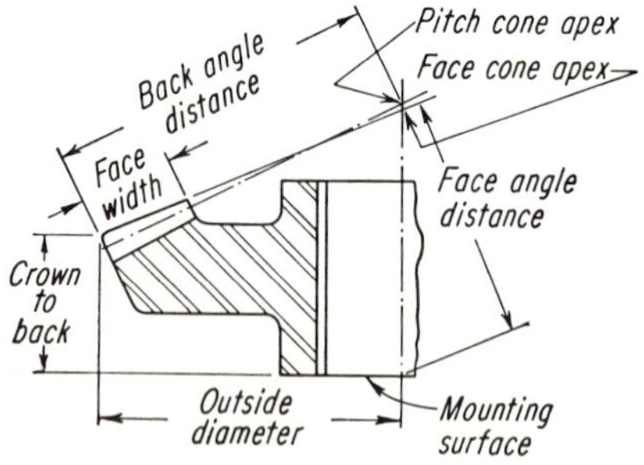

FIG. 5-15. Bevel-gear dimensions.

Table 5-23. Bevel Gear Addendum (for 1 Diametral Pitch)*

Ratio (m_g)		Addendum		Ratio (m_g)		Addendum	
From	To	Straight and Zerol	Spiral	From	To	Straight and Zerol	Spiral
1.00	1.00	1.000	0.850	1.52	1.56	0.730	0.620
1.00	1.02	0.990	0.840	1.56	1.57	0.720	0.620
1.02	1.03	0.980	0.830	1.57	1.60	0.720	0.610
1.03	1.04	0.970	0.820	1.60	1.63	0.710	0.610
1.04	1.05	0.960	0.820	1.63	1.65	0.710	0.600
1.05	1.06	0.950	0.810	1.65	1.68	0.700	0.600
1.06	1.08	0.940	0.800	1.68	1.70	0.700	0.590
1.08	1.09	0.930	0.790	1.70	1.75	0.690	0.590
1.09	1.11	0.920	0.780	1.75	1.76	0.690	0.580
1.11	1.12	0.910	0.770	1.76	1.82	0.680	0.580
1.12	1.13	0.900	0.770	1.82	1.89	0.670	0.570
1.13	1.14	0.900	0.760	1.89	1.90	0.660	0.570
1.14	1.15	0.890	0.760	1.90	1.97	0.660	0.560
1.15	1.17	0.880	0.750	1.97	1.99	0.650	0.560
1.17	1.19	0.870	0.740	1.99	2.06	0.650	0.550
1.19	1.21	0.860	0.730	2.06	2.10	0.640	0.550
1.21	1.23	0.850	0.720	2.10	2.16	0.640	0.540
1.23	1.25	0.840	0.710	2.16	2.23	0.630	0.540
1.25	1.26	0.830	0.710	2.23	2.27	0.630	0.530
1.26	1.27	0.830	0.700	2.27	2.38	0.620	0.530
1.27	1.28	0.820	0.700	2.38	2.41	0.620	0.520
1.28	1.29	0.820	0.690	2.41	2.58	0.610	0.520
1.29	1.31	0.810	0.690	2.58	2.78	0.600	0.510
1.31	1.33	0.800	0.680	2.78	2.82	0.590	0.510
1.33	1.34	0.790	0.680	2.82	3.05	0.590	0.500
1.34	1.36	0.790	0.670	3.05	3.17	0.580	0.500
1.36	1.37	0.780	0.670	3.17	3.41	0.580	0.490
1.37	1.39	0.780	0.660	3.41	3.67	0.570	0.490
1.39	1.41	0.770	0.660	3.67	3.94	0.570	0.480
1.41	1.42	0.770	0.650	3.94	4.56	0.560	0.480
1.42	1.44	0.760	0.650	4.56	4.82	0.560	0.470
1.44	1.45	0.760	0.640	4.82	6.81	0.550	0.470
1.45	1.48	0.750	0.640	6.81	7.00	0.540	0.470
1.48	1.52	0.740	0.630	7.00	∞	0.540	0.460

Table 5-24. Design Backlash for Bevel Gearing

Diametral pitch	Range of design backlash	Diametral pitch	Range of design backlash
1.00–1.25	0.020–0.030	3.50–4.00	0.007–0.009
1.25–1.50	0.018–0.026	4.0–5.0	0.006–0.008
1.50–1.75	0.016–0.022	5.0–6.0	0.005–0.007
1.75–2.00	0.014–0.018	6.0–8.0	0.004–0.006
2.00–2.50	0.012–0.016	8.0–10.0	0.003–0.005
2.50–3.00	0.010–0.013	10.0–12.0	0.002–0.004
3.00–3.50	0.008–0.011	12 and finer	0.001–0.003

* *Note:* In the earlier AGMA Standards and Gleason handbooks, the equations for addendum for all types of bevel gears were $a_{oG} = \dfrac{\text{Table 5-23}}{P}$. The values used in this table should be used only when checking calculations based on earlier standards.

5-9. Standard System for Worm Gears. The American Gear Manufacturers Association has published three standard systems covering the tooth proportions of worm gearing. The following data are based on information contained in the American Gear Manufacturers Association Standards, AGMA standard 341.01, "Design of General Industrial Coarse-pitch Cylindrical Worm Gearing," AGMA standard 374.03, "Design for Fine-pitch Worm Gearing," AGMA standard 342.01, "Design of General Industrial Double-enveloping Wormgears."

Calculating Procedure, General. The calculating procedure in Table 5-26 is arranged to provide a logical method of calculating worm gearing of standard tooth proportions. Such gears will be suitable for most applications. For a more complete treatment of the subject, the reader is referred to the original standards.

In applying the procedure outlined in Table 5-26, the reader is cautioned to observe the following limitations:

1. These data cover worms and wormgears with axes at right angles. The worm-gear member is hobbed for fully conjugate tooth surfaces.
2. These data are limited to cases wherein the hob used to produce the gear member is a counterpart of the mating worm in regard to its tooth profile, number of threads, and lead.
3. The tooth form of the fine-pitch worm is defined as that shape of thread which is produced by a symmetrical double-conical cutter or grinding wheel ground or dressed with straight-line elements having an included angle of 40°. This milling or grinding cutter is to be set at the lead angle of the worm when cutting at the standard worm pitch diameter.

   The coarse-pitch single-enveloping standard does not specify a standard tooth profile, stating only that the wormgear must be conjugate to the profile produced on the worm member.

   The double-enveloping standard states that the tooth form preferably should be straight-sided in the axial section and tangent to the base circle.

Basic Tooth Forms for Worm Gearing. The tooth forms of worm gearing have not been standardized to the same degree as tooth forms for spur gearing, for example. Since a special bob has to be made to cut the gears that will mesh with each design of worm and since the elements that constitute good worm gearing design are less well understood, there has been little incentive to standardize.

The most general practice in wormgear design and manufacture is to establish the shape of the worm thread and then to design a hob that will generate teeth on the gear that are conjugate to those on the worm. This practice is not followed in the case of double-enveloping gearing since special tooling based on the shape of the gear member is required to generate the worm.

The shape of the teeth of the worm member is dependent on the size of the tool and the method used to cut the threads. These threads may be cut with a straight-sided V-shaped tool in a lathe or milled with double conical milling cutters of up to 6″ diameter in a thread mill or hobbed with a special hob, or ground in a thread-grinding machine with a grinding wheel having a diameter up to 20″. In each case the worm thread profile will be noticeably different, especially in the case of the higher lead angles. Worms can also be rolled, in which case the thread profile will be still different. It is essential, therefore, for the designer of worm gearing either to specify the manufacturing method to be used to make the worm or to specify the coordinates of the worm profile.

Specific Calculation Procedure for Cylindrical Worm Gears. Table 5-26 shows a detail calculation procedure for cylindrical worm gears.

To start Table 5-26 an approximate center distance for the set can be obtained from Art. 13-3. Tables 5-28, 5-29, and 5-30 may be used to pick numbers of gear teeth and

Table 5-25. Tooth-depth Proportions, Single- and Double-enveloping Gearing, Standard Worm Gearing

Item	Symbol	Single-enveloping					Double-enveloping
		Fine pitch	Coarse pitch				
Lead angle	γ	....	Up to 30°	30°–34°59′	35′–39°59′	40–45	
Addendum	a	$0.3183p_n$	$0.3183p_x$	$0.2865p_x$	$0.2546p_x$	$0.2228p_x$	$0.225p_n$
Dedendum	b_G	$0.3820p_n + 0.002$	$0.3683p_x$	$0.3314p_x$	$0.2947p_x$	$0.2578p_x$	$0.611h_K$
Whole depth	h_t	$0.7003p_n + 0.002$	$0.6866p_x$	$0.6179p_x$	$0.5493p_x$	$0.4806p_x$	$0.5p_n$
Working depth	h_K	$0.6366p_n$	$0.6366p_x$	$0.5729p_x$	$0.5093p_x$	$0.4456p_x$	$0.9h_t = 0.45p_n$
Pressure angle, deg	ϕ_n	20 (on cutter)	20		25		20
% reduction in depth			0	10	20	30	

Table 5-26. Tooth Dimensions of Single-enveloping Worm Gearing

Item ref.*	Item	Member	Symbol	Equation or data reference
1	Gear ratio	.....	m_G	Required data $\left(m_G = \frac{N_G}{N_W}\right)$
2	Center distance	.....	C	Required data $\left(C = \frac{d_W + D_G}{2}\right)$
3	No. of teeth	Gear	N_G	
4	No. of teeth	Worm	N_W	$N_W = \frac{N_G}{m_G}$
5	Pitch diameter	Worm	d	$\frac{C^{0.875}}{2.2}$
6	Pitch diameter	Gear	D	$2\left(C - \frac{d}{2}\right)$
7	Circular pitch	Gear	p	$p = \frac{\pi D}{N_G}$
8	Axial pitch	Worm	p_x	$p_x = p$
9	Lead angle	.....	λ	$\lambda = \tan^{-1} \frac{D}{d m_G} = \tan^{-1} \frac{N_W p_{xW}}{\pi d}$
10	Lead	Worm	l	$l = \pi D \tan \lambda = N_W p_x$
11	Pressure angle cutter	Worm	ϕ_C	See Table 5-25
12	Cutter diameter	Worm	D_C	See Table 5-33
13	Normal pressure angle	Worm	ϕ_n	$\phi_n = \phi_C - \Delta\phi$ $\Delta\phi = \frac{5{,}400 d \sin^3 \lambda}{N_W(D_C \cos^2 \lambda + d)}$ (minutes)
14	Axial pressure angle	Worm	ϕ_x	$\phi_x = \tan^{-1} \frac{\tan \phi_n}{\cos \lambda_C}$
15	Addendum	Worm	a	See Table 5-25
16	Outside diameter	Worm	d_o	$d_o = d + 2a$
17	Face width	Gear	F_G	$F_G = (d_o^2 - d^2)^{1/2} \times 1.05$
18	Throat diameter	Gear	D_t	$D_t = 2a + D$
19	Outside diameter (throated)	Gear	D_o	$D_o = D_t + h$ $h = (d + 2a) - [(d - 2a)^2 - (0.6F_G)^2]^{1/2}$
	(Nonthroated)	Gear	D_o	$D_o = D_t$
20	Face width	Worm	F_W	$F_W = p_{xW}\left(4.5 + \frac{N_G}{50}\right)$
21	Root diameter	Worm	d_R	$d_R = d_o - 2h_t$ $d_R \text{ min.} = \frac{C^{0.875}}{3}$
22	Tooth thickness	Worm	t_W	$t_W = \frac{p_n}{2} - B$ See Table 5-32 for B
23	End radius of teeth	Gear	r_e	$r_e = 0.10F_G$
24	Edge radius at top land (tips of teeth)	Worm	r_t	$r_t = 0.050p_x$

* See instructions below.

numbers of worm threads. The normal pitch should, if possible, agree with one of the standard pitches in Table 5-31.

Detail instructions for each of the numbered items of Table 5-26 are:

1. Wormgear ratios are established by dividing the number of teeth in the worm gear by the number of "threads" or "starts" in the worm.

Table 5-27. Tooth Dimensions of Double-enveloping Worm Gearing

Item	Member	Item ref.*	Symbol	Equations
Gear ratio	Both	1	m_G	N_G/N_W
Center distance	Both	2	C	$(d_w + D_G)/2$
No. of teeth	Gear	3	N_G	
	Worm	4	N_w	N_G/m_G
Worm root diameter (min.)	Worm	5	d_R	$k_r(C^{0.875}/3)$
Pitch diameter (trial value)	Worm	6	d	$C^{0.875}/2.2$
Pitch diameter	Gear	7	D	$2c - d$
Lead angle (av.)	Worm	8	λ	$\tan^{-1}(0.87D/m_G d)$
Circular pitch:				
Axial	Worm	9	p_x	D/N_G
Normal	Worm	10	p_n	$p_x \cos \lambda$
Whole depth	Both	11	h_t	$p_n/2$
Working depth	Both	12	h_k	$0.9h_t$
Dedendum	Gear	13	b_G	$0.611h_k$
Worm root diameter	Worm	14	d_R	$d - 2b_G$†
Normal pressure angle	Both	15	ϕ_n	20° standard
Lead angle center of worm	Worm	16	λ_c	$\tan^{-1}(D/m_G d)$
Axial pressure angle	Worm	17	ϕ_x	$\tan^{-1}(\tan \phi_n/\cos \lambda_c)$
Face width	Gear	18	F_G	$d_R \pm \frac{1}{8}''$

* See instructions on page 5–70. † See item 14 on page 5–71.

Table 5-28. Recommended Number of Wormgear Teeth

Center distance, in.	Single-enveloping min. No. of teeth	Double-enveloping No. of teeth
2.000	20	24–40
3.000	25	24–50
4.000	25	30–50
5.000	25	30–50
8.000	25	40–60
10.000	29	40–60
15.000	35	50–60
20.000	40	50–70
24.000	45	60–80

2. The equations that follow assume that a specific center distance must be observed. If this is not the case, the designer is advised to utilize standard values of lead angle and normal pitch.
3. Experience has shown that the best worm gear designs result from the selection of numbers of wormgear teeth related to the operating center distance.
4. The number of teeth or threads in the worm is determined from the ratio and number of gear teeth. The number of teeth in gear and worm may have to be adjusted to get a suitable balance between lead per thread and root diameter. See items 9 and 21. Common factors between numbers of gear teeth and worm threads should be avoided.
5. Experience indicates that this equation gives the best values of worm pitch diameter for power gears. However, other values may be used to get a desired lead angle (see Fig. 5-16).
6. The pitch diameters of the wormgear and of the worm are of no kinematic significance. They are of aid only during the design of the gearing and as a

Table 5-29. Minimum Number of Wormgear Teeth for Standard Addendum

Pressure angle, deg	*Min. No. of teeth*
$14\frac{1}{2}$	32
20	25
25	20
30	15

Table 5-30. Suggested Limits on Lead Angle

No. of teeth	*Lead angle, deg*
1	Less than 6
1–2	Up to 12
3–4	10–24
5–6	15–36
7 or more	20 and higher

Table 5-31. Recommended Normal Pitches p_n

Fine pitch		Coarse pitch	
0.030	0.080	0.200	0.625
0.040	0.100	0.250	0.750
0.050	0.130	0.300	1.000
0.065	0.160	0.400	1.250
		0.500	1.500

result may be altered during design in order to get the best worm gear tooth proportions.

7. The normal circular pitch of the worm and of the wormgear must be the same. Table 5-31 shows commonly used standard values of normal circular pitch for worm gearing.
8. The axial pitch of the worm is made equal to the transverse circular pitch of the gear. Diametral pitches are not generally used for worm gearing.
9. The following should be observed when selecting the worm lead angle:
   a. In general, do not select angles in excess of 6° per thread (see Table 5-30).
   b. Low lead angles are less efficient. Self-locking may occur from 5 to 10°.
   c. For best efficiency, use from 15 to 30° lead angles. See Chap. 14 for detail data on worm gear efficiency.
10. The lead is the axial distance on the worm that a point will move when the worm is rotated through one revolution.
11. The pressure angle varies across the face of the wormgear. It is customary to use the pressure angle of the worm cutter in fine-pitch worm gear sets as the pressure angle for the worm gear sets. Table 5-25 gives nominal values of cutter pressure angle.
12. The tooth profile of the worm will vary depending upon the diameter of the cutter. In order to describe the desired worm profile completely by means of a cutter specification, it is necessary to specify the cutter thickness and outside diameters. Table 5-33 gives nominal values of cutter pitch diameters used for various types of worms.
13. The pressure angle of the worm measured in the normal section at the pitch diameter will be slightly different from the cutter pressure angle.

Table 5-32. Recommended Values of Backlash for Single-enveloping Worm Gearing and Double-enveloping Worm Gearing

Center distance	Backlash (amount worm should be reduced)	
2	0.003	0.008
6	0.006	0.012
12	0.012	0.020
24	0.018	0.030

Table 5-33. Recommended Values of Cutter Diameter (Single-enveloping Gearsets)

Type of worm	Suggested cutter diameter D_c, in.	Process
Low-speed power...........	4	Milled
High-speed power...........	20	Ground
Fine-pitch:		
Commercial quality.......	3	Milled
Precision quality.........	20	Ground

14. The axial pressure angle of the worm is the most commonly specified value when an attempt is made to specify the actual worm profile.
15. The addendum of the worm is given in Table 5-25. This value will vary depending upon the lead. In the case of small numbers of teeth, it will be necessary to use higher pressure angles to avoid undercut if standard values of addendum are to be employed (see Table 5-29).
16. The worm outside diameter should be reduced if the worm thread is too sharp at the outside diameter.
17. The face width of the wormgear tooth is limited by generating scallops produced by the hob. The equation gives about the largest face width (for maximum load capacity) which is still free of scallops.
18. The throat diameter of the wormgear can be turned in the blank before cutting or introduced by the hob in fine-pitch worm gears. It is not always present since the added tooth contact area produced by this feature may not warrant the extra cost to achieve it.
19. The outside diameter may be the same as the throat diameter in nonthroated gears.
20. The face width of the worm is the length of the threaded portion of the worm. It should be long enough to achieve the maximum contact ratio possible with the wormgear. Any length in excess of the value shown represents lost money in the manufacturing operation.
21. Root diameter should be sufficient to ensure that the worm has sufficient beam strength if the worm and shaft are integral. The root diameter must be sufficiently larger than the bore if of the bored type to provide ample stock over any keyway that is present.
22. The tooth thickness of the worm should be thinned to achieve the backlash desired in the wormgear mesh.
23. The ends of the wormgear may be chamfered or given an end radius. The latter is preferred.

24. The tips of the worm threads should be given a round to protect the hands of those working with the parts and also to reduce the danger of flaking during heat-treatment. Also, a sharp worm may cut its gear during break-in.

Specific Calculation Procedure for Double-enveloping Worm Gears. Table 5-27 shows a detail calculation procedure.

An approximate center distance may be obtained from Art. 13-3. The number of wormgear teeth may be picked from Table 5-28.

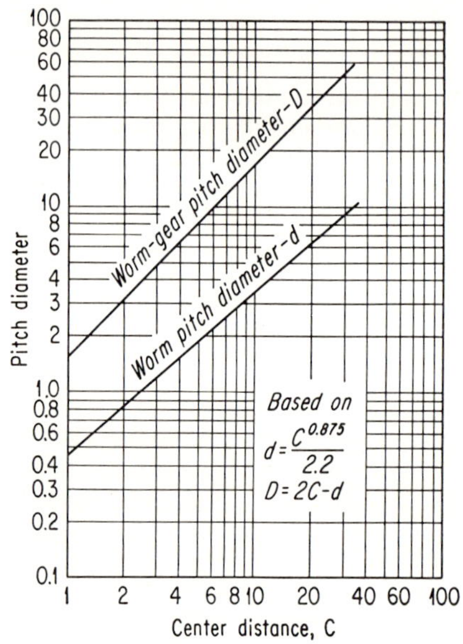

FIG. 5-16. Recommended pitch diameters for worms and wormgears.

Detail instructions for each item in the table are as follows:

1. Worm gear ratios are established by the number of teeth in the wormgear and the number of threads (starts) in the worm.
2. Center distance determines the load capacity.
3. The number of teeth in the gear member of multiple-thread worm gearsets should be selected from Table 5-28. The values shown yield gearing having the maximum horsepower capacity and greatest ease of manufacture. The number of teeth in the gear may exceed the numbers of teeth in Table 5-28.
4. The number of threads in the worm is based on the number of teeth in the wormgear and the required gear ratio.
5. The worm root-diameter equation gives the minimum usable value. Worms cut integral with the shaft should have a value of $k_r = 1.15$.
6. This equation gives a trial pitch diameter for the worm.
7. This gives a trial pitch diameter of the wormgear.
8. This is the average lead angle of the worm.

9 and 10. These equations give the pitch at the center of the worm in the axial and normal planes.

11 and 12. The whole depth applies to both the worm and the gear.

13. This is the dedendum of the gear tooth.

14. This value should equal or exceed the value from step 5. (Tolerance may be ±0.005″.) If it does not, repeat steps 6 through 14 until agreement is achieved.
15. Standard for double-enveloping worm gearing.
16. Lead angle at center section.
17. Engineering value.
18. This value gives maximum usable strength. For lightly loaded gearsets, this value may be decreased in proportion to the load.

5-10. Standard System for Face Gears. Fine-pitch (20 diametral pitch and finer) face gears have been standardized by the American Gear Manufacturers Association. This system is available as AGMA Standard 203.01, "Fine-pitch On-center Face Gears for 20-degree Involute Spur Pinions."

Coarse-pitch face gears have not been standardized. The following data are arranged to provide a logical method of calculating the tooth proportions of face gears. Such gears will be suitable for most applications. For a more complete treatment of the subject the reader is referred to the original standard.

Table 5-34. Tooth Proportions for Pinions Meshing with Face Gears

Item	Coarse pitch $P_d = 20$		Fine pitch $P_d = 20$	
No. of teeth in pinion N_p	$N_p < 18$	$N_p \geqq 18$	$N_p < 18$	$N_p \geqq 18$
Pressure angle ϕ, deg	20	20	20	20
Addendum a	See Table 5-36	$\frac{1}{P_d}$	See Table 5-36	$\frac{1}{P_d}$
Standard pitch diameter D_p	$\frac{N}{P_d}$	$\frac{N}{P_d}$	$\frac{N}{P_d}$	$\frac{N}{P_d}$
Working depth h_k	....	$\frac{2}{P_d}$	See Table 5-37	$\frac{2.0}{P_d}$
Whole depth h_t	....	$\frac{2.25}{2P_d}$	See Table 5-37	$\frac{2.2}{P_d} + 0.002$
Circular tooth thickness* t	....	$\frac{\pi}{2P_d}$	See Table 5-37	$\frac{\pi}{2P_d}$
Clearance c	....	$\frac{0.25}{P_d}$	$\frac{0.2}{P_d} + 0.002$	$\frac{0.2}{P_d} + 0.002$
Reference standard	None		AGMA 203.01	

*Thin teeth for backlash. See Art. 5-5 for general information on spur-gear backlash.

Caution should be exercised in using Table 5-34. The items shown apply only to gears that meet the following requirements:

1. The axes of the gear and the pinion must intersect at an angle of 90°.
2. The gear must be generated by means of a reciprocating pinion-shaped cutter having the same diametral pitch and pressure angle as the mating pinion and which must be of substantially the same size.
3. The pinion should be sized to meet Chap. 13 requirements on load-carrying capacity.
4. The minimum number of teeth in the pinion in this system is 12 and the minimum cutter pitch diameter is 0.250″.
5. The minimum gear ratio is 1.5:1 and the maximum ratio is 12.5:1.
6. The long- and short-addendum designs for the pinion with less than 18 teeth should not be generally used on speed-increasing drives.

Table 5-35. Minimum Numbers of Teeth in Pinion and Face Gear

Diametral pitch range	Min.* numbers of teeth	
	Pinion	Gear
20– 48	12	18
49– 52	13	20
53– 56	14	21
57– 60	15	23
61– 64	16	24
65– 68	17	26
69– 72	18	27
73– 76	19	29
77– 80	20	30
81– 84	21	32
85– 88	22	33
89– 92	23	35
93– 96	24	36
97–100	25	38

* The minimum numbers of teeth in the pinion is limited by the design requirements of the gear cutter. These requirements, in addition to the minimum-gear-ratio limitation, limit the numbers of teeth in the gear.

Table 5-36. Addendum of Face Gears and Pinions

No. of teeth in pinion	Coarse pitch		Fine pitch pinion addendum
	Pinion addendum	Gear addendum	
12	1.120	0.700	$\frac{1.1215}{P_d} - 0.002$
13	1.100	0.760	$\frac{1.1050}{P_d} - 0.002$
14	1.080	0.820	$\frac{1.0865}{P_d} - 0.002$
15	1.060	0.880	$\frac{1.0650}{P_d} - 0.002$
16	1.040	0.940	$\frac{1.0420}{P_d} - 0.002$
17	1.020	0.980	$\frac{1.0175}{P_d} - 0.002$
18 and 19	1.000 1.250	1.000 0.750	$\frac{1.0000}{P_d} - 0.002$ (for 18 and higher)
20 and 21	1.000 1.250	1.000 0.750	
22 and 29	1.000 1.200	1.000 0.800	
30 through 40	1.000 1.150	1.000 0.850	
40 and higher	1.000 1.100	1.000 0.900	

Table 5-37. Tooth Proportions for Fine-pitch Face Pinions

No. of teeth	Tooth thickness at standard pitch dia.	Whole depth	Working depth*
12	1.7878	2.0234	1.8234
13	1.7452	2.0654	1.8604
14	1.7027	2.1054	1.9054
15	1.6601	2.1424	1.9424
16	1.6175	2.1778	1.9778
17	1.5749	2.2118	2.0118
18	1.5708	2.2000	2.0000
19 and up	1.5708	2.2000	2.0000

* Divide by diametral pitch and subtract 0.002″.

Table 5-38. Tooth Proportions and Diameter Constants for 1 Diametral Pitch Face Gears, 20° Pressure Angle

No. of pinion teeth N_P	Gear dia. constants							
	$m_g = 1.5$		$m_g = 2$		$m_g = 4$		$m_g = 8$	
	m_o	m_i	m_o	m_i	m_o	m_i	m_o	m_i
12	.....	.....	1.221	1.020	1.221	0.960	1.221	0.945
13	1.202	1.064	1.202	1.015	1.202	0.959	1.202	0.945
14	1.187	1.062	1.187	1.011	1.187	0.958	1.187	0.944
15	1.174	1.052	1.174	1.007	1.174	0.957	1.174	0.944
16	1.161	1.051	1.161	1.004	1.161	0.956	1.161	0.944
17	1.156	1.041	1.156	1.000	1.156	0.955	1.156	0.944
18	1.150	1.039	1.150	0.997	1.150	0.954	1.150	0.943
	1.176	1.042	1.176	0.999	1.176	0.955	1.176	0.943
20	1.144	1.030	1.144	0.991	1.144	0.953	1.144	0.943
	1.166	1.032	1.166	0.993	1.166	0.953	1.166	0.943
22	1.140	1.022	1.140	0.987	1.140	0.952	1.140	0.943
	1.156	1.024	1.156	0.988	1.156	0.952	1.156	0.943
24	1.133	1.015	1.133	0.983	1.133	0.951	1.133	0.942
	1.150	1.017	1.150	0.984	1.150	0.951	1.150	0.943
30	1.121	1.001	1.121	0.975	1.121	0.949	1.121	0.942
	1.131	1.001	1.131	0.975	1.131	0.949	1.131	0.942
40	1.109	0.986	1.109	0.966	1.109	0.946	1.109	0.941
	1.113	0.986	1.113	0.966	1.113	0.946	1.113	0.941

Pinion Design. The pinion member in cases of numbers of teeth less than 18 has an enlarged tooth thickness to avoid undercut when cut with a standard cutter. Compared with the pinions designed in accordance with data in Art. 5-5 the outside diameter is somewhat less. This is to avoid the necessity of cutting the gear with shaper cutters that have excessively pointed teeth. The limit set is top land width $\geqq 0.40/P_d$. Table 5-37 gives the tooth proportions for face-gear pinions.

Face-gear Designs. The face-gear member is generated by a cutter having proportions based on the pinion with which the face gear will operate. The most important specification for the shape of the teeth of a face gear is a complete specification of the cutter to be used to cut it or, next best, a detailed specification of the mating pinion.

The face gear has two dimensions unique to face gears which control the face width of the teeth, the outer and inner diameter of the face gear.

The maximum usable face width may be estimated from Table 5-38. The outside diameter of the gear should not exceed m_oD. The inside diameter should not be less than m_iD. Thus,

$$\begin{aligned} \text{Outer diameter} &= m_oD = D_o \\ \text{Inner diameter} &= m_iD = D_i \\ \text{Pitch diameter} &= D \\ \text{Face width} &= \frac{D_o - D_i}{2} \end{aligned} \tag{5-42}$$

5-11. System for Spiroid and Helicon Gears. At the present time there are no AGMA standards covering Spiroid, Helicon, or Planoid®* gears. The following material is based upon information covered in "Spiroid Gearing," paper 57-A-162 of the ASME, and upon additional information supplied by its author, W. D. Nelson.

Although specialized designs can be developed which will best suit given applications, the standardized procedure presented here will yield designs that will meet the needs of most Spiroid and Helicon gear applications.

Planoid gears are often associated with Spiroid and Helicon gearing but are an entirely different form of gearing, used for ratios generally under 10:1 where maximum strength and efficiency are required. Their design is beyond the scope of this section.

The basic approach to the establishment of the Spiroid or Helicon tooth form is to establish the pinion-tooth form and then develop a gear-tooth form which is conjugate to it.

The tooth form of the Spiroid gear is comprised of teeth having different pressure angles on each side of the teeth; a "low" pressure-angle side and a "high" pressure-angle side. The choice of pressure angle will control the extent of the fields of conjugate action of the teeth. The values of pinion taper angle have been standardized by the Illinois Tool Works in order to achieve the best general-purpose gearing. These standard values are shown in Table 5-39.

The design of Helicon gearing is closely related to Spiroid gearing. Table 5-40 shows the general equations used in the design of both kinds of gearing. The general Spiroid formulas become Helicon formulas when the pinion taper angle τ is set equal to zero. Figure 5-17 shows a Spiroid gearset. A Helicon pinion and gear are similar except that the pinion taper angle τ and the gear face angle γ are both *zero* in Helicon gearing.

Spiroid Gearing. Spiroid gearing is suitable for gear ratios of 10:1 or higher. The numbers of teeth in the gear can range from 30 to 300.

The design procedure for Spiroid gearing differs somewhat from other types of gearing discussed in this chapter. The following are the basic steps:

1. Center distance is the basic starting point for the calculation of Spiroid tooth proportions. Center distance and gear ratio are first established based on information given in Chap. 13.
2. Lead and the pinion "zero" plane radius are then calculated.
3. The tooth proportions and the blank proportions are based on the pinion zero plane radius and are then calculated.

From this procedure, it can be seen that the pitch of the teeth is an end result of the calculating procedure for Spiroid gearing rather than the beginning as on other types of gearing.

Center distance governs the horsepower capacity of the gearset. Chapter 13 gives

* Spiroid, Helicon, and Planoid are trade-marks registered by the Illinois Tool Works, Chicago, Ill.

information on the center distances required for the various load ratings of Spiroid gears.

Table 5-39 shows the basic tooth proportions for Spiroid gearing.

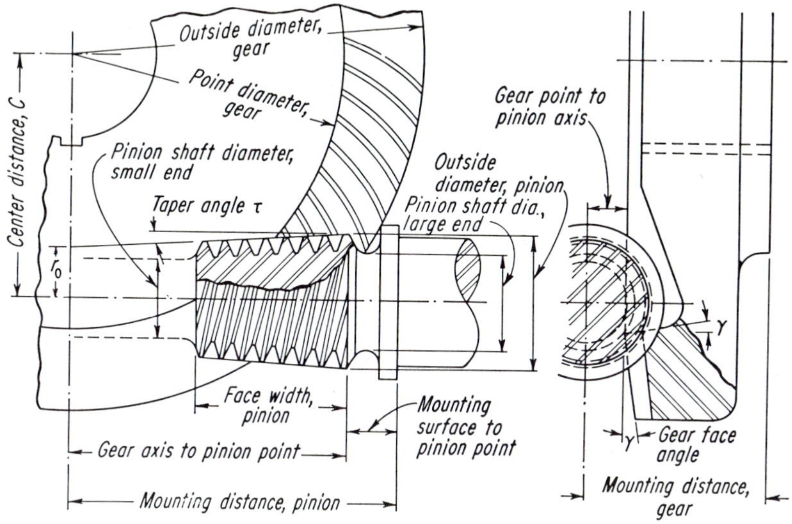

FIG. 5-17. Spiroid gearset. Mounting and gear-blank nomenclature.

Table 4-16 gives standard blank proportions for Spiroid gearing for specific values of center distance.

Table 5-40 gives the general equations used to determine Spiroid tooth proportions.

Table 5-41 gives a step-by-step procedure for calculating Spiroid gearing tooth and gear-blank proportions.

If optimum efficiency is desired, the size of the pinion relative to the gear should be kept small by using values of R_G/R_P which are larger than those shown in Fig. 5-18.

Table 5-39. Standard Tooth and Gear-blank Relationships, Spiroid and Helicon Gearing

Spiroid gearing:

1. "Sigma" angle, $\sigma_P = 40°$ (standard)
2. Pinion taper angle τ vs. gear face angle γ

Pinion taper angle, deg	*Gear face angle,* deg*	
τ	γ	
5	8	preferred
7	11	
10	14	

3. Gear ratio m_G = 10:1 higher
4. No. of teeth in gear:
   Varies with center distance (see Fig. 5-20)
   Hunting ratios are desirable with all multiple-thread pinions
5. Tentative pressure-angle selection

Ratio	*Pressure-angle selection*	
	Low side	*High side*
$M_G \leqq 16:1$	15°	35°
$M_G > 16:1$	10°	30°

* Approximate values, for $\sigma_P = 40°$.

This has the effect of producing an increased lead angle for a given center distance and gear ratio.

If control-gear applications are being considered, in which runout is critical, the driving side should be the low-pressure-angle side of the teeth. This minimizes the effects of runout in both members. A 10° taper angle τ may be used, since it

Table 5-40. General Equations for Spiroid and Helicon Gearing

Spiroid gearing:

$$\sin \phi_P = \frac{\tan \tau}{\tan \sigma_P} \tag{5-43}$$

$$R_P = \frac{C}{\sin \phi_P + (R_G/R_P) \cos \sigma_P} \tag{5-44}$$

$$R_G = \frac{R_G}{R_P} R_P \tag{5-45}$$

$$L = \frac{2\pi\, R_G \cos \sigma_P}{m_G - (R_G/R_P) \sin \sigma_P \cos \phi_P} \tag{5-46}$$

$$x_P = R_G \sin \sigma_P \tag{5-47}$$

$$r_o = R_P - x_P \tan \tau \tag{5-48}$$

$$D_W = \frac{0.6(L/N_P) \sec \tau}{[\sin \psi_1/\cos (\psi_1 + \tau)] + [\sin \psi_2/\cos (\psi_2 - \tau)]} \tag{5-49}$$

$$D_N = D_W \cos \tau \tag{5-50}$$

$$CLR = 0.07 \frac{L}{N_P} + 0.002 \tag{5-51}$$

$$\psi_{2L} = \tan^{-1} \left[\frac{R_P}{x_P} + \frac{CZ_P}{R_P(ky_P - x_P)} \right] \tag{5-52}$$

$$\tan \lambda_m = \frac{L \sec \tau}{2\pi r_m} \tag{5-53}$$

$$\psi_2 = \psi_{2L} + 5° \tag{5-54}$$

Helicon gearing:

$$R_P = \frac{C}{(R_G/R_P) \cos \sigma_P} \tag{5-55}$$

$$L = \frac{2\pi R_G \cos \sigma_P}{m_G - (R_G/R_P) \sin \sigma_P} \tag{5-56}$$

$$D_w = \frac{0.6(L/n)}{\tan \psi_1 + \tan \psi_2} \tag{5-57}$$

$$\psi_{2L} = \tan^{-1} \frac{R_P}{x_P} \tag{5-58}$$

$$\psi_{2L} = \tan^{-1} \frac{1}{(R_G/R_P) \sin \sigma_P} \tag{5-59}$$

$$\psi_2 = \psi_{2L} + 7° \tag{5-60}$$

requires only one-half the axial movement of the pinion to produce a given change in backlash.

For gearsets of the highest strength, the number of teeth in the gear should be kept low (30 to 40 for lower ratios) and a stub-tooth form employed. Such designs are beyond the scope of this section.

Straddle-mounted pinions may require a shaft of larger size (greater stiffness) than would result from the data shown in Fig. 5-17. In such cases, the values of R_G/R_P are decreased until a shaft of the desired size can be achieved. There is a definite limit to the maximum size of pinion for a given pressure angle. The size of the pinion may be increased until a "limiting pressure" angle condition is reached. Limiting pressure angle can be calculated by means of Eq. (5-54).

Helicon Gearing. Helicon gearing is suitable for gear ratios from 4:1 to 400:1 in light to medium loaded applications.

The design procedure for Helicon gears is somewhat similar to that of Spiroid gears but starts with the gear outside diameter as the basis of load capacity instead of the center distance as is used in Spiroid gearing.

The following are the basic steps:

1. The gear ratio m_G, pinion rpm, and horsepower are used to establish the necessary gear outside diameter as shown in Chap. 13.
2. Pinion pitch radius, center distance, lead, and working depth are next established.
3. From these values, the gear and pinion blank dimensions are established.

Specific Instruction for Calculation of Spiroid Gear Data in Table 5-41

1. *Center distance.* Spiroid gears are rated on the basis of their center distance (see Chap. 13). The tooth proportions are also based on center distance. As a result, the first value to be selected is the center distance, which is the basis of most of the remaining values.
2. *Gear ratio.* This value is selected on the basis of the service to be performed. In general, a single Spiroid mesh can have a ratio ranging from 10 to 110. Ratios larger than these can be obtained but are beyond the scope of this section. After the number of teeth in the pinion and in the gear have been selected, items 3 and 4, the exact ratio calculated from $N_G \div N_P$ should be entered in item 2 for use in all subsequent calculations involving ratio.
3. *Number of teeth, gear.* Figure 5-20 shows the ranges of tooth numbers available for various center distances. For high-strength designs, use lower numbers of teeth. For high rotational accuracy, use larger numbers of teeth. Whenever possible, use a hunting tooth in the gear.
4. *Number of teeth, pinion.* The number of teeth or threads in the pinion is a function of the number of teeth in the gear and the ratio.
5. *Taper angle.* A taper angle of 5° is standard. See Table 5-39 and also Figs. 5-17 and 5-18. Values between 5 and 10° may be used, although angles smaller tend to limit the extent of the conjugate action, and too large an angle (over 10°) tends to limit load capacity and also introduce manufacturing difficulties.

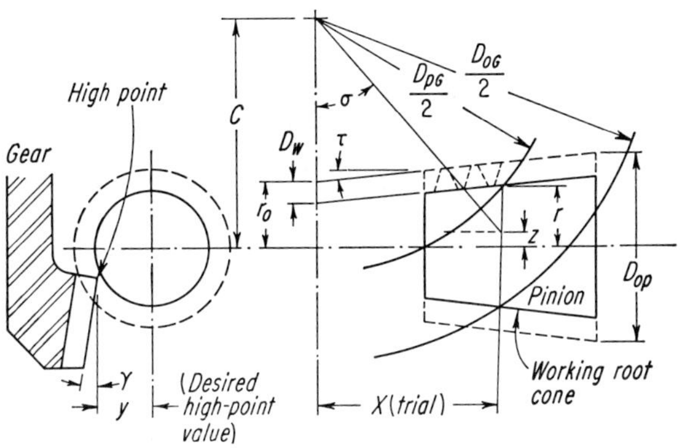

FIG. 5-18. Gear-tooth high point—Spiroid gearing.

Table 5-41. Tooth Proportions, Spiroid Gearing

Item	Member	Item ref.	Symbol	Equation	Example
Center distance	Both	1	C	....	3.75″
Gear ratio	Both	2	m_G	$\frac{N_G}{N_P}$	14.750
No. of teeth	Gear	3	N_G	....	59
	Pinion	4	N_P	N_G/m_G	4
Taper angle	Pinion	5	τ	....	5°
"Sigma" angle	Both	6	σ_P	....	40°
Gear to pinion pitch radius ratio	Both	7	R_G/R_P	From Fig. 5-20	4.15
		8	ϕ_P	$\sin \phi_P = \frac{\tan \tau}{\tan \sigma_P}$ if "standard proportions" are used $= 0.104265 = 5.9848°$	0.104264976
Pinion pitch radius	Pinion	9	R_P	$\frac{C}{\sin \phi_P + (R_G/R_P) \cos \sigma_P}$ if "standard proportions" are used $\frac{C}{0.104265 + (R_G/R_P)\ 0.766044}$	$\frac{3.75}{0.104265 + 4.15(0.766044)}$ $= 1.142127$
Gear pitch radius	Gear	10	R_G	$R_P(R_G/R_P)$	$= 1.142127 \times 4.15$ $= 4.7398272$
Conical lead	Pinion	11	L	$\frac{2\pi R_G \cos \sigma_P}{m_G - (R_G/R_P) \sin \sigma_P \cos \phi_P}$ if "standard proportions" are used $\frac{4.813196 R_G}{m_G - 0.639284(R_G/R_P)}$	$\frac{4.813196 \times 4.7398272}{14.75 - 0.639284(4.15)}$ $= 1.885903$

x coordinate of pitch point	Pinion	12	x_P	$R_G \sin \sigma_P$ if "standard proportions" are used $R_G \times 0.642788$	$4.7398272 \times 0.6427876$ $= 3.046702$
Pinion "zero plane" radius	Pinion	13	r_o	$R_P - x_P \tan \tau$ if "standard proportions" are used $R_P - 0.0874887 x_P$	$1.142127 - (0.0874887 \times 3.046702)$ $= 0.875575$
y coordinate of pitch point	Pinion	14	y_P	$R_P \cos \phi_P$	$1.142127 \times 0.9945496$ $= 1.13590195$
z coordinate of pitch point	Pinion	15	z_P	$R_P \sin \phi_P$	$1.142127 \times 0.1042650$ $= 0.1190839$
Limit pressure angle	Gear	16	ψ_{2L}	$\tan^{-1}\left[\frac{R_P}{x_P} + \frac{C z_P}{R_P(m_G y_P - x_P)}\right]$	$\frac{1.142127}{3.046702} + \frac{3.75 \times 0.1190839}{1.142127(14.75 \times 1.13590195 - 3.046702)}$ $= \tan^{-1} 0.4033965$ $\psi_{2L} = 21.965°$
Hi-side pressure angle (limiting)	Gear	17	ψ_2'	$\geqq \psi_{2L} + 5°$ if greater than 35° select a larger value of R_G/R_P. Recalculate steps 9 through 17	$21.965 + 5 = 26.96°$
Pressure angle	Hi-side gear	18	ψ_2	Select from Table 5-39 Note: ψ_2 must not be less than ψ_2'	35°
	Lo-side gear	19	ψ_1	Select from Table 5-39	15°
Working depth	Pinion	20	D_w	$0.6(L/N_P) \sec \tau$ ——— $\frac{\sin \psi_1}{\cos (\psi_1 + \tau)} + \frac{\sin \psi_2}{\cos (\psi_2 - \tau)}$ when $\tau = 5°$, $\psi_1 = 10°$, $\psi_2 = 30°$ $0.82341(L/N_p)$ when $\tau = 5°$, $\psi_1 = 15°$, $\psi_2 = 35°$ $0.64228(L/N_p)$	$0.64228 \frac{1.885903}{4}$ $= 0.3028194$

Table 5-41. Tooth Proportions, Spiroid Gearing *(Continued)*

Item	Member	Item ref.	Symbol	Equation	Example
Point diameter	Gear	21	D_{pG}	$2.2018C$	$2.2018 \times 3.75 = 8.2568$
Mid-plane distance	Pinion	22	x	$0.52C$	0.52×3.75 $x_1 = 1.950$ $x_3 = 1.96677$
Mid-plane root radius	Pinion	23	r	$r_o = D_w + x \tan \tau$	$(0.875575 - 0.3028194 + 1.950 \times 0.0874887)$ $r_1 = 0.7433586$ $r_3 = 0.744827$
	Pinion	24	z	$\dfrac{C}{1 + (x/r \tan \tau)}$	$\dfrac{3.75}{1 + \dfrac{1.950}{0.743359 \times 0.0874887}}$ $z_1 = 0.120974$ $z_3 = 0.120262$
	Pinion	25	k'	$(c - z)^2 + x^2$	$(3.75 - 120{,}974)^2 + (1.950)^2$ $k' = 16.9740$ $k_3 = 17.04318$
	Gear	26	k''	$(D_{pG}/2)^2$ k'' should equal k' within $\pm 0.005''$ if not, calc. x_2 by step 27 and use x_2 in place of x in steps 22 through 26. Repeat until $k'' = k' \pm 0.005''$	$(8.2568/2)^2$ $k'' = 17.04369$ $k'' - k' = 0.0697$ go to step 27 $k'' - k_3 = 0.0005$ hence go to step 28
		27	x_2	$(k'' - (C - z)^2)^{1/2}$	$(17.04369 - (3.75 - 0.12097)^2)^{1/2}$ $x_2 = 1.96821$ (to step 22) $x_3 = 1.96677$
Pinion axis to high point plane	Gear	28	y	Using final r and z steps 23 and 24 $[r^2 - z^2]^{1/2}$	$[(0.744827)^2 - (0.120262)^2]^{1/2}$ $y = 0.735054$

Mean radius	Pinion	29	r_m	Using final x steps 22 and 27 $r_o + x \tan \tau - D_w/2$	$0.875575 + (1.96677 \times 0.0874887) - 0.1514097$ $r_m = 0.896235$
Mean spiral angle	Pinion	30	λ_m	$\tan^{-1} \dfrac{L \sec \tau}{2\pi r_m}$	$\tan^{-1} \dfrac{1.885903 \times 1.0038198}{6.283185 \times 0.896235} = .336180$ $= 18.5817°$
Pressure angle check	Pinion	31	ψ_1	If $\lambda_m < 16°$, $\psi_1 = 10°$ If $\lambda_m \geq 16°$, $\psi_1 = 15°$ Correct ψ_1 in step 19 and ψ_2 in step 18 if necessary, and recalculate steps 20 through 30	$\lambda_m = 18.58° > 16°$ $\therefore \psi_1 = 15°$ *Checks*
Outside diameter	Pinion	32	D_{oP}	$2(r_o + 0.1043C)$	$2(0.875575 + 0.1043 \times 3.75) = 2.5334$
	Gear		D_{oG}	$3C$	$3 \times 3.75 = 11.25$
Face width	Pinion	33	F_P	$0.7308C$	$0.7308 \times 3.75 = 2.7405$
Gear axis to pinion point	Pinion	34		$1.1923C$	$1.1923 \times 3.75 = 4.4711$
Pinion shaft diameter — Large end		35		$0.632C^{0.833} - 0.034C$	$0.632 \times 3.75^{0.833} - 0.034 \times 3.75 = 1.7748$
Pinion shaft diameter — Small end		36		$0.382C^{0.833}$	$0.382 \times 3.75^{0.833} = 1.1498$
Normal working depth		37	D_N	$D_w \cos \tau$ if standard proportions are used D_w 0.996195	$0.3028194 \times 0.996195 = 0.3017$
Clearance		38	CLR	$0.07(L/N_p) + 0.002$	$0.07 \dfrac{1.885903}{4} + 0.002 = 0.035$
Land widths, tip and working root		39		$\dfrac{L}{5N_p}$	$\dfrac{1.885903}{5 \times 4}$ $= 0.0943$

6. *"Sigma" angle.* A standard value of 40° is recommended. This angle is a coordinate of the pitch point on the gear. The standard blank dimensions given in Table 5-42 are based on this standard value.

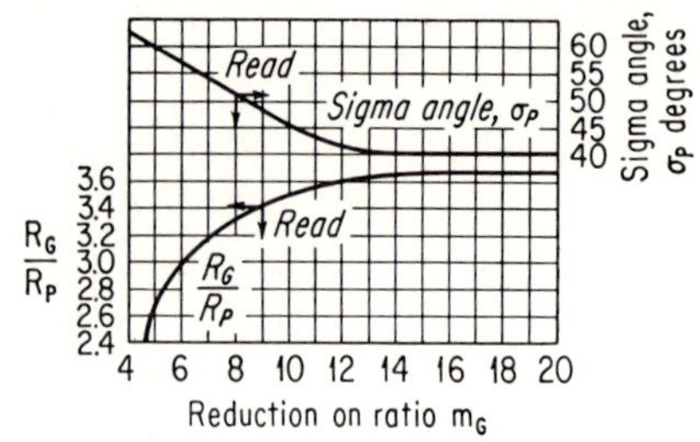

FIG. 5-19. Sigma angle for different ratios—Helicon gears.

7. *Gear to pinion pitch radius ratio.* This value controls the relative sizes of the pinion and gear. The values of R_G/R_P given in Fig. 5-20 are suitable for both cantilever and through shaft mountings.

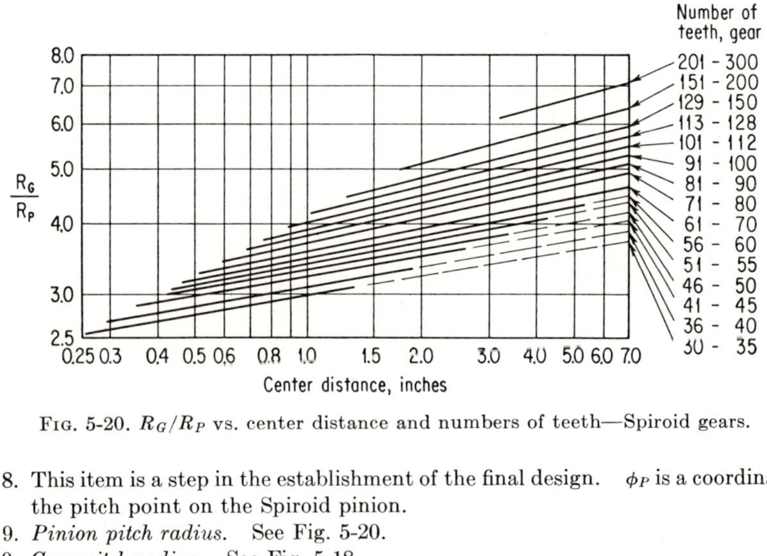

FIG. 5-20. R_G/R_P vs. center distance and numbers of teeth—Spiroid gears.

8. This item is a step in the establishment of the final design. ϕ_P is a coordinate of the pitch point on the Spiroid pinion.
9. *Pinion pitch radius.* See Fig. 5-20.
10. *Gear pitch radius.* See Fig. 5-18.
11. *Conical lead, pinion.* This is the lead measured along the pinion thread surface parallel to the axis.
12. x coordinate of pitch point. See Fig. 5-18.
13. Pinion zero-plane radius r_o. See Fig. 5-18.
14. y_P coordinate of pitch point. See Fig. 5-18.
15. z_P coordinate of pitch point. See Fig. 5-18.
16. A limiting condition of conjugate tooth action governs the maximum pressure angle (called limit pressure angle). This limit pressure angle is calculated, and the hi-side pressure angle is established from it.
17. Most designs employ a 30° pressure angle on the hi-side unless the limit pressure angle dictates the use of a higher value; 35° is the maximum value in any case. If ψ_2' exceeds 35° a larger R_G/R_P value should be selected in step 7, and items 9

through 17 recalculated. A larger value of R_G/R_P will yield a smaller pinion, thereby improving the undercut or limit pressure angle condition.

18. Select the "hi-side" pressure angle from Table 5-39. The value selected must be larger than the limiting hi-side pressure angle ψ_2' in step 17. The teeth of Spiroid pinions lie with their vertical axes roughly perpendicular to the tapered conical surface of the pinion blank. As a result, the pressure angle as seen by the gear member is different on each side of the pinion tooth. The "high" pressure-angle side is usually considered the nondriving side, although it can be used as satisfactorily as the low-pressure-angle side when the blanks are properly supported.
19. Select the "lo-side" pressure angle from Table 5-39. The low-pressure-angle side of the tooth is preferred as the driving side. It is the side of the teeth seen when looking along the pinion axis at the small end of the pinion taper.
20. *Working depth.* The effective height of the tooth.
21. *Point diameter gear.* See Fig. 5-18. This is the inner diameter of the gear blank. It is formed by the intersection of the top land cone and the inner cone of the gear blank. It is not critical and can be given a bilateral tolerance.

22 through 27. Steps 22 through 27 establish the exact position of the mid-plane of the pinion. See Fig. 5-18 (x trial). A tentative value is selected in step 22. A tentative value of r and of z is established. If the values do not agree with $(0.5D_{pG})^2$, step 27 is used to establish an improved value of x. Steps 23, 24, and 25 are repeated and k' compared with k''. The final agreement must be within $\pm 0.005''$.

28. *Pinion axis to high point plane y.* See Fig. 5-18.
29. *Mean radius.* In the mid-plane of the pinion a mean radius on the tooth where the spiral angle and the pressure angles can be examined is thus established.
30. *Mean spiral angle.* The spiral angle in the mean plane.
31. The pressure angle on the "lo-side" is checked in this step. A 10° pressure angle is used for mean spiral angles less than 16° and a 15° pressure angle is used for spiral angles of 16° and over.
32. *Outside diameter, pinion.* This is the diameter of the pinion blank at the large end (see Fig. 5-17). It is the theoretical diameter before any edge round is applied. It should be given a minus tolerance. *Outside diameter, gear.* The outside diameter of a spiroid gear is usually a cylindrical surface (see Fig. 5-17). It is not critical and can be given a bilateral tolerance.
33. *Face width, pinion.* This is the length of the thread portion of the pinion blank measured from the plane containing the largest diameter to the front face (see Fig. 5-17).
34. *Gear axis to pinion point.* This dimension controls the axial position of pinion member (see Fig. 5-17). The pinion mounting distance is established from this dimension.
35. *Pinion shaft diameter, large end.* This is the maximum diameter of the shaft at the large end of the pinion (see Fig. 5-17). It is based on sufficient cutter clearance to enable good manufacturing practices.
36. *Pinion shaft diameter, small end.* If a through-shaft or straddle-mounted pinion design is required, this dimension gives the largest permissible shaft extension at the small end of the pinion. See phantom lines (Fig. 5-17).
37. *Normal working depth.* This is the working depth of the tooth as measured perpendicular to the elements of the outside-diameter cone of the pinion or gear teeth.
38. *Clearance.* This is the clearance at the roots of the teeth. The whole depth of the tooth is the sum of the normal working depth and the clearance.
39. *Land widths, tip and working root.* This is the width of the top land and of the root land at the working root cone.

Table 5-42. Tooth Proportions for Helicon Gearing

Item	Member	Item ref.	Symbol	Equation
Outer diameter	Gear	1	D_{oG}	Given
Gear ratio	Both	2	m_G	Given N_G/N_P from items 18 and 19 (see explanation 2)
Sigma angle	Both	3	σ_P	Fig. 5-19
Gear to pinion pitch radius ratio	Both	4	R_G/R_P	Fig. 5-19
Outside diameter	Pinion	5	D_{oP}	$0.84D_{oG}/(R_G/R_P)$
Pitch radius	Pinion	6	R_P	$D_{oP}/2$
Center distance	Both	7	C	$R_P(R_G/R_P)\cos\sigma_P$
Lead		8	L	$\dfrac{2\pi C}{m_G - \left(\dfrac{R_G}{R_P}\right)\sin\sigma_P}$
Pitch radius	Gear	9	R_G	$(R_G/R_P)R_P$
x coordinate of pitch point	Pinion	10	x_P	$R_G\sin\sigma_P$
Limit pressure angle	Gear	11	ψ_{2L}	$\tan^{-1}\dfrac{R_P}{x_P}$
Hi-side pressure angle (limiting)	Gear	12	ψ_2'	$\geqq \psi_{2L} + 7°$ If greater than 35°, select a larger value of R_G/R_P and recalculate steps 4 through 12
Pressure angle . . . . . Hi-side	Gear	13	ψ_2	Select from Table 5-39. Note: ψ_2 must not be less than ψ_2'
Lo-side			ψ_1	Select from Table 5-39
Working depth	Pinion	14	D_W	$\dfrac{0.6(L/N_P)}{\tan\psi_1 + \tan\psi_2}$
Mean radius	Pinion	15	r_m	$R_P - 0.5D_W$
Mean spiral angle	Pinion	16	λ_m	$\tan^{-1}(L/2\pi r_m)$
Check of lo-side pressure angle	Pinion	17	ψ_1	$\psi_1 = 10°$ when $\lambda_m < 16°$* $\psi_1 = 15°$ when $\lambda_m > 16°$ Correct item 13 if necessary and recalculate items 14, 15, and 16
Number of teeth	Gear	18	N_G	Fig. 5-20†
		19	N_P	N_G/m_G Recalculate m_G in item 2
Hi-point value (pinion axis to top lands of gear teeth)		20	y	$R_P - D_W$

* From 15 to 17°, either pressure angle may be used.
† Do not use for Helicon R_G/R_P values.

Table 5-42. Tooth Proportions for Helicon Gearing (*Continued*)

Item	Member	Item ref.	Symbol	Equation
Clearance	Pinion	21	CLR	$0.07(L/N_P) + 0.002$
Inner diameter	Gear	22	D_{iG}	$0.734D_{oG}$
Gear axis to pinion inner end face	Pinion	23	s_i	$\sqrt{0.135D_{oG}^2 - C^2} + 0.017D_{oG}$
Gear axis to pinion outer end face	Pinion	24	s_o	$\sqrt{0.25D_{oG}^2 - C^2} + 0.033D_{oG}$
Face width	Pinion	25	F	$s_o - s_i$
Whole depth	Pinion	26	h_t	$D_w + CLR$

Specific Instruction for Calculation of Helicon-gear Data in Table 5-42

1. *Outer diameter, gear.* The basic dimension used to establish the capacity of Helicon gearsets is the outer diameter of the gear member. The designer should establish the outer diameter and the ratio before entering Table 5-42.
2. *Gear ratio.* This value is selected on the basis of the service to be performed by the gearset. In general, a single Helicon mesh can have a gear ratio ranging from 3 to 400 to 1. Because of problems in "cutting," ratios less than 7 to 1 are best made by such processes as molding or sintering. The information and procedures given in Table 5-42 are for Helicon gearsets of smaller sizes, suitable for manufacture by the hobbing process. Therefore, caution should be exercised in the design of ratios approaching 7 to 1 or lower.

   Items 2 through 19 should be based on the desired gear ratio, m_G, in a preliminary calculation. After specific numbers of teeth N_G and N_P in items 18 and 19 have been established, the exact ratio, N_G/N_P, item 2, should be calculated and items 3 through 19 modified if necessary.
3. *"Sigma" angle.* Enter Fig. 5-19 for value.
4. *Gear to pinion pitch radius.* Enter Fig. 5-19 for value.
5. *Outside diameter, pinion.* Calculate a "tentative" value. The "final" value, based on the final value of R_G/R_P is a pinion blank dimension.
6. *Pitch radius, pinion.* This is an engineering value used in subsequent calculations.
7. *Center distance.* The final value is a mounting dimension.
8. *Lead.* This is the axial advance of the pinion tooth per revolution.
9. *Pitch radius, gear.* This is an engineering value. $R_G = r_O$. See Fig. 5-18.
10. *x coordinate of pitch point.* This is an engineering value. This is x (trial) in Fig. 5-18.
11. *Limit pressure angle.* A limiting condition of conjugate tooth action governs the maximum pressure angle (called limit pressure angle). This limit pressure angle is calculated, and the hi-side pressure angle established from it.
12. *Hi-side pressure angle (limiting).* Most designs employ a 30° pressure angle on the hi-side unless the limit pressure angle dictates the use of a higher value. 35° is the maximum value in any case. If ψ_2^1 exceeds 35°, a larger R_G/R_P value should be selected in step 7, and items 4 through 12 should be recalculated. A larger value of R_G/R_P will yield a smaller pinion, thereby improving the undercut or limit pressure-angle condition.

13. *Pressure angle, hi-side.* Select the "hi-side" pressure angle from Table 5-39. The value selected must be larger than the limiting hi-side pressure angle $\psi_2{}^1$ from step 12. The pressure angles of the two sides of Helicon teeth differ, usual values being 10° and 30°. The "high" pressure-angle side is usually considered the nondriving side, although it can be used as satisfactorily as the low-pressure-angle side when the pinion blank is adequately mounted.

    Pressure angle, lo-side. Select the "lo-side" pressure angle from Table 5-39. The low-pressure-angle side of the tooth is preferred as the driving side.
14. *Working depth, pinion.* This is the effective height of the tooth.
15. *Mean radius, pinion.* This is an engineering dimension on which the mean spiral angle is based.
16. *Mean spiral angle, pinion.* This is an engineering value on which the "lo-side" pressure angle is based.
17. *Check of lo-side pressure angle.* The pressure angle on the lo-side is limited to values of 10° when the mean spiral angle is smaller than 16° and to a value of 15° in cases where the mean spiral angle is greater than 16°. After calculating the mean spiral angle, item 16, check that the lo-side pressure angle in item 13 is correct. If not, recalculate items 14, 15, 16, and 17.
18. *Number of teeth, gear.* Obtain recommended number from Fig. 5-20. Smaller numbers should be chosen for gearsets in which power transmission is of prime importance. Larger numbers should be used where rotational accuracy is important. Whenever possible, use a hunting tooth in the gear.
19. *Number of teeth, pinion.* The number of teeth in the pinion is a function of the number of teeth in the gear and the ratio.
20. *Hi-point value.* This is the distance from the axis of the pinion to the top lands of the gear teeth. This value, added to the thickness of the gear blank (top lands to mounting surface) gives the mounting distance of the gear.
21. *Clearance.* This is the distance from the working depth circle to the root diameter circle.
22. *Inner diameter, gear.* This diameter is the boundary of the inner ends of the gear teeth.
23. *Gear axis to pinion inner end face.* This is a mounting dimension of the pinion: it is the distance that the pinion teeth should extend toward the inner diameter of the gear in order to be fully effective. It locates the inner end of the face width of the pinion.
24. *Gear axis to pinion outer end face.* This distance may be used as the pinion mounting distance if desired.
25. *Face width.* This is the conventional face width of the pinion.
26. *Whole depth.* This is the cutting depth of the pinion and gear teeth.

REFERENCES

Standards of the American Gear Manufacturers Association and the American Standards Association

1. "Letter Symbols for Gear Engineering," AGMA standard 111.03, ASA standard B6.5-1954.
2. "Gear Nomenclature, Terms, Definitions, and Illustrations." AGMA standard 112.03, ASA standard B6.10-1954.
3. "Gear Cutting Tools, Single Thread Coarse-pitch Hobs," AGMA standard 121.02.
4. "Gear Cutting Tools, Single Thread Fine-pitch Hobs," AGMA standard 122.02.
5. "Gear Cutting Tools, Multiple Thread Coarse-pitch Hobs," AGMA standard 123.01.
6. "Gear Cutting Tools, Worm Gear Hobs," AGMA standard 124.01.
7. "Tooth Proportions for Coarse-pitch Involute Spur Gears," AGMA standard 201.02.
8. "Zerol Bevel Gears," AGMA standard 202.02.
9. "Fine-pitch On-center Face Gears for 20-degree Involute Spur Pinions," AGMA standard 203.01.

10. "20-degree Involute Fine-pitch System for Spur and Helical Gears," AGMA standard 207.04.
11. "Straight Bevel Gears," AGMA standard 208.01.
12. "Spiral Bevel Gears," AGMA standard 209.02.
13. "Design of General Industrial Double-enveloping Wormgears," AGMA standard 342.01.

Other References

14. Barber-Colman: "Hob Handbook," Barber-Colman Co., Rockford, Ill., 1954.
15. Baxter, M. L., Jr.: High Reduction Hypoids, *Machine Design*, Apr. 27, 1961.
16. Berner, G. A.: "Practical Notes for the Watchmaker," Editions Horlogeves Ch. Rohr Bienne (Switzerland), 1948.
17. Buckingham, Earle: "Analytical Mechanics of Gears," McGraw-Hill Book Company, Inc., New York, 1949.
18. Buckingham, Earle: "Spur Gears," McGraw-Hill Book Company, Inc., New York, 1928.
19. Davis, W. O.: "Gears for Small Mechanisms," NAG Press, Ltd., London, 1953.
20. Dudley, D. W.: "Practical Gear Design," McGraw-Hill Book Company, Inc., New York, 1954.
21. Dudley, D. W.: Modifications of Gear Tooth Profiles, *Prod. Eng.*, September, 1949.
22. Dudley, D. W.: Graphical Determination of Gear Contact Ratios, *Prod. Eng.*, March, 1948.
23. Fellows Gear Shaper Co.: "The Internal Gear," Fellows Gear Shaper Co., Springfield, Vt., 1956.
24. Fellows Gear Shaper Co.: "The Involute Curve" and "Involute Gearing," Fellows Gear Shaper Co., Springfield, Vt., 1950.
25. Grant, George B.: "A Treatise on Gear Wheels," 20th ed., Philadelphia Gear Works, Inc., Philadelphia, 1958.
26. Ham, C. W., E. J. Crane, and W. L. Rogers: "Mechanics of Machinery," 4th ed., McGraw-Hill Book Company, Inc., New York, 1958.
27. Illinois Tool Works: "Design of Gear Teeth."
28. Kinsman, F. W.: "Designing Non-standard Spur Gears for Standard Cutting Tools."
29. National Broach and Machine Co.: "Modern Methods of Gear Manufacture," National Broach and Machine Co., Detroit, Mich., 1950.
30. Nelson, W. D.: Spiroid Gearing, *Machine Design*, Feb. 16, Mar. 2, Mar. 16, 1961.
31. Spotts, M. F., and E. C. Varnum: "Contact Ratio of Hobbed Spur Gear Teeth," AGMA 109.11, October, 1957.
32. Tiplitz, C.: "Gear Contact Ratio," *Machine Design*, Nov. 13, 1958.
33. Van Keuren Co.: "Van Keuren Precision Measuring Tools Catalog and Handbook No. 36," The Van Keuren Co., Watertown, Mass., 1955.
34. Vogel, Michigan Tool Co.: "Involutometry and Trigonometry," Michigan Tool Co., Detroit, 1945.

Additional Standards

35. "Design of General Industrial Coarse-pitch Cylindrical Worm Gearing," AGMA standard 341.01.

Chapter 6

CENTER DISTANCE

By

PAUL M. DEAN, Jr., *Gear Technology Engineer, General Engineering Laboratory, General Electric Company, Schenectady, N.Y.*

This chapter deals with the distance between the shafts of meshing gears. This distance is called *center distance* in the case of gearing operating on nonintersecting shafts. Spur and helical gearing, both external and internal, wormgearing, hypoid gearing,* spiral gearing, Spiroid®† gearing, and Helicon®‡ gearing must all operate at specific center distances. Certain types of gears which operate on intersecting shafts, such as bevel gearing, do not have a center-distance dimension. However, their pitch surfaces must be maintained in the correct relationship; hence the axial position of these gears along their shafts is critical. Such gears, therefore, have a *mounting distance* which defines the axial position of the gears. Certain types of gears, such as hypoid gears, Spiroid gears, throated worms and wormgears, face gears, Helicon gears, and Planoid®§ gears, have both a center distance and a mounting distance.

When establishing the center distance for a set of gears, it is customary to first determine the theoretical center distance for the gears. The actual operating center distance is next determined. This center distance includes considerations of tolerance and the interrelationship of the various parts that may be included in the final assembly such as bearings, mounting brackets, and housings.

Since the function of the gear mountings is to maintain the required center distance and/or mounting distance, the gear designer should determine the actual degree to which the mountings will perform their required function. This will enable the designer to determine the amount of backlash or tooth modification that can be incorporated into the design of the teeth of some kinds of gearing to compensate, to some degree, for mounting deficiencies.

This chapter is divided into three sections. The first is a summary of equations useful in the determination of theoretical center distance. The second section covers gearing operating on nonstandard center distances. The third covers detailed considerations of individual elements that make up center distance such as tolerances, differential expansion of gears and their mountings, and general information on mounting distance.

* In the case of hypoid gearing, "offset" is the correct term for the distance between gear and pinion shafts.

† Spiroid is a trade-mark of the Illinois Tool Works, Chicago, Ill.

‡ Helicon is a trade-mark of the Illinois Tool Works, Chicago, Ill.

§ Planoid is a trade-mark of the Illinois Tool Works, Chicago, Ill.

GENERAL EQUATIONS RELATING TO CENTER DISTANCE

This section covers the calculation of the theoretical center distances at which various types of gearing will operate. The most generally used equations are summarized in Table 6-1. In the articles that follow, the special considerations upon which the equations are based are covered in detail.

6-1. Center-distance Equations. Table 6-1 is a summary of equations convenient to use to obtain the values of center distance for the various types of gearing shown in Fig. 6-1. Standard[1,*] letter symbols are used in Table 6-1.

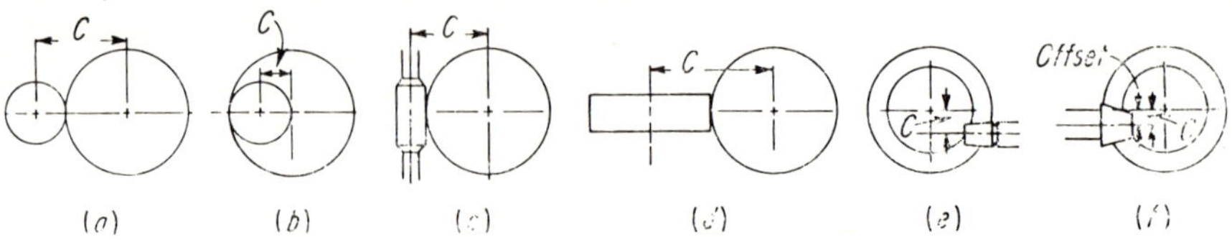

FIG. 6-1. Center distance, shown for various kinds of gears, is the distance between the axes of the shafts measured on a common normal. (*a*) Spur and helical gearing—external. (*b*) Spur and helical gearing—internal. (*c*) Wormgearing—single- and double-enveloping. (*d*) Spiral gearing. (*e*) Spiroid, Planoid, and Helicon gearing. (*f*) Hypoid gearing.

Center distance and tooth thickness are inseparable. In most cases, however, *standard* tooth proportions are used, and the simplified equations of center distance [Eqs. (6-1) through (6-4)] are adequate. *Standard* tooth proportions are defined in Art. 6-2. When the sum of the tooth thickness of pinion and gear does not equal the circular pitch, nonstandard center distances, as obtained from equations such as (6-5) and (6-8), will be required. Most problems involving center distance and tooth thickness fall into one of the following categories:

1. The thickness of the teeth of the gear and of the pinion is fixed. The center distance at which the gears will mesh properly is to be established.
   a. The sum of the tooth thickness of both members plus the design backlash is equal to the circular pitch. In this case, standard center distance is correct. Use Eqs. (6-1) through (6-4).
   b. The sum of the tooth thickness of both members plus the design backlash is *not* equal to the circular pitch. In this case, a nonstandard center distance is required. Use Eqs. (6-5) and (6-8), This problem is covered in greater detail in Art. 6-9.
2. The center distance is fixed. The tooth thicknesses for gears that will operate on the given center distance are to be established. If the center distance for the given diametral pitch and number of teeth is different from that obtained from Eq. (6-2), Eq. (6-5) must be solved and the operating pressure angle thus found used in Eq. (6-8), which is then solved for the term $(t_P + t_G + B)$. In Chap. 5, the ways in which this total tooth thickness can be divided between gear and pinion are discussed.
3. Neither the tooth thickness nor the center distance is fixed; the best values for both are to be established. In this happy case, the tooth thickness is usually established first on the basis of strength or some other appropriate consideration, and the center distance required is then based on Eqs. (6-8) and (6-5).
4. Both the center distance and the tooth thickness are fixed. The amount of backlash or the degree of tooth interference is to be established. This is a frequent "check problem" of an existing gear design. Equations (6-5) and (6-8) are used and solved for B. Minus values of B indicate tooth interference.

* Superscript numbers refer to references at the end of the chapter.

Table 6-1. Center-distance Equations

Standard center distance C

Spur, helical, and wormgears

$$C = \frac{d + D}{2} \tag{6-1}$$

$$C = \frac{n + N^*}{2P} \tag{6-2}$$

Internal gears

$$C = \frac{D - d}{2} \tag{6-3}$$

$$C = \frac{N - n}{2P} \tag{6-4}$$

Operating center distance C'

Spur, helical, and internal gears

$$C' = \frac{C \cos \phi}{\cos \phi'} \tag{6-5}$$

Spur, helical

$$C' = \frac{d' + D'}{2} \tag{6-6}$$

Internal

$$C' = \frac{D' - d'}{2} \tag{6-7}$$

General equations relating tooth thickness and center distance for parallel-axis gearing

External spur and helical gears

$$\text{inv } \phi' = \frac{n(t_P + t_G + B) - \pi d}{(n + N)d} + \text{inv } \phi \tag{6-8}$$

Internal gears

$$\text{inv } \phi' = \frac{\pi d - n(t_P + t_G + B)}{(N - n)d} + \text{inv } \phi \tag{6-9}$$

NOTE: To obtain C' use ϕ' from Eqs. (6-8) or (6-9) in Eq. (6-5).

General equations relating tooth thickness and center distance for nonparallel nonintersecting axis gearing

$$K_a = \frac{n}{N} = \frac{1}{m_G} \tag{6-10}$$

$$K_b = \text{inv } \phi_G + K_a \text{ inv } \phi_P - \frac{\pi(p_n - t_{nP} - t_{nG})}{N p_n} \tag{6-11}$$

$$K_d = \frac{\text{inv } \phi_G}{\text{inv } \phi_P} \tag{6-12}$$

$$K_e = K_a + K_d \tag{6-13}$$

$$\text{inv } \Phi_P = \frac{K_b}{K_e} \tag{6-14}$$

$$\text{inv } \Phi_G = K_b - K_a \text{ inv } \Phi_P \tag{6-15}$$

Check equations: To check equation series (6-10) through (6-15) above.

$$K_d' = \frac{\text{inv } \Phi_G}{\text{inv } \Phi_P} \tag{6-16}$$

$$K_d' = K_d \qquad \text{(check equation)} \tag{6-17}$$

$$D_P' = \frac{D_{bP}}{\cos \Phi_P} \tag{6-18}$$

$$D_G' = \frac{D_{bG}}{\cos \Phi_G} \tag{6-19}$$

* Does not apply to wormgearing.

Table 6-1. Center-distance Equations (*Continued*)

Use values from Eqs. (6-18) and (6-19) in Eq. (6-6) where

$$K_c = \frac{\sin \phi_G}{\sin \phi_P} = \frac{\sin \Phi_G}{\sin \Phi_P} \tag{6-20}$$

P = transverse diametral pitch
C = "standard" center distance [defined by Eqs. (6-1) and (6-2)]. See also Art. 6-2
C' = operating center distance
d = "standard" (n/P) pitch diameter of pinion. See also Art. 6-3
d' = operating pitch diameter of pinion. See also Art. 6-4
D = "standard" (N/P) pitch diameter of gear. See also Art. 6-3
D' = operating pitch diameter of gear. See also Art. 6-4
n = number of teeth in pinion
N = number of teeth in gear
ϕ = cutting pressure angle
B = design backlash, the total amount the teeth in both members are thinned for backlash considerations
ϕ' = operating pressure angle. See also Art. 6-5
t_P = circular (transverse) thickness of pinion tooth
t_G = circular (transverse) thickness of gear tooth
t_{nP} = circular (normal) thickness of pinion tooth
t_{nG} = circular (normal) thickness of gear tooth
ψ_P = helix angle of pinion
ψ_G = helix angle of gear
Φ_P = rolling pressure angle of pinion
Φ_G = rolling pressure angle of gear

6-2. Standard Center Distance. Most gear designs are based on *standard* tooth proportions. Such gears are intended to mesh on *standard* center distances. The equations which establish *standard* center distances are based on the following assumptions:

1. The sum of the circular tooth thicknesses (effective) equals the circular pitch minus the backlash.

$$t_P + t_G = p_t - B \tag{6-21}$$

This rule covers the case of long and short addendums in which the gear teeth are corrected for such things as undercut or balanced strength. In such cases, the tooth thickness of both members is altered sufficiently to permit the pair to mesh on standard center distances.

2. The tooth thickness of an individual gear is one-half the circular pitch (transverse) minus one-half the backlash.

$$t = \frac{p - B}{2} \tag{6-22}$$

This covers the more common case. It represents the approach used by most makers of "catalogue"-type gears. Such gears are all expected to operate on standard center distances regardless of the number of teeth in the pinion. The center distances on which such gears will operate may be calculated by means of Eqs. (6-1) and (6-2).

The tolerance on *standard* center distance can be bilateral (thus 5.000 ± 0.002″) if the value of B was chosen sufficiently large. This is the most convenient way from the standpoint of manufacturer. If B is not large enough, a unilateral tolerance (thus 5.000″ + 0.004″ − 000″) may be used.

6-3. Standard Pitch Diameters. The *standard* pitch diameter of a gear is a dimension of a theoretical circle. It is given for each type of gear by the following relations:

Spur gearing	$\frac{N}{P}$
Helical gearing	$\frac{N}{P}$
Bevel gearing	$\frac{N}{P}$
Wormgear	$\frac{Np}{\pi}$

Since the diametral pitch of a gear is fixed by the tool which is used to cut it (hob, shaper cutter, shaving cutter, etc.), and since the number of teeth in the gear is a whole number, the *standard* pitch diameter is an imaginary circle. It can have no tolerance, regardless of the variations in tooth thickness or in the center distance on which it is to operate.

Standard pitch diameter actually has little meaning; it is retained on gear drawings by convention, rather than by any engineering necessity.

6-4. Operating Pitch Diameters. As is shown elsewhere, it is entirely practical to operate involute gears of a specific diametral pitch and numbers of teeth at various center distances. It is convenient, in such cases, to calculate *operating* pitch diameters for such gears. See Eqs. (6-23), (6-24), (6-25), and (6-26).

External spur and helical pinions

$$d' = \frac{2C'}{m_G + 1} \tag{6-23}$$

Internal spur and helical pinions

$$d' = \frac{2C'}{m_G - 1} \tag{6-24}$$

External spur and helical gears

$$D' = \frac{2C'm_G}{m_G + 1} \tag{6-25}$$

Internal spur and helical gears

$$D' = \frac{2C'm_G}{m_G - 1} \tag{6-26}$$

where C' = operating center distance

m_G = gear ratio $\left(\frac{N}{n}\right)$

d' = operating pitch diameter of pinion

D' = operating pitch diameter of gear

These equations define the operating pitch diameters as being proportional to the transmitted ratio and the instantaneous center distance.

Since gears with thicker than standard teeth must operate on enlarged center distances, they will run on operating pitch diameters which are larger than their standard pitch diameters. The operating pitch diameter should not be specified as a drawing dimension on the detail drawing of the gear, since it will vary for every different gear and center distance.

6-5. Operating Pressure Angle. The pressure angle of an individual gear is based on the diameter of the base circle of the gear and upon the identification of the specific radius at which the pressure angle is to be considered. In standard gears, this radius is customarily the standard pitch radius. It is convenient to consider the pressure angle of gears operating on nonstandard center distances at the point of intersection of the line of action and line of centers. This is the definition of the operating pressure angle which may be calculated by means of Eq. (6-27)

$$\cos \phi' = \frac{C \cos \phi}{C'} \tag{6-27}$$

where ϕ' = operating pressure angle
C' = operating center distance
C = standard center distance [see Eq. (6-2)]
ϕ = standard pressure angle

It can be seen from the foregoing that a gear can be cut with a cutter of one pressure angle and operated at a different pressure angle. This flexibility causes much of the confusion in gear design. It is necessary to define accurately each of the standard elements of a nonstandard gear—pitch and pressure angle. A specification of base-circle diameter is highly desirable.

6-6. Operating Center Distance. The actual center distance at which a gear will operate will have a large influence on the way in which the gear will perform in service. The actual operating center distance is made up of the combined effects of manufacturing tolerances, the basic center distance, differential expansion between the gears and their mountings, and deflections in the mountings due to service loads.

The items that should be considered when determining the minimum and maximum operating center distance for any given gear design are discussed in detail in Arts. 6-9 and 6-10.

In any critical evaluation of a gear design, particularly in the field of control gearing, the minimum and the maximum operating center distance should be used in equations covering backlash, contact ratio, tooth-tip clearance, etc.

In this chapter the concept *operating center distance* is used in two ways. In one case, it is considered to be the center distance that results from the build-up of all tolerances that influence center distance in any one case. This concept is illustrated in case II in Art. 6-10. Thus it is the largest or smallest actual center distance that could be encountered in a given design. In the other case, the one covered by Eqs. (6-5) and (6-8), for example, it is the center distance at which gears of a specified tooth thickness and backlash will operate. Depending upon the problem encountered, the correct concept to use will have to be selected by the designer.

6-7. Center Distance for Gears Operating on Nonparallel Nonintersecting Shafts. The most frequently encountered example of this type of gear operation is in the shaving or lapping of gears. Here a cutter, usually a helical-toothed member, is run in tight mesh with a spur or helical gear. The operating center distance is dependent upon the tooth thickness of each member. Equations (6-10) through (6-19) should be solved in sequence, and the results used in Eq. (6-5) to obtain the operating center distance. Equation (6-20) gives the ratio constant for the transverse pressure angles for crossed-axis gearing.

6-8. Center Distance for Wormgearing. The center distance for wormgearing is based on the sum of the *standard* pitch diameters of the worm and wormgear [Eq. (6-28)]. The *standard* pitch diameter of the worm is obtained from Eq. (6-29). This circle has no kinematic significance but is the basis for worm-tooth proportions.

$$C = \frac{d_w + D}{2} \tag{6-28}$$

$$d_w = \frac{L}{\pi} \tan \psi \tag{6-29}$$

where d_w = pitch diameter of worm
C = center distance
D = pitch diameter of gear
L = lead
ψ = lead angle

CALCULATION OF GEARS TO OPERATE ON NONSTANDARD CENTER DISTANCES

On some occasions a set of gears must operate on a center distance which is not one-half the sum of the *standard* pitch diameter of the meshing gears. The designer is confronted with *nonstandard* center distance in several situations, the more important of which are

1. Gear trains in which the teeth are made to *standard* tooth thickness and backlash is introduced by increasing the standard center distance slightly.
2. Gear trains in which the number of teeth and pressure-angle relationship requires a long addendum (enlarged-tooth-thickness design) to avoid undercut and yet the number of teeth in the gear is so small that to make the gear sufficiently short-addendum to compensate for the pinion enlargement would cause undercut. In such cases, an enlarged center distance is usually indicated.
3. Gear trains in which the sum of the tooth thicknesses of pinion and gear is not equal to the circular pitch for reasons of tooth strength, wear, or scoring.
4. Gear trains in which a minor change in ratio (total number of teeth in mesh) has been made without a change in center distance.

In each of these cases, the calculation of operating center distance is performed using Eqs. (6-5) and (6-8) or (6-9).

6-9. Nonstandard Center Distance. By proper adjustments of the thickness of the teeth on each gear of a meshing pair, it is possible to achieve gear designs that will meet most *nonstandard* center distances. In Art. 5-3, the limitations governing tooth thickness are outlined. In cases where maximum tooth strength is not critical, it will be found that gears not exceeding these limitations will satisfy most nonstandard-center-distance problems encountered. This is discussed more fully in method 3 below.

Sometimes designers who do not realize the possibilities of gear design will select such things as fractional diametral pitches (10.9652, etc.) in order to make gears fit a nonstandard center distance. This is poor practice in that it necessitates special tooling, which may be costly. It is better to attempt to utilize standard-pitch tooling and, by proper adjustments of tooth thickness, to establish gear designs that will meet the nonstandard distances.

In cases wherein the center distance is *nonstandard,* the designer has four possible methods of designing gears to meet the given center distance.

1. A number of teeth in gear and pinion different from the numbers originally selected may be found which will more nearly meet the given center distance. A *Brocot* table is very useful in finding the numbers of teeth that will give different gear ratios ($1/m_G$). In this case, it is assumed that any of several ratios may be "good enough." If not, the following alternatives, which all assume that a specific ratio must be obtained, may be considered.
2. Equations (6-1), (6-2), (6-3), or (6-4) may be rearranged to solve for P_d. Thus a diametral pitch which will meet the gear ratio and center distance is found. This method will yield nonstandard diametral pitches for spur gears, which

will almost always lead to large manufacturing costs because of the need of nonstandard tooling, cutters, and master gears. This method should be used only if methods 1, 3, and 4 do not meet the needs of the application.

3. The gear and pinion can be made with teeth which are thicker or thinner than standard. The easiest approach when using this method is to follow method 2 above, and then use the results as the basis for a selection of the nearest *standard* diametral pitch to the one found. Equation (6-5) is then used, which will yield the operating pressure angle for the gears based on the new selection of diametral pitch and the desired center distance C'. The value of C in this equation is determined from Eq. (6-1) or (6-2). Equation (6-8) is then used and solved for the term $(t_P + t_G)$. The selection of values for each term, t_P and t_G, should be based somewhat on the ratios of their number of teeth. In general, pinions cannot be changed from standard as much as gears with large numbers of teeth. The selected value of tooth thickness should be checked for undercut in the event that t_P is less than $p/2$. If the thickness is appreciably greater than $p/2$ the pinion should be checked for pointed teeth. The same should be done for the gear, especially if the changed tooth thickness was not based on the gear ratio. Once the values of C', P, n, N, t_P, and t_G have been established, the remainder of the tooth proportions can be calculated from these values (see Chap. 5). If the addendum and whole depth are properly adjusted from the tooth thickness, standard tools can be used to cut and to inspect these gears.
4. If helical gears can be used, the lead angle can be adjusted to obtain the required tooth thickness–center distance relationship. In effect, the designer is following method 3 above but is adjusting the transverse tooth thickness by the selection of the required lead angle. The procedure in this case is to use Eq. (6-2) or (6-4) to establish the diametral pitch required in the transverse plane P to suit the given center distance. The nearest finer (smaller) diametral pitch for which tools are available is selected (P_n), and these values are used in Eq. (6-30) to establish the required lead angle.

$$\cos \psi = \frac{P}{P_n} \tag{6-30}$$

Although not an essential part of this problem, somewhat smoother-running gears will result if the designer can manage to pick diametral pitches which will result in a helical overlap of at least 2 (see Chap. 5). After the helix angle has been found, the designer can determine the remaining tooth proportions from data found in Chap. 5. This method can be used economically only if the hobbing process of cutting gears is to be used. Special guides and shaper cutters are required for every different lead of helical gear for gears cut by the shaping process. This then fixes limits on the values of helix angle that any given shop could cut.

ELEMENTS OF CENTER DISTANCE

Up to this point, this chapter has considered only the means of determining the theoretical center distance that is required by a set of gears. The theoretical center distance is either the *basic* center distance, as established by Eqs. (6-1), (6-2), (6-3), or (6-4), or the theoretical *tight-mesh* center distance, as established by Eqs. (6-5) through (6-20). In an actual gearset, however, the gears will be forced to operate on a real center distance which may be larger or smaller than the theoretical center distance by amounts which are based on the way that the part tolerances (bearings, casing, bores, etc.) will add up.

6-10. Effects of Tolerances on Center Distance. In all but the very simplest gear-casing designs, there are many tolerances that will govern the operating position of the gear and pinion shafts. Usually power gearing is not designed to small values of backlash. It is customary, therefore, to apply a generous amount of backlash to the gear teeth, a value which has been found by experience sufficient to prevent binding of the teeth. In control gearing, however, particularly in those applications which are operated in both directions and are intended to have a minimum of lost motion, a critical study of the effects of part tolerances is usually required. The tooth thickness can then be made as large as possible, ensuring a minimum of backlash with little risk of binding. The following two examples indicate a very simple case and a more complex case of center-distance calculations.

Case I

In Fig. 6-2 is shown a spur-gear pair mounted in a cast-iron gear case. The loads and speeds are such that the bearings are simply smooth bores in the cast iron with shafts running in them. Figure 6-3 is a vector diagram showing the forces produced on the bearings

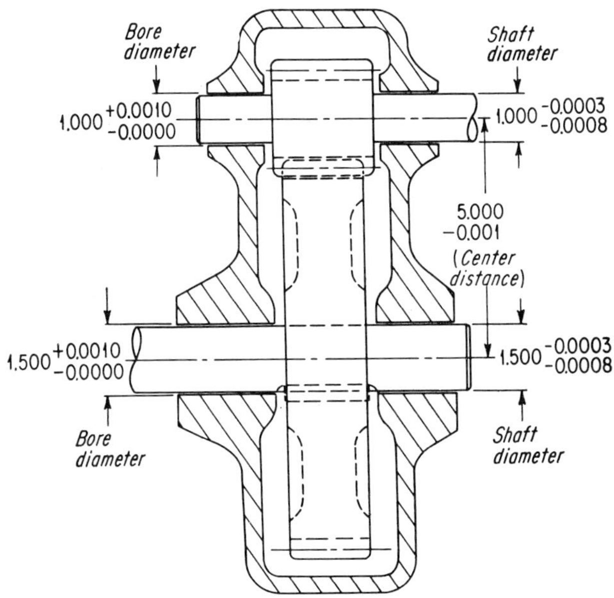

FIG. 6-2. Spur or helical gears operated in bearings bored in the gear casing exhibit the simplest case of tolerance build-up on center distance.

as a result of the gear-tooth loads. Figure 6-4 shows the displacements of the shafts in the bearings resulting from these loads.

In this example, it is desired to establish the largest and smallest center distance that the gears will ever experience in order to determine the range of backlash existing in this gearset. Table 6-2 outlines the calculations made to establish the various center distances at which the gears may be expected to operate in service.

Part A of Table 6-2 shows the calculation of the maximum distance that the axis of either shaft can move away from the axis of its bore. This is given as the eccentricity of the shaft within the bearing. It is assumed that the bearing is dry. For a very rough approximation of maximum and minimum possible center distances, these values of eccentricity can be added to and subtracted from the boring center distance plus its tolerance. This is shown in part B of Table 6-2. This assumes the forces acting on the shafts are tooth-load reactions which act along the line of centers. In the case of journal bearings which are properly lubricated, the calculated value of eccentricity of operation can be used. This

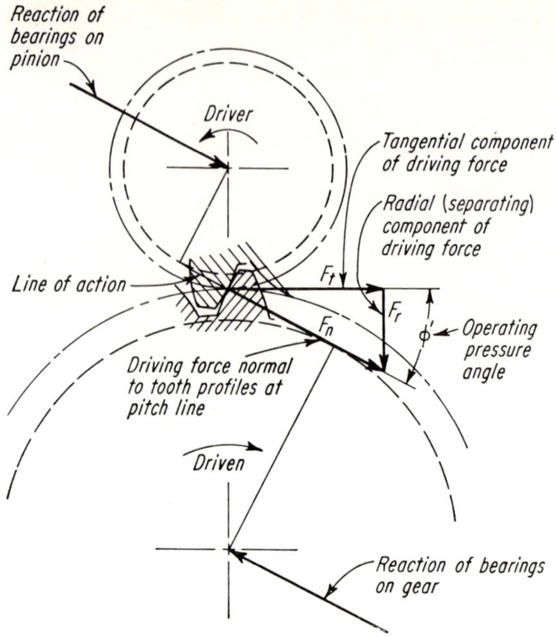

FIG. 6-3. The reaction forces developed by the bearings to overcome the gear-teeth driving forces are parallel to the line of action.

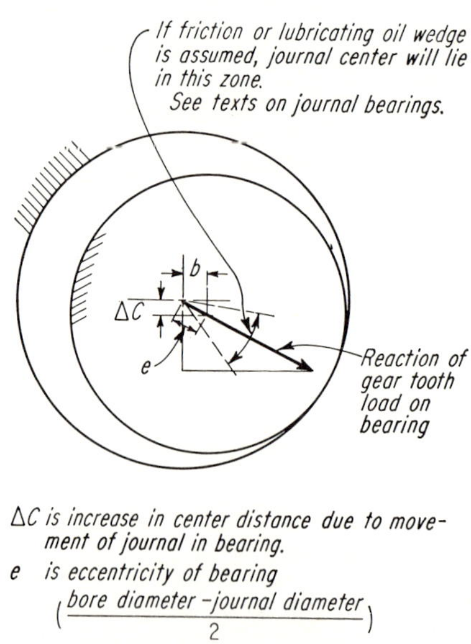

FIG. 6-4. The position occupied by the journal in a bearing is a function of the gear-tooth loads along the shaft and the effects of friction or of the oil wedge in the bearing.

calculation is discussed in texts on journal-bearing design. In most cases, if the designer can select the tolerances for the gear-tooth thickness and also the tolerances on the various parts of the casing and bearings that control the center distance, such that the backlash is not excessive when the calculations shown thus far are carried out, no further calculations need be made. However, when the design is to achieve an absolute minimum of lost motion, the more accurate evaluation of center distance should be carried out.

Part C shows a calculation assuming that the gear-tooth reactions will control the position of the journal in the bearing. The minimum center distance will occur when the minimum boring center distance is achieved and the minimum bearing clearance occurs. The tooth loads tend to increase the center distance in this example so that distances less than the minimum boring center distance cannot occur.

Table 6-2. Summary of Calculations for Case I

A. Calculation of Bearing Clearance and Shaft Eccentricity. See Fig. 6-2 for drawing dimensions

Bore for pinion shaft	Max. 1.0010	Min. 1.0000
Pinion-shaft diameter	Min. 0.9992	Max. 0.9997
Clearance in bearing (c)	Max. 0.0018	Min. 0.0003
Eccentricity of shaft position in bearing ($c/2$)	Max. 0.0009	Min. 0.00015
Bore for gear shaft	Max. 1.5010	Min. 1.5000
Gear-shaft diameter	Min. 1.4992	Max. 1.4997
Clearance in bearing (c)	Max. 0.0018	Min. 0.0003
Eccentricity of shaft position in bearing ($c/2$)	Max. 0.0009	Min. 0.00015

B. Calculation of Theoretical Maximum and Minimum Center Distance (assuming reaction forces acting along line of centers; values from above)

Center distance of bores (machining tolerance)	Max. 5.0010	Min. 5.000
Eccentricity of pinion bearing	0.0009	−0.0009
Eccentricity of gear-bearing center distance	0.0009	−0.0009
	Max. 5.0028	Min. 4.9982

C. Calculation of Theoretical Maximum and Minimum Center Distance (assuming reaction forces acting along line of action; values from above, part B)

Center distance of bores (machining tolerance)	Max. 5.0010	Min. 5.0000
Eccentricity, radial component of pinion bearing		
(0.0009) sin 20°	0.0003	
(0.00015) sin 20°	......	0.00005
Eccentricity, radial component of gear bearing		
(0.0009) sin 20°	0.0003	
(0.00015) sin 20°	......	0.00005
	Max. 5.0016	Min. 5.0001

In actual service, the maximum and minimum operating center distances will rarely reach these limits. In the case of the maximum center distance, a properly lubricated shaft will develop an oil wedge which will keep the journal away from the walls of the bore. Secondly, if the only separating loads on the shafts are those produced by gear-tooth reaction, the radial movement of the shaft along the line of centers will be e (sin ϕ). The proper evaluation of operating center distance will depend upon whether the application is intermittent in rotation and upon the direction of rotation.

Case II

A more complex case occurs when the pinion and gear shafts are mounted in separate units. The following example illustrates a case in which a motor having a pinion cut integral with the shaft is mounted by means of a rabbet in a gear case. This type of construction is often found in power hand tools and in aircraft actuators. This example also illustrates a critical evaluation of antifriction bearings. It is desired to determine the largest and smallest center distance that can occur in order to establish the maximum and minimum tooth thickness and the amounts of backlash that result (see Fig. 6-5).

Table 6-3, which summarizes the calculations, shows a total of 16 different elements which combine to control the maximum and minimum actual center distance. Since there

Table 6-3. Summary of Calculations for Case II*

Item discussed in footnotes	Surface designation (see Fig. 6-5) (1)	Tolerance (clearance), in. (2)	Equivalent change in center distance (3)	Min. center distance col. $(3)^c \times 10^{-6}$ (4a)	(4b)	Max. center distance col. $(3)^2 \times 10^{-6}$ (5a)	(5b)
1. Fit of motor shaft in inner race of ball bearing[a]	(1)	0.0000	0.0000	−0.00	−0.00	+0.00	+0.00
2. Eccentricity of inner ring of bearing[b]	(1-2)	0.0002	0.0001	−0.01	−0.01	+0.01	+0.01
3. Radial clearance in bearing[c]	(2-3)	0.0004–0.0000	0.0002	−0.04	+0.04	−0.04	+0.04
4. Eccentricity of outer ring of bearing[d]	(3-4)	0.0004	0.0002	−0.04	−0.04	+0.04	+0.04
5. Fit of outer race of bearing into motor end shield[e]	(4)	0.0008–0.0000	0.0004–0.0000	−0.16	−0.16	+0.16	+0.16
6. Concentricity of axis of bore with axis of rabbet on end shield[f]	(4-5)	0.0020	0.0010	−1.00	−1.00	+1.00	+1.00
7. Clearance between rabbet of motor end shield and bore of actuator end plate[g]	(5)	0.0025	0.00125	−1.56	−1.56	+1.56	+1.56
8. Concentricity of motor bore in actuator end plate and rabbet[h]	(5-6)	0.0010	0.0005	−0.25	−0.25	+0.25	+0.25
9. Distance between axis of gear-shaft bore and rabbet in actuator housing[i]	(6-7)	0.0020	0.0010	−1.00	−1.00	+1.00	+1.00
10. Clearance between bore for bearing and bearing outer race[e]	(7)	0.0008	0.0004	−0.16	−0.16	−0.16	+0.16
11. Eccentricity, outer race of bearing[d]	(7-8)	0.0004	0.0002	−0.04	−0.04	+0.04	+0.04
12. Radial clearance in bearing[c]	(8-9)	0.0004–0.0000	0.0002–0.0000	−0.04	+0.04	+0.04	+0.04
13. Eccentricity, inner race of bearing[b]	(9-10)	0.0002	0.0001	−0.01	−0.01	+0.01	+0.01
14. Fit of gear shaft in inner race of bearing[a]	(10)	0.0000	0.0000	−0.00	−0.00	+0.00	+0.00
15. Eccentricity, bearing journal and surface locating gear[j]	(10-11)	0.0002	0.0001	−0.01	−0.01	+0.01	+0.01
16. Fit gear on shaft	(11)	0.0002	0.0001	−0.01	−0.01	+0.01	+0.01
			Totals	−4.33	−4.17	+4.17	+4.33
			Probable values	$\sqrt{4.17} = -0.0020''$		$\sqrt{4.33} = +0.0021''$	

* Footnotes are at bottom of p. 6-13.

are so many tolerances, the use of the root mean square of these values is suggested as a reasonable approximation to operating conditions, and it is used here to establish the maximum and minimum probable center distance. This method gives over 99½ per cent assurance that the values shown will not be exceeded.

6-11. Machine Elements That Require Consideration in Critical Center-distance Applications. As was illustrated in cases I and II, the operating center distance for a pair of gears is made up of several elements, each of which contributes to the over-all center distance. The accumulation of tolerances on each of these elements must be considered in applications in which a minimum of backlash is to be established. The following is a consideration of the elements which have the largest contribution to variations in center distance:

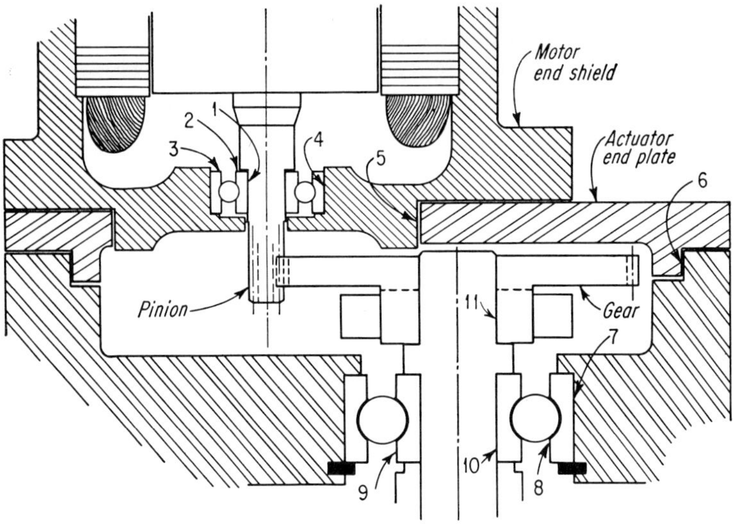

FIG. 6-5. Cross section of the first stage of gearing in an aircraft-type actuator. The numbers refer to the surfaces discussed in Table 6-3.

1. *Rolling-element bearings.* Ball and roller bearings consist of three major elements that contribute to backlash and to changes in center distance. Some of these are illustrated in case II. The outer race has machining tolerances that cause eccentricity of the axes of the inner raceway and the outer bore. Depend-

---

NOTE: The maximum values of center distance (Columns 4b and 5b) are based on the assumption that the separating forces between pinion and gear act to hold the shafts at their maximum separation within the limits of bearing clearance.

The minimum values of center distance (Columns 4a and 5a) are based on the assumption that the external forces on gear and pinion act to hold the shafts at their minimum separation within the limits of bearing clearance.

The minimum center distance (Column 4) is based on the assumption that all parts, made to their maximum clearances, are assembled to achieve the minimum possible center distance. The maximum center distance (Column 5) is based on the assumption that all parts, made to their maximum clearances, are assembled to achieve the maximum possible center distance.

Column (1) shows the surface designated on Fig. 6-5 by a number in a circle; (2) shows the tolerance (usually on a diameter); (3) shows the amount this tolerance can contribute to center distance ($\pm$); (4) shows the square of the tolerance contributing to the minimum center distance; and (5) shows the square of the tolerance contributing to the maximum center distance. The values in columns (4) and (5) are added algebraically and their square roots are shown at the foot of the column.

[a] Most bearing catalogues indicate bearing and shaft tolerances which will produce an interference fit. This fit cannot, therefore, contribute to the maximum or minimum center distance at which the gears will operate.

[b] Most bearing catalogues indicate a tolerance on runout between the bore and the raceway of the inner ring. This eccentricity contributes to the instantaneous center distance when the inner face is the moving portion of the bearing.

[c] Most bearings include a certain radial clearance between the inner and the outer race of a bearing. This clearance usually acts to allow an increase in the center distance due to the reaction between the two meshing gears.

[d] Most bearing catalogues indicate a tolerance on runout between the raceway and the outer diameter of the outer ring. This eccentricity contributes to the center distance at which the gears will operate. This value will change during running since the outer race is often permitted to creep.

[e] Most bearing catalogues indicate a bore tolerance on the hole that the bearing is fitted in. Usually the tolerance will provide clearance between the bearing and housing. When clearance exists, the gear reaction acts to increase the center distance to the limit allowed by the clearance.

[f] Many manufacturers of pilot-mounted motors specify a value of runout between the motor shaft and the pilot surface. It is customary to measure this by holding the motor shaft fixed (as between centers) and to rotate the motor about its shaft. A dial indicator running on this surface will show the runout (twice eccentricity).

[g] A clearance fit is usually specified on the hole into which the pilot of the motor is to fit. This clearance allows the motor to be shifted somewhat and, as a result, has a direct effect on center distance. The value shown assumes a tolerance of $\begin{smallmatrix}+0.0005''\\-0.0000''\end{smallmatrix}$ on the bore and a tolerance of $\begin{smallmatrix}+0.000''\\-0.002''\end{smallmatrix}$ on the diameter of the pilot on the motor.

[h] In an assembly, such as shown in this example, a tolerance must be placed on the location of the bore to receive the motor relative to the rabbet on the end plate.

[i] This, in effect, is the same type of tolerance as discussed under *g* above.

[j] Usual drafting practice is to specify a tolerance on the concentricity of the several cylindrical surfaces of a shaft. If these are all machined at one "setup," they will have little or no eccentricity relative to each other. In some cases, they must be machined in separate "setups."

ing upon how this member is installed, the center distance established by the bores in which this element is fitted will be increased or decreased. It is customary to let this element creep; thus the eccentricity will go through all positions. There is also a tolerance on the outer diameter of this member. This tolerance, plus the tolerance on the bore in which this element is fitted, can cause varying degrees of looseness and therefore changes in operating center distance.

The inner race has machining tolerances that cause eccentricity in the same manner as in the outer race member. In cases where the shaft rotates, it is customary to use a light interference fit between the inner race and the shaft. In such cases, the center of the gear will move in a path about the center of the ball or roller path of the bearing. The eccentricity of the gear will be that of the inner race of the bearing. This will cause a once per revolution change in center distance.

The clearance in the bearing is controlled by the selection of the diameter of the balls. This clearance allows greater or lesser changes in center distance of the gears supported by the bearings.

2. *Journal bearings.* Journal bearings can change the center distance at which the gear was intended to operate by an amount which is a function of the clearance designed into the bearing, its direction of rotation, the speed of rotation, and the lubricant used. This change from the distance actually established by the distance between the centers of the bores is due to the oil wedge developed. Handbooks give convenient methods of evaluating the eccentricity that can be expected of a given bearing design. When necessary, these refinements can be introduced into the calculations illustrated by cases I and II.
3. *Sleeve bearings.* The term "sleeve bearing" is used here to identify the journal-type bearing in which a sleeve of some bearing material is pushed into the gear casing or frames. The tolerance on the eccentricity of the outside diameter and the inside bearing surface is the element to consider in this type of bearing. In some cases, the clearance between the sleeve and the shaft is affected because of the interference fit between the sleeve and the bore in the casing.
4. *Casing bores.* The distance between bearing bores in a gear casing can be made up of several elements. In the simplest case, a single frame or casing has two holes drilled at a specified center distance. The tolerance is selected to give the machine operator the necessary working allowance (see Fig. 6-2). This tolerance has a direct effect on the operating center distance of the gears.

   In some designs, the casing is made up of several parts bolted together. One part may carry the bore for the gear shaft, and another may carry the bore for the pinion shaft. The center distance is then made up of a series of parts each having dimensions and tolerances which can add up in various ways to give maximum and minimun center distance. Case II illustrates this type of gearing.

6-12. Control of Backlash. In cases wherein it is necessary to control the amount of backlash introduced by the mountings, two courses of action may be employed. In the first, the center distance may be made adjustable. That is, provisions may be made so that, at assembly, the centers may be moved until the desired mesh is obtained. This entails a method of moving the parts through very small distances, and then being able to fix them securely when the desired distance has been reached. Provision must also be made for the assemblyman to see what he is doing when adjusting the mesh. In this approach, the difference between the smallest and the largest backlash in the adjusted mesh is only the effect of total composite error (run-out) in both meshing parts. Size tolerance on the teeth can be generous since this is one of the tolerances adjusted out of the mesh by the assemblyman.

After the center-distance adjustment is completed, the parts may be drilled and doweled, although this causes difficulties if replacement gears are ever to be used on these centers.

In the second approach, very close tolerances are held both in the mountings (boring center distance, bearings, etc.) and on the tooth thickness of the gears. This approach is used whenever interchangeability is specified and in mass production in which assembly costs are to be minimized.

6-13. Effects of Temperature on Center Distance. Many gear designs consist of gears made of one material operating in mountings made of another material having a markedly different coefficient of expansion. Since it is customary to manufacture and assemble gears at room temperature of about 68°F, an analytical check of the effective center distance present in the gears at their extremes of operating temperature is desirable.

Power gears may heat up because of the frictional losses in the mesh and may achieve an operating temperature above that of their casing. If made of the same materials as their mountings, such gears would have a larger apparent* pitch diameter relative to the center distance and would therefore have a smaller effective center distance at their running temperature. Some gears, particularly those in military applications, may be subjected to extreme cold for periods of time. Often these are steel gears operating in aluminum or magnesium casings. In this case, the center distance will shrink to a greater degree than the apparent pitch diameter of the gears, and the resulting backlash will be less than at room temperature.

In evaluating the effective center distance of a mesh at an operating condition other than room temperature, the temperature of the gear blanks as well as that of the mountings, which may be considerably different, must be considered as well as the coefficients of expansion of the gear and mounting materials. There are six operating conditions in which temperature can have an important effect on gear performances:

1. Gears operating at temperatures *higher* than their mountings:
   a. Operating temperatures *lower* than assembly-room temperature
   b. Operating temperatures *higher* than assembly-room temperatures
2. Gears operating at temperatures *lower* than their mountings:
   a. Operating temperatures *lower* than assembly-room temperatures
   b. Operating temperatures *higher* than assembly-room temperatures
3. Gears operating at temperatures essentially the *same* as their mountings:
   a. Operating temperatures *lower* than assembly-room temperatures
   b. Operating temperatures *higher* than assembly-room temperatures

Equations (6-31) and (6-32) evaluate each of these possibilities and establish the most critical extremes at which minimum or maximum effective center distance will occur.

The significance of these possibilities can be appreciated by considering the following service conditions. The first case is typical of power gearing under steady-state conditions. Condition 1*a* is a possibility at start-up under a cold operating environment. The second case is usually a transitory condition where the gears may have to perform adjacent to an external source of heat. The third condition is most typical of control gears which do not transmit enough energy to be at a temperature greatly different from their mountings.

* Apparent pitch diameter is the diameter at which a given value of tooth thickness is found.

The procedure recommended here is to calculate a minimum and a maximum effective center distance. These are based on the extremes of operating temperatures. From these values, the designer can calculate either a value of tooth thickness that will give the necessary operating backlash under the tightest mesh conditions, or the amounts of backlash that will be found in a given gearset.

Equations (6-31) and (6-32) give minimum and maximum effective center distances.

$$C_{e\,\text{min.}} = C_d + C_t' + \Delta C_{T_T} \tag{6-31}$$

$$C_{e\,\text{max.}} = C_d + C_t'' + \Delta C_{T_L} \tag{6-32}$$

where $C_{e\,\text{min.}}$ = minimum effective center distance that occurs under temperature extreme

$C_{e\,\text{max.}}$ = maximum effective center distance that occurs under temperature extreme

C_d = basic or nominal center distance

C_t' = minimum tolerance on center distance

C_t'' = maximum tolerance on center distance

ΔC_{T_L} = change in center distance; see note following Eqs. (6-33) and (6-34)

ΔC_{T_T} = change in center distance; see note following Eqs. (6-33) and (6-34)

Equations (6-33) and (6-34) indicate the amount that the center distance (effective) will change because the gears and their mountings shrink or expand at different rates.

$$\Delta C_T' = \frac{D}{2}(\Delta T_M' K_M - \Delta T_G' K_G) + \frac{d}{2}(\Delta T_M' K_M - \Delta T_G' K_P) \tag{6-33}$$

$$\Delta C_T'' = \frac{D}{2}(\Delta T_M'' K_M - \Delta T_G'' K_G) + \frac{d}{2}(\Delta T_M'' K_M - \Delta T_G'' K_P) \tag{6-34}$$

NOTE: Compare $\Delta C_T'$ and $\Delta C_T''$; assign the smallest positive number or the largest negative number to ΔC_{T_T}. Assign the largest positive number or the smallest negative number to ΔC_{T_L}. If signs are opposite, use this negative value for ΔC_{T_T} and use the positive value for ΔC_{T_L}. These values are used in Eqs. (6-31) and (6-32) to obtain effective center distance.

$$\Delta T_M' = T_M - T_R$$
$$\Delta T_M'' = T_M' - T_R$$
$$\Delta T_G' = T_G - T_R$$
$$\Delta T_G'' = T_G' - T_R$$

where T_R = assembly-room temperature (68°F)

T_M = minimum mounting temperature (operating)

T_M' = maximum mounting temperature (operating)

T_G = minimum gearing temperature (operating)

T_G' = maximum gearing temperature (operating)

K_P = coefficient of expansion of pinion material, in. per in. per °F

K_G = coefficient of expansion of gear material

K_M = coefficient of expansion of mounting material

6-14. Mounting Distance.* Bevel gearing, wormgearing, face gearing, and Spiroid gearing must be given close control on axial positioning (mounting distance) if good performance is to be achieved (see Fig. 6-6). Most of the elements that must be given control to achieve proper center distance must also be given close control to achieve proper mounting distance. If any of the above types of gearing can move along the axis of its shaft relative to the mating gear, the teeth will either bind or, in the opposite case, have excessive backlash.

* See Chap. 12 for general information on gear-mounting tolerances and practices.

Straight-toothed bevel gears are critical in respect to axial position, since backlash and tooth bearing pattern are affected by changes in axial position of the pinion and gear. Spiral bevel gears, hypoid gears, and Spiroid gears are also critical in that they tend to "screw" into mesh in one direction of rotation, which can cause binding, and will "unscrew" in the other direction of rotation, causing backlash, unless the mounting system is stiff enough to prevent axial shift. All properly designed mountings for gears of this type have bearings capable of withstanding all axial thrust loads imposed both by the gears and by possible external loads. Occasionally a design is

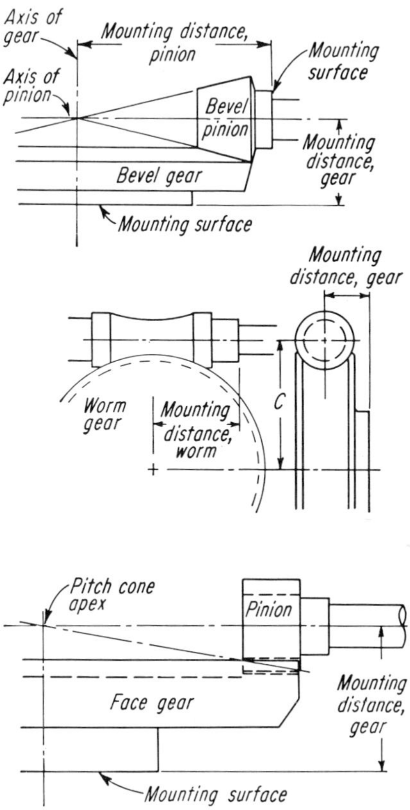

FIG. 6-6. Mounting distance, shown for various kinds of gears, is the distance from the axis of a gear to the mounting surface of its mating gear. This distance is measured parallel to the shaft of the subject gear.

attempted in which the designer relies on the separating forces produced in the mesh to keep the gears away from a binding condition. Only a few such designs are truly successful.

Wormgearing must also be accurately positioned if of the single- or double-enveloping design. If a throated member is permitted to move axially, the frictional force developed in the mesh will tend to move the member, and the geometry of the throated design will cause a reduction in backlash, sometimes to the point of binding.

Mounting distance is usually specified for each gear member as the distance from a specific mounting surface on the gear blank, a face of a hub, for example, to the axis of the mating gear. Bevel gears are usually stamped with the correct mounting distance for the pair. This distance was established in a bevel-gear test fixture at the time of manufacture.

The design of mountings for gearing requiring a mounting distance should include provisions for shimming or otherwise adjusting the position of the members at assembly. The calculation of mounting distance is accomplished in two steps. The design of the gear teeth (Chap. 5) includes a distance from the axes or pitch-cone apex to a given surface of the gear. To this distance is added the distance to the closest bearing face that is used to provide the axial location of gear and its shaft. Provision should be made in the design to secure the bearing seats that provide axial location at the specified mounting distances.

In the previous articles, radial clearance in rolling-element bearings was considered. In gear designs requiring a mounting distance, the axial clearance in one bearing on each shaft must be controlled. Reference to a bearing catalogue or handbook is recommended.

REFERENCES

Standards of the American Gear Manufacturers Association and the American Standards Association

1. "Letter Symbols for Gear Engineering," AGMA standard 111.03, ASA standard B6.5-1954.
2. "Gear Nomenclature, Terms, Definitions, and Illustrations," AGMA standard 112.03, ASA standard B6.10-1954.
3. "Basic Gear Geometry—Reference Information," AGMA 115.01.
4. "Tooth Proportions for Coarse-pitch Involute Spur Gears," AGMA standard 201.02.
5. "Zerol Bevel Gears," AGMA standard 202.02.
6. "Fine-pitch On-center Face Gears for 20-degree Involute Spur Pinions," AGMA standard 203.01.
7. "20-degree Involute Fine-pitch System for Spur and Helical Gears," AGMA standard 207.04, ASA standard B6.7-1956.
8. "Straight Bevel Gears," AGMA standard 208.01, ASA standard B6.13-1955.
9. "Spiral Bevel Gears," AGMA standard 209.02.
10. "Inspection of Fine Pitch Gears," AGMA standard 236.04, ASA standard B6.11-1956.
11. "Design of Coarse Pitch Cylindrical Worm Gearing," AGMA standard 341.01.
12. "Design of General Industrial Double-enveloping Wormgears," AGMA standard 342.01.
13. "Design for Fine Pitch Worm Gearing," AGMA standard 374.03, ASA standard B6.9-1956.

Other References

14. Buckingham, Earle: "Analytical Mechanics of Gears," McGraw-Hill Book Company, Inc., New York, 1949.
15. Buckingham, Earle: "Spur Gears," McGraw-Hill Book Company, Inc., New York, 1928.
16. Candee, Allan H.: Involute Gear Calculations Simplified, *Am. Machinist*, Sept. 13, 1945.
17. Dudley, D. W.: "Practical Gear Design," McGraw-Hill Book Company, Inc., New York, 1954.
18. Marshall, D.: Tables and Charts for Calculating Thermal Expansion of Metals, *Machine Design*, July, 1955.
19. Merritt, H. E.: "Gears," Sir Isaac Pitman & Sons, Ltd., London, 1955.
20. Thoen, R. L.: How to Find Exact Values of Backlash in Spur Gears, for Changes in Tooth Thickness or Center Distance, *Machine Design*, Jan. 23, 1958.
21. Vogel, Werner F.: "Involutometry and Trigonometry," Michigan Tool Co., Detroit, 1945.
22. Williamson, J. L.: Tables Simplify Calculations of Gear Testing Center Distances, *Machinery*, September, 1957.

Chapter 7

SPECIAL CALCULATIONS FOR SPUR AND HELICAL GEARS

By

M. F. SPOTTS, *Professor of Mechanical Engineering, Northwestern University, Evanston, Ill.*

This chapter covers a number of special calculations applicable to spur and helical gears. These gears have an *involute* form, and this permits an easy calculation of many useful quantities. The material given in this chapter should be considered as a supplement to the definitions of Chap. 4 and the basic tooth-design practices of Chaps. 5 and 6.

7-1. Notation for Chapter 7. Basic notation used in this chapter is given in Table 7-1.*

Table 7-1. Basic Notation for Chapter 7

a_1, a_2, addendum of pinion and gear respectively
B, backlash on pitch circle
C, center distance
d_1, d_2, pitch diameter of pinion and gear respectively
L, lead of helical gear or worm
m_p, contact ratio of profiles
N_1, N_2, number of teeth in pinion and gear respectively
P_d, diametral pitch for spur gear and for helical gear in plane of rotation
P_{nd}, normal diametral pitch of helical gear
p, circular pitch for spur gear and for helical gear in plane of rotation
p_n, normal circular pitch of helical gear
p_b, base pitch
r_1, r_2, pitch radius of pinion and gear respectively
r_{b1}, r_{b2}, radius of base circle of pinion and gear respectively
t, actual circular thickness of tooth on pitch circle in plane of rotation
t_n, normal circular tooth thickness of helical gear
z, length of contact between teeth
ϕ, pressure angle for spur gear, and for helical gear in plane of rotation
ϕ_n, pressure angle in normal section of helical gear
ψ, pitch helix angle for helical gear (between tooth and axis)
ψ_b, base helix angle for helical gear
ϵ, roll angle

7-2. Fundamental Gear Equations. A collection of fundamental equations used in gearing calculations is given in Table 7-2.

* For a list of notation used in gearing, see AGMA standard 111.03, or ASA standard B6.5-1954. See also Chap. 4.

Table 7-2. Fundamental Equations of Gearing

$p = \dfrac{2\pi r}{N}$	(7-1)	$t_n = t \cos \psi$	(7-7)
$p_n = \dfrac{2\pi r \cos \psi}{N} = p \cos \psi$	(7-2)	$m_p = \dfrac{z}{p_b}$	(7-8)
$P_d = \dfrac{N}{2r} = \dfrac{\pi}{p}$	(7-3)	$r_b = r \cos \phi$	(7-9)
$P_{nd} = \dfrac{N}{2r \cos \psi}$	(7-4)	$L = 2\pi r \cot \psi = 2\pi r_b \cot \psi_b$	(7-10)
$pP_d = p_n P_{nd}$	(7-5)	$\tan \phi_n = \tan \phi \cos \psi$	(7-11)
$p_b = \dfrac{2\pi r \cos \phi}{N} = p \cos \phi$	(7-6)	$\sin \phi_n = \sin \phi \cos \psi_b$	(7-12)

7-3. Intervals of Contact for Spur Gears. The length of contact between two mating spur gears, as shown in Fig. 7-1, extends from B_1 to B_2, the points where the

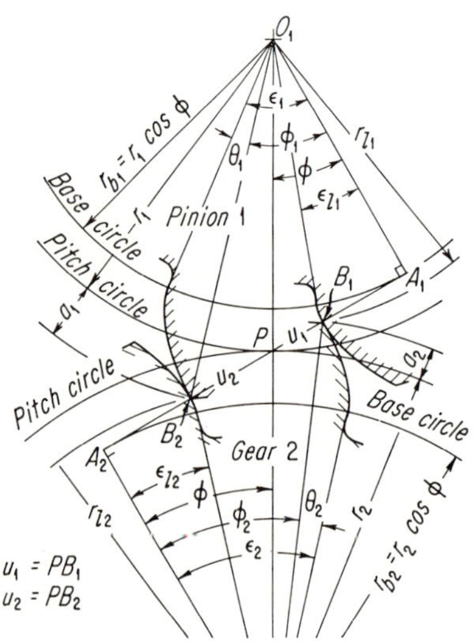

FIG. 7-1. Determination of interval of contact, roll angles, and form radii of spur gears.

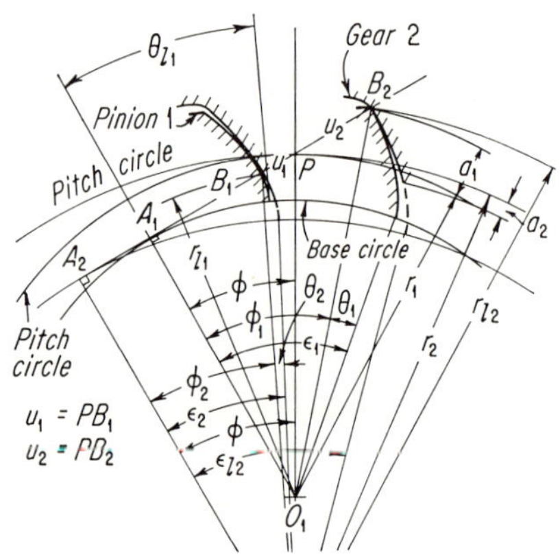

FIG. 7-2. Determination of interval of contact, roll angles, and form radii of internal spur gears.

two addendum circles cross the line of action. It is assumed that each gear has a sufficient number of teeth so that undercutting is not present. Equations for u_1 and u_2 are given in Table 7-3.

Base pitch p_b [Eq. (7-6)] is equal to the tooth-to-tooth distance between meshing pairs along the line of action. The contact ratio m_p, or number of normal pitches in the interval of contact, is

$$m_p = \frac{z}{p_b} = \frac{u_1 + u_2}{p_b} \tag{7-13}$$

For standard-addendum full-depth 20° gears the contact ratio can also be found from Table 7-4.

Table 7-3. Equations for Intervals of Contact, Roll Angles, and Form Radii

	Pinion	Gear
	External spur gears (Fig. 7-1)	
Auxiliary angle	$\cos \phi_1 = \dfrac{r_{b1}}{r_1 + a_1}$	$\cos \phi_2 = \dfrac{r_{b2}}{r_2 + a_2}$
Interval of contact	$u_1 = (r_2 + a_2) \sin \phi_2 - r_2 \sin \phi$	$u_2 = (r_1 + a_1) \sin \phi_1 - r_1 \sin \phi$
Roll angle at outside diameter, radians	$\epsilon_1 = \dfrac{r_1 \sin \phi + u_2}{r_{b1}}$	$\epsilon_2 = \dfrac{r_2 \sin \phi + u_1}{r_{b2}}$
Roll angle at theoretical limit radius, radians	$\epsilon_{l1} = \dfrac{r_1 \sin \phi - u_1}{r_{b1}} = \tan \phi_{l1}$	$\epsilon_{l2} = \dfrac{r_2 \sin \phi - u_2}{r_{b2}} = \tan \phi_{l2}$
Theoretical limit radius	$r_{l1} = \dfrac{r_{b1}}{\cos \phi_{l1}}$	$r_{l2} = \dfrac{r_{b2}}{\cos \phi_{l2}}$
Form diameter	$d_{f1} = 2\left(r_{l1} - \dfrac{0.025}{P_d}\right)$	$d_{f2} = 2\left(r_{l2} - \dfrac{0.025}{P_d}\right)$
	Internal spur gears (Fig. 7-2)	
Auxiliary angle	$\cos \phi_1 = \dfrac{r_{b1}}{r_1 + a_1}$	$\cos \phi_2 = \dfrac{r_{b2}}{r_2 - a_2}$
Interval of contact	$u_1 = -(r_2 - a_2) \sin \phi_2 + r_2 \sin \phi$	$u_2 = (r_1 + a_1) \sin \phi_1 - r_1 \sin \phi$
Roll angle at outside diameter, radians	$\epsilon_1 = \dfrac{r_1 \sin \phi + u_2}{r_{b1}}$	$\epsilon_2 = \dfrac{r_2 \sin \phi - u_1}{r_{b2}}$
Roll angle at theoretical limit radius, radians	$\epsilon_{l1} = \dfrac{r_1 \sin \phi - u_1}{r_{b1}} = \tan \phi_{l1}$	$\epsilon_{l2} = \dfrac{r_2 \sin \phi + u_2}{r_{b2}} = \tan \phi_{l2}$
Theoretical limit radius	$r_{l1} = \dfrac{r_{b1}}{\cos \phi_{l1}}$	$r_{l2} = \dfrac{r_{b2}}{\cos \phi_{l2}}$
Form diameter*	$d_{f1} = 2\left(r_{l1} - \dfrac{0.025}{P_d}\right)$	$d_{f2} = 2\left(r_{l2} + \dfrac{0.025}{P_d}\right)$

* The allowance of $0.025/P_d$ for excess of involute is purely arbitrary. A greater or less amount might be used. If the form diameter tends to come below the base circle, reduce the 0.025 as it is not possible to have any involute form below the base circle.

Table 7-4. Contact Ratio u_1/p_b or u_2/p_b for One 20° Full-depth Nonundercut Gear

(Based on a standard addendum of $a = 1/P_d$)

No. of teeth	Contact ratio	No. of teeth	Contact ratio	No. of teeth	Contact ratio	No. of teeth	Contact ratio	No. of teeth	Contact ratio	No. of teeth	Contact ratio
10	0.685	20	0.778	30	0.827	40	0.857	60	0.892	80	0.913
11	0.698	21	0.785	31	0.830	42	0.861	62	0.895	85	0.917
12	0.710	22	0.790	32	0.834	44	0.866	64	0.897	90	0.920
13	0.721	23	0.796	33	0.837	46	0.870	66	0.900	95	0.923
14	0.731	24	0.801	34	0.840	48	0.874	68	0.902	100	0.926
15	0.741	25	0.806	35	0.843	50	0.877	70	0.904	120	0.934
16	0.749	26	0.810	36	0.846	52	0.881	72	0.906	140	0.943
17	0.757	27	0.815	37	0.849	54	0.884	74	0.908	160	0.948
18	0.765	28	0.819	38	0.852	56	0.887	76	0.909	180	0.952
19	0.772	29	0.823	39	0.854	58	0.890	78	0.911	200	0.956

Teeth 10 to 18 are *normally* undercut. See Table 7-5 to get contact ratio of mating member. See Example 7-4.

To get contact ratio of the mesh add contact ratio of pinion to contact ratio of gear.

If made by a generating process, 20° full-depth gears, with fewer than 17 teeth, will be undercut as described in Art. 7-6. Should such gears be made by a process in which undercutting did not occur, the gears would be in danger of not meshing or rolling with other gears because of *interference* or overlap in the outlines of the teeth of two assembled gears.

Example 7-1. Find the contact ratio for a pair of 20° full-depth gears of 20 and 44 teeth.* $P_d = 4$.

Solution: $a_1 = a_2 = \dfrac{1}{P_d} = 0.25''$

$$r_1 = \frac{10}{4} = 2.5'' \qquad r_2 = \frac{22}{4} = 5.5''$$

By Eq. (7-9), $r_{b1} = 2.5 \times 0.939693 = 2.349232''$

By Eq. (7-9), $r_{b2} = 5.5 \times 0.939693 = 5.168309''$

Apply the following equations from Table 7-3:

$$\cos \phi_1 = \frac{2.349232}{2.75} = 0.854266 \qquad \phi_1 = 31.321°$$

$$\cos \phi_2 = \frac{5.168309}{5.75} = 0.898836 \qquad \phi_2 = 25.994°$$

$$u_1 = 5.75 \times 0.438285 - 5.5 \times 0.342020 = 0.639027''$$

$$u_2 = 2.75 \times 0.519837 - 2.5 \times 0.342020 = 0.574501''$$

$$\text{Normal pitch } p_n = p \cos \phi = \frac{\pi}{4} \times 0.939693 = 0.738035''$$

$$\text{Contact ratio } m_p = \frac{0.6390 + 0.5745}{0.738035} = 1.644$$

The same result is found by use of Table 7-4 by adding the entries for 20 and 44 teeth.

$$m_p = 0.778 + 0.866 = 1.644$$

To obtain smooth and even transfer of the load from one pair of teeth to the next, it is usually recommended that the value of m_p should not be less than 1.4.

7-4. Roll Angles and Form Diameters of Spur Gears. Roll angle is the angle subtended by the arc equal in length to the unwound string that is generating the involute. Thus in Fig. 7-1 ϵ_1 is the roll angle for the pinion at the outside diameter and ϵ_2 is the roll angle for the gear at the outside diameter. Equations for these angles can be written by inspection of Fig. 7-1 and are listed in Table 7-3.

The involute of a generated tooth must extend sufficiently far inwardly from the pitch circle to accommodate the tooth of the mating gear. Thus in Fig. 7-1 the involute of the pinion must extend at least to B_1 where the addendum circle of the gear crosses the line of action. Similarly the involute of the gear must extend at least to B_2. Point B_1 is located at the theoretical limit radius r_{l1} from center O_1, and B_2 is located at the theoretical limit radius r_{l2} from center O_2.

A line† from O_1 to B_1 divides angle ϵ_{l1} into two parts, ϕ_{l1} and inv ϕ_{l1}. Substitution of $\tan \phi_{l1} - \phi_{l1}$ for inv ϕ_{l1} shows that $\tan \phi_{l1}$ is equal to ϵ_{l1} in radians. An expression for theoretical limit radius r_{l1}, as shown in Table 7-3, is then easily written. Similar

* For a convenient table of natural trigonometric functions, at intervals of 0.001°, see J. Peters, "Seven-place Values of Trigonometric Functions," D. Van Nostrand Company, Inc., Princeton, N.J., 1942.

† To avoid complication of the drawing, this line is not shown in Fig. 7-1. For a discussion of the involute function, see Art. 7-8.

reasoning applies to the gear. A line from B_2 to O_2 divides ϵ_{l2} into ϕ_{l2} and inv ϕ_{l2}, from which it can be shown that ϵ_{l2} equals tan ϕ_{l2}.

Example 7-2. For the meshing gears of Example 7-1, find values for the various roll angles and theoretical limit radii.

Solution: From Example 7-1,

$$r_{b1} = 2.349232'' \qquad r_{b2} = 5.168309''$$
$$u_1 = 0.639027'' \qquad u_2 = 0.574501''$$

also

$$r_1 \sin \phi = 2.5 \times 0.342020 = 0.855050''$$
$$r_2 \sin \phi = 5.5 \times 0.342020 = 1.881111''$$

Apply the equations of Table 7-3 as follows:

$$\epsilon_1 = \frac{0.855050 + 0.574501}{2.349232} = 0.608518 \text{ radian} = 34.866°$$

$$\epsilon_{l1} = \frac{0.855050 - 0.639027}{2.349232} = 0.091955 \text{ radian} = 5.269°$$

$$\epsilon_2 = \frac{1.881111 + 0.639027}{5.168309} = 0.487614 \text{ radian} = 27.938°$$

$$\epsilon_{l2} = \frac{1.881111 - 0.574501}{5.168309} = 0.252812 \text{ radian} = 14.485°$$

$$\tan \phi_{l1} = \epsilon_{l1} = 0.091955 \qquad \phi_{l1} = 5.254°$$

$$r_{l1} = \frac{2.349232}{0.995799} = 2.35914''$$

$$\tan \phi_{l2} = \epsilon_{l2} = 0.252812 \qquad \phi_{l2} = 14.188°$$

$$r_{l2} = \frac{5.168309}{0.969498} = 5.33092''$$

In practice, the involutes should extend slightly farther toward the respective centers than points B_1 and B_2 in Fig. 7-1. This is to accommodate the effects of errors in the teeth and center distance of the mounting. A decrease in the radius equal to $0.025/P_d$ is sometimes specified. When this is deducted from the theoretical limit radii and the results are doubled, the form diameters d_{f1} and d_{f2} are obtained. Equations are shown in Table 7-3.

The form diameters for the gears of Example 7-1 are

$$d_{f1} = 2\left(2.35914 - \frac{0.025}{4}\right) = 4.7058''$$

$$d_{f2} = 2\left(5.33092 - \frac{0.025}{4}\right) = 10.6493''$$

The equations of Table 7-3 apply equally well to helical gears. It is obvious that a form diameter smaller than the base circle should never be specified.

7-5. Internal Gears. The geometric relationships corresponding to Fig. 7-1, but for internal gears, are shown in Fig. 7-2. The equations in Table 7-3 can be derived from inspection of the figure. For an internal gear, the correction $0.025/P_d$ must be added to the theoretical limit radius in forming the equation for the form diameter.

7-6. Undercutting and Interference. As a rule when gears are produced in large quantities, a generating process such as hobbing or shaping is used. Kinematically the action of a hob is similar to that of a rack cutter. Extensions must be provided on the ends of the teeth to cut the clearance f at the base of the teeth of the gear. Two typical forms of 20° rack-cutter teeth are shown in Fig. 7-3.

7-6 SPECIAL CALCULATIONS FOR SPUR AND HELICAL GEARS

Conjugate action between involute teeth takes place on the line of action and can have a maximum value of A_1A_2 (Fig. 7-1). These are called the *interference points* of the gears. When the number of teeth in a gear is small, the addendum of the rack or hob extends inside the interference point during the cutting operation. This results in a loss of a portion of the involute adjacent to the base circle, which materially shortens the duration of contact between a mating pair of teeth.

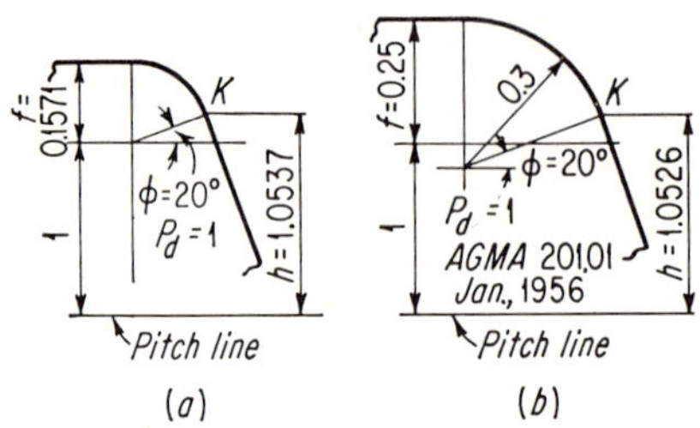

FIG. 7-3. Layout showing two methods of providing extensions on hob tooth for cutting clearance at base of teeth on the gear. $\phi = 20°$.

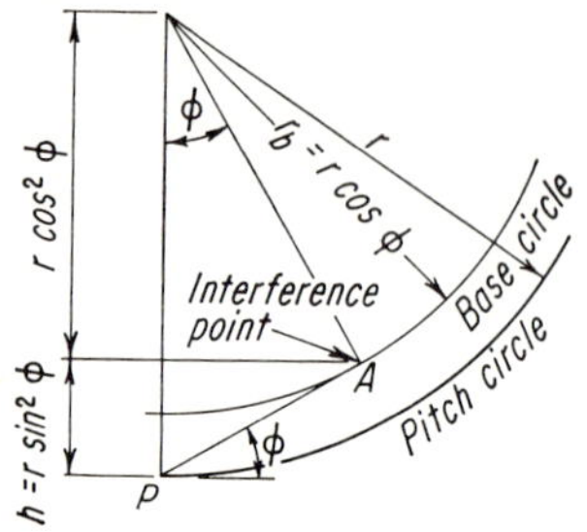

FIG. 7-4. Relationship required to prevent undercutting for gear of minimum number of teeth.

It is usually assumed that undercutting is due to the straight portion of the cutter tooth, which terminates at K in Fig. 7-3, and that the extension for clearance purposes beyond this point has little effect on the undercutting. Large-scale layouts of the cutting process have shown that this is a reasonably good assumption. Hence, in Fig. 7-4, if point K at height h passes through interference point A, the tooth will not be undercut. The minimum numbers of teeth to avoid undercutting, for full-depth teeth, are

$$\phi = 14\tfrac{1}{2}° \qquad N = 34$$
$$\phi = 20° \qquad N = 18$$
$$\phi = 25° \qquad N = 12$$

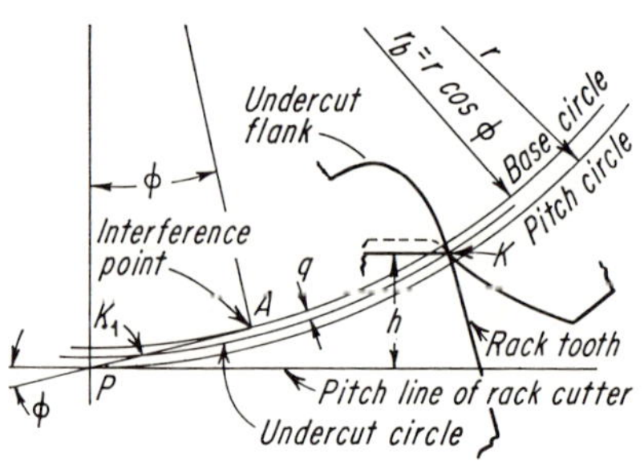

FIG. 7-5. Undercutting of spur-gear tooth.

When a gear has fewer teeth than the above, the situation will be shown in Fig. 7-5. Cutter tooth and gear are at the point of separating upon further motion to the right. The loss of involute in the radial direction is equal to q. The undercut circle thus determined crosses the line of action at K_1. Distance available for gear action is reduced to PK_1 instead of PA, as is theoretically possible. In practice, however, contact between teeth must stop some distance away from K_1 for undercut gears or from A for nonundercut gears to prevent the excessive contact stress that would otherwise result.

Undercutting and contact ratio for gears* produced by hobs with the teeth of Fig. 7-3 are given in Table 7-5.

Example 7-3. Find the value of the contact ratio for two 20° full-depth gears of 12 and 16 teeth when operating together. The hob used in cutting had teeth like Fig. 7-3.

Solution: Consider the 12-tooth pinion first. Table 7-5 indicates an available contact ratio of 0.545 in length PK_1 for 12 teeth. Table 7-4 and Fig. 7-1 indicate a contact ratio of 0.749 in interval u_1, as determined by the 16-tooth gear, so that the 0.545 figure holds.

* Methods used for the calculation of these values can be found in ref. 2.

Table 7-5. Undercutting and Contact Ratio PK_1/p_b for 20° Full-depth Hobbed Spur Gears

N	r*	q*	Contact ratio
10	5.0	0.0326	0.391
11	5.5	0.0243	0.467
12	6.0	0.0174	0.545
13	6.5	0.0119	0.624
14	7.0	0.0074	0.705
15	7.5	0.0041	0.787
16	8.0	0.0018	0.871
17	8.5	0.0004	0.956

* For $P_d = 1$.

Consider now the 16-tooth gear. Table 7-5 indicates sufficient involute for a contact ratio of 0.871. However, interval u_2 in Fig. 7-1 and Table 7-4 indicates a contact ratio of 0.710 required by the 12-tooth pinion. Hence the 0.710 figure holds and the undercutting of the 16-tooth gear has no effect.

$$\text{Total contact ratio} = 0.545 + 0.710 = 1.255$$

Example 7-4. Find the value of the contact ratio for two meshing 20° full-depth gears of 14 and 40 teeth.

Solution: The 14-tooth gear is undercut. Interval PK_1 of Fig. 7-5 has a contact ratio of 0.705 by Table 7-5.

Interval PB_2 or u_2 in Fig. 7-1 is also determined by the 14-tooth gear. By Table 7-4 the contact ratio is 0.731 Total contact ratio = 0.705 + 0.731 = 1.436.

In the power-transmission field, the operating characteristics of undercut gears are frequently not satisfactory.* When gears of few teeth must be used it is usually better either to increase the pressure angle or to make the design on the *extended center system* as described in Chap. 6.

7-7. Center Distance and Backlash. Calculations for gears are usually made on the assumption that operation will occur in metal-to-metal contact. The necessary clearance or backlash is then obtained by setting the cutting tool slightly deeper into the blank, which results in slightly thinner teeth. Backlash for involute gears can also be obtained by an appropriate increase in the center distance.

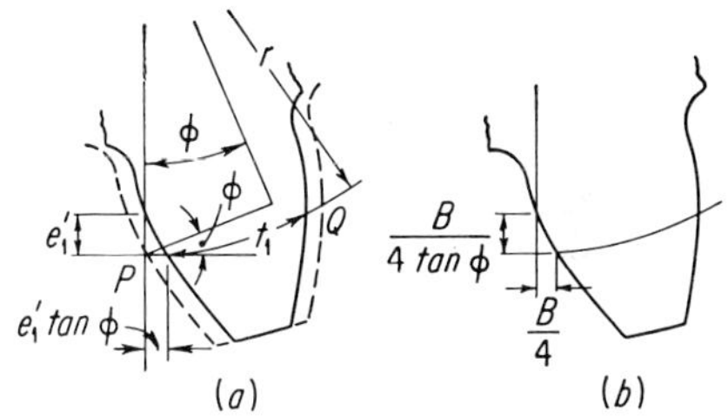

FIG. 7-6. Backlash determination in gears.

When the tool is set an amount e_1' deeper into the blank (Fig. 7-6*a*), then, to a good degree of approximation, arc PQ, or one-half the theoretical circular pitch, is reduced by $e_1' \tan \phi$ at both ends. The following equation can be written:

$$t_1 + 2e_1' \tan \phi = \tfrac{1}{2}p \tag{7-14}$$

or

$$e_1' = \frac{\tfrac{1}{2}p - t_1}{2 \tan \phi}$$

* For an extended discussion of interference phenomena of internal gears, see A. Fisher, The Involute Internal Gear Pair, *Machinery*, Vol. 44, pp. 521, 611, 1934; Vol. 45, pp. 37, 315, 421, 1934–1935.

The corresponding equation for the mating gear is

$$e_2' = \frac{\frac{1}{2}p - t_2}{2 \tan \phi} \tag{7-15}$$

Here t_1 and t_2 are actual tooth thicknesses on the standard pitch circles.

For small values of e_1' and e_2' as used for obtaining backlash, the actual metal-to-metal center distance for these gears is

$$C - \frac{p - t_1 - t_2}{2 \tan \phi}$$

If the gears are to operate with a backlash B, the center distance must be increased by $B/2 \tan \phi$ as shown in Fig. 7-6*b*. Actual center distance C_1 is then equal to

$$C_1 = C - \frac{p - t_1 - t_2 - B}{2 \tan \phi}$$

If $\Delta C = C - C_1$, the above equation reduces to

$$t_1 + t_2 + B = p - 2\,\Delta C \tan \phi \tag{7-16}$$

Example 7-5. A 20° master gear of 8 P_d, 36 teeth, has an actual tooth thickness of 0.1970″ on the standard pitch circle of 4.5000″ diameter.

(*a*) When meshed, metal-to-metal, with a production gear of 30 teeth, the average value for the center distance is 4.1240″. When a 50-tooth production gear is meshed with the master, the center distance is 5.3770″. Find the proper center distance for these two gears if they are to run with a backlash of 0.0030″.

(*b*) Find the backlash if the two production gears are assembled on a mount with a center distance of 4.9995″.

Solution: Circular pitch $p = \pi/P_d = \pi/8 = 0.3927''$.

(*a*) For 30-tooth gear and master:

$$\text{Standard center distance } C = \tfrac{1}{2} \times \frac{30 + 36}{8} = 4.1250''$$

$$\Delta C = C - C_1 = 4.1250 - 4.1240 = 0.0010''$$

By Eq. (7-16), $\quad B = 0 \qquad t_1 = p - 2\Delta C \tan \phi - t_2$

$$t_1 = 0.3927 - 2 \times 0.0010 \times 0.36397 - 0.1970 = 0.19497''$$

For 50-tooth gear and master:

$$\text{Standard center distance } C = \tfrac{1}{2} \times \frac{36 + 50}{8} = 5.3750''$$

$$\Delta C = C - C_1 = 5.3750 - 5.3770 = -0.0020''$$

$$t_2 = 0.39270 - 2(-0.002)0.36397 - 0.1970 = 0.19716''$$

For 30-tooth and 50-tooth: $\quad C = \tfrac{1}{2} \times \dfrac{30 + 50}{8} = 5''$

By Eq. (7-16),

$$\Delta C = \frac{0.39270 - 0.19497 - 0.19716 - 0.00300}{2 \times 0.36397} = -0.00334''$$

$$\text{Required center distance } C_1 = C - \Delta C = 5 - (-0.0033) = 5.0033''$$

(*b*) $\quad \Delta C = C - C_1 = 5 - 4.9995 = 0.0005''$

By Eq. (7-16), $\quad B = 0.39270 - 0.19497 - 0.19716 - 2 \times 0.0005 \times 0.36397 = 0.0002''$

The backlash for a pair of gears must be sufficient to permit the gears to turn freely under the worst combination of manufacturing tolerances and temperature variation. Values recommended* for backlash for assembled gears are shown in Table 7-6.

* See p. 83 of ref. 1.

Table 7-6. Recommended Backlash for Assembled Gears

P_d	Backlash, in.	P_d	Backlash, in.
1	0.025–0.040	5	0.006–0.009
1½	0.018–0.027	6	0.005–0.008
2	0.014–0.020	7	0.004–0.007
2½	0.011–0.016	8–9	0.004–0.006
3	0.009–0.014	10–13	0.003–0.005
4	0.007–0.011	14–32	0.002–0.004

It must be noted, however, for reasonable manufacturing costs, that, if the tolerances on the thickness of the teeth and the center distance are added, the result may be double or more the recommendation of the table.

Backlash B' measured perpendicular to the surfaces of adjacent teeth, as by a feeler gage, is given by

$$B' = B \cos \phi \tag{7-17}$$

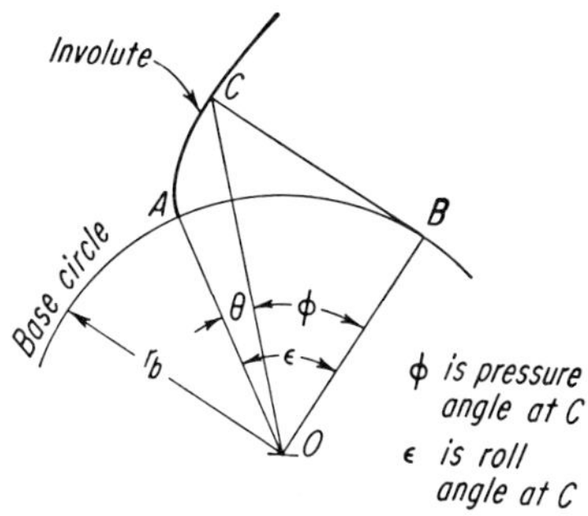

FIG. 7-7. Involute polar angle *θ*.

7-8. Involute Function. This function is widely used and very convenient in gearing calculations. In Fig. 7-7, BC is equal to arc AB. Then

$$\tan \phi = \frac{BC}{r_b} = \frac{\text{arc } AB}{r_b} = \epsilon$$

Let the angle AOC in radians be called the *involute function* of ϕ. Angle AOC is also the *involute polar angle* θ.

$$\text{inv } \phi + \phi = \frac{\text{arc } AB}{r_b} = \tan \phi \tag{7-18}$$

$$\text{inv } \phi = \tan \phi - \phi$$

or

$$\theta = \tan \phi - \phi$$

Values of the involute function at intervals of 0.1° are given* in Table 7-7.

7-9. Thickness of Teeth from Measurement over Rolls or Balls.† The thickness of the teeth on the pitch circle can be determined by measurements taken over rolls or balls placed between the teeth. Suitable equations are listed in Table 7-8.

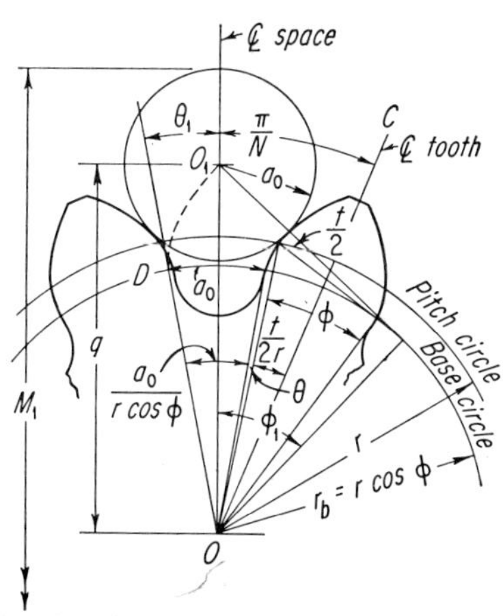

FIG. 7-8. Tooth thickness of spur gear by measurement over rolls. Even number of teeth.

Consider the spur gear of Fig. 7-8 with pressure angle ϕ and an even number of teeth N. With balls or rolls of radius a_o in diametrically opposite spaces, the equations for measurement M_1 and $\cos \phi_1$ of Table 7-8 are easily derived. Curve O_1D is an

* More extensive tables of involute functions can be found in refs. 5 (at intervals of 0.01°) and 6 (at intervals of 1 minute of arc) and in Earle Buckingham, "Manual of Gear Design," Sec. 1, The Industrial Press, New York, 1935 (at intervals of 0.01°).

† See also refs. 3 and 4.

Table 7-7.* Values of Involute Polar Angle $\theta = \tan\phi - \phi$

	0°	1°	2°	3°	4°	5°	6°	7°	8°	9°
0.0	0.000000	0.000002	0.000014	0.000048	0.000114	0.000222	0.000385	0.000612	0.000915	0.001305
0.1	0.000000	0.000002	0.000016	0.000053	0.000122	0.000236	0.000404	0.000638	0.000949	0.001349
0.2	0.000000	0.000003	0.000019	0.000058	0.000132	0.000250	0.000424	0.000666	0.000985	0.001394
0.3	0.000000	0.000004	0.000022	0.000064	0.000141	0.000265	0.000445	0.000694	0.001022	0.001441
0.4	0.000000	0.000005	0.000025	0.000070	0.000151	0.000280	0.000467	0.000723	0.001060	0.001488
0.5	0.000000	0.000006	0.000028	0.000076	0.000162	0.000296	0.000489	0.000753	0.001098	0.001536
0.6	0.000000	0.000007	0.000031	0.000083	0.000173	0.000312	0.000512	0.000784	0.001138	0.001586
0.7	0.000001	0.000009	0.000035	0.000090	0.000184	0.000330	0.000536	0.000815	0.001178	0.001636
0.8	0.000001	0.000010	0.000039	0.000097	0.000197	0.000347	0.000560	0.000847	0.001219	0.001688
0.9	0.000001	0.000012	0.000043	0.000105	0.000209	0.000366	0.000586	0.000881	0.001262	0.001740

	10°	11°	12°	13°	14°	15°	16°	17°	18°	19°
0.0	0.001794	0.002394	0.003117	0.003975	0.004982	0.006150	0.007493	0.009025	0.010760	0.012715
0.1	0.001849	0.002461	0.003197	0.004069	0.005091	0.006276	0.007637	0.009189	0.010946	0.012923
0.2	0.001905	0.002529	0.003278	0.004164	0.005202	0.006404	0.007784	0.009355	0.011133	0.013134
0.3	0.001962	0.002598	0.003360	0.004261	0.005315	0.006534	0.007932	0.009523	0.011323	0.013346
0.4	0.002020	0.002668	0.003443	0.004360	0.005429	0.006665	0.008082	0.009694	0.011515	0.013562
0.5	0.002080	0.002739	0.003529	0.004459	0.005545	0.006799	0.008234	0.009866	0.011709	0.013779
0.6	0.002140	0.002812	0.003615	0.004561	0.005662	0.006934	0.008388	0.010041	0.011906	0.013999
0.7	0.002202	0.002887	0.003703	0.004664	0.005782	0.007071	0.008544	0.010217	0.012105	0.014222
0.8	0.002265	0.002962	0.003792	0.004768	0.005903	0.007210	0.008702	0.010396	0.012306	0.014447
0.9	0.002329	0.003039	0.003883	0.004874	0.006025	0.007350	0.008863	0.010577	0.012509	0.014674

	20°	21°	22°	23°	24°	25°	26°	27°	28°	29°
0.0	0.014904	0.017345	0.020054	0.023049	0.026350	0.029975	0.033947	0.038287	0.043017	0.048164
0.1	0.015137	0.017603	0.020340	0.023365	0.026697	0.030357	0.034364	0.038742	0.043513	0.048702
0.2	0.015372	0.017865	0.020629	0.023684	0.027048	0.030741	0.034785	0.039201	0.044012	0.049245
0.3	0.015609	0.018129	0.020922	0.024006	0.027402	0.031130	0.035209	0.039664	0.044516	0.049792
0.4	0.015850	0.018395	0.021217	0.024332	0.027760	0.031521	0.035637	0.040131	0.045024	0.050344
0.5	0.016092	0.018665	0.021515	0.024660	0.028121	0.031917	0.036069	0.040602	0.045537	0.050901
0.6	0.016337	0.018937	0.021815	0.024992	0.028485	0.032315	0.036505	0.041076	0.046054	0.051462
0.7	0.016585	0.019212	0.022119	0.025326	0.028852	0.032718	0.036945	0.041556	0.046575	0.052027
0.8	0.016836	0.019490	0.022426	0.025664	0.029223	0.033124	0.037388	0.042039	0.047100	0.052597
0.9	0.017089	0.019770	0.022736	0.026005	0.029598	0.033534	0.037835	0.042526	0.047630	0.053172

	30°	31°	32°	33°	34°	35°	36°	37°	38°	39°
0.0	0.053752	0.059809	0.066364	0.073449	0.081097	0.089342	0.098224	0.107782	0.118060	0.129106
0.1	0.054336	0.060441	0.067048	0.074188	0.081894	0.090201	0.099149	0.108777	0.119130	0.130254
0.2	0.054924	0.061079	0.067738	0.074932	0.082697	0.091066	0.100080	0.109779	0.120207	0.131411
0.3	0.055518	0.061721	0.068432	0.075683	0.083506	0.091938	0.101018	0.110788	0.121291	0.132576
0.4	0.056116	0.062369	0.069133	0.076438	0.084321	0.092816	0.101964	0.111805	0.122384	0.133750
0.5	0.056720	0.063022	0.069838	0.077200	0.085142	0.093701	0.102916	0.112829	0.123484	0.134931
0.6	0.057328	0.063680	0.070549	0.077968	0.085970	0.094592	0.103875	0.113860	0.124592	0.136122
0.7	0.057940	0.064343	0.071266	0.078741	0.086804	0.095490	0.104841	0.114899	0.125709	0.137320
0.8	0.058558	0.065012	0.071988	0.079520	0.087644	0.096395	0.105814	0.115945	0.126833	0.138528
0.9	0.059181	0.065685	0.072716	0.080305	0.088490	0.097306	0.106795	0.116999	0.127965	0.139743

	40°	41°	42°	43°	44°	45°	46°	47°	48°	49°
0.0	0.140968	0.153702	0.167366	0.182024	0.197744	0.214602	0.232679	0.252064	0.272854	0.295157
0.1	0.142201	0.155025	0.168786	0.183546	0.199377	0.216353	0.234557	0.254078	0.275015	0.297475
0.2	0.143443	0.156358	0.170216	0.185080	0.201022	0.218117	0.236448	0.256106	0.277190	0.299809
0.3	0.144694	0.157700	0.171656	0.186625	0.202678	0.219893	0.238353	0.258149	0.279381	0.302160
0.4	0.145954	0.159052	0.173106	0.188180	0.204346	0.221682	0.240271	0.260206	0.281588	0.304527
0.5	0.147222	0.160414	0.174566	0.189746	0.206026	0.223483	0.242202	0.262277	0.283810	0.306912
0.6	0.148500	0.161785	0.176037	0.191324	0.207717	0.225296	0.244147	0.264363	0.286047	0.309313
0.7	0.149787	0.163165	0.177518	0.192912	0.209420	0.227123	0.246105	0.266464	0.288301	0.311731
0.8	0.151082	0.164556	0.179009	0.194511	0.211135	0.228962	0.248078	0.268579	0.290570	0.314166
0.9	0.152388	0.165956	0.180511	0.196122	0.212863	0.230814	0.250064	0.270709	0.292856	0.316619

* See Table 24-14 for seven-place values of involute polar angle.

Table 7-7. Values of Involute Polar Angle $\theta = \tan\phi - \phi$ (*Continued*)

	50°	51°	52°	53°	54°	55°	56°	57°	58°	59°
0.0	0.319089	0.344779	0.372370	0.402020	0.433904	0.468217	0.505177	0.545027	0.588044	0.634535
0.1	0.321577	0.347450	0.375240	0.405105	0.437223	0.471790	0.509027	0.549182	0.592531	0.639389
0.2	0.324082	0.350141	0.378130	0.408212	0.440566	0.475390	0.512907	0.553368	0.597053	0.644281
0.3	0.326605	0.352850	0.381042	0.411342	0.443934	0.479016	0.516816	0.557586	0.601611	0.649212
0.4	0.329146	0.355579	0.383974	0.414495	0.447326	0.482670	0.520755	0.561837	0.606205	0.654182
0.5	0.331706	0.358328	0.386928	0.417671	0.450744	0.486351	0.524724	0.566121	0.610834	0.659192
0.6	0.334283	0.361096	0.389902	0.420870	0.454187	0.490060	0.528723	0.570438	0.615500	0.664243
0.7	0.336879	0.363884	0.392899	0.424093	0.457656	0.493797	0.532753	0.574789	0.620203	0.669333
0.8	0.339494	0.366693	0.395918	0.427340	0.461150	0.497562	0.536813	0.579173	0.624943	0.674465
0.9	0.342127	0.369521	0.398958	0.430610	0.464670	0.501355	0.540905	0.583591	0.629720	0.679638

	60°	61°	62°	63°	64°	65°	66°	67°	68°	69°
0.0	0.684853	0.739397	0.798622	0.863053	0.933293	1.010043	1.094119	1.186482	1.288263	1.400812
0.1	0.690110	0.745101	0.804822	0.869805	0.940663	1.018106	1.102966	1.196216	1.299009	1.412719
0.2	0.695410	0.750852	0.811074	0.876615	0.948098	1.026244	1.111896	1.206045	1.309864	1.424751
0.3	0.700753	0.756651	0.817379	0.883485	0.955600	1.034456	1.120910	1.215970	1.320829	1.436910
0.4	0.706139	0.762498	0.823738	0.890415	0.963169	1.042744	1.130011	1.225994	1.331906	1.449198
0.5	0.711570	0.768393	0.830151	0.897406	0.970806	1.051109	1.139199	1.236116	1.343097	1.461618
0.6	0.717045	0.774338	0.836619	0.904458	0.978512	1.059552	1.148474	1.246339	1.354403	1.474170
0.7	0.722564	0.780333	0.843143	0.911572	0.986288	1.068073	1.157839	1.256664	1.365826	1.486857
0.8	0.728129	0.786379	0.849723	0.918748	0.994135	1.076674	1.167295	1.267092	1.377367	1.499681
0.9	0.733740	0.792475	0.856359	0.925989	1.002053	1.085356	1.176842	1.277624	1.389029	1.512643

	70°	71°	72°	73°	74°	75°	76°	77°	78°	79°
0.0	1.525747	1.665027	1.821046	1.996762	2.195871	2.423054	2.68433	2.98757	3.34327	3.76574
0.1	1.538994	1.679832	1.837677	2.015552	2.217238	2.447534	2.71262	3.02058	3.38224	3.81237
0.2	1.552386	1.694806	1.854508	2.034579	2.238891	2.472360	2.74133	3.05412	3.42188	3.85988
0.3	1.565925	1.709953	1.871541	2.053847	2.260834	2.497540	2.77048	3.08821	3.46222	3.90830
0.4	1.579614	1.725275	1.888781	2.073362	2.283073	2.523081	2.80007	3.12286	3.50328	3.95766
0.5	1.593456	1.740775	1.906231	2.093126	2.305613	2.54899	2.83012	3.15808	3.54507	4.00798
0.6	1.607451	1.756455	1.923895	2.113146	2.328461	2.57527	2.86064	3.19389	3.58762	4.05929
0.7	1.621604	1.772320	1.941776	2.133426	2.351623	2.60194	2.89163	3.23029	3.63094	4.11162
0.8	1.635916	1.788371	1.959878	2.153970	2.375105	2.62900	2.92311	3.26732	3.67505	4.16499
0.9	1.650389	1.804612	1.978206	2.174783	2.398913	2.65646	2.95509	3.30497	3.71998	4.21945

	80°	81°	82°	83°	84°	85°	86°	87°	88°	89°
0.0	4.27502	4.90003	5.68420	6.6957	8.0483	9.9465	12.7997	17.563	27.100	55.74
0.1	4.33173	4.97040	5.77370	6.8132	8.2090	10.1792	13.1658	18.220	28.607	62.10
0.2	4.38963	5.04240	5.86552	6.9341	8.3752	10.4217	13.5512	18.925	30.281	70.06
0.3	4.44874	5.11608	5.95975	7.0587	8.5474	10.6745	13.9576	19.681	32.152	80.29
0.4	4.50911	5.19149	6.05650	7.1871	8.7257	10.9383	14.3866	20.496	34.258	93.93
0.5	4.57077	5.26871	6.15586	7.3195	8.9106	11.2139	14.8401	21.377	36.644	113.03
0.6	4.63377	5.34780	6.25793	7.4561	9.1023	11.5022	15.3205	22.330	39.371	141.67
0.7	4.69816	5.42882	6.36283	7.5970	9.3014	11.8038	15.8300	23.367	42.518	189.42
0.8	4.76396	5.51184	6.47068	7.7426	9.5081	12.1199	16.3714	24.498	46.190	284.91
0.9	4.83124	5.59694	6.58160	7.8929	9.7230	12.4515	16.9478	25.737	50.529	571.39

involute determined by winding the string from O_1 upon the base circle. By writing the expression for angle DOC, the equation can be had for tooth thickness t on the pitch circle.

Example 7-6. A 20° spur gear has 12 teeth and $P_d = 4$. Measurement M_1 across rolls of diameter 0.6250″ is 4.1480″. Find tooth thickness t on the pitch circle.

Solution: Use equations from Table 7-8, case I.

$$q = \tfrac{1}{2}M_1 - a_o = \tfrac{1}{2} \times 4.1480 - 0.3125 = 1.7615''$$

$$\text{Radius of pitch circle } r = \frac{N}{2P_d} = \frac{12}{2 \times 4} = 1.5000''$$

Radius of base circle $r_b = 1.5 \times 0.939693 = 1.409539''$

$$\cos \phi_1 = \frac{1.409539}{1.7615} = 0.800192 \qquad \phi_1 = 36.852°$$

$$t = 2 \times 1.5 \left[\frac{\pi}{12} - 0.014904 + 0.106318 - \frac{0.3125}{1.409539} \right] = 0.39453''$$

Table 7-8. Equations for Finding Thickness of Teeth from Measurements over Rolls or Balls

(Valid only for involute teeth and true involute helicoids)

Case I. Spur Gear

$$M_1 = 2(q + a_o) \qquad \text{even number of teeth (Fig. 7-8)}$$

$$M_2 = 2\left(q \cos \frac{180°}{2N} + a_o\right) \qquad \text{odd number of teeth (Fig. 7-9)}$$

$$\cos \phi_1 = \frac{r \cos \phi}{q} = \frac{r_b}{q}$$

$$t = 2r\left(\frac{\pi}{N} - \theta + \theta_1 - \frac{a_o}{r_b}\right)$$

Case II. Internal Spur Gear

$$M_1 = 2(q - a_o) \qquad \text{even number of teeth (Fig. 7-10)}$$

$$M_2 = 2\left(q \cos \frac{180°}{2N} - a_o\right) \qquad \text{odd number of teeth}$$

$$\cos \phi_1 = \frac{r \cos \phi}{q} = \frac{r_b}{q}$$

$$t = 2r\left(\frac{\pi}{N} + \theta - \theta_1 - \frac{a_o}{r_b}\right)$$

Case III. Helical Gear

$$M_1 = 2(q + a_o) \qquad \text{even number of teeth (Fig. 7-11)}$$

$$M_2 = 2\left(q \cos \frac{180°}{2N} + a_o\right) \qquad \text{odd number of teeth}$$

$$\cos \phi_1 = \frac{r \cos \phi}{q} = \frac{r_b}{q}$$

$$t = 2r\left(\frac{\pi}{N} - \theta + \theta_1 - \frac{a_o}{r_b \cos \psi_b}\right)$$

Case IV. Internal Helical Gear

$$M_1 = 2(q - a_o) \qquad \text{even number of teeth (Fig. 7-12)}$$

$$M_2 = 2\left(q \cos \frac{180°}{2N} - a_o\right) \qquad \text{odd number of teeth}$$

$$\cos \phi_1 = \frac{r \cos \phi}{q} = \frac{r_b}{q}$$

$$t = 2r\left(\frac{\pi}{N} + \theta - \theta_1 - \frac{a_o}{r_b \cos \psi_b}\right) \qquad t_n = t \cos \psi$$

When the number of teeth is odd, the situation, as shown in Fig. 7-9, is modified to the extent that a different equation for measurement M_2 must be used.

Internal spur gears are handled by the equations of Table 7-8, case II. The geometry for an even number of teeth is illustrated in Fig. 7-10.

Case III pertains to helical gears. The geometry for an even number of teeth is given in Fig. 7-11. The helicoidal tooth surface for the right side of the groove is generated by the unwinding of triangle ACD from the base cylinder. Contact between the sphere and this surface is at P_2. This gives a distance from center O_1 to the involute at B in the transverse section of $a_o \sec \psi_b$. The equation for tooth thickness t is found by writing expressions for angle FOE. Equations for finding ϕ and ψ_b are given in Table 7-2.

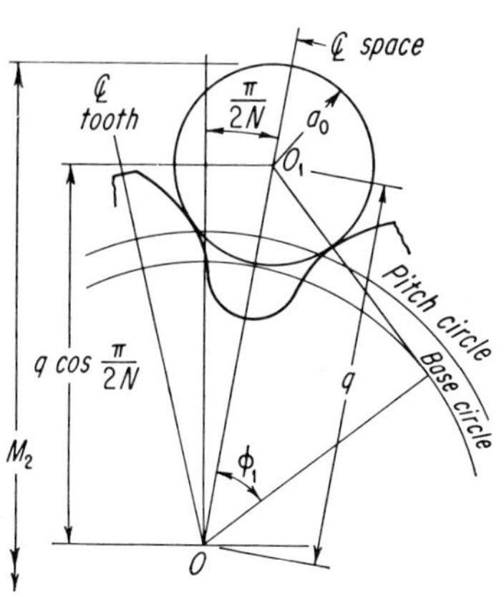

FIG. 7-9. Tooth thickness when number of teeth is odd.

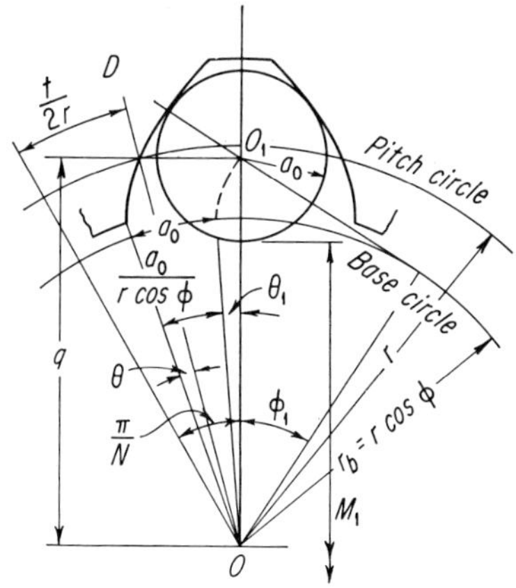

FIG. 7-10. Tooth thickness of internal spur gear by measurement between rolls or balls. Even number of teeth.

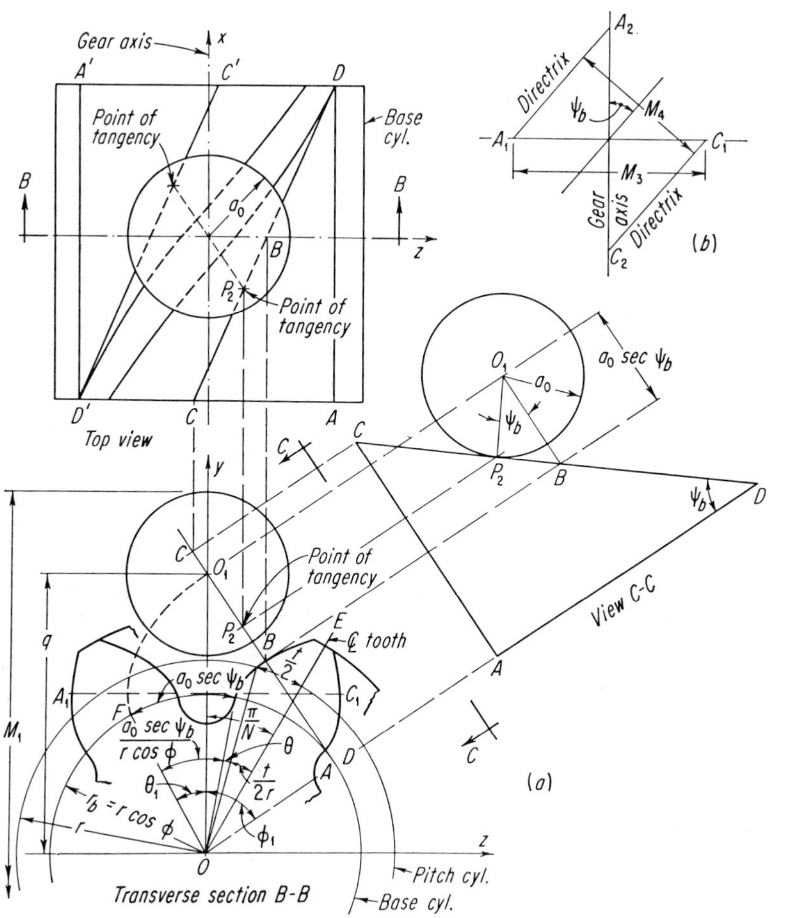

FIG. 7-11. Tooth thickness determined by balls for helical gear of even number of teeth.

Example 7-7. Find the tooth thickness t_n in the direction normal to the teeth for a helical gear of 19 teeth. $P_{dn} = 8$, $n = 20°$, and lead $L = 10''$. Measurement M_2 made over balls of diameter a_o of 0.1090″ is 3.8606″.

Solution:

$$\sin \psi = \frac{\pi N}{P_{dn} L} = \frac{19\pi}{80 \times 10} = 0.7461282 \qquad \psi = 48.2561°$$

By Eq. (7-10), $r = \dfrac{L \tan \psi}{2\pi} = \dfrac{10 \times 1.120646}{2\pi} = 1.783564''$

By Eq. (7-11), $\tan \varphi = \dfrac{\tan \varphi_n}{\cos \psi} = \dfrac{0.3639702}{0.6658048} = 0.5466621 \qquad \varphi = 28.6638°$

By Eq. (7-9), $r_b = r \cos \varphi = 1.783564 \times 0.8774494 = 1.564987''$

By Table 7-8, $q \cos \dfrac{\pi}{2N} = \dfrac{1}{2} M_2 - a_o = \dfrac{3.8606}{2} - 0.1090 = 1.8213$

$$\frac{\pi}{2N} = \frac{180}{38} = 4.7368° \qquad q = \frac{1.8213}{\cos 4.7368°} = 1.827542''$$

$$\cos \varphi_1 = \frac{r_b}{q} = \frac{1.564987}{1.827542} = 0.8563344 \qquad \varphi_1 = 31.0925°$$

Table 7-9. Equations for Finding Thickness of Teeth by Use of *m* and *μ* Functions

(Valid only for involute teeth and true involute helicoids)

Case I. Spur Gear

$$\mu_t = \frac{\pi}{2N} - \frac{a_o}{r \cos \phi}$$

$M_1 = 2(mr + a_o)$ even number of teeth

$M_2 = 2\left(mr \cos \dfrac{180°}{2N} + a_o\right)$ odd number of teeth

$$t = 2r\left(\frac{\pi}{N} - \mu - \frac{a_o}{r \cos \phi}\right)$$

Case II. Internal Spur Gear

$$\mu_t = -\frac{\pi}{2N} + \frac{a_o}{r \cos \phi}$$

$M_1 = 2(mr - a_o)$ even number of teeth

$M_2 = 2\left(mr \cos \dfrac{180°}{2N} - a_o\right)$ odd number of teeth

$$t = 2r\left(\frac{\pi}{N} + \mu - \frac{a_o}{r \cos \phi}\right)$$

Case III. Helical Gear

$$\mu_t = \frac{\pi}{2N} - \frac{a_o \sec \psi_b}{r \cos \phi}$$

$M_1 = 2(mr + a_o)$ even number of teeth

$M_2 = 2\left(mr \cos \dfrac{180°}{2N} + a_o\right)$ odd number of teeth

$t = 2r\left(\dfrac{\pi}{N} - \mu - \dfrac{a_o \sec \psi_b}{r \cos \phi}\right)$ $\qquad t_n = t \cos \psi$

Case IV. Internal Helical Gear

$$\mu_t = -\frac{\pi}{2N} + \frac{a_o \sec \psi_b}{r \cos \phi}$$

$M_1 = 2(mr - a_o)$ even number of teeth

$M_2 = 2\left(mr \cos \dfrac{180°}{2N} - a_o\right)$ odd number of teeth

$t = 2r\left(\dfrac{\pi}{N} + \mu - \dfrac{a_o \sec \psi_b}{r \cos \phi}\right)$ $\qquad t_n = t \cos \psi$

By Eq. (7-10), $\cot \psi_b = \dfrac{r \cot \psi}{r_b} = \dfrac{1.783564 \times 0.8923429}{1.564987} = 1.016974 \qquad \psi_b = 44.5178°$

By Table 7-8,

$$t = 2 \times 1.783564 \left[\frac{\pi}{19} - 0.0463854 + 0.0603936 - \frac{0.1090}{1.564987 \times 0.7130327} \right]$$

$$= 3.567128 \times 0.0816751 = 0.291346''$$

By Eq. (7-7), $t_n = t \cos \psi = 0.291346 \times 0.6658048 = 0.19398''$

Equations for the use of balls with internal helical gears are given in Table 7-8, case IV. The geometry for an even number of teeth is shown in Fig. 7-12.

Tables to assist in the calculations for measurements over rolls and balls are given in Chap. 23.

7-10. Measurement over Rolls or Balls by *m* and μ Functions. When tooth-thickness measurements must be made for a quantity of gears, it is convenient to prepare a graph which gives the tooth thickness t vs. measurement M_1 or M_2 for the range of values likely to be encountered in the lot of gears.

Equations suitable for preparing such a chart are given in Table 7-9. These are obtained from Table 7-8 by the following substitutions:

$$\mu = \text{inv}\ \phi - \text{inv}\ \phi_1 = \theta - \theta_1 \qquad (7\text{-}19)$$

$$m = \frac{q}{r} = \frac{\cos \phi}{\cos \phi_1} \qquad (7\text{-}20)$$

A trial value μ_t is obtained by taking tooth thickness t as equal to one-half the circular pitch, or $\pi r/N$. This gives a *target* value of μ to look for in the table.

Functions m and μ are tabulated in Chap. 24, Table 24-15. Use of the functions is illustrated by the following example.

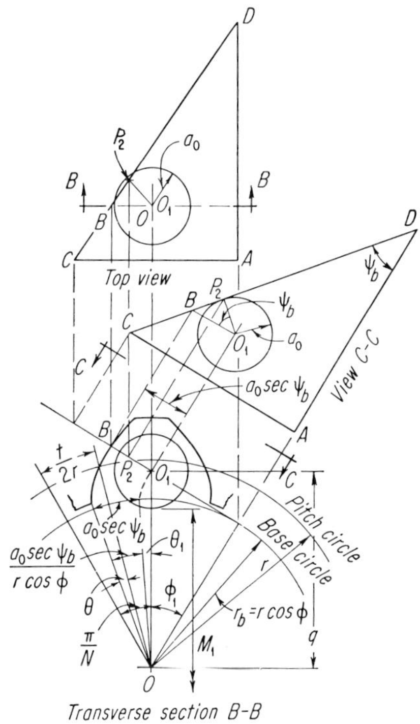

FIG. 7-12. Tooth thickness of internal helical gear by measurement between balls. Even number of teeth.

Example 7-8. A 20° spur gear has 18 teeth and diametral pitch equal to 4. Tooth-thickness measurements are to be made by means of rolls of diameter $2a_o$ equal to 0.4320″. Determine two pairs of coordinates for the curve of t vs. M_1.

Solution:

$$\text{Pitch radius } r = \frac{N}{2P_d} = \frac{18}{2 \times 4} = 2.2500''$$

By case I, Table 7-9, $\mu_t = \dfrac{\pi}{2 \times 18} - \dfrac{0.2160}{2.25 \times 0.93969} = -0.0149$

In Table 24-15 the trial value for μ_t lies between the two following entries for μ and m:

$\phi_1 = 25.0° \qquad \mu = -0.015071 \qquad m = 1.036836$

$\phi_1 = 24.8° \qquad \mu = -0.014319 \qquad m = 1.035158$

Use the first pair to compute values of t and M_1 as follows:

$$t = 2 \times 2.25 \left(\frac{\pi}{18} + 0.015071 - \frac{0.2160}{2.25 \times 0.93969} \right)$$
$$= 4.50(0.174533 + 0.015071 - 0.102161) = 0.39349''$$
$$M_1 = 2(1.036836 \times 2.25 + 0.2160) = 5.09776''$$

The second set of values of μ and m give the following:

$$t = 4.50(0.174533 + 0.014319 - 0.102161) = 0.39011''$$
$$M_1 = 2(1.035157 \times 2.25 + 0.2160) = 5.09021''$$

These are plotted in Fig. 7-13 and the curve drawn. Then, for any measurement M_1, the curve gives the corresponding pitch-line tooth thickness t. For example, for a measurement over the rolls of 5.0931″ the thickness is 0.3914″ to the nearest 0.0001″.

Calculation for μ_t need not have great accuracy and can be made with a slide rule if desired.

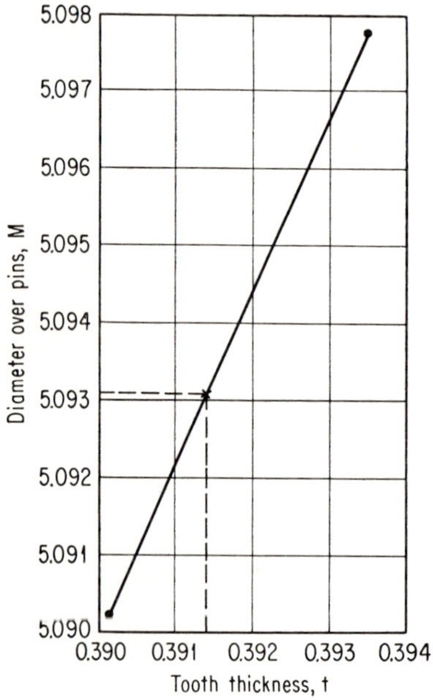

FIG. 7-13. Diameter over pins vs. tooth thickness.

7-11. Span Measurement by Vernier Calipers. Thickness of the teeth on the pitch circle can be found by taking a measurement* with a vernier calipers as shown in Fig. 7-14. Here OF and OG are the center lines between which N_1 teeth are located. Angle FOG then is equal to $2\pi N_1/N$. One-half of measurement M_3, or AB, is equal to distance BD when laid off around the base circle. One-half of tooth thickness t subtends an angle of $t/2r$. By writing the expression for angle DOE, the following equation is obtained:

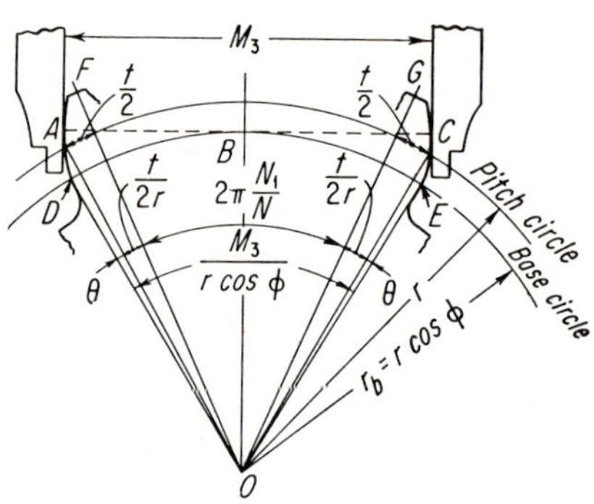

FIG. 7-14. Thickness of tooth determined by vernier calipers.

$$t = r\left(\frac{M_3}{r_b} - 2\pi\frac{N_1}{N} - 2\theta\right) \tag{7-21}$$

This equation can be easily adjusted for use on helical gears.† In Fig. 7-11, let a horizontal plane A_1C_1 be tangent to the base cylinder. Line A_1C_1 consists of the projection in the transverse section of directrixes A_1A_2 and C_1C_2 which are generating

* See Wildhaber, E., Measuring Tooth Thickness of Involute Gears, *Am. Machinist*, Vol. 59, p. 551, 1923.

† See Wildhaber, E., Measuring the Tooth Thickness of Involute Helical Gears, *Am. Machinist*, Vol. 59, p. 587, 1923.

the helicoidal surfaces. These particular directrixes, in the top view of Fig. 7-11*b*, are inclined to the gear axis at the base helix angle ψ_b. Measurement M_4, normal to the directrixes, is made by a vernier calipers. Then

$$M_4 = M_3 \cos \psi_b$$

When the value of M_3 from this equation is substituted the result is

$$t = 2r\left(\frac{M_4}{2r_b \cos \psi_b} - \pi \frac{N_1}{N} - \theta\right) \tag{7-22}$$

Example 7-9. An 18-tooth helical gear has $\phi_n = 14\frac{1}{2}°$ and $\psi = 22\frac{1}{2}°$. Normal diametral pitch P_{nd} is 5. Measurement M_3 when taken across two teeth ($N_1 = 1$) is 0.9368″. Find the value of normal tooth thickness t_n.

Solution:

By Eq. (7-11),

$$\tan \phi = \frac{\tan \phi_n}{\cos \psi} = \frac{0.258618}{0.923880} = 0.279926$$

$$\phi = 15.638°$$

By Eq. (7-12),

$$\cos \psi_b = \frac{\sin \phi_n}{\sin \phi} = \frac{0.250380}{0.269564} = 0.928835$$

By Eq. (7-4),

$$r = \frac{N}{2P_{nd} \cos \psi} = \frac{18}{2 \times 5 \times 0.923880} = 1.948306''$$

By Eq. (7-9),

$$r_b = r \cos \phi = 1.948306 \times 0.962983 = 1.876185''$$

By Eq. (7-22),

$$t = 2 \times 1.948306\left(\frac{0.9368}{2 \times 1.876185 \times 0.928835} - \frac{\pi}{18} - 0.006986\right)$$

$$= 3.896612(0.268784 - 0.174533 - 0.006986) = 0.340037'' \text{ in plane of rotation}$$

By Eq. (7-7),

$$t_n = t \cos \psi = 0.340038 \times 0.923880 = 0.31415''$$

7-12. Chordal Addendum and Chordal Thickness. The chordal thickness of a tooth can be measured by gear-tooth calipers especially made for the purpose. The measurement, as shown in Fig. 7-15*a*, is made at the pitch circle at a depth equal to $a + h_c$, the addendum plus the rise of the arc of the pitch circle. The central angle subtended by an arc equal to half the tooth thickness $t/2$ is $t/2r$ radians or $90t/\pi r$ degrees. The value of h_c used in setting the calipers is then given by

$$h_c = r\left(1 - \cos \frac{90t}{\pi r}\right) \tag{7-23}$$

Chordal thickness t_c for a tooth of thickness t is found by

$$t_c = 2r \sin \frac{90t}{\pi r} \tag{7-24}$$

In these equations angle $90t/\pi r$ is given in degrees with r equal to the pitch radius of the gear.

The above equations can be easily adapted for use on helical gears. Pitch radius r is replaced by R, the radius of curvature of the ellipse found by a plane intersecting normal to the tooth surface. Thickness t is replaced by t_n, tooth thickness in the normal plane. Then

$$h_c = R\left(1 - \cos\frac{90t_n}{\pi R}\right) \qquad (7\text{-}25)$$

$$t_c = 2R\sin\frac{90t_n}{\pi R} \qquad (7\text{-}26)$$

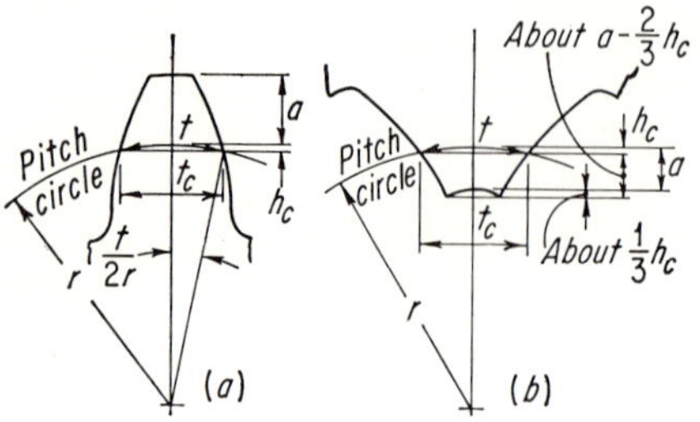

FIG. 7-15. Dimensions for measuring teeth by gear-tooth calipers.

By analytic geometry it can be shown that

$$R = \frac{r}{\cos^2\psi} \qquad (a)$$

where ψ is the helix angle, or angle between the direction of the teeth and the gear axis. Also

$$t_n = t\cos\psi \qquad (b)$$

$$p_n = p\cos\psi \qquad (c)$$

where t_n and p_n are, respectively, tooth thickness and circular pitch in the direction normal to the teeth. For spur gears $r = pN/2\pi$, and for helical gears, by Eq. (c),

$$r = \frac{p_nN}{2\pi\cos\psi} \qquad (d)$$

In Eq. (a),

$$R = \frac{p_nN}{2\pi\cos^3\psi} = \frac{N}{2P_{nd}\cos^3\psi} \qquad (e)$$

where $P_{nd} = \pi/p_n$, the normal diametral pitch.

Example 7-10. A helical gear of normal diametral pitch 3 and 45° helix angle has 15 teeth. Find values for h_c and t_c if the tooth thickness is exactly equal to one-half the circular pitch.

Solution:

By Eq. (e), $$R = \frac{15}{2 \times 3 \times 0.707107^3} = 7.07107''$$

As given, $$t_n = \tfrac{1}{2}p_n = \frac{\pi}{2P_{nd}} = \frac{\pi}{6}$$

In Eq. (7-25), $$h_c = R\left(1 - \cos\frac{90}{6R}\right) = 7.07107(1 - \cos 2.1213°) = 0.00484''$$

In Eq. (7-26), $$t_c = 2 \times 7.07107 \sin 2.1213° = 0.52348''$$

Values for h_c and t_c can be found directly from Table 24-18.

Corresponding measurements can be made on internal gears. As shown in Fig. 7-15b the rise of arc h_c is deducted from the top of addendum a. This is then adjusted by replacing at the bottom about one-third of h_c since the calipers ride on the more or less sharp corners at the addendum circle. The chordal-addendum setting of the calipers, to a good degree of approximation, is then equal to $a - \frac{2}{3}h_c$.

7-13. Angular Relationship between Teeth and Keyway. The angular position of the teeth with respect to a keyway can be found from the layout of Fig. 7-16.

Distances OB, OC, BC, and CE can be determined by measurement. The cosine theorem gives the value of angle δ.

$$\cos \delta = \frac{\overline{OB}^2 + \overline{OC}^2 - \overline{BC}^2}{2\,OB \times OC}$$

and

$$\sin \alpha = \frac{CE}{OC}$$

Angle γ is known from the spacing of the teeth. Hence angle β is given by

$$\beta = \gamma - \delta - \alpha \tag{7-27}$$

7-14. Variation of Tooth Thickness with Cutter Setting. For a hob or rack cutter, the thickness of the tooth and the width of the space on the pitch line are each equal to one-half the circular pitch. The thickness of the gear tooth which is

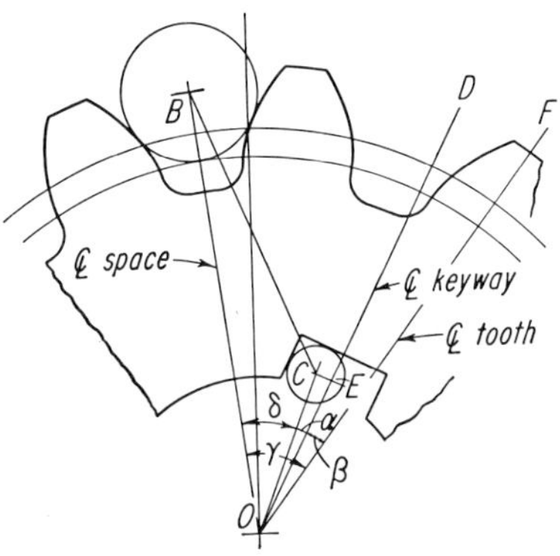

FIG. 7-16. Angular position of keyway with respect to teeth.

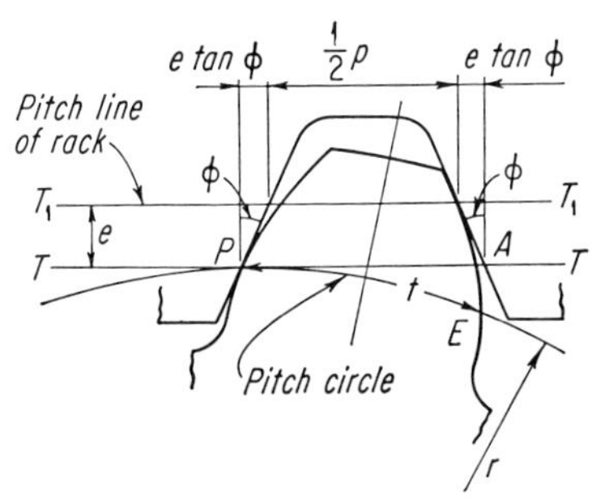

FIG. 7-17. Thickness of tooth on pitch circle as affected by cutter setting.

cut depends on the setting of the cutting tool with respect to the blank. In Fig. 7-17, the pitch line T_1T_1 of the rack cutter is located a distance e from TT, the tangent to the pitch circle, which gives a tooth thickness t of excessive width. During cutting, movement along the pitch circle is identical with movement along the pitch line of the rack. Distance PA is then equal to arc PE so that

$$t = 2e \tan \phi + \tfrac{1}{2}p \tag{7-28}$$

When distance e is taken to the inside of TT, the teeth become thinner than $\frac{1}{2}p$. Equation (7-28) applies with e substituted as a negative number. Such a setting is required to provide backlash in the assembly.*

7-15. Thickness of Tooth at Any Diameter. When the tooth thickness t on the pitch circle is known, its thickness t' at any other circle of radius r' can be found from the layout given in Fig. 7-18. Angle ϕ' can be found from

$$\cos \phi' = \frac{r \cos \phi}{r'} = \frac{r_b}{r'} \tag{7-29}$$

* Additional problems in the geometry of gearing are considered in ref. 6.

By writing expressions for angle DOE, the following equation for t' can be derived:

$$t' = 2q\left(\frac{t}{2r} + \text{inv } \phi - \text{inv } \phi_1\right) \tag{7-30}$$

$$= mt + 2q\mu \tag{7-31}$$

where $m = q/r$ and μ is read from Table 24-15 for the value of m

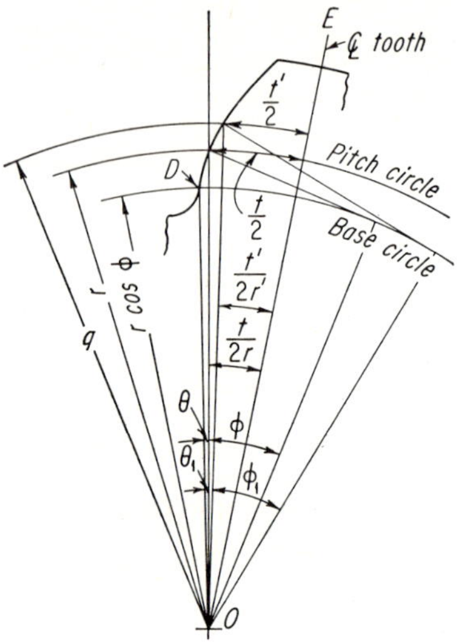

FIG. 7-18. Thickness of tooth at any radius q.

7-16. Radius at Which Teeth Become Pointed. The teeth become pointed when t' in Eq. (7-30) is reduced to zero. Then

$$\text{inv } \phi_1 = \frac{t}{2r} + \text{inv } \phi \tag{7-32}$$

or

$$\mu = -\frac{mt}{2q} \tag{7-33}$$

Angle ϕ_1 is thus determined. If R_1 is the radius to the tip of the pointed tooth,

$$R_1 = \frac{r \cos \phi}{\cos \phi_1} = rm \tag{7-34}$$

REFERENCES

1. Dudley, Darle W.: "Practical Gear Design," McGraw-Hill Book Company, Inc., New York, 1954.
2. Spotts, M. F.: Undercutting—Hobbed Spur Gear Teeth, *Machine Design*, Vol. 28, Apr. 19, 1956, p. 123.
3. Dalby, W. E.: "Spur and Helical Gear Problems," published by author, 600 Hidden Lane, Grosse Pointe 36, Mich., 1944.
4. LeGrand, R.: "The New American Machinists' Handbook," Sec. 4, McGraw-Hill Book Company, Inc., New York, 1955.
5. Vogel, W. F.: "Involutometry and Trigonometry," Michigan Tool Company, Detroit, Mich., 1945.
6. Buckingham, E.: "Spur Gears," McGraw-Hill Book Company, Inc., New York, 1928.
7. Spotts, M. F.: "Mechanical Design Analysis," Prentice-Hall, Inc., Englewood Cliffs, N.J., 1964.

Chapter 8

CALCULATION OF GEAR-TOOTH PROFILES

By

EDWARD C. VARNUM, *Head, Operations Research, Barber-Colman Company, Rockford, Ill.*

Gear-tooth profiles may be calculated by the use of conventional formulas and a desk calculating machine. Or they may be programmed and done on electronic computers such as the Bendix G-15. It is also possible to make certain desk calculations and then construct a graphical layout of the tooth on a drawing board.

Articles 8-1 to 8-8 show basic equations for tooth-profile calculations and indicate how electronic computers may be used on gear problems which are ordinarily rather time-consuming. Articles 8-9 and 8-10 show how to solve gear-profile problems by the more conventional calculation and layout procedure.

Articles 8-1 to 8-4 deal with involute gear profiles. Articles 8-5 to 8-8 deal with problems that have to do with milling or grinding single-enveloping worms. Articles 8-9 to 8-13 deal with bevel and hypoid gears.

Those who study this chapter closely will note that it is rather simple to apply the layout procedures of Arts. 8-9 and 8-10 to spur and helical gears. Likewise the programming techniques of the first articles can be applied to many other problems.

Electronic-computer Programs. Once a calculation is programmed, it takes only a few seconds to calculate each plot point. A 10-point profile may be determined in less time than it would take to get the first point using a desk calculator.

The formulas presented here have been programmed on a Bendix G-15 at Barber-Colman Company using an interpretive routine called "Intercom 1000 Double Precision."

An electronic-computer program contains several code sheets. These code sheets contain (1) notes, (2) command locations, (3) commands, and (4) storage locations. Table 8-1 is an example of a complete code sheet.

Because many readers will be using equipment other than Bendix or other interpretive routines, we shall exhibit only the notes of our programs. From these notes a computer program for IBM Royal-McBee, Burroughs, or other equipment can readily be constructed. The equations given can also be used directly to solve problems by conventional desk-calculation methods. Numerical values are given for key results so that the reader can check his solution of the problems given.

TROCHOIDS

A *trochoid* is a curve which is traced on a rotating plane by a designated point on a second plane. When the second plane moves in a straight line we have a *rack trochoid,* which is the curve on a gear traced by a point on a meshing rack.

8-1. Rack Trochoid. Consider a rack moving horizontally to the right in mesh with a gear whose center O is below the rack. The pitch line of the rack is tangent to the pitch circle of the gear at Q. Consider a point P lying on OQ at a distance b below Q. As the rack moves to the right the point on the rack which was at P will move to P'. During the time that P moves to P' on the rack, the point on the gear which was at P will rotate to point P'' and the point which was at Q will rotate to point Q''. As Q and Q'' are on the pitch circle the length of the arc QQ'' equals the length of the horizontal segment PP'. Let radius OP' be called r_t and angle POP' be called α (alpha). Calling the pitch radius of the gear R, we can determine

$$\cos \alpha = \frac{R - b}{r_t} \tag{8-1}$$

The common length of segment PP' and arc QQ'' is $(R - b) \tan \alpha$, and angle POP'' in radians will be $[(R - b) \tan \alpha]/R$. Let angle $P'OP''$ be denoted by θ_t (theta sub t),

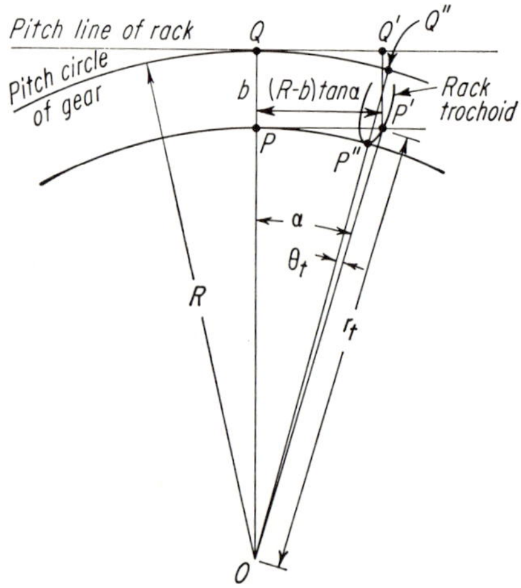

FIG. 8-1. Schematic drawing of rack trochoid.

which is the difference between angles POP' and POP''. By expanding and rearranging this difference we find

$$\theta_t = \frac{b}{R} \tan \alpha - \text{inv } \alpha \tag{8-2}$$

The polar coordinates of a rack trochoid are r_t and θ_t. The line $\theta_t = 0$ is a line of symmetry for the trochoid. On this line r_t reaches its minimum value of $R - b$.

Table 8-1 is a code sheet with which a point of a rack trochoid may be calculated on an electronic computer. Equation (8-1) is used to find α and Eq. (8-2) is evaluated to find θ_t for a given value of r_t. By changing r_t, new points on the trochoid may be found.

Equations (8-1) and (8-2) give the polar coordinates of a rack trochoid having its line of symmetry as its polar axis. In general this line of symmetry is not a convenient reference line. If we wish to use the gear-space center line for reference in polar coordinates we increase θ_t by δ_s (delta sub s) where the latter is given in radian measure by dividing B by R. B is the horizontal distance from the rack point to the vertical center line of the rack tooth.

Table 8-1. Code Sheet for Bendix G-15 Computer

INTERCOM PROGRAM

Prepared by E. C. Varnum Date 6-25-59

PROBLEM Rack Trochoid Page 1 of 1 Line 11

NOTES	Location	K	OP CODE	Address		S/D	(A) or (AA)
Clear and add R	1100		42	14	00		
Subtract b	1101		41	14	02		
Divide by r_t	1102		48	14	08		
Store cos $\alpha \rightarrow$ 1410	1103		49	14	10		
Clear and subtract cos α	1104		40	14	10		
Multiply by cos α	1105		44	14	10		
Add 1	1106		43	11	26		
Square root	1107		08	10	97		
Store sin $\alpha \rightarrow$ 1412	1108		49	14	12		
Divide by cos α	1109		48	14	10		
Store tan $\alpha \rightarrow$ 1414	1110		49	14	14		
Arctangent	1111		08	12	24		
Store $\alpha \rightarrow$ 1416	1112		49	14	16		
Subtract tan α	1113		41	14	14		
Store $-$inv $\alpha \rightarrow$ 1418	1114		49	14	18		
Clear and add b	1115		42	14	02		
Multiply by tan α	1116		44	14	14		
Divide by R	1117		48	14	00		
Add $-$inv α	1118		43	14	18		
Store $\theta_t \rightarrow$ 1420	1119		49	14	20		
Type r_t and tab	1120		33	14	08		
Clear and add θ_t	1121		42	14	20		
Divide by $\pi/180$	1122		48	11	28		
Type θ_t and carriage ret.	1123		38	21	00		
Ring bell	1124		63	00	00		
Halt	1125		67	00	00		

To find rectangular coordinates (X_t, Y_t) of the rack trochoid with the gear-space center line as the Y axis, we have

$$X_t = r_t \sin(\theta_t + \delta_s) \qquad Y_t = r_t \cos(\theta_t + \delta_s) \tag{8-3}$$

8-2. Trochoidal Fillet. In theory, each point on a rack creates its own trochoid on the meshing gear, and the final gear curve is the envelope of all these individual trochoids. In practice, only one of these many trochoids is of interest, namely, the trochoid for a corner point of the rack. Moreover, even this point exists only in theory because of wear; so we visualize a circle at the corner of the rack and consider the center of the circle as generating a rack trochoid. At each point of this rack trochoid we imagine a circle having its center at the point. The *trochoidal fillet* is then an envelope of this family of circles.

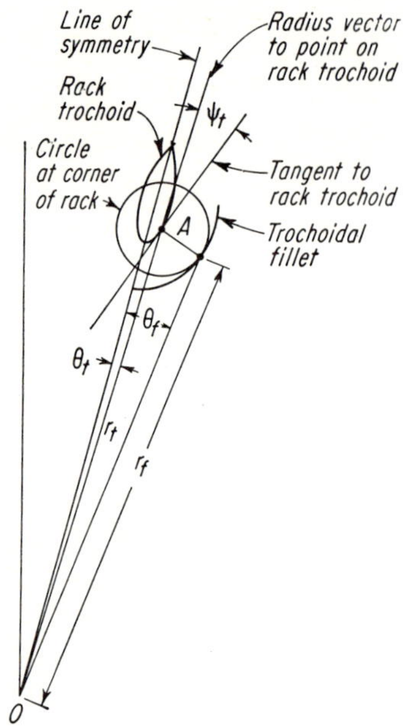

FIG. 8-2. Trochoidal fillet.

To find points on a trochoidal fillet we choose r_t and find θ_t from (8-1) and (8-2). These values of r_t and θ_t locate a point on the rack trochoid traced by the center of the circle. At this point on the trochoid we determine the angle ψ_t (psi sub t) between a radius to the point and a tangent to the trochoid at the point. The angle ψ_t is found by the relationship

$$\tan \psi_t = \frac{(R - b) - (r_t^2/R)}{\sqrt{r_t^2 - (R - b)^2}} \tag{8-4}$$

With angle ψ_t known and calling the radius of the circle A, we find the polar coordinates (r_f, θ_f) of a point on the trochoid fillet. This point lies on the normal to the rack trochoid at a distance A from the point (r_t, θ_t) of the rack trochoid.

$$r_f = \sqrt{r_t^2 + A(A - 2r_t \sin \psi_t)} \tag{8-5}$$
$$\theta_f = \theta_t + \alpha$$

$$\text{where } \tan \alpha = \frac{A \cos \psi_t}{r_t - A \sin \psi_t} \tag{8-6}$$

These equations are derived by considering a triangle having sides r_t, r_f, and A with the angle between r_t and A being the complement of ψ_t. Angle α (alpha) lies between r_t and r_f. The polar coordinates of a fillet point given in (8-5) and (8-6) have the line of symmetry as the polar axis. If δ_s is the angle between this line of symmetry and the gear-space center line, the rectangular coordinates of the fillet become

$$X_f = r_f \sin(\theta_f + \delta_s) \qquad Y_f = r_f \cos(\theta_f + \delta_s) \tag{8-7}$$

Program for Calculating Rectangular Coordinates of Trochoidal Fillet. In order to evaluate Eqs. (8-1) to (8-7), we need numerical values for the following dimensions, which we assume as shown:

R = pitch radius of gear = 10.000000″
A = radius of corner circle = 0.238600″
b = vertical coordinate of center of corner circle = 0.918400″
B = horizontal coordinate of center of corner circle = 0.197220″
r_t = radius to a point on the trochoid = 9.100000″

where b is measured from the pitch line of the rack and B is measured from the center line of the rack tooth.

Using these five values we perform the following 60 commands to find a trochoidal fillet point 0.366644″ to the right of the gear-space center line and 8.900140″ above a horizontal line through the center of the gear. As a guide in using the command list we tabulate the result of each command. The 15 stored numbers are placed in parentheses. For the remaining command lists in this chapter we shall give only the stored numerical values because the reader should be able to follow commands between storages without numerical values to check against.

No.	*Command*	*Result*
1.	Clear and add R	10.000000
2.	Subtract b	9.081600
3.	Store $R - b$	(9.081600)1
4.	Divide by r_t	0.997978
5.	Arccosine	0.063602
6.	Store α	(0.063602)2
7.	Tangent	0.063688
8.	Store tan α	(0.063688)3
9.	Subtract α	0.000086
10.	Store inv α	(0.000086)4
11.	Clear and add b	0.918400
12.	Multiply by tan α	0.058491
13.	Divide by R	0.005849
14.	Subtract inv α	0.005763
15.	Store θ_t	(0.005763)5
16.	Clear and add r_t	9.100000
17.	Multiply by r_t	82.810000
18.	Store $r^2{}_t$	(82.810000)6
19.	Clear and subtract $(R - b)$	−9.081600
20.	Multiply by $(R - b)$	−82.475458
21.	Add $r^2{}_t$	0.334542
22.	Square root	0.578396
23.	Store $\sqrt{r^2{}_t - (R - b)^2}$	(0.578396)7
24.	Clear and subtract $r^2{}_t$	−82.810000
25.	Divide by R	−8.281000
26.	Add $R - b$	0.800600
27.	Divide by $\sqrt{r_t{}^2 - (R - b)^2}$	1.384173
28.	Arctangent	0.945159
29.	Store ψ_t	(0.945159)8
30.	Sine	0.810590
31.	Store sin ψ_t	(0.810590)9
32.	Clear and subtract sin ψ_t	−0.810590
33.	Subtract sin ψ_t	−1.621180
34.	Multiply by r_t	−14.752738
35.	Add A	−14.514138
36.	Multiply by A	−3.463073
37.	Add $r_t{}^2$	79.346927
38.	Square root	8.907689
39.	Store r_f	(8.907689)10
40.	Clear and add r_t	9.100000
41.	Divide by A	38.139145
42.	Subtract sin ψ_t	37.328555

No.	*Command*	*Result*
43.	Store $(r_t/A) - \sin \psi_t$	(37.328555)11
44.	Clear and add ψ_t	0.945159
45.	Cosine	0.585613
46.	Divide by $(r_t/A) - \sin \psi_t$	0.015688
47.	Arctangent	0.015686
48.	Add θ_t	0.021450
49.	Store θ_f	(0.021450)12
50.	Clear and add B	0.197220
51.	Divide by R	0.019722
52.	Add θ_f	0.041172
53.	Store $\theta_f + \delta_s$	(0.041172)13
54.	Sine	0.041160
55.	Multiply by r_f	0.366644
56.	Store X_f	(0.366644)14
57.	Clear and add $\theta_f + \delta_s$	0.041172
58.	Cosine	0.999152
59.	Multiply by r_f	8.900140
60.	Store Y_f	(8.900140)15

Having found this trochoidal fillet point whose rectangular coordinates are 0.366644 and 8.900140, we may now change r_t only and repeat the same 60 commands to obtain another fillet point. In this way we may calculate enough plot points to draw the trochoidal fillet itself.

Note that all angles are measured in radians. In performing the arccosine command in step 5 and the arctangent commands in steps 28 and 47 the resulting angles must be changed to radians. This can be done by expressing the angle in degrees and decimals of a degree and then multiplying by 0.017453. This multiplier is pi divided by 180, which is carried as a constant on the Barber-Colman Bendix G-15 program tapes as 0.0174532925199.

INVOLUTES

An *involute* of a given curve C is a curve I such that C is the locus of the centers of curvature of I. In gearing theory, the curve C is always a circle, called the *base circle,* and an involute is visualized as being created by unwrapping a string from the base circle.

8-3. Tangent and Normal Coordinates of an Involute. Consider a point P on an involute curve. Draw the tangent at P which we call the x axis. Draw the normal at P and call it the y axis. We may find the x and y coordinates of a second point Q on the involute by use of the following equations, developed by Allan Candee:

$$x = r[\sin \phi \sin \theta + \cos \phi(\theta \sin \theta + \cos \theta - 1)] \quad (8\text{-}8)$$
$$y = r[\sin \phi(1 - \cos \theta) + \cos \phi(\sin \theta - \theta \cos \theta)] \quad (8\text{-}9)$$

where r = radius to point P

ϕ = pressure angle at point P

θ = angle between radii drawn to two points on the base circle, one on the normal through P and the other on the normal through Q

Theta (θ) and x are both positive when Q is above P and both negative when Q is below P. y is always positive and is measured toward the gear-tooth center line. At P, x and y are both zero.

Both r and phi (ϕ) are fixed for a given selection of P, so that additional points Q

on the involute can be calculated by varying θ only. The point P is usually taken to be the pitch point, in which case r is the pitch radius of the gear and ϕ is its pressure angle.

In order to facilitate computations using Eqs. (8-8) and (8-9), we have assumed three popular pressure angles (14.5, 20, 30°) and for each we have found the values of θ for which the bracket in (8-8) equals certain convenient values. With these solutions for θ we have evaluated the bracket in (8-9). From the resulting table we can readily obtain tangent and normal coordinates by multiplying the tabular values by r, the pitch radius.

For a given pressure angle ϕ, the values of x and y for the intersection point of the involute and the base circle are given by

$$x = r\{\cos\phi[\cos(\tan\phi) - 1]\} \qquad (8\text{-}10)$$

$$y = r[\sin\phi - \cos\phi \sin(\tan\phi)] \qquad (8\text{-}11)$$

Tangent and Normal Coordinates of an Approximating Circle. For purposes of approximation, the tangent and normal coordinates of the circle of curvature of the involute at point P may be found by

$$\frac{y}{r} = \tan\phi - \sqrt{\tan^2\phi - \left(\frac{x}{r}\right)^2} \qquad (8\text{-}12)$$

where r and ϕ are the radius and pressure angle at P and x and y are the tangent and normal coordinates of a second point S which lies on the circle of curvature for point P. This circle of curvature for point P has its center on the base circle and its radius is $r \tan\phi$, which is the length of the normal from the base circle to the point P.

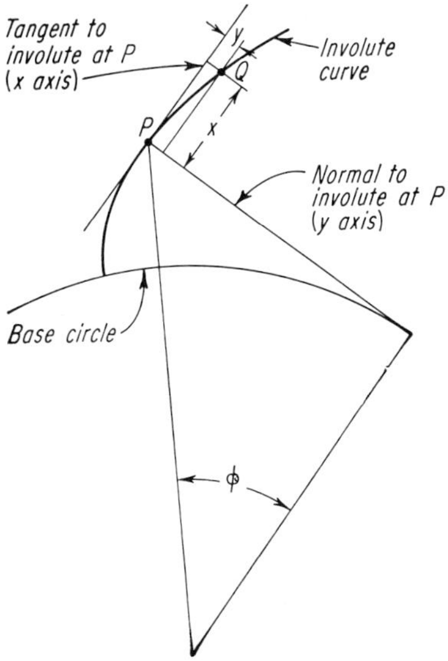

FIG. 8-3. Plotting tangent and normal coordinates of involute.

Table 8-3 has the same format as Table 8-2. For values of x/r near zero we note that the corresponding y/r values in the two tables are identical for practical purposes, but as Q and S move farther from P, the y/r values diverge appreciably. Remember that the differences observed in comparing respective entries in Tables 8-2 and 8-3 must always be multiplied by the pitch radius of the gear to determine their true magnitude.

Tangent and Normal Coordinates of an Involute at a Given Radius. Table 8-2 provides uniform steps along the tangent at the pitch point P. We now show how to calculate tangent and normal coordinates of points on an involute whose radii change uniformly. Let R be the radius of point Q and call its pressure angle ϕ_1. The theta value for point Q is equal to the difference between $\tan\phi_1$ and $\tan\phi$ when θ is measured in radians. However, the value of $\sec\phi_1$ can be expressed as R/r times the secant of ϕ, and thus we can find θ:

$$\theta = \sqrt{\left(\frac{R/r}{\cos\phi}\right)^2 - 1} - \tan\phi \qquad (8\text{-}13)$$

For negative values of x/r the tabular values of y/r are repeated, because the circle of curvature at P is symmetric about the normal drawn through P.

For a fixed value of ϕ, we calculate values of θ for various values of R/r and then use (8-8) to find x/r and (8-9) to find y/r. Note in (8-12) that R/r must never be less than $\cos\phi$; viz., the involute does not extend below the base circle.

Table 8-2. Tangent and Normal Coordinates of Involute Curve

x/r	Pressure angles, deg		
	14.5000000	20.0000000	30.0000000
0.2000000	0.0527684	0.0444761	0.0345812
0.1900000	0.0480781	0.0404334	0.0313500
0.1800000	0.0435875	0.0365694	0.0282697
0.1700000	0.0392971	0.0328847	0.0253405
0.1600000	0.0352076	0.0293802	0.0225632
0.1500000	0.0313201	0.0260568	0.0199385
0.1400000	0.0276362	0.0229160	0.0174673
0.1300000	0.0241577	0.0199595	0.0151506
0.1200000	0.0208872	0.0171894	0.0129897
0.1100000	0.0178276	0.0146080	0.0109861
0.1000000	0.0149826	0.0122183	0.0091414
0.0900000	0.0123567	0.0100236	0.0074574
0.0800000	0.0099551	0.0080279	0.0059365
0.0700000	0.0077842	0.0062359	0.0045808
0.0600000	0.0058518	0.0046528	0.0033933
0.0500000	0.0041672	0.0032852	0.0023769
0.0400000	0.0027421	0.0021404	0.0015350
0.0300000	0.0015909	0.0012275	0.0008717
0.0200000	0.0007322	0.0005571	0.0003914
0.0100000	0.0001905	0.0001425	0.0000989
0.0000000	0.0000000	0.0000000	0.0000000
−0.0100000	0.0002115	0.0001504	0.0001012
−0.0200000	0.0009117	0.0006212	0.0004099
−0.0300000	0.0023062	0.0014507	0.0009348
−0.0400000	0.0000000	0.0026962	0.0016860
−0.0500000	0.0000000	0.0044566	0.0026753
−0.0600000	0.0000000	0.0069773	0.0039169
−0.0700000	0.0000000	0.0000000	0.0054283
−0.0800000	0.0000000	0.0000000	0.0072309
−0.0900000	0.0000000	0.0000000	0.0093523
−0.1000000	0.0000000	0.0000000	0.0118290
−0.1100000	0.0000000	0.0000000	0.0147125
−0.1200000	0.0000000	0.0000000	0.0180810
−0.1300000	0.0000000	0.0000000	0.0220746
−0.1400000	0.0000000	0.0000000	0.0270870
0.0200000	0.0007322	0.0005571	0.0003914
0.0190000	0.0006633	0.0005039	0.0003536
0.0180000	0.0005975	0.0004533	0.0003177
0.0170000	0.0005351	0.0004052	0.0002836
0.0160000	0.0004758	0.0003598	0.0002515
0.0150000	0.0004199	0.0003169	0.0002213
0.0140000	0.0003672	0.0002767	0.0001930
0.0130000	0.0003179	0.0002392	0.0001666
0.0120000	0.0002720	0.0002043	0.0001421
0.0110000	0.0002295	0.0001720	0.0001195
0.0100000	0.0001905	0.0001425	0.0000989
0.0090000	0.0001550	0.0001157	0.0000802
0.0080000	0.0001230	0.0000917	0.0000634
0.0070000	0.0000946	0.0000704	0.0000486
0.0060000	0.0000698	0.0000518	0.0000358

Table 8-2. Tangent and Normal Coordinates of Involute Curve (*Continued*)

x/r	Pressure angles, deg		
	14.5000000	20.0000000	30.0000000
0.0050000	0.0000487	0.0000361	0.0000249
0.0040000	0.0000313	0.0000231	0.0000159
0.0030000	0.0000177	0.0000131	0.0000090
0.0020000	0.0000079	0.0000058	0.0000040
0.0010000	0.0000020	0.0000015	0.0000010
0.0000000	0.0000000	0.0000000	0.0000000
−0.0010000	0.0000020	0.0000015	0.0000010
−0.0020000	0.0000081	0.0000059	0.0000040
−0.0030000	0.0000183	0.0000133	0.0000090
−0.0040000	0.0000326	0.0000236	0.0000161
−0.0050000	0.0000513	0.0000371	0.0000251
−0.0060000	0.0000743	0.0000535	0.0000363
−0.0070000	0.0001017	0.0000730	0.0000494
−0.0080000	0.0001337	0.0000957	0.0000646
−0.0090000	0.0001702	0.0001215	0.0000819
−0.0100000	0.0002115	0.0001504	0.0001012
−0.0110000	0.0002576	0.0001825	0.0001226
−0.0120000	0.0003086	0.0002179	0.0001461
−0.0130000	0.0003647	0.0002565	0.0001717
−0.0140000	0.0004260	0.0002984	0.0001993
0.5000000	0.2962273	0.2544627	0.2043338
0.4900000	0.2842462	0.2442902	0.1961895
0.4800000	0.2726049	0.2343807	0.1882398
0.4700000	0.2612911	0.2247270	0.1804811
0.4600000	0.2502934	0.2153227	0.1729099
0.4500000	0.2396015	0.2061614	0.1655232
0.4400000	0.2292057	0.1972374	0.1583178
0.4300000	0.2190970	0.1885453	0.1512909
0.4200000	0.2092674	0.1800800	0.1444399
0.4100000	0.1997091	0.1718369	0.1377623
0.4000000	0.1904150	0.1638115	0.1312556
0.3900000	0.1813787	0.1559998	0.1249179
0.3800000	0.1725939	0.1483979	0.1187468
0.3700000	0.1640551	0.1410023	0.1127406
0.3600000	0.1557571	0.1338097	0.1068974
0.3500000	0.1476949	0.1268171	0.1012157
0.3400000	0.1398642	0.1200215	0.0956937
0.3300000	0.1322608	0.1134205	0.0903302
0.3200000	0.1248809	0.1070115	0.0851237
0.3100000	0.1177210	0.1007923	0.0800732
0.3000000	0.1107780	0.0947611	0.0751775
0.2900000	0.1040489	0.0889159	0.0704358
0.2800000	0.0975311	0.0832551	0.0658470
0.2700000	0.0912222	0.0777774	0.0614106
0.2600000	0.0851202	0.0724814	0.0571258
0.2500000	0.0792232	0.0673661	0.0529922
0.2400000	0.0735297	0.0624306	0.0490093
0.2300000	0.0680384	0.0576743	0.0451769
0.2200000	0.0627482	0.0530966	0.0414948
0.2100000	0.0576584	0.0486972	0.0379628

Table 8-3. Tangent and Normal Coordinates of Circle of Curvature at Pitch Point

x/r	Pressure angles, deg		
	14.5	20	30
0.2000000	0.0946571	0.0598743	0.0357477
0.1900000	0.0831666	0.0535281	0.0321592
0.1800000	0.0729214	0.0476250	0.0287764
0.1700000	0.0637252	0.0421406	0.0255955
0.1600000	0.0554353	0.0370538	0.0226131
0.1500000	0.0479447	0.0323465	0.0198260
0.1400000	0.0411709	0.0280025	0.0172312
0.1300000	0.0350487	0.0240080	0.0148262
0.1200000	0.0295258	0.0203508	0.0126084
0.1100000	0.0245598	0.0170202	0.0105758
0.1000000	0.0201159	0.0140069	0.0087262
0.0900000	0.0161654	0.0113028	0.0070579
0.0800000	0.0126846	0.0089008	0.0055694
0.0700000	0.0096536	0.0067947	0.0042592
0.0600000	0.0070563	0.0049795	0.0031262
0.0500000	0.0048794	0.0034507	0.0021691
0.0400000	0.0031121	0.0022047	0.0013873
0.0300000	0.0017459	0.0012385	0.0007799
0.0200000	0.0007745	0.0005499	0.0003465
0.0100000	0.0001934	0.0001374	0.0000866
0.0200000	0.0007745	0.0005499	0.0003465
0.0190000	0.0006989	0.0004963	0.0003127
0.0180000	0.0006272	0.0004454	0.0002807
0.0170000	0.0005593	0.0003972	0.0002503
0.0160000	0.0004954	0.0003518	0.0002217
0.0150000	0.0004354	0.0003092	0.0001949
0.0140000	0.0003792	0.0002694	0.0001698
0.0130000	0.0003269	0.0002322	0.0001464
0.0120000	0.0002786	0.0001979	0.0001247
0.0110000	0.0002340	0.0001663	0.0001048
0.0100000	0.0001934	0.0001374	0.0000866
0.0090000	0.0001566	0.0001113	0.0000702
0.0080000	0.0001238	0.0000879	0.0000554
0.0070000	0.0000948	0.0000673	0.0000424
0.0060000	0.0000696	0.0000495	0.0000312
0.0050000	0.0000483	0.0000343	0.0000217
0.0040000	0.0000309	0.0000220	0.0000139
0.0030000	0.0000174	0.0000124	0.0000078
0.0020000	0.0000077	0.0000055	0.0000035
0.0010000	0.0000019	0.0000014	0.0000009
0.0000000	0.0000000	0.0000000	0.0000000

In Table 8-4 we have assumed a pressure angle of 20° and a maximum R/r ratio of 1.2000 with a 0.01 decrement. The values of x/r and y/r in the second and third columns may be multiplied by the pitch radius of any 20° pressure angle involute spur gear to find tangent and normal coordinates of points on the gear. The values of theta given in the last column are expressed in degrees; so the results from (8-13) would need to be divided by 0.0174532925199 to yield the tabular values.

Program for Calculating Tangent and Normal Coordinates of an Involute at a Given Radius. The commands used in calculating x/r and y/r in Table 8-4 include the following 32 steps which were performed after determining the trigonometric functions of phi:

$\sin\phi = 0.3420201433$
$\cos\phi = 0.9396926208$
$\tan\phi = 0.3639702343$

1. Clear and add R/r
2. Store R/r: (1.100000)1
3. Divide by $\cos\phi$
4. Square
5. Subtract 1
6. Square root
7. Subtract $\tan\phi$
8. Store θ: (0.244548)2
9. Sine (radians)
10. Store $\sin\theta$: (0.242117)3
11. Multiply by $\sin\phi$
12. Store $\sin\theta \sin\phi$: (0.082809)4
13. Clear and add θ
14. Cosine (radians)
15. Store $\cos\theta$: (0.970247)5
16. Clear and add θ
17. Multiply by $\sin\theta$
18. Add $\cos\theta$
19. Subtract 1
20. Multiply by $\cos\phi$
21. Add $\sin\theta \sin\phi$
22. Store x/r: (0.110489)6
23. Clear and add 1
24. Subtract $\cos\theta$
25. Multiply by $\sin\phi$
26. Store $\sin\phi$ vers θ: (0.010176)7
27. Clear and subtract θ
28. Multiply by $\cos\theta$
29. Add $\sin\theta$
30. Multiply by $\cos\phi$
31. Add $\sin\phi$ vers θ
32. Store y/r: (0.014730)8

The eight numbers in parentheses were stored. The values used for $\sin\phi$, $\cos\phi$, and $\tan\phi$ were stored before the above steps were begun. These functions of ϕ may be either calculated or obtained from trigonometric tables. To prepare Table 8-4, storages 1, 6, and 8 were typed out and storage 2 was divided by $\pi/180$ and then typed. Finally, storage 1 was decremented and the process repeated.

Steps 9 to 21, inclusive, show how x/r can be calculated for an angle θ. The remaining steps show how y/r can be found. If tables of the trigonometric functions in radian measure are not available to desk-calculator operators, the θ value must be changed to decimals of a degree or to degrees, minutes, and seconds. For example, storage 2 is 0.244548 radian, which can be converted to 14.011568 degrees or to 14 degrees, 0 minute, and 41.6 seconds.

Tables for Small Values. For a gear with many teeth the addendum is so small relative to the pitch radius that only five or six entries in the first part of Table 8-2 would be useful. The values in the end of Table 8-2 are a refinement of those in the first part which will provide more plot points if necessary. The later parts of Tables 8-3 and 8-4 are corresponding refinements of their coarser tables.

8-4. Polar Coordinates of Involute Curve. By choosing a point P on an involute curve and passing a radius through P, we set up polar coordinates with radius OP as the polar axis. Consider a second point Q on the involute and let the ratio of OQ to OP be called m and the angle POQ be called by the Greek letter mu (μ) where μ is measured in radians and is considered positive if OP is greater than OQ, negative otherwise. Tables of m and μ have been prepared by Darle W. Dudley for various convenient values of the pressure angle of Q and for a fixed pressure angle of P.

Table 8-4. Tangent and Normal Coordinates at Convenient Radii (20° Pressure Angle)

R/r	x/r	y/r	θ, deg
1.2000000	0.2256379	0.0556555	24.6506887
1.1900000	0.2140350	0.0504509	23.6636788
1.1800000	0.2024413	0.0454902	22.6628278
1.1700000	0.1908608	0.0407744	21.6471577
1.1600000	0.1792971	0.0363046	20.6155798
1.1500000	0.1677541	0.0320819	19.5668762
1.1400000	0.1562358	0.0281078	18.4996778
1.1300000	0.1447460	0.0243838	17.4124372
1.1200000	0.1332888	0.0209114	16.3033951
1.1100000	0.1218683	0.0176927	15.1705380
1.1000000	0.1104888	0.0147297	14.0115447
1.0900000	0.0991545	0.0120251	12.8237168
1.0800000	0.0878699	0.0095817	11.6038889
1.0700000	0.0766396	0.0074028	10.3483080
1.0600000	0.0654686	0.0054923	9.0524715
1.0500000	0.0543618	0.0038548	7.7109013
1.0400000	0.0433248	0.0024958	6.3168238
1.0300000	0.0323633	0.0014218	4.8617010
1.0200000	0.0214838	0.0006408	3.3345185
1.0100000	0.0106933	0.0001627	1.7206563
1.0000000	0.0000000	0.0000000	0.0000000
0.9900000	−0.0105867	0.0001689	−1.8564461
0.9800000	−0.0210552	0.0006910	−3.8932125
0.9700000	−0.0313906	0.0015974	−6.1852206
0.9600000	−0.0415715	0.0029352	−8.8781337
0.9500000	−0.0515622	0.0047893	−12.3444096
0.9400000	−0.0612693	0.0073971	−19.3883514
0.9396926	−0.0615584	0.0075016	−20.0000000
1.0200000	0.0214838	0.0006408	3.3345185
1.0190000	0.0204006	0.0005792	3.1773136
1.0180000	0.0193184	0.0005206	3.0192294
1.0170000	0.0182370	0.0004650	2.8602480
1.0160000	0.0171565	0.0004126	2.7003515
1.0150000	0.0160770	0.0003632	2.5395210
1.0140000	0.0149984	0.0003169	2.3777371
1.0130000	0.0139207	0.0002736	2.2149798
1.0120000	0.0128439	0.0002335	2.0512284
1.0110000	0.0117682	0.0001965	1.8864613
1.0100000	0.0106933	0.0001627	1.7206563
1.0090000	0.0096195	0.0001320	1.5537903
1.0080000	0.0085466	0.0001045	1.3858396
1.0070000	0.0074747	0.0000801	1.2167793
1.0060000	0.0064039	0 0000590	1.0465837
1.0050000	0.0053340	0.0000410	0.8752261
1.0040000	0.0042651	0.0000263	0.7026789
1.0030000	0.0031973	0.0000148	0.5289133
1.0020000	0.0021305	0.0000066	0.3538992
1.0010000	0.0010647	0.0000017	0.1776056
1.0000000	0.0000000	0.0000000	0.0000000
0.9990000	−0.0010635	0.0000017	−0.1789514

Table 8-4. Tangent and Normal Coordinates at Convenient Radii (20° Pressure Angle) (*Continued*)

R/r	x/r	y/r	θ, deg
0.9980000	−0.0021261	0.0000066	−0.3592840
0.9970000	−0.0031876	0.0000150	−0.5410345
0.9960000	−0.0042480	0.0000267	−0.7242412
0.9950000	−0.0053073	0.0000418	−0.9089443
0.9940000	−0.0063654	0.0000603	−1.0951859
0.9930000	−0.0074225	0.0000822	−1.2830097
0.9920000	−0.0084784	0.0001076	−0.4724618
0.9910000	−0.0095331	0.0001365	−1.6635904
0.9900000	−0.0105867	0.0001689	−1.8564461
0.9890000	−0.0116391	0.0002048	−2.0510821
0.9880000	−0.0126903	0.0002442	−2.2475542
0.9870000	−0.0137402	0.0002872	−2.4459213
0.9860000	−0.0147890	0.0003339	−2.6462452
0.9850000	−0.0158366	0.0003841	−2.8485913
0.9840000	−0.0168829	0.0004381	−3.0530285
0.9830000	−0.0179279	0.0004957	−3.2596297
0.9820000	−0.0189716	0.0005570	−3.4684721
0.9810000	−0.0200141	0.0006221	−3.6796374
0.9800000	−0.0210552	0.0006910	−3.8932125
0.9790000	−0.0220951	0.0007638	−4.1092894
0.9780000	−0.0231335	0.0008404	−4.3279664
0.9770000	−0.0241707	0.0009209	−4.5493480
0.9760000	−0.0252064	0.0010053	−4.7735459
0.9750000	−0.0262407	0.0010938	−5.0006798
0.9740000	−0.0272736	0.0011862	−5.2308775
0.9730000	−0.0283051	0.0012828	−5.4642765
0.9720000	−0.0293351	0.0013834	−5.7010249
0.9710000	−0.0303636	0.0014883	−5.9412821
0.9700000	−0.0313906	0.0015974	−6.1852206
0.9690000	−0.0324161	0.0017107	−6.4330270
0.9680000	−0.0334400	0.0018285	−6.6849045
0.9670000	−0.0344623	0.0019506	−6.9410741
0.9660000	−0.0354830	0.0020772	−7.2017773
0.9650000	−0.0365021	0.0022084	−7.4672792
0.9640000	−0.0375195	0.0023442	−7.7378710
0.9630000	−0.0385351	0.0024846	−8.0138746
0.9620000	−0.0395490	0.0026299	−8.2956467
0.9610000	−0.0405612	0.0027801	−8.5835849
0.9600000	−0.0415715	0.0029352	−8.8781337
0.9590000	−0.0425799	0.0030955	−9.1797937
0.9580000	−0.0435864	0.0032609	−9.4891310
0.9570000	−0.0445909	0.0034316	−9.8067905
0.9560000	−0.0455935	0.0036079	−10.1335121
0.9550000	−0.0465939	0.0037897	−10.4701510
0.9540000	−0.0475922	0.0039773	−10.8177052
0.9530000	−0.0485882	0.0041708	−11.1773507
0.9520000	−0.0495820	0.0043705	−11.5504897
0.9510000	−0.0505734	0.0045766	−11.9388164

Table 8-5. Dudley Table of Involute Polar Coordinates m

μ	Pressure angles, deg		
	14.5	20	30
0.0015	0.9939122	0.9958116	0.9973917
0.0014	0.9943384	0.9960952	0.9975662
0.0013	0.9947613	0.9963781	0.9977406
0.0012	0.9951811	0.9966604	0.9979150
0.0011	0.9955977	0.9969421	0.9980892
0.0010	0.9960114	0.9972231	0.9982634
0.0009	0.9964221	0.9975035	0.9984375
0.0008	0.9968300	0.9977833	0.9986115
0.0007	0.9972352	0.9980625	0.9987853
0.0006	0.9976376	0.9983411	0.9989591
0.0005	0.9980375	0.9986190	0.9991328
0.0004	0.9984348	0.9988964	0.9993065
0.0003	0.9988297	0.9991732	0.9994800
0.0002	0.9992221	0.9994494	0.9996534
0.0001	0.9996122	0.9997250	0.9998267
0.0000	1.0000000	1.0000000	1.0000000
−0.0001	1.0003856	1.0002745	1.0001732
−0.0002	1.0007689	1.0005484	1.0003462
−0.0003	1.0011502	1.0008217	1.0005192
−0.0004	1.0015294	1.0010945	1.0006921
−0.0005	1.0019065	1.0013667	1.0008649
−0.0006	1.0022816	1.0016384	1.0010376
−0.0007	1.0026548	1.0019096	1.0012102
−0.0008	1.0030261	1.0021802	1.0013828
−0.0009	1.0033955	1.0024503	1.0015552
−0.0010	1.0037631	1.0027198	1.0017276
−0.0011	1.0041290	1.0029889	1.0018999
−0.0012	1.0044930	1.0032574	1.0020720
−0.0013	1.0048554	1.0035254	1.0022441
−0.0014	1.0052161	1.0037920	1.0024162
−0.0015	1.0055751	1.0040599	1.0025881

Table 8-5 is a Dudley table giving the radial ratios m for uniform 0.0001-radian steps of the polar angle μ where the fixed pressure angle of the reference point P for each column is one of the three popular angles of the preceding tables: 14.5, 20, or 30°.

Rectangular Coordinates of Involute Curve. With the μ and m values of a Dudley table, the rectangular coordinates of an involute curve may be found, with the y axis being the center line of the gear space and the x axis being a radius perpendicular to the y axis:

$$x = mr \sin (\delta - \mu) \qquad y = mr \cos (\delta - \mu) \tag{8-14}$$

where r is the pitch radius and delta (δ) is the angle between the space center line and the radius to the pitch point. To use Eq. (8-14) a Dudley table must be available for the pressure angle of the reference radius r and the angles δ and μ must both be in radians or degrees. If the gear at hand does not have a nominal pressure angle of 14.5, 20, or 30°, we can determine the radius of a point on the involute which does have one of these pressure angles and call it r. The delta (δ) in Eq. (8-14) must then be calculated from the space center line to the new reference radius.

To calculate a radius r having a standard pressure angle ϕ when a given radius R with a nonstandard pressure angle Φ is known, we use

$$r = R\frac{\cos\Phi}{\cos\phi} \tag{8-15}$$

Thus, for any involute with a known point of radius R and pressure angle Φ we can use (8-15) to find r and ϕ where ϕ is 14.5, 20, and 30° and then use the respective column of a Dudley table to provide m and μ values for use in (8-14).

The angle delta (δ) for any reference radial line is given in radians by the relationship

$$\delta = \frac{\pi}{N} - \frac{T}{D} \tag{8-16}$$

where N = number of teeth in gear
T = tooth thickness at reference radius
D = reference diameter, i.e., $2r$

By using Eq. (8-16) for δ, we can subtract from delta the values of μ directly from a Dudley table for use in Eq. (8-14).

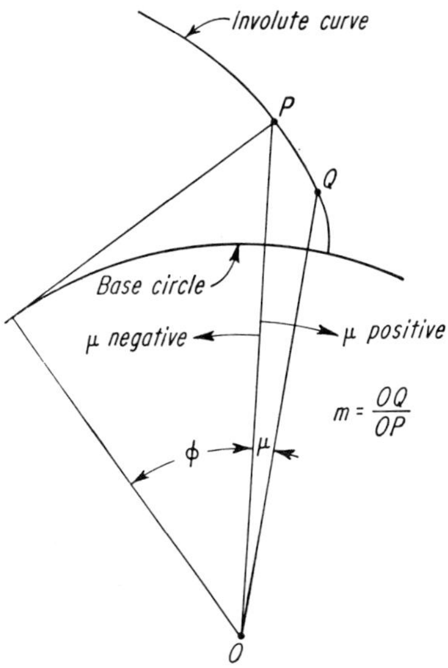

FIG. 8-4. Plotting involute with m and μ from Dudley table.

CUTTER PROFILE

When a rotating cutter is used to cut a form on a rotating workpiece, the axial section of the cutter would be identical with the axial section of the work if the axes of the work and cutter were parallel and there were no lead on the work. However, when the angle between the cutter and work axes is not zero, the axial section of the cutter is not identical with the axial section of the work. In fact, the axial cutter section is not exactly identical to the normal section of the work, because of interference.

8-5. Finding Axial Cutter Profile from Known Axial Section of Work. In determining the axial cutter profile, we assume that the following six numerical values are known:

r_0 = radius of a point P in axial section of work
z_0 = axial coordinate of point P
f' = slope of tangent line at P in axial work section
$R_p + r_p$ = sum of pitch radii of work and cutter
L = lead of work
Σ = setting angle between work and cutter axes

When these values are known, the following equation given by D. W. Dudley and H. Poritsky in a 1943 AGMA paper[3,*] contains only one unknown, θ:

$$\left(r\cos\theta - \frac{Lf'}{2\pi}\sin\theta\right)\left[-\sin\Sigma\left(z_0 + \frac{\theta L}{2\pi}\right) - r\sin\theta\cos\Sigma\right] + \left(r\sin\theta + \frac{Lf'}{2\pi}\cos\theta\right)\{\cos\Sigma\,[r\cos\theta - (R_p + r_p)]\} - rf'\sin\Sigma\,[r\cos\theta - (R_p + r_p)] = 0 \tag{8-17}$$

* Superscript numbers refer to references at the end of the chapter.

The solution of (8-17) for θ has been programmed on the Bendix G-15 at Barber-Colman Company using four code sheets similar to Table 8-1, i.e., in about 100 commands. When θ is found from (8-17), the axial coordinate z_2 of the axial section of the cutter may be found from (8-19) and the radial coordinate r_2 of the axial section of the cutter may be found from (8-22).

$$z = z_0 + \frac{\theta L}{2\pi} \qquad y = r \sin \theta \tag{8-18}$$

$$z_2 = (z - y \tan \Sigma) \cos \Sigma \tag{8-19}$$

$$y_2 = z_2 \tan \Sigma + y \sec \Sigma \tag{8-20}$$

$$x_2 = R_p + r_p - r \cos \theta \tag{8-21}$$

$$r_2 = \sqrt{x_2{}^2 + y_2{}^2} \tag{8-22}$$

The evaluation of (8-18) to (8-22) is straightforward once we have found θ from (8-17). After finding z_2 and r_2, we determine new values for r, z_0, and f' in the axial section of the work and repeat the solution of (8-17) for θ and then evaluate (8-18) to (8-22) until we have a sufficient number of points on the desired axial form of the cutter.

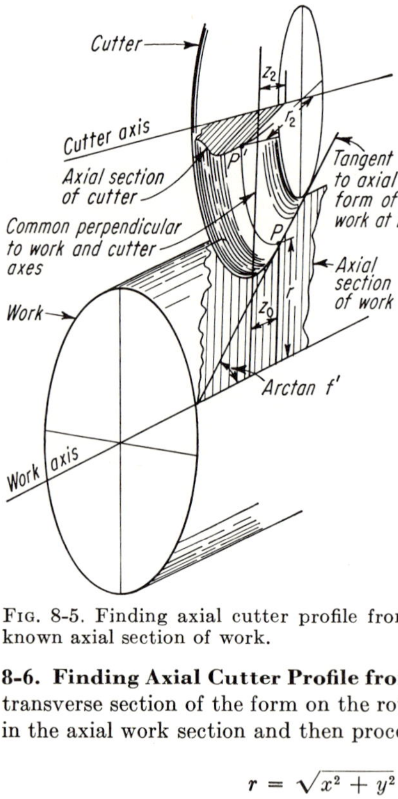

FIG. 8-5. Finding axial cutter profile from known axial section of work.

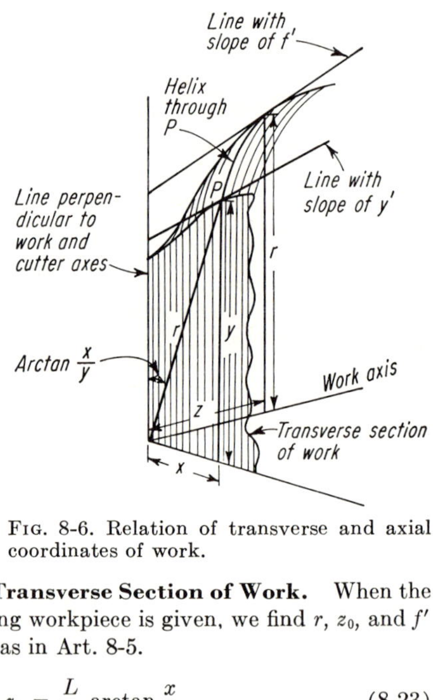

FIG. 8-6. Relation of transverse and axial coordinates of work.

8-6. Finding Axial Cutter Profile from Transverse Section of Work. When the transverse section of the form on the rotating workpiece is given, we find r, z_0, and f' in the axial work section and then proceed as in Art. 8-5.

$$r = \sqrt{x^2 + y^2} \qquad z_0 = \frac{L}{2\pi} \arctan \frac{x}{y} \tag{8-23}$$

$$f' = \frac{2\pi r(yy' + x)}{L(y - xy')} \tag{8-24}$$

where x and y are rectangular coordinates of a point P in the transverse section of the work and y' is the slope of a tangent line at P to the form in the transverse plane. The origin of the x and y axes lies on the work axis of rotation and the y axis passes through the cutter axis.

Table 8-6. Code Sheet from Cutter Profile Program

INTERCOM PROGRAM

Prepared by E. C. Varnum Date 6-23-59

PROBLEM Cutter Profile Page 9 of 20 Line 16

NOTES	Location	K	OP CODE	Address		S/T	(A) or (AA)
Clear and add x	1600		42	17	00		
Multiply by cos λ	1601		44	17	20		
Store x cos λ → 1044	1602		49	10	44		
Multiply by tan λ	1603		44	17	30		
Store x sin λ → 1736	1604		49	17	36		
Clear and add y'	1605		42	17	04		
Divide by sin λ	1606		48	17	22		
Store y' csc λ → 1738	1607		49	17	38		
Mark I: go to 1638	1608		26	16	38		
Multiply by r	1609		44	10	32		
Divide by y	1610		48	17	02		
Divide by tan λ	1611		48	17	30		
Store s → 1036	1612		49	10	36		
Clear and add x sin λ	1613		42	17	36		
Divide by y	1614		48	17	02		
Store (x/y) sin λ → 1014	1615		49	10	14		
Arctangent	1616		8	12	24		
Store arctan $[x \sin \lambda / y]$	1617		49	10	16		
Multiply by $(L/2\pi)$	1618		44	10	02		
Add x cos λ	1619		43	10	44		
Store z_0 → 1034	1620		49	10	34		
Clear and add y' csc λ	1621		42	17	38		
Add (x/y) sin λ	1622		43	10	14		
Multiply by r	1623		44	10	32		
Store num. → 1714	1624		49	17	14		
Clear and sub. y' csc λ	1625		40	17	38		

8-7. Finding Axial Cutter Profile from Known Normal Section of Work. When a specified form is desired in the normal plane of the work, rather than in the axial or transverse planes, we first find r, z_0, and f' in the axial work section and proceed as in Art. 8-5.

$$r = \sqrt{(x \sin \lambda)^2 + y^2} \tag{8-25}$$

$$z_0 = x \cos \lambda + \frac{L}{2\pi} \arctan \frac{x \sin \lambda}{y} \tag{8-26}$$

$$f' = \frac{2\pi r(yy' + x \sin^2 \lambda)}{L \sin \lambda(y - xy') + 2\pi r^2 \cos \lambda} \tag{8-27}$$

where x and y are rectangular coordinates of the normal form of the work and y' is the slope of a tangent to the normal form; the origin of the axes lying on the work axis and the y axis passing through the cutter axis as in Art. 8-6. The angle lambda (λ) in Eqs. (8-25) to (8-27) is the lead angle of the work. When this lead angle is a right angle its cosine is zero and its sine is unity, so that the equations in (8-7) reduce to those in (8-6) and, in truth, the normal and transverse sections are identical. Table 8-6 is a sample code sheet from the Barber-Colman cutter profile program, showing the evaluation of z_0 in (8-26) in 1613 to 1620.

WORK PROFILE

When a rotating cutter of known axial section is used to cut a rotating workpiece, the normal section of the work would be identical with the axial section of the cutter if the setting angle were zero and there were no lead on the work. When the setting angle is not zero, however, these above-mentioned sections are not exactly identical. The normal section of the work can be found from the axial section of the cutter in the manner indicated below.

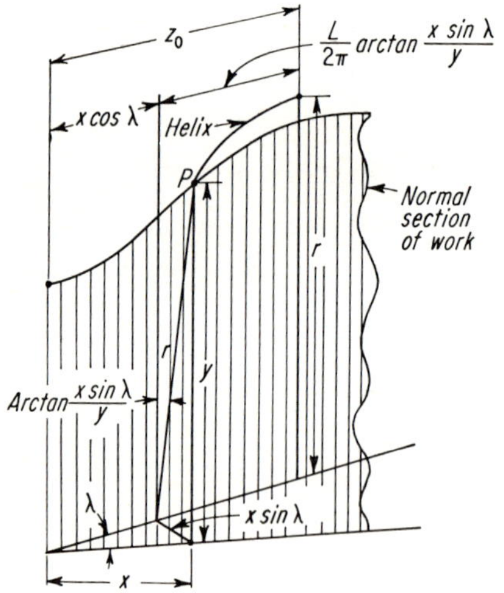

FIG. 8-7. Relation of normal and axial coordinates of work.

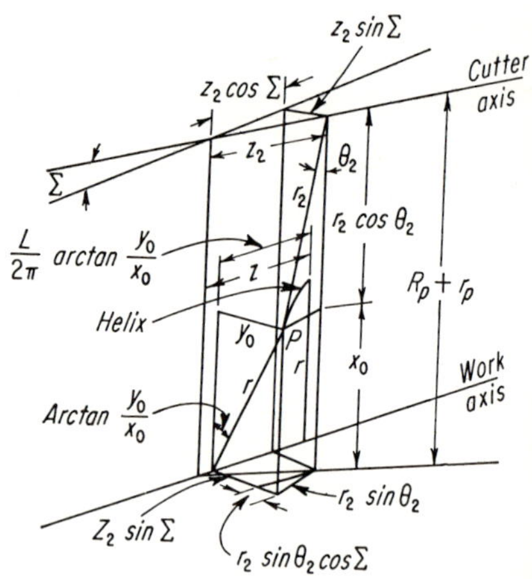

FIG. 8-8. Relationships between r_2, z_2 on cutter and x_0, y_0, r, and z on work.

8-8. Finding Normal Section of Work from Known Axial Section of Cutter. We assume that the following numerical values are known: r_2 = radius of a point P in axial section of cutter, z_2 = axial coordinate of point P in axial section of cutter, g' = slope of tangent line at P in axial cutter section, $R_p + r_p$ = sum of pitch radii of work and cutter. L = lead of work. Σ = setting angle between work and cutter axes.

For these values of a given point P on the cutter, we calculate an angle θ_2.

$$\theta_2 = \arctan \frac{B}{A} - \arccos \frac{C}{\sqrt{A^2 + B^2}} \tag{8-28}$$

where $A = \sin \Sigma (z_2 + r_2 g')$

$$B = (R_p + r_p) \cos \Sigma + \frac{L}{2\pi} \sin \Sigma$$

$$C = g' \left[(R_p + r_p) \sin \Sigma - \frac{L}{2\pi} \cos \Sigma \right]$$

Using the value of θ_2 found in (8-28) we solve the following transcendental equation in μ (mu):

$$\frac{2\pi R_p r}{L} \sin \mu + \frac{L}{2\pi} \mu - z = 0 \tag{8-29}$$

where $r = \sqrt{x_0{}^2 + y_0{}^2}$

$x_0 = (R_p + r_p) - r_2 \cos \theta_2$

$y_0 = z_2 \sin \Sigma + r_2 \sin \theta \cos \Sigma$

$$z = z_2 \cos \Sigma - r_2 \sin \theta_2 \sin \Sigma + \frac{L}{2\pi} \arctan \left(\frac{y_0}{x_0} \right)$$

With the value of μ found in (8-29), we now find the rectangular coordinates (x,y) in the normal section of the work:

$$x = \frac{r \mathit{sin} \mu}{\sin \lambda} \tag{8-30}$$

$$y = r \cos \mu \tag{8-31}$$

where x is measured in a direction making angle λ with the work axis, and y is measured in a direction parallel to the mutual perpendicular of the cutter and work axes.

Numerical Example of Finding Normal Section of Work from Known Axial Section of Cutter. To use the Barber-Colman work profile program prepared for the Bendix G-15 computer, the following inputs were typed in:

Lead of work = 3.9
Pitch diameter of work = 3.411
Pitch diameter of cutter = 5.224
Setting angle = 20°
Axial coordinate of first point = 0.140003
Radial coordinate of first point = 2.926419
Axial coordinate of last point = 0.310718
Radial coordinate of last point = 2.612000

With the above input values, the work profile program calculates the numerical values listed in the first paragraph of Sec. 8-8. Concentrating on the first point only, these values are as follows:

$r_2 = 2.926419$
$z_2 = 0.140003$
$g' = -1.84177723106$
$R_p + r_p = 4.3175$
$L = 3.9$
$\Sigma = 20$

To solve (8-28) with these numerical inputs, the following intermediate values are calculated:

$$\begin{aligned}\sin \Sigma &= 0.342020143326\\ \cos \Sigma &= 0.939692620787\\ A &= -1.79554038650\\ B &= 4.26941625639\\ C &= -1.64544513935\\ \arctan \frac{B}{A} &= -1.17269345957\\ \arccos \frac{C}{\sqrt{A^2+B^2}} &= -1.20760026056\end{aligned}$$

By subtracting $\arccos C/\sqrt{A^2+B^2}$ from $\arctan (B/A)$ we obtain

$$\theta_2 = 0.0349068099927 \text{ radian}$$

This is the solution of equation $A \cos \theta_2 + B \sin \theta_2 = C$. With the above value of θ_2, the following values are calculated:

$$\begin{aligned}r &= 1.40027276839\\ x_0 &= 1.39286371836\\ y_0 &= 0.143855788772\\ z &= 0.160509068848\end{aligned}$$

With these values of r and z, Eq. (8-29) is solved for μ and the resulting value used in (8-30) and (8-31) to obtain the coordinates

$$\begin{aligned}\mu &= 0.0359290848270 \text{ radian, from (8-29)}\\ \sin \mu &= 0.0359213551884\\ \cos \mu &= 0.999354655154\\ x &= 0.147076295174, \text{ from (8-30)}\\ y &= 1.39936910957, \text{ from (8-31)}\end{aligned}$$

The pitch point on the cutter is given as the last point in the data, for which we have

$$\begin{aligned}r_2 &= 2.612;\ z_2 = 0.310718\\ A &= -\ 1.53909205666\\ B &= 4.26941625639\\ C &= -1.64544513935\\ \arctan \frac{B}{A} &= -1.22480491978\\ \arccos [C(A^2+B^2)^{-1/2}] &= -1.19977897506\\ \theta_2 &= -0.0250259447140 \text{ radian}\\ x_0 &= 1.70631790238;\ y_0 = 0.0448526176372\\ r &= 1.70690730308;\ z = 0.33064373247\\ \mu &= 0.0622954218008 \text{ radian}\\ x &= 0.310715172609,\ y = 1.70359695323\end{aligned}$$

PROFILES OF BEVEL AND HYPOID GEARS

By C. B. King, *Associate Chief Research Engineer, Gleason Works, Rochester, N.Y.*

Profile layouts of bevel and hypoid gears are usually made for the following reasons:

1. To determine the geometry factor J used to determine the bending stress in the root fillet of a bevel or hypoid gear tooth. Formulas for load rating involving factor J are given in Chap. 13.

2. To obtain a visual inspection of the tooth profiles prior to manufacture. This provides a check on top lands and root fillets of the teeth.

The method here described gives tooth profiles in the mean normal section of the teeth. This is a section at right angles to the direction of the tooth at a point at mid-face or, in the case of spiral bevel and hypoid gears, at the point where spiral angles are specified.

8-9. Layout Procedure for Bevel Gears. Before proceeding with the layout, certain information must be obtained from the dimension sheet, gear drawing, or summary. These data should be transferred to a calculation format as shown in Table 8-7. On this sheet are determined the values of gear and pinion dimensions necessary for making an enlarged layout of the teeth in the normal plane at the middle of the face width.

The following numbers in parentheses refer to the items on the calculation format in Table 8-7. Subscripts R and L after the item numbers on the format refer to the right- and left-hand columns.

Table 8-7. Layout Data
(For sample combination 14/43)

Step	Operation	Pinion	Gear
1	$n,\ N$	14	43
2	$\gamma,\ \Gamma$	18°2′	71°58′
3	$\cos\gamma,\ \cos\Gamma$	0.95088	0.30957
4	$\gamma_o,\ \Gamma_o$	21°33′	73°43′
5	$\gamma_R,\ \Gamma_R$	16°17′	68°27′
6	$\alpha = (4) - (2)$	3°31′	1°45′
7	$\tan\alpha$	0.06145	0.03055
8	$\delta = (2) - (5)$	1°45′	3°31′
9	$\tan\delta$	0.03055	0.06145
10	$a_{Po},\ a_{Go}$	0.300	0.125
11	$b_{Po},\ b_{Go}$	0.172	0.347
12	F	1.625	1.625
13	P_d	4	
14	A_o	5.6527	
15	A	4.8402	
16	$(14) - (15)$	0.8125	
17	ϕ_c	20°	
18	$\cos\phi_c$	0.93969	
19	$\tan\phi_c$	0.36397	
20	ψ	35°	
21	$\sin\psi$	0.57358	
22	$\cos\psi$	0.81915	
23	$(22)^2$	0.67101	
24	$\dfrac{(15)}{(14)}$	0.85626	
25	$P_M = \dfrac{(13)}{(24)}$	4.6715	
26	$\dfrac{\pi(22)}{(25)}$	0.55088	
27	$a = (10) - (7)(16)$	0.2501	0.1002
28	$b = (11) - (9)(16)$	0.1472	0.2971
29	r_T	0.030	0.030
30	W_G		
31	t_{Po}	0.4974	
32	$t_P = (30) + 2.0(19)(28)_R$ or $(31)(22)(24)$	0.34888	
33	$t_G = (26) - (32)$	0.20200	
34	$a_1 = (13)(27)$	1.000	0.401
35	$b_1 = (13)(28)$	0.589	1.188
36	$t_1 = (13)(32) - 0.04;\ (13)(33)$	1.356	0.808

Table 8-7. Layout Data (*Continued*)

Step	Operation	Pinion	Gear
37	$r_1 = (13)(29)$	0.120	0.120
38	$W_{G1} = (13)(30)$		
39	$R_1 = \dfrac{(1)(24)}{2.0(3)}$	6.3034	59.4683
40	$R_{N1} = \dfrac{(39)}{(23)}$	9.3939	88.6251
41	$\Delta\phi = \dfrac{1720(36)_L}{(40)_R}$		
42	$\phi = (17) - (41)$ Formate $= (17)$ generated	20°	
43	$\cos\phi$	0.93969	
44	$\tan\phi$	0.36397	
45	$(23) + (44)^2$	0.80348	
46	$\dfrac{(13)(26)}{(43)}$	2.3449	
47	$p_2 = \dfrac{(46)}{(45)}$	2.9184	
48	$\Delta R_{bN1} = (40) - (40)(43)$	0.567	5.345
49	$F \times P_d = (12)_R(13)$	6.500	
50	m_F from Fig. 8-21	1.73	

(1) n,N numbers of teeth in pinion and gear, respectively. When making an analysis on a differential or planetary set where the two side gears are unequal, the layout should be made for both cases.

(2) γ,Γ pitch angles of pinion and gear, respectively.

(4) γ_o,Γ_o face angles of pinion and gear, respectively.

(5) γ_R,Γ_R root angles of pinion and gear, respectively.

(10) a_{P_o},a_{G_o} large-end addendums of pinion and gear, respectively.

(11) b_{P_o},b_{G_o} large-end dedendums of pinion and gear, respectively.

(12) F face width of pinion and gear, respectively. The gear face width should not exceed 25 per cent of the outside cone distance on Zerol®* bevels or 30 per cent of the outside cone distance on straight and spiral bevels.

(13) P_d diametral pitch at the outside cone distance.

(14) A_o outside cone distance.

(15) A mean cone distance. This is usually equal to $(14) - 0.5(12)_R$.

(17) ϕ_c nominal pressure angle of gear cutter.

(20) ψ nominal spiral angle at the center of the face width.

(25) P_M diametral pitch at the center of the face width.

(27) a addendum at the center of the face width.

(28) b dedendum at the center of the face width.

(29) r_T radius on the edge of the tools or cutters. Unless maximum radius is required, use a value of $0.120/P_d$ inches. When maximum radius is required, it must be determined according to the cutting method used for producing the gears. An approximation of maximum radius may be made by using a value of $0.240/P_d$ inches. If a cutting summary is available, the value may be taken directly from same.

(30) W_G point width of the spread-blade gear cutter. This item will be filled in only when the spread-blade point width is known.

(31) t_{P_o} large-end circular thickness of the pinion. If the point width of the spread-blade gear cutter [item (30)] is given, omit this item. If neither the gear cutter point width nor the pinion circular thickness is known, a value should be

* Registered trade-mark of the Gleason Works, Rochester, N.Y.

assumed. As a first approximation use

$$t_{Po} = \frac{1.5708}{P_d} + (a_{Po} - a_{Go})\frac{\tan \phi_C}{\cos \psi}$$

(32) t_P mean normal circular thickness of pinion. Use the first formula when W_G is given and the second formula when t_{P_o} is given.
(33) t_G mean normal circular thickness of gear.
(34)* a_1 addendum.
(35)* b_1 dedendum.
(36)* t_1 circular thickness. The constant 0.04 in the left-hand equation is an assumed average value of backlash for 1 diametral pitch. Except on straight bevel gears this is applied to the pinion only. On straight bevel gears subtract 0.02 from both equations.
(37)* r_1 tool edge radius.
(38)* W_{G_1} spread-blade gear point width.
(39)* R_1 transverse back-cone radius.
(40)* R_{N_1} normal back-cone radius.
(41)* $\Delta\phi$ pressure-angle correction on a Formate®† gear (in minutes).
(42) ϕ normal pressure angle. Use the upper formula for a Formate pair and the lower formula for a generated pair.
(48) ΔR_{bN1} layout distance between pitch circle and base circle in normal plane.
(50) m_F face contact ratio. This is obtained from Fig. 8-21. The product of face width and diametral pitch is given by item (49) and the spiral angle is given by item (20). The face contact ratio will be zero for straight bevel and Zero bevel gears. Figure 8-21 will not apply on gears with a face width less than three-tenths of the outside cone distance. In these cases use the formulas given on Fig. 8-21.

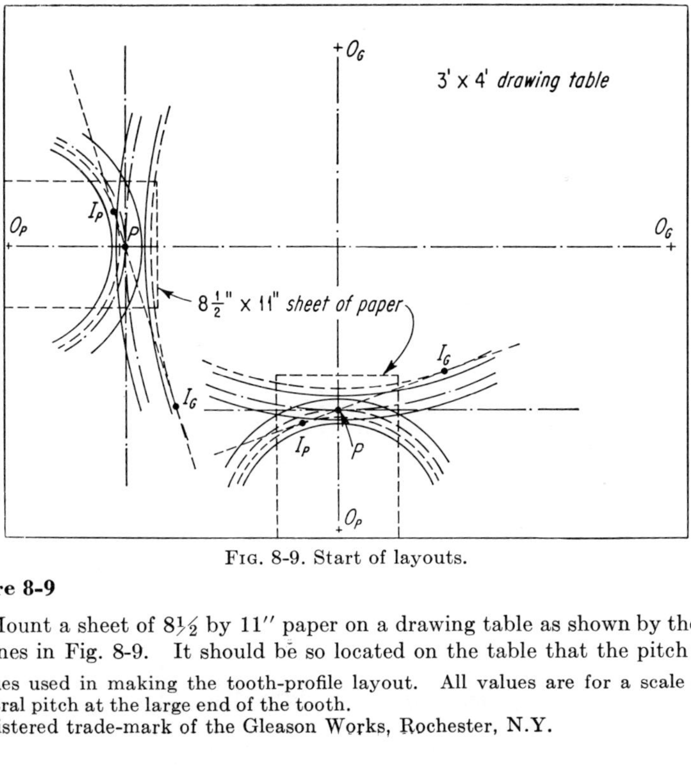

FIG. 8-9. Start of layouts.

Figure 8-9

1. Mount a sheet of 8½ by 11″ paper on a drawing table as shown by the broken lines in Fig. 8-9. It should be so located on the table that the pitch point P

* Values used in making the tooth-profile layout. All values are for a scale equal to 1 diametral pitch at the large end of the tooth.

† Registered trade-mark of the Gleason Works, Rochester, N.Y.

will lie about 3½″ from one end of the sheet of paper and the centers O_P and O_G of the pinion and gear, respectively, will lie on the table. If the center distance O_PO_G is greater than the length of the table, center O_G will have to be omitted.

$$O_PO_G = (40)_L + (40)_R \tag{8-32}$$

Figure 8-10

2. Draw the vertical center line through the center of the 8½ by 11″ sheet. In Fig. 8-9 this is the line of centers O_PO_G.
3. Draw a horizontal pitch line about 3 to 3½″ from the top of this sheet of paper.
4. The intersection of these two lines is the pitch point P.

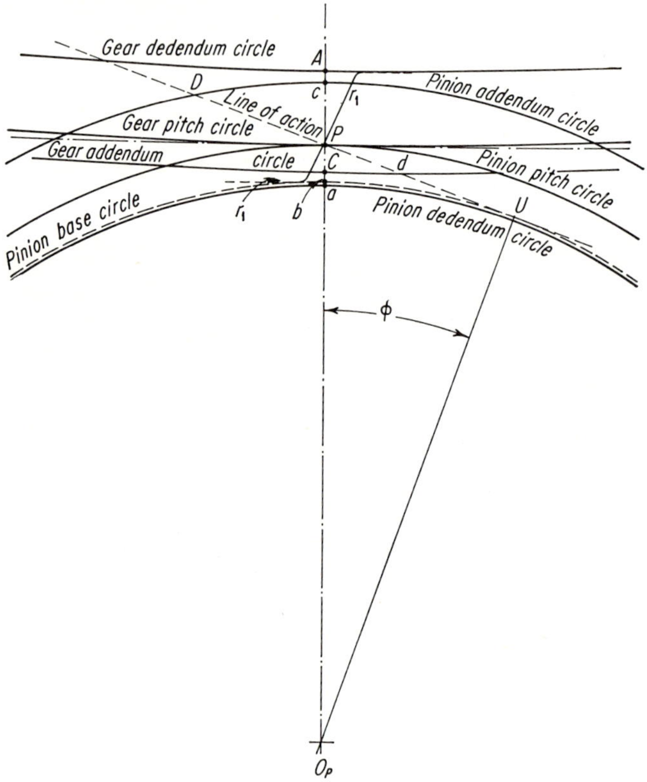

FIG. 8-10. Step 2 of layout.

5. From P measure off on the vertical center line the normal back-cone distances for pinion and gear [item (40)] giving points O_P and O_G. These are the centers of the pinion and gear, respectively. (Centers O_P and O_G may not lie on the 8½ by 11″ sheet.)
6. Next lay off the following dimensions:

Pinion	*Gear*
P_a = item $(35)_L$	PA = item $(35)_R$
P_b = item $(48)_L$	PB = item $(48)_R$
P_c = item $(34)_L$	PC = item $(34)_R$

7. Describe circles with radii O_{Pa}, O_{Pb}, O_PP and O_{Pc} with centers at O_P. These are the pinion dedendum circle, base circle, pitch circle, and addendum circle, respectively.
8. In like manner with radii O_GB, O_GA, O_GP, and O_GC describe circles with centers at O_G. These are the gear base circle (not shown on diagram), dedendum circle, pitch circle, and addendum circle, respectively. When the gear radius is greater than 100″, it will be satisfactory to draw straight horizontal lines through points B, A, P, and C to represent the gear arcs. When the gear radius is less than 100″ but center O_G cannot be located on the drawing table, it will be necessary to plot the points for these circles as explained below (see Fig. 8-11).
9. Through the pitch point P draw the common tangent I_PI_G to the two base circles. For generated gears this is the line of action.
10. Draw through the pitch point P a line perpendicular to the line I_PI_G. This line represents the position of the side of the generating tool for pitch-point contact. The sharp corners of the tool will be determined by the intersection of this cutting edge with two horizontal lines through points a and A.
11. On the sharp corners of the tool construct the arcs of the tool edge radii r_1 [item (37)] for the pinion and gear.
12. For Formate gears the path of contact will be a circular arc tangent to the line I_PI_G at the pitch point P. The radius of curvature of this circle will be

$$R_c = \frac{(40)_R(43)}{2.0} \qquad (8\text{-}33)$$

and its center will lie inside the gear pitch circle.

Figure 8-11

13. When the center of curvature of a circle cannot be located on the drawing board, it will be necessary to plot several points on the circle. This may be done as shown in Fig. 8-11 with the aid of the following curvature formula:

$$y = \frac{x^2}{2R} \qquad (8\text{-}34)$$

It is generally satisfactory to assume values of x equal to 2 and 4″.

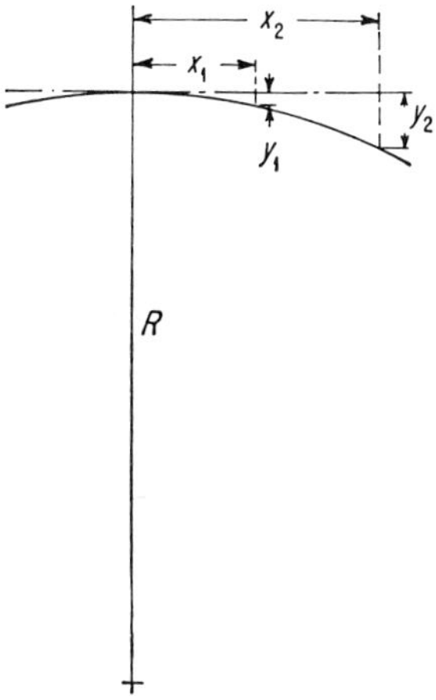

FIG. 8-11. Layout of circle by offsets.

Figure 8-12, Step A

14. Next, take a strip of sheet celluloid about 0.008″ thick, 3″ wide, and 12 to 15″ long and tack it over the drawing in the position shown in Fig. 8-12, step A.
15. For generated gears scribe on the celluloid a straight horizontal pitch line through point P. This is the pitch line of the rack-generating tool. For Formate gears scribe on the celluloid the pitch circle of the Formate gear.

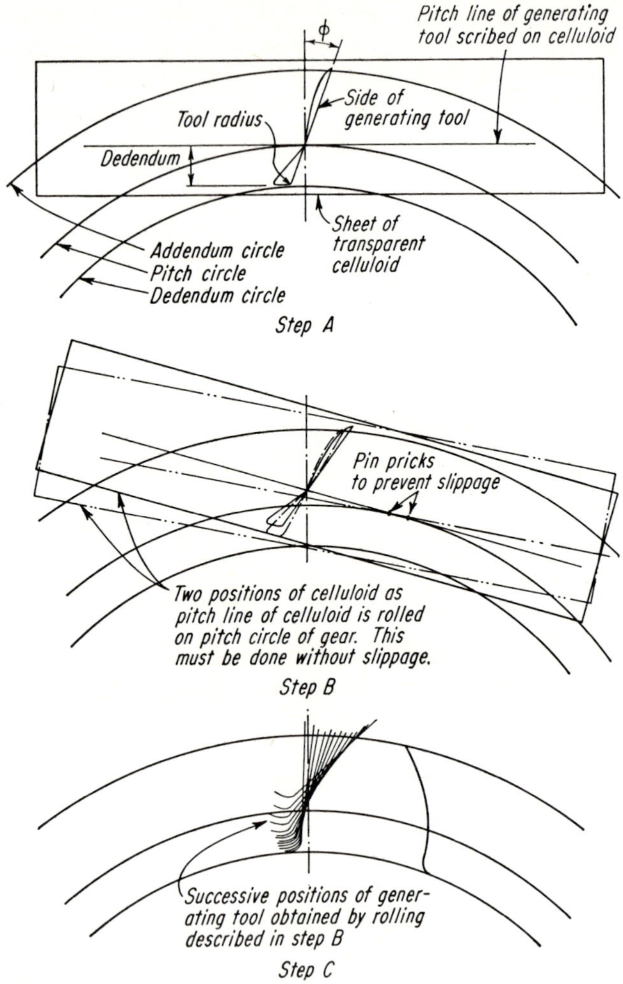

FIG. 8-12. Generating tooth profile by celluloid strip.

16. Now, scribe the side of the generating tool on the celluloid as described in step 10 above. The radius as explained in step 11 is also scribed. The celluloid is removed from the board and the section in the middle is removed, leaving a template of the side and corner of the tool as shown in Fig. 8-12, step A.

Figure 8-12, Step B

17. For generated gears roll the pitch line of the rack tool on the pitch circle of the pinion (or gear) in short intervals by pivoting the celluloid at adjacent points

of tangency during the rolling movement. At each successive position of the template trace the outline of the tool with a sharp pencil. For the Formate pinion roll the pitch circle of the Formate gear on the pitch circle of the pinion. Since the Formate gear is not generated, its profile form will be that of the generating tool described in step 10 with a fillet radius equal to the tool edge radius [item (37) r_1] as explained in step 11.

Figure 8-12, Step C

18. The traces formed by the above rolling will produce an envelope of the tooth form and the tooth fillet. If the gear is represented as a rack as explained in step 8 for gears with a radius greater than 100″ or if it is a Formate gear, the tooth will have the same form as the tool.

Before proceeding further with layout it will be necessary to compute values for locating the position of the load when the stress is a maximum.

First measure the distance along the line of action (path of contact for Formate gears) between points d and D, where the line crosses the addendum circles of gear and pinion, respectively (Fig. 8-10). This is the distance Z_{N1} and should be recorded on the calculation sheet of Table 8-8 as item (51).

Table 8-8, items (51) through (65):

Table 8-8. Layout Data

Step	Operation	Pinion	Gear
51	Z_{N1}	3.46	
52	$m_p = \dfrac{(51)}{(47)}$	1.1856	
53	$(52)^2$	1.4056	
54	$(50)^2 + (53)$	4.3985	
55	$\dfrac{(47)(53)}{(54)}$	0.93261	
56	$m_o = \sqrt{(54)}$	2.0973	
57	$\dfrac{0.5(54)}{(52)}$	1.8550	
58	$4.0(1)_L + 3.2(1)_R$	193.60	
59	$\dfrac{(1)_R - (1)_L}{(58)}$	0.14979	
60	$\dfrac{(59)(50)}{(52)}$	0.21857	
61	$1.0 - 0.5(56)$ or 0	0	
62	$(57) + (61)$	1.8550	
63	$2.0\sqrt{(56) - 1.0}$ or (56)	2.0973	
64	$(62) \pm (60)(63)$	2.3134	1.3966
65	$p_3 = (55)(64)$	2.1575	1.3025

(51) Z_{N1} length of action in the mean normal section. $Z_{N1} = dD$ on the layout.

(52) m_p profile contact ratio.

(56) m_o mismatch contact ratio. When m_o is greater than 2.0, there will always be at least two teeth in contact. When m_o is less than 2.0, there will be periods when only one tooth is carrying the load.

(61) When m_o is less than 2.0, use the left-hand formula, and when m_o is greater than 2.0, use zero.

(63) When m_o is less than 2.0 but greater than 1.0, use the left-hand formula, and when m_o is greater than 2.0, use item (56). When m_o is less than 1.0, item (63) will be zero.

(64) Use the plus (+) sign in the pinion column and the minus (−) sign in the gear column.

(65) p_3 distance in mean normal section measured along the line of action (path of contact) from the addendum circle of the mating member to the point of load application. If the value of p_3 calculated by this formula exceeds Z_{N1} [item (51)] make $p_3 = Z_{N1}$.

Figure 8-13

19. Measure off along the pinion circle from point P the distance $P_1 = (36)_L/2$. Then draw the pinion-tooth center line O_{P1}.

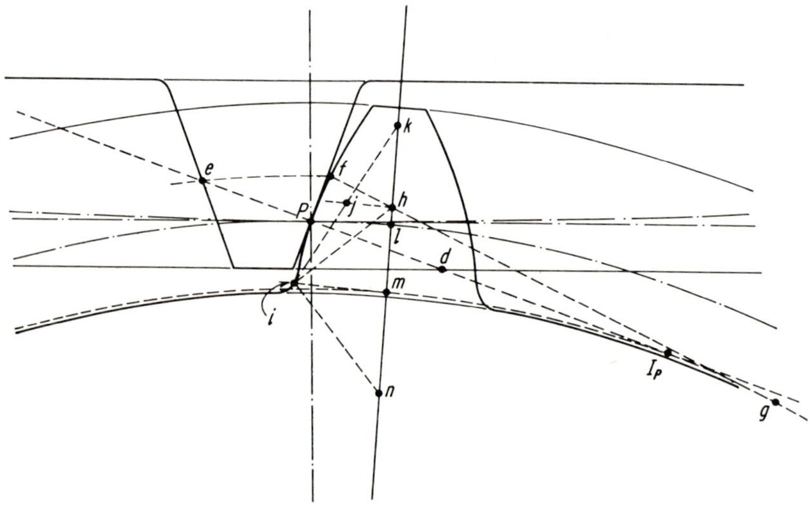

FIG. 8-13. Finding weakest section of pinion tooth.

20. Next, locate point e on the line of action (path of contact) for the pinion. The distance de is equal to the quantity p_3 for the pinion given on the calculation sheet by item $(65)_L$.
21. With O_P as a center and O_{Pe} as a radius, describe the arc of cutting the pinion-tooth profile at f. Point f is the point of load application on the pinion.
22. Draw line f_g through point f and tangent to the pinion base circle at g. This line intersects the center line of the pinion tooth at point h.
23. At point h erect a perpendicular hj to the pinion-tooth center line.

To find the weakest section it is necessary to locate the point of tangency between the inscribed parabola with vertex at point h and the tooth outline. This is done as follows:

24. With a straightedge and dividers find a position where the straightedge is tangent to the tooth fillet at some point i such that $ij = jk$. Point i is the point of tangency of the parabola with the tooth outline and therefore is the location of the weakest section.
25. Draw line ih.
26. Construct line in perpendicular to line ih at point i.
27. Construct line im perpendicular to the tooth center line O_p1.

28. The following values for the pinion may be taken from the layout by careful measurement:

$$\Delta R_{LN1} = h1$$

ΔR_{LN1} will be negative if point h lies inside the pinion pitch circle.

$$X_{LN1} = mm$$
$$t_{LN1} = im$$
$$h_{LN1} = hm$$
$$\phi_{LN} = \text{angle } fhj$$

Figure 8-14

29. The same procedure is used on the gear by first locating point E on the line of action and point F on the gear tooth at which the load is applied.

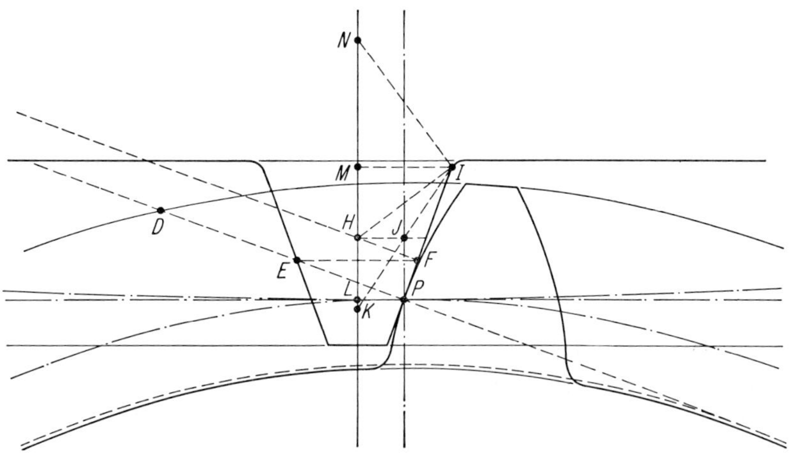

FIG. 8-14. Finding weakest section of gear tooth.

8-10. Layout Procedure for Formate Hypoid Gears. Before proceeding with the layout it is necessary that certain data and dimensions of the hypoid gears be known. These data should be entered on a calculation format shown as Table 8-9. On this sheet are determined the values of gear and pinion dimensions necessary for making an enlarged layout of the teeth in the normal plane at the middle of the face width.

The following numbers in parentheses refer to the items on the calculation format of Table 8-9.* Subscripts R and L after the item numbers on the calculation sheet refer to the right- and left-hand columns.

Table 8-9. Layout Data
(For sample combination 10/41)

Step	Operation	Pinion	Gear
1	n, N = ($H1$), ($H2$)*	10	41
2	γ_R, Γ_1 = ($H143$), ($H48$)	12°42′	74°52′
3	cos γ_R, cos Γ_1 = ($H144$), ($H50$)	0.97553	0.26107
4	R_p, R = (162), ($H12$)	1.2046	3.5220

* To enter data in Table 8-9, Gleason formats H_1, H_2 (or H_{1A}, H_{2A}) and H_3, H_4, and H_{FS} should be available. Consult Gleason Works, Rochester, N.Y.

Table 8-9. Layout Data *(Continued)*

Step	Operation	Pinion	Gear
5			
6			
7			
8			
9	$\tan \phi_0 = -(H176)$	−0.09423	
10	ϕ_0	−5°23′	
11	$\phi_1, \phi_2 = (H186), (H188)$	16°35′	−25°55′
12	F	1.500	1.250
13	$P_d = \dfrac{(H2)}{(H6)}$	4.9697	
14	$A_o = (H73)$	4.2732	
15	$A = (H72)$	3.6485	
16	$\Delta\phi' = (H91)(H170)$	2°52′	
17	$\phi_1', \phi_2' = (11) - (16)$	13°43′	−28°47′
18	$\phi_0' = (10) - (16)$	−8°15′	
19	$\dfrac{\phi_1}{2} = (H80)$	21°15′	
20	$\psi_P, \psi_G = (H172), (H167)$	50°33′	27°0′
21	$\sin \psi_P, \sin \psi_G$	0.77218	0.45399
22	$\cos \psi_P, \cos \psi_G$	0.63540	0.89101
23	$(22)^2$	0.40373	0.79390
24	$P_N = 2.0(H193)$	0.48094	
25	$P_M = \dfrac{\pi(22)}{(24)}$	4.1506	5.8203
26	$\dfrac{(23)_L + (23)_R}{2.0}$	0.59882	
27	$a = (H75) - (H87);\ H(87)$	0.2429	0.0498
28	$b = (27)_R + (28)_R - (27)_L;\ (H88)$	0.0884	0.2815
29	r_T	0.025	0.050
30	$W_G = (H199)$	0.110	
31			
32	$t_P = (30) + (H196)(28)_R$	0.33062	
33	$t_G = (24) - (32)$	0.15032	
34	$a_1 = (13)(27)$	1.207	0.247
35	$b_1 = (13)(28)$	0.439	1.399
36	$t_1 = (13)(32) - 0.04;\ (13)(33)$	1.603	0.747
37	$r_1 = (13)(29)$	0.124	0.248
38	$W_{G1} = (13)(30)$	0.547	
39	$R_1 = \dfrac{(13)(4)}{(3)}$	6.1367	67.0444
40	$R_{N1} = \dfrac{(39)}{(23)}$	15.2000	84.4494
41	$\Delta\phi = \dfrac{1{,}720(36)_L}{(40)_R}$	33′	
42	$\phi = (19) - (41)$ Formate	20°42′	
43	$\cos \phi$	0.93544	
44	$\tan \phi$	0.37787	
45	$(26) + (44)^2$	0.74161	
46	$\dfrac{(13)(24)}{(43)}$	2.5551	
47	$p_2 = \dfrac{(46)}{(45)}$	3.4453	
48	$\Delta\phi_2' = \dfrac{3{,}440[(36)_L + (36)_R]}{(40)_R}$	1°36′	
49	$\phi_2'' = (17)_R + (48)$	−27°11′	
50	$m_F = (H195)$	2.0046	

Table 8-10. Layout Data
(For sample combination 10/41)

Step	Operation	Pinion	Gear
51	Z_{N1}	3.365	
52	$m_p = \frac{(51)}{(47)}$	0.9767	
53	$(52)^2$	0.9539	
54	$(50)^2 + (53)$	5.3412	
55	$\frac{(47)(53)}{(54)}$	0.61531	
56	$m_o = \sqrt{(54)}$	2.3111	
57	$\frac{0.5(54)}{(52)}$	2.7343	
58	$4.0(1)_L + 3.2(1)_R$	171.2	
59	$\frac{(1)_R - (1)_L}{(58)}$	0.18107	
60	$\frac{(59)(50)}{(52)}$	0.38832	
61	$1.0 - 0.5(56)$ or 0	0	
62	$(57) + (61)$	2.7343	
63	$2.0\sqrt{(56) - 1.0}$ or (56)	2.3111	
64	$(62) \pm (60)(63)$	3.6317	1.8369
65	$p_3 = (55)(64)$	2.235	1.130

(1) n,N numbers of teeth in pinion and gear, respectively.
(2) γ_R,Γ_1 root angle of pinion and pitch angle of gear, respectively.
(4) R_P,R mean radius of pinion and gear, respectively.
(10) ϕ_0 root limit pressure angle.
(11) ϕ_1,ϕ_2 root pressure angle on the convex and concave side of the gear teeth, respectively.
(12) F face width of pinion and gear, respectively. The gear face width should not exceed 30 per cent of the outside cone distance given in item (14).
(13) P_d gear diametral pitch at outside cone distance.
(14) A_o gear outside cone distance.
(15) A gear mean cone distance.
(17) ϕ_1',ϕ_2' pitch-line pressure angle on the convex and concave side of the gear teeth, respectively.
(18) ϕ_0' pitch limit pressure angle.
(19) $\frac{\phi_1}{2}$ average pressure angle on the two sides of the teeth.
(20) ψ_P,ψ_G spiral angles of pinion and gear, respectively, at center of the face width.
(24) p_N mean normal circular pitch.
(25) P_M diametral pitch at the center of the face width.
(27) a addendum at the center of the face width.
(28) b dedendum at the center of the face width.
(29) r_T radius on the edge of the cutters.
(30) W_G point width of spread-blade gear cutter.
(32) t_P mean normal circular thickness of pinion.
(33) t_G mean normal circular thickness of gear.
(34)* a_1 addendum.

* Values used in making the tooth-profile layout. All values are for a scale equal to 1 diametral pitch at large end of the tooth.

(35)* b_1 dedendum.

(36)* t_1 circular thickness. The constant 0.04 in the left-hand equation is an assumed average value of backlash for 1 diametral pitch.

(37)* r_1 tool edge radius.

(38)* W_{G1} spread-blade gear point width.

(39)* R_1 transverse back-cone radius.

(40)* R_{N1} normal back-cone radius.

(41) $\Delta\phi$ pressure-angle correction on a Formate gear (in minutes).

(42) ϕ angle describing the direction of the path of contact in hypoid gears. This is not the pressure angle.

(48) $\Delta\phi_2'$ This formula gives the angles in minutes.

(50) m_F face contact ratio.

Figure 8-9

1. Mount a sheet of 8½ by 11″ paper on a drawing table as shown by the broken lines in Fig. 8-9. It should be so located on the table that the pitch point P will lie about 3½″ from one end of the sheet of paper and the centers O_P and O_G of the pinion and gear, respectively, will lie on the table. If the center distance O_PO_G is greater than the length of the table, center O_G will have to be omitted.

$$O_PO_G = (40)_L + (40)_R$$

Figure 8-15

2. Draw the vertical center line through the center of the 8½ by 11″ sheet. In Fig. 8-9 this is the line of centers O_PO_G.
3. Draw a horizontal line about 3 to 3½″ from the top of this sheet of paper.
4. The intersection of these two lines is the pitch point P.
5. From P measure off on the vertical center line the normal back-cone distances for pinion and gear [item (40)] giving points O_P and O_G. These are the centers of the pinions and gear, respectively. (Centers O_P and O_G may not lie on the 8½ by 11″ sheet.)
6. Next lay off the following dimensions:

Pinion	*Gear*
P_a = item $(35)_L$	PA = item $(35)_R$
P_b = item $(34)_L$	PB = item $(34)_R$

7. Describe circles with radii O_{Pa} and O_{Pb} with centers at O_P. These are the pinion dedendum circle and addendum circle, respectively.
8. In like manner, with radii O_GA and O_GB describe circles with centers at O_G. These are the gear dedendum circle and addendum circle, respectively. When the gear radius is greater than 100″, it will be satisfactory to draw straight horizontal lines through points A and B to represent the gear arcs. When the gear radius is less than 100″, but center O_G cannot be located on the drawing table, it will be necessary to plot the points for these circles as explained below (see Fig. 8-11).
9. Through the pitch point P draw line $O_P'P$ at the angle ϕ_0' [item (18)] to line of centers O_PO_G. Also construct line O_PO_P' perpendicular to line of centers O_PO_G. Point O_P', which is the intersection of these two lines, is the center of the pinion rolling circle. When angle ϕ_0' is negative (as in the example), point O_P' will lie to the right of point O_P.
10. With O_P' as a center and $O_P'P$ as a radius describe the pinion rolling circle.

* Values used in making the tooth-profile layout. All values are for a scale equal to 1 diametral pitch at the large end of the tooth.

11. In like manner, locate the center of the gear rolling circle O'_G, and draw the gear rolling circle through point P.
12. Also through point P draw line zz at an angle ϕ to the horizontal line through point P. [ϕ is given by item (42)].

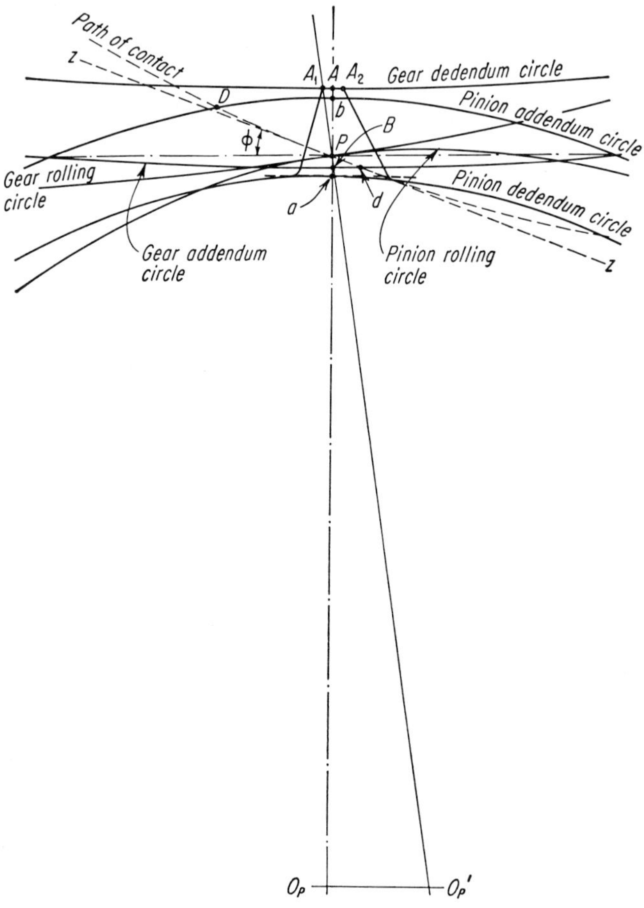

FIG. 8-15. Hypoid tooth.

13. For Formate hypoid gears the path of contact will be a circular arc tangent to the line zz at the pitch point P. The radius of curvature of this circle will be

$$R_c = \frac{(40)_R(43)}{2.0} \tag{8-35}$$

and its center will lie inside the gear rolling circle.
14. Through points A and a draw lines perpendicular to the line of centers O_PO_G. Measure off along the line through A one-half the distance W_{G1} [item (38)] on either side of point A, giving points A_1 and A_2.
15. Through point A_1 draw one side of the Formate gear tooth at an angle ϕ_1 [item $(17)_L$] to the line of centers O_PO_G, and through A_2 draw the other side of the Formate gear tooth at an angle ϕ_2 [item $(17)_R$]. The Formate gear is to be used as a tool for generating the pinion.

16. On the sharp corners of the tool construct the arcs of the tool edge radii r_1 [item (37)], for the pinion and gear.

Figure 8-11

17. When the center of curvature of a circle cannot be located on the drawing board, it will be necessary to plot several points on the circle. This may be done as shown in Fig. 8-11 with the aid of the following curvature formula:

$$y = \frac{x^2}{2R} \tag{8-36}$$

It is generally satisfactory to assume values of x equal to 2 and 4″.

Figure 8-16, Step A

18. Next, take a strip of sheet celluloid about 0.008″ thick, 3″ wide, and 12 to 15″ long and tack it over the drawing in the position shown in Fig. 8-16, step A.

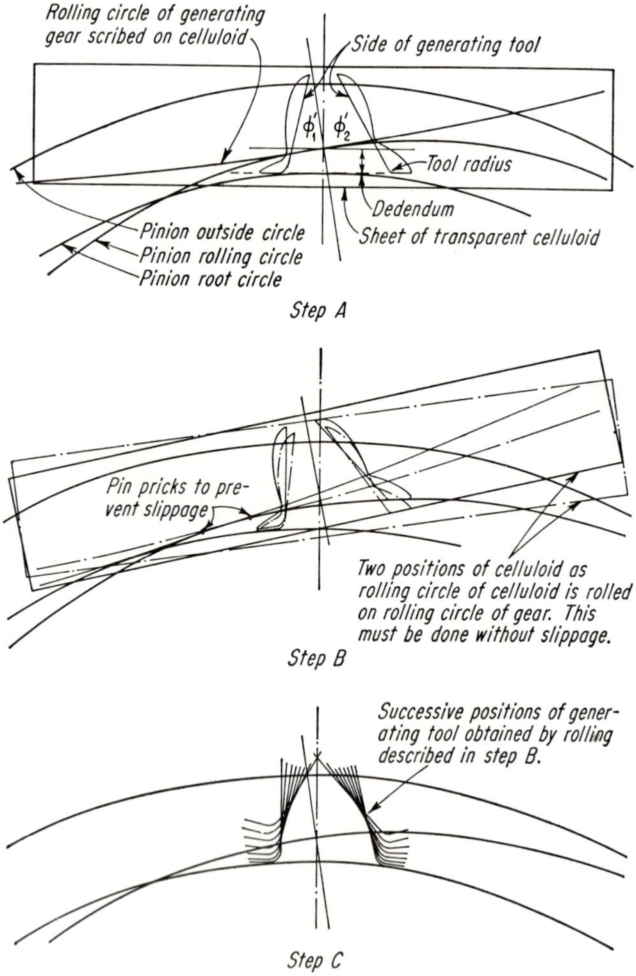

Fig. 8-16. Technique of making hypoid-tooth layout.

19. For Formate gears scribe on the celluloid the rolling circle of the Formate gear.
20. Now, scribe the sides of the Formate gear-tooth space on the celluloid as described in step 15 above, also the pinion edge radius as explained in step 16. The celluloid is removed from the board and the sections in the middle are removed, leaving a template of the sides and corner of the tool as shown in Fig. 8-16, step A.

Figure 8-16, Step B

21. For the Formate pinion roll the rolling circle of the Formate gear on the rolling circle of the pinion in short intervals by pivoting the celluloid at adjacent points of tangency during the rolling movement. Since the Formate gear is not generated, its profile form will be that of the generating tool described in step 15 with a fillet radius equal to the tool edge radius [item $(37)_R$] as explained in step 16.

Figure 8-16, Step C

22. The traces formed by the above rolling will produce an envelope of the pinion tooth form and tooth fillet.

Before proceeding further with the layout, it will be necessary to compute values for locating the position of the load when the stress is a maximum.

First, measure the distance along the path of contact between points d and D, where the path crosses the addendum circles of gear and pinion, respectively (Fig. 8-15). This is the distance Z_{N1} and should be recorded on the calculation sheet of Table 8-10 as item (51).

Table 8-10, items (51) through (65):

(51) Z_{N1} length of action in the mean normal section. $Z_{N1} = dD$ on the layout.

(52) m_p profile contact ratio.

(56) m_o mismatch contact ratio. When m_o is greater than 2.0, there will always be at least two teeth in contact. When m_o is less than 2.0, there will be periods when only one tooth is carrying the load.

(61) When m_o is less than 2.0, use the left-hand formula, and when m_o is greater than 2.0, use zero.

(63) When m_o is less than 2.0, use the left-hand formula, and when m_o is greater than 2.0, use item (56).

(64) Use the plus (+) sign in the pinion column and the minus (−) sign in the gear column.

(65) p_3 distance in mean normal section along the path of contact from the addendum circle of the mating member to the point of load application.

Figures 8-17 and 8-18

23. Measure along the horizontal line through P the distance QQ', the gear-tooth thickness given by item $(36)_R$.
24. Through Q' draw the side of the Formate gear tooth making an angle ϕ_2'' [item (49)] with the line of centers O_PO_G.
25. Construct the fillet radius [item $(37)_R$] at the root of the gear tooth. This completes the outline of the gear tooth.
26. In order to locate the pinion-tooth center line, it is first necessary to make an estimate of where the weakest section on the pinion will be. This will lie near the root of the tooth. By selecting some point i' on the fillet curve and drawing a line through it perpendicular to the line of centers O_PO_G we have our initial estimate. This line, which crosses one tooth profile at i' and the other tooth profile very close to i in Fig. 8-17, is now bisected to give a point very close to

m. The tooth center line may now be drawn by connecting this point with the pinion center O_P.

27. In like manner (Fig. 8-18), draw a line through point I' perpendicular to the line of centers O_PO_G, bisect this line to find the initial position of point M, and

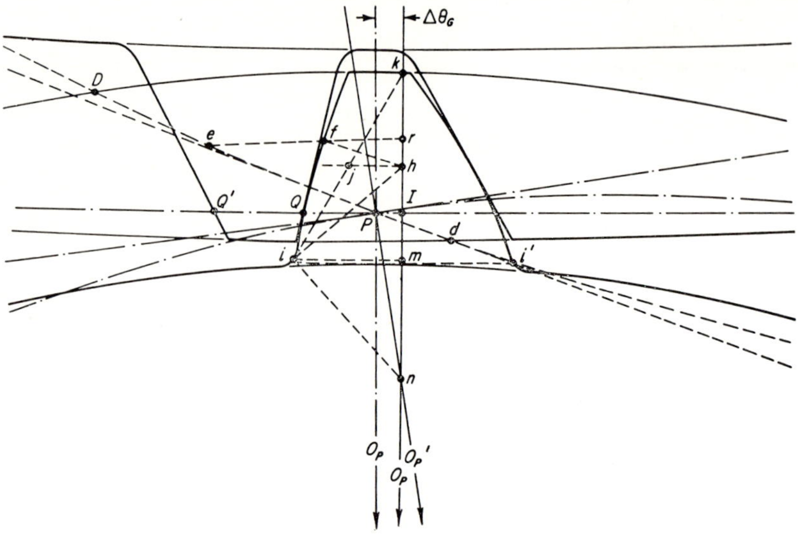

FIG. 8-17. Finding weakest section of hypoid pinion tooth.

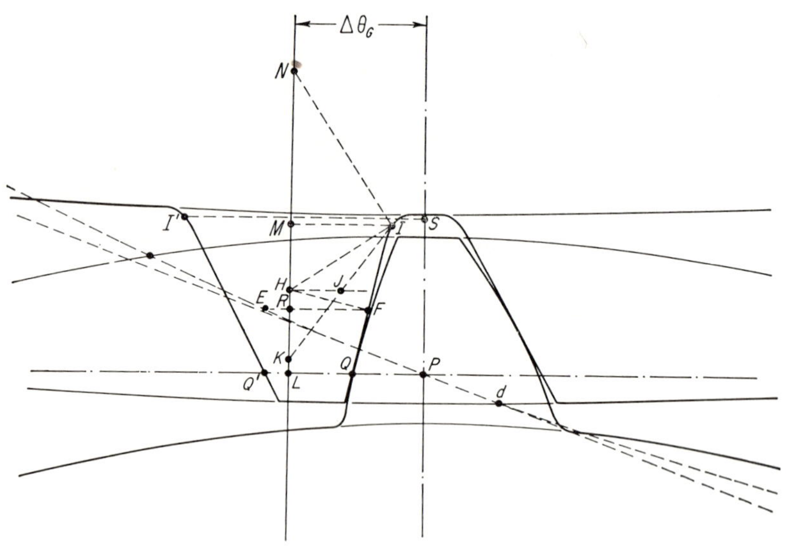

FIG. 8-18. Finding weakest section of hypoid gear tooth.

finally draw the gear-tooth center line by connecting point M with the gear center O_G. Because it is frequently impossible to locate the gear center O_G on the drawing table, the gear-tooth center line may be drawn through point M at an angle $\Delta\phi_G$ to the line of centers O_PO_G. Angle $\Delta\phi_G$ together with some of the other angles which are inconvenient to lay out on the drawing are calculated on the sheet of Table 8-11. Items (50a) through (50e) are dimensions taken from the layout (see Figs. 8-17 and 8-18). Angle $\Delta\phi_G$, given by item (50g), is in minutes.

Table 8-11. Layout Data

Step	Operation	Pinion	Gear
50a	MS (from layout)	1.15	
50b	PS (from layout)	1.365	
50c	$\Delta\theta_P$ (from layout)	1°0′	
50d	ef, EF (from layout)	0.94	0.91
50e	rl, RL (from layout)	0.635	0.56
50f	$(40)_R - (50b)$	83.0844	
50g	$\Delta\theta_G = \dfrac{3{,}440(50a)}{(50f)}$	48′	
50h	$\phi_{LNG} = (17)_L - (50g)$	12°55′	
50i	$(40) \pm (50e)$	15.8350	83.8894
50j	$\Delta\theta' = \dfrac{3{,}440(50d)}{(50i)}$	3°24′	37′
50k	$\phi_{LP'} = (17)_L + (50j)_L + (50j)_R$	17°44′	
50l	$\phi_{LNP} = (50K) - (50c)$	16°44′	

Table 8-11. Items on this sheet are to be calculated as required while making the hypoid layout.

(50a) MS distance (from layout) from point M determined in step 27 to the line of centers O_PO_G drawn parallel to the gear root circle (see Fig. 8-18).

(50b) PS distance (from layout) along line of centers from the pitch point P to point S (see Fig. 8-18).

(50c) $\Delta\theta_P$ angle between line of centers and pinion-tooth center line. This is taken from the layout (see Fig. 8-17).

(50d) ef, EF distance (from layout) from point e (E) as determined in step 28 (29) to point f (F) as determined in step 30 (31).

(50e) rl, RL distance (from layout) along tooth center line from point r (R) as determined in step 30 (31) to point l (L) where the tooth center line crosses the horizontal line through point P. rl will be positive when l lies between r and m, and RL will be positive when R lies between L and M (see Figs. 8-17 and 8-18).

(50g) $\Delta\theta_G$ angle between line of centers O_PO_G and gear-tooth center line. This angle is given in minutes.

(50h) ϕ_{LNG} angle FHJ.

(50i) Use the plus (+) sign in the pinion column and the minus (−) sign in the gear column.

(50j) $\Delta\theta'$ rotation angle. This angle is given in minutes by this formula.

(50k) $\phi_{LP'}$ angle between line fh and the horizontal line through point P. This is used in step 32 (see Fig. 8-17).

(50l) ϕ_{LNP} angle fhj.

28. Next, locate point e for the pinion on the path of contact. The distance de (Fig. 8-17) is equal to the quantity p_3 for the pinion and is given on the calculation sheet by item $(65)_L$.
29. In like manner locate point E for the gear. The distance DE (Fig. 8-18) is given by item $(65)_R$.

30. With O_P as a center and O_{P_e} as a radius, describe the arc *ef* cutting the pinion-tooth profile at *f* and the pinion-tooth center line at *r*. Point *f* is the point of load application on the pinion.
31. Similarly, describe arc *EF* about the gear center O_G cutting the gear-tooth profile at *F* and the gear-tooth center line at *R*. Point *F* is the point of load application on the gear.
32. Now, draw line *fh* normal to the pinion-tooth profile at *f*. Line *fh* will make an angle $\phi_{LP'}$ to the horizontal line through point *P*. [Angle $\phi_{LP'}$ is given by item (50*k*)].
33. Also draw line *FH* normal to the gear-tooth profile at *F*. Line *FH* will make an angle $\phi_{1'}$ to the horizontal line through point *P*. [Angle $\phi_{1'}$ is given by item $(17)_L$.]
34. At point *h* erect a perpendicular *hj* to the pinion-tooth center line.
35. In like manner, at point *H* erect a perpendicular *HJ* to the gear-tooth center line.

To find the weakest section it is necessary to locate the point of tangency between the inscribed parabola with vertex at *h*(or *H*) and the tooth outline. This is done as follows:

36. With a straightedge and dividers find a position where the straightedge is tangent to the tooth fillet at some point *i* (*I*) such that $ij = jk(IJ = JK)$. Point *i* (*I*) is the point of tangency of the parabola with the tooth outline and therefore is the location of the weakest section.
37. Draw line *ih* (*IH*).
38. Construct line *in* (*IN*) perpendicular to line *ih* (*IH*) at point *i* (*I*).
39. Construct line *im* (*IM*) perpendicular to the tooth center line O_Pl (O_GL).
40. The following values for the pinion may be taken from the layout by careful measurement (Fig. 8-17):

$$\Delta R_{LN1} = hl$$

ΔR_{LN1} will be negative if point *h* lies between points *l* and O_P.

$$X_{LN1} = mm$$
$$t_{LN1} = im$$
$$h_{LN1} = hm$$
$$\phi_{LNP} = \text{angle } fhj$$

This angle is calculated on the supplementary sheet at item (50*l*).

41. The following values for the gear may be taken from the layout by careful measurement (Fig. 8-18).

$$\Delta R_{LN1} = HL$$

ΔR_{LN1} will be negative if point *h* lies between points *L* and O_G.

$$X_{LN1} = MM$$
$$t_{LN1} = IM$$
$$h_{LN1} = HM$$
$$\phi_{LNG} = \text{angle } FHJ$$

This angle is calculated at item (50*h*).

Because the layout is enlarged, it is necessary to divide all linear values scaled from it by the diametral pitch, item (13) of Table 8-9, to obtain actual values.

Profile Contact Ratio. The meaning of *profile contact ratio* is shown on Fig. 8-19, which is similar to Fig. 8-14. The path of action is the line *DPd* intersecting the addendum circle of the pinion at *D*, passing through the pitch point *P*, and intersecting the addendum circle of the gear at d. As the pinion tooth turns from the beginning of contact at d to the end of contact at D, it turns through arc Q_1Q_2 on the pitch circle. The arc distance Q_1Q_2 divided by the circular pitch gives the profile contact ratio. In the case of generated gears, the path of action is a straight line, and the gear-tooth profiles as enveloped on the normal section layout are involutes of a circle. In this case the profile contact ratio is equal to distance dD divided by the normal base pitch. The profile contact ratio m_P is calculated at item (52) of Table 8-8.

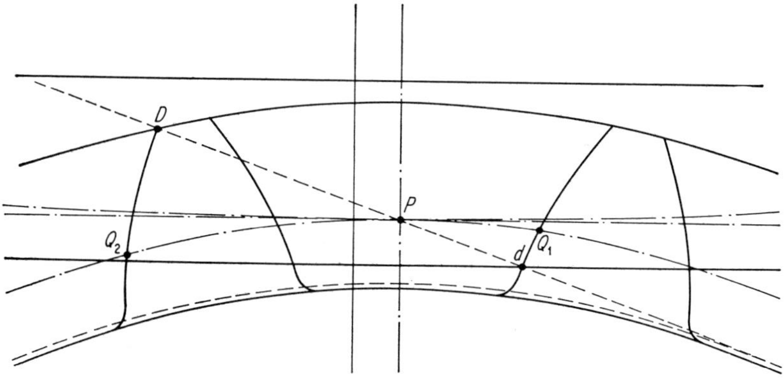

FIG. 8-19. Contact length is distance dD. This distance is used to determine the contact ratio.

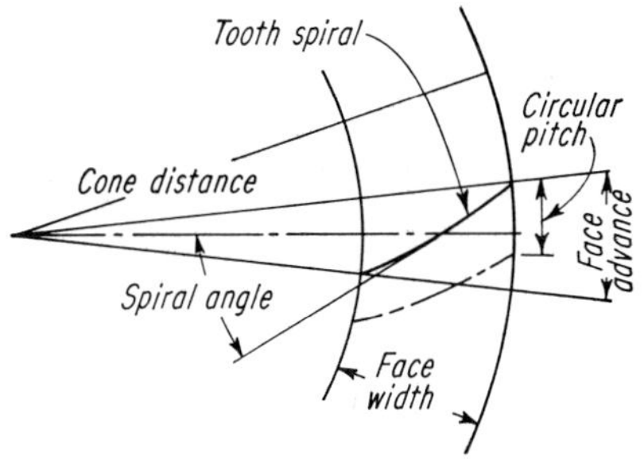

FIG. 8-20. Development of pitch cone.

Face Contact Ratio. Figure 8-20 shows the development of a spiral-bevel-gear pitch cone into a plane. The instantaneous axis is the central element (shown with center-line convention), and contact exists as the gear turns through the "face advance" angle. The face contact ratio $m_F = \dfrac{\text{face advance}}{\text{circular pitch}}$; it is calculated at line (50) of Tables 8-7 and 8-9.

The mismatch contact ratio m_o used in stress calculations $= \sqrt{(m_F)^2 + (m_P)^2}$.

Undercut. In Fig. 8-10, assume the straight portion of the rack to pass through point U. This point U is the limiting position of the straight rack profile, which may not extend beyond U without undercutting the pinion teeth. The limiting position of the generating profile is determined by projecting the center of the pitch circle to the path of action.

With spiral bevel and hypoid gears there exists a surface of action which is the locus (in space) of all points of contact. When a bevel or hypoid gear or pinion is generated, the cradle and cutter represent a generating gear which rolls with the member being generated, and a surface of action exists which represents the locus of points of contact between the cutter and the generated member. The projection of the axis of the workpiece to this surface of action defines a curve lying in the surface of action. The straight portion of the cutter profile must not extend beyond this limiting curve if undercut is to be avoided.

When laying out a spur-gear profile, if the rack used for generating the profile is sharp-cornered, the corner of the rack at the tip generates a curve which is one of the general class of curves called trochoids. Part of this trochoid is the tooth fillet. When cutter edge radii are used the tooth fillet is a generated curve parallel to the trochoid traced by the center of the edge round.

When no undercut exists, the tooth fillet and the tooth profile have a common tangent at their point of intersection.

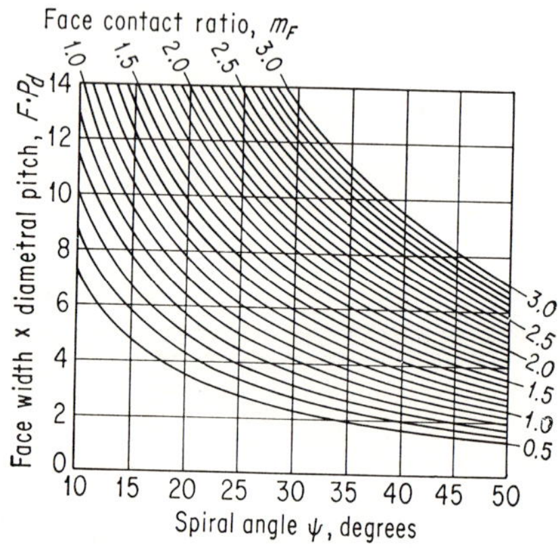

$$K_2 = \frac{F}{A_0}\left[\frac{2 - \dfrac{F}{A_0}}{2\left(1 - \dfrac{F}{A_0}\right)}\right]$$

$$m_F = \frac{\left(K_2 \tan \psi - \dfrac{K_2{}^3}{3} \tan^3 \psi\right) A_0 P_d}{\pi}$$

FIG. 8-21. Face contact ratio for spiral bevel gears.

With bevel gears, undercut may exist although none is shown by the normal section layout. This is because the layout is merely a close approximation—a two-dimensional representation of a three-dimensional situation. The systems of bevel-gear-tooth proportions discussed in Chap. 5 include graphs giving maximum values of dedendum angle for no undercut when sharp-pointed tools are used.

REFERENCES

1. Buckingham, E.: "Analytical Mechanics of Gears," pp. 48–52, 85–89, McGraw-Hill Book Company, Inc., New York, 1949.
2. Candee, A. H.: Involute Curve Calculations, *Prod. Eng.*, October, 1949, p. 135.
3. Dudley, D. W., and H. Poritsky: "Cutting and Hobbing Worms and Gears," American Gears Manufacturers Association paper, October, 1943, pp. 7–10.
4. Varnum, E. C.: Involute Curve Ratio Constants for Plotting Gear Tooth Profiles, *Prod. Eng.*, April, 1950, pp. 102–103.
5. Gleason Works: "Bevel and Hypoid Gear Design," Gleason Works, Rochester, N.Y.
6. Candee, Allan H.: "Introduction to the Kinematic Geometry of Gear Teeth," Chilton Co., New York, 1961.

Chapter 9

GEAR TOLERANCES

By

GEORGE W. MICHALEC, *Head of Special Products Department, GPL Division—General Precision, Inc., Pleasantville, N.Y.*

The contents of this chapter are devoted to the subject of gear tolerances, which are related to the information contained in all the previous and remaining chapters. Tolerances constitute a complex area of gear specification which directly affects gear performance, materials and finishes, fabrication and inspection techniques, and cost. Gear tolerances must be considered in the design and specification of gears and will also influence the design of associated elements such as bearings, shafts, and housing. Because of this broad effect, the gear engineer must comprehend and appreciate tolerances with a competent knowledge of proper specification. For this reason the subject of gear tolerances is given paramount significance with detailed coverage. In this chapter are a definition and complete review of tolerances for all types of gears and applications. Tolerance effect of function, fabrication, and cost is presented. As a design guide, numerous standardized tolerance values are presented.

TERMINOLOGY AND DEFINITIONS PERTAINING TO GEAR TOLERANCES

The subject of gear tolerances involves a number of special and general terms which require definition. These are set forth in the following section as a handy reference glossary.

9-1. Terms and Definitions

Active Profile. That portion of the tooth profile which encounters contact with the mating tooth. For standard gears it is nominally the addendum and the dedendum minus the clearance.

Allowance. The minimum specified difference between two dimensions.

Bellmouth. An extreme taper at the ends of the gear bore.

Body Dimension. A dimension of the gear pertaining to the hub or other portion of that which carries the teeth.

Concavity. Pertains to the degree of departure from flat to concave of a surface in the plane of rotation.

Convexity. Pertains to the degree of departure from flat to convex of a surface in the plane of rotation.

Fullness. An excessive amount of material beyond what is desired. Referring to a gear-tooth profile it is extra material on the space side of the desired profile.

Functional Element. A gear dimension or other criterion which directly bears upon gear performance. As an example, tooth thickness, which directly relates to backlash.

Gear-element Dimension. Refers to a dimension which pertains to the gear features, i.e., the teeth and profiles, as contrasted to a dimension pertaining to the hub or bore.

Gear Function. An operational feature of the gear, such as velocity ratio, linearity, or strength.

Gear Size. The precise tooth thickness or testing radius. In this use it is not meant as a general term for gear pitch or pitch diameter.

Generating Mounting Surface. The diameter and/or plane of rotation surface or shaft center which is used for locating the gear blank during fabrication of the gear teeth.

Hollow. A deficiency of material. Referring to a gear-tooth profile, it is the space or emptiness between the actual and desired profiles.

Index Tolerance. Allowable variation in position of the profile at the pitch line, in the middle of the face width and at each tooth.

Journal. The male member of a turning pair, such as the shaft length which engages a bearing.

Lateral Runout. The total indicator reading of an indicator placed against a surface in the plane of rotation as it is rotated.

Lay. The direction of the predominant pattern of a finished surface.

Microinch. One millionth of an inch (0.000001″).

Mounting Surface. The diameter and/or plane-of-rotation surface which is used for locating the gear in the application assembly.

Nonfunctional Element. A gear dimension or other criterion which does not affect gear performance, for example, hub diameter.

Non-gear-element Dimension. Refers to a dimension which does not pertain to the gear features of teeth or profiles. As examples, the hub or bore dimensions.

Over-rolls Measurement. Pertains to measurement, radial or diametrical, made across cylindrical rolls inserted between teeth. For fine teeth the associated small-diameter rolls are usually called pins and it is an over-pin measurement.

Parallelism. The degree of departure from parallel of two surfaces in the plane of rotation.

Position Error. The abbreviated equivalent term for tooth-profile position error, which is the degree of deviation of a gear's tooth profiles in space from the true positions they should have.

Profile. The complete contour of a gear tooth in a transverse plane.

Registering Surface. The surface in the plane of rotation which locates the gear blank axially in the generating machine and the gear in application. These are often identical surfaces, but not necessarily so.

Surface Finish. The degree of surface roughness and waviness pertaining to relative finely spaced irregularities produced by the cutting or abrading action of machine tools, or the result of a casting or forming process.

Surface Roughness. Pertains to the relatively finely spaced surface irregularities, produced by machining and abrasive operations. Roughness may be considered as superimposed on a "wavy" surface.

Surface Waviness. The surface irregularities which are of greater spacing than the roughness.

Testing Radius. The radial distance from the gear center to the theoretical pitch radius or pitch plane of the mating inspection master gear or rack.

Tolerance. The amount of variation permitted for a given dimension or other measured criterion.

Bilateral Tolerance. A variation value which affects the fixed dimension in two senses (positive and negative).

Unilateral Tolerance. A variation value which affects the fixed dimension in one sense (positive or negative).

Close Tolerance. A tolerance magnitude that is small for a particular dimension and therefore relatively difficult to achieve.

Open Tolerance. A tolerance magnitude that is large for a particular dimension and therefore relatively easy to achieve.

Top Generation. The gear-fabrication process which finishes the outside diameter of the gear teeth simultaneously with cutting of the tooth profiles.

Undercut. This is undesirable removal of tooth profile. Its magnitude is indicated by the hollowness between the profile and a radial line passing through the intersection of the base circle and the projected involute curve.

9-2. Symbols. The following symbols are used in this chapter:

a = addendum
B = backlash
b = dedendum
C = center distance
c = clearance
d = roll or pin diameter
M = over-roll measurement
N = number teeth, gear
n = number teeth, pinion
P = diametral pitch
p = circular pitch
R = pitch radius, gear
r = pitch radius, pinion
R_b = base-circle radius, gear
r_b = base-circle radius, pinion
R_c = radius to center of roll (or pin) inserted between teeth
t = tooth thickness
ϕ = pressure angle, standard
ϕ_o = pressure angle, operating
ψ = helix angle

INTRODUCTION TO GEAR TOLERANCE

We shall first consider the general philosophy of gear tolerances.

9-3. Definition of Gear Tolerance. As a general definition, a tolerance can be defined as "the magnitude of permissible variation of a dimension or control criterion from the specified value." Tolerances constitute an engineering legality for deviation from the ideal values. As with legal rulings, much good thought and planning must go into tolerance formulation.

Tolerances must be specified and used because of inevitable human failings and machine limitations which prevent ideal achievements during fabrication.

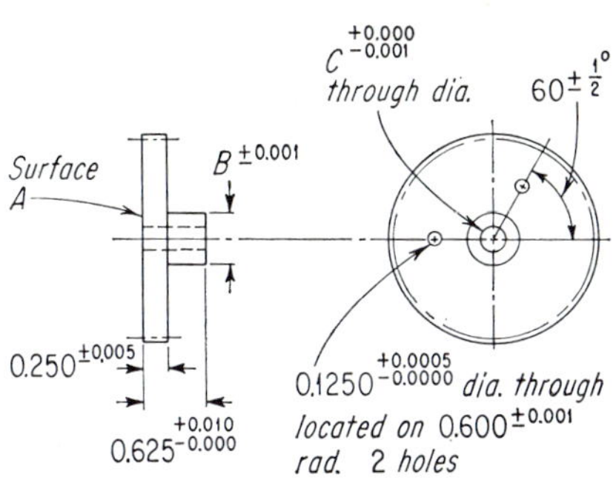

Notes:
1. *Surface A to be square with diameter* $B^{\pm 0.001}$ *within* 0.001 *per inch.*
2. *Bore* $C^{\pm 0.001}$ *to be concentric with diameter* $B^{\pm 0.001}$ *within* 0.0005 *TIR.*
3. *Material SAE* 4140 *hardened to* 30 *to* 34 *Rockwell C.*

FIG. 9-1. Examples of dimension and control tolerances.

9-4. Examples of Tolerance. Tolerances are applicable to measurements regardless of the nature of the measurement. Thus tolerances are to be considered in regard to the measures of length, weight, force, velocity, acceleration, etc.

In the case of gears the toleranced quantities are predominantly measures of length (or angle) and some measures of force, such as associated with material-hardness specification. The measures of length are either linear or angular. These in turn are toleranced in two general forms as shown in Fig. 9-1. Linear and angular dimensions directly located on the drawing are toleranced in an equally direct manner. However, some control measurements do not appear as direct dimensions but instead are control criteria or parameters in the form of notes. Examples are the runout, and squareness and hardness specifications shown in Fig. 9-1.

9-5. Gear-tolerance Magnitudes. No typical tolerance values can be stated as applying to the various gear dimensions. Each dimension has a unique tolerance which varies with many gear considerations such as size, precision, material, and application. As an indication of the general tolerance range, so as to fix the art of gearing in the full spectrum of measurements, pertinent gear tolerances range from several hundredths (0.010) down to 0.0001. This covers only the majority of gear tolerances, as there are exceptions. The several-hundred-fold variation affords a wide selection ranging from possibilities of wholly inadequate design to extreme overdesign and is the reason tolerances constitute an important subject.

9-6. Tolerances and Deviations. Dimensional variations within the band of allowable tolerances constitute *allowable deviations.* If the variation is beyond tolerance, the gear has not been made correctly and can be described quite rightfully as "in error." There has been some tendency in the gear trade to describe any deviation from theoretical perfection of profile, spacing, or lead as an *error.* The standard nomenclature of the American Gear Manufacturers Association (AGMA) considers an *error* to be an amount of deviation that is beyond the agreed-upon tolerance for the gear element.

When the amount of variation in a gear element is beyond limits, the amount of the variation that is over the limit may be described as *deviation beyond limits.* This permits a ready assessment of measured variations into the categories of allowable deviations and deviations beyond limits. There is then no confusion as to just what constitutes an error in a gear.

It should be pointed out that some variations in gear dimensions have no effect—or almost none—on gear performance and some most directly affect performance. The gear designer has the problem of constantly weighing the effect of any proposed tolerance on gear performance and of specifying proper values.

9-7. Allowances and Relation to Tolerances. *Allowance* has been defined in Art. 9-1 as the intentional clearance between mating parts. In the case of gears this would be

1. Clearance between gear bore and mating shaft
2. Clearance between integral gear shaft or mounting shoulder and mating bearing
3. Clearance between gear teeth, or "backlash allowance"

Allowance and tolerances are related in that the tolerance value adds to the allowance. This places a requirement for consistency between the two. For instance, if the allowance on backlash is intended to be very small, then a high order of accuracy is needed in the concentricity of the gears and in the center distance.*

9-8. Tolerances and Fabrication Cost. There is much significance to gear tolerances since they establish the practicability of the manufacturing process used to finish the gear parts. The cost of fabricating the whole gearset is a more or less direct function of the tolerances specified. The magnitude of the tolerances must be set so as to permit as low manufacturing costs as possible and still achieve the necessary gear functional performance. This requires a thorough understanding of critical and noncritical dimensions and controls and the interrelationship of tolerances.

The closer the tolerance, regardless of the particular dimension or control, the greater the fabrication effort required for achievement. This results in increased

* See Chap. 6 for a detail analysis of center-distance variations and other things that control the range of backlash in a gearset. Two sample problems are given.

manufacturing costs. The relationship between tolerance and cost is not linear, for as the tolerance approaches zero the task of achieving it becomes enormous and finally impossible. Similarly, at the other extreme, fabrication cost decreases with the relaxing of tolerance, but a value is soon reached beyond which little manufacturing relief is gained. In general tolerance vs. fabrication cost is a hyperbolic curve, as shown in Fig. 9-2. Note the curve is asymptotic to the two axes portraying the limits mentioned above. No typical numerical values can be stated because of the interdependence upon the dimension. As an example, the diameter of a shaft can be held to a closer tolerance than the length of shaft, for approximately the same effort and cost. Also, fabrication costs vary greatly among shops, preventing the derivation of any typical dollar costs. However, in general it can be said that the curve of Fig. 9-2 becomes asymptotic to 0.0001 for most critical gear tolerances.

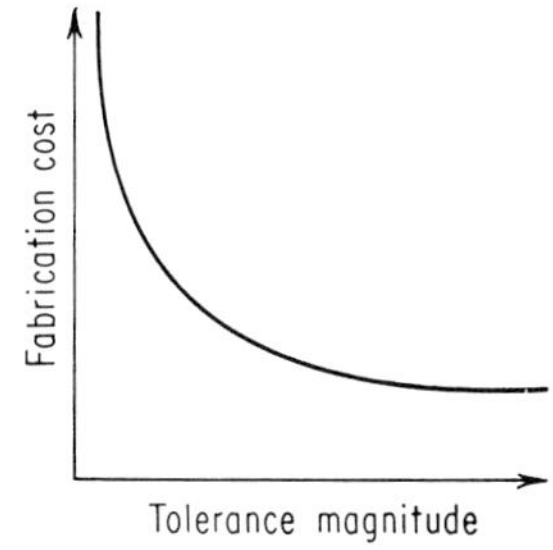

FIG. 9-2. The relationship of tolerance and fabrication cost.

TOLERANCE EFFECTS

A completely specified gear has so many dimensions and control criteria that it becomes a highly complex matter in regard to tolerance specification and its effect upon gear performance, fabrication, and inspection. Tolerances affect both the functional and nonfunctional characteristics of the gear.

9-9. Tolerance Effects on Function. Many toleranced gear dimensions bear directly upon functional operation of the gear. Gear function is defined as the intended action of the gear, i.e., the transmission of power at constant velocity, at specified speed, with other additional qualifications such as backlash magnitude, life, and noise. Since the function of a gear is its primary purpose, tolerances affecting it are the most critical. Typical functional tolerance examples are tooth thickness, profile deviation, hardness, surface finish, etc. The above tolerances have a very direct influence on gear function. In addition there are tolerances which have a secondary or negligible functional effect. An example is the tolerance on face width, which remotely affects the strength of the gear.

9-10. Nonfunctional Tolerance Effects. The functional elements of a gear are concentrated at the periphery with a relation to the center of rotation. To be practical these elements must be carried by a structure, or gear body. The dimensions of this gear body, i.e., web and hub design, mounting arrangements, etc., do not affect gear function, and associated tolerances are therefore noncritical in regard to gear function. Typical nonfunctional tolerances are those on hub diameter and length, keyway or equivalent mount design, etc.

9-11. Tolerance Effects on Fabrication. The specific magnitude of the critical gear tolerances has a determining effect upon the method of manufacture, tooling, jigs and fixtures, and inspection techniques. The basic fabrication method, machine, grind, mold, etc., will be established by the general gear quality. Specific dimension tolerances will influence methods, tools, jigs, and inspection.

TYPES OF TOLERANCES

Gear tolerances can be divided into three general classifications: functional gear-element tolerances, functional non-gear-element tolerances, and nonfunctional tolerances on body dimensions and other non-gear-element controls. In a complete gear

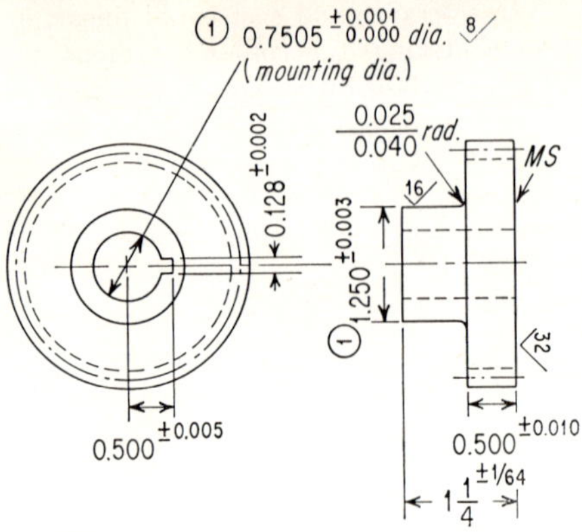

Notes:

1. *Diameters marked* ① *to be concentric within* 0.001 *TIR.*
2. *Lateral runout of MS not to exceed* 0.0015 *TIR.*
3. *Convexity of MS not to exceed* 0.0015.
4. *Concavity of MS not to exceed* 0.0015.

Material: AISI 416.
Hardness: 30 *to* 35 *Rockwell C.*
Finish: None.

Spur-gear Data
Full-depth Involute Form per ASA B6.7

No. of teeth	25
Diam. pitch	10
Pressure angle	20°
Standard pitch dia. (theoretical)	2.5000
Tooth form	*ASA* B6.1-1932
Addendum	0.100
Dedendum	0.157
Clearance	0.057
Testing radius	$1.250^{+0.000}_{-0.002}$
Measurement over wires (reference)	$2.7351^{+0.000}_{-0.0034}$
Dia. of wires	0.1728
Outside dia	$2.700^{+0.000}_{-0.005}$
Max. outside dia. runout	0.003 *TIR*
Max. total composite tolerance	0.0013
Max. tooth-to-tooth composite tolerance	0.0005
AGMA quality class	*No.* 11
Surface finish of active profile	16 √ *max. rms*
Max. lead tolerance	0.0004
Max. index tolerance	0.0015

FIG. 9-3*a*. Fully dimensioned and toleranced gear.

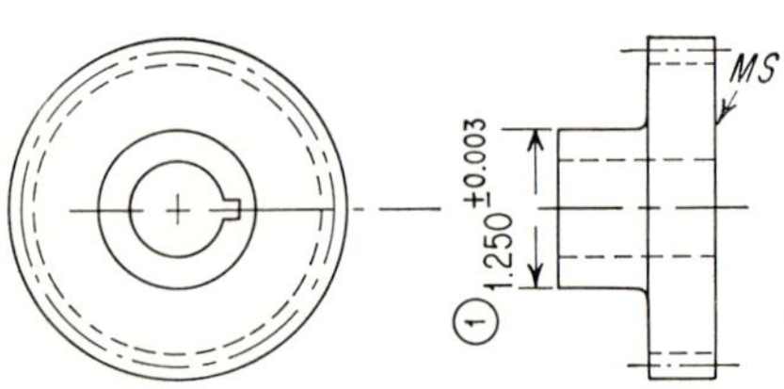

Notes:

1. *Diameters marked* ① *to be concentric within* 0.001 *TIR.*
2. *Lateral runout of MS not to exceed* 0.0015 *TIR.*
3. *Convexity of MS not to exceed* 0.0015.
4. *Concavity of MS not to exceed* 0.0015.

Material: AISI 416.
Hardness: 30 *to* 35 *Rockwell C.*
Finish: None.

Spur-gear Data
Full-depth Involute Form per ASA B6.7

No. of teeth	25
Diam. pitch	10
Pressure angle	20°
Standard pitch dia. (theoretical)	2.5000
Tooth form	*ASA* B6.1-1932
Addendum	0.100
Dedendum	0.157
Clearance	0.057
Testing radius	$1.250^{+0.000}_{-0.002}$
Measurement over wires (reference)	$2.7351^{+0.000}_{-0.0034}$
Dia. of wires	0.1728
Outside dia	$2.700^{+0.000}_{-0.005}$
Max. outside dia. runout	0.003 *TIR*
Max. total composite tolerance	0.0013
Max. tooth-to-tooth composite tolerance	0.0005
AGMA quality class	*No.* 11
Surface finish of active profile	16 √ *max. rms*
Max. lead tolerance	0.0004
Max. index tolerance	0.0015

FIG. 9-3*b*. Tolerances for functional gear elements.

FIG. 9-3. Various categories of toler-

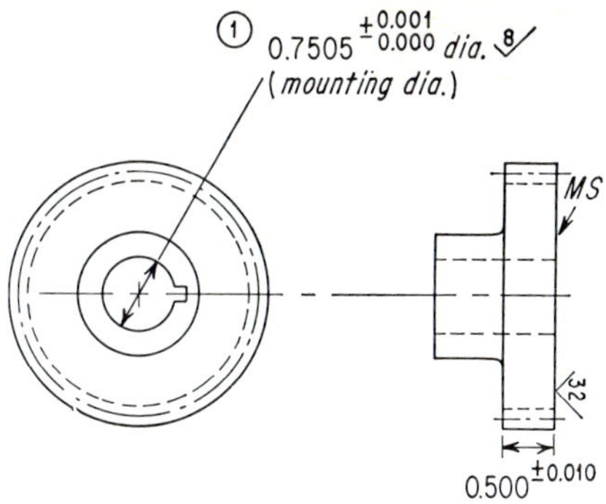

Notes:

1. *Diameters marked* ① *to be concentric within* 0.001 *TIR.*
2. *Lateral runout of MS not to exceed* 0.0015 *TIR.*
3. *Convexity of MS not to exceed* 0.0015.
4. *Concavity of MS not to exceed* 0.0015.

Material: AISI 416.
Hardness: 30 *to* 35 *Rockwell C.*
Finish: None.

Spur-gear Data
Full-depth Invalute Form per ASA B6.7

No. of teeth	25
Diam. pitch	10
Pressure angle	20°
Standard pitch dia. (theoretical)	2.5000
Tooth form	*ASA B*6.1-1932
Addendum	0.100
Dedendum	0.157
Clearance	0.057
Testing radius	—
Measurement over wires (reference)	—
Dia. of wires	—
Outside dia	—
Max. outside dia. runout	—
Max. total composite tolerance	—
Max. tooth-to-tooth composite tolerance	—
AGMA quality class	—
Surface finish of active profile	—
Max. lead tolerance	—
Max. index tolerance	—

FIG. 9-3*c*. Tolerances for functional nongear elements.

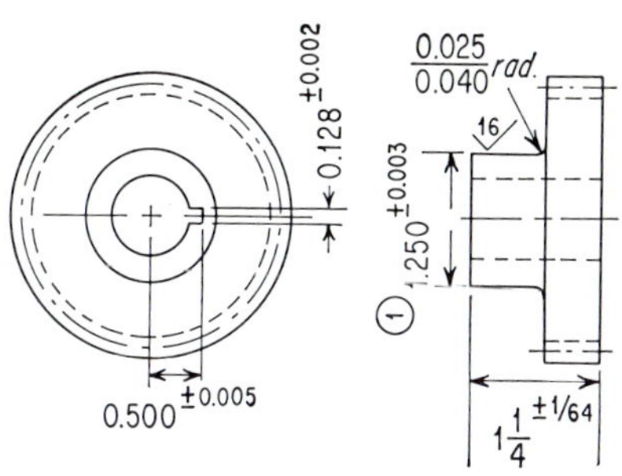

Notes:

1. *Diameters marked* ① *to be concentric within* 0.001 *TIR.*
2. *Lateral runout of MS not to exceed* 0.0015 *TIR.*
3. *Convexity of MS not to exceed* 0.0015.
4. *Concavity of MS not to exceed* 0.0015.

Material: AISI 416.
Hardness: 30 *to* 35 *Rockwell C.*
Finish: None.

Spur-gear Data
Full-depth Involute Form per ASA B6.7

No. of teeth	25
Diam. pitch	10
Pressure angle	20°
Standard pitch dia. (theoretical)	2.5000
Tooth form	*ASA B*6.1-1932
Addendum	0.100
Dedendum	0.157
Clearance	0.057
Testing radius	—
Measurement over wires (reference)	—
Dia. of wires	—
Outside dia	—
Max. outside dia. runout	—
Max. total composite tolerance	—
Max. tooth-to-tooth composite tolerance	—
AGMA quality class	—
Surface finish of active profile	—
Max. lead tolerance	—
Max. index tolerance	—

FIG. 9-3*d*. Nonfunctional tolerances.

anced gear dimensions and controls.

design all tolerances are interwoven to specify the gear. An example of a fully specified gear drawing is shown in Fig. 9-3*a*. Particulars of the three tolerance groupings and drawing examples are given below.

9-12. Functional Gear-element Tolerances. These tolerances pertain to the dimensions and control criteria associated with the gear teeth. They are

Tooth-form dimensions—addendum, dedendum, pressure angle, clearance
Tooth thickness
Tooth spacing
Tooth profile
Lead angle
Material hardness
Surface finish
Face width
Outside diameter
Outside-diameter runout
See Fig. 9-3*b*

9-13. Functional Non-gear-element Tolerances. These tolerances pertain to dimensions and control criteria not associated with the gear teeth but nevertheless having direct effect upon gear performance. They are

Mounting bore or journal diameter
Runout between gear-tooth elements and mounting diameter
Blank dimensions
  Outside-diameter runout
  Lateral runout
  Parallelism
  Convexity
  Concavity
See Fig. 9-3*c*.

9-14. Nonfunctional Dimension Tolerances. These tolerances pertain to dimensions relating to the tooth carrier or body and do not directly affect gear function. They are

Hub-design dimensions
Web-design dimensions
Mounting-fastening design—keyway, setscrew pinhole
Special body dimensions
See Fig. 9-3*d*

The nonfunctional and non-gear-element dimensions pertain to portions of the gear which are more properly termed general machine-design considerations rather than gear criteria. For this reason these dimensions and their tolerances will not be covered in this chapter.

FACTORS AFFECTING CHOICE OF TOLERANCE

Since the setting of tolerances is not an arbitrary matter, consideration must be given to the influences bearing on the choice. The most significant are the functional requirements. Often these primary considerations require compromises with standardization, methodizing, and manufacturing needs.

9-15. Functional Requirements. The primary tolerance purpose is to permit variation in dimensions and control parameters without degradation of gear performance beyond the limits established by the specification of the design. In high-performance designs the functional requirements will be the dominating factor in setting tolerances. When the functional performance is relatively open, the tolerance choice may be influenced and determined by the other factors such as standards, methods of tooling, and available fabrication equipment.

9-16. Standardization. Although functional requirements can be used to determine every tolerance it may not be practical because it would result in

1. Too many different tolerances—essentially a continuous spectrum
2. Unnecessarily excessive amounts of calculations and studies to establish tolerance values from the functional requirements
3. Excessive amount of special tooling
4. Too loose tolerances for broad functional requirements
5. Needless complication and prolonging of inspection

Thus, very often it is better to establish a limited number of standard tolerances and obtain the benefits of fewer varieties of tooling, fewer calculations, and increased unit quantities because of repeated uses of the same designs.

When standardized tolerances are used, the choice of specific tolerances must be such that the functional gear requirements are satisfied. Generally, the tolerances chosen will be slightly closer than function would dictate. The degree of excessiveness can be kept reasonable by establishing a rational number of standard tolerances.

9-17. Methodizing. Tolerance choices reflect upon machines, tools, fixtures, and operations. Achievement of tolerances depends upon the methodizing, which is centered about the tools and machines. The methodizing procedures will establish normal fabrication procedures and tolerance patterns.

9-18. Manufacturing Needs. Often a tolerance is not set by functional requirements but by manufacturing needs. One machine operation may affect the next. Jigs and fixtures may require tolerances for proper applications but otherwise the tolerances may not be needed. For example, lateral runout may be irrelevant regarding gear performance. However, in order to fabricate the gear and obtain a close functional tolerance on total composite variation, it may be necessary to control the lateral runout closely. In a similar manner many other tolerances may be dictated by fabrication needs.

9-19. Ranges of Numerical Tolerances. The numerical values of gear tolerances range across the entire spectrum of measurements. The broad open tolerances are in the order of $\frac{1}{32}''$, and the finest are down to millionths of an inch. In this broad area of choice all the influencing factors mentioned in the preceding section combine and interplay in setting tolerance values.

The gear elements and other dimensions affecting function utilize the more precise tolerances, generally in the order of ten-thousandths of an inch to one- or two-thousandths. Noncritical dimensions not affecting function are generally larger-valued tolerances, often in the order of several hundredths of an inch or greater.

NUMERICAL TOLERANCE VALUES

The tolerance values presented in the following sections represent recommended practice in the modern gear industry. Because gearing covers a vast field of applications and has many design variations, the tolerances are likewise numerous. The tolerances are presented for the many logical divisions of the gear field. This results in some duplication since many of the divisions overlap. Nevertheless, this is a worthwhile manner of presenting tolerances since the over-all application field is divided in these general divisions.

9-20. Basic Gear-blank Tolerances. The basic gear-blank tolerances pertain to a group of dimensions and control criteria that are of particular significance in the fabrication procedure and actually may or may not be of direct concern to the gear function. These blank dimensions and controls are

1. Bore diameter
2. Bellmouth of bore
3. Taper of bore
4. Journal or shoulder mounting diameter (if no bore)
5. Lateral runout
6. Face convexity
7. Face concavity
8. Face parallelism
9. Outside diameter
10. Outside-diameter runout

The significance of the first four items is evident since these dimensions can directly alter gear function. The importance of the last six items is relatively indirect since gears could perform satisfactorily in the presence of these malfunctions. However, gear blanks containing excessive convexity, concavity, and parallelism cannot be easily fabricated into good gears because of the ill effects upon machine and tool operation.

Specific numerical values of gear-blank tolerances are related to quality, size, application, fabrication method, etc. These will be presented in the appropriate following paragraphs.

9-21. Fundamental Tolerances for Spur Gears. Proper and complete dimension and control criteria tolerancing of a spur gear is shown in Fig. 9-3*a* and Fig. 11-6. This example represents the most fundamental type of gear upon which is based the dimensioning and tolerancing of other types.

The essential spur-gear tolerances affecting function are

Gear-element tolerances:

- Tooth thickness—over-roll measurement, center distance with master or other tooth-thickness control measurement
- Concentricity—runout of active sides of tooth with specified "reference" axis
- Profile—slope, fullness or hollow, irregularity, undercut allowable, tip modification, tip radius
- Spacing—tooth-to-tooth or "pitch tolerance," position error or "index tolerance"
- Alignment—parallelism of tooth sides with axis, convex form in a "crowned" tooth, allowable end radius
- Tooth surface finish—finish of active profile, finish of root fillet, allowable waviness or "lay" of surface, allowable "step" when active profile is ground and root fillet is unground

Non-gear-element tolerances:

- Outside diameter
- Outside-diameter runout
- Bore or journal diameter
- Lateral runout of mounting surface
- Face width
- Hardness
- Surface finish (roughness)
- Finish coating (if any)

Figure 11-6 shows an example of a spur-gear drawing with a tolerance specification. Figure 9-3*a* shows a spur gear toleranced by the composite system.

In the case of small gears where no profile modification is required and tooth loading is not extreme, the "composite" check with a master gear is widely favored. The composite check gives an indication of concentricity, profile, and spacing all at once. The diagram of a composite check is interpreted for the following:

Maximum tooth-to-tooth composite tolerance
Maximum total composite tolerance

Figure 9-4 shows an example of a composite check.

Figure 9-3*a* shows an example of a gear toleranced by the composite-check method.

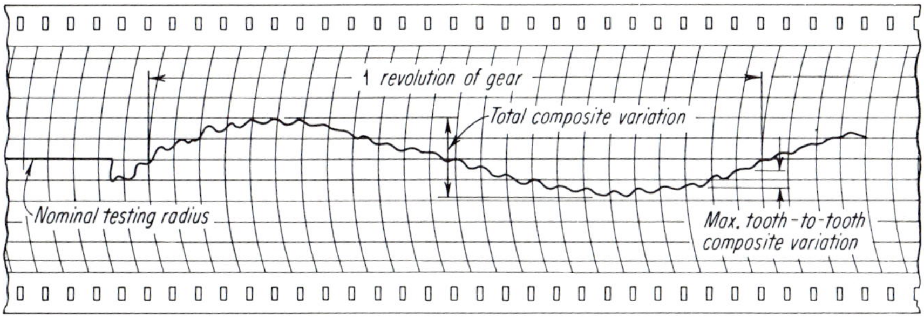

FIG. 9-4. Example of the graphical results of a composite check.

9-22. Gear-element Tolerance for Helical Gears. Tolerancing of helical gearing differs from spur gearing in that there is specific control of the helix angle and related parameters. Additional special tolerances beyond those listed for spur gears are

Helix angle
Normal pressure angle
Axial pitch

The helix angle and axial pitch are related, and it is redundant to specify tolerances on both. Either can be assigned the tolerance and the other dimensioned as a reference.

The helix angle of a helical gear is checked by using either a machine equipped to traverse an indicator along the helical spiral or a checking device equipped to move the indicator over an axial pitch at a time. In either case the *reading is taken normal to the tooth surface.* Thus deviations in either "lead" or "axial pitch" of a helical gear are in reality an excess or deficiency of material with respect to the true helical spiral of the tooth. The deviation reading of the helix of a helical gear is of the same nature as the deviation reading of the parallelism of a spur gear. The geometry of lead deviation is shown in Fig. 9-5.

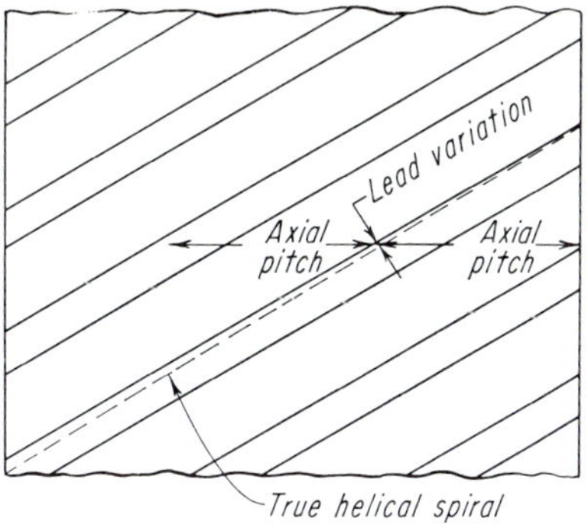

FIG. 9-5. Section of a helical gear to show lead deviation.

9-23. Fundamental Tolerances for Worms. The worm deviates from other gears in that it resembles a screw thread more and its teeth have cylindrical form rather than plane. In addition to the fundamental tooth-proportion tolerances, the worm specification must control the axial pitch and any special form factors such as in the case of globoidal worms. Pertinent tolerances for worm elements are given in Table 9-1.

Table 9-1. Tolerances for Coarse-pitch Cylindrical Worms*

(For worms ¾ to 12 in. diameter; ¼ to 4 in. axial pitch; in ten-thousandths of an inch)

Class	Axial pitch, in.	Runout measurement on pitch circle, total indicator reading					Pitch error in axial plane, multiple threads only					Profile-error total variation, not plus and minus			Lead-error total variation per convolution		Lead-error total variation per active length of worm		
		Pitch diam., in.					Pitch diam., in.					Lead angle, deg			Lead angle, deg		Lead angle, deg		
		¾	1½	3	6	12	¾	1½	3	6	12	0–15	15–30	30–45	0–15	15–30	0–15	15–30	30–45
Class 1 light duty	4	...	...	...	...	80	...	...	...	...	50	30	40	45	40	80	80	100	120
	2	...	...	...	50	60	...	...	...	35	40	13	15	17	25	35	35	40	45
	1	...	...	40	50	60	...	...	35	35	40	11	13	15	15	22	28	32	35
	½	...	30	40	50	60	...	25	35	35	40	10	12	14	15	20	25	28	32
	¼	30	30	40	...	...	20	25	25	...	...	10	11	12	12	18	22	25	28
Class 2 power transmission	2	...	...	...	30	30	...	...	...	20	25	10	12	14	17	22	23	25	30
	1	...	...	23	23	30	...	...	18	20	25	8	10	12	9	15	19	21	23
	½	...	18	23	23	30	...	12	18	20	25	7	9	11	9	14	17	19	21
	¼	15	18	23	...	...	12	12	15	...	...	7	8	9	8	12	15	17	19
Class 3 precision	2	...	...	...	15	...	...	...	...	15	...	8	10	12	15	20	18	20	24
	1	...	...	12	12	...	...	...	12	12	...	6	8	10	7	12	14	16	18
	½	...	8	12	12	...	...	10	12	12	...	5	7	9	7	11	12	14	16
	¼	8	8	12	...	...	10	10	12	...	...	5	6	7	6	9	10	12	14

* Taken from AGMA standard 234.01.

Table 9-2. Tolerances for Coarse-pitch Worm Gears*

(For worm gears ¾ to 100 in. diameter; ¼ to 4 in. axial pitch; in ten-thousandths of an inch)

Class / Pitch diam., in.	Circ. pitch, in.	Runout measurement on pitch circle, total indicator reading								Pitch variation measured on pitch circle in plane of rotation								Max. index error between any two teeth exclusive of runout effect								Required min. initial area of contact, %
		¾	1½	3	6	12	25	50	100	¾	1½	3	6	12	25	50	100	¾	1½	3	6	12	25	50	100	
Class 1 light duty	4	...	...	...	...	...	90	90	100	...	...	...	...	...	60	70	80	...	...	...	...	...	...	...	...	20
	2	...	...	...	...	60	80	80	90	...	...	...	...	35	40	50	60	...	...	...	...	...	...	...	...	20
	1	...	...	...	60	60	80	80	90	...	...	...	30	35	35	40	50	...	...	...	...	...	...	...	...	20
	½	...	...	50	60	60	80	80	...	...	...	25	25	30	30	35	...	...	...	...	...	...	...	...	...	20
	¼	...	30	50	60	60	80	...	...	...	20	20	20	25	25	...	...	...	...	...	...	...	...	...	...	20
Class 2 power transmission	2	...	...	...	...	25	30	35	40	...	...	...	...	15	20	25	30	...	...	...	...	...	...	...	...	30
	1	...	...	...	20	25	30	35	40	...	...	...	10	15	15	15	20	...	...	...	...	...	...	...	...	30
	½	...	...	20	20	25	30	35	...	...	...	10	10	10	12	15	...	...	...	...	...	...	...	...	...	30
	¼	...	15	20	20	25	30	...	...	...	9	9	9	10	10	...	...	...	...	...	...	...	...	...	...	30
Class 3 precision	2	...	...	...	...	20	25	...	...	...	...	...	...	9	9	...	...	...	...	...	...	20	27	...	...	40
	1	...	...	...	20	20	25	25	30	...	...	...	7	7	7	8	8	...	...	...	15	15	15	35	55	40
	½	...	...	10	15	20	25	25	30	...	...	6	6	6	6	7	7	...	...	15	15	15	15	30	45	40
	¼	...	10	10	15	20	25	...	...	...	3½	4	4	4	4	...	...	...	15	15	15	15	15	...	...	40

* Taken from AGMA standard 234.01.

9-24. Fundamental Tolerances for Worm Gears. Worm gears follow the basic tolerance form for spur gears except for throat design. Depending upon the geometric form, plain or throated, there will be tolerance considerations associated with the dimensions specifying the geometric form.

Typical tolerances for worm gears are presented in Table 9-2.

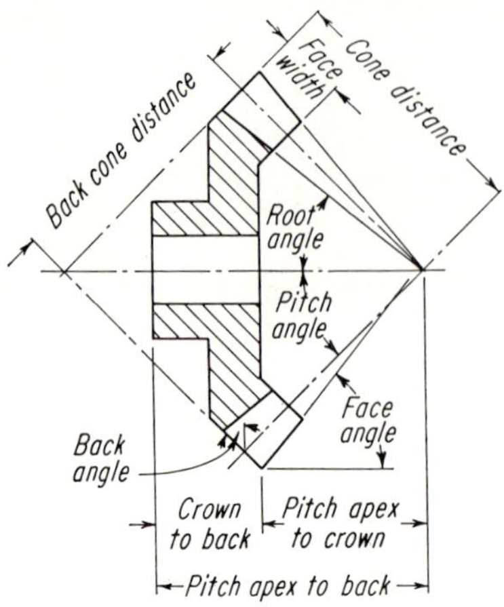

FIG. 9-6. Toleranced unique dimensions for bevel gears.

9-25. Fundamental Tolerances for Bevel Gears. Since bevel gears are radically different from spur gears, they require tolerance control of special parameters, which are

Cone distance (reference)
Crown to back
Pitch apex to back (reference)
Pitch apex to crown (reference)
Pitch angle
Face angle
Root angle
Back angle

These basic dimension parameters are shown in Fig. 9-6. Tolerances for the required close control of these dimensions are listed in Table 9-3.

Table 9-3. Fundamental Tolerances Pertaining to Bevel Gears

Dimension	Coarse pitch (19 diam. pitch and coarser) (up to 24″ dia.)		Fine pitch (20 diam. pitch and finer)	
	Commercial quality	Precision quality	Commercial quality	Precision quality
Cone distance (reference)	±0.010	±0.005	±0.005	±0.002
Crown to back	±0.005	±0.001	±0.001	±0.0003
Pitch apex to crown (reference)	±0.010	±0.005	±0.005	±0.002
Pitch apex to back (reference)	±0.010	±0.005	±0.005	±0.002
Pitch angle	±8 min	±2 min	±2 min	±30 sec
Face angle	±15 min	±5 min	±5 min	±1 min
Root angle	±15 min	±5 min	±3 min	±1 min
Back angle	±2°	±1°	±1°	±30 min
Face width	±0.015	±0.005	±0.010	±0.003

9-26. Tolerances for Special Gear Types. Tolerances for the more special gear types such as face gears, Spiroid®,* and Beveloid®† are established in a quite similar fashion to those for spur, helical, worm, and bevel types. The differences between these types is mainly in the geometry of the tooth. Allowable runouts, spacing errors, and the like follow a similar pattern for all gear types.

9-27. Tolerances for Quality Classes. A system of quality classes for all sizes of gears and a range of quality from a low order of commercial accuracy to the highest

* Registered trade-mark of the Spiroid Div. of Illinois Tool Works, Chicago, Ill.
† Registered trade-mark of Vinco Corp., Detroit, Mich.

order of precision have been established by AGMA. The gear-class system established is intended to be a "language bridge" for the designer, the user, and the manufacturer. Any given quality level represents a certain level of craftsmanship. The specified numerical tolerances for all the common elements of the gear clearly show what this level of quality amounts to for any particular pitch of gear tooth or diameter of gear blank or width of face.

The quality classes for spur, helical, and herringbone gears are given in a classification manual,[1,*] AGMA 390.01. In addition to quality classes, this manual lists the more common types of material and heat-treatments and establishes an "AGMA material number" for each one. The combination of the *AGMA quality number* and the *AGMA material number* makes an "AGMA class number." To illustrate this, an AGMA class number "8-H-14" means a gear finished to quality level 8 and made of material H-14. The H-14 material is a quenched and tempered alloy steel with a hardness of 285 to 325 BHN.

The gear-classification manual shows the quality levels and kinds of materials normally used in a wide variety of gearing applications. The system developed in the manual is flexible and allows for future advances and changes in the technology of gear manufacture. In any limited gear-product area it is possible to use the framework established by the manual as a building block and add to it additional tolerances and definitions to cover a limited and specific quality system for that particular gear-product area. Those involved in gear-quality specifications should obtain the manual and become familiar with the full detail of the system established.

Table 9-4 shows the tolerance values for AGMA quality classes 3 to 15 for gears in the pitch range of ½ to 19.99 diametral pitch. Tolerance values shown are in ten-thousandths of an inch. Thus a value of 80 in the table means 0.0080″.

It is interesting to note how the numerical values of the tolerances change with size. For instance, at a quality level of 10, an 8-pitch gear has a pitch tolerance of 5 for a 3″ diameter, 7 for a 50″ diameter, and 7½ for a 100″ diameter. The runout tolerance of the same gear changes more rapidly. At 3″ diameter it is 14, at 50″ it is 25, and at 100″ it is 30. At a quality level of 10, a 50″ diameter gear changes pitch tolerance from 8¾ to 7 if the pitch is changed from 2 to 8. The same 50″ gear would change runout tolerance from 50 to 25 in going from 2 to 8 pitch.

Table 9-5 shows the AGMA fine-pitch tolerance values. These go from 20 to 200 diametral pitch. Since most fine-pitch gears are checked by the composite system, tolerances are given in terms of this system rather than the more direct system of pitch tolerance, profile tolerance, etc., used for the coarser-pitch gears.

Broad Quality Levels. Words like "commercial," "precision," "high accuracy," and "aircraft quality" have been used rather loosely in the gear trade to designate gear quality. A more definitive and practical arrangement is the classification of all gear work into three broad quality ranges. These are

Commercial Quality Range. Gears cut or formed by the less precise methods. Gears are either lightly loaded for their size and material or have the capability of wearing in to a good fit on the job. Low cost is more important than excellence of performance. High reliability is not required.

Precision Quality Range. Gears are cut or formed on precision equipment. The teeth may have a finishing operation like shaving, grinding, or lapping. Gears are relatively heavily loaded in many cases. Some wearing in may occur on the job, but this is not counted on too strongly. Cost is important but it is balanced between the considerations for gear excellence of operation and gear reliability.

Ultraprecision Quality Range. Gears are cut or formed by the most reliable machines and then given the most accurate finishing the gear art can provide. Low cost is sacrificed to provide the ultimate in reliability or performance.

* Superscript numbers refer to references at the end of the chapter.

Table 9-4. Gear-tolerance Values for the Different AGMA Quality Classes*

Runout tolerance (Pitch dia., in.)

AGMA quality No.	Diam. pitch	¾	1½	3	6	12	25	50	100	200
3	½	...	...	...	...	...	770	1,010	1,360	
	1	...	...	...	...	540	710	930	1,250	
	2	...	...	...	382	498	660	860	1,150	
	4	...	...	280	355	460	608	800		
4	½	...	...	...	...	...	540	700	940	
	1	...	...	...	...	378	496	640	860	
	2	...	...	...	272	348	452	590	790	
	4	...	...	198	250	320	419	542	720	
5	½	...	...	...	...	...	396	510	665	880
	1	...	...	...	...	270	350	450	582	775
	2	...	...	...	184	233	302	390	510	680
	4	...	...	130	160	203	262	340	440	590
	8	...	91	112	140	177	228	290	380	
6	½	...	...	...	...	...	280	350	450	600
	1	...	...	...	...	188	235	295	378	508
	2	...	...	...	131	160	200	250	322	425
	4	...	...	92	110	135	170	210	270	360
	8	...	64	76	93	114	143	180	230	
	16–19.99	46	55	66	80	98	122	152	193	
7	½	...	...	...	...	...	209	260	335	445
	1	...	...	...	...	132	165	205	261	350
	2	...	...	...	84	103	130	163	210	280
	4	...	...	55	67	82	103	130	167	225
	8	...	37	44	54	66	82	103	132	
	16–19.99	26	31	37	45	55	69	86	110	

Pitch tolerance (Pitch dia., in.)

AGMA quality No.	Diam. pitch	¾	1½	3	6	12	25	50	100	200
6	½	...	....	...	....	....	50	55	62	70
	1	...	....	...	....	31	33	38	42	50
	2	...	....	...	26	27	29	33	37	42
	4	...	....	22	22	24	26	28	32	37
	8	...	18	18	19	21	23	25	28	
	16–19.99	16	16	16	17	18	19	21	23	
7	½	...	....	...	....	....	37	40	45	50
	1	...	....	...	....	23	25	28	32	37
	2	...	....	...	19	20	22	24	27	31
	4	...	....	16	16	17	19	21	24	28
	8	...	13½	14	14½	15½	17	19	21	
	16–19.99	11	11½	12	12½	13	14	15½	17½	

Profile tolerance (Pitch dia., in.)

AGMA quality No.	Diam. pitch	¾	1½	3	6	12	25	50
7	½	...	...	...	...	...	160	168
	1	...	...	...	...	100	106	111
	2	...	...	...	66	69	71	75
	4	...	...	44	45	48	50	52
	8	...	28	30	32	33	35	37
	16–19.99	22	22	22	23	24	25	28

* From AGMA "Gear Classification Manual for Spur, Helical and Herringbone Gears," AGMA 390.01, May, 1961.

NOTES:

1. Values for sizes other than shown can be found through direct interpolation.
2. Values shown are average. The gear manufacturer will make every effort to deliver gears that comply with tolerances shown. Therefore, minor variations from any tolerance that do not affect satisfactory operational use will not be cause for rejection.
3. Runout to be measured at the pitch line.
4. Pitch to be measured on the pitch circle in the plane of rotation.
5. Profile tolerance is exclusive of root and tip modification.
6. Unusual conditions may require that one or more of the individual quality tolerances be of a higher or lower number. In such cases it is possible to designate the quality number for each individual element. This should be done only when it results in a more satisfactory gear.

Table 9-4. Gear-tolerance Values for the Different AGMA Quality Classes *(Continued)*

AGMA quality No.	Diam. pitch	Runout tolerance, Pitch dia., in. ¾	1½	3	6	12	25	50	100	200	Pitch tolerance, Pitch dia., in. ¾	1½	3	6	12	25	50	100	200
8	½	...	...	...	...	...	160	200	255	350	...	...	...	...	...	26	28	31	35
	1	...	...	...	...	95	115	140	180	235	...	...	...	...	16	18	19	22	26
	2	...	...	...	58	68	82	100	125	160	...	...	...	14	14	15	17	19	22
	4	...	...	36	41	47	56	67	83	110	...	...	11	11	12	13	15	17	19
	8	...	25	28	32	36	42	50	60		...	9	10	10	11	12	13	14	
	16–19.99	19	21	23	26	30	35	42	51		8	8	8	9	9	10	11	12	
9	½	...	...	...	...	...	113	140	180	250	...	...	...	...	...	19	20	22	25
	1	...	...	...	...	68	81	100	125	165	...	...	...	...	11	12	14	16	19
	2	...	...	...	40	48	58	70	88	113	...	...	...	10	10	11	12	14	15½
	4	...	...	26	29	34	40	48	58	78	...	...	8	8	9	9½	11	12	14
	8	...	18	20	22	26	30	35	42		...	7	7	7½	8	8½	9½	10½	
	16–19.99	14	15	16	18	21	25	30	37		6	6	6	6¼	6¾	7¼	8	8¾	
10	½	...	...	...	...	...	82	100	130	175	...	...	...	...	...	13	14	15	18
	1	...	...	...	...	50	58	70	87	125	...	...	...	...	8¼	8¾	9¾	11	13
	2	...	...	...	30	34	40	50	63	82	...	...	...	7	7¼	8	8¾	10	11
	4	...	...	18	21	24	28	34	42	56	...	...	5¾	5¾	6¼	7	7½	8½	9¾
	8	...	13	14	16	18	21	25	30		...	5	5	5¼	5¾	6¼	7	7½	
	16–19.99	10	11	12	13	15	18	22	26		4¼	4¼	4½	4½	4¾	5	5½	6¼	
11	½	...	...	...	...	...	58	70	90	120	...	...	...	...	...	9½	10	11	13
	1	...	...	...	...	34	41	50	64	84	...	...	...	...	6	6	7	8	10
	2	...	...	...	21	24	28	35	44	57	...	...	...	5	5	6	6	7	8
	4	...	...	13	15	17	20	24	30	40	...	...	4	4	4½	5	5½	6	7
	8	...	9	10	11	13	15	18	22		...	3½	3½	4	4	4½	5	5½	
	16–19.99	7	8	8	9	11	13	15	18		3	3	3	3½	3½	3¾	4	4½	
12	2	...	...	...	15	18	21	25	32	42	...	...	...	3½	3¾	4	4½	5	6
	4	...	...	9	11	12	14	17	21	28	...	...	3	3	3¼	3½	3¾	4½	5
	8	...	6½	7	8	9	11	13	15		...	2½	2½	2¾	3	3¼	3½	3¾	...
	16–19.99	5	5½	6	7	8	9	11	13		2¼	2¼	2¼	2¼	2½	2½	2¾	3¼	

AGMA quality No.	Diam. pitch	Profile tolerance, Pitch dia., in. ¾	1½	3	6	12	25	50	100	200	Total composite tolerance, Pitch dia., in. 1	3	6	12	25	50	100	Lead tolerance, Face width 1" and less	2"	3"	4"	5"
8	½	...	...	...	...	...	110	118	123	130												
	1	...	...	...	...	70	75	78	81	85												
	2	...	...	...	46	48	50	52	55	58	...	...	...	80	94	111	135					
	4	...	...	30	32	34	35	37	39	41	...	46	52	58	68	79	93	7	12	17	19	24
	8	...	20	21	22	23	25	26	27		35	38	42	46	52	60	70					
	16–19.99	15	15	16	17	18	19	20	21		27	30	34	37	44	50	58					
9	½	...	...	...	...	...	80	83	88	90												
	1	...	...	...	...	50	52	56	59	60												
	2	...	...	...	32	33	35	37	39	40	...	...	...	57	66	78	95					
	4	...	...	21	22	23	25	25	27	29	...	33	37	42	48	54	66	6	10	14	16	20
	8	...	14½	15	16	17	17½	18	19		...	27	30	33	38	43	50					
	16–19.99	10¾	11	11½	12	12¾	13	14	15		19	22	24	27	31	36	42					
10	½	...	...	...	...	...	57	60	62	65												
	1	...	...	...	...	36	38	40	42	44												
	2	...	...	...	23	24	25	26	28	29	...	...	...	40	48	58	69					
	4	...	...	15	16	17	17½	18	19	20	...	23	26	29	34	39	47	5	8	11	13	16
	8	...	10½	11	11½	12	12½	13	14		...	19	21	24	26	30	35					
	16–19.99	7¾	8	8½	9	9¼	9½	10	10½		14	15	17	19	22	26	30					
11	½	...	...	...	...	...	40	42	46	48												
	1	...	...	...	...	25	27	28	30	31												
	2	...	...	...	17	18	18	19	20	21												
	4	...	...	11	11½	12	13	13½	14	15	...	17	18	21	24	28	34	4	6	9	10	12
	8	...	7½	8	8	8½	9	9½	9¾		...	14	15	17	19	22	26					
	16–19.99	5½	5½	6	6¼	6½	7	7¼	7½		10	11	12	14	16	18	21					
12	2	...	...	...	12	12½	13	14	14½	15												
	4	...	...	8	8¼	8½	9	9½	10	10½	...	12	13	15	17	20	25	3	5	7	8	10
	8	...	5¼	5½	6	6¼	6½	6¾	7		...	10	11	12	13	15	18					
	16–19.99	4	4	4¼	4½	4½	5	5¼	5½		7	8	9	10	11	13	16					

NOTES:

1. Values for sizes other than shown can be found through direct interpolation.
2. If total composite tolerance is specified, it should be in lieu of runout, pitch, and profile checks.
3. If runout, pitch, and profile tolerances are specified, they should be in lieu of total composite check.
4. Values shown are average. The gear manufacturer will make every effort to deliver gears that comply with the tolerances shown. Therefore, minor variations from any tolerance that do not affect satisfactory operational use will not be cause for rejection.
5. Runout to be measured at the pitch line.
6. Pitch to be measured on pitch circle in the plane of rotation.
7. Profile tolerance is exclusive of root and tip modification.
8. Lead tolerance is limited to 5" face and is measured normal to axis of tooth.
9. Applications requiring high-accuracy gearing necessitate the matching of pinion and gear profiles and pinion and gear leads so that the mismatch in a set does not exceed the tolerance on profile or the tolerance on lead for the applicable quality number. Matched sets can be provided at extra cost. In such a case, the manufacturer and user must agree on the details of the additional specifications covering how the matching is to be done and checked.
10. Unusual conditions may require that one or more of the individual quality tolerances be of a higher or lower number. In such cases it is possible to designate the quality number for each individual element. This should be done only when it results in a more satisfactory gear.

Table 9-4. Gear-tolerance Values for the Different AGMA Quality Classes (*Continued*)

AGMA quality No.	Diam. pitch	Runout tolerance									Pitch tolerance									Profile tolerance									Lead tolerance				
		Pitch dia., in.									Pitch dia., in.									Pitch dia., in.									Face width				
		¾	1½	3	6	12	25	50	100	200	¾	1½	3	6	12	25	50	100	200	¾	1½	3	6	12	25	50	100	200	1″ and less	2″	3″	4″	5″
13	2	...	...	...	11	12¾	15	18	23	30	...	...	...	2½	2¾	2¾	3¼	3½	4	...	...	...	8½	9	9¼	9¾	10¼	11	2½	3½	5	6	7
	4	...	...	6¾	7½	9	10½	12½	15½	20	...	...	2	2	2½	2½	2¾	3	3½	...	...	5¾	6	6	6½	6¾	7	7¼					
	8	...	5	5¼	5¾	6¾	7¾	9¼	11		...	1¾	1¾	2	2	2¼	2½	2¾		...	3¾	4	4¼	4¼	4½	4¾	5						
	16–19.99	5	5	5	5	5½	6½	7¾	9¾		1½	1½	1½	1¾	1¾	1¾	2	2¼		3	3	3	3¼	3¼	3½	3½	3¾						
14	2	...	...	...	7¾	9	10½	13	16	20	...	...	...	1¾	1¾	2	2¼	2½	3	...	...	...	6	6½	6¾	7	7¼	7¾	2	3	3½	4	5
	4	...	...	5	5½	6¼	7½	9	11	14	...	...	1½	1½	1½	1¾	2	2¼	2½	...	...	4	4¼	4¼	4½	4¾	5	5¼					
	8	...	5	5	5	5	5½	6½	7¾		...	1½	1½	1½	1½	1½	1¾	2		...	3	3	3	3¼	3¼	3½	3¾						
	16–19.99	5	5	5	5	5	5	5¾	6¾		1½	1½	1½	1½	1½	1½	1½	1¾		3	3	3	3	3	3	3	3						
15	2	...	...	...	5½	6½	7½	9	11½	15	...	...	...	1½	1½	1½	1½	1¾	2	...	...	...	4½	4½	4¾	5	5¼	5½	1½	2½	3	3½	4
	4	...	...	5	5	5	5½	6½	8	10	...	...	1½	1½	1½	1½	1½	1½	1¾	...	...	3	3	3¼	3¼	3½	3½	3¾					
	8	...	5	5	5	5	5	5	5½		...	1½	1½	1½	1½	1½	1½	1½		...	3	3	3	3	3	3	3						
	16–19.99	5	5	5	5	5	5	5	5		1½	1½	1½	1½	1½	1½	1½	1½		3	3	3	3	3	3	3	3						

NOTES:
1. The critical-accuracy requirements of quality classes 13, 14, and 15 require agreement between the manufacturer and user as to method of inspection.
2. Values for sizes other than shown can be found through direct interpolation.
3. Values shown are average. The gear manufacturer will make every effort to deliver gears that comply with the tolerances shown. Therefore, minor variations from any tolerance that do not affect satisfactory operational use will not be cause for rejection.
4. Runout to be measured at the pitch line.
5. Pitch to be measured on the pitch circle in the plane of rotation.
6. Profile tolerance is exclusive of root and tip modifications.
7. Lead tolerance is limited to 5″ face and is measured normal to axis of tooth.
8. Applications requiring high-accuracy gearing necessitate the matching of pinion and gear profiles and pinion and gear leads so that the mismatch in a set does not exceed the tolerance on profile or the tolerance on lead for the applicable quality number. Matched sets can be provided at extra cost. In such a case, the manufacturer and user must agree on the details of the additional specifications covering how the matching is to be done and checked.

Table 9-5. Fine-pitch-gear Tolerance Values for the Different AGMA Quality Classes*

AGMA quality No.	No. of teeth and pitch diameter	Diametral-pitch range	Tooth-to-tooth composite tolerance	Total composite tolerance
5	Up to 20 teeth incl.	20 to 80	0.0037	0.0052
	Over 20 teeth up to 1.999″	20 to 32	0.0027	0.0052
	Over 20 teeth 2 to 3.999″	20 to 24	0.0027	0.0061
	Over 20 teeth 4″ and over†	20	0.0027	0.0072
6	Up to 20 teeth incl.	20 to 200	0.0027	0.0037
	Over 20 teeth up to 1.999″	20 to 48	0.0019	0.0037
	Over 20 teeth 2 to 3.999″	20 to 32	0.0019	0.0044
	Over 20 teeth 4″ and over†	20 to 24	0.0019	0.0052
7	Up to 20 teeth incl.	20 to 200	0.0019	0.0027
	Over 20 teeth up to 1.999″	20 to 100	0.0014	0.0027
	Over 20 teeth 2 to 3.999″	20 to 48	0.0014	0.0032
	Over 20 teeth 4″ and over†	20 to 40	0.0014	0.0037
8	Up to 20 teeth incl.	20 to 200	0.0014	0.0019
	Over 20 teeth up to 1.999″	20 to 200	0.0010	0.0019
	Over 20 teeth 2 to 3.999″	20 to 100	0.0010	0.0023
	Over 20 teeth 4″ and over†	20 to 64	0.0010	0.0027
9	Up to 20 teeth incl.	20 to 200	0.0010	0.0014
	Over 20 teeth up to 1.999″	20 to 200	0.0007	0.0014
	Over 20 teeth 2 to 3.999″	20 to 200	0.0007	0.0016
	Over 20 teeth 4″ and over†	20 to 120	0.0007	0.0019
10	Up to 20 teeth incl.	20 to 200	0.0007	0.0010
	Over 20 teeth up to 1.999″	20 to 200	0.0005	0.0010
	Over 20 teeth 2 to 3.999″	20 to 200	0.0005	0.0012
	Over 20 teeth 4″ and over†	20 to 200	0.0005	0.0014
11	Up to 20 teeth incl.	20 to 200	0.0005	0.0007
	Over 20 teeth up to 1.999″	20 to 200	0.0004	0.0007
	Over 20 teeth 2 to 3.999″	20 to 200	0.0004	0.0009
	Over 20 teeth 4″ and over†	20 to 200	0.0004	0.0010
12	Up to 20 teeth incl.	20 to 200	0.0004	0.0005
	Over 20 teeth up to 1.999″	20 to 200	0.0003	0.0005
	Over 20 teeth 2 to 3.999″	20 to 200	0.0003	0.0006
	Over 20 teeth 4″ and over†	20 to 200	0.0003	0.0007
13	Up to 20 teeth incl.	20 to 200	0.0003	0.0004
	Over 20 teeth up to 1.999″	20 to 200	0.0002	0.0004
	Over 20 teeth 2 to 3.999″	20 to 200	0.0002	0.0004
	Over 20 teeth 4″ and over†	20 to 200	0.0002	0.0005
14	Up to 20 teeth incl.	20 to 200	0.00019	0.00027
	Over 20 teeth up to 1.999″	20 to 200	0.00014	0.00027
	Over 20 teeth 2 to 3.999″	20 to 200	0.00014	0.00032
	Over 20 teeth 4″ and over†	20 to 200	0.00014	0.00037
15	Up to 20 teeth incl.	20 to 200	0.00014	0.00019
	Over 20 teeth up to 1.999″	20 to 200	0.00010	0.00019
	Over 20 teeth 2 to 3.999″	20 to 200	0.00010	0.00023
	Over 20 teeth 4″ and over†	20 to 200	0.00010	0.00027

(*Table* 9-5 *continued on next page*)

Table 9-5. Fine-pitch-gear Tolerance Values for the Different AGMA Quality Classes* (*Continued*)

AGMA quality No.	No. of teeth and pitch diameter	Diametral-pitch range	Tooth-to-tooth composite tolerance	Total composite tolerance
16	Up to 20 teeth incl.	20 to 200	0.00010	0.00014
	Over 20 teeth up to 1.999″	20 to 200	0.00007	0.00014
	Over 20 teeth 2 to 3.999″	20 to 200	0.00007	0.00016
	Over 20 teeth 4″ and over†	20 to 200	0.00007	0.00019

* From AGMA "Gear Classification Manual for Spur, Helical and Herringbone Gears," AGMA 390.01, May, 1961.

† Table 9-5 is applicable to gears of 500 teeth maximum.

AGMA quality classes 3 to 7 cover the general range of commercial quality. AGMA quality classes 8 to 12 are precision quality. Class 8 is, of course, much less precise than class 12. AGMA quality classes 13, 14, and 15 are definitely of ultra-precision quality. Some would include class 12 in the ultraprecision grouping.

The bulk of gearing made falls into the commercial or precision range. Relatively few gears are made to the ultraprecision range of quality. The ultraprecision quality *should be asked for only when it is clearly needed.* Some examples of ultraprecision gears are master index gears for a machine tool, extremely high speed gears for a major power plant, radar gearing where the accuracy of sighting is a function of the gearing transmitting angular motion, missile-guidance gearing, and rocket-motor gearing.

In the fine-pitch field the obsolete standard AGMA 236.04 was used for many years to specify accuracy. Many drawings and gear technical papers are still in use that refer to AGMA 236.04. Table 9-6 shows an approximate comparison between the old practice and the current practice.

Table 9-6. Approximate Comparison of New and Previous Fine-pitch Quality Systems*

Previous fine-pitch system, AGMA 236.04			New fine-pitch system, AGMA 390.01		
AGMA quality No.	Tooth-to-tooth composite tolerance	Total composite tolerance	AGMA quality No.	Tooth-to-tooth composite tolerance	Total composite tolerance
Commercial 1	0.0020	0.0060	5 or 6	0.0027 or 0.0019	0.0052 or 0.0037
Commercial 2	0.0015	0.0040	6 or 7	0.0019 or 0.0014	0.0037 or 0.0027
Commercial 3	0.0010	0.0020	8	0.0010	0.0019
Commercial 4	0.0007	0.0015	9	0.0007	0.0014
Precision 1	0.0004	0.0010	10 or 11	0.0005 or 0.0004	0.0010 or 0.0007
Precision 2	0.0003	0.0005	12	0.0003	0.0005
Precision 3	0.0002	0.00025	13 or 14	0.0002 or 0.00014	0.0004 or 0.00027

* See Table 9-5 for exact values.

9-28. Complete Quality Tolerances. In the more precise gears, it is necessary to specify more tolerances than is the usual industry practice. Table 9-7 shows a complete tolerance listing for an 8-pitch aircraft gear of 6″ pitch diameter. This gear is basically AGMA quality class 12.

Table 9-7. Example of Complete Tolerance Listing for an Aircraft Gear of 8 Pitch, 6″ Pitch Diameter, and 2″ Face Width

Tolerance item	*Numerical amount*
Tooth thickness:	
Chordal	+0.0000 (for reference, check by pins) −0.0008
Over pins	+0.0000 −0.0019
Profile:	
Slope variation	±0.0003
Fullness or hollow	±0.0001
Irregularity	0.0001 (in any 10% of active roll)
Modification, amount	±0.0002
Start modification	±5% of active roll from design point
Spacing:	
Pitch tolerance	0.00028
Index tolerance	0.0006
Alignment:	
Lead tolerance	±0.00025
Max. % of face width that may be low because of radius or hollow	15
Total composite tolerance	0.0011
Surface finish:	
Active profile	20 rms
Root fillet	63 rms
Max. step at start of unground fillet	0.0015
Min. radius of curvature of root fillet	0.040
Hardness:	
Tooth surface	60 Rockwell C min.
Core	30–38 Rockwell C
Case depth (finished gears):	
Profile	0.025–0.040
Root fillet	0.030–0.045

NOTE: Gear is AGMA class 12-CH-15.

Note that the following things have been added to the basic accuracy limits for AGMA quality class 12:

Over-pins limit
Details of allowable profile variations
Index tolerance
Allowable low area in face width
Surface-finish limits
Allowable step
Root-fillet-radius limit
Hardness
Case depth

All the items mentioned above are important in the main reduction gears used to drive aircraft propellers through an epicyclic* gear train.

9-29. Tolerances per Application. In regard to application, gears divide into two general groupings based upon the fundamental functions of power or motion transmission. Power applications are the many cases of transmitting torque for the purpose of performing useful work. This grouping is divided into two types of applications: general power transmission, such as for household appliances, machine tools,

* See Chap. 3 for a discussion of epicyclic gear trains.

etc.; and high-power and high-speed applications as exemplified by turbine gearboxes. Motion transmission as a primary gear function usually is with negligible or light power transmission, although there are significant exceptions. Instruments, controls, and computers are examples of motion transmission. Tolerance practices associated with this scheme of classification are listed in Table 9-8.

Table 9-8. Tolerance Ranges per Gear Applications

Application	Tolerance level	Type of tolerancing
	Power	
Farm machinery	Low commercial	Tooth thickness and runout only
Household appliances	High commercial to precision	Composite sampling
Machine tools	High commercial to precision	Composite
Steam turbine	High to low precision	Check all gear elements
Aircraft engine	Precision	Check all gear elements
Rocket engine	Ultraprecision	Rigid and detailed check of all gear elements
	Motion	
Clock gears	Low to high commercial	Sampling check and final-product test
Industrial timers	Commercial	Sampling check and final-product test
Governor and valve gears on large engines	Precision	Composite check of all parts
Radar data gears	Ultraprecision	Composite check of all parts
Critical missile-guidance gears	Ultraprecision of highest order	Composite check plus additional checks

9-30. Tolerance per Speed of Operation. Another criterion for determining gear tolerances is the speed of operation. This is a condition which is somewhat independent of application, general quality class, transmitting power, and size. With increasing speed the dynamics of operation becomes paramount, and regardless of size, power, etc., critical qualities of the gear must be maintained. Comparative tolerances for basic gear criteria are given in Table 9-9.

Table 9-9. Comparative Tolerances for Various Pitch-line Velocities

Tolerance item	Pitch-line speed, fpm			
	500	5,000	20,000	40,000
Max. allowable runout	0.004	0.002	0.0012	0.0007
Max. allowable pitch tolerance	0.001	0.0004	0.0003	0.0002
Max. profile slope variation	±0.0005	±0.0004	±0.0003	±0.0002
Surface finish of active profile (rms microinches)	63	32	20	20
Composite tolerance:				
Max. tooth-to-tooth	0.002	0.0007	0.0004	0.0003
Max. total composite	0.005	0.0025	0.0015	0.001

NOTE: Comparison based on steel gears 5 to 20″ diameter, and 4 to 10 pitch.

9-31. Tolerances for Various Gear Materials. Gear-dimension tolerances are related to material in regard to practicability and limitations of achievement. Stable hard materials such as metallic alloys can be finished to a much higher degree of accuracy and surface finish compared with soft nonmetallic materials. The nonmetallic materials are subject to greater warpage, deflections, temperature variations, moisture absorption, and deterioration with aging. The nature of the material will also limit the possible techniques of fabrication. For example, nonmetallics cannot be conveniently ground, some ferrous alloys machine better than others, etc. Table 9-10 offers general tolerance ranges for various materials.

Table 9-10. General Tolerance Ranges for Various Gear Materials

Material	Limiting tolerance range*				
	Functional linear dimensions	Gear-body linear dimensions	Bore dia.	Journal dia.	Surface finish
Metallic:					
Ferrous, soft	0.0002–0.001	0.001–0.005	0.0002–0.0005	0.0001–0.0003	8–63
Ferrous, hardened	0.0001–0.001	0.0005–0.002	0.0002–0.0005	0.0001–0.0003	4–16
Nonferrous	0.0002–0.002	0.001–0.005	0.0002–0.0005	0.0002–0.0005	16–125
Nonmetallic	0.0005–0.003	0.002–0.015	0.001–0.005	0.001–0.003	32–125

* Tolerance limits given as a range, since exact limit is a function of the material and size of gear elements.

9-32. Tolerances for Various Fabrication Methods. The method of fabrication directly affects the tolerance limits that can be placed upon gear dimensions and controls. A comprehensive listing of fabrication methods is presented below.

Gear-fabrication Methods

1. General machining:
   Machine—generated
   Machine—form-milled or planed
2. Refinement methods:
   Grinding
   Shaving
   Burnishing
   Lapping
3. Forming—mechanical:
   Extruded
   Rolled
   Stamped
4. Forming—hot:
   Powdered sintered
   Molded—plastic
5. Casting:
   Sand
   Plaster
   Precision investment
   Die

For comparative purposes the above fabrication methods are relisted in Table 9-11 in descending order of gear-quality results. That is, the listing indicates the relative

Table 9-11. Quality Listing of Fabrication Methods in Descending Order

Fabrication methods	Comments	Achievable tolerance range, general degree of precision of final product	Order of use*
Grinding	Excellent finish and closest tolerances	0.0001–0.0005	SP
Shaving	Excellent finish, refinement on machine-cut accuracy	0.0001–0.0005	SP
Lapping	Fair finish. Some correction of dimensional variations	0.0002–0.0007	SP
Burnishing	Good finish. Gives uniformity of surface condition	0.0001–0.0002 dimensional variation on previous operation	R
Machine-cut generated	Good general precision	0.0005–0.002	N
Sintered	Good commercial quality	0.001–0.003	SP
Rolled	Commercial quality	0.002–0.005	SP
Extruded	Commercial quality	0.002–0.005	SP
Stamped	Commercial quality	0.002–0.005	HPL
Die cast	Commercial quality	0.002–0.005	HPL
Molded plastic	Low commercial quality, inexpensive	0.002–0.005	HPL
Form machine-cut	Good to poor quality	0.001–0.003	R
Precision investment	Fair quality, high cost	0.003–0.010	R
Plaster cast	Poor quality	$\frac{1}{64}$–$\frac{1}{32}$	R
Sand cast	Very poor quality	$\frac{1}{32}$–$\frac{1}{16}$	R

* N = normally used; SP = special applications; HPL = high-production low-cost applications; R = rarely used.

quality in terms of producible tolerance ranges that can be obtained by the various fabrication methods.

SPECIAL CONSIDERATIONS AND TOLERANCE VALUES

In addition to the specific areas and tolerance values covered in previous sections there are special considerations, worthy of study and review, which influence tolerance values. These are a miscellaneous grouping of items encompassing such things as materials and finishes, critical shapes and surface stresses, deburring, and surface coatings. The tolerance values presented in this section are to be selected, integrated, and evaluated along with the information given in previous sections.

9-33. Method of Gear-size Control and Tolerancing. Gear-size control can be accomplished by several methods which must be understood in order to relate and interchange tolerances. The several methods are

1. Tooth-thickness measurement (Fig. 9-7)
2. Over-rolls measurement (Fig. 9-8)
3. Testing radius (Fig. 9-9)

Methods 1 and 2 are analytical in that specific geometrical dimensions are directly measured. The tooth-thickness measurements may be direct on one tooth or across several as a chordal measure. Over-rolls measurements are limited to spur and helical gears and worms. The third method, testing radius, is functional in that the size measurement is done by rolling in intimate mesh with a master and noting the operating center distance.

Size by Pitch Diameter. The procedure of specifying gear size by pitch diameter and a tolerance is no longer widely practiced and accepted. Pitch diameter is somewhat meaningless since it does not exist until the gear is mated with another gear and then the pitch diameter depends upon exact center distance. The testing-radius specification achieves the objective of a pitch-diameter call-out. However, any of the previously mentioned size-call-out methods should be used in preference to pitch diameter.

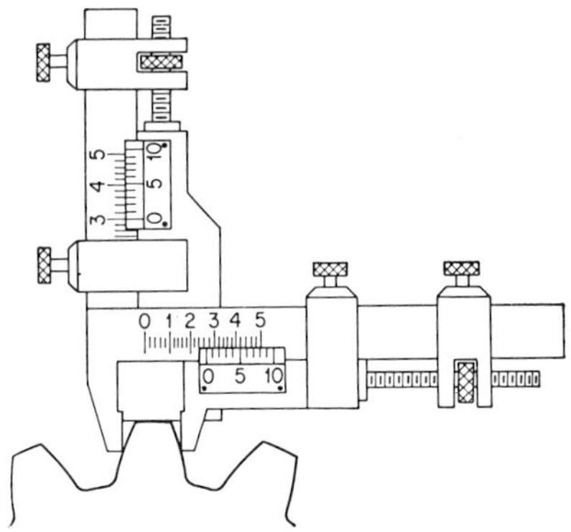

FIG. 9-7. Tooth-thickness measurement by calipers.

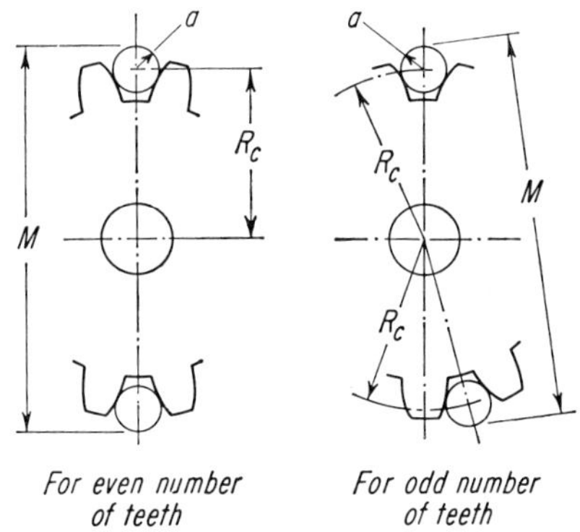

FIG. 9-8. Over-roll tooth-thickness measurement.

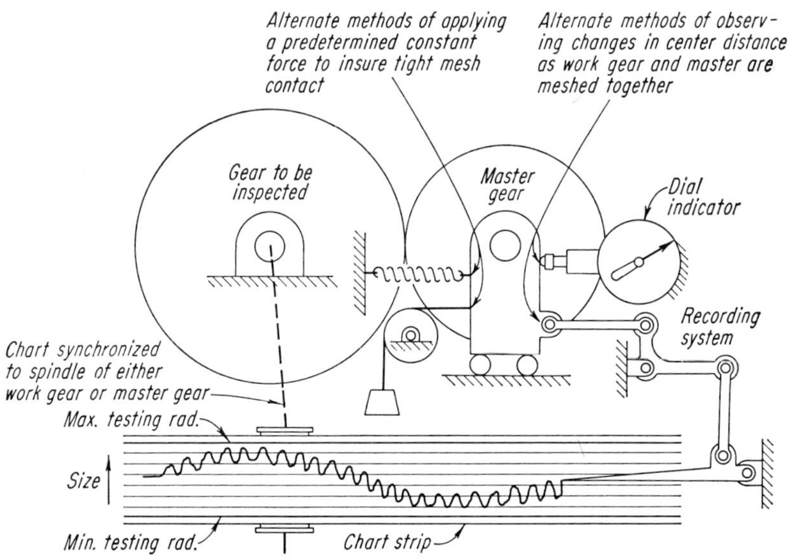

FIG. 9-9. Schematic diagram of a gear rolling fixture for measurement of testing radius.

9-34. Size Allowance, Tolerance, and Relation to Total Composite Variation. Gear size has three independent control criteria which are often misunderstood and confused, particularly when the size is designated in terms of testing radius. These are:

Size Allowance. This is the intended reduction (or oversize) from a standard or nominal value. This is shown in Fig. 9-10 as the region between the testing-radius value corresponding to standard size and maximum testing radius.

Size Tolerance. This is the allowable variation in gear size which extends downward in magnitude from the size allowance. In Fig. 9-10 it is the region bounded by the maximum and minimum testing radius.

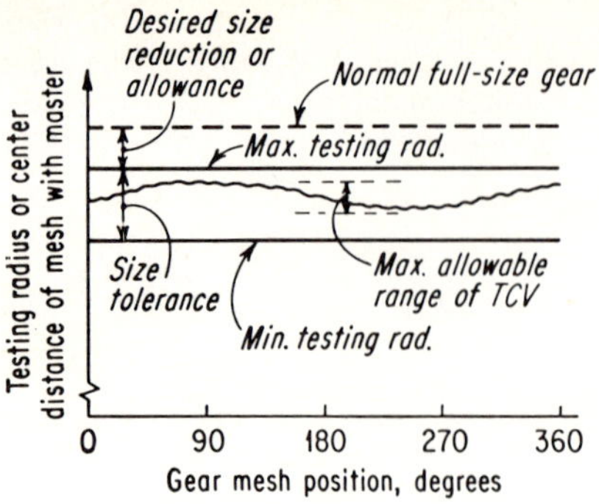

FIG. 9-10. Tooth-size allowance and tolerance in terms of testing radius.

Total Composite Variation (TCV). This is the range of variation of testing radius in a complete rotation with the master gear, as shown in Fig. 9-10.

It is significant to realize the above three items are independent and can be arbitrarily set at various values. The size allowance is a fixed value once chosen and can be zero or any finite value. However, the other two items are tolerances which cannot be zero or approach too closely without impossible results.

There are three general practices in establishing size tolerance and TCV in terms of testing radius, which are:

1. Set maximum and minimum testing-radius values with a maximum TCV value, all of which must fall within the maximum and minimum testing radii. This is shown in Fig. 9-11*a*.

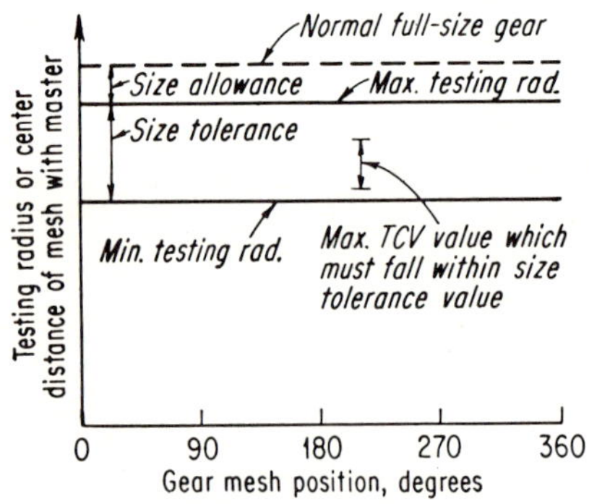

FIG. 9-11*a*. Minimum and maximum testing-radius values which must include maximum allowable TCV.

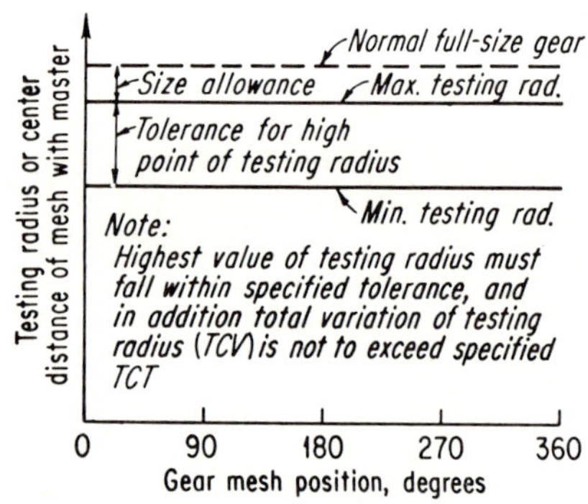

FIG. 9-11*b*. Minimum and maximum values of testing radius for high point of TCV. NOTE: Highest value of testing radius must fall within specified tolerance, and in addition total variation of testing radius (TCV) is not to exceed specified TCT.

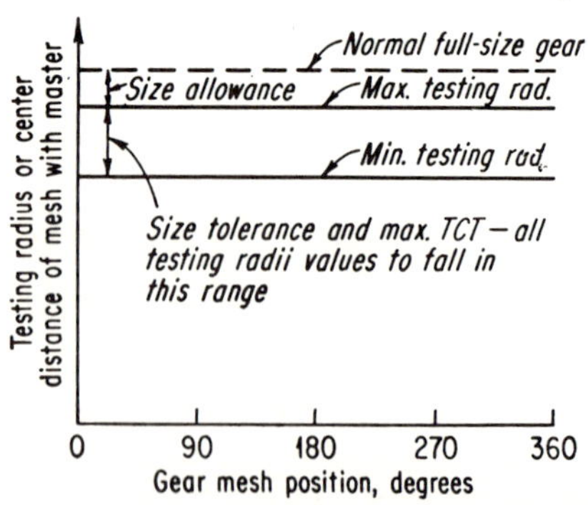

FIG. 9-11*c*. Minimum and maximum values of testing radius which also sets maximum allowable TCT.

FIG. 9-11. Various methods of testing radius-size control.

2. Set maximum and minimum testing-radius values for the high point of the total composite error. Separate from this the TCV has a maximum allowable value (see Fig. 9-11*b*).
3. Set maximum and minimum test radius, which also sets the maximum of TCT. In this case size and TCT are closely interrelated and are not independent (see Fig. 11*c*).

Procedure 1 is the simplest for execution but does not provide the best size control, since the size tolerance must be greater or at least equivalent to the maximum allowable TCV. Procedure 2 offers the greatest flexibility but is also more cumbersome to execute in the shop and inspection departments. Procedure 3 restricts independent consideration of size tolerance and TCV and therefore is of restricted use.

9-35. Values of Size Allowance and Total Composite Tolerance. The choice of size allowance is a function of many items, which are listed as

1. Size tolerance
2. Gear center tolerance
3. Gear quality (runout, tooth spacing, etc.)
4. Pitch diameter
5. Bearing runout
6. Mounting-surface runout
7. Thermal-expansion effects
8. Lubricant film space

Common denominators for most of the above considerations are diametral pitch and gear quality. For normal designs, allowance will directly vary as a function of pitch and inversely with quality. Table 9-12 lists typical values.

Table 9-12

Diam. pitch	Size allowance tooth thinning, in.	
	Commercial	Precision
½	0.075	0.025
1	0.025	0.010
2	0.015	0.005
3	0.010	0.004
4	0.007	0.0035
5	0.006	0.003
6	0.005	0.0025
7–9	0.004	0.002
10–13	0.003	0.0015
14–19	0.002	0.001
20–45	0.002	0.001
46–70	0.0015	0.001
71–120	0.0012	0.0007
121 and finer	0.0007	0.0005

NOTE: Size allowance is not necessarily limited to thinning of the teeth below nominal tooth thickness by the indicated values. It can also be accomplished by extending the center distance by an equivalent amount.

Total Composite Tolerance. The total composite tolerance and the subcomponent tooth-to-tooth composite tolerance are directly related to the gear quality.

Table 9-5 lists values per AGMA standard 390.01 for fine-pitch gears. These tolerances and classes are in extensive use in the gear trade. See Table 9-6 for listing of tolerances for commercial and precision fine-pitch gears and a comparison with the obsolete tolerances of AGMA 236.04.

9-36. Position Tolerance. Gears require a consideration which until recently has been almost neglected. Much emphasis has been placed on the total composite tolerance (TCT) control for achievement of linearity in angular transmission. This has been a subtle fallacy, particularly with high-accuracy gearing. The TCT, or composite-checking technique, is an adequate control of center-distance variation when rolling with a master but is incapable of detecting all angular variation. For example, as an extreme case, the TCT method cannot reveal the huge angular-transmission error which occurs when a gear has one more or less tooth than proper. It can only reveal tooth-to-tooth profile position variations but cannot show such accumulated variation. Absolute-angular-transmission accuracy can be revealed only by a tooth-profile position-variation check.

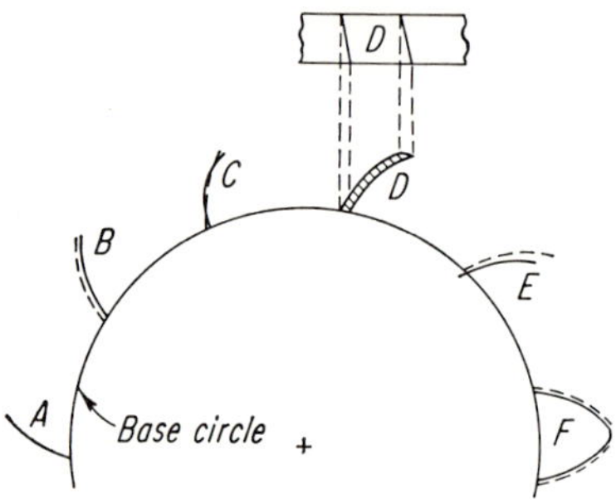

Notes: 1. The broken profiles are theoretically perfect profile positions in space, shown for comparison to the misplaced example profiles drawn solid.

2. Only one tooth profile drawn for each error example for simplicity of drawing, except tooth-thickness variation which must be shown as a relative shift in a pair of profiles.

3. Profile examples:

A. Reference profile—perfect form and position.
B. Profile-spacing variation.
C. Profile (deviation from true involute).
D. Profile lead variation.
E. Profile radial position (due to runout or wobble).
F. Tooth-thickness deviation.

FIG. 9-12. Examples of tooth-profile position variations.

Tooth-profile position variation or the shortened equivalent term "position variation" is defined as "the deviation of a gear's tooth profiles in space from the true positions they should have." By this is meant variations in the ρ or θ values for each point of the profile. The measure of position variation is a net sum of all variations affecting profile positions, which can be grouped as

1. Rotating variations
   a. Runout
   b. Lateral runout
2. Tooth-to-tooth variations
   a. Profile (deviation from true involute)
   b. Profile spacing (pitch deviation)
   c. Lead (zero for spurs)
   d. Tooth thickness

These variations are geometrically shown in Fig. 9-12.

Position variation may be further defined as either an angular or linear measure, and one value is convertible to the other. As a linear measure it would be the displacement from ideal position measured along the nominal pitch circle or any other convenient reference. As an angular measure, the displacement would be related to gear size. Thus the term "position variation" strictly interpreted requires the prefix angular or linear. Often the dimension will be inferred by the units of the specific data.

It is important to appreciate that, once a gear has been created as a single component, it contains position variation which is an inherent property. This does not become apparent until after it is mated in an application. Nevertheless, a simple lone gear can be evaluated for position variation by determining the degree of departure of its tooth profiles from perfect space positions. This makes it purposeful to have a term called "position variation" for a gear. It is also true, when a gear is assembled into an application, additional variations are encountered because of imperfections of the installation. These are generally runout of the bearings and shaft. These variations are not considered as part of the gear's intrinsic position variation and therefore are not included in this topic.

Specific numerical values of index tolerance for various quality types, etc., have been presented in previous tables. In addition, Table 9-13 offers a general list of values of maximum index error.

Table 9-13. Maximum-index-tolerance Classes

Class	Max. index tolerance, seconds of arc	Quality level		
		Commercial	Precision	Ultraprecision
		Pitch diameter, in.		
z	2,400	¼		
y	1,600	½		
x	1,060	1		
w	710	2	¼	
v	470	3	½	
u	315	6	1	
t	210	12	2	¼
s	140	24	3	½
r	93	48	6	1
p	62	96	12	2
n	41	...	24	3
m	27	...	48	6
l	18	...	96	12
i	12	...	...	24
h	8	...	...	48
g	5	...	...	96

9-37. Mounting Effects on Tolerances. The relationship of mounting surfaces to the functional gear teeth will have an important bearing on fabrication, inspection, and the practicability of close functional tolerances.

The mounting surfaces, usually designated MS on the drawing, are the critical dimensions upon which the gear is mounted in application. They can be bearing diameters, bore, shoulder diameters, etc. Concentricity and squareness (lateral runout) of the tooth elements are established by these same surfaces for mounting in the fabrication and inspection operations. Typical mounting surfaces are shown in Fig. 9-13. When the mounting surface is a diameter, it is often referred to as a "mounting diameter" and designated MD. Should it not be possible or convenient to use the mounting surface for the fabrication operation, a different reference surface must be chosen and closely controlled. This alternate mounting surface is termed "generating mounting surface," designated GMS on the drawing. The generating mounting surface must be closely controlled relative to the true mounting surface so as to provide the desired gear-function qualities relative to the mounting surfaces. Figure 9-14 shows typical generating mounting surfaces. Note that shaft centers are an example.

Tolerable amounts of discrepancy between mounting and generating mounting surfaces are related to gear quality. Suggested values are listed in Table 9-14.

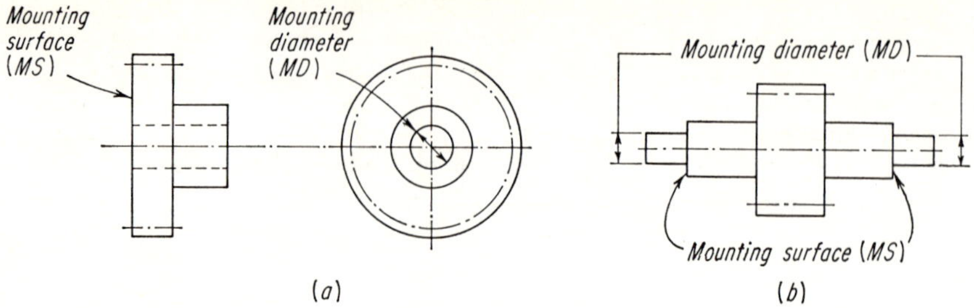

FIG. 9-13. Typical mounting surfaces (MS) or mounting diameters (MD). (*a*) Plain gear. (*b*) Gear shift.

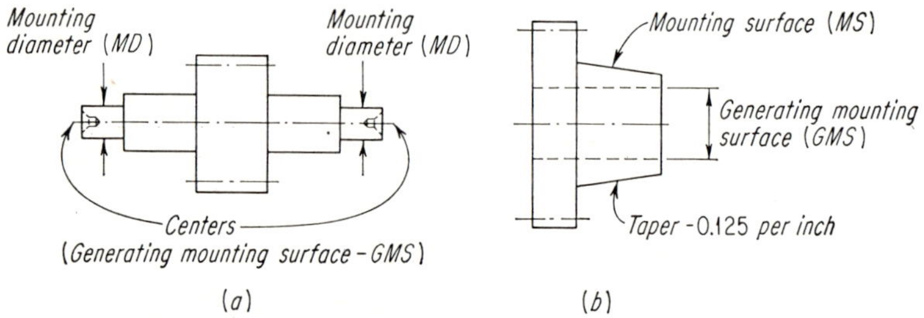

FIG. 9-14. Examples of generating mounting surfaces (GMS). (*a*) Gear shaft with centers. (*b*) Taper mounted gear.

Table 9-14. Tolerances for Relationship of Generating and Mounting Surfaces

Gear quality	Max. runout between concentric shoulders (TIR)	Parallelism, in. per in.	Runout between shaft centers and mounting surface (TIR)
Commercial gears (TCT 0.002 and greater)	0.001	0.002	0.0005
Precision gears (TCT 0.0005–0.001)	0.0002	0.0005	0.0001
Ultraprecision gears (TCT 0.0003 and less)	0.00005	0.0002	0.00005

9-38. Registering Surface Tolerances. The registering surface (RS) lies in the gear's plane of rotation and establishes axial position in the application. Figure 9-15 shows typical registering surfaces. This can be accomplished in the application by location of this surface against a shaft shoulder, collar, bearing race, etc. Squareness of location is affected as well as axial position. For most gears as exemplified by the spur type, the axial position is not highly critical. On the other hand, bevel, worm-gear, and face-gear types are very much subject to axial position.

Registering surfaces have critical tolerances regarding squareness and surface finish. Tolerance values vs. quality classes are given in Table 9-15. A completely designated gear may have the three types of critical surfaces, mounting surface (MS), generating mounting surface (GMS), and register surface (RS). Figure 9-16 illustrates a typical case.

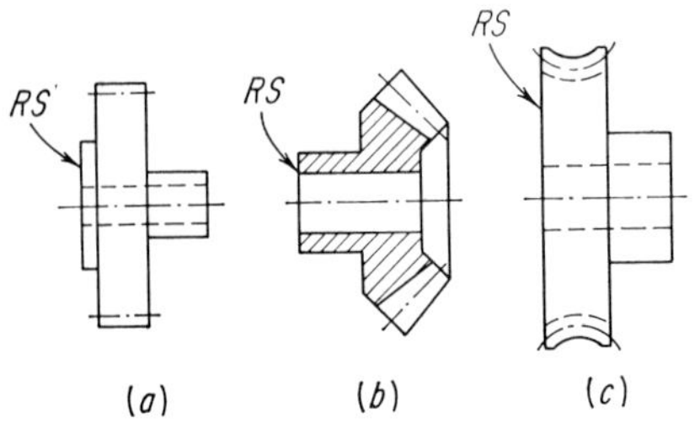

FIG. 9-15. Typical registering surfaces (RS). (*a*) Spur gear. (*b*) Bevel gear. (*c*) Worm gear.

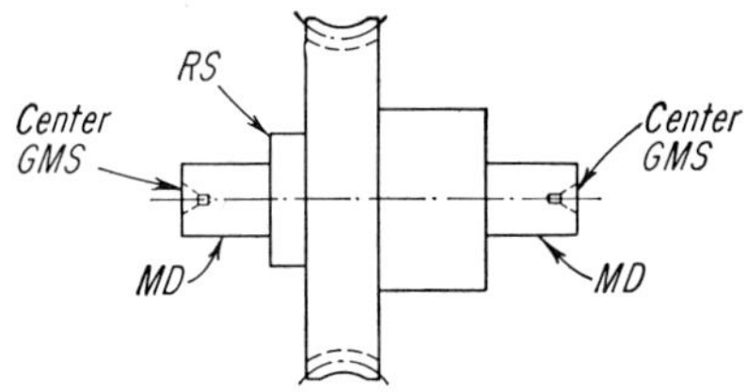

FIG. 9-16. Examples of designating, with accepted abbreviations, all critical generating mounting and registering surfaces.

Table 9-15. Tolerances for Registering Surfaces

Gear quality	Max. squareness deviation with axis of rotation, in. per in.	Surface finish rms, microin.
Commercial:		
Cut	0.002	63–125
Cast: Metallic	0.004	125–250
Plastic	0.006	63–250
Precision	0.0005	16–32
Ultraprecision	0.0002	8–16

9-39. Surface-finish Standards. Surface finish of a material is a criterion which unfortunately is sometimes neglected and often underappreciated and yet it can have extremely significant bearing upon gear performance. Often surface finish is not realized to be a control parameter, with tolerance, much like any other gear dimension.

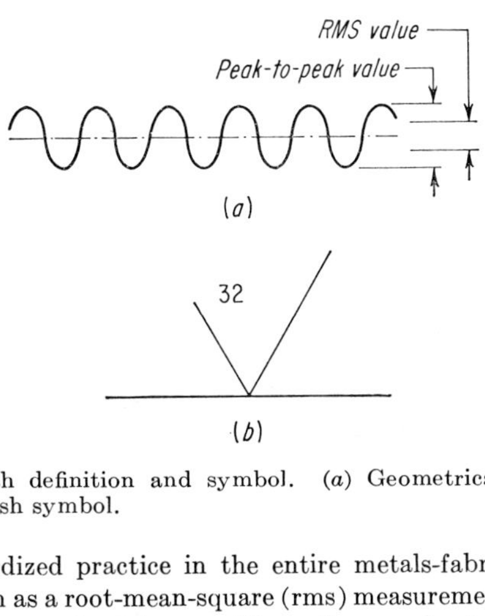

FIG. 9-17. Surface-finish definition and symbol. (*a*) Geometrical definition of surface finish. (*b*) Surface-finish symbol.

The accepted standardized practice in the entire metals-fabrication industry is to designate surface finish as a root-mean-square (rms) measurement of the normal variation of surface roughness and waviness. The units are microinches and the symbolic representation is the simple symbol ✓ with numerical value as shown in Fig. 9-17.

Standardized numerical values of maximum surface-finish variation have been established based upon a geometrical progression. These are listed and described in Table 9-16.

Table 9-16. Standardized Surface-finish Values

Symbol	Description
$^{1,000}\surd$	Indicates that the surface is very rough and uneven within the dimensional requirements
$^{500}\surd$	Indicates that the surface is rough and uneven within the dimensional requirements
$^{250}\surd$	Indicates that the surface must be smooth and even to a degree obtainable by tools removing large chips or shavings. Machining marks and grooves discernible to the eye and to touch are permitted, if the surface meets dimensional requirements
$^{125}\surd$	Indicates that the surface must be smooth and even to a degree obtainable by tools removing medium chips or shavings. Machining marks and grooves discernible to the eye are permitted, if the surface meets dimensional requirements
$^{63}\surd$	Indicates that the surface must be smooth and even to a degree obtainable by tools removing small chips. Machining marks and grooves discernible to the eye are permitted, if the surface meets the dimensional requirements
$^{32}\surd$	Indicates the surface must be smooth and even to a degree obtainable by tools removing small particles. Machining marks such as grooves must not be discernible to the eye or touch, if the surface meets the dimensional requirements
$^{16}\surd$	Indicates that the surface must be very smooth and even to a degree obtainable by tools removing very small particles. Machining marks such as grooves must not be discernible to the eye or touch, if the surface meets the dimensional requirements
$^{8}\surd$	Indicates that the surface must be even to a degree obtainable by tools removing minute particles, generally by grinding. Machining marks such as the fine patterns resulting from grinding must not be discernible to the eye or touch and the surface must have a polished appearance and meet dimensional requirements
$^{4}\surd$	Indicates that the surface must be even to a degree obtainable by tools removing very minute particles, generally by honing, lapping, or superfinishing. Machining marks such as very fine patterns must not be discernible to the eye or touch and the surface must have a highly polished appearance and meet dimensional requirements

Special finishes not covered by the standardized numerical values of Table 9-16 may be indicated by the symbol and numerical value. To a large degree the roughness of surface finish will be governed by the tolerance on the surface dimensions and the method of fabrication. For example, grinding will produce much finer finish than milling or turning. Also, if the dimension tolerance is in the order of several hundred millionths, crude surface finishes in the order of 50 microinches roughness and greater would be incompatible. Thus surface finish and tolerance are directly related, and in turn this places a requirement on fabrication method. Table 9-17 lists dimension tolerances and fabrication methods for the standard surface-finish values.

9-40. Surface-finish Tolerances. The practical achievement of close tolerances is very much influenced by the surface finish. Critical surfaces in this regard are tooth surfaces and mounting and registering surfaces.

Close tolerances require finer finishes in that otherwise surface roughness and waviness would become a significant portion of the tolerance and create difficulty of control and measurement. Also, the tolerance and finish combination influence the choice of fabrication, cutting speeds, and feed.

High-quality gears must have good tooth surface finish in order to meet precision and maintain quality over the life of the gear. Poor surface finish on functional surfaces, which might be adequate to pass initial inspection, will break down in early operation, resulting in falsely assumed gear quality. To maintain precision over a long life quality surface finish is necessary.

Table 9-18 relates surface finish and gear quality. Surface finish is toleranced and controlled by specifying only one number, which is the maximum value. Any lesser amount will also be satisfactory.

Table 9-17. Surface Finish and Applicable Dimension Tolerance

Symbol	Quality class	Description	Max. rms value, microin.	Suitable range of total tolerance	Typical fabrication methods	Approx. relative cost to produce
1,000 ✓	Extremely rough	Extremely crude surface produced by rapid removal of stock to nominal dimension	1,000	0.063–0.125	Rough sand casting, flame cutting	1
500 ✓	Very rough	Very rough surface unsuitable for mating surfaces	500	0.015–0.063	Sand casting, contour sawing	2
250 ✓	Rough	Heavy toolmarks	250	0.010–0.015	Very good sand casting, saw cutting, very rough machining	3
125 ✓	Fine	Machined appearance with consistent toolmarks	125	0.005–0.010	Average machining—turning, milling, drilling; rough hobbing and shaping; die casting, stamping, extruding	4
63 ✓	Fine	Semismooth without objectional toolmarks	63	0.002–0.005	Quality machining—turning, milling, reaming; hobbing, shaping; sintering, stamping, extruding, rolling	6
32 ✓	Smooth	Smooth, where toolmarks are barely discernible	32	0.0005–0.002	Careful machining; quality hobbing, and shaping; shaving; grinding; sintering	10
16 ✓	Ground	Highly smooth finish	16	0.0002–0.0005	Very best hobbing and shaping; shaving; grinding, burnishing	15
8 ✓	Polish	Semi-mirror-like finish without any discernible scratches or marks	8	0.0001–0.0002	Grinding, shaving, burnishing, lapping	20
4 ✓	Superfinish	Mirrorlike surface without tool, grinding, or scratch marks of any kind	4	0.00004–0.0001	Grinding, lapping, and polishing	25

Table 9-18. Surface Finish and Gear Quality Level

Gear quality class	Tooth surface finish rms, microin.
Commercial	64–250
Precision	8–32
Ultraprecision	4–16

NOTE: Based on 2″ diameter 20-pitch steel gear.

The above information is sufficient for most gear surface-finish requirements. For special cases requiring further detailed control, AGMA standard 236.04 provides helpful information. In these cases the over-all surface-finish requirement is divided into several additional aspects, which are

Roughness height, microinches
Roughness width, thousandths of an inch
Waviness height, thousandths of an inch
Lay, typical surface pattern
Zones of tooth profile, upper, lower, and root

A method of presenting this information on a gear drawing is shown in Fig. 9-18. Further information regarding numerical values of waviness height, roughness width, lay patterns, and zone specification is available in AGMA standard 236.04.

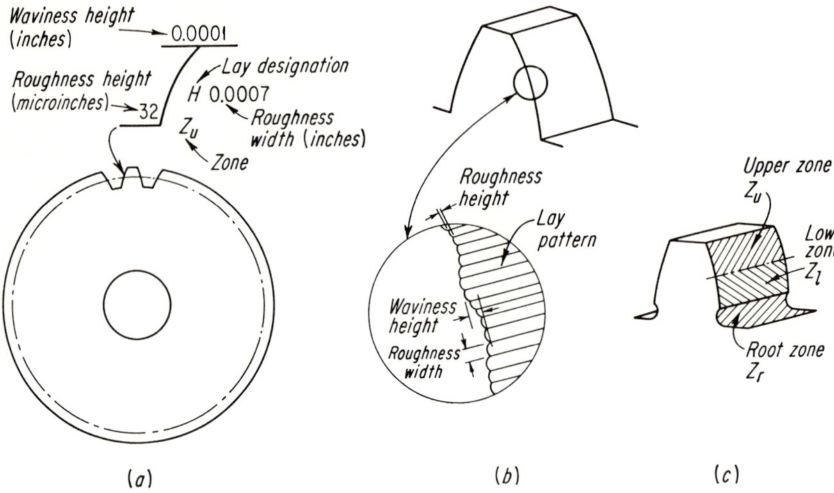

FIG. 9-18. Illustration of detailed control of tooth-profile surface finish. (*a*) Illustration of drawing specification. (*b*) Surface-finish breakdown. (*c*) Tooth-profile zones.

9-41. Materials and Tolerances. Achievement of gear quality is related to the material. Cut gears will be best when made from free-machining bronzes and steels. Ductile, soft, nonferrous alloys such as some aluminum alloys and tenacious steels such as stainless AISI 302 and 303 are comparatively difficult to cut. When grinding is the operation, the hard-steel alloys are best. Poorest achievement of tolerances occurs with nonmetallics, such as plastics. Table 9-19 is a comparative listing of materials and ease of achieving good surface finish and precision.

Table 9-19. Materials and Attainment of Relative Gear Quality (in Descending Order)

Type of Generating

Direct machine-cut	*Grinding*
Bronze alloys (free-cutting)	Hardened-tool-steel alloys
Brasses (free-cutting)	Hardened-carbon-steel alloys
Aluminum (hard types—2024 and 2075 types)	Hardened-stainless-steel alloys
Steels, mild	Semihard-steel alloys
Steels, alloy	Soft steels
Stainless steels (400 series)	Bronzes
Stainless steels (300 series)	Brasses
Soft aluminum alloys (2S, 3S types)	Aluminum alloys
Plastics—phenolic, nylon, etc.	

9-42. Finish-coating Effects on Tolerances. Coating finishes on gear-tooth surfaces and other critical parts, such as the mounting and registering surfaces, can have a drastic effect on tolerances. Most finishes result in dimension alterations, the seriousness depending upon the desired tolerance. For instance, precision gears are so accurate that changes of tenths of a thousandth of an inch will be objectionable.

Dimensional effects come from two causes. The coating has some measurable thickness, and various caustic and acid washes encountered during the processing remove material. For some processes only one effect is present or dominating. In other finishing processes both are present and the resultant effect on the gear is very dependent upon specific conditions. The preferred procedure is to leave tooth surfaces and other critical areas uncoated. If requirements compel application of a coating, careful control and choice are necessary. Various finish coatings and their effects are presented in Table 9-20.

To avoid finish coatings affecting critical gear tolerances, the teeth and other critical dimensions may be machined after the finishing coating is applied to the gear blank. Often the finish is not essential for the teeth but is needed for the gear body's appearance. Bare teeth of corrosive materials will obtain some protection from the operating lubricant and the constant rolling and sliding action. However, if there is no specified

Table 9-20. Finish-coating Effects on Critical Gear Dimensions

Finish type	Dimension changes			Recommended practice
	Degree	Magnitude	Type	
Paint	Great	0.002–0.010	Build-up	Avoid on critical surfaces by masking
Hard flash plate of chromium, nickel, etc.	Small	0.00005–0.0002	Build-up	Not for precision gears. When necessary, machine critical surfaces after finish
Anodized aluminum	Small	Zero to 0.0002	Build-up or reduction	Avoid for precision gears. When necessary, machine critical surfaces after finish
Iridite	Negligible	Zero to 0.00005	....	Acceptable except for ultraprecision tolerances
Electrofilms: electrolize	Small	0.00005–0.0001	Build-up	Acceptable except for ultraprecision tolerances
Passivation of stainless steels	None	Zero	....	Recommended

methodizing regarding the finish and critical dimensions, the gear drawing dimensions apply to the finish product. For this reason gear drawings, which have close tolerances affected by the called-for finish coat, should clearly specify if the finish is or is not to be applied to the critical surfaces.

9-43. Critical Geometrical Shapes and Tolerances. The ability to achieve good gear quality, and at a reasonable cost, is related to the critical geometrical shape of the gear and associated tolerances. Good quality will be more readily obtained if the

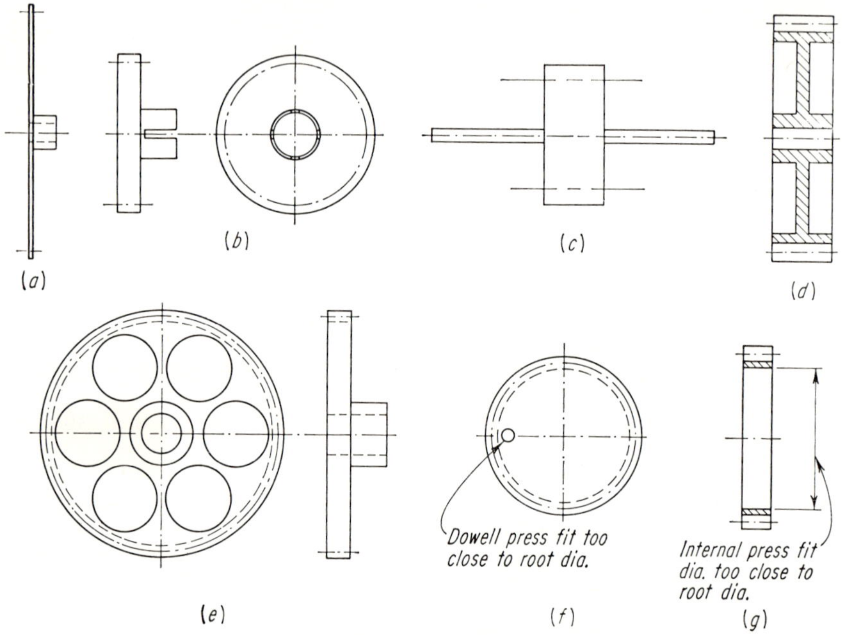

FIG. 9-19. Examples of critical dimensions affecting rigidity and distortion. (*a*) Gear body too thin for large diameter. (*b*) Split hub wall too thin. (*c*) Gear shaft too slender. (*d*) Body web too thin. (*e*) Lightening holes too near root diameter and web between holes too thin. (*f*) Dowel press fit too close to root diameter. (*g*) Internal press-fit diameter too close to root diameter.

design is physically rigid in all respects and of sufficient section to avoid distortions due to surface stresses. These considerations divide into four paramount areas of control:

1. Rigidity in fabrication and application
2. Distortions due to secondary operations
3. Stress distortions during fabrication
4. Tool clearances

Primary control will be obtained by selection of proper dimensions. A secondary but still significant influence comes from the tolerances associated with these dimensions.

Rigidity. Geometrical shapes which are dimensioned so as to be on the verge of excessive flexibility must be carefully toleranced to avoid dimensions drifting beyond the limiting conditions. This means tolerances must be unilateral and relatively close. Examples where the situation must be considered are narrow face widths of large-diameter gears; thin-walled bores; long slender integral shafts with undercuts; and holes, reliefs, and undercuts in the gear body located near the teeth. These examples are shown in Fig. 9-19.

Distortions Due to Secondary Operations. If a gear is of marginal rigidity, secondary operations such as press fitting and staking can cause distortions that affect function. Some examples are shown in Fig. 9-19*e*, *f*, and *g*.

Stress Distortions. Large thin sections will be subject to skin-stress distortions during fabrication. These can occur because of both the relief of skin stresses during machining cuts and the inducement of cold-working stresses during fabrication. The magnitude of tolerances on these dimensions is an indirect control of this difficulty,

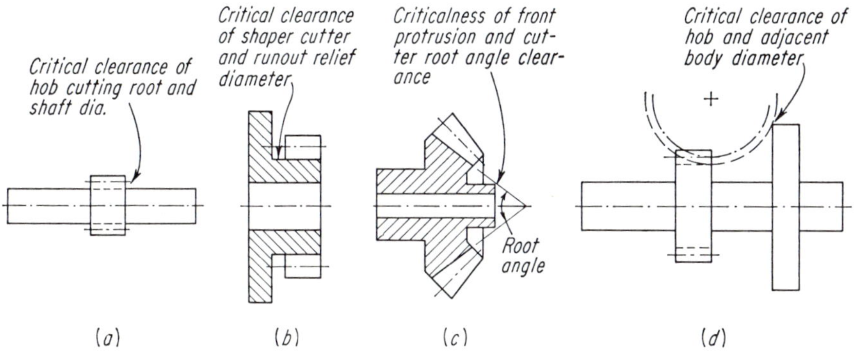

FIG. 9-20. Examples of critical gear tool clearances.

since it limits the fabrication technique. Wide-open tolerances will invite fast metal removal and consequent high skin stresses and distortions. A closer tolerance will compel more reasonable care, with resultant decreases in surface stress distortions.

Tool Clearances. Cutting-tool clearances and runouts are necessary fabrication considerations that cannot be compromised without resulting in tool damage, part damage, and gear-quality deterioration. Several types of clearances encountered are shown in Fig. 9-20. It is necessary to establish dimensions and set tolerances so as to avoid interference in all extreme cases.

9-44. Deburring and Tooth-edge Round-over Tolerances. The quality of tooth-profile deburring and edge breakage and round-over are related items that must

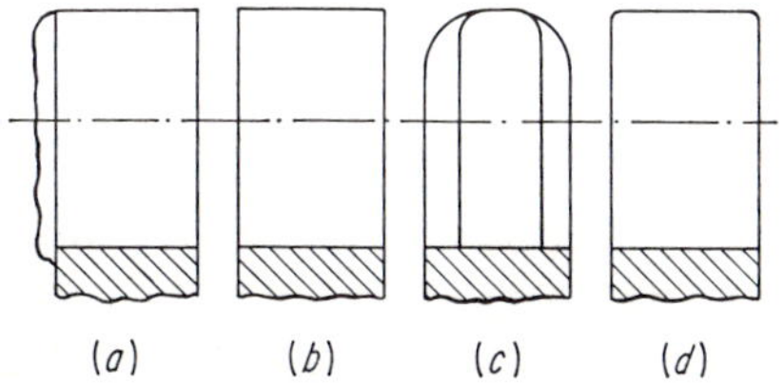

FIG. 9-21. Qualitative comparisons of tooth deburring. (*a*) Typical burred tooth on trailing edge. (*b*) Ideally deburred tooth. (*c*) Excessively deburred tooth with heavy round-over. (*d*) Acceptable deburring.

be controlled. Inadequate deburring can result in tooth-surface damage. Excessive deburring resulting in heavy tooth-edge round-over can cause significant loss of gear face and increased surface stresses. The possible variations of this problem are shown in Fig. 9-21. The occurrence of this problem is related to tooth width and material. Obviously wide gears, as compared with narrow gears, can afford more loss of width. Also, soft materials will be more difficult to deburr and subject to edge breakage and round-over. As a guide, the tolerance on deburring is absolute accomplishment. There are to be no discernible burrs to the eye or touch. The magnitude of edge breakage and tolerable amount of round-over are given in Table 9-21.

Table 9-21. Tooth-edge Breakage and Round-over

Quality and diam. pitch	Max. permissible edge round-over		
	Face width		
	Larger than ¼″	0.060 to ¼″	Less than 0.060″
Commercial:			
20 pitch and coarser	1/32	0.020	15% of face width
21 pitch and finer	1/64	0.010	15% of face width
Precision:			
20 pitch and coarser	0.015	0.010	10% of face width
21 pitch and finer	0.010	0.005	10% of face width

9-45. Tolerances and Noise. Gear noise is an indication of some degree of defective quality which has both technical and subjective significance. Noise directly results from clashing of loaded teeth and is related to geometrical defects in the tooth profiles due to cutting imperfections, tooth-load deflections, and vibrations. Additional noise can be emitted from forced and resonant vibration of gear and housing elements. This constitutes the technical aspects of noise and should make evident objections stemming from functional considerations.

As a subjective item, noise is objectionable if human beings are located within hearing distance. The high-pitch whine of high-speed gears is a detriment to users of the equipment.

Noise is to be controlled by specifying high accuracy of profile error, tooth-to-tooth spacing, pitch-line runout, and lateral runout. Maximum quietness will be obtained by specifying the best tolerance values listed for these quantities in previous tables. As a further guide, Table 9-22 lists typical tolerances.

Table 9-22. Typical Tolerances for Critical Noise-influencing Dimensions

Dimension or control item	Tolerance	
	Best obtainable quietness	Av. noise conditions
Profile tolerance, in	0.0001	0.0005
Tooth spacing, in	0.0001	0.0003
Pitch-line runout	0.0005 max.	0.001
Lateral runout	0.0005 per in., pitch radius 0.005 max.	0.002 per in., pitch radius 0.010 max.

ASSEMBLY AND INSTALLATION CONSIDERATIONS

The previous sections were concerned with tolerances related to individual gears. In addition, there are significant tolerances which apply to the assembly of gears into a housing and the meshing of gear pairs.

9-46. Center-distance Tolerances. Gear center-distance tolerance is of primary concern because it directly affects gear function in terms of backlash and contact ratio. Care must be exercised so as to dimension and tolerance center distance properly to avoid binding condition, excessive backlash, or so low a contact ratio that smooth operation and load-carrying capacity are affected. Center-distance tolerance is a function of

1. Backlash requirement
2. Gear quality
3. Pitch
4. Magnitude of center distance

These are further described as follows:

Backlash Requirement. The center distance directly affects backlash by the approximate equation

$$b = \Delta C \times 2 \tan \phi$$

where ΔC = change in center distance

The center-distance tolerance must be consistent with the backlash requirement, so as not to contribute backlash in excess of the stipulated maximum value or a disproportionate share.

Gear Quality. This is related to the center-distance tolerance in that there must be consistency between gear-size tolerance, runout, and center-distance tolerance. For example, it would be illogical to place a stringent requirement on center distance and simultaneously allow a liberal tolerance on gear size. Presumably good-quality gears are specified for a good reason and will also require good centers of equivalent quality tolerance.

Pitch. The diametral pitch bears on center-distance tolerance in that with fineness it demands closer control of center-distance variation so as to prevent low contact ratio.

Center-distance Magnitude. The center-distance magnitude influences the tolerance in that problems, negligible over short distances, become significant for large dimensions. This is due to temperature effects, work distortions and deflection, and magnified machine and tooling errors.

Methods of Gear-center Dimensioning. There are several different basic methods for dimensioning gear centers which are related to both gear function and fabrication method. They are

1. Center-to-center dimensioning
2. Coordinate dimensioning from fixed references
3. Coordinate dimensioning center-to-center
4. Polar-coordinate dimensioning

The features and peculiarities of these various methods can be stated as follows:

Center-to-center Dimensioning (See Fig. 9-22*a*). This is perhaps the dimensioning most analogous to the functional flow of power and motion through the gear train. It is point-to-point location using distance and reference angle. It has the feature of simplicity as related to the gearing but is limited in application because of build-up of accumulated tolerances from center to center. This is not bothersome if the housing gear centers are single bearing plate or pairs bored together. Interchangeability will be limited by the large accumulation of tolerances.

Coordinate Dimensioning from Fixed References (See Fig. 9-22*b*). This method provides good control on all centers without the fault of accumulating tolerances. For this reason it is suitable for use on double bearing plates and provides good

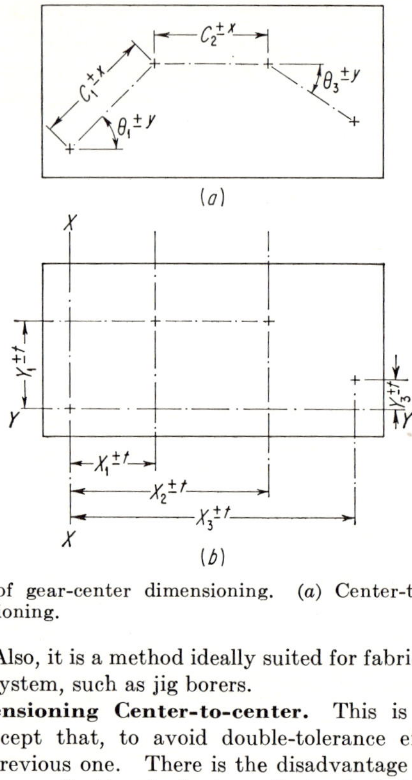

FIG. 9-22. Examples of gear-center dimensioning. (*a*) Center-to-center dimensioning. (*b*) Coordinate dimensioning.

interchangeability. Also, it is a method ideally suited for fabrication machines based upon the coordinate system, such as jig borers.

Coordinate Dimensioning Center-to-center. This is similar to the fixed reference method, except that, to avoid double-tolerance effects, each center is referenced from the previous one. There is the disadvantage of accumulated tolerances, which limits the application for best results to single-bearing-plate housings. Also, the nonaccumulative dimensioning makes it difficult for fabrication in the jig-borer-type machine.

Polar-coordinate Dimensioning. This method is most suited for a series of gears primarily related to a single fixed point in a polar relationship. It is also suitable for standard fabrication machines.

Table 9-23. Center-distance Tolerances for Various Magnitudes and Quality

Gear-train quality	Center distance				
	Under 1″	1–6″	6–12″	12–24″	Over 24″
Commercial	±0.002	±0.003	±0.005	±0.010	±0.010 per ft
Precision	±0.0005	±0.001	±0.002	±0.002	±0.002 per ft
High precision	±0.0001	±0.0002	±0.0002	±0.0003	±0.0005 per ft

The above represents basic methods of gear-center dimensioning. In practice one or more of the methods are used in even one design, and also often special combined systems are conceived, such as one coordinate and a center-to-center distance dimension.

Typical center-distance tolerances are listed in Table 9-23. In applying these tolerances it is necessary to give due consideration to the items listed in the earlier portion of this section as influential in determining tolerances of gear centers.

9-47. Tolerances on Mismatch of Gear Faces (See Fig. 9-23). The alignment of meshing gear faces is a control that can have importance if the gears are narrow-faced. In such instances the gears could conceivably lose enough engagement such that surface stresses could exceed design values and possible premature wear and

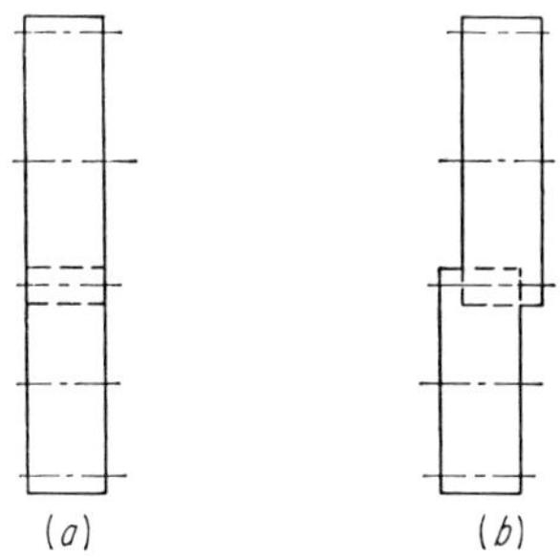

FIG. 9-23. Alignment of gear faces. (*a*) Ideal alignment. (*b*) Mismatched alignment.

failure could occur. The mismatch problem is dependent upon the relative magnitudes of tolerances causing mismatch and the gear face width. As a guide to controlling mismatch Table 9-24 lists practical values.

Table 9-24. Acceptable Values of Gear-face Mismatch

Nominal gear face width, in.	Mating gear or pinion face width	Max. % of gear-face mismatch
¼ or greater	Same as mate	20 but no greater than ⅛″
⅛–¼	Same as mate	20
0.060–⅛	Same as mate	15
Less than 0.060	Same as mate	10 but no less than 0.004″

For gear pairs in which one mate is significantly wider than the other, the match must be 100%.

9-48. Assembly Fits of Bores and Shafts. Fit of the gear bore and associated shaft is related to gear tolerances.* Precision gears must have a close fit to utilize the inherent gear quality. Lower-quality gears can have relaxed tolerances on gear fit. The fit of bore and shaft is a combination of allowance plus the tolerances on the bore and shaft. It is preferable to use standard sizes for the bore and vary the shaft diameter so as to require minimum-bore tooling and gages. Generally, shaft diameters can be fabricated to closer tolerances than bore diameters. Table 9-25 lists recommended values as a guide.

* See Art. 6-10 for a detail study of center-distance tolerance and other tolerances that effect backlash.

Table 9-25. Gear-bore and Mating-shaft Tolerances
Fine-pitch Gears (20 Diametral Pitch and Finer)

	AGMA class					
	Commercial			Precision		
	1	2	3 and 4	1	2	3
Bore dia.	+0.002	+0.001	+0.0007	+0.0005	+0.0002	+0.0001
Shaft dia.	−0.0005	−0.0005	−0.0003	−0.0002	−0.0002	−0.0001
Min. allowance	0.0005	0.0003	0.0002	0.0001	0.0001	0.0001

Coarse-pitch Gears (19 Diametral Pitch and Coarser)

	Nominal dia. in.					
	Commercial			Precision		
	Up to 1	1–4	Over 4	Up to 1	1–4	Over 4
Bore dia.	+0.001	+0.0015	+0.002	+0.0005	+0.0007	+0.001
Shaft dia.	−0.0005	−0.001	−0.002	−0.0002	−0.0003	−0.0005
Min. allowance	0.0003	0.001	0.0015	0.0001	0.0002	0.0005

REFERENCES

1. American Gear Manufacturers Assoc.: "Gear Classification Manual for Spur, Helical and Herringbone Gears," AGMA 390.01, May, 1961.
2. American GearManufacturers Assoc.: "20-degree Involute Fine-pitch System for Spur and Helical Gears (ASA B6.7-1956)," AGMA 207.04, October, 1960.
3. American Gear Manufacturers Assoc.: "Data Gears Accuracy Specification," AGMA 237.01, March, 1961.
4. Collins, L. J.: G. E. Reveals How It Makes Large Gears, *Am. Machinist*, Sept. 22, 1958.
5. Colvin, F. H., and F. A. Stanley: "Gear Cutting Practice," 3d ed., McGraw-Hill Book Company, Inc., New York, 1950.
6. Dean, Paul M.: Effects of Size on Gear Design Calculations, *Prod. Eng.*, April, 1954.
7. Dudley, Darle W.: "Practical Gear Design," McGraw-Hill Book Company, Inc., New York, 1954.
8. Hartman, M. A.: Mastering Master Gears, *Machinery*, January, 1957.
9. Merritt, H. E.: "Gears," 3d ed., Pitman Publishing Corporation, New York, 1954.
10. Michalec, G. W.: Critical Criteria for Precision Gears, *Elec. Mfg.*, Vol. 66, No. 1, July, 1960.
11. Michalec, G. W.: Methods of Specifying Precision of Spur Gears, *Prod. Eng.*, Vol. 27, No. 12, November, 1956.
12. Michalec, G. W.: "Gear Position Error Control," ASME Paper 59-1-21, December, 1959.
13. Michalec, G. W.: Precision Gearing, *Machine Design*, Vol. 27, pp. 1–4, January–April, 1955.
14. Michalec, G. W.: Tolerances for Over-pin Gear Measurements, *Prod. Eng.*, Vol. 28, No. 5, May, 1957.
15. Parkinson, A. C., and H. W. Downey: "Gears, Gear Production and Measurement," Pitman Publishing Corporation, New York.

Chapter 10

GEAR MATERIALS

By

DARLE W. DUDLEY, *Manager, Advance Gear Engineering, General Electric Company, Lynn, Mass.*

The gear trade uses a wide variety[1,*] of steels, cast irons, bronzes, aluminum, phenolic laminates, plastics, and other materials for gears. This chapter will cover the basic information on all the more commonly used gear materials and their processing treatments. Data on the relative load-carrying capacities of different gear materials are covered in Chap. 13.

In some cases things like trade practices, shop equipment available to make gears, or specific design requirements leave the designer little choice as to which gear material to use. In other cases it may be possible to consider a wide variety of gear materials. When the designer has a wide choice, things like cost of raw stock, relative load-carrying capacity for a given size, adaptability of material to mass-production processes, and resistance to corrosion enter into the choice of a material. As an example, many toys and gadgets use stamped-bronze gears. This is not a low-cost raw material, but low-cost gears can be made by stamping and this material resists corrosion.

FERROUS GEAR MATERIALS†

The "ferrous" gear materials are those which contain mainly the element iron. Iron with percentages of carbon in the range of 0.15 to about 1.5 per cent is customarily called "steel." Iron with larger percentages of carbon such as 2 to 4 per cent is customarily called "cast iron." In a steel the carbon and iron are mostly in chemical combination with each other. In a cast iron the carbon is mostly in the free state. The whole family of ferrous materials used in gear work comprises steels, gray cast iron, Meehanite®,‡ ductile iron, malleable iron, and sintered iron.

The ferrous materials are of great importance in the gear trade. Generally speaking, they represent the lowest-cost raw material in terms of cents per pound. The aircraft-type hardened steels can carry the greatest load in terms of horsepower per unit volume. The ferrous materials represent the greatest tonnage of any kind of material used to make either toothed gears or gear casings.

* Superscript numbers refer to references at the end of the chapter.

†A considerable amount of material for this section was prepared by Antoni Tabak, Gear Metallurgist, General Electric Co., Lynn, Mass.

‡ Registered trade-mark of the Meehanite Metals Corp., New Rochelle, N.Y.

10-1. General Characteristics of Steels and Cast Irons. The properties of steels and cast irons vary rather widely depending on composition and heat-treatment. Table 10-1 shows in outline form the general range of properties obtainable.

Some interesting general principles can be noted from a study of Table 10-1. The steels shown vary from 0.2 to 0.6 per cent carbon. All these steels could be heat-treated to obtain about 200 BHN. All except AISI 1020 can be treated to 350 BHN. The hardness of a steel is generally a function not of its composition but rather of its heat-treatment. It is true that the *maximum* obtainable hardness for a given steel is determined by its composition.

Table 10-1. General Properties of Steels and Cast Irons

Material	Approx. % carbon	Hardness, BHN	Modulus of elasticity, psi	Density, lb per cu in.	Yield strength, psi	Tensile strength, psi
Steels:						
AISI 1020......	0.2	180	28.5×10^6	0.283	54,000	89,000
AISI 1040......	0.4	200	28.5×10^6	0.283	63,825	99,000
		350	28.5×10^6	0.283	93,000	123,000
AISI 4140......	0.4	200	28.5×10^6	0.283	60,500	95,000
		350	28.5×10^6	0.283	155,000	170,000
AISI 4340......	0.4	220	28.5×10^6	0.283	68,500	108,000
		350	28.5×10^6	0.283	166,000	175,000
AISI 1060......	0.6	350	28.5×10^6	0.283	168,000	175,000
		550	28.5×10^6	0.283	265,000	275,000
Cast iron:						
AGMA 20......	3.5	150	12×10^6	0.255	....	20,000
AGMA 40......	3.5	175	16×10^6	0.260	....	40,000
AGMA 60......	3.5	200	20×10^6	0.265	....	60,000
Ductile iron:						
ASTM 80-60-3..	3.5	200	24×10^6	0.265	60,000	80,000

Most of the materials used for gears, other than the steels, are not subject to hardness control by heat-treatment. Composition rather than heat-treatment determines the hardness of most bronzes, zincs, plastics, and laminates.

The ultimate tensile strength of steel is a very direct function of the Brinell hardness. Note that the values shown are approximately 500 times the Brinell hardness number. The only case where this rule fails is where a steel piece is not uniformly hardened. Then the ultimate strength of the part is not so great as the surface hardness would indicate.

The steels are all about the same density. Cast iron is less dense and tends to vary over a range of as much as from 0.255 to about 0.275 lb per cu in. The modulus of elasticity of cast iron is less than that of steel, and it also tends to vary with different irons. In steel the modulus is quite constant.

The yield strength of steels tends to range from about 60 per cent of the tensile strength in a low-hardness steel to about 90 per cent of the tensile strength for a high-hardness steel.

Gray cast iron is a brittle material that really does not have a yield point like steel. A very slight yielding may occur at around 25 per cent or more of the tensile strength. Ductile iron does have a yield point somewhat like steel.

Significance of the Hardness of Steel or Iron. Since the steels and irons vary so widely in hardness, it is desirable to form some mental impressions of the significance of different hardnesses. Table 10-2 shows some general levels of hardness and their meaning in the gear trade. Table 24-19 lists the wide variety of hardness-testing instruments that are used to measure gear hardness. Table 24-20 shows approximate numerical conversions from one scale to another. In general discussions of hardness it is common practice to speak of Brinell hardness number (BHN) or Rockwell C hardness. These two scales are the best known in the trade.

Table 10-2 skips the region of 400 to 500 BHN, which is seldom used. The steel is too hard to cut; and if one is going to the trouble of grinding, the part might just as well be made still harder so that more capacity is obtained.

Table 10-2. Relative-hardness Levels for Steel or Iron Gears

Hardness		Machinability	Comments
Brinell	Rockwell C		
150–200	...	Very easy	Very low hardness. Minimum load-carrying capacity
200 250	... 24	Easy	Low hardness, moderate load capacity. Widely used in industrial gear work
250 300	24 32	Moderately hard to cut	Medium hardness. Good load capacity. Widely used in industrial work
300 350	32 38	Hard to cut, often considered limit of machinability	High hardness. Excellent load capacity. Used in lightweight, high-performance jobs
350 400	38 43	Very hard to cut. Many shops cannot handle	High hardness. Load capacity excellent provided heat-treatment develops proper structure
500 550	51 55	Requires grinding to finish	Very high hardness. Wear capacity good. May lack beam strength
587	58 63	Requires grinding	Full hardness. Usually obtained as a surface hardness by case carburizing. Very high load capacity for aircraft gears, automobile gears, trucks, tanks, etc.
	65 70	May be surface-hardened after final machining	Superhardness. Generally obtained by nitriding. Very high load capacity

10-2. Heat-treatment of Steels and Cast Irons. The basic process for hardening a steel material consists of heating it to a red-hot heat, quenching in oil or water, and then reheating to temper the steel. The steel has to be heated hot enough to "austenitize" the steel. Generally this is in the range of 1450 to 1600°F. The quench must be fast enough to avoid the formation of undesirable intermediate products. A well-quenched gear will have a maximum amount of "martensite." This makes the gear very hard and somewhat brittle. The reheat for tempering brings the part up to some temperature that is below the critical temperature at which the part austenitizes. After tempering the part has a lower hardness and the martensite will have undergone more or less transformation. The resulting material is much tougher and more ductile than the original quenched material. By controlling the tempering temperature, it is possible to get a wide range of hardness values. For instance, a part with 0.60 carbon might quench out at 600 BHN. A temper at 350°F would toughen the steel but cause almost no drop in hardness. A temper at 1000°F might drop the hardness to something like 400 BHN. A temper at 1200°F might bring the hardness down to around 300 BHN. It can thus be seen that any desired hardness *less than the as-quenched hardness* can be obtained by controlling the tempering temperature.

Steels that are lean in carbon will not develop full hardness even in the as-quenched condition. Figure 10-1 shows the approximate *maximum* hardness that can be obtained for different carbon contents. To obtain maximum hardness the part must either be small and quenched drastically or it must contain enough alloys so that a slower quench will work and thus permit larger sizes or milder quenching. In straight carbon steels a rod ½″ in diameter quenched in water will develop all the hardness that is possible. In a highly alloyed steel a rod as large as 3″ in diameter quenched in oil will develop the full hardness potential of the carbon.

Figure 10-1 shows that it takes about 0.60 per cent carbon to get up to a full hardness of 600 BHN. Greater amounts of carbon put more carbides in the gear surface

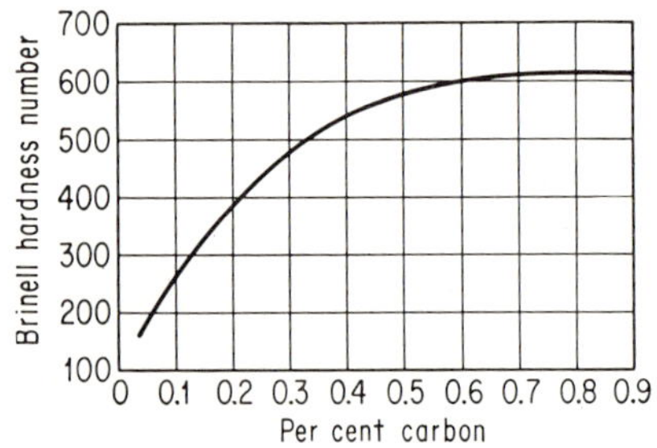

FIG. 10-1. Maximum hardness of steel.

and make the part more wear-resistant but they do not appreciably increase the maximum hardness. The "eutectoid" of carbon and iron is at 0.85 per cent carbon. This amount of carbon is considered an optimum for the carbon content of a case-carburized part.

Carbon contents in the range of 0.10 to 0.20 per cent considerably reduce the maximum hardness that can be obtained. A gear made to a low carbon content like 0.15 per cent might be held to as much as 300 BHN but it would not be a very good gear. The wear resistance would be poor because of not having enough carbides in the surface. Generally gears made in the 200 to 350 BHN range have 0.40 to 0.60 per cent carbon. A low carbon content like 0.10 or 0.20 per cent carbon is useful, though, as the *core* material for a gear that is case-carburized after the teeth are cut. The case is enriched with carbon up to around 0.85 per cent. After quenching and tempering, the carbon-rich case will come out at around 600 BHN and the core material—because of the lean amount of carbon—will develop a hardness of only around 250 BHN. The core is much tougher and more ductile than the case. The cased tooth has better endurance strength and impact strength and less sensitivity to notches than a gear tooth would have if it were made out of a homogeneous material having 0.85 per cent carbon all the way through the tooth.

Figure 10-2 shows what can be obtained by water-quenching and tempering[2] an AISI* 1045 steel. This is a plain-carbon steel having 0.45 per cent carbon. Figure 10-3 shows the properties that can be obtained by oil-quenching and tempering an AISI 4340 steel. AISI 4340 is a 0.40 per cent carbon-nickel steel relatively rich in

* The term "AISI" refers to the American Iron and Steel Institute. Their standards for steel compositions are widely used in industry.

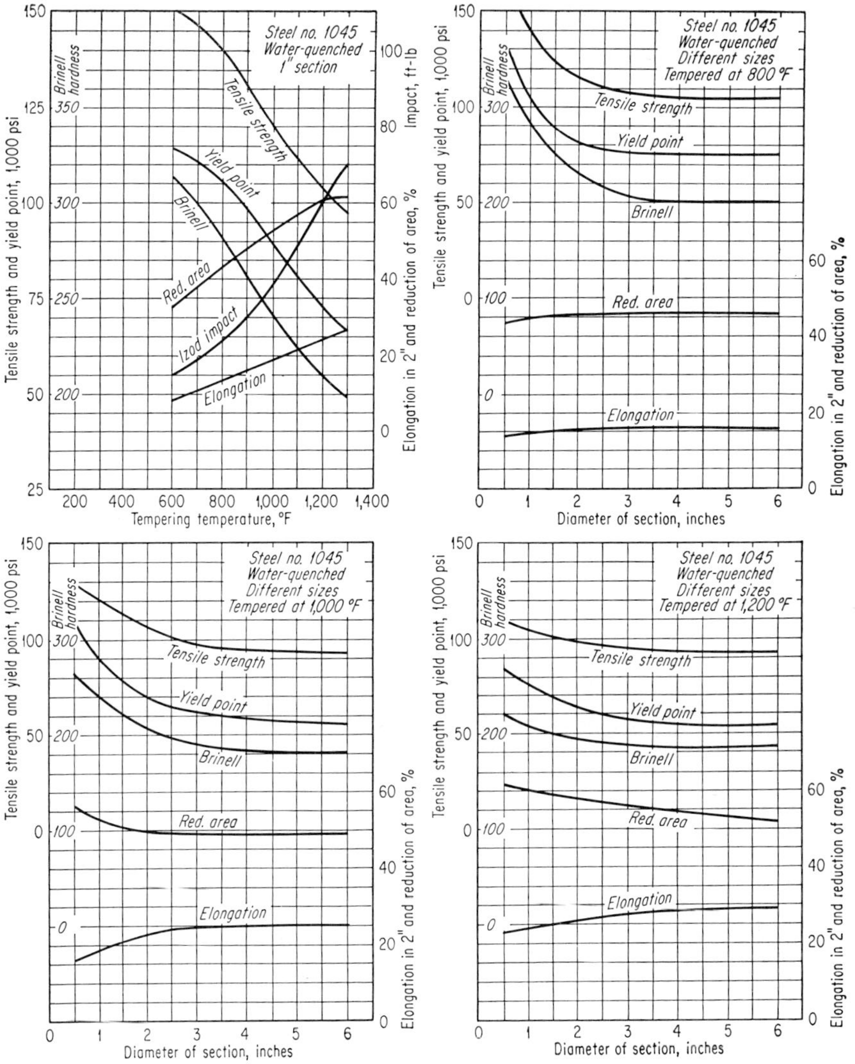

FIG. 10-2. Properties of water-quenched and tempered carbon steel 1045 in different sizes. (In sizes over 1″ the properties represent those at the section midway between the surface and the axis on samples cut longitudinally.) In section ½ to 2″, inclusive, quenched from 1475 to 1525°F; over 2 to 4″, inclusive, from 1500 to 1550°F; over 4″, from 1525 to 1575°F. (*Courtesy of International Nickel Co., New York.*)

alloy content. Figures 10-2 and 10-3 show some interesting things about how the alloy content affects the properties obtained after heat-treatment. A 1″-diameter section of the AISI 1045 developed only 310 BHN after a temper at 600°F. The same size piece of AISI 4340 developed 500 BHN after a 600°F temper. The reason for this is that a 1″ rod is too big a piece of AISI 1045 steel to quench out to the full hardness that might be obtained from the carbon content.

The difference is quite significant for a 3″ rod tempered at 800°F. The AISI 1045 piece makes only 200 BHN while the AISI 4340 piece comes out at 410 BHN.

Composition, % *

Carbon	Manganese	Nickel	Chromium	Molybdenum
0.38–0.43	0.60–0.80	1.65–2.00	0.70–0.90	0.20–0.30

* Phosphorus and sulfur each 0.040 max.; silicon 0.20–0.35

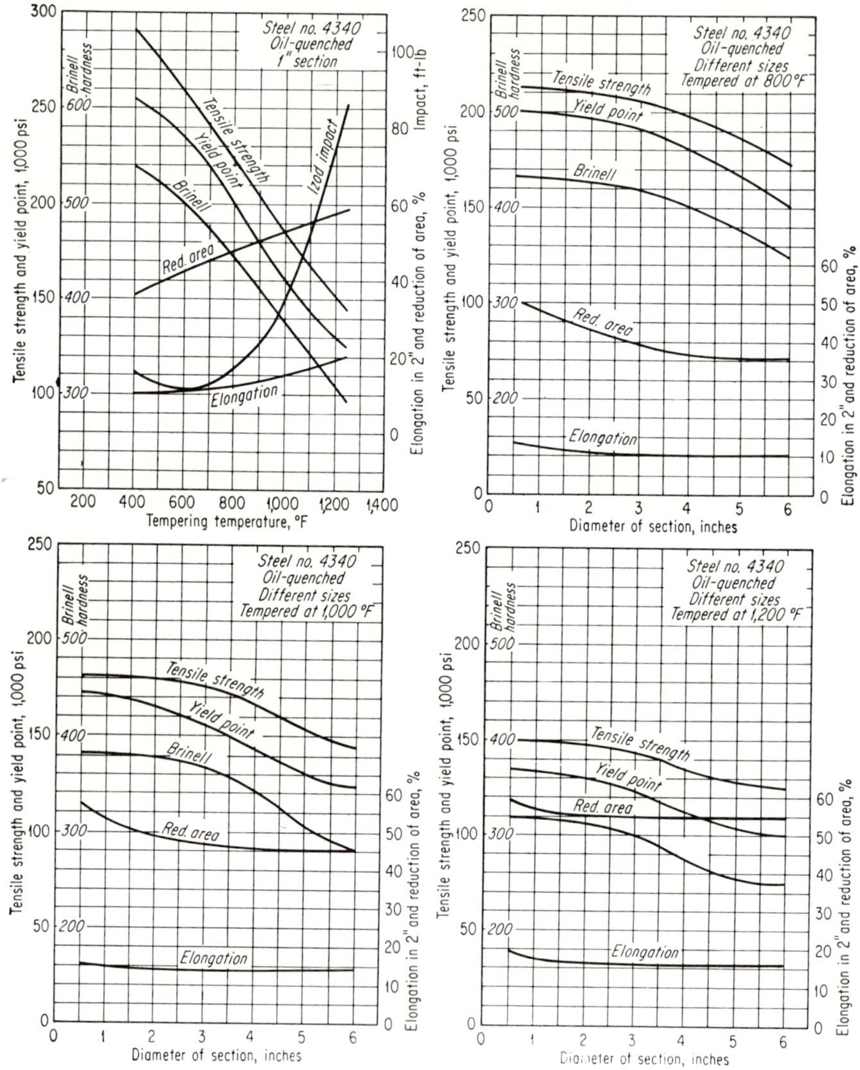

FIG. 10-3. Properties of oil-quenched and tempered nickel-chromium-molybdenum steel 4340 in different sizes. (In sizes over 1″ the properties represent those at the section midway between the surface and the axis on samples cut longitudinally.) In sections ½ to 2″, inclusive, quenched from 1500 to 1550°F; over 2 to 4″, inclusive, from 1525 to 1575°F; over 4″, from 1550 to 1600°F. (*Courtesy of International Nickel Co., New York.*)

Some further insight on the affect of alloy content on hardenability can be gained by studying "Jominy" test results. The standard *Jominy* test is made by quenching one end of a 1" rod in water under closely controlled conditions. After the part has cooled, hardness readings are taken every ⅟₁₆" back from the spot where the quench ended. Figure 10-4 shows Jominy data for AISI 1045 and 4340. Figure 10-5 shows Jominy data for AISI 1020 and 3310. In each case the highly alloyed steels show a good hardness for a considerable distance back from the end of the quench water line. The plain-carbon steels develop about the same hardness right at the quench line, but they drop off in hardness rapidly back of the quench line. This drop-off is caused by the fact that they did not cool fast enough.

Figure 10-6 shows the effect of several different alloying elements on the size of rod that will quench out to give the desired structure. The curves for "100 per cent martensite" mean that all the carbon in the steel transformed into martensite on quenching. The 50 per cent martensite curves mean that only half the carbon present formed martensite. The numbers on the curves indicate percentages of the elements shown. The curves indicate that the amount of alloy needed to develop best results

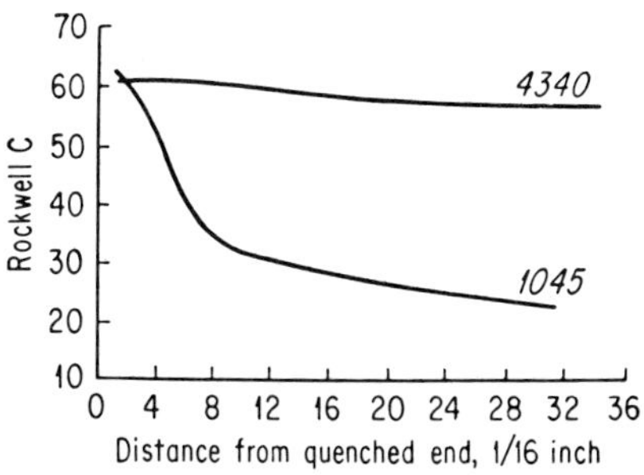

FIG. 10-4. Jominy end-quench hardenability for typical through-hardening steels. (*Courtesy of International Nickel Co., New York.*)

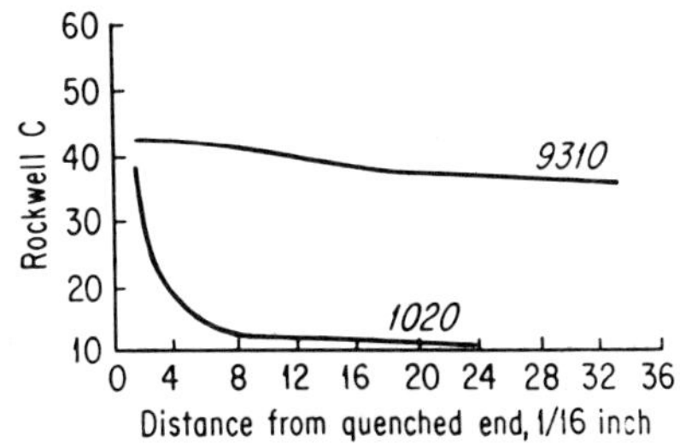

FIG. 10-5. Joining end-quench hardenability for typical carburizing steels. (*Courtesy of International Nickel Co., New York.*)

for a given size of rod depends quite a bit on the carbon content. For instance, 3 per cent nickel is not enough to get all the carbon into martensite in a 1" rod if the carbon content is 0.30 per cent. However, if the carbon content is up to about 0.50 per cent, 3 per cent nickel will do the job.

The data presented in this article so far show that a wide range of properties can be obtained by proper heat-treatment of a steel. It is also clear that the alloy content and the carbon content of a steel profoundly affect the response of the steel to a given heat-treatment. The different compositions of gear steels shown later in this chapter have become popular mainly because each has the composition to respond properly to a reasonable heat-treating cycle and then fill a given gear application well.

Gear metallurgists use a variety of special terms in their work. Some of these terms have already been used in this chapter. Others will be needed in the remaining part. In view of this a glossary of terms will be given.

Glossary of Metallurgical and Heat-treating Terms

Aging. A change in an alloy by which the structure recovers from an unstable or metastable condition produced by quenching or by cold working. The degree of stable equilibrium obtained for any given grade of steel is a function of time and temperature. The change in structure consists of precipitation and is marked by a change in physical and mechanical properties. Aging which takes place slowly at room temperature may be accelerated by an increase in temperature.

Annealing. A broad term used to describe the heating and cooling of steel in the solid form. The term annealing usually implies relatively slow cooling. In annealing the size, shape, and composition of the steel product and the purpose of the treatment determine the temperature of the operation, the rate of temperature change, and the time at heat.

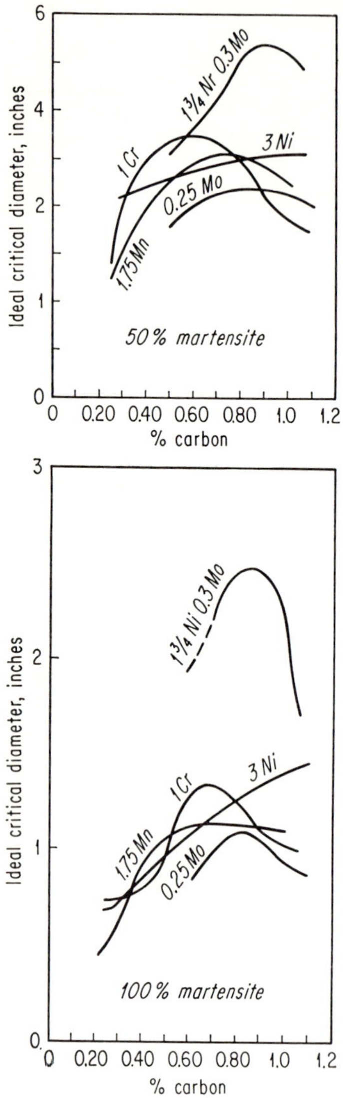

FIG. 10-6. Influence of carbon content on hardenability of several alloy steels. (*Courtesy of International Nickel Co., New York.*)

Annealing is used to induce softness, to remove stresses, to alter physical and mechanical properties, to remove gases, to change the crystalline structure, and to produce a desired microstructure.

Austempering. A process for the transformation of austenite by quenching in a special bath, as a molten-salt bath or molten lead, to allow transformation at a constant temperature to produce a structure other than martensite.

Austenite. In steels the gamma form of iron with carbon in solid solution. Austenite is tough and nonmagnetic and tends to harden rapidly when worked below the critical temperature.

Austenitic Steels. Steels which are austenitic at room temperature.

Basic Steel. Steel melted under a slag having a basic reaction and in a furnace having a basic bottom and lining.

Bessemer Process. A method for making steel by blowing air through molten pig iron to reduce the carbon content and to remove certain elements by oxidation.

Blue Brittleness. Brittleness occurring in steel when in the temperature range of 400 to 700°F, or when cold after being worked within this temperature range.

Bright Annealing. A process of annealing usually carried out in a controlled atmosphere to protect the surface of the material from oxidation or discoloration.

Brinell Hardness. A hardness number determined by applying a known load to the surface of the material to be tested through a hardened-steel ball of known diameter. The diameter of the resulting permanent indentation is measured. This method is unsuitable for measuring the hardness of sheet or strip metal.

Carburizing. Diffusing carbon into the surface of iron-base alloys by heating such alloys in the presence of carbonaceous materials at high temperatures. Such treatments, followed by appropriate quenching and tempering, harden the surface of the metal to a depth proportional to the time of carburizing.

Case Depth. That portion of a carburized, nitrided, or cyanided iron-base-alloy part in which the carbon or nitrogen content has been noticeably increased.

Casehardening. A process of surface hardening involving a change in the composition of the outer layer of an iron-base alloy by inward diffusion from a gas or liquid followed by appropriate thermal treatment. Included are carburizing, nitriding, carbonitriding, and cyaniding.

Cementite. A microconstituent of steel consisting of Fe3C (iron carbide, containing 6.69 per cent C). It is hard, brittle, and weakly magnetic.

Cold Working. Permanent deformation of a metal below its recrystallization temperature. Some metals can be hot-worked at room temperature while others can be cold-worked at temperatures in excess of 1000°F.

Controlled Cooling. A process by which a steel object is cooled from an elevated temperature in a predetermined manner to avoid hardening, cracking, or internal damage, or to produce a desired metallurgical structure.

Creep. The slow deformation of steel under stress at elevated temperatures.

Critical Range. The range of temperature required for completion of phase transformations in alloys.

Crystal. A physically homogeneous solid in which the atoms are arranged in a three-dimensional repetitive pattern.

Decarburization. The loss of carbon from the surface of solid steel during heating.

Deoxidizing. To remove oxygen from molten metal; or to reduce scale on metal.

Ductility. The property of a metal which allows it to be permanently deformed before final rupture. It is commonly evaluated by tensile testing in which the amount of elongation or reduction of area of the broken test specimen, as compared with the original, is measured.

Elastic Limit. Maximum stress to which a metal can be subjected without permanent deformation.

Endurance Limit. The maximum stress to which material may be subjected an infinite number of times without failure.

Ferrite. Iron in the alpha form in which alloying constituents may be dissolved. Ferrite is magnetic, soft, and acts as solvent for manganese, nickel, and silicon.

Flakes. Internal fissures in steel with a bright scaly appearance. Flakes can be minimized in forgings and rolled products by thermal treatment.

Full Annealing. A softening process in which metal is heated to a temperature above the transformation range and after being held for a proper time is cooled slowly to a temperature below the transformation range. The rates of cooling are very slow.

Grain Size. The grain-size number is determined by a count of a definite microscopic area, usually at 100 magnification. The larger the grain-size number, the smaller the grains.

Grains. Individual crystals in metals.

Hardenability. The property which determines the depth and distribution of hardness induced by quenching. The higher the hardenability value the greater the depth to which the material can be hardened and the slower the quench which can be used.

Heat-treatment. One or more operations involving the heating and cooling of a metal in the solid state for the purpose of obtaining certain desired conditions or properties. Heating and cooling for the prime purpose of mechanical working are not included in the meaning of the definition.

Impact Test. A test in which one or more blows are suddenly applied to a specimen. The results are usually expressed in terms of energy absorbed or number of blows of a given intensity required to break the specimen.

Inclusions. Particles of nonmetallic materials, usually silicates, oxides, or sulfides which are mechanically entrapped or are formed during solidification or by subsequent reaction within the solid metal. Impurities in metals.

Isothermal Annealing. A method of annealing which consists of heating to the proper temperature above the critical range for austenitizing, followed by rapid cooling to a suitable temperature and holding for a sufficient time at that temperature to complete the transformation.

Isothermal Transformation. A process of transforming austenite in any given steel to ferrite or a ferrite-carbide aggregate at any constant temperature within the transformation range.

Jominy Test. The Jominy test is used for determining end-quench hardenability of steel. It consists of water-quenching one end of a 1″ diameter bar under closely controlled conditions and measuring the degree of hardness at regular intervals along the side of the bar from the quenched end up.

Killed Steel. Steel deoxidized with a strong deoxidizer like silicon or aluminum to reduce the oxygen content of the steel to a minimum to prevent any gaseous reaction between carbon and oxygen during solidification.

Martempering. A hardening process in which the steel part is quenched from the austenitizing temperature in a salt or molten bath held at a temperature above or slightly below the temperature at which martensite begins to form. The part is held at this temperature long enough to equalize in temperature throughout but not long enough to permit any transformation to occur. The steel is then cooled in air through the martensite range to accomplish full transformation.

Martensite. A microconstituent in quenched steel characterized by an acicular, or needle-type, structure. It has the maximum hardness of any structure obtained from the decomposition of austenite.

Modulus of Elasticity. The ratio, within the elastic limit, of the stress to the corresponding strain. The stress in pounds per square inch is divided by the elongation in inches per inch of the original gage length of the specimen. Also known as Young's modulus.

Nitriding. Adding nitrogen to solid iron-base alloys by heating the steel in contact with ammonia gas or other suitable nitrogenous material.

Normalizing. A process in which a steel is heated to a temperature above the transformation range and is subsequently cooled in still air to room temperature.

Pearlite. The lamellar aggregate of ferrite and cementite resulting from the direct transformation of austenite at the lower critical point.

Proof Stress. The greatest static stress at which a material exhibits an arbitrarily permanent deformation. No standard has been adopted and published values may not be comparable.

Proportional Limit. The greatest load per square inch of original cross section for which the elongation is proportional to the load applied.

Residual Stress. Stresses remaining in a part after the completion of working, heat-treating, welding, etc., due to phase changes, expansion, contraction, and other phenomena.

Rockwell Hardness. A hardness number as determined by a Rockwell hardness tester, a direct-reading machine which may use a steel ball or a diamond penetrator.

Secondary Hardening. Increase in hardness developed by tempering high-alloy steels after quenching, usually associated with precipitation reactions.

Stress Relief. A process of reducing internal residual stresses in a metal part by heating to a suitable temperature and holding the part at that temperature for a proper time. Stress relieving may be applied to parts which have been welded, machined, cast, heat-treated, or worked.

Temper Brittleness. Brittleness resulting from the slow cooling of certain steels through certain temperature ranges (below transformation). Notched-bar impact tests are used to reveal brittleness.

Tempering. A process of reheating quench-hardened or normalized steels to a temperature below the transformation range.

Tensile Strength. The maximum load per unit of original cross-sectional area carried by a material during a tensile test.

Yield Point. The load per unit of original cross section at which a marked increase in deformation occurs without any increase in load.

10-3. Carbon and Alloy Steels. The plain-carbon steels have only small amounts of alloy. In the AISI series these steels have "10" for the first two digits like "AISI 1020" or "AISI 1045." The last two digits signify the carbon content. AISI 1020 has 0.20 per cent carbon.

Table 10-3 shows the compositions of the more popular gear steels. Those having numbers starting with 2 or higher have relatively high alloy content and are considered *alloy* steels.

Many steel companies sell nonstandard steels under their own brand names or composition numbers. Frequently these nonstandard steels will make excellent gears for a given job and the nonstandard material can be obtained at lower price and with quicker delivery than the standard AISI steels. Usually the nonstandard steel compares quite closely with the specification of some standard steel. The behavior of the nonstandard steel can usually be predicted quite well by comparison with the behavior of the standard steels having the nearest composition.

For critical gear work it is usually advisable to work to recognized AISI steel specifications. The limits on composition are set as closely as is practical. The standard steel can be used on a job and the precise heat-treating cycle established to give the desired results. The user can be confident that each batch of standard steel will behave almost exactly like the last batch. The nonstandard steels are usually made to much looser limits. The user may have trouble from batch to batch with changes in the heat-treat distortion, hardnesses not coming out right with the same heat-treating temperature, or an occasional batch having enough impurities in it to reduce its strength significantly.

Table 10-3. Composition of Typical Gear Steels

AISI No.	Chemical-composition limits, %				
	C	Mn	Ni	Cr	Mo
1015	0.13/0.18	0.30/0.60			
1025	0.22/0.28	0.30/0.60			
1045	0.43/0.50	0.60/0.90			
1060	0.55/0.65	0.60/0.90			
1118	0.14/0.20	1.30/1.60	(0.08/0.13 sulfur)		
1137	0.32/0.39	1.35/1.65	(0.08/0.13 sulfur)		
1320	0.18/0.23	1.60/1.90			
1335	0.33/0.38	1.60/1.90			
2317	0.15/0.20	0.40/0.60	3.25/3.75		
2340	0.38/0.43	0.70/0.90	3.25/3.75		
3140	0.38/0.43	0.70/0.90	1.10/1.40	0.55/0.75	
3250	0.48/0.53	0.40/0.60	1.65/2.00	0.90/1.20	0.20/0.30
4047	0.45/0.50	0.70/0.90	....	....	0.20/0.30
4023	0.20/0.25	0.70/0.90	....	....	0.20/0.30
4130	0.28/0.33	0.40/0.60	....	0.80/1.10	0.15/0.25
4140	0.38/0.43	0.75/1.00	....	0.80/1.10	0.15/0.25
4320	0.17/0.22	0.45/0.65	1.65/2.00	0.40/0.60	0.20/0.30
4340	0.38/0.43	0.60/0.80	1.65/2.00	0.70/0.90	0.20/0.30
4620	0.17/0.22	0.45/0.65	1.65/2.00	....	0.20/0.30
4640	0.38/0.43	0.60/0.80	1.65/2.00	....	0.20/0.30
4820	0.18/0.23	0.50/0.70	3.25/3.75	....	0.20/0.30
5132	0.30/0.35	0.60/0.80	....	0.75/1.00	
5145	0.43/0.48	0.70/0.90	....	0.70/0.90	
E52100	0.95/1.10	0.25/0.45	....	1.30/1.60	
6120	0.17/0.22	0.70/0.90	....	0.70/0.90	(Va 0.10 min.)
6150	0.48/0.53	0.70/0.90	....	0.80/1.10	(Va 0.15 min.)
8620	0.18/0.23	0.70/0.90	0.40/0.70	0.40/0.60	0.15/0.25
8640	0.38/0.43	0.75/1.00	0.40/0.70	0.40/0.60	0.15/0.25
E9310	0.08/0.13	0.45/0.65	3.00/3.50	1.00/1.40	0.08/0.15
9840	0.38/0.43	0.70/0.90	0.85/1.15	0.70/0.90	0.20/0.30

NOTES: The prefix E means a grade of steel made by the basic electric-furnace process. The phosphorus and sulfur are limited to 0.040 maximum for all the steels except AISI 1118 and 1137—which are resulfurized for free machining—and E9310 and E52100—which are held to a maximum of 0.025 phosphorus and sulfur.

Table 10-4 shows hardenability data for different gear steels. The effect of alloy[2] on hardenability is clearly shown in this table. In a water quench AISI 1045 will through-harden only in a rod of ½″ or less. AISI 4340 will through-harden in a rod as large as 2.80″ diameter even with a slower oil quench.

Table 10-5 shows nominal heat-treating temperatures for the different gear steels. The "reheat temperature" column applies only to those steels which are ordinarily used for case carburizing. The "M_s temperature" is significant when a "mar-quench" is used.

Table 10-4. Hardenability Data for Typical Gear Steels
(Data based on minimum composition range except where noted)

AISI No.	Hardness of 90% martensite, Rockwell C	Critical cooling rate, °F per sec at 1300°F	Size of round that will through-harden	Mildly agitated quenching medium
Through-hardening steels				
1045	50.5	400*	0.50†	Water
1060	54	125*	1.20†	Water
1137	48*	....	1.4*	Water
1335	46	195	1.00	Water
2340	49	125	0.60	Oil
3140	49	125	0.60	Oil
3250	55*	....	3.0*	Oil
4047	52	195	1.00	Water
			0.40	Oil
4130	44	305	0.70	Water
4140	49	56	1.00	Oil
4340	49	10	2.80	Oil
5132	45*	....	1.6*	Oil
5145	51	125	0.60	Oil
E52100	60	30	1.30	Oil
6150	53	77	0.80	Oil
8640	50*	60*	2.0*	Oil
9840	49			
Data for core of case-carburizing steels				
	Approx. core hardness, Rockwell C			
1015	22	400†	0.50†	Water
1025	27	400†	0.50†	Water
1118	33	400†	0.50†	Water
1320	35	305	0.70	Water
2317	30	150†	0.50	Oil
4023	...	....	0.70*	Water
4320	35	195	0.10	Oil
4620	35	305	0.20	Oil
			0.70	Water
4820	35	77	0.80	Oil
6120	35	5	3.80	Oil
8620	35	250*	0.80	Water
			0.30	Oil
E9310	30	30*	3.0*	Oil

* Estimated values.
† Obtained from nominal compositions rather than from minimum of specification range.

How to Choose a Gear Steel.* With so many gear steels available, the designer is often quite uncertain as to what steel he should be using. Here are some general guides:

1. Use a steel with no more alloy content than is necessary for the size of part to harden properly.

* See ref. 9 for AGMA standards on steel.

2. If wear is a problem, a higher carbon content may be helpful (AISI 1060 will outwear AISI 1025).
3. If machinability is a problem a lower carbon content may be helpful (AISI 1025 will machine better than AISI 1060 at the same hardness).
4. Consider the price of the raw steel and the price of processing it into gears and select the material that will give the lowest-priced gear that is adequate to do the job at hand.

Table 10-5. Heat-treating Data for Typical Gear Steels

AISI No.	Normalizing temp., °F	Annealing temp., °F	Hardening temp., °F	Carburizing temp., °F	Reheat temp., °F	M_s temp., °F
1015	1700	1600	....	1650–1700	1400–1450	
1025	1650–1750	1600	1575–1650	1500–1650		
1040	1650–1750	1450	1525–1575			
1045	1600–1700	1450	1450–1550			
1060	1550–1650	1400–1500	1450–1550	....	....	555
1118	1700	1450	....	1650–1700	1650–1700	
1320	1600–1650	1500–1700	....	1650–1700	1450–1500	740
1335	1600–1700	1500–1600	1500–1550	....	....	640
2317	1560–1750	1575	....	1650–1700	1450–1500	725
2340	1600–1700	1400–1500	1425–1475	....	....	555
3140	1600–1700	1450–1550	1500–1550	....	....	590
4047	1550–1750	1525–1575	1475–1550			
4130	1600–1700	1450–1550	1550–1650	....	....	685
4140	1600–1700	1450–1550	1525–1625	....	....	595
4320	1600–1800	1575	....	1650–1700	1425–1475	720
4340	1600–1700	1100–1225	1475–1525	....	....	545
4620	1700–1800	1575	....	1650–1700	1475–1525	555
4640	1600–1700	1450–1550	1450–1550	....	....	605
4820	1650–1750	1575	....	1650–1700	1450–1500	685
5145	1600–1700	1450–1550	1475–1525			
E52100	....	1350–1450	1425–1600	....	....	485
6120	1700–1800	1600	....	1700	1475–1550	760
6150	1650–1750	1550–1650	1550–1650	....	....	545
8620	1600–1800	1575	....	1700	1425–1550	745
E9310	1650–1750	1575	....	1650–1700	1425–1550	650
9840	1575–1725	1475–1700	1475–1525	....	....	575

Table 10-6 shows in a general fashion what steels might be considered for different size and hardness of gear parts. It is assumed that the teeth are to be through-hardened to a relatively uniform* hardness. Table 10-6 is limited in showing only a sampling of steels that might be used for the gear sizes listed. Many other compositions not shown have been used with good results in certain applications.

How to Process the Steel for a Gear. Raw steel obtained from the mill may be in an as-forged condition with hard and soft spots. It may be "hot-rolled" or it may be "cold-rolled." Quite often the material must be annealed to bring the hardness down to a uniform low value to permit fast machining. Several other operations may be needed. The following list shows the whole gamut of heat-treat operations that might be needed on a tough job.

* See Art. 10-4 for a discussion of shallow-hardening steels.

Table 10-6. General Recommendations on Choice of Steel for Different Gear Sizes

Pitch of teeth	Wall thickness,* in.	Hardness†	Gear steels
Through-hardening			
10–30	½	200 BHN	1045, 1137, 1335, 4047
		300 BHN	1045, 1060, 3140, 4047
5–15	1	200 BHN	1045, 1060, 3140, 4047
		300 BHN	2340, 3140, 3250, 4140, 4340, 4640
2½–8	2	200 BHN	1060, 2340, 3250, 4340, 5145, E52100
		300 BHN	2340, 3250, 4340, 4640, 8640, 9840
1¼–4	4	200 BHN	2340, 3250, 4140, 4340, 4640, 9840
		300 BHN	3250, 4340
Carburizing			
10–30	½	58 Rockwell C	1015, 1025, 1118, 1320, 4023, E9310
5–15	1	58 Rockwell C	2137, 4620, 6120, 8260, E9310
2½–8	2	58 Rockwell C	4620, E9310

* "Wall thickness" is based on thickest section of rim, web, or shaft that must develop the minimum hardness throughout.

† Hardnesses are minimum; 300 BHN would normally be a range of 300 to 350 BHN.

1. Anneal the forging to a low hardness. Rough-machine away excess stock.
2. Normalize the part. (If the gear is contour forged, omit step 1 and normalize the forging.) Machine the part except for teeth.
3. Heat the part to hardening temperature, quench, and draw to design hardness. Finish journals and cut teeth.
4. Stress-relief temper at a temperature just slightly below that used for the draw in step 3. Refinish journals or centers (if necessary). Shave or lap gear teeth.

Sometimes the stress-relief temper may be used after rough-cutting the teeth and before finish-cutting. Quite often the pitch of the teeth is small enough to not require this operation. The normalize treatment is not too often needed. Its main value lies in cutting down the distortion after a quench-and-draw treatment. Gears that are to be carburized after cutting the teeth are critical from the standpoint of distortion after quenching since the warpage may be enough to cause most of the case to be ground away in certain spots. A normalize treatment after the gear is in its approximate final shape (web and rim roughed out) has often been very helpful on a critical carburizing job.

In many cases it is possible to buy steel from a mill where the bar is already heat-treated and then make finished gears without doing any of the four heat-treating steps listed above. A manufacturer making 2″ pitch diameter gears of 10 pitch at 200 BHN should be able to buy heat-treated stock and do no heat-treatment. Conversely, a manufacturer making 24″ gears of 3 pitch at 350 BHN will probably have to do all four of the heat-treating steps listed.

When gears are made by welding it is common practice to weld a quenched and tempered ring to low-hardness web plates and shaft. After welding the complete weldment is given a stress-relief temper. If the weldment is quite large and complicated, a second stress-relief treatment may be needed after the weldment is machined and the teeth are rough-cut. The stress-relief treatment takes out localized machining and welding stresses and prevents the part from distorting later on as the part ages.

10-4. Localized Hardening of Steel Gears. The surface durability of gear teeth is roughly proportional to the square of the surface hardness. This means that a gear tooth with 600 BHN hardness on the surface may be able to carry as much as 9 times the power of a gear with only a surface hardness of 200 BHN! Generally speaking, the limit of machinability is at about 350 BHN. In addition to this the beam strength of through-hardened gear teeth is a maximum in the 350 to 400 BHN range and slightly decreases at higher hardness. This situation poses the problem of how to take advantage of the high surface durability of gears in hardness ranges above 350 BHN and at the same time have good beam strength and a means of making the gears with accurate teeth. The answer to this problem lies in *localized* hardening of gear teeth.

Table 10-7 shows five commonly used methods for the localized hardening of gear teeth. In each case the surface of the tooth is made very hard and the center of the tooth, the gear rim, and the gear web are at a medium or low hardness. Generally speaking, teeth that are given a localized hardening treatment are brought up to a full hardness in the neighborhood of 600 BHN (or 60 Rockwell C). A gear might be hardened to only 500 BHN, but there is no point in stopping at 500 when you can go higher.

Table 10-7. Methods for Localized Hardening of Gear Teeth

Method	*Description of method*
Carburizing	The teeth are usually finish-cut, carburized, and ground. Sometimes gears less than 12″ in diameter are finished before carburizing and distortion is low enough to permit using the gears without grinding
Nitriding	The teeth are usually finished before nitriding. If properly done, nitrided gears do not distort much. In addition the nitride case is so thin it may be damaged by grinding
Induction hardening	The teeth are usually finished before hardening. If properly done and the pitch is not too coarse, distortion is very low. These gears have a deep hardened layer and can be ground without trouble except in the case of critically loaded aircraft gears
Shallow hardening	The teeth may or may not be ground
Carbonitriding	The teeth are usually finished before carbonitriding

The details of the different methods of localized hardening are given in Arts. 10-5 to 10-8.

Figure 10-7 shows the general nature of the endurance-strength curves for all the different methods of gear-tooth hardening. For through-hardened teeth the endurance strength reaches a peak at around 400 BHN and tends to drop off at higher hardnesses. With very carefully developed and controlled localized hardening, the endurance strength can be brought up to about 150,000 psi in the region of 550 BHN surface hardness. All the localized treatments are quite fussy to control properly. With poor control the strength drops off quickly and is no better than that of a low-hardness gear tooth. Some of the things that tend to spoil the strength of teeth that have been given a localized treatment are

Lack of case depth in root fillet
High residual tensile stress from an improper quenching technique
Lack of hardness in root fillet
Grinding burns in root fillet
Decarburization of root fillet
Notches, tears, or scratches in root fillet

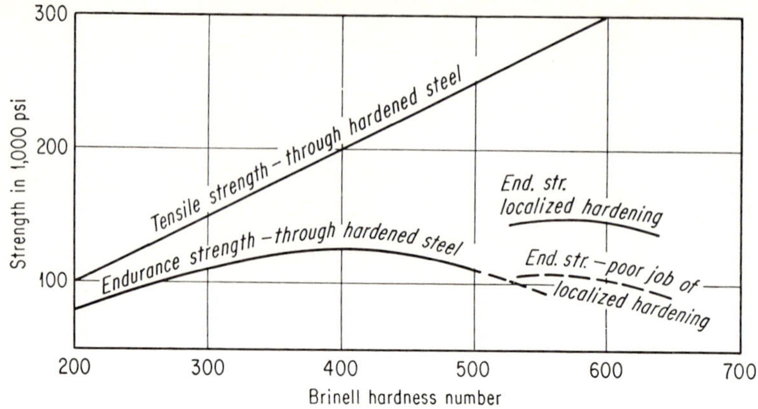

FIG. 10-7. Endurance strength of gears as a function of hardness.

The endurance strength plotted in Fig. 10-7 is for a "unidirectional" stress. For fully reversed stress the values are about 70 per cent of those shown.

Figure 10-7 should not be used directly to design gears. The proper load rating of gears requires that a proper margin of safety be used and that consideration be given to tooth errors and the roughness of the driving and driven apparatus plus many other factors. Chapter 13 covers the subject of load rating and lists appropriate design stresses for the different gear applications.

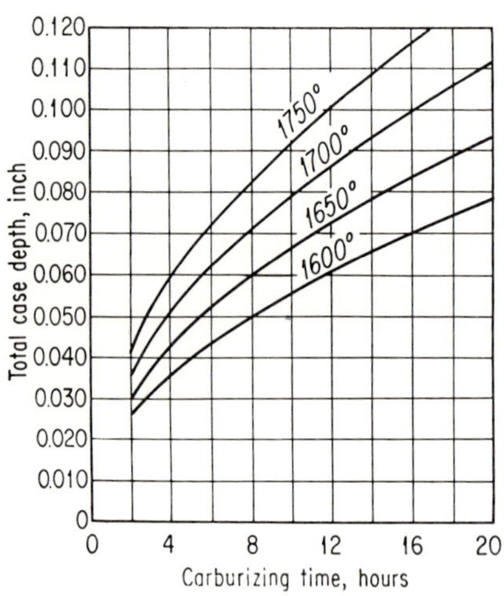

FIG. 10-8. Nominal carburizing time and temperatures for different total case depths gas-carburized. (*Courtesy of International Nickel Co., New York.*)

10-5. Carburizing Gears. Carburizing is one of the most widely used methods for hardening gear teeth. The gear to be carburized is hobbed, shaped, or cut by some other method. A low-carbon steel having a carbon content in the range of 0.10 to 0.25 per cent is used for the gear blank. The cut gear is placed in a carburizing medium and heated above its critical range. Carbon is absorbed into the surface layer of the gear tooth, and after one or more hours at temperature, the carbon has penetrated to give the required case depth. Rim areas, web areas, and hub areas that are not to be carburized are masked off during the carburizing cycle. (In some cases an area may be carburized and then the carbon case removed by machining or grinding after carburizing.)

The carburized gear may be quenched immediately after removal from the carburizing chamber; or it may be slow-cooled so as to not develop its hardness and then, after some further processing, reheated and quenched.

The carburizing medium may be a gas atmosphere or a liquid like a salt containing free carbon. In many simple carburizing setups the gear is packed with charcoal* and heated in a chamber to carburizing temperature.

* Certain chemicals are added to the charcoal.

The concentration of carbon in the case may go as high as 1.40 per cent. For the best hardness and fatigue strength, the carbon content should be in the 0.80 to 0.90 per cent range. The control of carbon content is achieved by regulating the richness of the carburizing medium. Special equipment has been developed by builders of heat-treating furnaces to control carbon potential.

Figure 10-8 shows the normal carburizing time to obtain different case depths. The curves are based on data obtained from gas-carburizing tests.

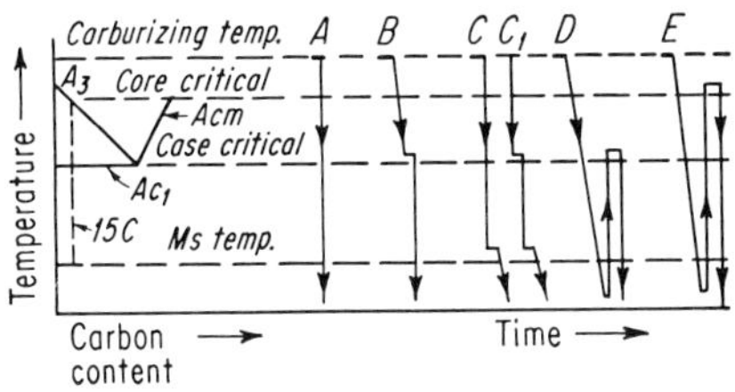

Treatment	Case	Core
A. Adapted to fine-grained steels	Unrefined—carbides Dispersed—austenite Retained—reduced distortion	Unrefined—hardened Case supported
B. Adapted to fine-grained steels	Unrefined—slight carbide network—less distortion than A	Unrefined—less hardening than A, depending on alloy content
C. Adapted to fine-grained steels	Unrefined—carbide Dispersed—max. hardness, less distortion than A	Unrefined—hardened Improved toughness Case supported
C_1	Least distortion	Hardened and toughened
D. Adapted to fine-grained steels	Refined—tendency to carbide-network distortion as in B	Unrefined—soft and machinable Min. case support
E. Adapted to coarse-grained steels	Solution of excess carbide—austenite retained less than A, medium distortion	Refined—max. Strength—tough Case supported

FIG. 10-9. Diagrammatic representation of modern treatments after carburizing and case and core characteristics. (*Courtesy of International Nickel Co., New York.*)

Carburizing rates vary rather appreciably. It is common practice for the heat-treater to put several test coupons in with the gears to be carburized. Coupons are pulled from the furnace at regular intervals and checked immediately for case depth. The caruburizing is stopped as soon as the coupons indicate the case depth is within the required limits.

Figure 10-9 shows several heat-treatment techniques that may be used on carburized gears. Some treatments favor the case and some the core. The choice of treatment depends on grain size and carbon content as well as the end use of the gears.

Table 10-3 lists the compositions of carburizing steels as well as other steels. Table 10-4 segregates the carburizing steels and shows the normal minimum core hardness. Table 10-5 shows normal carburizing temperatures for different steels.

There is fairly general agreement among gear experts that a casehardened gear should have a core hardness of about 35 Rockwell C. If the core is very hard, such as 50 Rockwell C, the teeth tend to be too brittle. If the core is of low hardness like 20 Rockwell C the teeth lack beam strength and the case may tear loose from the core. For most high-capacity jobs, the core hardness can be held to a range of 30 to 40 Rockwell C with good results. A close control on core hardness is difficult because the core is heat-treated purely as a secondary operation to the heat-treatment for the case. Also, the core is very sensitive to small differences in carbon content. (Figure 10-1 showed the sensitive relation between maximum hardness and carbon content.)

An optimum casehardness is 60 Rockwell C. For different gear applications, it is advisable to consider several specifications on casehardness. Table 10-8 shows a recommended practice. This agrees quite well with the practices of a majority of gear manufacturers.

Table 10-8. Recommended Casehardness

Application	*Drawing specification*
General-purpose industrial gearing	55 Rockwell C min.
High-capacity industrial gearing	58 Rockwell C min.
Aircraft gearing	58–63 Rockwell C
Max. capacity aircraft or rocket gearing	60–63 Rockwell C

The maximum hardness specification of 63 Rockwell C is of some question. A carburized case that has been given a proper draw after quenching will not be harder than this value. The specification is of some value in spotting cases where the draw was forgotten. However, it is quite possible to forget the draw and still not exceed 63 Rockwell C. Generally when 63 Rockwell C is exceeded, it indicates only that the hardness-checking device is a point or two out of calibration.

The case depth of a carburized tooth is primarily a function of pitch. The bigger the tooth, the more case is needed to carry the loads that will be imposed on the tooth. For each size of tooth there is an optimum case depth. Too much case makes the tooth brittle with a tendency for the top of the tooth to shatter off. Too thin a case reduces tooth strength and resistance to pitting. Table 10-9 shows the general practice on case depth for different diametral pitches.

Table 10-9. Recommended Case Depths

Normal diametral pitch	*Case-depth specification, in.*
20	0.010–0.018
16	0.012–0.023
10	0.020–0.035
8	0.025–0.040
6	0.030–0.050
4	0.040–0.060
2	0.070–0.100

For extremely critical gears a smaller range of case depth may be used than that shown in Table 10-9. The change is made by raising the minimum case depth. For instance, a critical 10-pitch job might be held to 0.025 to 0.035 case depth.

The case depth specified is the total case depth as noted in an etched specimen. Some specify case depth by considering as "effective case" only the part of the case whose hardness exceeds 50 Rockwell C. The *effective* case is about 75 per cent of the total case.

10-6. Nitriding Gears. Nitriding[4] is a process for casehardening alloy-steel gears. The gear to be nitrided is given a quench-and-draw treatment to establish a medium hardness in the *sorbitic* range. The gear blank is finish-machined to size and the teeth are cut and finished. Then the parts of the gear that are not to be nitrided are masked off with copper plate or some other suitable medium. Then the gear is put in a nitriding furnace and heated to about 1000°F. The nitriding is done by ammonia gas which breaks down into atomic nitrogen and hydrogen at the surface of the steel.

The atomic nitrogen slowly penetrates the steel surface and combines with elements like aluminum, chromium, molybdenum, tungsten, and vanadium to form very hard nitrides. The useful case on a gear is formed by the nitrogen and *alloying elements* present in the gear. A straight-carbon steel cannot be successfully nitrided.

The nitriding is carried out at a temperature that is well below the critical temperature of the steel. This means that the gear blank is not red-hot and no molecular changes are being made in the body of the gear—provided the gear blank was drawn at a temperature at least as hot as the nitriding temperature prior to nitriding. A gear properly nitrided will exhibit little or no distortion. This makes it possible to finish nitrided gears prior to nitriding. Generally the only work done after nitriding is to strip the copper plate and polish journals or bores to exact size.

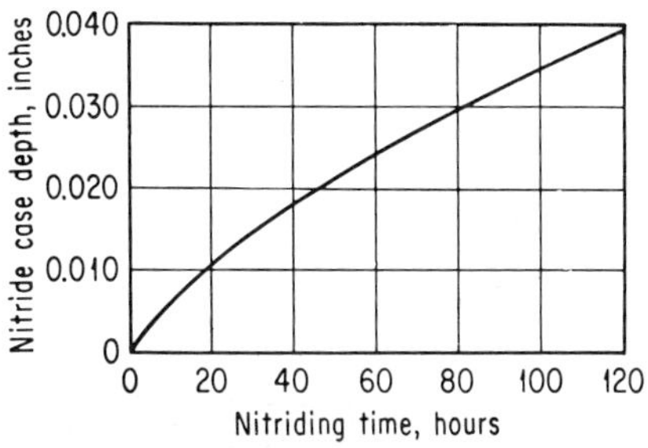

FIG. 10-10. Nominal time required for different nitride case depths.

Nitrided parts have a surface layer of superrich nitrides which are weak and brittle. This surface layer is called a "white layer" because it etches out white in a micrograph. With a special two-cycle nitriding process, often called a "diffusion" cycle, it is possible to hold the white layer to 0.0005″ or less. If just a simple nitriding cycle is used with a rich ammonia gas, the white layer may build up to 0.002 or 0.003″. A thin white layer is generally not harmful to a gear. A thick layer of 0.001 or more may cause trouble. Quite often gears with a thick white layer are lapped or ground to remove the white layer.

If nitrided gears are ground, only a very slight amount of metal should be taken off. The hardness of the nitride case drops off much more rapidly than that of a carburized case. An amount of grinding that would be quite permissible on a carburized case can ruin a nitrided part by removing a substantial amount of the high-hardness material.

Table 10-10 shows the compositions of some typical nitriding steels. Table 10-10 also shows typical nitriding temperatures and hardnesses for these steels.

Nitride cases are formed at a much slower rate than carburized cases. Figure 10-10 shows the nominal time required for different nitride-case depths.

Generally speaking, nitrided gears do not require so much case depth as casehardened gears. Nitrided gears with rather thin cases have performed excellently on many critical aircraft jobs. It appears that the lack of case depth can be compensated for by the extra high hardness of the outer case and a harder core underneath the case. Table 10-11 shows some nominal case-depth specifications for nitrided parts.

10-7. Induction-hardening Gears. Gear teeth may be surface-hardened by high-frequency alternating currents. The process involves wrapping an induction coil around the piece. Generally the part is rotated inside the coil. In a few seconds the teeth are brought up to a red heat (above critical temperature). The gear is withdrawn from the coil and given a controlled spray quench by a special quench ring or agitated bath.

Table 10-10. Nominal Compositions of Nitriding Steels

Steel	Carbon	Manganese	Silicon	Chromium	Aluminum	Molybdenum	Nickel
Nitralloy®* 135	0.35	0.50	0.30	1.20	1.15	0.20	
Nitralloy 135 modified	0.41	0.55	0.30	1.60	1.0	0.35	
Nitralloy N	0.23	0.55	0.30	1.25	1.15	0.25	3.5
Nitralloy N 5-2	0.23	0.35	0.25	0.5	2.0	0.25	5(Va 0.10)
AISI 4340	0.40	0.70	0.30	0.80	....	0.25	1
AISI 4140	0.40	0.90	0.30	0.95	....	0.20	

Nominal Temperatures Used in Nitriding and Hardnesses Obtained

Steel	Temp. before nitriding, °F	Nitriding, °F	Hardness, Rockwell C	
			Case	Core
Nitralloy 135	1250	975	65–70	30–34
Nitralloy 135 modified	1250	975	65–70	32–36
Nitralloy N	1200	975	65–70	40–44
Nitralloy N 5-2	1275	975	65–70	45–50
AISI 4340	1100	975	48–53	27–33
AISI 4140	1100	975	50–55	23–33

* Nitralloy is a trade name of the Nitralloy Corp., New York.

Table 10-11. Recommended Nitride Case Depths

Normal diametral pitch	*Case-depth specification, in.*
20	0.005–0.010
16	0.008–0.013
10	0.012–0.018
8	0.014–0.020
6	0.016–0.022
4	0.020–0.028

Before induction hardening the gear blank is heat-treated to a medium hardness such as 33 to 38 Rockwell C. The teeth are finish-cut and finish-shaved or lapped for final accuracy. If the induction-hardening cycle has been properly developed for the part, almost no distortion occurs. Since only the surface layers of the teeth are heated, the bulk of the gear blank serves as a cold, strong fixture to maintain the accuracy of the portion of the part that is heated to a red heat.

After induction hardening the gear is given a low-temperature draw (in the 300 to 400°F range). The teeth may be brushed lightly to remove the slight amount of scale that develops. Since a slight dimensional change may occur in bores or journals, it may be desirable to finish the bores or journals to size after induction hardening.

If there is a very high intensity of electric power per square inch of piece, it is possible to get a fairly uniform case somewhat analogous to that obtained by carburizing or nitriding. Normally, though, an induction-hardened piece is hardened almost all the way through the tooth at the pitch line. At the bottom of the tooth, there may be an unhardened area with the coarser-pitch teeth.

Figure 10-11 shows a comparison of typical hardness patterns obtained by different surface-hardening treatments. Note views *b* and *c* showing typical induction-hardening patterns.

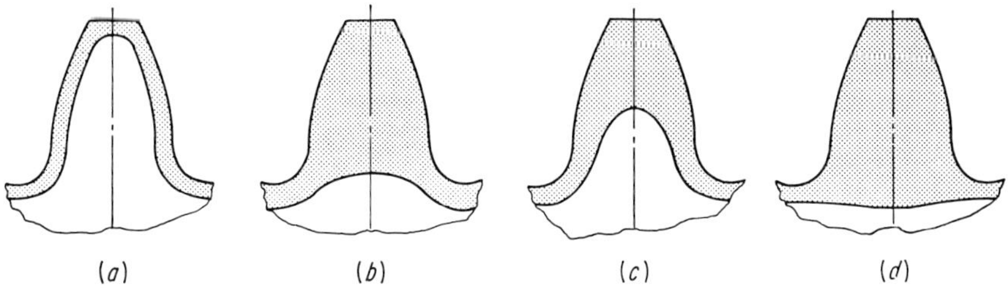

FIG. 10-11. Typical hardness patterns. (*a*) Carburized or nitrided. (*b*) Induction-hardened or shallow-hardened. (*c*) Induction-hardened or shallow-hardened. (*d*) Flame-hardened.

In an induction-hardened tooth that requires high beam strength, it is necessary to get a reasonable depth of hardened stock in the center of the root fillet. Table 10-12 shows how much is needed.

Table 10-12. Recommended Depth* of Hardness in Root Region for Induction-hardened Teeth

Normal diametral pitch	*Recommended hardness depth, in.*
20	0.010–0.040
16	0.015–0.060
10	0.020–0.100
8	0.030–0.125
6	0.045–0.150
4	0.060–0.175

* Depth reading taken in center of root fillet. Depth should be greater at sides of root fillet.

Even though the right amount of hardened material is obtained in the root region, it is difficult to obtain high beam strength with an induction-hardened tooth. As the outer parts of the gear tooth are induction-heated they expand rapidly and at high temperature become plastic. The net result is that the material is upset. Upon cooling the layers contract and they now tend to want to crack because of having insufficient volume to match the rest of the gear blank. However, the transformation of sorbite to martensite involves an *increase* in volume. This can compensate for the plastic upsetting.

With a clever development program it is usually possible to work out an induction-hardening cycle so that the timing on the heating, the delay before quench, and the quenching are such as to obtain a part that is free from dangerous residual tensile stresses and has high fatigue strength. Induction hardening is a good process for low-cost uniform-quality high-production gear jobs where a sound development program can be used before the gear part is put into production. It is a risky process for job-shop work where a few parts are made without benefit of several development samples and fatigue tests on some of these samples.

The frequency of electric power has a relation to the pitch of the teeth. Table 10-13 shows the general relation.

Frequencies in the order of 500,000 or higher cycles per sec are obtained by vacuum-

Table 10-13. Frequencies Generally Recommended for Induction Hardening

Normal diametral pitch	*Frequency, cycles per sec*
20	500,000–1,000,000
10	300,000– 500,000
8	300,000– 500,000
6	10,000– 500,000
4	6,000– 10,000
2	6,000– 10,000

tube oscillators. Spark-gap oscillators work in the 10,000 to 300,000 range. Motor generators are used in the 5,000 to 10,000 range.

A rather large amount of power is required to induction-heat a whole gear. A concept of power requirements can be obtained by multiplying the pitch diameter of the gear by the face width (in inches). This quantity can be divided into the kilowatts of power of the machine used to get a measure of the intensity of the power. Table 10-14 shows a general guide for induction-hardening power requirements. Equipment now on the market will do gears up to about a 30″ pitch diameter and a 5″ face width.

Table 10-14. Power Requirements for Induction-hardening Gears by the Whole-gear-at-a-time Method

Power $\frac{\text{kw}}{\text{pitch dia.} \times \text{face width}} = \text{kw per sq in.}$	*Remarks*
100	Very high power; heating time under 10 sec, may get a "case" pattern
50	High power; heating time may be in 5- to 15-sec range
20	Medium power; heating time may be in 8- to 20-sec range
10	Low power; heating time may be in 10- to 40-sec range
5	Very low power; heating time may be in 20- to 100-sec range

The heating time depends somewhat on pitch and the efficiency of the equipment used as well as the intensity of power. To escape all serious distortion, it is usually necessary to get the part up to temperature in 20 sec or less. Parts heated 50 sec or longer generally require grinding after induction hardening.

The problem of getting enough power to induction-harden gears has led to extensive work on the problem of induction hardening a tooth at a time. This can be done by wrapping a coil around a single tooth or even by using a short moving coil that "scans" a tooth across the face width. The *tooth-at-a-time* method has the promise that any size of gear can be done with only a moderate amount of power. The method has the problem that the mechanics of the hardening make it even more difficult to get high fatigue strength than in the gear-at-a-time method. Also there is some distortion trouble because some teeth must get hardened under the *discontinuity* of already having the neighbor tooth hardened. So far this method has been helpful in improving wear capacity of gears, but the method has not been widely used in the gear trade to produce high-fatigue-strength gears. Much work is being done to develop this process, and there may be significant developments in this area in the future.

Medium-carbon steels are used for induction hardening. Some of the popular steels are AISI 1040, 1050, 4340, and 4350.

10-8. Shallow-hardening Gears. Gears may be flame-hardened by oxyacetylene flames using special burner heads. To achieve uniform hardening the gear is usually rotated in the flame. Like induction hardening, the gear blank is made medium hard and the teeth are cut and finished prior to hardening. With a good flame-hardening setup the distortion is low enough to permit the gear to be used without grinding after hardening.

In a flame-hardened gear, the heat travels inward by conduction. By comparison, the induction-hardening process develops heat by electronic action *inside* the surface of the tooth. The hardness pattern cannot be controlled in flame hardening the same as it is controlled in induction heating. See Fig. 10-11 for a typical pattern.

Medium-carbon steels are used for flame hardening. The depth requirements for flame hardening in the root fillet are similar to induction hardening. To get high fatigue strength it is necessary to do a development job on each part and establish burner design and intensity, heating time, and quenching procedure.

In large sizes, gear teeth may be shallow-hardened by using a steel with limited hardenability and a special *heat-quench-and-draw* cycle. For instance, an AISI steel of 1040 composition will not harden all the way through a 3-pitch tooth. By using a controlled spray quench it is possible to shallow-harden coarse-pitch teeth and get a hardness pattern similar to that obtained in induction hardening (see Fig. 10-11, view *c*).

Large gear teeth which have been properly shallow-hardened have good wear resistance and fatigue strength. This process requires a considerable amount of development work and fatigue testing to get the cycle just right for each gear. It is a good low-cost production method, but it is not handy to use in job-lot production of a few gears. For very critical applications, shallow-hardened gears require grinding. Generally speaking, their distortion on hardening is less than that of a carburized gear but more than that of a good nitrided or induction-hardened gear.

Still another method of shallow-hardening gears is the one called "cyaniding." One way of using this process is to harden gears whose teeth have already been finished by putting them in a bath of molten sodium cyanide. A shallow case is formed which contains both nitrides and carbides.[3]

A variety of treatments may be used to produce a combined carburized and nitrided type of case. More properly these treatments are called "carbonitriding" treatments.

The carbonitriding treatments have the advantages that distortion is low and cycle time is much shorter than in nitriding. The case produced is not so hard as a nitride case. The depth is shallow, being in the order of 0.010 to 0.025″. If this case is backed up with a very hard core and the pitch is not too coarse (6 pitch or finer), the teeth tend to have good wear resistance and good fatigue strength.

Gears to be cyanided are made with more carbon in the core than case-carburizing gears. They do not require aluminum or chromium like a nitriding gear. AISI 4640 is typical of one of the popular steels for cyanided parts. AISI 5132 is popular in automotive work.

The case on cyanided gears runs in the order of 57 to 63 Rockwell C. Generally, it is possible to get good cyanided gears without extensive development work. This process has been used widely in both job-lot and production-lot applications.

10-9. Gray-cast-iron Gears. Gray cast iron has long been used as a gear material. Cast iron is low in cost and it can be cast easily into any desired shape for the rim, web, and hub of a gear. Cast iron machines easily.

Cast-iron gears generally show a good resistance to wear and are often less sensitive to lubrication inadequacies than steel gears. Cast iron has good dampening qualities. It is generally felt that cast-iron gears run more quietly than steel gears of the same size and accuracy.

Cast-iron gear teeth have about three-quarters of the surface load-carrying capacity of steel gears of the same pitch diameter and face width. Their beam strength capacity is about one-third that of steel gears of the same normal diametral pitch. By making cast-iron gears somewhat coarser in pitch for the same pitch diameter, it is possible to design cast-iron gearsets about the same size as steel gearsets for a given application.

Gray cast iron has low impact strength and should not be used where severe shock loads occur.

Gray cast iron has free carbon in the form of flakes. The matrix of the material is a combined carbon and iron that is very similar to steel. Gray cast iron may be heat-treated like steel. Even the cooling of the casting in the mold can be considered as a heat-treatment with a mild quench. In many gear applications this is enough of a heat-treatment and the gears are used with only the "as-cast" heat-treatment.

AGMA standard 242.02 lists six classes of gray irons according to tensile strength. The amount of alloy in the composition is not specified, but it is assumed that enough alloy will be used to meet the requirements specified for standard test bars and also to meet the minimum hardness specified for the teeth. Table 10-15 shows the standard AGMA classes for gray cast iron. Those planning to use gray cast iron should study AGMA 242.02 for full detail requirements of the standard.

Table 10-15. AGMA Standard[10] Classes of Gray Cast Iron

Class No.	Min. tensile strength, psi	Min. Brinell hardness on toothed portion
20	20,000	
30	30,000	175
35	35,000	185
40	40,000	200
50	50,000	215
60	60,000	220

Meehanite. This is a special kind of gray iron made in licensed[6] foundries. The proprietary process used tends to control the graphite flakes so that they are smaller and more uniformly dispersed. The process tends to produce a matrix in the pearlitic or sorbitic condition. Several grades of Meehanite are available. These cover a range of tensile strengths similar to that of regular gray cast iron. Although low in impact strength, Meehanite can be treated to have rather good impact strength for a cast iron. A slight amount of ductility can also be obtained with Meehanite.

10-10. Ductile Iron. Ductile iron is a cast iron having the free carbon in the form of spheroidal nodules instead of flakes. Ductile iron is often called "nodular" iron even though the official designation is *ductile* iron.

The fact that the carbon is in little round balls instead of flakes explains why this cast iron behaves a lot like steel. The flakes in gray cast iron act as stress risers and make the impact strength and fatigue strength low for gray cast iron. The nodules in ductile iron do not have a serious stress-riser effect. Ductile iron has good impact strength and enough ductility to have elongations in the range of 2 to 15 per cent, depending on the class. The fatigue strength of ductile iron tends to be almost as good as that of steel of equal hardness.

Ductile iron may be used as-cast or it may be quenched and drawn to a wide range of hardness levels. Gears may be cut from ductile-iron blanks with the minimum hardness specified all the way from 165 to 350 BHN. Ductile-iron gears may be surface-hardened by flame- or induction-hardening methods up to a surface hardness of 450 BHN minimum.

Ductile iron is a relatively new material. It was first developed[5] in 1948. There has been much use of the material[11] in gear work, but it is still too early to have fully evaluated and standardized the material for all gear applications. AGMA standard 244.01 gives some tentative data on nodular-iron gear materials.

Ductile iron for gears is often specified by specifications developed by the American Society of Testing Materials. Table 10-16 shows some of the more popular specifications for gear use.

Table 10-16. Ductile-iron Specifications

Specifying body	Spec. No.	Class	Min. tensile strength, psi	Yield strength, psi	Elongation, %
ASTM	A-339-55	80-60-03	80,000	60,000	3
		60-45-10	60,000	45,000	10
ASTM	A-396-55T	120-90-02	120,000	90,000	2
		100-70-03	100,000	70,000	3

A ductile-iron gear made with a blank material specified at 120,000 psi minimum tensile strength should have a minimum hardness on the gear teeth of 255 BHN. At a minimum tensile strength of 60,000 psi, the minimum tooth hardness should be 155 BHN.

NONFERROUS GEAR MATERIALS

By Clarence Underwood, *Gear Consultant, Raytheon Co., Wayland, Mass.*

A wide variety of bronzes, aluminum alloys, zinc alloys, and nonmetallic plastics and laminates are used to make gears. In many cases the nonferrous gear is driven by a steel pinion. When loads are light and the parts are small, it is quite often possible to make both members of the pair from nonferrous material.

The nonferrous materials are used for a variety of reasons. Certain bronzes will stand high sliding velocity with a steel worm better than any other material. The nonferrous materials tend to be more corrosion-resistant than the ferrous materials. Some of the nonferrous materials have less mass per unit of volume than ferrous material. This is important in certain instrument gear work where inertia of rotating parts must be kept extremely low. Some nonferrous materials lend themselves to very low cost mass-production techniques. For instance, stamped bronze gears and injection-molded plastic gears are about as low in cost as is possible for a toothed gear.

Table 10-17 shows some general characteristics of the nonferrous gear materials. Detail data are given in the articles that follow.

10-11. Gear Bronzes. A family of four bronzes accounts for much of the nonferrous gear materials chiefly because of their "wear-resistance" characteristics. These characteristics result from the alloying of small amounts of tin, zinc, lead, and traces of other elements and the manner in which they mix or position themselves with the base element copper. The purpose of this section is to emphasize the important advantages contributed by the different alloys and to identify the resulting materials by recognized specifications.

The four gear bronzes are:

Phosphor bronze
Manganese bronze
Aluminum bronze
Silicon bronze

Table 10-17. General Characteristics* of Non-ferrous Gear Materials

Material	Modulus of elasticity, psi × 10^6	Density, lb per cu in.	Ultimate strength, psi	Yield strength, psi
Phosphor bronze	15	0.32	40,000	18,000
Manganese bronze	14–18	0.31	65,000	30,000
Aluminum bronze, heat-treated	19.4	0.27	100,000	50,000
Silicon bronze	15	0.31	45,000	20,000
Aluminum alloy	10.6	0.10	68,000	48,000
Zinc alloy	14	0.24	40,000	26,000
Magnesium alloy	6.5	0.065	44,000	32,000
Brass alloy	15	0.306	76,000	45,000
Nylon (73°F)	0.4	0.036	11,200	8,500
Phenolic laminates:				
Paper	1.5 (length) 1.0 (crosswise)	0.042	13,500	10,000
Fabric	1.0 (length) 0.8 (crosswise)	0.042	13,500	10,000

* The values shown in this table are typical. Individual compositions may vary appreciably.

The metallurgy of bronze[7] is based on the ability of copper to form solid solutions with the various alloy elements. The extent of solubility varies widely, usually forming more than one type of solid solution with copper. This accounts for the effect on the physical properties produced by varying the amount of alloy element.

When a bronze gear is worn to a good contact, the hard constituent of the microstructure has become aligned to the rubbing surface. The softer constituent, it is then believed, allows this aligning process.[1] The major alloying elements are described with brief explanations of their contributions:

Zinc (Zn) is added to the basic copper-tin alloys to increase strength properties but to some extent at the sacrifice of bearing properties.[2]

Tin (Sn) is the principal alloying agent in phosphor bronze. It is a major contributor used to affect strength, but if improper melting is made the tin may oxidize, resulting in very hard crystals of tin which could be very abrasive.

Lead (Pb) is used to improve machinability, and as with each contributing element the compromise must be recognized. Although the limited amount of lead has little effect on tensile strength, it does cause a lowering of shear strength and ductility. Lead does not combine with the base alloy, but it does act like a "lubricant."

Iron (Fe) is another major contribution for strength even in the materials called "nonferrous."

Phosphor Bronze. These bronzes are tough and hard and capable of high fatigue resistance. This group covers a wide range of copper-tin and copper-tin-lead bronzes all of which are deoxidized with phosphorus. A long-time gear material of this group is 89-11 *gear bronze* used particularly for worm wheels that operated with hardened-steel worms for moderate to high sliding velocities and moderate tooth loads. Adding a trace of nickel provides fine grain structure. Another of this group is 88-10-2 *G bronze,* or *gun metal,* having good physical qualities and good resistance to wear. Both these phosphor bronzes have good resistance to the corrosive action of sea water, have good machinability, and make excellent castings.

Manganese Bronze. This is tough, having about the tensile strength and ductility of ordinary cast steel, one of the toughest without benefit of heat-treatment. Its high physical characteristics are due to the hardening and deoxidizing agents iron, manganese, aluminum, and tin. It makes excellent close-grain castings that result in a smooth surface quality.

Aluminum Bronze. Another tough strong casting material like manganese bronze, it is lighter in weight and capable of higher physical characteristics because of added elements of iron, manganese, and nickel, plus the fact that it can be heat-treated. The attainment of highest strength will be at the sacrifice of the most important property of the gear tooth, ductility.

It would be best to produce gear blanks as centrifugally cast parts.

Silicon Bronze. This casting material produces smooth golden-yellow gears capable of accurate thin sections but requiring special foundry technique. For gears of fine pitch, it is advisable to cast blanks oversize to assure the removal of surface defects. Centrifugal castings will help in this respect also. This material has importance in many electrical applications because of its nonmagnetic quality.

Trade Standards in Bronzes. Many organizations have standardized bronze compositions. The gear designer engaged in all-round gear work will find that he needs to be familiar with and use specifications set up by several technical organizations and government agencies.

AGMA 243.01 shows classes[12] of bronze casting for use in worm-gear speed reducers. These are all of the phosphor-bronze type. The kinds listed are as follows:

Class 1. Nickel-tin bronze
Class 1c. Chill cast-nickel bronze
Class 2. Tin bronze
Class 2c. Chill cast-tin bronze
Class 3. Leaded bronze

The AGMA bronzes all have about 10 per cent tin. They are similar to the "gun metal" bronze.

Table 10-18 shows the compositions of several kinds of bronzes. Table 10-19 shows a cross referencing of different bronze specifications and general data on the properties and uses of the various bronzes.

Table 10-18. Gear Bronzes

Common name and trade term	Principal elements in composition, %				
	Cu	Sn	Zn	Fe	Others for total of 100%
A. Gun metal, cast G bronze 88-10-2	88	10	2	0.15	Pb 0.3, Ni 1.0
B. Phosphor bronze, wrought	88	10	0.75	....	Pb 1–2.5, Ni 1.0
C. G bronze	88	8	4	0.15	Pb 0.3, Ni 1.0
D. Phosphor bronze, cast 89-11 gear bronze	89	11			
E. Nickel-tin bronze, cast	87	10	....	....	Ni 2
F. Manganese bronze, cast	60	1	Remain.	2	Mn 1.5, Al 1.0, Pb 0.4
G. Manganese bronze, cast	68	0.2	Remain.	4.5	Mn 5, Al 7.5
H. Silicon bronze, cast	88	2	5	2.5	Mn 1.5, Si 4
J. Aluminum bronze, cast	86	....	....	1	Al 10
K. Aluminum bronze, cast	83	....	....	4	Al 11, Ni 2.5
L. Aluminum bronze, cast	78	....	....	4	Al 11, Ni 5, Mn 3.5

NOTES: This table in conjunction with Table 10-19 lists gear bronzes as diminishing in order of preference for machinability and resistance to wear, but increasing in order of toughness. It is important to note that the physical properties can be met only by careful proportioning of elements and the practice of good foundry technique.

Table 10-19. Gear Bronzes, Physical Properties

	Ultimate tensile strength, 1,000 psi	Yield, 1,000 psi	Specification Nos.				
			MIL	FED	ASTM	SAE	AGMA
A	40	18	....	QQ-L-225 comp. 16	B143	62	234.01 class 2
B	35	..	....	....	....	63	
C	40	..	B-16540 grade A	QQ-L-225 comp. 5			
For gears of low to moderate service, free machining, resistant to corrosive action of sea water							
D	35	18	....	....	....	65	
Predominantly for moderate strength and toughness of worm wheels used with hard-steel worms and moderate to high pitch-line velocity							
E	50	..	....	....	....	....	243.51 class 1
F	65	25	....	QQ-B-726 class A	B147-8A	43	243.71 class 2
G	90	45	....	QQ-B-726 class B	B147-8B	430A	243.71 class 3
Sound castings, good machinability, moderate hardness, good ductility							
H	45	20	....	QQ-C-593	B198		
Good strength, toughness, corrosion resistance, nonmagnetic							
J	65	25	B-16033 class 2	QQ-B-671 class 2	B148-9B	68B	243.61 class 2
J, heat-treated	80	40	B-16033 class 2	QQ-B-671 class 2	B148-9B-HT		243.61 class 3
K	75	30	B-16033 class 3	QQ-B-671 class 3			
K, heat-treated	90	45	B-16033 class 3	QQ-B-671 class 3			
L	90	40	B-16033 class 4	QQ-B-671 class 4			
L, heat-treated	110	60	B-16033 class 4	QQ-B-671 class 4			
High strength, good fatigue resistance for gears under repeated stresses, good strength at higher than usual temperatures							

10-12. Lightweight Metal Alloys for Gears. Lightweight gears contribute not only to weight reduction in gearboxes but to the low-inertia effect of rotating parts. The smoothest-running gear trains consist of well-balanced, low-inertia gears.

Alloyed aluminum is the most common light metal for these gears. Three prominent grades are 6061-T6, 2024-T4, and 7075. Occasionally blanks are made from 356-T6 castings. Centrifugal casting reduces the problem of porosity.

Operating tests have shown that gears made of these materials having light loads and suitable lubrication can be used in the as-machined condition giving a few million cycles of operating life.

Greater wear life can be obtained by having the gears anodized. When this is done with the proper control, the dimensional build-up may be as small as 0.0001 per surface. The result is a hard aluminum-oxided surface resistant to salt-atmosphere corrosion and wear. Lubrication is important as with all applications involving gears and particularly here to help prevent the oxide coating from breaking into small abrasive particles (see Fig. 10-12).

Even greater resistance to wear is possible when the 6061-T6 material is given the anodic hard-coat treatment. Tests have shown the resulting surface to resist wear better than casehardened steels (WADD).[16] The coating is brittle but completely bonded to the base material. It will not withstand extended or heavy shock loads, but where medium-loaded gears must provide long life, the low-inertia aluminum (6061-T6) gears with Martin hard coat, even with scant lubrication, will perform better than hardened-steel gears for wear resistance.

Minimum backlash conditions are not advised because of difficulties in controlling close tolerances on the surface build-up. The build-up is uniform, and unlike plating it does not form into a ball at the tips of the gear teeth. A magnified examination of a section will show cracks at the fillets of teeth. This is not detrimental. The design for tooth strength must be with respect to the base material.

Magnesium alloys, lighter in weight than the aluminum alloys, will require testing and usage additional to what has been accomplished to date before definite "pro or con" evaluation can be published.

Thus far it appears, on the basis of tests with rolls under combined sliding and rolling action, that magnesium with its hard coating will withstand applied loads amounting to 50 per cent of the loads permissible with Martin hard-coat aluminum.

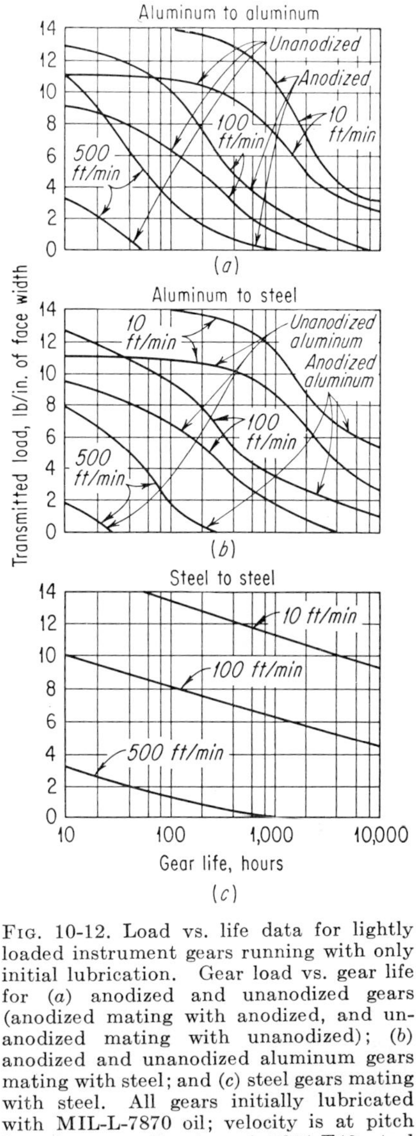

FIG. 10-12. Load vs. life data for lightly loaded instrument gears running with only initial lubrication. Gear load vs. gear life for (*a*) anodized and unanodized gears (anodized mating with anodized, and unanodized mating with unanodized); (*b*) anodized and unanodized aluminum gears mating with steel; and (*c*) steel gears mating with steel. All gears initially lubricated with MIL-L-7870 oil; velocity is at pitch line of gears. Aluminum is 2024-T42; steel is stainless steel QQ-S-763a class 6. (*From Product Engineering*, p. 165, *September*, 1956.)

Die-cast Materials.[8] Die-cast gears provide great savings when large quantities are involved because they suddenly become the finished gear integral with keys, shoulders, and ratchets and at times with the only machining needed the removal of flash on one edge. It is well not to add a finishing operation to the teeth because the outer surface is harder than the subsurface. Improvements in the alloying of materials and processes have resulted in smooth surfaces and accuracies well suited for commercial-class gears when produced by die casting.

Table 10-20. Coefficient of Thermal Linear Expansion, Inches per Degree Fahrenheit* ($\times 10^{-6}$)

Aluminum alloy, °F	2024-T4(1)	6061-T6(2)	7075-T6
−58 to 68	11.9	12.1	12.1
68 to 212	12.6	12.0	12.9
68 to 392	13.2	13.5	13.5
68 to 572	13.7	14.0	14.4

* Data from "The Aluminum Data Book," Reynolds Metals Co., Louisville, Ky., 1950.
NOTE:
1. T4 is solution heat-treat.
2. T6 is solution heat-treat followed by artificial aging resulting in strength increase.

When it has been decided to specify die-cast gears, unless the designer has had considerable experience in this area, the optimum in design consistent with the application can be obtained only after discussions with die-casting manufacturers. The conditions of influence are that

Quantities warrant the making of the die.

Nonferrous metals meet requirements.

Cast teeth have requisite strength and can be used as cast.

If the gear teeth forms involve undercuts, parts of the teeth may have to be finished machine-cut. Helical gears can be cast and must be formed by part of the die that can be turned to clear at the time of ejection, or the gear must be turned a few degrees to make ejection possible.

"When parts are die cast in brass or bronze, cost is much higher than for zinc or aluminum although strength, hardness and wear resistance are higher than for alloys of lower melting point."[8] Zinc alloys at 82 to 100 BHN, not offering much resistance to wear, have excellent casting prerequisities.

Zinc Die Castings. The majority of die castings are made from zinc alloys because of the advantages of mechanical properties and casting characteristics: higher production speeds, greater variation in section thickness, and closer dimensional tolerances than are obtainable with any other commonly used die-casting alloys. Because of lower pressures and temperatures zinc die castings afford less die maintenance. Only brass die castings exceed the impact strength of zinc die castings. Two principal alloys have the following typical values:

Alloy	Tensile strength,† psi	Elongation in 2″, %	Charpy impact on square specimens, ft-lb	BHN on square specimens, 500-kg load, 10-mm ball, 30 sec
AG-40A (XXIII)	41,000	10	43	82
AC-41A* (XXV)	47,000	7	48	91

* Die casting of alloy AC-41A shall not be used in applications where they will be subjected to prolonged temperatures above 200°F.

† Tension tests on round specimens.

Table 10-21. Alloy Cross-reference Charts, ADCI-M1-55T*

(These charts provide a means of readily identifying the commercial alloy designation of a die casting from the alloy designation appearing in government or other standard specifications)

1. Aluminum-base-alloy Die Castings

Commercial designation			13†	380†	360†	218	39-11	43
Nominal composition			Si 12%	Cu 3.5% Si 9.0%	Mg 0.5% Si 9.5%	Mg 8%	Cu 3.8% Si 11%	Si 5%
Symbol	Specifying organization	General Spec. No.	Designations used in specifications					
ASTM	American Society for Testing Materials	B-85	S12A S12B	SC84A SC84B	SG100A SG100B	G8A	SC114A	S5C
SAE	Society of Automotive Engineers		305	306 308	309		303	304
FED	Federal Specifications	QQA-591a-1	Comp. No. 1 Comp. No. 2	Comp. No. 10 Comp. No. 11	Comp. No. 12 Comp. No. 12A	Comp. No. 7		Comp. No. 3
MIL	Military Specifications	MIL-A-15153a	Class 1	Class 10	Class 9	Class 7		Class 3
AMS	Aeronautical Materiel Spec.		4292	4291A-2	4290E			
AF	Air Force	MIL-A-8294	Comp. No. 1		Comp. No. 2			

* Reprinted by permission of the American Die Casting Institute, Inc., 366 Madison Ave., New York 17, N.Y.

† These alloys available in two grades with different iron-analysis levels. See ADCI M2 and M3. The lower iron grades are commercially designated 13x, A380, and A360.

Table 10-21. Alloy Cross-reference Charts, ADCI-M1-55T (*Continued*)

2. Magnesium-base-alloy Die Castings

Commercial designation			AZ91A	AZ91B
Nominal composition			Al 9% Zn 0.7%	Al 9% Zn 0.7%
Symbol	Specifying organization	General Spec. No.	Designation used in specifications	
ASTM	American Society for Testing Materials	B94	AZ91A	AZ91B
SAE	Society of Automotive Engineers		501	501A
FED	Federal Specifications	QQ-M-38	Only	
AMS	Aeronautical Materiel Spec.	4490D	Only	

3. Copper-base-alloy Die Castings

Commercial designation	Z30A	ZS331A	ZS144A
Nominal composition	Cu 60 %	Cu 65 % Si 1 %	Cu 81 % Si 4 %
General Spec. No.	Designation used in specifications		
B-176	Z30A	ZS331A	ZS144A

4. Zinc-base-alloy Die Castings

Commercial designation			
Nominal composition			Al 4%
Symbol	Specifying organization	General Spec. No.	Designation used in specifications
ASTM	American Society for Testing Materials	B86	AC41B
SAE	Society of Automotive Engineers		903 925
FED	Federal Specifications	QQZ363	A B
AMS	Aeronautical Materiel Spec.	4803	

Table 10-22. Zinc-base-alloy Die Castings, ADCI-M9-57T*

1. Alloy Section

The zinc-base alloy shown below is the composition generally used. The alloy is available with more rigid copper limits for special applications where the nature of the requirement justifies the higher metal and processing costs involved

2. Chemical Composition

The chemical composition of zinc-base-alloy die castings, as shown in the table, is assured under the continuing analyses provided in the Certified Zinc Alloy Plan for Die Castings; see ADCI-M11

Chemical Composition: Zinc-alloy Die Castings

Designation	*Standard alloy*
Composition, %:[a]	
Copper[b]	0.0 –1.25
Aluminum	3.5 –4.3
Magnesium	0.03–0.08
Iron	0.10
Lead	0.007
Cadmium	0.005
Tin	0.005
Zinc	Remainder[c]

3. Properties and Constants

Typical properties and constants are shown in the table below. The values indicated are for separately die-cast test bars and do not represent values for specimens cut from die castings

Typical Properties: Zinc-alloy Die Castings

Properties and constants:[d]	[e]
Tensile strength, psi	41,000–47,000
Elongation, % in 2″	10–7
Impact strength, Charpy, ft-lb	43–48
Shear strength, psi	31,000–38,000
Specific gravity	6.6
Weight per cu in., lb	0.24
Melting point (liquidus), °F	727.9
Thermal expansion, in. per in. per °F[f]	15.2
Electrical conductivity, % of copper standard	26

* Reprinted by permission of the American Die Casting Institute, Inc., 366 Madison Ave., New York 17, N.Y.

[a] Composition in per cent, maximum unless shown as a range.

[b] For special applications the alloy may be obtained with less than 0.25 per cent copper (ASTM designation AG40A) or with copper content of 0.75 to 1.25 per cent (ASTM designation AC41A). For the majority of commercial applications, a copper content in the range of 0.25 to 0.75 per cent will not adversely affect the serviceability of die castings and should not serve as a basis for rejection.

[c] Zinc-alloy die castings may contain nickel, chromium, silicon, and manganese in amounts up to their solubility at freezing temperature of 0.02, 0.02, 0.035, and 0.5, respectively. No harmful effects have ever been noted from the presence of these elements in these concentrations, and therefore analyses are not required for these elements.

[d] Data furnished by New Jersey Zinc Co., ASTM B86, and ADCI.

[e] Values are typical, and where double figures are shown they represent the properties for copper content at minimum and maximum of composition. See ADCI M10 for characteristics of zinc-base die castings and M11 for certified zinc die castings.

[f] To be multiplied by 10^{-6}.

Table 10-23. Characteristics of Aluminum Alloys, ADCI-M4-55T*

Selection of an aluminum-base-alloy die casting requires evaluation not only of physical properties, constants, and chemical composition of the die casting but also of the alloy itself and its effect on production of the die casting and in finishing

The table includes certain casting and other outstanding characteristics which are usually considered in selecting a die-casting alloy for a specific application. The characteristics are rated from 1 to 5, 1 being the best and 5 being the least desirable alloy. In applying these ratings, it should be noted that all the alloys have sufficiently good characteristics to be accepted by users and producers of die castings. Hence a ratings of 5 indicates a commercial alloy, although in certain cases its application may be limited or its manufacture may be restricted to relatively simple die castings

Alloy	Die-casting characteristics					Other characteristics[k]						
	Approx. melting range, °F	Resistance to hot cracking[a]	Pressure-tightness	Die-filling capacity[b]	Anti-soldering to the die[c]	Resistance to corrosion[d]	Machining[e]	Polishing[f]	Electroplating[g]	Anodizing[h] (appearance)	Chemical oxide coating[i] (protection)	Strength at elevated temp.[j]
Standard:												
13, S12B 13X, S12A	1065–1080	1	1	1	1 3	3	4	5	3	5	3	3
380, SC84B A380, SC84A	1000–1100	2	2	2	1 2	4	3	3	1	3	5	2
Special:												
360, SG100B A360, SG100A	1035–1105	1	1	1	2 3	2	3	3	1	3	3	2 1
39-11, SC114A	960–1080	2	2	1	3	4	3	3	2	4	4	2
218, G8A	995–1150	5	5	4	5	1	1	1	5	1	1	4
43, S5C	1065–1170	2	3	3	4	3	5	4	2	2	3	5

NOTE: These data should be used in connection with ADCI-M2 and M3 in determining selection of an aluminum-base-alloy die casting.

* Reprinted by permission of the American Die Casting Institute, Inc., 366 Madison Ave., New York 17, N.Y.

[a] Ability of alloy to withstand stresses from contraction while cooling through hot-short or brittle temperature range.

[b] Ability of molten alloy to flow readily in die and fill thin sections.

[c] Ability of molten alloy to flow without sticking to the die surfaces. Ratings are given for both iron compositions for 13, 13X; 380, A380; and 360, A360.

[d] Based on resistance of alloy in standard-type salt-spray test.

[e] Composite rating based on ease of cutting, chip characteristics, quality of finish, and tool life.

[f] Composite rating based on ease and speed of polishing and quality of finish provided by typical polishing procedure.

[g] Ability of the die casting to take and hold an electroplate applied by present standard methods.

[h] Rated on lightness of color, brightness, and uniformity of clear anodized coating applied in sulfuric acid electrolyte.

[i] Rated on combined resistance of coating and base alloy to corrosion.

[j] Rating based on tensile and yield strengths at temperatures up to 500°F, after prolonged heating at testing temperature.

[k] Die castings are not usually solution heat-treated. Low-temperature aging treatments may be used for stress relief or degrowthing. Die castings are not generally gas- or arc-welded or brazed.

Table 10-24. Aluminum-base-alloy Die Castings, ADCI-M2-55T*

1. Alloy Selection—Standard Alloys

The aluminum alloys most generally used are shown below. Die-casting, and other processing, characteristics data involved in alloy selection are covered in ADCI M4. Special aluminum alloys are shown in ADCI M3. Special alloys should be selected only when the nature of the requirement justifies the higher metal and processing costs involved

2. Composition and Properties

The chemical compositions of standard aluminum-base-alloy die castings and related typical values for physical properties and constants are shown in the table. The typical values indicated are for separately die-cast test bars and do not represent values for specimens cut from die castings

Chemical Composition and Properties: Aluminum-alloy Die Castings

Designation	Standard alloys[a]	
Commercial ASTM	13 S12B	380 SC84B
Composition, %:[b]		
Copper	0.6	3.0–4.0
Iron	2.0	2.0
Silicon	11.0–13.0	7.5–9.5
Manganese	0.35	0.50
Magnesium	0.10	0.10
Zinc	0.50	1.0
Nickel	0.50	0.50
Tin	0.15	0.35
Total others	0.25	0.50
Aluminum	Remainder	Remainder
Properties and constants:[c]		
Tensile strength, psi	36,000	43,000
Yield strength (0.2% offset), psi	21,000	25,000
Elongation, % in 2″	2.0	2.0
Shear strength, psi	25,000	28,000
Fatigue strength, psi[d]	19,000	20,000
Specific gravity	2.65	2.71
Weight per cu in., lb	0.096	0.098
Melting point (liquid), °F	1080	1100
Thermal conductivity, cgs	0.29	0.23
Thermal expansion, in. per in. per °F[e]	11.9	12.1
Electrical conductivity, % of copper standard	31	23

* Reprinted by permission of the American Die Casting Institute, Inc., 366 Madison Ave., New York 17, N.Y.

[a] These standard alloys are also available with 1.3 per cent maximum iron content (designated 13X and A380) having 3 per cent elongation.

[b] Composition in per cent, maximum unless shown as range. Same as ASTM B85.

[c] Data furnished by Aluminum Company of America and ASTM B85. Revisions are in progress and will be made from time to time.

[d] R. R. Moore rotating-beam test data at 500,000,000 cycles.

[e] To be multiplied by 10^{-6}. Temperature range 68 to 392°F.

Brass Die Castings (Copper-base). Although these possess the highest mechanical and corrosion-resistant properties, die life is comparatively short, resulting in castings higher in costs than other die-cast metals.

The following give brief résumés, the first two being general-purpose copper-base alloys:

Alloy Z30A. Yellow brass with high tin and lead contents for good machinability

Alloy ZS331A. Higher strength, easier to die-cast but more difficult to machine

Alloy ZS144A. Used generally for high-strength and wear-resistance applications and is the most difficult to machine

Table 10-25. Aluminum-base-alloy Die Castings, ADCI-M3-55T*

1. Alloy Selection—Special Alloys

Standard aluminum-base-alloy die castings (ADCI M2) provide commercially satisfactory combinations of properties and characteristics (ADCI M4) at lowest cost. Special aluminum-base-alloy die castings are available at higher costs of applications involving certain specific requirements and should be specified *only* when necessary for the end use

For example:

Alloy 360, SG100B, provides somewhat better ductility and corrosion resistance than either of the standard alloys

Alloy, 39-11, SC114A, has excellent fluidity or die-filling capacity which is of advantage in producing large thin-walled intricate die castings

Alloy 218, G8A, provides the best combination of strength, ductility, corrosion resistance, and finishing qualities but is more difficult to die-cast

Alloy 43, S5C, is sufficiently ductile for special applications requiring forming or upsetting during assembly operations but is difficult to die-cast

NOTE: Reference to ADCI M2 and M4 is advised before specifying special aluminum-base-alloy die castings.

2. Composition and Properties

The chemical compositions of special aluminum-base-alloy die castings and related typical values for physical properties and constants are shown in the table. The typical values indicated are for separately die-cast test bars and do not represent values for specimens cut from die castings

Chemical Composition and Properties: Aluminum-alloy Die Castings

Designation	Special alloys			
Commercial ASTM	360[a] SG100B[a]	39-11 SC114A	218 G8A	43 S5C
Composition, %:[b]				
Copper	0.6	3.0–4.5	0.25	0.6
Iron	2.0	1.3	1.8	2.0
Silicon	9.0–10.0	10.5–12.0	0.35	4.5–6.0
Manganese	0.35	0.50	0.35	0.35
Magnesium	0.40–0.6	0.10	7.5–9.5	0.10
Zinc	0.50	1.0	0.15	0.50
Nickel	0.50	0.50	0.15	0.50
Tin	0.15	0.35	0.15	0.15
Total others	0.25	0.50	0.25	0.25
Aluminum	Remainder	Remainder	Remainder	Remainder
Properties and constants:[c]				
Tensile strength, psi	43,000	43,000	41,000	30,000
Yield strength (0.2% offset), psi	25,000	27,000	27,000	16,000
Elongation, % in 2″	3.0	2.0	7.5	5.0
Shear strength, psi	27,000	28,000	28,000	19,000
Fatigue strength, psi[d]	19,000	19,000	23,000	17,000
Specific gravity	2.63	2.70	2.57	2.65
Weight per cu in., lb	0.095	0.098	0.093	0.096
Melting point (liquid), °F	1105	1080	1150	1170
Thermal conductivity, cgs	0.27	0.23	0.23	0.34
Thermal expansion, in. per in. per °F[e]	12.2	11.7	13.9	12.9
Electrical conductivity, % of copper standard	29	23	24	37

The alloys listed herein are "Special." For standard alloys, see ADCI-M2-55T.

* Reprinted by permission of the American Die Casting Institute, Inc., 366 Madison Ave., New York 17, N.Y.

[a] This alloy is also available in lower iron composition, A360.

[b] Composition in per cent, maximum unless shown as range. Same as ASTM B85.

[c] Data furnished by Aluminum Company of America and ASTM B85. Revisions are in progress and will be made from time to time.

[d] R. R. Moore rotating-beam test data at 500,000,000 cycles.

[e] To be multiplied by 10^{-6}. Temperature range 68 to 392°F.

Table 10-26. Magnesium-base-alloy Die Castings, ADCI-M7-56T*

1. Alloy Selection

The magnesium alloy shown below is the standard composition generally used. This standard alloy is available in a high-purity composition with 0.10 maximum copper content (designated AZ 91A) having improved corrosion resistance

2. Composition and Properties

The chemical compositions of magnesium-base-alloy die castings and related typical values for physical properties and constants are shown in the table. The typical values indicated are for separately die-cast test bars and do not represent values for specimens cut from die castings

Chemical Composition and Properties: Magnesium-alloy Die Castings

Designation Commercial and ASTM	Standard alloy AZ91B
Composition, %:[a]	
Aluminum	8.3–9.7
Zinc	0.4–1.0
Manganese, min	0.13
Silicon, max	0.5
Copper, max	0.30
Nickel, max	0.03
Others, total, max	0.3
Magnesium	Remainder
Properties and constants:[b]	
Tensile strength, psi	34,000
Tensile yield strength (0.2%) offset), psi	23,000
Elongation, % in 2″	3
Shear strength, psi	20,000
Compressive yield strength, psi	22,000
Ultimate compressive strength, psi	58,000
Fatigue strength, psi[c]	14,000
Specific gravity	1.80
Weight per cu in., lb	0.066
Thermal conductivity, cgs	0.17
Thermal expansion, in. per in. per °F[d]	15.2
Electrical conductivity, % of copper standard	10

See ADCI M8 for characteristics of magnesim-alloy die castings.

* Reprinted by permission of the American Die Casting Institute, Inc., 366 Madison Ave., New York 17, N.Y.

[a] Composition in per cent as shown. *Cf.* ASTM B94.

[b] Data furnished by Dow Chemical Co.; ADCI; ASTM.

[c] R. R. Moore rotating-beam test at 500,000,000 cycles.

[d] To be multiplied by 10^{-6}. Temperature range 68 to 392°F.

Magnesium Die Castings. These, the lightest in weight, may get treated by the die caster for protection against tarnishing of the surface occurring during storage. The strength-to-weight ratio of magnesium vs. aluminum die castings favors magnesium. Comparing the values for yield strength, a very close relationship exists. In this regard, as ambient temperatures rise, magnesium alloys are more stable than aluminum alloys.

Trade Standards in Die-casting Materials. The American Die Casting Institute has standardized the commonly used zinc-alloy, aluminum-alloy, magnesium-alloy, and copper-base-alloy materials. Tables 10-21 to 10-27 show the data for these materials. Chemical compositions and physical property data are both given.

10-13. Nonmetallic Gears. Nonmetallic or "plastic" gears are frequently used in order to reduce or eliminate sounds objectionable to a particular environment. The

Table 10-27. Copper-base (Brass) -alloy Die Castings, ADCI-M5-56T*

1. Alloy Selection

The copper-base alloys most generally used are shown below. Die-casting and machining characteristics involved in alloy selection and the basis for specifying brass die castings are covered in ADCI M6

2. Composition and Properties

The chemical compositions of copper-base-alloy die castings and related typical values for physical properties and constants are shown in the table. The typical values indicated are for separately die-cast test bars and do not represent values for specimens cut from die castings

Chemical Composition and Properties: Brass Die Castings

Designation	Standard copper-base alloys		
Commercial and ASTM	Z30A	ZS331A	ZS144A
Composition, %:[a]			
Copper	57.0 min.	63.0 –67.0	80.0 –83.0
Silicon	0.25 max.	0.75– 1.25	3.75– 4.25
Lead, max	1.50	0.25	0.15
Tin, max	1.50	0.25	0.25
Manganese, max	0.25	0.15	0.15
Aluminum, max	0.25	0.15	0.15
Iron, max	0.25	0.15	0.15
Magnesium, max	....	....	0.01
Other elements, max	0.50	0.50	0.25
Zinc	30.0 min.	Remainder	Remainder
Properties and constants:[b]			
Tensile strength, psi	55,000	70,000	85,000
Yield strength (0.2% offset), psi	30,000	35,000	50,000
Elongation, % in 2″	15	25	25
Impact strength, Charpy, ft-lb	40	50	70
Hardness, Rockwell B scale	55–60	68–72	85–90
Modulus of elasticity[c]	15	15	20

See ADCI M6 for information relating to alloy selection and basis for specifying brass die castings.

* Reprinted by permission of the American Die Casting Institute, Inc., 366 Madison Ave., New York 17, N.Y.

[a] Composition in per cent as indicated. Same as ASTM B176.

[b] Data from ADCI—average typical properties for separately die-cast test bars.

[c] To be multiplied by 10^{-6}.

same reason that these gears are quiet also explains their further usefulness and teaches us what is to be done with all gear meshes: the damping-out properties permit gear-tooth irregularities to roll or slide through mesh by what may be described as conformability. This occurs at the instant of contact, and for the material to be of practical usage it must have good "memory" characteristics.

Nylon is a thermoplastic material with many favorable qualities as a gear material. It has excellent resistance to the usual wear phenomena. In many applications the only lubricant was the original as applied. Formulations of nylon are being improved upon; so it will be necessary to check manufacturers' reports and literature for up-to-date mechanical and physical characteristics.

Each design application should have prototype models made since various conditions can affect "plastic" materials. They are more sensitive than metals to moisture absorption and temperature changes. For critical applications or where sections of a

Table 10-28. Physical Properties of Nonmetallic Gears

Property	Poly-carbonate	Polyamide	Acetal	Phenolic fabric, LE: Cross-wise	Phenolic fabric, LE: Length-wise
Tensile strength, psi × 10^3	9–10.5	8.5–11	10	9.5	13.5
Flexural strength, psi × 10^3	11–13	14.6	14	13.5	15
Elongation, %	60–100	60–300	15–75		
Impact strength, ft-lb per in.	12–16	0.9–2.0	1.4–2.3	1	1.25
				Thickness: 1/8″ 1/4″ 1/2″ 1″ up	
Water absorption, % per 24 hr	0.3	1.5	0.4	1.3 0.95 0.70 0.55	
Coefficient of thermal linear expansion, °F, in. per in. per °F	3.9	5.5	4.5	1.1	
Heat resistance, °F (continuous)	250–275	250	....	250	
Representative trade names	Lexan	Nylon, Zytel	Delrin	Phenolite, Textolite	

* This table is courtesy of General Electric Company, Plastics Division, Pittsfield, Mass. The tabulated data are average values and should not be used for specifications.

Table 10-29. Suggestions for Machining Gears of Laminated Plastics

Sawing and Fly Cutting

Gear blanks are formed by either band sawing or fly cutting. Fly cutting is generally limited to thicknesses less than 1/2″ and a diameter less than 5″. This permits the drilling of a center hole in the same operation and produces a blank of sufficiently accurate outside diameter so that a "topping" hob can be used for cutting teeth without a turning operation.

When band sawing, leave at least 1/8″ stock for finishing.

Drilling

To obtain best results, use high-speed-steel drills and Carboloy-tipped boring tools. Certain drills are available with steeper twists and wider flutes than standard, and these will help chip clearance although it is still desirable to lift the drill from the work and remove chips when drilling deep holes. It is very desirable to use very sharp-edged tools on all machining operations.

Turning

For turning Textolite blanks, use high-speed-steel or Carboloy-tipped tools. In using the steel, speeds are usually limited to approximately 750 fpm. With the Carboloy, speeds well above 1,500 fpm can be used with rigid lathes. In general use a very high speed with a light cut.

You will obtain the smoothest finish with a round-nose tool with a radius of 7/32″ or more. Grind tools with very little top rake and considerable end clearance.

Keyseating

Key slots can be broached or milled. In general, the technique should be the same as in machining brass.

Tooth Cutting

Best results will be obtained with the highest speed available from the work arbor and a rather light feed. It is generally desirable to back up the blank so that the emerging cutter does not raise a burr or turn over the laminations.

gear may provide stress raisers it may be advisable to stress-relieve or anneal the part. This applies to gears machined from nylon stock and molded gears. Residual stresses in the latter can be severe.

Delrin, an acetal resin, is a material used for gears showing some improvements over nylon. The same kinds of considerations for dimensional stability must be given as in cases when designing with nylon. Table 10-28 shows some general properties of nonmetallic gear materials.

Laminates as gear materials provide several advantages: quietness, little lubrication, and manufacturing ease. Sheet stock made from phenolic resins that are combined with fine-weave linen provide the best condition.

Blanks may be punched out or circles may be obtained to fit into quantity manufacturing. Whether one blank or several are used, a metal backing-up disk is required to obtain a "clean" edge on the end piece.

Gears from laminate stock are usually driven members of a gear mesh.

If pitch-line velocities are high, a steel pinion driving a phenolic gear should be hardened to prevent scoring and premature wear on the pinion.

A precaution frequently omitted is consideration for the high elasticity of nonmetallic gears. Too much deflection because of this allows the *entering driven* tooth to scrape over the flank of the driver and scuff the surface. To avoid this where the pinion is metal and the driven tooth nonmetallic, it is always advisable to use the long- and short-addendum system for teeth proportions, except if undercutting would result.

The laminated materials offer some problems in machining. Table 10-29 is a digest of tips on how to machine laminated material.

REFERENCES

1. Dudley, Darle W.: "Practical Gear Design," pp. 152–189, McGraw-Hill Book Company, Inc., New York, 1954.
2. The International Nickel Co., Inc.: "Nickel Alloy Steels," The International Nickel Co., Inc., New York, 1949.
3. National Broach and Machine Co.: "Modern Methods of Gear Manufacture," 3d ed., National Broach and Machine Co., Detroit, Mich., 1950.
4. The Nitralloy Corporation: "Nitralloy and the Nitriding Process," The Nitralloy Corp., New York, 1954.
5. The International Nickel Co., Inc.: "Ductile Irons," The International Nickel Co., Inc., New York, 1957.
6. Meehanite Metals Corp.: "Handbook of Meehanite Metals," Meehanite Metals Corp., New Rochelle, N.Y., 1948.
7. American Manganese Bronze Co.: "Reference Book Bronze Casting Alloys," American Manganese Bronze Co., Philadelphia, 1953.
8. The New Jersey Zinc Co.: "Practical Considerations in Die Casting Design," The New Jersey Zinc Co., New York, 1955.
9. American Gear Manufacturers Assoc.: "Gear Materials—Steel," AGMA 241.01, June, 1954.
10. American Gear Manufacturers Assoc.: "Cast Iron Gear Blanks," AGMA 242.02, September, 1946.
11. American Gear Manufacturers Assoc.: "Nodular Iron Gear Materials," AGMA 244.01, June, 1959.
12. American Gear Manufacturers Assoc.: "Cast Bronze Gear Blanks," AGMA 243.01, June, 1954.
13. Wulff, John, Howard F. Taylor, and Amos J. Shayler: "Metallurgy for Engineers," John Wiley & Sons, Inc., New York, 1956.
14. American Gear Manufacturers Assoc.: "Gear Classification Manual," AGMA 390.01, August, 1960.
15. American Gear Manufacturers Assoc.: "Recommended Practice for Carburized Gearing," AGMA 246.01, October, 1960.
16. Gillig, F. G. (Cornell Aeronautical Laboratory, Inc.): "Study of Hard Coat for Aluminum Alloys," WADC Technical Report 53–151, Wright Air Development Center, Dayton, Ohio, 1953.

Chapter 11

GEAR DRAWINGS

By

GEORGE L. McCAIN, *Consulting Engineer, Birmingham, Mich.*

This chapter shows how to make gear drawings. Gears often include complicated structures which are dictated by design but which have no relation to the proper functioning of the gear. Therefore, the examples will show only the dimensioning for the teeth and the mounting features. The data for gear-tooth dimensions will vary from a bare minimum required for manufacture, to the complete specification for quality control of processes by inspection.

Illustrations of drawing practice in this chapter cover a wide variation of gear applications. Selection of the data is based on the service required of the gear and the processing equipment to be used by the manufacturer. The use of *inclined block lettering* in the illustrations indicates dimensions, notes, or data which belong on the drawing. Instructions for the draftsmen are in *lower case.*

11-1. General Drawing Practice. The features of the gear may be shown in a single view or may require several views, depending on the complication. Generally a section across the axis, or a side view and an end view (along the axis) as shown in Fig. 11-3 are the preferred illustrations; but where feature relations, such as shown in Fig. 11-6, must be specified, additional views may be used.

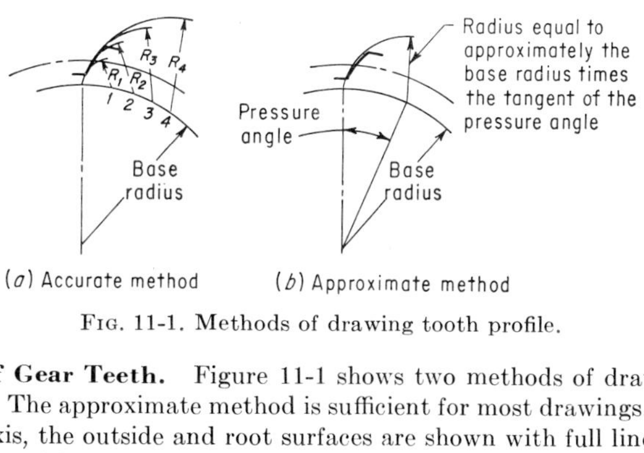

FIG. 11-1. Methods of drawing tooth profile.

Showing of Gear Teeth. Figure 11-1 shows two methods of drawing the gear-tooth profile. The approximate method is sufficient for most drawings. In a section through the axis, the outside and root surfaces are shown with full lines. The pitch lines are shown with center lines (short and long dashes). In a side view the root surfaces are shown as hidden lines (short dashes). The side view is usually used to show the angle of the tooth, as in Fig. 11-3. In the end view (along the axis) one or two teeth may be shown if desired, and the remainder indicated by using phantom lines (two short dashes, one long dash), as in Fig. 11-3, for the outside and root circles. If other features of a gear, such as a keyway, spline tooth, or bolthole, must be related to a gear tooth or space, the relation must be shown.

Dimensioning. Generally, the axial section dimensions relate to the design of the blank. Accuracy of mounting must be the basis of accuracy of the gear teeth, and the drawing should show the finished gear with suitable limits on which to base final inspection processes. *The axis of reference should always be indicated.* It could be the axis of the center hole, the axis of the pitch circle of a splined hub, the axis of a pilot on which the gear is mounted, or the center of a group of holes for mounting bolts. Dimension values are indicated on the drawing examples by x's placed in recommended locations. The use of inclined x's suggests also the number of decimal places usually given for each dimension value. Some drawings show limits with two decimal places, while others show untoleranced two-place decimals. If a general tolerance note is included on the drawing, no limits with two decimal places need be used.

Dimensions are in three classifications, for reference, manufacturing, and inspection. Dimensions without tolerances which are used to define features of the gear, or are the basis for other toleranced dimension values, are reference dimensions, and should be marked "REF," meaning for reference only. Values marked "min" or "max" are working dimensions used also for inspection. The "min" or "max" indicates one limit, where the other limit is subject to control by some other dimension value. Where some of the dimension values are grouped in data form, such as in Figs. 11-3, 11-4, and 11-5, each reference item may be marked "REF," as in Fig. 11-3, or they may be grouped together, as in Fig. 11-5. Sometimes a toleranced dimension, such as the value for the circular tooth thickness, may become a reference dimension if it is the basis for the limits for the chordal tooth thickness or measurements with pins. Dimension limits define the blank and indicate features such as the depth of cut and the amount of chamfer. They indicate the tolerance allowed the workman in setting the machine tool which performs the operation.

Method of Manufacture. Wherever possible, the method of making the gear should be left to the manufacturer; however, the method of finishing may be specified, as cut, ground, shaved, or lapped. Requirements for interchangeability, replacement, or functioning may necessitate the use of a particular manufacturing method. This is true, for example, for certain types of bevel gears.

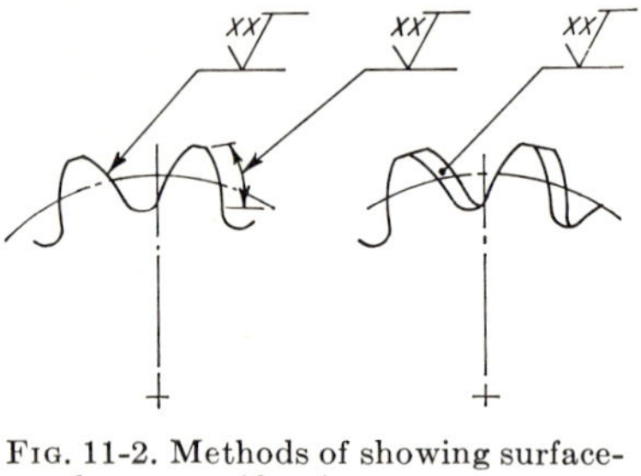

FIG. 11-2. Methods of showing surface-roughness specifications.

Surface Finish. Specifications for maximum roughness allowable are indicated by symbols, as in Fig. 11-2, in accordance with American standards. Figure 11-2 is given as a guide for the application of surface-roughness values for the tooth profiles.

Material Specifications. Specification of material and heat-treatment are usually included in a separate block of data on the lower part of the drawing, as follows:

MATERIAL XXX STEEL (forging, casting, bar stock, stamping)
FURNACE HARDEN AND DRAW AFTER ROUGH MACHINING AND BEFORE FINISH MACHINING TO OBTAIN XX BRINELL HARDNESS. CHECK AT X (AS INDICATED ON DRAWING)
INSPECT MAGNETICALLY TO XXX SPECIFICATION

In addition, it may sometimes be desirable to list forging number, casting or pattern number, other part number (if remachined), or rough-stock designation.

11-2. Spur Gears. The minimum data required for cutting a spur gear are shown in Fig. 11-3 for an external gear, and in Fig. 11-4 for an internal gear. The working depth in both cases is determined by data from the mating gear, and it must be sufficient to ensure clearance over the maximum outside diameter of that gear.

Specifications for Chordal Tooth Measurements. The circular-tooth-thickness limits are the basis for measurements of chordal tooth thickness with vernier calipers, as in Fig. 11-3. In case the desired final inspection is with verniers, the

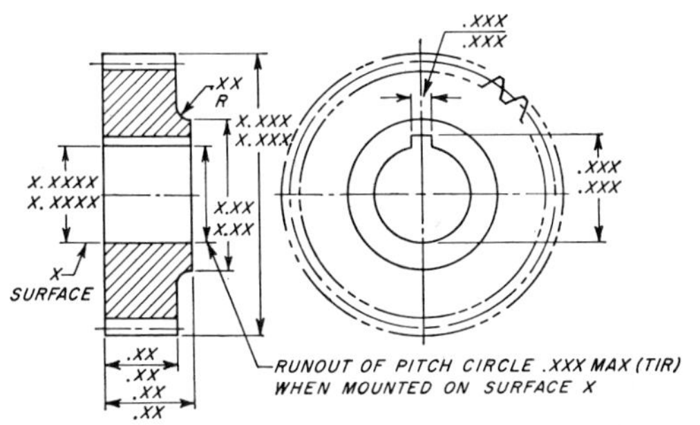

SPUR-GEAR TOOTH DATA	
NUMBER OF TEETH	XX
DIAMETRAL PITCH	XX
PRESSURE ANGLE	XX°
PITCH DIAMETER	X.XXXX
BASE DIAMETER	X.XXXX
CIRCULAR THICKNESS (REF)	.XXX-.XXX
WHOLE DEPTH (MIN)	.XXX
WORKING DEPTH (MAX)	.XXX
CHORDAL ADDENDUM (MAX OD)	.XXX
CHORDAL THICKNESS	.XXX-.XXX

FIG. 11-3. External spur gear.

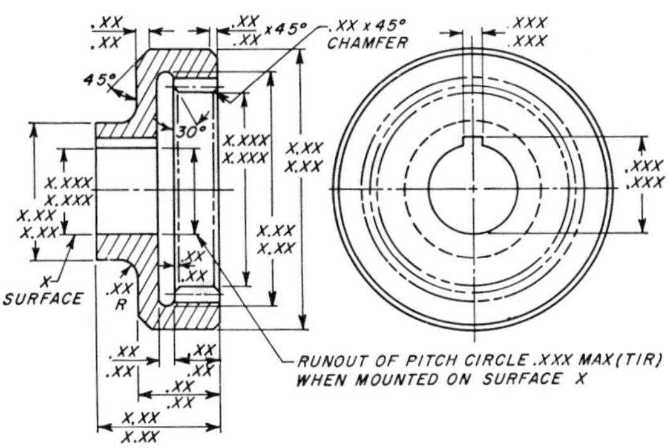

SPUR-GEAR TOOTH DATA	
NUMBER OF TEETH	XX
DIAMETRAL PITCH	XX
PRESSURE ANGLE	XX°
PITCH DIAMETER	X.XXXX
BASE DIAMETER	X.XXXX
CIRCULAR THICKNESS (REF)	.XXX-.XXX
WHOLE DEPTH (MIN)	.XXX
WORKING DEPTH (MAX)	.XXX
MEASURING PIN DIA	.XXXX
MEASUREMENT BETWEEN PINS	X.XXX-X.XXX

FIG. 11-4. Internal spur gear.

drawing data should be as shown in Fig. 11-3. The measuring caliper rests on the top of the tooth; therefore, a correction in the chordal addendum may be required to compensate for variations of the outside diameter. The chordal addendum is usually given for the maximum outside diameter.

Specifications for Inspection with Pins. Measuring pins contact the working surfaces of the gear teeth; therefore, measurements over pins are accurate checks for either intermediate (process) inspection or final inspection. The drawing should show only the final inspection measurements, but intermediate process drawings or operation sheets may include such measurements for a gear before shaving or grinding, in order to control the amount of material to be removed in the final operation. Where such data are to be used, they should be given as follows:

MEASURING-PIN DIAMETER	*.XXXX*
MEASUREMENTS OVER PINS	*X.XXX–X.XXX*

The pins for such measurements are placed in exactly opposite spaces for gears with even numbers of teeth. For gears with odd numbers of teeth the spaces used are away from opposite position by half the circular pitch. For internal gears, as shown in Fig. 11-4, measurements are given between pins.

Where desired, measurements with pins may be specified by a drawing dimension, as shown in Fig. 11-5. If it is desirable to show also the change factor for convenience in process control, it may be added to the data as follows:

MEASUREMENT CHANGE FACTOR X.XXX

Where measurements over (or between) pins are given as final inspection specifications, the toleranced dimensions for tooth thickness should be marked "REF" as shown in Fig. 11-3.

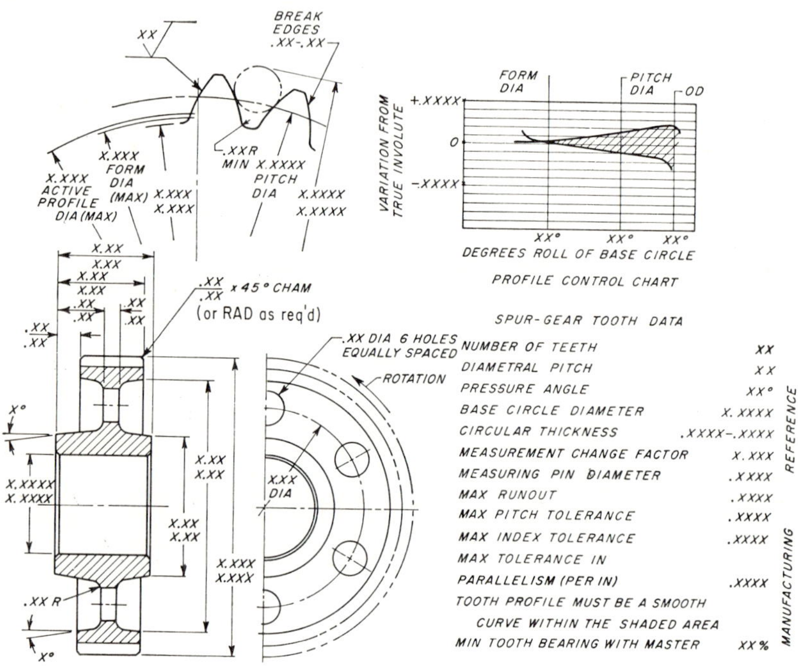

FIG. 11-5. Precision spur gear.

Inspection with Master Gears or Racks. Master gears or racks are widely used in gear inspection. A spring-loaded master is rolled in tight mesh with the gear being inspected, and variation in center distance is measured. The resulting check is called a "composite" check because it measures a combined effect of runout, pitch error, tooth thickness, profile error, and lateral runout.

In cases where a check with a master is to be used for tooth accuracy the drawing specification should read

MAX. TOOTH-TO-TOOTH COMPOSITE TOL. .XXXX
MAX. TOTAL COMPOSITE TOL. .XXXX

In high-volume manufacture of small precision gears it is quite expedient to set up the master gear check with maximum and minimum limits on center distance so that the master check measures tooth thickness as well as the accuracy of the tooth meshing. In this case the drawing specification should include the following in addition to the composite limits:

CENTER DISTANCE WHEN IN TIGHT MESH WITH STANDARD MASTER GEAR (OR RACK) *X.XXXX–X.XXXX*

In some cases there will be high-volume production of gears that are not too critical on the tooth-meshing accuracy. In this case appropriate center-distance limits with the master may be adequate to control quality and the tooth-to-tooth and total composite limits may be omitted.

Special Inspection. Special methods of inspection may be used where desired, and the devices to be used should be shown on the drawing with dimension values. For example, special gage blocks may be used.

Inspection in Sets. Inspection in pairs or trains is sometimes required where gears are not individually interchangeable, and especially where quiet operation is essential and backlash must be controlled. In such cases the data to be given are as follows:

PART NUMBER OF MATING GEAR	*XXXXX*
NUMBER OF TEETH IN MATING GEAR	*XX*
BACKLASH (AT CENTER DISTANCE OF X.XXX)	*.XXX–.XXX*

Where the gears are shown on an assembly drawing with the center distance, the "backlash" note only may be used.

Geometric Tolerancing of Gears. This relates to the accuracy of the tooth profile, the spacing of the teeth, the lead or parallelism, and concentricity of the teeth in relation to a datum surface (see Chaps. 9 and 23). Tolerances are usually given as maximum permissible amounts. Individual variations due to inaccuracies in the producing equipment must be recognized and provision made for their control by specifying tolerances. Where required, permissible variations (errors) may be specified as follows:

MAXIMUM RUNOUT	*.XXX*
MAXIMUM PROFILE TOLERANCE	*+.XXXX* to *–.XXXX*
MAXIMUM PITCH TOLERANCE	*.XXXX*
MAXIMUM INDEX TOLERANCE	*.XXXX*
MAXIMUM TOLERANCE IN PARALLELISM	*.XXXX*

Except for runout and profile errors, detailed inspection is usually applied only to gears of highest-precision quality because of the cost of inspection.

Control of Tooth Form. Where close control of profile is required, the drawing note may be expanded to include the specification of both positive (outside) and negative (inside) variations from the specified profile. The chart, as shown in Fig. 11-6, is one way of specifying the allowable variations and may be added to the drawing where required. Where such control is included, the following dimensions should be added, usually in note form:

FORM DIAMETER MAX (*or MIN*)	*X.XXX*
MINIMUM RADIUS OF FILLET	*.XXX*
ROOT DIAMETER	*X.XXX – X.XXX*

Figure 11-6*A* shows the profile chart in the form used on the gear checker, and this form may be used where desired.

Tooth-form Modification. Design requirements sometimes make it necessary to relieve the tips of the teeth of the driven gear. This prevents interference at the active form diameter of the mating gear which might otherwise occur because of

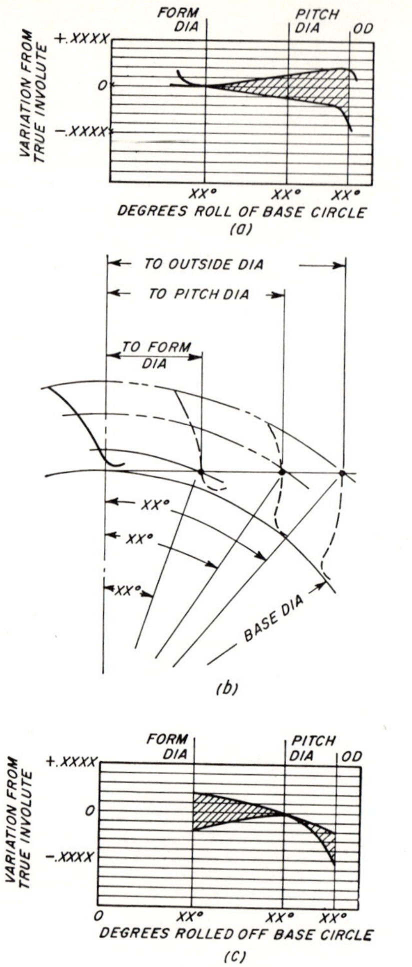

FIG. 11-6. Profile-control chart. (*a*) Profile-control chart. (*b*) Establishing of basic check points. (*c*) Example of profile-control chart based on pitch diameter. This type of chart may be used as an alternate to the one shown in (*a*) above.

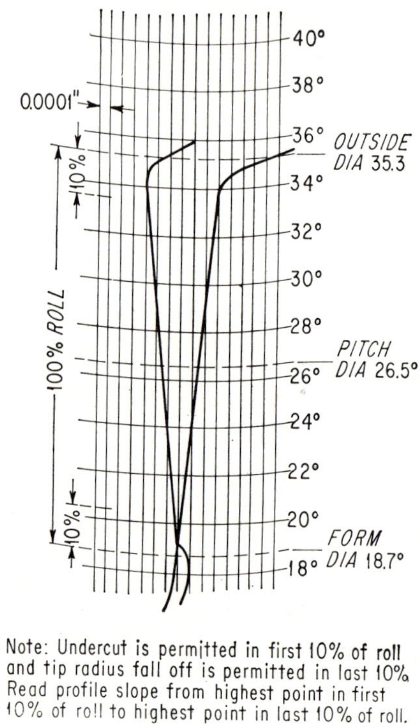

FIG. 11-6*A*. Sample profile-control chart drawn for +0.0003 to −0.0003 limits on involute slope.

manufacturing tolerance or tooth deflection under load. The chart (Fig. 11-6) may be used to show desired modifications for either spur or helical gears; the variation from the true involute is given in the plane of rotation. It is desirable to have the modification follow a smooth curve within the shaded area, and a note may be added as follows:

TOOTH PROFILE MUST BE A SMOOTH CURVE WITHIN THE SHADED AREA

The explanation (*b*) shows the location of the basic check points, starting at zero where the involute starts at the base circle. As the gear is rotated through the designated angles, measurements are made along a tangent to the base circle. This explanation may be included on the drawing if desired. Some gears require profile control or modification for both drive and coast sides of the teeth. In such cases, two control charts may be used.

Control of Tooth Contact. Where rigid operating conditions demand such inspection, the area of contact may be defined by a note, such as

MINIMUM TOOTH BEARING WITH MASTER XX%

Application of Specifications (Combinations). For many precision gears, various control specifications are desirable and necessary. However, care should be exercised to avoid double dimensioning, or dimensions (specifications) which encroach upon each other. For example, if measurements with pins are specified, vernier-caliper measurements are not used; if measurements with master gears are specified, measurements with pins are not used for final inspection.

Use of Abbreviations. Drawing data shown on these pages are mostly spelled out completely, but abbreviations* may be used if desired, according to ASA standard "Abbreviations for Use on Drawings." Since the drawing is usually an important part of a manufacturing contract, there should never be any doubt about the meaning of a specification. *"When in doubt, spell out."*

Crowned Teeth. Crowning is specified as the amount on one side that the end is relieved, and is measured as a lead variation of the tooth. Where the crown is offset, as shown in Fig. 11-7, a dimension indicating the center of the crown should be given and the amount of the crown shown at the thinnest end of the tooth.

Rack Teeth. Rack teeth are dimensioned with the desired location from the end of the rack blank, and from some datum surface or line convenient for manufacture and inspection. The datum line may be the back face of the blank or a line through reamed holes or other accurate features of the rack (see Fig. 11-8).

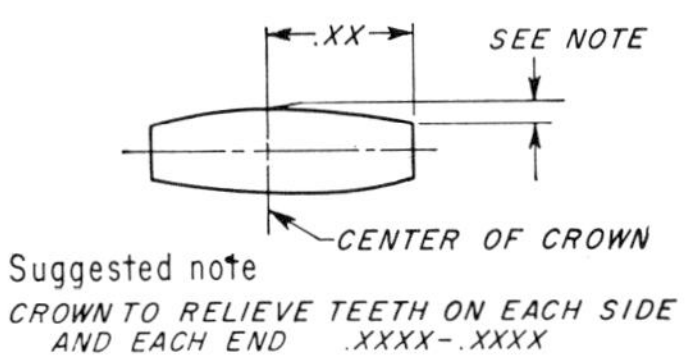

FIG. 11-7. Dimensioning of crowned teeth.

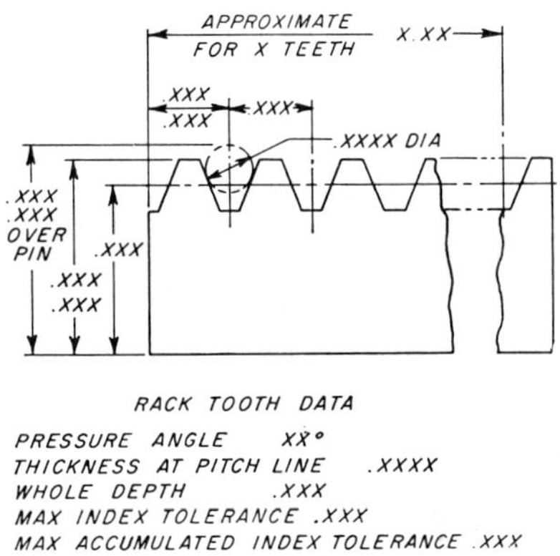

FIG. 11-8. Dimensioning of rack teeth.

11-3. Helical Gears.

Most of the practices outlined for spur gears apply also to helical gears. However, measurements are usually made with balls rather than with pins.

Normal and Transverse Planes. The features of helical teeth may be considered and dimensioned in a transverse plane (a plane perpendicular to the axis), the same as in the case of spur gears, or a normal plane (perpendicular to the teeth). In a spur gear the transverse and normal plane are the same. It is necessary to use designations "transverse" and "normal" for helical gears, for various tooth dimensions and angles. A helical gear tooth has involute form only in its transverse plane.

Gear-cutting Practices. In the design and dimensioning of helical gears, the choice between normal and transverse values depends on the practice to be adopted

* See Art. 4-17 for detail information on abbreviations for gear terms.

in the selection and use of the gear-cutting tools. It is common practice to use a spur-gear hob to generate helical gears. The standard tooth dimensions, for which the hob is made, would apply in the normal plane of the helical gear; the resulting transverse values, such as circular pitch and diametral pitch, will not be standard.

There are other practices in hobbing and other generating methods in which the cutting tools are designed and made for a standard transverse diametral pitch and pressure angle. The tooth specifications are given in transverse values; even so, some values will probably be given in the normal plane. The dimensions and data on the drawing should be in accordance with the practice of the manufacturer and the designer of the gear.

Helical-gear-tooth Data. The data for helical gears are similar to those used for spur gears, with additional data required for the helix angle of the teeth. The required dimensions for helical gear teeth are shown in Fig. 11-9. The shaft angle,

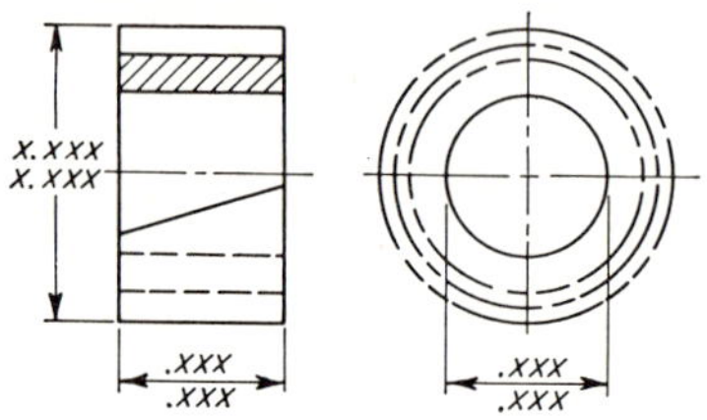

HELICAL-GEAR TOOTH DATA	
NUMBER OF TEETH	XX
NORMAL DIAMETRAL PITCH	XX
TRANSVERSE DIAMETRAL PITCH	XX
NORMAL PRESSURE ANGLE	XX°
TRANSVERSE PRESSURE ANGLE	XX°
PITCH DIAMETER	X.XXXX
NORMAL CIRCULAR THICKNESS	.XXX-.XXX
NORMAL CHORDAL ADDENDUM	.XXX
NORMAL CHORDAL THICKNESS	.XXX
WHOLE DEPTH	.XXX
HELIX ANGLE LEFT (or right) HAND	XX°
LEAD	X.XXX

FIG. 11-9. Helical gear, minimum dimensioning.

which also must be known, is given only when the shafts of mating gears are not parallel or at 90°.

Helical gears are used where quiet and smooth operation is required. In order to maintain close limits, inspection by measurements over balls is the usual method chosen. Measurements with two pins can be used only for even numbers of teeth; so that balls are usually used. Figure 11-10 shows a precision helical gear. The profile-control chart, shown in Fig. 11-6*A*, may also be used with this method of dimensioning tooth profile. Where it is not desired to show the enlarged teeth, profile tolerances, runout, pitch diameter, and measurement over pins may be added to the other tabulated data.

For gear-tooth data in the normal plane, the principal items are

NORMAL DIAMETRAL PITCH	*XX.XXX*
NORMAL PRESSURE ANGLE	*XX°XX.X′*
NORMAL CIRCULAR PITCH	*.XXXX*
NORMAL CIRCULAR THICKNESS	*.XXXX*
NORMAL CHORDAL ADDENDUM	*.XXX*
NORMAL CHORDAL THICKNESS	*.XXX–.XXX*

Specifications for Inspection. Inspection of chordal thickness, measurement over balls, measurements with master gears or racks, special inspection or inspection in sets or trains applies to helical as well as to spur gears. Inspection for unavoidable deviations is the same as that described for spur gears, except that the term "lead deviations" is used for helical gears, instead of "deviation in parallelism." Profile control and modification are applied as for spur gears.

11-4. Bevel Gears and Hypoid Gears. In bevel gears the axes intersect, and usually the shaft angle is 90°. In hypoid gears, the axes are offset. The general dimensions of bevel gears are the outside diameter, pitch angle, cone distance, and face width. These all appear in a layout made in a plane which contains the gear and pinion axes.

Tooth Proportions. In a cross section of bevel gear teeth the proportions are similar to those of an equivalent spur gear which has the same curvature of the pitch surface and the same circular pitch.

It is customary to use long-addendum pinion and short-addendum gear teeth, and to design the tooth thickness of pinion and gear with consideration of their respective strength. The diametral pitch and tooth dimensions apply to the outer ends of the teeth unless otherwise specified.

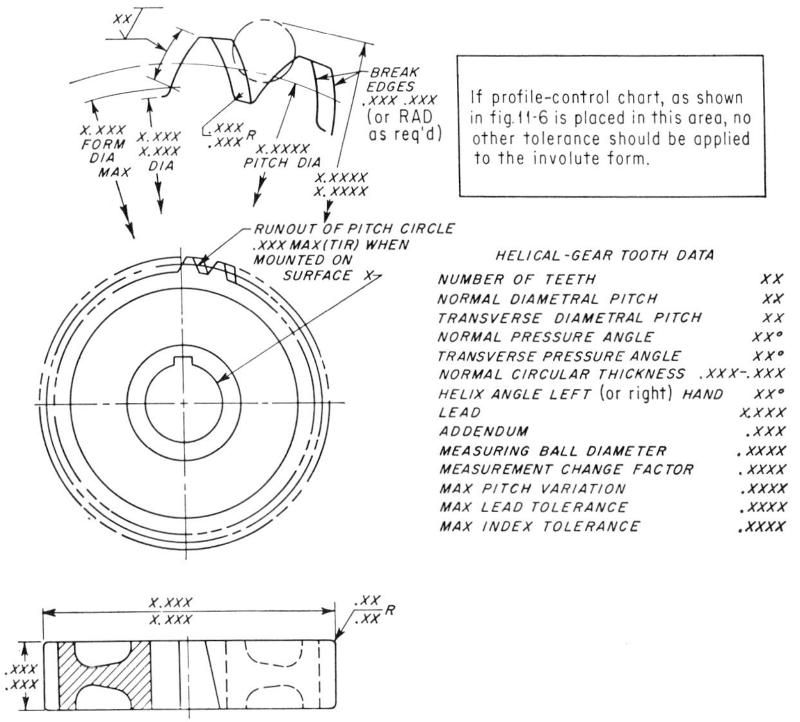

FIG. 11-10. Precision helical gear.

Method of Cutting. In the production of bevel gears and hypoid gears, important differences may exist in such features as the following:

1. Form of tooth profile
2. Kind and form of cutting and generating tools
2. Cutting and generating methods
4. Types of cutting and generating machines
5. Methods of setting up and operating machines

Spiral Bevel Gears. In spiral bevel gears the pitch diameter and diametral pitch also are customarily given at the outer ends of the teeth. The specified spiral angle is at the mean cone distance. The spiral angle varies along a tooth. The pressure angle is a value in a normal plane and more specifically is nominally the pressure angle of the cutter (see Fig. 11-12).

Detailed tooth specifications vary according to the cutting equipment to be used and

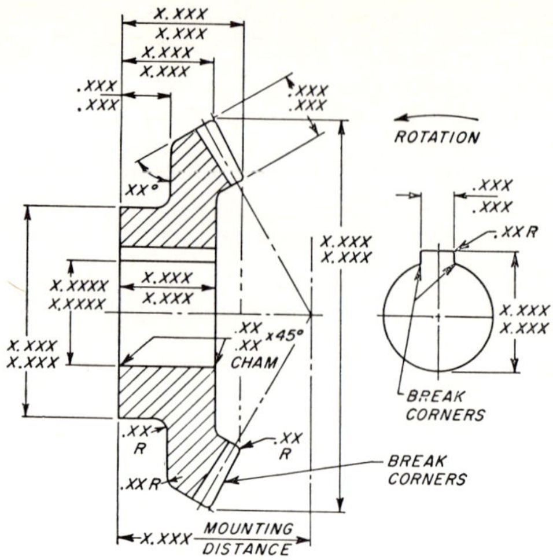

STRAIGHT BEVEL TOOTH DATA
NUMBER OF TEETH XX
DIAMETRAL PITCH XX
PITCH DIAMETER X.XXX
PITCH ANGLE XX° X'
SHAFT ANGLE XX°
PRESSURE ANGLE XX°
LIMIT POINT WIDTH .XXX
ADDENDUM .XXX
WHOLE DEPTH .XXX
CONE DISTANCE X.XXX
TOOL ADVANCE .XXX
ROOT ANGLE XX° X'
TOOTH ANGLE XX° X'
CHORDAL ADDENDUM .XXX
CHORDAL THICKNESS .XXX
DRIVER (or driven)
SPEED XXX RPM (or variable)
BACKLASH IN ASSEMBLY XXX-.XXX
MATING GEAR NO. XXXXX (XX TEETH)
MAX RUNOUT .XXX

FIG. 11-11. Straight bevel gear, hub type.

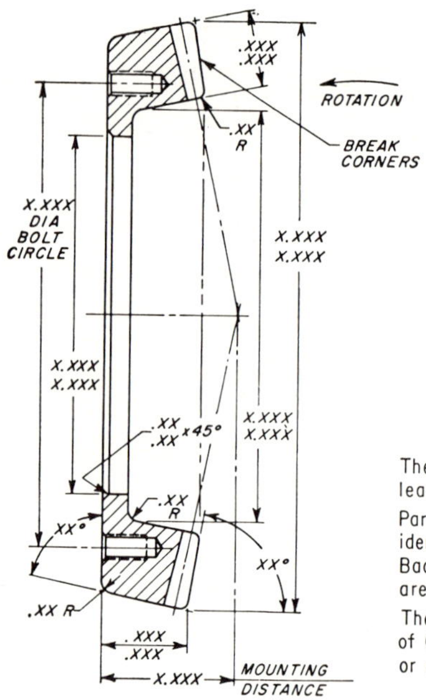

SPIRAL (or Zerol) BEVEL TOOTH DATA
NUMBER OF TEETH XX
DIAMETRAL PITCH XX
PITCH DIAMETER X.XXX
PITCH ANGLE XX° X'
SHAFT ANGLE XX°
PRESSURE ANGLE XX°
ADDENDUM .XXX
WHOLE DEPTH .XXX
SPIRAL ANGLE XX° X'
HAND OF SPIRAL RIGHT (or left)
DRIVER (or driven)
SPEED XXX RPM (or variable)
BACKLASH IN ASSEMBLY .XXX-.XXX
MATING GEAR NO. XXXX (XX TEETH)
MAX RUNOUT .XXX
SUMMARY NO. XXXX

The diameter of the throat, inside of ring, must be at least 0.38 greater than the inside bore diameter.

Part number, serial number, backlash, date, and other identification should be marked on the outside of the ring. Back face should be ground only if the tooth profiles are ground.

The runout (TIR) should be specified as a maximum of 0.003 for cut teeth and 0.001 for ground teeth or nonhardened precision gears.

FIG. 11-12. Spiral bevel gear.

generally will be determined by the manufacturer. Because of the variations in the methods of cutting and grinding, satisfactory duplication of tooth contact can be obtained only by using the proper cutter specifications and machine settings. These are customarily recorded on a summary (of machine settings). See Chap. 20. The number (or other identification) of the summary should be recorded on the drawing data, or if desired, the complete summary may be included on the drawing.

Tolerances. Tolerances are applied to the measurements of the blank size and shape, tooth thickness, and sometimes the depth, but there are no simple or direct methods for checking the tooth surfaces, as there are for spur and helical gears. Tolerances are indicated for the blank dimensions in Figs. 11-11 through 11-16.

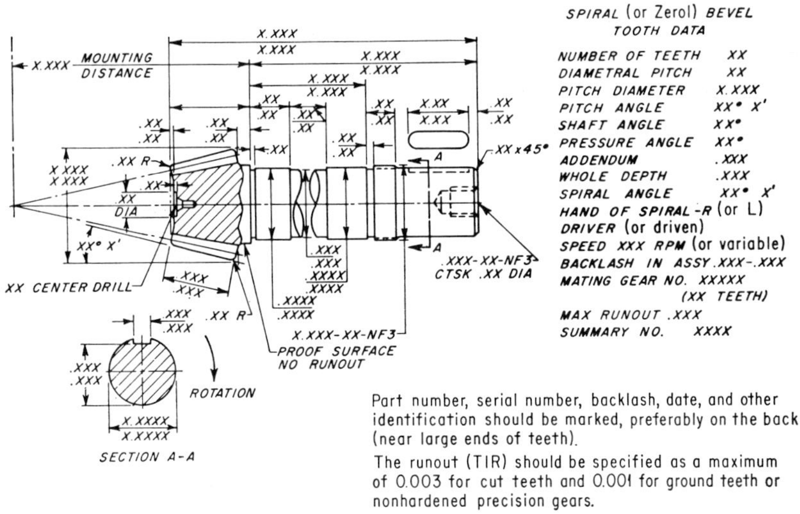

FIG. 11-13. Spiral bevel pinion.

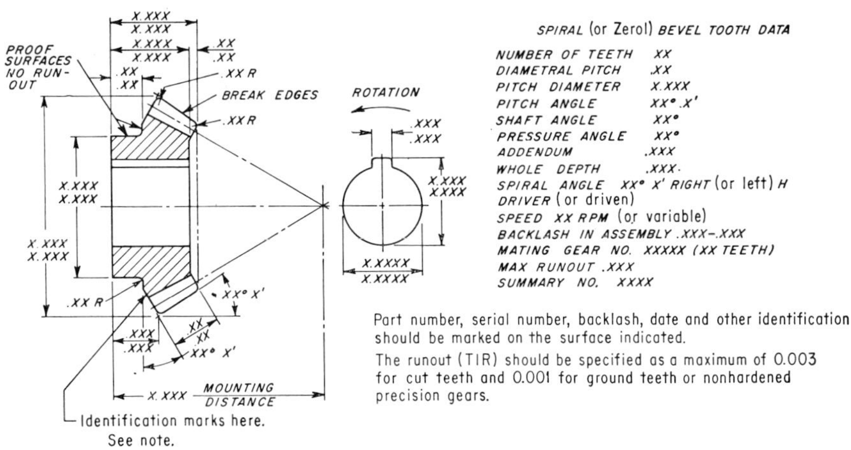

FIG. 11-14. Spiral bevel pinion, hub type.

An alternative method of applying tolerances to the bevel-gear blank is shown in Fig. 11-17. This is based on the principles of zone tolerancing. The desired basic contour is defined by untoleranced dimensions and the permissible deviations are indicated by phantom lines inside and outside the basic contour.

Inspection. Final inspection is made by running a gear-and-pinion pair in a test stand in accurately correct relationship. Acceptance is determined on the basis of sound and the location, extent, and shape of the tooth contact. In Figs. 11-11 through 11-16, the backlash and runout notes refer to such inspection.

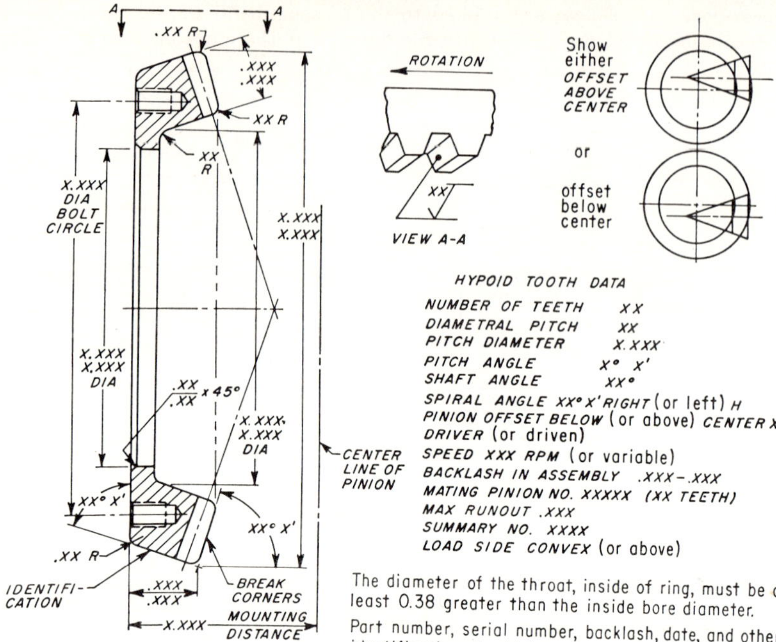

HYPOID TOOTH DATA

NUMBER OF TEETH XX
DIAMETRAL PITCH XX
PITCH DIAMETER X.XXX
PITCH ANGLE X° X'
SHAFT ANGLE XX°
SPIRAL ANGLE XX° X' RIGHT (or left) H
PINION OFFSET BELOW (or above) CENTER X.XXX
DRIVER (or driven)
SPEED XXX RPM (or variable)
BACKLASH IN ASSEMBLY .XXX–.XXX
MATING PINION NO. XXXXX (XX TEETH)
MAX RUNOUT .XXX
SUMMARY NO. XXXX
LOAD SIDE CONVEX (or above)

The diameter of the throat, inside of ring, must be at least 0.38 greater than the inside bore diameter.

Part number, serial number, backlash, date, and other identification should be marked on the outside of the ring. Back face should be ground only if the tooth profiles are ground.

The runout (TIR) should be specified as a maximum of 0.003 for cut teeth and 0.001 for ground teeth or nonhardened precision gears.

Fig. 11-15. Hypoid gear.

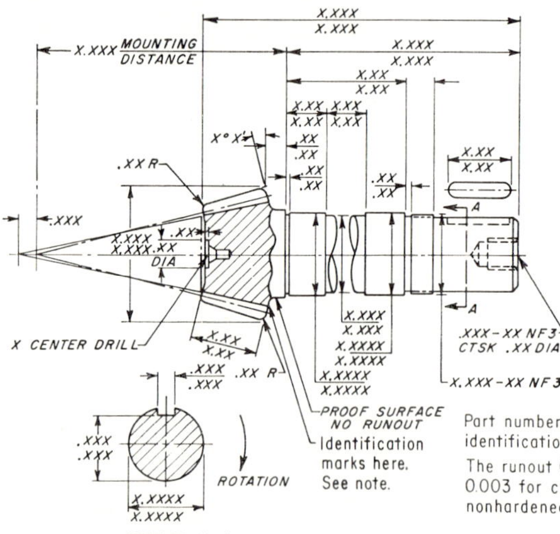

HYPOID TOOTH DATA

NUMBER OF TEETH XX
DIAMETRAL PITCH XX
PRESSURE ANGLE AVERAGE XX°
SHAFT ANGLE XX°
SPIRAL ANGLE XX RH (or LH)
OFFSET BELOW (or above)
CENTER X.XXX
DRIVER (or driven)
SPEED XXX RPM (or variable)
BACKLASH IN ASSY .XXX–.XXX
MATING GEAR NO. XXXXX
(XX TEETH)
MAX RUNOUT .XXX
SUMMARY NO. XXXX

Part number, serial number, backlash, date, and other identification should be marked on the surface indicated

The runout (TIR) should be specified as a maximum of 0.003 for cut teeth and 0.001 for ground teeth or nonhardened precision gears.

Fig. 11-16. Hypoid pinion.

Drawing Data. Essential dimensions for the gear teeth are indicated in Figs. 11-11 through 11-16. These dimensions are mostly for the gear blanks, and of course, design will dictate other shapes than shown on these examples. Some of the data in note form, such as "pitch angle" and "root angle," may, if desired, be placed on the drawing as dimensions.

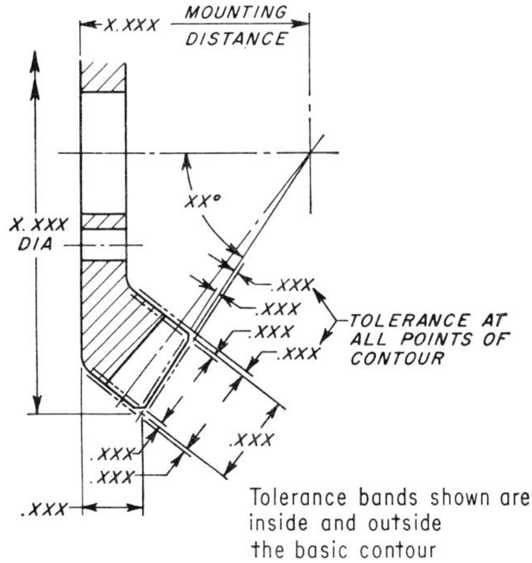

FIG. 11-17. Alternative method of tolerancing bevel gears.

11-5. Wormgears. In all types of wormgears, the nominal pitch circle of the worm is generally assumed to be at the mean of the working depth. In view of the fact that not all worm-thread forms are standardized, it is recommended that the same manufacturer make both the worm and the mating wormgear.

Characteristics. The cylindrical worm, as shown in Fig. 11-18, may be used with either the straight or the throated type of wormgear, as shown in Fig. 11-19. The straight (nonthroated) wormgear and worm (also called "crossed-helical gears") may

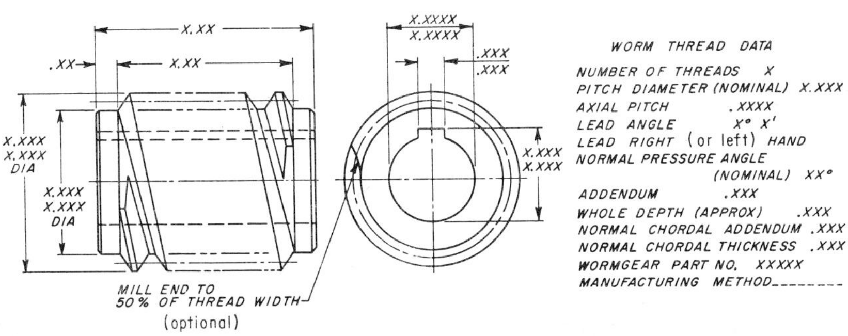

FIG. 11-18. Worm, cylindrical type.

have involute form. The throated-type wormgear, used with a cylindrical worm, usually has a tooth form which matches the form of a worm thread produced by a straight-sided cutter.

Optional Data. Reference to special processes or tools may be desirable, because of the closely related manufacturing of the worm and wormgear. The method of finishing may be specified, such as the use of double-conical grinding wheels or milling cutters, and in such cases the tool data may be included in the specifications. Where this is done, some values, such as the actual pressure angle, may be permitted to vary from the nominal value.

Tolerances. The geometrical tolerances which are used for spur and helical gears are mostly not applicable for wormgears. Wormgears are made in sets, to run only

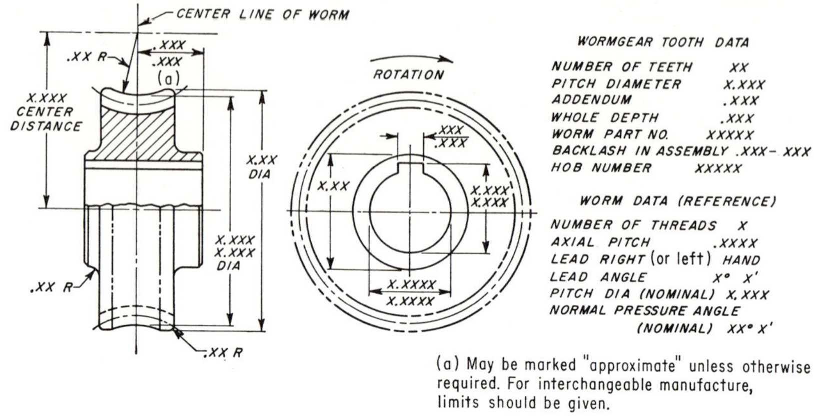

Fig. 11-19. Wormgear, throated type.

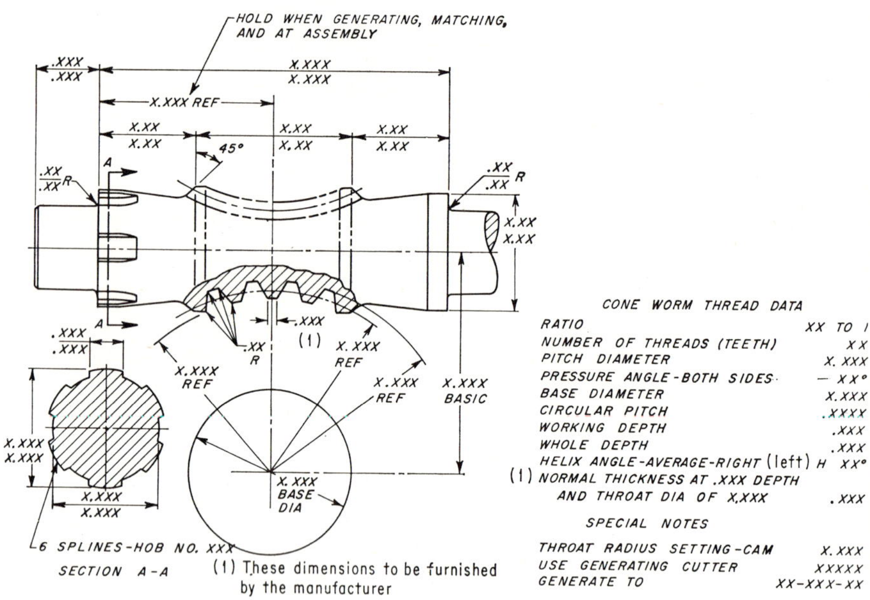

Fig. 11-20. Cone worm.

with each other; therefore, the important tolerances are those for the blank, center distance, and other mounting dimensions. The component parts in the design must be checked to determine that the right amount of clearance and end play will be present throughout the range of tolerances.

Cone-Drive®* Wormgears. These double-enveloping double-throated-type wormgears, as shown in Figs. 11-20 and 11-21, both have straight-sided teeth, all

* Registered trade-mark, Michigan Tool Co., Detroit, Mich.

profiles being tangent to a common base circle, and contact the full depth. The spline shown, near the end in Fig. 11-20, is used for the purpose of driving the part in the hobbing machine; therefore the dimensions and tolerances should be supplied by the gear manufacturer. The axial tooth tip and normal thickness dimensions shown in both Figs. 11-20 and 11-21 are determined as the result of hob design and therefore should be furnished by the gear manufacturer.

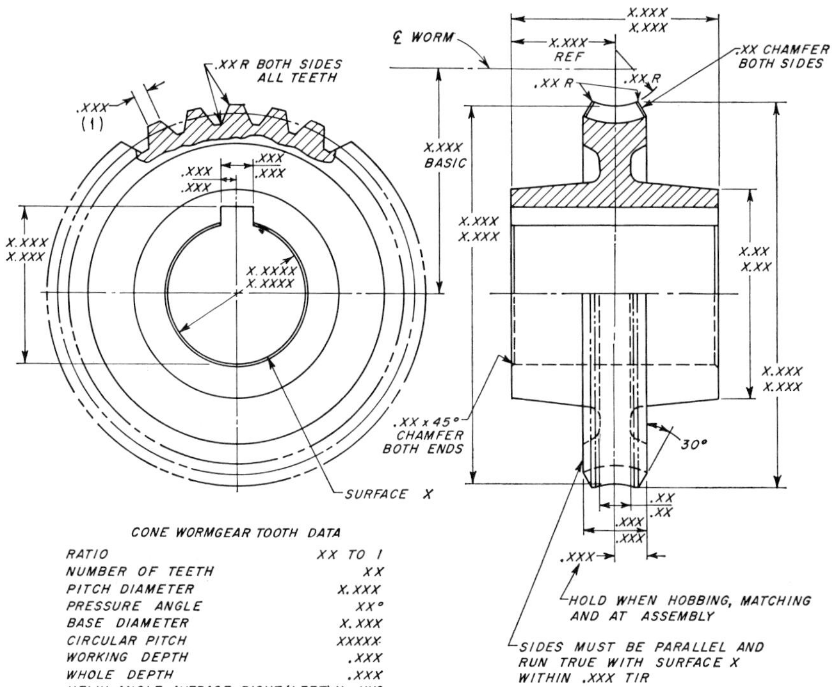

CONE WORMGEAR TOOTH DATA

RATIO	XX TO 1
NUMBER OF TEETH	XX
PITCH DIAMETER	X.XXX
PRESSURE ANGLE	XX°
BASE DIAMETER	X.XXX
CIRCULAR PITCH	X.XXX
WORKING DEPTH	.XXX
WHOLE DEPTH	.XXX
HELIX ANGLE–AVERAGE-RIGHT (LEFT) H	XX°
(1) NORMAL THICKNESS AT .XXX DEPTH AND THROAT DIA OF X.XXX	.XXX

SPECIAL NOTES

USE HOB NO.	XXXXX
DO NOT TURN OVER WHEN HOBBING	
MAKE FROM STANDARD BLANK	XXXXX

(1) These dimensions to be furnished by the manufacturer.

Fig. 11-21. Cone wormgear.

11-6. Special Gear Types. Examples of some special gears will be considered next.

Spiroid®* Gears. The design of these skew-axis gears is explained in other chapters, and Figs. 11-22 and 11-23 show the drawing arrangement and dimensioning recommended. The backlash variation, pinion-and-gear runout, and tooth bearing may all be checked in the operating position. The summary sheets (furnished by the manufacturer), include much of the data on the drawing and in addition the data on cutting tools with their setting angles, driving gears, machine identification, heat-treatment, and inspection data. The pinion shown in Fig. 11-22 is a cantilever (overhung) type of mounting; however, pinions with straddle-type mounting may also be used.

Planoid®* Gears. The Planoid is a right-angle gear for ratios from 1.5 to 10. It is a skew-axis drive which is designed with a pinion offset of approximately one-sixth

* Registered trade-marks, Illinois Tool Works, Chicago, Ill.

to one-third of the gear diameter. The tooth flanks of the gear are plane surfaces, as shown in Fig. 11-25. Tools for cutting these gears are special for each particular ratio and center distance. The basic member of a gearset is the gear, because all tooth-design specifications are applied to it.

The gear teeth are straight and may be milled, ground, or broached. Both sides of the teeth are cut in one pass (see Fig. 11-25).

The pinion teeth are cut in a continuous generating operation similar to hobbing (see Fig. 11-24).

Face Gears. The face gear is one which is produced on the face of a ring by the generating action of a gear-shaper cutter. The axes of the cutter and blank during the cutting process are usually located at a 90° angle to each other. The inclination of the

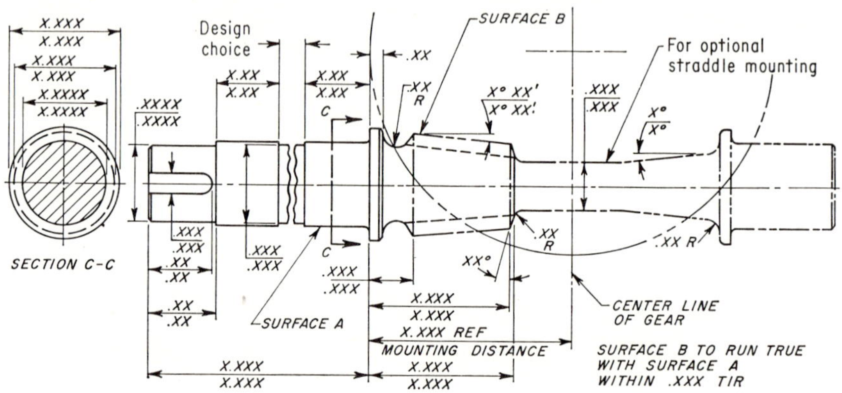

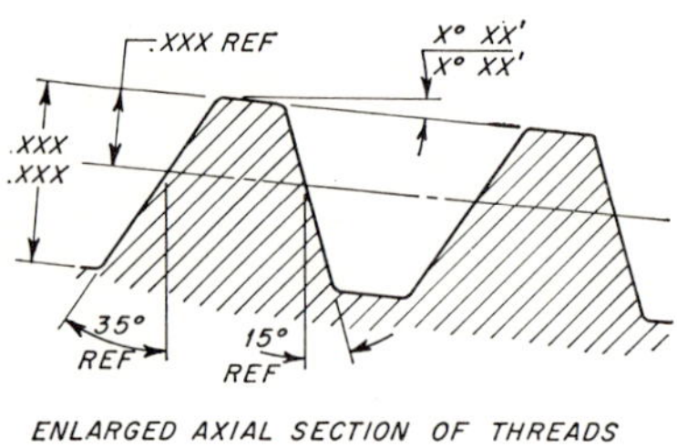

SPIROID PINION DATA

RATIO	XX TO 1
NO. OF THREADS LH (RH)	XX
CENTER DISTANCE	X.XXX
MATING GEAR NUMBER	XXXXX
LEAD	X.XXXX
ZERO PLANE RADIUS	.XXX
THREAD ANGLE	XX° XX'
TOLERANCE CLASS	X
BACKLASH	.XXX-.XXX
SUMMARY SHEET NO.	XXXX

FIG. 11-22. Spiroid pinion.

sides of the teeth varies from the inside to outside extremities, being less at the inner extremity. The tooth sides are substantially straight. Where the axes of the face gear and pinion intersect, the face gear is called "on-center," and where the axes do not intersect, it is called an "offset" face gear. The mating pinion may have either spur or helical teeth, and the cutter used for the gear should usually have the same number of teeth as the mating pinion and generate on the design pinion center distance (see Figs. 11-26 and 11-27).

The length of the active gear face is limited by two conditions:

1. Trimming of the tooth flank at the inner end. The cutter diameter extends on an arc on both sides of the line of travel and will eventually trim the gear teeth at the inner end, since these gear teeth converge toward the center.
2. Top land coming to a point at the outer end. The increasing inclination of the sides of the teeth, as they recede from the center, will eventually eliminate the top

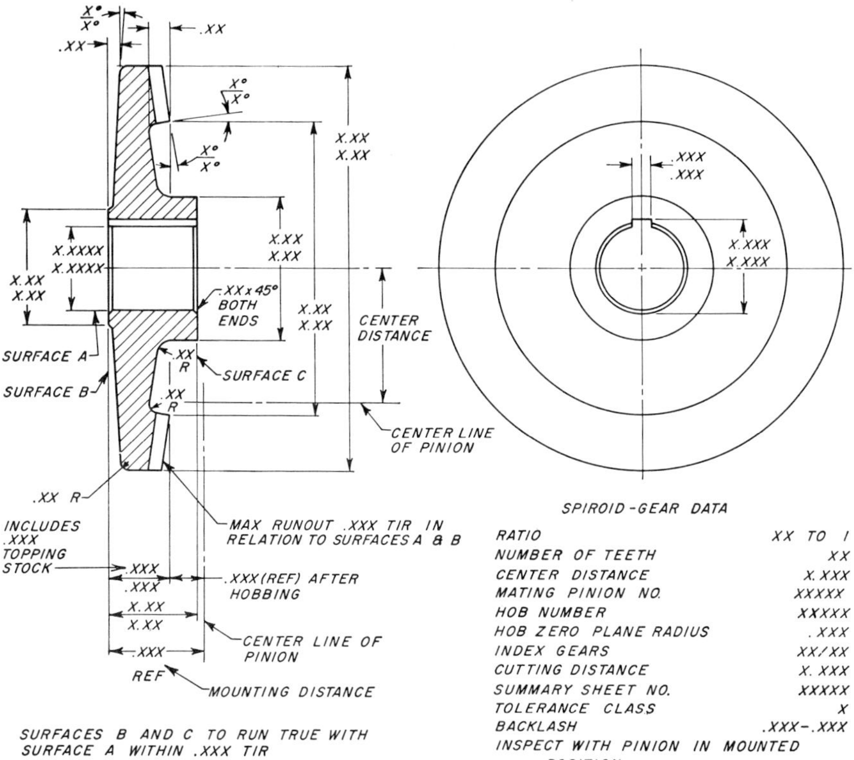

SPIROID-GEAR DATA

RATIO	XX TO 1
NUMBER OF TEETH	XX
CENTER DISTANCE	X.XXX
MATING PINION NO.	XXXXX
HOB NUMBER	XXXXX
HOB ZERO PLANE RADIUS	.XXX
INDEX GEARS	XX/XX
CUTTING DISTANCE	X.XXX
SUMMARY SHEET NO.	XXXXX
TOLERANCE CLASS	X
BACKLASH	.XXX–.XXX
INSPECT WITH PINION IN MOUNTED POSITION	

FIG. 11-23. Spiroid gear.

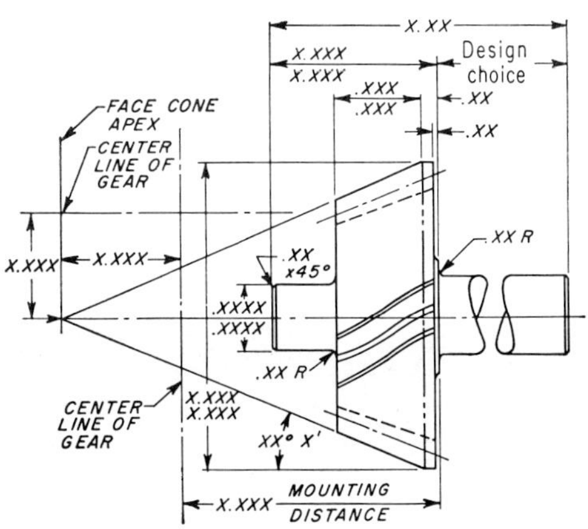

PLANOID PINION TOOTH DATA

NUMBER OF TEETH	XX
RATIO	XX/XX
CIRCULAR THICKNESS	.XXX
WHOLE DEPTH	.XXX
PITCH ANGLE	XX° X′
ROOT ANGLE	XX° X′
SPIRAL ANGLE	XX° X′ LH (or RH)
BACKLASH IN ASSY	.XXX–.XXX

FIG. 11-24. Planoid pinion.

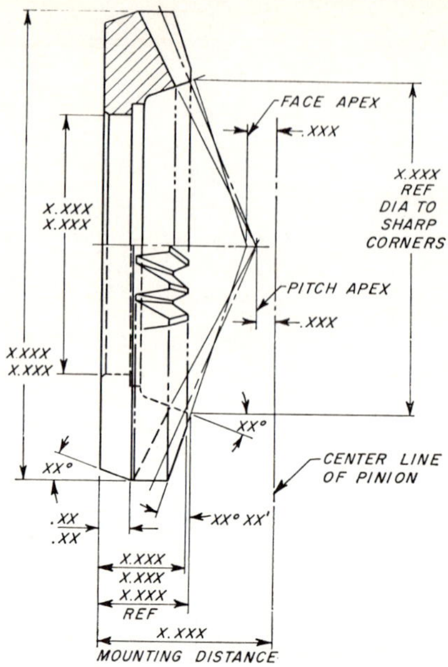

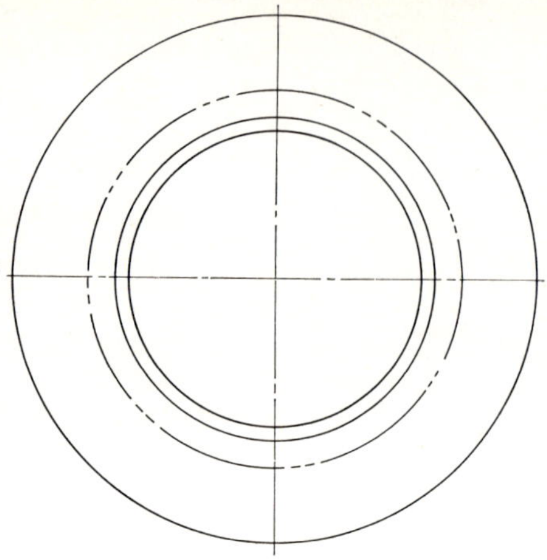

PLANOID-GEAR TOOTH DATA

NUMBER OF TEETH	XX
RATIO	XX/XX
OFFSET OF PINION	X.XXX
CIRCULAR THICKNESS	.XXXX
WHOLE DEPTH	.XXX
PITCH ANGLE	XX°
ROOT ANGLE	XX°
SPIRAL ANGLE LH (or RH)	XX°
PRESSURE ANGLE-DRIVE	XX°
PRESSURE ANGLE-COAST	XX°
BACKLASH IN ASSY	.XXX-.XXX

Fig. 11-25. Planoid gear.

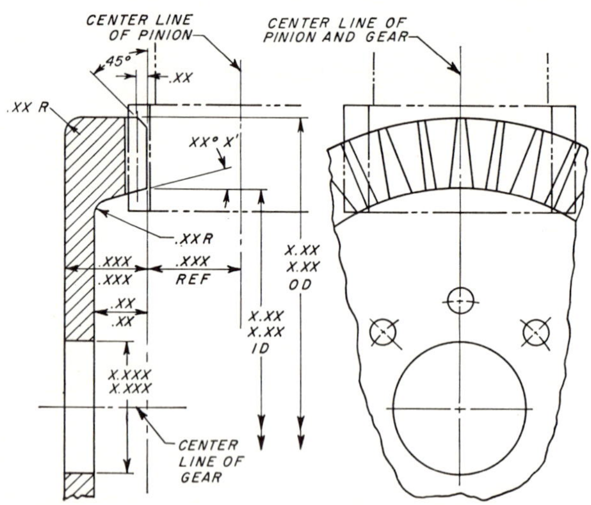

FACE-GEAR TOOTH DATA

NUMBER OF TEETH	XX
ADDENDUM	.XXX
WHOLE DEPTH	.XXX
BACKLASH IN ASSY	.XXX-.XXX

REFERENCE DATA-PINION

NUMBER OF TEETH	XX
OUTSIDE DIAMETER	X.XXX
DIAMETRAL PITCH	XX
PRESSURE ANGLE	XX°
PITCH DIAMETER	X.XXX
ADDENDUM	.XXX
WHOLE DEPTH	.XXX
TOOTH THICKNESS (CIRC) AT PITCH CIRCLE	.XXX

Fig. 11-26. Face gear, "on-center" type.

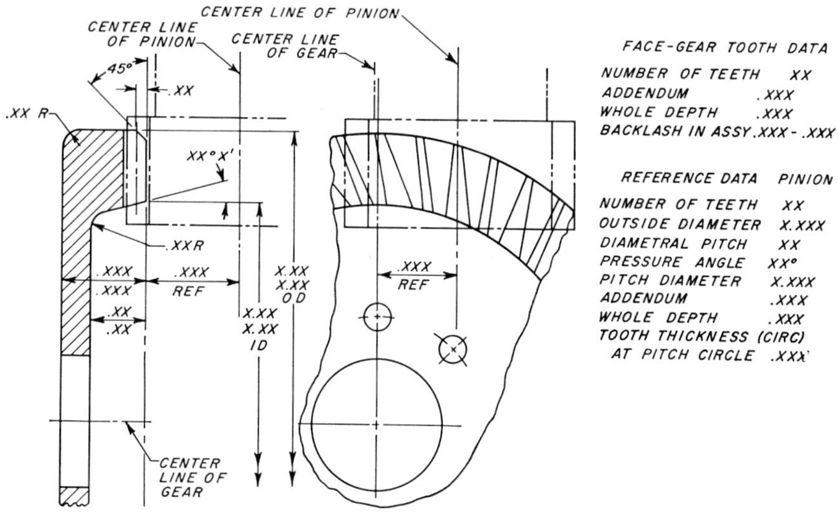

FIG. 11-27. Face gear, "offset" type.

land until it becomes a point. For "offset" gears, the amount of offset affects both conditions 1 and 2. See Chap. 5 for design data.

The "on-center" gear is shown in Fig. 11-26 and the "offset" gear in Fig. 11-27. In both cases the reference data for the pinion teeth define the shaper cutter to be used.

REFERENCE

1. American Gear Manufacturers Assoc.: "Formats for Gear Specification Data," AGMA 114.01, October, 1960.

Chapter 12

GEAR REACTIONS AND MOUNTINGS

By

MARTIN A. HARTMAN, *Research Specialist, Rocketdyne Division of North American Aviation, Canoga Park, Calif.*

No matter how perfectly a gear is designed and made, it must be properly mounted with all its reactions and moments completely contained if it is to function without failures.

In this chapter the methods for calculating gear reactions for each type of gear will be given, and the principles of proper gear mounting will be explained and examples shown for each gear type.

12-1. Mechanics of Gear Reactions. The main function of a gear is to transmit motion and/or power. The main function of the supporting body is to neutralize or create a state of equilibrium. Since a gear is a rotating or moving body, a state of "dynamic equilibrium" must be obtained.

To be in dynamic equilibrium, all the reactions from the rotating gear must be neutralized by equal and opposite forces supporting the gear shaft. All couples and moments from the gear or power source must also be contained. In essence, the total work forces and moments in must be equaled by the total forces and work out. Only the most basic form of dynamics will be considered. It is realized that shaft unbalance, accelerating and decelerating changes, and various other dynamic forces exist. These are normally not an important factor. Gears generally rotate about a near uniform center axis which always tends to pass through its center of gravity and percussion; gears and the connecting bodies reach a state of relative constant velocity. Only short time periods of extreme acceleration or deceleration exist. However, these conditions must be considered and taken into account when they form the major loading conditions. They will not be considered further here.

In mechanics there are basic quantities which must be defined and understood.

Force is any action causing variation of velocity or motion in either magnitude or direction. Force will be measured in pounds. Distance is linear units measured in inches or feet. Time is generally expressed in minutes. Work is the product of distance and force and is in inch-pounds. Power is work divided by time or inch-pounds per minute; a common rating is horsepower, which is the equivalent of 396,000 inch-pounds per minute. Velocity is unit of length traversed in a given time interval and is expressed as feet per minute. Acceleration is the rate of change of velocity for a given time or feet per minute per minute. Inertia is that property of a body at rest to stay at rest and a body in motion to stay in motion, unless acted upon by an external force.

To have dynamic equilibrium, all the laws of dynamics must be obeyed and all forces must have direction, magnitude, and a point of application. This will be shown by a graphical or pictorial representation by means of a line with an arrowhead. The arrow denotes direction, the length of line its magnitude, and the tail the point of origin.

Summation of Forces and Moments. Any number of lines of force in three planes can be resolved into one resultant.

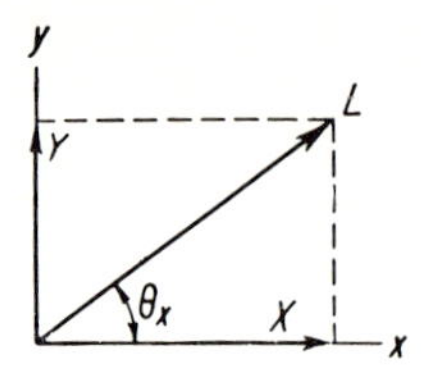

FIG. 12-1. Vector diagram.

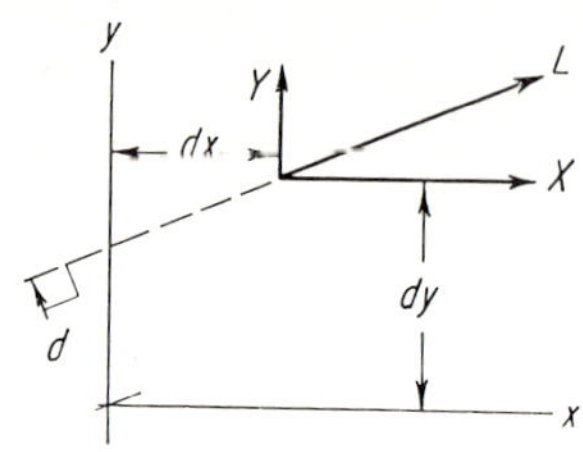

FIG. 12-2. Moment diagram. $M = L \cdot d = X\,dy + Y\,dx$.

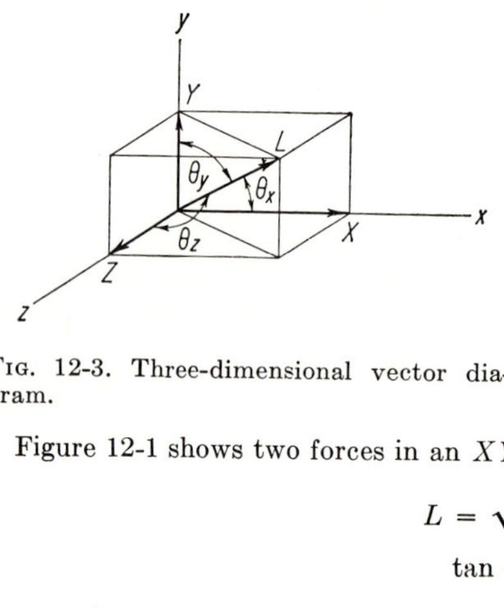

FIG. 12-3. Three-dimensional vector diagram.

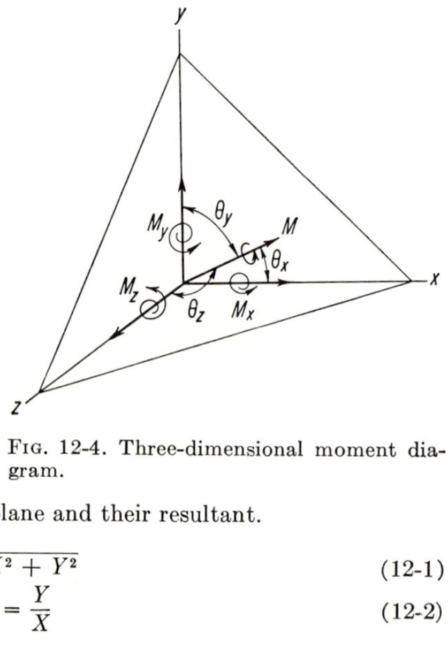

FIG. 12-4. Three-dimensional moment diagram.

Figure 12-1 shows two forces in an XY plane and their resultant.

$$L = \sqrt{X^2 + Y^2} \tag{12-1}$$

$$\tan \Theta_x = \frac{Y}{X} \tag{12-2}$$

conversely

$$X = L \cos \Theta_x \tag{12-3}$$

$$Y = L \sin \Theta_x \tag{12-4}$$

A moment of a force is equivalent to the sum of the individual moments of its components. This is true in either coplanar or three planes.

In Fig. 12-2 a moment is shown in the XY plane:

$$M = L \cdot d \tag{12-5}$$

$$M = X(dy) + Y(dx) \tag{12-5a}$$

These same laws hold for three planes, and for dynamic equilibrium their total sum must equal zero except for the power transmitted or lost.

In Fig. 12-3 for forces

$$L = \sqrt{X^2 + Y^2 + Z^2} \tag{12-6}$$

$$\begin{aligned} X &= L \cos \Theta_x \\ Y &= L \cos \Theta_y \\ Z &= L \cos \Theta_z \end{aligned} \tag{12-7}$$

Summation of moments is as shown in Fig. 12-4.

$$M = M_x \mathbin{+\!\!\!\!\rightarrow} M_y \mathbin{+\!\!\!\!\rightarrow} M_z \tag{12-8}$$

or

$$M = \sqrt{M_x^2 + M_y^2 + M_z^2} \tag{12-8a}$$

$$\cos \Theta_x = \frac{M_x}{M} \qquad \cos \Theta_y = \frac{M_y}{M} \qquad \cos \Theta_z = \frac{M_z}{M} \tag{12-9}$$

It is also important to know two other facts pertaining to couples and moments.

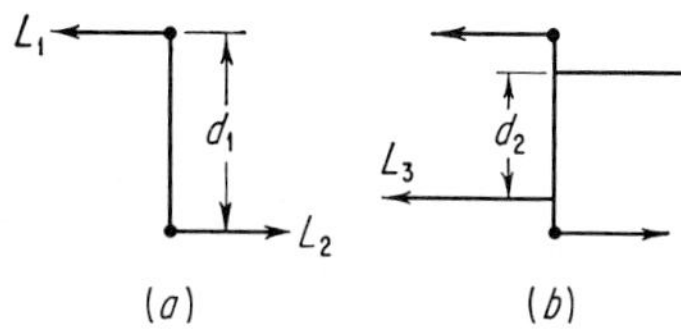

FIG. 12-5. A couple. (a) Couple. (b) Couple in equilibrium.

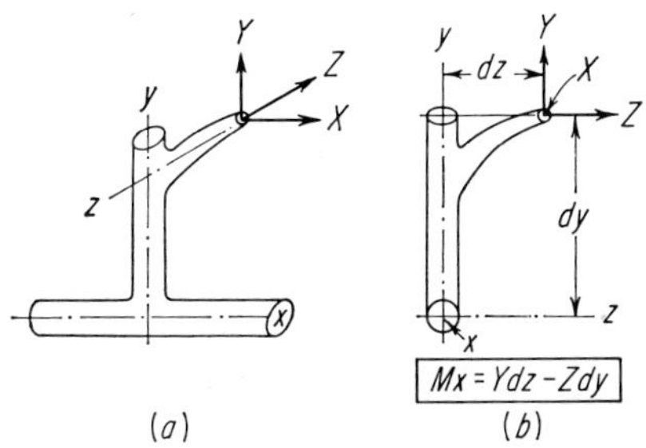

FIG. 12-6. Force vectors at a point in space.

Figure 12-5 shows a couple which has no resultant; the two forces are equal but act in opposite directions. This produces rotation, and the moment is the product of one force multiplied by the entire distance between them. Only an equal and opposite couple can balance another couple. It cannot be balanced or held stationary by a single force or reaction (see Fig. 12-5b).

All gears rotate about an axis, and it is here that the application of moments in space can be simplified. No force has a moment about a parallel axis.[2,*]

Figure 12-6a and b shows how the moments in space are made coplanar and the X axis is a point. Therefore, the moment of force about the X axis is due only to the sum of the Y and Z forces and arms.

To have equilibrium,

$$\Sigma X = 0 \qquad \Sigma Y = 0 \qquad \Sigma Z = 0 \tag{12-10}$$

$$\Sigma M_x = 0 \qquad \Sigma M_y = 0 \qquad \Sigma M_z = 0 \tag{12-11}$$

Since this shows only three independent variables, there can exist a maximum of three unknown values. To have dynamic equilibrium there cannot be more than six unknown quantities to be solved, and both sets of equations (12-10) and (12-11) must be completely satisfied. These rules always hold true for gears rotating about any axis.

To reduce gear reactions and their resultant bearing-load calculations to the

* Superscript numbers refer to references at the end of the chapter.

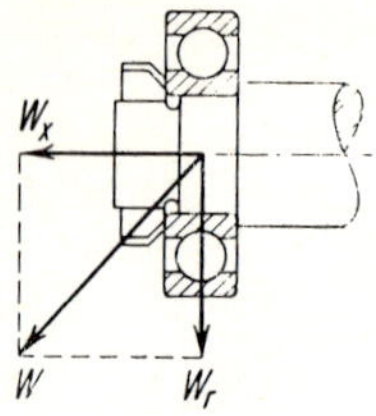

FIG. 12-7. Load reactions at a bearing. W = combined load. W_r = radial load at 90° to axis of rotation. W_x = thrust load directly along axis of rotation.

simplest form, no matter how many moments or forces there are on a gear, always resolves itself to two basic bearing loads, one of which is axial and the other radial, as shown in Fig. 12-7.

The one basic rule that must be applied to all gear mounting and mechanics analysis is "summation of all forces must equal zero and summation of all moments must equal zero."

12-2. Basic Gear Reactions, Bearing Loads, and Mounting Types. There are various sources of gear reactions and bearing loads. These are further complicated by the manner in which the gear and its shaft are mounted.

The main sources of loads are torque, reactions, weight, centrifugal forces, and vibrations. In most cases the gear torque is the main applied load and is usually caused by power input and work being done at the output.

To determine torque or twisting moment and loads from horsepower, the amount of horsepower, the revolutions per minute, the pitch diameters of gear and pinion, and the reduction ratio must be known.

Let P_H = horsepower
T = torque, lb-in.
n = rpm
r = pitch radius
d = pitch diameter
M_g = reduction ratio or gear teeth divided by pinion teeth

$$T = \frac{P_H \times 63{,}025}{n} \tag{12-12}$$

If gear reduction or increase ratios are known and torque is known for one pinion, any other torque and load for meshes in a gear train may be computed. A decrease in speed increases torque and vice versa.

$$T_2 = T_1 M_g \tag{12-13}$$

The tangential load at the gear pitch diameter is designated W_t, the gear pitch radius is r, and the tangential driving load is calculated below.

$$W_t = \frac{T}{r} \tag{12-14}$$

Gear Reactions to Bearings. With W_t known it is possible to calculate the separating force W_r' and the axial force W_x. Once these are determined, the values may be added vectorially so that W_t and W_r' determine the total radial load W_r on the gear and bearing and the axial forces W_x.

As stated in Art. 12-1 and shown in Fig. 12-8 the total radial (W_r) and axial loads (W_x) determined in space are all that is necessary to determine total loads W on the gear teeth and reactions on bearings.

$$W_r = W_r' \leftrightarrow W_t \tag{12-15}$$
$$W = W_r \leftrightarrow W_x \tag{12-16}$$

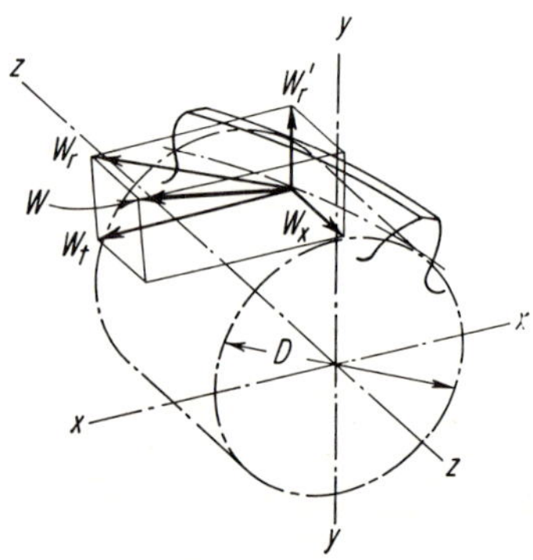

FIG. 12-8. Tooth reactions on a gear loaded in three planes.

Directions of Loads. The method of finding the calculating W_r', W_r, W_x will be found later on and is shown in spur and helical gears.

There are also general rules which hold true for directions of loads on driving and driven members.

"The radial load on the bearing supporting a gear tooth that is driving is always opposite to its direction of rotation." "The radial load on the bearing supporting a gear tooth that is being driven is always the same as its direction of rotation."

The rules for determining axial loads tend to be more complicated. "When viewed from the axis of rotation and the hand of spiral is left on a counterclockwise rotating driving gear its axial bearing thrust is away from the viewer." "When viewed from the axis of rotation and the hand of spiral is right on a counterclockwise rotating driving gear the axial bearing thrust is toward the viewer." "When viewed from the axis of rotation and the hand of spiral is left on a clockwise rotating driving gear the axial bearing thrust is toward the viewer." "When viewed from the axis of rotation and the hand of spiral is right on a clockwise rotating driving gear the direction of axial bearing thrust is away from the viewer." "The direction of the driven gears axial thrust will always be directly opposite the axial thrust of the driving gear." This can all best be shown in graph form in Table 12-1.

Table 12-1. Direction of Axial Thrusts on Driving and Driven Gears*

Hand of spiral	Direction of rotation	Driving	Driven
Left	Clockwise	Toward viewer	Away from viewer
	Counterclockwise	Away from viewer	Toward viewer
Right	Clockwise	Away from viewer	Toward viewer
	Counterclockwise	Toward viewer	Away from viewer

* For thrust directions on bevel and hypoid gears, see Art. 12-7.

It should always be kept in mind that the tangential, radial, axial, and total vectorial sums of any of these same components on a driving gear are always directly opposite to those on the driven gear.

Additional Considerations. Consideration must be given not only for the axial and radial bearing loads generated from the geometry of the gears but for the other external forces which must be supported by the shaft and bearings. For example, spur gears should not normally generate any axial thrusts, but on a common shaft a pump, fan or propeller could be attached, generating tremendous axial forces or centrifugal forces, and these would have to be restrained as well as the gear forces.

Transfer of power in any mechanism always results in losses due to inefficiency. In the calculations presented here, 100 per cent efficiency is assumed. In types of gearing with high rubbing rates, such as worms, the power should be reduced because of the higher losses incurred.

Weights of actual gears usually are so small they are of no consequence. Direction of gravity and weight must be considered when large gears or when large masses are attached to the gear shaft. Care must also be taken in allowing for the effect of g, or gravity loadings. In some cases values as high as 15 g's are encountered.

Forces caused by the rotation of a mass around a center different from its own center of gravity are called centrifugal forces. Parts which are out of balance, have large external masses, or rotate in a planetary field have centrifugal forces which must be considered. The most common formula for calculation is

$$W_{cf} = \frac{\bar{r}wn^2}{35,200} \qquad (12\text{-}17)$$

where W_{cf} = centrifugal force in pounds
$\bar{r}$ = radius from rotational center to center of gravity
n = rpm
w = weight, lb

or, where inch-ounces of unbalance are known,

$$W_{cf} = 1.75 \left(\frac{n}{1{,}000}\right)^2 \times \text{in.-oz} \tag{12-18}$$

All gear reactions can be resolved into tangential force, separating force, and axial thrust. The relative values of these will vary in intensity as to the types of gears, and the loads applied to the bearings are extremely dependent on the type of mounting used.

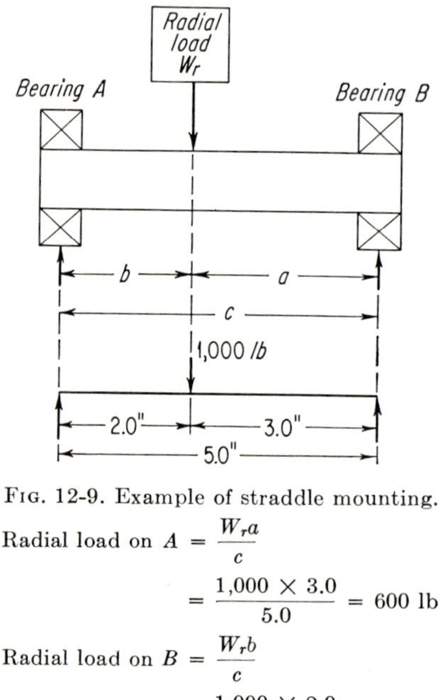

FIG. 12-9. Example of straddle mounting.

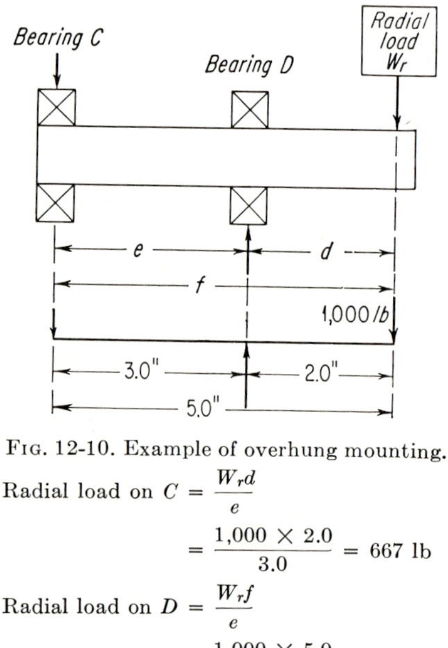

FIG. 12-10. Example of overhung mounting.

$$\text{Radial load on } A = \frac{W_r a}{c} = \frac{1{,}000 \times 3.0}{5.0} = 600 \text{ lb}$$

$$\text{Radial load on } B = \frac{W_r b}{c} = \frac{1{,}000 \times 2.0}{5.0} = 400 \text{ lb}$$

(600 + 400 = 1,000. Check.)

$$\text{Radial load on } C = \frac{W_r d}{e} = \frac{1{,}000 \times 2.0}{3.0} = 667 \text{ lb}$$

$$\text{Radial load on } D = \frac{W_r f}{e} = \frac{1{,}000 \times 5.0}{3.0} = 1{,}667 \text{ lb}$$

(1,667 − 667 = 1,000. Check.)

Types of Mountings. There are really only two basic types of mountings—straddle and overhung.

In straddle mounting, as shown in Fig. 12-9, the radial load is divided in inverse proportion to the distance from points of application to the total distance between shaft supports. The bearing reactions act opposite to the direction of the load from the gearing.

In an overhung mounting as in Fig. 12-10, the load is applied outside the two supports. This produces a greater reaction or load on the bearing nearest the applied load. The load on the bearing farther away is smaller. The bearing loads are not in the same directions. The bearing reaction nearest the applied load is opposite to the direction of the load. The reaction on the other bearing, farthest away, is in the same direction as the applied load.

Since, in an overhung-mounted gear the largest bearing load is nearest the mesh, this assumes greater importance in heavily loaded optimum-designed bearings mountings. To compensate for this, generally the bearing nearest the mesh is a type which has the greatest radial load-carrying ability. The bearing farthest away is often a ball or axial resistance bearing and constrains both the axial and the smaller radial loads. This method is used when there are no great temperature or deflection problems. In some cases two different-sized bearings are used with the greater-capacity bearing nearest the overhung member. It is wise to locate the axial-thrust retaining bearing next to the gear when large temperature variations or deflections exist. Change of rotation and potential thrust-direction changes must be taken into account.

While it is necessary to balance out reactions loads, from now on whenever a load is calculated it will be the actual load applied to the bearing or support in the correct direction. It will not be the bearing reaction, which opposes the load and is in the opposite direction of the bearing load.

There are relative merits between the overhung- and straddle-mounted gears and shafts. The most commonly used and preferred method is the straddle-mounted. It balances loads, more efficiently utilizes the bearings, and in most cases reduces misalignment and deflection problems.

When there are extreme or unusual axial-thrust loads or special mounting problems, the overhung mounting has its merits.

Efficiencies. Gear efficiencies and resultant power losses can cause great variation in the actual power delivered and the load transmitted. These loss values vary from ½ per cent per mesh to values as high as 80 per cent per mesh depending on the type of gear, lubrication, bearings, and degree of accuracy of manufacturing. It is a poor gear design or application which has an efficiency less than 50 per cent. It is generally a nonplanar application which has great losses. Some designers allow a set value for each mesh or increase the effective angle several degrees off angle. While 100 per cent efficiency is assumed in all these calculations, it is always necessary to at least consider the effect of efficiency. If the power loss is low, generally a few percentage points or less, it is best to ignore this effect on bearings and gears. Coplanar gears can be from 97 to 80 per cent efficient. When critical, experience or actual test data should be used to determine losses and reference should be made to Chap. 14.

As a quick guide for efficiencies or if useful data are not available, the values in Table 12-2 are suggested for good conditions.

Table 12-2. Approximate Gear Losses for Various Types

Different types of gear	*Total range of losses per mesh, %*
Spur and helical external	½–3
Spur and helical internal	½–3
Bevel gears	½–3
Worms	2–50
Spiroids®*	2–50
Hypoids	2–50
Crossed helicals	5–50

It should be noted that all nonplanar gears have great variation due to ratios, sizes, materials, lubricants, and relative sliding.

* Registered trade-mark of Spiroid Div., Illinois Tool Works, Chicago, Ill.

12-3. Basic Mounting Arrangements and Recommendations. Once the bearing loads are calculated for any gear, experience, art, and common sense have given some pointed recommendations and rules. These rules apply to all simply mounted gears no matter what type or shape they take. Various other rules for each specific type of gear will be shown later under their appropriate heading.

Keep bearing mountings for gears as close as possible to their faces allowing reasonable space for lubrication and arrangements. This eliminates large moments and reduces vibration problems.

Whenever possible, use only two bearings for each gear shaft. When making a center-distance layout for gear proportions, also consider shaft, spline, and bearing sizes and load-carrying abilities. These often are limiting conditions.

Proper support must be given to the gear tooth by both the rim and the web. The following applies to most properly stressed gear designs. The rim, that part of the gear blank directly under the teeth, should have a minimum metal thickness at least equal to the whole depth, extending below the root of the gear for its entire tooth length. Commercial gears always exceed this value. The web thickness supporting the rim should be a value of 100 to 70 per cent of the whole depth for aircraft and rocket applications and two to three times the whole depth for commercial applications. Quite often the web contains lightening holes. The space of the metal between the holes should not be less than the width of the lightening hole. When gears are mounted in a train, it is wise to select the number of lightening holes in a manner to ensure a hunting action. This reduces the chance of torsional and vibration build-up in the gear train.

Whenever possible, straddle-mount both members of a mesh. Straddle-mounted gears should have the spread between bearing of approximately 70 per cent of the pitch diameter as a minimum. When only one gear can be straddle-mounted it should be the gear with the highest radial load.

Overhung gears should have a spread between bearing of 70 per cent of the pitch diameter and the spread between bearings should be at least twice the overhang. The shaft supporting the overhung gear should be greater in diameter than the overhang (see Fig. 12-14).

Idlers can often be positioned to offset and minimize bearing loads. Consideration must be given to the two meshes on the gear train setting up undesirable torsional vibration and overloads.

In general, for all types of gears, the tangential forces applied to the bearing are opposite to the direction of rotation of the driving member. The separating forces are away from the tooth surfaces. The driven gear has its tangential forces and separating forces always opposite to the driving pinion or driving force. Reverse direction of rotation and there is a reversal of tangential driving loads but the separating forces tend to remain in the same direction away from the teeth (toward center).

Axial thrusts vary for all types of gears, but they also reverse direction with a reversal of rotation except in the case of Zerol®* and straight bevel gears.

If the total force on any pinion or any gear of a single mesh is found, the force on the mating gear of that mesh is directly opposite to it. When calculations are laborious, this fact can be used to establish the direction and magnitude of the forces on the mating gear for that individual meshing point in space.

All shafting should be checked for torsional and moment load-carrying ability. The distance between supports and shaft diameter should preferably be designed so that operating speed is below the first critical speed. The shaft should not operate near any critical vibration excitation factors.

Bearing and Shaft Alignment. It is extremely important to minimize gear-tooth mounting displacements. The bearing sizes and shaft dimensions are determined only in part by strength and life considerations. The above-mentioned rules are good examples which should be followed besides strength and stress knowledge. The bearing spacing and shaft stiffness as well as the housing and methods of attachments between gear, shaft, bearing, and housing must be considered. In most applications the deflections should not exceed 0.001 at the working surfaces between the gear and pinion. Total amount of allowable misalignment between gear surfaces

* Registered trade-mark of Gleason Works, Rochester, N.Y.

should be further investigated and calculated. The method of calculating increased stresses in spur and helical gears due to misalignment should be investigated by the formulas shown in Chap. 13. The total contributing factors to misalignment should not exceed the calculated or known allowable value.

Bearings. From the tentative gear layout calculate the bearing loads and establish the preferred types and sizes of bearings. Actual calculations should be based on data that follow in detail each type of gearing and mounting that is involved. It is an accepted practice to preload the bearings carrying the thrust loads in order to obtain mountings with minimum total axial displacements under operating loads. The amount of preload depends upon the mounting, load, and operating speed and should be established in collaboration with the bearing manufacturer. At this time the design of the gearbox with complete speed, load, and operating data should be submitted to the bearing manufacturer for approval of bearing sizes, types, bearing mountings, lubrication, and bearing life. The calculation of the bearing-load capacities and life is not included here, but such data and information are readily available from any first-rate bearing manufacturer. Many bearing manufacturers have engineering-design books which are given to users of their products. Selection of dimensional fits can also be learned. Generally, the most important rule is for the tight or interference fit to be between the inner race and shaft and a loose or controlled slippage to be between the outer race and the housing or bearing liner.

Mounting Gears to Shaft. Whenever possible the gear should be made integral with the shaft. However, because of complex shapes and bearing and mounting arrangements, and with several gears and components mounted on a single shaft, this is not always possible. When gears must be assembled onto the shaft, involute splines are often used. In mounting gears with splined bores, a cylindrical centering fit is recommended to avoid excessive eccentricity in the final mounting. The most satisfactory method is to provide a suitable length of cylindrical fit at one or both ends of the gear hub for centering purposes. The splines are then used for driving only. This method is particularly applicable in the case of involute side-fit splines which cannot be adequately centered from the major or minor diameters (see Fig. 12-35). When space or length allowable is critical, the major-diameter-fit involute spline is often used. The internal spline is often broached and the major diameter of the external member is ground to fit the broached part. When used this way the spline is stressed from both tangential driving torques and possible tip loading from interference fits caused by assembly or temperature contractional affects.

Present practice is to avoid the use of minor-diameter-fit involute splines whenever possible. In many applications straight-sided splines and serrations are used. Serrations and straight-sided splines are generally treated in the same manner and their rules of usage are interchangeable. Hardened gears with straight-sided splines in the bore should be centered in assembly by the bore or minor diameter of the splines, which should be ground concentric with the teeth after hardening. Unhardened gears with straight-sided splines should be centered in cutting and in assembly by the major diameter of the splines. Since heat-treatment may introduce distortion in all splines, it is important that the splines be of no greater length than is actually required for load transmission.* In blanks with long hubs, the splines should be located as nearly as possible to the gear teeth and, if possible, directly under the gear teeth.

Keys, pins, and press-fit diameters are used in many gear applications. In data gearing loads normally are light and the mounting method is picked to give the least amount of running-index error. The integral gear and shaft is preferred with the cylinder fit of gear to shaft next. When heavy loads and extreme accuracy requirements are needed, ingenious combinations of various basic mounting methods must be employed.

* Generally the length of a spline is equal to or less than the spline pitch diameter.

Housing. The function of the housing is to form a strong base in which to mount the bearings and support the gears and shafts and to create an environment and space where a satisfactory lubricant may be introduced to lubricate and cool the gears. The housing is also used to mount and support various other components such as accessories and parts near or common to the gearing. In many installations the lubrication system is contained entirely within the housing; in some oil systems, lines into and out of the housing must be supplied. The actual design and calculation of most gearboxes can be so complicated that a general practice is to employ experience or test data or actually to submit the gear housing to loads and measure the results by dial or strain gages. In simple-type shapes basic stress analysis is used. With the bearings selected and the general arrangement established it is possible to proceed with the design of the supporting structure. The design should ensure, whenever possible, that loads from external sources such as belts, propellers, and pumps are carried entirely independently of the gear mountings; otherwise the design must be made sufficiently rigid to carry these external loads without affecting the operating position of the gears. Flexible-type couplings are preferred to connect the assembly to the input and output loads in order to protect the gears from outside forces because of shaft deflections or misalignments.

The housing and supports should be of adequate section and of suitable shape, with ribbing to support the bearings rigidly against the forces, as indicated by the analysis of their magnitude and direction. The choice of material for the housing depends to a great extent upon the application. If weight is not a factor ferrous castings or a combination of ferrous castings with weldments may be used. Where weight is very important, aluminum or magnesium castings may be used for the housing. When light metal housings are used, special design considerations must be given to minimize the effect of temperature changes on the mountings and to provide adequate rigidity under all operating conditions. Best practice calls for mounting the bearings in steel liners either fitted and pinned in the housing bores or cast in and machined in place. Also the thrust bearings which control the position of the gears on their axes should be located close to the gear teeth. Two opposed bearings should be arranged so that temperature changes will not seriously affect preloading. It should be noted that in order to obtain equal rigidity in a light metal casting it may be necessary to use sections of two to four times the thickness of those used in steel or iron.

Inspection Hole. As an aid to assembly and for periodic inspection in service, an opening should be provided in the housing to permit the inspection of the tooth surfaces of at least one member of the pair without disassembly. This is especially desirous in nonplanar-axis gears or in bevel and hypoid gears.

Break-in. While gears and bearings are designed, mounted, and manufactured to carry their rated loads without trouble, the initial period of operation is most critical, and preferably a "break-in" run at lighter loads or slower speed should be made. The load should gradually be increased until full operating load and speed are reached to check the complete gearbox.

12-4. Bearing-load Calculations for Spur Gears. The most common gear used is the spur type. This is true not only because of the simplicity of stress calculations, design, and manufacturing but largely because of the ease of mounting, calculations, and retention of bearing loads. Spur gears are almost always devoid of any large self-generated axial thrusts. Their only components are the tangential driving load and its separating components which results in a radial load only, which is the vectorial sum of these two forces [see Eq. (12-15)].

In a simple spur gear, only a radial load is present and W_r can be used interchangeably with the total load W. Since this fact assumes greater importance in other types of gears, such as helical, W_r will be used here.

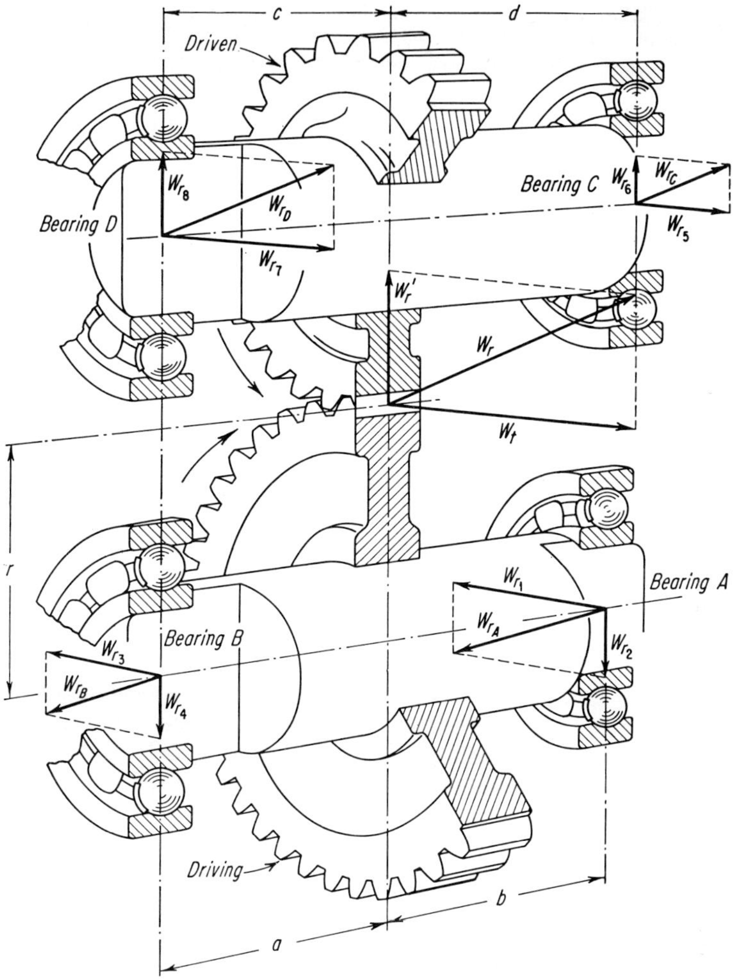

FIG. 12-11. Spur-gear bearing loads.

Forces	Bearing A	Bearing B
Tan force W_t	$W_{r_1} = \dfrac{W_t a}{a + b}$	$W_{r_3} = \dfrac{W_t b}{a + b}$
Separating force W_r'	$W_{r_2} = \dfrac{W_r' a}{a + b}$	$W_{r_4} = \dfrac{W_r' b}{a + b}$
Total radial load	$W_{r_A} = \sqrt{(W_{r_1})^2 + (W_{r_2})^2}$	$W_{r_B} = \sqrt{(W_{r_3})^2 + (W_{r_4})^2}$
	Bearing C	Bearing D
Tan force W_t	$W_{r_5} = \dfrac{W_t c}{c + d}$	$W_{r_7} = \dfrac{W_t d}{c + d}$
Separating force W_r'	$W_{r_6} = \dfrac{W_r' c}{c + d}$	$W_{r_8} = \dfrac{W_r' d}{c + d}$
Total radial load	$W_{r_C} = \sqrt{(W_{r_5})^2 + (W_{r_6})^2}$	$W_{r_D} = \sqrt{(W_{r_7})^2 + (W_{r_8})^2}$

The tangential driving load can be calculated by Eq. (12-14). The separating load is designated W_r'. For universal usage, the separating load is a function of the transverse pressure angle or that pressure angle formed by a plane slicing through the gear tooth at 90° to the axis of rotation. In a spur gear, the normal pressure angle ϕ and the transverse pressure angle ϕ_t are one and the same. To simplify matters later on, the transverse pressure angle ϕ_t will always be used in separating and radial-load calculations.

$$W_r' = W_t \tan \phi_t \tag{12-19}$$

Let W_r designate the total radial load

$$W_r = \sqrt{W_t^2 + W_r'^2} \tag{12-20}$$

If W_t is designated in the X plane and W_r in the Y plane, Eq. (12-20) agrees exactly with Eq. (12-1), which is for forces in one plane.

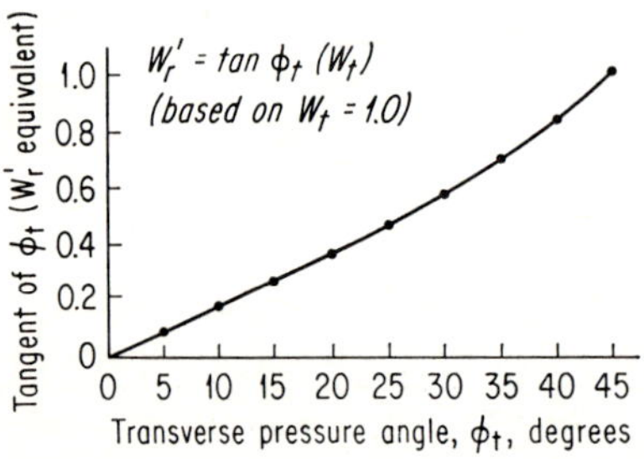

FIG. 12-12. Chart of relation W_r' for W_t = 1.0.

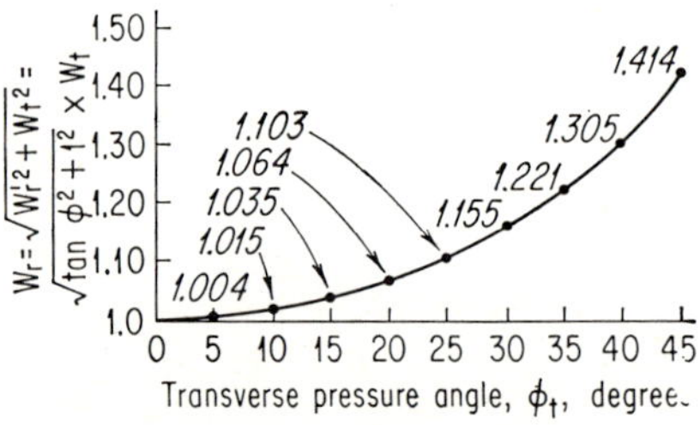

FIG. 12-13. Chart of total radial load vs. ϕ_t.

The values of W_r, W_t, and W_r' will all be considered as originating from the pitch diameter of the pinion or gear and in the center of its contact face, as will W and W_x later on.

Figure 12-11 shows a perspective of two spur gears and their respective driving or driven load.

The pressure angle of the gear controls the effect of the separating component. The smaller pressure angle gives lesser separating and total forces, as shown in Fig. 12-12. Equation (12-20) is plotted against the tangent of the pressure angle to calculate W_r proportions shown in Fig. 12-13. A pressure angle of 0 to 45° represents one unit of tangential driving load in pounds.

Figures 12-9 and 12-10 and sample problems show how to calculate overhung- and straddle-mounted gear resultant bearing reactions.

Spur gears have no calculable axial thrust; however, they tend to walk or to be displaced by torque and other components. It is wise to allow for small amounts of axial movements or to restrain the gear. In heavily loaded spurs, or when spurs are common to shafts carrying other axial loads, it is a necessity to restrain or position them.

Internal Gears. All data shown so far have applied to external gears. However, there are many applications where internal gears are used. In external gears, the direction of rotation is changed in the mesh. In internal gears, the rotation between the two gears is in the same direction. In most cases one or both of the gears are overhung. While it is possible to arrange straddle mounting, normally space and length do not permit it. Internal gears offer the advantage of larger reduction or step up changes in smaller center distances, but at the expense of allowable space for bearing-mounting arrangements.

Procedures for calculating bearing loads are shown in Fig. 12-14; W_r', W_t, and W_r are calculated as before.

Gears in Trains. Many times gears are in trains either in one continuous plane with idlers or else in offset planes on intermediate shafts and in planetary drives. Planetary gears will be considered as a separate type of gear train.

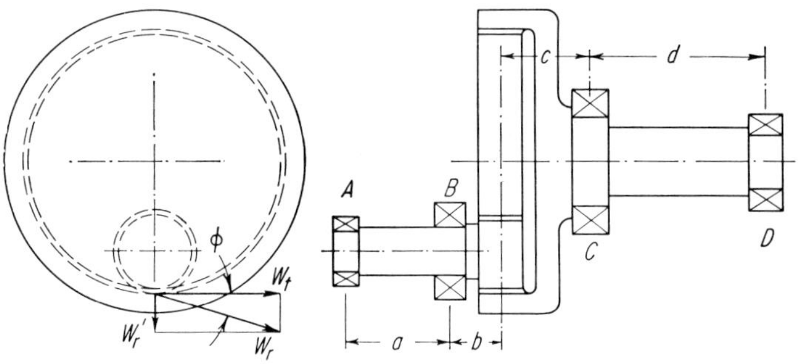

FIG. 12-14. Internal spur-gear bearing loads.

Forces	Pinion loads	
	Bearing A	Bearing B
Tan force W_t	$W_{r_1} = \dfrac{W_t b}{a}$	$W_{r_3} = \dfrac{W_t(a+b)}{a}$
Separating force W_r'	$W_{r_2} = \dfrac{W_r' b}{a}$	$W_{r_4} = \dfrac{W_r'(a+b)}{a}$
Total load	$W_{r_A} = \sqrt{(W_{r_1})^2 + (W_{r_2})^2}$	$W_{r_B} = \sqrt{(W_{r_3})^2 + (W_{r_4})^2}$
	Gear loads	
	Bearing C	Bearing D
Tan force W_t	$W_{r_5} = \dfrac{W_t(c+d)}{d}$	$W_{r_7} = \dfrac{W_t c}{d}$
Separating force W_r'	$W_{r_6} = \dfrac{W_r'(c+d)}{d}$	$W_{r_8} = \dfrac{W_r' c}{d}$
Total load	$W_{r_C} = \sqrt{(W_{r_5})^2 + (W_{r_6})^2}$	$W_{r_D} = \sqrt{(W_{r_7})^2 + (W_{r_8})^2}$

Idlers. The idlers may be used to change direction of rotation, to gain additional distance between centers with smaller-sized and greater quantity of parts, or to provide additional mounting pads at various rpm's for driven components, shafts, or accessories.

If only the input and output shafts are used and a constant horsepower is driven through the train, the load on all the gear teeth in the train is the same. When the idlers are all arranged in one straight line the two separating components W_r' cancel out but the tangential driving loads are added directly. Under the aforementioned conditions, the resultant bearing load is twice W_t.

When the gear train is offset at an angle between centers, the total forces tend to combine or cancel out. For accuracy of results a graphical or vector analysis should be made.

Figure 12-15 shows a vector analysis of idler loads.

If in Fig. 12-15 the angle on BC in relation to AB had been acute instead of obtuse, the reaction W_B would have been much smaller in magnitude. If it is possible, proper selection of rotation and angles of intersection can be adjusted to reduce the total bearing loads generated in an idler gear train.[4]

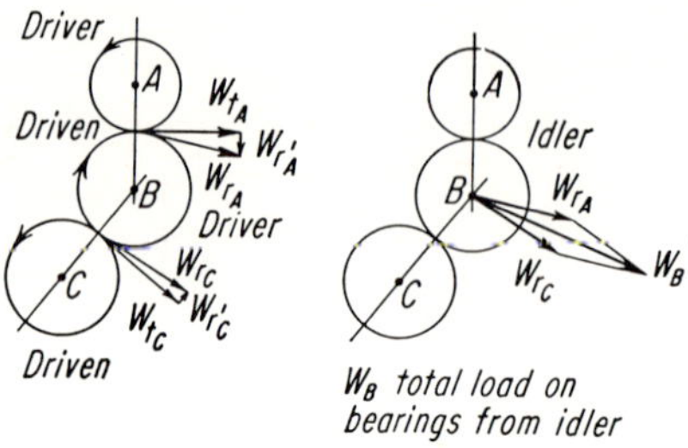

FIG. 12-15. Vector analysis of idler loads.

Intermediate Gears. The other most common gear train in spur gears is the intermediate-shaft arrangement. Its main advantage is the combination of greater gear-ratio reductions and a gain in center distances for mounting various combinations.

There are basically two gear meshes but only three shafts instead of four. Also the planes of mesh are parallel but separated (see Fig. 12-16).

The total loads that are applied to and must be reacted against by bearing D are found by a summation of moments about bearing C. The same is true as to finding the forces on bearing C. Summation of forces, however, is taken about bearing D. When contact is at an angle or several meshes are concerned, both vertical and hori-

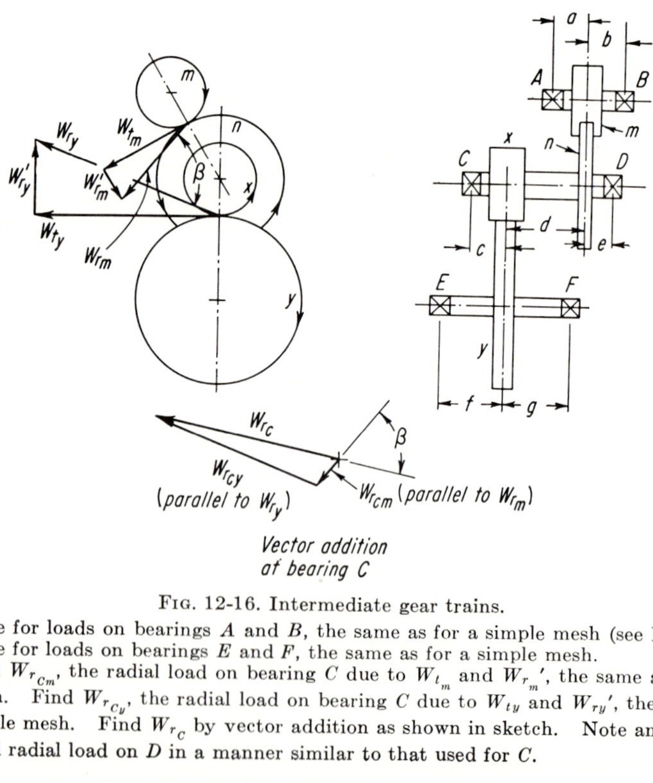

FIG. 12-16. Intermediate gear trains.

1 Solve for loads on bearings A and B, the same as for a simple mesh (see Fig. 12-11).
2 Solve for loads on bearings E and F, the same as for a simple mesh.
3. Find $W_{r_{C_m}}$, the radial load on bearing C due to W_{t_m} and W_{r_m}', the same as a simple mesh. Find $W_{r_{C_y}}$, the radial load on bearing C due to W_{ty} and W_{ry}', the same as a simple mesh. Find W_{r_C} by vector addition as shown in sketch. Note angle β.
4. Find radial load on D in a manner similar to that used for C.

zontal planes of moments must be taken. The bearings must resist a load to make the total moment summation equal zero. The proportions of spacing between bearings are very critical. The distances or moment arms should be selected to favor the F bearing with the lesser load-carrying capacity. For example, bearing F in Fig. 12-16 would in most gearboxes be the smaller because of lack of space. Therefore, the distance g on the y gear was intentionally made longer to reduce the bearing load on F and put the greater load on E, as it should have more space available for a larger bearing with greater load-carrying capacity. Also, whenever possible, the distance between gears and bearings is as small as is feasible to eliminate large moment loads in the shafts. The shaft on x-n gears will have a coupling tendency to cock or walk, and consideration should be given to axial thrust bearings to resist the coupling effect. To aid the positioning of planes of intersection on the gear teeth, avoid large axial moments and overhangs.

Planetary Gears. Planetary gears are used most often for maximum horsepower transmitted in the smallest space or when large changes in speed ratios are needed. The increased carrying capacity is generally possible because the gear reactions tend to equalize each other, reducing bearing loads, size required, and total weight. Helical, bevel, and spur gears are used in planetary drives, but spurs are often preferred because of ease of manufacturing and elimination or reduction of axial-thrust loads.

For simplicity's sake, spur gears has been assumed in these discussions but helical and other types of gears can just as easily be used.

In a simple planetary arrangement there is generally a sun gear, a planet gear, an internal ring gear, and a cage or arm (see Fig. 12-17).

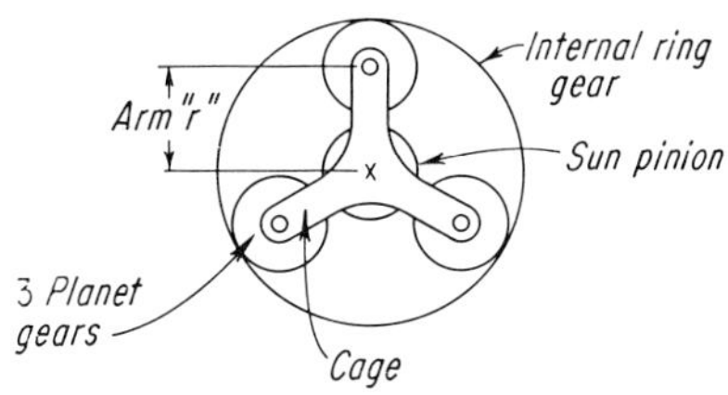

FIG. 12-17. Simple planetary arrangement.

One Planet. Load reactions are different if the planetary drive has one planet gear from those if there are two or more planet gears. The one-planet-gear calculations are basically the same as the combination of an external idler and an internal gear and external pinion. For example, the sun and one mesh planetary drive are the same as two external gears meshing. The reaction on the internal gear is the same as in a simple one meshing with an external pinion. However, on the planet gear, consideration must be given for the reactions on its center of rotation from the cage or arm.

Several Planets. If two or more planets are used, preferably three or more, the reactions are different because three planets mate with one sun gear and one ring gear; and the bearing reactions tend to balance out. There are basically zero resultant separating bearing forces on the sun pinion and the internal ring gear. The planet gear bearing has the loads from the two meshes and the cage driving arm plus the weight and centrifugal force from the planet and its bearings. When there are high-speed planets, the effect of centrifugal force must be considered. If speeds and weights are low, centrifugal force may be ignored. If they are high, compute them by means of Eq. (12-17) or (12-18).

Centrifugal force may become the prime factor in the final decision of bearing design or type of planetary gear drive used.

When the total gear reaction W and W_{cf} (centrifugal force) are found, they are added vectorially or computed as below:

$$L = \sqrt{(W_{cf})^2 + (W)^2} \tag{12-21}$$

where L = total load of all forces
W = total gear reactions
W_{cf} = centrifugal force

However, helix forces, unbalance, and driven propeller or fan reactions must be considered when they are tied into the planetary-drive system. When two or three planets are used, the loads should be considered shared equally by all planets. When there are more than three planets, it is not always possible to share the load equally and some allowance should be made for unequal loading of all the individual planets.

The following special recommendations apply to idlers. Preferably use two or more planets to carry the load. When possible use three planets. If more than three are needed many designers feel it is wise to go to five or more because of a lack of load-sharing tendencies.

Always mount the entire system as flexibly as possible. For example, always mount the planets on spherical floating bearings or mounts. Make all mountings or actual holding surfaces for bearings as rugged as is allowable. Keep planets and outer bearing cages or surfaces as light in weight as can be utilized to reduce centrifugal forces.

12-5. Bearing-load Calculations for Helicals. So far all data have been based on spur gears and all radial values calculated are directly applicable to helical gears if the transverse pressure angle is considered.

Helical gears have thrust and reactions in all three planes. Besides the tangential driving and separating forces, there is also an axial-thrust load. This thrust is due to the helix angle and is normally considered emanating from the tooth contact at the pitch diameter.

Since the thrust is applied at the distance of the pitch radius from the axis of rotation a radial moment is produced. In spur gears no bearing moments had to be considered until the gear was mounted. However, in helicals the moment is produced within the gear.

This information on helicals can be transcribed into the laws of mechanics. Three forces are shown in Fig. 12-8.

$$
\begin{aligned}
W_t &= \text{tangential driving load}\\
W_r' &= \text{separating force}\\
W_x &= \text{axial thrust}\\
W &= \text{total load}\\
M_a &= \text{moment produced by axial thrust}
\end{aligned}
$$

W_t and W_r' are found as before. It is important, though, to consider a new effect on the pressure angle.

$$
\tan \phi_t = \frac{\tan \phi_n}{\cos \psi} \tag{12-22}
$$

where ϕ_n = normal to tooth-face pressure angle

ϕ_t = pressure angle normal to axis of rotation of gear

ψ = helix angle or the acute angle between the tooth-face surfaces and the axis of rotation measured at the pitch diameter. When ψ is 0, $\cos \psi$ is 1.0 and the part is the same as a spur gear

The separating load is a function of transverse pressure angle, which in a helical gear is always greater than ϕ_n, the normal pressure angle. Therefore, W_r' will always be a slight amount larger than in a spur gear but is calculated as before in Eq. (12-19).

The helical angle ψ assumes great importance in calculating the axial and separating load, and in the equations presented, the helix angle will be used for all calculations.

$$
\tan \psi = \frac{\pi \times \text{pitch dia.}}{\text{lead}} = \frac{\pi d}{L} \tag{12-23}
$$

Lead is the distance a point would advance axially across the face of the gear for one revolution measured at the pitch diameter. A small helix angle will have a high lead and a large helix angle will have a low lead.

W_x increases with a higher helix angle and whenever possible should be kept to smallest angle. However, for a helix angle to be of any value it should have a face overlap of a little over 1. This means the contacting distance of face advance along the pitch-diameter periphery of a helical tooth is equal to or greater than the transverse circular pitch (see Fig. 12-18).

Normally for high-speed gears a face advance of over 2 is preferred.

$$\text{Face contact ratio} = m_f = \frac{\text{face advance}}{\text{circ. pitch}} = \frac{Q_f}{p_t} \tag{12-24}$$

As gears become more heavily loaded with higher speeds and quietness is a requirement, a greater overlap ratio is needed. Therefore, the selection of the helix angle becomes a direct function of face advance

$$\tan \psi = \frac{Qf}{F} \tag{12-25}$$

With the selection of ψ the helical-gear forces can be determined as shown in Fig. 12-8.

In its simplest form,

$$W_x = W_t \tan \psi \tag{12-26}$$

The total load W is the vectorial summation of $W_t \nrightarrow W_x$ and W_r'.

$$W = \sqrt{W_t^2 + W_r'^2 + W_x^2} \qquad \text{for } X, Y, \text{ and } Z \text{ axes} \tag{12-27}$$

More consideration must now be given to summation of forces in various planes. In some cases, values will be additive and in other cases they will be subtractive.

The moment produced from the helix is now calculated:

$$M_a = W_x r \tag{12-28}$$

where M_a = axial moment
W_x = axial thrust
r = pitch radius of gear

M_a produces an additional moment and, in turn, an additional radial load which must be contained by the bearings and supports.

Also notice that, if the vectors representing W_t, W_r', and W_x are assumed to be, respectively, the X, Y, and Z axes, Eq. (12-27) conforms to Eq. (12-6).

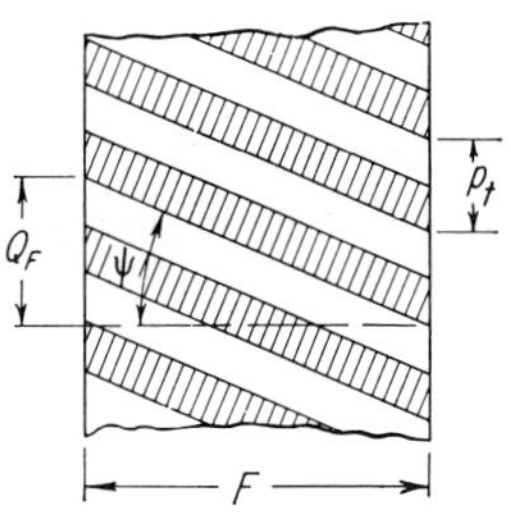

FIG. 12-18. Helical-gear-face advance. Face advance = $Q_F = F \tan \psi$. Contact ratio of faces $Q_F/p_t \geqq 1$.

From the standpoint of load produced axially, it is always desired to keep it to the absolute minimum value. Since a helical gear always requires a bearing to resist an axial force this can be a disadvantage in some installations. The forces can become so large as to be detrimental in large helix angles. See Fig. 12-19 and Table 12-3 for details of bearing-load calculations.

Double Helical. To overcome axial thrust the forces are sometimes counteracted or nullified by double-helical gears. The total face is made into two equal halves both with the same helix angle but opposite hands. The axial thrusts oppose one another and the forces are contained in the gear and not transmitted to the bearing. The double-helical gears are often made of two separate halves of helical gears bolted or joined together, or else a space between the gears is allowed in the middle for tool clearances. There are some gear machine tools which permit a continuous herringbone when face width is limited.

Only one gear in a double-helical train is positioned or restrained axially by a bearing and the rest of the gears then get their location from this one part. For speed-increasing gears it is often wise to use or consider either helical or double-helical gears. As axial leads and helix angles increase so also do the merits of double-helical gears increase.

Skewed or Crossed Helicals. Besides helical gears on parallel axes, it is possible to have the axes of rotation skewed or at an angle with each other. The latter gears

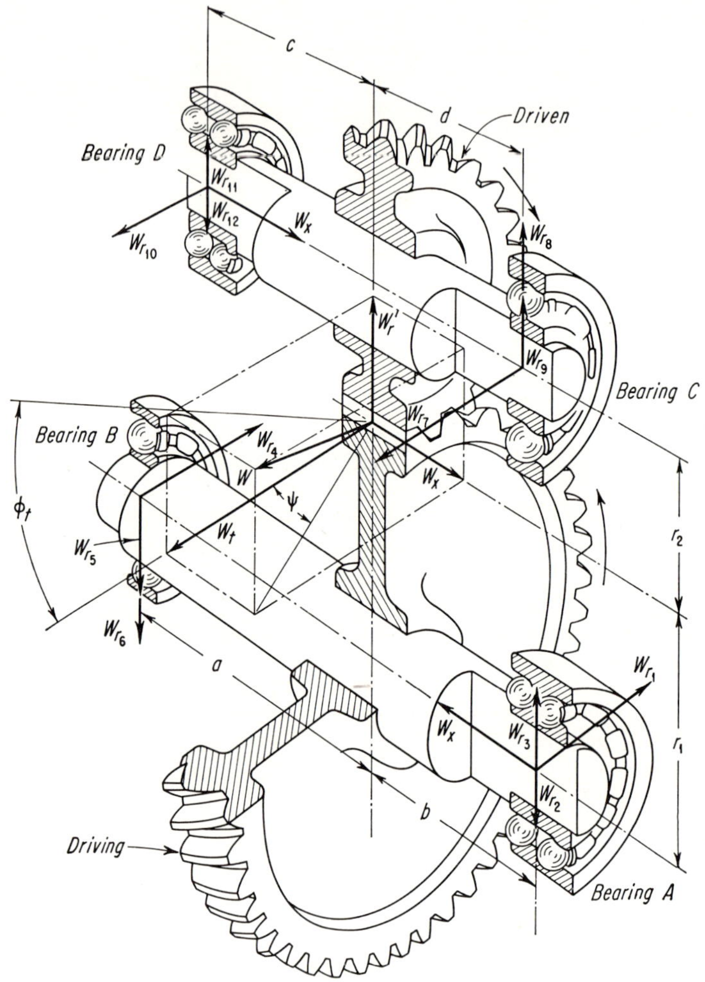

FIG. 12-19. Helical-gear bearing loads.

are called crossed-helical or skewed-helical gears. When the individual gear reactions are found, the same methods are used. In helical gears, the total reaction for the pinion is equal and opposite to the mating gear. In a pair of crossed helicals, each gear and pinion must be considered and calculated separately.

The direction of W_x is generally obvious.

In crossed-helical gears one of the mating parts could be a helical or spur gear and it would mate with a helical gear. All spiral helicals tend to have only point contact and are used to transmit light loads and have small bearing reactions. Allowance

Table 12-3. Equations for Helical-gear Bearing Loads with Straddle Mounting

Force	Gear bearing loads		Pinion bearing loads	
	Bearing A	Bearing B	Bearing C	Bearing D
Tan force W_t	$W_{r_1} = \frac{W_t a}{a + b}$	$W_{r_4} = \frac{W_t b}{a + b}$	$W_{r_7} = \frac{W_t c}{c + d}$	$W_{r_{10}} = \frac{W_t d}{c + d}$
Separating force W_r'	$W_{r_2} = \frac{W_r' a}{a + b}$	$W_{r_5} = \frac{W_r' b}{a + b}$	$W_{r_8} = \frac{W_r' c}{c + d}$	$W_{r_{11}} = \frac{W_r' d}{c + d}$
Thrust W_x	$W_{r_3} = \frac{W_x r_1}{a + b}$	$W_{r_6} = \frac{W_x r_1}{a + b} = W_{r_3}$	$W_{r_9} = \frac{W_x r_2}{c + d}$	$W_{r_{12}} = \frac{W_x r_2}{c + d} = W_{r_9}$
Total radial load	$W_{r_A} = \sqrt{(W_{r_1})^2 + (W_{r_2} - W_{r_3})^2}$	$W_{r_B} = \sqrt{(W_{r_4})^2 + (W_{r_5} + W_{r_6})^2}$	$W_{r_C} = \sqrt{(W_{r_7})^2 + (W_{r_8} + W_{r_9})^2}$	$W_{r_D} = \sqrt{(W_{r_{10}})^2 + (W_{r_{11}} - W_{r_{12}})^2}$
Total thrust	W_x (may be applied to either bearing A or B)		W_x (may be applied to either bearing C or D)	

NOTE: See Fig. 12-19 for definition of terms and bearing locations.

must be made for the helix angle in relation to the axis of rotation of each gear. In external helical gears the hands of the mating parts are equal and opposite, but in crossed helical all types of hands and angles of helix can and must be considered. Calculations of the bearing loads from the gear reactions are the same as for helical gears. More care and consideration must be given to the angles of helix in relation to the shaft angles. The hand of the helices and shaft angles can be varied so that direction of rotation is reversed. Once the true helix angle and the actual direction of rotation between the driver and driven gears are determined, the reactions are easily found. Figure 12-20 shows a pair of crossed-helical gears. The light lines on the driving gear show the teeth as they would appear to the viewer. The meshing point considered in the diagram is *underneath* the driver. After allowing for the direction of rotation, shaft angle, and hand of spirals, the direction of the forces can be found using Table 12-1. The angles ψ_P and ψ_G can be found and knowing W_t the axial thrusts for the gear and pinion are determined as in any helical gear. See Fig. 12-19 for details of bearing-load calculations.

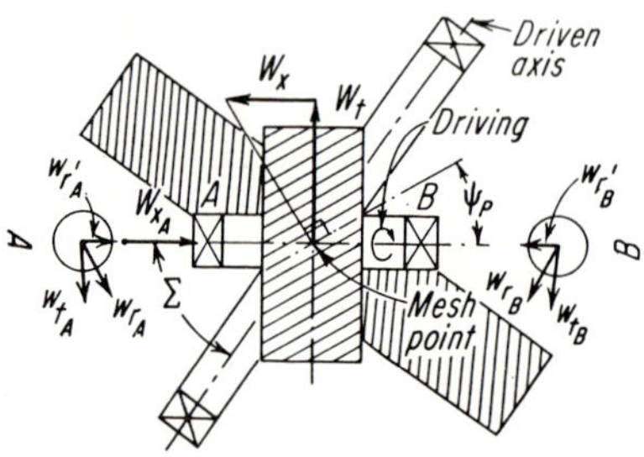

FIG. 12-20. Bearing loads on crossed-helical gears. Regardless of shaft angle between cross-helical pinion and crossed-helical gear, the directions of the pinion tangential force W_t and the pinion thrust force W_x are as shown above for the mesh point.

12-6. Mounting Practice for Bevel and Hypoid Gears.* All gears operate best when their axes are maintained in correct alignment, and although bevel and hypoid gears have the ability to absorb reasonable displacements without detriment to the tooth action, excessive misalignments reduce the load capacity and complicate the manufacture and assembly of the gears. Accurate alignment of the gears under all operating conditions may be accomplished only by good design, accompanied by accurate manufacture and assembly of both mountings and gears. The following recommendations are given to aid in the design of bevel and hypoid gearbox assemblies in order to obtain the best performance of the gears.

Analysis of Forces. In the design of the mountings for bevel and hypoid gears, the first step is to establish the magnitude and directions of the axial and separating forces, from Figs. 12-21 and 12-22, and draw a diagram of the resultant force on the gears as shown in Figs. 12-23 to 12-26. These provide the basis for the design of the gear blank and the mountings and for the selection and arrangement of the bearings. In case the gear must operate in both directions of rotation or if there is reversal of torque, the force diagram should include both conditions of operation. Clockwise and counterclockwise rotation are as viewed from the back of the gear looking toward cone center. Values shown are for shaft angles at right angles but values and determination of directions also apply to shaft angles other than 90°.

Rigid Mountings. Gear mountings should be designed for maximum rigidity since the problems involved in producing satisfactory gears multiply rapidly when deflections in the mountings cause excessive gear displacements. It is necessary to modify the standard cutting in order to narrow and shorten the tooth contacts to suit the flexible mounting. The decrease in bearing area raises the unit tooth pressure and reduces the number of teeth in contact, which results in problems of noise and increases the danger of surface failure and tooth breakage.

* The original material for Arts. 12-6 and 12-7 was prepared by M. L. Fallon, Senior Application Design Analyst and Warner T. Cowell, Chief Engineer, Gear Engineering Test Center, Gleason Works, Rochester, N.Y.

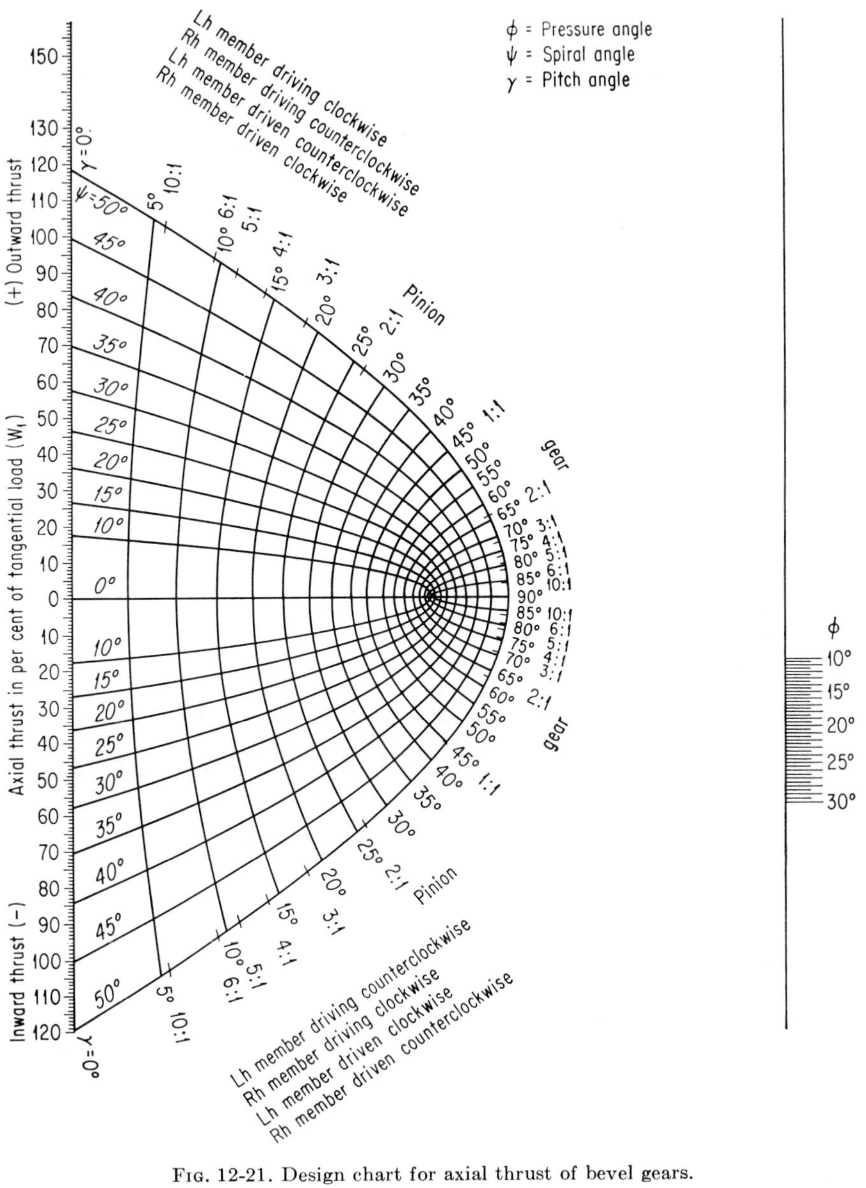

FIG. 12-21. Design chart for axial thrust of bevel gears.

Maximum Displacements. For gears from 6 to 15″ diameter the desirable limits of gear and pinion displacements at maximum load are listed below:

1. The pinion offset should not exceed ±0.003″ from correct position.
2. The pinion should not yield axially more than 0.003″ in either direction.
3. The gear offset should not exceed ±0.003″ from correct position.
4. The gear should not yield axially more than 0.003″ in either direction on miters or near miters or more than 0.010″ away from the pinion on higher ratios.

The above limits are for average applications in which the gears operate over a range of loads and at maximum load approximately 10 per cent of the time. When maximum load occurs for longer periods reduce the above accordingly. Somewhat wider tolerances are allowable for larger gears.

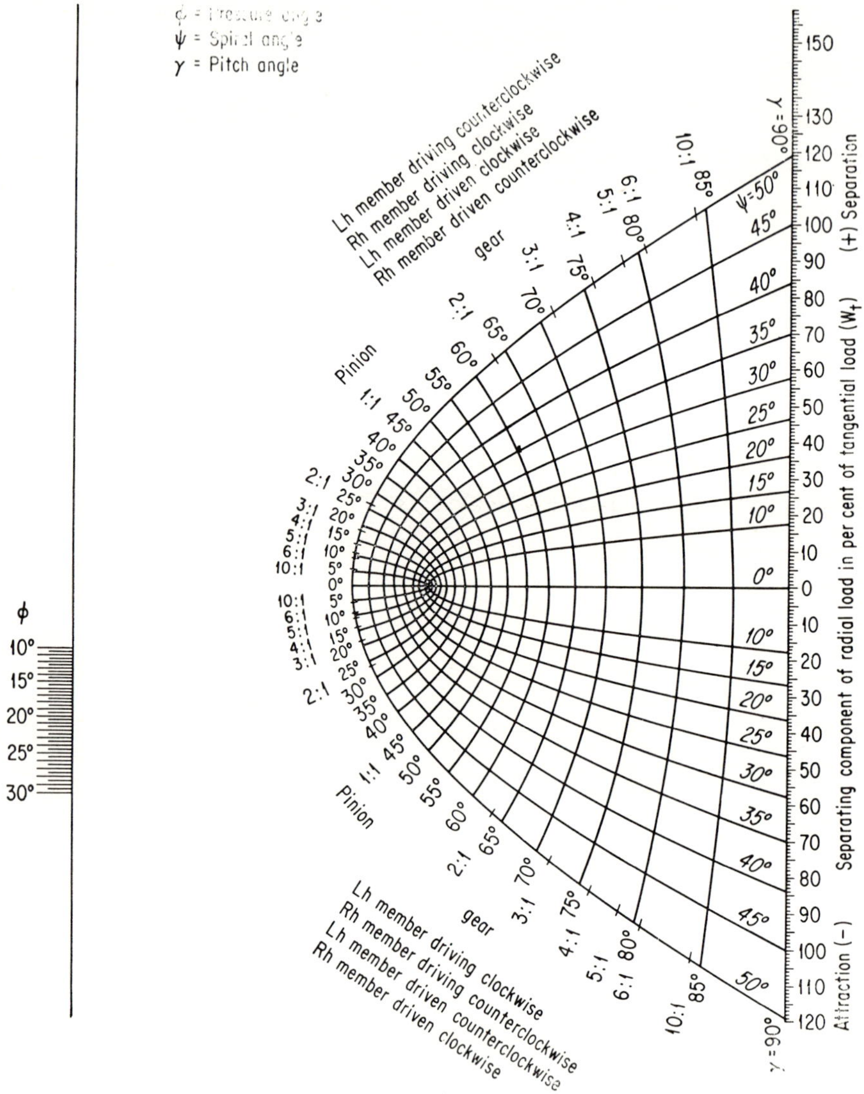

FIG. 12-22. Design chart for separating component of radial load for bevel gears.

Rolling-element Bearings. While plain bearings may be used in bevel- or hypoid-gear mountings, it is usually much easier to maintain the gears in satisfactory alignment with rolling-element bearings. Either ball or roller bearings are satisfactory. However, it is important that the type and size of the bearings be carefully selected to suit the particular application, considering especially the operating speeds.

loads, the desired life of the bearings, and the rigidity required for the best operation of the gears.

Straddle Mounting. The preferred design of a bevel or hypoid gearbox provides straddle mountings for both gear and pinion, and this arrangement is generally used for industrial and other heavily loaded applications. Usually, the desired bearing life and mounting rigidity may be obtained more economically with a straddle mounting. When it is not feasible to use this arrangement for both members of a pair the gear member which usually has the higher radial load should be straddle-mounted.

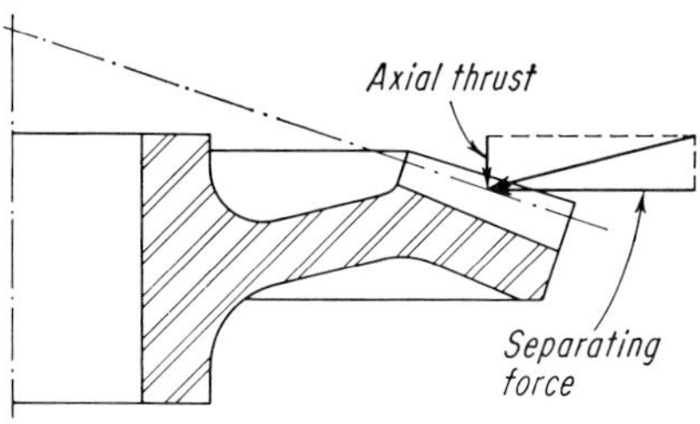

FIG. 12-23. Resultant force in axial plane due to right-hand gear being driven counterclockwise or driving clockwise, also left-hand gear being driven clockwise or driving counterclockwise.

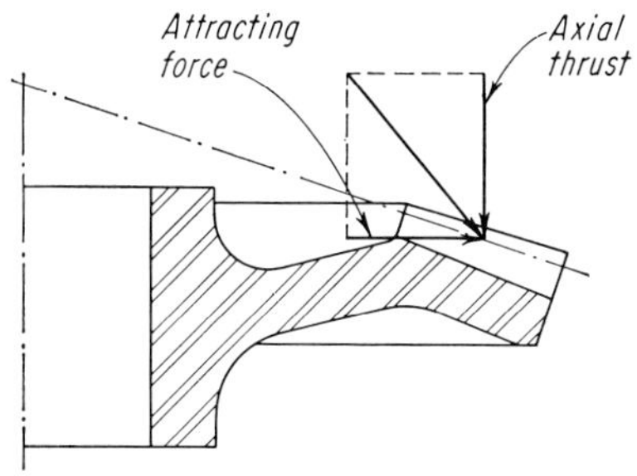

FIG. 12-24. Resultant force in axial plane due to right-hand gear being driven clockwise or driving counterclockwise, also left-hand gear being driven counterclockwise or driving clockwise.

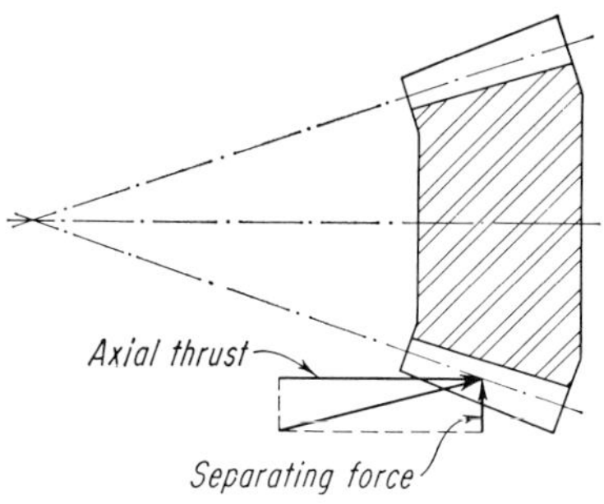

FIG. 12-25. Resultant force in axial plane due to left-hand pinion driving clockwise or being driven counterclockwise, also right-hand pinion driving counterclockwise or being driven clockwise.

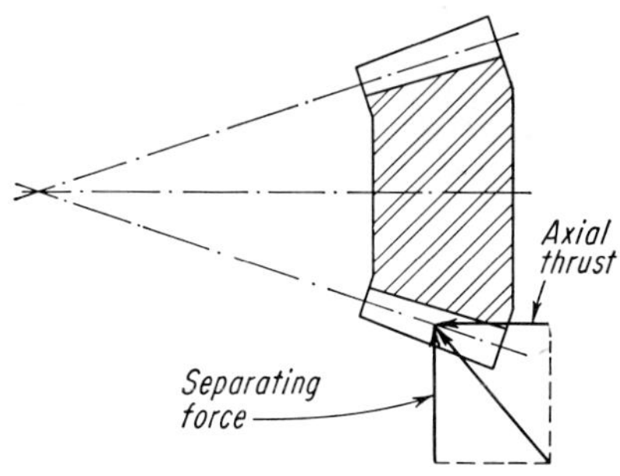

FIG. 12-26. Resultant force in axial plane due to left-hand pinion driving counterclockwise or being driven clockwise, also right-hand pinion driving clockwise or being driven counterclockwise.

In straddle mounting the gear or larger member of a pair, usually two tapered roller bearings or angular-contact ball bearings are mounted opposed with sufficient spacing between them to provide adequate control of the gear (Fig. 12-27).

A second straddle-mounting arrangement (Fig. 12-28), used mostly for pinions, provides an inboard or pilot bearing for pure radial support of the one end of the pinion shaft, and on the opposite end bearings suitable for carrying both radial and thrust loads. In this location several different types of bearings are used: a single double-row deep-groove angular-contact ball bearing or two single-row angular-contact bearings mounted *DB* or *DF*, or two tapered roller bearings either direct- or indirect-mounted.

A similar arrangement is used also for straddle-mounting the gear or larger member of a pair in large applications, and in an aluminum or magnesium housing, where temperature changes would seriously affect the preloading or thrust bearings if spaced one on either end of the gear shaft (Fig. 12-29).

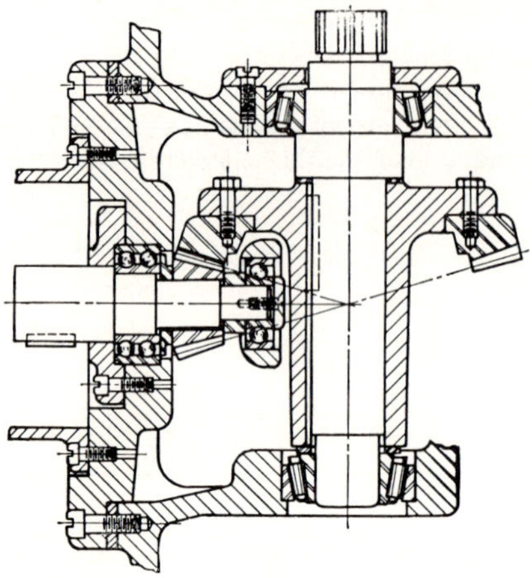

Fig. 12-27. Webless-type gear member straddle-mounted using two opposed tapered roller bearings widely spaced to provide good control of the gear.

Overhung Mounting. When the pinion is mounted overhung on tapered roller bearings, the indirect mounting should be used (Fig. 12-30). This arrangement provides greater stability to the mounting for a given spacing of the bearings, and the thrust load for normal operation is thus carried by the bearing adjacent to the pinion, adding further to its stability.

The bearing adjacent to the pinion should be located as closely as possible to the pinion teeth to reduce the overhang. See Fig. 12-31.

All spiral bevel and hypoid gears should be locked against thrust in both directions. Provision against inward thrust on straight and Zerol bevel gears is often omitted provided the conditions of operation are such that inward thrust cannot occur. One of the advantages of Zerol and straight bevel gears is that change of rotation does not change direction of axial thrust.

Since the contact pattern of spiral bevel and hypoid gears is controlled by their axial position, special attention must be given to the selection of the type and to the preloading of the bearings.

The lubrication of bevel and hypoid gears is most important; for specific data refer to Chap. 15 or to instructions of the Gleason Works.

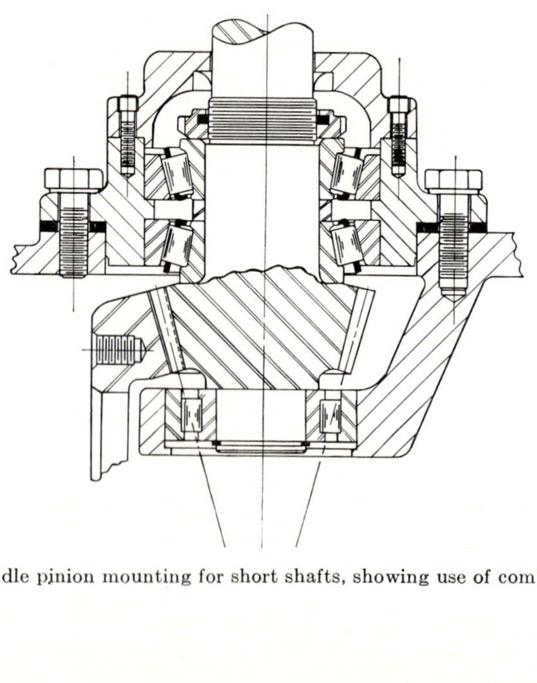

Fig. 12-28. Straddle pinion mounting for short shafts, showing use of combined thrust and radial bearings.

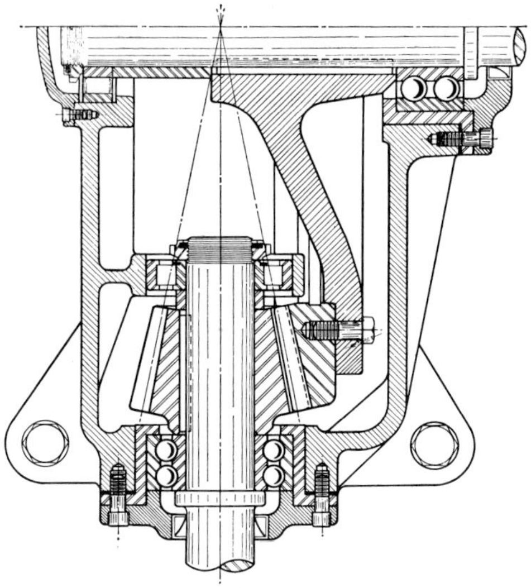

FIG. 12-29. Straddle mounting for both members of a spiral bevel pair, using a straight roller radial bearing on one end of the shaft and combined thrust and radial bearing on other end. Shim-type adjustments for positioning both gear and pinion.

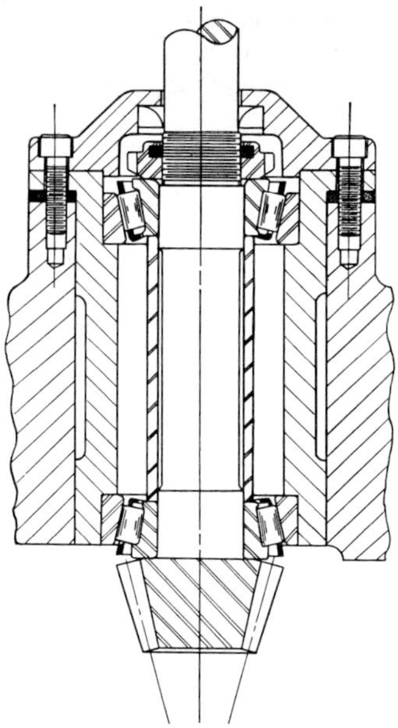

FIG. 12-30. Typical overhung pinion mounting with shim adjustment for positioning pinion and selected spacer for preloading bearings.

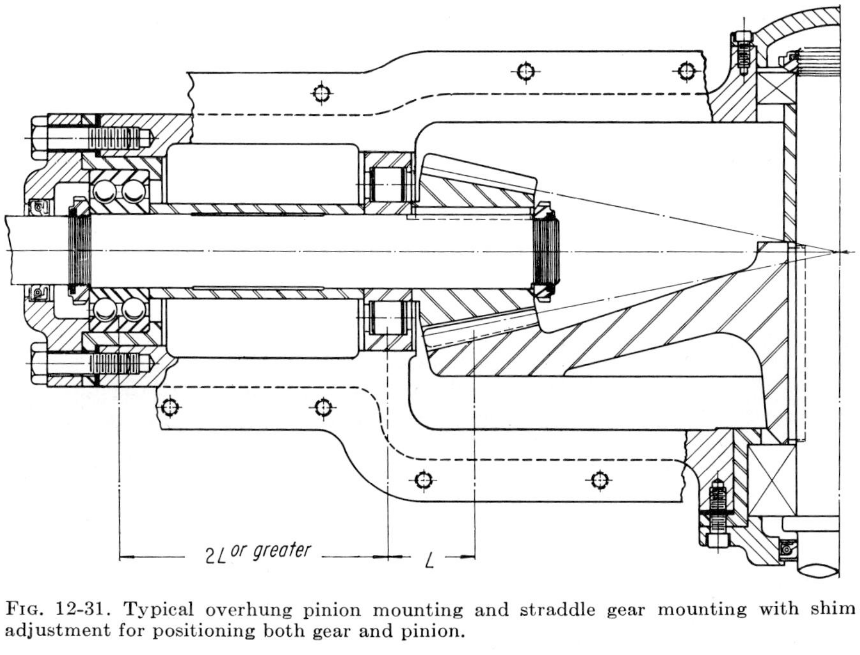

FIG. 12-31. Typical overhung pinion mounting and straddle gear mounting with shim adjustment for positioning both gear and pinion.

Gear-blank Design. The gear blank should be designed to avoid excessive localized stresses and serious deflections within itself. The direction of the resultant force should be considered in the design of the blank and in the method employed in mounting it on its shaft or centering hub (see Figs. 12-32 to 12-37).

Ring gears are made in three general types: the webless ring, web-type ring, and counterbored type. Of these the bolted-on webless ring design shown in Fig. 12-27 is best for hardened gears larger than 7″ in diameter. The fit of the gear on its

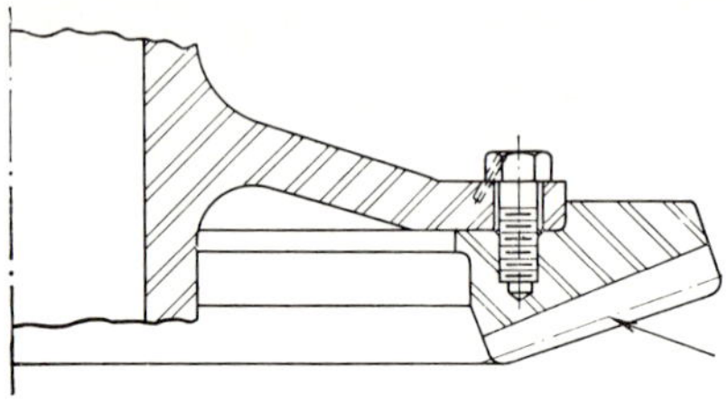

FIG. 12-32. Method of mounting gear to absorb the thrust with minimum amount of deflection.

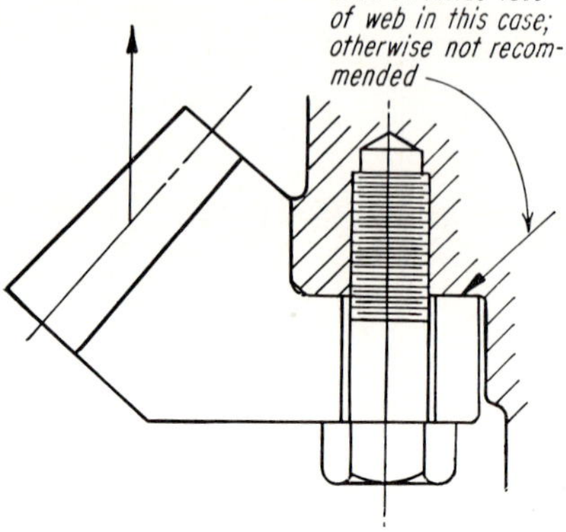

FIG. 12-33. Method of mounting gear when thrust is inward.

centering hub should be either size-to-size or a slight interference. These gears should be clamped to the centering hub with fine-thread cap screws as shown in Figs. 12-27, 12-33, and 12-37 or with bolts. Several methods of locking screws and nuts in place are indicated in the illustrations. Self-locking-type nuts and cap screws are widely used for this purpose. For severe operating conditions cap screws should be locked more positively than is possible by wiring. The method of centering the gear on the counterbore, shown in Fig. 12-37, is recommended for large gears, especially near miters. Figure 12-33 illustrates mounting for a gear which operates with an inward thrust. Designs in which screws or bolts are subjected to added tension from gear forces should be avoided. On severe reversing or vibrating installations separate dowel drives may be used with the designs or splines as in Figs. 12-35 and 12-39.

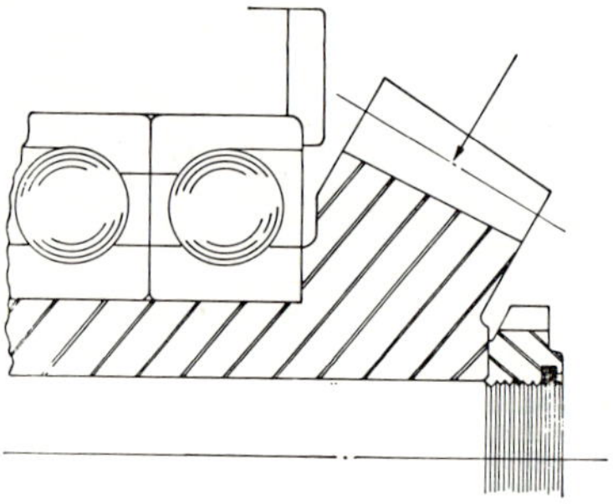

FIG. 12-34. Blank sections beneath the teeth should be sufficiently rigid in the direction of thrust so that deflections will be minimized.

The use of dowels or dowel-fit bolts has been found unnecessary in most automotive and industrial drives for, with bolts or cap screws drawn tightly, the friction of the ring gear on its hub prevents shearing of the screws.

Hardened gears smaller than 7″ in diameter may best be designed with integral hubs. A front hub on a Zerol bevel, spiral bevel, or hypoid gear should in no case intersect the extended root line so as to interfere with the path of the cutter toward the root cone apex as shown in Fig. 12-38. On hub-type gears the length of the hub should be at least equal to the bore diameter, and the end of the hub should be securely clamped against a shoulder on the shaft.

Whether the gear is mounted on a flanged hub or is made integral with the hub, the supporting flange should be made sufficiently rigid to prevent serious deflections in the direction of the gear axis at the mesh point. For larger gears the web preferably should be made conical and without ribbing to permit machining for balance, to eliminate oil churning when dip lubrication is used, and to lessen the danger of

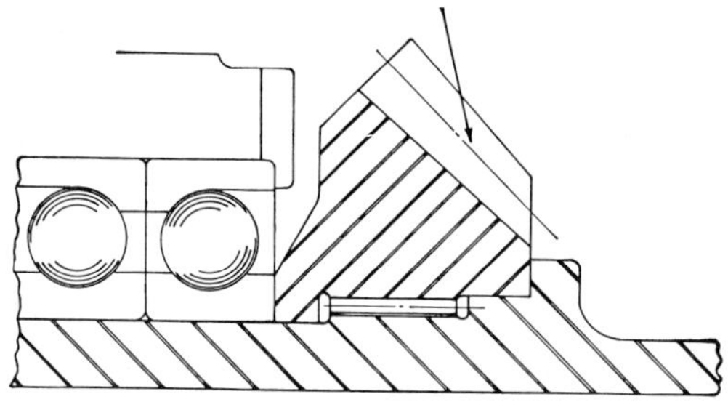

FIG. 12-35. Gear blank designed to take care of inward thrust. Also a preferred mounting with spline drive.

stress concentrations being set up within the castings. The gear-hub and bearing arrangement should be designed with sufficient rigidity that the use of a thrust button behind the gear mesh will be unnecessary. The area of a thrust button is too small to provide durable support.

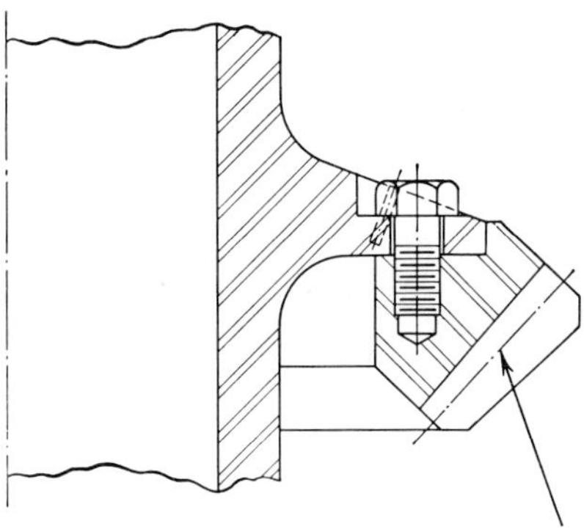

FIG. 12-36. Method of mounting gear to hub to absorb high outward thrust.

For hardened gears, the blanks should be designed with sections and shape suitable for hardening with minimum distortion.

Gear and Pinion Adjustments. Provision should be made for adjusting both gear and pinion axially at assembly. Shim-type adjustments as shown in Figs. 12-28 and 12-29 or threads and nuts with locks as shown in Fig. 12-40 may be used. Shims less than 0.015″ thick may "pound out" in service. When used adjacent to a bearing, the shims should be placed next to the stationary member. If an assortment of shims varying in thickness in steps of 0.003″ is used (such as 0.020, 0.023, 0.026) the member to be adjusted may be positioned within ±0.0015″.

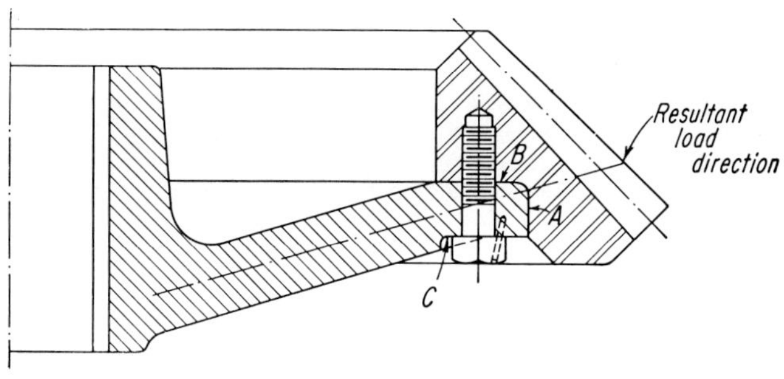

FIG. 12-37. Webless-type miter gear—counterbored type.

Assembly Procedure. The following is a recommended procedure for the assembly of bevel and hypoid gearbox assemblies:

Thoroughly clean housing and parts, following the bearing manufacturer's instructions in handling the rolling-element bearings.

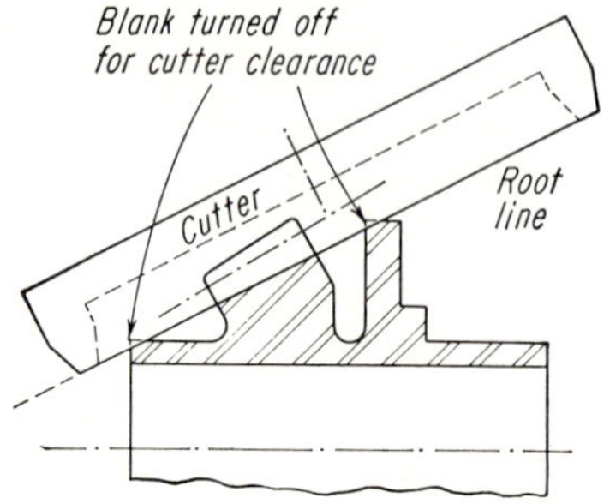

FIG. 12-38. When using Zerol bevel, spiral bevel, and hypoid gears, clearance must be provided for the cutter as shown.

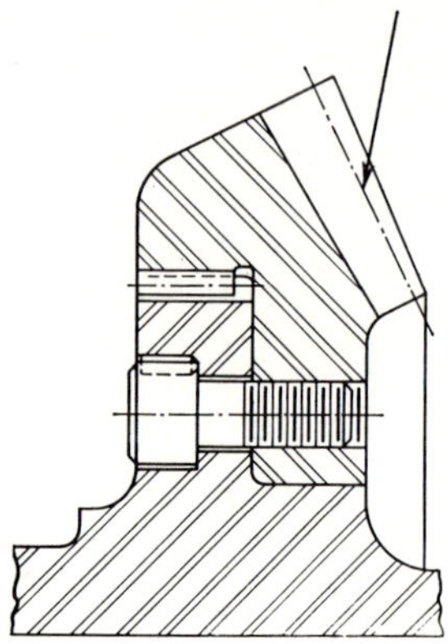

FIG. 12-39. Positive drive for severe reversing or shock loads.

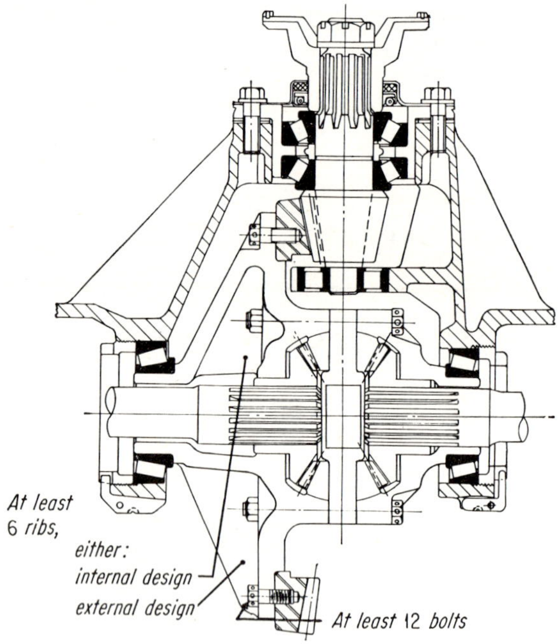

FIG. 12-40. Truck axle—straddle-mounted pinion, and with threaded-nut adjustments for the gear bearings.

Keep lapped or matched gears in their original sets.

Assemble even-ratio pairs with the teeth mating as lapped. Tooth and mating spaces are usually designated with X markings.

Set pinion on its correct axial position by measurement.

Adjust gear along its axis until the specified backlash is obtained.

Securely lock gears and thrust bearings in this position. Note, for spiral bevel and hypoid gears, both members should be locked against thrust in both directions.

As a final check, paint the teeth with marking compound and observe the tooth-contact pattern under light load. This may indicate the need for further adjustment.

Check to ensure that there is an adequate supply of the specified lubricant to the mesh and to the bearings.

12-7. Calculation of Bevel and Hypoid Bearing Loads. The normal load on the tooth surfaces of bevel or hypoid gears may be resolved into two components: one in the direction along the axis of the gear and the other perpendicular to the axis. The axial force produces an axial thrust on the bearings, while the force perpendicular to the axis produces a radial load on the bearings. The direction and magnitude of the normal load depends upon the ratio, pressure angle, spiral angle, hand of spiral, direction of rotation, and whether the gear is the driving or driven member.

Hand of Spiral. The hand of spiral on spiral bevel and hypoid gears is denoted by the direction in which the teeth curve; that is, left-hand teeth incline away from the axis in the counterclockwise direction when an observer looks at the face of the gear and right-hand teeth incline away from the axis in the clockwise direction. The hand of spiral of one member of a pair is always opposite to that of its mate. It is customary to use the hand of spiral of the pinion to identify the combination; that is, a left-hand combination is one with a left-hand spiral on the pinion and a right-hand spiral on the gear. The hand of spiral has no effect on the smoothness and quietness of operation or on the efficiency. Attention, however, is called to the difference in the effect of the thrust loads as stated in the following paragraph.

A left-hand spiral pinion driving clockwise (viewed from the back) tends to move axially away from the cone center; while a right-hand pinion tends to move toward the center because of the oblique direction of the curved teeth. If there is excessive end play in the pinion shaft because of faulty assembly or bearing failure, the movement of a right-hand pinion driving clockwise will take up the backlash under load, and the teeth of the gear and pinion may wedge together, while a left-hand spiral pinion under the same conditions would back away and introduce additional backlash between the teeth, a condition which would not prevent the gears from functioning. When the ratio, pressure angle, and spiral angle are such that it is possible, the hand of spiral should be selected to give an axial thrust that tends to move both the gear and the pinion out of mesh. Otherwise, the hand of spiral should be selected to give an axial thrust that tends to move the pinion out of mesh. Often the mounting conditions will dictate the hand of spiral to be selected.

In a reversible drive there is, of course, no choice unless the pair performs a heavier duty in one direction a greater part of the time.

In the calculation of reactions, Zerol and straight bevel gears may be treated as a special case of spiral bevel gears in which the spiral angle is zero degrees. In these, the direction of the axial thrust is always outward on both members regardless of the direction of rotation and whether the pinion is the driving or driven member.

On hypoids when the pinion is below center and to the right when facing the front of the gear, the pinion hand of spiral should always be left. With the pinion above center and to the right, the pinion hand should always be right (see Fig. 12-41).

Spiral Angle. The spiral angle, if possible, should be selected to give a face contact ratio of at least 1.25. For maximum smoothness, the face contact ratio should be between 1.50 and 2.0. On straight or Zerol bevel gears the spiral angle will, of course, be zero.

Tangential Load.* The tangential load on a bevel or hypoid gear is given by

$$W_t = \frac{126{,}050P}{d_{mG}n} \tag{12-29}$$

* In all calculations herein, the tangential load is calculated at the mean pitch radius.

where W_t = tangential load, lb on gear
P = hp transmitted
$d_{mG} = D - F \sin \Gamma$ = mean pitch diameter of gear, in.
D = gear pitch diameter, in.
F = gear face width, in.
Γ = gear pitch angle
n = speed of the gear, rpm

The tangential load on the mating bevel pinion will be equal to the tangential load on the gear. The tangential load on the mating hypoid pinion, however, must be determined as follows:

$$W_{tP} = W_{tG} \frac{\cos \psi_P}{\cos \psi_G} \tag{12-30}$$

where W_{tP} = tangential load on pinion, lb
W_{tG} = tangential load on gear as determined above, lb
ψ_P = pinion spiral angle
ψ_G = gear spiral angle

Axial Thrust. The value of the axial thrust is dependent upon the tangential tooth load, spiral angle, pressure angle, and pitch angle. This may be determined

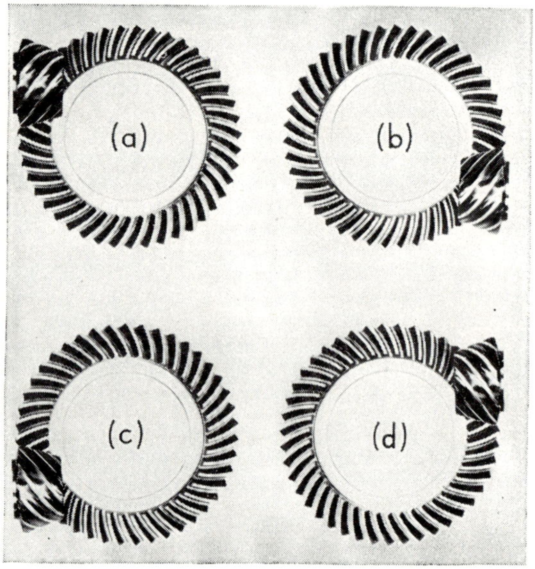

FIG. 12-41. Both pinions shown in (*a*) and (*b*) are referred to as having an offset "below center," while those in (*c*) and (*d*) have an offset "above center." In determining the direction of offset, it is customary to look at the face of the gear with the pinion at the right.

with the aid of Fig. 12-21. Figure 12-21 is symmetrical about a horizontal center line; therefore there are two points of intersection of the pitch-angle γ and the spiral-angle ψ curves. Selection of the proper point is dependent upon the hand of spiral, the direction of rotation, and whether the member is the driving or driven one. This is given in Fig. 12-21. Note that the intersection points for the pinion and its mating gear will always lie on opposite sides of the horizontal center line. The pressure angle ϕ is given on the scale at the right. A straight line connecting the pressure angle with the intersection point of the spiral-angle and pitch-angle curves can now be extended to the scale on the left. This scale gives the axial thrust W_x in per cent of the tangential tooth load W_t. The spiral angle on straight and Zerol gears is zero.

For determining the axial thrust in hypoid gears, the pressure angle of the driving side should be used. Instead of using the pitch angles, the face angle of the pinion and the root angle of the gear should be used.

Note that the spiral angle, pitch angle, and pressure angle for the corresponding member must be selected in each case.

The actual axial-thrust load may be determined by multiplying the tangential load in pounds for the corresponding member by the per cent given in Fig. 12-21.

$$W_x = (\text{per cent from Fig. 12-21}) \times W_t \tag{12-31}$$

Figure 12-43 illustrates the direction of rotation and direction of thrust.

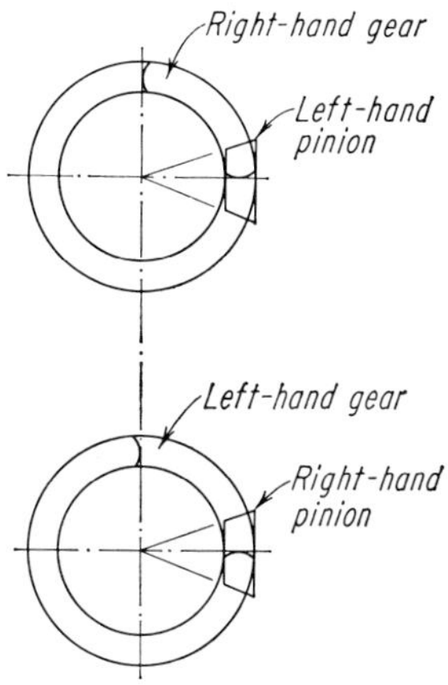

FIG. 12-42. Diagram illustrating hand of Zerol. The upper drawing shows a left-hand and the lower a right-hand Zerol-bevel-gear combination.

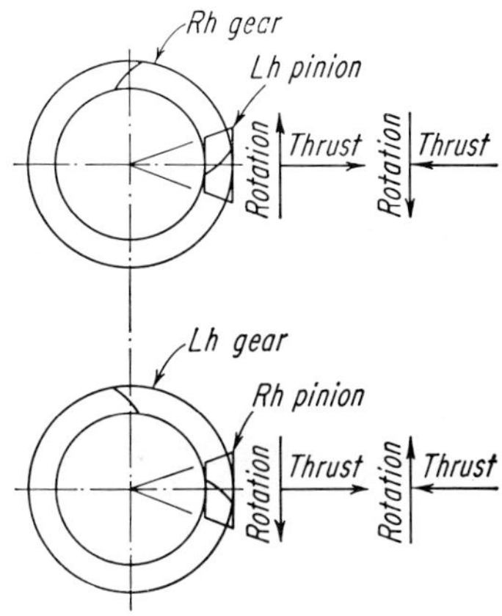

FIG. 12-43. Diagram illustrating hand of spiral and its effect on the direction of axial thrust in spiral bevel gears. The upper drawing shows a left-hand and the lower a right-hand spiral bevel combination.

Radial Load. The radial load caused by the separating force between the two members may be determined in a similar manner to that used for the axial thrust with the aid of Fig. 12-22. The same procedure is followed here as outlined above under Axial Thrust. The scale on the right gives the component part of the radial load W_r in per cent of the tangential tooth load W_t.

This component of the actual radial load may be determined by multiplying the tangential load in pounds for the corresponding member by the per cent given in Fig. 12-22.

$$W_r' = (\text{per cent from Fig. 12-22}) \times W_t \tag{12-32}$$

In addition to the above there is a radial component caused by the tangential load itself, and a third component caused by the couple which the axial thrust on the given member produces.

The total radial load produced on the bearings will therefore depend on the resultant of these three components.

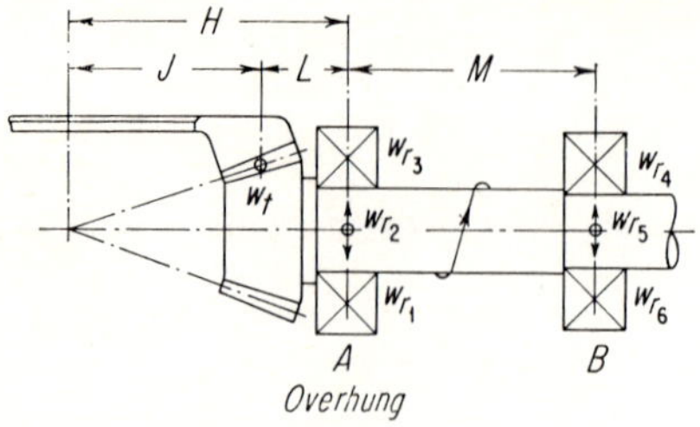

FIG. 12-44. Diagram showing load components on a pinion and bearings—overhung mounting.

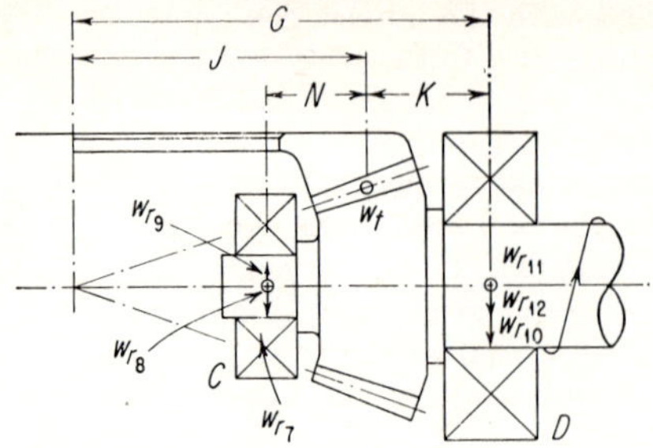

FIG. 12-45. Diagram showing load components on a pinion and bearings—straddle mounting.

For the overhung mounting the radial components on bearing A will be (see Fig. 12-44)

$$W_{r_1} = W_r' \frac{L + M}{M} \tag{12-33}$$

$$W_{r_2} = W_t \frac{L + M}{M} \tag{12-34}$$

where W_r' = separating component from Fig. 12-22
W_t = tangential tooth load

$$W_{r_3} = W_x \frac{d_m}{2M} \tag{12-35}$$

where W_x = axial thrust load
L = distance along the axis from center of the face width to center of bearing A*
M = spacing between centers of bearings A and B
$d_m = d - F \sin \gamma$ = mean pitch diameter†
d = pitch diameter
F = face width
γ = pitch angle

The total radial component on bearing A will be

$$W_{r_A} = \sqrt{W_{r_2}{}^2 + (W_{r_1} - W_{r_3})^2} \tag{12-36}$$

In like manner, the radial components on bearing B will be

$$W_{r_4} = W_r' \frac{L}{M} \tag{12-37}$$

$$W_{r_5} = W_t \frac{L}{M} \tag{12-38}$$

$$W_{r_6} = W_x \frac{d_m}{2M} \tag{12-39}$$

The total radial component on bearing B will be

$$W_{r_B} = \sqrt{W_{r_5}{}^2 + (W_{r_4} - W_{r_6})^2} \tag{12-40}$$

* For a hypoid pair use Eq. (12-52).
† For a hypoid pair use Eq. (12-49) for d_m and find N by directions given for Eq. (12-52).

For the straddle mounting shown in Fig. 12-45 the radial components on bearing C will be

$$W_{r_7} = W_r' \frac{K}{N + K} \tag{12-41}$$

$$W_{r_8} = W_t \frac{K}{N + K} \tag{12-42}$$

$$W_{r_9} = W_x \frac{d_m}{2(N + K)} \tag{12-43}$$

where N = distance along axis from center of face width to center of bearing C*
K = distance along axis from center† of face width to center of bearing D

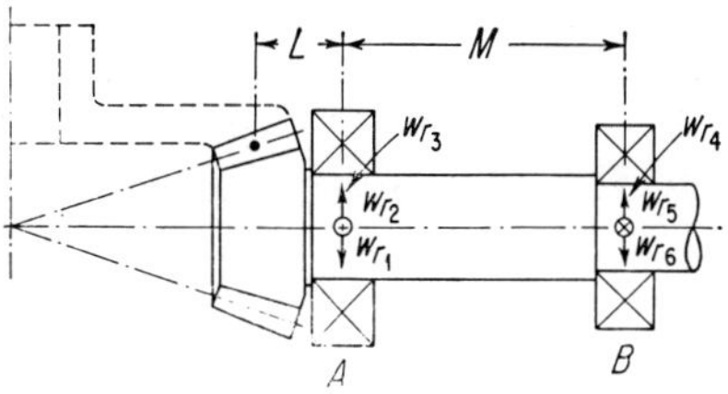

FIG. 12-46. Overhung mounting.

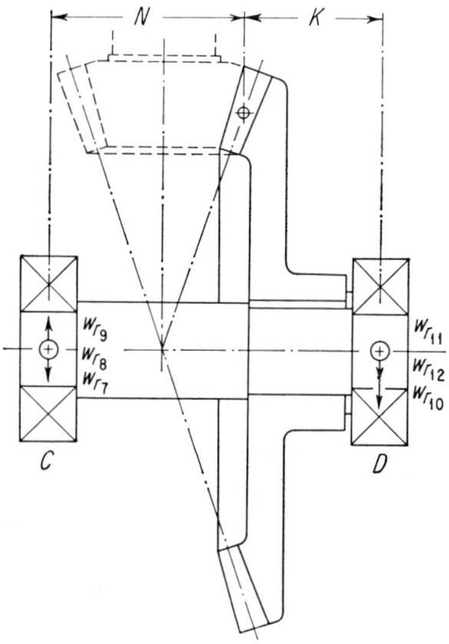

FIG. 12-47. Straddle mounting.

The total radial component on bearing C will be

$$W_{r_c} = \sqrt{W_{r_8}^2 + (W_{r_7} - W_{r_9})^2} \tag{12-44}$$

In like manner, the radial components on bearing D will be

$$W_{r_{10}} = W_r' \frac{N}{N + K} \tag{12-45}$$

$$W_{r_{11}} = W_t \frac{N}{N + K} \tag{12-46}$$

$$W_{r_{12}} = W_x \frac{d_m}{2(N + K)} \tag{12-47}$$

The total radial component on bearing D will be

$$W_{r_D} = \sqrt{W_{r_{11}}^2 + (W_{r_{10}} + W_{r_{12}})^2} \tag{12-48}$$

* For a hypoid pair use Eq. (12-49) for d_m and find N by directions given for Eq. (12-52).
† For a hypoid pair find K by Eq. (12-51). See Fig. 12-45.

For a hypoid pair, change the preceding equations as follows:

$$d_{mG} = D - F \sin \Gamma_R = \text{mean pitch diameter of hypoid gear} \qquad (12\text{-}49)$$

$$d_{mP} = d_{mG} \frac{n}{N} \frac{\cos \psi_G}{\cos \psi_P} = \text{mean pitch diameter of hypoid pinion} \qquad (12\text{-}50)$$

where D = gear pitch diameter
F = gear face width
Γ_R = gear root angle
n = number of teeth in hypoid pinion
N = number of teeth in hypoid gear
ψ_G = gear spiral angle
ψ_P = pinion spiral angle

$$K = G - J \qquad (12\text{-}51)$$

$$L = H - J \qquad (12\text{-}52)$$

where G = distance along axis from center line of mate to center of bearing D
H = distance along axis from center line of mate to center of bearing A

$$J_P = \frac{d_{mG}}{2} \cos \epsilon \qquad \text{for hypoid pinion}$$

$$J_G = \frac{d_{mP}}{2} \qquad \text{for hypoid gear}$$

where $\epsilon = \tan^{-1} [\tan (\psi_P - \psi_G) \sin \Gamma_R]$
M = spacing between centers of bearings A and B
N = spacing between centers of bearings C and D minus dimension K
See Figs. 12-44 and 12-45.

Example of Spiral-bevel-gear Bearing-load Calculation. A sample problem for a 16-tooth pinion driving a 49-tooth gear will demonstrate how the equations given in this article are used.

Given a 16/49 combination, 5 diametral pitch, 1.500″ face width, 20° pressure angle, 35° spiral angle.

Load data: 71 hp at 1,800 rpm of pinion, left-hand pinion driving clockwise, gear straddle mounted—pinion overhung.

Table 12-4. Spiral-bevel-gear Sample Data and Bearing Arrangement

	Pinion		Gear	
Pitch dia.	d	3.200″	D	9.800″
Pitch angle	γ	18°5′	Γ	71°55′
	$\sin \gamma$	0.3104	$\sin \Gamma$	0.9506
Mean pitch dia.	$d_{m_P} = d - F \sin \gamma$	2.734″	$d_{m_G} = D - F \sin \Gamma$	8.374″
Overhung mounting:				
L		1.500″		
M		5.000″		
Straddle mounting:				
K				2.500″
N				3.500″

The tangential load on the spiral bevel gear above will be

$$W_{tG} = \frac{126{,}050P}{d_{mG}n} = \frac{126{,}050 \times 71}{8.374 \times 588} = 1{,}818 \text{ lb}$$

Table 12-5. Solution to Sample Problem on Spiral Bevel Bearing Loads

Forces	Load calculations	Bearing loads			
		Overhung mounting		Straddle mounting	
		Bearing A	Bearing B	Bearing C	Bearing D
Tangential W_t	$W_t = \frac{126{,}050P}{d_{m_G} n}$	$W_{r_2} = W_t \frac{L+M}{M}$	$W_{r_5} = W_t \frac{L}{M}$	$W_{r_8} = W_t \frac{K}{N+K}$	$W_{r_{11}} = W_t \frac{N}{N+K}$
		$W_{r_2} = 1{,}818 \frac{1.5+5.0}{5.0}$ $= 2{,}363$	$W_{r_5} = 1{,}818 \frac{1.5}{5.0}$ $= 545$	$W_{r_8} = 1{,}818 \frac{2.5}{3.5+2.5}$ $= 758$	$W_{r_{11}} = 1{,}818 \frac{3.5}{3.5+2.5}$ $1{,}060$
Thrust W_x	$W_x = W_t \times$ thrust factor See Fig. 12-21	....	$W_x =$ thrust factor $\times W_t$	....	$W_x =$ thrust factor $\times W_t$
			$W_x = +0.80 \times 1{,}818$ $= 1{,}454$		$W_x = +0.20 \times 1{,}818$ $= 364$
Separating W_r'	$W_r' = W_t \times$ separating factor See Fig. 12-22	$W_{r_1} = W_r' \frac{L+M}{M}$	$W_{r_4} = W_r' \frac{L}{M}$	$W_{r_7} = W_r' \frac{K}{N+K}$	$W_{r_{10}} = W_r' \frac{N}{N+K}$
		$W_{r_1} = 364 \frac{1.5+5.0}{5.0}$ $= 473$	$W_{r_4} = 364 \frac{1.5}{5.0}$ $= 109$	$W_{r_7} = 1{,}454 \frac{2.5}{3.5+2.5}$ $= 606$	$W_{r_{10}} = 1{,}454 \frac{3.5}{3.5+2.5}$ $= 848$
Thrust couple	$W_x \frac{d_m}{2}$	$W_{r_3} = W_x \frac{d_m}{2M}$	$W_{r_6} = W_x \frac{d_m}{2M}$	$W_{r_9} = W_x \frac{d_m}{2(N+K)}$	$W_{r_{12}} = W_x \frac{d_m}{2(N+K)}$
		$W_{r_3} = 1{,}454 \frac{2.734}{2 \times 5.0}$ $= 398$	$W_{r_6} = 1{,}454 \frac{2.734}{2 \times 5.0}$ $= 398$	$W_{r_9} = 364 \frac{8.374}{2(3.5+2.5)}$ $= 254$	$W_{r_{12}} = 364 \frac{8.374}{2(3.5+2.5)}$ $= 254$
Total radial load	$W_r = \sqrt{\Sigma(\text{loads})^2}$	$W_{r_A} = \sqrt{(W_{r_2}{}^2 + (W_{r_1} - W_{r_3})^2}$	$W_{r_B} = \sqrt{(W_{r_5}{}^2 + (W_{r_4} - W_{r_6})^2}$	$W_{r_C} = \sqrt{(W_{r_8}{}^2 + (W_{r_7} - W_{r_9})^2}$	$W_{r_D} = \sqrt{W_{r_{11}}{}^2 + (W_{r_{10}} + W_{r_{12}})^2}$
		$W_{r_A} = \sqrt{(2{,}363)^2 + (473-398)^2}$ $= 2{,}364$	$W_{r_B} = \sqrt{(545)^2 + (109-398)^2}$ $= 617$	$W_{r_C} = \sqrt{(758)^2 + (606-254)^2}$ $= 836$	$W_{r_D} = \sqrt{(1{,}060)^2 + (848+254)^2}$ $= 1{,}536$

W_t = tangential tooth load, lb
P = power transmitted, hp
$d_{m_G} = D - F \sin \Gamma$ = mean pitch diameter of gear, in.
D = gear pitch diameter, in.
F = gear face width, in.
Γ = gear pitch angle
n = speed of gear, rpm
W_r' = separating component
W_x = axial-thrust component
L = axial distance from center of face width to center line of bearing A
M = spacing between center line of bearings A and B
N = axial distance from center of face width to center line of bearing C
K = axial distance from center of face width to center line of bearing D
$d_m = d - F \sin \gamma$
d = pitch diameter, in.
γ = pitch angle

where $P = 71$ hp

$d_{m_G} = 8.374''$

$$n = \frac{16}{49} \times 1{,}800 = 588 \text{ rpm of gear}$$

The tangential load on the spiral bevel pinion above will be equal to the tangential load on the gear:

$$W_{tP} = W_{tG} = 1{,}818 \text{ lb}$$

The axial-thrust loads for both gear and pinion are obtained from Fig. 12-21.

$$\begin{aligned} W_x &= (\text{per cent from Fig. 12-21}) \times W_t \\ W_{xP} &= +0.80 \times 1{,}818 = 1{,}454 \text{ lb (outward)} \\ W_{xG} &= +0.20 \times 1{,}818 = 364 \text{ lb (outward)} \end{aligned}$$

The separating force for both gear and pinion is obtained from Fig. 12-22.

$$\begin{aligned} W_r' &= (\text{per cent from Fig. 12-22}) \times W_t \\ W'_{rP} &= +0.20 \times 1{,}818 = 364 \text{ lb (separating)} \\ W'_{rG} &= +0.80 \times 1{,}818 = 1{,}454 \text{ lb (separating)} \end{aligned}$$

In Table 12-5 are tabulated the formulas and sample calculations for the resultant bearing loads on the above pair of gears.

NOTE: Since an overhung mounting is used on the pinion, the formulas for bearings A and B are used, and since a straddle mounting is used on the gear, the formulas for bearings C and D are used. If both members had been overhung-mounted, the formulas for bearings A and B would have been used for both gear and pinion, or if both members had been straddle-mounted, the formulas for bearings C and D would have been used for both gear and pinion. In any case, the dimensions and loads for the corresponding member must be used.

The thrust is always taken at one mounting point only. This may be taken by a single bearing as shown, or by a set of duplex bearings mounted in tandem.

Table 12-6. Hypoid-gear Sample Data and Bearing Arrangement

	Pinion		Gear	
Pitch dia.	. . . .	. . . . .	$D = \dfrac{N}{\text{diam. pitch}}$	9.375″
Pitch angle:				
Pinion face angle	γ_o	21°47′		
Gear root angle	. . . .	. . . . .	Γ_R	66°27′
	$\sin \gamma_o$	0.37110	$\sin \Gamma_R$	0.91671
Mean pitch dia	$d_{m_P} = d_{m_G} \dfrac{n \cos \psi_G}{N \cos \psi_P}$	3.161″	$d_{m_G} = D - F \sin \Gamma_R$	8.057″
Distance from center line of mate to center of bearing	H (from layout)	5.615″	G (from layout)	4.581″
	$J_P = \dfrac{d_{m_G}}{2} \cos \varepsilon$	3.740″	$J_G = \dfrac{d_{m_P}}{2}$	1.581″
Overhung mounting (see Figs. 12-44 and 12-46):				
$L = H - J_P$	. . . .	1.875″		
$M =$	. . . .	4.750″		
Straddle mounting (see Figs. 12-45 and 12-47):				
$K = G - J_G$	. . . .	. . . . .	. . . .	3.000″
$N =$	. . . .	. . . . .	. . . .	6.250″

Example of Hypoid-gear Bearing-load Calculation. A sample problem of a 14-tooth hypoid pinion driving a 47-tooth hypoid gear is given in the following:

Given a 14/47 combination, 5.013 gear diametral pitch, 1.438″ face width, 1.750″ pinion offset below center. Pressure angles 15°20′ on forward drive, 27°10′ on reverse drive. Spiral angles 21°28′ on the gear, 45°3′ on the pinion. Pinion face angle = 21°47′. Gear root angle = 66°27′.

Load data: 60 hp at 1,500 rpm of pinion, left-hand pinion driving clockwise, gear straddle-mounted—pinion overhung.

The tangential load on the hypoid gear above will be

$$W_{tG} = \frac{126{,}050 P_H}{d_{mG}n} = \frac{126{,}050 \times 60}{8.057 \times 447} = 2{,}100 \text{ lb}$$

where $P_H = 60$ hp
$d_{mG} = 8.057$
$n = {}^{14}\!/_{47} \times 1{,}500 = 447$ rpm of gear

The tangential load on the hypoid pinion above will be

$$W_{tP} = W_{tG}\frac{\cos \psi_P}{\cos \psi_G} = 2{,}100 \times \frac{0.70649}{0.93063} = 1{,}594 \text{ lb}$$

$$\psi_P = 45°3' \quad \cos \psi_P = 0.70649$$
$$\psi_G = 21°28' \quad \cos \psi_G = 0.93063$$

The axial-thrust loads for both gear and pinion for forward drive are obtained from Fig. 12-21.

$$W_x = (\text{per cent from Fig. 12-21}) \times W_t$$
$$W_{xG} = +0.11 \times 2{,}100 = 231 \text{ lb (outward)}$$
$$W_{xP} = +1.08 \times 1{,}594 = 1{,}722 \text{ lb (outward)}$$

The separating forces for both gear and pinion are obtained from Fig. 12-22.

$$W_r' = (\text{per cent from Fig. 12-22}) \times W_t$$
$$W_{r_G}' = +0.48 \times 2{,}100 = 1{,}008 \text{ lb (separating)}$$
$$W_{r_P}' = -0.01 \times 1{,}594 = -16 \text{ lb (attracting)}$$

In Table 12-7 are tabulated the formulas and sample calculations for the resultant bearing loads on the hypoid pair.

Required Data for Bearing-load Calculations. To calculate the bearing loads, the following data must be available:

Bevel Gears

Gear ratio (number of teeth in gear and pinion)
Gear pitch diameter or diametral pitch
Pitch angles (both gear and pinion)
Face width
Pressure angle
Spiral angle
Horsepower or gear torque to be transmitted
Gear rpm
Hand of spiral
Direction of rotation (viewed from the back)
Sketch showing method of mounting and bearing spacing
Shaft angle

Table 12-7. Solution to Sample Problem on Hypoid-gear Bearing Loads

Forces	Load calculations	Bearing loads: Overhung mounting, Bearing A	Bearing loads: Overhung mounting, Bearing B	Bearing loads: Straddle mounting, Bearing C	Bearing loads: Straddle mounting, Bearing D
Tangential W_t	$W_t = \dfrac{126{,}050P}{d_m n}$	$W_{r_2} = W_t \dfrac{L+M}{M}$ $W_{r_2} = 1{,}594 \dfrac{1.875 + 4.750}{4.750}$ $= 2{,}223$	$W_{r_5} = W_t \dfrac{L}{M}$ $W_{r_5} = 1{,}594 \dfrac{1.875}{4.750}$ $= 629$	$W_{r_8} = W_t \dfrac{K}{N+K}$ $W_{r_8} = 2{,}100 \dfrac{3.0}{6.25 + 3.00}$ $= 681$	$W_{r_{11}} = W_t \dfrac{N}{N+K}$ $W_{r_{11}} = 2{,}100 \dfrac{6.25}{6.25 + 3.00}$ $= 1{,}419$
Thrust W_x	$W_x = W_t \times$ thrust factor See Fig. 12-21	$W_x =$ thrust factor $\times W_t$ $W_x = +1.08 \times 1{,}594$ $= 1{,}722$	....	....	$W_x =$ thrust factor $\times W_t$ $W_x = +0.11 \times 2{,}100$ $= 231$
Separating W_r'	$W_r' \times W_t \times$ separating factor See Fig. 12-22	$W_{r_1} = W_r' \dfrac{L+M}{M}$ $W_{r_1} = -16 \dfrac{1.875 + 4.750}{4.750}$ $= -22$	$W_{r_4} = W_r' \dfrac{L}{M}$ $W_{r_4} = -16 \dfrac{1.875}{4.750}$ $= -6$	$W_{r_7} = W_r' \dfrac{K}{N+K}$ $W_{r_7} = 1{,}008 \dfrac{3.0}{6.25 + 3.00}$ $= 327$	$W_{r_{10}} = W_r' \dfrac{N}{N+K}$ $W_{r_{10}} = 1{,}008 \dfrac{6.25}{6.25 + 3.00}$ $= 681$
Thrust couple	$W_x \dfrac{d_m}{2}$	$W_{r_3} = W_x \dfrac{d_m}{2M}$ $W_{r_3} = 1{,}722 \dfrac{3.161}{2 \times 4.750}$ $= 573$	$W_{r_6} = W_x \dfrac{d_m}{2M}$ $W_{r_6} = 1{,}722 \dfrac{3.161}{2 \times 4.750}$ $= 573$	$W_{r_9} = W_x \dfrac{d_m}{2(N+K)}$ $W_{r_9} = 231 \dfrac{8.057}{2(6.25 + 8.00)}$ $= 101$	$W_{r_{12}} = W_x \dfrac{d_m}{2(N+K)}$ $W_{r_{12}} = 231 \dfrac{8.057}{2(6.25 + 3.00)}$ $= 101$
Total radial load	$W_r = \sqrt{\Sigma(\text{loads})^2}$	$W_{r_A} = \sqrt{W_{r_2}{}^2 + (W_{r_1} - W_{r_3})^2}$ $W_{r_A} = \sqrt{(2{,}223)^2 + (-22 - 573)^2}$ $= 2{,}301$	$W_{r_B} = \sqrt{W_{r_5}{}^2 + (W_{r_4} - W_{r_6})^2}$ $W_{r_B} = \sqrt{(629)^2 + (-6 - 573)^2}$ $= 855$	$W_{r_C} = \sqrt{W_{r_8}{}^2 + (W_{r_7} - W_{r_9})^2}$ $W_{r_C} = \sqrt{(681)^2 + (327 - 101)^2}$ $= 718$	$W_{r_D} = \sqrt{W_{r_{11}}{}^2 + (W_{r_{10}} + W_{r_{12}})^2}$ $W_{r_D} = \sqrt{(1{,}419)^2 + (681 + 101)^2}$ $= 1{,}620$

W_t = tangential tooth load, lb
P = power transmitted, hp
n = speed, rpm (values for corresponding member)
W_r' = separating component
W_x = axial-thrust component
L = axial distance from center of gear face width at point of contact to center line of bearing A
M = spacing between center line of bearings A and B
N = axial distance from center of gear face width at point of contact to center line of bearing C
K = axial distance from center of gear face width at point of contact to center line of bearing D
$d_m = D - F \sin \Gamma_R = d_{m_G}$ = mean pitch diameter of hypoid gear
$= d_{m_G} \dfrac{n}{N} \dfrac{\cos \psi_G}{\cos \psi_P}$ = mean pitch diameter of hypoid pinion
D = gear pitch diameter, in.
F = gear face width, in.
Γ_R = gear root angle
n = number of teeth in hypoid pinion
N = number of teeth in hypoid gear
ψ_G = gear spiral angle
ψ_P = pinion spiral angle

Hypoid Gears

Gear ratio (number of teeth in gear and pinion)
Gear pitch diameter or diametral pitch
Gear root angle
Pinion face angle
Gear face width
Pressure angles (both sides of tooth)
Spiral angles (both gear and pinion)
Horsepower or gear torque to be transmitted
Gear rpm
Hand of spiral
Direction of rotation (viewed from the back)
Sketch showing method of mounting and bearing spacing
Shaft angle

12-8. Bearing-load Calculations for Worms. Basically the worm and gear bearing-load calculations are similar to helical-gear reactions except for three main

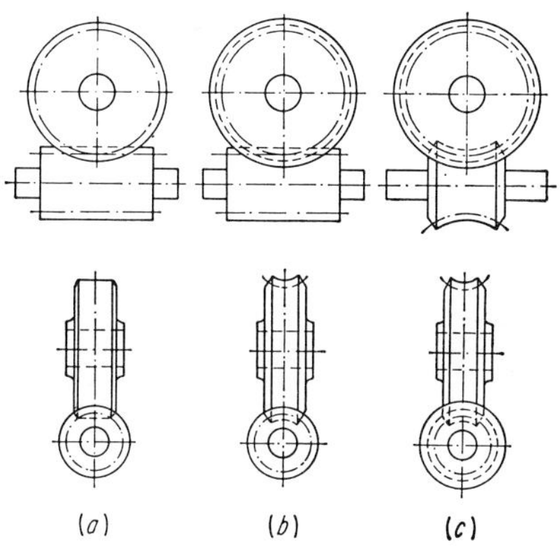

FIG. 12-48. Types of worm gears. (*a*) Nonthroated crossed-helical. (*b*) Single-throated cylindrical worm gear. (*c*) Double-throated double-enveloping worm gear. (*a*) Point contact. (*b*) Line contact. (*c*) Line contact or area contact.

differences. All the members of the worm-gear family run on *crossed axes*. Generally the axes are crossed at 90°.

Besides the shaft arrangement, worm gears have *overlap*. There may be no overlap, overlap in one plane, or overlap in two planes. "Overlap" is that characteristic where one part tries to envelop the mating part by being so curved as to tend to wrap around the mating part. The overlap gives a greater area of tooth engagement. It also makes the mounting more sensitive. See Fig. 12-48 for worm types.

The third unusual feature of worm gears is that of ratio. The ratio is defined as the number of gear teeth divided by the number of worm threads (or worm "starts"). A single-start worm may have several turns of thread, but if it meshes with a gear with 100 teeth the *ratio* is 100:1. If the worm had two starts and the gear had 100 teeth, the ratio would be 50:1.

Calculation of Forces in Worm Gears. Figure 12-3 showed the basics of a force that acted in three planes. This concept is helpful in understanding how the force at the worm-gear mesh acts.

The efficiency of worm gears is quite variable. A single-thread worm may have an efficiency as low as 50 per cent. A multiple-thread worm with at least 15° lead angle and a reasonable rubbing speed will usually have an efficiency in the range of 90 to

95 per cent. Since worm-gear efficiency is so variable it cannot be neglected in calculating load reactions. Article 14-3 may be used to estimate a relatively accurate efficiency. In particular see Eq. (14.27).

The worm-gear overlap gives a relatively wide zone of contact. The pressure angle and the thread angle vary quite appreciably throughout this area. It is

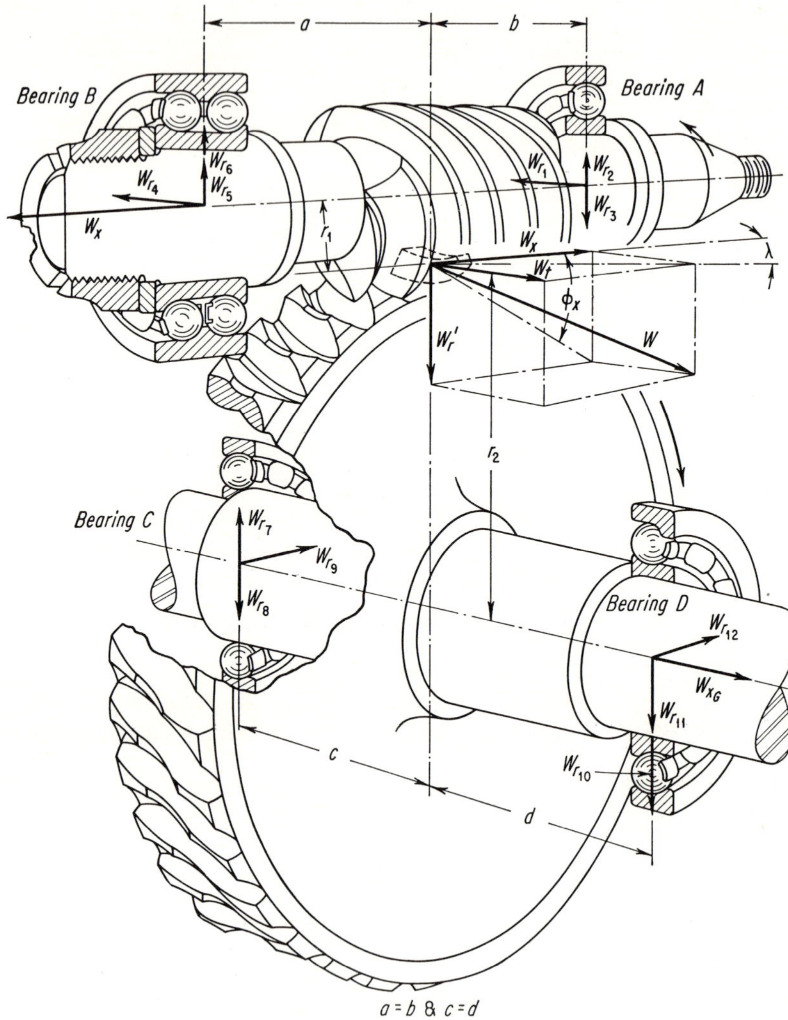

FIG. 12-49. Worm-gear bearing loads.

customary to calculate the worm-gear load reactions as if all the contact were at the theoretical pitch point. This permits use of the theoretical pressure angle, the theoretical lead angle, etc., as variables in the equations. Although this method is not really exact it is good enough to get load reactions to size bearings and design casing structures.

Figure 12-49 shows the case of a worm and gear mounted on 90° axes and each straddle-mounted. Load reactions at the mesh and at each of the bearings are shown.

Table 12-8. Equations for Worm-gear Bearing Loads with Straddle Mounting

Forces	Worm bearing loads		Worm-gear bearing loads	
	Bearing A	Bearing B	Bearing C	Bearing D
Worm tan force W_t	$W_{r_1} = \frac{W_t a}{a+b}$	$W_{r_4} = \frac{W_t b}{a+b}$	$W_{r_7} = \frac{W_t r_2}{c+d}$	$W_{r_{10}} = \frac{W_t r_2}{c+d} = W_{r_7}$
Separating force W_r'	$W_{r_2} = \frac{W_r' a}{a+b}$	$W_{r_5} = \frac{W_r' b}{a+b}$	$W_{r_8} = \frac{W_r' d}{c+d}$	$W_{r_{11}} = \frac{W_r' c}{c+d}$
Thrust W_x	$W_{r_3} = \frac{W_x r_1}{a+b}$	$W_{r_6} = \frac{W_x r_1}{a+b} = W_{r_3}$	$W_{r_9} = \frac{W_x d}{c+d}$	$W_{r_{12}} = \frac{W_x c}{c+d}$
Total radial load........	$W_{r_A} = \sqrt{(W_{r_1})^2 + (W_{r_2} - W_{r_3})^2}$	$W_{r_B} = \sqrt{(W_{r_4})^2 + (W_{r_5} + W_{r_6})^2}$	$W_{r_C} = \sqrt{(W_{r_9})^2 + (W_{r_7} - W_{r_8})^2}$	$W_{r_D} = \sqrt{(W_{r_{12}})^2 + (W_{r_{10}} + W_{r_{11}})^2}$
Total thrust...	W_x (may be applied to either bearing A or B)		$W_{x_G} = W_t$ (may be applied to either bearing C or D)	

NOTE: See Fig. 12-49 for definition of terms and bearing locations.

The tangential force on the worm may be calculated by

$$W_t = \frac{126{,}050 P_H}{d n_W} \tag{12-53}$$

where P_H = input hp
d = worm pitch diameter
n_W = rpm of worm

The separating force is

$$W_r' = \frac{W_t \tan \phi_n}{\tan \lambda} = W_t \tan \phi_x \tag{12-54}$$

where ϕ_x = axial pressure angle of worm
= ϕ_t, transverse pressure angle of gear
λ = lead angle of worm (also helix angle of gear)
$\tan \lambda = \dfrac{\text{lead}}{\pi d}$ (see Art. 4-18 for basic gear equations)
d = worm pitch diameter

The worm thrust force is

$$W_x = \frac{W_t}{\tan \lambda} \tag{12-55}$$

The details for worm-gear bearing-load calculations are shown in Table 12-8. Figure 12-49 illustrates the bearing loads in a worm gearset.

The method of worm-gear bearing-load calculation just described may be used for crossed-helical gears on right-angle axes, for cylindrical worm gears, and for double-enveloping worm gears.

Mounting Tolerances. The worm-gear types vary all the way from the non-enveloping crossed-helical to the double-enveloping type. In the crossed helical, alignment is not critical in any direction. Axial movement merely shifts the contact area along the length of the part. Center-distance error changes backlash and the amount of tooth contact, but neither of these things is highly critical in most applications.

The cylindrical worm gearset is critical on center distance, shaft angle, and axial position of the gear. Slight errors in any of these three things will tend to change a full-face-width contact pattern to only a small amount of contact at the end of the tooth. The contact area in a worm gearset may drop from 100 to 15 per cent with an error in alignment of as little as 0.010″ in a foot.

The double-enveloping worm gearset is critical in the three directions of the cylindrical set plus a fourth direction of the axial position of the worm.

Table 12-9. Typical Mounting Tolerances for High-capacity Worm Gearsets, in.

Center distance, in.	Tolerance on center distance	Tolerance on axial position	Tolerance in alignment per ft
0–3	±0.0005	±0.001	±0.008
3–6	±0.001	±0.002	±0.005
6–15	±0.002	±0.003	±0.003
15 up	±0.003	±0.004	±0.002

Table 12-9 shows some typical mounting tolerances for high-capacity worm gearsets. The most precise worm gears used in timing devices, radar work, etc., are mounted with about half the error shown in Table 12-9. On the other hand, worm gearsets for commercial service and relatively light load service are mounted with about twice the error shown in Table 12-9.

Worm-gear-blank Considerations. The worm member is usually made of steel. Generally this member offers no problem in design. In a few cases the worm may be long and slender with a wide span between its bearings. In this case there may be danger of serious "bowing" of the worm between supports. Calculations should be made in such a case to make sure that the worm shaft is not stressed too high. Also, the deflection should be calculated. This deflection acts like a change in center distance and should be judged on this basis.

The worm gear is generally made of bronze in the toothed area. Quite often the hub is made of steel or cast iron. The hub should have a large enough hole in it to assure a strong shaft. The length of the hub should be about one-third the gear pitch diameter to assure a stable mount.

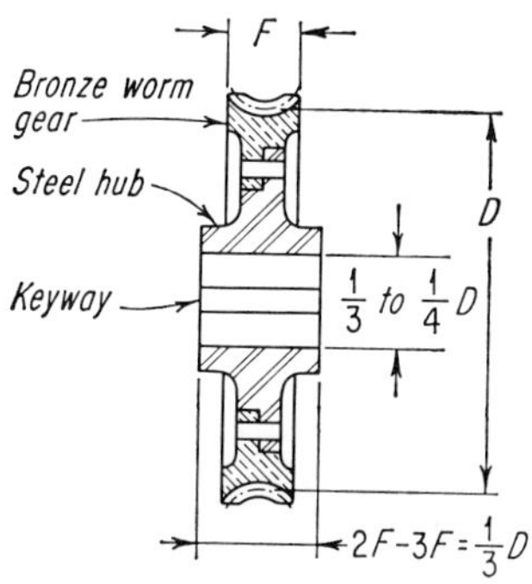

FIG. 12-50. Worm-gear blank proportions.

Figure 12-50 shows a typical worm gear with a bronze ring bolted to a steel hub.

Run-in of Worm Gears. All types of worm gears benefit by a "run-in." This is particularly true when the gear member is made of bronze.

The as-cut worm gear generally has generating flats on its surface and the surface itself is somewhat imperfect because of "oversize" effects of the hob. As the gear is broken in its surface wears a slight amount to conform to that of the worm. In addition the surface of the gear becomes smooth and polished. Tests on bronze gears have shown that the amount of load which will cause seizure of the surface may go up as much as tenfold after a careful break-in!

In breaking in worm gears, the load is gradually increased and the speed is kept low. The surfaces are watched carefully. If small amounts of bronze start adhering to the worm surface, it is necessary to stop the break-in and dress down the worm threads with fine abrasive paper to remove the bronze. The break-in may then continue.

After the break-in is complete and the tooth surfaces polished and the contact pattern has been obtained to a satisfactory level of quality, the gears are stamped as a *matched* set. Exact axial-distance settings used in the break-in are stamped on the set to aid in field installation.

12-9. Bearing-load Calculations for Spiroids. The Spiroid gear is a crossed-axes type of gear somewhat intermediate between a worm gear and a hypoid gear. Its originators, the Illinois Tool Works of Chicago, Ill., consider it to be a screw type of gear. The pinion of the Spiroid set is like a tapered worm. The pinion meshes on the side of the gear rather than the outside diameter of the gear. The gear does not wrap around the worm as a conventional worm gear does.

The Spiroid pinion has two effective pressure angles which are called the "high" pressure angle and the "low" pressure angle. Figure 12-51 shows the high side and the low side of a typical Spiroid pinion. The low side of the Spiroid pinion is preferred for driving because it has the largest amount of tooth contact area and it exerts the least separating force on the gear.

Before bearing-load calculations can be made for a Spiroid set, it is necessary to establish design details like ratio, cone angle, offset, and pressure angle. The pinion has a constant lead, but because of the taper the lead angle is variable. Figure 12-52 shows how the lead angle varies along the length of the cone.

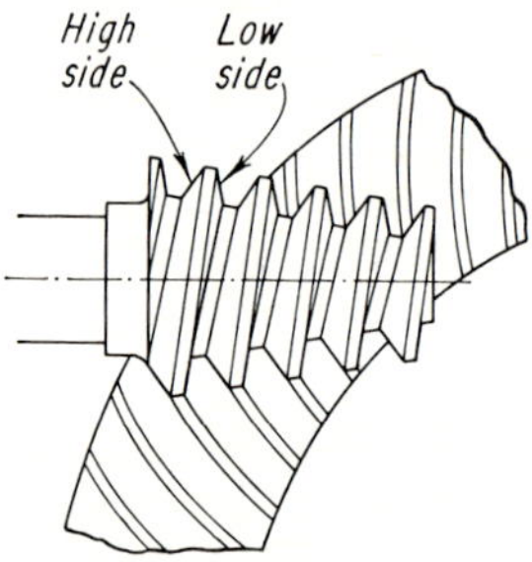

Fig. 12-51. High and low side of spiroid gear.

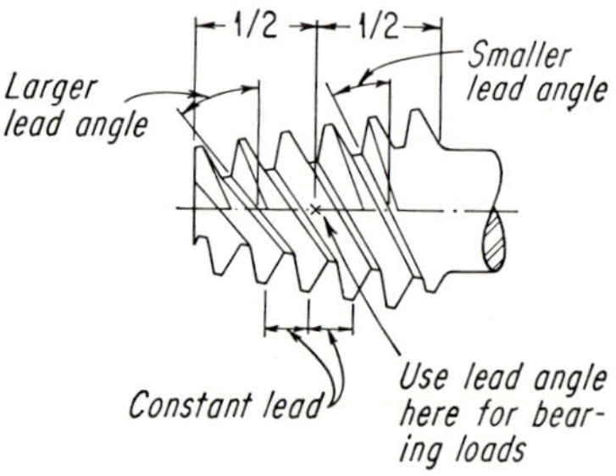

Fig. 12-52. Spiroid-pinion lead angle.

Bearing-load calculations are based on the lead angle, the pressure angles, the pitch diameters, and the center distance at the mean length of the pinion.

The exact calculation of Spiroid bearing loads is a complicated process. Generally, it is satisfactory to calculate approximate bearing loads using special constants. The error is usually less than 10 per cent. If exact load data are needed, the Illinois Tool Works is willing to furnish assistance on a specific request.

The approximate tangential driving force may be obtained from the output horsepower or output torque as follows:

$$W_t = \frac{63{,}025P}{n_g R} = \frac{\text{output torque}}{R} \tag{12-56}$$

where P = output hp
n_G = rpm of gear
R = gear pitch radius

The approximate radial force on the gear is

$$W_{rG} = 0.91W_t \tag{12-57}$$

The radial force on the pinion is

$$W_{rP} = fW_t \tag{12-58}$$

where f = 0.60 for 35° pressure angle on high side
= 0.52 for 30° pressure angle on high side
= 0.35 for 15° pressure angle on low side
= 0.21 for 10° pressure angle on low side

The thrust force on the pinion is equal to the radial force of the gear. Likewise the thrust force on the gear is equal to the radial force on the pinion. Thus

$$W_{xP} = W_{rG} \qquad (12\text{-}59)$$
$$W_{xG} = W_{rP} \qquad (12\text{-}60)$$

The point of application of these forces should be taken at the mid-face of the gear and the mid-depth of the tooth.

For a straddle-mounted Spiroid set, the bearing loads may be determined by a procedure similar to that shown for worm gears in Table 12-8. Substitute W_{rP} for the worm tan force W_t. Assume zero for the separating force. Substitute W_{xP} for the worm thrust W_x. In the Spiroid-gear calculation, use zero for the separating force and substitute W_{rG} for the W_x term in the worm-gear bearing-load part of the table. The radii r_1 and r_2 should be taken to the mid-point of the Spiroid mesh.

Figure 12-53 shows the Spiroid-gear meshing point and the direction of the tangential driving force.

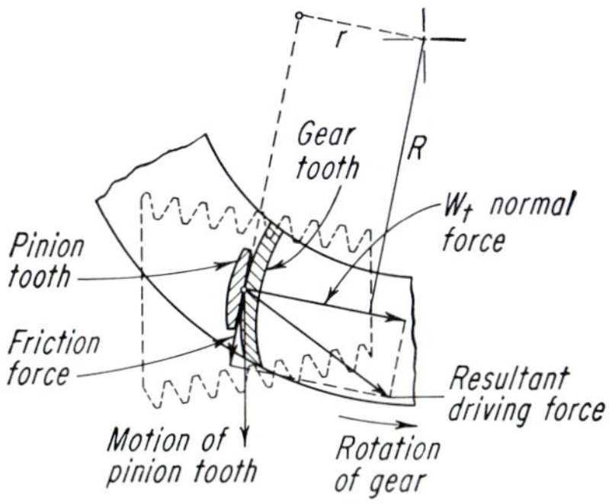

FIG. 12-53. Spiroid-gear bearing loads.

12-10. Bearing-load Calculations for Other Gear Types. Space does not permit detail bearing-load equations for several additional types of gears. The Planoid ®,* Helicon ®,* face gear, Beveloid ®,† and others are not covered. However, the principles used in this chapter will make it relatively easy for the designer to set up equations for any special type. The first step in solving the problem is to find the radial, separating, and thrust forces at some center point of the contact area. Then calculations are made to determine the final radial and axial reaction at each bearing location. If there is more than one load point on the same shaft, proper vector addition must be made to get the over-all resultant.

In solving complex bearing-load problems, the following rules should be kept in mind:

1. When all the gearing forces have been transferred to a given bearing mount, the final result is one resultant radial load and (depending on the bearing type) one resultant thrust (or axial) load.
2. The summation of all forces must equal zero.
3. The summation of all moments must equal zero.

In the case of new gear types, the manufacturer of the type or the inventor of the type will usually be glad to furnish potential customers with detail calculations for the forces at the meshing point.

* Registered trade-marks of the Spiroid Div. of Illinois Tool Works, Chicago, Ill.
† Registered trade-mark of Vinco Corp., Detroit, Mich.

REFERENCES

1. New Departure Division of General Motors Corp.: "New Departure Handbook," Vol. 2, pp. 27–73, 145–158, Bristol, Conn.
2. Singer, Ferdinand L.: "Engineering Mechanics," Harper & Brothers, New York, 1943.
3. Marlin Rockwell Corp.: "Ball Bearing Load Computation Manual," Jamestown, N.Y., 1952.
4. Barish, Thomas: Idler Gears, *Machine Design*, Cleveland, Ohio, Nov. 28, 1957.
5. Gleason Works: "Bevel and Hypoid Gear Design," pp. 1–104, Gleason Works, Rochester, N.Y., 1956.
6. Merritt, H. E.: "Gears," Chaps. 20, 23, Pitman Publishing Corporation, New York, 1955.
7. Michigan Tool Co.: "Cone Drive Gears," Catalog 60, pp. 1–16, 38–52, Cone Drive Gear Div., Michigan Tool Co., Detroit, Mich., 1953.
8. Illinois Tool Works: "Spiroid Gears," Illinois Tool Works, Chicago, Ill., 1959.
9. American Gear Manufacturers Assoc.: "Nomenclature—Letter Symbols for Gear Engineering," AGMA standard 111.03, 1954.
10. Marks, Lionel S.: "Mechanical Engineers' Handbook," 6th ed., Sec. 8, Gearing, McGraw-Hill Book Company, Inc., New York, 1958.

Chapter 13

LOAD RATING OF GEARS

By

E. J. WELLAUER, *Director of Research and Development, The Falk Corporation, Milwaukee, Wis.*

It is the purpose of this chapter to present useful design data for estimating the load-carrying capacity of gears. The precise computation of gear capacity is an extremely difficult process since a large number of variables are involved such as gear-cutting accuracy, mounting errors, elastic deflections, material quality, tooth stiffness, characteristics of connected machinery, and many others.

To simplify the estimation of load capacity, the major information required is listed, and then tables which can be used for preliminary or approximate ratings of various types of gears are presented. Finally, for those problems requiring refined analyses of gear capacity, additional authoritative formulas are provided.

GENERAL CONSIDERATIONS FOR LOAD-RATING GEARS

As a start on the problem of load rating of gears, it is desirable to consider the kinds of failures that may limit gear capacity and the application characteristics that enter into the estimate of load capacity.

13-1. Limitation on Load Rating of Gears. The maximum rating of a gearset is normally determined by one of the following:

Strength Rating. This is the load the tooth can carry without permanent deformation or fracture. Strength failure normally occurs with a fatigue fracture initiated at or near the root fillet.

Profile-durability Rating. This is the load that can be carried without damaging the profile. Normally, the limiting factor is the pitting of the profile at or below the pitch point. Occasionally, when high overloads are experienced, plastic profile deformation is encountered.

Scoring Rating. Loads exceeding this rating break down the oil film and allow the contacting surfaces to weld together, abrade, and otherwise destroy themselves. This rating is particularly to be considered for highly loaded case-carburized and casehardened gears or gears operating at high speeds, high temperatures, and with thin oils.

APPLICATION DATA REQUIRED FOR GEAR RATING

13-2. Data Required before Gears Can Be Rated. Before an attempt is made to rate gears as much accurate information as possible should be obtained about the following:

Type of Gearing. The axial and angular relations required for the input and output determine the type of gear which can be used. Table 13-1 shows the types of gears conventionally used for various shaft arrangements.

Table 13-1. Types of Gears
(Shaft axes)

Parallel	Intersecting	Nonintersecting nonparallel
Spur external Spur internal Helical external Helical internal	Straight bevel Zerol®* bevel Spiral bevel Face gear	Crossed helical Single-enveloping worm Double-enveloping worm Hypoid Spiroid®† Planoid®† Helicon®† Beveloid®‡

* Registered trade-mark of Gleason Works, Rochester, N.Y.
† Registered trade-marks of Illinois Tool Works, Chicago, Ill.
‡ Registered trade-mark of Vinco Corp., Detroit, Mich.

Character of Power Source and Driven Load. In addition to the transmitted load, dynamic forces are imposed upon gear teeth. These are partially determined by the torque characteristics of both the driving and the driven machine. Table 13-2 shows values for typical commercial drives which can be used as a guide.

Table 13-2. Typical Examples of Drive Characteristics

Power source:	
Uniform	Electric motor, turbine
Light shock	Multicylinder internal-combustion engine
Medium shock	Single-cylinder internal-combustion engine
Load on driven machine:	
Uniform	Centrifugal blower Pure liquid agitator Belt conveyor (uniform feed)
Moderate shock	Lobe-type blower Liquid and solid agitator Belt conveyor (nonuniform feed)
Heavy shock	Ore crushers Single-cylinder compressor Reciprocating conveyor

The magnitude of transient torque peaks during starting or operation should be determined as well as whether or not torque or directional reversals are present. If the load cycle varies, the number of cycles or operating time for each load should be obtained.

Operational Facts. Information should be gathered about input and output speeds, speed variations, expected life cycles, and the degree of reliability required to protect human safety or guard against operational shutdown hazards.

Gear-manufacturing Facilities. The size range and accuracy of gear-cutting equipment as well as gear-manufacturing and -finishing operations have an important effect on load capacity. For example, a high-speed gearset operating at 15,000 fpm pitch-line velocity must be manufactured to special tolerances and manufacturing techniques. These are not always available in every machine shop or gear-cutting plant. The designer must be certain that proper facilities are available to manufacture gears expected to operate at high speeds or in unusual conditions before selecting gear proportions or attempting load ratings.

Mounting.* The load capacity is dependent upon the operating accuracy of the gear mesh. The accuracy of bearing bore size and location, shaft and tooth misalignment due to elastic deflections of the shafts, teeth, mountings, or case, bearing clearances, etc., should be calculated. This information determines whether the gears are to be considered as high precision or commercial precision, or not specified as to accuracy. Normally, these data are recorded as the lead deviation across the face for spur or helical gears and as the lift of bevel pinions or gears.

Materials and Physical Properties. The ultimate tensile, yield strength, and fatigue properties of probable materials† to be used should be known, and so should abnormal design configurations which might affect internal stress or response to heat-treatment. If unusual temperatures or atmospheric conditions are expected to be encountered, the ability of the material to withstand these must be considered. It is desirable to know the modulus of elasticity when surface durability is to be computed.

Rating Recommendations. If the data secured in these categories are limited or seem inaccurate, an approximate estimate of the gear capacity is normally the best that can be expected. To simplify these estimates, the series of tables in Art. 13-3 should be used. These tables are also useful during preliminary design phases when an approximate estimate of gear size or load capacity can allow more detailed design investigations to proceed before it is necessary to make precise evaluations. Many times these tabulated capacities entirely satisfy the needs.

If considerable accurate data are available to evaluate the previously mentioned categories and the designer requires precise size or load-capacity calculations, pertinent information and reference can be found in Arts. 13-5 to 13-8.

SIMPLIFIED ESTIMATES FOR GEAR CAPACITY

In most cases a quick estimate of gear capacity can be made by using appropriate tables and modifying factors.

13-3. Tabulation of Gear Capacity. The surface contact stress usually limits the rating of spur, helical, and bevel teeth except for casehardened gears for which tooth strength is sometimes the limit of load capacity. Wormgears, hypoids, Spiroids, Planoids, and Helicons are usually rated by surface durability because the pitch is made coarse enough to eliminate the possibilities of breakage. The following tabulations are for rated gear capacity, which is usually converted into a service rating by dividing this rated capacity by the service or reliability factors from Tables 13-15 and 13-18 listed in Art. 13-4. Article 13-4 also contains data with which it is possible to estimate the capacity for hardnesses other than those listed as well as to adjust for different numbers of load cycles.

Spur and Helical Gears. For preliminary estimates of spur- or helical-gear size use the formula

$$C^2F = \frac{31{,}500}{K}\,\frac{P}{n_P}\,\frac{(m_G + 1)^3}{m_G} \tag{13-1}$$

* Full rating capacity of bevel gears can only be obtained by adjusting the mounting dimensions to develop a proper tooth contact pattern at full load (see Art. 23-9).

† Full load capacity of gears requires a proper choice of material and processing of material. See Chap. 10 for general information on materials. Note case-depth specifications in Tables 10-9 and 10-11.

Table 13-3. *K* Factor for Preliminary Estimates of Helical- and Spur-gear Size

Application	Service characteristics		BHN of material		Pitch-line speed, fpm	Accuracy	*K* factor
	Driver	Driven	Pinion	Gear			
Turbine driving generator	....	....	225	180	Over 4,000	High precision	110
Internal-combustion engine driving compressor	....	....	225	180	Over 4,000	High precision	60
Motor driving compressor	....	....	225	180	Over 4,000	High precision	55
General-purpose industrial drive	Uniform	Uniform	575 350 210	575 300 180	1,000 1,000 1,000	Commercial Commercial Commercial	500–1,000 350–450 170–250
	Uniform	Uniform	575 300 210	575 300 180	3,000 3,000 3,000	Commercial Commercial Commercial	475–750 275–375 125–200
Large industrial gears—hoists, kilns, mills	Uniform	Moderate shock	225	180	1,000 max.	Generated	80–100
	Uniform	Moderate shock	260	210	....	Generated	130–170
Aircraft (single pair)	Engine	Auxiliary drive	58 Rockwell C min.	58 Rockwell C min.	10,000	High ground	1,000 (at take-off)
Aircraft planetary	Engine	Propeller	58 Rockwell C min.	58 Rockwell C min.	3,000–10,000	Ground	600 (at take-off)
Automotive transmission	....	In low gear	58 Rockwell C min.	58 Rockwell C min.	....	High	1.500
Small commercial	Uniform	Uniform	Steel 350 BHN	Phenolic laminate	Under 1,000	Commercial	75
Small commercial	Uniform	Uniform	Steel 350 BHN	Nylon	Under 1,000	Commercial	50
Small gadget	Uniform	Uniform	Steel 200 BHN	Zinc-alloy die casting	Under 1,000	Commercial	25
Small gadget	Uniform	Uniform	Steel 200 BHN	Brass or aluminum	Under 500	Commercial	25
Small gadget	Uniform	Uniform	Brass or aluminum	Brass or aluminum	Under 500	Commercial	15

NOTES:
1. For spur gears reduce loading 25 per cent—make certain they will operate at required pitch-line velocity.
2. These values for 10 hr per day—for continuous service divide by 1.25.
3. For heavy shock conditions, divide values by at least 1.5.
4. For other hardness values, see Table 13-16.

Table 13-4. Nominal Capacity of Steel Spur Gears
(For pinion of 300 BHN min. and gear 255 BHN min.)

Ratio, m_G	Center distance, C	Pinion pitch dia., d	Face width, F	Rpm of pinion: 100	720	1,750	3,600
				Hp at different pinion speeds			
1	0.500	0.500	¼	0.009	0.062	0.140	0.260
2	0.750	0.500	¼	0.012	0.078	0.176	0.328
4	1.250	0.500	¼	0.013	0.089	0.201	0.373
8	2.250	0.500	¼	0.014	0.092	0.206	0.384
1	1.000	1.000	½	0.064	0.410	0.888	1.55
2	1.500	1.000	½	0.081	0.518	1.12	1.96
4	2.500	1.000	½	0.092	0.588	1.28	2.23
8	4.500	1.000	½	0.095	0.607	1.31	2.30
1	2.000	2.000	1	0.462	2.80	5.66	8.80
2	3.000	2.000	1	0.584	3.54	7.18	11.1
4	5.000	2.000	1	0.667	4.02	8.17	12.6
8	9.000	2.000	1	0.681	4.14	8.37	13.0
1	4.000	4.000	2	3.50	19.3	34.2	38.6
2	6.000	4.000	2	4.42	24.4	43.2	48.8
4	10.000	4.000	2	5.03	27.8	49.2	55.5
8	18.000	4.000	2	5.17	28.6	50.6	57.1
1	8.000	8.000	4	25.2	119.0	149.0	
2	12.000	8.000	4	31.9	150.0	188.0	
4	20.000	8.000	4	36.3	170.0	215.0	
8	36.000	8.000	4	37.3	175.0	221.0	
1	16.000	16.000	8	169.0	560.0		
2	24.000	16.000	8	214.0	707.0		
4	40.000	16.000	8	244.0	803.0		
8	72.000	16.000	8	250.0	827.0		
1	32.000	32.000	16	931.0			
2	48.000	32.000	16	1,176.0			
4	80.000	32.000	16	1,337.0			

NOTES:
1. Class 1 AGMA service for industrial applications.
2. Pitch-line speed not over 4,000 fpm.
3. No allowance for shock loads, $K_o = 1.0$.
4. Data based on 32 pinion teeth for $m_G = 1.0$, 28 teeth for $m_G = 2$, 24 for $m_G = 4$, and 20 for $m_G = 8$.
5. Based on 20° pressure angle, full-depth tooth.
6. See Art. 13-4 for other modifying factors.

where $K = \frac{W}{Fd}\left(\frac{m_G + 1}{m_G}\right)$

C = center distance, in.
F = net face width, in.
P = horsepower
n_P = pinion, rpm
m_G = ratio = N_G/N_P
W = tangential driving load, lb = $\frac{2T_P}{d}$
T_P = pinion torque, lb-in.
d = pinion pitch diameter, in. (take at large end of bevel gear)

Typical values for K are listed in Table 13-3. When the C^2F value has been obtained, C and F dimensions can be studied and a selection made to best fit the design requirements. For best practice the ratio F/d should not exceed 1. If it does, the deflections must be checked to make certain they do not impair contact. Ratios of F/d exceeding 2 are seldom used in commercial practice.

For ready reference Table 13-4 lists the nominal rated capacity of steel spur gears for a range of centers and face widths. Table 13-5 contains similar information for helical gears.

Table 13-5. Nominal Capacity of High-speed Helical Gears
(For pinion of 245 BHN min. and gear 210 BHN min.)

Ratio, m_G	Center distance, C	Pinion pitch dia., d	Face width, F	Rpm of pinion: 100	720	1,750	3,600	6,000	10,000
				Hp at different pinion speeds					
1	2.000	2.000	2.0	0.906	5.71	12.5	14.8	24.7	41.2
2	3.000	2.000	2.0	1.22	7.66	16.8	19.8	33.0	55.0
4	5.000	2.000	2.0	1.46	9.18	20.1	23.7	39.6	65.9
8	9.000	2.000	2.0	1.62	10.2	22.3	26.3	43.9	73.2
1	4.000	4.000	4.0	6.61	39.7	84.1	119.0	198.0	330.0
2	6.000	4.000	4.0	8.86	53.3	113.0	158.0	264.0	440.0
4	10.000	4.000	4.0	10.6	63.9	135.0	190.0	317.0	528.0
8	18.000	4.000	4.0	11.8	71.1	150.0	211.0	351.0	585.0
1	8.000	8.000	8.0	46.2	265.0	544.0	949.0	1,580	2,640
2	12.000	8.000	8.0	62.0	355.0	730.0	1,270	2,110	3,520
4	20.000	8.000	8.0	74.2	426.0	874.0	1,520	2,530	4,220
8	36.000	8.000	8.0	82.7	473.0	971.0	1,690	2,810	4,680
1	16.000	16.000	16.0	281	1,500	3,690	7,600	12,700	
2	24.000	16.000	16.0	376	2,000	4,920	10,100	16,900	
4	40.000	16.000	16.0	451	2,400	5,910	12,200	20,300	
8	72.000	16.000	16.0	503	2,670	6,560	13,500	22,500	
1	32.000	32.000	32.0	1,940	12,200	29,500			
2	48.000	32.000	32.0	2,590	16,200	39,400			
4	80.000	32.000	32.0	3,100	19,500	47,300			

NOTES:
1. Class 1 AGMA service.
2. No allowance for shock loads, K_o = 1.0.
3. Based on full-depth teeth, 20° transverse pressure angle.
4. See Art. 13-4 for other modifying factors.
5. Values at left of heavy line rated by AGMA 211.01.
6. Values at right of heavy line rated by AGMA 421.04.

Table 13-6. Nominal Capacity of Straight or Zerol Bevel Gears
(For case-carburized pinion and gear, 55 Rockwell C min.)

Ratio, m_G	Pinion pitch dia., d	Gear pitch dia., D	No. pinion teeth, N_P	Face width, F	Rpm of pinion: 100	720	1,750	3,600
					Hp at different pinion speeds			
1	1.000	1.000	22	3/16	0.094	0.611	1.34	2.38
2	1.000	2.000	20	5/16	0.156	1.01	2.22	3.95
4	1.000	4.000	17	9/16	0.277	1.80	3.96	7.04
8	1.000	8.000	14	1 1/16	0.334	2.18	4.78	8.49
1	2.000	2.000	23	3/8	0.523	3.14	6.73	
2	2.000	4.000	20	5/8	0.864	5.18	11.1	
4	2.000	8.000	17	1 1/8	1.54	9.24	19.8	
8	2.000	16.000	14	1 3/8	1.86	11.1	23.9	
1	4.000	4.000	24	3/4	2.93	16.1		
2	4.000	8.000	21	1 1/4	4.85	26.7		
4	4.000	16.000	18	2 3/16	8.41	46.4		
8	4.000	32.000	15	2 5/8	10.0	55.1		
1	8.000	8.000	30	1 9/16	16.7			
2	8.000	16.000	26	2 1/2	26.5			
4	8.000	32.000	23	3 7/16	36.2			

NOTES.
1. Class 1 AGMA service for industrial applications.
2. Pitch-line speed not over 1,000 fpm.
3. No allowance for shock loads, $K_o = 1.0$.
4. Based on 20° pressure angle, full-depth tooth.
5. Theoretical face width does not exceed 0.28 cone distance.

Table 13-7. Nominal Capacity of Large Spiral Bevel and Hypoid Gears

Ratio, m_G	Pinion* pitch dia., d	Gear pitch dia., D	No. gear teeth, N_G	Face width, F	Rpm of pinion: 100	720	1,750	3,600
					Hp at different speeds			
1	12.00	12.0	36	2 1/2	59.0	265.0	516.0	
1	16.00	16.0	36	3	101.0	437.0		
1	20.00	20.0	36	3 3/4	175.0	758.0		
1	24.00	24.0	40	4 3/4	276.0	1,135.0		
1	30.00	30.0	45	6	451.0	1,998.0		
1	36.00	36.0	45	7 1/2	733.0			
2.54	5.50	14.0	56	2 1/4	17.0	93.6	175.0	288.0
2.50	8.00	20.0	50	3	34.1	168.0	335.0	
2.50	11.20	28.0	65	4	82.2	370.0	719.0	
2.75	13.09	36.0	66	5	131.0	590.0	1,148.0	
4.53	3.08	14.0	68	2 1/4	7.46	44.2	92.2	158.0
4.68	4.27	20.0	75	3	15.7	86.7	178.0	300.0
4.50	6.67	30.0	77	4 1/2	42.1	214.0	390.0	714.0
5.00	7.20	36.0	90	5	53.0	271.0	494.0	
5.00	9.60	48.0	95	6 1/2	108.2	535.0	1,064.0	
7.00	5.14	36.0	98	5	34.5	190.0	355.0	584.0
7.00	6.86	48.0	105	6 1/2	66.3	337.0	614.0	1,123.0

Spiral angle and pressure angle are variable to fit application.
Based on case-carburized pinion and gear. Pitch-line velocity not over 6,000 fpm.
NOTES:
1. Class 1 AGMA service for industrial applications.
2. No allowance for shock loads, $K_o = 1.0$.
3. Do not increase face widths, except with more detailed study.

* Pinion pitch diameters shown are for *spiral bevel sets*. For *hypoid* the pinion to go with the same gear would be about 10 per cent larger.

Table 13-8. Nominal Capacity of Spiral Bevel Gears
(For case-carburized pinion and gear, 55 Rockwell C min.)

Ratio, m_G	Pinion pitch dia., d	Gear pitch dia., D	No. pinion teeth, N_P	Face width, F	Rpm of pinion: 100	720	1,750	3,600	10,000
					Hp at different pinion speeds				
1	1.000	1.000	20	3/16	0.139	0.906	1.99	3.53	7.73
2	1.000	2.000	15	5/16	0.239	1.55	3.41	6.06	13.3
4	1.000	4.000	11	9/16	0.501	3.26	7.15	12.7	27.8
8	1.000	8.000	8	11/16	0.598	3.89	8.54	15.2	30.4
1	2.000	2.000	21	13/32	0.725	4.35	9.32	15.7	
2	2.000	4.000	16	21/32	1.21	7.28	15.6	26.4	
4	2.000	8.000	12	1 5/32	2.50	15.0	32.2	54.3	
8	2.000	16.000	8	1 13/32	3.13	18.8	40.2	67.9	
1	4.000	4.000	23	27/32	4.11	22.6	41.9	73.8	
2	4.000	8.000	18	1 5/16	7.23	39.8	73.8	130.0	
4	4.000	16.000	14	2 7/32	13.4	73.8	137.0	240.0	
8	4.000	32.000	10	2 21/32	16.0	88.1	163.0	287.0	
1	8.000	8.000	29	1 11/16	24.4	120.0	225.0		
2	8.000	16.000	24	2 21/32	42.5	209.0	393.0		
4	8.000	32.000	19	3 15/32	56.8	280.0	525.0		

NOTES:
1. Class 1 AGMA service for industrial applications.
2. Pitch-line speed not over 4,000 fpm.
3. No allowance for shock loads, $K_o = 1.0$.
4. Based on 20° pressure angle, full-depth tooth.
5. Assume 35° spiral angle.
6. Theoretical face width does not exceed 0.30 cone distance.
7. Face width rounded to nearest lower 1/32″.

Table 13-9. Nominal Capacity of Straight Face Gears
(For pinion of 300 BHN min. and gear 255 BHN min.)

Ratio, m_G	Pinion pitch dia., d	Gear pitch dia., D	No. pinion teeth, N_P	Face width, F	Rpm of pinion: 100	720	1,750	3,600
					Hp at different pinion speeds			
1.5	1.000	1.500	20	0.085	0.018	0.116	0.251	0.440
	1.400	2.100	22	1/8	0.040	0.249	0.523	0.875
	2.000	3.000	24	11/64	0.105	0.637	1.29	2.00
	3.000	4.500	30	1/4	0.365	2.10	3.98	5.38
	4.000	6.000	36	23/64	0.899	4.96	8.79	9.92
	6.000	9.000	36	35/64	2.66	13.5	20.6	
2.0	1.000	2.000	20	9/64	0.024	0.154	0.334	0.586
	1.400	2.800	22	13/64	0.078	0.488	1.03	1.72
	2.000	4.000	24	19/64	0.194	1.18	2.38	3.70
	3.000	6.000	30	7/16	0.599	3.45	6.54	8.84
	4.000	8.000	36	37/64	1.38	7.62	13.5	15.2
	6.000	12.000	36	55/64	4.24	21.5	32.8	
4.0	1.000	4.000	20	3/8	0.073	0.466	1.01	1.77
	1.400	5.600	22	33/64	0.174	1.09	2.30	3.84
	2.000	8.000	24	23/32	0.491	2.97	6.01	9.34
	3.000	12.000	30	1 1/32	1.59	9.17	17.4	23.5

NOTES:
1. Class 1 AGMA service.
2. Pitch-line velocity not over 4,000 fpm.
3. No allowance for shock loads, $K_o = 1.0$.
4. Based on full-depth, 20° pressure angle tooth.

Bevel Gears. The nominal capacity of straight bevel and Zerol gears is listed in Table 13-6, large spiral bevels in Table 13-7,[1,*] and smaller spiral bevels in Table 13-8.[1] Straight bevel gears sometimes are used other than casehardened, and the tabulated capacity can be adjusted for hardness by the data in Art. 13-4, Table 13-16. Spiral bevel gears are usually used case-carburized and hardened.

Face Gears. Table 13-9 is for face gears of medium hardness. The pinion of a face gearset has about the same capacity per inch of face as a spur-gear pinion of equal face width. Face gears cannot be ground for high accuracy, but if manufacturing tolerances and heat-treating distortion can be maintained accurately, case-carburized face gears will have a capacity 150 to 200 per cent of that shown in Table 13-9.

Table 13-10. Nominal Capacity of Crossed-helical Gears
(For driver of 60 Rockwell C and driven phosphorus bronze)

Ratio, m_G	Center distance, C	Helix angle of pinion, ψ_1	Pinion pitch dia., d	Gear pitch dia., D	Rpm of pinion: 100	720	1,750	3,600	10,000
					Hp at different pinion speeds				
1	1.000	45	1.000	1.000	0.003	0.015	0.030	0.051	0.085
3	1.366	60	1.000	1.732	0.006	0.039	0.082	0.137	0.228
5	1.9435	60	1.000	2.887	0.010	0.066	0.139	0.231	0.385
8	1.572	75	1.000	2.144	0.019	0.123	0.260	0.433	0.720
1	2.000	45	2.000	2.000	0.018	0.117	0.245	0.408	
3	2.732	60	2.000	3.464	0.047	0.312	0.658	1.10	
5	3.887	60	2.000	5.774	0.081	0.528	1.11	1.85	
8	3.1435	75	2.000	4.287	0.152	0.988	2.08	3.46	
1	4.000	45	4.000	4.000	0.143	0.932	1.96	3.26	
3	5.464	60	4.000	6.928	0.382	2.50	5.27	8.75	
5	7.7735	60	4.000	11.547	0.646	4.22	8.90	14.8	
8	6.287	75	4.000	8.574	1.21	7.92	16.7		
1	8.000	45	8.000	8.000	1.14	7.46	15.7		
3	10.928	60	8.000	13.856	3.05	20.0	42.2		
5	15.547	60	8.000	23.094	5.16	33.8	71.2		

NOTES:
1. Class 1 AGMA service for industrial applications.
2. Pitch-line speed not over 4,000 fpm.
3. No allowance for shock load, $K_o = 1.0$.
4. Units given a short "break-in" run to polish contact area.

Crossed-helical Gears. Crossed-helical gears have point contact. The nature of this contact as obtained by the geometric proportions and "wearing in" largely determines the capacity. Table 13-10 shows the capacity which can be obtained with a casehardened driver and a bronze driven gear carefully broken in to polish the surfaces and broaden the contact area.

At slow speeds cast-iron gears might have double the capacity shown in the table provided that no sand or other abrasive inclusions are present in the metal.

Cylindrical Wormgearing. The limiting rating on the input power of wormgears depends upon both the contact stress and the sliding velocity. For purposes of preliminary estimates, it is best to compare the capacity of a range of sizes and pro-

* Superscript numbers refer to references at the end of the chapter.

portions such as those in Table 13-11. These gears have good contact bearing, good finish, and high accuracy. The worms are casehardened and ground meshing with chill-cast phosphorus-bronze wheels. Approximately 20 per cent higher ratings can be achieved by using centrifugally cast bronze. The tabulated gearsets are operating under 6,000 fpm sliding velocity and are for a shock-free system operating 10 hr or less per day. If it is desired to operate 24 hr per day in a system having some shock, the capacities should be reduced to 75 per cent of the values shown. For infrequent or short operating cycles the capacities may be multiplied by 1.5.

Table 13-11. Nominal Capacity of Cylindrical Wormgearing

Ratio, m_G	Center distance, C	Worm pitch dia., d	Effective face width, F_e	Rpm of worm: 100	720	1,750	3,600	10,000
				Hp at different worm speeds				
5	2	0.825	$\frac{15}{32}$	0.30	1.6	2.8	3.5	5.0
8	2		$\frac{1}{2}$	0.20	0.9	2.1	2.6	3.9
15	2		$\frac{15}{32}$	0.15	0.7	1.4	1.9	3.4
25	2		$\frac{15}{32}$	0.10	0.5	1.0	1.4	2.9
50	2		$\frac{15}{32}$	0.03	0.3	0.5	0.7	1.0
5	4	1.525	$\frac{7}{8}$	2.1	9.7	14.5	18.5	23.0
8	4		$\frac{15}{16}$	1.8	6.8	10.6	14.2	19.0
15	4		$\frac{15}{16}$	0.9	4.6	7.0	8.8	11.3
25	4		$\frac{15}{16}$	0.6	3.0	4.8	6.7	8.6
50	4		$\frac{15}{16}$	0.4	1.6	2.6	4.2	5.2
5	8	2.800	$1\frac{11}{16}$	12.0	44.5	66.6	86.0	
8	8		$1\frac{13}{16}$	8.0	32.5	48.0	65.0	
15	8		$1\frac{13}{16}$	5.5	23.0	33.5	47.0	
25	8		$1\frac{13}{16}$	4.0	15.2	22.0	34.0	
50	8		$1\frac{13}{16}$	2.1	8.0	11.3	20.0	
5	16	5.100	$3\frac{1}{8}$	65	190	265	350	
8	16		$3\frac{3}{8}$	50	140	210	265	
15	16		$3\frac{3}{8}$	36	100	145	185	
25	16		$3\frac{3}{8}$	23	66	100	130	
50	16		$3\frac{3}{8}$	12.5	30	50	75	

NOTES:
1. Class 1 AGMA service.
2. Sliding velocity not over 6,000 fpm.
3. No allowance for shock loads, $K_o = 1.0$.

The output torque rating can be determined as follows:*

$$T = \frac{P}{n_p} 63{,}000\, m_G \frac{\eta}{100} \quad \text{lb per in.} \tag{13-2}$$

where m_G = number of teeth in gear divided by number of threads in worm

$$\eta = \text{efficiency} = \left(100 - \frac{m_G}{2}\right) \text{per cent}$$

With special attention to materials selection and quality, using the maximum in accuracy refinements and suitable lubrication, the tabulated capacities can be

* This formula is quite approximate. See Art. 14-3 for more precise methods of determining wormgear efficiency.

increased 50 to 100 per cent. These high capacities would not be considered normal machine-design practice for the average wormgear set.

When other materials or limited sliding velocities are used, multiply the tabulated capacities by the values in Table 13-12.

Table 13-12. Wormgear Rating Modifications for Different Materials and Sliding Velocities

Material and hardness worm	Wormgear	Sliding velocity, fpm	Multiply rating by
Steel, 53 Rockwell C min.	Bronze-phosphor	6,000–10,000	1.0
Steel, 35 Rockwell C min.	Bronze-phosphor	Under 3,000	1.0
Steel, 53 Rockwell C min.	Bronze-manganese	Under 400	2.0
Steel, 53 Rockwell C min.	Bronze-manganese forged	Under 3,000	1.4
Cast iron	Cast iron	Under 400	1.6
Cast iron	Bronze-phosphor	Under 3,000	1.2

The above comments and capacities refer to the *mechanical* rating. Sufficient natural heat dissipation or coolers must be available to keep the lubricating oil at a reasonable temperature, usually less than 190°F.

Double-enveloping Wormgears. The capacity of double-enveloping wormgears is shown in Table 13-13. The worms must be heat-treated to a hardness of 32

Table 13-13. Double-enveloping Wormgearing

Ratio, m_G	Center distance, C	Worm pitch dia., d	Rpm of worm: 100	720	1,750	3,600	10,000
			Hp at different worm speeds				
5	2	0.830	0.35	1.72	2.96	4.20	6.20
15	2	0.830	0.15	0.77	1.40	1.86	2.65
50	2	0.850	0.05	0.25	0.45	0.66	0.98
5	4	1.730	3.0	12.9	19.8	25.6	34.5
15	4	1.550	1.27	5.95	9.64	13.0	17.8
50	4	1.660	0.39	1.82	3.07	4.22	5.60
5	8	3.450	24.4	84.4	119.2	148	
15	8	2.940	10.4	40.9	61.2	74.5	
50	8	2.900	3.28	12.8	19.5	24.5	
5	16	5.200	172	480	620		
15	16	5.100	78	256	350	435	
50	16	5.100	25	85	119	139	
5	24	7.350	550	1,420	1,840		
15	24	7.350	252	715	934		
50	24	7.350	77.5	226	292		

Notes:

1. Class 1 AGMA service.
2. Sliding velocity not over 6,000 fpm.
3. No allowance for shock loads, $K_o = 1.0$.

Rockwell C minimum or casehardened. The service rating should be computed as suggested for cylindrical wormgears. Equation (13-2) for output-torque rating also applies.

Hypoid Gears. A hypoid gear can be considered as having the same rating as a spiral bevel gear of the same size. The mating hypoid pinion will then be about 10 per cent larger. Tables 13-7 and 13-8 can then be used for preliminary estimates of hypoid-gear capacity.

Spiroid Gears. Table 13-14 shows the capacity of Spiroid gears when both members are carburized. Final finishing by grinding or lapping is desirable to achieve these capacities. Operation is expected to be with an adequate supply of extreme-pressure lubricant.

Table 13-14. Nominal Capacity of Spiroid Gears
(Pinion and gear casehardened, 60 Rockwell C min.)

Ratio, m_G	Center distance, C	Pinion OD, d_o	Gear OD, D_o	Rpm of pinion: 100	720	1,750	3,600	10,000
				Hp at different pinion speeds				
10.250	0.500	0.437	1.500	0.0142	0.0697	0.129	0.1948	0.3354
14.667	0.500	0.421	1.500	0.0120	0.0589	0.109	0.1645	0.2834
25.500	0.500	0.423	1.500	0.0088	0.0432	0.080	0.1208	0.2080
47.000	0.500	0.427	1.500	0.0064	0.0313	0.058	0.0876	0.1508
10.250	1.000	0.853	3.000	0.0937	0.4601	0.852	1.287	
14.667	1.000	0.821	3.000	0.0759	0.3726	0.690	1.042	
25.500	1.000	0.827	3.000	0.0550	0.2700	0.500	0.7550	
47.000	1.000	0.837	3.000	0.0377	0.1852	0.343	0.5179	
71.000	1.000	0.761	3.000	0.0299	0.1469	0.272	0.4107	
10.250	1.875	1.507	5.625	0.4961	2.435	4.51	6.810	
14.667	1.875	1.463	5.625	0.4125	2.025	3.75	5.663	
25.500	1.875	1.435	5.625	0.2915	1.431	2.65	4.002	
47.000	1.875	1.448	5.625	0.1804	0.8856	1.64	2.476	
71.000	1.875	1.308	5.625	0.1441	0.7074	1.31	1.978	
106.000	1.875	1.215	5.625	0.1188	0.5832	1.08	1.631	
10.200	3.250	2.395	9.750	2.101	10.31	19.0		
14.667	3.250	2.465	9.750	1.760	8.64	16.0		
25.500	3.250	2.319	9.750	1.21	5.94	11.0		
47.000	3.250	2.344	9.750	0.7205	3.537	6.55		
71.000	3.250	2.090	9.750	0.5544	2.722	5.04		
106.000	3.250	1.933	9.750	0.4455	2.187	4.05		
10.167	5.125	3.342	15.375	7.007	34.40	63.7		
14.250	5.125	3.342	15.375	5.918	29.05	53.8		
25.333	5.125	3.092	15.375	3.949	19.39	35.9		
47.500	5.125	2.910	15.375	2.255	11.07	20.5		
71.000	5.125	3.097	15.375	1.727	8.478	15.7		
106.000	5.125	2.841	15.375	1.386	6.804	12.6		

NOTES:
1. Class 1 AGMA service.
2. Pitch-line speed not over 1,700 fpm.
3. No allowance for shock loads, $K_o = 1.0$.
4. Based on tooth proportions recommended by Spiroid Division of Illinois Tool Works, Chicago, Ill.

Planoid Gears. These gears should rate similarly to hypoid gears of the same physical size.

Helicon. This type of gear can be rated at 50 per cent of the capacity of a similarly sized Spiroid gear.

Beveloid. Beveloids can be designed to mesh with contacts varying from a point as secured with crossed helicals to almost a line contact such as is present in a bevel gearset. Hence the capacity can best be estimated on the basis of the type of gearset the particular beveloid gear contact best simulates. The capacity will therefore

Table 13-15. Service Factors for Electric-motor Drives

Application	10-hr* service	24-hr service
Agitators:		
Pure liquids	1.00	1.25
Liquids and solids	1.25	1.50
Liquids—variable density	1.25	1.50
Blowers:		
Centrifugal	1.00	1.25
Lobe	1.25	1.50
Vane	1.00	1.25
Compressors:		
Centrifugal	1.00	1.25
Lobe, rotary	1.25	1.50
Reciprocating:		
Single cylinder	1.75	1.75
Multicylinder	1.25	1.50
Conveyors—uniformly loaded or fed: apron, assembly, belt, bucket, chain, flight, oven, screw	1.00	1.25
Conveyors—heavy duty, not uniformly fed: apron, assembly, belt, bucket, chain, flight, oven, screw	1.25	1.50
Conveyors—severe duty:		
Live roll	†	†
Reciprocating or shaker	1.75	1.75
Crushers:		
Ore or stone	1.75	1.75
Sugar	....	1.50
Elevators:		
Bucket—uniform load	1.00	1.25
Bucket—heavy load	1.25	1.50
Bucket—continuous	1.00	1.25
Centrifugal discharge	1.00	1.25
Escalators	1.00	1.25
Freight	1.25	1.50
Gravity discharge	1.00	1.25
Fans:		
Centrifugal	1.00	1.25
Cooling towers, induced and forced draft	†	†
Induced draft	1.25	1.50
Large (mine, etc.)	1.25	1.50
Large industrial	1.25	1.50
Light (small-diameter)	1.00	1.25
Feeders:		
Apron or belt	1.25	1.50
Disk	1.00	1.25
Reciprocating	1.75	1.75
Screw	1.25	1.50
Machine tools:		
Auxiliary drives	1.00	1.25
Main drives	1.25	1.50
Punch press (geared)	1.75	1.75
Tapping machines	1.75	1.75
Mills, rotary:		
Ball, dryers, coolers, cement kilns, pebble, rod	....	1.50
Kiln	1.25	1.50
Tumbling barrels	1.75	1.75
Pumps:		
Centrifugal	1.00	1.25
Proportioning	1.25	1.50
Reciprocating:		
Single-acting, 3 or more cylinders	1.25	1.50
Double-acting, 2 or more cylinders	1.25	1.50
Single-acting, 1 or 2 cylinders	†	†
Double-acting, single cylinder	†	†
Rotary: gear, lobe, vane	1.00	1.25
Screens:		
Air washing	1.00	1.25
Rotary—stone or gravel	1.25	1.50
Traveling water intake	1.00	1.25

* For intermittent duty (3 hr per day) use next lower service factor than shown in table. For occasional duty (½ hr per day) make individual study. Nominal service factors: 0.80, 1.0, 1.25, 1.50, 1.75, and 2.0.

† Service factor not established. Make individual study.

NOTE: Engine drives. For multicylinder engines use the next higher service factor than shown in table. Single-cylinder engine-drive applications require individual study. Service factors usually will be higher than multicylinder engines. These service factors assume that the system is free from serious critical and torsional vibrations.

normally vary from 20 to 70 per cent of the capacity of an equivalent-sized straight bevel gearset.

13-4. Modifying Factors for Tabulated Gear Capacities. The rating of the gearsets tabulated in Art. 13-3 can be modified for different service conditions, materials, required life cycles, etc., by means of the following data.

Service Factors. By means of field experience gear users and manufacturers have accumulated a list of service factors which are used to determine the service capacity. These factors measure such intangible items as shock loads, periods of overload, required reliability, and hours of operation. The rated capacities tabulated in

Table 13-16. Multiplying Factor for Relative Horsepower Durability Rating

Material and BHN	Steel									
	180	225	245	285	310	360	400	Case-car-bu-rized	Cast iron	Bronze gear, steel pinion
Steel:										
180	1.00	1.30	1.46	1.82	2.09	2.73	3.30	3.36	0.84	0.67
225	0.77	1.00	1.12	1.40	1.62	2.11	2.57	2.62	0.65	0.52
245	0.69	0.89	1.00	1.25	1.45	1.88	2.29	2.35	0.59	0.47
285	0.55	0.71	0.80	1.00	1.14	1.50	1.85	1.87	0.47	0.37
310	0.48	0.62	0.70	0.87	1.00	1.31	1.58	1.63	0.41	0.32
360	0.37	0.48	0.54	0.67	0.77	1.00	1.22	1.23	0.31	0.24
400	0.30	0.39	0.43	0.55	0.63	0.83	1.00	1.02	0.25	0.20
Case-carburized	0.29	0.38	0.42	0.53	0.61	0.81	0.98	1.00	0.25	0.20
Cast iron	1.19	1.53	1.70	2.14	2.46	3.26	3.92	4.00	1.00	0.80
Bronze gear, steel pinion	1.50	1.93	2.14	2.70	3.09	4.10	4.93	5.03	1.26	1.00

Read across for gear-material specification of present rating.
Read down for gear-material specification for which new rating is desired.
Rating present material × multiplying factor = rating for new material.
Case-carburized 55 to 65 Rockwell C (distortion factor = 0.90).

Table 13-17. Life Factors—Spur, Helical, and Bevel

Total life cycles required*	Durability rating, life factor C_L, spur, helical, bevel	Strength rating, life factor K_L				
		Spur and helical				Bevel
		160 BHN†	250 BHN	450 BHN	Case-car-burized‡	Case-car-burized‡
1,000	. . .	1.6	2.4	3.4	2.7	4.6
10,000	1.5	1.4	1.9	2.4	2.0	3.1
100,000	1.3	1.2	1.4	1.7	1.5	2.1
1 million	1.1	1.1	1.1	1.2	1.1	1.4
10 million and over	1.0	1.0	1.0	1.0	1.0	1.0

* Check for pinion and gear-tooth contact cycles.
† Brinell hardness of tooth.
‡ Carburized 55 Rockwell C minimum.

Table 13-18. Factors of Safety

Requirements for application	For gears rated for surface durability	For gears rated by fatigue strength
High reliability	1.25 and higher	1.50 and higher
Fewer than 1 failure in 100	1.00	1.00
Fewer than 1 failure in 3	0.80*	0.70

* At this load plastic deformation of the profile might occur.

Art. 13-3 should be divided by these service factors to obtain the allowable service rating. The estimated transmitted load also can be multiplied by the service factor to obtain the rated load of the gears selected from Art. 13-3. The service factors listed in Table 13-15 are typical of those used for general industrial purposes. Special applications as for vehicular transmissions, some machine tools, etc., might use service factors of 1 or less.

Gear Hardness. When so indicated in Art. 13-3 the tabulated rated capacities can be corrected for other hardness or materials by Table 13-16. These hardness and material corrections are applicable only when scoring or wear do not require case-hardened materials or special gear bronzes.

Life Cycles. The ratings tabulated in Art. 13-3 are for 50 million cycles and over. For lesser numbers of cycles the rating can be multiplied by the life factors shown in

Table 13-19. Recommended Lubricants
Spur, Helical, Bevel Gears

Low-speed centers	Ambient temp., °F	
	15–60, use AGMA No.	50–125, use AGMA No.
Parallel shaft (single reduction), up to 8″	2	3
Over 8″ and up to 20″	2	4
Over 20″	3	4
Parallel shaft (double reduction), up to 8″	2	3
Over 8″ and up to 20″	3	4
Over 20″	3	4
Parallel shaft (triple reduction), up to 8″	2	3
Over 8″ and up to 20″	3	4
Over 20″	4	5
Spiral or straight bevel-gear units		
Cone distance up to 12″	2	4
Cone distance over 12″	3	5
Gear motors	2	4
High-speed units*	1	2

* For speeds over 3,600 rpm or pitch-line velocities over 4,000 fpm. It may be desirable to use a lubricant one grade higher or lower than those specified, depending on operating conditions.

Table 13-19. Recommended Lubricants (*Continued*)
Cylindrical and Double-enveloping Wormgear Units

Worm centers	Worm speed up to rpm	Ambient temp., °F		Worm speed† above rpm	Ambient temp., °F	
		15–60*	50–125		15–60*	50–125
Up to 6″ inclusive:						
Cylindrical worms.	700	7 comp.	8 comp.	700	7 comp.	8 comp.
Double-enveloping worm. . . .	700	8 comp.	8A comp.	700	8 comp.	8 comp.
Over 6″ centers up to 12″ centers:						
Cylindrical worms.	450	7 comp.	8 comp.	450	7 comp.	7 comp.
Double-enveloping worms. . . .	450	8 comp.	8A comp.	450	8 comp.	8 comp.
Over 12″ centers up to 18″ centers:						
Cylindrical worms.	300	7 comp.	8 comp.	300	7 comp.	7 comp.
Double-enveloping worms. . . .	300	8 comp.	8A comp.	300	8 comp.	8 comp.
Over 18″ centers up to 24″ centers:						
Cylindrical worms.	250	7 comp.	8 comp.	250	7 comp.	7 comp.
Double-enveloping worms. . . .	250	8 comp.	8A comp.	250	8 comp.	8 comp.
Over 24″ centers:						
Cylindrical worms.	200	7 comp.	8 comp.	200	7 comp.	7 comp.
Double-enveloping worms. . . .	200	8 comp.	8A comp.	200	8 comp.	8 comp.

* Pour point of the oil used should be less than the minimum ambient temperature expected.

† Wormgears of either type operating at speeds above 2,400 rpm or 2,000 fpm rubbing speed may require force-feed lubrication.

Viscosity Range for AGMA Lubricants

AGMA lubricant No.	Viscosity range, SUV sec	
	At 100°F	At 210°F
1	180–240	
2	280–360	
3	490–700	
4	700–1,000	
5	. . . .	80–105
6	. . . .	105–125
7 comp.*	. . . .	125–150
8 comp.*	. . . .	150–190
8A comp.*	. . . .	190–250

* Compounded with 3 to 10 per cent acidless tallow.

Table 13-17. Service factors, when used for estimating capacity, normally include an evaluation of the life factor.

Reliability. Present engineering practice sometimes demands that the designer secure maximum reliability or in some instances design for a calculated risk. Failure in Table 13-17 does not mean an immediate failure under applied load but rather a shorter life than the minimum specified. Table 13-18 is a list of suggested factors of safety for certain reliabilities. A service factor normally includes a consideration of reliability requirements.

Lubrication. It is necessary that gears be lubricated with oils having proper characteristics. Worm gears and similar teeth which operate with a high sliding velocity must use specially compounded oils. Sometimes these special oils are required only during the running-in period. The oil viscosity must accommodate the bearings and remain liquid at lowest operating temperature and not thin out too much at the highest operating temperature. The load capacity of oils is dependent upon the surface conditions of the teeth and normally increases slightly with an increase in viscosity. Extreme-pressure additives are normally helpful to protect or smooth over profiles which might have been partially damaged by overloads. Table 13-19[2] is a list of lubricants* compatible with the ratings shown in Art. 13-3.

Thermal Rating. It is necessary to limit the horsepower transmitted by a gearset below that which causes an excessive temperature rise. Normally, gearsets can be operated safely with a temperature rise of between 70 and 90°F with maximum temperatures of 200°F. More frequent oil changes should be made when temperatures continually exceed 180°F. Table 13-20 shows the heat dissipation which can be expected for gears enclosed in conventionally designed cases. If more specific data are not available, Table 13-21 can be used to estimate gear losses including those due to mesh and bearing friction, oil churning, and windage.

Table 13-20. Heat Dissipation

Area exposed to moving air, sq ft	*Heat dissipation, hp/sq ft/°F rise*
10	0.002
20	0.0019
50	0.0017
100	0.0013
140	0.0012

The Table 13-20 data are intended for use with enclosed speed reducers having no external oil cooling system. When the oil is pumped through a cooling system, it is possible to design the system to take away almost any amount of heat. See Art. 15-6 for information on oil cooling systems.

Table 13-21. Gear Unit Efficiencies*

Type of gear	*Losses per each reduction, %*
Spur	2
Helical	2
Bevel	2
Worms and other gears having considerable sliding such as crossed-axis helical, hypoid, and Spiroid	$m/2$ (m = reduction ratio)

* See Chap. 14 for more complete data on gear efficiency.

DETAILED ANALYSIS OF GEAR CAPACITY

For more precise rating of gears or for special cases departing from conventional procedures a more detailed analysis is required. In some cases pilot-model or full-sized unit tests are desirable. The following rating methods have been used for these

* See Chap. 15 for more complete information on gear lubrication.

special cases. In most instances the basic data are from the various standards and practices of the American Gear Manufacturers Association. These specialized practices or those issued by other similar organizations should be consulted for additional detailed data which might prove helpful.

13-5. Surface Durability of Spur, Helical, Herringbone, and Bevel Gear Teeth. Surface durability is measured as the resistance of the profile to the fatigue phenomenon known as "pitting." A general formula applicable to spur, helical, herringbone, and bevel gears can be used to determine the limiting load to prevent destructive pitting.[3] The data presented apply to external gears, unless otherwise noted.

Fundamental Surface-durability Formula. The fundamental surface-durability formula for gear teeth is:

$$s_c = C_p \sqrt{\frac{W_t C_o}{C_v} \frac{C_s}{dF} \frac{C_m C_f}{I}} \tag{13-3}$$

where s_c = calculated contact stress number

The relation of calculated contact stress number to allowable contact stress number is

$$s_c \leqq s_{ac} \frac{C_L C_H}{C_T C_R}$$

where s_{ac} = allowable contact stress number. This depends on:

1. Material composition
2. Mechanical properties
3. Number of cycles
4. Temperature
5. Size
6. Residual stress
7. Work hardness

Table 13-22. Allowable Contact Stress Number s_{ac}

Material	Surface hardness, BHN	s_{ac}
Steel	180 min.	95,000
	240 min.	115,000
	300 min.	135,000
	360 min.	160,000
	440 min.	190,000
	55–63 Rockwell C (case-carburized)	200,000
Cast iron:		
AGMA grade 20	. . . .	30,000
AGMA grade 30	175 min.	50,000
AGMA grade 40	200 min.	65,000
	Tensile strength min., psi	**s_{ac}**
Bronze:		
AGMA 2c (10–12% tin)	40,000	30,000
Aluminum bronze ASTM-B-148-52 Alloy 9C, heat-treated	90,000	65,000

NOTE: These values are suggested for general design purposes. Higher values may be used when high-quality material is used, when section size and design allow maximum response to heat-treatment, and when proper quality control is effected by adequate inspection.

Table 13-22 shows typical values for s_{ac}.

C_L = life factor. The life factor C_L adjusts the allowable loading for a required number of cycles of operation wherein the gear pitting will not exceed the boundary between corrective and progressive pitting.

C_H = hardness ratio factor. The hardness ratio factor, C_H, depends on:

1. Gear ratio
2. Hardness of pinion and gear

C_T = temperature factor. The temperature factor, C_T, is generally taken as 1.0 when gears operate with oil or gear blank temperatures not exceeding 250°F. In some instances it is necessary to use a C_T value greater than 1.0 for carburized gears operating at oil temperatures above 160°F. When oil temperatures above 180°F are encountered, more frequent oil changes are necessary.

C_R = factor of safety. The factor of safety is introduced to offer the designer an opportunity to design for high reliability, or in some instances, to design for a calculated risk. Failure does not mean an immediate failure under the applied load, but rather a shortening of life.

The definition of the items in Eq. (13-3) continues as follows:

C_p = coefficient depending upon elastic properties. From the theory of cylindrical contacting surfaces, it is

$$C_p = \sqrt{\frac{k}{\pi\left(\frac{1-\mu_P^2}{E_P} + \frac{1-\mu_G^2}{E_G}\right)}}$$

where μ_P and μ_G = Poisson's ratio for pinion and gear, respectively = 0.30

E_P and E_G = modulus of elasticity for pinion and gear

k = 1.0 for gears with nonlocalized contact (most spur, helical, and herringbone gears)

k = $\frac{3}{2}$ for gears with localized contact (most bevel gears)

W_t = transmitted tangential load, lb, at the operating pitch diameter. The transmitted tangential load is calculated directly from the power transmitted by the gearset. (When operating near a critical speed of the drive, a careful analysis of conditions must be made.) When the transmitted load is not uniform, consideration should be given not only to the peak load and its anticipated number of cycles, but also to intermediate loads and their number of cycles. Frequently, a gearset is required to carry a peak load for a limited number of cycles, and then a much smaller load for a much larger number of cycles. In such cases the gear rating should be adjusted so that each load and number of cycles carried by the gearset is within allowable stress limits. The transmitted tangential load is

$$W_t = \frac{P \times 33{,}000}{v_t} = \frac{2T}{d} = \frac{P \times 126{,}000}{n_P d}$$

where P = transmitted power, hp

T = pinion torque, in.-lb

v_t = pitch-line velocity, fpm

C_o = overload factor. The overload factor makes allowances for the roughness or smoothness of operation of both the driving and driven apparatus. Specific overload factors can be established only after considerable field experience is gained in a particular application. In selecting the overload factor, consideration should be given to the fact that many prime movers develop momentary overload torques appreciably greater than those determined by the nameplate ratings of either the prime mover or the driven apparatus. Service factors have been established where field data are available for specific applications. These service factors include not only the overload factor, but also the life factor and factor of safety. Service factors* for many applications are listed in AGMA standards and should be used whenever applicable. If a specific service factor is used in place of the overload factor C_o, use a value of 1.0 for C_R and C_L. Table 13-23 shows recommended

Table 13-23. Overload Factors C_o

Power source	Character of load on driven machine		
	Uniform	Moderate shock	Heavy shock
Uniform	1.00	1.25	1.75
Light shock	1.25	1.50	2.00
Medium shock	1.50	1.75	2.25

This table is for speed-decreasing drives only. For speed-increasing drives of spur and bevel gearing add $0.01(N_G/N_P)^2$ to the factors in Table 13-23, where N_P = number of teeth in the pinion, N_G = number of teeth in the gear. For speed-increasing helical and herringbone gears, further consideration is required.

values of overload factors. See Table 13-2 for typical characteristics of driving and driven machines.

C_v = dynamic factor. The dynamic factor C_v depends on:

1. Effects of tooth spacing and profile error
2. Effects of pitch-line speed and rpm
3. Inertia and stiffness of all rotating elements
4. Transmitted load per inch of face
5. Tooth stiffness
6. Lubricant properties

Commonly used C_v factors are shown in Fig. 13-1.

C_s = size factor. The size factor reflects nonuniformity of material properties. It depends primarily on:

1. Tooth size
2. Diameter of parts
3. Ratio of tooth size to diameter of part
4. Face width
5. Area of stress pattern
6. Ratio of case depth to tooth size
7. Hardenability and heat-treatment of materials

The size factor may be taken as unity for most gears, provided a proper choice of steel is made for the size of the parts and the case depth or hardness pattern is adequate.

d = pinion operating pitch diameter, in. (taken at outer end of tooth for bevels)

F = face width, in.

C_m = load-distribution factor. The load-distribution factor C_m evaluates the effects of nonuniform distribution of load across the face and depends upon:

1. Misalignment of axes
2. Lead deviations
3. Elastic deflection of shafts, bearings, and housings

* Some typical service factors are shown in Table 13-15.

C_f = surface-condition factor. The surface-condition factor C_f depends on:

1. Surface finish as affected by cutting, shaving, lapping, grinding, shot peening, etc.
2. Residual stress
3. Plasticity effects (work hardening)

The surface condition factor C_f may be taken as unity, unless tests or field experiences indicate otherwise.

I = geometry factor. The geometry factor accounts for the effect of dimensional proportions such as the profile curvature at the contact, the pressure angle and the load sharing.

$$I = \frac{C_c}{m_N}$$

where C_c = curvature factor at pitch line

m_N = load-sharing ratio which depends on profile contact ratio, face contact ratio, and the crowning effect.

The following formula may be used to calculate the power directly:

$$P_{ac} = \frac{n_P F}{126{,}000} \frac{I C_v}{C_s C_m C_f C_o} \left(\frac{s_{ac} d}{C_p} \frac{C_L C_H}{C_T C_R} \right)^2 \tag{13-4}$$

where P_{ac} = allowable power, hp.

n_p = pinion speed, rpm

When service factors are available, use these instead of the term C_o. The terms C_L, C_T, and C_R will then have a value of 1.0.

Spur-gear Surface-durability Rating. Using Eq. (13-3) or (13-4) the surface durability of spur gears can be computed. The following values are generally used for the various factors:

C_o = overload factor. See Table 13-23.

C_v = dynamic factor. See Fig. 13-1. Use a value of 1.0, curve 1, for high-precision shaved or ground spur gears when no appreciable dynamic load can be developed. Use curve 3, Fig. 13-1, for high-precision shaved or ground spur gears when it is possible to develop a dynamic load. Use curve 4, Fig. 13-1, for spur gears finished by hobbing or shaving. When milling cutters are used to cut teeth or when inaccurate teeth are generated, lower dynamic factors than the above must be used.

C_s = size factor. Use a value of 1.0 for general application. For case-carburized gears of 2 diametral pitch or coarser a size factor of 1.25 might well be included. Face widths over 10 in. should be derated 10 to 20 per cent for the size effect.

C_m = load-distribution factor. Use the values in Table 13-24.

Table 13-24. Load-distribution Factor—Spur Gears

Condition of support	Face width			
	2″ and under	6″	9″	16″ and over
Accurate mountings, low bearing clearances, minimum elastic deflection, precision gears	1.3	1.4	1.5	1.8
Less rigid mountings, less accurate gears, contact across full face	1.6	1.7	1.8	2.2
Accuracy and mounting such that less than full face contact exists	Over 2.2			

C_f = surface-condition factor. Use a value of 1.0.

I = geometry factor. The spur-gear geometry factors are shown in Fig. 13-2. The values are for the worst condition of tooth contact considered as being at the point where single tooth contact begins.

s_{ac} = allowable stress. Use Table 13-22.

C_L = life factor from Table 13-17.

C_T = temperature factor. Use 1.0 unless operating temperature exceeds 250°F above which some derating might be necessary.

C_H = hardness ratio factor. Use 1.0.

C_R = factor of safety. Use Table 13-18. If the service factors in Table 13-15 are used, C_T, C_H, and C_R should each be 1.0.

C_p = elastic coefficient. Use Table 13-25.

Table 13-25. Spur and Helical Gears Elastic Coefficient C_p, Nonlocalized Contact

Pinion material	Modulus of elasticity	Gear material and modulus of elasticity			
		Steel 30×10^6	Cast iron 19×10^6	Aluminum bronze 17.5×10^6	Tin bronze 16×10^6
Steel	30×10^6	2,300	2,000	1,950	1,900
Cast iron	19×10^6	2,000	1,800	1,800	1,750
Aluminum bronze	17.5×10^6	1,950	1,800	1,750	1,700
Tin bronze	16×10^6	1,900	1,750	1,700	1,650

Poisson's ratio = 0.30.

Helical- and Herringbone-gear Surface Durability. Equation (13-3) or (13-4) can be used, with the following as suggested values:

C_o = overload factor. Use Table 13-23.

C_v = dynamic factor. See Fig. 13-1. Use curve 1, Fig. 13-1, a value of 1.0, for high-precision shaved or ground helical or herringbone gears when no appreciable dynamic load can be developed. Use curve 2, Fig. 13-1, for high-precision helical gears. Use curve 3, Fig. 13-1, for high-precision shaved or ground helical and herringbone gears when it is possible to develop a dynamic load. This value is recommended for commercial helical and herringbone gears used for general industrial applications. When milling cutters are used to cut teeth or when inaccurate teeth are generated, lower dynamic factors than the above must be used.

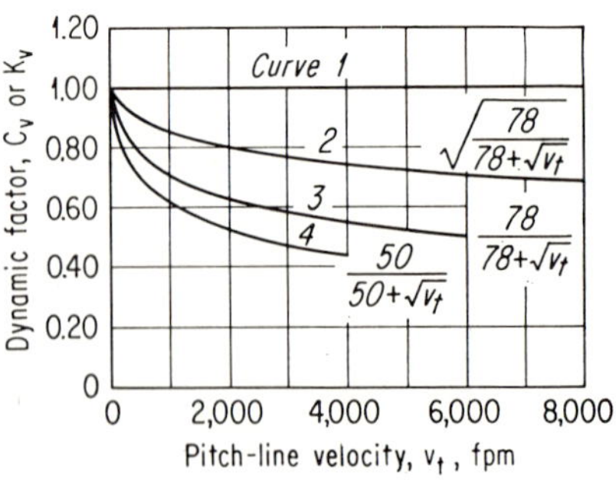

FIG. 13-1. Dynamic factor, C_v and K_v.

Table 13-26. Load-distribution Factor, Helical Gears

Condition of support	Face width			
	2″ and under	6″	9″	16″ and over
Accurate mountings, low bearing clearances, minimum elastic deflection, precision gears	1.2	1.3	1.4	1.7
Less rigid mountings, less accurate gears, contact across full face	1.5	1.6	1.7	2.0
Accuracy and mounting such that less than full face contact exists	Over 2			

Table 13-27. Load-distribution Factor for Spur, Helical, and Herringbone Gears, C_m and K_m

Conditions ratio of F/d	Contact	C_m and K_m
1.0 or less	95% face-width contact obtained at ⅓ torque 95% face-width contact obtained at full torque	1.4 at ⅓ torque 1.1 at full torque
	75% face-width contact obtained at ⅓ torque 95% face-width contact obtained at full torque	1.8 at ⅓ torque 1.3 at full torque
	35% face-width contact obtained at ⅓ torque 95% face-width contact obtained at full torque	2.5 at ⅓ torque 1.9 at full torque
	20% face-width contact obtained at ⅓ torque 75% face-width contact obtained at full torque	4.0 at ⅓ torque 2.5 at full torque
	Teeth are crowned 35% face-width contact obtained at ⅓ torque 85% face-width contact obtained at full torque	2.5 at ⅓ torque 1.7 at full torque
Over 1 less than 2	Calculated combined twist and bending of pinion not over 0.001″ over entire face Pinion not over 250 BHN hardness 75% contact obtained at ⅓ torque 95% contact obtained at full torque	2.0 at ⅓ torque 1.4 at full torque
	Calculated combined twist and bending of pinion not over 0.0007″ over entire face Pinion not over 350 BHN hardness 75% contact obtained at ⅓ torque 95% contact obtained at full torque	2.0 at ⅓ torque 1.4 at full torque
	30% contact obtained at ⅓ torque 75% contact obtained at full torque	4.0 at ⅓ torque 3.0 at full torque
	Twist and bending exceeds 0.001″ over entire face	Calculate effects of deflection and either adjust helix angle to compensate for deflection or increase C_m to allow for both alignment errors and deflection

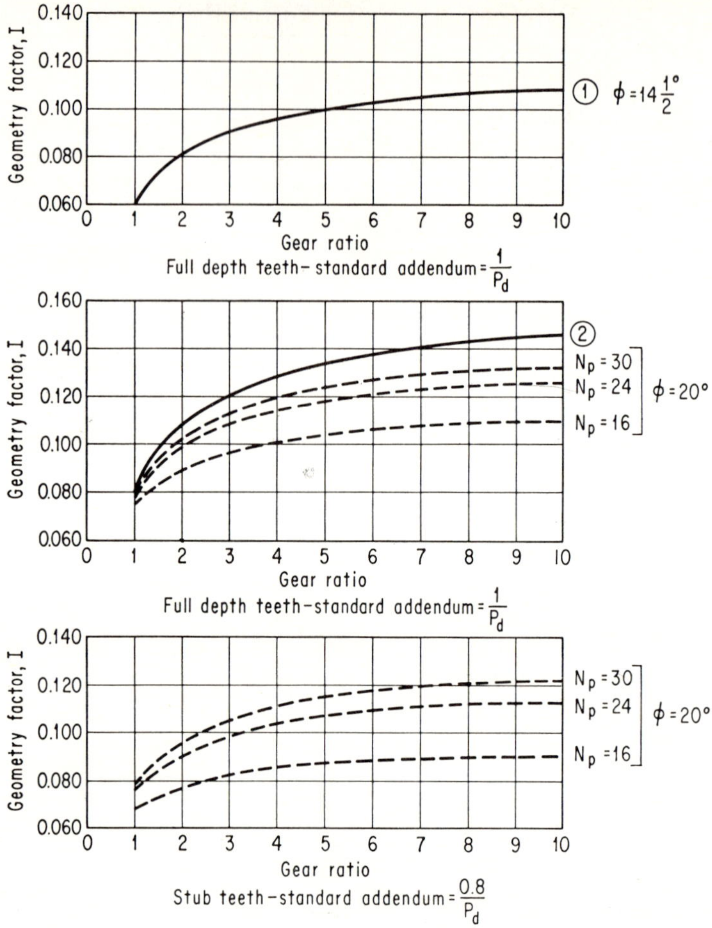

FIG. 13-2. External spur gear, geometry factor I. Curves 1 and 2 pitch-line loading. All others modified for the beginning of single-tooth contact.

C_s = size factor. Use a value of 1.0 for general applications. For case-carburized gears of 2 diametral pitch or coarser, a factor of 1.25 might well be used. The size factor for helical gears is partially included when the C_m factors in Fig. 13-3 are used.

C_m = load-distribution factor. Use the values in Table 13-26. Manufacturers of wide-face helical and herringbone gears generally control the magnitude of misalignment and load distribution by design restrictions and special manufacturing procedures. For these cases the values of C_m shown in Table 13-27 are appropriate.

For enclosed helical gears equivalent to commercial gear units in accuracy and mounting rigidity, the load-distribution factor C_m shown in Fig. 13-3 can be used as a guide. The load-distribution factor shown includes other in-built factors affecting gear durability such as size effect and limited alignment errors. When F/d ratio exceeds 2, a more detailed analysis is suggested.

C_f = surface-condition factor. Use a value of 1.0.

I = geometry factor

$$I = \frac{C_c}{m_N}$$

where C_c = curvature factor at pitch line

$$= \frac{\sin \phi_t \cos \phi_t}{2}\left(\frac{m_G}{m_G + 1}\right) \qquad \text{external gears}$$

$$= \frac{\sin \phi_t \cos \phi_t}{2}\left(\frac{m_G}{m_G - 1}\right) \qquad \text{internal gears}$$

$$m_N = \frac{p_N}{0.95Z} \qquad \text{(external and internal gears)}$$

p_N = normal base pitch

Z = length of line of action, in.

s_{ac} = allowable stress. Use Table 13-22.

C_L = life factor from Table 13-17.

C_T = temperature factor. Use a value of 1.0, unless operating temperatures exceed 250°F, above which some derating might be necessary.

C_H = hardness-ratio factor. Use Fig. 13-4.

C_R = factor of safety. Use Table 13-18. If the service factors in Table 13-15 are used, C_T, C_H, and C_R should each be 1.0.

C_p = elastic coefficient. Use Table 13-25.

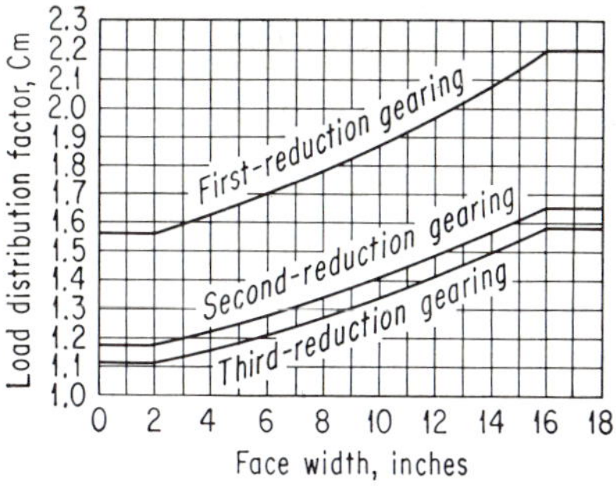

FIG. 13-3. Helical-gear load, distribution factor C_m. For fourth and additional reductions use third-reduction values.

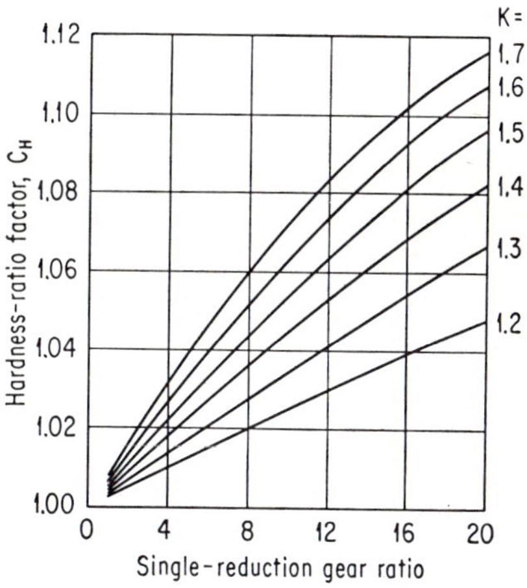

FIG. 13-4. Hardness-ratio factor C_H.

$$K = \frac{\text{BHN of pinion}}{\text{BHN of gear}}$$

When $K < 1.2$ use $C_H = 1.00$.

Bevel-gear Durability Rating. Using Eqs. (13-3) or (13-4) the surface durability of bevel gears can be computed using values for the various factors as follows:

C_o = overload factor. See Table 13-23.

C_v = dynamic factor. See Fig. 13-1. Use curve 1, a value of 1.0, for generated bevel gears having the preferred pattern of tooth contact and accurate spacing and concentricity. The system should be such that no appreciable dynamic load can be developed. Use curve 2, Fig. 13-1, for large-planed spiral bevel gears. Use curve 4, Fig. 13-1, for large-planed straight bevel gears. When milling cutters are used to cut teeth or when inaccurate teeth are generated, lower dynamic factors than the above must be used.

C_s = size factor. Use a value of 1.0.

C_m = load-distribution factor. When the deflections or displacements of the gear and pinion from their theoretical position are known, the C_m value can be obtained from Fig. 13-5. When actual displacements are not known, the C_m

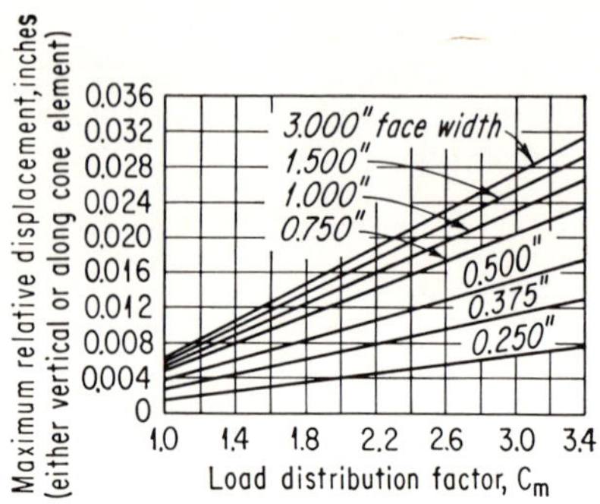

FIG. 13-5. Load-distribution factor for bevel gears C_m.

values can be obtained from Table 13-28. Where mountings are carefully designed for maximum rigidity, use the smaller factor from the table. Where mounting rigidity is questionable, use the larger factor.

Table 13-28. Bevel-Gear Load-distribution Factors C_m

Application	Both members straddle-mounted	One member straddle-mounted	Neither member straddle-mounted
General industrial	1.00–1.10	1.10–1.25	1.25–1.40
Automotive	1.00–1.10	1.10–1.25	
Aircraft	1.00–1.25	1.10–1.40	1.25–1.50

NOTE: Because of the smaller bearing loads for straddle mountings, the over-all designs are sometimes smaller and lighter for equal rigidity. Frequently space limitations dictate the type of mounting for a particular installation. Usually one member of the pair can be straddle-mounted, but it is not always feasible to use a straddle mounting for both members. An overhung mounting might be just as good as, or better than, a straddle-mounting arrangement. However, rigid supports as well as adequate capacity are required for both types of mountings.

C_f = surface-condition factor. Use a value of 1.0.

I = geometry factor. Figure 13-6 gives geometry-factor values for generated spiral bevel gears with 20° pressure angle, 35° spiral angle, and 90° shaft angle when designed according to "System for Spiral Bevel Gears," AGMA standard 209.02. These values are valid provided the face width does not exceed three-tenths of the cone distance or $10/P_d$ and is not less than one-quarter of the cone distance. Figure 13-7 gives geometry-factor values for generated Coniflex®* (crowned) straight bevel gears and Zerol bevel gears with 20° pressure angle and 90° shaft angle when designed according to AGMA standard 208.01, "System for Straight Bevel Gears," or AGMA standard 202.02, "Zerol Bevel Gear System," provided the face width does not exceed one-third of the cone distance (on Zerols one-fourth of the cone dis-

* Coniflex is a registered trade-mark of the Gleason Works, Rochester, N.Y.

tance) or $10/P_d$. For other tooth proportions, other pressure angles, or a shaft-angle different from 90° the geometry factor I for bevel gears can be determined by the method outlined in Appendix A of AGMA 215.01.[3]

s_{ac} = allowable stress. Use Table 13-22.

C_L = life factor from Table 13-17.

C_H = hardness-ratio factor. Use a value of 1.0.

C_T = temperature factor. Use a value of $K_T = (460 + T_F)/620$ where T_F is the peak operating oil temperature in degrees Fahrenheit.

C_R = factor of safety. Use Table 13-18. If service factors in Table 13-15 are used, C_T, C_H, and C_R should each be 1.0.

As a possible aid to the selection and use of allowable stress values the following is nserted pertinent to special applications.

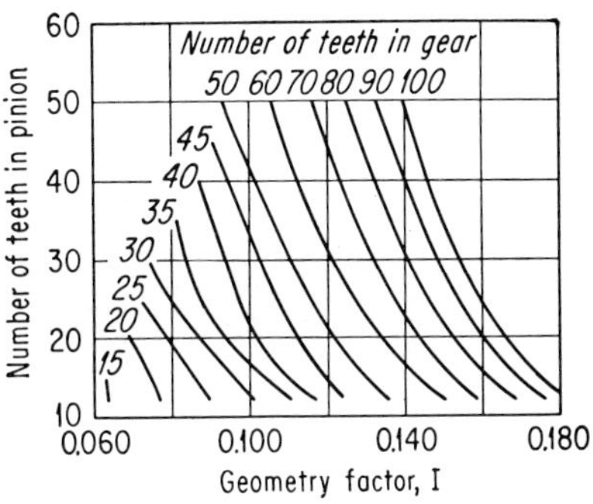

FIG. 13-6. Geometry factors I for spiral bevel gears (AGMA 209.02). Shaft angle 90°. Pressure angle 20°. Spiral angle 35°.

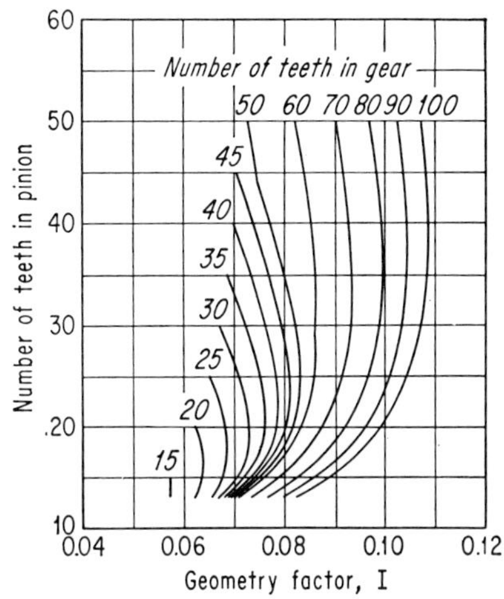

FIG. 13-7. Geometry factors I for straight and Zerol bevel gears (AGMA 208.01 and 202.02). Shaft angle 90°. Pressure angle 20°.

Special Applications of Bevel Gears. 1. *Aircraft Power Drives.* These applications usually involve continuous loads and peak loads for large numbers of cycles. Nearly all such gears are made from casehardened steel with ground tooth surfaces. The allowable contact stress in this case is 240,000 psi.

2. *Automotive Drive Gears.* Passenger-car gears of casehardened steel are designed for a maximum contact stress of 375,000 psi. Truck gears of casehardened steel are designed for a maximum contact stress of 500,000 psi. The above stresses are based on maximum torque through low gear (on truck, a transmission ratio not exceeding 8:1) or maximum tractive-effort torque, whichever is the lesser. Tractive-effort torque should be based on a coefficient of friction between wheels and pavement of 0.85.

3. *Marine Gears.* In general, marine gears of casehardened steel are designed for a maximum contact stress of 200,000 psi. Special application may be designed for higher stress values, but these generally involve limited-life considerations.

4. *Railway Gears.* The design of railway drives is based largely on tractive-effort torque using a coefficient of friction between wheels and rails of 0.25. Based on maximum torque, the contact stress for casehardened steel should not exceed 300,000 psi.

5. *Tractors.* Both farm and industrial tractors should be designed for a maximum contact stress for casehardened steel of 325,000 psi. Tractive-effort torque should be

based on a coefficient of friction between wheels and ground of 0.65, except on lug-type wheels where a value of 1.0 should be used.

C_p = elastic coefficients. Use Table 13-29.

Table 13-29. Elastic Coefficient C_p, Bevel or Other Gears with Localized Contact

Pinion material	Modulus of elasticity	Gear material and modulus of elasticity			
		Steel 30×10^6	Cast iron 19×10^6	Aluminum bronze 17.5×10^6	Tin bronze 16×10^6
Steel	30×10^6	2,800	2,450	2,400	2,350
Cast iron	19×10^6	2,450	2,250	2,200	2,150
Aluminum bronze	17.5×10^6	2,400	2,200	2,150	2,100
Tin bronze	16×10^6	2,350	2,150	2,100	2,050

Poisson's ratio = 0.30.

13-6. Strength Rating of Spur, Helical, Herringbone, and Bevel Gear Teeth. The strength of spur, helical, herringbone, and bevel gear teeth can be determined by using a basic bending-stress equation (13-5)[4] applicable to each type.

Basic Bending-stress Equation. The basic equation for the bending stress in a gear is calculated as follows:

$$s_t = \frac{W_t K_o}{K_v} \frac{P_d}{F} \frac{K_s K_m}{J} \tag{13-5}$$

where s_t = calculated tensile stress at the root of the tooth, psi

Load:

W_t = transmitted tangential load, lb, at operating pitch diameter

K_o = overload factor

K_v = dynamic factor

Tooth size:

P_d = transverse diametral pitch (use at large end of teeth for bevel gears)

F = face width, in.

Stress distribution:

K_s = size factor

K_m = load-distribution factor

J = geometry factor

Note that the above equation is divided into three groups of terms, the first of which is concerned with the load, the second with tooth size, and the third with stress distribution.

The relation of calculated stress to allowable stress is

$$s_t \leqq \frac{s_{at} K_L}{K_T K_R}$$

where s_{at} = allowable stress for material, psi

s_t = calculated stress, psi

K_L = life factor

K_T = temperature factor

K_R = factor of safety

Fundamental Power Formula. In preparing handbook data, for gear designs already developed, the following formula can be used to calculate the power directly:

$$P_{at} = \frac{n_P d K_v}{126{,}000 K_o} \frac{F}{K_m} \frac{J}{K_s P_d} \frac{s_{at} K_L}{K_R K_T} \tag{13-6}$$

where P_{at} = allowable power of gearset, hp
n_P = pinion speed, rpm
d = operating pitch diameter of pinion, in. (at large end for bevels)

Tangential Transmitted Load. The tangential transmitted load is calculated directly from the power transmitted by the gearset. (When operating near a critical speed of the drive, a careful analysis of conditions must be made.) When the transmitted load is not uniform, consideration should be given not only to the peak load and its anticipated number of cycles, but also to intermediate loads and their number of cycles.

The tangential transmitted load is

$$W_t = \frac{P \times 33{,}000}{v_t} = \frac{2T}{d} = \frac{P \times 126{,}000}{n_P d}$$

where P = power transmitted, hp
T = pinion torque, lb-in.
v_t = pitch-line velocity, fpm
n_P = pinion speed, rpm

Geometry Factor *J*. The geometry factor evaluates the shape of the tooth, the position at which the most damaging load is applied, stress concentration and correction due to geometric shape, and the sharing of load between one or more pairs of teeth.

Load-distribution Factor K_m. The evaluation of this factor is critical to gear rating. The load-distribution factor K_m depends upon:

1. Misalignment of axes of rotation
2. Alignment errors due to tooth inaccuracies
3. Elastic deflections of shafts, bearings, and blanks

The foregoing errors can combine so that less than the full face-width contacts with the mating element, or full contact but nonuniform loading can result. For large lead deviations, corrections are sometimes made by tapering the teeth so that a more uniform contact condition is secured under load. The deflections of the teeth themselves compensate for misalignment by a calculable amount:

With

$$F_m = \left(\frac{2W_t}{eG}\right)^{1/2} \qquad \text{spur gears}$$

$$F_m = 2\left(\frac{W_t}{eG}\frac{p_b}{ZK}\right)^{1/2} \qquad \text{helical gears}$$

(A value of G which has been used is 1.2×10^6. It can vary up to 2.5×10^6.)

where F_m = maximum theoretical face width which would contact with load varying from zero at one side of face to a maximum at other side
e = lead deviation between mating teeth, in. per in.
G = stiffness constant, lb per in. lead deviation per in. of face
Z = total length of line of action in plane of rotation, in.
p_b = base pitch, in.
K = inclination factor compensating tooth stiffness for helix angle

then for

$$F_m < F\colon K_m = \frac{2F}{F_m}$$

$$F_m > F\colon K_m = \frac{F_m}{F_m - (F/2)}$$

Cognizance should be had of the fact that the operating alignment errors are not always easily evaluated. In these instances tabulations of typical values for commercial gears will be presented.

Size Factor K_s. The size factor depends primarily upon:

1. Tooth pitch
2. Diameter of parts
3. Ratio of tooth size to diameter of part
4. Face width
5. Dimensions of maximum stress pattern and gradient of stress therein
6. Ratio of casehardened depth to tooth thickness
7. Hardenability and heat-treatment of materials

That a size factor exists in the performance of materials seems evident from numerous service experiences.

The size factor may be taken as unity for most spur and helical gears, provided that a proper choice of steel composition and heat-treatment procedures is made. When coarse-pitch casehardened teeth are used, it might be desirable to derate to reflect the lesser strengthening effect of the surface compressive stresses developed by the case. With K_s equal to unity for a 10-diametral-pitch gear, a K_s of 1.25 might be suitable for a 1-diametral-pitch casehardened tooth.

Standardized size factors are available for bevel gears.

Dynamic Factor K_v. The dynamic factor depends on:

1. Effect of tooth spacing and profile errors
2. Effect of pitch-line speed and rpm
3. Inertia and stiffness of all rotating elements
4. Transmitted loads per inch of face
5. Tooth stiffness

A complex procedure is required to evaluate accurately the effect of dynamic loadings caused by tooth error and elastic flexibility. The dynamic factors, as determined by recent tests, seemingly show that the dynamic-load increment increases linearly with the velocity unless resonant conditions occur but becomes a lesser percentage of the transmitted load as the load increases. The effect of the variation in tooth deflection as determined by changes in contact length or number of teeth in contact can be as important as the dynamic increment caused by the errors conventionally found in commercial gearing. It is for this reason that the design of uniform contact conditions needs attention and explains why spur gears with their 1 to 2 tooth-contact variation usually develop higher dynamic stresses than helical gears in which only a minor contact variation normally exists.

Overload Factor K_o. The overload factor makes allowances for the roughness or smoothness of operation of both the driving and driven apparatus. Specific overload factors can be established only after considerable experience is gained in a particular application.

In determining the overload factor, consideration should be given to the fact that many prime movers develop momentary overload torques appreciably greater than those determined by the nameplate ratings of either the prime mover or the driven apparatus.

Factor of Safety K_R. An increasing awareness of the spread or scatter of material properties and the similar performance of actual machine elements subjected to fatigue stress has brought about a wider scope of meaning to the term "factor of safety." The old concept of factor of safety to guard against any possible failure, or to cover ignorance, must accommodate modern notions regarding risk or reliability as related to percentages of failure. The values in Table 13-18 seem reasonable. Failure in this table does not mean immediate fracture by the applied load, but rather a shorter life than expected.

When yield strength is under consideration, K_R values of 1.33 for "conventional industrial applications" and 3.00 for "high reliability" have been used provided the maximum peak torque is accurately known.

Life Factor K_L. The stress developed in gears is not directly proportional to the load and even the stress concentration varies with the number of cycles; therefore, it is evident that determining an accurate life factor is difficult. When long life is required, consideration must be given to the normal wear of the gear and bearings and the effect this might have on contact conditions.

Table 13-17 shows values which have been used to adjust the allowable stress for the required number of life cycles. The slope of the curve in the finite region will vary

Table 13-30. Allowable Fatigue Stress s_{at}

Material	Material hardness, BHN	s_{at}		
		Spur	Helical	Bevel
Steel	140 min.	20,000–22,000	20,000–22,000	11,000
	180 min.	25,000–28,000	25,000–28,000	14,000
	300 min.	35,000–40,000	35,000–45,000	19,000
	450 min.	45,000–50,000	45,000–60,000	25,000
	55–63 Rockwell C (case-carburized)	55,000–65,000	55,000–65,000	30,000
Cast iron:				
AGMA grade 20	....	5,000	5,000	2,700
AGMA grade 30	175 min.	8,500	8,500	4,600
AGMA grade 40	200 min.	13,000	13,000	7,000

with the material, hardness, and heat-treatment, the lower values being applicable to the lower hardnesses.

When a gearset is to be operated at varying loads and number of cycles, the author's experience seems to indicate that the loadings beneath the endurance limit do not contribute to a decreased life. The exact point at which accumulative fatigue damage to gears occurs is not yet well established.

Temperature Factor K_T. When gears operate at oil or gear blank temperatures not exceeding 160°F, K_T is generally taken at 1.0. In some instances, it is necessary to use a K_T value greater than 1.0 for case-carburized gears in a temperature range above 160°F. One accepted basis of correction for bevel gears, which is sometimes also used for spur and helical gears, is

$$K_T = \frac{460 + T_F}{620}$$

where T_F = the peak operating oil temperature, °F

Allowable Stress s_{at}. An allowable design stress for unity application factor and 10 million cycles of load application is determined by field experience for each material and condition of that material. This stress is designated s_{at}.

The allowable stress for gear materials varies considerably with heat-treatment, forging or casting practice, material composition, and such treatments as shot peening.

The allowable fatigue design stresses given in Table 13-30 are nominal values intended for general application. Even lower values may be required if the steel composition and heat-treatment are poorly suited to the application. Higher stress values may be permitted in some cases by careful gear design, manufacturing pro-

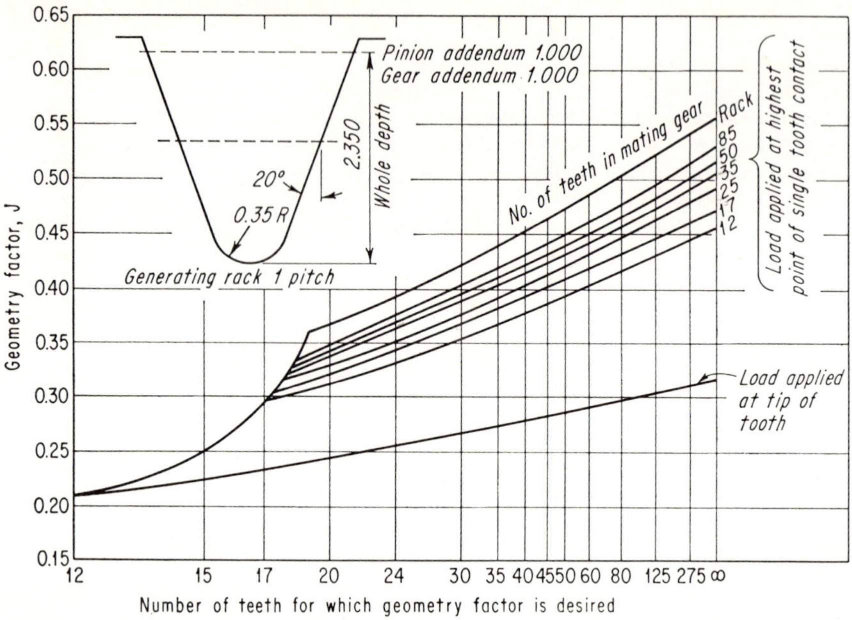

FIG. 13-8. Geometry factors J 20° spur, standard addendum.

cedures, choice of steels, surface peening, and the use of the best heat-treating procedure. When the higher allowable stress values are used, the K_L values in Table 13-17 must be modified.

It will be noted in Table 13-30 that the values for allowable stress are different for bevel gears than for spur and helical gears. This apparent difference is compensated for by the bevel-gear-size factor; so that a spur gear and bevel gear of 10 diametral pitch will have substantially the same load capacity.

The designer might wisely consult other AGMA rating practices and use the more certain stress values given there whenever his application falls into one of these.

Use 70 per cent of these values for idler gears and other gears where the teeth are loaded in both directions.

When the gear is subject to infrequent momentary high overloads, the maximum allowable stress is determined by the allowable yield properties rather than the fatigue strength of the material. This stress is designated as s_{ay}. Table 13-31 shows sug-

Table 13-31. Allowable Yield Stress s_{ay}

Heat-treatment	BHN	s_{ay}
Annealed or normalized	150	30,000
	200	50,000
	250	75,000
Quenched and tempered	200	60,000
	250	85,000
	300	110,000
	350	135,000
	400	160,000

gested values for allowable yield strength. When yield is the governing stress, the stress-concentration factor is sometimes considered ineffective.

Spur-gear Strength Rating. The spur-gear strength rating can be computed by using Eq. (13-5) or (13-6) and the following values for the various factors.

J = geometry factor. Accurate spur gears develop the most critical stress when load is applied at the highest point of the tooth where a single pair of teeth is carrying all the load. Less accurate spur gears having errors preventing two pairs of teeth from sharing the load may be stressed most heavily when load is applied at the tip. Figure 13-8 shows the geometry factor for equal-addendum involute spur gears of 20° pressure angle. In these curves, it is assumed that the theoretical stress-concentration factor is not seriously affected by surface finish, plasticity, residual stresses, or other factors. Table 13-32 shows the variation in base pitch between the gear and pinion which determines whether or not load sharing exists in 20° pressure angle spur gears. For special cases the geometry factor can be determined by the procedures shown in Appendix A of AGMA information sheet 225.01.

Table 13-32. Limiting Error in Action for Steel Spur Gears (Variation in Base Pitch)

No. of pinion teeth	Allowable error when teeth share load*			Amount of error when teeth fail to share load†		
	Load per in. of face			Load per in. of face		
	500 lb	1,000 lb	2,000 lb	500 lb	1,000 lb	2,000 lb
15	0.0004	0.0007	0.0014	0.0006	0.0011	0.0023
20	0.0003	0.0006	0.0011	0.0006	0.0011	0.0023
25	0.0002	0.0005	0.0009	0.0006	0.0011	0.0023

* Use upper curves of Fig. 13-8, highest point of single tooth loading.
† Use lower curve of Fig. 13-8, tip loading.

K_m = load-distribution factor. Use Table 13-24 for general design purposes. Manufacturers of wider-face-width gears generally find it necessary to control misalignment by other means than by allowed rates of misalignment. To handle such cases, Table 13-27 shows appropriate values of K_m.

K_s = size factor. Standard size factors are not yet established for cases where there is a detrimental size effect. Sometimes a value greater than unity must be used, particularly for coarse-pitch carburized teeth. For a 1-diametrial-pitch tooth a K_s value of 1.25 might be used.

K_v = dynamic factor. See Fig. 13-1 and: Use curve 1 ($K_v = 1$) for high-precision shaved or ground spur gears when the drive system is such that no appreciable dynamic load is developed. Use curve 2 for high-precision shaved or ground spur gears when the drive system allows dynamic loads to be developed. Use curve 4 for spur gears finished by hobbing or shaping. When milling cutters are used to cut the teeth or inaccurate teeth are generated, lower dynamic factors than shown must be used since the dynamic factor reflects the effect of inaccuracies in profile, tooth spacing, and runout.

K_o = overload factor. Use values from Table 13-23 for conditions listed in Table 13-2. Use the service factors from Table 13-15 if they are applicable. Then terms K_L, K_R, and K_T will each be 1.0.

K_R = factor of safety. Use the values shown in Table 13-18.

K_L = life factor. Obtain from Table 13-17.

K_T = temperature factor, commonly used as 1.0 if the oil or gear blank temperature does not exceed 250°F.

s_{at} or s_{ay} = allowable stress. Use values in Table 13-30 or 13-31.

Helical- and Herringbone-gear Strength Rating. The strength rating of helical and herringbone gears can be computed by using Eq. (13-5) or (13-6) and the following values for the various factors.

J = geometry factor. For helical gears the critical point of load application occurs when the oblique line of contact intersects the tip. Figures 13-9 and 13-10 show geometry factors for equal-addendum involute gears. For special cases or other tooth proportions, the geometry factor can be computed by the procedures shown in AGMA Information Sheet 225.01.

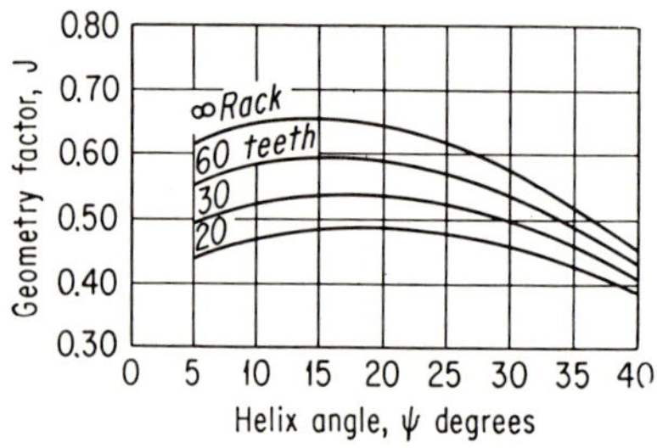

FIG. 13-9. Geometry factors J, 22° normal pressure angle helical gear.

$$\text{Addendum} = \frac{1}{P_{nd}} \qquad \text{Hob tip radius} = \frac{0.28}{P_{nd}}$$

$$d = \frac{N_P}{P_d} \qquad m_N = \frac{p_N}{0.95Z}$$

Value for Z is for an element of indicated numbers of teeth and a 75-tooth mate.

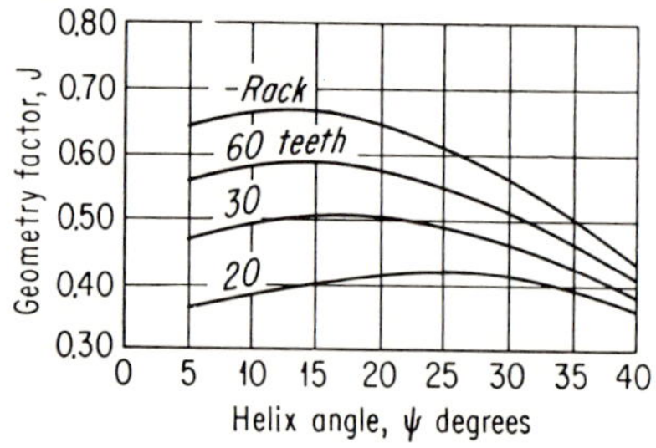

FIG. 13-10. Geometry factors J, 15° normal pressure angle helical gear.

$$\text{Addendum} = \frac{1.155}{P_{nd}} \qquad \text{Hob tip radius} = \frac{0.44}{P_{nd}}$$

$$d = \frac{N_p}{P_d} \qquad m_N = \frac{p_N}{0.95Z}$$

Value for Z is for an element of indicated numbers of teeth and a 90-tooth mate.

K_m = load-distribution factor. Use Table 13-26 or, for wide-face-width gears manufactured with the necessary control, use Table 13-27.

K_s = size factor. The size factor may be taken as unity for most helical and herringbone gears provided a proper choice of steel is made for the size of the parts and the case depth or hardness pattern is adequate. For case-carburized gears K_s might be used with values varying up to 1.25 for a 1-diametral-pitch tooth.

K_v = dynamic factor. See Fig. 13-1. Use curve 1 (K_v = 1.0) for high-precision helical and herringbone gears when the drive system is such that no appreciable dynamic load is developed. Use curve 2 for high-precision helical and herringbone gears when the drive system allows dynamic loads to be developed. This curve is recommended for commercial helical gears. When milling cutters are used to cut the teeth or inaccurate teeth are generated, lower dynamic factors than shown must be used since the dynamic factor reflects the effect of inaccuracies in profile, tooth spacing, and runout.

K_o = overload factor. Use values from Table 13-23 for conditions listed in Table 13-2. Use the service factors from Table 13-15, if they are applicable. Then terms K_L, K_R, and K_T will each be 1.0.

K_R = factor of safety. Use the values shown in Table 13-18.

K_L = life factor. Obtain from Table 13-17.

K_T = temperature factor, commonly used as 1.0 if the oil or gear blank temperature does not exceed 250°F.

s_{at} or s_{ay} = allowable stress. Use Table 13-30 or 13-31.

Bevel-gear Strength Rating. The strength rating of straight and spiral bevel gears can be computed using Eq. (13-5) or (13-6) and the following values for the various factors.

J = geometry factor.* In bevel gears the position of load application and the amount of sharing of the load are determined by integrating the contact area.

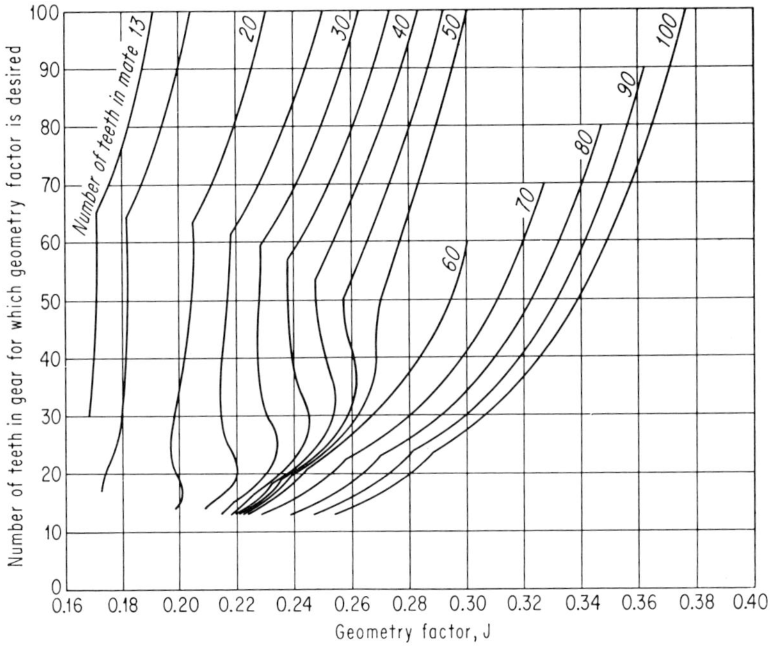

FIG. 13-11. Geometry factors J, straight bevel gears (AGMA 208.01). Shaft angle 90°. Pressure angle 20°.

Figures 13-11 and 13-12 show the geometry factors for typical bevel gears. Consult AGMA Information Sheet 225.01 if geometry factors for special cases or other tooth proportions are desired.

K_m = load distribution. In bevel gearing, when the deflections or displacements of the gear and pinion from their theoretical position are known, the load-distribution factor K_m can be obtained from Fig. 13-5. When the estimated or actual displacements are not known, an estimate of the K_m factor may be obtained from Table 13-28. Where mountings are carefully designed for maximum rigidity, use the smaller factor from the table. Where mounting rigidity is questionable, use the larger factor.

* Additional data on geometry factors may be obtained from publications[13] of the Gleason Works, Rochester, N.Y.

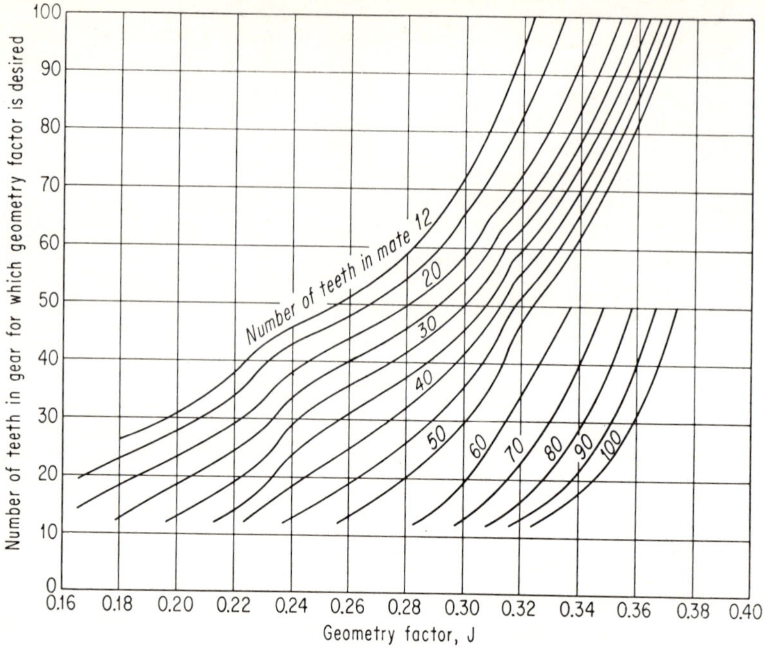

FIG. 13-12. Geometry factors J, spiral bevel gears (AGMA 209.02). Shaft angle 90°. Pressure angle 20°. Spiral angle 35°.

K_s = size factor. The size factor for bevel gears is given as a function of the diametral pitch and may be obtained from Table 13-33. For gears of 16 diametral pitch and finer use a size factor of 0.5.

Table 13-33. Size Factor for Bevel Gears K_s

Diametral pitch P_d	*Size factor* K_s
1	1.00
2	0.84
3	0.76
4	0.71
6	0.64
8	0.59
10	0.56
16 and finer	0.50

K_v = dynamic factor. See Fig. 13-1. Use curve 1 for generated bevel gears having the preferred pattern of tooth contact and accurate tooth spacing and concentricity when the drive system is such that no appreciable dynamic load is developed. Use curve 2 for large-planed spiral bevel gears when the drive system allows dynamic loads to be developed. Use curve 4 for large-planed straight bevel gears. When milling cutters are used to cut the teeth or inaccurate teeth are generated, lower dynamic factors than shown must be used since the dynamic factor reflects the effect of inaccuracies in profile, tooth spacing, and runout.

K_o = overload factor. Use values from Table 13-23 for conditions listed in Table 13-2. Use the service factors from Table 13-15 if they are applicable. The terms K_L, K_R, and K_T will then each be 1.0.

K_R = factor of safety. Use the values shown in Table 13-18.

K_L = life factor. Obtain from Table 13-17.

K_T = temperature factor. When gears operate at oil or gear blank temperatures not exceeding 160°F, K_T is generally taken at 1.0. In some instances, it is necessary to use a K_T value greater than 1.0 for case-carburized gears in a temperature range above 160°F. One accepted basis of correction for bevel gear is the following:

$$K_T = \frac{460 + T_F}{620}$$

where T_F is the peak operating oil temperature in degrees Fahrenheit.

s_{at} = allowable stress. Use Table 13-30. As a guide to the selection of s_{at} values for special applications the following information might prove useful.

1. *Aircraft Power Drives.* These applications usually involve continuous loads and peak loads in excess of 6 million cycles. Nearly all such gears are made from case-hardened steel with ground tooth surfaces. The allowable bending stress in this case is 30,000 psi. Aircraft gears should, in general, be designed for equal stress on gear and pinion.

2. *Automotive-drive Gears.* Passenger-car gears of casehardened steel are designed for a maximum pinion bending stress of 60,000 psi. Truck gears of casehardened steel are designed for a maximum pinion bending stress of 80,000 psi. The above stresses are based on maximum torque through low gear (on truck, a transmission ratio not exceeding 8:1) or maximum tractive-effort torque, whichever is the lesser. Tractive-effort torque should be based on a coefficient of friction between wheels and pavement of 0.85. In general, automotive gears are designed for equal life on gear and pinion.

3. *Marine Gears.* In general, marine gears of casehardened steel are designed for equal stress with a maximum bending stress of 30,000 psi. Special applications may be designed for higher stress values, but these generally involve limited-life considerations.

4. *Railway Gears.* The design of railway drives is based largely on tractive-effort torque using a coefficient of friction between wheels and rails of 0.25. Based on maximum torque, the bending stress for casehardened steel should not exceed 30,000 psi. Gears should be designed for equal stress.

5. *Tractors.* Both farm and industrial tractors are designed for a maximum pinion bending stress for casehardened steel of 40,000 psi. Industrial tractors are frequently designed for equal stress, although some designers attempt to compromise between equal life and equal stress. Farm tractors are usually designed for equal life. Tractive-effort torque should be based on a coefficient of friction between wheels and ground of 0.65, except on lug-type wheels where a value of 1.0 should be used.

13-7. Power Rating of Cylindrical Wormgears. The following data are based primarily upon AGMA Standard Practice 440.03.[5] The information is limited to worm speeds of 3,600 rpm and less and sliding velocities of 6,000 fpm and less. The power-rating formula applies to ratios from 3:1 to 100:1, inclusive.

Input Power Rating of Cylindrical Wormgearing. The formula for *input* horsepower rating of the gearing is

$$P = \frac{W_t D n}{126{,}000 m_G} + \frac{v W_f}{33{,}000} \tag{13-7}$$

The first expression provides the output horsepower, the second the power loss at the tooth mesh.

where W_t = tangential load on wormgear teeth, lb

$$W_t = K_s D^{0.8} F_e K_m K_v \qquad (13\text{-}8)$$

where K_s = materials factor. See Table 13-34. This factor contains a size correction by reducing the K_s value as the face width increases.

Table 13-34. Materials Factor K_s, Cylindrical Wormgears

Gear face width, actual	Sand cast	Static chill cast	Centrifugal cast
3″ and less	700	800	1,000
4	665	780	975
5	640	760	940
6	600	720	900
7	570	680	850
8	530	640	800
9	500	600	750

NOTE:
1. These factors are for copper-tin and copper-tin-nickel bronze gears operated with a worm having a minimum surface hardness of 58 Rockwell C.
2. The K_s factor shall not exceed 800 where center distance of worm and gear is less than 3.00″.

F_e = effective face width of gear, in.
= actual gear face width, excepting it shall not exceed $2d/3$ in which d is the mean worm diameter (diameter at mid-point of working depth of worm thread), in.

K_m = ratio correction factor. See Table 13-35.

Table 13-35. Ratio Correction Factor K_m

Ratio m_G	K_m	Ratio m_G	K_m
3.0	0.500	14.0	0.799
3.5	0.554	16.0	0.809
4.0	0.593	20.0	0.820
4.5	0.620	30.0	0.825
5.0	0.645	40.0	0.815
		50.0	0.785
6.0	0.679		
7.0	0.706	60.0	0.745
8.0	0.724	70.0	0.687
		80.0	0.622
9.0	0.744		
10.0	0.760	90.0	0.555
12.0	0.783	100	0.490

K_v = velocity factor. See Table 13-36.

The above are empirical factors applicable to worm drives designed and manufactured to the high quality standards established for the normal commercial enclosed speed reducers. Any departure from these standards requires a reevaluation of rating. The lubricant must be satisfactory for wormgears operating under the applied conditions.

Definition of the items in Eq. (13-7) continues as follows:

D = pitch diameter of gear, in.

n = rotational speed of worm, rpm

m_G = gear ratio = N_G/N_W

v = sliding velocity at mean worm diameter, fpm

= $\pi nd/12 \cos \lambda$

where λ = lead angle of threads at mean worm diameter, deg.

Table 13-36. Velocity Factor K_v

Sliding velocity, fpm	Velocity factor K_v	Sliding velocity, fpm	Velocity factor K_v
1	0.649	550	0.358
1.5	0.647	600	0.340
10	0.644	700	0.310
20	0.638	800	0.289
30	0.631	900	0.289
40	0.625	1,000	0.258
60	0.613	1,200	0.235
80	0.600	1,400	0.216
100	0.588	1,600	0.200
150	0.558	1,800	0.187
200	0.528	2,000	0.175
250	0.500	2,200	0.165
300	0.472	2,400	0.156
350	0.446	2,600	0.148
400	0.421	2,800	0.140
450	0.398	3,000	0.134
500	0.378	4,000	0.106
		5,000	0.089
		6,000	0.079

W_f = frictional force, lb

$$= \frac{\mu W_t}{\cos \lambda \cos \phi_n}$$

where μ = coefficient of friction. See Table 13-37 (also see Chap. 14)

ϕ_n = normal pressure angle of worm thread, deg

Input Power Rating of Complete Wormgear Drive. Because of possible higher losses in wormgear reducers, it is sometimes desirable to rate the input horsepower of the complete assembly, accounting for power lost in oil churning, bearings, and seals.

Input power rating of reducer P_r is

$$P_r = P_l + P \tag{13-9}$$

where P_l = power loss determined experimentally at no-load power input or by calculation

P = input power to gear mesh [Eq. (13-7)]

The thermal rise for a worm-type speed reducer should not exceed 100°F. To determine the probable temperature rise for the natural heat dissipation or forced cooling it is necessary to know the efficiency, which is

$$\eta = \frac{P_r - (vW_f/33{,}000) + P_l}{P_r} \times 100 \text{ per cent}$$

Another generally used formula gives the relation

$$\eta = \left(100 - \frac{m_G}{2}\right) \quad \text{per cent}$$

Service Rating. The rating determined by Eq. (13-7) is for continuous service free from shock loading and a total duration up to 10 hr per day.

Table 13-37. Coefficient of Friction*μ, Cylindrical Wormgearing

Sliding velocity v, fpm	Coefficient of friction μ	Sliding velocity v, fpm	Coefficient of friction μ
1	0.115	200	0.044
3	0.107	250	0.040
5	0.099	300	0.037
10	0.090	350	0.035
15	0.085	400	0.032
20	0.080	500	0.030
25	0.077	700	0.026
30	0.073	900	0.023
40	0.069	1,200	0.020
50	0.065	1,500	0.018
60	0.062	2,000	0.016
70	0.060	3,000	0.014
80	0.058	.....	.....
100	0.054	.....	.....
120	0.052	.....	.....
150	0.049	.....	.....

* Values of coefficient of friction shown above are nominal values for standard industrial designs. See Chap. 14 for more detail information on coefficient of friction.

The service rating P_s is

$$P_s = \frac{P_r}{K_o}$$

where P_r = input horsepower rating [Eq. (13-7)]

K_o = overload or service factor. See Table 13-38. Table 13-2 can be used as a guide to the driven-machine-load classification.

13-8. Power Rating of Double-enveloping Wormgears. The following data are abstracted from AGMA standards 214.02[6] and AGMA 441.02[7] pertinent to double-enveloping wormgearing and double-enveloping wormgear speed reducers. The information is applicable to wormgearing operating at a 90° shaft angle and sliding velocities up to 4,000 fpm or 3,600 rpm of the worm, whichever is limiting.

Table 13-38. Service Factors in Single- and Double-reduction Cylindrical-worm and Helical-worm Speed Reducers

Prime mover	Duration of service per day	Driven-machine-load classifications		
		Uniform	Moderate shock	Heavy shock
Electric motor	Occasional ½ hr	0.80	0.90	1.00
	Intermittent 2 hr	0.90	1.00	1.25
	10 hr	1.00	1.25	1.50
	24 hr	1.25	1.50	1.75
Multicylinder internal-combustion engine	Occasional ½ hr	0.90	1.00	1.25
	Intermittent 2 hr	1.00	1.25	1.50
	10 hr	1.25	1.50	1.75
	24 hr	1.50	1.75	2.00
Single-cylinder internal-combustion engine	Occasional ½ hr	1.00	1.25	1.50
	Intermittent 2 hr	1.25	1.50	1.75
	10 hr	1.50	1.75	2.00
	24 hr	1.75	2.00	2.25
Following Service Factors Apply for Applications Involving Frequent Starts and Stops				
Electric motor	Occasional ½ hr	0.90	1.00	1.25
	Intermittent 2 hr	1.00	1.25	1.50
	10 hr	1.25	1.50	1.75
	24 hr	1.50	1.75	2.00

Explanatory Notes:

1. Time specified for intermittent and occasional service refers to total operating time per day.
2. Term "frequent starts and stops" refers to more than 10 starts per hour.
3. Normal starting or momentary peak loads up to 300 per cent of the basic rating are permissible.

Table 13-39. Basic Pressure Constant K_s, Double-enveloping Wormgearing

Center distance	*Basic pressure constant* K_s
1.25	0.007
1.5	0.013
2	0.033
2.5	0.065
3	0.120
4	0.290
5	0.60
6	1.05
7	1.72
8	2.60
9	3.75
10	5.20
15	18.5
20	44.5
25	90.0
30	155
35	255
40	480

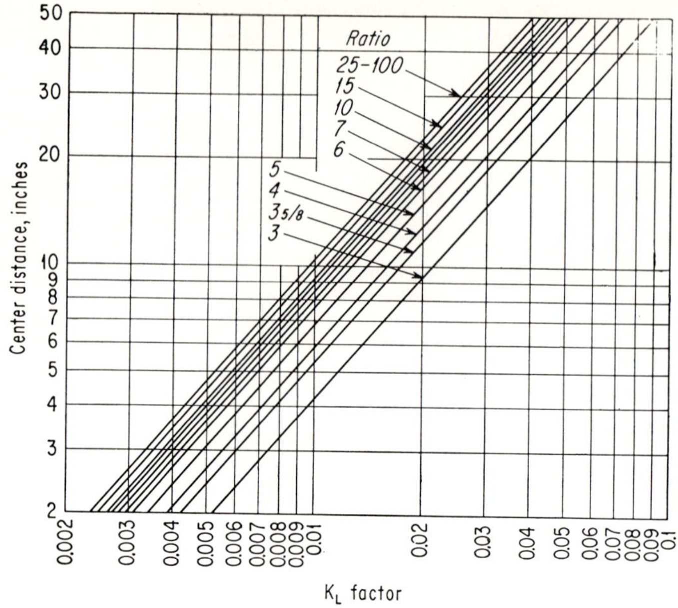

FIG. 13-13. Thread length factor, double-enveloping wormgears.

Input Power Rating of Double-enveloping Wormgearing. The formula for *input* horsepower rating of the gearing is

$$P = K_s K_m K_v \frac{n}{m_G} \tag{13-10}$$

where K_s = basic pressure factor. See Table 13-39.
K_m = ratio correction factor. See Table 13-40.
K_v = velocity factor. Use Figs. 13-13 and 13-14.
n = input speed of worm, rpm
K_L = thread length factor. Use Fig. 13-13.

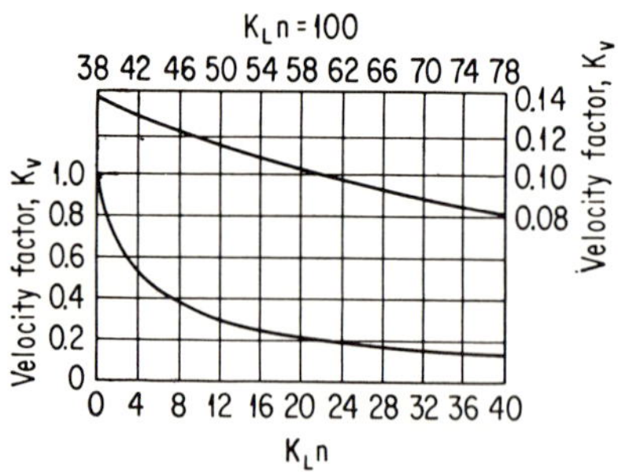

FIG. 13-14. Velocity factor, double-enveloping wormgears. (Multiply K_L times n and read curve.)

Table 13-40. Ratio Correction Factor K_m, Double-enveloping Wormgearing

Ratio	*Ratio correction factor* K_m
3	0.41
3.5	0.46
4	0.51
4.5	0.55
5	0.58
6	0.62
7	0.65
8	0.66
9	0.68
10	0.69
15	0.72
20	0.74
30 and above	0.75

This formula is subjected to the following restrictions:

1. The worm root diameter should not be less than

$$d_R = \frac{C^{0.875}}{3}$$

where d_R = root diameter of worm, in.
C = center distance, in.
As the ratio decreases, the worm root diameter may be increased as follows: from no increase for 8:1 ratio, the worm root should be increased gradually to plus 15 per cent for a 3:1 ratio. This root diameter is intended for use on worms where threads are cut integral with the shaft.
2. The worm threads must be cut integral with the worm shaft of an alloy steel heat-treated to a minimum hardness of 32 Rockwell C.
3. The gears should be a 10 per cent tin bronze having a minimum tensile strength of 40,000 psi.
4. The lubricant must be satisfactory for the operating conditions.
5. Normal starting or peak momentary loads should not exceed 300 per cent of this rating.

For ratings below 100 rpm of the worm the input horsepower may be calculated by reducing the rating in the proportion that the given speed bears to 100.

Service Rating. To obtain the service rating of the gearset the capacity as determined by Eq. (13-10) must be divided by the applicable service factors obtained from Table 13-41.

Table 13-41. Service Factors

Class 1. Normal 8- to 10-hr service, free from recurrent shocks (i.e., shock loads that recur at approximately even and frequent intervals). Hp ratings taken from the charts are to be applied directly in this classification provided only that the thermal rating is not exceeded.

Class 2. 8- to 10-hr service where recurrent shock loading is encountered, or 24-hr service where no shock loading is experienced. Class 1 rating must be reduced by dividing by 1.2 for this service. Thermal rating must not be exceeded.

Class 3. 24-hr shock-load service. Class 1 rating must be reduced by dividing by 1.3 for this service, and thermal rating must not be exceeded.

Class 4. For intermittent service where worm is operated at 100 rpm or more, the class 1 rating must be divided by the factor indicated below, depending upon the frequency and duration of the load. Thermal rating need not be considered for this class of service.

Class 5. Low-speed service, where the worm speed is less than 100 rpm, will require output-torque ratings in inch-pounds, calculated as in Eq. (13-11) independent of the load except that where such service is intermittent the factors below will apply.

Total min operation allowed			Service factor	Service-factor designation	
Per hr	Per cycle	Per cycle			
Multiple cycles per hr	1 cycle per hr	1 cycle per 2 hr or more		Class 4 service	Class 5 service
	5	10	0.6	4/0.6	5/0.6
2	10	20	0.7	4/0.7	5/0.7
5	15	30	0.8	4/0.8	5/0.8
10	20	40	0.9	4/0.9	5/0.9

Output-torque Rating. The output-torque rating is derived from the input power equation (13-10) by

$$T = \frac{P}{n} 63{,}000 m_G \frac{\eta}{100} \qquad (13\text{-}11)$$

where η = gear efficiency, per cent
= 97 per cent for 100 rpm output and below for ratios 6:1 or less
= $100 - (m_G/2)$ for other ratios

For more precise estimations of over-all efficiencies for output-torque or thermal-dissipation calculations, the data in Table 13-42 can be used. This table is representative of the over-all efficiency of a complete speed reducer including bearing, oil churning, seal, and mesh losses.

Table 13-42. Typical Over-all Efficiencies, Double-enveloping Wormgear Speed Reducers

Lead angle of worm	*Efficiency, 1,500–2,400 rpm**
1	50
2	62
3	70
4	76
5	79
10	87
15	91
20 and above	92

* Subtract the following per cent for other speeds.

Rpm	*Subtract %*
1,300–1,500	1
800–1,300	3
400–800	8
200–400	12
50–200	18
10–50	20

SCORING

Besides rating a set of gears for strength and durability, attention sometimes must be given to the problem of scoring. Scoring occurs when the oil film breaks down, allowing metal-to-metal contact resulting in welding and tearing apart until the tooth-profile surface is deteriorated and tooth action is affected. Scoring is identified as radial* scratch patterns of varying severity located near the tip of the pinion or gear.

Factors which affect scoring are the oil-film strength and chemical activity, contact pressure, sliding velocity, material hardness and heat-transfer rate, surface finish, lubricant temperatures, and probably additional similar factors.

13-9. Scoring Probability. Scoring is most likely to occur if maximum loading is carried by casehardened gears or through-hardened gears above 450 BHN. For such gears the probability of scoring increases as the pitch-line velocities exceed the limit shown in Table 13.43.

13-10. Scoring Criterion. The designer usually rates a set of gears for profile durability or strength capacity. An estimate is then made of the probability of encountering scoring trouble. There is no simple handbook method of scoring rating.

* The scoring marks are not radial on gears with offset axes like worms, hypoids, crossed-axes helicals, etc.

Table 13-43. Scoring Probabilities—Heavily Loaded Gears

P_d	Pitch-line velocity	Remarks
20	10,000	Probability of scoring increases as pitch-line velocity exceeds these values
10	7,000	
5	6,000	

Much attention is being given the scoring problem at the present time, but the theory and data are beyond the scope of this handbook. Equation (13-12) provides a scoring criterion which might prove helpful for spur, helical, and bevel gears. If the scoring-criterion number is above the values shown in Table 13-44, a possibility exists that scoring will be encountered. Better oils, extreme-pressure additives, improved surface finish, lower oil temperatures, increased hardness, and profile modifications can be used to prevent scoring. More elaborate analysis might prove helpful using data and details from Dudley,[1] AGMA,[8] Kelley,[9] or Shipley.[17]

Table 13-44. Critical Scoring-criterion Numbers

Blank temp., °F	100°	150°	200°	250°	300°
Kind of oil*	Critical scoring-index numbers				
AGMA 1	9,000	6,000	3,000		
AGMA 3	11,000	8,000	5,000	2,000	
AGMA 5	13,000	10,000	7,000	4,000	
AGMA 7	15,000	12,000	9,000	6,000	
AGMA 8A	17,000	14,000	11,000	8,000	
Grade 1065, Mil-0-6082B	15,000	12,000	9,000	6,000	
Grade 1010, Mil-0-6082B	12,000	9,000	6,000	2,000	
Synthetic (Turbo 35)	17,000	14,000	11,000	8,000	5,000
Synthetic Mil-L-7808D	15,000	12,000	9,000	6,000	3,000

NOTES:
1. See Chap. 15 for listings of gear oils and their properties.
2. Generally, the gear blank temperature is only slightly higher than the oil inlet temperature.
* See AGMA 250.02 for details of standard gear oils.

$$\text{Scoring-criterion number} = \left(\frac{W_t}{F}\right)^{3/4} \frac{n_P^{1/2}}{P_d^{1/4}} \tag{13-12}$$

where W_t = tangential driving load, lb
F = contacting face width, in.
n_P = pinion, rpm
P_d = diametral pitch

Rating Formula for Industrial Bevel Gears with Allowance for Scoring. The generally used formula for surface-load capacity of industrial bevel gears contains factors which protect against scoring as well as pitting. A number of formulas have been proposed for the calculation of bevel-gear capacity strictly on the basis of scoring resistance. However, none of these have been applied successfully at this date.

Table 13-45. Material Factors

Gear			Pinion			Material factor C_M
Material	BHN	Rockwell C	Material	BHN	Rockwell C	
Casehardened steel. . . .	. . . .	58*	Casehardened steel. . . .	. . . .	60*	1.33†
Casehardened steel. . . .	. . . .	55*	Casehardened steel. . . .	. . . .	55*	1.00
Flame-hardened steel.	. . . .	50*	Casehardened steel. . . .	. . . .	55*	1.00
Flame-hardened steel.	. . . .	50*	Flame-hardened steel. .	. . . .	50*	1.00
Oil-hardened steel. . . .	375–425	. . .	Oil-hardened steel. . . .	375–425	. . .	0.70
Heat-treated steel. . . .	250–300	. . .	Casehardened steel. . . .	. . . .	55*	0.50
Heat-treated steel. . . .	210–245	. . .	Heat-treated steel.	245–280	. . .	0.35
Cast iron.	. . . .	. . .	Casehardened steel. . . .	. . . .	55*	0.40
Cast iron.	. . . .	. . .	Flame-hardened steel. .	. . . .	50*	0.40
Cast iron.	. . . .	. . .	Annealed steel.	160–200	. . .	0.30
Cast iron.	. . . .	. . .	Cast iron.	. . . .	. . .	0.30

* Minimum value.
† Gears must be file hard.

The following empirical formula is based on many years of successful experience with all types of bevel gears. It is suitable for use in designing industrial bevel gears of 8 pitch or coarser. The formula was developed by the Gleason Works of Rochester, New York.

$$T_P = FC_MC_SK_5 \tag{13-13}$$

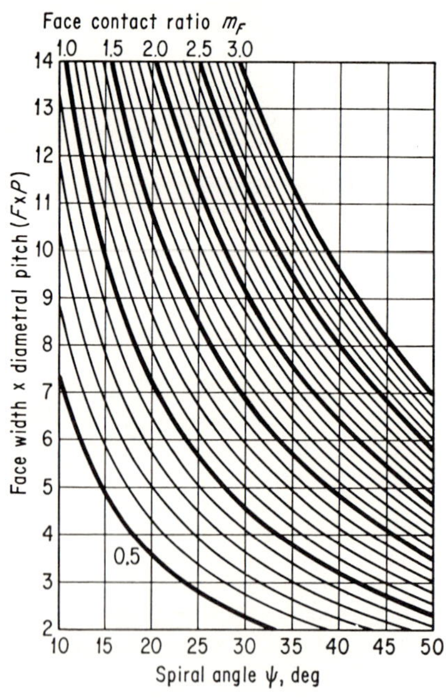

FIG. 13-15. Face contact ratio m_F.

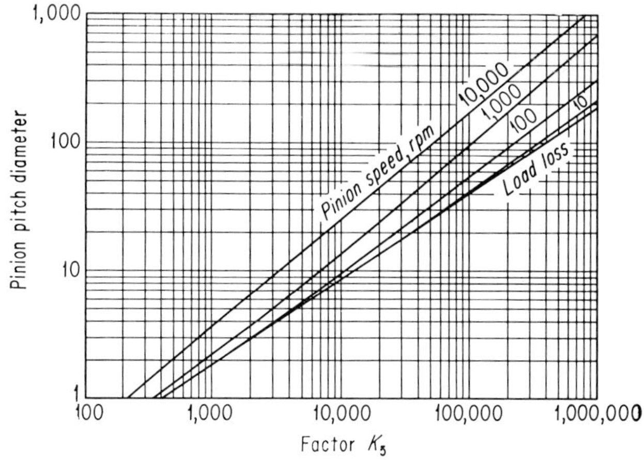

FIG. 13-16. Pinion diameter vs. factor K_5—surface capacity.

Table 13-46. Profile Contact Ratio—m_p
(20° pressure angle)

Number of teeth in pinion	Spiral bevel and hypoid gears	Straight bevel and Zerol bevel gears
6	0.87	
7	0.95	
8	1.02	
9	1.07	
10	1.11	
11	1.15	
12	1.19	
13	1.20	1.48
14	1.21	1.49
15	1.22	1.51
16	1.23	1.52
17	1.24	1.53
18	1.25	1.54
19	1.25	1.55
20	1.26	1.56
21	1.26	1.57
22	1.27	1.58
23 and 24	1.28	1.59
25	1.29	1.60
26	1.29	1.61
27 and 28	1.30	1.62
29	1.30	1.63
30 and 31	1.31	1.64
32 and 33	1.32	1.65
34 and 35	1.32	1.66
36 and 37	1.33	1.67
38, 39, and 40	1.34	1.68

where T_P = maximum allowable pinion torque, lb-in. ($P_H = \frac{T_P n_P}{63{,}025}$, where P_H is the maximum continous horsepower and n_P is the pinion rpm.)

F = net face width, in.

C_M = material factor for surface capacity (see Table 13-45).

C_S = factor for contact ratio. $C_S = \sqrt{0.4(m_F + m_p)}$.

m_F = face contact ratio (see Fig. 13-15). If m_F is greater than 2.00, use a value of 2.00 in this formula. For straight bevel and Zerol bevel gears, use $m_F = 0$.

m_p = profile contact ratio (see Table 13-46).

K_5 = combined factor for pinion diameter, velocity factor, pinion speed, and allowable unit load. Values are given on the curve sheet (Fig. 13-16).

The allowable horsepower obtained from this formula should be considered as a "service horsepower," or the allowable horsepower when the service factor is unity. Tables 13-6 to 13-8 are based on this formula.

REFERENCES

1. Dudley, D. W., and R. C. Bryant: How to Choose a Right Angle Gear, *Prod. Eng.*, 1960.
2. American Gear Manufacturers Assoc.: "Lubrication of Industrial Enclosed Gearing, Includes Appendix Sheet 250.02A, Typical Manufacturers' Oils Meeting Standard AGMA 250.02," AGMA 250.02.
3. American Gear Manufacturers Assoc.: "Surface Durability of Spur, Helical, Herringbone and Bevel Gear Teeth," AGMA 215.01.
4. American Gear Manufacturers Assoc.: "Strength of Spur, Helical, Herringbone and Bevel Gear Teeth," AGMA 225.01.
5. American Gear Manufacturers Assoc.: "Practice for Single and Double Reduction Cylindrical-worm and Helical-worm Speed Reducers," AGMA 440.03.
6. American Gear Manufacturers Assoc.: "Surface Durability of Double-enveloping Wormgearing," AGMA 214.02.
7. American Gear Manufacturers Assoc.: "Practice for Double-enveloping Wormgear Speed Reducers," AGMA 441.02.
8. American Gear Manufacturers Assoc.: "Progress Report by Automotive Gearing Committee Sec. II, PVT Values of Gear Teeth," AGMA 101.02, October, 1951.
9. Kelley, B. W.: A New Look at the Scoring Phenomena of Gears, *Trans. SAE*, 1952.
10. Gleason Works: "Gleason 20° Straight Bevel Gear System," Gleason Works, Rochester, N.Y., 1959.
11. Gleason Works: "Gleason Zerol Bevel Gear System," Gleason Works, Rochester, N.Y., 1956.
12. Gleason Works: "Gleason Spiral Bevel Gear System," Gleason Works, Rochester, N.Y., 1959.
13. Gleason Works: "Bending Stresses in Bevel Gear Teeth," Gleason Works, Rochester, N.Y., 1955.
14. Gleason Works: "Strength of Bevel and Hypoid Gears," Gleason Works, Rochester, N.Y., 1955.
15. Gleason Works: "Surface Durability—Pitting Formulas for Bevel Gear Teeth," Gleason Works, Rochester, N.Y., 1960.
16. Gleason Works: "Bevel and Hypoid Gear Design," Gleason Works, Rochester, N.Y., 1956.
17. Shipley, Eugene E.: Design and Testing Considerations of Lubricants for Gear Applications, *Lubrication Engineering*, pp. 148–152, April, 1958.

Chapter 14

LOADED GEARS IN ACTION

By

EUGENE E. SHIPLEY, *Project Engineer—Advance Development, General Electric Company, Lynn, Mass.*

Gear analysis and design are no more than an engineering guess until the gears go to work. They must be fully loaded and subjected to dynamic conditions to be sure they will perform as predicted. It is only then that the efficiency of the gear mesh, the dynamic-load effects, and the over-all performance of the gearbox can be noted and measured.

Those who design gears are constantly surprised that some gears run better and last longer than would be expected by the design formulas while others fail prematurely, even when operated well within the design limits of transmitted horsepower. New gear designs must be based on both theoretical and practical experience to assure that the very best gearsets are manufactured and installed in customer locations. The gear designer must be able to evaluate the friction of the gear mesh, bearing and windage losses, dynamic loads, and kinds and causes of gear failures in order to do the best possible job. In addition, he will be required to know the various types and kinds of gear-performance testing that can be done at the factory to simulate and supplement field conditions.

EFFICIENCY OF GEARS

One of the chief problems of any gear designer is that of obtaining reliable efficiency information on which to base his gear designs. Most gear designers, however, realize that gears are a very efficient method of transmitting power. In fact, in many applications gears are the only practical way to perform the task. Gears have the advantage not only of transmitting power, but also of transmitting this power at any speed ratio desired. Gears have efficiencies in the range of 98 per cent or more. On the surface it appears that this small loss could be neglected by gear designers and often it is. In many applications, however, this small friction loss can cause considerable concern, since it must be dissipated as heat throughout the gear system.

Gears with nonintersecting nonparallel axes, such as wormgears or Spiroids,®* have a meshing action consisting primarily of a sliding motion and consequently higher friction losses than parallel-axes gears such as spur or helical gears which have a combination of a high degree of rolling motion with a somewhat lesser degree of

* Registered trade-mark of the Spiroid Div. of Illinois Tool Works, Chicago, Ill.

sliding motion. It is quite evident that the efficiency of a given type of gear is tied in with the amount of sliding that takes place in the meshing action.

14-1. Sliding Velocity. Often the efficiency of various types of gears is expressed in terms of a sliding velocity. This being the case, it becomes quite important to understand how to calculate the sliding velocity of different gear drives.

Spur Gears. In the case of conjugated gear teeth such as spur or helical gears, the sliding velocity can be determined by the following equations:[1,*]

$$v_s = v_1 - v_2 \tag{14-1}$$

where v_1 = rolling velocity of the point in question on the pinion
v_2 = rolling velocity of the corresponding point on the gear

$$v_s = s(\omega_P + \omega_G) \tag{14-2}$$

where ω_P = angular velocity of pinion
ω_G = angular velocity of gear
s = distance along line of action to the point in question from the pitch point

Since the sliding velocity is expressed as the algebraic difference of the two rolling velocities of the point in question, it is often more convenient to calculate v_s in terms of v_1 and v_2 as indicated below.

$$v_1 = \frac{n_P \pi \rho_1}{360} \tag{14-3}$$

$$v_2 = \frac{n_G \pi \rho_2}{360} \tag{14-4}$$

where n_P = rpm of pinion
ρ_1 = radius of curvature of pinion
n_G = rpm of gear
ρ_2 = radius of curvature of gear

Sometimes sliding velocities are required at specific points such as the sliding velocity at the pinion tip or the sliding velocity at the gear tip. In this case,

$$v_{sP} = \frac{n_P}{114.59} \times Q_r \frac{m_R + 1}{m_R} \tag{14-5}$$

where v_{sP} = sliding velocity at pinion tip
Q_r = arc of recession of driving gears, radians

$$v_{sG} = \frac{n_G}{114.59} \times Q_a \frac{m_R + 1}{m_R} \tag{14-6}$$

where v_{sG} = sliding velocity of gear tip
Q_a = arc of approach, radians

Bevel Gears. As in the case of spur and helical gears, the sliding velocity of a point on a bevel gear, a distance s from the pitch point, is approximately

$$v_s = s(\omega_P{}^2 + \omega_G{}^2 + 2\omega_P\omega_G \cos \Sigma)(\sin^2 \phi + \cos^2 \psi \cos^2 \phi)^{1/2} \tag{14-7}$$

where ω_P = angular velocity of pinion radians per sec
ω_G = angular velocity of gear radians per sec
Σ = shaft angle

* Superscript numbers refer to references at the end of the chapter. See p. 141 of reference.

In the case of right-angle bevel drives, $\Sigma = 90°$ and

$$v_s = s(\omega_P \cos\gamma + \omega_G \cos\Gamma)(\sin^2\phi + \cos^2\psi\cos^2\phi)^{1/2} \tag{14-8}$$

where γ = pitch-cone angle of pinion, $\gamma + \Gamma = 90°$
Γ = pitch-cone angle of gear
ψ = spiral angle

Wormgears. The sliding velocity, or rubbing velocity, as it is generally referred to for wormgears, can be calculated as follows:[2]

$$v_r = \frac{0.262 n_W d}{\cos\lambda} \tag{14-9}$$

where v_r = rubbing velocity, fpm
λ = lead angle
n_W = rpm, worm
d = pitch diameter of worm

Crossed-axes Helical. Crossed-axes helical gears can be calculated in a similar manner.[3,*]

$$v_r = \frac{0.262 nD \sin\Sigma}{\cos\psi_2} \tag{14-10}$$

where $\Sigma = \psi_1 + \psi_2$
D = diameter of driver
n = rpm of driver
ψ_2 = helix angle of driven
ψ_1 = helix angle of driver

When the shaft angle = 90° sin of Σ equals 1.00 and the above equation becomes

$$v_r = \frac{0.262 nD}{\cos\psi_2} \tag{14-11}$$

Hypoids. The sliding velocity for a hypoid gear can be found by using a type of equation similar to that for wormgears or crossed-axes helical gears. In fact, a hypoid gear is an offset-axis bevel gear and has much in common with an offset-axis helical, or "crossed-axis" helical, as it is more properly called.

$$v_s = \frac{0.262 Dn}{\sin\gamma} \tag{14-12}$$

where v_s = sliding velocity, fpm
D = diameter of base cylinder of pinion hyperboloids
n = rpm of pinion
γ = angle of generatrix with axis of base cylinder of pinion

14-2. Efficiency of Coplanar Gears (*Parallel Axes and Intersecting Axes*). In applications where large amounts of power are being transmitted, the efficiency of gears becomes very important. The most efficient gear designs are required from the point of view of conservation of energy as well as a reduction in the amount of heat rejected to the geared system. In general terms, the nonparallel- nonintersecting-axis drives have higher sliding or rubbing velocities, resulting in higher losses than parallel-axis and intersecting-axis drives. Consequently, the latter types are most efficient. For this reason, spur, helical, or bevel gears are usually desired for high-power-transmission applications unless the design conditions are such that other types of gear drives have more attractive features.

Spur Gears. The over-all efficiency of spur gears, all gears for that matter, is dependent on three separate and distinct types of losses. These three types are

* See p. 191 of reference.

commonly known as (1) windage and churning losses, (2) bearing losses, and (3) gear-mesh losses. The efficiency information presented here is based only upon the losses of the gear mesh.[4]

The per cent efficiency for spur gears is

$$E = 100 - P_l \tag{14-13}$$

where P_l = per cent power loss

The power-loss equation is

$$P_l = \frac{50f}{\cos\phi}\left(\frac{H_s^2 + H_t^2}{H_s + H_t}\right) \tag{14-14}$$

where f = average coefficient of friction, dimensionless
ϕ = pressure angle, deg
H_s = specific sliding velocity at start of approach action
H_t = specific sliding velocity at end of recess action

H_s and H_t can be determined as follows:

$$H_t = \frac{m_G + 1}{m_G}\left[\sqrt{\left(\frac{r_o}{r}\right)^2 - \cos^2\phi} - \sin\phi\right] \tag{14-15}$$

and

$$H_s = (m_G + 1)\left[\sqrt{\left(\frac{R_o}{R}\right)^2 - \cos^2\phi} - \sin\phi\right] \tag{14-16}$$

where m_G = gear ratio, a whole number
r_o = outside radius of pinion, in.
r = pitch radius of pinion, in.
R_o = outside radius of gear, in.
R = pitch radius of gear, in.

H_s and H_t can also be determined from the following relationships:

$$H_s = \frac{v_{sG}}{v} \tag{14-17}$$

where v_{sG} can be determined from Eq. (14-6) and

$$v = 0.262 \times n \times d \tag{14-18}$$

where d = pitch diameter of pinion, in.
n = rpm of pinion

$$H_t = \frac{v_{sP}}{v}$$

where V_{sP} can be determined from Eq. (14-5).

Figures 14-1 and 14-2 give values for the average coefficient of friction f vs. pitch-line velocity for various K-factor gear loadings.

A short-cut method has been developed to analyze standard spur-gear trains which is quite helpful, particularly in the early design stages. The power loss of a given set of gears is a function of the coefficient of friction of the gear mesh and the so-called mechanical advantage of the gear mesh, that is:

$$P_l = \frac{f}{M} \times 100 \tag{14-19}$$

where P_l = per cent power loss
f = coefficient of friction
M = mechanical advantage of the mesh

The mechanical advantage M, for 20 and 25° pressure angle standard spur gears, has been plotted for various combinations of number of gear and pinion teeth and can be found by referring to Figs. 14-3 and 14-4. The same coefficient of friction shown in Figs. 14-1 and 14-2 is used in the above equation. The curves shown in Fig. 14-5 give the nominal efficiency of spur gears.

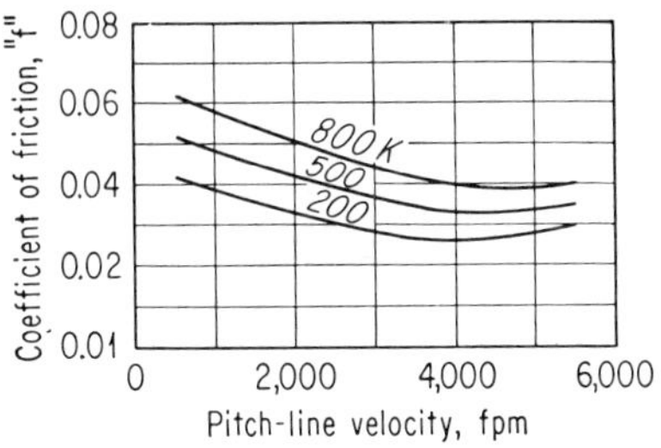

FIG. 14-1. Average coefficient of friction of lightweight petroleum oil. 120°F oil inlet hardened gears 45 SUS at 100°F.

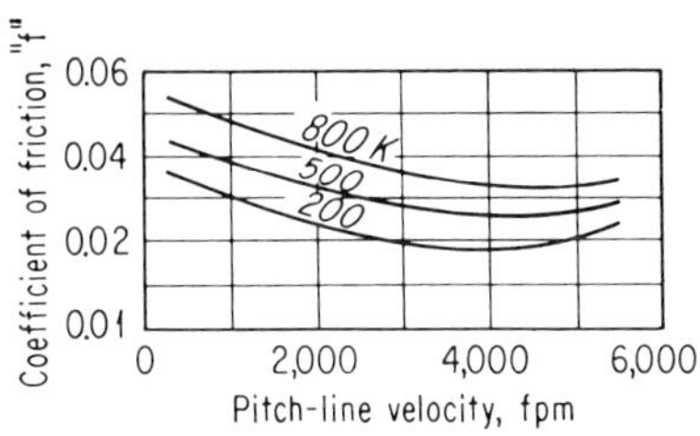

FIG. 14-2. Average coefficient of friction of medium-weight petroleum oil. 120°F oil inlet hardened gears 300 SUS at 100°F.

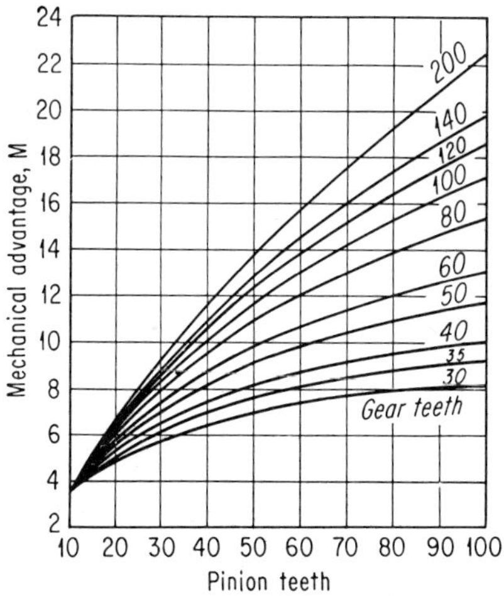

FIG. 14-3. Mechanical advantage 20° pressure angle spur gears. Pressure angle = 20°. Per cent power loss = P. Coefficient of friction = f. Mechanical advantage = M. $P = f/M \times 100$.

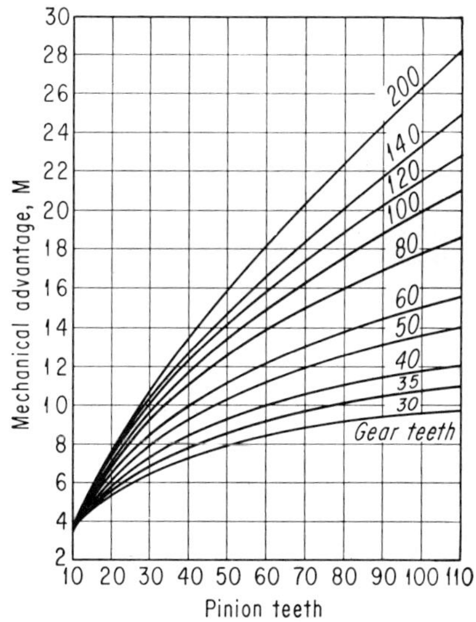

FIG. 14-4. Mechanical advantage 25° pressure angle spur gears. Pressure angle = 25°. Per cent power loss = P. Coefficient of friction = f. Mechanical advantage = M. $P = f/M \times 100$.

Internal Spur Gears. The efficiency of internal spur gears is very similar in nature to that of external spur gears and can be determined by the same general equations, namely:

$$E = 100 - P_l \tag{14-13}$$

and

$$P_l = \frac{50f}{\cos\phi}\left(\frac{H_s^2 + H_t^2}{H_s + H_t}\right) \tag{14-14}$$

except that H_s and H_t are determined in a slightly different manner, as specified by the following equations:

$$H_t = \frac{m_G - 1}{m_G}\left[\sqrt{\left(\frac{r_o}{r}\right)^2 - \cos^2 \phi} - \sin \phi\right] \tag{14-20}$$

and

$$H_s = (m_G - 1)\left[\sqrt{\left(\frac{R_i}{R}\right)^2 - \cos^2 \phi} - \sin \phi\right] \tag{14-21}$$

where R_i = inside radius of internal gear

It is quite obvious from these equations that the power loss of internal spur gears will be lower than that of external gears and consequently more efficient for a given coefficient of friction. Since both internal and external spur gears are very similar in nature except for the differences in specific sliding velocities, the curves of coefficient of friction vs. rpm as shown in Figs. 14-1 and 14-2 can be used for internal spur gears as well as external spur gears.

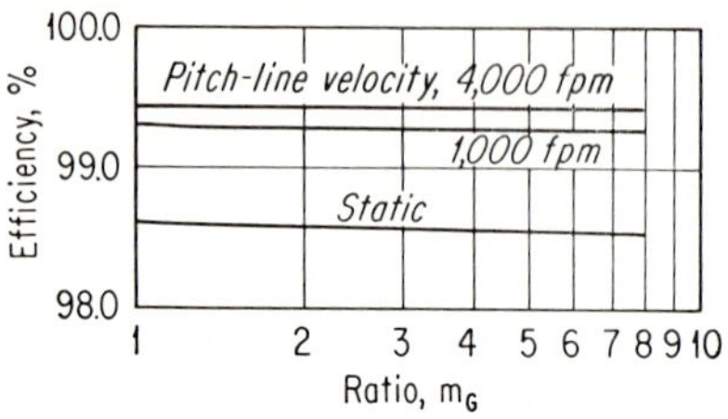

FIG. 14-5. Nominal efficiency of spur gears. Based on a good grade of petroleum oil. Hardened and ground gears 60 Rockwell C.

Single-helical Gears. The efficiency of helical gears is somewhat similar in nature to spur and internal gears. In fact, if a helical system is analyzed in terms of an equivalent spur gear, the equation for the per cent power loss can be written as follows:[5,*]

$$P_l = \frac{50f \cos^2 \psi}{\cos \phi_n}\left(\frac{H_s^2 + H_t^2}{H_s + H_t}\right) \tag{14-22}$$

Where H_s and H_t are determined by solving Eqs. (14-15) and (14-16), it must be kept in mind, however, that

ϕ_n = normal pressure angle of helical gear
ψ = helix angle
f = coefficient of friction

The resultant efficiency is determined by the same general efficiency equation $E = 100 - P_l$.

It will become obvious from these equations that a helical gear is somewhat more efficient than a similar spur gear. Often this is an important consideration when designing gears for minimum power loss.

Helical gears are similar in nature to spur gears; therefore, the equations of coefficient of friction as shown in Figs. 14-1 and 14-2 can be used for calculation purposes.

Double-helical and Herringbone Gears. Power loss or the efficiency of double-helical gearing and herringbone gearing is determined in exactly the same manner as for single-helical gears. Since the power being transmitted by a set of double-helical gearing splits exactly in two, the total power loss is just double that for one side. The per cent power loss calculation need be determined for only one side. This per cent loss can then be multiplied by the total transmitted horsepower of the double-helical set to obtain total power losses.

Straight Bevel Gears. Straight bevel gears can be likened to spur gears except, of course, that they mesh at right angles. Bevel gears are often analyzed as equivalent spur gears by using Tredgold's approximation to determine the spur-gear cross

* See p. 345 of reference.

sections. Fundamentally, the number of teeth in virtual spur gears can be found as follows:[6]

$$N_{vG} = \frac{N_G}{\cos \Gamma} \tag{14-23}$$

$$n_{vP} = \frac{N_P}{\cos \gamma} \tag{14-24}$$

where N_{vG} = number of teeth in virtual spur gear
N_G = number of teeth in actual bevel gear
N_{vP} = number of teeth in virtual spur pinion
N_P = number of teeth in actual bevel pinion
Γ = pitch-cone angle of bevel gear
γ = pitch-cone angle of bevel pinion

When the above information is completely integrated into the power-loss equations for spur gears, the following equation can be derived for determining the per cent power loss for straight bevel gears:

$$P_l = 50f\left(\frac{\cos \Gamma + \cos \gamma}{\cos \phi_n}\right)\left(\frac{H_s^2 + H_t^2}{H_s + H_t}\right) \tag{14-25}$$

H_s and H_t are determined by solving Eqs. (14-15) and (14-16). It must be kept in mind that

ϕ_n = normal pressure angle of the bevel gear
r_o = outside radius of large end of bevel pinion
r = pitch radius of large end of bevel pinion
R_o = outside radius of large end of bevel gear
R = pitch radius of large end of bevel gear

The efficiency for straight bevel gears is again determined by the same general efficiency equation.

$$E = 100 - P_l \tag{14-13}$$

Since bevel gears have so much in common with spur gears,[7] it is quite proper to use the curves of Figs. 14-1 and 14-2 to determine the coefficient of friction of bevel gears. The curve shown in Fig. 14-6 gives the nominal efficiency of straight bevel gears. It is interesting to note from these curves that the straight bevel gear mesh is quite efficient, being in the range of 99 per cent or better.

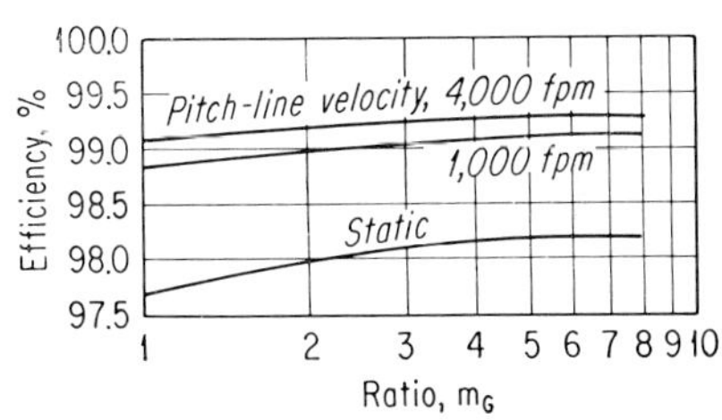

FIG. 14-6. Nominal efficiency of straight bevel gears. Based on a good grade of petroleum oil. 33 to 40 Rockwell C.

Zerol®* Bevel Gears. Zerol bevel gears are made in a different type of machine than the straight bevel gears.[8] The difference is that the straight bevels are cut on a bevel-gear planer which moves the cutting tool back and forth in straight lines, whereas the Zerol bevel gears are cut with a rotary cutter. This rotary motion produces a curved tooth section in the length of the tooth; however, the tooth has a zero-degree spiral angle and consequently is very similar to a straight bevel gear from an efficiency point of view. Actually, the efficiency of a Zerol bevel gear can be calculated using the information and equations for straight bevel gears as described in the above article.

Figures 14-1 and 14-2 can also be used to determine the mesh efficiency of Zerol bevel gears. One of the chief advantages of Zerol bevel gears is that they can be hardened and ground quite accurately. They are often made this way and con-

* Registered trade-mark of the Gleason Works, Rochester, N.Y.

sequently can be loaded higher than straight bevel gears. This being the case, then the curves in Fig. 14-6 would be slightly higher for hardened and ground Zerol bevel gears.

Spiral Bevel Gears. Spiral bevel gears have often been referred to as helical bevel gears.[9] In practice, however, the spiral bevel gear does not have a true helical spiral. They are cut on or ground on the same machine that produces Zerol bevel gears, except that the cutting tool is set at an angle to the axis of the gear, thus producing the helical-gear effect.

From an efficiency point of view, a spiral bevel gear is very good and can be compared with the results of helical gearsets. Generally, spiral bevels are used where high load and high speeds are encountered, since they can be hardened and ground and produced quite accurately.

The per cent power loss for a spiral bevel gear can be determined by solving the following equation:

$$P_l = 50f(\cos \Gamma + \cos \gamma) \frac{\cos \psi^2}{\cos \phi_n} \left(\frac{H_s^2 + H_t^2}{H_s + H_t} \right) \tag{14-26}$$

where Γ = pitch-cone angle of spiral bevel gear
γ = pitch-cone angle of spiral bevel pinion
ψ = spiral angle
ϕ_n = normal pressure angle
f = coefficient of friction

H_s and H_t must be calculated by using Eqs. (14-15) and (14-16); however, the general comment made in the discussion of straight bevel gears applies equally to spiral bevel gears. The coefficient of friction as shown in Figs. 14-1 and 14-2 can be used for calculation purposes.

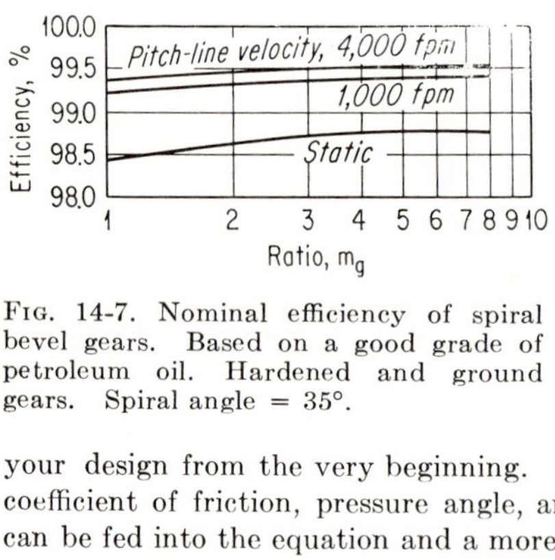

FIG. 14-7. Nominal efficiency of spiral bevel gears. Based on a good grade of petroleum oil. Hardened and ground gears. Spiral angle = 35°.

Figure 14-7 gives the *nominal* efficiency of spiral bevel gears vs. gear ratio. Note that the efficiency has been plotted for various values of pitch-line velocity. Actually, these curves have been based on full rated power being transmitted by the gearset and are only representative of what can be done. They are quite useful as a guide, but if close efficiency estimates must be made, it would be wise to calculate your design from the very beginning. By using the proper equations, the values of coefficient of friction, pressure angle, and gear geometry for your particular design can be fed into the equation and a more precise answer will be obtained.

14-3. Efficiency of Nonplanar Gears (*Nonintersecting and Nonparallel Axes*). Many gear applications require the features of this general class of gearing. Although coplanar gears are generally more efficient, this class of gearing can be designed with high-performance characteristics.

Wormgears. The per cent efficiency for wormgear drives can be determined as follows for the case where the worm drives the worm wheel:

$$E = 100 \frac{\cos \phi_n - f \tan \lambda}{\cos \phi_n + f \cot \lambda} \tag{14-27}$$

where ϕ_n = normal pressure angle
f = coefficient of friction
λ = pinion lead angle

When the worm wheel drives the worm, Eq. (14-27) becomes

$$E = 100\frac{\cos\phi_n + f\tan\lambda}{\cos\phi_n - f\cot\lambda} \tag{14-28}$$

In many instances, not all the information pertaining to a given set of wormgears is known. In this case a simplified equation for wormgear efficiency can be used.[5,*]

$$E = 100\frac{\tan\lambda}{\tan(\lambda + \eta)} \tag{14-29}$$

where η = friction angle
$\tan\eta = f$ approximately

Since generally the coefficient of friction has been determined with some experimental error, Eq. (14-29) is a fair approximation.

It is quite helpful to analyze the relationship of lead angle and the efficiency of various combinations of wormgear drives by studying Tables 14-1 and 14-2. Tables

Table 14-1. Efficiency of Worm Gearset, Worm Driving

Worm lead angle, deg	f = 0.015 eff., %	f = 0.02 eff., %	f = 0.03 eff., %	f = 0.04 eff., %	f = 0.05 eff., %	f = 0.07 eff., %	f = 0.10 eff., %
3	77.5	72.0	65.0	55.0	48.0	42.0	35.0
5	86.0	81.5	75.1	70.0	63.0	55.0	48.0
10	92.0	89.5	84.5	82.0	77.0	72.0	63.0
15	94.5	92.5	89.5	87.0	83.2	78.0	72.0
25	96.2	95.1	92.6	90.5	88.5	84.0	78.0
35	96.8	95.9	93.9	92.1	90.2	86.8	81.5
45	97.1	96.2	94.2	92.4	90.5	87.2	82.0
55*	96.8	95.8	93.8	91.8	89.8	86.0	80.5
65	96.0	94.6	92.1	89.3	87.3	82.0	75.0
75	93.9	91.9	88.1	84.0	80.0	72.0	61.0
80	91.0	88.0	82.1	75.3	71.0	59.0	40.0
85	82.5	75.3	63.0	50.3	40.0	20.0	

* Generally worms are not designed for more than 45° lead angle. An angle of 55° is usually the very upper limit for a practical design.

Table 14-2. Efficiency of Worm Gearset, Wormgear Driving

Worm lead angle, deg	f = 0.015 eff., %	f = 0.02 eff., %	f = 0.03 eff., %	f = 0.04 eff., %	f = 0.05 eff., %	f = 0.07 eff., %	f = 0.10 eff., %
3	69.0	58.0	40.0	20.0	0		
5	82.5	77.0	63.0	53.0	40.0	20.0	
10	91.0	88.0	82.5	75.3	71.0	58.0	40.0
15	93.9	91.9	88.2	84.0	80.0	72.0	61.0
25	96.0	94.6	92.1	89.8	87.0	81.9	75.0
35	96.8	95.7	93.8	91.8	89.8	86.0	80.5
45	97.1	96.2	94.3	92.5	90.6	87.1	82.2
55*	96.8	96.0	93.9	92.0	90.2	86.8	81.5
65	96.3	95.1	92.6	90.7	88.5	84.0	78.0
75	94.3	92.8	89.5	87.0	83.3	78.0	72.0
80	92.0	89.5	85.3	82.0	78.0	72.0	64.0
85	86.0	81.9	81.3	70.0	63.0	56.0	47.0

* Generally worms are not designed for more than 45° lead angle. An angle of 55° is usually the very upper limit for a practical design.

* See p. 347 of reference.

14-1 and 14-2 were determined by assuming various values for the coefficient of friction and solving the efficiency equations. In order to use Tables 14-1 and 14-2 profitably, values of the coefficient of friction for your given conditions must be known.

Figure 14-8 gives some experimental values of f based on using a good grade of mineral oil after the gearset has been well broken in.

Figure 14-9 shows the nominal efficiency of cylindrical wormgears as a function of lead angle and rubbing velocity. It is understood that double-enveloping wormgears also follow such a pattern as outlined in Fig. 14-9, although it has been said that double-enveloping wormgears are more efficient under starting conditions than cylindrical wormgears.

Crossed-axes Helical. Crossed-axes helical gears are actually nonenveloping wormgears and as such can be calculated with efficiency equations very similar in nature to equations for cylindrical wormgears.

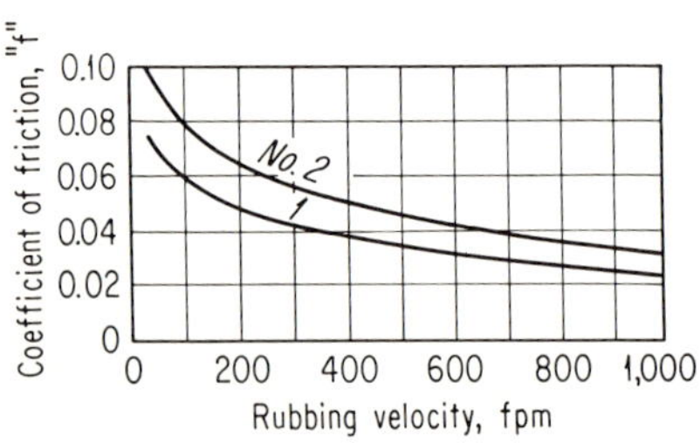

FIG. 14-8. Coefficient of friction for wormgears. Based on a good grade of petroleum oil. No. 1 = case-carburized and ground worm and phosphor-bronze gear. No. 2 = cast-iron worm and wormgear.

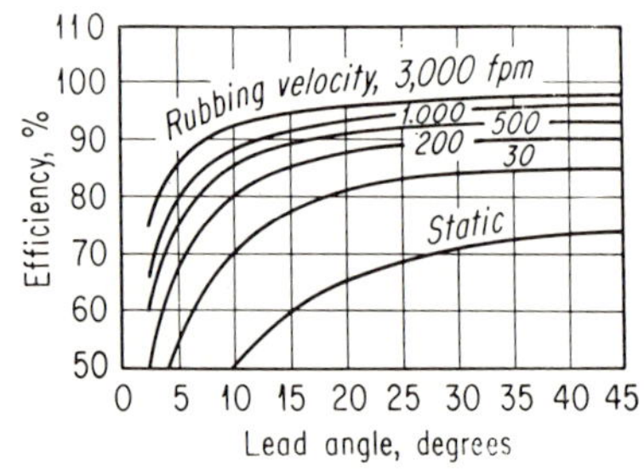

FIG. 14-9. Nominal efficiency of wormgears. Based on case-carburized and ground worm and phosphor-bronze gear.

Since crossed-axes helical gears can be of the same hand or opposite hands and the spiral angle or helix angle can be different for either member, the equations for the efficiency are a little tricky and must be used with caution. For the case where the gears are of the same hand, then,[5,*]

$$E = 100 \frac{1 - f \tan \psi_f}{1 + f \tan \psi_d} \tag{14-30}$$

where f = coefficient of friction
ψ_d = helix angle of driver
ψ_f = helix angle of follower

When the gears are of opposite hand and ψ_d is less than ψ_f,

$$E = 100 \frac{1 - f \tan \psi_f}{1 - f \tan \psi_d} \tag{14-31}$$

When the gears are of opposite hand and ψ_d is greater than ψ_f,

$$E = 100 \frac{1 + f \tan \psi_f}{1 + f \tan \psi_d} \tag{14-32}$$

The coefficient of friction f for spiral gears is practically the same as that for wormgears; therefore, the same curves of f vs. rubbing velocity can be used (see Fig. 14-8).

Spiroid Gears. Spiroid gears are comparable in performance and efficiency with wormgears and in general can be used in many applications where heretofore worm-

* See p. 355 of reference.

gear drives were specified.[10] The efficiency of Spiroid gears can be determined from an equation which has much in common with the basic wormgear-efficiency equation. Efficiency in per cent for Spiroid gears can be expressed as follows:

$$E = 100 \frac{\cos \phi_n + f \cot \alpha_g}{\cos \phi_n + f \cot \lambda} \tag{14-33}$$

where ϕ_n = normal pressure angle
f = coefficient of friction
α_g = gear spiral angle (with respect to the plane of rotation)
NOTE: This is 90° minus the spiral angle as used for spiral bevel gears.
λ = pinion spiral angle

$$\tan \lambda = \frac{L}{2\pi r_m} \tag{14-34}$$

and L = conical lead
r_m = mean cone pinion radius at a point half way across the face

In order to evaluate Eq. (14-33) fully, however, additional information is required, such as

$$\alpha_g = 90° - (\sigma - \lambda) \tag{14-35}$$

where σ = normally standard at 40°
λ = lead angle as indicated above

The normal pressure angle is taken as 10° except when the gearset is run with the contact on the opposite side of the pinion teeth. Then it is driving on the high-pressure-angle side, which usually has a pressure angle of 30°. In order to understand better the geometry relationship of the above equations, refer to Fig. 14-10, which shows a contacting segment of a Spiroid gear and a Spiroid pinion.

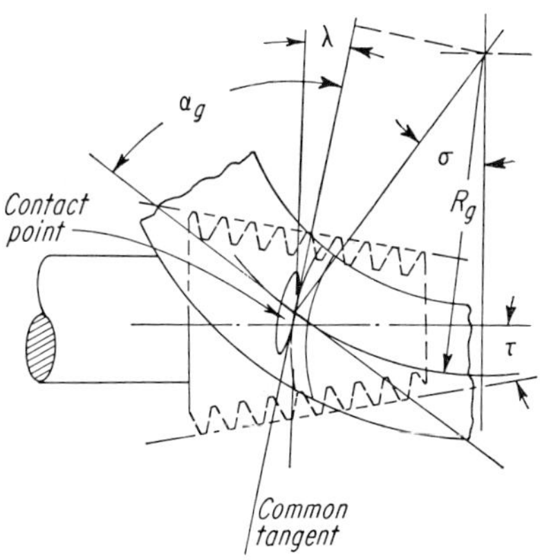

FIG. 14-10. Contacting segment of a Spiroid gear and pinion.

The relationship of the coefficient of friction vs. sliding velocity can be seen in Fig. 14-11. Curves have been plotted for the efficiency of Spiroid gears vs. pinion Spiroid angle (see Fig. 14-12). Note from Fig. 14-12 that the gears used were hardened and ground. Little difference has been observed when bronze gears have been used instead of hardened gears. At very low speeds, though, a bronze gear and a hardened pinion combination does appear to be somewhat more efficient.

Helicon®* Gears. Helicon gears are very similar in nature to Spiroid gears in that they are skew-axis gears. They differ by virtue of the fact that the pinions have a zero degree taper angle, whereas standard Spiroid pinions have a 5° taper angle.[11] Helicon gears also are flat-faced instead of conical-faced like Spiroid gears. Other than this, the gear types are very much alike and can be calculated using exactly the same equations as outlined above for Spiroid gears. The standard σ "sigma" angle is 40°. The 10° pressure angle applies to the Helicon gear as well as the Spiroid gear. Figure 14-11 can be used to determine the coefficient of friction for Helicon gears. Figure 14-12 also depicts typical Helicon gear efficiency vs. rubbing velocity. It must be kept in mind, however, that Helicon gears by design have less face width in contact and will carry less load than their brothers the Spiroid gears. Consequently, the

* Registered trade-mark of Spiroid Div., Illinois Tool Works, Chicago, Ill.

over-all efficiency of Helicon drives will be less efficient as compared with the over-all efficiency of Spiroid gear drives.

Hypoid Gears. Hypoid gears have much in common with plain or spiral bevel gears.[12] In fact, hypoid gears are actually offset-axis bevel gears, and as such, they are subjected to a high degree of sliding as are most of the other types of offset or "nonplanar" gears. In comparison with a wormgear, however, the rubbing speed of a hypoid relative to its own pitch-line velocity is not so high. The per cent efficiency for hypoid gears can be determined as follows:

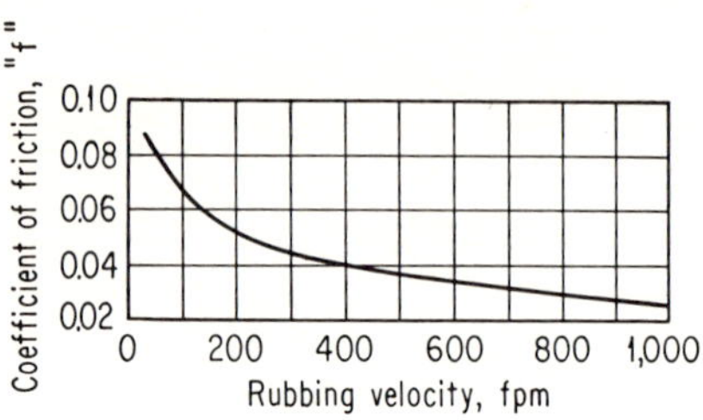

FIG. 14-11. Coefficient of friction vs. sliding velocity of Spiroids. Based on carburized pinion and gear, 60 Rockwell C, and a good grade of petroleum oil.

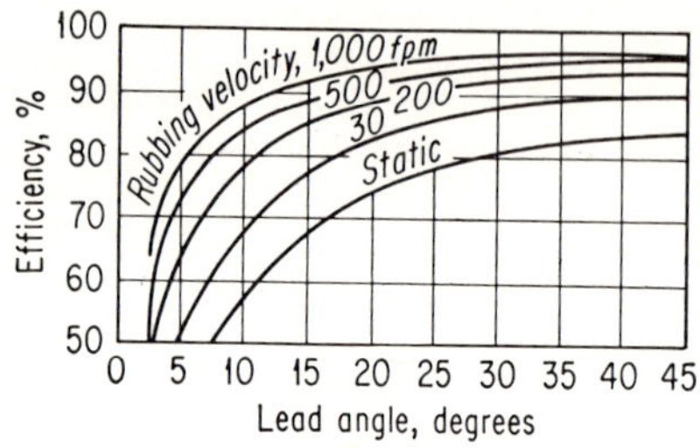

FIG. 14-12. Nominal efficiency of Spiroid gears. Based on case-carburized pinion and gear, 60 Rockwell C, and a good grade of petroleum oil.

$$E = 100 \frac{\cos \phi_n + f \tan \psi_g}{\cos \phi_n + f \tan \psi_p} \tag{14-36}$$

where ϕ_n = normal pressure angle
f = coefficient of friction
ψ_g = gear spiral angle
ψ_p = pinion spiral angle

The nominal efficiency for hypoid gears can be found in Fig. 14-13. These curves were plotted for the conditions where the sum of the gear and pinion spiral angle equals 75° and the normal pressure angle is 22°30′. It is quite interesting to note that at low ratios the hypoid gears are quite efficient. Generally speaking, the coefficient of friction for hypoid gears is similar to friction values of wormgears, and consequently the curves shown in Fig. 14-8 can be used for hypoid-gear calculations.

Planoid®* Gears. Planoid gears perform a function similar to that of hypoid gears. They are right-angle gears with offset axes. The tooth surfaces of the gears are planes and the pinions are slightly tapered with helical teeth.[13] The recommended ratio range for Planoid gears is from 15:1 to 10:1. The efficiency of Planoid gears compares quite favorably with that of hypoid gears.

The efficiency of Planoid gears can be determined from the following equation:

$$E = 100 \frac{\cos \phi_n + f \cot \alpha_g}{\cos \phi_n + f \cot \lambda} \tag{14-37}$$

where ϕ_n = normal pressure angle
α_g = gear spiral angle
λ = pinion spiral angle
f = coefficient of friction

* Registered trade-mark of Spiroid Div., Illinois Tool Works, Chicago, Ill.

The efficiency of Planoid gears compares favorably with that of hypoid gears. Since Planoids are usually designed for low ratio, they generally are quite efficient, being in the range of 96 per cent or better. In reference to Fig. 14-13, Planoid gears fit nicely in the range of industrial and automotive gears from an efficiency point of view. Like hypoid gears, the efficiency of Planoid gears can be calculated using the coefficient-of-friction data of wormgears as outlined in Fig. 14-8.

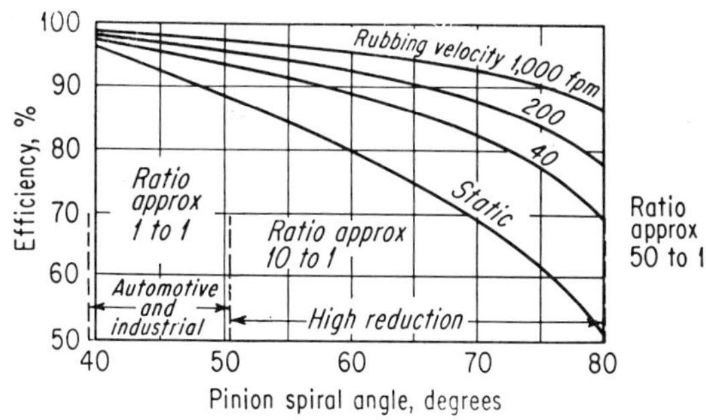

FIG. 14-13. Nominal efficiency of hypoid gears. Based on the sum of gear and pinion spiral angles equal to 75°, and normal pressure angle of 22°30′.

14-4. Gear Trains. In order to analyze a complete gear train, information pertaining to all the losses of a gearbox must be considered, not just the gear mesh itself. As stated previously, windage and churning losses and bearing losses are very important factors and at times contribute considerably to the over-all efficiency of a complete gear train. Even the gear arrangement itself, in some cases, must be carefully analyzed to be sure that a complete loss picture is obtained. For example, a simple planetary is inherently more efficient than a simple star arrangement, even though they are generally composed of very similar gear elements. It is the load transmitted and the meshing velocity, among other things, that in the final analysis determine the amount of power being dissipated in the mesh. Sometimes both the load and the meshing velocities are difficult to determine in complex gear arrangements.

Bearing Losses. In order for gear elements to transmit power, they must be supported in the gear casing by some sort of bearing which will permit free and accurate rotation with a minimum amount of heat generation. Generally, of course, the bearings are lubricated with an ample supply of oil both to minimize the coefficient of friction in the bearing and to act as a cooling medium to carry away the heat that has been generated.

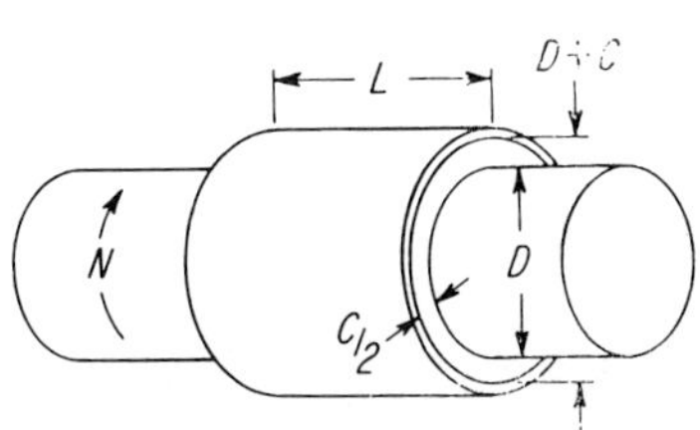

FIG. 14-14. Elements of a sleeve bearing.

These bearings usually fall into two main categories, namely, rolling contact bearings such as ball and roller bearings, or hydrodynamic bearings commonly known as "sleeve"-type bearings. Usually, a given gearbox design is equipped with either one type of bearing or the other. In a few applications, however, a mixture of the two types has been used.

The horsepower loss of the hydrodynamic-type sleeve bearing can be calculated from the laws of viscosity since the power consumption of this type of bearing results from the shearing of the oil film.

If the shaft is assumed to be in the center of the bearing, the losses can be calculated using the well-known Petroff equation[14,*] with reference to Fig. 14-14.

$$P_b = 3.79 \times 10^{-13} \frac{Z_2 n^2 D^3 L}{C} \tag{14-38}$$

where P_b = horsepower loss in bearing
Z_2 = outlet oil viscosity (absolute), cp
n = rpm
D = diameter, in.
L = length of bearing, in.
C = clearance diametrical, in.

This equation serves as a very good approximation for lightly loaded bearing since the shaft on lightly loaded bearings tends to ride in the center.

In practice, however, and particularly with bearings that carry an appreciable load, the shaft usually runs eccentric in the bearing. In this case, the above calculated horsepower loss (P_b) must be modified by a power-loss coefficient which is a numerical representation of the bearing characteristics.

$$P'_b = JP_b \tag{14-39}$$

where J = power-loss coefficient

The power-loss coefficient can be calculated[14,†] by solving the following rather complex equation with reference to Fig. 14-15:

$$J = \frac{pCe \sin \varphi \times 10^7}{2.90\pi ZnD^2} \frac{1}{2\pi} \int_0^{2\pi} \frac{BC}{2Lh} d\Theta \tag{14-40}$$

where C, Z, D, L, and n are as previously defined
p = load per unit projected area, psi
e = eccentricities, in.
Θ = angle to point on bearing surface measured from load line
B = width of reduced bearing oil film in expanding region of bearing, in.
h = film thickness, in.
φ = attitude angle between line of center and load line

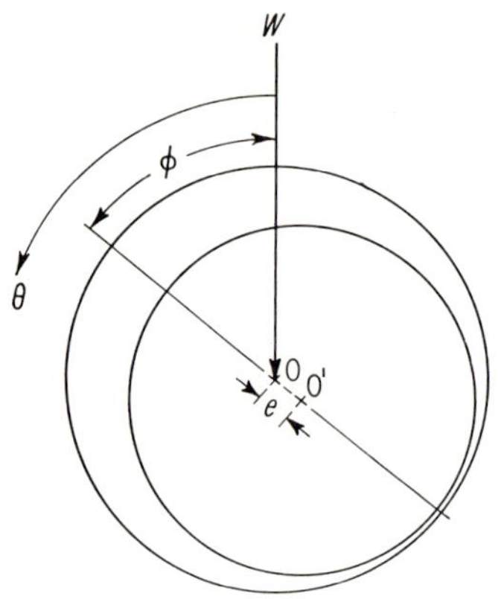

FIG. 14-15. Definition of a loaded sleeve bearing.

The above equation for J is somewhat difficult to solve; however, J has been determined for many applications and plotted in Fig. 14-16. You will note from Fig. 14-16 that J has been plotted for various L/D ratios vs. the Summerfield number S.

$$S = 2.42 \times 10^{-9} \times \left(\frac{D}{C}\right)^2 \times \frac{Zn}{p} \tag{14-41}$$

In order to determine J, it is a simple matter to calculate the Summerfield number and the L/D ratio, then go to Fig. 14-16 and "pick off" the proper value for J.

If the particular bearing under consideration is an elliptical-type sleeve bearing, then the J values found in Fig. 14-17 should be used. The P_b in Eqs. (14-38) and (14-39) can be used to calculate the basic horsepower loss. For solving these equations for an elliptical bearing design, the average diameter is used for D.

* See p. 192 of reference.
† See p. 220 of reference.

Fig. 14-18 can be used for determining the proper value of J for four axial-groove sleeve bearings. Again, Eqs. (14-38) and (14-39) can be used to calculate the horsepower loss.

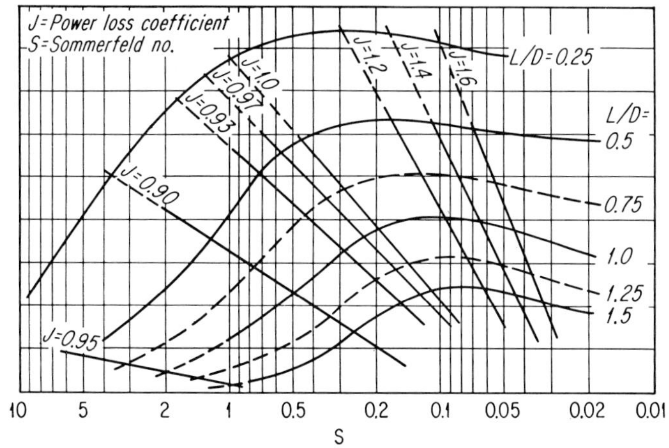

FIG. 14-16. Plot of J cylindrical bearings.

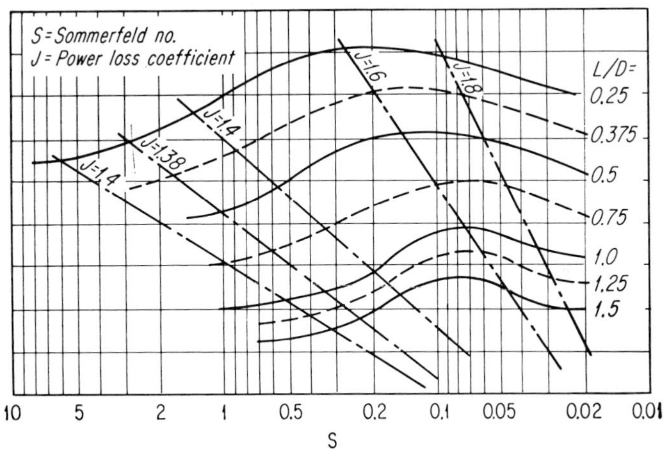

FIG. 14-17. Plot of J elliptical bearings.

The power loss of simple thrust bearings can also be determined in similar fashion by using the laws of viscous flow.

$$P_t = 3.79 \times 10^{-13} \times \frac{Z_2 n^2}{t} (R_2^4 - R_1^4) \tag{14-42}$$

where P_t = horsepower loss of a simple thrust bearing
Z_2 = outlet viscosity, cp
t = oil-film thickness
R_1 = inside radius of thrust bearing
R_2 = outside radius of thrust plate

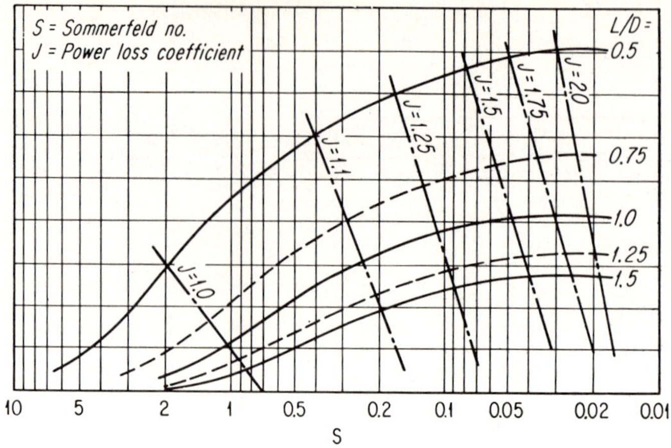

FIG. 14-18. Plot of J four-groove bearings.

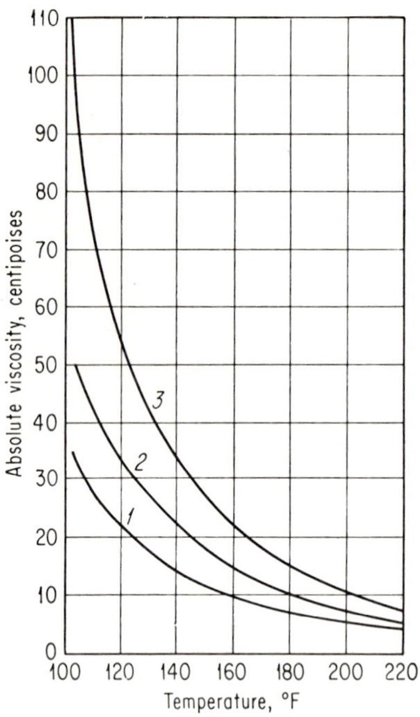

FIG. 14-19. Viscosity of petroleum oil. (1) AGMA 1—200 SSU at 100°F. (2) AGMA 2—300 SSU at 100°F. (3) AGMA 3—600 SSU at 100°F.

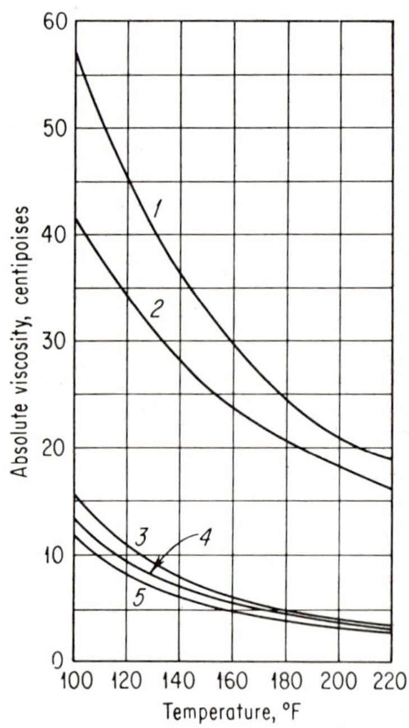

FIG. 14-20. Viscosity of synthetic oils. (1) Dow Corning XF-258 (Silicone). (2) GE Versalube F-50 (Silicone). (3) MIL-L-9236B (Cellutherm 2505A). (4) MIL-7808D. (5) MIL-L-25336 (Sinclair L-743).

Figure 14-19 gives the values of absolute viscosity in centipoises for various grades of petroleum oil vs. temperature of the oil. This chart is quite useful when calculating bearing losses. Figure 14-20 gives similar information on other types and grades of oils, some of which are synthetics.

It is extremely difficult to calculate bearing losses and have them coincide exactly with measured bearing losses. There are many sound reasons why variations between actual and predicted bearing losses exist. Some of these reasons are:

1. Bearing misalignment at assembly
2. Variation in viscosity-temperature behavior of the lubricating oil from that used in computation
3. Presence of phenomena such as oil turbulence or recirculation
4. Variation in operating variables such as shaft speed and bearing loads from those used in computations

The losses of rolling-contact bearings depend to a large degree on the type and quantity of lubricant used in a given application. Exact predictions of these losses are extremely difficult. However, a considerable amount of experimental work has been done. Empirical equations were determined from these investigations. Like most experimental work, the results must be used with extreme care and understanding.

One of the simplest approaches to the bearing-loss problem is to determine the torque loss in the bearing and then calculate the horsepower loss as follows:[14,*]

$$P_b = \frac{Tn}{63{,}000} \tag{14-43}$$

where P_b = horsepower loss in bearing
T = torque loss per bearing, lb-in.
n = rpm of bearing

The torque loss T can be found simply by the use of a constant coefficient of friction for each type of bearing.

$$T = f\frac{D_1}{2}W \tag{14-44}$$

where T = loss torque lb-in.
D_1 = bore diameter of bearing, in.
W = load on bearing, lb

This approach offers a fast and reasonable approximation based upon normal operating loads and speeds with favorable conditions.

The above equation can be used for a variety of bearing types as long as the proper coefficients of friction are used[15,†] (see Table 14-3). It must be kept in mind, however, that, under starting conditions or if higher-viscosity lubricants are used, higher values of coefficient of friction will be obtained.

Table 14-3

Type of bearing	*Coefficient of friction f*
Radial ball bearing (single-row deep groove)	0.0015
Self-aligning ball bearing	0.0010
Angular-contact bearing	0.0013
Thrust ball bearing	0.0013
Cylindrical roller bearing	0.0011
Spherical roller bearing	0.0018
Tapered roller bearing	0.0018

* See p. 83 of reference.
† See p. 21 of reference.

Figure 14-21 can be used as a guide to determine the exact type of bearing used in your particular design.

Some gear designers prefer to determine the approximate torque loss of rolling-contact bearings by using the pitch diameter of the rolling elements for the appropriate diameter in the torque equation[16] or

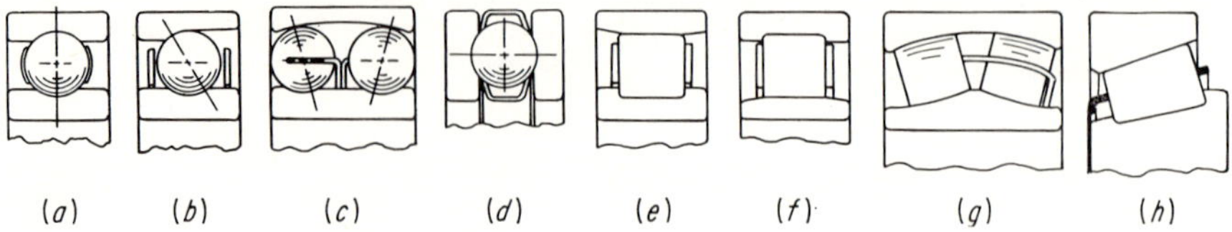

FIG. 14-21. Types of bearings. (*a*) Radial ball bearing. (*b*) Angular-contact bearing. (*c*) Self-aligning ball bearing. (*d*) Thrust ball bearing. (*e*) Cylindrical roller bearing (outside race free). (*f*) Cylindrical roller bearing (inside race free). (*g*) Spherical roller bearing. (*h*) Tapered roller bearing.

$$T = f\frac{D_p}{2}W \tag{14-45}$$

where D_p = pitch diameter of rolling elements

This equation is perhaps theoretically more correct in that the rolling-contact area is where the actual loss takes place, not in the bore of the bearing. This equation also permits some consideration for the differences between bearings of the same bore diameter but with larger balls or rollers. Of course, the larger ball or roller operates a larger pitch diameter. It is somewhat more difficult to obtain the operating pitch diameter of a rolling-contact bearing; however, it can be approximated by the following equation:

$$D_p = \frac{D_1 + D_2}{2} \tag{14-46}$$

where D_p = pitch diameter of elements, approximately, in.
D_1 = bore diameter
D_2 = outside diameter

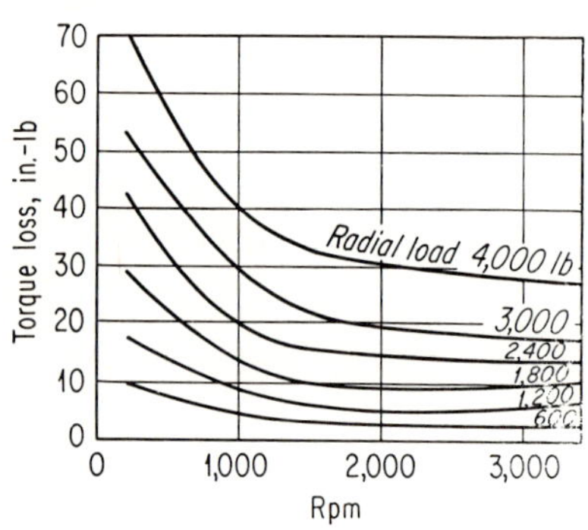

FIG. 14-22. Losses of 22,308 spherical roller bearings. Bearing torque loss vs. rpm and load 40-mm bore × 33-mm width. Double-row spherical roller bearing. Light aircraft oil at 120°F inlet.

As an example of how the actual bearing loss varies for a spherical roller bearing, some 22,308 bearings were tested by the author. Figure 14-22 shows these results. Note that the loss torque at slow speeds is quite high compared with that at the faster speeds. It is also quite interesting to note that, as the speed is increased the losses in the bearing tend to increase, undoubtedly because of the churning effect of the oil.

A very complete work on friction torque of ball bearings was carried out by Manlio Muzzoli.[14,*] The equation developed for the friction torque of deep-groove ball bearings under radial load is as follows:

$$T = 0.0868KV\left[n^{0.5}(2.6d^{0.2} - 3.33)n^{0.375d-0.2} + \frac{(0.454W)^m}{72n^{0.73} + 0.150N_b(d - 6)^{0.055}}\right] \tag{14-47}$$

* See p. 84 of reference.

where d = ball diameter, mm
N_b = no. of balls
n = rpm
W = load, lb
K = constant, lubricant
V = constant, conformity
m = determined by Eq. (14-48)

$$m = 1.65 + 0.004n^{0.54} + 0.005dN_b \tag{14-48}$$

The constant V can be evaluated by referring to Fig. 14-23. Note that V is plotted against conformity, which is the race curvature divided by ball diameter. Bearings are generally manufactured to a conformity of 52 per cent, which results in a value of V equal 1.1.

The constant K, which takes into consideration the type of lubricant, can be found in Table 14-4.

Table 14-4

Fluid grease	$K = 1.0$
Dense grease	$K = 1.2$
Very dense grease	$K = 2.0$
Very light oil	$K = 0.4$
Light oils	$K = 0.6$

Dr. Booser and Dr. Wilcock[14,*] state they have found that modern good ball-bearing greases will give a K value of 0.10 to 0.15, which is somewhat better than those given in Table 14-4. They also indicate from the results of many of their tests that the Muzzoli equation (14-47) is fundamentally correct.

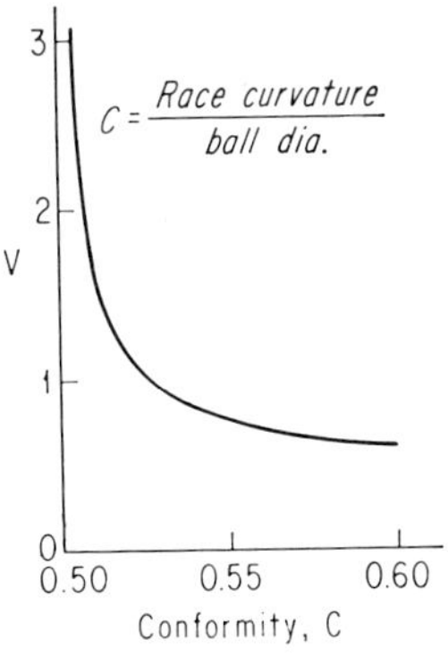

FIG. 14-23. Plot of V vs. conformity.

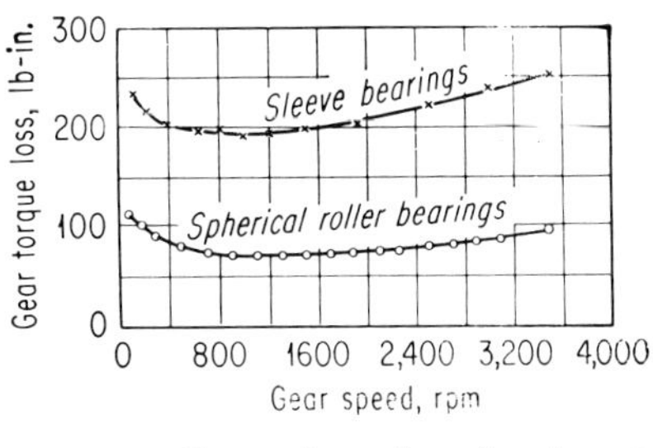

FIG. 14-24. Comparison of gearbox torque losses (sleeve bearings vs. rolling-contact bearings).

The curves as indicated in Fig. 14-24 give a good graphical comparison of the losses of a given gearbox when mounted on first sleeve-type bearings and then rolling-contact bearings. In both cases, the gearbox was loaded with exactly the same amount of torque. In many respects the curves are just what would be expected in that the sleeve-bearing design has higher breakaway torque. The losses of the sleeve bearing at speed are still higher in that a considerable amount of oil is required to keep the

* See p. 85 of reference.

bearing functioning properly and this generally results in a high viscosity-loss effect. However, at really high speeds and loads, these two curves tend to cross, thus making the sleeve-bearing performance more comparable with the rolling-contact bearing.

Windage Losses. The windage losses of a gearbox are very difficult to measure or calculate. Investigations have been carried out on this subject, but very little good information has been published. In general terms, windage-loss determination is still pretty much based on individual experience of the gear designer and experimental measurements on specific gearboxes in question.

Actually the windage losses for a given gear design depend on many things such as:

1. The diameter of the rotating elements
2. The length of the rotating elements
3. The speed of rotation
4. The web or gear-blank design
5. The over-all casing design
6. The type of oil-feed system
7. The operating temperature and viscosity of the oil
8. The pressurization of the casing

Some equations have been written that give a fair approximation of the windage losses of small-diameter gears, say up to approximately 20″ in diameter with an L/D ratio of approximately 0.5.

$$P_w = \frac{n^3 D^5 L^{0.7}}{100 \times 10^{15}} \qquad (14\text{-}49)$$

where P_w = horsepower loss due to windage
n = rpm
D = diameter of rotating element
L = length of rotating element

Another interesting equation that sometimes is quite useful is the equation that was empirically derived for a rotating smooth body in air. The rotating element was not enclosed by a gear casing, nor was it exposed to any lubricant.

$$P_w = \frac{15}{0.746}\left(\frac{n}{1{,}000}\right)^3\left(\frac{D}{100}\right)^4\left(\frac{5L}{100} + \frac{D}{100}\right) \qquad (14\text{-}50)$$

The answer obtained by using this formula will undoubtedly give the minimum value of windage obtainable.

Sometimes it is quite helpful to realize just when you should start to be concerned about windage losses, even if the methods of evaluating it leave something to be desired.

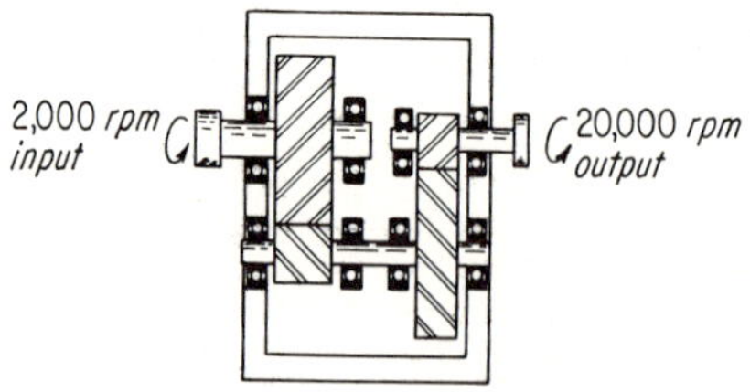

FIG. 14-25. High-pitch-line-velocity gearing.

Often windage losses are combined with other extraneous losses in the system. Stated in another manner, when the mesh losses and bearing losses are subtracted out of the total losses, windage losses are what remain.[17] A gear designer should concern himself about windage losses when his design starts to suffer from any of the troubles indicated schematically in the following sketches, namely, Figs. 14-25, 14-27, 14-28, 14-29, 14-31, and 14-33.

Figure 14-25 indicates a high-speed gearbox with pitch-line velocities higher than 10,000 fpm. Generally, when the meshing velocity reaches 10,000 fpm, the windage losses begin to increase to the point where a sophisticated design should take them into consideration. There is not much that can be done about high pitch velocities, except to keep the diameters of the rotors as small as possible.

As an indication of the effect windage losses may have on a high-speed gear train, refer to Fig. 14-26. This information was obtained experimentally from efficiency tests on a single-offset aircraft gearbox. It is quite interesting to note that in a well-designed gearbox little, if any, windage loss is present at the lower speeds.

Figure 14-27 indicates a tight-gear-casing arrangement whereby the gear casing fits tightly around the gearing. This is done because of weight and space consideration; however, it usually leads to windage troubles. Relief can often be found by shrouding the gears with oil shields. The oil-discharge port, in particular, must be protected from the "hurricane" within the casing in order to assure a free flow of oil from the casing.

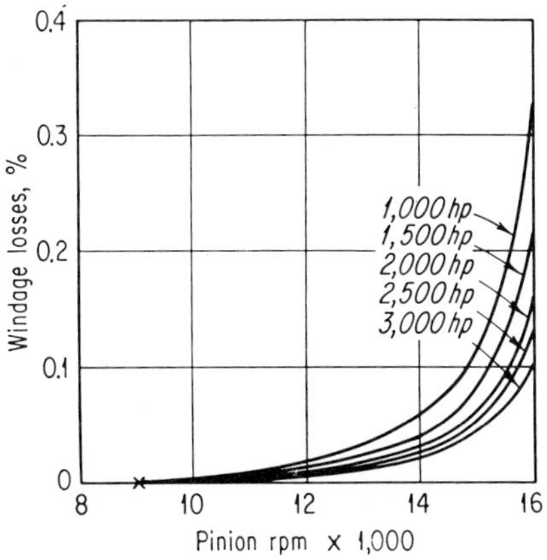

FIG. 14-26. Windage losses—aircraft gear. Single offset. MIL-L 7808D oil. 250°F inlet oil.

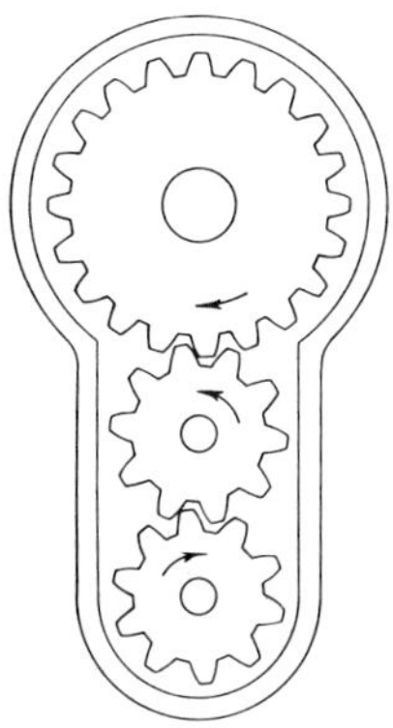

FIG. 14-27. Tight-gear-casing arrangement.

Not only does this type of arrangement lead to excessive heat being generated in the oil because of churning, but sometimes more drastic troubles are encountered, such as pumping of the oil within the casing, which generally results in a hydraulic braking action. If this trouble is encountered, the gear unit must be shut down immediately or the unit may suffer a catastrophic failure. In addition to shrouding, a unit of this nature should be scavenged with a pump that has from 2½ to 3 scavenge ratio.

Figure 14-28 illustrates a compact-gear-train arrangement. Usually a gear train of this sort is an accessory-drive package, whereby the gears themselves do not have excessively large face widths, but they generally are quite high speed. Some of the gearing is not too highly loaded in that often the center distance required dictates the size of the gearing, not the power being transmitted. Any free oil in a gearbox of this nature gets "kicked" and battered about before it finally leaves the box to be recirculated again. In a gear design of this nature, it is the best policy to keep the oil flow at an absolute minimum to the gear mesh, as well as to the bearings. Generally, by carefully observing the early test results of a given design, oil can be redirected and redistributed to lubricate the complex gear train adequately without having to resort to shrouding. In some instances, however, a well-placed oil deflector can be quite beneficial. Again, removing the oil from the gear casing and preventing any build-up of surplus oil is a must; thus usually a scavenge pump is required, but a 2:1 scavenge ratio generally is high enough.

Figure 14-29 illustrates a poor or inadequate oil-scavenge arrangement. This type of trouble sometimes accompanies a tight-gear-casing arrangement or a compact-gear-train arrangement. Often enough, it occurs in more simple types of gearboxes such as the one illustrated in Fig. 14-29, primarily because of engineering oversight. When inadequate drainage and scavenge capacity are present, the gearbox will fill up with oil, causing extreme foaming and hydraulic braking. Shrouding the gears in areas where the oil has a tendency to pile up helps. The discharge ports must be

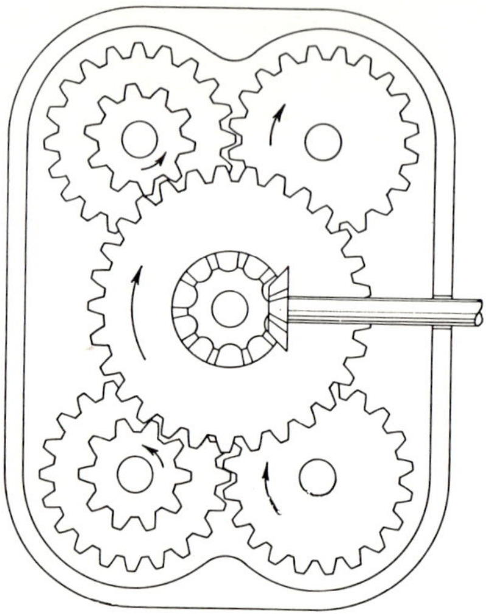

FIG. 14-28. Compact-gear-train arrangement.

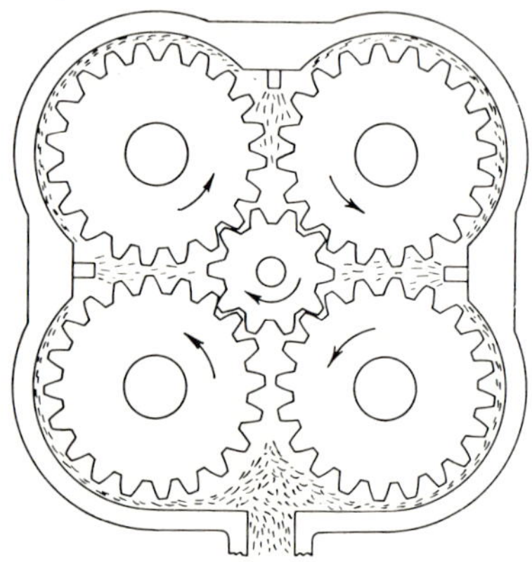

FIG. 14–29. Inadequate oil scavenge.

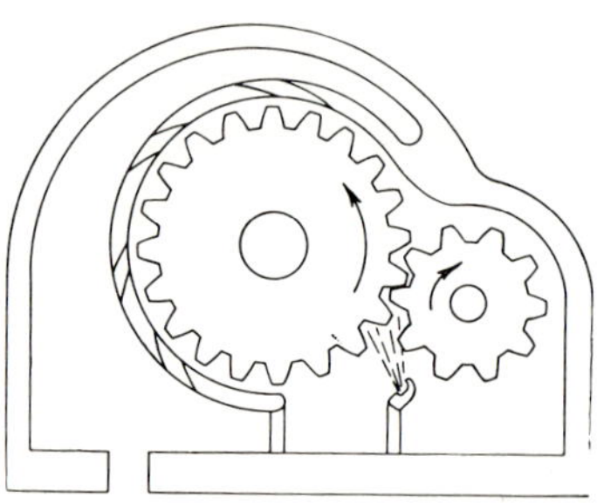

FIG. 14-30. Oil strippers.

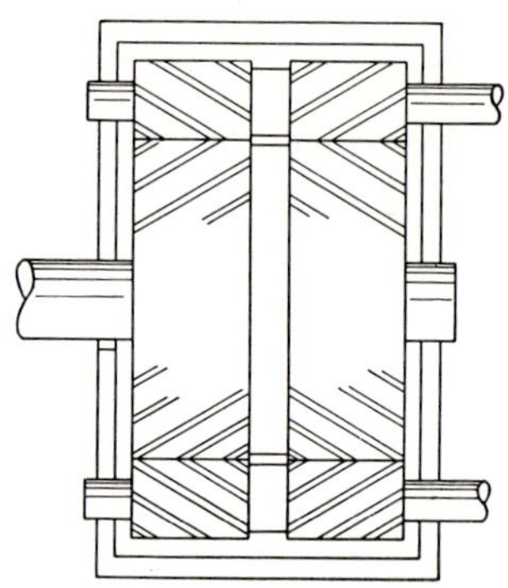

Fig. 14-31. Wide-face-width gearing.

shielded in order to permit a free flow of oil from the casing. Sometimes, if possible, oil strippers can be used to advantage (see Fig. 14-30). Of course, in any gear casing where excessive oil becomes a problem, it is always wise to cut down to the bare minimum the amount of oil that is initially put into the gear casing.

Figure 14-31 shows a wide-face-width gear design. In this sketch, the L/D ratio of the gearing is quite high. Because there is considerable face width, the windage loss tends to build up in relation to the length of the rotating bodies. If the gear design was such that the face width was only ½″ or so, the windage loss might be quite small, but since the gearing could be up to several feet in length, the losses due to windage increase considerably. Another factor that contributes to additional losses is that the oil sometimes becomes trapped in the mesh, since it has a much greater distance to go before it can be expelled from the ends. Actually, there is very little that can be done to help a design of this nature, except to allow for plenty of room within the casing. If the gearing is double-helical, as most likely it would be with large-face-width gearing, the helix angle could be so arranged that the helix angle would tend to force the oil toward the ends of the gear teeth. This arrangement would prevent the oil from being stacked up at the gap between helices.

From a windage-loss point of view, the inlet-oil temperature should be kept as high as practical within good over-all operational characteristics of both the oil and the equipment. As an example of temperature effects on the windage losses of a given gearbox, refer to Fig. 14-32. This figure shows the no-load losses of a given gearbox with respect to inlet-oil temperature. Except for the churning losses of the bearings at no load and the churning losses of the gear mesh itself, the losses shown are basically windage losses and depict to a great extent the effect of oil temperature on over-all windage losses.

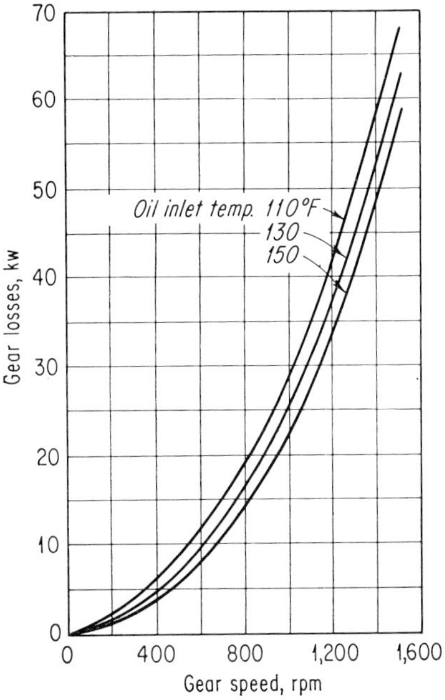

FIG. 14-32. Oil temperature effects on windage losses. Oil-inlet pressure 8 psi, Navy 2190 T oil.

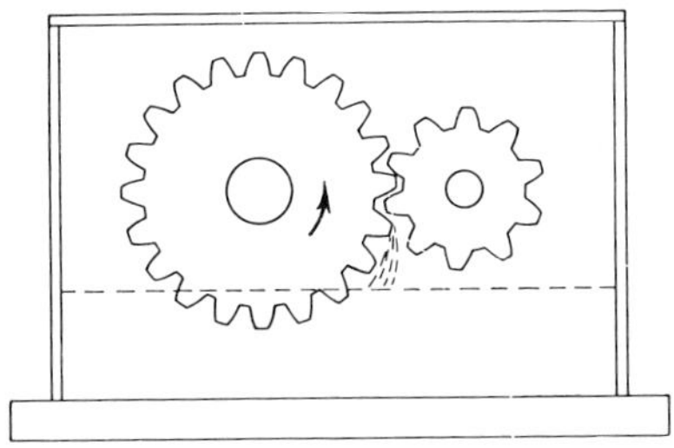

FIG. 14-33. Dip- or splash-feed lubrication.

Figure 14-33 shows a simple gearbox that is lubricated by the dip or splash feed system. Any time that a design of this nature is considered, the designer must be alert to the possibility of high losses due to churning of the oil. Not only must the designer be aware of this problem, but he also must be aware of the tendency of the lubricant to channel, thus rendering the lubrication system ineffective. This arrangement is generally utilized for low-speed drives; in fact, many wormgear applications use this system.

Single-offset Gears. A single-reduction two-element gearbox is the simplest type of gear train, and consequently the determination of the over-all efficiency of this arrangement is quite straightforward. First, the mesh loss is determined by solving Eq. (14-13) or (14-14) with the help of Fig. 14-1. Second, the bearing losses are determined by solving Eq. (14-38) for sliding-contact bearings. This horsepower loss should then be converted to the per cent loss basis, so that it may be added to the mesh loss. If rolling-contact bearings are present, then Eq. (14-43) should be used, and the loss torque determined should also be converted to a per cent loss basis, and again added to the mesh loss. Third, the windage losses for the gearbox should be determined by solving Eq. (14-49), first for the pinion and then for the gear. The results should be converted to a per cent loss, then added to the total loss already accumulated. This new total represents the over-all loss of the complete gear train. Caution should be exercised to assure that there are no bad seal rubs or oil trapping (see Figs. 14-27, 14-28, and 14-33). Factors of this nature cause additional losses and

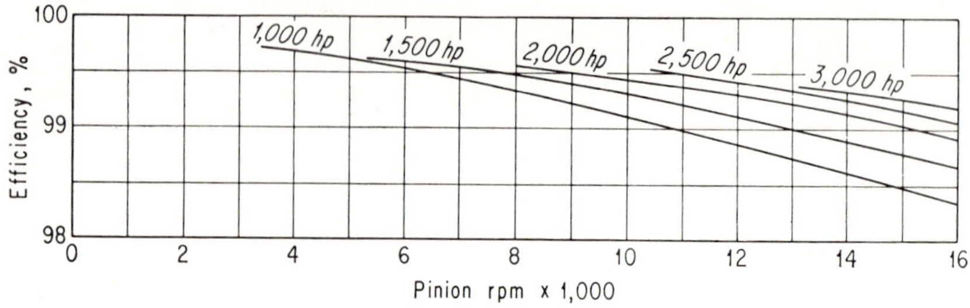

FIG. 14-34. Over-all efficiency—single-offset gear. Constant horsepower. Inlet-oil temperature 250°F; oil—MIL-L-7808C.

are not indicative of good design practice. Gear trains of this type are very efficient, particularly if high gear-tooth loading is present. Figure 14-34 shows the results of recent tests on an aircraft single-offset gearbox with rolling-contact bearings. Note from Fig. 14-34, that the gearbox is of a relatively high-speed and high-load design.

Double-reduction Gears. A double-reduction gear design is actually two single-offset gear trains in series. This being the case, the over-all gearbox efficiency is determined exactly like the single-offset gear train. Each mesh, each bearing, and each rotor is calculated for the appropriate loss. The individual losses are then totaled up and the over-all gearbox efficiency is determined. Triple-reduction gear trains, etc., are also calculated in a similar manner, as well as any combination of spur, helical, bevel, or internal gears.

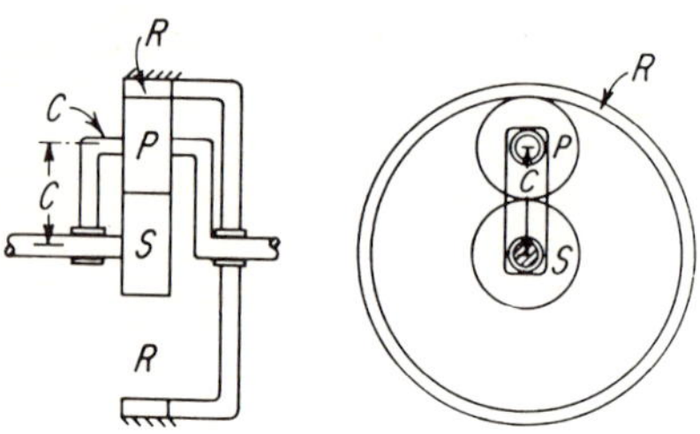

FIG. 14-35. Simple planetary drive.

Simple Planetary Drive. A simple planetary-gear system consists of three major parts; a fixed member, a driving member, and a driven member.[18,*] A common arrangement of this combination is shown in Fig. 14-35. In this train of gears, R is the fixed member, S is the driving member, and C is the driven member. In order for the cage C to function, an idler P is required in the system. The efficiency of this drive depends on the tooth loads and the velocity of the tooth arrangements. The product of tooth load and engagement velocity gives a measure of the equivalent power being transmitted by the mesh. Generally, the equivalent power being transmitted is less than the actual power being transmitted.

Actual power transmitted can be determined as follows:

$$P_a = W_t \times v \tag{14-51}$$

where W_t = actual load being transmitted

v = pitch-line velocity of actual transmitted load

* See p. 129 of reference.

$$W_t = \frac{2T}{d_S} \tag{14-52}$$

$$v = 0.2618 \times d_S \times n_S \tag{14-53}$$

where T = torque transmitted
d_S = pitch diameter of sun pinion
n_S = rpm of sun pinion

The equivalent power being transmitted through the sun gear mesh is

$$P_s = W_t \times v_S \tag{14-54}$$

where v_S = pitch-line velocity of tooth engagement of sun pinion

$$v_S = v + \frac{d_R}{d_R + d_S} \tag{14-55}$$

where v = from Eq. (14-53)
d_R = pitch diameter of ring gear
d_S = pitch diameter of sun pinion

The equivalent power being transmitted through the ring gear mesh is

$$P_R = W_t \times v_R$$

However, $v_S = v_R$ (14-56)

Therefore, $P_R = P_S$

The efficiency of the drive can be determined by calculating the per cent loss for each mesh. The power loss for each mesh is the product of the per cent loss and the equivalent power being transmitted. The over-all efficiency then is found by dividing output power by the input power or

$$E = \frac{W_t \times v - (\text{per cent loss sun} \times W_t \times v_S + \text{per cent loss ring} \times W_t \times v_R)^1\!/_{100}}{W_t \times v} \tag{14-57}$$

Note the per cent loss of the sun mesh and the per cent loss of the ring mesh can be determined by solving Eqs. (14-13) and (14-14) modified for internal gears.

To better understand the above efficiency calculations, and with reference to Fig. 14-35, a sample problem will illustrate its usage.

Given:

10 hp
Input speed, 1,000 rpm
10 diametral pitch
Sun pinion driving
$d_S = 2'' d_R = 8'' d_P = 3'' C = 3.5''$

Solution:

$$\text{Reduction ratio} = 1 + \frac{d_R}{d_S} = 1 + \frac{8}{2} = 5$$

Rpm of driven member = 1,000/5 = 200
Pitch-line velocity of transmitted load

$$v = 0.2618 \times d_S \times n_S$$
$$= 0.2618 \times 2 \times 1{,}000 = 523.6 \text{ fpm}$$

$$W_t = \frac{2 \times T}{d_S}$$

$$W_t = \frac{2 \times 630}{2} = 630 \text{ lb}$$

Actual power transmitted

$$P_a = W_t \times v$$
$$P_a = 630 \times 523.6 = 330{,}000 \text{ ft-lb per min}$$

Equivalent power transmitted by sun gear mesh

$$P_S = W_t \times v_S$$
$$v_S = v \times \frac{8}{8+2} = \frac{d_R}{d_R + d_S}$$
$$v_S = 523.6 \times \frac{8}{(8+2)} = 418 \text{ fpm}$$
$$P_S = 630 \times 418 = 263{,}000 \text{ ft-lb per min}$$

Equivalent power transmitted by ring gear mesh

$$P_R = W_t \times v_R$$
$$P_R = W_t \times v_S = P_S$$
$$P_R = 263{,}000 \text{ ft-lb per min}$$

From Eq. (14-14) and (14-14) modified, the per cent power loss for the sun gear mesh and the ring gear mesh has been determined to be 1.05 and 0.324, respectively. Then the efficiency of the over-all gearbox is

$$\sum = \frac{330{,}000 - (1.05 \times 263{,}000 + 0.324 \times 263{,}000)^{1}\!/_{100}}{330{,}000}$$
$$\sum = 98.9 \text{ per cent}$$

It is quite evident, however, from studying the problem, that

$$\frac{P_R}{P_a} = \frac{P_S}{P_a} = \frac{(m_R - 1)}{m_R} \tag{14-58}$$

where m_R = over-all ratio

Then Eq. (14-57) can be reduced to the following:

$$E = 100 - (\text{per cent loss sun} + \text{per cent loss ring})\frac{m_R - 1}{m_R} \tag{14-59}$$

Simple Star Gears. A star-gear arrangement consists of the three basic parts of the simple planetary, except that the cage C is stationary and the ring gear is allowed to rotate (see Fig. 14-36). From this diagram it is obvious that the driven ring gear R will rotate in the opposite direction from the driver S. Actually, the star gear train is simply an internal gear train with an idler P, which changes the direction of rotation of the driven ring gear R. This being the case, the over-all efficiency can be determined by first calculating the losses of mesh S to P and then

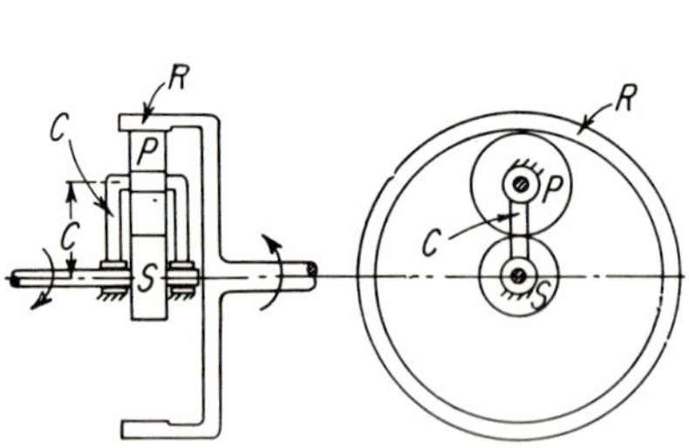

FIG. 14-36. Simple star gear.

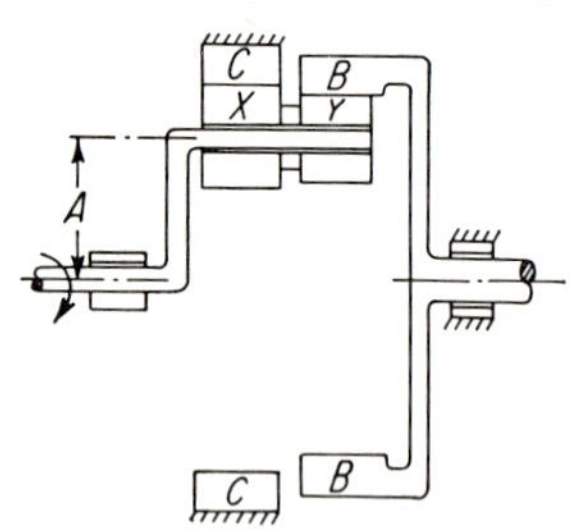

FIG. 14-37. A simple four-gear differential.

the losses of mesh P to R. It must be remembered, however, that the bearing and windage losses of the idler P should be added into the total losses only once.

Differential Gears. A simple four-gear differential gear train is shown in Fig. 14-37. A is the arm through which the input torque must act, and gear B is the output gear. Generally, in practice, the fixed gear C and the rotating gear B are nearly equal in diameter, and this, of course, also applies to both the pinions that must mesh with gears C and B. A differential gear train permits a very high speed reduction with only a few gears and in a minimum amount of space.[18,*] The amount of power that can be transmitted is somewhat limited by virtue of the high gear-tooth loads and high engagement velocities of the mesh. In order to determine the efficiency of a differential drive, the gear-tooth loads on both the fixed and rotating ring gear must be determined as well as the meshing velocities of both members.

The power being transmitted can be calculated as follows:

$$P_a = W_t \times v \tag{14-60}$$

where $W_t = \dfrac{T}{A}$ = load at center pinion, lb

v = $0.2618 \times 2A \times n$ = velocity of center of pinion, fpm

T = torque input, lb-in.

A = torque arm, in.

n = rpm of input torque arm A

The equivalent power on the fixed ring gear C can be determined as follows:

$$P_C = v_C \times W_C \tag{14-61}$$

where v_C = meshing velocity of ring gear C, fpm

W_C = load applied to ring gear C, lb

$$v_C = V \frac{d_C}{d_C - d_X} \tag{14-62}$$

$$W_C = W_t \times Y \frac{d_C - d_X}{d_C d_Y - d_B d_X} \tag{14-63}$$

where d_X, d_B, d_C and d_Y are pitch diameters of respective parts.

The equivalent power on the rotating ring gear B can be determined as follows:

$$P_B = v_B \times W_B \tag{14-64}$$

where v_B = meshing velocity of ring gear B, fpm

W_B = load applied to ring gear B, lb

$$v_B = v \frac{d_C d_Y}{d_X(d_B - d_Y)} \quad \text{fpm} \tag{14-65}$$

$$W_B = W_t d_X \frac{d_B d_Y}{d_C d_Y - d_B d_X} \quad \text{lb} \tag{14-66}$$

The efficiency of the drive can be determined by calculating the per cent loss for each mesh. The power loss for each mesh is the product of the per cent loss and the equivalent power being transmitted. The over-all efficiency is found by dividing output power by the input power or

$$E = \frac{W_t \times v - (\text{per cent loss}_C \times W_C \times v_C + \text{per cent loss}_B\ W_B v_B)^1\!/_{100}}{W \times v} \tag{14-67}$$

Note the per cent loss of mesh C and the per cent loss of mesh B can be calculated by using Eqs. (14-13) and (14-14).

* See p. 128 of reference.

A sample problem will illustrate the usage of the above equations for a given four-element differential. In reference to Fig. 14-37, the following information is given:

Output torque = 840,000 lb-in.
Pitch diameter of fixed ring d_C = 31.2″
Pitch diameter of pinion d_X = 8.8
Pitch diameter of rotating ring d_B = 29.2″
Pitch diameter of pinion d_Y = 7.6″
Diametral pitch = 2½
Output rpm = 1 rpm = n
Diameter of the torque arm A = 11.2

Solution: The over-all ratio is found by solving the following:

$$m_R = \frac{1}{1 - (d_C d_Y / d_B d_X)} \qquad (14\text{-}68)$$

Since the output speed is equal to 1, this is also equal to the input speed of torque arm A.

$$m_R = \frac{1}{1 - (31.2 \times 7.6/29.2 \times 8.8)} = 12.9515 = n$$

Velocity of the center of the pinion, fpm

$$v = 0.2618 \times 2A \times n \qquad (14\text{-}69)$$
$$v = 0.2618 \times 2 \times 11.2 \times 12.9515 = 75.937 \text{ fpm}$$

Load at the center of pinions

$$W_t = \frac{T}{A} \qquad (14\text{-}70)$$
$$W_t = \frac{840{,}000/12.9515}{11.2} = 5{,}791 \text{ lb}$$

Input power—or power being transmitted

$$P_a = W_t \times v \qquad (14\text{-}71)$$
$$P_a = 5{,}791 \times 75.937 = 439{,}753 \text{ ft-lb per min}$$

Meshing velocity of fixed ring gear C

$$v_C = v\frac{d_C}{d_C - d_X} \qquad (14\text{-}72)$$
$$v_C = 75.937\frac{31.2}{31.2 - 8.8} = 105.769 \text{ fpm}$$

Load applied to ring gear C

$$W_C = W_t d_Y \frac{d_C - d_X}{D_c d_Y - d_B d_X} \qquad (14\text{-}73)$$
$$W_C = 5{,}791 \times 7.6 \frac{31.2 - 8.8}{31.2 \times 7.6 - 29.2 \times 8.8} = 49{,}691 \text{ lb}$$

The equivalent power on the fixed ring gear C

$$P_a = W_C v_C \qquad (14\text{-}74)$$
$$P_a = 105.769 \times 49{,}691 = 5{,}256{,}000 \text{ ft-lb per min}$$

Meshing velocity of rotating ring gear B

$$v_B = v \frac{d_C d_Y}{d_X(d_B - d_Y)} \tag{14-75}$$

$$v_B = 75.937 \frac{31.2 \times 7.6}{8.8(29.2 - 7.6)} = 94.729 \text{ fpm}$$

Load applied to rotating ring gear B

$$W_B = W_t \times d_X \frac{d_B - d_Y}{d_C d_Y - d_B d_X} \tag{14-76}$$

$$W_B = 5{,}791 \times 8.8 \frac{29.2 - 7.6}{31.2 \times 7.6 - 29.2 \times 8.8} = 55{,}481 \text{ lb}$$

The equivalent power on rotating ring gear B

$$P_B = W_B \times v_B \tag{14-64}$$

$$P_B = 94.729 \times 55{,}481 = 5{,}256{,}000$$

From Eq. (14-14) modified, the per cent power loss for ring gear mesh C and for ring gear mesh B has been calculated to be 0.503 and 0.601 per cent, respectively. The efficiency for the over-all gear train is

$$E = \frac{439{,}753 - (0.503 \times 5{,}256{,}000 + 0.601 \times 5{,}256{,}000)\frac{1}{100}}{439{,}753}$$

$$E = 86.8 \text{ per cent}$$

It is quite evident, however, from studying the problem that

$$\frac{P_C}{P_a} = \frac{P_B}{P_a} = \frac{m_R - 1}{1} \tag{14-77}$$

where R = over-all differential ratio

Equation (14-67) can be reduced to the following:

$$E = 100 - (\text{per cent loss}_C + \text{per cent loss}_B)(m_R - 1) \tag{14-78}$$

An equation of this nature is quite helpful in determining the over-all efficiency of a given design. It can also be used as a check on the more detailed calculations that must be carried on when a device such as this is designed for a particular application.

Often on a design problem of this nature, a quick approximation of the efficiency which could be determined quite quickly on a slide rule is good enough. An approximate equation for the efficiency of a four-element differential is as shown:[19]

$$E = \frac{1}{1 + \frac{1}{5}[(1/N_X) + (1/N_Y) - (1/N_C) - (1/N_B)](m_R - 1)} \tag{14-79}$$

where R = over-all ratio of train

N_X = number of teeth on gear X

N_Y = number teeth on gear Y

N_C = number teeth on gear C

N_B = number teeth on gear B

The efficiency calculated in this manner for the above-mentioned sample problem would give an approximate efficiency of

$$E = \frac{1}{1 + \frac{1}{5}(\frac{1}{22} + \frac{1}{19} - \frac{1}{78} - \frac{1}{73})(12.9515 - 1)}$$

$$= 85.4 \text{ per cent approximately}$$

This approximation when compared with the efficiency obtained from Eq. (14-67) or (14-78) shows good correlations.

DYNAMIC LOADS ON GEAR TEETH

When gears are run on a test stand or in the field, they are subject to all the forces that can be applied to rotating equipment. In addition to these forces, gears are unique in that they are used in pairs or trains to transmit power and motion, and geometric errors in the gears themselves can cause considerable overloading while the gears are in operation.

14-5. Dynamic Loads on Spur Gears. When gears are operating under dynamic conditions, there are many factors that affect their over-all operation and performance.[20] These many factors affect the loading conditions on a given gear mesh by combining and producing a critical error in action which in the final analysis may produce gear-tooth overloads in excess of safe operational limits. The dynamic load is affected by such items as:

Elasticity of the gear materials as well as the gear mounting
Torsional deflection and beam bending of the gear shafting
Gear-tooth loading and deflection
Gear-tooth spacing and profile errors
Speed and pitch-line runout of the rotating bodies
Mounting arrangement and alignment
The dynamic balance of rotating elements
Temperature differentials throughout the gearbox
The mass of the gears and shafting
The mass of the connected bodies

The critical error in action produced and maintained by a combination of the above factors is similar in nature to the total error noted on a "red-liner" chart of a gear meshing with a "master" gear. The "red-liner" chart depicts the error in action of a production gear meshing with a very accurate "master" gear with little or no amount of error. Generally, the error shown on a "red-liner" chart is considered to be the result of various manufacturing errors of the production gear. The critical error in action along with the natural vibrations of the geared system cause sudden accelerations of the gears, followed by impacts when the mating gear teeth come back into mesh. This intermittent gear-tooth action sets up forced vibrations in the gear and pinion and does not permit smooth continuous action as would generally be expected from a constant applied load.

Considerable work has been done by various investigations on dynamic loads of gear teeth such as Buckingham, 1931,[20] Buckingham, 1949,[3] Tuplin, 1950,[21] Tuplin, 1953,[22] Strauch, 1953,[23] Reswick, 1954,[24] Attia, 1956,[25] Niemann and Rettig, 1958,[26] Zeman, 1957,[27] Harris, 1957,[28] and Kohler, 1959.[29] In addition, there has been unpublished work done by Caterpillar Tractor Co., the General Electric Company, and undoubtedly other concerns. These many investigators, to say the least, are not in full agreement on how to evaluate the dynamic-load effects. Their work points to the fact that no one best method for determining dynamic loads has been found, and the gear designer is left pretty much on his own. Actually, the existence of dynamic tooth loadings has been known and taken into consideration for some time, except that it has been on a less formal basis. Designers have had to rely on past experience, reduced allowable-stress levels, service safety factors, experimental values applicable to narrow specific fields, etc., in order to put their designs into successful operation. Upon occasion, even the more experienced gear designer finds himself in trouble because he has not taken into consideration the dynamic-loading effect of gears in action.

14-6. Brief Summary of Dynamic-load Investigation. In order for one to take advantage of what has been accomplished to date in the field of dynamic loads, a quick look at the results of this melting pot is beneficial. Zeman[27] * has done a fair job of summing this up from a practical point of view. Figure 14-38 shows a curve taken from Zeman's paper which has the dynamic load plotted against peripheral

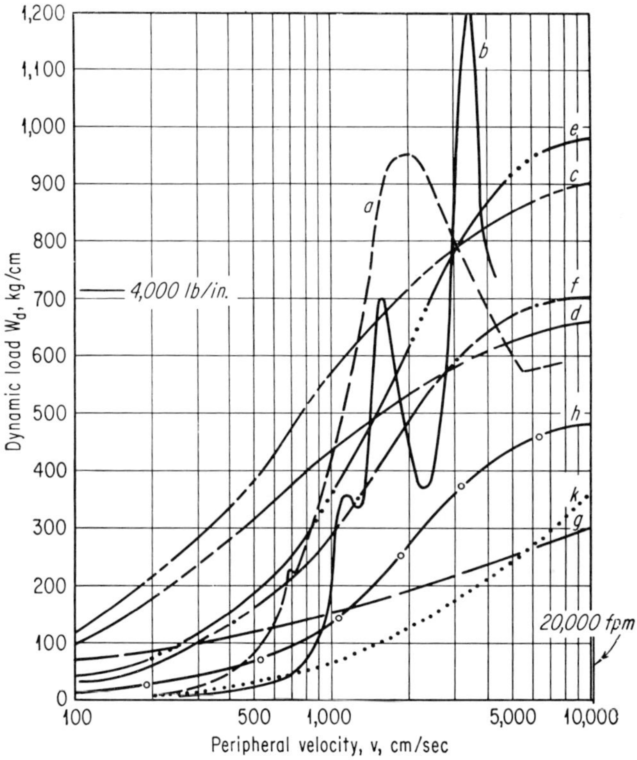

FIG. 14-38. Dynamic-load comparisons. (*a*) Zeman's formula—single disturbance. (*b*) Zeman's formula—continuous error. (*c*) Buckingham's old formulas, 420 kg. (*d*) Buckingham's old formulas, 120 kg. (*e*) Buckingham's new formulas, 420 kg. (*f*) Buckingham's new formulas, 120 kg. (*g*) Niemann's new formula. (*h*) Niemann's old formula. (*k*) Tuplin's data.

speed for a given set of spur gears. He has calculated the dynamic load using his own method as well as the methods put forth by other investigators for the gear data given in Table 14-5.

Table 14-5

Ratio	1:1
No. of teeth	30
Pitch, diametral	2.54
Center distance	11.8″
Error, magnitude (peak to peak)	0.002″
Material	Steel
Load	420 and 120 kg (921 and 263 lb)

* See pp. 244–254 of reference.

One thing is quite obvious. All the curves follow the general trend of increasing dynamic load as the peripheral speed is increased. Using the word loosely, there seems to be some correlation in the results of the various investigators. Most of the results also indicate that a maximum value of dynamic load is reached and then the curves start to level out. Zeman's results[27] seem to peak faster than the rest; however, his maximum values are somewhat in line with Buckingham's. Both Niemann's and Tuplin's results are somewhat lower than Buckingham's. Data from Harris's investigation[28] are not presented in Fig. 14-38. However, Harris states, based on test results, that perhaps the maximum load the gear teeth will see is twice the load that gives zero error in velocity ratio. The load that gives zero error in velocity ratio is the design load with the gear teeth correctly modified. Kohler states[29] in the conclusion of his thesis that his test results revealed, irrespective of error, that the dynamic load never exceeded three times the design load. Again, in reference to Fig. 14-38, the top curve of Buckingham's results shows that the dynamic load is 2.38 times the design load. To further back up this general trend of Buckingham's work, Table 14-6 gives the results of many investigations using the Buckingham method of calculation.

Table 14-6

Design load W	Dynamic load Wd	Ratio Wd/W
4,160	9,770	2.35
5,270	12,195	2.31
625	1,567	2.52
725	1,465	2.02
2,080	4,580	2.20
2,939	8,326	2.84

From a practical point of view, it makes little difference which theory is the proper one, as long as the results obtained are correct, and if they are not correct, they should be at least on the conservative side.

Strictly from a design point of view, the author feels that the Buckingham theory of dynamic loads is the proper approach to use, considering the present state of the art. In reference to Fig. 14-38, it is somewhat on the conservative side. In comparison with Harris and Kohler, loaded gear trains calculated by the Buckingham method seem to take a middle-of-the-road trend. Professor Buckingham was consulted personally, and the results of several lengthy discussions indicated that he felt his method would produce dynamic loads somewhat on the high or conservative side. Although undoubtedly new theories and more physical investigations may reveal that Buckingham's approach is outdated, it seems at the present the most logical and most direct method of assuring the customer that the gear trains being designed, tested, and shipped will work as specified in service.

At this point it must be pointed out that, in special cases, when the design load is quite small compared with the mass effect of the gears and pinions, the connected shafts, as well as the connected bodies, the ratio of calculated dynamic load to design load is much higher[30]—in the range of 8.

The author has some practical experience that will verify this high ratio of dynamic load to design load. For example, in applications where very small amounts of power are being transmitted from a high-speed high-horsepower source (Fig. 14-41, a governor drive), gears often fail from dynamic-load effect. Generally, from a torque-transmitted basis, the gears are very lightly loaded; in fact, often the face widths are sized more from a practical manufacturing point of view instead of the torque being transmitted. When gears of this nature fail, the design engineer is hard pressed for technical reasons to explain the failure. When the problem is analyzed from a dynamic-load-effect approach, usually ratios of 6 to 10 are apparent. This readily explains the premature gear failure.

14-7. When Dynamic Loads Should Be Considered. Actually the dynamic load should be determined for each and every new gear application. This way the design engineer will have a backlog of information with which to compare a given dynamic-load calculation. If a diligent procedure is set up whereby the apparatus is calculated in the design stage and observed in service in the field, and maintenance and failure records are carefully analyzed, the dynamic-load effect will soon have as much meaning for the average gear designer as does the transmitted load.

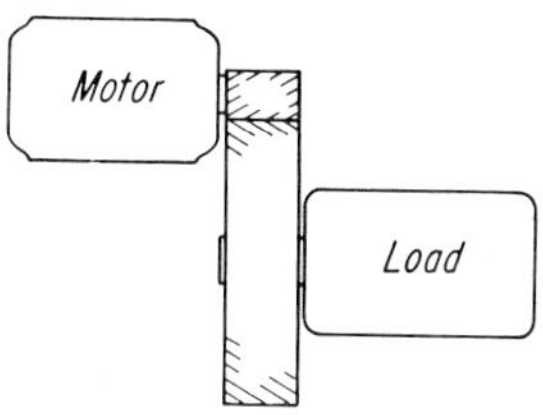

FIG. 14-39. Close-coupled stiff shafting.

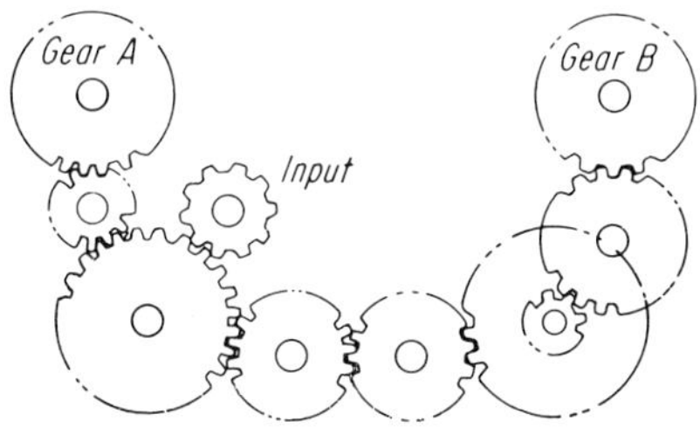

FIG. 14-40. Accessory gearbox.

There are, of course, many well-established applications where it is not necessary that dynamic-load calculations be made, since past experience has already established the amount of dynamic load present. There are, however, several types of gear arrangements where it is quite important that very careful consideration be given to the dynamic-load effects during the design phase. In these arrangements, the omission of a dynamic-load evaluation would be termed "poor design practice."

Figure 14-39 shows a very close coupled gear train. In this arrangement the shafts that join the mass of the rotor to the pinion and the mass of the load to the gear are very heavy and stiff; consequently, a considerable amount of this connected mass will influence the gear mesh. A dynamic-load calculation will determine the magnitude of this mass effect as well as perhaps point out ways in which this effect could be reduced.

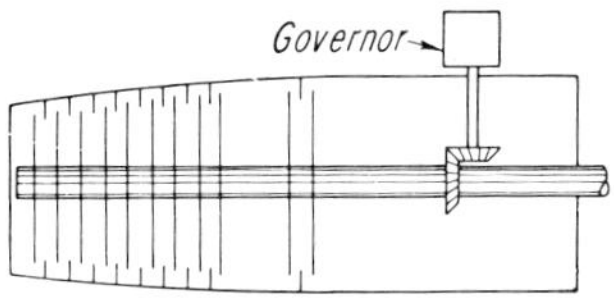

FIG. 14-41. Power take-off, light loads.

Figure 14-40 shows a gear train of an aircraft accessory gearbox. Because of the arrangement, the mass effect of each and every gear has an additive effect on the last gear in the train. If the design of this box is not carefully analyzed for dynamic-load effects, the whole arrangement could fail on test, particularly the two end gears A and B.

When very light loads are driven from a high-speed transmission as outlined in Fig. 14-41, the gearing is subjected to some of the worst dynamic-load conditions. Generally the high-speed power source has its own vibration characteristics because of its speed and relative high mass. The take-off gears are quite small because of the very light transmitted load. Usually, when this arrangement is calculated, the dynamic load is considerably larger than the rather small transmitted load, perhaps something in the order of 8 to 10.

Often it is desirable to have a multi-input drive whereby the prime movers feed into a gear train which may drive several loads but in the final analysis are cross-linked to permit operation or sharing of the load while one prime mover is down for repairs. Such an arrangement is shown in Fig. 14-42. From a dynamic-load point of view, this arrangement is quite risky and should be given every consideration.

Another very tough and difficult problem is that of a high-speed quill drive where, for example (see Fig. 14-43), a high-pressure lube pump is driven through a quill from a relatively high-speed high-power gear train. The quill is intended to compensate for misalignment due to temperature effects and the pinion shifting due to applied load. In many applications, the quill will lock up on one end and the other

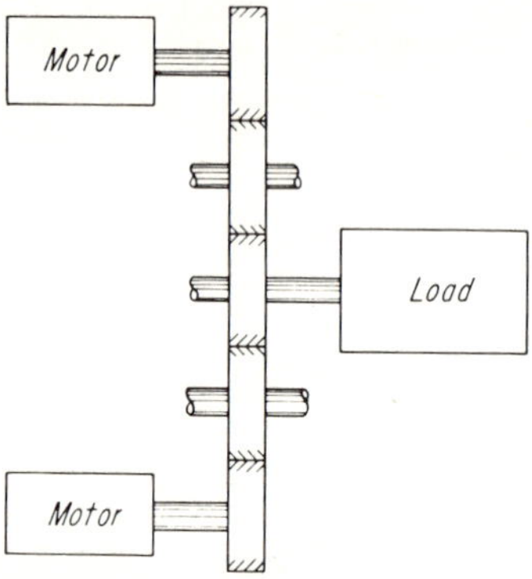

FIG. 14-42. Multiple power inputs.

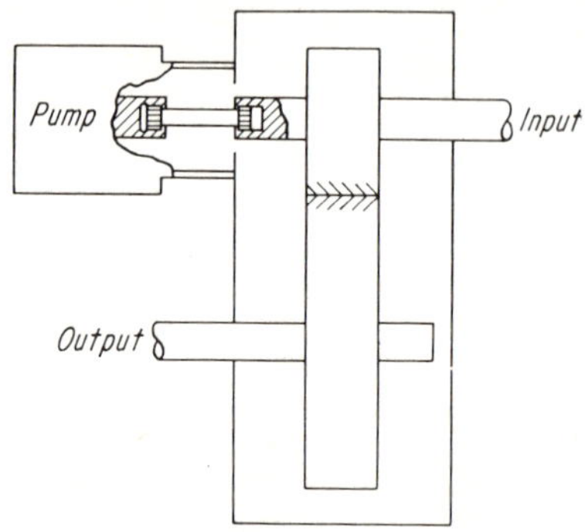

FIG. 14-43. High-speed quill drives.

end will take a beating because of a combination of mass effects of the main drive gears and wear due to misalignment.

14-8. Buckingham's[3,]* Solution to the Dynamic-load Effect on Gear Teeth. The load that causes the driven gear to move forward or ahead to make up for a distance lost due to error in a preceding mesh is known as the "acceleration load." The resultant impact when the teeth come together again in the mesh is known as the "dynamic load." Actually, there are two load surges for every tooth engagement; the "acceleration load," which is set up by the first phase of engagement, followed by the impact or "dynamic load," which is really the reaction. The impact load or "dynamic load" is always the maximum load value of the tooth engagement.

The acceleration load is a function of f_1, the force needed to accelerate rigid bodies through the critical error, and f_2, the force required to do the work of deforming the tooth by elastic deformation equal to the amount of the critical mesh error. The acceleration load f_a then is the resultant of the two where each does its part during the meshing cycle:

$$\frac{1}{f_a} = \frac{1}{f_1} + \frac{1}{f_2} \tag{14-80}$$

The dynamic load is then expressed as a function of the acceleration load f_a and f_2:

$$W_d = W_t + \sqrt{f_a(2f_2 - f_a)} \tag{14-81}$$

where W_d = dynamic load, lb
W_t = transmitted load, lb
f_a = acceleration load, lb
f_2 = force required to do the work of elastic deformation, lb

* See pp. 426–452 of reference.

In order to make a dynamic-load calculation using the Buckingham method for a given gearset, some quantities must be known and some preliminary calculation must be made:

M_p = pinion mass effect at pitch radius of pinion, lb-sec² per ft
M_{a1} = mass effect of bodies connected to pinion at pitch radius of pinion, lb-sec² per ft
M_G = gear mass effect at radius of gear, lb-sec² per ft
M_{a2} = mass effect of bodies connected to gear at pitch radius of gear, lb-sec² per ft
Z_1 = elasticity of pinion shafting, lb per in.
Z_2 = elasticity of gear shafting, lb per in.
e = effective error assumed or known, in.
E_1 = modulus of elasticity of pinion
E_2 = modulus of elasticity of gear

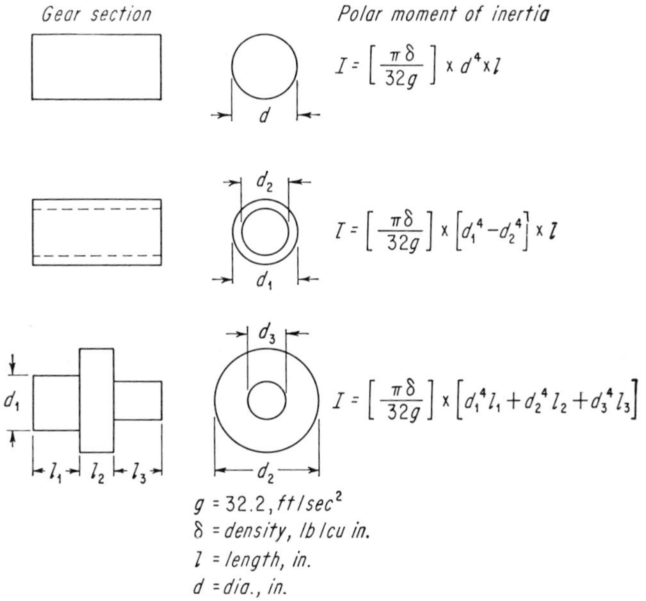

$$I = \left[\frac{\pi\delta}{32g}\right] \times d^4 \times l$$

$$I = \left[\frac{\pi\delta}{32g}\right] \times \left[d_1^4 - d_2^4\right] \times l$$

$$I = \left[\frac{\pi\delta}{32g}\right] \times \left[d_1^4 l_1 + d_2^4 l_2 + d_3^4 l_3\right]$$

FIG. 14-44. Method used to calculate polar moment of inertia.

The above items can be determined in the following manner. To simplify the determination of moments of inertia for various shapes of gears, refer to Fig. 14-44.

$$M_p = \frac{I}{R_1^2} \tag{14-82}$$

where I = polar moment of inertia of pinion, lb-in.² sec² per ft
R_1 = pitch radius of pinion, in.

$$M_{a1} = \frac{I}{R_1^2} \tag{14-83}$$

where I = polar moment of inertia of connected bodies to pinion shaft, lb-in.² sec² per ft
R_1 = pitch radius of pinion, in.

$$M_G = \frac{I}{R_2^2} \tag{14-84}$$

where I = polar moment of inertia of gear, lb-in.^{2} sec^2 per ft
R_2 = pitch radius of gear

$$M_{a2} = \frac{I}{R_2^2} \tag{14-85}$$

where I = polar moment of inertia of connected bodies to gear shaft, lb-in.^{2} sec^2 per ft
R_2 = pitch radius of gear, in.

$$Z_1 = \frac{K_1}{R_1^2} \tag{14-86}$$

where K_1 = stiffness of shaft—pinion, lb-in.
R_1 = pitch radius of pinion, in.

$$Z_2 = \frac{K_2}{R_2^2} \tag{14-87}$$

where K_2 = stiffness of shaft—gear, lb-in.
R_2 = pitch radius of gear, in.

$$K_1 \text{ or } K_2 = \frac{\pi G}{32} \times \frac{1}{\Sigma l/(d_1^4 - d_2^4)} \tag{14-88}$$

where G = shear modulus = 11.8×10^6 for steel
l = any length, in.
d_1 = outside diameter, any shaft
d_2 = inside diameter, any shaft

In high-speed apparatus, there is a limited time in which the meshing error is permitted to act. When this critical speed is reached, the dynamic load will maximize and then decrease from this point with increasing speeds. Often it is convenient to check this critical speed for your particular application to see if your design permits time for the full error in action to take place.

$$n_c = \frac{30}{N_P}\sqrt{\frac{6W_t}{eM}} \tag{14-89}$$

where n_c = rpm of pinion where dynamic load is maximum
N_P = number of teeth in pinion
W_t = transmittal load, lb
e = error in action, in.
M = mass effect at pitch line of gears

If N_c is smaller than the design speed of the apparatus under question, then the dynamic load should be determined with a new "effective error" that is less than the original assumed error. The effective error can be determined as follows:

$$e^1 = \frac{6Wt^2}{M} \tag{14-90}$$

where e^1 = effective error
W = transmittal load
M = mass effect at pitch line of gears
t = time, sec, $= \dfrac{30}{nN_1}$
N_P = number of teeth in pinion
n = rpm of pinion

Table 14-7. Calculation of Dynamic Load by Dudley Method

1. Design hp	865 given
2. No. of pinion teeth N_P	37 given
3. No. of gear teeth N_G	309 given
4. Pitch dia. of pinion d_P	3.700 given
5. Pitch dia. of gear D_G	30.900 given
6. Face width F	$8\frac{1}{8}$ given
7. Pinion rpm n_P	10,022 given
8. Pitch-line velocity, 0.2618 × (4) × (7)	9,707.9
9. (8) × (8)	94.244×10^6
10. Pinion torque = 63,000 × (1) ÷ (7)	5,437.5
11. tan driving load W_t = 2.00 × (10) ÷ (4)	2,939.2
12. Helix angle ψ	23° given
13. cos (12) × cos (12)	0.84732
14. Constant C_1 (read Fig. 14-45)	9.094
15. Constant C_2 (read Fig. 14-45)	0.00113
16. Mass effect of pinion M_P	0.550 [Eq. (14-82)] given
17. Mass effect of body connected to pinion M_{a1}	117.3 [Eq. (14-83)] given
18. Mass effect of gear M_g	11.000 [Eq. (14-84)] given
19. Mass effect of body connected to gear M_{a2}	8.9042 [Eq. (14-85)] given
20. Elasticity of pinion shaft z_1	0.20×10^6 [Eq. (14-86)] given
21. Elasticity of gear shaft z_2	9.3×10^4 [Eq. (14-87)] given
22. Effective error e	0.00025 (assume value)
23. Modulus of elasticity of pinion E_1	30×10^6
24. Modulus of elasticity of gear E_2	30×10^6
25. 1.0 ÷ (4) + 1.0 ÷ (5)	0.302632
26. 2.0 × (15) × (25)	0.000684
27. (26) × (17) × (9) × (13)	6,406,992
28. [(16) + (M_2)] (27)	74,000,758 [assume M_2 = (18)]
29. M_2 × (22) × (20)	550.0 [assume M_2 = (18)]
30. (28) + (29)	74,001,308
31. M_2 × (22) × (17) × (20)	64,515 [assume M_2 = (18)]
32. $\sqrt{(30)^2 + 4.0(27) \times (31)}$	74,012,479
33. (32) − (30)	11,171
34. (33) ÷ 2.0(27)	0.000871
35. (16) + (34)	0.550871
36. (26) × (19) × (9) × (13)	486,316
37. (18) + (35) × (36)	5,617,373
38. (35) × (22) × (21)	12.8077
39. (37) + (38)	5,617,386
40. (22) × (19) × (35) × (21)	114.0180
41. $\sqrt{(39)^2 + 4.0(36) \times (40)}$	5,617,406
42. (41) − (39)	20
43. (42) ÷ 2.0 × (36)	0.000020
44. M_2 = (18) + (43)	11.000020

Check how close $M_2 = M_g$; (44) = (18). If necessary substitute calculated M_2 back into (28), (29), and (31) and recalculate until assumed and calculated values of M_2 check to first five places.

45. (35) × (44)	6.059592
46. (35) + (44)	11.550891
47. (45) ÷ (46)	0.524599
48. (47) × (26) × (9) × (13)	28,668
49. 1.0 ÷ (23) + 1.0 ÷ (24)	666.7×10^{-10}
50. (49) × (14)	0.606292×10^{-6}
51. (50) × (11) ÷ (6)	0.000219
52. (51) ÷ (13)	0.000258
53. (22) ÷ (52) + 1.0	1.968992
54. (11) × (53)	5,787.2613
55. (48) × (54)	165.90921×10^6
56. (48) + (54)	34,455.26
57. (55) ÷ (56)	4,815.9057
58. 2.0 × (54) − (57)	6,758.6169
59. (57) × (58)	32.54886×10^6

Table 14-7. Calculation of Dynamic Load by Dudley Method *(Continued)*

60. $\sqrt{(59)}$	5,705.161
61. $W_d = (11) + (60)$	8,644

Item 61 is the dynamic load which will occur unless the gear is running so fast that the full reaction cannot occur.

62. $5{,}400 \times (11)$	15.8717×10^6
63. $(7) \times (7) \times (2) \times (2) \times (47)$	7.1818×10^{10}
64. $e = (62) \div (63)$	0.000221
65. $(64) \div (52) + 1.0$	1.856589
66. $(65) \times (11)$	5,456.886
67. $(48) \times (66)$	156.4380×10^6
68. $(48) + (66)$	34,124.886
69. $(67) \div (68)$	4,584.2791
70. $2.0 \times (66) - (69)$	6,329.4928
71. $(69) + (70)$	29.01616×10^6
72. $\sqrt{(71)}$	5,386.67
73. $W_d = (11) + (72)$	8,326

Item 73 is the maximum dynamic load that can occur in the time while a tooth goes through mesh. Item 64 is the maximum error that can be effective in creating dynamic load. Use whichever is smaller, item 61 or item 73, as the calculated dynamic load.

Once this information has been collected, the actual calculation of the dynamic load can get under way. Space will not permit the repeat of the many and various steps in making a calculation as outlined by Buckingham;[3] however, Dudley[1,*] in "Practical Gear Design" has a short-cut procedure for making the final dynamic-load calculation for a single pinion-and-gear combination, either spur or helical. Table 14-7 gives this method step by step. In order to use the Dudley[1] short-cut method, the deflection constants C_1 and the acceleration constants C_2 found in Fig. 14-45 are required.

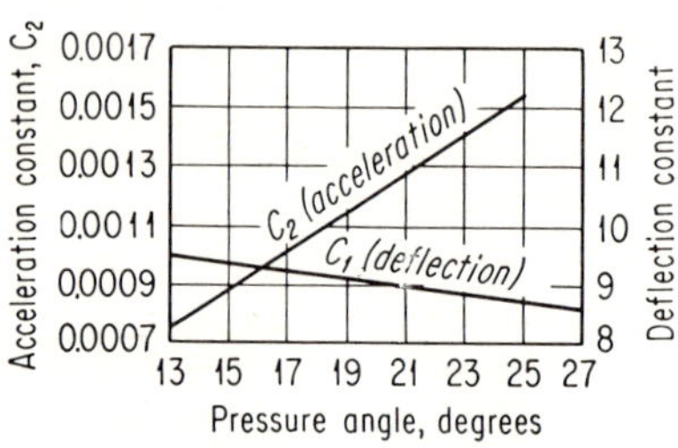

FIG. 14-45. Constant curves.

14-9. Dynamic-load Effect on Gear Trains. The analysis of trains of gears is very similar to that of a single pair of gears, except that, when a given mesh is under consideration, the mass effect of the other gears in the train must be taken into account.

Simple Train. To better understand how a simple train of gears is handled, refer to the diagram of Fig. 14-46.

Let I_1 = polar moment of inertia of driving member—pinion
I_2 = polar moment of inertia of idler
I_3 = polar moment of inertia of driven gear

For mesh—pinion to idler,

$$M_p = \frac{I_1}{R_1^2} \tag{14-91}$$

$$M_G = \frac{I_2}{R_2^2} + \frac{I_3}{R_3^2} \tag{14-92}$$

For mesh—idler to gear,

$$M_p = \frac{I_1}{R_1^2} + \frac{I_2}{R_2^2} \tag{14-93}$$

$$M_G = \frac{I_3}{R_3^2} \tag{14-94}$$

* See pp. 329–331 of reference.

The dynamic load would then be determined for each mesh by substituting the appropriate values of M_G and M_P into the calculation sheet of Table 14-7.

Compound Trains. An example of a compound train[30] is shown in Fig. 14-47.

Let I_1 = polar moment of inertia of driving pinion
$I_{2,3}$ = polar moment of inertia of cluster gear
I_4 = polar moment of inertia of driven gear

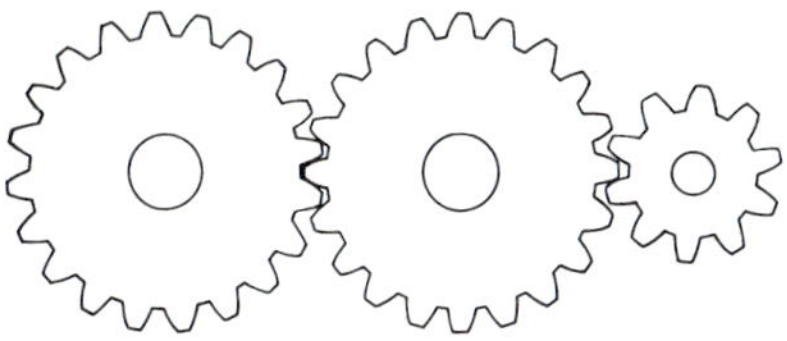

FIG. 14-46. Simple train.

FIG. 14-47. Compound train.

For mesh—pinion to cluster gear,

$$M_p = \frac{I_1}{R_1^2} \tag{14-95}$$

$$M_G = \frac{I_{2,3}}{R_2^2} + \frac{R_3^2}{R_2^2} \times \frac{I_4}{R_4^2} \tag{14-96}$$

For mesh—cluster gear to driven gear,

$$M_p = \frac{I_{2,3}}{R_3^2} + \frac{R_2^2}{R_3^2} \times \frac{I_1}{R_1^2} \tag{14-97}$$

$$M_G = \frac{I_4}{R_4^2} \tag{14-98}$$

This same procedure can be used for more complex gear trains. In the final analysis, however, each mesh must be calculated using the appropriate values of M_G and M_P in the calculation sheet of Table 14-7.

GEAR PERFORMANCE

Gear trains or gear drives to be considered as satisfactory must meet a standard of performance as interpreted by the ultimate customer. Actually the desired performance of a given gear drive depends to a large measure on the end use of the apparatus. A good-performance gear means different things to different customers. For example, a good-performance gear to a marine engineer means extreme reliability. He envisions the difficulty of cutting through the decks and compartments of ships to replace faulty gear equipment. To a submarine captain, a good-performance gear should be ultraquiet. This would give a measure of protection against detection by the enemy. A fighter pilot wants a high horsepower-to-weight ratio to give him a speed advantage, yet also he wants a degree of reliability to prevent catastrophic failures. A controls engineer wants extreme accuracy in transmitted motion, since a few thousandths in backlash may cause him to miss the target area.

14-10. Factors That Affect the Performance of Gear Units. A good-performance gear should be described as one that operates as well as or even better than the prediction of the design engineer. From an engineering point of view, however, there are several basic fundamental criteria for judging equipment of this nature such as the efficiency of the geared drive, the reliability as compared with the design life, or the size and weight of the drive vs. the horsepower transmitted, and the over-all job that the gear must do. Table 14-8 shows a very interesting and revealing study of various types of gear drives. It must be remembered, however, that this table must be used only as a rough guide for ball-park comparisons.

Table 14-8. Relative Performance of Gear Drives

Type of gear transmission	Relative wt. per total job done	Hp per lb	Input speed	Over-all ratio
Marine	105	0.075	6,500	64
Navy	36	0.5	8,000	35
Industrial	160	1.0	8,000	4
Industrial (lightweight)	32	2.0	19,000	16
Railway	74	0.875	2,100	4.1
Aircraft (power)	16	12–18	17,000	13
Rockets and missile	12	25–50	30,000	5

From a classical point of view, perhaps the ultimate in gear performance should be synonymous with the ultimate in gear capacity. The best-performing gear drive would then be the drive that could carry the most horsepower with the least weight. This being the case, in reference to Table 14-8, the aircraft and rocket geared systems would be way out front. There are several factors other than the transmitted load and the speed of operation that affect the over-all performance of gear applications such as:

Power Loss. The amount of heat being dissipated within the gearbox is a good indication of how the over-all gearbox is operating. This known heat loss can then be compared with the calculated power loss of the unit under consideration.

The calculated power loss should be greater than the measured power loss. If this is not the case, then the gear unit must be carefully investigated. Often when there is an excessive heat loss within the gearbox, the unit is not scavenging properly and the design may be of the questionable type, as indicated by Figs. 14-27, 14-28, and 14-29. Oil* that is not immediately removed from high-speed gear units will store up and cause considerable heat loss because of churning of the oil. In many cases, improper scavenging will cause a hydraulic-braking effect, which sometimes will overload the gear unit to the point of failure. At times, hot bearings will cause an increase in the temperature differential across the gear casing, and often this is an indication of incipient failure. When a gear unit is first started up, however, it is difficult to tell if the bearings are running hot or not, unless the unit is fully instrumented. If the unit has been previously run, then it is quite obvious when the temperature differential increases.

Vibrations. In order for a gearbox to operate satisfactorily, it must run within safe vibration limits. Normally, if the vibrations of a given gearbox run more than 5 mils under steady-state conditions, you should suspect that something is wrong. Several things can cause undesirable vibrations in a gearbox. Dynamic balance of the rotating elements should be first on the list of suspects. All rotating elements should be checked balanced as soon as trouble is encountered. Next, consideration should be given to connecting bodies, such as couplings and shafting. All shafting should be rechecked for critical-speed indications; this includes the gearing within the gearbox as well as the outside shafting. Often a flexible coupling will stick in an unbalanced position and then cause considerable vibration when the unit is up to speed. Runout on all bearing surfaces and the pitch line of the gears should be checked to see if they are within drawing limits. Sometimes the face of the coupling

* See Chap. 15 for detail recommendations on gear-lubrication systems.

flanges will be out of square with respect to the rotating axis, causing bolted-on equipment to run somewhat out of balance. Profile errors of the gear-tooth elements often can cause the mesh to run rough, which in turn can set up bad vibrations. Sometimes a bad rolling-contact bearing which has been damaged at assembly can cause a gear unit to vibrate and, of course, this usually leads to premature failures. Another possible cause of rough running is a bent or damaged gear tooth. This sometimes happens at assembly and it is usually quite difficult to detect unless the gears are pin-checked.

Alignment and Mounting.* Proper or improper alignment can affect the operation of a gearbox in several ways.[31] For example, if the external alignment of the gearbox to the connected load is not made carefully, the external coupling may fail because of excessive heat generation in the coupling engagement. If this happens, usually the coupling becomes quite "sticky" and then transmits bending moments back into the gearbox. If the unit continues to run, the shaft extension to which the coupling is attached often fails back near the first firm support area, which is usually the bearing just within the gearbox. Shafts have been known to fail in this manner where the shaft-extension loading normally would be considered quite conservative. Internal alignment is also very important. The mounting arrangement plays an important part in determining the degree of alignment which will ultimately exist within the gearbox. Sometimes at manufacturing, the casing bores are not machined true with respect to each other or parallel with respect to the set of bores in which the mating gear must run. This obvious mismatch of bores causes alignment problems in the mesh which are often compounded by additional misalignment due to internal clearance in the bearings. The alignment of the mesh is also affected by shaft deflections and gear blank deflections, as well as casing distortions. If the pinion is wide-faced with a high length-to-diameter ratio, torsional windup of the pinion can also contribute to the alignment condition of the mesh. It makes little difference what causes the misalignment in the mesh. The net result is a load concentration on one end or the other of the mesh, which usually results in premature failures of the gear or pinion teeth by pitting or scoring or breakage, or a combination of all three. Generally, misalignment conditions within a gearbox must be eliminated if at all possible. If elimination is not possible, it must be corrected for by such means as crowning, dubbing off the ends of the teeth, or helix-angle correction. If the mismatch is such that the ends of the teeth are being forced into tight mesh, perhaps additional backlash is required as well as some generous profile modification.

Lubrication. In order for a gearbox to operate at all, it must be supplied with an adequate lubricant. The lubricant permits the bearings, as well as the gear mesh, to operate with a minimum of power loss.[32] In addition to the job of lubricating these rubbing surfaces to cut down on friction, the lubricant usually acts as a cooling medium to keep the rubbing parts within satisfactory temperature limits.[33] Often in high-load high-speed applications, the lubricant also acts as a carrying medium for antiwear and antiscoring agents. On wide-face-width gearing, it is very important that the distribution of the oil across the mesh is uniform to eliminate any uneven heating effects. The lubricant also helps in eliminating temperature differential throughout the gear-casing assembly. It tends to wash out high-temperature areas and heat up low-temperature areas.

Normally, oil is thought of as a necessary and beneficial fluid; however, at times it can be very troublesome and detrimental. If oil is allowed to fill up within the gearbox, it will cause excessive power loss and ultimate failure due to the braking effect. Because oil is a fluid, it often causes undue concern with respect to seal leakage and proper ventilation of the gearbox. If the complete lubrication system has not been

* See Chap. 12 for detail recommendations on gear-mounting practice.

carefully designed with due regard to oil turnover and proper settling-out time, considerable trouble can be encountered with respect to oil foaming. Some antifoaming agents can be utilized in the oil; however, a healthy solution to the problem is to allow plenty of oil for the over-all gear box with adequate scavenge capacity and allow the foam to settle out in the storage tank. Oil foaming can also be a problem at reduced pressures.[35] This trouble is encountered particularly at high altitudes, and it makes the use of a circulation oil system very difficult (see Fig. 14-48).[35] At high-altitude operation where a recirculation system is required, design engineers have had to pressurize the complete lube system. Because the gearbox depends on the lube oil for its very existence, the supply of oil must not be interrupted. Often a catastrophic gear failure can be traced back to a lube-pump failure or a lube-line break or even to plain carelessness on the part of an operator in failing to turn on the lube system in the proper manner.

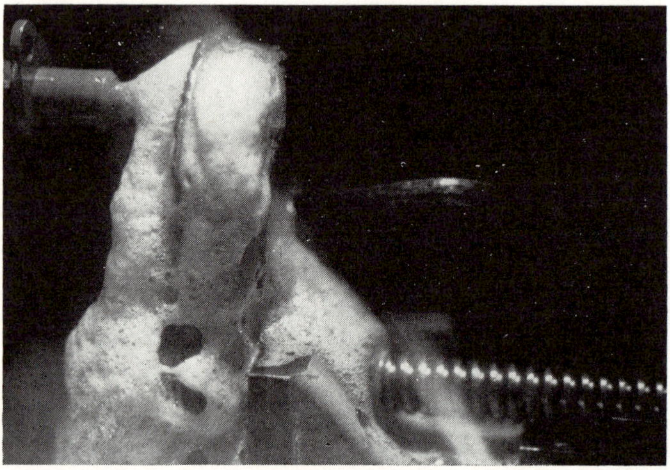

FIG. 14-48. Oil foaming from a nozzle. *(From Hartman, M. A., Gears for Outer Space, Machinery, August,* 1959, *pp.* 115–121.)

Environmental Conditions. Gears must operate in many different environments and under various adverse conditions. Gears have been designed, to name a few, to operate under water, water lubricated, poorly lubricated, and not lubricated, in high-temperature fields,[34] in low-pressure atmospheres,[35] in contaminated atmospheres of all kinds, and under a variety of high speed and high load conditions. The environments in which gears must operate are not only important from the point of view of the harm that the environment does to the gearbox, but also from the viewpoint of what the gear may do to the specific areas or atmosphere in which they are working. For example, a gear must be extremely quiet if it is located in an area where people must rest or sleep. A gear that may contaminate food will be required to operate without lubricants. Radio waves emitted by gears may interfere with various control equipment. Oil vapors escaping from geared equipment may be objectionable to personnel who must work in the immediate vicinity.

It is not the purpose of the author to go into the details of how to cope with these many environmental conditions; however, the design engineer must be aware of these conditions since, in most cases, they will affect the basic concept of his whole gear design.

14-11. Gear Failures. Gears fail in service from such causes as pitting of gear teeth, breakage of the gear teeth, long-range wear, plastic deformation, scoring, and other types of destructive wear not so common. The AGMA standard 110.03,[36] "Nomenclature of Gear-tooth Wear and Failure," breaks down all types of gear failures into four general classes:

1. Wear
2. Plastic flow
3. Surface fatigue
4. Breakage

Each class is then divided into several forms and types with definitions. It is highly recommended that design engineers obtain copies of this standard for handy reference.

A good gear designer needs to know how to recognize and analyze gear failures. In order for past experience to be of any benefit, failures must be properly investigated and appropriate corrective action taken.

Wear. Wear is a very important phenomenon, particularly in high-speed gearing that must be in operation for an unlimited length of time. Wear has been defined from a gear engineer's point of view as that kind of tooth damage whereby layers of metal have been more or less uniformly removed from the surface.[1,*] The most common causes of gear-tooth wear are metal-to-metal contact due to an inadequate oil film, abrasive particles in the oil supply, failure of the oil film in the contact area causing rapid wear or scoring, and chemical wear due to the composition of the oil and its additives.[33]

Figure 14-49 shows a general plot of pitch-line velocity vs. torque capacity of a gearset.[1,†] The various areas indicate the type and regions where gear failures are most likely to occur. Areas 1, 2, and 3 are the failure regions associated with the wear phenomenon. In area 1, the gear is not running fast enough to develop an oil film. In area 2, the speed is fast enough to develop a film and the gear will run an indefinite period provided the lubricant is free of foreign material and is noncorrosive. In area 3, rapid wear or scoring will take place since the load and speed are high enough to break down the existing oil film. Ideally, of course, gear designs should be such that they operate wholly within area 2.

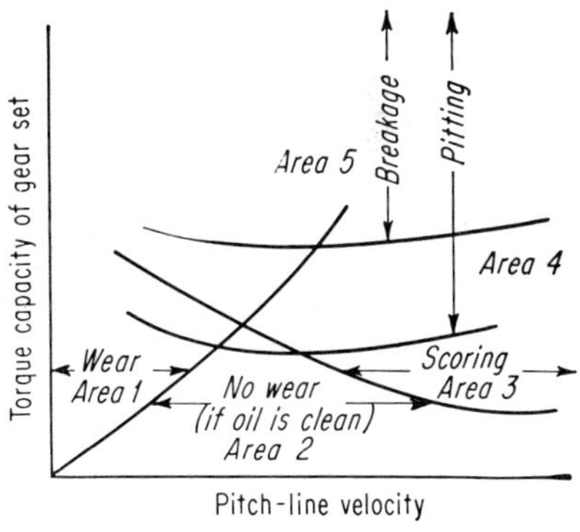

FIG. 14-49. Regions of gear failures.

It is a general practice in the gear trade, and a good one, to break in gearboxes and gradually build up to full load and full speed over a period of time. This procedure permits the contacting surfaces to polish in. Normal wear or polishing in is actually a very slow wear process. However, the amount of metal removed should be well within the design life of the equipment. Figure 14-50 shows an excellent example of a well-polished spiral bevel gear. This gear has seen many cycles of operation, and although some metal has been removed, the gear is in excellent condition.

Figure 14-51 shows a helical gear with a moderate amount of wear. The pitch line is clearly visible, indicating some high metal. Unless the load is reduced or the lubrication improved this gear will ultimately fail from pitch-line pitting.

Abrasive wear can be of considerable concern to the successful operation of a gear train. It is only logical that abrasive particles must be kept from the lube system.

* See p. 294 of reference.
† See p. 295 of reference.

However, at times, this is a very difficult task to perform. Abrasive wear is self-perpetuating in that a few particles can remove metal from the tooth surfaces, which in turn act as a lapping compound which tends to increase the rate of metal removal.

Figure 14-52 shows what can happen if abrasive wear is permitted to continue. Note from Fig. 14-52 that approximately one-third of the tooth thickness has been removed and yet the tooth surface still shows deep radial grooves in the direction of sliding.

Scoring is very rapid wear of the tooth surfaces initiated by virtue of the fact that the oil film has failed, permitting metal-to-metal contact.[33] This intimate contact

FIG. 14-50. Polishing in spiral bevels.

FIG. 14-51. Moderate wear.

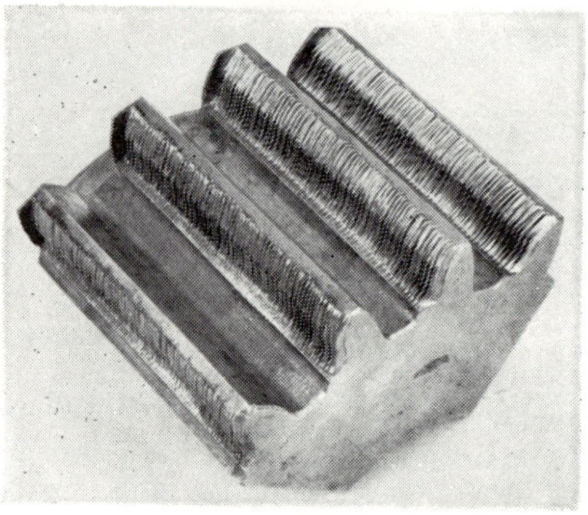

FIG. 14-52. Abrasive wear.

FIG. 14-53. Moderate scoring.

produces a welding and tearing action which removes metal rapidly from the tooth surfaces. Although scoring is triggered off by a lubrication failure, there are many factors other than the oil that influence the final scoring resistance of the gear mesh such as tooth-surface pressure, properties of the material, surface finish, surface treatments, and surface rubbing velocities. Although scoring failures are quite easy to recognize, they are quite hard to analyze. Scoring can be broken down into several categories, which often helps to ferret out the real cause of the failure. The several types of scorings are as follows: *initial* or *slight scoring, moderate scoring, severe scoring, misalignment scoring, scoring due to load concentrations.* Initial or slight scoring usually tends to heal itself unless the load or speed of the apparatus is increased,

Often scoring of this nature occurs if the gearbox has not been properly broken in. Figure 14-53 shows a typical case of moderate scoring. Note that there are many areas on the tooth surfaces that have not been detrimentally affected. Gear of this nature could be put back into service if something is done to improve the scoring conditions. For example, reduce the transmitted load slightly, increase the viscosity of the lubricant, decrease the speed of operation, or add antiscoring additives to the

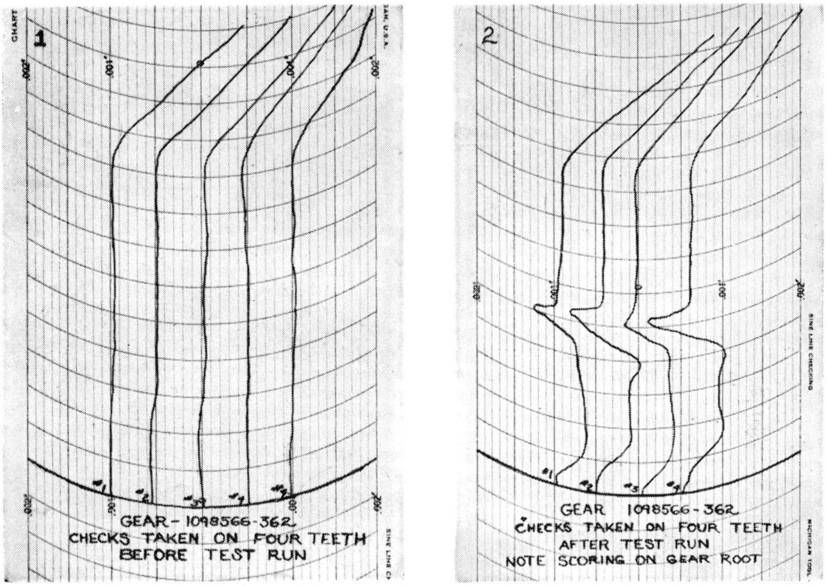

FIG. 14-54. Scoring failure—involute charts.

FIG. 14-55. Severe scoring.

lubricant. Figure 14-54 shows a chart of the involute profile of a gear before and after a scoring failure. The amount of profile destruction is quite evident. It is also quite interesting to note that material has been scooped out of the dedendum and built up on the pitch line of the gear. This scoring failure appeared slightly worse than Fig. 14-53 but not anywhere near so severe as that of Fig. 14-55, which depicts a really bad case of severe scoring. Observe from Fig. 14-55 the welding and tearing action that has been in progress, particularly the metal that has been dragged radially over the tips of the gear. Often the transmitted load is not high enough to cause scoring in the mesh. However, if the gears become misaligned because of deflections or manufacturing errors, the load may concentrate on the ends of the gear teeth. This concentrated load may now be high enough to break down the oil film and permit end scoring or misalignment scoring. Figure 14-56 is a good example of this type of failure. A failure of this nature at times is a blessing in that usually scoring shows up early, often on the test stand, indicating a misaligned condition. The engineer,

FIG. 14-56. Misalignment scoring.

FIG. 14-57. Load-concentration failure.

realizing this condition exists, can do something about it before the unit is put into service. If a misaligned condition goes undetected, trouble usually shows up many hours later in the form of broken ends of the teeth or at best heavy pitting on the ends of the teeth.

Generally, misaligned conditions are corrected by helix-angle corrections or by relieving the ends of the teeth from contact or, if conditions warrant, putting a slight crown on the tooth surface in the axial direction. Crowning must be done with extreme care and careful analysis of the over-all situation since crowning in effect concentrates the load in the center of the gear teeth and can cause a load-concentration failure, as indicated in Fig. 14-57.

Plastic Flow. Gears fail from plastic flow when the surface yields and deforms under heavy loads. Usually this occurs with soft and medium-hard materials, but it can happen on harder gear-tooth surfaces. Generally this plastic flow is characterized by finned material overhanging the tips of gear teeth as indicated by Fig. 14-55. Actually Fig. 14-55 started out as a scoring failure, but because of the destruction of the profile, the load increase due to dynamic effects caused, in addition to the severe scoring, considerable plastic flow of the rolling and peening type.

Rippling and *ridging* are also failures associated with plastic flow. Generally rippling shows up as a wavy surface which may or may not lead to ultimate failure. It generally can be improved by extreme-pressure additives in the oil and better oil distribution. Decreasing the transmitted load of course helps, but this cannot always

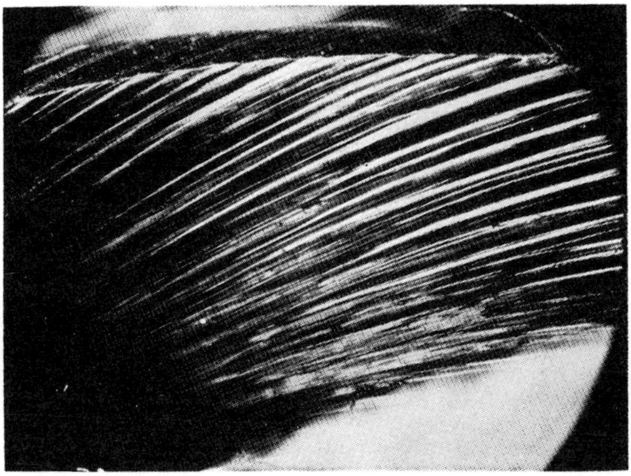

FIG. 14-58. Ridging. (*Courtesy of Socony Mobil Oil Co., New York.*)

be accomplished. Ridging is quite similar in nature to rippling; however, instead of just a surface imperfection rather deep ridges appear and by their very nature are more detrimental to the gear surface than rippling. Ridging usually shows up on worm-gears and hypoid gears (see Fig. 14-58). It is the author's belief that ridging failures are quite closely associated with scoring failures, but because of the high rubbing velocity, a considerable amount of polishing takes place in the mesh, and the net result is ridging.

Surface Fatigue. Surface fatigue, or pitting as it is normally called, is a surface failure due to exceeding the endurance limit of the surface material. Gears under load produce repeated surface and subsurface stresses, and if the loads are high enough and the stress cycle repeated often enough, chunks of metal will fatigue from the surface producing pits. Pitting can be characterized by three different types such as *initial pitting* or *corrective pitting, destructive pitting,* and *spalling.* Generally, pits of the corrective-pitting type are quite small in diameter in the range of $\frac{1}{64}$ to $\frac{1}{32}''$ in diameter. They occur in localized overstressed areas and tend to distribute the load by progressively removing high contact spots. When the load is more evenly distributed, the pitting action is reduced and eventually polishes over. Destructive pitting is more severe and the pits are almost always much larger in size. Figure 14-59 shows a typical pinion with destructive pitting. Pitting of this nature is always fatal in that the profile is destroyed beyond any hope of recovery and, if the transmitted load remains the same, the pitting will get progressively worse until the gear unit runs so rough it must be taken out of service or the gear teeth break off. Spalling is practically the same as destructive pitting except the term is more commonly used when the pits are quite wide in diameter and cover a considerable area. The spiral bevel pinion shown in Fig. 14-60 is an excellent example of spalling. Sometimes

FIG. 14-59. Destructive pitting.

spalling occurs when the pits originate in a casehardened pinion at or near the transition line between the hard case and the softer core material. Spalling also can be formed by destructive pittings breaking into each other, leaving wide voids.

When gears fail from destructive pitting or spalling, the gear designer must do a considerable amount of research. Unless the mesh shows evidence of heavy misalignment, which usually can be corrected, pitting is evidence that the design may not have the capacity* to carry the transmitted load. If this is the case, then something must be done to improve the load-carrying capacity. Medium-hardened gears can be fully hardened. This will increase the endurance limit of the surface material perhaps high enough to prevent pitting. A substitute gear material can be used such as a high-capacity material nitrided or maybe a material case-carburized.† Either of these materials is excellent for increased pitting capacity. Often changes in materials and heat-treatments are not enough and a redesign of the gear must be made. Increasing the face width of the gear will reduce the load per inch of face and increase pitting resistance. The center distance of the gearbox could also be increased. This would reduce the load per inch of face, with the net result of improving the surface pitting resistance. There are undoubtedly other factors that will increase the effective endurance limit of a given gear design. However, they require a more complete understanding of the finer points of the art of gear design, and perhaps a good text should be consulted such as "Practical Gear Design."[1]

FIG. 14-60. Spalling. (*Courtesy of Lubrizol Corp., Cleveland, Ohio.*)

* See Chap. 13 for methods to calculate the load-carrying capacity of a gearset.

† See Chap. 10 for detail information on hardening gears.

Gear-tooth Breakage. Breakage is a failure by breaking a whole tooth or a substantial portion of a tooth by either overload, shock, or the more common fatigue phenomena of repeated stressing above the endurance limit of the gear teeth in beam bending. Figure 14-61 depicts a typical fatigue failure. Note from this failure that the focal point, or characteristic "eye" of the fatigue break, is clearly visible.

Failures of this nature are caused by such things as overload due to design oversights, hob tears or notches in the root fillets, inclusions, heat-treatment cracks, and misalignments.[37] A *fatigue* failure actually takes many cycles of operation to develop. This is why the "eye" is quite often evident.[1] The break initially starts with a slight crack followed by oil seeping into the initial crack. The oil and slight motion develop "fretting corrosion," which is clearly visible in the fatigue "eye" and is an indication of the time necessary for a failure. Generally the break is quite smooth, which also indicates a considerable amount of working of the cracked surfaces before actual failure.

FIG. 14-61. Typical fatigue failure.

FIG. 14-62. Helical pinion failed from overload breakage.

Overload breakage causes a stringy fibrous-appearing break. In the harder materials this break has a more silky appearance, but the break still shows evidence of being wrenched apart. Overload failures are caused by sudden misalignment, bearing failure which tends to freeze up the gear transmission, or large foreign particles passing through the mesh. Figure 14-62 shows a casehardened helical pinion that has failed from overload breakage. It is quite interesting to note that several teeth have chunks broken out corresponding to the number of teeth in the contact zone. When gear teeth break or when foreign material passes through the tooth engagement, often quite catastrophic failure results. Figure 14-63 shows what can happen when something wedges the mesh apart causing a complete mismatch of tooth engagement.

Often gear-tooth breakage is a secondary effect from excessive wear or pitting failures. Figure 14-64 shows a spur gear with destructive pitting and spalling. This photograph clearly indicates fatigue cracks originating from stress risers set up by excessive pitting. If this gear is left in service, a typical pitch-line fatigue break will occur.

When gear teeth break, the investigator can often piece together the wreckage and quite accurately determine the tooth that failed first. If an end breaks off a given gear tooth, there may be some special reason. Misalignment is quite often to blame, and a careful analysis of the contact marking will point this out. Sometimes gears are damaged at assembly or during manufacture. If, perchance, the ends or the gear teeth are upset above the tooth surface, severe overloads will occur, and this section of the gear tooth fails from fatigue.

Other Types and Causes of Gear Failure. There are many cases of gear failure other than the four classical examples previously outlined. A few of the widely different kind will be considered.

FIG. 14-63. Failed gearbox.

FIG. 14-64. Spur gear with destructive pitting and spalling.

FIG. 14-65. Lubrication failure.

Quite often, a simple lubrication failure will cause a catastrophic failure of an entire gearbox. If, for example, a lubrication line ruptures or an operator fails to turn on the lubrication system, the complete gearbox will be starved for oil. The net result is extreme overheating, failed bearings, and badly scored and pitted gear teeth. Figure 14-65 is an example of a gear failure of this nature. The failure will show evidence of excessive heating, a deep blue color, the hardness of the material will be drawn back, and premature pitting due to lack of hardness will be under way. The lubricant can cause other troubles that are not quite so readily detectable as a starved gearbox such as: the viscosity of the lubricant may not be high enough to produce a good oil film; the "oiliness" of the lubricant may be below a safe level to prevent high mesh friction; the lubricant may not have adequate "wetting" properties; the lubricant may collect contaminants or break down and form harmful chemical by-products; the lubricant may oxidize and sludge up the filter causing either a restriction in flow or, by forcing the filter to by-pass,[38] permit unfiltered oil access to the bearings and gear mesh.

FIG. 14-66. "Fretting corrosion," ball bearing.

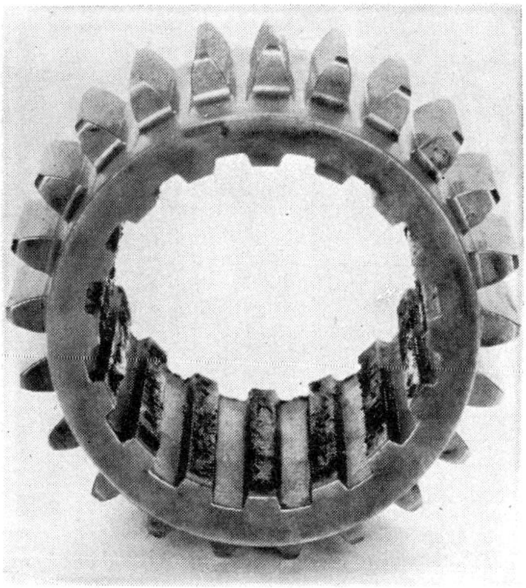

FIG. 14-67. "Fretting corrosion," spline.

"Fretting corrosion" can often lead to a failure of the complete gearbox. "Fretting corrosion" is most often found in areas where there are high loads, high vibrations, and little if any lubrication, such as ball-bearing fits, spline couplings, bushings, and grease-packed applications where there is no fluid present to wash away wear particles. Figure 14-66 shows what can happen to a gear journal when the ball-bearing fit starts to fail from "fretting corrosion." Many mils of material have been "fretted" away causing a high misalignment of the gear mesh. The by-products of "fretting corrosion" are clearly visible in Fig. 14-67. This build-up of sludge and wear particles indicates that considerable metal has been removed from the spline fit. Continuing to operate a gear in this condition would ultimately lead to gear failures.

At times the location and design of oil nozzles can influence the operation of a gearbox. It is very important to deliver oil to the bearing and the gear meshes in the proper quantities to lubricate and to carry the heat away. If the nozzles are not doing a proper job of wetting down the gear and bearing surfaces, adjustments must be made. Perhaps more oil nozzles are required or a relocation of existing nozzles must be made. Often, because of the windage of the mesh, oil nozzles do not have the ability to break down this barrier. A higher back pressure must be used to create the proper jet velocity to penetrate the windage of the gear mesh.

Another cause of gear failures can be traced to the various types of flexible couplings used to connect a given gearbox either to the prime mover or to the driven load.

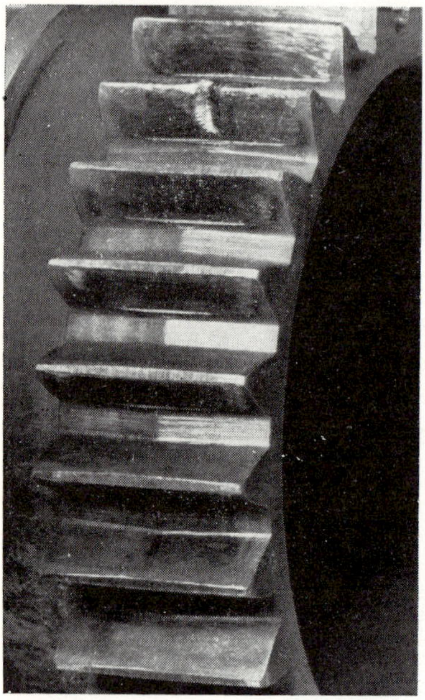

FIG. 14-68. Dental-tooth coupling.

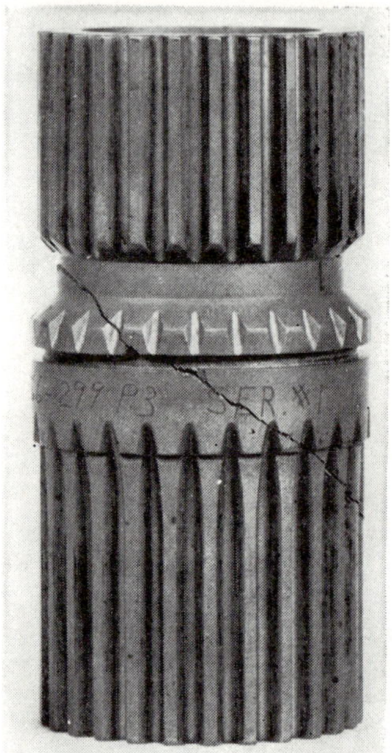

FIG. 14-69. Quill drive—failed.

Quite often, flexible couplings of the dental-tooth type are used such as illustrated in Fig. 14-68. This coupling has failed from the so-called "worm-tracking" effect as indicated by the metal that has been removed in the center of the contact zone.[39] The coupling also shows sign of heavy abrasive wear and "pickup." Generally, when couplings fail in this manner, they become sticky and seize. When this happens, bending moments are transferred into the gear casing, causing considerable trouble to the bearings and contributing to misalignment of the mesh. Couplings used for slightly different purposes can also cause considerable concern. Figure 14-69 shows a failed quill drive used to operate a scavenge pump. This quill failed from normal fatigue set up by stress concentration caused by hob nicks and abrupt section changes. Once a lube or scavenge pump fails, the failure of the complete gear unit is imminent.

14-12. Performance Testing of Gears. To assure the customer that the production gearbox will meet all the requirements as outlined by the specifications, the gear unit should be tested in the factory. Its performance should be observed and adjustments made, if necessary.

Full-power Testing. In order to measure gear performance, the unit must be operated under full-load and full-speed conditions. Ideally, the unit should be tested in exactly the same manner in which it is to be used in the field. It should be driven by its own prime mover and it should drive the required load. In practice at times it is impractical to duplicate exact field conditions in the shop. The practice of transmitting full horsepower through a given gearbox into the actual field equipment or some other load device is known as *power-absorption testing*. When the speed and horsepower rating of a given gear train is low, full-power-absorption testing is practical. Either the actual driven load can be used to absorb the power or a water brake or an electric dynamometer can absorb the power.

When the horsepower rating of the gear unit is large, power-absorption testing becomes more difficult. High-horsepower driving equipment is required as well as high-horsepower absorption equipment. Not only does the equipment become expensive, but when the tests are under way, a large amount of power must be dissipated as losses in the water brake or dynamometer. To simplify and to reduce the cost of full-load-testing high-horsepower gearboxes, a technique known as *torque*

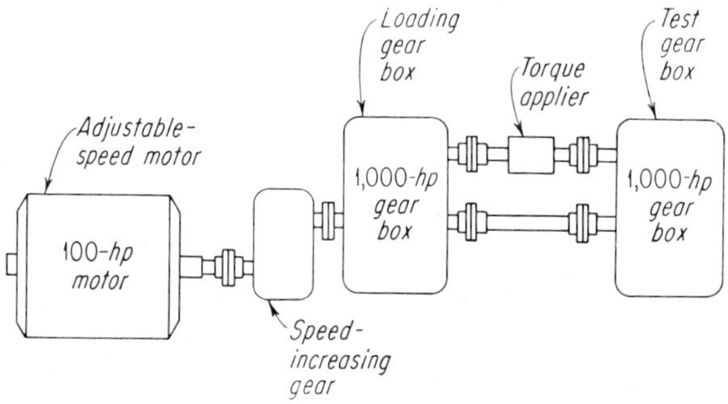

FIG. 14-70. Torque-testing setup.

testing[40] is widely used in the gear trade. Torque testing is based on the "four-square" principle whereby two gearsets are joined back to back and locked together after the full twisting couple is applied to one of the connecting shafts. Now that the gearsets are torqued together, the driving motor needs only to supply the static and dynamic friction losses of the bearing and gears as both units are brought up to speed. When the gearsets are up to speed, they are operating at an equivalent horsepower, equal to that horsepower required for a full-power-absorption test.[41] A schematic diagram of a torque-testing setup can be found in Fig. 14-70. Although torque testing does not fully duplicate exact field conditions, it does permit the factory to full-load-test even the highest-horsepower gear units manufactured.

Accessory Gearbox Testing. Accessory gears are usually designed with a single input or driver with multiple output or driven members. Accessories consist of such things as alternators, lube pumps, hydraulic pumps, power take-offs, and governors. It becomes almost impossible to make an accessory-box torque test; therefore, most accessory gear drives are tested using the power-absorption method. Since the accessories themselves are so many and varied, water brakes are usually used for the power-absorption device. Generally, accessory boxes are rated in the vicinity of 200 hp as a maximum. This being the case, the amount of power being absorbed and converted into losses is not prohibitive from a cost point of view. Figure 14-71 indicates an aircraft accessory gear being loaded with three water brakes to simulate various accessories. It is usually the practice to load accessory gears higher than the design load in order to wring out the gear-tooth loadings on the test stand. Water brakes permit extra power absorption that might not be available if the actual accessories were used instead of the brakes. Accessory-gearbox testing can also be programmed with certain prescribed load and speed cycles if water brakes are used simulating as nearly as possible certain expected service conditions.

Component Testing. Although in the final analysis the production gearbox should be run under field conditions, it has been beneficial in many respects to run components of gearboxes on various types of laboratory test equipment. Component testing can be accomplished at a lower cost and, perhaps, under more controlled testing conditions. Theoretical investigations can be undertaken such as a comprehensive program on the scoring resistance of various material-lubricant combinations or a study on the wear characteristics of certain types of gear steels.[42]

FIG. 14-71. Accessory-box testing. (*Courtesy of General Electric Co., Lynn, Mass.*)

Bearings and spline couplings can also be investigated on various component-test equipment. In fact, component testing has been an excellent source of information for design purposes. It is quite easy to try out new things on a component-test stand and observe how they operate and function. Based on this information, new equipment can be designed with a high degree of assurance for successful operation on the first production unit.

The component-test equipment shown in Fig. 14-72 is typical of the many gear-testing rigs that are in operation throughout industry. Equipment of this nature can be used to investigate pitting, wear, scoring, tooth breakage, lubrication, dynamic loads, and many other gear characteristics.

Efficiency Testing. In many gear applications, the efficiency of the gear train must be known within very close limits. Some customers, for example, request that the efficiency of a given production gearbox be measured at the factory and demand that this measured efficiency fall within the guaranteed limit. Of course, efficiencies can be calculated with a fair degree of precision;[4] however, when extreme accuracy is required, there is no substitute for efficiency testing.

If the gear train under consideration is of a design that lends itself to torque testing, then the over-all efficiency of a unit may be determined on a loss-torque basis. Two identical units are set up back to back with the prescribed amount of locked-in torque. A very accurate torque meter must be installed in the drive motor shaft next to the "four-square" torque loop. This meter will then measure the amount of torque that is required to drive the two locked-up gear trains. Since the motor supplies only the

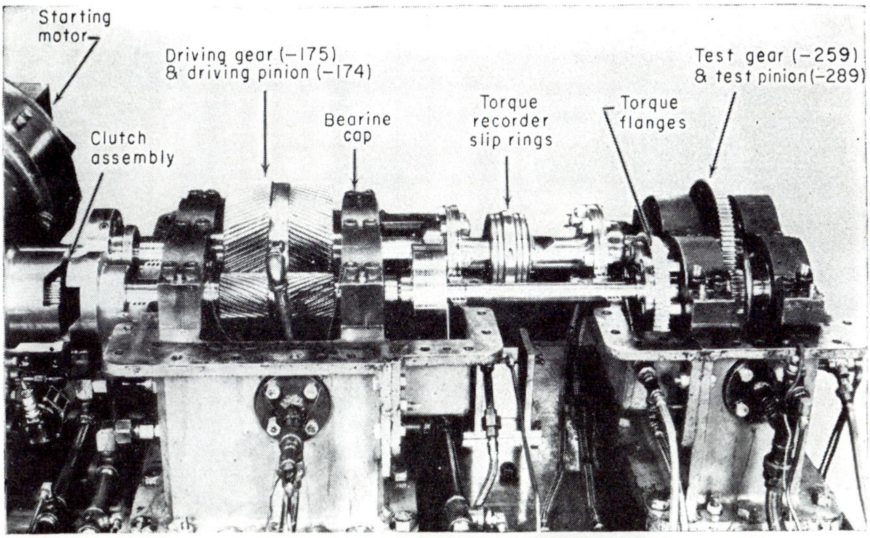

FIG. 14-72. Component-gear tester. (*Courtesy of General Electric Co., Lynn, Mass.*)

losses to the system, in effect, the inefficiency of two identical gearboxes has been measured. The efficiency can then be calculated on a torque basis.

Namely, efficiency in per cent equals

$$E = \frac{T_t - T_{L/2}}{T_t} \times 100 \tag{14-99}$$

where E = efficiency, per cent
T_t = transmitted torque
T_L = loss torque

Often a given gearbox does not lend itself to a simple back-to-back torque test. Therefore, other means of efficiency testing must be devised. A simple heat balance on a loaded gearbox transmitting full horsepower will give the over-all efficiency of the geared transmission. In effect, all the losses of a gearset are reduced to heat, which is dissipated to the atmosphere and carried away by the lubrication system. If this heat loss can be measured accurately, then the basic efficiency of the drive has been measured.

The heat carried away by the lubrication system can be measured very accurately by carefully determining the oil flow through the gear casing and observing the temperature rise of the oil as it goes into and comes out of the gearbox. The heat loss by radiation and convection is very hard to determine; however, this problem can be by-passed by "lagging" the gearbox to keep heat loss of this nature to an absolute minimum. Of course, the heat that normally would escape will now show up as an increase in oil temperature across the gearbox. The "lagging" job on the gear casing

FIG. 14-73. Aircraft unit before lagging. (*Courtesy of General Electric Co., Lynn, Mass.*)

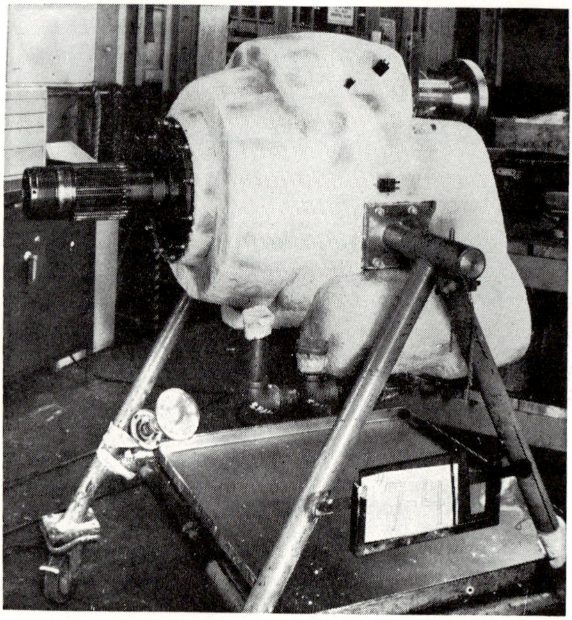

FIG. 14-74. Aircraft unit after lagging. (*Courtesy of General Electric Co., Lynn, Mass.*)

must be of high quality; in fact it should be done by an expert using the very best insulating materials. The aircraft gear casing in Fig. 14-73 and 14-74 shows a before and after "lagging" job. Note from these photographs that an ample amount of lagging has been used and that a professional job of "lagging" has been accomplished. The efficiency of a gearbox from a heat-balance test can be determined by the following equations:

$$Q = MC_p \, \Delta T \qquad (14\text{-}100)$$

where Q = heat loss, Btu per min
M = lb of oil per min
C_p = specific heat of oil
ΔT = temperature rise, °F (inlet to outlet)

$$P_L = \frac{Q}{42.4} \tag{14-101}$$

where P_L = horsepower loss
42.4 = Btu per hp

$$E = \frac{P_T - P_L}{P_T} \times 100 \tag{14-102}$$

where E = efficiency, per cent
P_T = transmitted horsepower
P_L = horsepower loss

Lubrication Testing. In order to get the best performance from a gearbox, the design engineer must not only have a good design, but he must use the best lubricant for his application. The complete lubrication system, in fact, must meet all the performance requirements of the gearbox or else the gear itself will never make it. We know from experience that the two major failures relating to the lubricant are scoring and normal or long-range wear.[33] To evaluate a new gear oil is rather a time-consuming and costly project. The procedure may vary somewhat, depending upon the ultimate use of the oil, but in general, the following steps should be taken: All available information pertaining to the oil should be obtained from the manufacturer and the various laboratories that have been working with the oil. Laboratory checks should be made on an oil sample, such as viscosity, pour point, flash point, and oxidation stability. Comparison tests should be run on bench tests such as the Shell four-ball wear tester and the Timken wear tester. Lubricated compression-roll tests should be run under simulated service conditions and the results compared with previous tests. Scoring tests using test gears on a four-square gear-testing apparatus must be run to determine the "critical scoring temperature" of the oil. Wear tests should be run on a radioactive gear tester to determine the wear characteristics of the new oil.

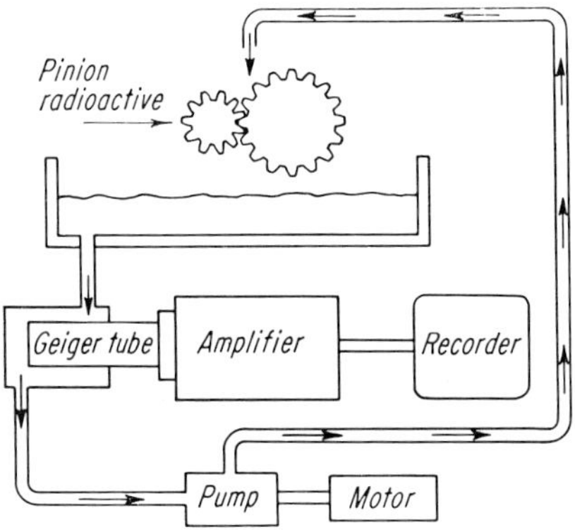

FIG. 14-75. Radioactive wear test.

Since wear as such is very hard to detect and measure, the radioactive technique of gear testing has been adapted for studying the wear phenomenon.[33] Minute quantities of wear can be detected in short periods of time and several oils may be tested with the same sets of gears.

The schematic test setup of Fig. 14-75 gives a good flow picture of how the radioactive technique is used for testing gears and gear lubricants. The pinion to be tested is sent to the nearest atomic pile, where it is irradiated with neutrons. This makes the pinion radioactive. The pinion and gear are then mounted in a test stand. The gears are loaded and the test machine put into motion. The lubricant is circulated through the gearbox, past the counter tube, and back into the sump, as indicated by Fig. 14-75. When the active gears commence to wear, the lubricating oil transports the worn iron particles past the counter where they are detected. It is a rather simple laboratory technique to convert the counting rates as recorded to milligrams of wear per unit of time.

In most cases environmental conditions affect the lubricant more than any other characteristic of a gear setup.[43] Because of this, most environmental testing has been centered around the lubricant as one of the main variables. A large per cent of environmental testing has been done under close laboratory supervision on component-gear-test equipment.[38]

At the present time, work is under way on both high-temperature and low-temperature gear operation for the rocket and missile field. High-altitude atmospheres are under investigation.[35] Scoring and wear tests are in progress under all kinds of adverse conditions. Gear-lubrication problems of all sorts are under investigation, such as grease lubrication, solid-film lubrication,[49] exotic-liquids lubrication, lubricant additives, and many other new and different environmental conditions.

Field Testing. Field operation is the best proving ground an engineer can find for evaluation of gear equipment. The various gear designs must operate successfully in the field, and much can be learned from obtaining and observing operational records of the gear arrangements that are in service.

In many instances, service operating conditions are much different from the original design conditions, and it is not uncommon to find a customer using the equipment far in excess of what has been recommended. The reverse is quite often true, however, and a design engineer can easily be fooled by the seemingly successful operation of gearsets in the field. The apparent successful operation may be the result of not operating the gear unit up to design conditions.

By firmly following the policy of keeping close tabs on field operations, the gear engineer not only obtains good information upon which to base his new designs but also creates a good working relationship with the customer.

REFERENCES

1. Dudley, Darle W.: "Practical Gear Design," McGraw-Hill Book Company, Inc., New York, 1954.
2. Philadelphia Gear Works: "Philadelphia Gears," Philadelphia Gear Works, Philadelphia, Pa., 1958.
3. Buckingham, Earle: "Analytical Mechanics of Gears," McGraw-Hill Book Company, Inc., New York, 1949.
4. Shipley, Eugene E.: How to Predict Efficiency of Gear Trains, *Prod. Eng.*, August 1958, pp. 44–45.
5. Merritt, H. E.: "Gears," 3d ed., Pitman Publishing Corporation, New York, 1955.
6. Candee, Allan H.: Bevel Gears in Aircraft, *Trans. ASME*, May, 1943, pp. 267–277.
7. Gleason Works: "20° Straight Bevel Gear System," Gleason Works, Rochester, N.Y., 1949.
8. Gleason Works: "Zerol Bevel Gear System," Gleason Works, Rochester, N.Y., 1950.
9. Gleason Works: "Spiral Bevel Gear System," Gleason Works, Rochester, N.Y., 1951.
10. Illinois Tool Works: "Spiroid Gears," Illinois Tool Works, Chicago, Ill., 1959.
11. Illinois Tool Works: "Helicon Gears," Illinois Tool Works, Chicago, Ill., 1959.
12. Gleason Works: "Bevel and Hypoid Gear Design," Gleason Works, Rochester, N.Y., 1947.
13. Illinois Tool Works: "Planoid Gears," Illinois Tool Works, Chicago, Ill., 1958.

14. Wilcock, Donald F., and E. R. Booser: "Bearing Design and Application," McGraw-Hill Book Company, Inc., New York, 1957.
15. Palmgren, Arvid: "Ball and Roller Bearing Engineering," S. J. Burbank & Co., Inc., Philadelphia, Pa.
16. Barish, Thomas: "Fundamentals of Ball Bearing Design," unpublished paper, March, 1957.
17. Watt, Hamilton R.: Gear Tooth Efficiencies, I, II, III, *Power Transmission*, March, April, June, 1951.
18. Buckingham, Earle: Manual of Gear Design, Sec. 2, *Machinery*, 1935.
19. Tuplin, W. A.: Compound Epicyclic Gear Trains, *Machine Design*, April, 1957, pp. 100–104.
20. Buckingham, Earle: Dynamic Loads on Gear Teeth (Report of the ASME Special Research Committee on the Strength of Gear Teeth), American Society of Mechanical Engineers, New York, 1931.
21. Tuplin, W. A.: Gear Tooth Stresses at High Speed, *Proc. Inst. Mech. Engrs. (London)*, Vol. 163, pp. 162–175, 1950.
22. Tuplin, W. A.: Dynamic Loads on Gear Teeth, *Machine Design*, October, 1953, pp. 203–211.
23. Strauch, H.: Zahnradschwingungen, *VDI Zeitschrift*, Vol. 95, No. 6, pp. 159–163, 1953.
24. Reswick, J. B.: Dynamic Loads on Spur and Helical Gear Teeth, *Trans. ASME*, Vol. 77, pp. 645–654, 1954.
25. Attia, A. Y.: "Dynamic Loading of Spur Gear Teeth," Doctoral Thesis, University of Sheffield, 1956.
26. Niemann, G., and H. Rettig: "Errors Induced by Dynamic Gear Tooth Loads," International Conference on Gearing, London, September, 1958.
27. Zeman, J.: Dynamic Forces in Gear Drives, *VDI Zeitschrift*, Vol. 99, No. 6, pp. 244–254, February, 1957.
28. Harris, Stephen L.: "Dynamic Loads on the Teeth of Spur Gears," The Institution of Mechanical Engineers, December, 1957.
29. Kohler, H. K.: "The Mechanism and Measurement of Dynamic Loading in Spur Gears," Ph.D. Thesis Submitted to the University of Sheffield, May, 1959.
30. Buckingham, Earle: "Gear Tooth Loads," Massachusetts Gear & Tool Company, 1953.
31. Cleare, G. V.: "Gear Tooth Failures in Highly Stressed Transmission Gears," International Conference on Gearing, The Institute of Mechanical Engineers, September, 1958.
32. Tuplin, W. A.: Gear Tooth Lubrication, *Machine Design*, August, 1954, pp. 125–131.
33. Shipley, Eugene E.: Design and Testing Considerations of Lubricants for Gear Applications, *Lubrication Eng.*, April, 1958, pp. 148–152.
34. Shipley, Eugene E.: Investigation of Factors Affecting High Temperature Gear Operation, *Lubrication Eng.*, March, 1959, pp. 98–104.
35. Hartman, M. A.: Gears for Outer Space, *Machinery*, August, 1959, pp. 115–121.
36. American Gear Manufacturers Assoc.: "Standard Nomenclature of Gear-tooth Wear and Failure," AGMA 110.03, 1960.
37. Frederick, S. H., and A. D. Newman: "Gear Failures," International Conference on Gearing, The Institution of Mechanical Engineers, September, 1958.
38. Shipley, Eugene E.: Will Gears Operate at 600°F?, *Am. Machinist*, February, 1958, pp. 101–103.
39. Batter, J. F., and F. E. Reed: "Wear in Heavily Loaded Lubricated Surfaces Subjected to Oscillating Motion," American Society of Lubrication Engineers, Paper 59GS-12, January, 1959.
40. Shipley, Eugene E.: 12 Ways to Load Test Gears, *Prod. Eng.*, January, 1958, pp. 77–82.
41. Newman, A. D.: "Load-carrying Tests of Admiralty Gearing," International Conference on Gearing, The Institution of Mechanical Engineers, London, September, 1958.
42. Crook, A. W.: The Lubrication of Rollers, *Phil. Trans. Roy. Soc. London*, A, Vol. 250, pp. 387–409, January, 1958.
43. Research Report, Inert Gas Boosts Oil's Ability, *Chem. Eng. News*, October, 1958, pp. 34–35.
44. Hirst, W., and J. K. Lancaster: The Influence of Oxide and Lubricant Films on the Friction and Surface Damage of Metals, *Proc. Roy. Soc. London*, A, Vol. 223, pp. 324–338, 1954.
45. Buckingham, Earle: "Spur Gears," McGraw-Hill Book Company, Inc., New York, 1928.
46. Radzimovsky, Eugene I.: Planetary Gear Drives, *Machine Design*, June, 1959, pp. 144–153.

47. Dudley, Darle W.: Modification of Gear Tooth Profiles, *Prod. Eng.*, September, 1949.
48. Barish, Thomas: Idler Gears, *Machine Design*, November, 1957, pp. 121–125.
49. Moore, Wayne L.: "Reliability of Gear Applications for Aero-space Vehicles," American Gear Manufacturers Association, March, 1960.
50. Editor: Full Load Efficiency Testing of Turbine Gear Units, *Engineering*, March, 1959, pp. 306–307.
51. Shipley, Eugene E.: "Efficiency of Involute Spur Gears," American Society of Lubrication Engineers, January, 1959.
52. Barish, Thomas: "On Hertz Contact with Sliding," American Gear Manufacturers Association, March, 1960.
53. Crook, A. W.: Simulated Gear-tooth Contacts. Some Experiments upon Their Lubrication and Subsurface Deformations, *Proc. Inst. Mech. Engs. London*, Vol. 171, No. 5, 1957.
54. Buckingham, Earle: Dynamic Loads on Gear Teeth, *Machinery*, October, 1958, pp. 117–121.
55. Attia, A. Y.: "Dynamic Loading of Spur Gear Teeth," ASME Paper 58-SA32, 1958.
56. Gatcombe, Ernest K., and Roy W. Prowell: "Rocket Motor—Gear Tooth Stress Analysis . . . Hertzian Contact Stresses and Times," ASME Paper 59-A-256, 1959.
57. Michigan Tool Company: "Cone-drive Gears," Michigan Tool Company, Detroit, Mich. 1950.

Chapter 15

LUBRICATION OF GEARS

By

FOLKE RICHARDZ, *Consulting Specialist, Gearing Products, Westinghouse Electric Corporation, Buffalo, N.Y.*

All gears, regardless of type and material, will gain in life expectancy by being properly lubricated. The lubrication of gears should therefore be as much a part of the design requirements for successful operation as is the calculation of gear-tooth stresses or surface durability. The art of proper lubrication is not so well established as the methods of gear-tooth calculations, and it is the purpose of this chapter to explore and evaluate not only the generally accepted methods of lubrication and types and viscosity of oil for various operating conditions, but also methods for retaining the lubricant within prescribed boundaries and means for keeping the viscosity of the lubricant at temperatures which will assure maximum protection to gears and bearings.

15-1. Systems and Methods of Gear Lubrication.* The methods by which lubricant is applied to gear teeth vary with type of gear, speed in terms of pitch-line velocity, surface finish, hardness of the material, and combination of materials. In open gearing or gearing mounted on pedestal supports and not enclosed, a method of applying lubricant by a paddle or brush applies almost exclusively to slow-speed coarse-pitch unprotected gears. Since this method is tedious and expensive, it is fast disappearing and is called to the designer's attention only as a method to be avoided whenever possible.

Figure 15-1 illustrates splash lubrication commonly used in speed reducers. The illustration depicts double-reduction gearing, and it is rather obvious that the second-reduction or slow-speed gearset should lubricate satisfactorily as the gear is dipping into the lubricant reservoir.

The high-speed gearset, since it does not dip into the lubricant, may be questionable in regard to satisfactory lubrication. Because of lack of experimental data, adequate lubrication is assured by the method indicated in Fig. 15-2, where a splash pan is introduced and the oil level is increased to permit the high-speed gear to dip. To assure adequate and satisfactory lubrication at low speed, a check should be made by calculating the critical speed.

To attain the desirable end result, the speed must be such that the centrifugal force exceeds gravity.[1,†] Viscosity will of course enter the problem to some extent but

* See Chap. 14 and particularly Art. 14-4 for further information on gear performance and lubrication.

† Superscript numbers refer to references at the end of the chapter.

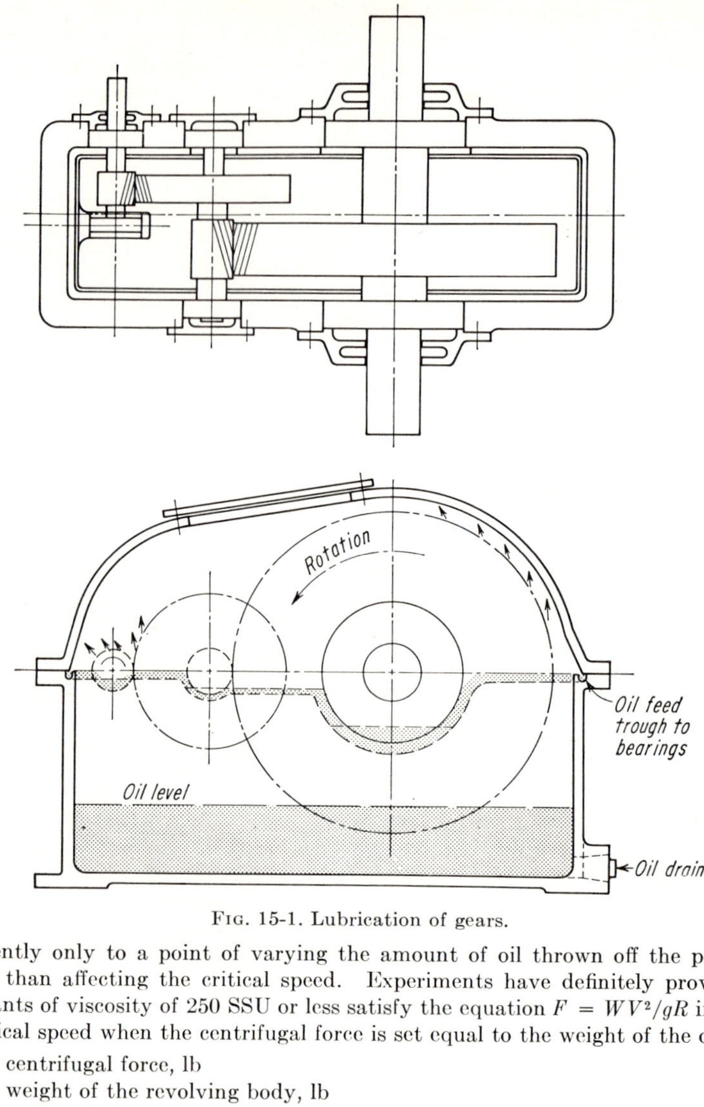

FIG. 15-1. Lubrication of gears.

apparently only to a point of varying the amount of oil thrown off the periphery rather than affecting the critical speed. Experiments have definitely proved that lubricants of viscosity of 250 SSU or less satisfy the equation $F = WV^2/gR$ in regard to critical speed when the centrifugal force is set equal to the weight of the object.

F = centrifugal force, lb
W = weight of the revolving body, lb
R = radius of the gear, ft
V = velocity, fps
N = rpm
g = acceleration due to gravity = 32.16. Since we have assumed $F = W$ our equation becomes $1 = V^2/gR = 0.000341RN^2$ or $N^2 = 2{,}935/R$.

As either of the two terms may be variable, we can solve for one with the other fixed. As an example, consider the radius of the gear as 2 ft, and it is desired to know how slowly it may run with safety for good splash lubrication. We substitute R in the equation and solve for N,

$$N^2 = \frac{2{,}935}{2} = 1{,}467.5 \qquad N = 38 + \text{rpm}$$

For unusually slow speeds, where the splash from the gears does not reach the oil troughs feeding oil to the bearings, scrapers as illustrated in Fig. 15-3 are effective. The arm of the scraper extending from the case to the side of the gear should be made adjustable so that the scraper may be brought within approximately a $\frac{1}{32}''$ from the side of the gear. The tip of the scraper may be made from either metal or a material such as synthetic rubber.

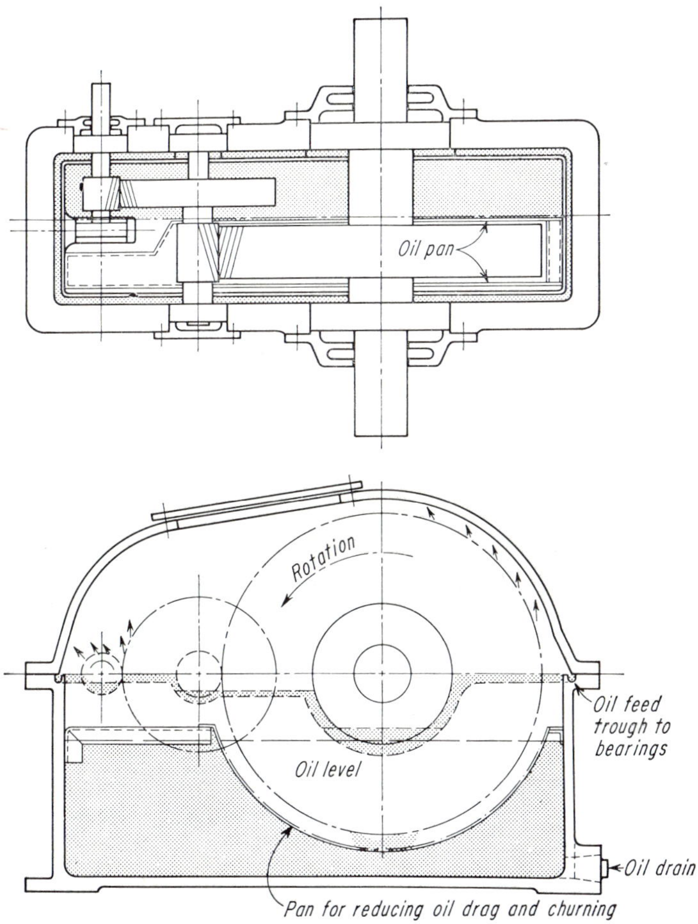

FIG. 15-2. Restricted splash lubrication.

Figure 15-4 illustrates typical splash lubrication as employed in a double-reduction gear motor. Figure 15-5 illustrates a forced-feed lubrication system frequently used either from a central system or an integral pump driven from one of the shafts in the gear drive or from a separate motor-driven pump. When the forced-feed method of gear lubrication is employed in gear reducers having antifriction bearings and no provision for oil-feed troughs to the bearings, a fine stream from the force-feed system should be directed onto the bearings. When sleeve-type bearings are employed, adequate feed lines to the bearings must always be provided.

Figure 15-6 shows a typical forced-feed system for high-speed industrial gear units employing sleeve bearings. The amount of oil required to lubricate sleeve bearings satisfactorily is dependent on the specified bearing clearance and the rpm involved. A satisfactory basis for selecting an oil pump to provide an adequate amount of lubricant to gears and bearings is to allow 0.03 gal of oil per square inch of the projected total bearing area involved and 0.5 gal* for each inch of gear face width.

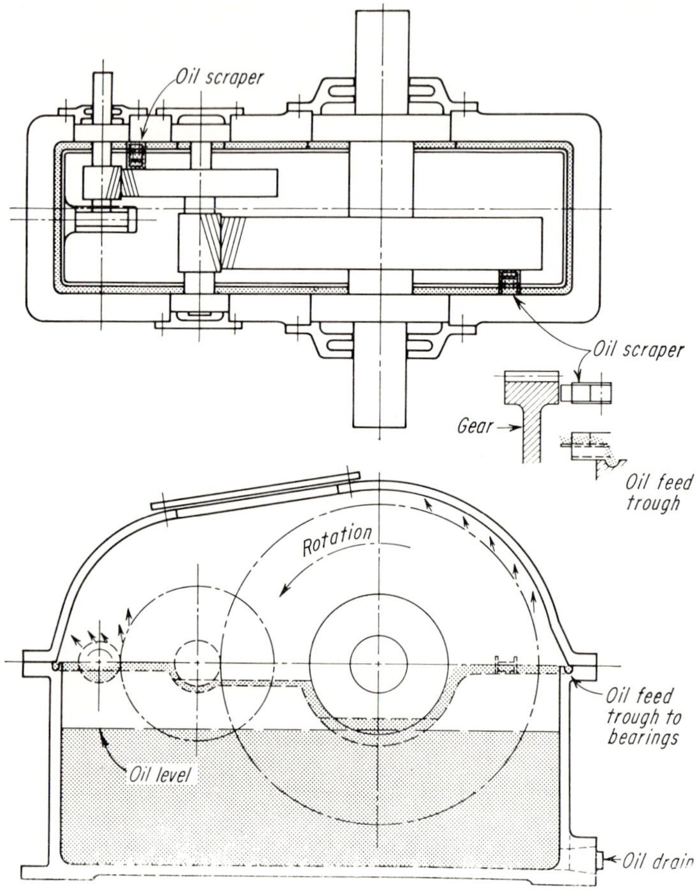

FIG. 15-3. Method of feeding oil to bearings at very slow speed without forced feed.

In general, high-speed gear drives are unidirectional; however, at times dual direction of rotation is required. In addition, on applications such as pipelines, where diesel or gas engines are used as prime movers, a back pressure in the line may develop causing the gear unit to run in the direction for which it was not designed. If a unidirectional pump is used without safeguards for such an event, the oil piping may burst. Since the bearings would be starved for lubrication, an expensive repair would result.

Figure 15-7 shows a simple check-valve arrangement acting as a safeguard for occurrences of this type and also satisfying the requirement for dual direction of rotation.

* See Art. 15-6. In some cases this is not enough oil.

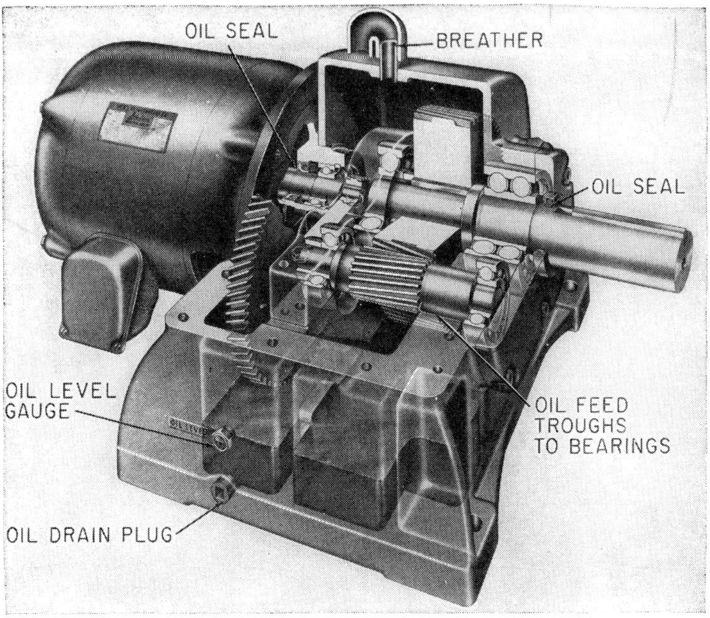

FIG. 15-4. Gear motor with splash-lubricated double-reduction gearing. (*Courtesy of Westinghouse Electric Corporation, Pittsburgh, Pa.*)

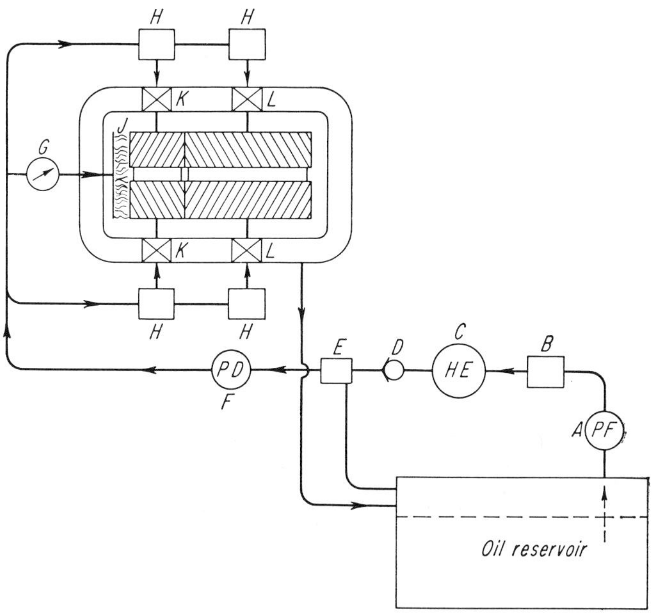

FIG. 15-5. Schematic diagram of a forced-feed gear-lubrication system. A = lube oil pump. B = lube oil strainer. C = heat exchanger. D = check valve. E = pressure-relief valve. F = protective device. G = pressure gage. H = lube oil adj. valves. J = oil spray to gearing. K = pinion bearings. L = gear bearings.

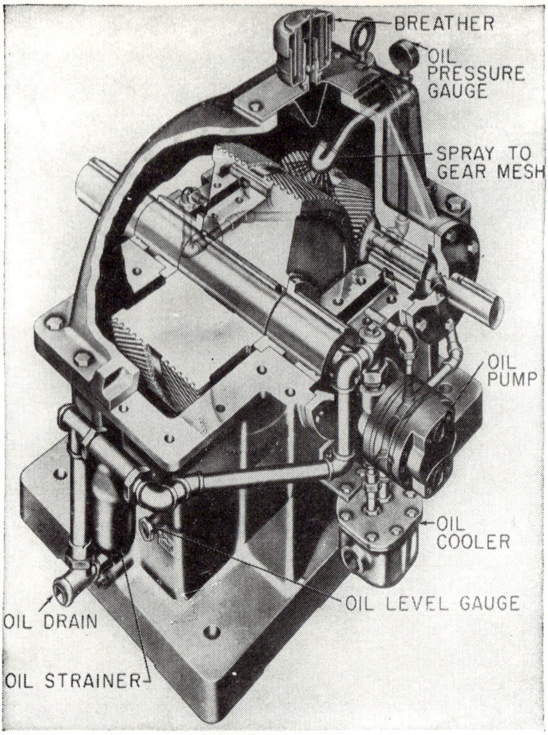

FIG. 15-6. High-speed industrial gearset with forced-feed lubrication system. (*Courtesy of Westinghouse Electric Corporation, Pittsburgh, Pa.*)

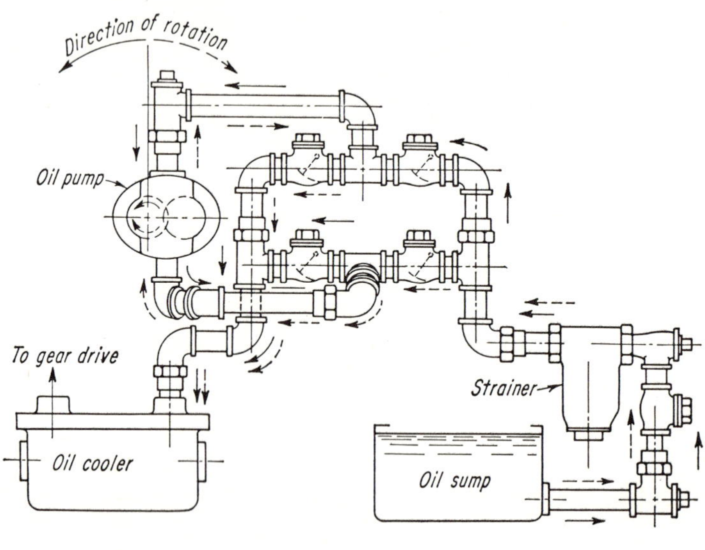

FIG. 15-7. Check-valve arrangement for dual direction of rotation.

On some aircraft and special applications of high-speed gearing where the weight is critical, the rim of the gear as well as the central web is made exceptionally thin and therefore presents a different lubrication problem. The lubricant in this case is generally directed to the underside of the rim and the sides of the web primarily to keep the metal cool and the gear mesh* lubricated essentially by the mist created by the spray on rim and web (Fig. 15-8).

Spiral bevel and hypoid gears at peripheral velocity up to approximately 2,000 fpm are generally lubricated by dipping into the oil supply. At higher velocities forced-feed lubrication is generally employed. The direction of spray should be such that it covers the entire width of the gear face.

Vertical gear drives are most frequently forced-feed lubricated and particularly where more than one gear reduction is employed. Figure 15-9 shows a typical vertical double-reduction gear drive forced-feed lubricated.

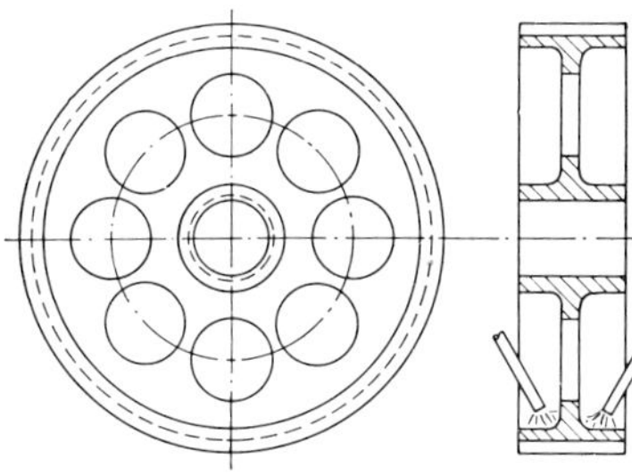

FIG. 15-8. Lubrication of aviation gearing.

In certain applications where it is not convenient or practical to immerse the pinion in the oil reservoir or employ forced feed, an oil flinger is mounted on the pinion shaft of sufficient diameter to dip into the oil reservoir as illustrated in Fig. 15-10. The oil flinger is designed in such a manner that it throws a sufficient amount of lubricant into the gear mesh and bearings.

In summary, systems and methods of gear lubrication are

1. Paddle, brush, or drip cup, used primarily for slow-speed, open, or semiprotected gearing
2. Dip or splash, used for medium pitch-line velocity up to approximately 2,500 fpm
3. Restricted dip, using oil splash pan. Employed for pitch-line velocities up to approximately 3,500 fpm
4. Pressure lubrication by
   a. Attached lube oil pump
   b. Separate motor-driven lube oil pump
   c. Central lube oil pressure system; pressure lubrication may be employed for low as well as high pitch-line velocities, provided proper control of volume and pressure is provided
5. Mist; this method is restricted to the more unusual conditions such as high rpm's or applications where build-up of lubricant is not tolerated

15-2. Types and Viscosity of Gear Lubricants. The most common types of gear lubricants used to hold gear-tooth wear to a minimum and minimize rubbing and sliding friction between metallic gear teeth are generally a product of vegetable, animal, or mineral origin and, recently, synthetic lubricants such as silicon compounds. The mineral oils, be they of the paraffin series (generally referred to as Pennsylvania oil) or

* See Art. 15-6 for other methods of lubricating aircraft gears.

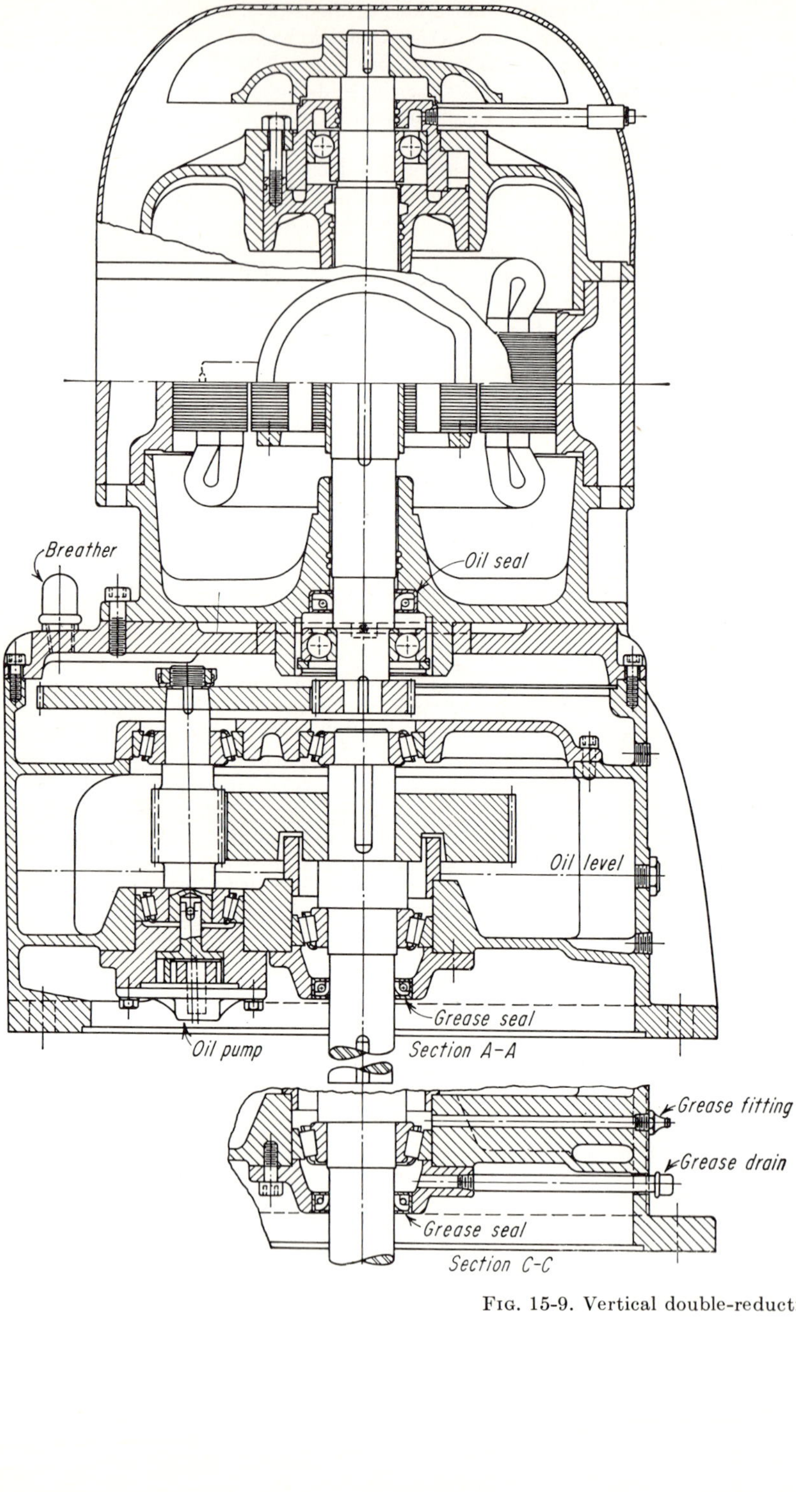

FIG. 15-9. Vertical double-reduction

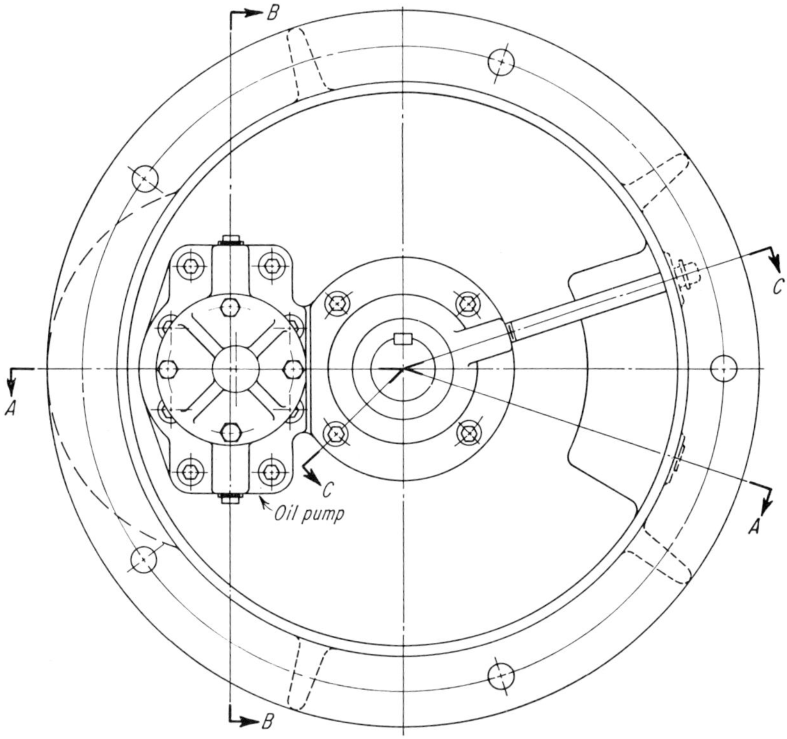

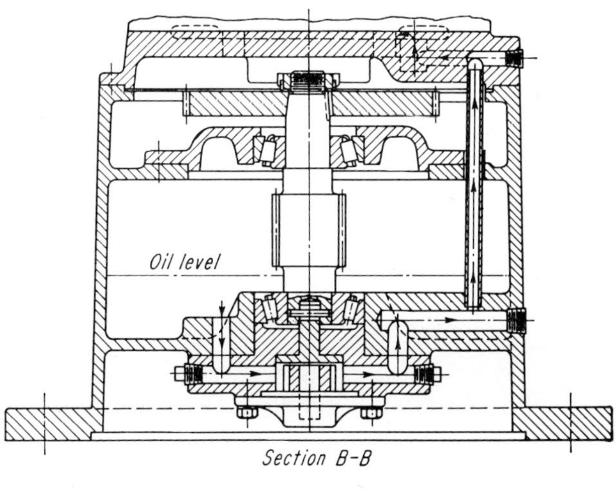

Section B-B

gear with forced-feed lubrication.

the asphaltic series, occupy at least for the present the number-one place in gear lubrication. The mineral oils in many of today's highly loaded gears are in many instances not capable of satisfactorily carrying the load imposed and must therefore be fortified with other elements such as sulfur, chlorine, and lead naphthanates and combinations of these elements and as such they are generally referred to as non-corrosive extreme-pressure lubricants. In gear-lubrication research and testing engineers continue to explore the reasons for gear-tooth wear, pitting, and failures of various types. The American Gear Manufacturers Association has issued standard nomenclature[2] identifying the various and most common types of gear-tooth surface deterioration. Gear lubricants of high viscosity at light load show higher friction loss, but as the load increases, the heavy-viscosity lubricants show less frictional losses, indicating that viscosity is an important factor and the gear-tooth loading should therefore be given due consideration by the designer in the selection of lubricants. Based on present gear-tooth design limitations by AGMA standards for strength and

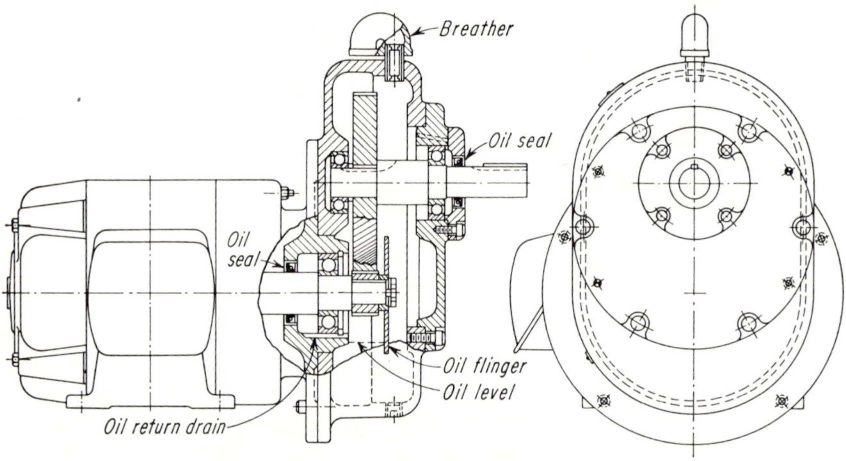

FIG. 15-10. Oil flinger used as a lubrication means.

durability, industrial gear-lubrication standards have been developed by the gear manufacturers.[3,4,5]

Oil-film-strength measurements and evaluations are to some extent questionable, but in general the Timken Roller Bearing Company's extreme-pressure testing machine has provided a comparatively reliable evaluation in regard to load rating and comparison of lubricants. An extreme-pressure oil should not have a value of less than 35 lb on the Timken extreme-pressure test machine. Presently extreme-pressure gear oils will measure 35 to 70 lb for viscosities of 44 to 900 sec at 210°F. Straight mineral oils are always less than 35. In addition to the Timken extreme-pressure testing machine there are other test equipments fairly widely used such as the Almen tester and the Shell four-ball extreme-pressure tester. The use of extreme-pressure lubricants for the initial "running-in" period of gearing has by actual experience been found to have definite advantages, as they will to a great extent prevent incipient or initial scoring, galling, and seizure. This feature is attributed to the greater film strength and antiweld properties. Application of extreme-pressure lubricants to gearing where heavy shock and vibration loads are encountered in many instances has improved the performance and minimized maintenance.

Under heavy load conditions, a heavy mineral lubricant would ordinarily be required. The operating temperatures may exceed good practice and cause rapid oxidation and general deterioration of the lubricant. By changing to a lighter-viscosity extreme-pressure lubricant, equal or better results will be found in regard to load-carrying capacity and a decided drop in operating temperatures, tending to minimize the frequency of lubricant change due to less oxidation deterioration.

The gear lubricant, be it straight mineral oil, extreme-pressure lubricant, etc., will be of additional value to the user if it contains protective additives.[6] The most commonly used and important protective additives are rust inhibitors, defoamants, and oxidation inhibitors. The rust inhibitor is one of the most important when the equipment is subjected to condensation due to fluctuating temperature and where the gearing may be idle for extended periods of time and operating outdoors with little or no protection. The oxidation inhibitor is of great value when the gearing is operating in the higher-temperature range. The defoamant is of particular value when the lubricant is subject to agitation due to splash lubrication at high pitch-line velocities or forced-feed lubrication where the oil pump capacity is in excess of the oil reservoir capacity.

Viscosity and viscosity index are important factors in the selection of gear lubricants. Gear-teeth contact areas are relatively small; therefore, the contact unit pressures are comparatively high. It is therefore important to provide a film of lubricant of sufficient strength to withstand the localized pressure during the period of tooth contact. The pitch-line velocity of the gearset governs the length of time the oil film must withstand the pressures. When speeds are high, the gear-tooth contact time is short and the loads usually not so heavy, and as a result, a lubricant of a relatively low viscosity may be used. At the lower speeds the loads are generally high and the tooth contact time longer, and consequently a lubricant of a higher viscosity should be used. Where high-speed and low-speed gearing are contained in the same gearbox and lubricated from the same source of lubricant, the lubricant viscosity required for the lower-speed gearing should be selected.

The viscosity index was established by Dean and Davis in 1929.[7,8] Essentially it is an arbitrary system evaluating the relationship between viscosity and temperature. The standard of comparison is Pennsylvania oil, rated 100, and coastal naphthinic, rated 0 when refined by the same sulfuric acid process.

Present modern refining methods make it possible to obtain higher values for both Pennsylvania and coastal oils. The significance of viscosity index (V.I.) in gear lubrication is that, when a gear drive is subjected to extreme variations in temperatures, it is possible that the oil of a low V.I. may become so viscous that it presents a problem in starting by the prime mover. For example, an oil of zero V.I. having a viscosity of 100 SSU at 210°F would have approximately 2,640 SSU at 100°F; an oil of 100 V.I. with the same viscosity at 210°F, namely, 100 SSU, would have only approximately 1,200 SSU at 100°F, or a ratio in excess of 2:1 in favor of the high V.I. oil. Generally, a V.I. of 60 and above should be selected where temperatures vary over a wide range. When gear drives are subjected to low temperatures where starting becomes a problem because of too viscous an oil, it may be advantageous to insert an electric-immersion-type heater in the oil reservoir of sufficient capacity to bring up the temperature of the volume of oil to provide a viscosity that will permit easy starting. The general requirement and range of gear lubricant for enclosed gearing according to the American Gear Manufacturers may be covered by eight AGMA lubricant numbers of viscosities as indicated in Table 15-1.[3]

There is no infallible rule as to the exact viscosity of lubricant to be used for the various sizes or ratings of gear units, but if the gearing is designed in line with AGMA standards, it is generally adequate to use the lubricant recommendations shown in Table 15-2.

Lubrication of Wormgearing. Since wormgearing is subjected to all sliding friction as against sliding and rolling in involute spur, helical, and herringbone gear-

Table 15-1. Viscosity Range for Various AGMA Lubricants

AGMA lubricant No.	Viscosity range, SSU	
	At 100°F	At 210°F
1	180–240	
2	280–360	
3	490–700	
4	700–1,000	
5		80–105
6		105–125
7 comp.*		125–150
8 comp.*		150–190
8A comp.*		190–250

* Oils marked "comp." are those compounded with 3 to 10 per cent of acidless tallow or other suitable and accepted animal fats. The oils marked "comp." are essentially intended for wormgearing but may also be used for any gearing where the viscosity is considered suitable for the application.

Table 15-2

Type and size of unit, main gear low-speed centers	Ambient temp., °F	
	15–60°F, use AGMA No.	50–125°F, use AGMA No.
Parallel shaft (single reduction):		
Up to 8″	2	3
Over 8 up to 20″	2	4
Over 20″	3	4
Parallel shaft (double reduction):		
Up to 8″	2	3
Over 8 and up to 20″	3	4
Over 20″	3	4
Parallel shaft (triple reduction):		
Up to 8″	2	3
Over 8 and up to 20″	3	4
Over 20″	4	5
Planetary-gear units:		
OD housing up to 16″ dia	2	3
OD housing over 16″ dia	3	4
Spiral or straight bevel gear units:		
Cone distance up to 12″	2	4
Cone distance over 12″	3	5
Gear motors	2	4
High-speed units*	1	2

* For speeds over 3,600 rpm or pitch-line velocities over 4,000 fpm. It may be desirable to use a lubricant one grade higher or lower than those specified, depending on operating conditions.

ing, the lubrication offers a different problem. Also, since the worm is generally made from steel and the worm wheel from bronze, an extreme-pressure lubricant may not prove to have the same beneficial effect since welding action between steel and bronze is not apparent. In general the best result in wormgear lubrication is obtained with a steam cylinder oil or an oil containing 3 to 10 per cent acidless tallow. The American Gear Manufacturers Association makes recommendations as indicated in Table 15-3.

Table 15-3. Cylindrical- and Double-enveloping Wormgear Units

Worm centers	Worm speed up to rpm	Ambient temp., °F		Worm speed above rpm*	Ambient temp., °F	
		15–60	50–125		15–60	50–125
Up to 6″ inclusive:						
Cylindrical worms..........	700	7 comp.	8 comp.	700	7 comp.	8 comp.
Double-enveloping worms...	700	7 comp.	8A comp.	700	8 comp.	8 comp.
Over 6″ centers up to 12″:						
Cylindrical worms..........	450	7 comp.	8 comp.	450	7 comp.	7 comp.
Double-enveloping worms...	450	8 comp.	8A comp.	450	8 comp.	8 comp.
Over 12″ centers up to 18″:						
Cylindrical worms..........	300	7 comp.	8 comp.	300	7 comp.	7 comp.
Double-enveloping worms...	300	8 comp.	8A comp.	300	8 comp.	8 comp.
Over 18″ centers up to 24″:						
Cylindrical worms..........	250	7 comp.	8 comp.	250	7 comp.	7 comp.
Double-enveloping worms...	250	8 comp.	8A comp.	250	8 comp.	8 comp.
Over 24″ centers:						
Cylindrical worms..........	200	7 comp.	8 comp.	200	7 comp.	7 comp.
Double-enveloping worms...	200	8 comp.	8A comp.	200	8 comp.	8 comp.

* For wormgears of either type operating at speeds above 2,400 rpm or 2,000 fpm, rubbing speed may require forced-feed lubrication. In general, a lubricant of lower viscosity than shown in Table 15-3 may be used with a forced-feed system.

The pour point of any lubricant selected should be below the lowest operating temperatures expected.

Where heavy loads are encountered in all types of gears except wormgearing and hypoid gears, the so-called mild extreme-pressure lubricant should be employed as an additional insurance against scoring and incipient or initial pitting. The viscosity of the extreme-pressure lubricant should generally be equal to that recommended for straight mineral oils. AGMA standard 252.01 deals with mild extreme-pressure lubricants for industrial enclosed gearing.

Hypoid gears, Spiroid®* gears, Planoid®* gears, and Helicon®* gears all have a greater sliding action than any other gearing except wormgearing, and in addition both members are generally made from steel; therefore, an extreme-pressure lubricant should be used for best results at least during the "break-in" or "run-in" period, after which a regular mild extreme-pressure lubricant will often be satisfactory.

15-3. Gear Wear and Failures Caused by Inadequate Lubrication. All gears are to some extent dependent on lubrication. Metallic gears are particularly dependent on lubrication, and the life of nonmetallic gears is increased considerable by the use of the proper type and volume of lubricant.

Metallic gears operating dry will wear out quickly. The life of nylon gears or gears of comparable material is governed more by breaking strength than by wear; however, the life of the gear teeth prior to breaking is considerably reduced when no lubricant is provided.

* Registered trade-marks of the Spiroid Div., Illinois Tool Works, Chicago, Ill.

Wear is a relative term and must be used in our terminology with a great deal of discretion. The American Gear Manufacturers Association[3] defines *gear-tooth wear* as "loss of material from the contacting tooth surfaces resulting from the friction of metal-to-metal contact or the abrasive action of foreign matter passing thru the gear mesh." It has been said: "A gear wears out until it wears in after which it never wears out." After a gear has worn in, it will have an indefinite life, provided it is properly lubricated with a clean lubricant of the proper viscosity and film strength, and operating within its rated capacity.

Destructive wear is a deterioration of the tooth surface or a change in the tooth shape that may appreciably shorten the life or impair the smoothness of transition from one tooth to the other. The types of wear or gear-tooth failures for which lubrication or lack of lubrication is generally considered as the main cause of trouble are:

1. *Abrasive wear*, a surface injury caused by too thin or low-viscosity lubricant; a lubricant of a satisfactory viscosity at normal operating temperatures but containing abrasives in suspension from scale and sand not properly removed from the castings or gear blanks, etc.; impurities of the oil from the surrounding atmosphere or burrs detached from tooth surfaces or other machined surfaces.

2. *Scratching*, an aggravated form of abrasive wear. It shows up as scratch lines or tears of the metal in the direction of sliding. It may be caused by foreign particles brought up into the mesh by the lubricant or material embedded in the tooth faces or weld splatter, etc., loosened by the lubricant and vibration and carried into the mesh by the lubricant.

3. *Scoring*, scuffing, seizing, galling denoting essentially the same type of failure and generally caused by tearing out of small particles welded together as a result of metal-to-metal contact, and may also be caused by loading beyond the design limit for the type of lubricant used. A mild extreme-pressure lubricant will often prevent this type of failure because of its ability to prevent welding.

4. *Corrosi e wear*, a surface deterioration caused by moisture or other contaminants such as excessive acid in the lubricant. Rusting due to contamination with water from condensation in localities where the temperature fluctuates widely. Some extreme-pressure lubricants are corrosive in action and may cause corrosive wear.

Where gear-tooth loads are high, the best insurance against abrasive wear is to select a heavy-bodied lubricant. Tests have proved that the use of a 600 SSU lubricant as against a 400 SSU lubricant has increased the spalling resistance by 100 per cent and reduced the efficiency by only 0.4 per cent.

Under load conditions requiring lubricants to have powerful antiwear additives, the character of these additives can influence metallurgical failures. Under moderate loads, and with the use of mild antiwear oils, surface deterioration and distortion due to friction are relatively insignificant. Laboratory and field experience have proved that improper selection of the type of additive as well as quantity of the additive employed fór specific operating conditions can shorten fatigue life and cause premature surface failure such as spalling.

Certain types of tooth design make lubrication more of a problem. Enlarged-addendum or all-addendum pinions are subject to higher sliding velocities for the same rpm when compared with equal-addendum gearing, and since greater sliding is involved more attention must be given to proper lubrication.

Under condition of heavy load, employing high-hardness gear teeth of, for example, the carburized type in combination with high sliding velocity such as experienced in hypoid gears, lack of effective lubrication can result in high burnish and eventually may develop into a large number of short parallel ridges across the tooth surface in the direction of the sliding and in a relatively short time will make the gears inoperative.

15-4. Methods of Retaining Lubricants in Gearboxes. The design engineer should consider the method of retaining the lubricant inside the gear enclosure just as much a part of the lubrication as the lubrication itself, since, if adequate and effective means are not provided for retaining the lubricant, the gear drive will not function to the satisfaction of the ultimate user.

There are many types of oil seals and oil-seal arrangements; therefore, the designer should select the type most suitable for the application.

Radial Seal. Figure 15-11 shows two types of widely used and effective mechanical-type oil seals. The sealing element is generally made of a pliable material such as impregnated tanned leather, silicones, synthetic rubber such as neoprene, Buna N, etc. The sealing element requires some means of "hugging" the shaft to prevent oil leakage due to wear of this element, etc., generally accomplished by the use of a carefully selected garter-type spring designed to provide the right tension to seal effectively and yet not create high enough friction to cause the shaft and the sealing element to overheat and become brittle. An important point for the designer to consider when employing this type of seal is that the shaft portion on which the seal rides must be smooth. To specify a ground surface finish is not enough. It has been found that for best results the shaft surface at the seal should have a surface finish of 16 microinches or less. The seals must be handled with care. It is important that the sealing

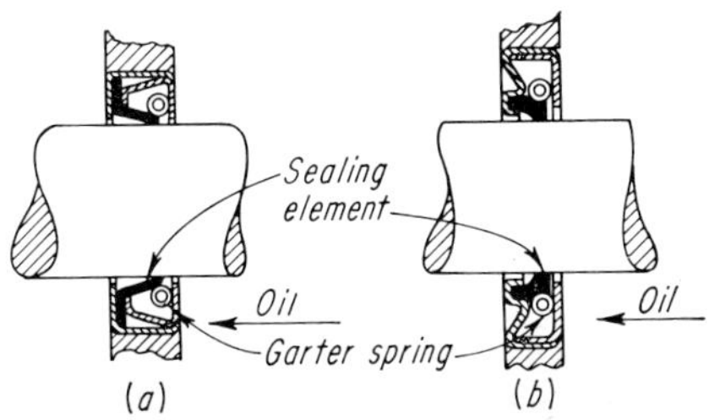

FIG. 15-11. Radial-type oil seal.

element be free of nicks and surface imperfections at the sealing surface, as otherwise it cannot prevent oil leakage. Normal care should be exercised when assembling these seals on their respective shaft. Where they are to be pushed "down" over keyways or sharp edges, it is important to protect the element by the use of a smooth sleeve or cover the sharp edges of the keyway by a material such as "scotch tape" or its equivalent. The sealing lip contacting the shaft should for best result be coated with a good grade of grease.

Face-contact General-purpose Seal. Numerous variations are available in regard to spring arrangements, means of driving the rotary part of the seal, as well as methods of attaching the seal to the shaft. The one main feature in common among all face-type seals is the two lapped sealing surfaces, as they must, in order to seal effectively, be held to an unusually high degree of flatness. Seals of this type will seal effectively up to 200 psi (see Fig. 15-12).

Concentric Radial Oil-groove Seal. This seal is extensively used in both geared speed reducers and increasers and other apparatus where design conditions justify. It is of simple, yet effective, design. It is almost invariably made as an integral part of an end cap. The radial clearance between shaft and seal should be relatively small and in the order of 0.006 to 0.012″, depending on the shaft diameter. The seal may be made from practically any material, but in order to prevent possible scoring of the shaft, in case of bearing failure, a material such as aluminum or bronze may be preferred. The radial grooves should be of sufficient depth and width to carry off any oil leakage along the shaft and provided with oil-drain back holes of sufficient size and number to provide free flow back into the gear case or oil reservoir (see Fig. 15-13).

15-5. Methods of Cooling Gear Lubricant. As indicated in Art. 15-2, the lubricant viscosity is to a considerable extent influenced by temperature; therefore, the designer must give proper consideration to the design of gearboxes in regard to thermal capacity and methods by which the lubricant may be kept at a suitable temperature in order to lubricate gears and bearings effectively and economically. In speed-reduction gear drives where the pitch-line velocity is relatively low, the design surface

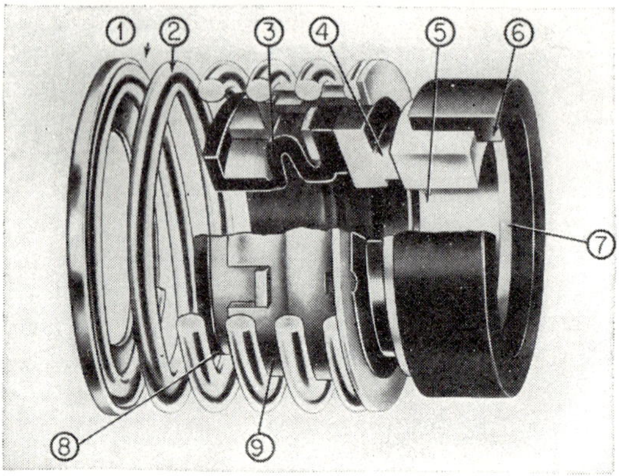

FIG. 15-12. Face-contact general-purpose seal. (1) Spring retainer. (2) Spring. (3) Synthetic rubber bellows. (4) Sealing washer. (5) Lapped sealing faces. (6) Seat ring. (7) Floating seat. (8) Driving band. (9) Driving-band pin.

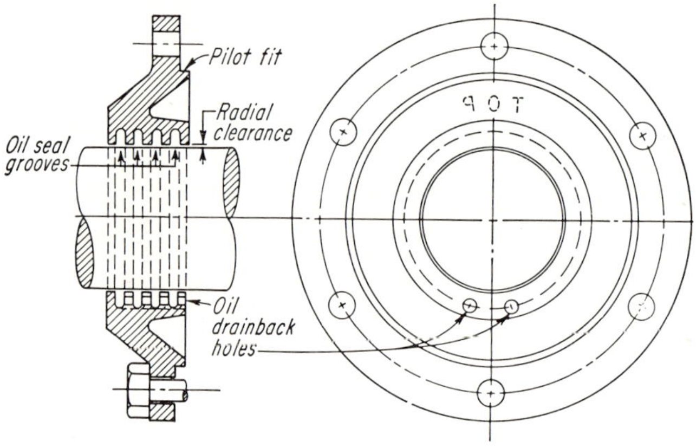

FIG. 15-13. Combination end cap and radial oil-groove seal.

area may be entirely adequate to dissipate the heat generated by the churning of the oil, the friction losses in the bearings, together with the sliding and rolling losses at the gear mesh. Chapter 14 deals with gear efficiency and may be consulted in determining the required heat removal where really close values are required. Chapter 13 discusses *thermal ratings* of gears.

Oil coolers or heat exchangers should in general be selected with a liberal safety margin. It is recommended that they be based on the maximum expected ambient temperature as well as the highest anticipated load condition and in addition a factor to compensate for deposit of impurities in the tubing and/or plate-type cooling grids. For the purpose of selecting the heat exchanger or cooler it is generally sufficient to assume an efficiency loss of 2 per cent for each gear reduction when spur, helicals, herringbone, bevels, and spiral bevels are employed. For wormgearing the efficiency loss for the selection of a suitable cooler may be assumed to be 1 − [100 − (gear ratio/2)]; where closer determinations are required refer to AGMA standards.[9,10] The American Gear Manufacturers Association[11] gives formulas and experience curves for determining the natural thermal capacity of gear drives.

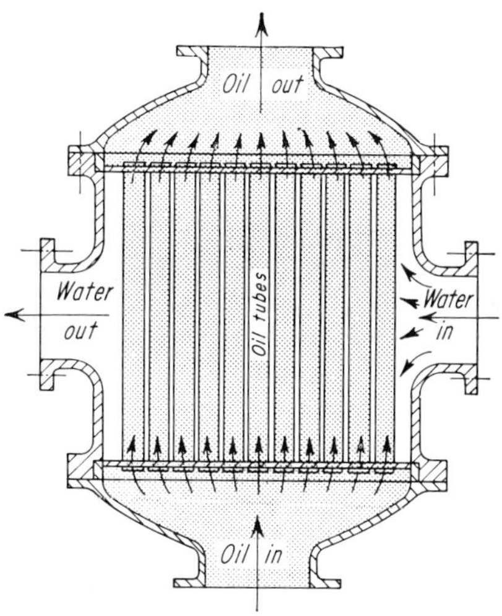

FIG. 15-14. Tube-type oil cooler.

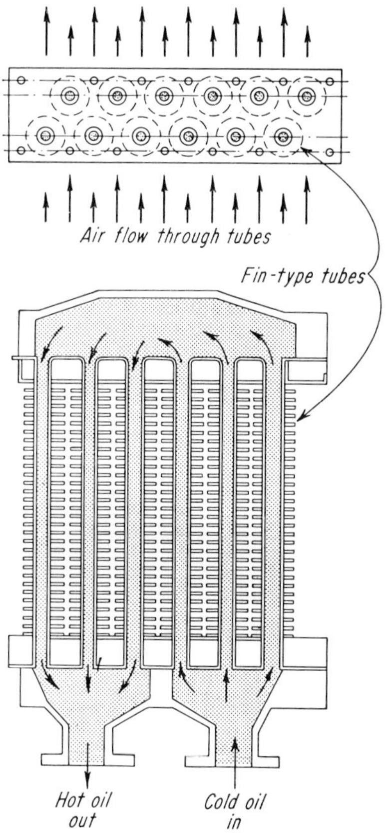

FIG. 15-15. Air-to-oil cooler.

There are many makes of oil coolers or heat exchangers. The basic principle of all types of cooling devices is the same. Their purpose is to act as a self-contained unit through which two liquids may flow or one liquid and an air stream separated from one another by metal tubing or walls. One liquid is hot and the air or the other liquid cool. While the hot liquid is going through the exchanger the cool liquid or air surrounds the hot liquid, tending to create an equilibrium of temperatures; in other words the hot liquids become cooler and the cool liquid or air becomes warmer.

The cooling medium in gear-unit cooling systems is generally water; however, where water is not generally available or is scarce, air is circulated by fans over and around the cooling tubes and/or plates. Air coolers in general take on larger proportions when compared with "water-to-oil" coolers.

Figure 15-14 illustrates schematically the principle of a water-to-oil cooler.

A typical schematic general arrangement of a cooling system for gear drives is illustrated in Art. 15-1 as Fig. 15-5.

Figure 15-15 illustrates a typical general-purpose type of air-to-oil cooler.

This chapter has dealt with gear-unit lubrication in general, and it is intended to be of aid to gear-unit designers as well as users.

Research on gears and gear lubrication is continually being pursued by gear manufacturers and major lubricant suppliers; therefore, for full information as to new developments, it is recommended that the literature of the American Gear Manufacturers Association, One Thomas Circle, Washington 5, D.C., and the American Society of Lubrication Engineers, 343 South Dearborn Street, Chicago 4, Ill., be referred to.

15-6. Aircraft-gear Lubrication. There is a steady increase in the amount of gearing used in aircraft and space-vehicle designs. Extreme lubrication problems exist in this field, and in view of this, some general design criteria are included to cover some of the peculiarities in this field.

Amount of Oil. In a small aircraft gear unit, the amount of oil is the least amount that can be pumped through an orifice in the gearbox and maintain a reasonable pressure. Because of the hazard of clogging with lint or small specks of scale or other foreign materials, a nozzle of 0.030″ diameter is generally considered a minimum. At the maximum predetermined operating temperature, a minimum pump pressure of 25 psi is generally considered acceptable. These two limits can be used to obtain the minimum pump capacity and oil quantity required.

In larger gearboxes, the amount of oil is a function of the power transmitted. For spur, helical, and bevel gears (all coplanar axes) Table 15-4 may be used as an approximate guide for the amount required provided the amount is not below the minimum discussed above.

Table 15-4. Amount of Oil per Gear Reduction for Aircraft Gears

*Ratio of hp to gal per min**	*Comment*
400 hp per gpm	A generous supply. Temp. rise should be under 50°F
800 hp per gpm	Probably OK if gears and bearings have good efficiency. Temp. rise should be under 75°F
1,200 hp per gpm	A lean supply. Use only for a high-efficiency design. Temp. rise may be under 100°F
1,600 hp per gpm	A starvation supply. Must be a very high efficiency unit. Temp. rise may exceed 100°F

* In *double*-reduction gearing, it is assumed that losses double; therefore, double the transmitted horsepower before dividing by the total oil flow.

Although Table 15-4 is intended primarily for aircraft, it indicates a good principle to use in checking high-horsepower marine and industrial units. The general rule of thumb of 0.5 gal per in. of face for industrial gearing may be too lean for many high-capacity gear units.

When wormgears, hypoids, Spiroids, and other offset-axis gears are used in aircraft, the efficiency is quite variable (depending on ratio and geometry of drive). In these cases, it is necessary to make a heat-loss calculation for the gears and bearings and then evaluate or calculate the oil flow to keep the temperature within allowable limits.

Application of Oil. The smaller aircraft gearboxes (100 hp or less) may get by with a splash system and air cooling when employing a "finned" box even with pitch-line speeds up to as high as 5,000 fpm and rotational velocities up to 8,000 rpm.

Somewhat larger gear units (up to about 300 hp) may operate satisfactorily using oil nozzles directed at the gear webs or the gear mesh, or both. An oil mist is violently circulated in the gearbox and several gears and bearings having no direct nozzle will still get adequate lubrication and cooling.

The more powerful gear drives (500 to 5,000 hp) tend to require a nozzle to each mesh and each bearing. The teeth will often heat up to as much as 500°F even with an incoming oil temperature of 200°F and with a generous oil flow directed to the web, hub, and/or gear mesh.

The subject of where to place the oil nozzle on a critical gear mesh has been debated for years by experts in the field. In the opinion of Darle W. Dudley,[12] based on experience gained from numerous experiments and tests in aircraft work and from discussion with engineers from various aircraft concerns throughout the country as well as abroad:

1. In most cases, a nozzle can be developed so that its spray pattern and trajectory are satisfactory for location on either the *incoming* or *outgoing* sides of the mesh.
2. *Cooling* is often somewhat more effective on the outgoing side. *Lubrication* is somewhat more effective on the incoming side.
3. Closely controlled scoring tests with lean oil flows tend to show the highest load capacity with oil on the incoming side.
4. Spur gears with 1″ or more of face width and 10,000 fpm or more pitch-line velocity are sensitive to oil trapping when the nozzle is on the incoming side and may fail because of undue vibration caused by oil "wedging" and rough running.

Table 15-5. Aircraft Gear Oils

Type	Military Spec.	Viscosity, cs		Top limit temp., °F	Pour point (max.), °F	Application
		100°F	210°F			
Grade 1005. .	Mil-0-6081B	5.0	1.7	175	−75	Jet-engine accessory gears. Moderate power
Grade 1010. .	Mil-C-6081B	10.0	2.5	175	−70	Jet-engine accessory gears. Moderate power
Grade 1065. .	Mil-0-6082B	105	12	175	0	Aircraft reciprocating engines. Accessory and power gears
Grade 1100. .	Mil-0-6082B	260	21	175	+10	Same as above
Synthetic. . . .	Mil-L-7808D	14.2	3.6	300	−75	Accessory gears, main power gears
Synthetic. . . .	(Turbo 35) none	38	7.9	300	−65	Heavier diester than 7808 used in turbo-prop engine and accessory and power gears
Synthetic. . . .	(Target) Mil-L-9236A	300 max.	10 min.*	400	−75	Would be used as a 400°F jet engine and gear lubricant
Synthetic. . . .	Mil-L-25336	12.7	3.3	300	−75	Same as 7808. Used where more load-carrying ability is required
	Mil-0-6086					
	Grade L	31	5	175	−40	Gearboxes where extreme-pressure properties are required. Helicopter gearboxes, rocket gears, radar gears
Gear oil.	Grade M	70	8	175	−20	

* 3.0 min at 400°F.

Kinds of Oil. Well-engineered aircraft gears continue to amaze gear experts by their ability to run on thin oils having relatively poor load-carrying capacity. Thin oils are needed in aircraft work to permit starting at temperatures as low as −70°F and to keep oil churning losses and oil pumping losses to a minimum.

The ability of well-engineered aircraft gearing to tolerate marginal lubricant stems from several things:

1. The tooth surface finish is exceptionally smooth and often as fine as 20 microinches (rms) or better. Surface hardness is frequently 60 Rockwell C or higher.
2. Accuracy of spacing, lead, and profile is held very close. (Tooth-to-tooth spacing is usually within 0.0002″.)
3. Profile and helix modifications are used wherever appreciable deflections occur.
4. Gearboxes and oil systems are kept extremely clean and free from foreign material. (Filters as fine as 10 microns or finer are often used.)

The kinds of oil most frequently used in aircraft gear work are shown in Table 15-5 These oils are generally made to meet a military specification. The specification covers properties of the fluid and certain performance tests that the fluid must meet. The exact molecular composition is not specified. This makes it possible for different oil companies to supply fluids of somewhat different chemical composition to meet a given specification. Users are sometimes puzzled that a drum of oil bought to a given military specification from one oil company will not behave quite the same as a drum of oil from another company.

15-7. Grease Lubrication of Aircraft Gears. Most of the gears used in aircraft actuators operate intermittently and can generally get by using grease lubrication. Small instrument and radar gears also use grease lubrication. Amounts of power are low. The gears may or may not run intermittently.

When very low speed gears are lubricated, the grease will stick to the ends and back sides of the teeth. As the grease is rubbed off the working side of the tooth, a tiny amount of oil will bleed out of the grease mixture and seep around to keep the working side of the tooth wet. In high-speed jobs that tend to fling the grease clear of the teeth, it is necessary to have the gear casing so constructed that the grease that is flung off eventually shakes back into the path of some gear and is then tracked back into the meshes so as to keep some oil on the teeth.

Technically a *grease* is a more or less solid mixture of an oil in a soap. The lubrication job is done by the small amounts of oil that bleed out of the soap and wet the rubbing surface.

Table 15-6. Aircraft Greases

Military Spec.	Oil type	Thickener	Temp. range, °F	Applications
Mil-G-3278A......	Diester	Lithium soap	−65 to 250	General-purpose lightly loaded gear drives
Mil-G-7118.......	Diester	Lithium soap	−65 to 250	Aircraft gears and actuator screws (more load-carrying ability than Mil-G-3278A)
Mil-G-7187.......	Petroleum or diester	Lithium soap plus graphite	−40 to 200	Starter gear and selector-valve gears when frequent lubrication not possible
Mil-L-3545........	Petroleum	Sodium soap	−20 to 300	High-speed high-temp. application
Mil-G-25013A	Silicone	Nonsoap	−65 to 400	High-temp. aircraft actuator, gearboxes, couplings

Grease lubrication of gears works when the rubbing surfaces can manage to keep wet with oil and when there is no particular cooling problem.

Table 15-6 shows some typical aircraft greases. The ones shown have been widely used on aircraft gears.

REFERENCES

1. Himes, W. H.: "Splash Lubrication Fails at Subcritical Speeds," *Machine Design*, December, 1932, pp. 27, 28, 53.
2. American Gear Manufacturers Assoc.: "Standard Nomenclature—Gear Tooth Wear and Failure," AGMA 110.02 (ASA B6.12—1954).
3. American Gear Manufacturers Assoc.: "Lubrication of Industrial Enclosed Gearing," AGMA 250.02.
4. American Gear Manufacturers Assoc.: "Data on Extreme Pressure Lubrication," AGMA 254.04.
5. Richardz, F.: Lubrication of Industrial Gear Drives, *Lubrication Eng.*, Vol. 4, pp. 25-27, February, 1948.
6. Shipley, Eugene E.: Design and Testing Consideration of Lubricants for Gear Application, *Lubrication Eng.* April, 1958, pp. 148-151, 191.
7. *Chem. & Met. Eng.*, Vol. 36, p. 618, 1929.
8. The Texas Company: Viscosity and Viscosity Index Publication, *Lubrication*, Vol. 31, No. 4, April, 1945.
9. American Gear Manufacturers Assoc.: "Practice for Single- and Double-reduction Cylindrical-worm and Helical-worm Reducers," AGMA 440.03.
10. American Gear Manufacturers Assoc.: "Practice for Double-enveloping Worm Gear Speed Reducers," AGMA 441.02.
11. American Gear Manufacturers Assoc.: "Thermal Capacity for Helical and Herringbone Gear Speed Reducers," AGMA 423.01.
12. Dudley, Darle W.: "Practical Gear Design," pp. 305-306, McGraw-Hill Book Company, Inc., New York, 1954.

Chapter 16

GEAR CUTTING

By

STUART J. JOHNSON, *Chief Engineer, Small Tools Division, Barber-Colman Company, Rockford, Ill.*

Toothed gear wheels may be made by cutting teeth, die casting or forming teeth, and by grinding. Teeth may be finished by grinding, shaving, or lapping or they may be used as cut. In this chapter the gear-tooth-cutting methods of milling, hobbing, shaping, shear cutting, and broaching will be covered.

GEAR MILLING

Gear teeth are produced by milling with a form-milling cutter. After each tooth space is milled, the gear blank is indexed to the next cutting position. Gear milling can be applied to the roughing and finishing of spur, helical, and straight bevel gear teeth.

16-1. Application of Gear Milling. Gear milling is universal in application. In practice, however, it is usually confined to producing replacement gears or small-lot production, roughing and finishing coarse-pitch gears, and finish-milling fine-pitch gears having special tooth forms. Depending upon the size of the gear and the capacity of the machine, standard milling machines are used with either manual or automatic indexing mechanisms. On a production basis, where milling is used to produce large coarse-pitch gears, special automatic indexing gear-cutting machines are used. Small automatic gear-cutting machines are applied to the volume production of fine-pitch gears having special tooth forms.

Gear milling finds its widest application in roughing and finishing external spur gears and in roughing external helical gears. Straight bevel gears are roughed out by milling prior to generating; and in some cases these gears are finished by milling. In addition to gears, milling is applied to sprockets, splines, racks, ratchets, and other forms. Although it is usually applied to external tooth forms, internal tooth forms can also be produced by milling.

Milling can be used to machine almost any tooth form. The teeth can be either equally or unequally spaced and the tooth forms can be either symmetrical or unsymmetrical. In milling the small gear of a two-gear cluster or teeth adjacent to a shoulder, adequate clearance is required to allow the cutter to produce full-depth teeth for the specified face width.

Standard-type milling machines are available in various sizes for application to gear milling. The size and pitch of the gear that can be produced will largely depend upon the machine capacity available.

In addition to standard milling machines, special machines are available for the primary function of milling gears and other tooth forms. Usually referred to as "gear-cutting machines," they are available to cut gears of 1 diametral pitch* and up to 9 ft diameter. For fine-pitch gears, small machines with automatic features provide a method applicable to volume production.

FIG. 16-1. Large gear after finish milling of teeth. *(Courtesy of Gould and Eberhardt Div., Norton Co., Worcester, Mass.)*

Although milling can produce gears of accuracy equivalent to that which is produced by other gear-cutting methods, it is normally not used where accurate tooth spacing is required. Whether it is hand-operated or automatic, the accuracy of the tooth spacing is directly dependent upon the indexing mechanism. Since milling gear teeth is a form-cutting operation in which each cutter tooth produces the complete gear-tooth profiles, fine surface finishes can be obtained. Depending upon conditions, such as gear material, milling-cutter design, and milling setup, surface finishes in the range of 15 to 20 microinches can be produced.

FIG. 16-2. Gear-milling cutter and spur gear.

16-2. Principles of Gear Milling. Gear milling is a machining process in which the gear teeth are produced by a form cutter. In the case of spur gears, the form on the cutter teeth is reproduced on the gear. When helical gears are milled, the cutter-tooth form is not reproduced on the gear.

Gear cutters are not universal for a complete range of teeth. To produce theoretically correct gear teeth, the cutter-tooth form must be designed for the specific number of teeth. However, where a small error in tooth form is acceptable, cutters which are designed for a range of teeth can be used. Standard cutters which are discussed in the chapter on gear-cutting tools are available for established ranges of teeth. The form on the cutter is made correct for the lowest number of teeth in that particular range. Thus, all teeth within the range are provided with sufficient tip relief. The same form is produced on all tooth spaces within that range.

* Coarser-pitch gears may be cut with special cutters such as an end-mill cutter.

The tooth form of a standard 14½° involute gear cutter is actually a composite of the involute and cycloidal forms. The purpose of this form is to allow gears with a 14½° pressure angle and a small number of teeth to be produced by a cutter. This requires a modified involute form to allow gears and pinions to be meshed without interference. Consequently, gears cut by this type of cutter will not mesh properly with those produced by other types of standard involute gear-cutting tools. They will, however, mesh properly with other gears of the same pitch which are cut by a standard gear cutter. Gear cutters with a standard involute form can be ordered special, and gears produced by these cutters will roll properly with gears produced by other types of standard involute gear-cutter tools.

For spur gears, a gear cutter will produce tooth spaces which have the same form as the cutter teeth. Since the cutter axis is at 90° to the gear axis, the axis of the cutter is contained in the transverse plane of the gear. Under these conditions, the gear-tooth profiles are finished when the cutter teeth pass through this plane. Therefore, the form on the cutter teeth is reproduced in the gear.

The form on helical-gear-tooth spaces will not be the same as the cutter-tooth form. In milling helical teeth, the cutter axis is usually set at the helix angle of the gear. At this setting, the axis of the cutter is contained in the normal plane through the center of the gear-tooth space. Under these conditions, only one point on the finished profile is produced in this normal plane. All other points on the finished profile are produced in different planes. Therefore, the form on the cutter teeth is not reproduced in the gear. In addition to the setting angle, the diameter of the cutter affects the gear-tooth form and must be considered in designing the proper cutter.

In practice, standard cutters are sometimes used to cut helical gears. The range of the cutter to be used can be determined by finding the equivalent number of spur teeth. For small helix angles, a standard cutter for the proper range of teeth will produce a close approximation to the desired gear-tooth form. However, gear cutters for gears with large helix angles are usually designed specifically to produce the desired form.

16-3. Milling-machine Operation. Milling gear teeth is primarily a form-milling operation to which the general techniques of milling apply. Except for the addition of an indexing mechanism, and the determination of lead gears and setup for helical teeth, machine operation is comparable with conventional milling practice.

With milling considered as a standard-type machining operation for which ample information is available, the following discussion is confined to some of the more important aspects of milling gear teeth.

In selecting the proper cutter to be used, attention should be given to the number of teeth in the gear and the tooth form specified. For spur gears, standard range cutters as discussed in Chap. 22, "Gear-cutting Tools," can be used to produce gears having the standard gear-cutter form. It is only necessary to select the proper range of cutter for the number of teeth specified in the gear. Where the tooth form produced by a standard range cutter is not applicable, a special cutter designed to produce the specified form is used.

Theoretically, a special cutter is required to mill a helical gear of a specified tooth form and helix angle. Determination of the equivalent number of spur teeth provides an approximation which is often used to find the standard range cutter required. The equivalent number of spur teeth is found by dividing the actual number of gear teeth by the cosine of the helix angle cubed.

Cutters for straight bevel gears are made as standard range cutters. Cutter selection depends upon the number of teeth in both the gear and the pinion. The number of teeth for which the cutter should be selected can be found by

$$\tan C = \frac{N_G}{N_p} \tag{16-1}$$

Number of teeth to select cutter for gear

$$\frac{N_G}{\cos C} \qquad (16\text{-}2)$$

Number of teeth to select cutter for pinion

$$\frac{N_p}{\sin C} \qquad (16\text{-}3)$$

where N_G = number of teeth in gear
N_p = number of teeth in pinion

For miter gears, only one cutter is required.

In using a gear cutter, the correct gear-tooth form is produced only when the cutter is set to its proper depth of cut. A reduced depth of cut increases the thickness of the gear tooth, but the form is in error. A reduction in gear-tooth thickness to provide backlash is a design feature which should be incorporated into the cutter as an increase in cutter-tooth thickness. A more involved and time-consuming method of thinning the gear teeth requires that an additional cut be taken on each profile of each gear-tooth space. After all the gear-tooth spaces have been cut, the gear is indexed through

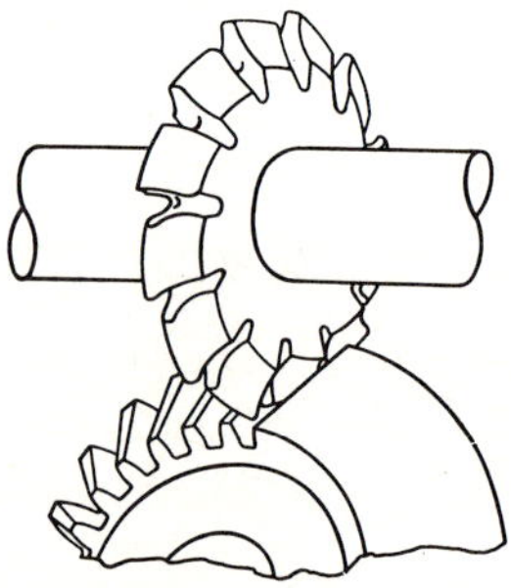

FIG. 16-3. Gear-milling cutter and bevel gear.

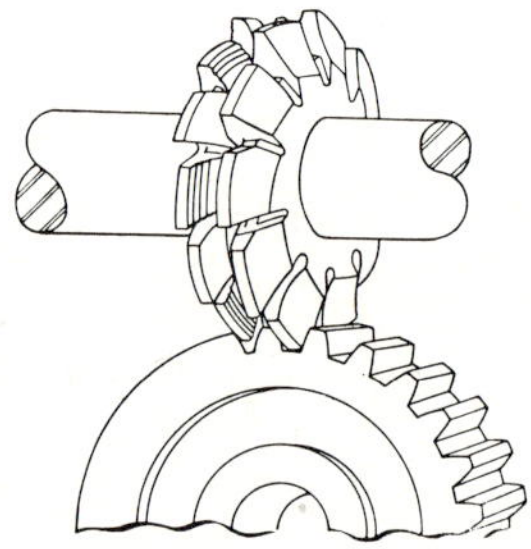

FIG. 16-4. Roughing and finishing cutters ganged together.

a slight angle based upon one-half the amount of backlash. A cut is then taken on one side of each tooth space. An identical operation is performed on the opposite side of each tooth space.

With a single cutter, the cutter-tooth form is centered correctly with the gear axis so that a symmetrical tooth space is produced. More than one tooth space can be cut by using a gang of single cutters or a multiple tooth cutter. Normally, a roughing and a finishing cutter are ganged, with the finishing cutter being centered with respect to the gear. Multiple-tooth cutters are usually used to mill rack teeth. Gang-type cutters and multiple-tooth cutters are designed specifically for the job.

The feed for finishing gear teeth is normally in the range of 0.002 to 0.004″ per cutter tooth. Higher feeds can be used for roughing cuts. To improve the finish, reduced feeds may be required. Other factors which affect the feed include cutter design, tooling rigidity, machine capacity, and work material.

Cutter speed is primarily dependent upon the machinability of the gear-blank material. Steel at 100 BHN can be milled at approximately 150 sfm with high-speed steel cutters. If the hardness is increased to 300 BHN the speed should be reduced to approximately 50 sfm. Normally, reduced speeds are required for lower machinability rating of the work material.

16-4. Advantages and Limitations of Gear Milling. Gear milling can be used to advantage in producing replacement gears or a small quantity of gears on standard milling machines with tools which are usually less expensive than other types of gear-cutting tools. Milling can be applied as a roughing or finishing operation on gears, racks, splines, and many special tooth forms. In a number of cases, milling can produce tooth forms which would not be applicable to generating methods.

Gear-milling cutters are not universal for a complete range of teeth, as are standard gear hobs and shaper cutters. Helical gears usually require cutters which are designed for the specific part. Although milling can produce accurate gears, tooth-spacing accuracy is limited by the accuracy of the indexing mechanism, and most indexing mechanisms do not produce spacing accuracy comparable with generating methods.

GEAR HOBBING

Almost any external tooth form which is uniformly spaced about a center can be hobbed. It is not necessary that the form be symmetrical about an individual axis, but each tooth or group of teeth must be a duplicate of any other tooth or group. The kinds of work which can be hobbed include spur gears, helical gears, splines, serrations,

FIG. 16-5. Hobbing a helical pinion. (*Courtesy of Barber-Colman Co., Rockford, Ill.*)

worms, wormgears, and special forms. Many gears are finished-hobbed in one cut, others are hobbed with a semifinishing hob prior to shaving or grinding, and some are rough-hobbed before the finish-hobbing operation. Hobbing is applicable to mass production or to jobbing operations. Automatic machines can be designed for the mass production of a specific part. The ease of setup makes hobbing machines especially adaptable to small-lot production.

16-5. Application of Gear Hobbing. Although the hobbing process is often associated only with the manufacture of gears and splines, many other forms can be cut. It is also recognized as an economical means of producing ratchets, sprockets, and other special forms. Standard hobs have been established for some of these forms.

The versatility of hobbing makes it an economical method of cutting gears. One hob of a given pitch will cut the teeth of all involute spur and helical gears of the same normal pitch and pressure angle, including all numbers of teeth and helix angles.

The size of the teeth and the size of the work are limited only by the capacity of the hobbing machine on which the part is to be cut. Hobbing machines are available for cutting any diameter up to 200″. Face width is limited by the length of hob travel. Pitches from less than 1 to 300 diametral pitch have been hobbed.

The accuracy of parts produced by hobbing is determined by the accuracy of the machine, the rigidity of the tooling, the accuracy of the blanks, the care in mounting the hob and the work, and the accuracy of the hob. With an accurate machine and proper attention to the tooling, hobbing can produce AGMA quality 12 gears in the coarse pitches and AGMA quality 12 precision gears in the fine pitches. Precision machines are available for hobbing closer tolerance when necessary.

As an example, 8 diametral pitch gears of 3″ pitch diameter and with a face width of 1⅜″ are cut within AGMA quality 12* tolerances. The runout of the pitch diameter is held within 0.0007, the pitch tolerance is within 0.00025, and the profile tolerance is less than 0.00055. The lead variation on this gear with a 20° helix angle is held within 0.0005″.

As an example of fine-pitch accuracy, 64 diametral pitch spur gears with a pitch diameter of 0.7500″ and a face width of 0.093″are hobbed within AGMA quality 12 precision tolerances. Fine-pitch gears are normally inspected on a variable-center-distance fixture. The tolerances for this quality of gear are 0.0003″ for tooth-to-tooth composite tolerance and 0.0005″ for total composite tolerance. The tolerance for runout of the blank on this class of gear is 0.0015″.

The finish on the gear teeth is dependent upon accuracy of the hob, the rigidity of the tooling, and the amount of feed. With adequate tooling and a small amount of feed, a very fine finish can be obtained. Although it is difficult to inspect the finish on gear-tooth profiles, it is possible to obtain a finish of 15 microinches. On roughing operations, a coarse feed would be used, and the finish might be as rough as 150 microinches.

Hobbing is applicable to all types of gear materials. High-speed steel hobs are used to cut ferrous, nonferrous, and nonmetallic materials. Carbide-tipped hobs can be used for the mass production of nonferrous metals and nonmetallic materials if the hobbing machine can utilize the high speeds required by carbides. Steel can sometimes be hobbed in the hardness range of 40 to 48 Rockwell C with high-speed steel hobs. In this case, the part is normally rough-hobbed, hardened, and finish-hobbed.

Hobbing machines can also be adapted to perform special operations. Gear teeth can be chamfered with a special hob used on a standard hobbing machine. A crowned tooth form can be hobbed on a machine which continually changes the center distance between the hob and work according to the desired configuration of the crown. Taper root splines are cut on a machine which feeds the hob at an angle to the axis of the spline. Machines which have a work spindle set at an angle to the plane of the hob slide travel will cut tapered serrations. Special worm-gear machines are designed to feed the hob at 90° to the work axis. Worm gears are also hobbed on machines equipped with power feed in a radial direction. Worms and threads can be either hobbed or milled on a machine which allows the proper angular setting of the tool. Many other types of special parts can be cut on hobbing machines which are designed specifically for that part.

16-6. Theory of the Hobbing Process. The generation of a gear tooth is a continuous indexing process in which both the cutting tool and the workpiece rotate in a constant relationship while the hob is being fed into the work. As the hob is fed across the work once, all the teeth in the work are completely formed.

In most instances, the unusual feature of the hobbing process is that the form on the hob is not reproduced on the work. The largest application of hobbing is the cutting of involute gears with a hob having teeth with essentially straight sides at a given pressure angle. Hobs can also be designed with other than straight sides to produce the desired form on the work.

* See Chap. 9 for detail information on gear tolerances and Arts. 9-27 and 9-28 for data on AGMA quality numbers.

The hob is basically a worm which has been fluted and has form-relieved teeth. The flutes provide the cutting edges. Each tooth is relieved radially to form clearance behind the cutting edge, allowing the faces of the teeth to be sharpened while retaining the original tooth profile.

Figures 16-7 and 16-8 are schematic diagrams showing the relationship between the hob and the gear. The teeth of the hob cut into the gear blank in successive order and each in a slightly different position. Each hob tooth cuts its own profile, which in the case of an involute hob is straight-sided, but the accumulation of these straight cuts produces a curved form on the gear teeth. Therefore, instead of being formed in one profile cut as in milling, the gear teeth are formed a little at a time in a series of cuts. This is known as the *generating process* of cutting gears.

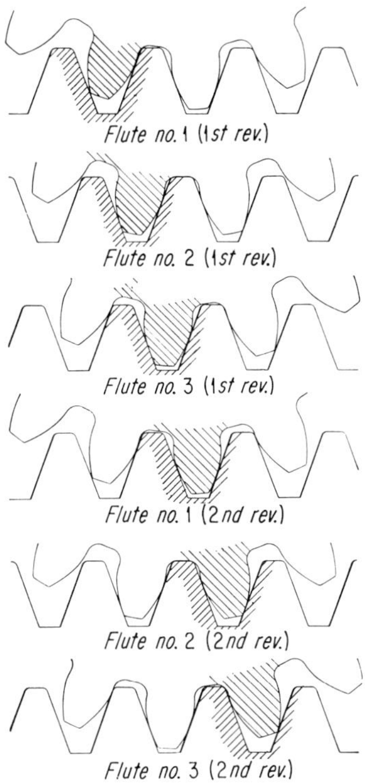

FIG. 16-6. How different hob flutes form the gear tooth.

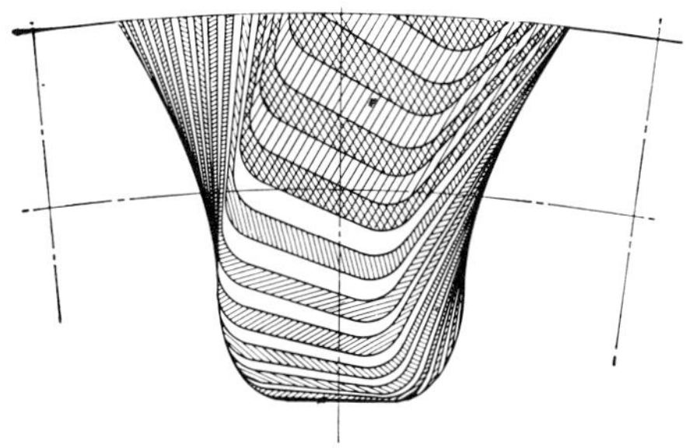

FIG. 16-7. Chip loads cut by successive hob teeth.

16-7. Hobbing-machine Relationships. The generating process requires accurate relationships between various elements of the machine in order to produce the desired results. The hob rotates in a definite relationship with the rotation of the work. For spur gears with a single-thread hob, the blank moves one tooth space while the hob rotates once. The rotation is timed by means of change gears. For helical gears, the rotation of the work is slightly retarded or advanced in relationship to the rotation of the hob, and the feed is also held in a definite relationship with the work and the hob.

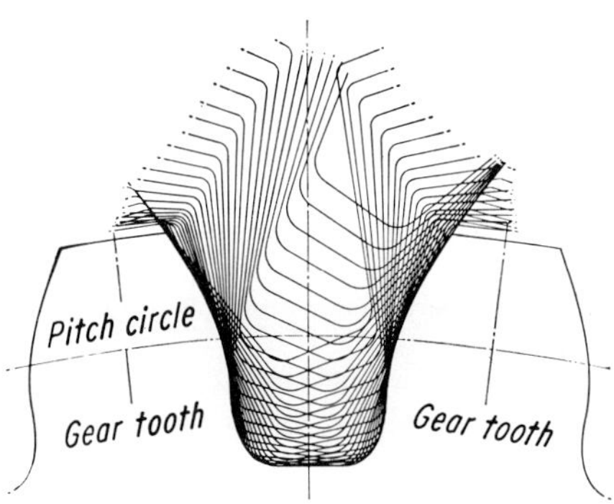

FIG. 16-8. Complete generating action of the hob.

Figure 16-9 is a schematic drawing of a typical mechanical hobbing-machine gear train. The drive to the hob and work spindles is obtained from a main drive motor through speed-change gears. From the driven shaft of the speed-change gears, the speed of the hob spindle is fixed by the gearing in the machine. The drive to the work spindle from the speed-change gears goes through index-change gears to the index worm and wormgear. Therefore, with a given set of index-change gears, a definite relationship between the rotation of the hob and the rotation of the work is maintained. Since the drive that feeds the hob into the work is taken from the index-change gears through feed-change gears to the feed screw, a relationship is established between the work and the feed. It is the relationship between the rotation of the hob, the rotation of the work, and the amount and direction of feed that enables the gear to be cut. Index- and feed-change gears establish the definite relationship for a specific gear. Speed-change gears can be changed without affecting the relationship between the hob, work, and feed.

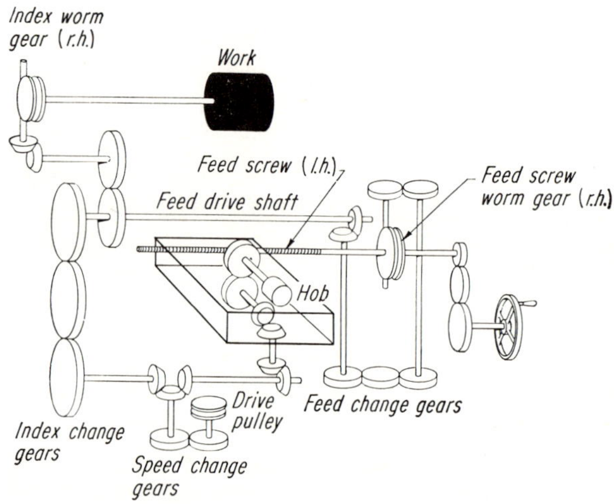

Fig. 16-9. Schematic drawing of a hobbing machine.

Some hobbing machines are equipped with a differential which holds a constant relationship between the hob, the work, and the feed even during independent movement of the hob slide. This is accomplished by holding the feed screw, the index worm, and the hob spindle in a constant geared relationship at all times, even when the feed is disengaged. A differential is recommended for hobbing helical work when two cuts are required or when a tooth must be held in relation to a feature on the part. It can also save a considerable amount of time in setting up the machine on short runs of helical parts.

The following relationships are typical for some makes of hobbing machines. To maintain the required relationship between the index and feed, there are two constants which are established by the gearing in the machine and which are used to determine the correct change gears. The machine-index constant is the number of revolutions of the hob spindle during one revolution of the work spindle when the index-change gears have a 1:1 ratio. The machine-feed constant is the distance in inches which the hob slide will advance during one revolution of the work spindle when the ratio of the feed-change gears is 1:1. These constants are shown on the machine.

For spur gears, the formula for determining the index-change gear ratio is

$$\text{Index gear ratio} = \frac{KT}{N} \tag{16-4}$$

where K = machine index constant
T = number of threads on hob
N = number of teeth on gear

The feed gear ratio for cutting spur gears is determined by dividing the feed by the machine-feed constant. The formula is

$$\text{Feed gear ratio} = \frac{F}{M} \tag{16-5}$$

where F = feed, in. per revolution of work
M = machine-feed constant

For helical gears, the basic formula for figuring index change gear is

$$\text{Index gear ratio} = \frac{KCT}{(CN) \pm 1} \tag{16-6}$$

where K = machine index constant
C = constant
T = number of threads on hob
N = number of teeth on gear

The selection of the plus or minus sign is determined by the hand of the hob, the hand of the gear, and the feed direction. C is a different number for each different set of conditions, depending upon the diametral pitch, the desired feed, and the helix angle. It is actually a constant whose primary purpose is to establish a relationship between the index- and feed-change gears so that the correct helix angle will be cut on a specific gear. It can be defined as the number of revolutions of the work as the hob is fed one axial pitch of the work. Mathematically, this can be shown by the following equation:

$$C = \frac{p_n}{F \sin \psi} \tag{16-7}$$

where p_n = normal circular pitch
F = desired feed, in. per revolution of work
ψ = helix angle of work

It is not necessary to use the theoretical value of C as determined above, since changing its value will affect only the amount of feed. Therefore, C can be assumed to be any value which is fairly close to the theoretical value. For convenience, C is usually assumed to be a whole number which, when substituted in the basic formula, will produce a ratio that is factorable exactly into suitable numbers of gear teeth. Of course, the farther the actual C is from its theoretical value, the farther the actual feed will be from the desired feed.

Actually, neither the index nor the feed gears have to be held to an exact ratio as long as the resulting combination of these gears will produce the desired lead. Normally, however, it is easier to obtain a satisfactory setup if the index gears represent the exact ratio. The feed can then vary slightly—usually not more than 0.000015″—from its theoretical value and still produce a lead which is satisfactory.

In order to establish the proper relationship between the index and the feed, the exact value of C which is used in the index formula must be used in figuring the feed.

$$\text{Feed} = \frac{p_n}{C \sin \psi} \quad \text{in. per revolution} \tag{16-8}$$

where p_n = normal circular pitch
C = exact C used in figuring index gears
ψ = helix angle of work

With the feed determined, the feed gear ratio is found as given for spur gears.

To obtain the tooth profile, cutting clearance, and tooth thickness for which the hob is designed, the hob must be accurately positioned. For spur gears, the hob swivel setting is equal to the lead angle of the hob. The actual procedure for determining the setting angle for helical gears is to add the lead angle of the hob to the helix angle of the gear if they are of opposite hands, and to subtract them if they are of the same hand.

16-8. Hobbing-machine Operation. As with most machining operations, a number of factors influence the optimum speed-feed combination. In hobbing, these factors include material being hobbed, finish, accuracy, machine capacity, hob specifications, and hob life. In any specific hobbing department, finish and accuracy are normally established items for the particular hobbing operation and the part to be cut. The problem is to determine a hob speed and feed for a given hob and machine capacity that will produce the established finish and accuracy at a given production rate with justifiable tool life.

Hob Speeds and Feeds. Hob speed (surface feet per minute) is primarily a function of the machinability of the material being hobbed, the hob specifications, the machine capacity, and the hob life desired. The machinability rating of a material varies with its structure and its hardness. It is commonly compared with the machinability of AISI B-1112 cold-drawn steel, which is rated at 100 per cent. The following table provides a starting point in determining a satisfactory sfm.

Diam. pitch	*Sfm* (100% *machinability*)
1–16	180
16–32	200
32 and finer	220

With other things being equal, an increase in machinability rating would warrant an increase in hob speed. However, the hob material and hob tooth form should also be considered. Where carbide-tipped hobs have been applied to nonferrous metals and nonmetallic materials, hob speeds have been increased beyond the operating speed range for high-speed steel hobs. In many cases, hob speed has been limited by the maximum machine speed available. In addition, the type of high-speed steel in the hob influences the speed which can be used and the wear that is obtained. Pointed hob teeth or forms with only a small corner radius may require reduced hob speeds to eliminate excessive wear. The top corner of a tooth on a gear hob does a great deal of work in forming the root fillet of a gear tooth. With a small portion of the cutting edge doing a big cutting job, too high a hob speed may cause premature cutting-edge failure.

The amount of feed applied to a specific job varies with the finish and accuracy required and with the machine capacity and tooling employed. Fine-pitch precision gears are usually hobbed within the range of 0.005 to 0.020″ per revolution. Gears coarser than 20 pitch are usually finish hobbed in the range of 0.010 to 0.030″ and are roughed or semifinished in the range of 0.040 to 0.100″ per revolution.

For single-thread hobs and even thread-tooth ratios, the size of the feed marks or scallops increases with an increase in feed rate. With a prime thread-tooth ratio and some hob runout, the feed marks are broken up, and it is often difficult to find a feed-mark pattern. In axial feeding, the feed marks follow a spiral path around the circumference of the part, and they can usually best be seen on the root surface. A high feed rate produces long, deep feed marks, resulting in a rough finish having peaks and valleys, which affects the accuracy of the part. The feed rate applied must be within the capacity of the machine and tooling used to clamp and drive the part. Lack of rigidity enhances deflections which often cause chatter marks and gouges on the hobbed surfaces. For a preshave hobbing operation, these surface irregularities are often too deep to be removed by shaving when the amount of shaving stock allowed is in the normal range.

Although improved designs and techniques which may promote equivalent or even better results are investigated and often successfully applied, the initial hob speed and initial feed for a new job are usually based upon past performance records for similar parts. For essentially the same kind of part being produced in several different hobbing departments, past experience varies for reasons which include the acceptance level of finish and accuracy, the conditions and capacity of the hobbing machines available, and the tooling designs employed. As a result, it is difficult to recommend a hob speed and feed for a specific job.

Estimating Hobbing Production. An estimate in advance of production is of prime interest in any machining operation. As applied to hobbing, the production estimate is required to provide production figures from which the cost of the part may be determined, and to ascertain the number of hobbing machines required for the desired amount of production.

The actual computation connected with finding the time required to hob a given part is relatively simple. However, there are many variables that must be taken into consideration. If the existence of these variables is appreciated and they are properly applied, the estimate should be fairly reliable. To determine the numerical values for these variables requires broad experience tempered with careful consideration for the particular job.

Basically, the expression for determining the hobbing time is the linear distance traversed by the hob carriage on the bed ways divided by the velocity of the hob carriage. The mathematical expression is

$$\text{Hobbing time} = \frac{NL}{T \times \text{rpm} \times F} \quad \text{min} \tag{16-9}$$

where F = feed, in. per revolution of work
L = total travel of hob carriage, in.
N = number of teeth on gear
T = number of threads on hob
rpm = hob speed

To receive a reliable solution to this expression, each item should be investigated individually. Feeds and speeds should be determined by past experience on similar jobs or by a test under actual job conditions. They are discussed in the preceding article.

The one element of this equation which must be calculated is the total length of hob-carriage travel. The total length of travel to complete the generation of the part is divided into three elements, i.e., approach, face width, and overrun. The face width is indicated on the part print, but approach and overrun must be calculated. When more than one blank per load is to be hobbed, the total face width of all the blanks must be considered.

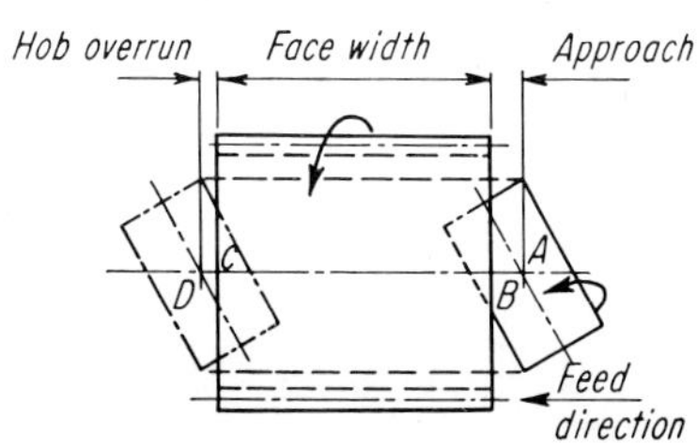

FIG. 16.10. Total length of hob travel is from A to D.

The hob approach is the distance from the point of initial contact between the hob and blank to the point where the hob first cuts at full depth. This distance varies with the hob diameter, part diameter, depth of cut, and hob swivel-setting angle. These points have a geometric relationship that has been plotted in charts A and B (see Figs. 16-11 and 16-12).

Where the swivel-setting angle is small, as often is the case, chart *A* is sufficient. Find the approach distance by following the curve corresponding to the hob diameter until it intersects the tooth depth, and read the approach by following the horizontal line which contains that intersection point.

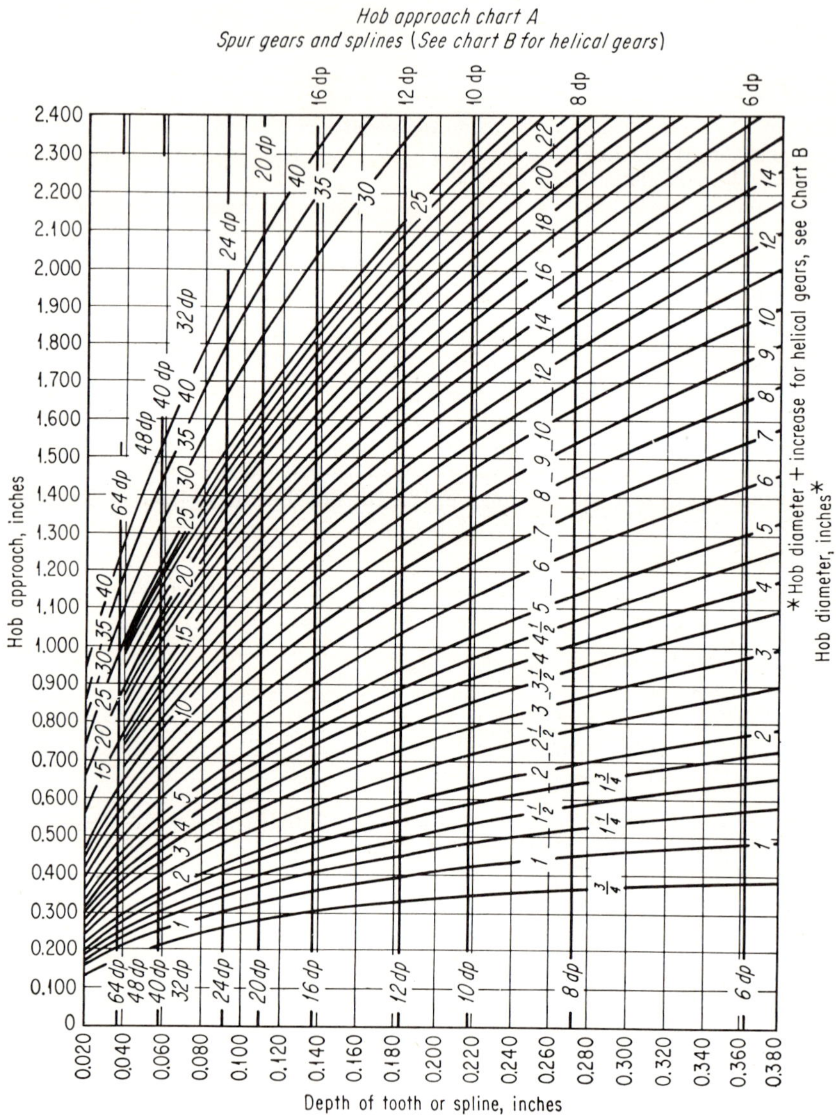

FIG. 16-11. Chart to determine hob approach distance.

Setting the hob at an angle has the same effect as increasing the hob diameter, since the hob has the form of an ellipse in the axial plane of the part. For the larger swivel-angle settings which usually exist when hobbing helical parts, the actual hob diameter must be increased by a certain number prior to applying chart *A*. Use chart *B* to find the number required for a specific job. Determine the point at which the sum of the hob and part diameters intersects the line representing the swivel-angle setting, and read the number at the left of the chart. Add this number to the hob diameter, and use chart A. If the sum of the hob and part diameters is greater than the largest sum

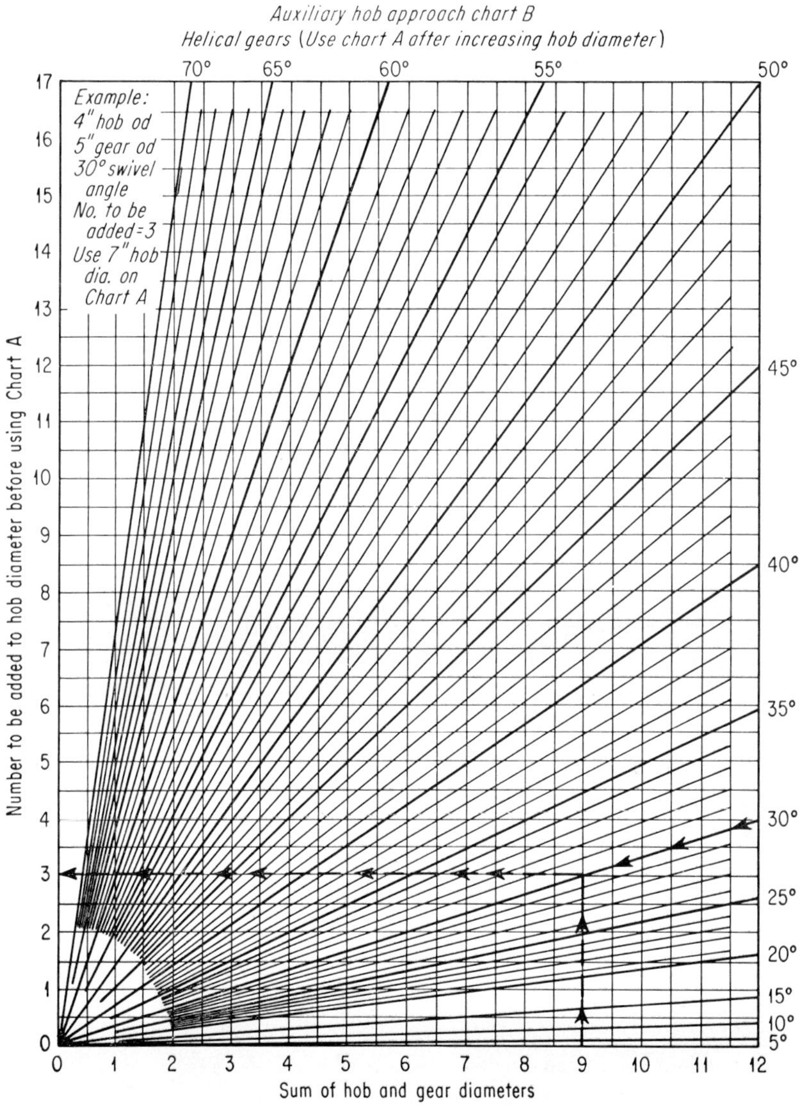

FIG. 16-12. Chart to correct hob size for helical-gear effect.

shown in chart B, take half the sum of the hob and part diameters, find the appropriate number on the chart, and double the answer.

Hob overrun is the linear hob-carriage travel beyond full-depth cutting that is required to complete generation of the teeth. For helical-gear hobbing where the hob swivel-angle settings are usually large, the overrun should be calculated and added to the approach and face width to obtain the total hob-carriage travel. A practical formula for involute forms that has been checked by test and gives safe, full-generating results is

$$\text{Overrun} = \frac{b \cos \psi \tan S}{\tan \phi_n} \quad \text{in.} \tag{16-10}$$

where b = dedendum of gear
ψ = helix angle of gear
S = swivel-setting angle
ϕ_n = normal pressure angle

The sum of the approach, face width, and overrun provides the total travel of the hob carriage, which can be substituted in the equation. The answer found from the equation gives the time required for cutting the part. Added to this cutting time should be a handling time which is influenced by the work-holding device and the factors affecting material handling. This sum gives the total time per cycle.

Locating the Hob. A conventional-type multiposition generating hob can be used across its entire active face width. In each location relative to the part axis,

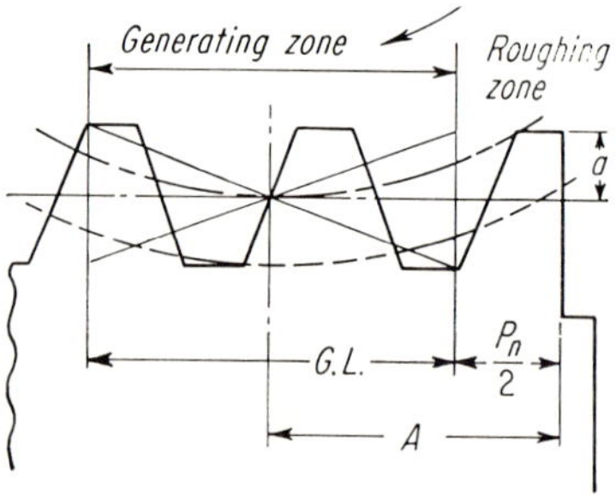

FIG. 16-13. Distance A from hob end.

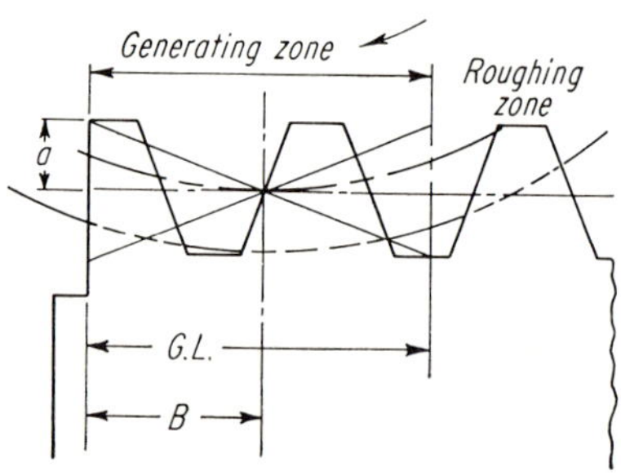

FIG. 16-14. Distance B from hob end.

however, there must be sufficient hob teeth in the generating zone to produce completely the required gear-tooth form. This restricts the location of the hob when the teeth at either end of the hob are being used. Also, roughing teeth are required when the part rotates into the end of the hob.

The blank must pass across the roughing zone prior to reaching the generating zone. Consequently, when the blank for an involute gear rotates into the end of the hob, the roughing zone should contain adequate hob teeth so that the cutting load on the first roughing tooth is not excessive. The number of roughing teeth required varies with the diameter of the blank, the hob swivel-setting angle, and the feed. A minimum setting dimension can be determined for spur and helical involute gears by the following approximate expression. In this expression, we have assumed a maximum generating length so that safe settings will result regardless of the helix angle or diameter of the blank.

$$A = \frac{a}{\tan \phi_n} + \frac{p_n}{2} \quad \text{in.} \tag{16-11}$$

where A = minimum setting dimension measured along hob axis
a = addendum of gear
p_n = normal circular pitch
ϕ_n = normal pressure angle

When the opposite end of the hob is used, ample roughing teeth are usually available. To ensure that generation is complete, however, the minimum end-setting dimension should be at least

$$B = \frac{a}{\tan \phi_n} \quad \text{in.} \tag{16-12}$$

where B = minimum setting dimension measured along hob axis

In hobbing other than involute forms, similar restrictions on the end settings of the hob are applicable. Because the geometry of the involute curve is most familiar to all users of hobbing machines, only application to the involute form is discussed. With the end settings restricted as indicated, the total axial hob shift is limited to its active face width minus the sum of $(A + B)$. In shifting the hob across the part, the total amount of available shift should be known.

Shifting the Hob. The purpose of hob shifting is to increase hob life by uniformly distributing the wear so that all the hob teeth have approximately the same amount of wear. The importance of proper hob shifting becomes apparent when the hob is removed from the hobbing machine for sharpening. If some of the teeth are dulled more than others, useful material will have to be ground away from the teeth that are not dulled before the badly dulled teeth are sharpened. This uneven wear can be avoided by shifting the hob regularly in small amounts.

Where the hob is shifted manually, it remains at a particular setting until the resulting wear reaches the allowed maximum. The hob is then shifted to a new setting where sharp teeth are available, and wear is allowed to reach the predetermined maximum amount before the hob is shifted to the next setting. At each setting, the maximum amount of wear can normally be found on a group of teeth located in the roughing zone adjacent to the generating zone. In shifting to a new setting so that these teeth are no longer doing any cutting, a number of teeth having less than the allowable wear are shifted past the zone where maximum wear occurs. Unless a predetermined amount of shift is made frequently, a nonuniform wear pattern is the result.

The amount of shift should be based upon the maximum allowable hob wear. Primarily, the amount allowed varies depending upon the pitch of the hob and whether the operation is finishing or semifinishing. After a maximum wear limit for a particular operation has been set, the amount of shift can be determined by testing various amounts of shifts. The initial shift increment can usually be found from past-performance records. In arriving at the initial shift, new jobs should be compared with old jobs with consideration of such variables as machinability, numbers of teeth, length of cut, and type of tooth form. Because of the many variables which exist, no attempt has been made to recommend amounts of shift. Instead, a table of recommended amounts of maximum wear has been included.

Frequent hob sharpening pays dividends in terms of uniform accuracy, consistent finish, and longer hob life. Therefore, it is advantageous to limit the maximum amount of wear. The recommended maximum wear of Table 16-1 can be applied as a guide to establish the amount of wear which should be allowed.

Table 16-1. Recommended Maximum Hob Wear
(All readings in thousandths of an inch)

Diam. pitch	Operation	
	Semifinishing	Finishing
4–10	15–20	10–15
11–19	10–15	8–12
20–48	6–10	4–8
48 and finer	5–7	3–5

The tabulated wear is the maximum amount of dullness which extends behind the tooth face. For involute gear hobs, it can usually be found on the side of the tooth near the top corner. The table values are representative of current practice using high-speed-steel hobs.

The direction of shift is determined by whether you desire to move the hob teeth from the generating zone to the roughing zone, or vice versa. Normally, the former method is used. The hob teeth having sharp cutting edges are moved progressively from a light-cutting-load zone into the roughing zone where the cutting load is heaviest. Teeth which generate the part tooth form have been exposed to relatively light cutting loads. Consequently, compared with using duller teeth in the generating zone, tooth form and finish are improved. For spur gears or small hob swivel-setting angles, sharp teeth can be moved into the generating zone by shifting in a direction which is opposed to part rotation. Shifting with the rotation moves dulled teeth into the generating zone. For helical forms or large hob swivel-setting angles, a portion of the hob at one end enters the blank before the portion at the other end; and, consequently, it does the roughing work. Sharp teeth can be moved into the generating zone by shifting the hob toward the part.

Climb and Conventional Hobbing. For spur-tooth forms, where the hob swivel-setting angle is relatively small, the type of cutting action can be either "conventional" or "climb," depending upon the rotation of the hob and the feed direction. For helical tooth forms, where the hob swivel-setting angle is usually large, the type of

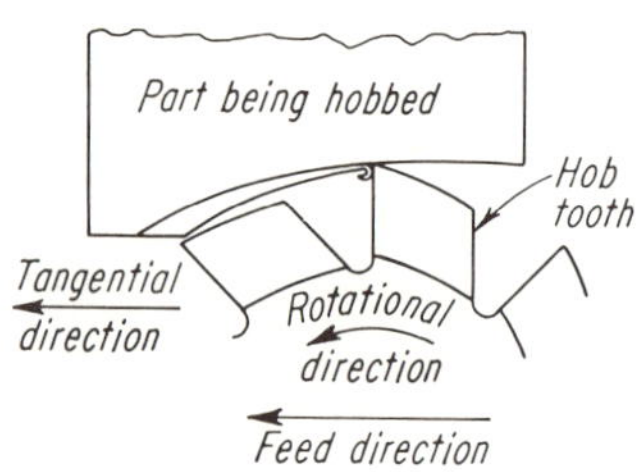

Fig. 16-15. Conventional hobbing.

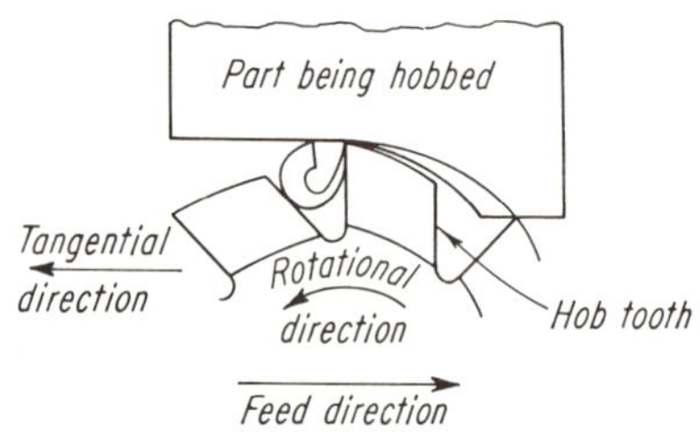

Fig. 16-16. Climb hobbing.

cutting action is determined by the rotational direction of the hob in relation to the direction of gear rotation.

In conventional-type cutting when the hob swivel-setting angle is small, the hob is fed into the part in a direction which is in agreement with the tangential vector denoting the direction of hob rotation. Although the chips removed vary in size and shape, the centered hob tooth removes a chip which is comparable with those obtained in conventional milling. The chip starts out thin and becomes increasingly thicker as the hob tooth sweeps through the cut.

For parts having large helix angles, conventional-type cutting exists when a component of hob rotation is opposed to the part rotation. This holds true when the hand of the hob and the hand of the part are alike. With a large part helix angle, the hob setting angle is usually large, tending to position the axis of the hob parallel with the axis of the part. Since the direction of part rotation depends solely upon the rotation and hand of the hob, the hand of the part determines whether hob rotation opposes part rotation. This kind of conventional-type cutting may be considered as tangential and independent of the axial-feed direction.

With small hob swivel-setting angles, climb-type cutting takes place when the hob is fed into the part in a direction which opposes the tangential vector denoting the direction of hob rotation. As with conventional-type cutting, the chips vary in size and shape. However, unlike conventional-type cutting, the chip starts out thick and becomes increasingly thinner as the hob tooth sweeps through the cut.

When the hand of the hob and the hand of the part are unlike for a large hob swivel-setting angle, a component of hob rotation agrees with the direction of part rotation. This combination of unlike hands produces a "roll-in" effect which is essentially climb cutting, but in a tangential direction rather than in an axial direction. Under these conditions, the hob tends to drive the part, reducing the torsional load in the index drive.

The selection of climb cutting over conventional-type cutting is usually based upon a comparison of surface finish obtained from each. There is no definite rule which can be used to determine which will produce the best finish for the specific job under consideration. When conditions permit, the most satisfactory type of cutting can be determined from test runs on the specific part. In many cases, climb-type cutting has been applied to replace conventional-type cutting because a better finish was obtained with equivalent hob life and accuracy.

16-9. Advantages and Disadvantages of Hobbing. Two of the primary requisites of a gear are that it must be quiet and long-wearing. Accuracy of the gear is required if it is to meet these rigid standards. With ground hobs, a machine in good condition, and reasonable care in setting up and operating the machine, hobbing will consistently produce gears to very close tolerances. Accuracy of tooth spacing, lead, and profile can be closely controlled under mass-production conditions.

The constant indexing of the work makes hobbing rapid as well as accurate. Multiple-thread hobs have greatly increased production on many jobs. Carbide-tipped hobs have also increased production on phenolic, aluminum, and other nonferrous materials.

The versatility of hobbing makes it an economical method for producing gears. One hob will cut all numbers of teeth of a given pitch and pressure angle, and many different tooth forms can be cut on the same machine by changing the form on the hob.

The limitations of the hobbing process are very few. Two types of gears which are outside the range of hobbing are bevel gears and internal gears. Also, gears with an integral feature which is so close that it causes interference with the hob are not applicable to the hobbing process.

GEAR SHAPING*

Like hobbing, gear shaping is a generating process. The tool used is a "pinion"-type tool while hobbing uses a "worm"-like tool. Ordinarily gear teeth are shaped on circular blanks but the process lends itself to producing teeth on cam surfaces curved to almost any mathematical function.

Shaping is used both as a jobbing operation and as a basis for automatic machines mass-producing gears.

16-10. Application of Gear Shaping. In addition to the cutting of external and internal spur, helical, and herringbone gears, the gear shaper can also be applied to the cutting of roller and silent chain sprockets, external and internal ratchets, external and internal splines, interrupted-tooth gears, segment gears, elliptical gears, face gears, worm threads, and racks.

There are a number of models of gear shapers, ranging in size from a machine which will cut gears as small as $\frac{1}{16}''$ in diameter and as fine as 200 pitch, to one which has a capacity of 23 ft and less than 1 diametral pitch. Machines are available with a capacity up to 60″ face width.

As in any gear-cutting process, the accuracy of gears produced by gear shapers depends upon many factors, such as condition of the machine, tooling rigidity, accuracy of the cutter, the material to be cut, blank accuracy, and the skill and care of

* A considerable amount of material for this section was furnished by Ed Kitchen, Gear Engineer, Farrel-Birmingham Co., Buffalo, N.Y,

the operator. In the coarse-pitch range, gears can be produced with AGMA quality 12 tolerances.

The material being cut has an important influence on the finish which can be obtained. For most of the commonly used alloy-steel gear materials, a finish of 40 to 60 microinches can be obtained. Finishes of 20 microinches can be produced as a finish cut.

Gears can be shaper-cut at comparatively high hardness. In properly heat-treated alloy-steel forgings, such as AISI 4140 and 4340, gears have been cut at 43 Rockwell C. Steel castings of about 0.40 carbon can be cut at a maximum hardness of approximately 25 Rockwell C.

The gear shaper is suited to the addition of fixtures and attachments to increase its range of application. Other attachments can be added to some types which will make them semiautomatic or fully automatic in operation.

FIG. 16-17. Shaping internal gear. (*Courtesy of Farrel-Birmingham Co., Buffalo, N.Y.*)

When internal gears are to be cut close to a shoulder, a clearance groove or recess must be provided for the cutter to run into at the end of the cutting stroke. As a general rule it is advisable to have the groove as wide as it can be in order to provide plenty of chip space and eliminate the possibility that the cutter teeth may hit the lower face of the groove.

Some gear shapers generate the two helices of a herringbone gear simultaneously, without a center groove. In addition these machines will cut single-helical gears with helical angles up to 45°, herringbone gears with a center groove, and double-helical gears having staggered teeth, either with or without a center groove.

16-11. Theory of Shaping Process. One gear of an interchangeable series will rotate with another gear of the same series with proper tooth action. It is therefore evident that a tooth curve driven in unison with a mating blank will "generate" in the latter the proper tooth curve to mesh with itself. Similarly, a gear, which is made up of a series of tooth curves, is capable of "generating" in a blank a corresponding series of tooth curves, suitable for meshing with itself. This method of tooth "generation" is known as the "molding-generating" process.

One gear may serve as a cutting tool to produce another gear. To work as a tool, the cutting gear must be hardened and have cutting clearances on the tooth sides.

FIG. 16-18. Shaping external gear on a gear shaper. (*Courtesy of Fellows Gear Shaper Co., Springfield, Vt.*)

FIG. 16-19. Shaping continuous-tooth herringbone pinion. (*Courtesy of Farrel-Birmingham Co., Buffalo, N.Y.*)

The cutting gear is rotated in correct ratio with the gear blank. By reciprocating the cutting gear, metal is removed from the gear blank and teeth are developed. This, basically, is the principle upon which the gear shaper operates, as is indicated diagrammatically in Fig. 16-20.

That portion of a gear tooth lying inside the base circle from which the involute is developed is of noninvolute shape. How this shape is produced by the gear-shaper cutter is shown in Fig. 16-21. In the tooth space to the right of the illustration are shown the successive positions taken by the cutter tooth as it "rolls" into the tooth space. The cutter teeth, of course, are longer than the gear teeth, so as to provide the necessary clearance at the bottom of the tooth spaces for the mating gear.

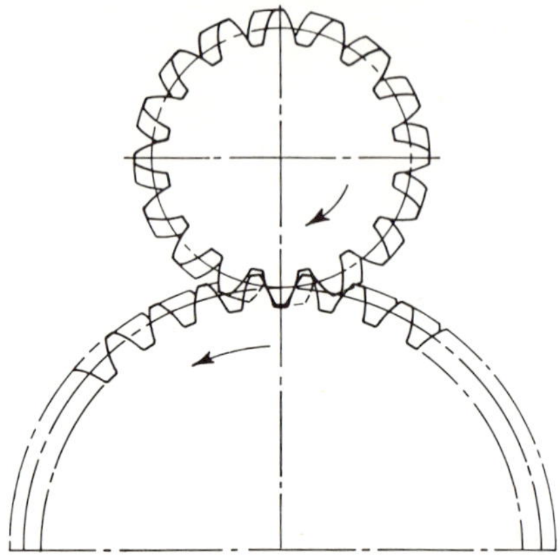

FIG. 16-20. Shaper cutter and work.

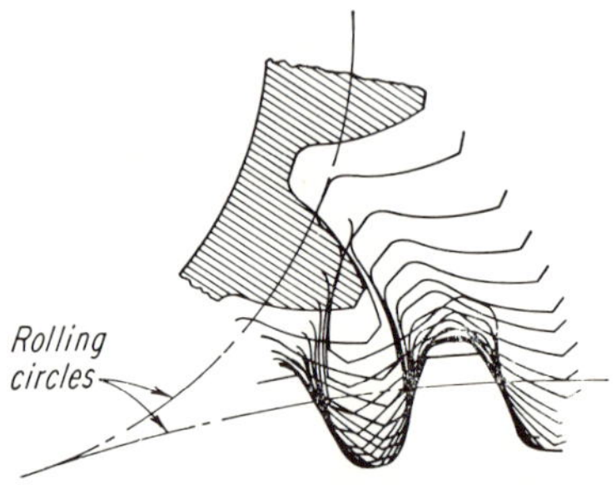

FIG. 16-21. Generating action of shaper cutter.

This illustration also indicates the nature of the chip taken by the gear-shaper cutter, the heaviest portion of the chip, after reaching full depth, being in the flank and fillet, and the lightest chip on the involute portion of the tooth, thus assuring a fine finish.

16-12. Gear-shaper Operation. When a spur gear is being machined, the reciprocating motion of the cutter spindle is restrained to operate in a straight path by a positive straight guide.

If a gear having helical teeth is being machined, a helical guide is used. As the cutter spindle reciprocates, the helical guide produces a helical motion to the cutter spindle. It is evident that the reciprocating control is independent of any other function of the machine. For external helical gears, cutters of the opposite hand are used. Internal helical gears require cutters of the same hand.

The cutter spindle is reciprocated through a crank-arm mechanism. The eccentric position of the crankpin determines the length of stroke, and the crank arm is adjustable so as to position the cutter relative to the work.

The work spindle is held in an apron, which is provided with a positive type of relieving mechanism. The function of this relieving mechanism is to withdraw the work from the cutter on the return or noncutting stroke of the cutter to avoid cutter rub. The rotation of the work spindle is held in relationship to the rotation of the cutter spindle through change gears and auxiliary mechanism. Thus the correct relation is maintained between the number of teeth on the cutter and work. The indexing mechanism is of the continuous-revolution type.

After the proper guide and cutter have been selected, determine the work-change gears according to the following formula:

$$\text{Change gear ratio} = \frac{K \times N}{T} \tag{16-13}$$

where K = machine index constant
N = number of teeth on gear
T = number of teeth on cutter

A helical guide causes the cutter to trace the same lead as the lead of the guide. For the same guide and cutter, a fixed helix angle is produced on the work regardless of the number of teeth on the work. This can be expressed in terms of lead, according to the following formula:

$$\frac{L}{T} = \frac{l}{N} \tag{16-14}$$

where L = lead of helical guide
T = number of teeth on cutter
l = lead of gear
N = number of teeth on gear

The principal factors controlling the cutting speed on the gear shaper are the length of the cutting stroke required, machinability, and the degree of quality and cutter life that is desired. Table 16-2 indicates cutting speeds for various materials. These speeds may be used as a guide in the selection of the initial speed. A more favorable speed can be determined by trial cuts.

Table 16-2. Cutting Speeds for Use on Gear Shapers*
(Based on a maximum cutting stroke of 1" and the maximum degree of machinability for the analyses specified)

Material	Surface speed, fpm	Remarks
Cast iron	50–60	Ordinary gray-iron castings
Steel (mild)	75–85	0.15–0.20% carbon, no hardening alloys
Steel (high-carbon)	50–60	0.50% carbon, 1.15% manganese, 0.55% chromium
Steel (tool)	40–50	0.90–1% carbon, 0.25% manganese
Steel (chrome-nickel)	40–50	SAE N3250
Steel (stainless)	25–35	
Brass (soft)	100–110	
Bronze (naval)	45–55	
Aluminum	190–200	With coolant
Fibrous materials	75–85	

* From Fred H. Colvin and Frank A. Stanley, "Gear Cutting Practice," 3d ed., McGraw-Hill Book Company, Inc., 1950.

When converting cutting or surface speed into strokes per minute, it is necessary to take into consideration the fact that the actual distance traveled by the cutter spindle, per revolution of the crankshaft, is twice the length of the stroke. The following formula is used for determining the number of strokes per minute:

$$s = \frac{12 \times S}{2L} = \frac{6S}{L} \tag{16-15}$$

where s = number of strokes per minute
L = length of cutting stroke, in.
S = surface speed, fpm

Based upon a cutting speed of 60 fpm and cutting steel at about 200 BHN, the following table gives the number of strokes per minute.[4]

Gear face width, in.	*Strokes per min*
½	600
1	350
3	110
8	40

Because of the twisting motion imparted to a helical cutter, its actual cutting speed is greater than that of a spur cutter used at the same stroke length and number of strokes per minute. The difference depends upon the helix angle of the cutter used. The actual cutting speed can be found by dividing the selected cutting speed by the cosine of the helix angle of the cutter.

The selection of the rotary feed depends largely upon the finish and degree of accuracy desired. It is evident that a fine-pitch gear requires a finer feed than a coarse-pitch gear and also that a coarser feed can be used for roughing than for finishing cuts, particularly when the gear is roughed on one machine and finished on another. The most suitable combination of feed and speed to meet requirements is best determined by trial to find the combination giving the highest production with the longest cutter life.

Rotary feed is the number of strokes of the cutter per revolution of the cutter. Table 16-3 shows some representative values for the number of strokes per revolution of the cutter.[4] These are average values which can be used for cutting good commerical gears.

Table 16-3. Strokes per Revolution of Cutter

Cutter data	No. of strokes					
	Roughing			Finishing		
	No. of gear teeth			No. of gear teeth		
	12	35	100	12	35	100
4 pitch, 16 teeth	1,000	800	500	1,200	900	600
6 pitch, 18 teeth	800	500	300	1,000	700	500
24 teeth	1,100	700	400	1,300	900	700
10 pitch, 30 teeth	800	500	300	1,000	700	500
40 teeth	1,000	700	400	1,300	900	700
20 pitch, 20 teeth	600	400	300	600	400	300
40 teeth	800	550	400	800	550	400

The change gear or pulley ratios provided on the machine may not permit the table values to be obtained. Usually the effect of number of strokes is not too critical. Roughing and finishing are frequently done using the same number of strokes.

The number of cuts required to complete a gear depends primarily upon the pitch, machinability, and finish required. On large steel gears having a 200 to 260 BHN, the number of cuts in the 2 diametral pitch range may vary from 7 to 9. For the roughing cuts, the successive depth of cut would begin at about 0.300″, decreasing to about 0.100″. The finishing-cut depth increments would be from 0.015 to 0.005″ for each gear revolution. In the 4 to 10 diametral pitch range, the number of cuts would be in the range between 7 and 3, with the coarser pitches requiring the higher number of cuts. For pitches finer than 10 diametral pitch, the number of cuts can be reduced to 3 or less. The production time in minutes to cut a gear using a pinion-type cutter may be estimated by the following formula:

$$\text{Time} = \frac{N \times r \times n}{T \times s} \tag{16-16}$$

where N = number of teeth on gear
r = strokes per revolution of cutter
n = number of cuts
T = number of teeth on cutter
s = number of strokes per minute

16-13. Advantages and Disadvantages of Shaping. Gear shaping is a versatile and accurate method for producing gears. It is used to cut spur and helical teeth, and external and internal gears. In addition, herringbone and face gears can be produced. Shaping is the method for cutting cluster gears or gear teeth which are adjacent to a shoulder. From a production standpoint, shaping may be advantageous on gears of narrow face width because of the relatively short approach and overrun.

A limitation of gear shaping is the length of cut. Gear parts which are integral with long shafts may be difficult to mount in the machine. Another limitation is the requirement for a separate helical guide for each helix angle to be cut.

SHEAR CUTTING OF GEARS

Form-tool blades are used to shear-cut each space between teeth at the same time. The speed of shear cutting and the nature of the operation are quite similar to broaching. The patented machine used for this process is called the Shear-Speed®* gear shaper.

The Shear-Speed process is a high-production method of producing gear teeth.

16-14. Application of Shear-Speed Cutting. The Shear-Speed cuts all teeth of external spur forms simultaneously at a high rate of production. Helical forms are not applicable to this method. While the majority of forms shaped by the Shear-Speed are external, there are cases where this method is most advantageous in the shaping of internal forms. There are, however, certain limitations on shaping internal forms on the Shear-Speed. Shear-Speed shaper flexibility is demonstrated by the broad range of parts that can be produced and the wide size and shape variations. The product range includes gears, splines, sprockets, ratchets, and miscellaneous external shapes. Part tooth forms may be involute, straight-sided, or contoured. The forms can be either symmetrical or unsymmetrical.

Although some clearance on both sides of the part is required for overtravel of the form-cutting tool, parts which are integral with closely spaced flanges or shoulders such as cluster gears can be readily cut by the Shear-Speed process.

In the Shear-Speed process, it is possible to form-cut involute spur gear teeth with practically any required tooth modification. This includes preshave root relief, semitopping, full fillet modifications, and modifications in the involute form. Normally this method is applied to semifinishing or finishing of gear teeth.

* Registered trade-mark of the Michigan Tool Co. of Detroit, Mich.

Shear-Speed machines are designed for mass production of gears. Therefore, they are most efficiently applied to high-volume production. The Shear-Speed process can also be applied to short runs when they are repetitive at frequent intervals.

Machines are available to cut a maximum diameter of 20″ and a maximum face width of 6″. Gears up to 2 diametral pitch can be produced. The finest pitch which

Fig. 16-22. Shear-Speed cutter above work. (*Courtesy of Michigan Tool Co.*)

Fig. 16-23. Shear-Speed gear shaper and stack of three gears. (*Courtesy of Michigan Tool Co., Detroit, Mich.*)

is recommended on these machines is 12. Gear materials which can be cut on these machines include steel, cast iron, bronze, brass, aluminum, fiber, plastic, and others.

16-15. Shear-Speed Principles. All the teeth in a gear are cut simultaneously in the Shear-Speed gear-cutting process. The same number of gear-cutting tools as there are teeth in the gear to be cut are assembled radially in the cutting toolholder mounted in the head of the machine.

The number of tools depends on the number of spaces, slots, or forms to be cut around the periphery of the work blank. All the tools may have the same form, or the forms may vary depending upon the part specifications. The tools may be equally or unequally spaced.

The form of the cutting edge of the tool is an exact duplicate of the space, slot, or form. All tools are form-relieved to maintain correct form after sharpening.

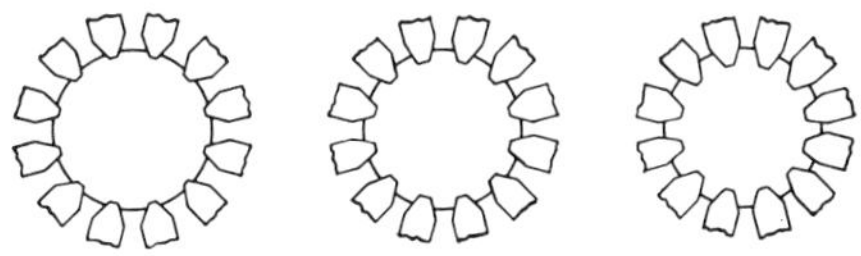

FIG. 16-24. Progressive cutting of gear by shear process.

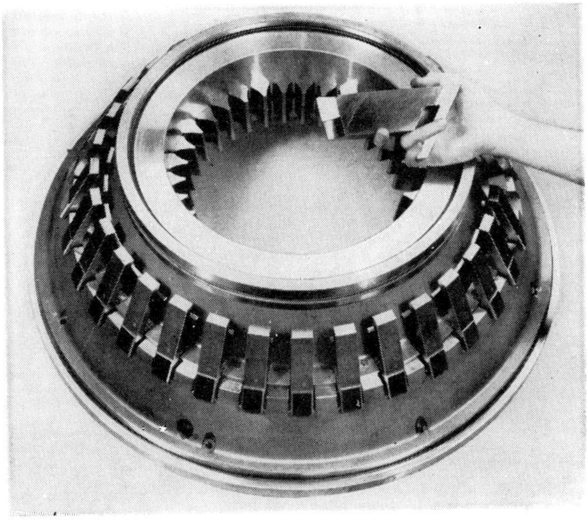

FIG. 16-25. Shear-Speed cutting tools and toolholder. (*Courtesy of Michigan Tool Co., Detroit, Mich.*)

With the toolhead stationary, the part is reciprocated past the cutting tools. Each tool is fed radially a predetermined distance with each stroke, until the full tooth depth is cut. All tools are retracted on each return stroke of the gear to avoid drag and are returned to their initial position at the end of each cycle.

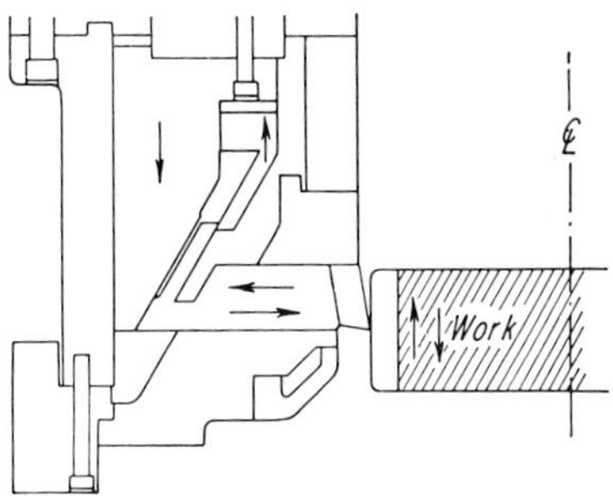

FIG. 16-26. Feed control of shear cutters.

The toolhead assembly consists of three parts, a retainer housing, a radially slotted member which controls tool alignment (Fig. 16-25), and a movable double cone-shaped guiding unit (Fig. 16-26) which controls the feed. All tools are fed radially into the work at the same time by use of the double cone-shaped toolhead.

The cutting tools can be compared with the teeth on a broach. Since each tool must have the same form as the tooth space on the gear, each set of tools and the toolhead are normally designed to produce a specific part.

16-16. Machine Operation, Shear-Speed Cutting. Cutting tools are assembled in slots of the toolholder, with the cutting edges inside and facing the bottom of the toolholder. The angular projections on the cutting tools are at the outside of the toolholder with the projections pointing toward the top of the toolholder. When the toolholder is raised into position by means of jacks in the head of the machine, the angular projections fit between two cones. The outer cone controls infeed of the cutting tools on the upstroke; the inner cone controls retraction of the cutting tools on the downstroke. Thus fastening the toolholder in place in the head of the machine

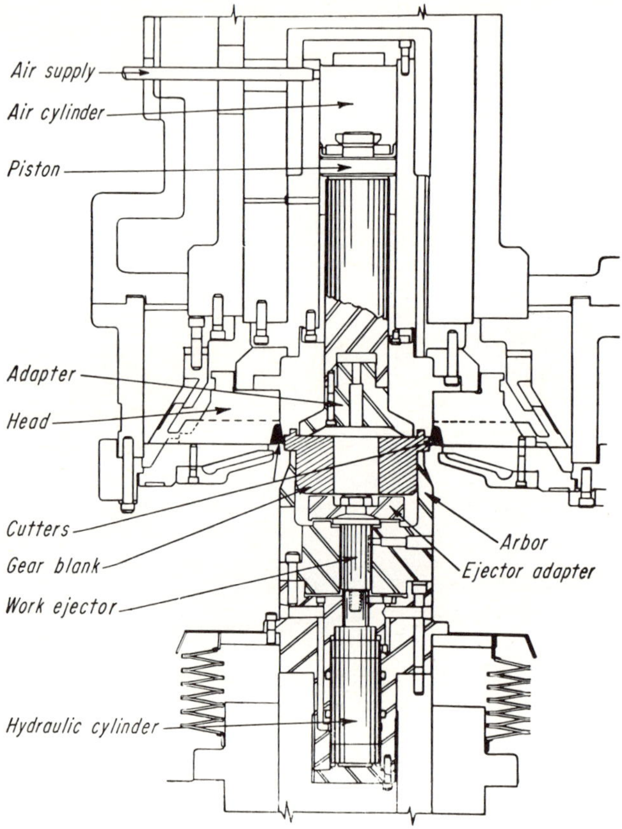

FIG. 16-27. Shear-Speed machine. (*Courtesy of Michigan Tool Co., Detroit, Mich.*)

places the cutting tools in correct cutting position in relation to the vertically reciprocating gear blank. Provision is made for centering the gear blank in relation to the cutting tools.

The cutter-head assembly is mounted in the head slide, which also carries the feed and relief cams and their drives. This slide travels on scraped ways from loading to cutting positions and is changed from one position to the other by means of a hydraulic cylinder. When the slide lowers the head into cutting position, it is automatically locked rigidly in place by two wedges, one at either side of the head assembly. These wedges are actuated by individual hydraulic cylinders. Lock action is automatic. The locking wedges are retracted automatically at the end of the cutting cycle to allow the slide to return to loading position.

The actual feed functions only when the head slide is locked in "cutting" position. Downfeed of the cones is controlled by a cam which gradually reduces the feed on each cutting stroke. The rate of feed is controlled by a ratchet mechanism which turns the cam. The pawl of this mechanism is adjustable to move the cam an amount equal to either one or two teeth on the ratchet wheel for each reciprocation of the work, thereby regulating the rate of feed of the tools. The rate of infeed decreases as the tools are fed to depth. A ratchet wheel with the correct number of teeth is selected to give the desired infeed of the tools.

The cones are raised a definite amount for each return stroke by an eccentric relief cam. As the work starts on its downstroke, this rotating cam raises the cone assembly slightly and holds it up until the downstroke is completed. This causes the inner cone to pull the tools away from the work. The cam returns the cones to a cutting position for the cutting stroke. This relief action is completely independent of the feed and acts identically on each return stroke regardless of the position of the cones during the machine cycle.

The gear blank is reciprocated vertically through the cutter assembly while the cutting tools are fed in a predetermined amount for each stroke of the gear, until all teeth are formed to full depth. Provision is made to retract the blades on each return stroke to avoid drag, and to return the blades to starting position at the end of the cycle.

Length of reciprocating stroke of the work is adjustable. Recommended length of stroke is the face width of the gear being cut plus 1/4″. Speed of stroke is regulated by change gears on the machine. It is recommended that change gears selected give a speed of 20 fpm cutting stroke.

The work is mounted on an adaptor and is clamped hydraulically or pneumatically. Mechanical clamping may also be used. Clamping and release may be either automatic or semiautomatic. Parts with shafts as well as with holes can be cut by the use of proper adaptors. The work-holding adaptor or fixture is mounted on the top of a ram.

The following examples indicate typical production data. In both cases, a finish of approximately 50 microinches can be obtained.

Example 1

Number of teeth	13 spur
Diametral pitch	12
Pressure angle	14°30′
Outside diameter	1.290″
Whole depth of tooth	0.1645″
Face width	0.786″
Strokes per minute	126
Surface feet per minute	21
Infeed per stroke, average	0.0036″
Length of stroke	1.0″
Teeth in ratchet	50
Pieces per load, manual loading	1
Pieces per hour at 80 per cent efficiency	77

Example 2

Number of teeth	56 spur
Diametral pitch	12.7–16.9
Pressure angle	20°
Outside diameter	4.527″
Whole depth of tooth	0.167″
Face width	0.480″
Strokes per minute	81.1
Surface feet per minute	21.1
Infeed per stroke, average	0.0037″
Length of stroke	1.56″
Teeth in ratchet	50
Pieces per load, manual loading	2
Pieces per hour at 80 per cent efficiency	102

16-17. Advantages and Disadvantages of Shear-Speed Cutting. Shear-Speed form cutting is applicable to cutting gears, splines, sprockets, and other spur-tooth forms which may be unsymmetrical or unequally spaced. Since all the spaces between the teeth are cut at one time, gears can be cut at a high rate of production. The production advantage is greater for gears with large numbers of teeth. Each machine is usually put on the production of one high-volume part.

The cutting tools are easily removed for sharpening and are sharpened all at one time on a surface grinder. The process is limited to coarse-pitch gears having spur teeth. In addition, each model is limited to a small range of diameters and pitches.

GEAR BROACHING

Broaching is a widely used process for making keyways, finishing holes, cutting splines, etc. The process is also applicable to making gear teeth.

Fig. 16-28. Broaching an internal gear. (*Courtesy of Colonial Broach Co., Detroit, Mich.*)

The form of the space between broached gear teeth corresponds to the form of the broach tooth. Broaching is normally a high-production process. Occasionally it may be used in fairly small lot production to get accuracy on a critical part that cannot readily be made to a high degree of accuracy by other methods.

16-18. Application of Broaching. Broaching is a machining process that removes metal from a part by pulling or pushing a multiple-toothed tool called a "broach" through or along the surface of the part. It is a fast and accurate process that produces fine surface finishes. This section is limited to the discussion of broaching as it applies to gears and similar parts.

Gear broaching is usually applied to the cutting of internal spur and helical parts. Normally, all teeth are cut in one pass of the broach. In addition to gears, other parts which can be broached include splines, serrations, and other internal tooth forms. Although broaching is generally applied to internal tooth forms, it is sometimes advantageous to broach external tooth forms.

When all the teeth are cut simultaneously, broaching is a high-production method of cutting gears and splines and is normally applied to high-volume production. Roughing and finishing are normally combined in one pass of the broach. However, some parts require two operations, rough broaching and finish broaching. In some special applications, broaching is done one tooth at a time.

Broaching machines can be designed to cut gears of almost any size and pitch. Each application must be considered in terms of the major and minor diameters, the length of the part, and the material. Some of the large vertical broaching machines have a capacity of 75 tons and a stroke of 72″.

A broach can be used to machine almost any internal form. The teeth can be either equally or unequally spaced, and tooth forms can be either symmetrical or unsymmetrical. The tooth form must be uniform in the direction of the broach axis, and the wall of the part must be strong enough to withstand the broaching pressure. Gear broaching usually requires a part with a through hole. However, blind holes can be broached if there is sufficient clearance beyond the broached section and if a series of punch-type broaches are used.

Broaching can produce close tolerances on profile, spacing, lead, and size. As an example, an internal pump gear with 35 teeth, 10 diametral pitch, 24° pressure angle, and $^{11}/_{16}$″ face width is broached within AGMA quality 10 tolerances on profile, parallelism, and spacing. Print specifications call for involute profile errors to be within ±0.00025. Spacing is held to 0.0005″, lead within 0.0003″, and lead variation within 0.0005″. Both faces must be parallel and square with the pitch diameter and the outside diameter within 0.001″ indicator reading. In addition, the pitch diameter must be concentric with the outside diameter within 0.001″.

Most types of ferrous and nonferrous materials can be broached. In steels, the hardness range for broaching is normally between 12 and 22 Rockwell C. However, a considerable amount of broaching is done at hardnesses up to 35 Rockwell C. In some cases, the finish on the hard gear may be superior to that on a gear made of softer material. Steels below 12 Rockwell C are difficult to broach because of their tendency to pick up metal on the sides of the broach teeth and on the lands, resulting in tears and poor surface finish. Parts harder than 35 Rockwell C can be broached, but it is seldom economical because of the high cost of broach maintenance.

To broach work-hardening materials, it is necessary that there should be sufficient step or cut per tooth so that each tooth will bite under the surface of the metal—including the first tooth. Otherwise, the material being broached may tend to break down the cutting edges of the tooth.

Steel forgings and cast iron can be broached, but the material should be of as uniform density or hardness as possible. Commercial grades of brass, bronze, aluminum, and plastic can normally be broached at high speeds.

16-19. Broaching Principle. Broaching is a machining method by which the cutting tool is pulled or pushed over the surface to be machined. The method of operation affects the construction of a broach. *Push* broaches are normally short and therefore less expensive than the longer pull broaches. However, one *pull* broach can sometimes remove as much stock as several push broaches. Most internal broaching is done with pull broaches. Not only do they remove more stock than push broaches, but they are also capable of taking much longer cuts.

The actual cutting action of any single broach tooth is very much like that of a single form tool. Each section of the broach has as many teeth as there are tooth spaces on the gear. The diameter of each section increases progressively to the major diameter which completes the tooth form on the part. In some cases, broaches can be designed with a removable finishing section which finishes the complete tooth form.

Assume an internal gear blank with an inside diameter finish reamed to a 3.8000″ diameter in which internal gear teeth are to be broached to a major diameter of 4.2314″. The broach will produce the gear teeth by the simple method of each succeeding section of the broach increasing in diameter by increments from 3.8000 to 4.2314″. Each broach tooth contributes some portion of the internal form. As shown in Fig. 16-29, the gear-tooth space is produced by successive broach teeth whose form depth increases progressively. The broach teeth in the finishing sections complete the gear-tooth form by producing the major diameter.

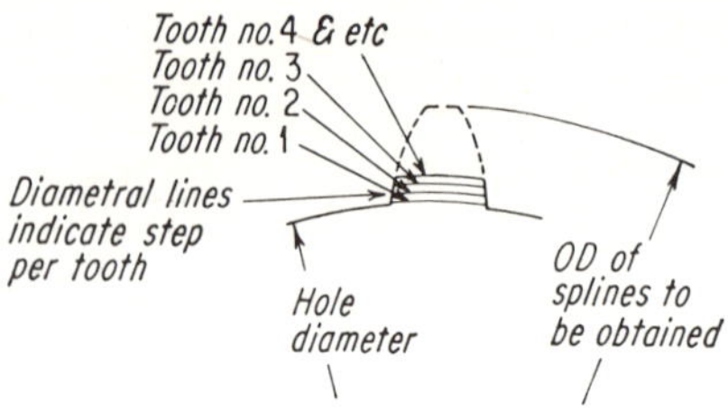

FIG. 16-29. Cutting action of a broach.

Differences which must be considered in broaching arise more in chip disposal than in chip separation. In ordinary cutting on a lathe, milling machine, shaper, or planer, chips are removed almost as soon as formed. However, chips from a broach are of necessity confined in the space between the teeth for the length of the cut. Therefore, it is absolutely necessary to provide adequate chip space when designing the broach. Unless this is done, there is a chance of dulling or even breaking the tool, and roughing or tearing the surface of the work being broached. Chip space is limited by the pitch of the teeth and in small broaches by the diameter of the broach. Broaching in general requires the use of chip breakers. Otherwise too wide a chip may be produced.

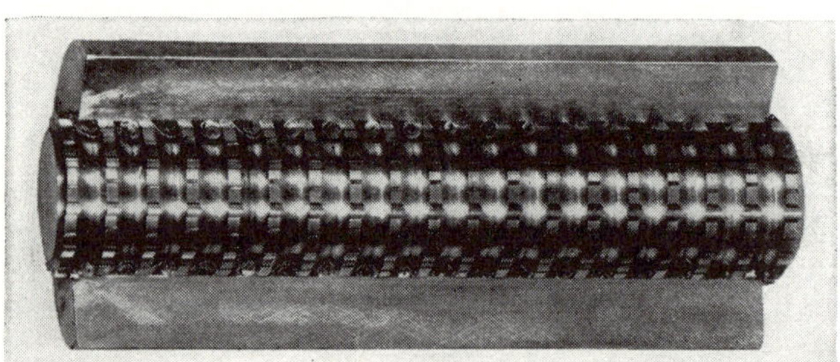

FIG. 16-30. Broach chips revealed by cutting through a workpiece. (*Courtesy of Colonial Broach Co., Detroit, Mich.*)

Such chips are hard to control and are usually taken care of by providing staggered chip breakers on successive broach teeth.

16-20. Broaching Operation. Machine selection plays a vital part in the relative success of the broaching operation. A primary factor to be considered is the broach itself since, in good broaching practice, broach design to some extent precedes machine selection. Parts which require heavy stock removal should be applied to machines with heavy tonnage. However, lighter machines can frequently do a good job by using a finer step per tooth with more passes.

Gears are broached on both horizontal and vertical machines using pull-type broaches. Horizontal machines are generally considered to be adaptable to a wide variety of broaching operations. On some models, provisions are made for accessibility of a broach drive head to facilitate the broaching of internal helical gears. Both pull-up and pull-down vertical machines are applicable to broaching gears. The pull-down type is more convenient on heavy parts and where it is necessary to locate an internal tooth in relation to a feature on the part, such as an external gear tooth. Both types can be equipped with automatic broach-handling mechanisms, making it unnecessary for the operator to handle the broach. Other types of machines are available for special applications. For large-volume work, machines can be equipped with automatic handling devices.

In addition to the size and type of broaching machine required, there are several general considerations which must be given to the gear specifications.

Broaching of several narrow-face gears at one pass of the broach will result in an off-lead condition if the parts are not processed with all faces parallel. Most internal-gear plants grind either one or both sides of the gear-locating faces to ensure parallelism. With wide-face gears, a broach of excessive length would be required for a one-pass operation. Consequently, more than one broach would be applied with each one removing a portion of the stock in its pass through the gear.

Some internal gears require a pitch circle which is concentric with an external contour of the part. Since the broach has a tendency to follow the hole through which it is pulled or pushed, the internal gear teeth should be broached prior to finishing the

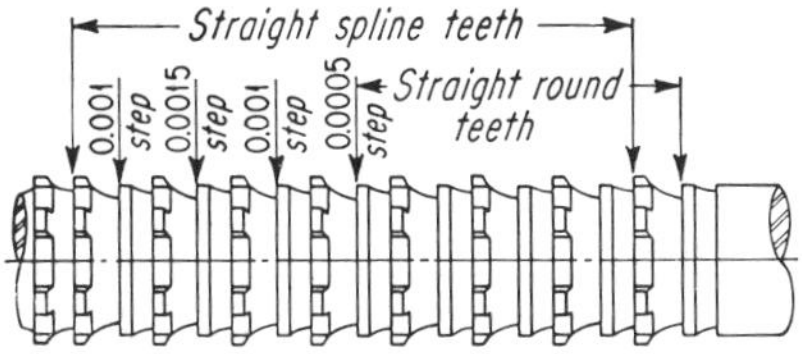

FIG. 16-31. Dwell tooth broach.

external contour. The part can then be located on the pitch circle for finishing the external contour. When the part is to be located on the minor diameter, a dwell tooth broach, as shown in Fig. 16-31, will provide concentricity between the minor diameter and the pitch diameter of the internal gear.

Broaching of internal spur gears with the desired zero helix angle is dependent upon the alignment of the puller end and the follower mechanism of the broaching machine. To achieve accuracy, it is necessary to align the puller and retriever mechanisms properly. In addition, the axis of the broach must be perpendicular to the face of the fixture which locates the gear and absorbs the broaching forces.

Internal helical gears are broached in much the same manner as spur gears except that the broach must rotate as it passes through the blank. Rotation of the broach is accomplished by means of a drive head on the machine. Helical gears are produced on both vertical and horizontal broaching machines equipped with special attachments. It may not be practical to broach gears with a large helix angle.

What would be feed in other types of machining is designed into the broach and is called chip per tooth. In a gear broach, the teeth increase in diameter from the first roughing tooth to the first finishing tooth. This increase in diameter or cut per tooth need not be uniform. Normally, finishing teeth are of uniform diameter to take off the final increment of stock, completing the form to the major diameter. Cut per tooth in the broach has a bearing on finish. Usually, a smaller step helps to give a good finish by producing a thin chip which curls up in the chip space and keeps clear of the cutting edge. The cut per tooth must be decreased as hardness or toughness increases. A step of 0.0005″ on broach diameter is about the minimum for a clean cut.

Speed is an important consideration in broaching. Cutting speeds for gear broaching are normally in the range of 8 to 30 fpm. Broaching speeds as low as 2 or 3 fpm are used effectively to broach extremely hard-to-cut materials.

Cutting speed can be varied for an individual broach application. Actually, considerations of power requirements, tooth strength, broach strength, finish desired, and choice of coolant offer considerable limitations. Free-cutting material may be broached faster than tough stringy material. Other considerations, however, usually make it advisable to use a different broach for a different material.

In general steels are broached with high-speed-steel broaches at surface speeds ranging from 10 to 30 fpm. Cutting speeds range from 8 fpm for hard stainless to 20 or 25 fpm for free-machining types.

Cast iron and malleable iron can be broached with high-speed-steel broaches at maximum speeds up to 30 fpm in normal operation and at 120 fpm with carbide broaches. Cast iron may be cut with a greater step or cut per tooth than free-cutting steel, but break-out is a possibility and may limit the cut.

Fig. 16-32. Drive head for helical internal broaches. (*Courtesy of Colonial Broach Co., Detroit, Mich.*)

Cutting speeds up to 30 fpm are practicable in broaching commercial grades of brass and bronze. Aluminum and magnesium materials can be broached at speeds of over 30 fpm.

It is not difficult to estimate the production possible from a broaching setup if a few simple factors are known. These include machine speed, return speed, length of stroke, and time to load and unload. Below is an example of how the production rate can be determined.

Cutting speed. . . . . 30 fpm (360″ per min)
Return speed. . . . . 60 fpm (720″ per min)
Stroke. . . . . 48″

$$\text{Cutting time} = \frac{48'' \times 60 \text{ sec per min}}{360'' \text{ per min}} = 8 \text{ sec}$$

$$\text{Return time} = \frac{48'' \times 60 \text{ sec per min}}{720'' \text{ per min}} = 4 \text{ sec}$$

Starting and stopping	=	2 sec
Loading and unloading	=	5 sec
Total time per piece	=	19 sec

16-21. Advantages and Limitations. Gear broaching is primarily applicable to high-volume jobs which require high production rates. Gears can be broached accurately, since a number of the elements are controlled by the accuracy of the broach. In addition, broaching can produce a fine finish on the gear-tooth profiles.

Although gear broaching can be applied to external gears, it is confined primarily to the production of internal spur and helical gears. A gear broach will cut only parts which have identical gear specifications. Features which prevent the broach from passing through the part may limit the application of broaching.

REFERENCES

1. Barber-Colman: "Hob Handbook," Barber-Colman Company, Rockford, Ill.
2. Barber-Colman: "Operators Manual," Barber-Colman Company, Rockford, Ill.
3. Colonial Broach: "Guide to Better Broaching," Manual CB-58, Colonial Broach and Machine Company, Detroit, Mich.
4. Dudley, Darle W.: "Practical Gear Design," McGraw-Hill Book Company, Inc., New York, 1954.
5. Farrel-Sykes: "Gear Generators," *Bull.* 460, Farrel-Birmingham Company, Inc., Ansonia, Conn.
6. Fellows: "Gear Production Equipment," The Fellows Gear Shaper Company, Springfield, Vt.
7. Fellows: "The Art of Generating with a Reciprocating Tool," The Fellows Gear Shaper Company, Springfield, Vt.
8. Fellows: "The Internal Gear," The Fellows Gear Shaper Company, Springfield, Vt.
9. Gould and Eberhardt: "Gear Cutting Machines," *Bull.* 250, Gould and Eberhardt, Irvington, N.J.
10. Michigan Tool: "The Michigan Shear-Speed Shaper Line," *Bull.* 1800-54, Michigan Tool Company, Detroit, Mich.
11. Michigan Tool: "Shear-Speed," *Bull.* SS-55, Michigan Tool Company, Detroit, Mich.
12. National Broach: "Broaching Practice," National Broach and Machine Company, Detroit, Mich.
13. The Metal Cutting Tool Institute: "How to Use and Maintain Gear Cutting Tools," Gear Generating Tools Subdivision of the Metal Cutting Tool Institute, New York, N.Y.
14. Colvin, Fred H., and Frank A. Stanley: "Gear Cutting Practice," 3d ed., McGraw-Hill Book Company, Inc., New York, 1950.

Chapter 17

GEARS PRODUCED BY DIES*

By

E. G. CHARTIER, *President, Tekniks Engineering Company, Elmhurst, Ill.*

Various methods to produce toothed gear wheels by rotating machinery were discussed in Chap. 16. In this chapter, we shall deal with the various methods to produce a toothed gear wheel, segment, or any flat or crowned irregularly shaped toothed part completely in one or a sequence of continuous operations. When gears are produced by the die process partially or entirely, the initial tool costs are usually high, and consequently this type of gearing is generally restricted to large-volume production, where inexpensive parts are of paramount importance and the surface specifications of the tooth profile and finish are not the major requirement.

17-1. Outline of Die Processes. The most commonly used techniques to produce toothed gear elements by the die processes are vertical and horizontal punch presses, cold drawing or extrusion, injection and compression molding, die casting, rolling and

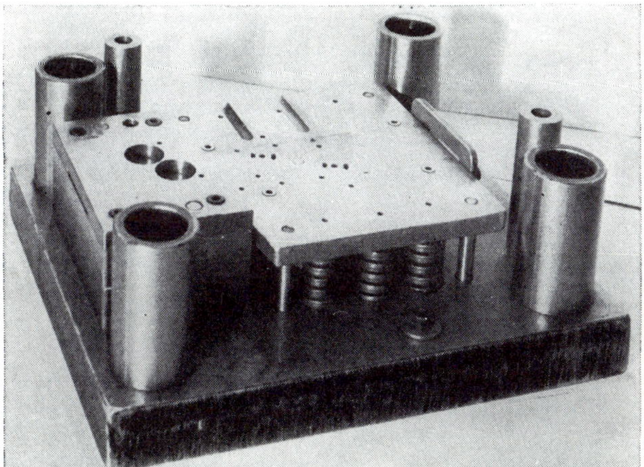

FIG. 17-1. Bottom portion of press die, 64 pitch, 64 teeth with lancing. (*Courtesy of Sunbeam Corp., Chicago, Ill.*)

* A considerable amount of material for this chapter was prepared by A. J. Lemanski, Supervisor of Gear Engineering, Lycoming-Avco, Stratford, Conn.

FIG. 17-2. Top portion of press die, 64 pitch, 64 teeth with lancing. (*Courtesy of Sunbeam Corp.*)

FIG. 17-3. Punch section of press die shown in Figs. 17-1 and 17-2. (*Courtesy of Sunbeam Corp.*)

forming, sintering or powder metallurgy, and special procedures for specific purposes such as for thin steel laminated pinions, and the punching of difficult materials such as laminated fibers.

Gear-stamping Process. This process requires the use of a press equipped with a set of tools called "dies," which have the master form of the toothed gear element built in and will produce exactly the shape that is so constructed. A typical set of dies is shown in Figs. 17-1 and 17-2, and an enlarged view of the punch section of this die is shown in Fig. 17-3. The stock for producing these gears usually comes in rolls, although 12-ft-long strips are occasionally used. The stock is usually passed through a device called a roll straightener, which irons out the stock and makes it flat, and then passes through the dies. As the top portion of the die comes into contact with the stock, the pressure forces the stock into the bottom portion of the die, where the punches can enter, and a shearing action takes place, which produces a part to the shape of the punch that entered the die. This shearing action causes the teeth and outside periphery to be somewhat rough and rounded at one edge. This rounding is called *drawdown,* and a typical example is shown in Fig. 17-4. This shearing and rounding characteristic increases as the stock thickness increases, and thicknesses over $\frac{1}{16}''$ are not recommended. If punched gears are desired with less drawdown and roughness, an additional operation can be performed called *shaving,* which does exactly what the name implies, in that an additional die station is provided with closer clearance and the gear is finely trimmed or shaved to remove some of the drawdown and roughness. A wide choice of materials can be used, including all the low- and medium-carbon steels; all brass alloys, which incidentally give the best finish; half- and full-hard aluminum alloys; and nonmetallic materials. When high-carbon and stainless steels are used, the material for the die and the construction are special, and each application should be considered individually. Gears manufactured by this process have a wide variety of applications, among which are toys; clock and timer mechanisms; watches; counter gear trains; water and electric meters; small appliances such as mixers, irons, blenders, toasters, can openers, dryers, washers; and similiar devices.

FIG. 17-4. Gear of 50 teeth, 64 pitch, 0.020" thickness, blanked two at a time, 150 strokes per minute, No. 1010 steel. (*Courtesy of Sunbeam Corp.*)

Cold-drawing or Extrusion Process. This process requires the least tool expenditure for mass production of toothed gear elements and is extremely versatile, in that almost any shape that can be desired can be reproduced. As the name of the process implies, the material is passed through several block dies with the final die the shape of the desired toothed element, and as the material is run through these dies it is actually squeezed into the shape of the die. Since the material is displaced by pressure the outside surface of the material is quite hard and smooth. The bars usually are from 10 to 12 ft in length, and this allows manufacturers to operate screw machines in the most efficient manner to complete the final shape and size of the toothed gear element, usually called a pinion. It is a recognized fact that small gears, pinions, and special shapes can be sliced from the bar more economically than they can be cut from solid. In some instances it would be impossible to produce the desired shape of pinion in lengths in any other way. Any material that has good drawing properties may be used for drawn pinion rod, such as high-carbon steels, brass, bronze, Monel, aluminum, and stainless steel. Gears and pinions manufactured by this process have a large variety of application and have been used on watches, electric clocks, spring-wound clocks, typewriters, multigraph machines, motion-picture machines, carburetors, magnetos, small motors, switch apparatus, taximeters, cameras, slot machines, all types of mechanical toys, and many other parts for machinery of all kinds.

Injection-molded Gear Processes. Injection molding embraces the use of a thermoplastic material, that is, a material that can be heated and made into a liquid

FIG. 17-5. Machine used to injection-mold nonmetallic gears. *(Courtesy of Fellows Gear Shaper Co., Springfield, Vt.)*

state, and by pressure forced into a closed chamber whose shape has the desired form of the toothed element to be produced. Although many types of thermoplastic material can be readily molded, the most common type for producing gears is nylon, because it has the properties that are best suited to the usual requirements demanded of toothed gear elements.

Typical molding machines are shown in Figs. 17-5 and 17-6. The plastic powder or pellets are placed into the hopper where they are preheated to the correct temperature and are then forced through the machine where they are heated additionally to their liquid state and forced by pressure into the mold, which has two parts. The mold remains closed for several minutes or seconds depending upon the volume of the part being molded, and then the rear part of the mold is opened, or pulled back, and the parts are removed. Depending upon the number of parts molded at one time (usually from two to eight), the mold is designated as a two- or eight-cavity mold.

The parts are connected by runners, which are in reality feeders to each part so that the mold will fill completely as quickly as possible. When the parts are removed these runners must be taken from the parts, and there is usually a small area on the part where this gate was removed. Under ordinary conditions no further work on the gate is necessary; however, in critical cases this extra material might cause interference, and the excessive portion must be removed by hand with a sharp tool.

Once the tooling is complete, large volume can be made economically, and the parts will be identical as long as the setup conditions remain unchanged. Almost any

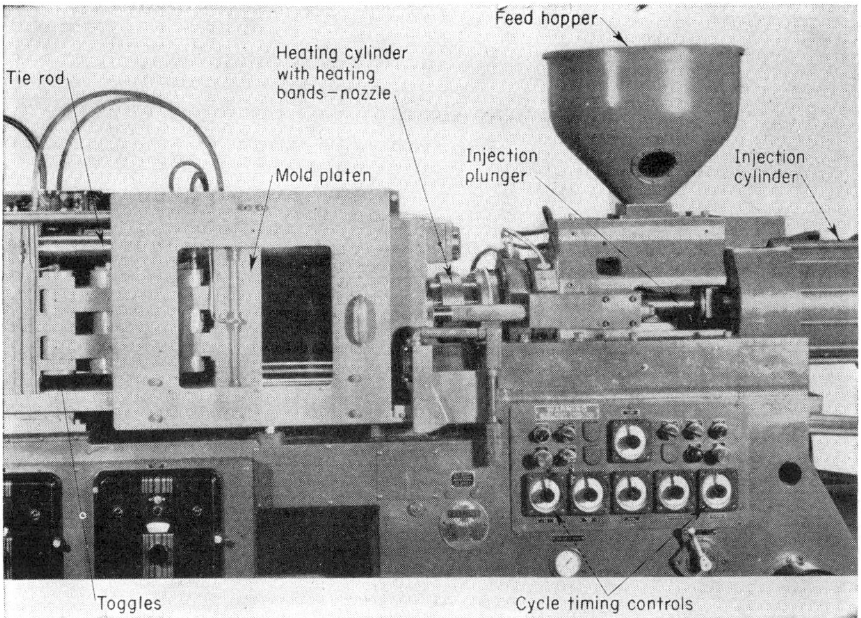

FIG. 17-6. Close-up view of the mechanism of an injection-molding machine. (*Courtesy of Fellows Gear Shaper Co.*)

FIG. 17-7. 56-tooth 64-pitch nylon molded gear with two projections 0.040″ thick. Marks from knockout pins can be seen. (*Courtesy of Sunbeam Corp.*)

desired shape can be produced; however, care must be exercised to maintain fairly constant sections. Diametral pitches from 20 to 120 can be molded very well, and combination or clusters of gears and pinions can be made as one piece. Where large volume can offset the high initial cost of tooling, nylon plastic gears and pinions offer a wide selection of design and configuration. A typical example of a molded gear with a projection included is shown in Fig. 17-7.

While nylon is used almost exclusively in gear molding, other thermoplastic materials can be used, and it is advisable to consult the plastic molder to assist in selecting the proper material for each specific case. Applications for molded gears include a wide variety of products, among which are clocks, timers, small motors, instruments, toys of all types, small appliances such as mixers, blenders, vacuum cleaners, and small tools, and many other similar items. Each application should be thoroughly investigated, however, with respect to load, temperature, tolerances, and assembly design specifications.

Compression-molded gears are made from a thermosetting material such as phenolic compounds. They are used in various applications as mentioned for the thermosetting materials; however, they are subject to more variation in dimensions, are restricted in design, and are not so popular. Applications for use of gears from thermosetting material should be studied with the molder before attempting to specify them. Wide-face gears that do not require close tolerances, and gears of unusual shapes, or gears with tooth profile and additional surfaces are applications.

Die-casting Process. Die casting, like injection molding, embraces the use of molten metal instead of plastic. The metal is forced under pressure into a metal-cavity mold or die and is maintained in that position until the molten metal has solidified. The pressure is then relieved and the mold is opened and the parts removed. Since the metal is very hot all parts must be handled with asbestos gloves. The die-casting machine is similar to the plastic-molding machine in operation and appearance. Like plastics, only certain specific metals are suitable for die casting; these include zinc, brass, magnesium, and aluminum. Each is best suited for a specific purpose, and the exact material specification should be reviewed with the die-casting manufacturer before using.

Die-cast gears are made by two different methods, the hot chamber and the cold-chamber methods. For the hot-chamber type, the machine has a melting pot which permits the injection cylinder to be immersed into the molten metal at all times. The injection cylinder, which is actuated by power, forces the metal into the mold to complete the cycle, while the cold-chamber method provides for a separate melting furnace. The metal is introduced into the cylinder by a hand or mechanical ladle, and the cylinder forces the metal into the mold. The hot-chamber process is limited with regard to types of metal and pressure applied to molten metal. Alloys that have a tendency to dissolve iron or steel are not used, since the proper clearances between the plunger and the cylinder are difficult to maintain. Since zinc does not require the forces of the other materials this is generally the only material that is used for the hot process. Aluminum, magnesium, and brass require very high pressures to obtain desired densities; therefore they are usually cast by the cold-chamber method.

Die-cast gears can be used for high-volume low-duty applications, where low-cost gears of intricate blank shape are desired. Spur, helical, worm, clusters, bevel, face, and other unusual shapes of gears can be readily cast. Applications cover a wide area of use such as toys, vending machines, small gear trains, counters, hand tools, control devices, gardening equipment, washers, dryers, cameras, small speed reducers, business machines, and similiar items.

Rolling or Forming Process. This process, as the name implies, consists of passing bars through sets of rolls which have shapes that will produce the thread or worm desired. Since this is a rotating type of operation, the toothed element produced is usually a worm thread as applied to gearing. Rolling the material between dies leaves the worm thread smooth and burnished, free from chatter or cutter marks, and increases the hardness of the material. When the worm is rolled the material is re-formed in a continuous pattern of the shape rolled, and no waste is incurred because the material is displaced into the desired shape and the original bar is elongated by the amount of metal displaced. In practice it is usual to make the bar diameter somewhat smaller than the finished outside diameter of the worm to be rolled, and the displacement of material fills the areas of increased diameter.

Any ductile material can be rolled, but straight-carbon steels, high-speed steels, nonleaded brass, and copper constitute the majority of materials used in the production of rolled worms. Many variations in rolling characteristics and finishes are encountered with the different materials. Rolled worms are used in motion-picture projectors, slide projectors, toys, appliances, small motors, business machines, and similar devices.

Sintering or Powder-metallurgy Process. This type of process derives its name from the fact that the gears are made from a compound of powder metal and must be processed through a furnace to obtain final homogeneous structure. This furnace operation is called *sintering*. The raw material is composed of finely ground particles of iron, brass, or a combination of each and is forced under high pressure into metal cavities or dies with the shape of the toothed element desired. When the part is removed from this die it is in a state known as "green" and has no strength. If

FIG. 17-8. Powdered-metal gear and cam. (*Courtesy of U.S. Graphite Co., Div. Wickes Corp., Saginaw, Mich.*)

FIG. 17-9. Powdered-metal pinion, 18 teeth, 40 pitch, $\frac{3}{8}''$ face width. (*Courtesy of U.S. Graphite Co.*)

handled it will crumble into small particles. The green gears are then processed through furnaces and are heat-treated. Parts processed in this manner are fairly accurate; however, if more accuracy is desired, an additional operation called "coining" may be performed. This consists in passing the part through another die and repressing it to make it conform to the die more accurately and to remove the slight distortion that takes place in heat-treating. Coining also tends to smooth the surface and make it more dense. Powdered-metal toothed elements that are not coined can be vacuum-impregnated with lubricating oil, as the metal after sintering is usually about 18 per cent porous. These porous areas are filled with oil. Coining destroys the porous structure on the outside surfaces of the parts, and they cannot therefore be oil-impregnated.

Powdered-metal toothed elements can be molded in clusters, or with cams attached as shown in Fig. 17-8, or with irregular holes as shown in Fig. 17-9. They cannot be made thin, however, and the pitch must be coarse enough so that the chordal tooth thickness is sufficient to maintain the cross section of the part. Gears as fine as 64 diametral pitch have been produced; however, each application should be reviewed with the manufacturer for specific recommendations for each specific case. Typical applications are for business machines, toys, power tools, appliances, small motor drives, instruments, and similar items. Any use for impact should be carefully investigated. Tooling for this type of process is moderately expensive; however, if the volume is sufficiently large, it will be more than offset by the economical cost of the parts.

Special Processes. For special applications that require low noise characteristics, low load capacity, minimum vibration, and little shock load, gears made from thermosetting laminated plastics and fibers may be considered. They are most effective at pitch-line velocities of 400 fpm and over and have a high degree of elastic deformability and a low modulus of elasticity. Of the industrial thermosetting laminates, cloth-base grades are the most commonly used for the greater load capacities, and the paper-based laminates are used for the small gears where motion transmission at low noise level is the predominating factor.

Laminated-plastic gears can be punched up to thicknesses of ⅛″, although 3⁄32″ is more commonly specified. Toothed gear elements punched in this manner must

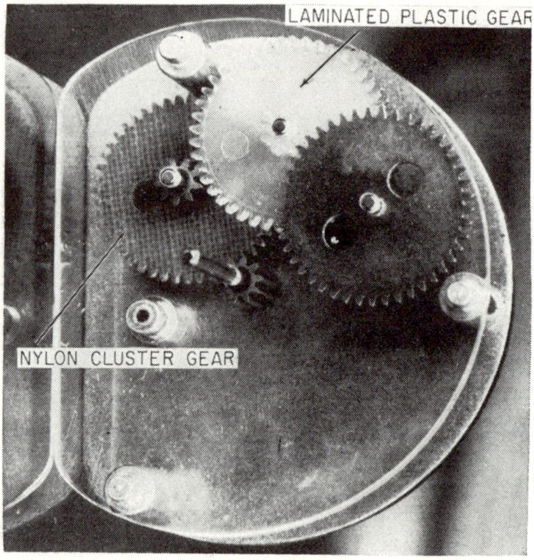

FIG. 17-10. Laminated-plastic and nylon cluster gear. (*Courtesy of Sunbeam Corp.*)

usually be heated prior to punching. While experimentation is in process to punch 80 diametral pitch, 0.025″ thick, it has not yet been solved for production runs, but 64 diametral pitch has been successfully punched. Since the material is resilient, some trial and error is necessary to arrive at the proper tool size to make the exact size of gear desired. Additionally, shrink is involved as the material is warm when punched and will shrink as it returns to its original state. Each application for this type of gear should be presented to the plastic manufacture to ensure selection of the best material for the specific application. Gears of this type are used in electric clocks, small motor gear trains, instruments, and machines.

While punching laminated-plastic gears has been described it is not generally used for punched gears, since the cost is relatively expensive. This material is more readily hobbed in the fine-pitch area. Figure 17-10 shows an experimental laminated cloth-based plastic gear that has been punched. It is 80 diametral pitch with 45 teeth, is very effective in noiseless operation, and absorbs a substantial amount of oil because of the open surfaces around the periphery of the tooth profile. Considerable trials were made to produce this gear to the size desired, and in this application of high-speed gear reduction units it is operating very effectively in respect to noise, wear, and transmission of low torque.

17-2. Gear Stamping.* Low-cost lightweight gears for no-load to medium-duty applications can be designed for stamping from sheet metal. In designing gears for stamping, the backlash must be taken care of in the tooth-form design, as once the die is made no alterations are practical as in the cutting methods. The tooth form is designed, therefore, drawn very accurately fifty times size, and reproduced on a glass master, which has the provisions for backlash, whole depth, and corrected tooth thickness provided. Additionally to compensate for wear it is usually good practice to provide an additional amount to the whole depth, so that, when the punch wears, the tooth will not become too shallow.

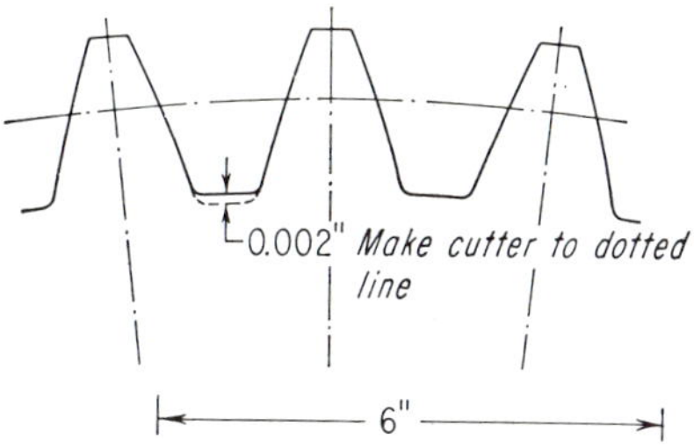

FIG. 17-11. Typical 50-times-size tooth form. (*Courtesy of Sunbeam Corp.*)

Figure 17-11 shows a representative tooth form for a 60-tooth 60-pitch gear. Note that 0.002″ has been added to the whole depth in this case. In preparing the tooth form the thickness of the tooth at the pitch diameter is usually reduced by a standard amount. The recommended reduction in tooth thickness for gears of this type is shown in Table 17-1.

Table 17-1

Pitch range	*Reduction of* 1.5708/*DP*, *in.*
150–180	−0.0008
121–149	−0.0012
100–120	−0.0015
65–99	−0.0020
20–64	−0.0025

It is also well to keep in mind that the full face width will be reduced by approximately 10 per cent because of drawdown, and the corners should have at least 0.003″ radius to prevent excessive die maintenance. Typical gears produced at a high rate from commercial dies are shown in Figs. 17-12 and 17-13. Generally there are no blank requirements except where a lancing or special tab forming is specified as part of the gear. In these cases, the tool engineer will usually provide the correct sequences and stock widths. Compound dies are usually employed to produce this type of gearing as it maintains a better concentricity between the center hole and the outside diameter and produces the entire gear at one stroke. Where lancing or forming is required a progressive die is usually made, and the tolerances of the part depend upon the exactness of the location of pilots in each die station. Typical pilot pins are shown in Fig. 17-2.

Lightweight gears are usually made from No. 1010 steel, quarter hard; leaded brass; full-hard aluminum; and sometimes steels with higher carbon content without any special die-design requirements. When high-carbon steels and stainless steels are used, special carbide inserts are necessary. Stamped gears range in size from 20 through 120 diametral pitch, and from 0.010 to 0.078″ thick. The finer the pitch the thinner the material that should be specified. Table 17-2 shows recommended thick-

* Part of the data used in this article was furnished by Harold Winzeler of Winzeler Manufacturing Co., Chicago, Ill.

ness for various pitches that are commonly used and require no special care in die maintenance. With special setups and/or shaving, the thicknesses in Table 17-2 can be modified; however, die life and tolerance deviation will be greater.

Table 17-2

Pitch range	*Thickness, in.*
20–36	0.020–0.078
38–60	0.015–0.062
62–72	0.010–0.040
74–90	0.010–0.035
92–120	0.010–0.025

Tolerances for stamped gears are good, and AGMA[1,*] quality class 9 can be produced without difficulty or excess cost. Profile can be made to within 0.002″, and

Fig. 17-12. 60-tooth 60-pitch stamped gear, 0.025″ thickness. (*Courtesy of Sunbeam Corp.*)

Fig. 17-13. 50 tooth, 60 pitch, 0.046″ thickness, stamped two at a time, 220 pieces per min.

concentricity to within 0.001″ TIR. Surface finish usually depends upon raw material used and can be as good as economy warrants. Flatness will be acceptable provided a roll straightener is used when stamping the gears. Finishing costs are low, and barrel tumbling will usually remove the burrs that are existing. No protective coating is required on the nonferrous materials and on steel a rust preventative such as nickel flash or black oxide will preserve the part.

In running stamped gears, a first-piece inspection is usually made when setting up the part for the first time; that is, a few parts are run, and the press is stopped until these first few pieces have been checked. This may be done by comparing the part with the tooth form on a comparator or by rolling with a master gear or by measuring over wires, or all three, depending upon the accuracy desired. If the gear is satisfactory, the press will then be started and many thousands can be made. Barring unusual acts, all the gears will be the same as the first pieces. A check must be made every 2 hr approximately to ensure that the die does not break a tooth or that excessive burrs are not developing because of some imperfection in the material or that some foreign matter has not embedded itself in the die portion.

* Superscript numbers refer to references at the end of the chapter.

Production of gears in this manner is a function of the tooling and the type of press that is used. Sometimes two gears are made at the same time, so that on every stroke of the press two gears are produced. On high-speed presses, 200 strokes per minute is common, and consequently 400 gears per minute are produced. It can be easily

(a)

(b)

FIG. 17-14. (a) Stamped gears with lightening holes, cutouts, pockets, serrated center holes, and similar features. (b) Formed gears may be used to provide right-angle drive for noncritical designs. (*Courtesy of Winzeler Manufacturing Co., Chicago, Ill.*)

recognized that substantial quantities can be made in a very short time, and therefore gears in this category are relatively inexpensive, costing as low as $3 per thousand in large quantities. As an example of actual costs, a 45-tooth 48-pitch gear, 0.032″ thick, made of No. 1010 steel and nickel flashed 0.0002″ thick, has a tool charge of $890 and a piece price of $11.39 per thousand in 1,000 lots, $5.39 per thousand in 5,000 lots, $3.98 per thousand in 10,000 lots, and $2.09 per thousand in 50,000 lots. There is no saving in amounts greater than 50,000.

Summarizing, stamped gears offer the designer a number of advantages which are valuable in the consideration of economical lightweight, light-duty gear trains.

1. Sizes are relatively what is desired within the range of common pitches and thicknesses. A great many stock gears can be purchased from existing manufacturers, thus eliminating tool costs, and catalogues listing items available are easily obtained.
2. Shapes that can be made in a punch press are obtainable, such as lancings, irregular holes, tabs, partial toothed profiles, bosses, formings, and cutouts (see Fig. 17-14).
3. Material choice is large. All ferrous and nonferrous materials which can be stamped are available. Finish is obtainable in many grades and is specified by the raw stock.
4. Final finishing is economical and is usually done in large quantities by tumbling. Protective coatings are also economical, being performed by the barrel method.
5. While initial tool cost is high, volume production results in very low piece-part costs.
6. Lead times to tool new gear designs are long and expensive.
7. Once a tool is made, alterations are not practical.
8. No preliminary tryout for new parts is possible. Once a die is set up, parts will be made as designed.
9. The entire face width of the gear is not available for load because of loss of surface due to drawdown.

17-3. Cold Drawing or Extrusion.* Low-cost pinions for large-volume production for light- to medium-duty loads can be designed for manufacture from cold-drawn pinion rod. In designing pinions to be made from drawn pinion-rod stock, the backlash and undercut must be provided for in the tooth design, as once the several dies are completed, alterations are not practical as with cutting methods. The tooth form is designed, therefore, drawn very accurately at fifty times size, and reproduced on a glass master. This tooth form has provisions for backlash, whole depth, and corrected tooth thickness provided. Since the original die is made with a milling cutter, no undercut can be obtained as with hobbing since the cutter could not be retracted; therefore, the generated tooth form is modified by making a line tangent to the bisector of the space and connecting to the involute portion of the tooth above the base circle to the root radius. Additionally, in order to compensate for wear of tools at the root, additional depth is provided. This is of greater consideration than the allowance specified in stamped gears as the wear in drawn stock tools is much greater. Figure 17-15 shows a representative tooth form for a 10-tooth 68-pitch pinion. Note that 0.0015″ has been added to the whole depth. Table 17-3 shows the recommended amount to be added to the whole tooth depth for various diametral pitches. The tooth thickness at the pitch diameter should be reduced by the same amount as recommended for stamped gears and shown in Table 17-1. Actually the pinion rod is the blank; the rods come in 10- to 12-ft lengths, and the shape of the pinion detail is completed on the screw machine.

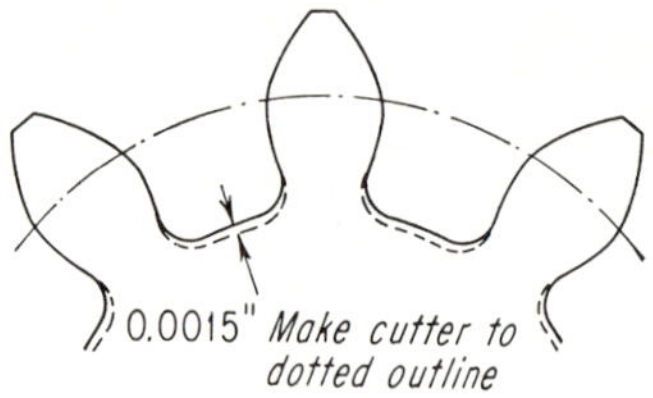

Fig. 17-15. Typical 50-times-size pinion tooth form. (*Courtesy of Tekniks Engineering Co., Elmhurst, Ill.*)

* Part of the data used in this article was furnished by Clarence Johnson of Utility Manufacturing Co., North Wilbraham, Mass.

Usually from four to seven dies are made, the first-pass die being a shape similar to the pinion desired but with large radii to allow easier metal displacement. After each pass the material is coiled and annealed for the next pass, and with each succeeding pass the stock assumes a shape nearer the final design until the last pass, which is exactly like the tooth form specified.

Any material that has good cold-working properties can be used for drawn pinion rod. While steels, stainless steels, aluminum, brass, bronze, and monel can be drawn, the preferable choice is brass and free-machining steel. One of the outstanding properties of pinion rod is that the surface is smooth and hard, and since the material is cold-worked, the outside surface is several Rockwell points harder than the core. Therefore, in cases where close-tolerance holes are required, care must be exercised in drilling so the tool will not drift because of the soft center.

Pinion rod from 16 to 100 diametral pitch can be obtained, but as the pitch becomes finer, it becomes more difficult to obtain the close tolerances that are sometimes desired on fine-pitch pinions. In addition, this is only the blank, and the final tolerance of the pinion will depend to a great extent upon the collet design and the accuracy of the screw-machine operation. In this regard concentricity will depend on the setup of the machine. Concentricities of less than 0.003″ TIR are difficult, and 0.005″ TIR is more generally practical. A tolerance of plus or minus 0.002″ on all pinion dimensions is considered standard, but all dimensions can be held to a total of 0.002″ if required, without excessive difficulty. There are some so-called standard pitches available, on which pinions are available without tool charge. These are 24, 32, 40, 48, 64, and 80. Many are available in both 14½ and 20° pressure angles. It is well to consult the manufacturer for each specific tooth pinion desired to ascertain if it is available.

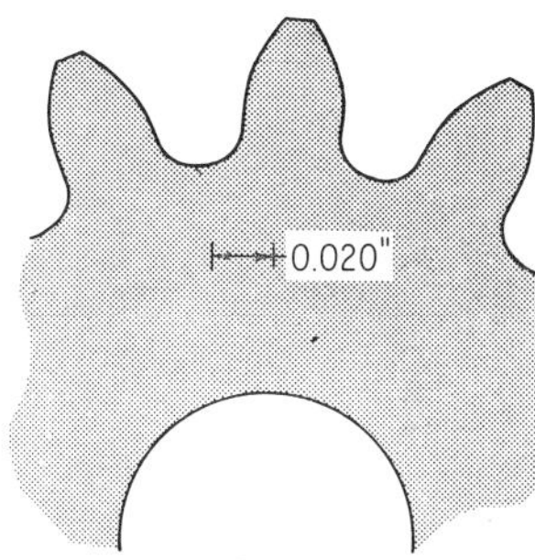

Fig. 17-16. 10-tooth drawn pinion rod showing photograph of actual tooth form. (*Courtesy of Tekniks Engineering Co.*)

Inspection of pinion rod usually consists of cutting a piece from the bar approximately 1/32″ thick, sanding to remove burrs, and comparing the tooth form with the master on a comparator, checking hardness with a Rockwell tester, measuring with appropriate tools, and inspecting with a glass visually for defects on the surface.

To summarize, drawn pinion rod

1. Is economical for volume production.
2. Has low initial tool costs.
3. Provides excellent and durable surface finish.
4. Has uniformity of material structure.
5. Is extremely versatile for unusual or odd shapes.
6. Is extruded; therefore sharp corners, extra-deep thin teeth, and undercut should be avoided.

Table 17-3. Whole-depth Correction for Drawn Pinion Rod

Pitch range	*Amount to be added, in.*
16–24	0.0030
26–38	0.0025
40–62	0.0020
64–80	0.0015

A print of a 10-tooth drawn pinion enlarged from a comparator chart is shown in Fig. 17-16. This is an actual photograph of a shadow as it appears on the comparator window.

17-4. Injection Molding. Low-cost pinions and gears for large-volume production in many unusual and intricate shapes may be designed for manufacture from nylon by injection molding. In the design of gears and pinions for this method of manufacture the tooth form must not only provide for final shape and modifications to eliminate backlash but also for shrinkage, swelling, deformation under load, and moisture absorption. A metal gear cannot be exactly replaced by a nylon gear, unless modifications are made. The geometry of injection-molded gear-tooth forms is not too important. Any tooth form that can be made by a cutting process in metal can be reproduced by molding, and many times a more intricate shape can be made by the molding process than by any other method. Since nylon has a thermal coefficient of expansion approximately ten times that of steel, additional clearances in the tooth form must be provided so that the gears will run properly at the maximum and minimum temperatures dictated by the design requirements. Success has been achieved by reducing the tooth thickness in accordance with the dimensions given in Table 17-4, and enlarging the center distance for standard-pitch-diameter gears and

Table 17-4

Pitch range	*Reduction in tooth thickness of 1.5708/DP, in.*
100–120	−0.002
65–99	−0.003
42–64	−0.0035
20–40	−0.004

pinions by the amounts recommended in Table 17-5. While this reduction in thickness and increase in center distance may seem to be excessive it is well to remember that a 48-tooth 48-pitch gear will increase in size approximately 0.006″ from zero to 100°.

Table 17-5

Pitch range	*Increase in center distance, in.*
100–120	+0.002
65–99	+0.003
42–64	+0.0035
20–40	+0.0045

A fifty-times-size tooth form should be generated as described in the other articles to determine the best shape, and it is often necessary to employ a larger diametral pitch than originally contemplated because of the tooth reduction necessary. A great deal of this shrinkage and dimensional variation can be eliminated by using metal inserts for the center of the gear. The use of an insert reduces the amount of nylon and consequently the shrinkage. The use of inserts offers a compromise between cost and design objectives. In general, the design of the tooth form should, where possible, conform to the American standard 20° involute fine-pitch system (ASA B6) with the added modifications recommended here.

Mold or die design of nylon-gear toothed elements should follow the same general procedure as for other commercial thermoplastics. Nylon is fast-setting so that, in spite of its fluidity in the molten state, it will not flow so far in a thin groove as some other slower-setting compounds. Flash is always a problem on pinions and gears, and care must be exercised in the mold and gate design to eliminate the necessity of removing flash by hand if practical. Designing the molds[14] is a task that should be left to those skilled in the art, since each design presents a separate and distinct problem and must be analyzed individually. To realize the fullest advantages of nylon gears, close cooperation between the molder and the engineer is very desirable. The molder will usually make valuable contributions to best adapt the part to injection molding. Among the variables the molder will consider are the size and type of molding machine, design and shape of part, and best flow pattern to obtain the best part with reference to residual stresses, cycle, and temperature. Only after discussions of all these variables can the final gear design be specified and the proper diametral pitch, tooth shape, and configuration be determined.[11]

Almost any desired shape can be produced, provided the sections are fairly constant. A cluster gear and pinion of 120 diametral pitch is being molded successfully, and the large size is limited only to the amount of variation that can be tolerated in dimensional tolerance.

The accuracy and tolerances of nylon molded gears are generally as good as for molded metal gears; however, nylon will slowly absorb moisture from the surrounding air until equilibrium is reached, which varies with the relative humidity of the surroundings. Therefore, in production parts it is usually well to recommend annealing and/or moisture conditioning.[9] Exact tolerance specifications for each part should be reviewed with the molder.

The inspection of nylon gears introduces conditions that are not prevalent with metal gears. A great deal of care must be exercised in inspection, since nylon has the property of resiliency, which causes it to resist deformation by springing back when the applied load is removed. It is difficult, therefore, to measure nylon gears by pin measurements.[15] The projection method is used a great deal to check the gears, but difficulty arises when cluster designs are used, since the pinion must be viewed by reflection, and nylon does not have good reflective properties. With thin gears, the translucency of the material causes a halo around the profile, which is another objection. Since inspection costs for large-volume production for low-cost gears must be in line many manufacturers use a fifty-times-size tooth form as described in previous articles and check parts at random on an optical comparator. A projection inspection such as this considers several kinds of errors that are usually measured by a master gear. If additional control is desired, a simple master-gear fixture with variable center distance can be made up and used on a sampling-plan basis. Since the cost of each gear is low, the inspection obviously must be done at intervals to keep the costs proportional, and inspection of these gears run at high production rates should follow the accepted standards of items run in volume.

Production rates for nylon gears and pinions vary from 200 to 4,500 per hour, depending upon size and number of cavities. For example, a cluster gear and pinion, with a gear of 48 teeth 120 pitch, pinion of 12 teeth 100 pitch, and mold of 8 cavities, produces 1,240 units per hour, and the cost is approximately $13 per thousand pieces.

Summarizing, the injection-molded gear and pinion offer the designer a number of advantages for lightweight and medium-load applications.

1. Sizes are almost unlimited.
2. Shapes that are relatively uniform in cross section although irregular are easily obtainable.
3. No finishing is usually required.
4. While initial tool costs are high, volume production results in low-cost unit parts.
5. In many applications no lubrication is necessary.
6. Lead times to tool new designs are long and expensive. Once the mold is made, changes are impractical and expensive.
7. No preliminary tryouts for new parts are possible; parts must come from the final mold. Material choice is limited. Nylon is used almost exclusively, although other materials have been used.

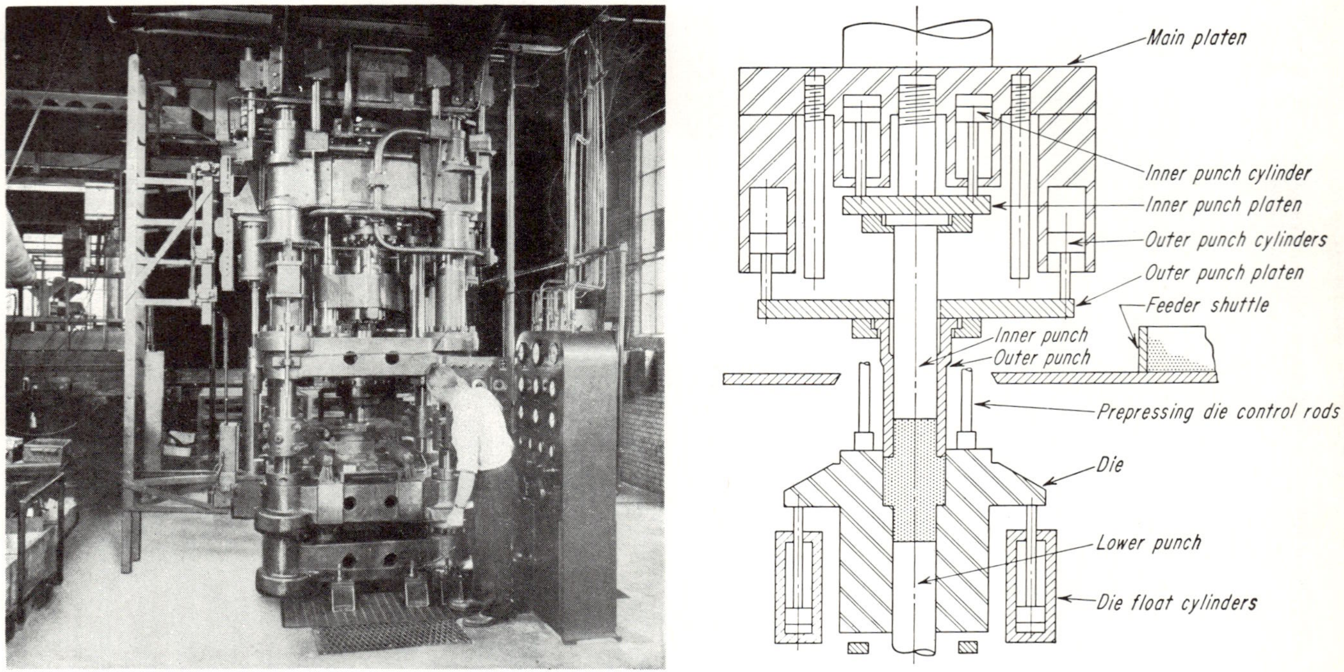

FIG. 17-17. Details of a die design and photograph of powdered-metal press. (*Courtesy of Industrial Equipment Div., Baldwin-Lima-Hamilton Co., Philadelphia, Pa.*)

8. Surface finish is extremely good and has excellent wearing characteristics. Final dimensions must be carefully analyzed with respect to variations from temperature, moisture, and load.

17-5. Powdered-metal Sintered Gears. Parts produced by the powder-metallurgy process have a much wider range of characteristics than other metals, and designs for this type of gearing must be given special consideration. Tooth form, backlash, and other design modifications must be taken care of in tooling. Since the parts are made from finely sifted powder, there are no blanks. Tooth form can usually follow standard proportions, with the exception that, on pinions, no undercut is permissible because of tooling methods. Material thickness between root diameter and hole must be of sufficient amount to prevent cracking of the pinion if it is to be pressed onto a shaft in a subsequent operation.

The die design is of a general nature in accordance with Fig. 17-17; however, each particular application may have special arrangements to produce a given part. The powdered-metal manufacturer should be consulted for each individual part, as the density, configuration, and tolerances affect the die design. When closer tolerances are required an additional die is prepared to coin the part after it is sintered. This is in reality a sizing operation that squeezes the part under heavy pressure to conform to the die shape more exactly, and no further distortion will take place after this operation, since the part is not usually subjected to additional heat-treatment.

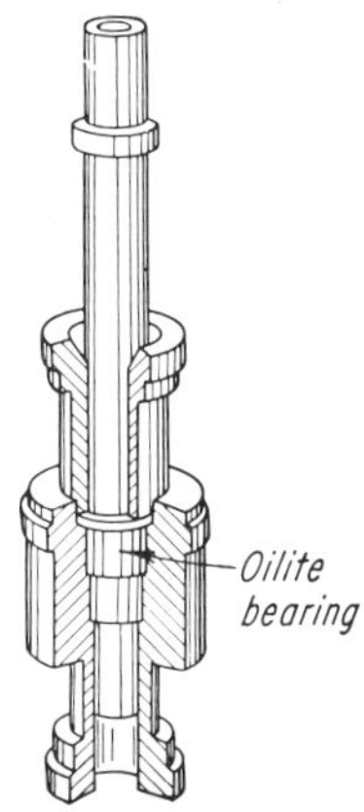

FIG. 17-18. Typical sizing tool. Tool shown is sizing an Oilite bearing. (*Courtesy of Amplex Co., Redwood City, Calif.*)

Material choices are varied, and the base irons, steels, brass, and numerous alloys of each are employed. In choosing materials an unusual factor must be considered that is not present in any other method, and that is that the raw material may be the best possible type but, if it is sintered incorrectly, will not have the properties desired. A good material improperly heat-treated will make a very poor part. The final product is therefore controlled to a great extent by the ability of the manufacturer to sinter properly.

Advances in the art of powder metallurgy with respect to gears have been swift during the past several years, and acceptance by manufacturers has been equally prominent, so that today the use of this type of gearing is rather common in appliances,

FIG. 17-19. Powdered-metal gear assembled to a punched gear. (*Courtesy of Sunbeam Corp.*)

many instruments, hardware, and outdoor equipment. The ability to have collars, bosses, and other appendages as part of the gear has been instrumental in cost reduction on this type of equipment. Typical gears of this type are shown in Figs. 17-19 and 17-20.

In general, iron of the common variety is used, but for special applications any number of special alloys may be employed. A typical alloy of iron-copper that has been used with great success for electric tools, garden equipment, and other devices of this type is listed in Table 17-6. This is by no means the ultimate, as there are as

FIG. 17-20. Typical powdered-metal gear with a hub. (*Courtesy of U.S. Graphite Co.*)

FIG. 17-21. Powdered-metal gear with a driving key integral with the hub. (*Courtesy of U.S. Graphite Co.*)

many alloys as there are manufacturers, each type being suited best for a particular application. When a specific design has been decided upon it is recommended that the manufacturer be called in to consult on the final characteristics. For detail design procedures and articles the references are good sources, and there are many excellent and reliable manufacturers who will give a great deal of assistance in the finalization of gear design from powdered metal.

The use of powdered metal in tools has been through several concentrated experiments, and a particular example is in the use in ¼-hp portable electric drill. The gear just ahead of the final reduction is subject to heavy shock loads, and ordinary

Table 17-6. Typical Physical Properties for Alloyed Powdered Metal
(Iron-copper with controlled combined carbon)

Density dry, gm per cu cm	Young's modulus in tension, psi	Tensile strength, psi	Transverse rupture strength, psi	Radial crushing strength	Hardness, 15T scale
5.8	12.5×10^6	39,800	70,500	80,100	70–76
6.0*	14.0	47,400	80,300	93,000	74–79
6.2	15.4	55,100	89,700	......	78–84
6.4†	16.9	63,000	99,800	......	80–85
6.6†	18.6	69,200	110,000	......	83–88

Applications: Gears and other structural parts where loading is moderately high.

Advantages: The greater strength in conjunction with combined carbon (pearlitic structure) on the part's surface results in higher hardness to give maximum wearability with oil impregnation.

* Standard production density.

† Not all parts can be manufactured to these densities as a result of their complicated shape; it may require special tooling to reach these densities because of higher than normal pressures needed to obtain these values. Heat-treating may be considered to improve physical strength if the part does not lend itself to high density.

Typical Properties Obtained after Heat-treatment

Density	Tensile strength	Radial crush	Hardness, 30N	Hardness, Rockwell C
6.0	58,500	90,000	35–45	15–25
6.2	66,000	105,000	45–55	25–35

The hardness values obtained cannot be correlated to those hardness values of solid metals of similar composition. Superficial-hardness testers are normally used to obtain readings in order to try to compensate for the inherent seating of the tester's ball on the surface pores common to all powder-metallurgy parts, the results of which give readings below the actual particle hardness.

powdered metal of the common iron type was tried. After several cycles the gears fractured under stall load, and after exhaustive tests an alloy of 5 per cent copper and iron was used and vacuum-impregnated with a good grade of turbine oil of approximately 700 sec viscosity. The parts were not sized because sizing prevents the vacuum impregnation of oil. Conventional copper infiltration, that is, placing a slug of copper on the gear and sintering, was tried and produced a "washing effect." When the copper was placed on the back section of the gear proper flow was not achieved, and the gear-tooth portion therefore did not have adequate strength. When copper was placed on the gear-tooth portion, the "washing" effect created a spongy porosity to a depth of 0.010″ and provided weak gear teeth and poor wear resistance. Consequently the alloy of iron and copper was used and the as-sintered porosity was about 10 per cent. Heat-treatment was also tried but added no factors for improvement to operation and was eliminated. Conclusions from these tests and subsequent data from field tests of over one year indicate that presintering alloys of iron-copper compounds are considerably better than infiltrated copper parts for shock loads of the intermittent type.

Powdered-metal gears have been produced as fine as 80 and experimentally as fine as 96 diametral pitch. However, 64 diametral pitch is usually the finest that is obtainable on a commercial basis and without special considerations. Tolerances are very good, and AGMA quality 9 can be obtained without difficulty. Finer tolerances may be obtained in special cases.

Inspection techniques are similar to those used for other types of metal gears.

Production of these gears is large, and several hundred pieces per hour can be produced without difficulty.

In summarizing:

1. Powdered-metal gears are economical for high-volume production.
2. Initial tool costs are moderate.
3. Surface finish is good and wears in.
4. The material is porous and can be vacuum-impregnated with lubricant, thus providing long life.
5. Clusters and unusual shapes may be made easily.

17-6. Die Casting. Low-cost pinions and gears for large-volume production, in many unusual and intricate shapes, may be designed for manufacture from aluminum, brass, zinc, and magnesium. Brass is sometimes used but is expensive from both tool and material viewpoints. In designing gears and pinions for die casting, the tooth form must provide for not only final shape and modifications to eliminate backlash but also shrinkage, parting-line effects, and the draft which is necessary to remove the piece from the mold.

The geometry of die-cast molten-metal gears and pinions is not too important, as any tooth form that can be produced by the cutting process, and many that cannot, can be reproduced by die casting. Also in many cases, a gear, pinion, and shaft can be molded as one integral piece. Since the parts are made from metal which is in a molten state and solidified in a die under pressure, the final shape cannot be held to such close tolerances as cut gears, unless special considerations are taken into account. For all practical purposes the final results are the same as for injection-molded nylon gears, and success has been achieved using the reduced tooth thickness specified in Table 17-4, and enlarging the standard center distance as recommended in Table 17-5.

In general the tolerances for die-cast gears are governed by the weight of the part, and plus or minus 0.001″ is common on castings not over 1 oz in weight. As the weight increases the permissible tolerance increases until plus or minus 0.005″ is necessary on castings of 6 oz. Closer tolerances may be achieved without too much difficulty; however, the cost will be substantially higher. A fifty-times-size tooth form should be generated as described in the preceding article to determine exact tooth shape and to provide specifications for the mold maker. In general the design should conform to the AGMA standard fine-pitch involute system.

Mold design, as stated in the nylon article, should be left to those skilled in the art, and there are several excellent manufacturers who will assist the engineer in this endeavor. In general, when gears and pinions are designed for die casting, the parting-line locations should be studied, because they will affect appearance; size must be considered, and if too heavy, lightening areas must be provided; and section thicknesses are most important and should be maintained as uniform as possible, although abrupt section changes should never be specified. Long unbroken surfaces should be avoided, since they may be wavy or tapered. Holes should have as large a taper as possible, and where straight holes are required, a secondary reaming operation will be required. Undercuts should be avoided as they cause the tools to be made with removable pieces, slow down production, and increase costs. Radii should be provided at all intersections to add strength and uniform flow of molten metal. Since the parts have gates and runners their location should be indicated, as they may interfere with the final performance of the part. In some cases external threads and knurls can be cast as part of the gear; however, they are not in any sense replacements for the machined threads and knurls, and the manufacturer should be consulted before such items are specified. Gears and pinions can be produced in this manner from 1 oz to several pounds. However, as the parts become larger it is usually necessary to perform secondary operations to meet the specifications required.

Inspection of this type of gearing can be made in the conventional manner with pins, micrometers, and optical-comparator methods. Since the quantity involved is usually large, a quality-control sampling plan is usually employed. Parts are made at many pieces per hour similar to production methods of nylon, and in some cases stock gears and pinions can be procured from manufacturer's inventory.

This type of gear and pinion offers the designer the same advantages as those of nylon, except that metal cast gears can carry more load per unit than nylon. It is possible to cast bevel, spur, worm, helical, stepped, and shouldered gears. Most die-cast gears are small and for low loads; however, gears of 5″ diameter are not uncommon.

17-7. Gear Teeth and Worm Threads Formed by Rolling. Parts produced by this method are used only in mass-production quantities. It is a generally accepted fact that, if a part can be rolled as designed or be redesigned slightly to be rolled, it will cost less than other methods and the resulting part will have superior strength and finish because of the material-displacement process. When designing toothed elements for this process it is usually good practice to have the thread depth about 18 per cent of the major diameter. Greater depths are difficult to make and maintain

1. Design worm as 2 start if possible.
2. Keep thread depth less than one-fifth of major diameter. However, in 2 start worms, good straightness can be maintained with deeper threads.
3. High helix angles increase end-feeding tendency.
4. Always design worm shaft like this if possible,

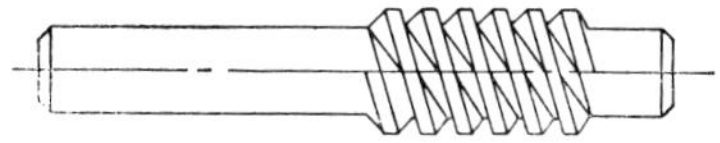

so that minor diameter is larger than adjacent shaft diameter.
Or like this,

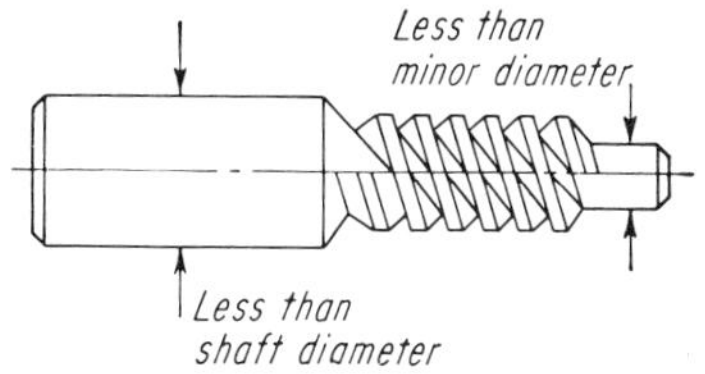

to avoid rolling between shoulders.
5. Make thread runout as long as possible—2 pitches long.
6.

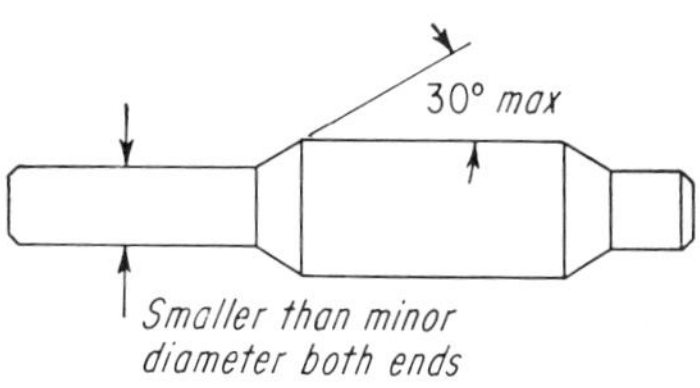

7. Always design to 20° pressure angle normal. Higher pressure angles desirable on worms.
8. On gears for infeed rolling try to get helix angle and pitch so that two teeth are in simultaneous contact with rolling dies.

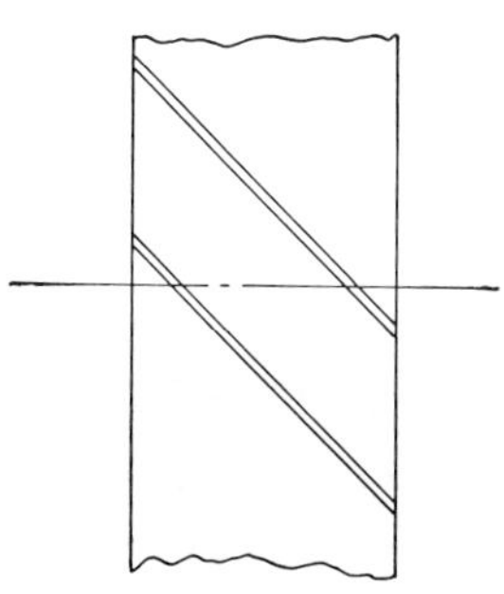

FIG. 17-22. Design hints for rolled worms. (*Courtesy of Frank Gareri.*)

straightness of shaft. Worm threads should not be specified between shoulders, and the runout of the worm onto the main shaft should be equal to at least two threads. See Fig. 17-22 for recommendations on design characteristics for worm threads.

Blanks for rolled worms are usually prepared by screw-machine process, extrusion, and/or cold heading. The blanks should be accurate as the finished worm accuracy depends upon the blank size. Since the material is designed to pass through two or more rolls which are closed together, a large blank will not allow the rolls to close properly and the worm will be oversize; also a small blank will not fill the void area of the roll when closed and the resulting worm will be undersize. If good accuracy is desired it is sometimes necessary to have the blank pass through a secondary operation for size control. Exact blank diameters may be found from the manufacturer; however, an approximation can be made by subtracting one whole depth of tooth from the finished outside diameter of the worm.

Worms are rolled by several methods, and excellent results are claimed for a patented process[29] now in use. Another method in the developed stage is called "chipless[26] forming" and is done on a special machine. Generally, however, the rolling is performed, as the name implies, between two sets of rolls with the rolls designed to produce the desired shape. Most thread rolling follows the direction of the threads. The behavior of materials during the rolling operation is related to the material properties such as hardness, ductility, and toughness. In this process the dies are progressively forced into the blank as they roll. The penetration of the die is being resisted by the material being rolled, and the displacement is actually a sliding movement within the internal structure of the material.

Straight-carbon steels, structural alloy steels, nonleaded brasses, and copper constitute a majority of rolled toothed elements. Many variations in characteristics and finishes are obtained, and final selection usually depends upon compromise of rollability and machinability. Worms from 48 to 8 diametral pitch have been rolled with success, and production rates are high but vary considerably with material and accuracy required. Summarizing, rolling toothed-form elements is desirable when mass-production quantities are required, low cost is important, and design can accommodate rolling.

Rolled Splines. In recent years the rolling of involute splines[30] has become well established in the automotive trade. It is common practice to roll the splines on the main drive shaft of an automobile. The process is extremely fast. The finished teeth are very smooth and accurate and have a work-hardened surface.

Rolled splines usually have a 30 or 45° pressure angle. The spline teeth are shallow in depth, being about 50 per cent of the depth of a full-depth gear tooth. Medium-carbon-steel shafts are used at a medium hardness. Production rates may be as fast 10 sec floor-to-floor time per part.

Rolled Gear Teeth. The rolling of gear teeth has long been considered but is still used only to a limited extent. The rolling of gear teeth presents some fairly difficult problems. A normal 20° pressure angle gear tooth of full depth has too low a pressure angle and too much depth for an easy rolling job. Normally gear teeth have fairly narrow face widths in proportion to their diameter. The ends of gear teeth that are rolled tend to be imperfectly formed and need to be machined away. In some cases the addendum of the rolled tooth is extruded out in such a manner as to make the tooth weak.

Some rather good gear-tooth rolling has been accomplished, and it is likely that gear-tooth rolling will continue to develop as a promising gear-tooth production process.

REFERENCES

1. American Gear Manufacturers Assoc.: "Gear Classification Manual," AGMA 390.01, May, 1960.

2. Strasser, Frederic: Stamped Gears, *Machine Design*, September, 1957.
3. Maschmeyer: Wear Life Aluminum Gears, *Prod. Eng.*, September, 1956.
4. Talmage, Robert: Sintered Steel Parts, *Machine Design*, June, 1957.
5. Powder Metallurgy Precision Metal Moulding, Republic Steel Co., *Publ.*, June, 1957.
6. High Strength Parts, *Business Week*, January, 1955.
7. Martin, L. D.: Small Gears for Appliances, *Elec. Mfg.*, August, 1956.
8. Everhart: Design for Powder Metal, *Material Design Eng.*, April, 1959.
9. Cheney: Designing with Nylon, *Machine Design*, March, 1958.
10. Martin, L. D.: Moulded Nylon Gears, *Machine Design*, November, 1955.
11. Martin, L. D.: Tooling for Moulded Nylon Gears, *Tool Engr.*, July, 1956.
12. Hall, K. W., and H. H. Alvord: Nylon Spur Gears, *Mech. Engr.*, May, 1959.
13. Cheney, A. J.: Nylon Gear Capacity, Dupont Co., *Publ.*
14. Hall, K. W., and H. H. Alvord: Nylon Mould Design, September, 1957.
15. Martin, L. D.: Nylon Gear Inspection, *Machinery*, December, 1955.
16. Powder Metallurgy Gears, Metal Powder Assoc. *Publ.*, spring issue, 1958.
17. Powder Metallurgy Today, F. J. Stokes Machine Co. *Publ.*
18. "Die Casting," American Die Casting Institute,
19. Zinc Die Castings, Gries Reproducer Corp. *Publ. Bull.* 1-C-8-57.
20. Die Cast Gear and Pinions, Gries Reproducer Corp. *Publ. Bull.* 3001B.
21. Die Casting, "American Machinists' Handbook," McGraw-Hill Book Company, Inc., New York, 1955.
22. Carson, Gordon B.: Powder Metal Dies, "Production Handbook."
23. Carson, Gordon B.: Die Casting, "Production Handbook."
24. Small Parts—Zinc Die Castings, *Prod. Eng.*, April, 1960.
25. Thread and Form Rolling, *Machine Design*, January, 1958.
26. Pelphery, H., "Chipless Forming," ASME Paper 59-Prod-12.
27. Pelphery, H.: "Rolled Forming Toothed Parts," ASME Paper 23T9.
28. Cold Forming Gears, *Am. Machinist*, June, 1958.
29. "Cold Forming Gears," Grob, Inc., Paper DM-59.
30. Dudley, Darle W.: When Splines Need Stress Control, *Prod. Eng.*, December, 1957.

Chapter 18

GEAR FINISHING BY SHAVING, HONING, AND LAPPING

By

BEN F. BREGI, *Vice President—Manufacturing and Research, National Broach and Machine Company, Detroit, Mich.*

Several types of gear-finishing methods are available to improve the quality of gears following conventional gear-tooth roughing and semifinishing operations such as hobbing and shaping. The gear-shaving process can be carried out on gears having hardnesses up to 40 Rockwell C scale. On hardened gears that are subject to heat-treatment distortion, a widely used processing sequence is to shave their teeth before heat-treatment with shaving cutters that have modified tooth forms. These cutters provide tooth surfaces that compensate for fire-change movements.

There are three methods of finishing gear teeth after they are hardened. These include honing, grinding (see Chap. 19), and lapping. Grinding is a process that can be used to remove errors resulting from heat-treatment distortion for which no compensation has been made. Honing and lapping are methods of removing minor amounts of stock to improve surface finish, make minor tooth-form corrections, and generally improve the sound characteristics of the gear.

Gear-tooth honing is a particularly effective method of removing nicks and burrs from the active profiles of the teeth after heat-treatment. Combined with its ability to improve surface finish and make minor form corrections, the honing process is rapidly being accepted as an operation through which all gears are processed following heat-treatment.

GEAR SHAVING

Gear shaving is a free-cutting gear-finishing operation which removes small amounts of metal from the working surfaces of the gear teeth. Its purpose is to correct errors in index, helical angle, tooth profile, and eccentricity—also to improve tooth surface smoothness and to eliminate the danger of tooth end-load concentrations in service. Shaving reduces gear noise. It also increases the gear's load-carrying capacity, its factor of safety, and its service life.

Gear finishing (shaving) and gear cutting (roughing) are essentially different. Any machine designed primarily for one cannot be expected to do both with equal effectiveness or with equal economy.

Gear shaving can correct the inaccuracies inherent in gear cutting. It is equally effective as a control for troublesome fire distortions caused by heat-treatment.

If the metallurgical structure of gear-tooth surfaces were not disturbed during the gear-cutting operation, normal heat-treat distortion might be predicted and compensation made for it. However, roughing cuts of the forming tool cold-work the metal leaving unequal strains in it. The heavier these cuts, the greater the strains. Release of these strains during the hardening operation makes it impractical to predict either the amount or direction of the resulting distortion.

By shaving, the overly stressed surface metal is removed and thus fire distortion is brought under control.

18-1. Rotary Crossed-axes Shaving. This method of shaving is widely used for spur and helical external gears of all sizes and types. Because the process permits a hunting-tooth action between work gear and cutter, it can effectively correct spacing errors. It makes use of a serrated rotary cutter in the form of a helical gear (Fig. 18-1).

FIG. 18-1. A typical gashed helical shaving cutter.

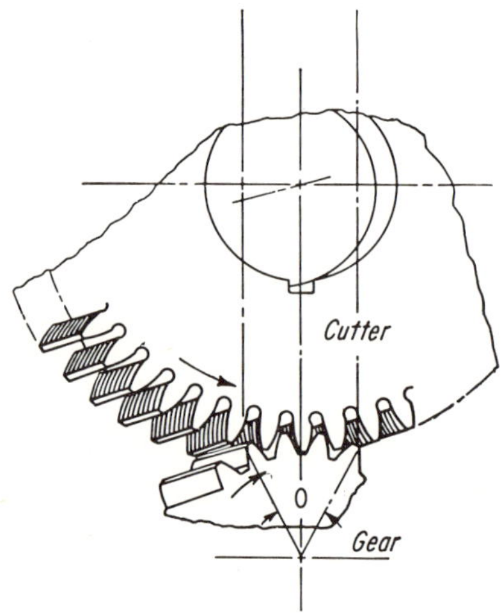

FIG. 18-2. Work gear in mesh with shaving cutter.

This gear has a helical angle different from that of the gear to be shaved. The axes of cutter and work gear are crossed at a predetermined angle during the shaving operation.

When cutter and work gear are thus rotated in close mesh (Fig. 18-2) the edge of each cutter gash shaves a fine hairlike chip as it moves over the surface of a work-gear tooth. The chip is somewhat like that produced by a diamond boring tool.

The finer the cut, the less the pressure required between tool and work, eliminating the tendency to cold-work the surface metal of the work-gear teeth.

The rotary gear-shaving process can be applied to gears having teeth as fine as 180 diametral pitch ranging up to those having 2 diametral pitch. A variety of standard shaving machines are available for shaving gears in a size range from ¼ to 220″ pitch diameter. Shaving cutters are made in pitch diameters from 2 to 13″ to accommodate the range of diametral pitches. Typical production rates for shaving operations on different gears are shown in Table 18-1.

The rotary shaving process is carried out on a shaving machine that has a motor-driven cutter head and a reciprocating work table (Figs. 18-3 through 18-9). The cutter head is adjustable to obtain the desired crossed-axes relationship with the work. The work, carried between live centers, is driven by the cutter.

FIG. 18-3. Work gear on shaving-machine table below shaving cutter.

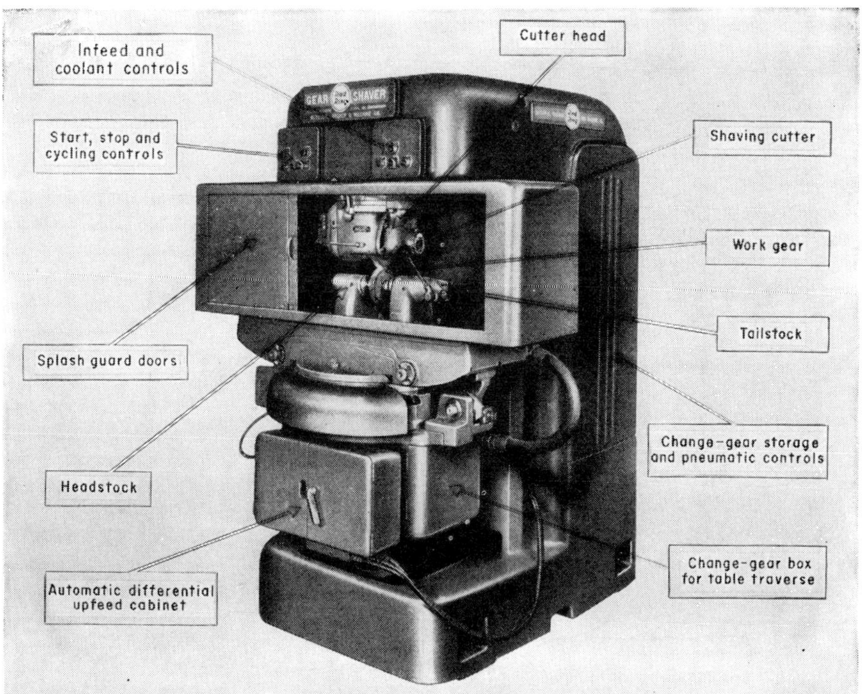

FIG. 18-4. A universal rotary shaving machine for gears from 1 to 12″ pitch diameter. Uses **axial-traverse**, right-angle-traverse, or angular-traverse method. (*Courtesy of National Broach and Machine Co., Detroit, Mich.*)

During the shaving cycle, the work is traversed back and forth across the face of the cutter in one of several selected methods and fed an increment into the cutter at each work traverse.

Rotary Gear-shaving Cutters. These are high-precision hardened and ground generating tools held to class A and AA tolerances in all principal elements.

The characteristics of the cutter—its tooth profile, helix angle, index, and eccentricity—are all transferred to the work gear. The gear can surpass the cutter in precision. This lies in the fact that the "hunting-tooth" action of shaving may produce a better tooth-to-tooth spacing accuracy on the work than that of the cutter.

FIG. 18-5. A rotary shaving machine that uses the axial-traverse method for finishing gears from 1 to 12″ pitch diameter. (*Courtesy of National Broach and Machine Co.*)

Serrations in the cutter extend the full length of the tooth, terminating in a clearance space at the bottom. These clearance spaces provide unrestricted channels for a constant flow of coolant to dispose of chips promptly. They also permit uniform depth of penetration and increase cutter life.

The shaving cutter is operated at high speed, often 400 or more surface feet per minute. Feed is fine and tool contact zone is restricted.

Cutter life depends on several factors: operating speed, feed, material and hardness of the work gear, required work-gear tolerances, type of coolant, and the size ratio of cutter to work gear.

In some cases a cutter will shave up to 80,000 gears before it needs sharpening. It may generally be resharpened from four to eight times.

Crossed-axes Principle. To visualize the crossed-axes principle, consider two parallel cylinders of the same length and diameter (Fig. 18-10). When brought

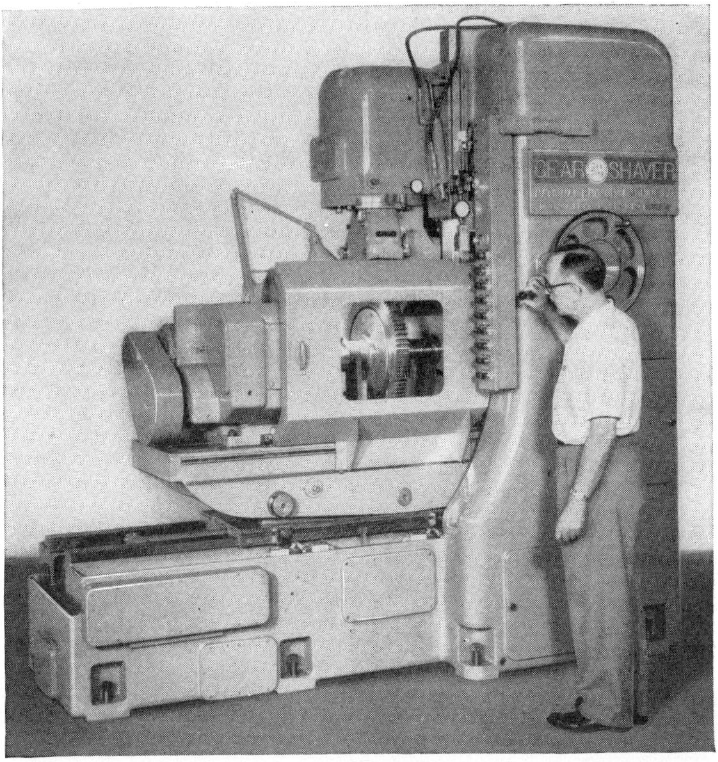

FIG. 18-6. A rotary shaving machine for gears from 3 to 24″ pitch diameter. Work table travels to the left to clear work from cutter for easy loading and unloading. (*Courtesy of National Broach and Machine Co.*)

FIG. 18-7. A rotary gear-shaving machine for gears from 4 to 36″ pitch diameter. (*Courtesy of National Broach and Machine Co.*)

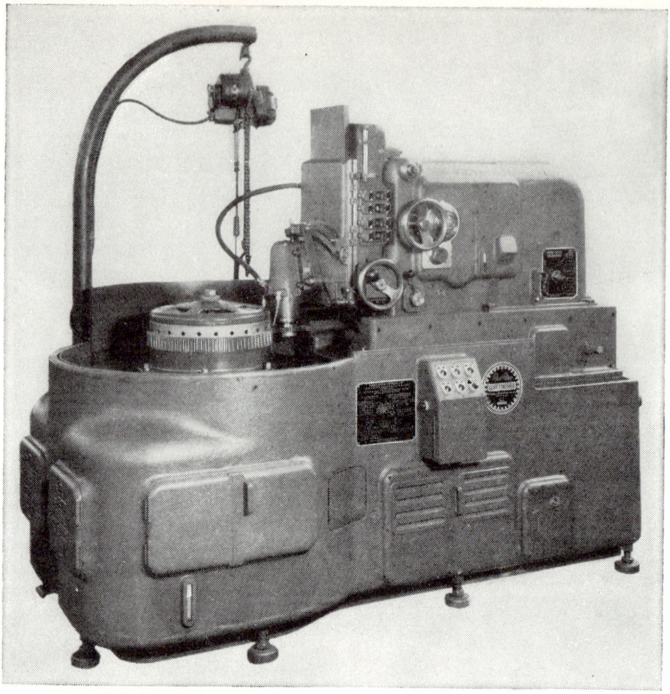

FIG. 18-8. A rotary gear shaver for finishing gears from 12 to 36″ pitch diameter. (*Courtesy of National Broach and Machine Co.*)

FIG. 18-9. A rotary gear-shaving machine for gears from 30 to 180″ pitch diameter. (*Courtesy of National Broach and Machine Co.*)

together under pressure, their common contact surface is a rectangle having the length of a cylinder and a width which varies with contact pressure and cylinder diameter.

When one of these cylinders is swung around so that the angle between its axis and that of the other cylinder is increased up to 90°, their common contact plane remains a parallelogram but its area steadily decreases as the axial angle increases.

The same conditions prevail when, instead of the two plain cylinders, a shaving cutter and a work gear are brought together. When the angle between their axes is of from 10 to 15°, tooth-surface contact is reduced and pressure required for cutting is small.

As the work gear is moved away axially from the point of intersection of the axes, backlash develops. Conversely, as it is returned to the point of axial intersection, backlash decreases until the two members engage in tight mesh with the teeth of the cutter wedging between those of the work gear. Thus each succeeding cutting edge sinks deeper into the work-gear tooth until the point of axial intersection is reached.

For shaving, the axes of cutter and work gear are crossed at an angle usually in the range of 10 to 15° or approximately equal to the difference between their helical angles.

Crossing the axes compensates for the variation in sliding action due to involute roll. As gear and cutter teeth come together, cutting proceeds diagonally toward the pitch line and in the reverse direction as the pitch line is crossed and the teeth begin to separate. This assures uniform cutting over the entire tooth profile.

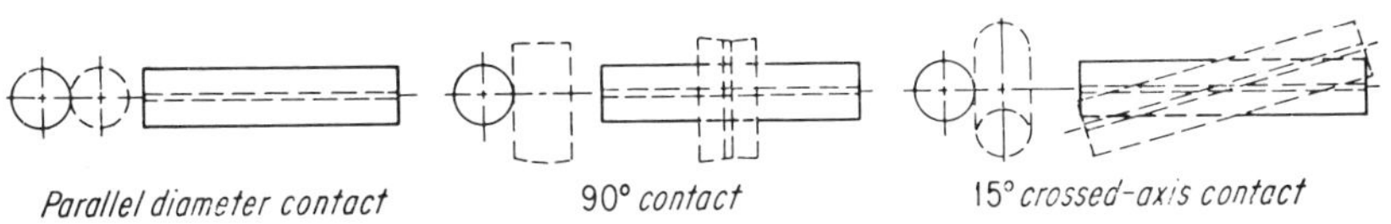

FIG. 18-10. Change in contact area as angle between axes is varied.

The location of the contact area between the cutter teeth and the gear teeth depends upon the location of the center line of crossed axes. The pivot point of the cutting tool must pass beyond the ends of the gear teeth to finish the teeth properly their full length. When gears having no interference from adjacent gears are shaved, the cutter is positioned on the spindle with the center of the cutter, the pivot point, and the intersection of the axes lying in a common plane. The length of the reciprocating stroke is set to allow the point of intersection of the axes to pass the end of the gear teeth at each extremity, finishing the gear teeth their entire length.

To pass the pivot point of the cutter beyond both faces of a gear having an interference preventing free lateral travel (reciprocating motion), it is sometimes necessary to position the cutter on the spindle with the pivot point off center. This is done by placing a spacer on the spindle between the locating shoulder and the cutter. In this way, the cutter is positioned with its edge (next to the interference) nearer the intersection point of the axes. The width of the spacer required depends upon the diameter of the interference portion, the difference in helical angles between the gear and the cutter, and the clearance between the edge of the gear and the interference.

In selecting cutters for shaving shoulder gears, it is necessary to determine the crossed-axes angle that can be used without interference between the cutter and the shoulder. By constructing the outside diameter of the shoulder and the outside diameter of the cutter (Fig. 18-11) with their centers on the XX axis, the point A is established. This is the distance from the XX axis by which the cutter would interfere with the shoulder. As a safety factor, the angle should be small enough to allow a clearance between the cutter and the shoulder at the extremity of the reciprocating stroke. By drawing in the line of action from the pressure angle, the point B is established. This is the distance from the XX axis by which the outside diameter of the cutter intersects the line of action or goes out of contact with the gear. Points A and B lie in a plane (containing one side of the cutter) that is at an angle to the side of the gear. This is the maximum crossed-axes angle. The width of recess should be great enough to permit a minimum of 3° and preferably 5° crossed-axes angle.

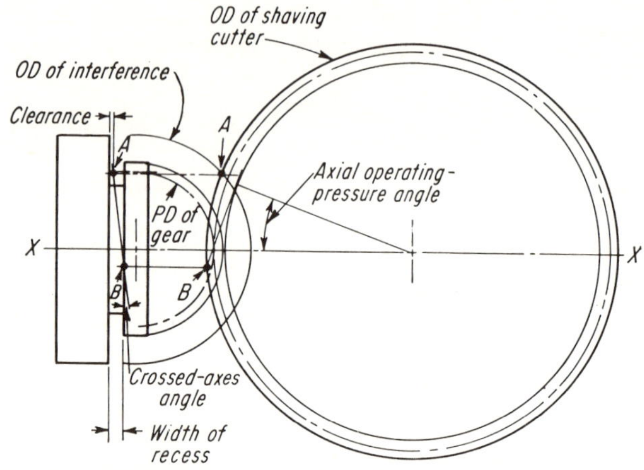

FIG. 18-11. Determination of crossed-axes angle for rotary-shaved shoulder gears.

Relationship between Cutting and Guiding Action. Increasing the angle between cutter and work-gear axes increases cutting action, but as this reduces the width of the contact zone, guiding action is sacrificed. Conversely, guiding action can be increased by reducing the angle of crossed axes but at the expense of cutting action. At zero angle there is practically no cutting.

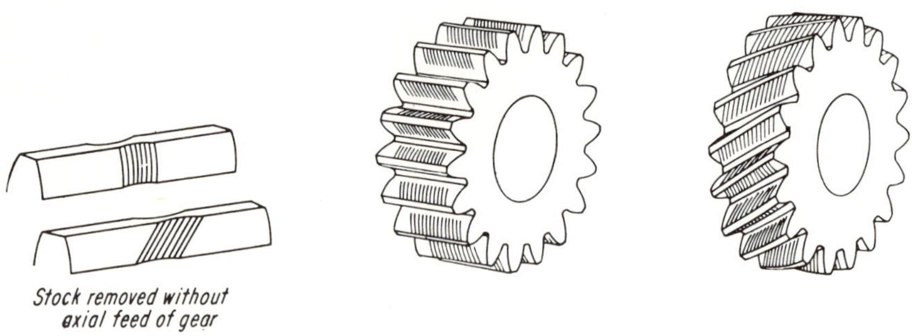

FIG. 18-12. Gears shaved without table reciprocation.

Gears shaved without table reciprocation (Fig. 18-12) show the band of cutter contact less than the gear face width. They also exhibit a deeper cut in the middle of the band than on either side.

Gears shaved with about ¼″ of table reciprocation (Fig. 18-13) have teeth whose profiles are perfect over the distance of reciprocation but fade out at each end. The chordal thickness at the center of the contact band is less than that at the ends.

Gears shaved with full table reciprocation (Fig. 18-14) are finished to full depth

FIG. 18-13. Gears shaved with only ¼″ of table reciprocation.

FIG. 18-14. Gears shaved with full reciprocation.

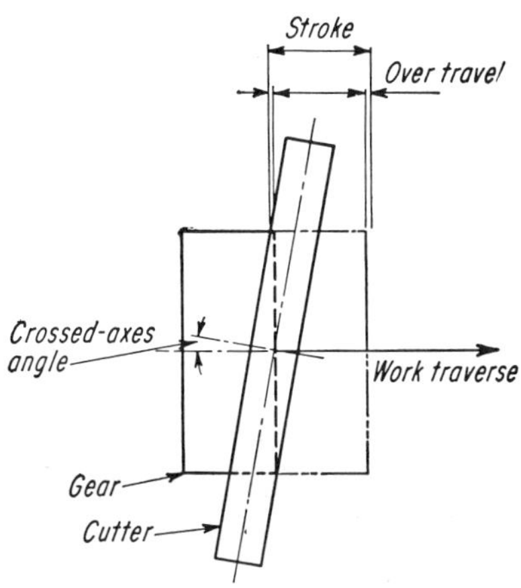

FIG. 18-15. Relationships between cutter and gear with axial-traverse shaving.

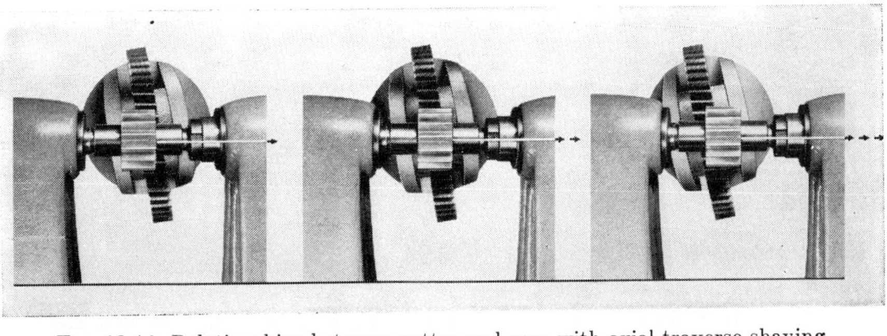

FIG. 18-16. Relationships between cutter and gear with axial-traverse shaving.

over the entire face. Tooth thickness may be decreased by increasing the feed of work toward the cutter.

Axial-traverse Shaving. This is one of several methods of moving the pivot point of the shaving cutter across the work gear in rotary crossed-axes shaving. The methods vary according to the direction of the reciprocating stroke. In the original method (Figs. 18-15 and 18-16) the work gear is reciprocated in mesh with the cutter in a path parallel to the work-gear axis. It can be applied to all types of spur and helical internal or external gears where the cutter can pass each end of the gear to provide adequate pivot-point travel. It is particularly adapted to shaving extremely wide-faced gears and long involute splines as well as narrow-faced gears.

The cutter meshes with the work gear in crossed-axes relationship. The work gear is fed vertically into the cutter at the end of each cutting stroke until the desired tooth size is obtained on the gear. The direction of rotation of the work is reversed at the end of each stroke. The number of passes required in axial-traverse shaving depends on the amount of total stock removal as well as the upfeed rate. Usually the last few reciprocations of the cutter are made without upfeed to provide finishing stroke.

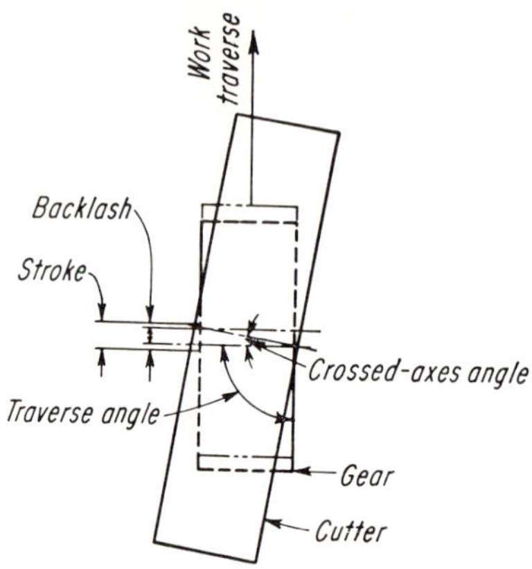

Fig. 18-17. Relationships between cutter and gear with right-angle-traverse shaving.

Right-angle-traverse Shaving. In this method, pivot-point travel is provided by reciprocating the cutter across the axis of the work gear at 90° (Figs. 18-17 and 18-18). It is particularly adaptable to shaving relatively narrow-faced gears, shoulder gears, and herringbone gears. The danger of damaging the shaving cutter when shoulder gears are shaved is reduced by the absence of the lateral reciprocating motion. The shaving cutter for this process is wider than the gear being shaved. Shaving action is accomplished by a right-angle feed between the shave cutter and the work gear. Usually one pass and return of the cutter to starting position finish the gear.

Fig. 18-18. Relationships between cutter and gear with right-angle-traverse shaving.

The cutter used for right-angle-traverse shaving may be similar to that used for other rotary shaving methods and the crossed-axes relationship may be identical. The shaving cutter and the work gear are rotated as intermeshing gears with their axes at an angle. The shaving cycle is started with the cutter past center (but in mesh with the work gear) and consists of one pass to the opposite side of center and return to starting position. The direction of the cutter rotation is reversed at the end of each stroke. The surface finish of the gear teeth is sometimes improved by one or more finishing strokes after the gear has been sized.

Angular-traverse Shaving. The principal difference between axial-traverse and angular traverse shaving is in the direction of reciprocation of the work through and under the cutter (Figs. 18-19 and 18-20).

Although axial-traverse shaving requires a number of table strokes, each with its increment of upfeed, angular-traverse shaving may be done in just two strokes with no upfeed while fixed center distance between cutter and work is maintained. Angular-traverse shaving, however, is not restricted to a two-stroke cycle.

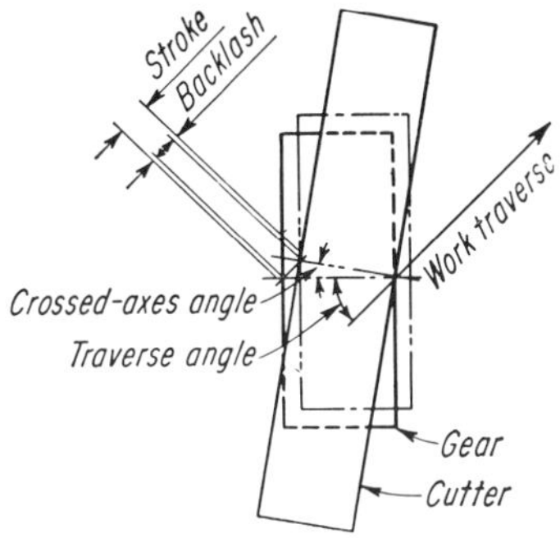

FIG. 18-19. Relationships between cutter and gear with angular-traverse shaving.

The traverse angle for angular-traverse shaving is the angle between the direction of traverse and the work-gear axis. Theoretically this angle is such that, when the cutting stroke starts, the planes of the near sides of work gear and cutter intersect on the work-gear axis. At the end of the cutting stroke, the planes of their far sides intersect. The maximum theoretical traverse angle is determined as follows:

$$\text{Tangent max. traverse angle} = \frac{\text{cutter face width} \times \text{sine crossed-axes angle}}{\text{gear face} - (\text{cutter face} \times \text{cosine crossed-axes angle})} \tag{18-1}$$

The traverse angle may be set at any point from 0 to 90°. Normally it will vary from 30 to 60° in order to obtain optimum conditions of cutting speed and tooth surface finish for any specific work gear and cutter combination.

A full 90° traverse may be used provided the cutter has special (differential) serrations and its face width exceeds that of the work gear. In the range of 30 to 60°, the

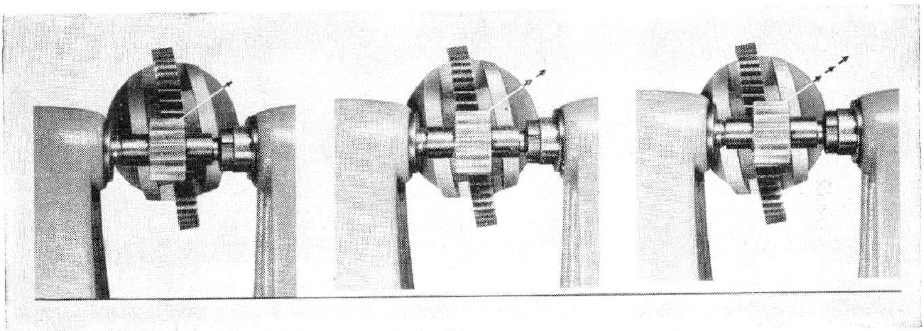

FIG. 18-20. Relationships between cutter and gear with angular-traverse shaving.

standard angular-traverse cutter without specially designed serrations will suffice. In many cases, the axial-traverse type of shaving cutter may be used, provided a slight amount of crown in the gear is permissible.

Angular-traverse shaving is much faster than axial-traverse shaving. In many cases it is up to 50 per cent faster. It also offers greater latitude in handling shoulder gears with critical clearance between the gear and the shoulder.

With angular-traverse angles of 30 to 60°, cutting is not restricted to a small zone of the cutter as it is in axial-traverse shaving but is migrated across the entire cutter face. Consequently cutter life is extended.

When center distance is fixed, each stroke begins and ends with a slight amount of backlash. Thus, as the stroke ends, pressure between cutter and work fades out—the best possible condition for maximum tooth-surface smoothness. Furthermore, when loading with fixed center distance, the operator can more easily detect oversize work gears which might damage the cutter.

Multistroke Angular-traverse Shaving. The addition of an automatic upfeed mechanism (Fig. 18-21) enlarges the scope of angular-traverse shaving by making it feasible for multistroke operations. This device feeds the work into the cutter in a series of small increments, synchronized with table reciprocation.

Removing stock from the work gear, in a series of small increments instead of two larger increments, further increases cutter life. It also makes the process adaptable to gears requiring more stock removal than can be handled on a two-stroke cycle.

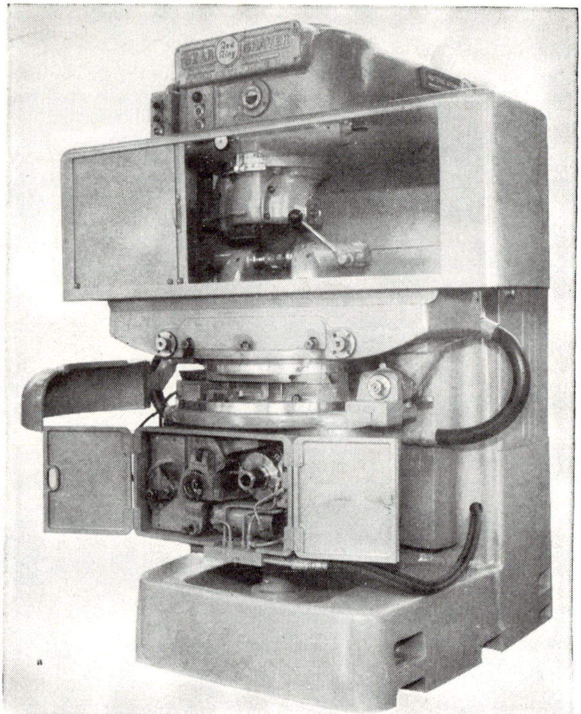

FIG. 18-21. A universal rotary-shaving machine with cover for the differential upfeed mechanism opened to show details. This machine finishes gears from 1 to 8″ pitch diameter. (*Courtesy of National Broach and Machine Co.*)

When upfeed is completely automatic, there can be no danger of an error in selecting feed rates. Since the cycle starts and stops in a position of maximum backlash, loading and unloading can be very fast.

Crown Shaving. Crowned tooth forms avoid the danger of concentrated loads on the ends of gear teeth where they are most vulnerable.

Conventional gear teeth are designed on the assumption that mating teeth will always remain parallel and that the load transmitted will always be equally distributed across the face width of the teeth. This is an ideal condition that is seldom attained.

Several factors influence the parallelism of operating gear teeth and the resulting distribution of load on individual teeth. These include the dimensional and elastic characteristics of the gear housings, shafts, and bearings and the gear material as well as the accumulated dimensional errors in the teeth themselves.

If, by chance, perfect parallelism is achieved at some specific load, it is lost when that load changes, because the curve of elastic deflection in supporting parts is a straight line.

Any deviation from the parallel relationship of conventional teeth concentrates the load on their ends where they are most vulnerable to failure.

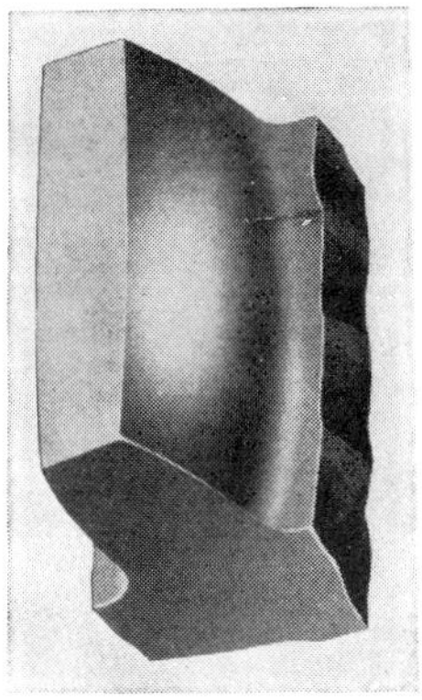

FIG. 18-22. Appearance of a crowned gear tooth showing contact area.

Tooth forms that completely avoid this danger (Figs. 18-22 and 18-23) have a modification of both horizontal section and vertical profile. This is often referred to as "crowned lead" and "crowned involute" and is called an elliptoid tooth form. The sides of a pitch-line section through a crowned tooth are elliptical rather than straight and parallel as they are for a conventional tooth. Thus load concentrations are avoided at the tooth ends. Usually a crown of 0.0003" per in. of face width is sufficient. This is well within allowable limits of good gear practice for even conventional teeth, but the crowned tooth is always thinner at its ends. This may restrict tooth load to about 85 per cent of face width, but it protects vulnerable tooth ends.

A crowned tooth has involute profile with a plus curvature of 0.0002 to 0.0003" at the pitch line to accommodate deflection and tip interference between mating gears. The involute at each end of the gear is modified by a fall-off or a low curve to avoid tip contact.

Crowned tooth forms can be produced on shaving machines utilizing the axial-traverse method by rocking the work table as it is reciprocated. This rocking action causes the cutter teeth to sink deeper into the work-gear teeth at their ends than it does in the middle, thus thinning the teeth progressively toward their ends. Both the point of maximum tooth thickness and the amount of crowning are readily controlled. For angular-traverse high-production shaving, the cutter is modified to provide crowning.

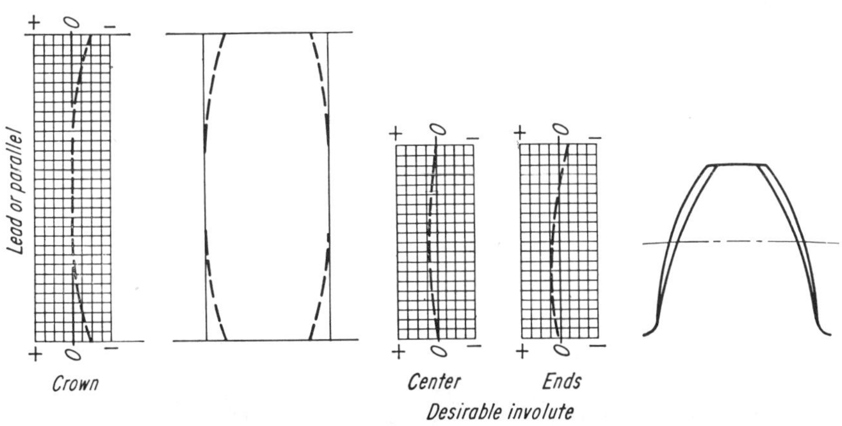

FIG. 18-23. Tooth modifications in both profile and at pitch-line section of a crowned gear tooth.

Shaving Internal Gears. When an internal gear is shaved, the operation is either carried out on an external gear shaver with a special cutter-head attachment or on machines especially designed for handling internal gears (Figs. 18-24 and 18-25).

Fig. 18-24. A rotary shaver for internal gears. This machine is equipped with a fully automatic loader arrangement. Such a machine will handle gears from 3 to 12″ pitch diameter. (*Courtesy of National Broach and Machine Co.*)

Fig. 18-25. A rotary gear shaver finishing internal gears from 22 to 48″ pitch diameter (*Courtesy of National Broach and Machine Co.*)

Internal gears are shaved of necessity by the axial-traverse process. The diameter and number of teeth of the cutter are limited by the nature of the meshing action between an internal gear and external gear-type cutter. If the gear is supported at both ends during the shaving process, a further limitation is placed on cutter diameter. The largest cutter is possible when a pot-chuck type of support is provided for the internal work gear.

Automating Gear-shaving Processes. The gear-shaving process is carried out on machines of various basic standard types. These machines can be adapted to serve a variety of production requirements ranging from low production using manual-loading means to high-production fully automated loading arrangements.

FIG. 18-26. A rotary gear shaver equipped for semiautomatic loading with approximate work locator, automatic air cylinder, door closer, and air-operated tailstock.

Semiautomatic loaders (Fig. 18-26) can be used for medium-production schedules. With these accessories, the work gear is placed on an approximate locator, and a cycle button is pressed. The head and tailstocks automatically engage the work gear, the coolant doors close, the shaving cycle is carried out, the doors open, and the gear is released for unloading and placement of another part in the locator.

Full-automation requirements for shaving processes are provided by magazine-type automatic loaders. Usually of a rocker-type compact design (Fig. 18-27), these loaders include size-checking gages that prevent entry of oversized parts. Magazines can be filled by an operator or receive gears from a factory conveyor line.

Automatic Gaging for Shaving Automation. When full-automation techniques are applied to gear-shaving equipment, in-process gages are a requirement to maintain product quality and protect the high-production equipment. Such gear variables as size, helix angle, involute profile, and spacing can be checked with in-process gaging equipment. These gages (Fig. 18-28) can inspect, sort, and reject gears, either before or after the shaving operation or both as required.

Applying the Rotary Gear-shaving Process. For maximum effectiveness, it is necessary to understand the fundamental purpose of gear shaving and to adhere to good shop practice.

FIG. 18-27. A rocker-type automatic loader for cluster gears installed on a rotary gear shaver.

FIG. 18-28. An automatic gage for checking accuracy of shaved gears.

Table 18-1. Typical Shaving Production Rates for Different Gears

Name of gear	Material	No. of teeth	Diam. pitch	Pitch dia., in.	Pressure angle, deg	Helix angle, deg	Face width, in.	Stock removal, in. over pins	Production rates, loading method, and shaving process
Planet pinion	Steel	12	13.5	1⅛	20°	27°16′2″	1	0.005	240 per hr, auto. load, angular traverse
Timing gear	Cast iron	28	10	2 5/16	14°30′	23°17′49″	1 1/16	0.010	130 per hr, semiauto. load, angular traverse
Shoulder gear (small gear on two-gear cluster)	Steel	14	6/8	2 9/16	20°	Spur	1 1/16	0.006	160 per hr, semiauto. load, angular traverse
Cluster gear (one of large gears on a four-gear cluster)	Steel	29	9.25	4	16°30′	35°56′14″	1 1/16	0.010	95 per hr, semiauto. load, angular traverse
Timing gear	Aluminum	56	10	6½	14°30′	23°17′49″	1	0.010	95 per hr, semiauto. load, angular traverse
Bull gear	Steel	158	7	24	20°	20°	4½	0.010	2 per hr, manual load, axial traverse
Bull gear	Steel	400	6	72	14°30′	22°30′	12	0.001–0.002″ total on tooth thickness	5 hr each, manual load, axial traverse
Marine propulsion gear	Steel	1,128	6	200	14°30′	30°	36 (herringbone)	0.001–0.002″ total on tooth thickness	130 hr, manual load axial traverse

It must be remembered that shaving is a precision operation. As such, it must not be expected to remove large amounts of stock. Consequently, hobbing and shaping tolerances should be held within reasonable commercial limits. The closer these limits, the faster and easier it is to attain the desired precision in the shaved gear.

This is especially true in the case of stub-tooth gears because the short tooth overlap tends to cause the shaving cutter to follow the form of the hobbed tooth.

Shaving readily reduces tooth-profile error 66⅔ per cent and more. For best results, the rough gear-cutting operation should leave the teeth in a form similar to that desired in order that the remaining stock may be removed evenly over the tooth surfaces by the shaving operation. Thus involute tolerance of the shaved gear can be held to within 0.0002″.

Tooth-to-tooth spacing or pitch error can be held to 0.0003″ or less when that of the preshaving operation does not exceed 0.001″, depending upon individual conditions.

Shaving reduces eccentricity error 66⅔ per cent or more. The error in lead for helical gears or tooth parallelism for spur gears can be reduced from 0.001″ per in. of face to 0.0002″ by shaving.

The success of the shaving operation is also affected by the metallurgy and heat-treatment of the gears before and after shaving.

The metallurgical structure of gear steel for shaving is preferably a laminated pearlitic or sorbitic microstructure with a McQuaid-Ehn grain size of from 6 to 8 and a machinability rating of not less than 25 per cent. Forged gears should have uniform density; gears should be clean and free from scale.

While the majority of gears are shaved in the green, they may be shaved after heat-treatment if the Rockwell C hardness does not exceed 40.

In shaving before heat-treatment, allowance should be made in both profile and helix for heat-treat distortion. This allowance is based on the metallurgy of the gear material and the hardening process selected.

It is preferable to shave from the same locating points or surfaces that were used in hobbing or shaping. Faces, hole diameter, and centers should be qualified. The hole should be square with the locating face.

Locating points on arbors and fixtures should be held to a tolerance of not more than 0.0002″. If centers are used, they should be free of nicks and burrs and qualified for true center angle.

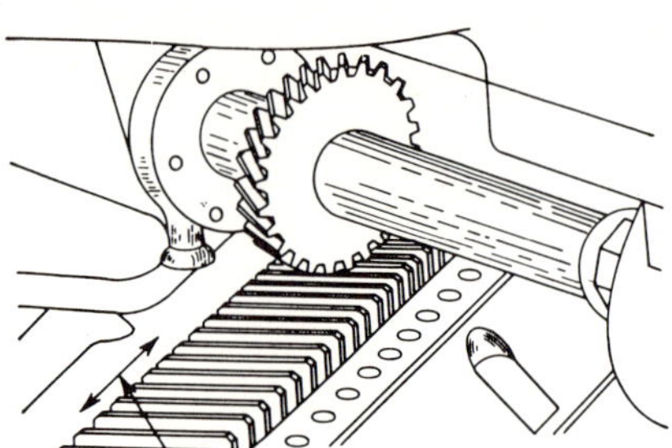

FIG. 18-29. Relationship between cutter and work with rack shaving. (*Courtesy of Michigan Tool Co., Detroit, Mich.*)

Gears with splined holes are located from major diameter, pitch diameter, or spline bore. There should be no interference of cutter at root fillet.

Table 18-1 gives some typical production rates for rotary shaving.

18-2. Rack Shaving. In rack shaving (Fig. 18-29) the work-gear teeth are finished by rolling the work in tight mesh with a cutting tool that has the general shape of a rack. The rack is reciprocated back and forth and fed into depth at the end of each stroke. The crossed-axes angle depends on the relative helix angles of the rack teeth and the work gear. Rack shaving is applicable to gears under 6″ diameter.

The rack-shaving tool has a series of serrated blades. The edges of the serrations form parallel cutting edges that remove metal from the gear teeth. The thickness of each blade is equal to the circular pitch of the gear. Racks with blades parallel to each other and perpendicular to the sides of the rack holder (straight racks) are used for finishing helical gears up to 30° helix angle. Racks with blades set in angular location to the side of the holder finish spur gears as well as helical gears whose helix angle is too small or too large to be shaved on a straight rack.

18-3. Basic Design Considerations for Shaved Gears. Shaving-cutter design is a complex process that relies on a broad background of experience. The myriad of design details of the cutters are usually specified by the tool manufacturers. There are, however, several basic considerations that users of shaving processes should know and understand in order to make maximum use of this finishing process.

Fillets. The tips of the shaving-cutter teeth must not contact the gear-root fillet during the shaving operation (Fig. 18-30). If such contact does occur, excessive wear of the cutter results and the accuracy of the involute profile is affected.

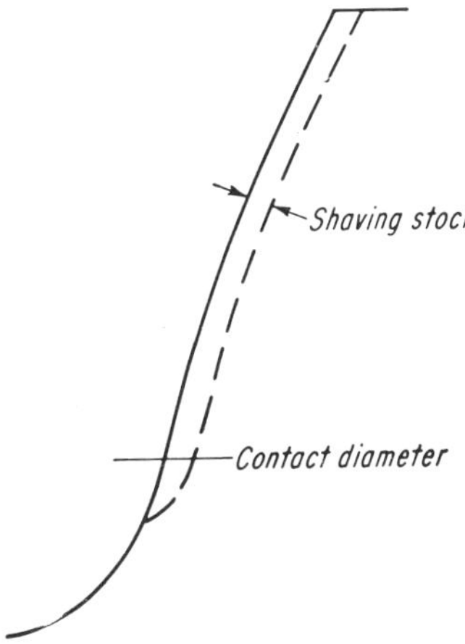

Fig. 18-30. Condition present when shaving cutter contacts gear-root fillet.

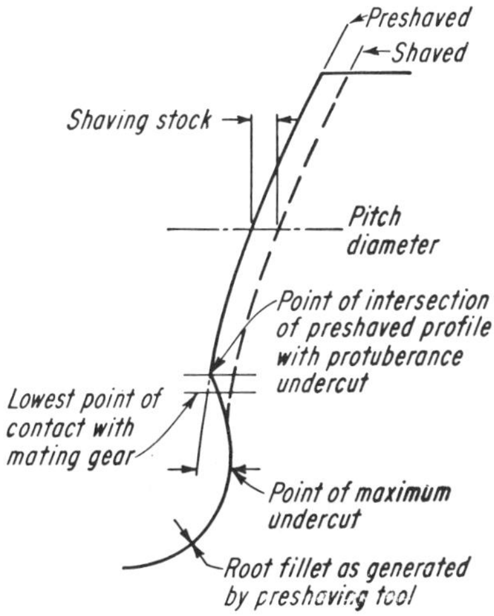

Fig. 18-31. Condition present when shaving a gear tooth produced by a protuberance-type hob or shaper cutter.

The shaving cutter must finish the gear tooth below its active profile. Thus the height of the fillet should not exceed the lowest point of contact between the shaving-cutter teeth and the teeth on the work gear.

Shaper cutters generate higher fillets than hobs for the same depth of cut. As a result, shaper-cut gears are cut deeper than hobbed gears. Tooth depth for 1 to 19 diametral pitch shaper-cut gears, inclusive, should be 2.35/diametral pitch. For 20 diametral pitch and finer, a depth of 2.40/diametral pitch plus 0.002″ allows ample clearance for the shaving cutter under ordinary conditions.

When full-round root fillets are used, additional whole depth must be provided. It is advisable to maintain the same length of involute (straight portion) on the pre-shaving tool and determine the size of tip radius and extra depth from that point as shown in Fig. 18-31. The shape of the fillets is a generated form determined by the type of preshave tool. It is usually specified as the form produced by a hob or shaper cutter with a given tip radius or shape.

Protuberance. Protuberance-type (high-point) hobs and shaper cutters are often used prior to shaving to produce a slight undercut or relief near the base of the tooth. This method assures a smooth blending of the shaved tooth profile and the unshaved tooth fillet, as well as reducing shaving-cutter tooth-tip wear (Fig. 18-31).

Standard hob and shaper-cutter designs usually generate a natural undercut on gears having small numbers of teeth. In many cases, however, this undercut is excessive from the standpoint of good gear design. For this reason, protuberance-type tools are seldom recommended for gears having fewer than 20 teeth with a 20° pressure angle, or 30 teeth with a $14\frac{1}{2}$° pressure angle.

The following rules should be followed in designing gears for protuberance-type tools:

1. The amount of protuberance should be determined from mating-gear conditions.
2. The amount of protuberance should not be excessive to a point where the teeth are weakened.
3. The protuberance must be properly positioned. A hob or shaper cutter designed to produce the proper undercut on a 30-tooth gear will produce too high an undercut on the active profile of a 20-tooth gear. The undercut produced by the same tool on a 100-tooth gear will be too low for any useful purpose. A preshave protuberance-type hob or shaper cutter for crowned gear teeth may require some modification in protuberance design to avoid fillet interference with the cutter at both extremes of the gear face.

Protuberance-type tools serve a useful purpose on shaved gears that have restricted tooth clearances (shallow tooth depth) as well as on large gears. In both cases, the fillet is moved out of the path of the shaving-cutter tip, a desirable condition.

Indiscriminate use of protuberance-type preshave tools on all shaved gears regardless of the number of teeth is a questionable practice. Theoretically individual protuberance-type tools are required for each gear in accordance with its number of teeth. This is not a practical method, however, and one tool is used for a range of gears with varying numbers of teeth. Often a hob or shaper cutter with no protuberance can be used for gears with small numbers of teeth, a second protuberance-type tool being used for the gears with larger numbers of teeth.

On long- and short-addendum gears, the amount and position of protuberance must be carefully specified, so that the final tooth design is not impaired. Enlarged tooth-form and mating-gear layouts can be used to advantage in studying protuberance problems.

The amount of undercut produced by the protuberance-type tool should be from 0.0005 to 0.001″ greater than the amount of shaving stock for each side of the gear tooth. On crowned gears this calculation should be made for the thin end. The position of the undercut should be such that its upper margin meets the involute profile at a point below its contact diameter.

On small pinions, it is practically impossible to maintain this objective. Under these conditions, the junction point may come slightly above the contact diameter. The undercut should be of a shape to allow at least 0.0005″ of stock at the mating-gear contact diameter to be removed at the shaving operation.

A small amount of involute profile is generated in the region of contact diameter. The point of maximum undercut is determined by the amount of high point and radius on the tool.

Root fillets can often be increased slightly on protuberance-cut teeth if they remain sufficiently low to clear the tips of the shaving-cutter teeth. Properly designed protuberance forms on preshaved gear teeth can be most beneficial in the shaving process.

Shaving Stock. The amount of stock removed by the shaving process is a key to its successful application. Sufficient stock should be removed to permit correction of errors in the preshaved teeth. However, if too much stock is removed, cutter life and work accuracy are effectively reduced.

Table 18-2 shows normal amounts of stock to be removed in shaving.

Table 18-2. Recommended Amount of Stock to Leave for Shaving

Normal diam. pitch	Shaving stock on tooth thickness, in.	Stock allowed over two pins, in. Pressure angle		
		14½°	20°	25°
2–3.999	0.0045	0.013	0.011	0.009
4–5.999	0.0040	0.011	0.099	0.007
6–7.999	0.0035	0.009	0.007	0.006
8–12.999	0.0030	0.007	0.006	0.005
13–16.999	0.0025	0.005	0.004	0.004
17–25.999	0.0020	0.004	0.003	0.003
26–32.999	0.0015	0.002	0.002	0.002
33–47.999	0.0010	0.002	0.001	0.001
48–95.999	0.0005	0.001	0.001	0.001
96 and up	As little stock as possible			

GEAR HONING

The gear-tooth-honing process is a hard-gear-finishing method that was developed to improve the sound characteristics of hardened gears. Three basic factors influence the operating sound level of running gears. These are (1) nicks and burrs, (2) surface finish, and (3) minor errors in tooth shape.

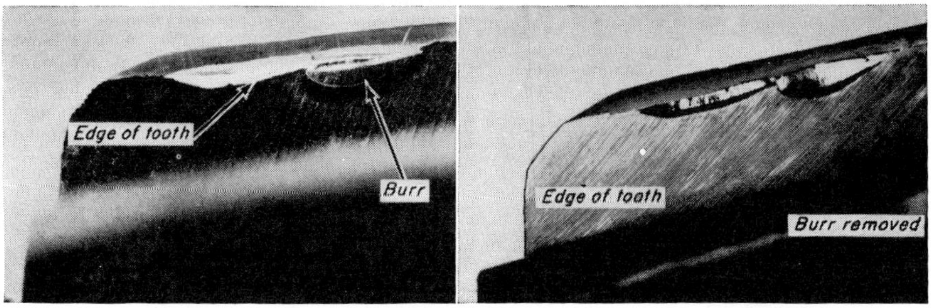

Fig. 18-32. Appearance of nicked tooth before and after honing.

The honing process was originally developed to remove nicks and burrs (Fig. 18-32) that are unavoidably encountered on production gears because of handling. Further development work with the process revealed that minor corrections in tooth shape and improved surface finish could be achieved.

The honing process uses an abrasive-impregnated plastic helical gear-shaped tool (Fig. 18-33). This tool is run in mesh with the hardened gear in crossed-axes relationship. The work gear is driven by the honing tool at high speeds (up to 1,000 sfpm) while being traversed back and forth across the honing tool in a path parallel to the work-gear axis. The work gear is driven in both directions during the honing cycle.

18-4. Gear-tooth-honing Methods and Machines. The honing process can be applied to both external and internal spur and helical gears. Specialized machines (Figs. 18-34 and 18-35) have been developed to serve these applications. Both taper- and crown-honing operations can be carried out if desired.

FIG. 18-33. Typical honing tools for external gears (left) and internal gears (right). (*Courtesy of National Broach and Machine Co., Detroit, Mich.*)

FIG. 18-34. A gear-honing machine for external gears from 1 to 12″ pitch diameter. (*Courtesy of National Broach and Machine Co.*)

Two basic methods are used to hone gear teeth. One is called the zero-backlash method and the other is called the constant-pressure method. The zero-backlash method is used for hardened gears made within commercial tolerance specifications. The constant-pressure method is used to hone teeth on hardened gears produced to dimensions outside commercial tolerance ranges.

With the zero-backlash method, the process is carried out with the tool placed into initial controlled tight mesh with the work. The toolhead is locked in this fixed-center-distance location throughout the honing cycle.

In the constant-pressure method, the honing tool is placed into an initial precision pressure-controlled tight mesh. The tool and cutter are kept in pressure-controlled tight-mesh relationship throughout the honing cycle.

Table 18-3 shows some typical production rates for honing.

FIG. 18-35. A gear-honing machine for internal gears from 3 to 12″ pitch diameter. (*Courtesy of National Broach and Machine Co.*)

18-5. Honing Tools. The honing tool is a throw-away type that is discarded at the end of its useful life. A hone may process from 5,000 to 8,000 pieces before it is replaced. It is made of a type of plastic material that is impregnated with abrasive grains such as silicon carbide and formed in a precision mold.

Honing tools are supplied with various abrasive grits, the coarsest being a 60-grit tool. Tools up to 280 grit can be supplied. The grit of the tool depends upon the gear material, tooth specifications, and condition of tooth surface finish after hardening.

The tool does not load up during the process. Conventional honing oil is utilized as the coolant for gear-tooth honing. The honing tool retains its form as it wears. Theoretically it can be used until the teeth are of insufficient thickness to carry the honing tool load. From 0.001 to 0.002″ of stock, measured over pins, is the recommended amount of stock removal for the honing process.

18-6. Applying the Honing Process. Considerable savings can be effected in production by applying the honing process to remove nicks and burrs. Typical production rates for honing various gears are shown in Table 18-3.

Table 18-3. Typical Production Rates for Honing Different Gears

Name of gear	Material	No. of teeth	Diam. pitch	Pitch dia., in.	Pressure angle, deg	Helix angle, deg	Face width, in.	Production rate, loading method, and honing process
Sun gear	Steel	35	14.35	1.500	17°30′	22°11	1.205	135 per hr, auto. loading, zero backlash
Stem gear	Steel	15	9.25	2.003	16°30′	35°56′	0.800	110 per hr, manual loading, zero backlash
Sun gear	Steel	44	18	2.578	20°	18°30′	0.687	120 per hr, manual loading, zero backlash
Sun gear	Steel	44	18	2.578	20°	18°30′	1.125	95 per hr, manual loading, zero backlash
Third-speed gear	Steel	25	7.769	3.046	17°30′	32°51′	0.620	120 per hr, semi-auto. loading, zero backlash
Drive gear	Steel	41	10.16	4.042	20°	Spur	0.315	140 per hr, semi-auto. loading, zero backlash
Countershaft drive gear	Steel	48	8.082	6.743	20°	28°16′	0.973	60 per hr, semi-auto. loading, constant pressure
Drive gear	Steel	76	10	8.000	20°	20°	2½	20 per hr, manual loading, constant pressure
Drive gear	Steel	84	8	12.000	20°	30°	3	12 per hr, manual loading, zero backlash
Bull gear	Steel	76	4	19.000	25°	Spur	3	3 per hr, manual loading, zero backlash
Bull gear	Steel	140	6	24.000	14°30′	15°	4	2 per hr, manual loading, zero backlash
Bull gear	Steel	216	6	36.000	14°30′	Spur	6½	1½ hr each, manual loading, constant pressure

The usual method of removing nicks and burrs from hardened gear teeth is to sound-test the gears against masters on sound-testing machines. Any nicks indicated by sound are searched for visually and removed by a pencil-type hand grinder, and the gear is respeeded to make certain that the nick is removed. By passing all hardened gears over the honing machine, the sound test, nick detection, nick removal, and second sound test are avoided.

Some shape correction can be achieved in removing the recommended stock of 0.002″ over pins. A helical gear in the 5″ diameter range may show lead correction of 0.0004″, involute-profile corrections of 0.0003″, and eccentricity correction of 0.0004″.

The question of using the honing process as a method of salvaging hardened gears is a matter of economics. If gears to be salvaged have a relatively large tooth-shape error, honing time is increased and tool life is decreased. When such gears represent a large investment in production time and material, honing may be the most economical method to retain this investment.

The surface finishes commonly associated with rotary gear-shaving processes are in the range from 10 to 40 microinches rms. The honing process, because it is not a heavy-stock-removal or tooth-correction process, cannot be substituted for the shaving operation, which is usually performed on the soft gear following generation of the teeth by hobbing or shaping.

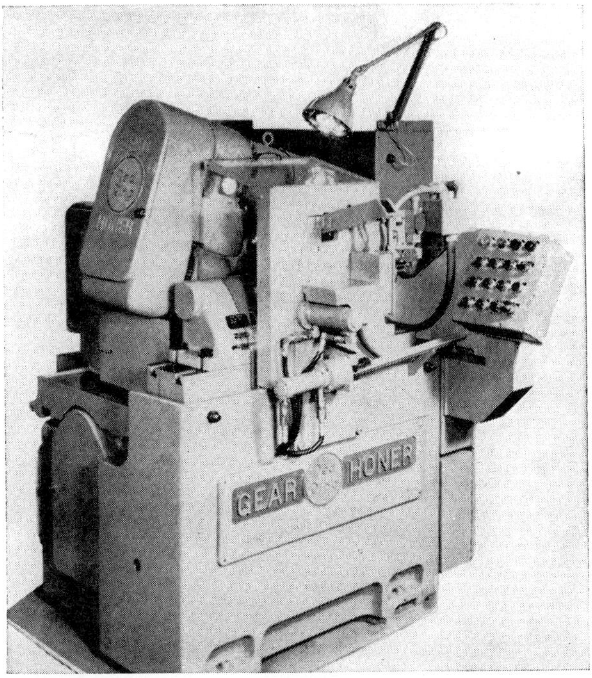

FIG. 18-36. A gear-honing machine equipped for fully automatic loading. (*Courtesy of National Broach and Machine Co.*)

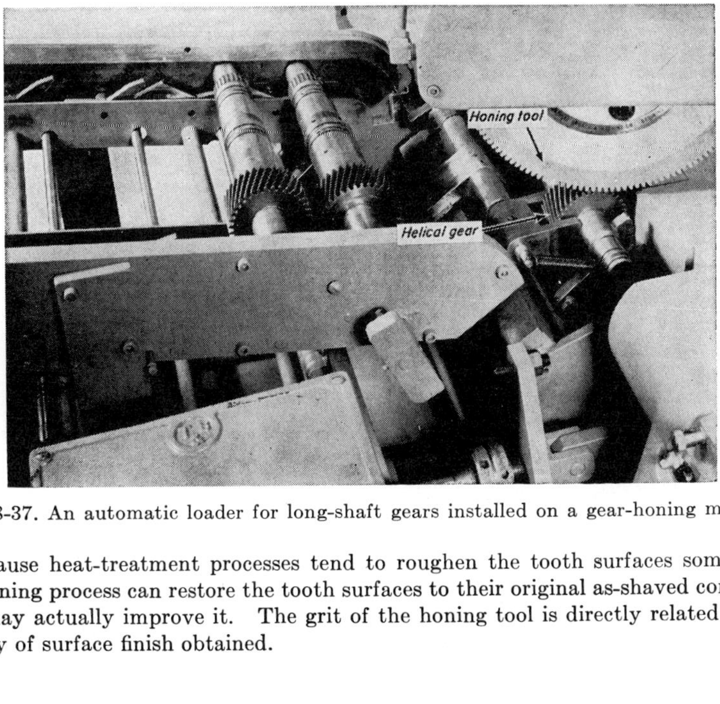

FIG. 18-37. An automatic loader for long-shaft gears installed on a gear-honing machine.

Because heat-treatment processes tend to roughen the tooth surfaces somewhat, the honing process can restore the tooth surfaces to their original as-shaved condition and may actually improve it. The grit of the honing tool is directly related to the quality of surface finish obtained.

To hone production gears economically requires that one grit of tool and a relatively short honing cycle are provided. What is produced in the way of surface finish thus represents a compromise.

In applying the honing process it should be remembered that the tool must first remove nicks and burrs; then it should make the minor improvements in tooth shape and surface finish. The resulting improvement in surface finish is an added feature that can help to promote long gear wear life and improve sound characteristics.

18-7. Automating Gear-tooth-honing Processes. Three types of loading arrangements are available on honing machines to meet different production requirements. Honing machines can be equipped with manual, semiautomatic, or fully automatic loading devices. On the manual models, the operator loads the gear, actuates a control, closes the coolant door, pushes a button for initiating the honing cycle, opens the door, and removes the gear.

In semiautomatic loading arrangements, the operator loads the gear on an approximate locator and pushes a button. The coolant door closes automatically, and the chucking of the work gear, honing cycle, unchucking, and door opening follow in an automatic sequence.

Fully automatic gear honers (Figs. 18-36 and 18-37) have magazine-type loaders. The operator merely keeps the magazine full of gears and the machine is loaded, cycled, and unloaded automatically. With the fully automatic type, the honer can receive parts from a factory conveyor system and put honed parts back on the line to provide full automation.

GEAR LAPPING

Gear lapping is a method of correcting minute heat-treatment distortion errors of involute profile, helix angle, spacing, and eccentricity in hardened gears. Lapping is done by either of two general methods: (1) by running the gear in mesh with a gear-shaped lapping tool or (2) by running two mating gears together. The latter method is normally used for bevel, spiral bevel, and hypoid gears. Spur and helical gears are usually lapped by the first method. In both methods an abrasive lapping compound is the medium that accomplishes removal of metal.

When gears run together, there is a variation in sliding velocity which decreases from a maximum value at the root of the teeth to zero at the pitch line and then increases to another maximum value at the tip of the teeth. Therefore, when gears are run together for excessive periods of time with an abrasive medium, the accuracy of the involute profile is affected in a detrimental manner. To make up for the lack of sliding in the vicinity of the pitch line of spur and helical gears, an auxiliary sliding action in the direction of the gear axis has to be supplied for the lapping process. Thus it is possible to approximate a uniform lapping action over the entire generated tooth profile.

Gear laps are usually made of an open-structured material such as cast iron which will imbed and carry the abrasive that performs the metal-removal function while the work gear and lap are run together. Laps are made as large in diameter as possible and then recut to a smaller number of teeth when worn. Usually a reduction in tooth number by five makes a satisfactory recut lap.

The time required to lap a gear depends upon the amount and types of errors to be corrected. An average gear can be lapped from ½ to 2 min per side of gear teeth. If lapping times are longer than this, the gear is not being held within reasonable tolerances for distortion. Long lapping cycles* are not desirable, since the accuracy of the gear can be destroyed while an attempt is being made to improve it.

* See Art. 18-13 for information on how wide-face helical gears may be lapped for several hours with good results.

18-8. Rotary Parallel-axis Reciprocating Lapping. As pointed out before, the lapping process for spur and helical gears requires that some method of axial reciprocation be provided to provide additional sliding in the area of the pitch line. One method that has been used successfully makes use of three lapping tools, each of which reciprocates simultaneously in parallel-axis mesh with the work gear. Each of the laps has different tooth-addendum characteristics which enables them to contact the work gear at different operating pitch diameters. Thus the sliding action is more evenly distributed across the tooth surface.

Another method makes use of two laps. Internal gears are also lapped by the rotary reciprocating method. An internal gear-type lapping tool can also be applied in a similar manner to lap an external gear.

The reciprocating motion in all these applications has to operate faster than the rotary motion to strive to achieve the same sliding-action balance conditions.

18-9. Rotary Crossed-axes Reciprocating Lapping. In this method a single lap and work gear rotate in mesh with their axes at an angle. The work gear is moved across the face of the lap with an axial reciprocating motion that can be slower than that required in parallel-axes lapping because of the additional sliding action introduced by crossing the axes. The lap must reciprocate across the entire face of the work gear in this process to make certain that proper contact occurs between the lap and gear teeth.

18-10. Increasing Lapping Abrasion. Lapping abrasion can be increased either by cramping (decreasing the center distance between work gear and lap) or by braking the driven shaft. Cramping causes the lap teeth to wedge in the space between adjacent work-gear teeth. When the center distance is reduced more than 0.010″ with crossed-axes lapping, a change in crossed-axes angle is necessary. Runout and spacing errors are reduced by cramp-lapping processes.

Power tailstock braking (braking of the driven shaft) keeps a constant center distance between work gear and lap. A braking action is accomplished by having the work gear drive a gear-type hydraulic pump in the tailstock of the machine while the work gear is being driven by the lap. The load can be varied by regulating a relief valve in the flow line of the pump. Braking action is thus the same for both directions of driving. When fixed-center-distance lapping methods are used, it is extremely important that the lap tooth extend into the work-gear tooth as much or slightly more than the mating gear.

18-11. Lapping Compounds. The uniformity of the abrasive medium has a direct effect on the results achieved with gear lapping. Lapping compounds can be made with either an oil- or water-soluble base. They should also contain a rust retardant. Grain size of the abrasive is dependent on the nature of the work-gear material and hardness. Lapping compounds should spread easily to a uniform thickness on the lap tooth surfaces.

18-12. Lapping Bevel, Spiral Bevel, and Hypoid Gears. These gears are lapped together in sets to remove the effects of heat-treatment growth and distortion. Lapping operations are carried out on machines that duplicate functional mating conditions. The lapping machine is provided with a means of oscillating the pair of gears both axially and tangentially. The lapping operation is carried out until a suitable contact pattern is achieved by regulating the position of the gear axes on the machine.

18-13. Lapping Wide-face Helical Gears. As was pointed out earlier in this chapter, lapping tends to destroy the accuracy of the involute profile. When the lapping is done on parallel axes, the sliding velocity decreases to zero at the pitch line. There is a lapping action above and below the pitch line, but no lapping action on the pitch line. In the case of wide-face helical gears, there is usually an axial contact ratio of 2 or more. The axial overlap has a tendency to maintain the trueness of the gear-tooth action *even when a slight amount of metal is worn away above and below the pitch line.*

It has been found by experience that fairly good profile accuracy can be maintained on wide-face helical gears when they are lapped by rotation of the lap and work only (no axial reciprocating motion). Wide-face double-helical gears may be lapped for several hours with a cast-iron lap or with their mating pinions. On teeth as large as 3 diametral pitch, lapping may remove as much as 0.002″ from a gear and 0.005″ from a mating pinion with profile errors being only in the order of 0.001″. Even though there is no sliding at the pitch line, the metal is abraded away by the cramping action of the lap or pinion.

Where substantial correction is needed in spacing or helix, a gear may be lapped with a pair of laps. One lap may be made *long* addendum while the other may be made *short* addendum. Each lap will have a "rolling" pitch line with the gear that is above or below the pitch line that will be established by the mating pinion. The net result of this is that one lap will slide on the pitch line established by the other lap. This helps to lap the whole gear-tooth profile evenly and makes the lapping go faster than it would with laps cut to roll on the same pitch line as the pinion.

Where a gear is lapped with two or more cast-iron laps to correct spacing and helix errors, it is usually given a final lap with its mating pinion. This tends to match the helices of the pinion and gear and also matches their profiles.

REFERENCES

1. ASTME: "Tool Engineers Handbook," 2d ed., Gear Finishing, Sec. 44—Gear Manufacturing, McGraw-Hill Book Company, Inc., New York, 1959.
2. "Modern Methods of Gear Manufacture," 3d ed., National Broach & Machine Co.
3. Dudley, Darle W.: "Practical Gear Design," McGraw-Hill Book Company, Inc., New York, 1954.
4. Ashton, H.: "Gear Measurement Analysis," Machine & Tool Blue Book, September, 1958.
5. Motz, C.: Stop Operating Noise with Shaved Gears, *Power Transmission Design*, March, 1959.
6. Bregi, B. F.: Specify Honing on Drawings and Gear Noise Will Practically Vanish, *Power Transmission Design*, February, 1960.
7. Mentley, Max B.: How to Improve Gear Shaving Operations, *Tool Engr.*, October, 1956.
8. American Gear Manufacturers Assoc.: "Lapping Procedures for Gears," AGMA 254.07, December, 1955.

Chapter 19

GRINDING SPUR AND HELICAL GEARS

By

WAYNE H. GLOVER, *Vice President and General Manager, Indiana Gear Works Division, Buehler Corporation, Indianapolis, Ind.*

Grinding is generally used to finish gear teeth, but in some cases fine-pitch teeth are cut from the solid by grinding.

This chapter covers the several methods of grinding spur and helical gears. The grinding of bevel gears is covered in Chap. 20. The grinding of worms is covered in Chap. 21.

19-1. Reasons for Gear Grinding. In general gear grinding is a finishing operation performed during the manufacturing cycle to achieve certain desired characteristics in the finished gear. Grinding is frequently used to finish the gear-tooth profiles on parts which have been heat-treated to a high hardness level after the gear has been cut. Case-carburized gears have a surface hardness of 60 to 65 Rockwell C when fully hardened and are commonly tempered to 58 to 63 Rockwell C prior to finishing. Nitrided gears are commonly produced with a surface hardness of 91 to 92 Rockwell 15N (equivalent to 63 to 70 Rockwell C). Carburizing and subsequent hardening require heating to relatively high temperatures, and sufficient distortion to exceed acceptable limits may be present after heat-treatment. While nitriding is performed at considerably lower temperatures than carburizing the time cycle is much longer and size changes take place which usually require correction. Since casehardening is required in many gear applications to obtain desired wear and strength characteristics it is often desirable to utilize grinding as a finishing process.

In many cases through-hardened gears are finish-ground. Even though through-hardened gears in the range of 44 to 50 Rockwell C might be finish-cut to desired tolerances it is usually standard practice to cut, heat-treat, and finish-grind where hardness values above 40 Rockwell C are used. Factors which must be considered in selecting the best manufacturing method for gears of this type are the cost of gear-cutting tools, cutter-sharpening time, and increased cutting time vs. the cost of grinding and the ability to obtain the required accuracy and surface finish with available equipment.

Gears which are in the medium-hardness range of 25 to 40 Rockwell C are also ground in some cases. This may be done to save the cost of expensive cutting tools such as hobs, shaper cutters, broaches, or shaving cutters; or it may be done to produce a desired surface finish or accuracy on a gear difficult to manufacture, to incorporate profile modifications or changes to other elements, or to prepare the part for further processing such as induction hardening.

Fine-pitch gears may be finish-ground from the solid in numerous cases. Since the volume of material removed is quite small it is often more economical to grind from the solid, particularly in cases where cutting and final grinding are indicated.

A number of other factors must be considered in determining whether gear grinding should be used on a particular part for a given application. Most of these items are controversial, and the true answer can be obtained only by testing the finished article. Gear noise can sometimes be reduced by grinding even though the dimensional tolerances are no more precise. Lubrication retention is sometimes related to surface finish on the tooth flank. Parts with light cutter marks have yielded longer life than gears with finer finishes in some critical applications. Claims for superior lubrication retention have been made for gears manufactured by hobbing or grinding methods which produce a number of small flats on the active profile of the tooth. Accuracy is certainly an important factor, and in many cases grinding or other finishing methods must be employed to produce a finished product to the desired tolerances.

19-2. Form Grinding. Form grinders use a disk wheel to grind both sides of the space between two adjacent gear teeth while the gear or spline is maintained in a fixed radial position. Since this type of grinding is similar to milling gear teeth the tooth profile is not generated and the involute form is dressed into the side of the wheel. Figure 19-1 shows a comparison of a form-grinding wheel which finishes the root and a wheel which finishes the profile only.

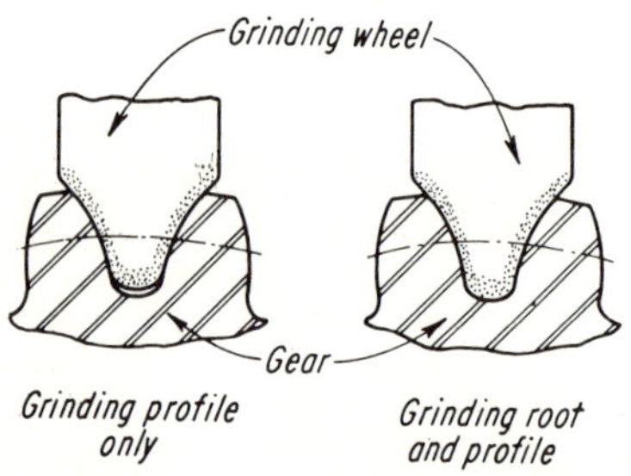

FIG. 19-1. Form-grinding wheel and work.

The involute form is usually dressed into the side of the wheel by means of a pantograph-type dresser utilizing diamonds to form the wheel. Figure 19-9 shows the construction of a typical wheel trimmer. The desired involute form is produced on the forms *A*. These forms are made from master templates utilizing pantograph equipment to reduce the errors in layout and manufacturing to a minimum. The trimmer has two reduction pantographs with pins which follow the form profile. These pantograph arms carry diamond-pointed tools *B* which produce the desired contour on the wheel *C*. Care must be exercised to locate the trimming attachment exactly on center to produce the proper form on both sides of the tooth space. Desired modifications are produced on the forms, and involute modification is relatively simple on this type of dresser. Some late models of form grinders use multiple-ribbed wheels which grind more than one tooth space simultaneously and thus reduce grinding time.

Tooth-spacing errors are controlled to a great extent by the accuracy of the indexing mechanism. A multiple-tooth index plate having the same number of spaces as the part being ground is utilized on most machines. After the grinding wheel passes through the work and returns, the work is indexed to the next tooth and so on around the part. Therefore, the spacing error in the finished part is dependent upon the accuracy of the index plate used. This error can be minimized by using an index plate which is larger than the part being ground. Larger grinders index by a change-gear mechanism and use a plate with only one gash for locking the work arbor or fixture after indexing.

External form grinders usually mount the work on centers and internal grinders usually provide for a workholding fixture to be mounted to the indexing head. In each case sufficient clearance is required in the part design to allow the wheel to pass through or run out at each end of the work. Figure 19-7 illustrates the requirements for clearance. Splines have a finished-length requirement and a hob radius which allows for grinding-wheel die-out.

External helical gears can be ground on at least one type of form grinder. The helix is produced by rotating the gear as the wheel passes through the work. The rotation is controlled by a lead bar which has the same lead as the helical gear at the pitch diameter. This has the obvious disadvantage of requiring a number of lead bars when a variety of parts are ground on a common machine.

Estimating Form-grinding Time. The time to form-grind a gear consists of four parts. These are (1) handling time, (2) time to rough-grind, (3) time to finish-grind, and (4) time to dress the grinding wheel.

Handling time consists of loading time, time required for equalizing of grinding stock, unloading time, and time required for inspection. This depends upon the

FIG. 19-2. Aircraft gear being ground on a form grinder. *(Courtesy of Gear Grinding Div., Michigan Tool Co., Detroit, Mich.)*

productivity of the tooling, the accuracy of the work received, and the type of inspection equipment provided.

All but the smallest gears require more than one wheel dressing per gear. The number of dressings required depends upon the number of teeth, the face width, the grade of grinding wheel, the diameter of the grinding wheel, the accuracy required in the finished part, the hardness of the part, and the metallurgical characteristics of the material being ground.

Table 19-1 shows some typical data for dressing time and number of wheel dressings required. It is assumed that that material is carburized and hardened to 60 Rockwell C and that moderate accuracy is required. Late-model production machines utilizing fixed-diamond-wheel trimmers may reduce these times considerably.

Rough- and finish-grinding times will vary considerably and will depend upon the amount of stock left for grinding, inaccuracies in the gear to be ground, metallurgical characteristics of the part, type of coolant, accuracy required in the finished gear, and productivity of the grinding machine.

Sufficient stock must be left to allow complete clean-up of the surface to be ground.

Excessive amounts of stock over and above that required for clean-up result in increased grinding time. Figure 19-3 shows the amount of stock normally required for grinding.

Excessive inaccuracies in the gear to be ground are caused by inferior manufacturing practices. Enough stock must be removed to eliminate the errors produced in rough-cutting, errors produced by heat-treating distortion, and variations in finish machining relative to the gear to be ground. Therefore, the maintenance of good surface finish and accurate dimensions in cutting, painstaking control of heat-treating practices, and care in maintaining the proper relation of other dimensions to the gear elements in finish machining will yield the lowest gear-grinding times.

Metallurgical characteristics can adversely affect grinding times. Carburized cases containing carbide network are more susceptible to cracking during grinding than are those which are free from network. Gears having a light decarburized skin after carburizing and hardening can also cause difficulty. The soft surface tends to load up the wheel and contribute to surface tempering.

Table 19-1. Form-wheel Dressing
(Wheel hardness I, J, K; grit 60 through 120)

Size of work		Wheel dia., in.	No. of dressings	Sec per dress
No. of teeth	Pitch			
20	3–6	6	3	70
60	3–6	6	6	70
20	7–12	12	2	50
75	7–12	6	5	50
20	16 and finer	12	2	35
150	16 and finer	12	4	35

Selection of the proper coolant is an important factor. There are a vast number of good coolants available, and a wise selection will contribute to productivity.

The time required to perform a rough- or finish-grinding operation by form grinding excluding setup, handling time, and dressing time can be estimated by Eq. (19-1). In this formula a "cycle" is the cutting action which occurs on a single tooth space between downfeeds of the grinding wheel. Early-model machines feed down only once per revolution of the work and the wheel makes a stroke across and back on each tooth. A cycle for this type machine is two strokes.

$$\text{Grinding time, min} = \frac{\text{No. of teeth} \times \text{No. of cuts}}{\text{cycles per min}} \tag{19-1}$$

Many late-model machines are equipped with automatic downfeed features. These machines feed down at each end of the stroke and grind each tooth space to a predetermined size during roughing. In this case a roughing cycle is one stroke and a finishing cycle is two strokes. Grinding time for this type of grinder is expressed in Eq. (19-2).

$$\text{Total grinding time, min} = \frac{\text{No. of teeth} \times \text{No. of rough cuts}}{\text{roughing cycles per min}} + \frac{\text{No. of teeth} \times \text{No. of finish cuts}}{\text{finishing cycles per min}} \tag{19-2}$$

The number of roughing cuts required depends upon the amount of stock left for grinding and the amount removed per cut. As stated previously, sufficient stock must be allowed to compensate for cutting inaccuracies, heat-treating distortion, and errors introduced in manufacturing between cutting and finish grinding. Assuming that the total error is within the stock allowed, the maximum number of cuts is

$$\text{No. of cuts} = \frac{\text{stock left for grinding}}{\tfrac{1}{2}\ \text{stock normally removed per cut}} \tag{19-3}$$

In Eq. (19-3) the stock left for grinding is in terms of tooth thickness or two times the excess stock per surface. The normal amount of stock removed is the amount that

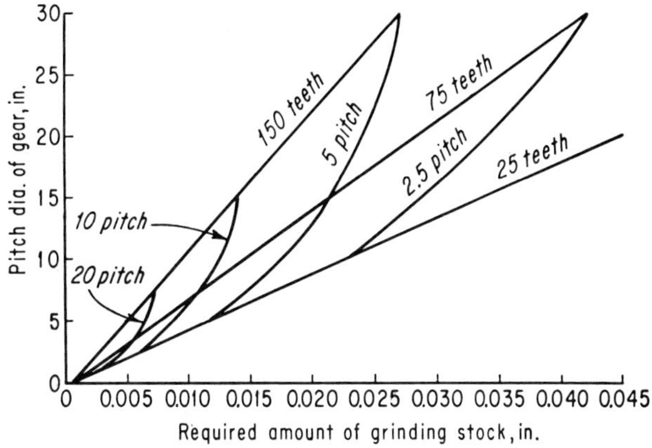

FIG. 19-3. Amount of grinding stock (on tooth thickness). (1) Chart is based on standard aircraft quality practice for carburizing or through-hardening. (2) With special heat-treating techniques it is possible to reduce stock to about 60 per cent. (3) Commercial heat-treating practice may require a stock increase to about 160 per cent.

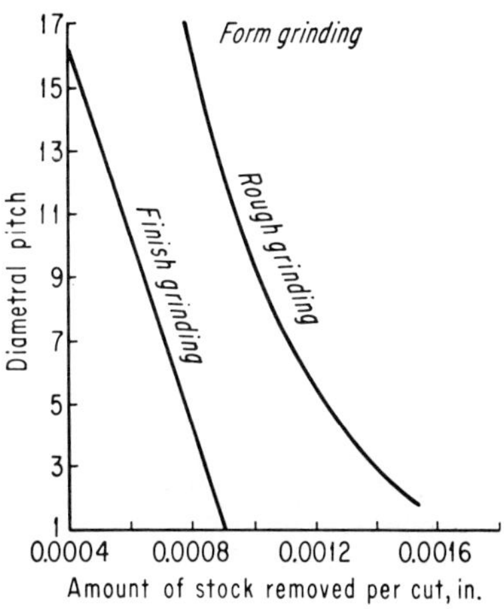

FIG. 19-4. Amount of stock (on tooth thickness) removed per cut of a form grinder.

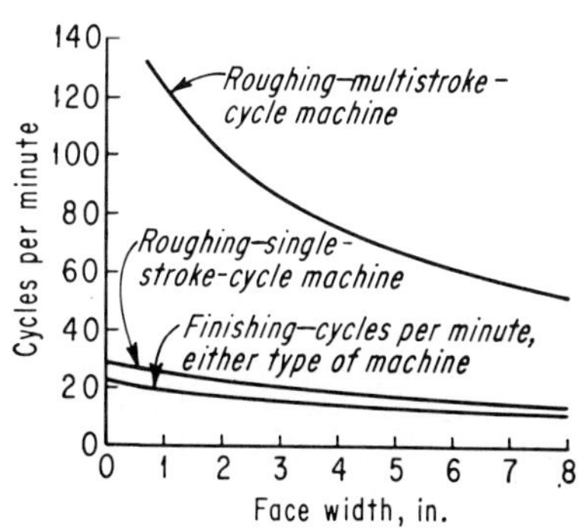

FIG. 19-5. Form-grinding cycles per minute relative to face width.

would be removed from the tooth thickness if the grinding wheel were cutting on both sides. The minimum number of cuts would be one-half of that given by Eq. (19-3) and the average number of cuts would be approximately three-fourths of that given by Eq. (19-3).

Figure 19-3 shows the amount of stock normally needed for grinding. The values shown are average values and may be decreased if the anticipated errors are small or may be increased where the design of the part makes control of errors difficult.

Rough-grinding cuts usually remove the maximum amount of stock which can be removed without causing surface tempering, rehardening, or grinding cracks. Finishing cuts take a smaller amount of stock. This will vary with the surface finish and profile accuracy required. Figure 19-4 shows the average amount of stock removed per cut by form grinding.

FIG. 19-6. Helical aircraft gear being form-ground. (*Courtesy of National Broach and Machine Co., Detroit, Mich.*)

The rate which a form-grinding machine is set to stroke (see Fig. 19-5) depends upon the face width being ground, the amount of overtravel of the wheel, the type and condition of material being ground, the rigidity of the work, and the range of speeds on the machine. Machines which rough-grind one tooth space complete before indexing can have considerably less overtravel than would be required if the wheel had to clear the work completely to allow the work to index.

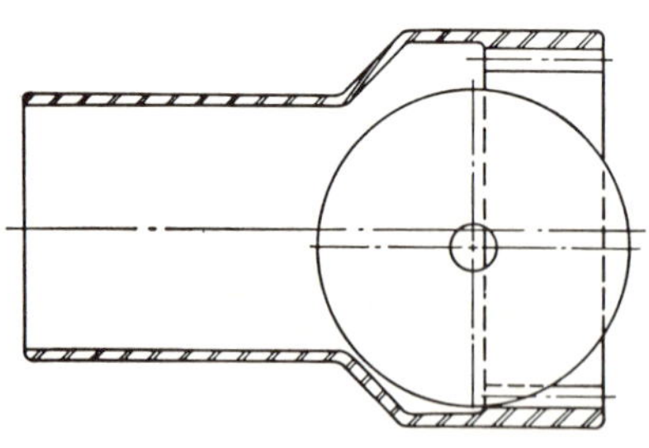

FIG. 19-7. Schematic arrangement to form-grind internal gear.

Form-grinding machines for external gears have a wide range of capacities. Present machines will grind a $\frac{3}{4}''$ minimum root diameter to a 36″ outside diameter and will accommodate 80″ between centers. Special machines have been designed to grind gears up to 6 ft in diameter. Internal form grinders are available to accommodate parts with internal diameters of $2\frac{1}{2}$ to 30″. Machines are available to form-grind external helical gears up to 18″ pitch diameter, and additional machines are being developed. Machines utilizing two

FIG. 19-8. Form grinding internal ring gear for aircraft engine. (*Courtesy of Indiana Gear Works, Div., Buehler Corp., Indianapolis, Ind.*)

wheels to permit form grinding of gears with undercut are now being used. New models with multiple grinding wheels and multiribbed wheels to increase productivity are also being manufactured. Figure 19-10 shows diagrams of multiple-wheel applications. Crowning attachments can be utilized on some current models.

19-3. Generating Grinding with a Disk Wheel. A gear is ground on a generating-type grinding machine by rolling the gear in relation to the grinding wheel. As wheel reciprocates back and forth across the face width of the part, the part rotates slowly about its own center. The reciprocating grinding wheel acts the same as a straight-sided rack in mesh with the gear being ground. The grinding wheel makes several strokes across the face width per increment of roll and in reality produces a series of flats on the tooth profile; however, these flats are quite small and the finished tooth appears to be a continuous involute curve when measured on conventional checking equipment. The ground gear performs as if the profile were a continuous involute also.

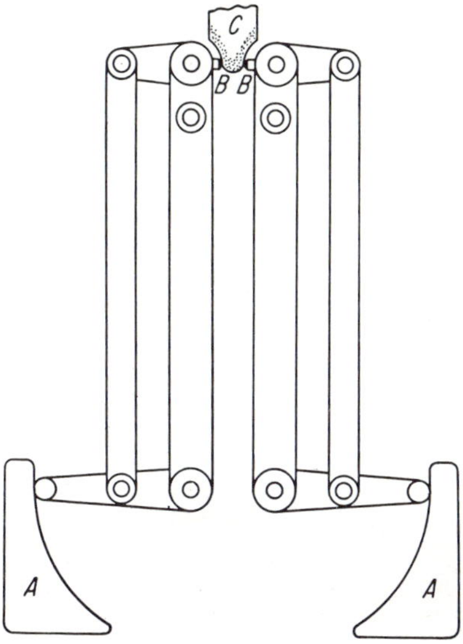

FIG. 19-9. Typical wheel-dressing arrangement for a form grinder.

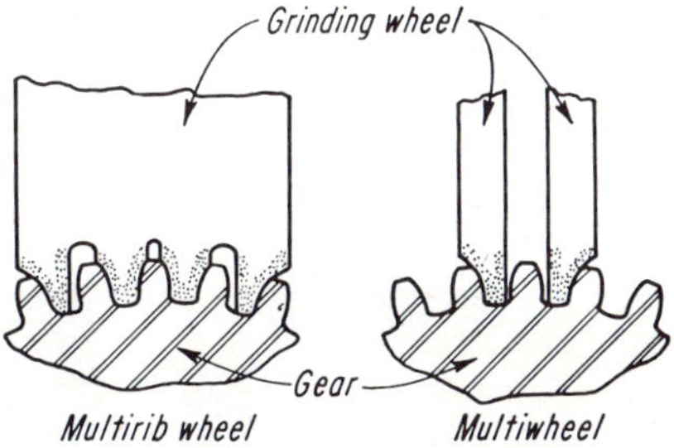

FIG. 19-10. Form grinding with ribbed wheels and multiple wheels.

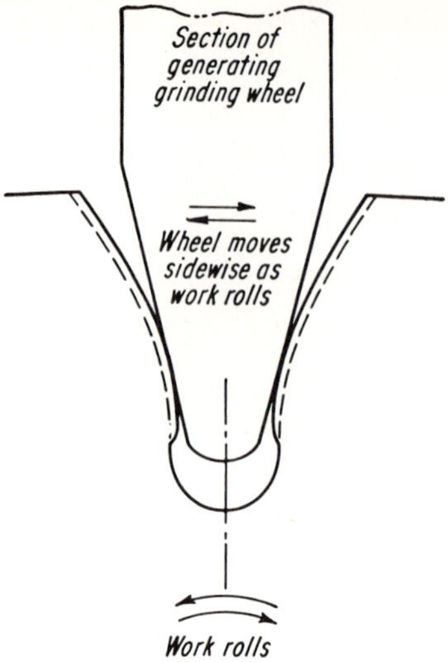

FIG. 19-11. Generating grinding wheel and work.

Generating disk-wheel grinders are manufactured which utilize conical dressed disk wheels and saucer-shaped wheels.

Conical Wheel Grinder. The conical-shaped wheel has an angle dressed on the sides of the wheel which is equal to the pressure angle of the part being ground. When external helical gears are ground the wheel is dressed to the normal pressure angle of the gear. This is performed by using dressing diamonds which are set on the proper angle. Wheel-dressing mechanisms for this type of grinding are relatively simple since no special form is required on the wheel.

The generating motion and indexing are controlled by means of a master gear and master rack. The master gear generally is of the same pitch and usually contains the same number of teeth as the gear being ground. However, when grinding gears with a small number of teeth it is not uncommon to use a master gear having a multiple number of teeth. The master gear and rack are commonly $22\frac{1}{2}°$ pressure angle because of the good rolling action with that pressure angle. A linkage mechanism actuates the rack, which causes the master gear to rotate about its axis. This in turn rotates the arbor containing the work, and the involute form is generated by the relative motion of the part and wheel. The

FIG. 19-12. Grinding railroad pinion on a generating grinder with a conical wheel. *(Courtesy of General Electric Co., Lynn, Mass.)*

FIG. 19-13. Grinding helical aircraft gear with a saucer-shaped wheel. (*Courtesy of General Electric Co., Lynn, Mass.*)

master gear and work are indexed one tooth by means of a ratchet type of index and the next tooth is ground.

This type of grinder can, in general, generate both external spur and helical gears. Helical master gears and mating racks must be used for helical gears. In grinding a helical gear the rack and master gear have the pitch-in plane of rotation and the wheel is dressed with the normal pressure angle. Internal gears cannot be ground on this type of machine. It is also possible to grind an undercut, but it is usually

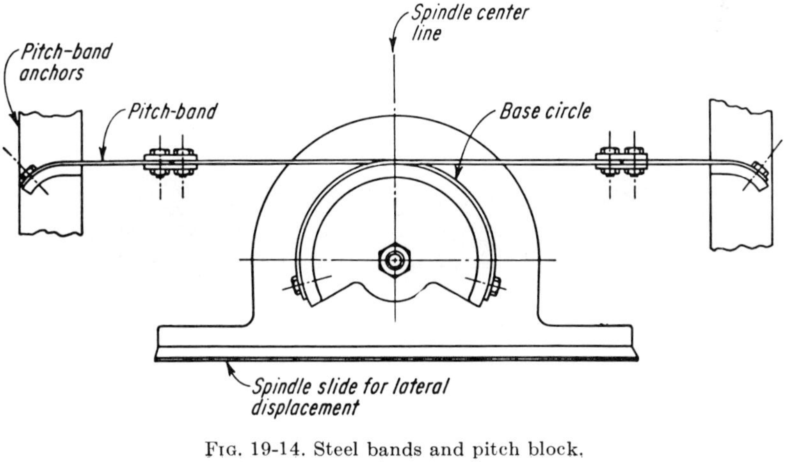

FIG. 19-14. Steel bands and pitch block.

impractical to generate profiles with tip relief or other modifications. Tip modification can be approximated by making a second grind with a wheel having a greater pressure angle. This requires timing the second grind with the first very accurately and is generally avoided in critical applications.

Grinding machines of this type are available for gears up to 18″ diameter. No new machines are being manufactured which utilize conical wheels.

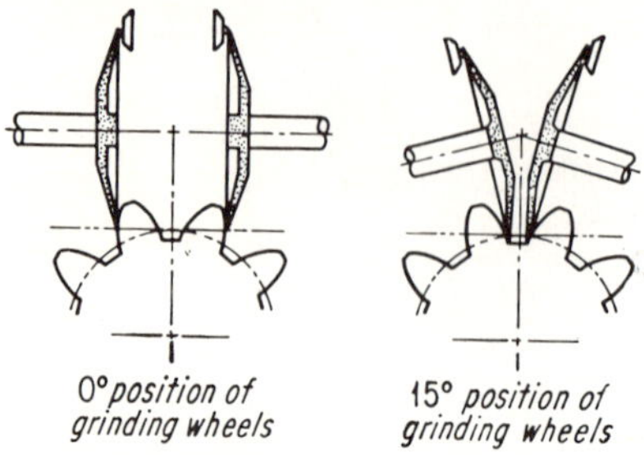

Fig. 19-15. Different wheel settings for saucer-wheel machine.

Saucer-shaped-wheel Machines. The basic principles for grinding with this type of machine are quite similar to those described above. The generating motion is usually produced by steel bands which are actuated over a pitch block which is the same diameter as the base circle of the gear being ground (see Fig. 19-14).

Two wheels are used and they may be set at 0 or 15° (see Fig. 19-15). When the wheels are set at 0° only one point of each wheel is in contact with the gear and when set at 15° each wheel contacts the gear profile at two points.

Indexing is controlled by means of an index plate similar to that used in most form-grinding machines. Worm gearing is used for indexing in some of the larger models of these machines.

Gears can be produced with tip or root modifications by changing the normal generating motion with a cam. Modifications can be controlled accurately and are readily reproducible.

Fig. 19-16. Grinding a large helical gear with a saucer-wheel-type machine. (*Courtesy of the Maag Gear Wheel Co., Zurich, Switzerland.*)

Machines are available for grinding external spur and helical gears from 1 to 142″ pitch diameter. Internal gears can be ground utilizing this type of grinding principle and machines are available for grinding approximately 9 to 20″ pitch diameter spur and helical gears.

Estimating Grinding Time with Disk-wheel Method. The grinding time for gears finished on a generating machine is composed of the same basic elements that are required in form grinding. The amount of stock left for grinding by this process is essentially the same, and metallurgical characteristics, type of coolant, and inaccuracies produced during manufacturing affect the grinding time as in form grinding.

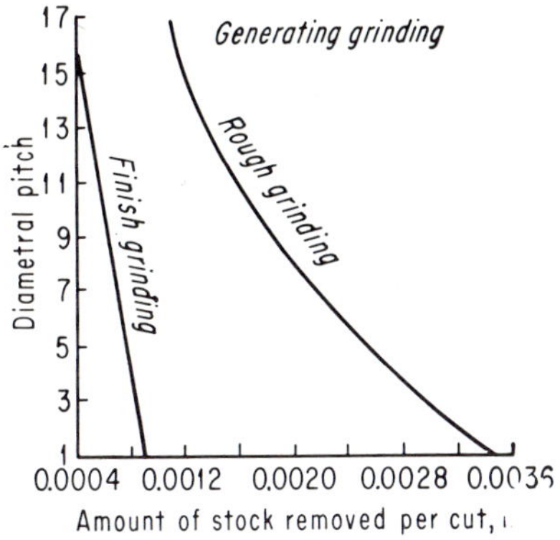

FIG. 19-17. Amount of stock (on tooth thickness) removed per cut of a generating grinder as a function of pitch.

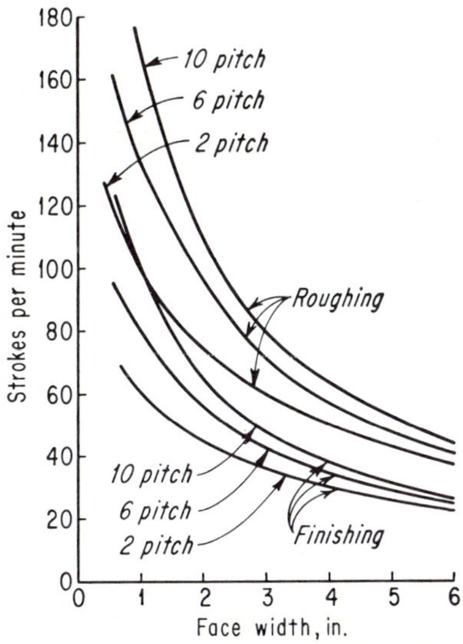

FIG. 19-18. Strokes per minute of a generating grinder based on face width.

The time required to rough- or finish-grind a gear may be calculated by Eq. (19-4). This is the actual grinding time to which must be added handling time, wheel dressing, and inspection time by the operator to obtain the productivity.

$$\text{Grinding time} = \frac{\text{No. of gear teeth} \times \text{No. of cuts} \times X}{\text{No. of wheels}} \tag{19-4}$$

where $X = \text{index time} + \dfrac{\text{strokes per tooth}}{\text{strokes per min}}$

An allowance of approximately 0.04 to 0.06 min is made for index time of generating grinders for gears up to 24" diameter. Indexing time for larger machines will generally be longer and will vary considerably with the part being ground. Figure 19-18 shows typical stroking rates for generating grinders.

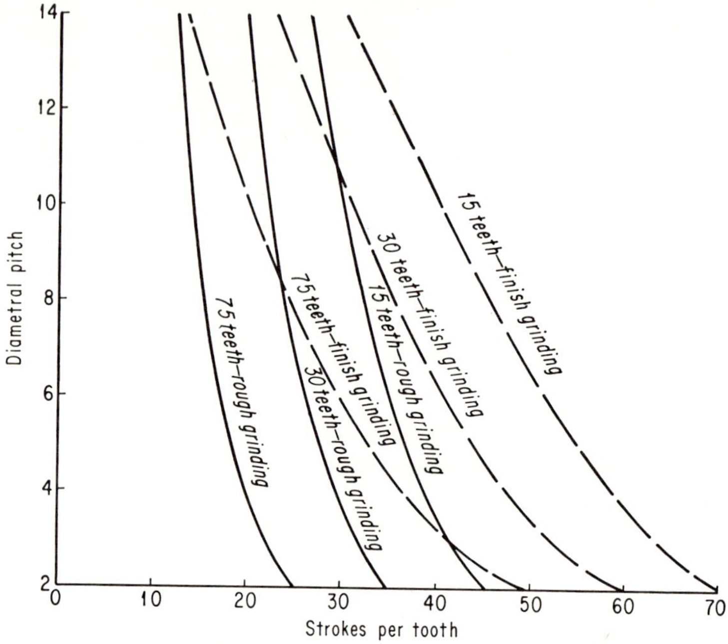

FIG. 19-19. Strokes per tooth, per cut for generating grinding.

The amount of stock removed per cut can be estimated by using the values given in Fig. 19-17. The number of strokes per tooth are estimated by referring to Fig. 19-19. The type of machine, machine condition, metallurgical characteristics, and type of gear will have some effect on these values.

Table 19-2 lists typical wheel-dressing times for generating grinding.

Table 19-2. Time to Dress Generating Grinding Wheels

Size of work		Wheel dia., in.	No. of dressings	Sec per dress
No. of teeth	Diam., pitch			
20	3–6	12	4	70
60	3–6	9	8	70
20	7–12	12	3	50
75	7–12	12	6	50
20	16 and finer	12	2	35
150	16 and finer	12	4	35

In general, generating machines, with the exception of the threaded-wheel-type generating grinders, are less productive than the other types of grinding machines. While more stock can be removed per cut without tempering than in form grinding, the wheel or wheels make point contact with the work and several strokes are required for each tooth.

Accuracy. Generating grinding has a number of advantages. The form used on the wheel is quite simple to produce, and dressing mechanisms are relatively simple. The machine is quite straightforward in principle and can be operated by less experienced mechanics. One operator can run a number of machines depending on the grinding cycle of the parts being run. Profiles can be modified on *certain models.* Some machine manufacturers claim superior lubrication-retention qualities due to the series of flats that are produced.

One disadvantage of these machines is productivity. Modifications cannot be produced on most machines. The accuracy is limited by the accuracy of the master gear and master rack on the disk-wheel machines. The saucer-wheel machines are excellent from the standpoint of accuracy.

19-4. Generating Grinding—Threaded Wheel. Threaded-wheel grinding machines operate on the same basic principles as gear-hobbing machines. Both spur

FIG. 19-20. Grinding a helical gear with a threaded wheel (*Courtesy of Indiana Gear Works Div., Buehler Corp., Indianapolis, Ind.*)

and helical external gears can be ground on this equipment, which will grind parts up to 20″ pitch diameter.

The threaded wheel is rough-formed by crushing and finish-dressed with diamonds in most cases. Some machine builders recommend crushing without any final forming with diamonds. Tip modifications can be lapped on the diamonds to produce almost any type of modification on the gear being ground.

Crushing a new wheel requires about 4 hr. Therefore, most users normally stock wheels for each pitch and pressure angle which can be dressed with diamonds for each part to be ground.

Dressing a precrushed wheel usually requires about 20 min. More time may be required if modifications are made since lapping the diamonds is a trial-and-error method. Two diamonds are used to dress the two sides of the thread in the wheel.

The work is mounted vertically on centers, on an arbor, or in a fixture to fit the machine centers. When grinding helical gears, the workholding slide is set at the helix angle of the gear to be ground. Change gears are utilized which are calculated to produce the required lead and fine adjustment is made by use of a sine bar. The work is driven by a hydraulic motor which is synchronized with another hydraulic motor driving the wheel. As the work and wheel rotate the work is oscillated (see Fig. 19-20) to permit grinding across the face width. The wheel feed is controlled hydraulically to control size of the finished part. When the work is mounted between centers hydraulic pressure is applied to the top center for driving.

FIG. 19-21. Threaded-wheel grinding machine. (*Courtesy of Fellows Gear Shaper Co., Springfield, Vt.*)

Estimating Grinding Time by Threaded-wheel Method. Several gears can be ground between dressings. After the gear is rough-ground the wheel is moved axially and a new portion of the wheel generates the finish profile. For very precise gears the cycle may be as follows:

1. Rough-grind part 1.
2. Move axially and dress wheel.
3. Finish-grind part 1.
4. Rough-grind part 2.
5. Move wheel axially and redress.
6. Finish-grind part 2.

These machines are very productive and are quite accurate. The time to grind a gear by this process may be estimated by Eq. (19-5).

$$\text{Grinding time} = \frac{\text{No. of teeth} \times (\text{face width} + \text{overtravel}) \times \text{No. of cuts}}{\text{No. of threads} \times \text{feed} \times \text{rpm of wheel}} \tag{19-5}$$

Table 19-3 shows the amount of stock removed per grind pass and the feed per revolution of the work for different classes of work. These values are frequently modified depending upon the condition of the part to be ground.

Productivity, ease of modification, and accuracy are the principal advantages of these machines. Late-model machines are available with attachments for crowning which allow this operation to be performed during the actual grinding cycle. This saves a considerable amount of time and is a decided advantage.

This type of grinder is quite complicated and sensitive and thus requires a considerable amount of operator skill.

Table 19-3. Typical Values for Feed and Stock Removal When Grinding with a Threaded Wheel

Diam. pitch	Feed per revolution of work		Stock removed per grind pass	
	Roughing	Finishing	Roughing	Finishing
24	0.055	0.025	0.0015	0.0004
16	0.055	0.025	0.0015	0.0005
10	0.055	0.025	0.0015	0.0006
6	0.055	0.025	0.001	0.0006

The amount of stock removed with an average grinding pass and average rates of feed is shown.

The values are tabulated for precision gears. It may be necessary to deviate markedly from the values shown, depending upon the design of the gear, grinding fixture being used, material being ground, and its hardness.

A typical wheel speed is 1,800 rpm for a 13″ diameter wheel. The amount of overtravel for grinding with a threaded wheel need not be large like the overtravel allowance of a hob. Typical overtravel allowances for threaded-wheel grinding are ½″ on 10 pitch and ¾″ on 6 pitch.

19-5. Control of the Grinding Process. Gear grinding is normally one of the last operations performed on a ground gear. To prevent scrapping costly parts rigid control of the process should be maintained.

Control of gear grinding begins with the heat-treating operations. Excessive case-carbon content can result in the presence of retained austenite on the surface. During grinding the retained austenite transforms because of heat and pressure from the grinding wheel and surface tempering and/or grinding cracks frequently occur.

Insufficient surface and/or atmosphere protection during heat-treatment can allow oxidation on the surface, which results in a thin layer of decarburized material on the surface. This soft material causes the wheel to "load up" and overheat, causing surface tempering of the finished part. Removal of this decarburized layer prior to grinding can result in a considerable increase in production with a decrease in rejection rate.

Excessive case carbon can also contribute to the formation of a carbide network or excessive free carbides in the case structure. Since this material is extremely hard, local overheating tends to occur and surface tempering takes place.

Excessive distortion in heat-treating causes many problems in the gear-grinding operation. If the machine operator does not start the grinding cycle on the largest tooth or at the point of maximum runout, abnormal amounts of stock per cut will be removed at those points and cracking, tempering, or dimensional errors will be present. It should be mentioned here that inadequate control of the gear-cutting operations or intermediate operations between heat-treatment and grinding will contribute to or cause this same problem.

The importance of adequate stress relieving during heat-treatment cannot be overstressed. Long tempering cycles increase the ductility of the casehardened surface without decreasing the hardness appreciably and balance and/or decrease residual stresses. This produces a more desirable surface stress pattern which allows the grinding operation to be performed faster and decreases the probability of rejection. Stress relieving after grinding is usually practiced to decrease the possibility of latent grinder cracks developing during storage or in the early cycles of use.

Coolant control is also an important factor in grinding control. This consists of adequate filtering to remove dirt and metal particles, the use of a sufficient volume to

FIG. 19-22. Grinding cracks on an improperly ground gear.

FIG. 19-23. Nital etch indications of surface tempering.

prevent excessive foaming or aeration, and a means to prevent an undue temperature rise or drop in the coolant. Temperature can usually be controlled sufficiently by controlling the volume in the system and the room temperature, but heat exchangers may be employed for critical situations.

The type of grinding wheel used is another important factor in grinding control. Wheels are available in various hardnesses, structure, grain size, and bond. Experimentation and experience are often necessary to determine the proper combination to be used for each given application. After the proper wheel has been selected, it is quite important that the wheel be dressed frequently to provide the best grinding action and to eliminate damage to the gear being ground. Many factors, including the cleanliness of the coolant and the cleanliness and type of material being ground, influence the frequency of dressing.

Cracking (frequently referred to as grinding checks because of their appearance) is one of the signs of improper control. Frequent magnaflux inspection during the grinding operation is recommended to assure proper grinding technique. One hundred per cent inspection by magnaflux should be made after grinding.

Surface tempering is another sign of malpractice in grinding. The most widely used inspection method for detecting tempering is the nital etch process. This consists of the following steps:

1. Thoroughly clean. Vapor degreasing or solvent degreasing may be used.
2. Rinse thoroughly.
3. Dip in a 4 per cent by volume solution of nitric acid for 1 min.
4. Rinse.
5. Dip in a 5 per cent by volume solution of hydrochloric acid for approximately 2 min. Acetone, water, or alcohol may be employed here.
6. Rinse.
7. Dip in solution of ammonia and water or other neutralizing solution.
8. Rinse in cold water.
9. Rinse in hot water.
10. Blow dry.
11. Inspect visually.
12. Oil to prevent rust.

There are a number of variations of this procedure which are used, but the vast majority deviate only slightly from this method. The interpretation of the results is straightforward if adequate criteria for acceptance and rejection are provided to the inspector. Tempered areas etch black, overheated areas containing rehardened spots are black with a white spot, and the background area containing no tempering is gray. Rehardened areas are almost always cause for rejection. The seriousness of tempered areas depends upon their location and the hardness as determined by a superficial-hardness test.

REFERENCES

1. Dudley, Darle W.: "Practical Gear Design," McGraw-Hill Book Company, Inc. New York, 1954.
2. Tarasov, L. P.: "Injury in Ground Surfaces," Norton Company, 1947.
3. Buckingham, Earle: "Spur Gears," McGraw-Hill Book Company, Inc., New York, 1928.
4. Colvin, F. H., and F. A. Stanley: "Gear Cutting Practice," 3d ed., McGraw-Hill Book Company, Inc., New York, 1950.

Chapter 20

BEVEL- AND HYPOID-GEAR MANUFACTURING

By

HARRY J. HART, *Liason Engineer, Gleason Works, Rochester, N.Y.*

CHARLES B. KING, *Associate Chief Research Engineer, Gleason Works, Rochester, N.Y.*

HARRY PEDERSEN, *Chief Design Analyst, Gleason Works, Rochester, N.Y.*

GILMORE M. SPEAR, *Research Engineer, Gleason Works, Rochester, N.Y.*

The various types of bevel and hypoid gears are described in Chap. 2, and the theory for producing conjugate pairs is outlined in Chap. 1. Information concerning tooth proportions and the design of bevel and hypoid gears is given in other chapters. The material in this chapter deals primarily with the machinery and equipment used to produce the teeth of such gears—their basic principles of operation, application, approximate range, and relative productivity. It also briefly touches upon other aspects of the manufacturing process which directly affect the production of the teeth, including processing of the gear blank and heat-treating.

The tools, methods, and machines used for producing bevel and hypoid gears are quite different from those used for other types of gears. There are two general classes of these machines: those used for gears with straight teeth, and those for gears with curved teeth. As is the case with all other gear machines, there are various models to accommodate differing requirements of size, range, and productivity. In order to understand these specific machines, it is first necessary to understand the basic principle of a generating machine, and also of a nongenerating machine. In order to understand the application of the differing machines, it is important to know of the different processes used to produce the teeth in gearsets for various requirements.

20-1. The Basic Process. Generation can be called the basic process in bevel- and hypoid-gear making, because at least one member of every pair must be generated; this is usually the member having the lesser number of teeth, called the pinion. The theory of generation, as applied to these gears, involves the following concepts:

1. An imaginary nondeformable bevel or hypoid gear called the generating gear. This may be conceived of as a crown gear, a mating gear, or some other bevel or hypoid gear.
2. A gear blank or workpiece made of plastic, deformable, nonelastic material on which teeth are to be produced.
3. The positioning of generating gear and workpiece so that the teeth of the generating gear are in mesh with the teeth of the workpiece.

4. The turning of generating gear and workpiece on their respective axes according to a prescribed motion. (With bevel gears the turning motion is defined by the imaginary nonslip rolling of the pitch surfaces of generating gear and workpiece.)
5. The envelopment of teeth of the workpiece by the teeth of the generating gear.

20-2. The Basic Generator. The actual production of gear teeth by the generating process requires:

1. That at least one tooth of the generating gear be described by the motion of a cutting tool or a grinding wheel.
2. That the workpiece be positioned in relation to the cutting tool or grinding wheel so as to mesh the teeth of the workpiece with the teeth of the generating gear as they are represented by cutter or tool.
3. That the cutting tool or grinding wheel be carried on a rotatable machine member called a cradle, the axis of which is identical with the axis of the generating gear.
4. That cradle and workpiece rotate on their respective axes, *or roll together,* exactly as would the workpiece and the imaginary generating gear.
5. That a means of indexing be provided.

Thus the rotating cradle carrying a cutter, tool, or grinding wheel represents a generating gear rolling with the workpiece, according to a prescribed relative motion. The motion of the tools or rotation of the cutter or wheel must sweep out, or describe, the teeth of the generating gear. As the tool, cutter, or wheel is carried about the cradle axis, teeth are enveloped on the rotating workpiece.

The concept of the imaginary generating gear is the real key to understanding of the generating process. Figure 20-1 illustrates the representation of a generating gear by a face-mill cutter on a Gleason hypoid-gear generator. Fundamental design differences that exist among different types of gear generators have originated because of different concepts of the imaginary generating gear. For example:

1. The generating gear might be one of a pair of complementary gears which fit one another like a mold casting. If such gears have 90° pitch angle they are referred to as complementary crown gears. A series of interchangeable bevel gears could be based on these generating gears. Straight bevel and some spiral bevel gears are derived from generating gears having pitch angles close to 90°.
2. The generating gear for a spiral bevel pinion could be either a spiral bevel or a hypoid gear.
3. The generating gear for a hypoid pinion could be a hypoid gear having a pitch angle and offset different from the mating gear.
4. The generating gear for a spiral bevel or hypoid pinion could be a gear having a helicoidal instead of a conical pitch surface; this is used for the generation of two tooth sides simultaneously.

Tooth surfaces of the above generating gears are described by tools, cutters, or wheels which are adjustable to some degree relative to the axis of the generating gear or cradle. Most of the machines with which such generating gears are represented use face-mill cutters or comparable grinding wheels. Spindles of cutter or wheel are adjustable radially from the cradle axis and are tiltable or inclinable with respect to the cradle axis; they can also be swiveled about a point in the cutter or wheel axis. This permits the representation of a variety of generating gear teeth. Machines having tiltable cutter axes (known as tilting-spindle machines) usually operate so as to maintain a constant ratio of turning between cradle and work during the generating roll.

Some of the larger generators and generating grinders, and the older types of pinion roughers, have wheel or cutter spindles which are not tiltable. That is, they are fixed in direction parallel to the cradle axis but may be adjusted radially with respect to it. With this type of generator the ratio of angular velocities between cradle and work spindle may be varied during generation. The ratio of turning velocity between work spindle and cradle is called the *ratio of roll,* and methods employing a nonconstant ratio of roll are called *modified-roll methods.* All machines on which the cutter or wheel spindle is fixed in direction parallel to the cradle axis have provision for modified roll and are referred to as *modified-roll machines.*

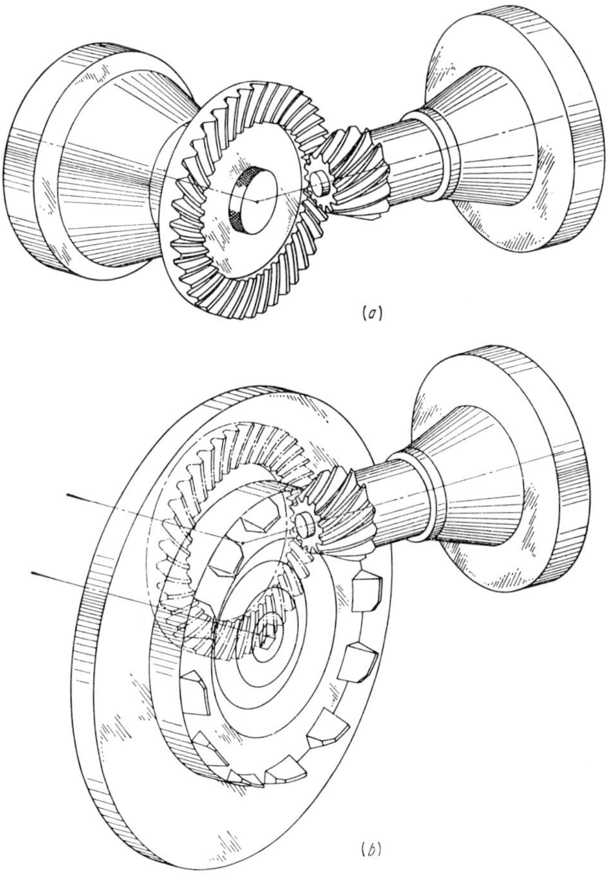

FIG. 20-1. How a face-mill cutter represents a generating gear. (*a*) Pinion in mesh with mating gear. (*b*) Face-mill cutter representing generating gear.

The generating gear must be replaced by a rotatable member which carries some type of cutting tool for sweeping out one or more of its tooth surfaces. This member is known as the cradle (see Fig. 20-2). The gear to be cut is held by a rotatable spindle which is part of a work head. The work head is adjustable axially so that gears of different mounting distance can be accommodated. For the simplest generating methods, the work head is positioned so that the pitch-cone apex of the workpiece falls on the cradle axis.

The work head is supported by a swinging base which is rotationally adjustable about a vertical axis. The swinging feature of this base is required in order to accommodate gears of different pitch angles. Beneath the swinging base is the sliding base. This member is movable and adjustable in a direction parallel to the cradle axis. Adjustment is made in the position of this base in order to establish the correct depthwise relationship between the cutting tool and the workpiece. It can also be moved back from the cutting position in order to obtain clearance for mounting and dismounting the workpiece. The sliding base is supported by the main frame of the machine, to which is attached a housing that supports the cradle member.

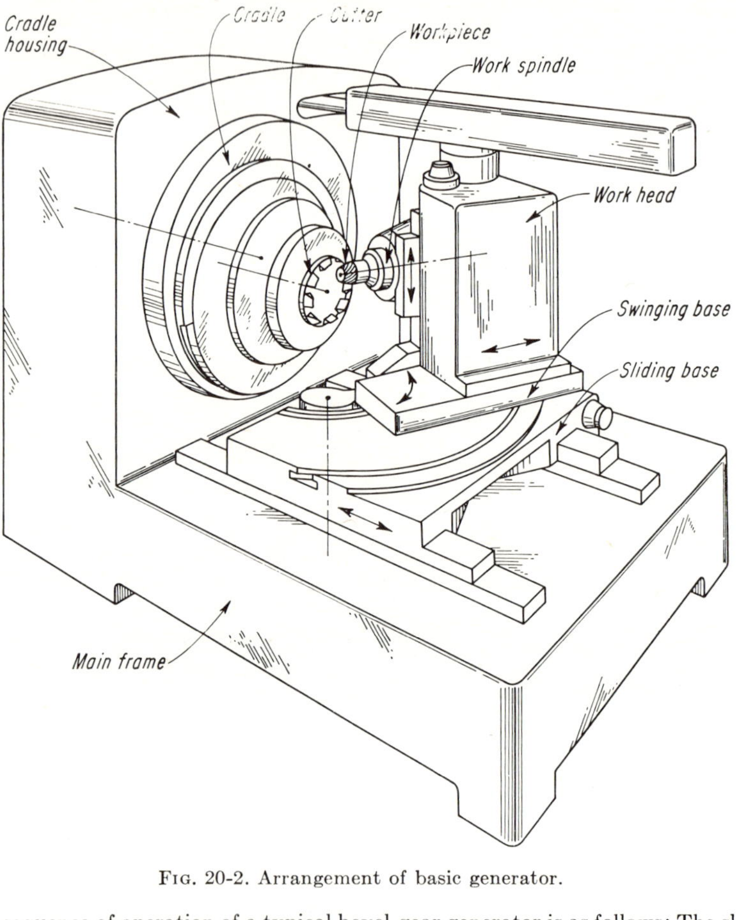

FIG. 20-2. Arrangement of basic generator.

The sequence of operation of a typical bevel-gear generator is as follows: The sliding base moves forward to bring the work into engagement with the cutter, and the generating motion of the cradle and work begins (Fig. 20-3*a* and *b*). After the operation is completed on a tooth space, the sliding base is moved back (Fig. 20-3*c*), and the cradle and work spindle reverse their rotation. The cradle returns to its original position, but the work is indexed one tooth during its return roll (Fig. 20-3*d*). The sliding base then moves forward for the start of the next cycle.

The rotating elements of the machine, i.e., cradle, work spindle, and cutter, are connected by means of gear trains. The block diagram in Fig. 20-4 illustrates how these elements are connected and shows the general location of the change gears used to control the various machine functions. The drive starts at the motor and goes through a double set of change gears. One of these sets controls the speed of the cutting tool while the other set, known as the feed gears, controls the time interval for a machine cycle. A fixed number of shaft revolutions enter the reverse mechanism from the feed gears during each cycle and are converted into the two-directional rotation needed to drive the generating train. Normally, the reverse mechanism is such

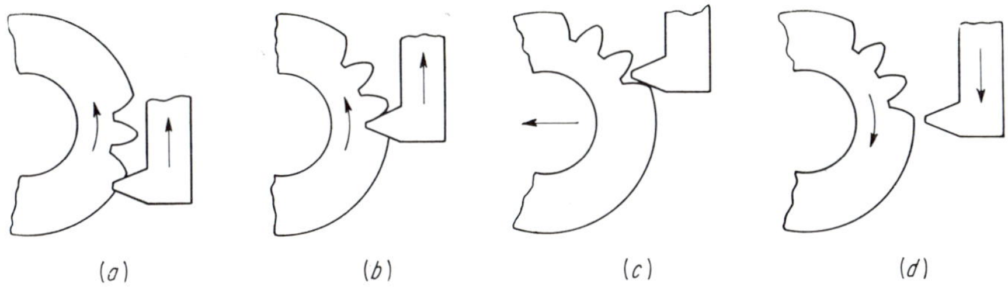

FIG. 20-3. Four stages of generating cycle. (*a*) Start of roll. (*b*) Center of roll. (*c*) Withdraw. (*d*) Return.

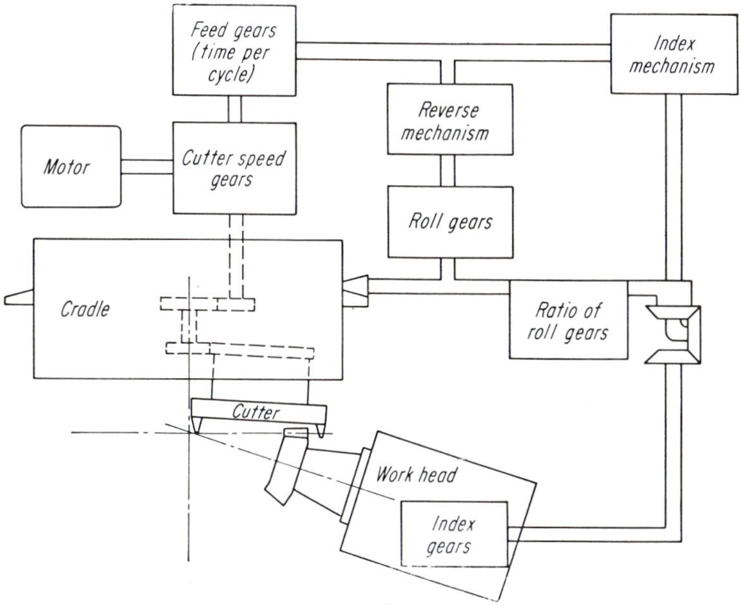

FIG. 20-4. Basic-generator-drive diagram.

that the generating train is rotated slowly during generation and rapidly during the return roll.

The roll gears between the reverse mechanism and the generating train control the amount of rotation imparted to the cradle and work. This is always expressed as "cradle roll," that is, the number of degrees through which the cradle member rotates in each direction during a cycle. The amount used in any case must not be less than that needed to sweep out the surface of action which exists kinematically between the imaginary generating gear and the workpiece.

The ratio-of-roll gears, in conjunction with the index gears, cause the work spindle to roll with the cradle in the same way that the work would roll with an imaginary generating gear. These gears establish the imaginary rolling pitch surfaces.

The indexing is accomplished by adding a fixed amount of rotation to the work end of the generating train by means of a differential. The index gears divide this rotation so that the proper number of teeth is produced in the workpiece.

20-3. The Basic Nongenerator. In order to achieve fastest production rates, the gear member of spiral bevel pairs, approximately 3:1 ratio and higher, and of hypoid pairs, approximately 2.25:1 and higher, can be made with nongenerated tooth surfaces, which are straight in the normal section. These tooth surfaces, formed by a single sweep of a cutting tool, can be produced much more rapidly than the tooth surfaces of generated gears. (Generated profiles, because they are curved, require a series of generating cuts to envelop the profile shape.) (See articles on nongenerated gears,

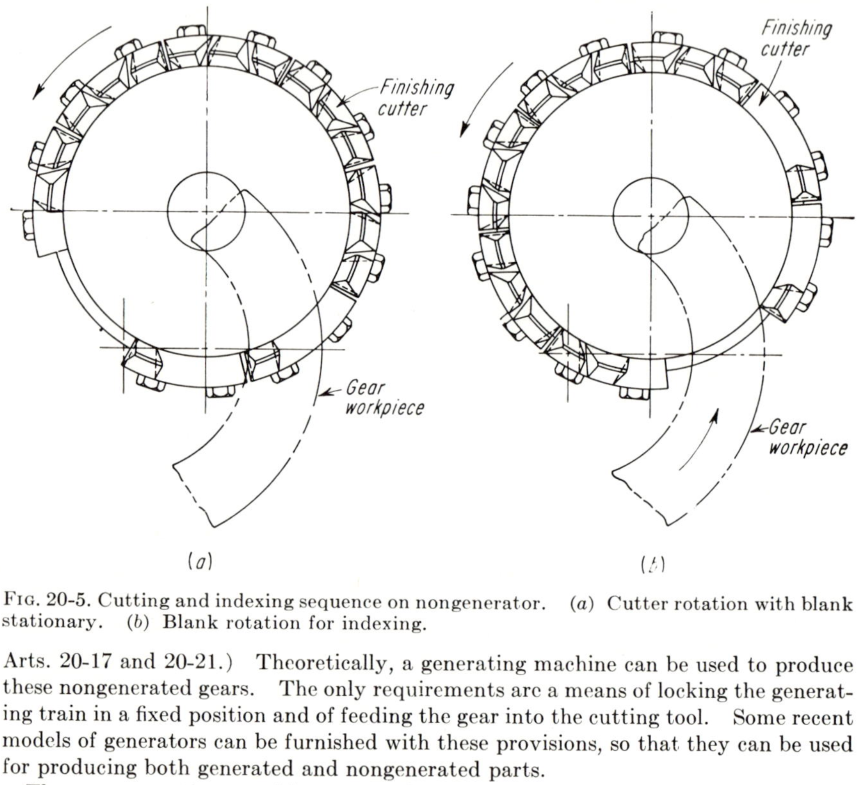

FIG. 20-5. Cutting and indexing sequence on nongenerator. (*a*) Cutter rotation with blank stationary. (*b*) Blank rotation for indexing.

Arts. 20-17 and 20-21.) Theoretically, a generating machine can be used to produce these nongenerated gears. The only requirements are a means of locking the generating train in a fixed position and of feeding the gear into the cutting tool. Some recent models of generators can be furnished with these provisions, so that they can be used for producing both generated and nongenerated parts.

The nongenerating machines are relatively simple compared with generating machines; fewer machine components are required, since, except for the indexing of the workpiece, the only motion involved is that of the cutting tool. Nongenerated gears, which include both Formate®* and Helixform®* gears, are cut with a conical face-mill-type cutter. For Formate gears a Single Cycle®* cutter makes the finishing cut on both sides of one tooth space with each revolution. The inside and outside cutting blades are stepped radially, causing the cutter to act like a broach. The last two blades are an inside and an outside finishing blade, which are spaced so that they cut one at a time. After they pass through the tooth space, the work is indexed quickly by means of an index-plate mechanism while a gap in the cutter passes the work. This sequence is indicated in Fig. 20-5.

* Registered trade-marks of the Gleason Works, Rochester, N.Y.

The position of the Single Cycle cutter relative to the nongenerated gear is indicated in Fig. 20-6. The cutter axis is located with respect to the gear apex by the distances H and V, which are functions of the mean cone distance, the spiral angle, and the cutter radius. Figure 20-7 is a schematic picture of a machine for cutting these gears. The cutter is mounted in a spindle which is adjustable upward for the V setting and across for the H setting. The work spindle is adjustable for root angle and also in the direction of the cutter axis for depth setting. These two components are, of course, mounted on the machine frame.

20-4. Gear-blank Requirements Prior to Tooth Processing. The sequence of operations and the type of equipment used for production of bevel gear blanks vary widely, depending upon size, quantity, quality, and shape. In general the machining processes and tooling used are similar to those used in the manufacture of any other accurate machine part. The purpose of this article is not to specify in detail methods of blank preparation or tolerances but to list the general requirements which must be met, so that the gear machines can produce parts of desired quality.

The primary requirement is that the blank, as delivered to the gear machine, must be in proper condition for producing the teeth in accurate relation to certain important surfaces on the blank. The most important of these surfaces are those used for locating the gear *radially* and *axially* on the gear machine. When the size and shape of the gear permit, it is preferable to use these same surfaces for cutting, testing, die quenching, and grinding, to avoid an accumulation of errors, and to eliminate the need for machining numerous surfaces to close tolerances. Equally important, in the case of hardened gears, are the reference or locating surfaces used for the above-mentioned operations which follow the tooth cutting.

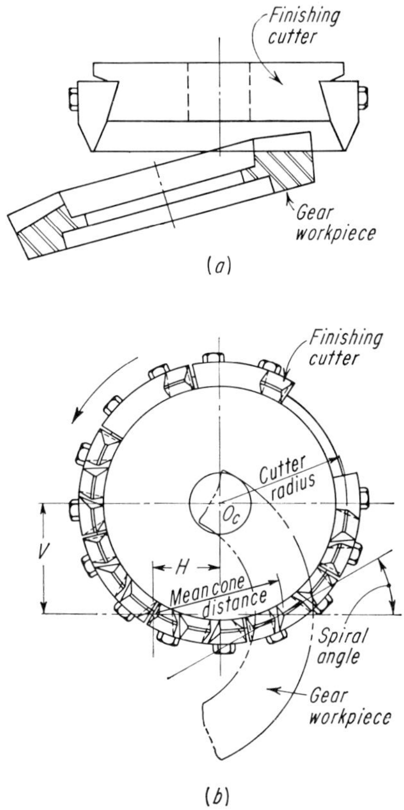

Fig. 20-6. Relative position of Single Cycle cutter and Formate gear. (*a*) Plan view. (*b*) Elevation view.

Wherever possible, the surfaces used for locating the blank in the gear machine should be the same surfaces used for mounting the gears in the final assembly. Figure 20-8 shows the various surfaces and dimensions which should carry specific tolerances, in the case of a bevel gear blank of conventional bored-type design.

As a general rule in small and medium sizes of blanks for industrial gears of good quality, locating bores or shanks are held to a total tolerance of 0.0005″. This is reduced to 0.0002″ for precision gears, and to a "wring fit" on workholding equipment if extreme precision is required.

The axial locating surface must be flat and square with the locating bore or shank within a total axial runout of 0.0005″ for good-quality industrial gears, and a smaller tolerance may be required for precision gears.

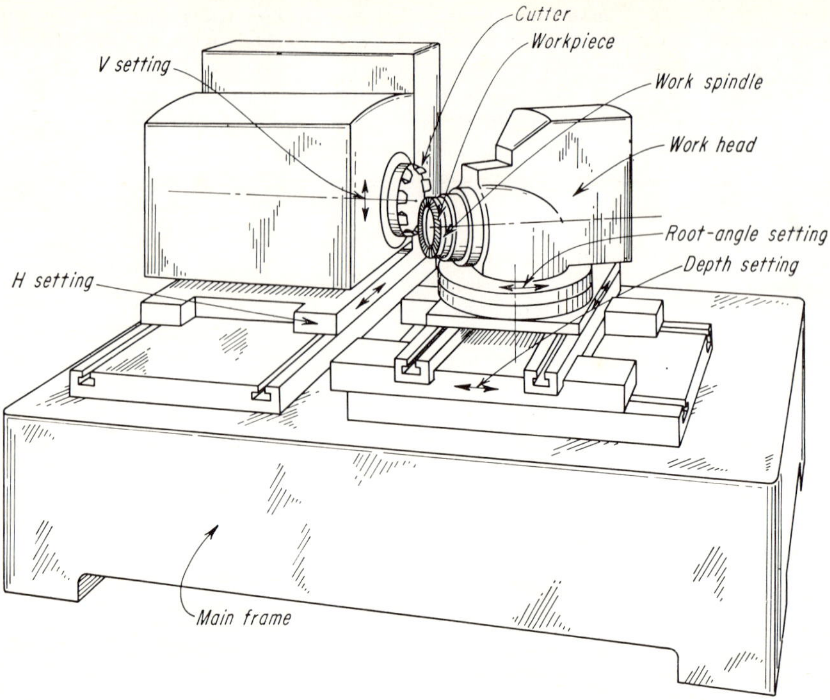

FIG. 20-7. Arrangement of basic nongenerator.

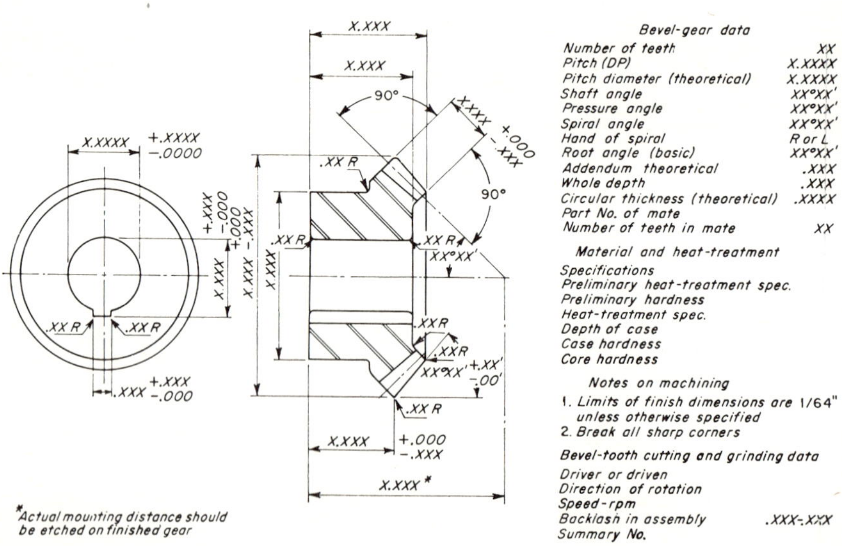

FIG. 20-8. Toleranced bevel gear-blank dimensions.

The axial and radial location of the crown point, as well as the face angle, must be controlled within specific limits, which depend generally upon diametral pitch and gear size. Otherwise tooth depth and operating clearance will vary from gear to gear, because the gear machine produces teeth in fixed relative position to the blank locating surfaces.

20-5. Bevel- and Hypoid-gear Manufacturing Procedures. Tables 20-1 and 20-2 show the sequence of processes directly related to tooth formation of gears for various types of applications.

Table 20-1. Straight-tooth Bevel Gears Applications and Manufacturing Procedures

Type	Precision, fine-pitch (unhardened)	Precision load-carrying (hardened)	Commercial	Commercial coarse-pitch, also gears with hub or flange projecting above root plane
Typical application	Instruments	Instruments and aircraft accessory drives	General industrial, automotive differential	General industrial
Procedure	Cut teeth in one operation, test	Cut teeth, soft test, harden, finish-grind teeth, hard test	Cut teeth in one operation, soft test, harden, hard test	Rough-cut teeth, finish-cut teeth, soft test, harden, hard test

Table 20-2. Spiral Bevel, Zerol®* Bevel, and Hypoid Gears Applications and Manufacturing Procedures

Type	Precision, fine-pitch (unhardened)	Precision, load-carrying (hardened)	Commercial	Automotive axle
Typical application	Instruments	Instruments and aircraft accessory drives	General industrial	Passenger cars, trucks, and busses
Procedure	Cut teeth in one operation, test	Cut teeth,† soft test,† harden, finish-grind teeth, hard test	Rough-cut teeth, finish-cut teeth, soft test, harden, lap, hard test	Rough-cut teeth, finish-cut teeth, soft test, burnish teeth (pinions only), harden, match pairs, lap in pairs, hard test

* Registered trade-mark of the Gleason Works, Rochester, N.Y.

† May be omitted on gears of through-hardening material approximately 20 diametral pitch and finer.

20-6. Bevel-gear Workholding Equipment. In the cutting, testing, lapping, and grinding of bevel and hypoid gears, the chucking equipment plays a vital part. The important points to consider in connection with chucking equipment are rigidity, concentricity, accuracy of size, and an ample, uniform force for holding the gear in

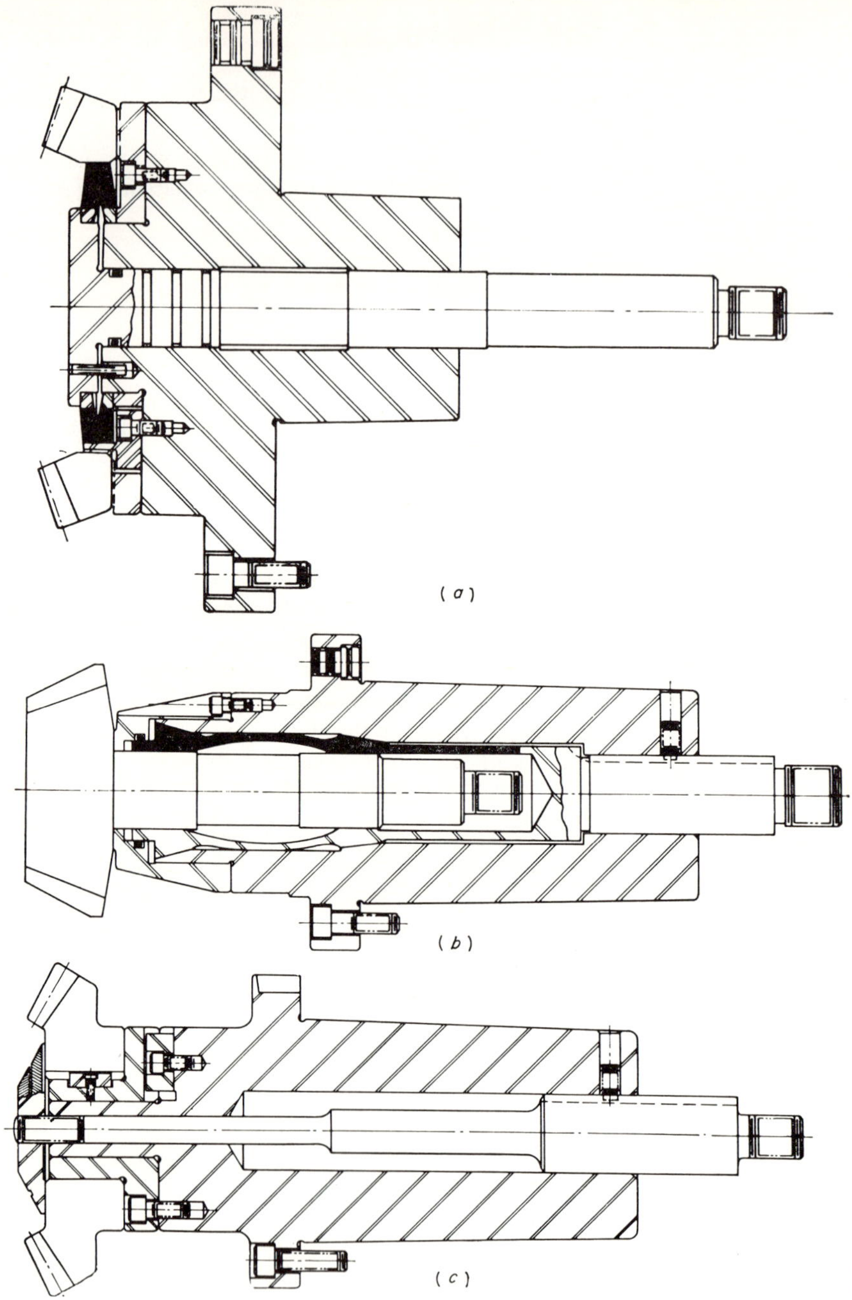

FIG. 20-9. Typical workholding equipment. (*a*) Production-type arbor for cutting gear. (*b*) Production-type chuck for cutting pinion. (*c*) Adaptable-type arbor for bored gear.

place. Excessive overhang, loose bushings, a multiplicity of fits, and equipment which must be trued up on the machine should be avoided. Chucks and arbors should, in general, be maintained within 0.0002″ of the correct size at all times. When mounted in the cutting or testing machine the radial and axial locating surfaces should run true within 0.0002″ full indicator reading. Even closer tolerances are required for gears of extreme precision.

Bevel-gear cutting and testing machines are usually equipped with a work spindle having a taper bore which receives the taper shank of the arbor. In manufacture the large end of the bore is held very accurately to size, so that the shoulder of a correctly made arbor will seat on the face of the work spindle, when the arbor is pulled snugly into the taper bore.

To seat properly in the taper bore and on the face of the spindle, the arbor must be made with the correct "draw." The draw is the amount that the shoulder of the arbor is wrung into the spindle by hand. This amount should be from 0.004 to 0.012″, depending on the size of the taper.

Nearly all modern bevel-gear-cutting machines are equipped with hydraulic chucks to which nearly all types of workholding equipment may be readily adapted. Some typical methods of chucking blanks for gear cutting are shown in Fig. 20-9*a*, *b*, and *c*.

STRAIGHT-BEVEL-GEAR MACHINES

This section deals with the four basic types of straight-bevel-gear-cutting machines, as well as straight-bevel-gear-grinding equipment. The subjects covered include description of machine operating cycles, required setup data, tooling requirements, and capacity and production data of the various machines.

20-7. Two-tool Generators and Roughers. As the name implies, these machines employ two reciprocating tools, each of which cuts on opposite sides of a tooth. These are the most universal of the straight bevel machines, and initial tooling cost is the least expensive, but production rates are appreciably slower than on other types of straight bevel generators. The two-tool machines are used for (1) work beyond the size range of interlocking cutter machines (Art. 20-8) and the Revacycle®* machine (Art. 20-10); (2) parts which have integral hubs or flanges projecting above the root line, precluding use of the above-mentioned machines; and (3) installations for minimum-quantity requirements over a wide range of work which cannot be accommodated by one model of the other types of straight bevel machines.

Roughing. Gears which are finish-cut on these machines must first be rough-cut. Where the production requirements warrant, the roughing is done on separate machines which are similar in appearance to the generators but which have no provision for generating roll. Otherwise the rough-cutting operation is performed on the generators. On both the roughing machines and the generators, the roughing cut is made by feeding the blank straight into the tool without roll.

A method of roughing which is used to good advantage on flat gears having a relatively large number of teeth and a face width not exceeding one-quarter of the cone distance, is the "double-index" method. By means of this method, two slots are roughed during each cutting cycle, one roughing tool operating in each of the tooth slots. The index mechanism is arranged to index the blank for half the number of teeth in the gear being roughed.

Required Setup Data. In order to make the machine setups for producing a pair of right-angle gears, the operator must be provided with the gear data and one calculated machine setting called the tooth angle (Fig. 20-10). The remainder of the setup data are taken from tables furnished with the machine. When the shaft angle

* Registered trade-mark of the Gleason Works, Rochester, N.Y.

of the pair is not 90°, the ratio-of-roll gears and data for checking the roll must be calculated, in addition to the usual setup information.

Tools. Figure 20-11 shows typical roughing and finishing tools which are specified by diametral pitch and pressure angle. These are mounted in holders which are attached to the front of reciprocating slides on the face of the cradle. Each tool can be adjusted relative to the holder so that the end of the tool is in the gear-root plane and so that the tool edge travels along a line which intersects the cradle axis. This tool positioning is checked by means of gaging fixtures which are set on a proof block. The position of the tool-holder on the face of the tool slide is adjusted to accommodate the cone distance of the work, and the length of stroke is adjusted so that the tool overtravels the face width by a small amount.

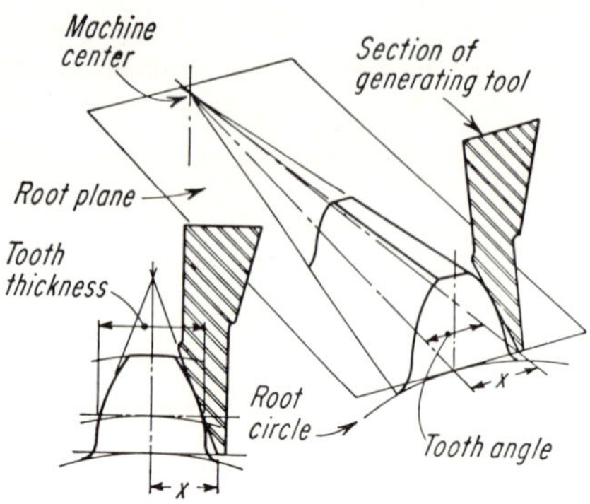

FIG. 20-10. Straight bevel gear-tooth angle.

Several types of tools are available for the roughing operation, the simplest being straight-sided V-shaped tools used for gears of 3 diametral pitch and finer, as in Fig. 20-11. Tools with curved profiles, available under the trade name Acruf®,* however, can be designed and furnished for specific parts, and these rough the teeth much closer to the finished shape than the V tools. This results in a saving of time and tool life in the finishing operation. A third type of roughing-tool set, Tanruf®,* which is used for heavier pitches (usually 3 diametral pitch and coarser) consists of a slotting tool, followed by a side-cutting tool. The first tool cuts only on the bottom

FIG. 20-11. Roughing tools (left) and finishing tools.

of the tooth space while the second tool cuts only on the side. The second tool is usually made with a straight cutting edge but can be furnished with a curved profile for closer roughing.

Coniflex®* Provision. Coniflex gears are Gleason straight bevel gears whose teeth are crowned lengthwise, to localize the tooth contact. In order to produce such gears, the generators have two angularly adjustable guides on the back of each slide

* Registered trade-mark of the Gleason Works, Rochester, N.Y.

which ride on a pair of fixed rollers. When these guides are in line, the tool is stroked along a straight line. When the guides are set out of line, as in Fig. 20-12, the tool is stroked along a curved path. The amount and position of the curvature can be varied by the setting of the guides. A table furnished with the machine lists the guide settings which are used to make the tooth bearing length approximately one-half the face width. The machine operator can readily vary these settings to make the tooth contact longer or shorter.

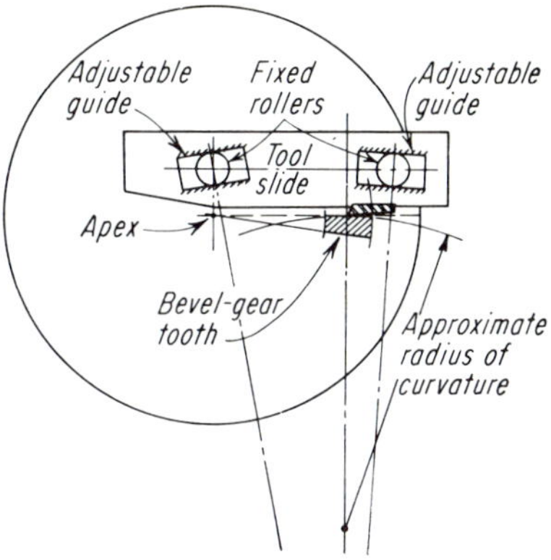

FIG. 20-12. Coniflex provision on two-tool generators.

Finishing Cycle. The completely automatic cutting cycle of the finishing operation consists of three phases. The first is a semifinishing operation which is done with the cradle turning clockwise and the sliding base held back from the full-depth position by a small amount. Next, the sliding base is automatically advanced to full depth and the cradle and work rotations are reversed to make the finishing cut. At the end of the generation, the work is withdrawn and indexed to the next tooth. Semifinishing the tooth surface prior to finishing leaves a uniform thickness of material for the finishing, and thereby ensures the best possible surface finish.

Table 20-3. Two-tool Generators and Roughers Capacity and Production Data

Capacity range	No. 14 Coniflex generator	No. 24A Coniflex generator and No. 24 rougher
Max. diam. pitch	3	1¼
Max. cone distance, in.	12	18
Max. face width, in.	3½	6
Max. ratio	10:1	10:1

Representative Production Rates, AISI 4620 Material, 160 to 190 BHN

	Roughing, min	Finishing, min
8 diam. pitch miter gear, 32 teeth, 0.750″ face	10.3	12.1
1½ diam. pitch miter gear, 32 teeth, 4″ face	128.0	101.0

Average Machine Setup Time, Hr

Roughing	½
Finishing	1

20-8. Interlocking Cutter Machines. The most recent type of straight bevel generator is that which employs a pair of interlocking disk-mill cutters having a multiplicity of blades which cut on the two sides of a tooth space. Figure 20-13 is a close-up view of a finished gear and the interlocking cutters. These cutters are not stroked in the direction of the teeth; consequently slightly concave root land surfaces are produced.

In nearly all cases, gears cut on these machines are completed from the solid blank in one operation which combines the roughing and finishing cuts. The size range of each of these machines (Table 20-4) is not so great as that of a two-tool generator, but the

FIG. 20-13. Close-up of Coniflex generator cutters and workpiece.

productivity is as much as four to five times greater. These machines are well suited for the economical production of straight bevel gears in widely varying quantities.

Coniflex Provision. As seen in Fig. 20-14, the disk cutters rotate on axes which are inclined to the face of the cradle. The cutting edges do not lie in the planes of rotation of the cutters but instead sweep out slightly internal conical surfaces. These dished cutter surfaces produce teeth which are slightly crowned in the lengthwise

Table 20-4. Interlocking-cutter Machines, Capacity and Production Data

Capacity range	No. 102A	No. 104	No. 114
Diam. pitch	16 and finer	3–20	2½–16
Max. cone distance, in.	2¼	4¼	7
Max. face width, in.	7/16	1⅜	2½
Max. ratio	Any practical ratio	10:1	10:1

Representative Production Rates, AISI 4620 Material, 160 to 190 BHN, 32-tooth Miter Gear

Machine	Gear size	Cutting time, min.
No. 102A	24 diam. pitch, ¼″ face	2
No. 104	6 diam. pitch, ⅞″ face	5
No. 114	3 diam. pitch, 1⅝ face	14

Average Machine Setup Time, Hr

No. 102A generator	1¼
No. 114 generator	2½–3

direction. The amount of crowning, or tooth-bearing localization, varies with the dish angle, which is determined by the blade angle. The cutter blade angle is selected according to the length of contact desired. The pressure angle of the gears being cut is obtained by means of machine settings.

Cutters. The cutters for the smallest machine are available with one of three different blade angles, but nearly all users find that one angle, 20°, is satisfactory for all gears. Seven pairs of these cutters with different blade-point widths cover the pitch range from 16 to 64 diametral pitch. The cutters for the larger machines are made with five different blade angles. A series of nine tooth-bearing lengths can be obtained with these cutters by using different blade angles (in some cases, for the two members of a gearset). Normally, a selection can be made from two or three bearing lengths which are between 35 and 65 per cent of the face width.

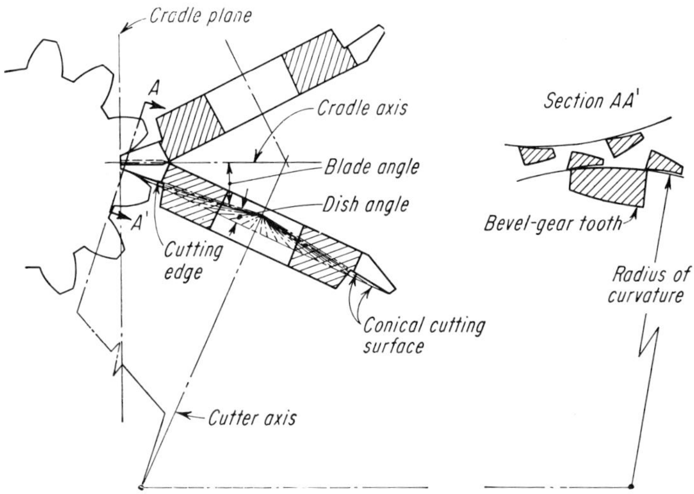

FIG. 20-14. How disk-mill cutters produce Coniflex teeth.

The theoretical blade-point width is calculated directly from the gear-tooth proportions. In actual use the next smaller blade point is selected from the series of standard widths furnished by the manufacturer.

The blades of Coniflex cutters are relief-ground when manufactured and thus require sharpening only on the front face. Blade-to-blade spacing and cutter diameter, as well as angle and surface finish of the front face, must be closely controlled in sharpening. Special-purpose sharpeners, offered by the gear-machine manufacturer, are recommended to assure consistency in sharpening, as well as longest cutter life.

Machine Cycles. The small generator uses a continuous-index type of cycle in which the work spindle turns continuously in one direction at a constant rate, while the cradle is oscillated in timed relationship to produce the generating roll. The cutters roll into a new tooth space at the start of each generating cycle and skip over some predetermined number of teeth on the return roll. The cradle rotation is produced by a generating cam which actuates the cradle through an adjustable linkage. A series of cams is furnished to cover the required range of rotation.

On the larger machines, the generating train is driven by a cam which permits the use of the type of cycle illustrated in Fig. 20-15. This cycle begins with the work and cradle stationary—the cutters situated somewhat above the center of the generating roll. A cam feeds the gear blank into the cutters, stopping just short of full depth. After this infeed roughing, the work and cradle roll down, rough-generating the tooth space. At the bottom of the roll, the base sets in to the full-depth position and the tooth space is finish-generated during the up-roll. At the top of the roll, the work is withdrawn and the cradle and work roll down to the infeed roughing position. During this brief down-roll, the blank is indexed to the next tooth space.

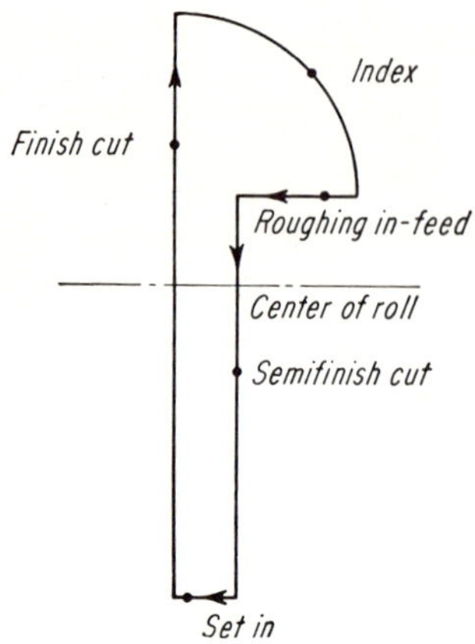

FIG. 20-15. Completing cycle diagram for Coniflex generator.

Required Setup Data. In order to make the machine setups for producing a pair of gears, the operator must be provided with the gear data, a summary of the machine settings, and cutter specifications, and the necessary work holding equipment and cutters. The machine settings can be obtained from the machine manufacturer or can be determined readily by means of a relatively short calculation procedure.

20-9. Straight Bevel Coniflex Grinder (See Art. 20-20—Grinding Considerations). The straight bevel Coniflex grinder is an outgrowth of the interlocking-cutter machines (Art. 20-8), which employ a pair of inclined cutters whose blades interlock in a tooth space. This cutting arrangement is not adaptable to a grinding machine unless the cutters are first rotated apart by one pitch of the generating gear so that the blades no longer interlock. Then the cutter can be replaced by solid grinding wheels as seen in Fig. 20-16. This arrangement requires somewhat more cradle and work roll to complete the generation of the two profiles being ground, but there is no noticeable loss of efficiency, owing to the fast cycle times used in grinding. The two wheels grind the tooth flanks, the fillet surfaces, and the root land to produce a smooth, blended tooth space.

The standard version of this machine, which uses wheels from 10½ to 8″ in diameter, can grind gears as fine as 24 diametral pitch. The machine can be equipped (at time of manufacture) with an additional set of auxiliary wheel spindles and a dresser

Table 20-5. No. 105 Straight Bevel Coniflex Grinder Capacity and Production Data

	Capacity range	
	Min.	Max.
Diam. pitch	48*	3
Cone distance, in	¾	4¼
Face width, in		1⅜
Ratio	1:1	6:1

* Auxiliary wheel spindles required for gears 24 diametral pitch and finer.

Representative Production Rates
(0.003 to 0.0035″ Finishing Stock per Side)

4 diametral pitch miter gear, 32 teeth, sec per tooth	30
10 diametral pitch miter gear, 32 teeth, sec per tooth	20
Average machine setup time, hr	4–6

modification for grinding gears as fine as 48 diametral pitch. These spindles, which are mounted on the standard spindles and driven by the same motors, accommodate wheels from 4 to $2\frac{3}{4}''$ in diameter.

Coniflex Provision. Lengthwise crowning of the tooth surfaces for tooth-bearing localization is obtained on this machine in the same manner in which it is obtained on

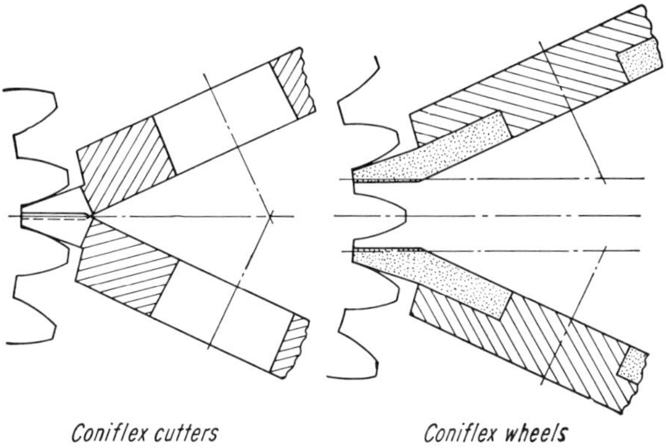

FIG. 20-16. Positioning of Coniflex grinding wheels.

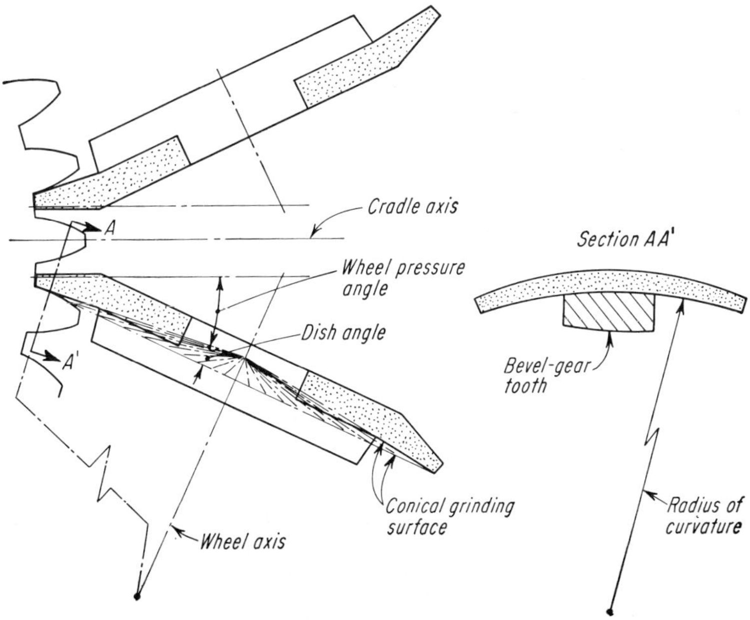

FIG. 20-17. How grinding wheels produce Coniflex teeth.

the interlocking-cutter machines. The wheel surfaces which grind the tooth flanks are disposed at a slight internal angle to the plane of rotation and thus remove more metal at the ends of the teeth than at the center (Fig. 20-17). The amount of this ease-off is a function of the dish angle and the wheel pressure angle. These angles are selected according to the length of contact desired, and the required gear pressure angle is then produced by means of machine settings.

Required Prior Processing. All gears, regardless of diametral pitch, must be cut in a semifinishing operation prior to grinding on this machine. The recommended stock allowance is 0.003″ in tooth bottom, and from 0.003 to 0.005″ on each tooth side. When standard-diameter wheels are used gears with face width up to ½″ can be semifinished on a two-tool generator or one of the larger interlocking-cutter machines. Gears with face widths over ½″ should be semifinished on the No. 104 Coniflex generator so that the root land surfaces will match. The No. 102A Coniflex generator is recommended for semifinishing where the special small-diameter wheels are used.

Machine Cycle. In order to assure uniformity of the gears produced on this machine, an automatic-cycle unit is used to control the entire grinding operation after the work is mounted and aligned with the wheels. This unit contains a drum with a series of cam surfaces which determine the sequence of the cycle events, including automatic wheel dressing, work feed, and generating roll. A typical cycle consists of two dressing operations—one roughing and one finishing—and five grinding passes on all teeth. The generating motion stops each time a grinding pass is completed on all teeth and begins again after the necessary wheel dressing and feed (work advance) are completed for the next pass. The variable feed is produced by a disk cam which is controlled from the automatic-cycle drum. As the wheel diameter is decreased by dressing, the wheel motor speed is automatically controlled to maintain a constant speed on the grinding surfaces.

Required Setup Data. The operator requires a summary of machine settings and a recommended grinding sequence. These can be obtained from the machine manufacturer or can be calculated by the user's engineering department, following detailed calculating instructions furnished by the manufacturer.

20-10. Revacycle Machine. The Revacycle process is the fastest method of cutting straight bevel gears of commercial quality. It is used primarily for large-volume production of gears for such applications as automotive-type differentials and agricultural implements. The Revacycle machine is usually furnished with an automatic loading and unloading unit. Although initial tooling cost is greater than for other types of straight-bevel-gear machines, the extremely fast production rate results in lowest over-all cost for large-scale production. Revacycle-cut gears are truly conjugate but are not interchangeable with gears produced by other methods. Revacycle tooth proportions differ from those of other straight bevel gears, and thus gear blanks must be designed specifically for this process.

Cutter and Completing Machine Cycle. Most Revacycle-cut gears are produced with one completing operation, using a large-diameter (21 or 25″) cutter, continuously rotating in a horizontal plane at a uniform rate. The cutter blades, which extend radially outward from the cutter head, have concave edges which produce convex profiles on the gear teeth. During the cutting operation the workpiece is held motionless while the cutter is moved by means of a cam in a straight line across the face of the gear and substantially parallel to its root line. This motion makes it possible to produce a straight tooth bottom, while the desired tooth shape is produced by the combined effect of the motion of the cutter and the shapes of the cutter blades. There is no depthwise feed of the cutter into the work, the effective feed being obtained by making each cutter blade progressively longer than the one before it.

The completing cutter contains three kinds of blades: roughing, semifinishing, and finishing. One revolution of the cutter completes each tooth space, and the work is indexed in the gap between the last finishing blade and first roughing blade.

Figure 20-18*a* shows the Revacycle cutter in position at the beginning of the cut. As the cutter rotates counterclockwise, blades of gradually increasing length contact the workpiece until the root line of the tooth is reached. Figure 20-18*b* shows a transverse view of the tooth space as roughed to full depth. Each chip extends the full width of the slot, except for the amount of stock left for finishing. Figure 20-18*c* is an axial section of the same tooth showing that the roughing chips extend the full length of the tooth. During the first part of the roughing the cutter moves from *A* to *B* (Fig. 20-18*a*). It then remains stationary at *B* until full depth is reached.

The roughed tooth does not have the proper taper. It is substantially correct along the diagonal line in Fig. 20-18*c*, but the portions of tooth surface lying to the

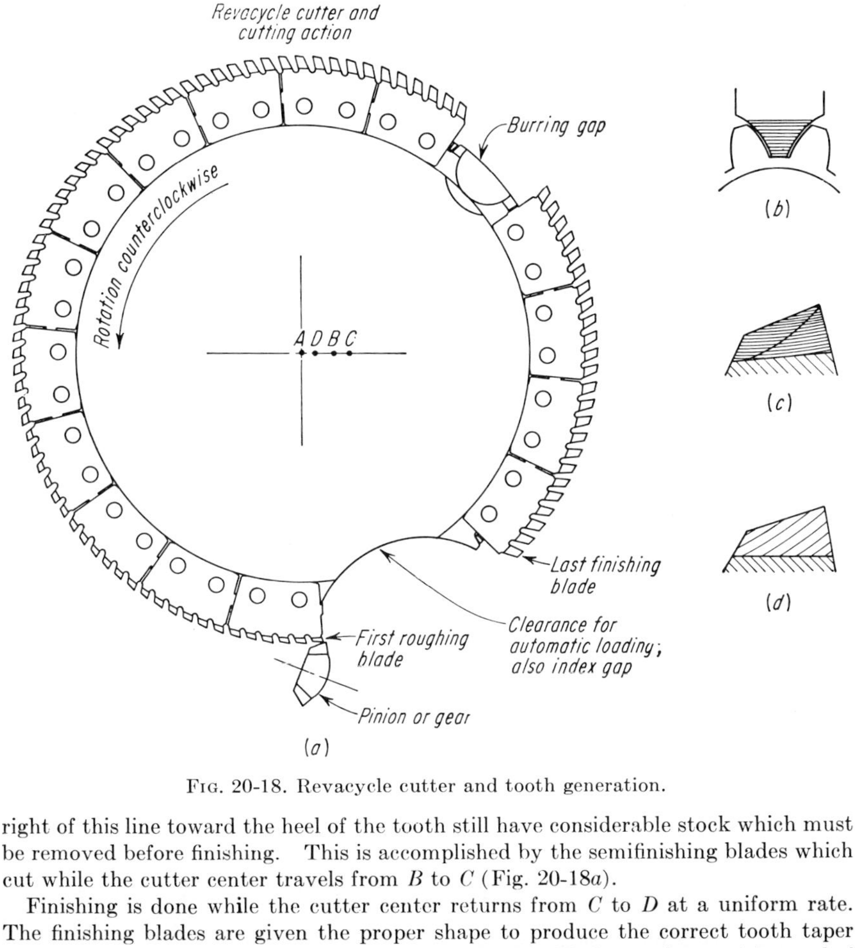

Fig. 20-18. Revacycle cutter and tooth generation.

right of this line toward the heel of the tooth still have considerable stock which must be removed before finishing. This is accomplished by the semifinishing blades which cut while the cutter center travels from *B* to *C* (Fig. 20-18*a*).

Finishing is done while the cutter center returns from *C* to *D* at a uniform rate. The finishing blades are given the proper shape to produce the correct tooth taper and proper profile at every point along the tooth. The blades cutting at either end of the tooth space are made slightly wider than necessary for correct taper in order to ease off the ends of the tooth and produce a crowning which results in localized tooth bearing.

The resulting finished tooth surface is generated, being made up of a series of inclined cuts similar to those shown in Fig. 20-18*d*. In general, a different cutter is required for each different gear specification.

Two-cut Operation. For the small percentage of Revacycle work which is too deep for completing in one cut (Table 20-6), separate roughing and finishing operations are used. Separate cutters and separate machine setups are required for each operation. The cutters and machine cycle are similar to those described for completing, with some slight differences. The roughing cutter has no semifinishing or finishing blades, and there is no translation of the cutter in the roughing operation. The finishing cutter, however, translates as in the completing operation, with semifinishing blades cutting on the first translation and finishing blades cutting during the return stroke.

Table 20-6. No. 109 Revacycle Machine, Capacity and Production Data

Capacity

Max. cone distance, in.	5
Max. ratio	5:1

	21″ dia. cutter	25″ dia. cutter
Max. diam. pitch	5	$3\frac{1}{2}$
Max. root face width, in.	1.000	1.250
Max. whole depth for completing in one cut	0.420″	0.500″
Max. whole depth for two-cut operation	0.585″	0.650″

Production Rates

Average cutting time per tooth per operation AISI 4620 material, 160 to 190 BHN

21″ cutter	2.65 sec
25″ cutter	2.90 sec
Average setup time (exclusive of loader setup)	5–6 hr

Required Setup Data. In order to set up for Revacycle cutting, the operator requires a summary, or list, of cutter and cam specifications and machine settings. These data have been calculated and developed for standard series of ratios which cover nearly all usual automotive differential combinations and are available from the machine manufacturer. For other ratios, however, a special calculation and development must be made by the machine manufacturer.

Automatic Deburring. A deburring attachment is provided so that burrs may be removed during the actual cutting of the gear teeth. The attachment is designed to remove the profile and root burrs from the heel of the teeth. A gap in the cutter between the semifinishing and finishing blades (see Fig. 20-18*a*) provides the space for mounting the burring tools. The shape of the burring tool is determined by the design of the gear, and generally the shape of the tool must be changed for each different gear specification.

Cutter Maintenance. Revacycle cutter blades are relief-ground when manufactured and thus require sharpening on the front face only. Blade-to-blade spacing, the angle of the plane of the front face, and surface finish of the front face must all be closely controlled in sharpening. In addition, when new segments are assembled in heads, cleanliness of assembly, accuracy of position of segment-locating keys, and close control of segment-holding bolt tension are all necessities for proper cutting results. The gear-machine manufacturer recommends use of a special cutter-assembly stand and a special-purpose automatic wet-type sharpener for all Revacycle production units.

20-11. Straight-bevel-gear Planers. The bevel-gear planer is the original type of machine for producing cut straight bevel gears. It is still used to cut large gears, and, because it can be set up easily with a minimum of tooling, it is very useful for repair work where a wide variety of coarse-pitch gears is encountered. Of the two machines listed in Table 20-7, the No. 154 planer is the more frequently used.

The workpiece is mounted in a work spindle which is held stationary except for indexing, and the tooth surfaces are produced by a forming process using a tool which

Table 20-7. Straight Bevel Planers, Capacity and Production Data

Capacity range	No. 154	No. 192*
Max. cone distance, in.	38¼	101
Max. diam. (1:1 ratio), in.	54	140
Max. face width, in.	14	34
Ratio	1:1–8:1	1:1–5:1

* This is a combination-type machine which also cuts spur gears.

Representative Production Rates, No. 154 Planer, AISI 4140 or 4340 Material, 269 to 310 BHN

Gear data			Approx. total cutting time, hr per pair
Ratio	Depth, in.	Face width, in.	
27:27	1.75	5	15.4
16:74	1.75	8	32.5
28:28	2.96	13	57

is reciprocated on a guided tool arm. Since the surface of a tooth (Fig. 20-19) is made up of converging straight-line elements, it can be swept out by a line which is pivoted at an apex and guided along the necessary profile at the heel. Figure 20-19 shows how this is accomplished in the machine with the rounded corner of a tool which is stroked along a line and guided by a former.

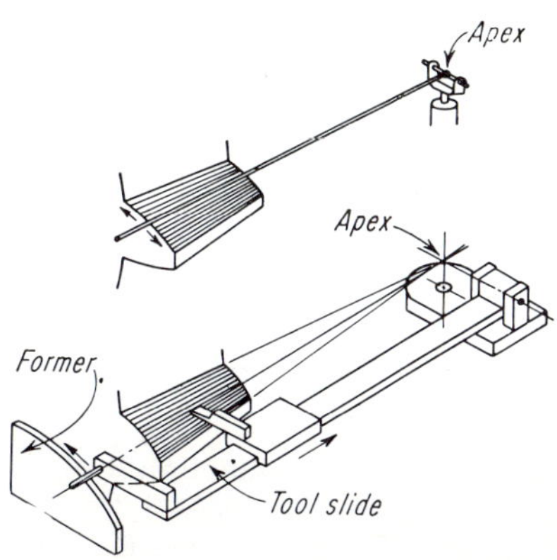

FIG. 20-19. Principle of planing method.

Formers. Two formers are required for each gear—one for each side of the tooth. Theoretically a pair of formers is required for each gear ratio, but in practice a pair is designed to cover a small range. A set of 25 pairs of formers can be furnished to cover all 90° shaft angle ratios from 1:1 to 8:1, for either 14½ or 20° pressure angle. This system of formers is based on the use of equal-addendum tooth proportions for all ratios. (Use of the current Gleason long- and short-addendum system, as applied to generated gears, would require a special former for each different tooth specification.)

Coniflex Provision. The No. 154 planer is equipped with a device for cutting Coniflex gears. The tooth-bearing localization is produced by a slight motion of the tool arm as the tool is stroked along the tooth. The length and position of the tooth contact can be determined and controlled by the machine operator.

Cutting Sequence. Normally five or six cutting operations are required to produce a finished gear. The first is a roughing operation which is made by feeding a slotting tool or a corrugated V tool to full depth. If the first cut is made with a slotting tool, a second cut is required with a V tool or a corrugated V tool. These cuts are made with the form roller resting on a straight guide. After roughing, the tooth formers are installed and the teeth finished by making two cuts on each side.

Tools. Slotting tools are specified by point width and depth of cut; corrugated roughing tools by point width, depth of cut, and pressure angle; finishing tools by point width only.

Required Setup Data. The operator needs only to know the tooth proportions of the gear to be cut, plus the former list and index gear list furnished with the machine, to set up the planer.

PLANING GENERATORS—FOR STRAIGHT BEVEL, ZEROL BEVEL, SPIRAL BEVEL, AND HYPOID GEARS

Planing generators are unique among bevel- and hypoid-gear machines, in that they cut both straight-tooth and curved-tooth gears. Their use is generally confined to gears of large size. This section describes the machine operating cycles, required setup data, tooling requirements, and capacity and production data of these machines.

20-12. Planing Generators. As the name implies, these machines employ planer-type tools but at the same time produce gears with generated tooth surfaces. They are used for generating bevel and hypoid gears which are larger than about 35″ in diameter or coarser than 1½ diametral pitch. The standard machines can cut spiral, Zerol, and straight bevel gears and, when equipped with the proper work head, can cut hypoids as well. Table 20-8 gives an outline of the capacity of these machines; more detailed data are available from the manufacturer.

Table 20-8. Planing Generators, Capacity and Production Data

Capacity range	No. 70	No. 91
Cone distance, in.	9–47	9–47
Max. dia. at 90° pitch angle, in.	72	91
At 45° pitch angle, in.	68	84
Max. face width, in.	10	12
Coarsest diam. pitch	¾	¾
Extreme ratio	Any practical ratio	

Representative Production Rates, AISI 4140 or 4340 Material, 269 to 310 BHN

Gear data			Cutting time, hr per pair
Ratio	Diam. pitch	Face width, in.	
60:60	1.4	9	45
18:130	2	5½	25
34:75	1.25	7½	34.5

The finishing tools have straight cutting edges and are used singly to represent the side of a generating gear tooth. They are mounted in a reciprocating slide which is carried on the face of the cradle and connected to a rotating crank by a connecting rod (see Fig. 20-20). The required tooth profiles are obtained on the work by the usual method of rolling the work with the generating gear. The lengthwise shape of the teeth is determined by a unique combination of several motions. These are (1) the stroke of the tool, (2) the continuous uniform rotation of the work, and (3) an angular oscillation of the cradle, produced by an eccentric. The latter motion is needed to modify the effect which the first two motions have on the lengthwise shape of the

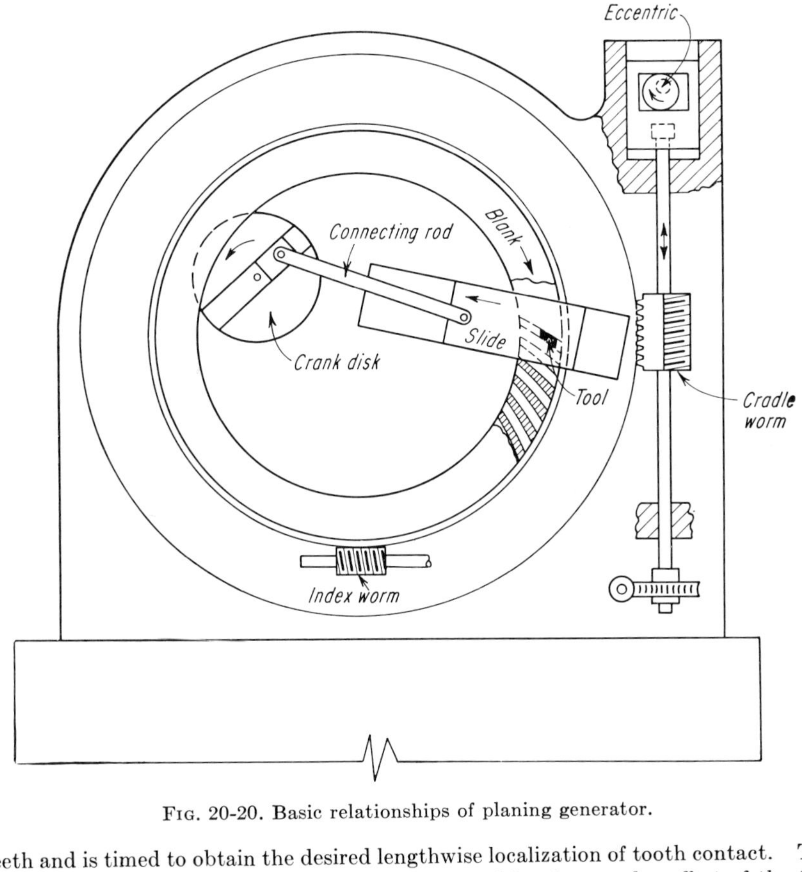

FIG. 20-20. Basic relationships of planing generator.

teeth and is timed to obtain the desired lengthwise localization of tooth contact. The direction, or spiral angle, of the teeth is determined by the angular offset of the tool slide from the cradle axis.

The continuous work rotation is principally for indexing purposes. In effect, the tool makes a cut on all the teeth in succession in one generating position, and then the cradle and work roll together a slight distance before a cut is taken again on all the teeth in the new generating position. Actually, however, the rolling is continuous and occurs gradually until all teeth are completely generated on the last pass around the gear. As a consequence of this continuous cutting action, accurate tooth spacing and uniform tooth shape are obtained, if the blank is reasonably homogeneous.

Cutting Sequence. Several separate cutting operations are required to complete a part, the number depending upon tooth depth and profile shape. Flat gears generally are roughed without roll, using first a corrugated tool, followed by a cut with a V roughing tool. This is followed by at least two side-cutting operations on each side of the tooth, including the generating cuts with roll. Pinions are also roughed first with a corrugated tool and then a V rougher, but with roll. The roughing is followed by size-cutting, semifinishing, and finishing cuts on each side of the tooth.

Tools. The tools are of simple form and consequently are inexpensive. Corrugated roughing tools are universally furnished with $14\frac{1}{2}°$ pressure angle, regardless of the pressure angle of the part being cut, but point width and depth of cut must be specified. Both V roughing tools and finishing tools are specified by point width, pressure angle, and depth of cut.

Required Setup Data. The operator requires a summary, which lists calculated machine settings for both roughing and finishing and also specifies limit point widths. The summary can be obtained directly from the machine manufacturer or can be prepared by the user's engineering department, following calculation instructions furnished by the manufacturer.

FACE-MILL-TYPE CUTTING MACHINES AND GRINDERS—SPIRAL BEVEL, ZEROL BEVEL, AND HYPOID GEARS

With the exception of the large gears cut on planing generators (Art. 20-12), spiral bevel, Zerol bevel, and hypoid gears are cut with face-mill cutters or ground with comparable cup-type wheels. A wide variety of (1) cutters, (2) methods, and (3) machines is available to cover the entire range of this work—from the smallest instrument-type gears up to approximately 35″ diameter.

This section explains these three factors in detail, so that the best choice can be made for producing a specific part at a given production level.

After the type of cutter, cutting or grinding method, and machines have been chosen, it is necessary to provide the operator with a summary, which lists in detail the cutter or wheel specifications, machines to be used, and their settings. This summary can be obtained from the machine manufacturer or can be prepared by the user's engineering department, following detailed calculating instructions furnished by the manufacturer.

20-13. Face-mill Cutters; Types and Uses. Face-mill cutters have three general types of blades, integral, segmental, and inserted. They are used for both rough and finish cutting.

Integral-blade or solid cutters are made from a single piece of tool steel. That is, the blades are part of the cutter head. Usually these cutters are used for fine-pitch work and are 6″ or less in diameter.

Segmental cutters are made up of segments which have two, three, or four blades per segment. The segments are bolted to the cutter head around the periphery.

Inserted-blade cutters have individual blades bolted to slotted heads.

The three types of face-mill cutters are shown in Fig. 20-21. Inserted-blade cutters usually have parallels for changing diameter and adjusting wedges for truing as shown. Some designs, however, such as the Helixform cutter, have neither parallels nor wedges, while other designs, such as the Triplex®* rougher, have only parallels.

For finishing, each of the three types of cutters can be furnished with all outside-cutting blades, all inside-cutting blades, or alternate outside and inside blades. Roughing cutters may have either alternate inside and outside blades or may be the Triplex type, which has end- or bottom-cutting blades alternately spaced with inside and outside blades.

Table 20-9 shows the various categories and uses of face-mill cutters of the segmental and inserted-blade types. Some of the preceding categories merit special explanation.

* Registered trade-mark of the Gleason Works, Rochester, N.Y.

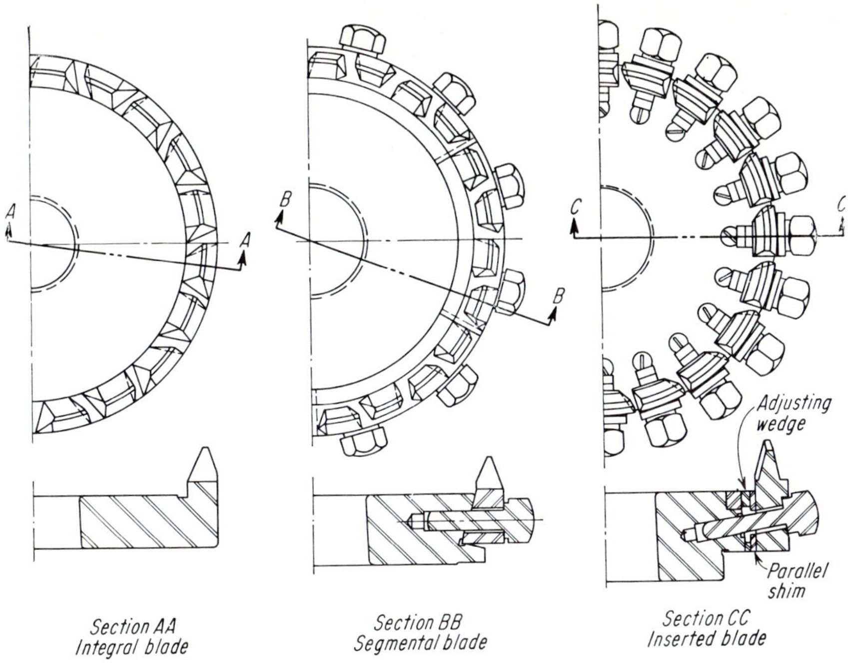

FIG. 20-21. The three general types of face-mill cutter.

Table 20-9. Types and Uses of Face-mill Cutters

Type of cutter	Type of blade	Kind of blade		Adjusting wedges for truing	Parallels for diameter changes	Used for
		Inside	Outside			
Helixform.	Inserted	Both		No	No	Nongenerated gear finishing
Single Cycle	Segmental	Both		No	No	Nongenerated gear finishing
Cyclex....	Segmental	Both		No	No	Nongenerated gear completing
Triplex....	Inserted	Both plus bottom cutting blades		No	Yes	Gear roughing
Inside blade...	Inserted	All	None	Yes	Yes	Pinion finishing
Outside blade...	Inserted	None	All	Yes	Yes	Pinion finishing
Alternate blade	Inserted	Both		Yes	Yes	Gear and pinion finishing and roughing, also jobbing and Duplex* processes

REMARKS: Alternate-blade, inside-blade, and outside-blade Gleason finishing cutters having special hardened heads are called Hardac® cutters. Alternate-blade cutters designed for roughing, or Triplex cutters, using special hardened heads, are referred to by the trade name Roughac®. Both are registered trade-marks of the Gleason Works, Rochester, N.Y.

* Fine-pitch gears, completed by the Duplex method, use integral-blade cutters.

Cutters Used for Nongenerated Gears. Helixform, Single Cycle, and Cyclex®* cutters are used for nongenerated gears. The gears produced are referred to as nongenerated because the teeth are not enveloped by successive positions of the cutter blades (see Art. 20-3, The Basic Nongenerator, and Art 20-17).

Helixform Gear-finishing Cutter. This inserted-alternate-blade cutter has eight blades spaced at 36° intervals with the remaining space being used as the index gap. The first two inside and first two outside blades shown in Fig. 20-22 are semifinishing blades, while the last four blades are finishing blades. The blades are stepped in both a radial and axial direction to take a precalculated chip load, and one cutter revolution finishes each tooth slot. The semifinishing blades remove from 0.004 to 0.007″ total stock depending on the design and tooth size of the particular gear. The finishing blades remove a total of 0.003″ stock in steps of 0.002 and 0.001″,

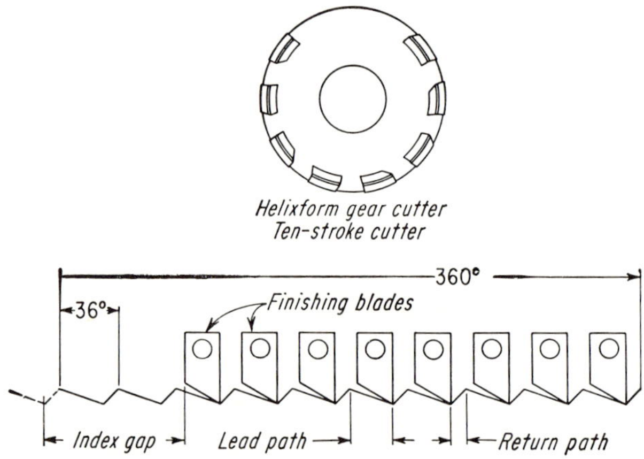

FIG. 20-22. Helixform gear-cutter diagram.

respectively. Each blade of the Helixform cutter is positioned so that only one blade is cutting at any given time.

Single Cycle Gear-finishing Cutter. This cutter is a segmental alternate-blade cutter with from 10 to 16 blades, depending on the cutter size. A gap between the last finishing blade and first semifinishing blade provides the space for indexing the work (see Fig. 20-5). Blades are stepped in both a radial and axial direction to take a precalculated chip load. Figure 20-5 shows the blade arrangement for a 9.0″ mean diameter segmental Single Cycle cutter. Between 0.015 and 0.030″ total finishing stock is removed by Single Cycle cutters, depending on the tooth size of the gear. The finishing blades are spaced so that there is never more than one blade cutting in the tooth slot at any given time.

Cyclex Completing Cutter. The Cyclex cutter is an alternate-blade segmental completing cutter that roughs and finishes both sides of a tooth slot in one operation. The cutter is composed of two sections, roughing and finishing, with the two finishing blades set lower than the highest roughing blade so that they do not engage the work during the roughing portion of the cutting cycle.

There are two blade arrangements for the roughing section as shown in Fig. 20-23. The Cyclex cutter for use on the Formate machine has all the roughing blades set at the same height, and corresponding points on the cutting edges are at the same radius from the center of the cutter. The Cyclex cutter for use on a generator specially

* Registered trade-mark of the Gleason Works, Rochester, N.Y.

arranged for Formate cutting differs from that described above in that the roughing blades are stepped in an axial direction to take a precalculated chip load.

The two finishing blades on these cutters are set at the same height but with space between them so that they are not in engagement with the workpiece at the same time. They are also separated from the roughing blades so that a roughing and finishing blade will not engage the gear blank at the same time.

Triplex Gear-roughing Cutter. The Triplex cutter has bottom- or end-cutting blades alternately spaced with inside and outside cutting blades. The end-cutting blades are generally set 0.010″ higher than the side-cutting blades to relieve the amount of end cutting necessary in the finishing operation. Triplex cutters are most

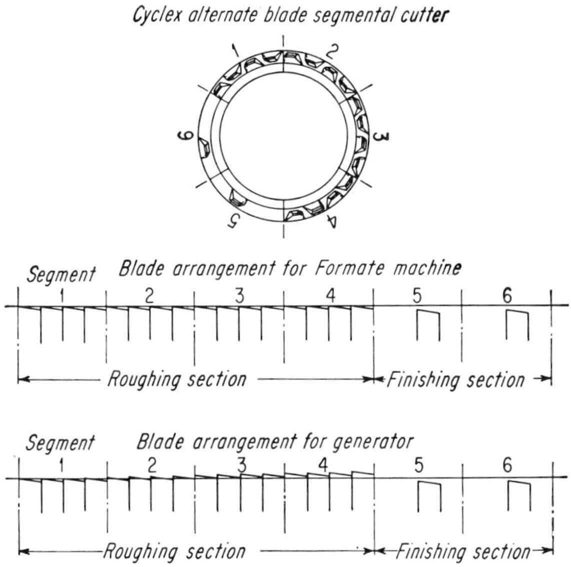

FIG. 20-23. Cyclex completing cutter diagram.

generally used when the slot width of the gear tooth in the root plane is 0.065″ or greater.

Hardac Finishing Cutter. These finishing cutters may be one of two types: alternate inside and outside blades, or cutters with all inside or all outside blades. The blades of these cutters are mounted in hardened and ground heads that have adjusting wedges for truing. See Fig. 20-21 for a cross section of this type of cutter.

Maintenance. All face-mill cutter blades are sharpened on the front face so that the primary cutting edge lies in a plane containing the cutter axis. The blades are relieved in such a manner as to maintain pressure angle and diameter with successive sharpenings. Blade-to-blade spacing and angle of front face must be closely controlled in sharpening, specific tolerances varying with type and size of cutter. The finish of the sharpened surface should be within a limit of 10 microinches.

The gear-machine manufacturer offers special-purpose sharpeners to maintain the required tolerances. Several models are available, varying as to size, range of cutters accommodated, and degree of automatic operation.

A high degree of precision is essential in the assembly of segments, or blades, in heads, for purposes of either refilling used cutters or changing cutter specifications. Duplication of blade angles and concentricity of blades must be held to close tolerances—the latter within a fraction of a ten-thousandth of an inch on some finishing cutters, for example. This necessitates care, insistence on cleanliness, and use of torque wrenches to maintain uniform cutter bolt tension.

Stand-type fixtures are offered by the gear-machine manufacturer to facilitate cutter reassembly. These fixtures include the equipment necessary for measuring trueness of angle and concentricity of blades.

20-14. Face-mill Cutters; Specifications. The important cutter specifications which directly affect the tooth form of the gears are the diameter, pressure angle, and point width. Hand of cutter must also be specified, depending upon cutting conditions.

Diameter. The diameter referred to in a cutter having alternate inside- and outside-cutting blades is the average diameter (see Fig. 20-24). Blade-point diameter is specified for cutters having all inside or all outside blades.

Table 20-10. Range Covered by Available Face-mill Cutter Diameter

Cutter dia.	No. of blades	Approx. mean cone distance, in.	Approx. max. face width, in.	Max. standard whole depth,* in.
1.1	8	0.5–0.75	0.25	0.125
1.5	12	0.75–1.0	0.3125	0.1875
2.0	16	1.0–1.5	0.4375	0.1875
2.75	20	1.25–1.75	0.5625	0.250
3.5	20	1.5–2.75	0.750	0.350
4.5	20	2.5–3.0	1.0	0.375
6.0	16, 20	2.75–3.5	1.25	0.375
7.5	16, 20	3.5–4.0	1.5	0.50
9.0	16, 20, 24	4.0–5.25	1.875	0.5625
12.0	12, 20, 28, 32	5.25–7.5	2.5	0.750
16.0	24, 40	7.5–15.0	3.5	0.750
18.0	28	7.5–15.0	4.0	0.750
18.0	24	7.5–15.0	4.0	14½° 1.3125 17½° 1.125 20° 1.000

* Can be exceeded by use of special-length blades.

The cutter diameter for a particular pair of gears is chosen from the 12 available standard diameters, by considering cone distance, face width, whole depth of tooth, and spiral angle. Table 20-10 is given as a general guide for cutter selection. A good rule is to select a cutter diameter equal to or slightly less than twice the cone distance. Generally, if there is a choice between two diameters, the smaller is preferred.

Pressure Angle. Generally the nominal pressure angle of the cutter blade is the same as the normal pressure angle of the gear to be cut. The standard is 20°, although 16, 22½, and 25° are also commonly used. Since the cutter-blade tips travel along the root cone instead of the pitch cone, the inside and outside blade angles must be made respectively greater and less than the normal pressure angle of the gear teeth. This difference between cutter-blade angle and normal pressure angle is denoted by a system of cutter numbers.

Point Width. For cutters having alternate inside and outside blades the point width indicates the distance between the inside and outside cutting edges at the tip of the blades as shown in Fig. 20-24. For cutters having all outside or all inside blades the blade point indicates the distance between the cutting edge and clearance side at the tip of the blade.

Alternate-blade cutters must have blade points less than the cutter point width in order to provide clearance for chips.

Toprem®* Feature. Lapping a pair of gears above 2:1 ratio may produce interference between the flank of the pinion tooth and the top of the gear tooth. Toprem blades in the pinion-finishing cutters prevent this interference.

Near the tip of a Toprem blade there is a small straight segment of the blade profile where the pressure angle is slightly less than the pressure angle of the rest of the blade. The amount of this modification can be selected from the following general rule: depth of modification equals one-third the gear addendum, plus the clearance between the top of the gear tooth and the bottom of the pinion-tooth space at the large end.

Hand of Cutter. Another specification for cutters relates to the direction in which the cutter rotates when it is cutting. A right-hand cutter is one which rotates clockwise when viewed from the back, while a left-hand cutter rotates counterclockwise.

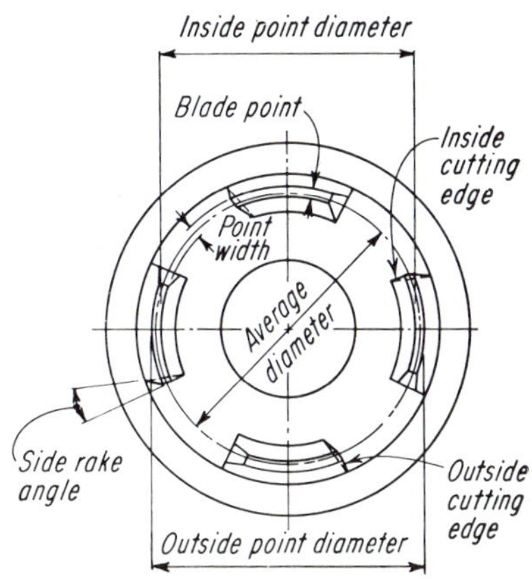

FIG. 20-24. Face-mill-cutter specifications.

When the hand of cutter is the same as the hand of the spiral of the gear or pinion, the direction of cut is from the small end to the large end of the teeth. Therefore, flat gears (greater than 3:1 ratio) which are roughed without generation use cutters having the same hand of spiral as the gear, so that the thrust of the cut forces the work against the chuck.

On generating cuts, however, another factor must be taken into account. When the machine cradle is rotating counterclockwise as viewed from the front of the machine, the machine is said to be on the up-roll; clockwise rotation is called down-roll. Right-hand cutters climb cut on the up-roll, whereas left-hand cutters climb cut on the down-roll. Better finish and cutter life are obtained by climb cutting during both the roughing and finishing operations, and the hand of cutter is usually specified accordingly.

20-15. General Classification of Methods. There are four basic cutting or grinding methods, namely, spread blade, single side, fixed setting, and single setting.

The rotating cutting edges of a face-mill cutter generate an imaginary cutter surface; when grinding is the finishing operation, these cutter surfaces may be described by grinding wheels. In general, the methods defined below apply to both cutting and grinding.

Spread-Blade Method. This refers to the process where the member being produced is generated with a circular face-mill cutter with alternate inside and outside blades which cut the tooth surfaces on both sides of a tooth space simultaneously. The width of the tooth space is controlled by the point width of the cutter. Uses of the Spread-Blade method are:

1. For the gear member of Formate or Helixform pairs
2. With the jobbing system—for the gear member
3. For either or both members of generated pairs
4. For roughing

Single-Side Method. This is the process where the member being produced is finished by a circular face-mill cutter with alternate inside and outside blades which cut the tooth surface on each side of a tooth space in separate operations with inde-

* Registered trade-mark of the Gleason Works, Rochester, N.Y.

pendent machine settings. The cutter-point width is considerably less than the space width of the finished gear. The width of the tooth space is controlled by rotating the blank about its axis through a measured distance, or by means of a stock-dividing gage if sufficient production is involved. This method is used for pinion finishing and sometimes for the gear member of generated pairs.

Fixed-Setting Method. This refers to the process where the member being produced is finished by two circular face-mill cutters—one with inside blades only for cutting the convex side of the tooth, the other with outside blades only for cutting the concave side of the tooth. The two sides of the tooth are produced separately with two entirely different machine setups. For volume production, pairs of machines are used—one machine of the pair for one side of the tooth, the other machine for the other side of the tooth. Width of the tooth space is usually controlled with the stock-dividing gage. The term *Fixed-Setting* is applied to the finish cutting of pinions, only, which may be pinions of a generated pair or Formate or Helixform pinions.

Single-Setting Method. This is a variation of the Spread-Blade method and is used when available cutters have smaller point widths than required for spread-blade cutting. Both sides are cut with the same machine settings, and the blank is rotated about its axis in either or both directions to remove the amount of stock necessary to produce the correct tooth thickness. After the first cut, only one blade at a time is cutting in the tooth slot. This method is used with the Unitool* system.

Combinations of cutting methods, calculating methods, and special machining procedures make up the various *gear-cutting processes* that are applicable to Helixform, Formate, and generated pairs. In general, all processes are theoretically adaptable to grinding.

20-16. Rough Cutting. The first roughing cut on all work is a spread-blade operation. For most work, this is the only roughing operation. However, it is occasionally desirable to perform a second roughing cut on a pinion, in order to conform more closely to the finished taper of the tooth. If this is done, the second cut is performed with the same cutter, but with a set-over of the machine settings. This second cut is therefore a single-side operation. Current generating-type roughing machines include provisions for automatically changing these settings between up-roll and down-roll of the cradle, so that the two cuts are performed in one machine operation.

Generating roll is omitted when the pitch angle of the gear to be roughed is approximately 65° or greater. Gears having pitch angles less than 65°, and all pinions, are roughed with generating roll.

20-17. Processes for Nongenerated Gears. The gear member of spiral bevel, Zerol bevel, or hypoid pairs has relatively little profile curvature when the ratio is high, and a substantial gain in productivity can be made by cutting or grinding the gear without generating roll. Such gears are referred to as *nongenerated* gears, the classification including both Helixform and Formate gears. In volume production generation of the gear is usually omitted for spiral bevel pairs of ratio 3:1 and higher, and for hypoid pairs having ratios of 2.25 and higher. Helixform and Formate *pairs* consist of a nongenerated gear produced by the named method, and a pinion with teeth generated to match. Most automotive rear-axle-drive gears are made by these processes.

Figure 20-25 illustrates the basic difference between tooth-profile formation by generation and nongeneration. In the former type of operation, the cutter contacts the tooth profile at only one point at any one time, whereas in the latter, the cutter makes contact over the entire profile at once.

* Trade-mark of the Gleason Works, Rochester, N.Y.

Helixform. Gears called by this trade name are of a nongenerated type having helicoidal tooth surfaces cut by the Spread-Blade method. Helixform pinions are generated, generally using the Fixed-Setting method, to match the Helixform gear. The process can be applied to spiral bevel, hypoid, and Zerol gears.

The advantages offered by this process are ease of tooth-bearing development, improved quality and adjustability, and speed of production, which is comparable with the Formate process. Helixform gears, like Formate gears, are straight-sided in the normal section. As each blade of the Helixform cutter passes through a tooth space, the cutter is advanced axially and then returned before entry of the subsequent blade. The cutter thus has combined rotational and reciprocating motion, and the work is indexed in a gap between the blades.

In Fig. 20-26 the gear-cutter axis has been placed perpendicular to the gear-face plane. This is possible because, as the cutter rotates, it is given an axial advance so

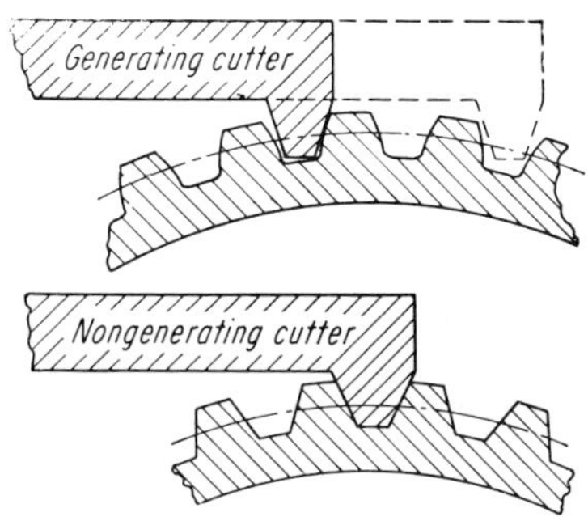

FIG. 20-25. Difference between generating and nongenerating.

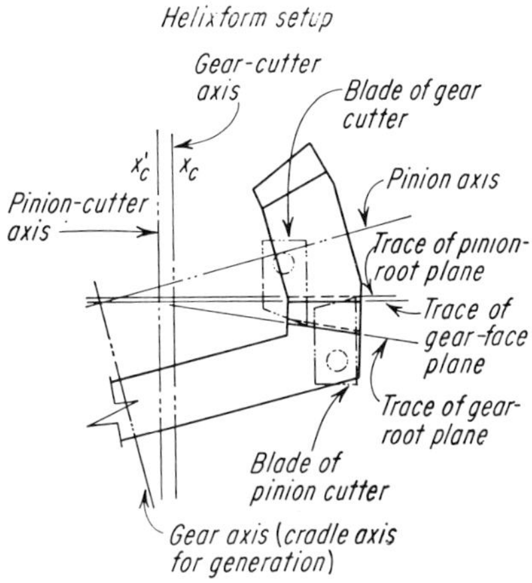

FIG. 20-26. Relationships of cutters and workpieces in Helixform cutting.

that the path of the cutter-blade tips is tangent to the gear-root plane. Since gears are designed with constant clearance, the gear-face plane and pinion-root plane are parallel. If the cutter-blade angles are made equal to the pinion-root plane pressure angles, the gear-cutter axis, which is perpendicular to both the gear-face plane and pinion-root plane, is parallel to the pinion-cutter axis.

Helixform gears are roughed on a companion machine to the Helixform gear-finishing machine by a method similar to Formate roughing. The cutter is fed in the direction of the cutter axis until the full-depth position is reached. The difference between the conical surface of the roughed gear and the helical surface of the finished gear is small enough to be corrected by the semifinishing blades of the Helixform gear-finishing cutter.

Formate. The gears to which this trade-mark is applied are nongenerated gears produced with face-mill cutters. They have tooth surfaces which are surfaces of revolution and are usually straight in the normal section. The Formate process can be applied to hypoid, spiral bevel, and Zerol gears.

Because of tooth-design considerations, it can be used only on flat gears having pitch angles greater than approximately 70°.

All Formate gears are cut by the Spread-Blade method, except Unitool gears, which are cut by the Single-Setting method.

Formate pinions are generated to mate with Formate gears. For most Formate pinions, the Fixed-Setting method is used; the Unitool process, however, uses the Single-Side method, while duplex helical pinions are cut by the Spread-Blade method.

Figure 20-27 shows the positions for the gear- and pinion-cutter axes in the Formate setup. If the cone angles of the gear cutter are made equal to the root plane pressure angles of the gear, the gear-cutter axis x_c is perpendicular to the gear-root plane. In this position, the tips of the gear-cutter blades always lie in the gear-root plane as the cutter rotates. The conical surfaces of the gear teeth are thus defined by the position of the gear cutter and its cone angles.

The pinion-root plane is inclined to the gear-root plane by the angles. To produce teeth having depthwise taper, the peripheral direction of the pinion-cutter blade tips must be tangent to the pinion-root plane near the middle of the face, as shown.

Use of Single Cycle Cutter (Art. 20-13). The use of this cutter requires that gears be roughed without generation before finishing by feeding the cutter in an axial direction into the full-depth position. The conventional alternate-blade face-mill roughing cutter may be used or, more commonly, a Triplex roughing cutter (Art. 20-12), which contains alternate end- and side-cutting blades.

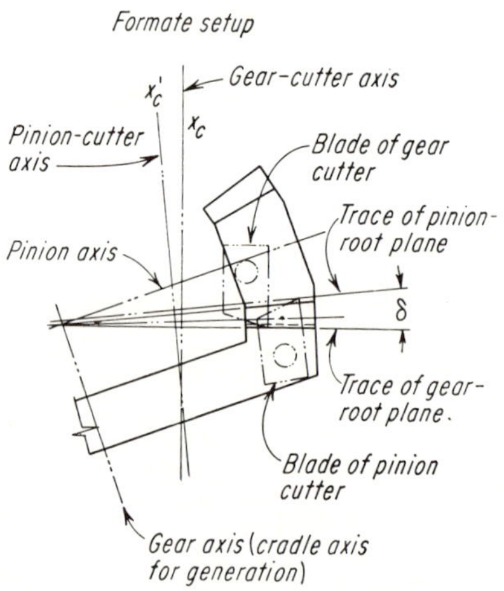

FIG. 20-27. Relationships of cutters and workpieces in Formate cutting.

Gear finishing is accomplished with a single revolution of the Single Cycle cutter in each tooth space (Fig. 20-5).

Use of Cyclex Cutter (Art. 20-12). This cutter provides a means of completing, both roughing and finishing, in one cut from the solid blank. It is used where the quantity of production does not warrant separate roughing and Single Cycle finishing equipment. The major advantages of Formate gears in regard to accuracy of tooth spacing, quality of finish, and fast production rates are closely maintained by the Cyclex method.

During the roughing portion of the cutting cycle, the cutter makes a number of turns, and feed is provided by a cam which delivers uniform motion in the direction of the cutter axis.

After the last roughing blade has passed through the tooth slot, and as full depth is reached, the cutter is rapidly advanced axially by the cutter feed cam to a fixed position. The cutter is held in this advanced position until both finishing blades pass through the tooth slot, finishing the tooth shape. The cutter is then rapidly withdrawn to prevent the roughing blades from contacting the work. Additional withdraw of the cutter provides the necessary clearance to index to the next tooth.

Use of Alternate-blade Cutters. Fine-pitch gears of sets produced by the Formate helical Duplex process may be cut by the Spread-Blade method with simple alternate-blade cutters on generators having tilting cutter spindles. A special setup places the cutter in a position for correct contact when the gear and cradle are coaxial. The cradle and gear are rotated together as one unit during infeed to whole depth. Details and advantages of the Duplex helical method are given below (Art. 20-18).

Formate gears cut by the Unitool process using alternate-blade cutters depend upon the Single-Setting method. The gear member is cut on a generator arranged for Formate cutting, that is, with both the cradle and work spindle held fixed by clamps during the cut. The work-spindle clamp is released automatically for indexing. The Unitool cutter is fed to whole depth with no relative generating motion so that the surface on the tooth is a counterpart of the surface of revolution of the cutter. See Table 20-15 correlating cutters, methods, processes, and machines for Helixform and Formate pairs.

20-18. Processes Used for Either Generation or Nongeneration. There are two processes which can be used for either generation or nongeneration; Duplex Spread-Blade and Unitool.

Duplex Spread-Blade Method. This is a method in which both the gear and pinion are cut Spread-Blade. The gear may or may not be generated, while the pinion must always be generated. This method is suited to large- or small-scale production of small spiral bevel, Zerol bevel, and hypoid gears, such as those used in instruments and drives for outboard motors, power tools, and aircraft accessories.

The setup for generating the pinion member of Formate duplex cut pairs represents an imaginary generating gear having a helicoidal pitch surface. This requires relative feed of the cutter along the cradle axis during the generating roll, which is referred to as *helical generation.* This is required for both members of hypoid pairs and pinions, only, on spiral bevels. When semifinishing is required prior to Duplex helical grinding, it is usually done with helical generation.

Duplex grinding on machines without tilting spindles or helical movement is done with wheels having surfaces that are spherical, or which are curved surfaces of revolution. Semifinishing prior to grinding on this type of machine must be done by a comparable process without cutter-spindle tilt.

Refer to Tables 20-22, 20-23, and 20-24 for relationships of methods, processes, and machines.

In order to use the Duplex Spread-Blade method, special tooth proportions are required. The blanks must be designed for the particular method of cutting to be employed, following calculating instructions available from the machine manufacturer.

Although control over the tooth bearing is not quite so complete with this method as with the Fixed-Setting method, control over tooth size, concentricity between the two sides of the teeth, and smoothness of the fillets and tooth bottoms is excellent.

Unitool Method. The purpose of the Unitool method is to make practical and economical the production of good-quality spiral bevel, Zerol bevel, and hypoid gears in very small quantities. It is not intended for mass production, where the Fixed-Setting method is used for pinion finishing. It is slower than the Jobbing System (Art. 20-19) because both members of the pair are cut Single-Side. In brief, the Unitool method covers the following:

1. Spiral bevel and Zerol bevel gears of any ratio
2. Hypoid gears of 3:1 ratio and higher
3. All reasonable tooth numbers, pitches, spiral angles, and pressure angles
4. All reasonable tooth designs including equal-addendum teeth as well as standard long- and short-addendum designs, stub teeth, and special root-line direction.

Its features are the following:

1. The same cutter is used for both members of the pair.
2. Each part is completed before removing from the cutting machine.
3. One cutter covers a wide range of work.
4. Six cutters cut practically all jobs within the range of the machines used by this method.
5. Roughing is generally performed with the finishing cutter.
6. Calculations for machine settings are comparatively short.

20-19. Processes Used Only for Generation. Gleason 20° Spiral Bevel Gear Jobbing System. The purpose of the jobbing system is to make practical and economical the production of high-quality spiral bevel gears in moderate quantities. It is not intended for mass production where the Fixed-Setting method is used for pinion finishing. This system is applicable to the cutting of a wide range of spiral bevel gear ratios:

1. With mean cone distances between 2.5 and 7.0″
2. With 35° spiral angle
3. With 20° pressure angle

This system features a simplified method for obtaining tooth proportions, blank dimensions, and machine settings. Only a minimum of calculation work is necessary, together with a minimum of cutters.

Cutting of the gear is accomplished by the Spread-Blade method, while the mating pinion is finished by the Single-Side method. Both members are roughed prior to finishing. All cutters used with this system contain alternate inside- and outside-cutting blades.

Quality of the tooth bearing produced by this method is good, since direction and width of contact are controlled, and a choice in the length of the tooth bearing is obtained by selection of the proper pinion cutter. Set-over from one side of the pinion tooth to the other requires a change in only four machine settings.

A few roughing cutters are provided, but roughing can also be done with the finishing cutters for small quantities.

Special Jobbing. This is similar to the 20° Jobbing System explained above but can be used for cutting other spiral angles using the 20° jobbing cutters. This process is also applicable to cutting spiral bevel gears with other pressure angles provided the necessary cutters are used.

Calculations for the Special-Jobbing method are somewhat longer, since it is not practical to tabulate settings for every pressure angle and spiral angle.

Versacut.* This is a process which requires only a few cutters to accommodate a wide variety of spiral bevel gears on the largest face-mill generator, covering the following range:

Outer cone distance	3–13″
Diametral pitch	10–1.6
Pressure angle	20°
Spiral angle	0–40°
Ratios	1:1–10:1

The cutters can be used for both roughing and finishing, and each gear can be completed before being removed from the machine. Where production quantities merit, however, a roughing cutter can be used for a separate roughing operation. The machine settings for roughing are also the finishing settings for one side of the tooth. Each side of each member is finish-cut separately.

Arco.* This is a special process of cutting and grinding angular spiral bevel gears having cone distances which are beyond machine limits for conventional generation. Gears made by this method are usually designed with constant depth, and both members are generated. Modified roll is used when generating the pinion.

Allcone.* This term has been applied to a special process for generating gears of long cone distance. With this process the workpiece has motion of translation with respect to the cradle axis in addition to the generating roll. The method is used only on machines with linear vertical motion of the workhead during generation.

See Tables 20-22, 20-23, and 20-24 correlating cutters, methods, processes, and machines for generated spiral bevel, Zerol, and hypoid gears.

* Trade-marks of the Gleason Works, Rochester, N.Y.

20-20. Grinding Considerations. The requirement for ground bevel and hypoid gears for certain types of power transmission has increased greatly in recent years. A knowledge of the merits of ground bevel and hypoid gears, of the machines employed, and of the machine cycles used is needed before the decision to grind is made.

Purpose. The primary reason for grinding bevel and hypoid gears is to obtain optimum accuracy in the teeth of gears which require a heat-treatment or hardening operation. To obtain maximum life, sometimes under heavily loaded conditions, the use of hardened gears is mandatory. These gears are normally cut in a gear-cutting machine prior to the heat-treatment operation. Some distortion of the teeth is unavoidable in the heat-treatment process, and this distortion destroys the accuracy of shape and dimensions originally produced in the cut member.

There are procedures other than grinding which are used to minimize or partially correct the heat-treatment distortion. Generally, quenching dies are employed in the heat-treatment operation to decrease the amount of distortion. A subsequent lapping operation, which is a partially corrective measure only, often provides gears which are quite satisfactory for their particular application. However, when optimum smoothness in motion transmission is required because of high speed, heavy loads, or accuracy requirements, a grinding operation must be used on hardened gears in lieu of lapping, to assure satisfactory results.

Typical Applications Requiring Ground Gears. It is the application of the gear, rather than the type of gear, which dictates the need for grinding. The following outline indicates the general fields of power transmission in which ground bevel and hypoid gears are employed, and the predominating reasons for the requirement of ground teeth.

1. *Aircraft.* Aircraft gears are designed to provide maximum load capacity with minimum weight, necessitating the use of casehardened steel. The same design restrictions usually result in considerable section variations which induce excessive heat-treatment distortions. Grinding the teeth to remove hardening distortion and to improve accuracy of the gears has permitted higher operating speeds, longer operation, and heavier loading.
2. *Machine Tools.* The requirement for accuracy (gear trains, dividing heads, etc.) and long life under heavily loaded conditions has increased the demand for casehardened ground bevel and hypoid gears.
3. *Instrument.* The prime requirement is accuracy. Many fine-pitch instrument gears are ground from the solid in a single operation.
4. *Marine Propulsion.* Long life under heavily loaded conditions requires the use of hardened and ground gears.
5. *High-speed Hand Tools.* The high-speed application requires the use of ground gears which can operate at speeds considered impractical with unground gears.
6. *High-speed Mills and Presses.* Hardened and ground gears have proved advantageous for the requirements of high-speed smooth power transmission under heavily loaded conditions.

Accuracy of Ground Bevel and Hypoid Gears. The following values are indicative of the accuracy and quality which can be maintained in production when using current gear-grinding machines: tooth-to-tooth spacing as close as 0.0002″; eccentricity of the teeth relative to gear bore or shaft within 0.00025″ if proper regard is given to the gear blank processing and the workholding equipment; tooth size (measured in terms of backlash) to within 0.001 to 0.002″ between mating members; and surface finish within a range of 15 to 30 microinches. This uniformity of quality and dimensional accuracy, as provided by grinding machines, can be obtained by no other process on hardened bevel and hypoid gears.

Portions of Teeth Which Are Ground. Bevel and hypoid gear-grinding machines grind the entire configuration of the tooth slot—that is, the profile, fillet radius, and root or bottom land. It is primarily due to this fact, combined with the accuracy and uniformity of ground teeth, that longer life is obtained with ground bevel gears even under operation at high speeds or in heavily loaded applications.

The grinding wheel is dressed so that the fillet radius, top land, and profile are all tangent at their junction points, creating a smooth blend without an abrupt change in curvature. The rate of dress is controlled so as to permit a uniform dressing action at all points of the wheel. This is particularly important in the area of the edge radius where the proper dressing speed helps to prevent burning in the fillet.

Table 20-11. Recommended Semifinishing Practices and Grind Stock Allowances for Spiral Bevel, Zerol Bevel, and Hypoid Gears

Diam. pitch	Stock allowance per tooth side, in.	Stock allowance in tooth bottom, in.	Remarks
20 and finer	....	.....	All gears 20 diam. pitch and finer can be ground from the solid by the Duplex method
10–20	0.003–0.004	0.003	1. Gears of 12 to 20 diam. pitch should be semifinished using conical (straight-profile) cutters in all cases
Coarser than 10	0.004–0.005	0.003	2. Gears coarser than 12 diam. pitch should be semifinished with cutter shapes similar to shape employed in grinding wheel; i.e., if spherical (curved-profile) grinding wheels are to be used, spherical (curved-profile) semifinishing cutters are required. If conical (straight-profile) grinding wheels are to be used, conical (straight-profile) semifinishing cutters are required

1. Exceeding the above limits for stock allowance will (*a*) reduce the load-carrying capacity of casehardened gears; (*b*) increase the grinding time in order to eliminate the risk of burning or checking; and (*c*) increase wheel and diamond wear.

2. To maintain the recommended minimum stock allowance, it is essential that related control surfaces and dimensions of the gear blank be held to close tolerances.

3. It is essential that hardening distortion be kept to a minimum by means of the necessary quenching dies and presses.

Recommended Semifinishing Practices and Grind Stock Allowances. Minimum amount and uniform distribution of stock left by the semifinishing operation are extremely important. These are the bases for economical production in the grinding operation, and a necessity for obtaining casehardened gears with high load-carrying capacity. Table 20-11 lists the recommended semifinishing practices and stock allowances for spiral bevel, Zerol bevel, and hypoid gears of various diametral pitches. (Similar values are applicable when grinding straight bevel gears.)

Gears of diametral pitch coarser than 20 should be semifinished before grinding. When the Duplex method of grinding is used, a single Duplex semifinishing cut from the solid will suffice, but the same calculating method must be used for both cutting and grinding. This is important for the following reason: since there are two types of grinding machines—one having a grinding-wheel axis fixed parallel to the cradle axis, and the other permitting inclination of the wheel axis with respect to the cradle axis—there are, perforce, two different methods of representing the generating gear. With modified-roll machines the angle between work axis and cradle axis is always 90° plus the root angle; with the "tilting" type of machines the angle between cradle axis and work axis is not so restricted, and imaginary generating gears of many different pitch angles may be represented by tilting the grinding-wheel axis and work axis so that the surface of the generating gear tooth is tangibly represented. Gears, and particularly pinions, have substantially different root surfaces when manufactured on different types of machines.

In order to hold the stock allowance to the limit established in Table 20-11, semi-finishing and grinding must be based on the same calculation method.

The above requirements in method, which are based upon uniform stock removal, do not preclude interchangeability, to a practical degree, of gears ground on both types of machines. For optimum interchangeability, however, it is best to follow the recommendation of the machine manufacturer.

Nital Etch Inspection. The subject of nital inspection is covered as a whole in Art. 19-5. However, it should be especially noted that production procedures should include a periodic nital etch inspection to check for burning. Consult the grinding-machine manufacturer for details.

Coolants. Preferably, bevel-gear grinding is done with an oil coolant, and without water, a typical oil coolant having specifications as follows:

Viscosity 150 to 180 SSU
Mineral base, 2 per cent sulfur contained

Water-soluble oil coolants can be used, however, but when grinding casehardened steel, results will be less satisfactory. Provision should be made for keeping the coolant clean; the latest types of bevel- and hypoid-gear grinders are equipped with a centrifugal separator.

The Grinding Cycle. Prior to production grinding, the most efficient grinding cycle for a particular gear is established. This depends upon the material to be ground. The grinding cycle, which is fully automatic, consists of sequential operations of dressing, wheel feed, and grinding. Repetition of identical operations for every gear of a production lot ensures uniformity of the product. For typical cycles, see Tables 20-18 and 20-19.

20-21. Machines for Nongenerated Gears. A Helixform machine can be used to cut both Helixform and Formate gears, whereas Formate gear finishers can be used only for Formate gears.

The manufacture of exactly duplicate gears on both Helixform and Formate finish-cutting machines is accomplished by the use of setup gages. These gages are furnished for specific jobs and are designed to reproduce exactly the machine setup to which the gage was originally fitted. In this way a definite reference is established and the original gear may be exactly duplicated at any time by setting the machine to this gage. Moreover, the use of the gage permits setting up any number of machines alike so as to produce identical gears.

Helixform Machines. Helixform gear-finishing machines have a stroke cam which transmits axial motion to the cutter spindle by means of an adjustable pivoted arm. A no-stroke cam is substituted for the Helixform stroke cam in order to cut Formate gears.

The stroke cam is designed with two lead and return paths and rotates five times for each revolution of the cutter. The proportion of lead to return path is approximately 3:1, which means that 27° is the maximum angular face width of the gear which can be cut. Since the stroke cam has a timed relationship with the cutter, one lead path is always used to cut the coast side of the gear and the other lead path the drive side. A range of modified lead-path cams permits independent drive and coast-side tooth-bearing developments to suit individual assembly deflections and hardening distortions.

The work is mounted on an arbor (Fig. 20-9*a*), which is actuated hydraulically. The work spindle in which the arbor is mounted is locked in place during cutting and

Table 20-12. Capacities of Machines for Nongenerated Gears

	No. 607 Helixform Gear Finisher and No. 606 Gear Rougher	No. 22 Formate Gear Finisher	No. 22 Gear Rougher	No. 18 Formate Gear Grinder
Cone distance max., in.	5¼	11	11	12⅞
Pitch angle max	90°	90°	90°	90°
Pitch angle min	45°	71°34′	45°	68°12′
Min. ratio	2½:1	3:1	3:1	2½:1
Pitch dia. max., in.				
At 90° shaft angle	10½	22 for 12:1 ratio	22 for 12:1 ratio	24
At 30° spiral angle		21 for 3:1 ratio	21 for 3:1 ratio	
Max. diam. pitch	2½	2½	2½	2
Face width max., in	....	4	4	4
Angular face max	27.5°			
Index range	20–150	31–78	31–78	20–78
Cutter or grinding-wheel diameters, in	6, 7½, 9	9, 12, 16, 18	9, 12, 16, 18	6, 7½, 9, 12

Table 20-13. Production Data, Cutting Machines for Nongenerated Gears
(Average feeds and speeds, based on AISI 4620 steel at 160 to 190 BHN)

Operation and machine	Cutter dia., in.	Feed, sec per tooth	Speed, fpm
Roughing:			
No. 606	6, 7½, 9	3.5–4.5	145–165
No. 22	12, 16, 18	10–12	110–120
Single Cycle finishing:			
No. 607	6½, 7½, 9	3.0–4.0	20–25
No. 22	12, 16, 18	10–12	20–25
Helixform finishing:			
No. 607	6, 7½, 9	3.5–4.5	25–30
Completing Cyclex:			
No. 607	6	7–8	130–150
	7½	9–11	roughing blades
	9	10–12	
No. 22	12	16–17	45–50
	16	25–30	finishing blades
	18	30	

automatically released for indexing. In dechucking action the gear is pushed off the centering portion of the centering ring so that it needs only to be lifted off the arbor.

Cutter speed and feed are controlled by change gears. Table 20-13 shows average speeds and feeds.

Rapid indexing is accomplished with an index-plate mechanism.

On the gear-roughing machine, the cutter is advanced toward the work in the direction of its axis by a cam motion. A linkage adjustment permits from 60 to 100 per cent of the cam throw to be delivered to the spindle. Workholding, index, and cutter speed and feed are controlled the same as on the finisher.

The gear is chamfered automatically on the outside diameter by a tool which cuts on both the bottom and concave side of the tooth slot.

Formate Cutting Machines. With the Formate process the rotation of the cutter on its axis is the only motion used while finishing the gear. The cutter, which is mounted on a support integral with the cutter-spindle housing, is driven through a minimum number of gear meshes.

Workholding equipment, speed and feed control, and indexing are essentially the same as with the Helixform machine.

Table 20-13 shows average speeds and feeds for Formate machines. Since machines using Cyclex cutters are doing both roughing and finishing, these machines are equipped with two-speed motors.

Formate roughing and Cyclex machines have a feed cam that advances the cutter in the direction of its axis until the full-depth position is reached. A change in depth of tooth requires a different cam, for efficient operation.

Automatic chamfering attachments may be used on the roughing machines. They chamfer the acute angles (i.e., heel and toe edges) of the gear teeth after the tooth has been cut. A burring cutter set on the outside diameter of the gear rides the back angle and removes all burrs.

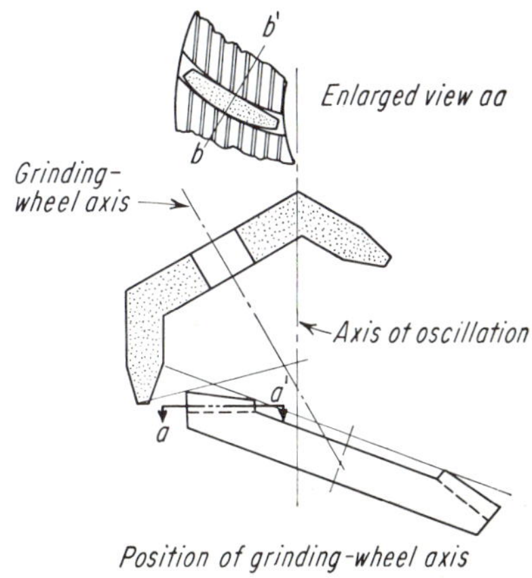

FIG. 20-28. Relationship of wheel and workpiece on Formate grinder.

Formate Grinding Machines. The Formate grinder differs from face-mill cutting machines in the shape, position, and motion of the grinding wheel itself. A flared cup-type grinding wheel is positioned so as to have line contact with the sides of the tooth and is oscillated from end to end of the tooth while grinding. Both sides of the tooth space are ground simultaneously. Figure 20-28 shows the position of the grinding-wheel axis and the equivalent cutter axis.

Table 20-14. Production Data, No. 18 Formate Grinder

Av. wheel speed, fpm	3,800–4,500
Av. cycle time per tooth, sec	2.6–4.1
Av. No. of passes per tooth based on 0.003–0.004" stock per side of tooth	3–4

Representative Production Rate for Finish Grinding AISI 4820 Material, Carburized 60 Rockwell C Hardness

Gear data	*Finish-grinding time*
57 teeth 2.85 diam. pitch 3.00" face	22.5 min

The grinding-wheel axis is inclined at an angle with the cutter axis, or axis of oscilla tion. The normal radius of curvature of the grinding-wheel surface must remain smaller than the radius of curvature of the equivalent cutter on the concave side of the tooth and larger on the convex side of the tooth. This makes it possible at any given

Table 20-15. Correlation of Machines, Methods, and Processes for Nongenerated Gears and Mating Pinions, Spiral Bevel, Zerol Bevel, and Hypoid Gears

Gear rough-cutting machine No.	Gear finish-cutting machine No.	Gear finish-grinding machine No.	Gear method	Gear-finishing cutter type	Pinion rough-cutting machine No.	Pinion finish-cutting machine No.	Pinion finish-grinding machine	Required pinion finish-cutting machine features	Pinion method	Process and application
Formate Pairs										
	102	. . .	Spread-Blade	Alternate-blade	. . . .	102	. . . . . .	T, H	Spread-Blade	Duplex helical for 16 diam. pitch and finer
108, 606R, 118	108, 607, 118	18	Spread-Blade	Cyclex† or Single Cycle	106, 106R, 116, 116R, 108, 118	106, 108, 116, 118	7	T	Fixed-Setting	Highest-quality work, large-volume production
108, 606R, 118	108, 607, 118	18	Spread-Blade	Cyclex† or Single Cycle	116, 116R, 118	116, 118	27	M*	Fixed-Setting	Highest quality work, large-volume production
108, 606R, 118	108, 118	18	Spread-Blade	Cyclex† or Single Cycle	106, 108	106, 108	7	T, H*	Spread-Blade	Duplex helical for fine pitches
108, 118	108, 118	. . .	Single-Setting	Alternate-blade	106, 108, 116, 118	106, 108, 116, 118	. . . . . .	T	Single-Side	Unitool for small quantities
22R, 28	22, 28	18	Spread-Blade	Cyclex† or Single Cycle	26, 26R, 28	26, 28	27, 137	M	Fixed-Setting	Highest-quality work, large-volume production
Helixform Pairs										
606R	607	. . .	Spread-Blade	Helixform	106, 106R, 116, 116R, 108, 118	106, 108, 116, 118	. . . . . .	T	Fixed-Setting	Highest-quality work, large-volume production

The postscript R after a machine designates a model used for roughing only.
H = helical motion.
T = tilting cutter spindle.
M = modified roll.
* Asterisk denotes that feature is special equipment.
† Cyclex cutter completes gear; no previous roughing cut required.

moment to have the grinding wheel in contact at only one section *b-b′* along the tooth, as shown in section *a-a′* of Fig. 20-28.

Wheel Dressing. The two sides and end of the grinding wheel are dressed by three diamonds which are hydraulically operated and automatically controlled.

The hydraulically operated diamond dressers are adjustable for pressure angle, radius, wheel shape, wheel diameter, point width, and speed and feed of dress.

Grinding Wheels. Grinding wheels for Formate grinding are generally operated between 3,800 and 4,500 sfpm. A commonly used wheel specification in the 2 to 10 diametral pitch range is A-46H-8V.

Machine Cycle. The most efficient grinding cycle for a particular gear is governed by the material to be ground and is established during the calculation of the machine settings.

A typical cycle for the Formate machine is as follows: rough-dress, grind, finish-dress, grind, grind. The machine cycle time is generally 2.6 to 4.1 sec per tooth, with a 10-stroke (six effective) cycle. For each additional 0.001″ stock per side allowed in cutting, over the average recommended 0.0035″, one rough dress and one grinding pass (10-stroke cycle) is allowed.

After the machine has been set up and the grinding sequence selected, the work is advanced to the grinding position, the wheel is dressed to the correct shape, and grinding begins. A series of feed cams advances the oscillating wheel into the work, and at the completion of the grind, the wheel is withdrawn and the work indexed to the next tooth slot. Indexing is done by a notch-type index plate. A predetermined number of grinds is required to complete the part, each grind removing a preset amount of stock. At preselected intervals between grinds the wheel is automatically dressed.

20-22. Machines for Generated Gears. Generators using face-mill cutters, and grinders using cup-type wheels, are made in two types—*tilting-spindle machines* and *modified-roll machines*. The basic differences and relation to the means of representing the imaginary generating gear are explained in Art. 20-2, The Basic Generator.

Cutting-machine Cycles. Operation on both types of machines is entirely automatic. On a typical cycle, as soon as the workpiece is chucked, the feed cam

Table 20-16. Capacities of Machines for Cutting Generated Gears

	No. 102	Nos. 106, 108*	Nos. 116, 118*	Nos. 26, 28*
Cone distance max., in.	$2\frac{5}{16}$	$4\frac{1}{4}$	9	$16\frac{1}{2}$
Machine root angle:				
Max.	90°	90°	90°	Pitch angle 84°18′
Min.	−12°	−12°	−12°	Pitch angle 5°42′
Extreme ratio	Any practical ratio	Any practical ratio	Any practical ratio	10:1
Pitch dia. max., in. (30° spiral angle)	$3\frac{3}{4}$ for 10:1 ratio, $2\frac{5}{8}$ for 1:1 ratio	$8\frac{1}{2}$ for 10:1 ratio, 6 for 1:1 ratio	18 for 10:1 ratio, $12\frac{3}{4}$ for 1:1 ratio	33 for 10:1 ratio, $23\frac{1}{4}$ for 1:1 ratio
Max. diam. pitch	16	4	2	$1\frac{1}{2}$
Face width max., in.	$\frac{7}{16}$	$1\frac{1}{4}$	$2\frac{3}{4}$	4
Index range	5–200	5–97 and most numbers up to 175	5–150	5–100 and most numbers up to 200
Cutter dia., in.	1.1, $1\frac{1}{2}$, 2, $2\frac{3}{4}$, $3\frac{1}{2}$	$1\frac{1}{2}$, 2, $2\frac{3}{4}$, $3\frac{1}{2}$, $4\frac{1}{2}$, 6, $7\frac{1}{2}$	6, $7\frac{1}{2}$, 9, 12	9, 12, 16, 18

* Machine has features required for cutting Formate gears.

moves the workhead and gear into cutting position and generating roll begins. At the completion of one tooth space, the cradle and work spindle reverse direction, the work is withdrawn, and indexing takes place. The workhead moves forward and the sequence of motion is repeated until all teeth are cut; the machine then stops automatically.

Tilting-spindle Cutting Machines. This is the type of machine used extensively for automotive pinion production. It is made in a series of sizes ranging from the smallest machine for fine-pitch work, which mounts face-mill cutters of from 1.1 to 3½″ diameter, up to the largest size, which mounts cutters from 6 to 12″ diameter.

Table 20-17. Production Data, Machines for Cutting Generated Gears Average Feeds for Face-mill Generators Based on AISI 4620 Steel at 160 to 190 BHN

Automotive and Truck Pinions

Machine	*Feed, sec per tooth*
No. 106	
Rougher	15
Finisher	30*
No. 116	
Rougher	19
Finisher	35*
No. 26	
Rougher	30
Finisher	80*

Representative Production Rates

Machine	Diam. pitch	Ratio	Face, in.	Roughing feed, sec per tooth	Finishing feed, sec per tooth	
No. 102	40	18:36	5/32	Completing feed, sec per tooth 2.6		
No. 106	9	18:36	5/8	Completing feed, sec per tooth 10		
					Gear	Pinion
No. 106	9	18:36	5/8	14	14	24*
No. 116	4	18:36	1½	30	30	40*
No. 26	2	18:36	3	80	80	80*

* Pinions finished by the Fixed-Setting method. Time includes both convex and concave sides—a total of two cuts.

Some tilting-spindle machines are optionally equipped with modified roll. The variation from constant angular motion is made on the work spindle and results from the action of an eccentric roller against a flat plate. The linear movement so produced is fed into the generating train by means of a rack, pinion, and differential.

The generating train—a train of gears connecting cradle and work spindle—is driven by a drum cam. This allows the gearing to turn at a varying rate so as to produce optimum chip loads throughout the cutting cycle. Intermittent indexing is used on the larger machines, accomplished by a modified Geneva mechanism, and cutting is done on either up-roll (counterclockwise turning of cradle) or down-roll. Continuous indexing is used on the smallest machine.

Feed is cam-controlled, and on the larger machines, a variety of drum cams having more than one path permits either gradual or rapid advance to full depth. Completing cams are designed for finish cutting from solid blanks by using first a straight infeed and then generation during both down- and up-roll, similar to the cycle for Coniflex generators depicted in Fig. 20-15.

When finish-cutting pinions from the solid blank, a helical generation is required; this necessitates a special infeed during the generating roll (in addition to the advance to whole-depth position).

On the small tilting-spindle generators Formate gears may be produced using simple alternate-blade cutters as described in Art. 20-17. Some of the larger generators permit Formate cutting with Single Cycle or Cyclex cutters.

For the larger models there are companion roughing machines which cut on either up-roll or down-roll. In addition, they can be set up for double-roll operation, a set-over from one side of the tooth to the other taking place between up-roll and down-roll.

Table 20-18. Capacities of Generating Grinders

	No. 7	No. 27	No. 137
Cone distance max., in.	5½	15*	†
Pitch angle max.	84°17′	71°34′	71°34′
Pitch angle min.	5°42′	5°42′	5°42′
Extreme ratio	Any practical ratio	3:1‡	3:1‡
Pitch dia. max., in.:			
90° shaft angle	6 for 1:1 ratio	21 for 1:1 ratio	21 for 1:1 ratio
30° spiral angle	8½ for 10:1 ratio	31½ for 3:1 ratio	31½ for 3:1 ratio
Wheel dia., in.	2-6	5½-13½	6-21
Diam. pitch (coarsest)	4	2	1½
Diam. pitch (finest)	Limited only by min. practical point width		
Face width max., in.	1¼	3	4

* With 12″ wheels.

† The No. 137 grinder has a special feature permitting generation of gears of long cone distance. Refer to Gleason Works.

‡ Refers to gears only, which should be Formate, above 3:1 ratio. Pinions of any practical ratio can be ground.

With this type of cut the teeth are taper-roughed to leave a minimum of stock for finishing.

Modified-roll Cutting Machine. The larger machines, mounting face-mill cutters of 9 to 18″ in diameter, are made without tilt adjustment of the cutter spindle. That is, the direction of the cutter axis always remains parallel to the cradle axis. As explained in Art. 20-2, The Basic Generator, the ratio of angular velocity between cradle and work varies during generation. On the larger machines the motion of the cradle is modified by means of an eccentric roller acting against a plate. The harmonic linear motion so derived causes an axial displacement of the cradle drive worm; control of the modifying motion is through a set of change gears. Some of the large generators are arranged for Formate cutting with Single Cycle and Cyclex cutters. Intermittent indexing is used, and cutting may be done on either up- or down-roll.

Roughing machines of this type can be used either for single-roll cutting or for double-roll operation in which a set-over from one side of the tooth to the other takes place between up- and down-roll, leaving a minimum of stock for finishing.

Grinding-machine Cycles. Although the generating cycles of the two types of grinders are different, as explained below, the complete grinding sequence is automatically controlled in similar manner on both types. The number of grinding passes and dresses is determined prior to production grinding, depending upon the material and amount of stock to be removed. The complete grinding cycle consists of the operations of wheel feed, dressing, grinding, and work advance in predetermined sequences. All these are actuated in desired order by movement of a variable stop drum, which rotates to successive positions after each complete grinding pass around the workpiece. Once the machine cycle is started, the grinder automatically proceeds through all the above operations in sequence until final size is reached.

Typical cycles and grinding times are shown in Tables 20-19 and 20-20.

Table 20-19. Typical Cycle for Generating Grinder, Finish Grinding
(Teeth semifinished prior to grinding)

Operation	Feed (advance into work), in.	Stock removed (per side), in.
Rough-dress and grind	0.003 approx.	0.0008–0.001
Finish-dress and grind	None (float)	0.0002–0.0006
Grind	0.0025	0.0009
Grind	0.0015	0.0006
Grind	0.0015	0.0006
Total	....	0.0031–0.0037

Typical time per index 1.9 to 3.2 sec.

Representative Production Rates, SAE 9316 Material, Carburized, 60 Rockwell C Hardness

Gear data	*Finish-grinding time, min*
40 teeth, 10 diam. pitch, 0.405″ face	10.5
27 teeth, 3.5 diam. pitch, 1.500″ face	12.8

Table 20-20. Typical Cycle for Generating Grinder—Teeth Ground from a Solid Blank—Whole Depth 0.088″

Operation	Feed (advance into work), in.	Stock removed (per side), in.
Rough-dress:		
Grind	0.010	0.0036
Grind	0.010	0.0036
The above steps are repeated three times sequentially until wheel is advanced to 0.080″ depth.		
Finish-dress:		
Grind	No feed	
Grind	0.004	0.0014
Grind	0.002	0.0007
Grind	0.002	0.0007
Total	0.088	0.032 per side

Typical time per index 1.9 to 2.7 sec.

Representative Production Rate, AMS 6415 Material, 55 Rockwell C Hardness

Gear data	*Completing grinding time, min*
31 teeth, 20 diam. pitch, 0.295″ face	29.4

Grinding Wheels. Cup-type wheels for generated grinding are generally operated between 3,800 and 4,500 fpm. Typical wheel specifications are A-60H-7V and A60M-B.

Tilting-spindle Grinding Machines. These machines use a cup-shaped grinding wheel instead of a face-mill cutter, and in general, all cutting methods and processes not requiring modified roll are equally applicable to grinding. The generating process has been described in Art. 20-2, The Basic Generator.

Intermittent indexing of the workpiece is accomplished by a modified Geneva mechanism. Grinding takes place while the cradle and the work turn about their respective axes far enough to complete the generation of the side of a tooth or the adjacent sides of a tooth space, depending on the method used. When generation is complete, the work is withdrawn from engagement with the wheel, and indexing of the workpiece takes place while cradle and work spindle are returning to their starting positions.

Table 20-21. Features of Generating Cutting Machines and Grinders for Spiral Bevel, Zerol Bevel, and Hypoid Gears

Machine	Features	Range of cutter dia., in.
No. 102 generator	C, T, FR	1.1–3½
No. 106 generator	I, T, (H)	1½–7½
No. 108 generator	I, T, (H), F	3½–7½
No. 116 generator	I, T, (M)	6–12
No. 118 generator	I, T, (M), F	6–12
No. 26 generator	I, M	9–18
No. 28 generator	I, M, F	9–18
No. 7 grinder	I, T, H	2–6
No. 27 grinder	C, M	5½–13½
No. 137 grinder	C, M, A	6–21
No. 106 rougher	I, T, S	3½–7½
No. 116 rougher	I, T, S	6–12
No. 26 rougher	I, M, S	9–18

Optional features are coded in parentheses.
A = arranged for Allcone process.
C = continuous indexing.
F = arranged for Formate cutting.
FR = will cut Formate gears with 1:1 roll.
H = helical generation.
I = intermittent indexing.
M = modified roll.
T = tilting cutter spindle.
S = set-over provision for double-roll roughing.

Wheel Dressing. Wheels are dressed by a cam-controlled hydraulically operated dresser. A series of standard cams is used to produce all straight-sided wheel profiles and most of the curved profiles that are required by the various grinding methods. Two lapped diamonds, optically preset before mounting in the dresser, are used for small and medium-sized work, while wheels for the finest pitches are dressed with natural stones.

Modified-roll Grinding Machines. These machines use cup-shaped wheels in place of face-mill cutters and, like some of the larger cutting machines, have wheel spindles that are always parallel to the cradle axis. Operation is different from the larger cutting machines, however, in that the work spindle turns continuously at a constant rate, but when generation of a tooth side or sides is complete, the wheel is withdrawn from engagement with the work. During the time that the wheel is disengaged, the work continues to turn while the cradle motion reverses and the cradle returns to its starting position. The wheel then moves forward and engagement with the work is resumed in a different tooth space. The number of tooth spaces included between successive engagements is controlled by change gears and depends on the number of teeth in the workpiece and on the amount of turning necessary to complete the generation. Turning of the cradle is referred to as *cradle roll;* it is derived from a cam and is either uniform or nonuniform, depending on the distance set between the cam geometrical center and the center of cam rotation; this setting provides the modified roll.

Wheel Dressing. Grinders without tilting-wheel spindles use three natural-dressing diamonds mounted in rotatable nibs which are inclined to the surface normal of the wheel. The diamond nibs are traversed across the wheel sides and end by a hydraulically actuated mechanism. This mechanism rotates the side-dressing diamonds about axes substantially parallel to the surface normals of the wheel sides and

Table 20-22. Correlation of Machines, Methods, and Processes, Generated Spiral Bevel Gear Pairs

Rough-cutting machine No.	Finish-cutting machine No.	Finish-grinding machine No.	Method		Required finish-cutting machine features	Process and application
			Gear	Pinion		
	102	7	Spread-Blade	Spread-Blade	H, T	Duplex helical 16 diam. pitch and finer
106, 106R, 108	106, 108	7, 27	Spread-Blade	Fixed-Setting	.....	Highest-quality work, large-volume production
106, 106R, 108	106, 108	7	Spread-Blade	Spread-Blade	H,* T	Duplex helical, fine pitches
106, 106R, 108	106, 108	.....	Spread-Blade	Single-Side	T	Jobbing, for moderate quantities
106, 108	106, 108	.....	Single-Setting	Single-Side	T	Unitool, for small quantities
116, 116R, 118	116, 118	27, 137	Spread-Blade	Fixed-Setting	.....	Highest-quality work, large-volume production
116, 116R, 118	116, 118	.....	Spread-Blade	Single-Side	T	Jobbing, for moderate quantities
116, 118	116, 118	.....	Single-Setting	Single-Side	T	Unitool, for small quantities
26, 26R, 28	26, 28	27, 137	Spread-Blade	Fixed-Setting	.....	Highest-quality work, large-volume production
26, 26R, 28	26, 28	...	Single-Side	Single-Side	M	Versacut, for small and moderate quantities

The postscript R after a machine number designates a model used for roughing only.
H = helical motion.
T = tilting cutter spindle.
M = modified roll.
* Feature is special equipment.

also rotates the end diamond about an axis substantially perpendicular to the surface normals. An adjustment of the dresser which offsets the axes of diamond rotation from the wheel axis permits either conical or curved-profile wheels to be dressed.

SPIRAL BEVEL AND HYPOID PINION BURNISHING

Burnishing, which is a plastic smearing, is accomplished by rubbing pressure which flattens and spreads minute surface irregularities to a perfectly even surface. The process, which is very rapid, is applied to spiral bevel and hypoid pinions, in the soft state, primarily for the purpose of minimizing subsequent lapping time. It is used in large-volume production installations, such as automotive-axle manufacturing.

Table 20-23. Correlation of Machines, Methods, and Processes, Generated Zerol Bevel Gear Pairs

Rough-cutting machine No.	Finish-cutting machine No.	Finish-grinding machine No.	Method		Required finish-cutting machine features	Process and application
			Gear	Pinion		
	102	7	Spread-Blade	Spread-Blade	T, H	Duplex helical 16 diam. pitch and finer
106, 106R, 108	106, 108	7	Spread-Blade	Fixed-Setting	T	Highest-quality work, large-volume production
106, 106R, 108	106, 108	7, 27	Spread-Blade	Fixed-Setting	.....	Highest-quality work, large-volume production
106, 106R, 108	106, 108	7	Spread-Blade	Spread-Blade	T, H*	Duplex helical for fine pitches
106, 108	106, 108	.....	Single-Setting	Single-Side	T	Unitool for small quantities
116, 116R, 118	116, 118	.....	Spread-Blade	Fixed-Setting	T	Highest-quality work, large-volume production
116, 116R, 118	116, 118	27, 137	Spread-Blade	Fixed-Setting	.....	Highest-quality work, large-volume production
116, 116R, 118	116, 118	27, 137	Spread-Blade	Spread-Blade	.....	Duplex spherical, large quantities, fine pitches
116, 118	116, 118	.....	Single-Setting	Single-Side	T	Unitool for small quantities
26, 26R, 28	26, 28	27, 137	Spread-Blade	Fixed-Setting	.....	Highest-quality work, large-volume production
26, 26R, 28	26, 28	.....	Single-Side	Single-Side	M	Versacut for small and moderate quantities

The postscript R after a machine number designates a model used for roughing only.
H = helical motion.
T = tilting cutter spindle.
M = modified roll.
* Feature is special equipment.

Pinions are burnished by running in a burnishing machine under light pressure, in metal-to-metal contact with a special hardened and etched burnishing gear. This produces a very fine lustrous finish on the pinion teeth. It should be emphasized that the sole purpose of the burnishing operation is to produce a smooth uniform surface, and it cannot be used to correct results of improper cutting.

20-23. Burnishing-machine Operation. The pinion-burnishing machine uses the pinion to drive the burnishing gear, with Stuart's Thread-Kut Base Cutting Oil, or equivalent, as a lubricant. As the pair is rotated, the position of the gear is changed continuously and automatically to effect a combined horizontal and vertical movement relative to the pinion. This movement results in spreading the burnishing action over the entire tooth surface, and at the same time preserves the localized tooth bearing which is obtained when the gears are run on correct centers. As the pinion is meshed with the gear, the sleeve in which the gear spindle is mounted is moved back against a spring. The spring pressure, which can be varied to suit the particular job, holds the gear against the pinion throughout the range of automatic movement.

The automatic cycle of the burnishing machine provides for running the pair equal amounts of time in both the forward and reverse directions. Time per side may be varied, by change gears, from 6 to 30 sec. An average time is about 20 sec per side, or 40 sec per pinion.

The burnishing machine accommodates pinions over the normal range of automotive passenger-car applications.

Table 20-24. Correlation of Machines, Methods, and Processes, Generated Hypoid Gear Pairs

Rough-cutting machine No.	Finish-cutting machine No.	Finish-grinding machine No.	Method		Required finish-cutting machine features	Process and application
			Gear	Pinion		
	102	7	Spread-Blade	Spread-Blade	T, H	Duplex helical 16 diam. pitch and finer
106, 106R, 108	106, 108	7	Spread-Blade	Fixed-Setting	T	Highest-quality work, large-volume production
106, 106R, 108	106, 108	7, 27	Spread-Blade	Fixed-Setting	....	Highest-quality work, large-volume production
106, 106R, 108	106, 108	7	Spread-Blade	Spread-Blade	T, H	Duplex helical for fine pitches
116, 116R, 118	116, 118	....	Spread-Blade	Fixed-Setting	T	Highest-quality work, large-volume production
116, 116R, 118	116, 118	27, 137	Spread-Blade	Fixed-Setting	....	Highest-quality work, large-volume production
26, 26R, 28	26, 28	27, 137	Spread-Blade	Fixed-Setting	M	Highest-quality work, large-volume production

The postscript R after a machine number designates a model used for roughing only.
H = helical motion.
T = tilting cutter spindle.
M = modified roll.

20-24. Burnishing Gear. Burnishing gears for Formate or Helixform pairs are preferably finished on Formate or Helixform machines, and are cut single-side with special cutters. If these machines are not available, however, a generator can be used.

Treatment of the burnishing gear-tooth surfaces by an acid etch after hardening serves two purposes. First, it roughens the tooth surface sufficiently to permit removal of small pinion surface blemishes. Second, it lengthens burnishing-gear life, since the gear can wear a small amount before becoming perfectly smooth. The gear can be re-etched several times, to restore the surface after wear. Etching is accomplished by immersing the gear 3 to 7 min in a 50 per cent solution of hydrochloric acid (18°Bé) and water at 180°F, followed by a rinse and neutralizing in hot alkaline solution.

20-25. Burnishing Methods. There are two methods for distributing the burnishing action over the desired portion of the pinion tooth. With the Top-and-Flank method, preferred for better coverage and control of contact, the burnishing gear is developed to have a full-length contact with the pinion. The gear is given the proper movement during the burnishing cycle to move the contact completely from top to flank of tooth. Occasionally such a burnishing-gear development is not possible, and then the "Toe-and-Heel" method is used, in which the burnishing gear is developed to have a contact only slightly longer than a production gear. The gear is given a movement during the burnishing cycle to move the contact completely from toe to heel.

SPIRAL BEVEL AND HYPOID GEAR LAPPING

Lapping is a polishing operation used to refine the tooth surface and to improve the tooth bearing. The mating gear and pinion are run together under a controlled light brake load while a mixture of abrasive compound and suitable vehicle is pumped onto the pair.

Hardened spiral bevel and hypoid gears which are required to transmit appreciable loads smoothly and quietly are nearly always lapped, unless the teeth are to be ground after hardening. Zerol bevel and straight bevel gears are not lapped because they do not have the progressive lengthwise tooth contact required for effective lapping action.

After a gear and pinion have been lapped together, they must remain as a pair; they cannot be used as interchangeable single pieces.

Lapping is a routine production operation and should not be considered a salvage operation for the correction of excessive hardening distortion, improper gear cutting, or runout.

To ensure best results and to minimize lapping time, it is necessary to make allowances, in the soft tooth-bearing development, for heat-treatment distortion. Therefore, control of both cutting and heat-treatment is essential for production lapping. Variations in either of these processes will result in changes in tooth bearing, which require special lapping-machine settings or extra lapping time, thus losing the production features of the lapping operation.

20-26. Lapping-machine Cycles and Rates. The pair to be lapped is mounted in the lapping machine in much the same manner as for testing (Art. 23-9). Setup fixtures may be used to position the machine heads at correct mounting distances, and the desired backlash is automatically established by a machine setting. As the pinion drives the gear during the lapping operation, it is necessary that their relative position be changed, so that contact will move progressively from one end of the tooth to the other, thus lapping the entire working surface of the tooth. This is done automatically by the lapping machine.

After the "start" button is pushed, to start the lapping-compound pump and driving-spindle rotation, the gears are run for a specific time cycle on one side. This cycle includes a predetermined number of passes, during each of which the point of lapping action is moved to "wipe out" entirely the desired portion of the tooth surface.

Spindle rotation then momentarily stops, settings are automatically changed to predetermined independent positions for the other side, and the second side is then lapped. At the conclusion of the complete lapping cycle, the compound pump shuts off before the gears stop rotating, so that they are free of excessive compound before handling.

Average lapping time for conventional automotive passenger-car hypoid gears varies from 4 to 6 min per pair.

20-27. Lapping-machine Types. There are two general types of lapping machines, the primary difference between them lying in the motions which they impart to the mating gear and pinion during the lapping cycle.

Machines with All-linear Movements. Machines of this type (see Table 20-25) have a lapping cycle in which the position of the gear is changed automatically in three linear directions, as shown in Fig. 20-29. The movements are accomplished by a cam-controlled primary vertical motion of the gear spindle, and any required change of

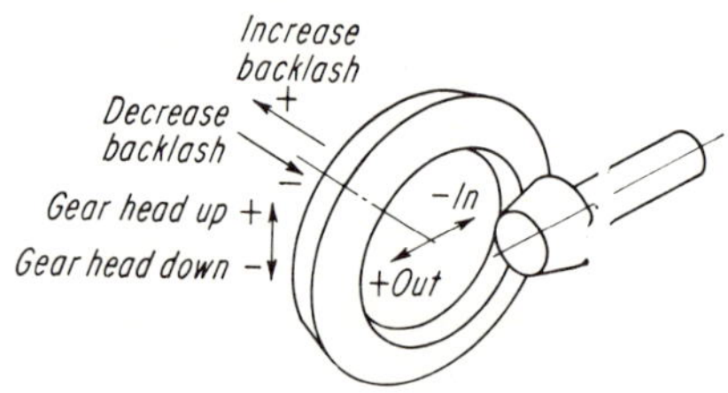

Fig. 20-29. Gear movements on lapping machines with all-linear motions.

throw in any of the three directions is accomplished by a simple adjustment of dial-operated controls. In order to determine the movements required, a vertical and horizontal (V and H) check of tooth contact is made for both forward and reverse sides of the gear (Art. 23-9, on testing). The vertical movement is determined directly from the vertical check thus obtained. Horizontal motion is controlled by a dial setting, and backlash control is obtained by an automatic axial movement of the gear spindle. The number of passes is set by a counter, and the speed of the movement can be varied between the two ends of the teeth.

Table 20-25. Capacities of Lappers with All Linear Motions

No. 15 Angular Hypoid Lapper	
Max. dia. gear, in.	17
Range of shaft angles, deg	0–180
Max. offset for drive spindle above or below center, in.	3
Min. and max. position of nose of drive or pinion spindle from center of machine, in.	4 and 14
Min. and max. position of nose of brake or gear spindle from center of machine, in.	5–13
Drive-head horizontal adjustment, in., at right angles to spindle axis (for long-cone-distance angular gears):	
Away from operator	5
Toward operator	0
No. 119 Hypoid Lapper—for 90° Shaft Angle, Only	
Max. dia. gear, in.	17
Max. offset of gear spindle above or below center of pinion spindle, in.	3
Min. and max. distance from center of gear spindle to nose of pinion spindle, in.	6¼–10½
Min. and max. distance from center of pinion spindle to nose of gear spindle, in.	5–10¾

When the machine automatically changes from reverse to forward side lapping, a set-over mechanism provides a means of automatically shifting the relative position and movement of gear and pinion to obtain the correct lapping positions for each side of the tooth, independent of the other.

An adjustable dynamometer-type brake on the driven spindle provides the load applied during the lapping operation.

Machines with Swinging and Linear Movements. Machines of this type (see Table 20-26) use three motions in a horizontal plane to accomplish the lapping action; a swinging motion of the pinion head, a motion in the direction of the pinion axis, and a motion in the direction of the gear axis. The directions of the machine motions are

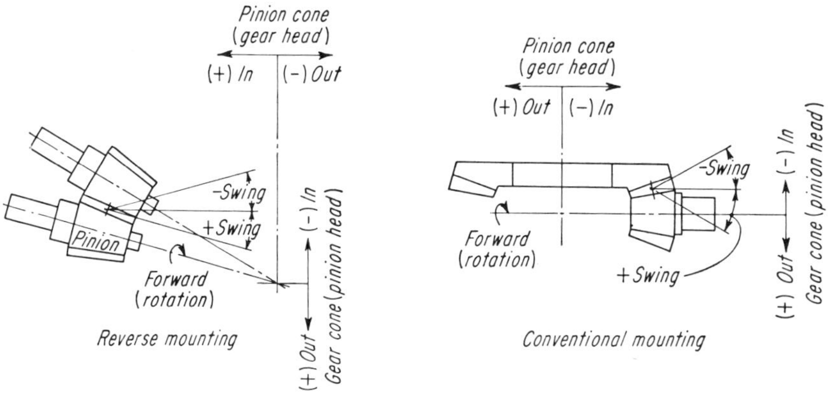

FIG. 20-30. Movements on lapping machines with swinging and linear motions.

shown in Fig. 20-30. The swinging motion is obtained by rotating the pinion head about an axis which is perpendicular to both the gear and pinion axes and passes through the teeth of the two mating members at a central point of mesh. This is the command motion which causes pinion and gear head slides to reciprocate simultaneously and proportionately.

Table 20-26. Capacities of Lappers with Swinging and Linear Motion

No. 503 Hypoid Lapper—for 90° Shaft Angle	
Dia. of gear (max.), in	8½
Dia. of gear (max. for 1:1 ratio), in	6
Offset of gear spindle above or below center of pinion spindle (max.), in	3
No. 504 Hypoid Lapper	
Dia. of gear (max.), in	36
Dia. of gear (max. for 1:1 ratio), in	24
Offset of gear spindle above or below center of pinion spindle (max.), in	7½
Angular adjustment, deg	50–110

All motions are independently controlled so machine setup is simplified. The gears are mounted in the machine at standard mounting distances, and by means of dial settings, trial amounts of the three motions are activated. When the desired extremity of lapping position is reached, the corresponding dial settings are "locked." This procedure is followed to determine motion limits for toe and heel position on both drive and reverse sides.

During the lapping cycle, the machine automatically moves to the above predetermined limits, thus lapping the entire desired tooth area. Movements and length of time for each side are independently set, and the lapping rate can be made either uniform or variable from one end of the tooth to the other.

An automatic backlash-control mechanism assures that each pair of a production lot is lapped with the same amount of backlash throughout the entire lapping cycle.

Load is provided by an adjustable dynamometer-type brake on the driven spindle.

20-28. Lapping Compounds. The compounds used in lapping machines should consist of a mixture of oil and abrasive. The compound must flow readily and still be heavy enough to hold the abrasive grains in suspension. For lapping spiral bevel gears, a compound such as Carborundum C-280-V20S (RI medium) or its equivalent is suitable. Because of the sliding action present, a finer grade, such as Carborundum C-400-V20S (RI fine), is used for lapping hypoid teeth.

20-29. Lapping by Manual Positioning. This method of lapping is necessarily slow and is generally used only on large gears beyond the range of the lapping machines, for which production quantities are small. It is often referred to as brush lapping in fixed position. The gears are lapped in central, toe, and heel positions on both sides, each position requiring an independent vertical and horizontal setting, as in testing. A lapping paste, which can be obtained with several different grades of abrasive, is applied to the gear teeth with a brush.

REFERENCES

1. Wildhaber, E.: Precision Bevel Gears Cut Quickly, *Am. Machinist*, June 7, 25, 1945.
2. King, C. B., and G. M. Spear: "Helixform: A Recent Development in Spiral Bevel and Hypoid Gears," Institution of Mechanical Engineers, Paper 11, London, September, 1958.
3. Spear, G., C. B. King, and M. L. Baxter, Jr.: Helixform Bevel and Hypoid Gears, *J. Eng. for Industry*, ASME, August, 1960.
4. Wildhaber, E.: Basic Relationship of Hypoid Gears, *Am. Machinist*, 1946.
5. Gleason Works: "How to Test Bevel Gears," Gleason Works, Rochester, N.Y.
6. Gleason Works: "Gleason Bevel Gear Manufacture," Gleason Works, Rochester, N.Y.
7. Candee, A. H.: "Large Spiral Bevel and Hypoid Gears," ASME Paper MSP-51-9, 1929
8. Wildhaber, E.: Relationships of Bevel Gears, *Am. Machinist*, Aug, 30, Sept. 27, Oct. 11 25, 1945.

Table 20-27. Outline of Bevel and Hypoid Gear Materials and Heat-treat Processing

Part I. Precision-type Gears

Type	Precision fine-pitch (light load)	Precision, load-carrying							
		Fine pitch				Medium pitch			
Typical application	Instruments	Instruments, high-speed tools				Aircraft accessory drives			
						Medium stress		High stress	
Materials	1. .40 % C steel (AISI 1040 or 1141) 2. Bronze (SAE 65) 3. Stainless steel (type 416)	1. Nitralloy 2. AISI 4340	Medium-carbon alloy steel (AISI 4140, 4340)			Low-carbon medium-alloy steel (AISI 4620, 8617)		Low-carbon high-alloy stee (AISI 3310, 9310)	
Heat-treatment and associated processing		1. Quench and draw rough blank 2. Finish machine 3. Cut teeth 4. Nitride	1. Heat-treat blank (34 Rockwell C+) 2. Finish blank 3. Grind teeth from solid	1. Heat-treat blank (34 Rockwell C max.) 2. Finish blank 3. Grind teeth from solid	1. Heat-treat blank (34 Rockwell C max.) 2. Finish blank 3. Cut teeth 4. Flame-harden teeth	Carburize and harden	1. Carburize and harden 2. Grind teeth	1. Carburize and harden	1. Carburize and harden 2. Grind teeth
Machine for hardening operation	....	....	....	....	Gear flame hardener (Gleason No. 1)	Quenching press (Gleason No. 16 and No. 529) Roller-quenching machine (Gleason No. 140)			

Table 20-27. Outline of Bevel and Hypoid Gear Materials and Heat-treat Processing *(Continued)*

Part II. Commercial Gears

Type	Commercial			
	Medium pitch		Coarse pitch	
Typical application	General industrial and automotive		General industrial	
	Medium stress	High stress	Medium stress	High stress
Materials	Low-carbon medium-alloy steel (AISI 4027, 8617, 4620)	Low-carbon high-alloy steel (AISI 4820, 3310)	Medium-carbon alloy steel (AISI 4140, 4340)	Low-carbon alloy steel (AISI 4620, 4820, 3310)
Heat-treatment and associated processing	Carburize and harden	Carburize and harden	1. Heat-treat blank 2. Finish blank 3. Cut teeth 4. Flame-harden teeth	1. Carburize and harden
Machine for hardening operation	Quenching press (Gleason Nos. 16 and 26) Pinion-quenching machine (Gleason No. 16) Roller-quenching machine (Gleason No. 140)		Gear flame hardener (Gleason No. 2)	Quenching press (Gleason Nos. 26, 36, and 529) Roller-quenching machine (Gleason No. 140)

Chapter 21

WORM AND WORMGEAR CUTTING AND GRINDING PROCESSES

By

T. W. KHIRALLA, *Staff Engineer, Curtiss-Wright Corporation, Curtiss Division, Caldwell, N.J.*

The manufacture of worms and wormgears is different in many respects from the manufacture of spur and helical gears or the manufacture of bevel gears. This chapter considers the many special things involved in the manufacture of worms and wormgears.

The Chap. 16 material on hobbing and milling is somewhat related to this chapter. The material on hobs and cutters in Chap. 22 is also related to the discussion of wormgear hobs in this chapter.

The rating of wormgear sets for power-transmitting capacity is handled in Chap. 13. Chapter 23 covers wormgear inspection.

In general, the production of commercial-grade wormgears poses very few problems. Gears of higher accuracy demand very close attention. Each manufacturing facility may develop its own peculiar methods of production consistent with the type of work it is called upon to do. To cover these various approaches, in their many forms, is beyond the scope of the present work. Here, we shall concern ourselves only with those processes which are both common and basic within the wormgear industry.

21-1. The Removal of Material. A glance at a pile of chips usually generates very little interest, but we must recognize the fact that the methods used to produce these chips determine the quality of the finished components. The removal of material, to produce the desired geometry in a gear, is a subject that requires constant and careful study. The ultimate cost of a completely machined gear is a function of the speed with which the excess material has been removed. It is necessary, therefore, that:

1. The equipment be in the best of condition
2. The cutting tools be properly designed
3. The cutting edges of the tools be properly sharpened and maintained
4. The cutting speeds and feeds be properly selected
5. Cutters and work be securely and properly mounted

It should be emphasized that the cutting speeds and feeds used must be determined and based on the material to be cut, how this material responds to cutting, and the surface finish desired on the gear-tooth or worm-thread profiles.

The processes used to remove material may involve rotary motion such as is obtained with a hob or milling cutter, or rectilinear motion such as is obtained with a reciprocating tool. In the manufacture of worms and wormgears, the rotary-cutting process is almost exclusively used.

21-2. Mounting of Hob or Cutter. Except for the different types of gears that they may cut, the mounting of a hob or cutter in the machine requires the utmost in attention, care, and precision, if the end product is to meet the required blueprint specifications. It will be noted that all hobs and cutters have ground reference diameters that were used by the manufacturer to produce the tool. These reference diameters may be the hubs extending out from each end of the tool body, or a ground shank on each side of the cutting section, if the cutter is of the shank type. This is true except for the pinion-type gear-shaper cutter. In the latter instance, the trueness of the cutter is checked by indicating the trueness of the tip lands of the cutter teeth with a dial indicator.

In all cases, the shafts, arbors, or cutter spindles must be checked for trueness before the cutting tools are mounted in place. After trueness has been established, the cutting tool is mounted on the arbor or spindle and securely locked in place.

21-3. Mounting of the Work Arbor or Fixture. The work spindle on the cutting machine must also be checked for its trueness of rotation, prior to mounting the work arbor or cutting fixture. When this has been done and the work spindle has been adjudged in proper working order, the work arbor or cutting fixture is mounted in place.

The mounting surfaces that will support the work blank must then be checked to make certain that they are true with respect to the rotation of the work spindle about the axis of rotation. After this has been achieved, the machine is ready for the details of setting up the other components and systems such as the feed gears, the index gears, and the differential gear train, if the machine is equipped with this last feature.

21-4. The Thread-milling Process. Under this heading we shall discuss the milling of worm threads as differentiated from hobbing. A worm-thread-milling machine is basically a screw-cutting lathe and is designed specially for the production of machine components having thread forms. Its basic elements are a work spindle, cutter head, lead screw, tail stock, and change-gear train. The lead screw actuates the cutter head and provides the latter with a controlled linear motion, parallel to the axis of the worm blank. The gear train provides the proper time relation between the rotation of the work spindle and the linear motion of the cutter head. The cutter head is geared, through the change gears, in the proper time relation with the rotation of the worm blank so that, for each revolution of the work, the cutter advances one lead parallel to the axis of the worm.

The cutter head is mounted on a rotating member so that its axis may be set to any desired angle with respect to the axis of the worm blank. The angle between the axis of the cutter and the axis of the worm blank is usually equal to the lead angle of the worm being cut. This positions the plane of rotation of the cutter in proper relationship with the thread helix.

Setting Up the Thread Miller. Each thread-milling machine has its own special features, depending upon the design. The operator's manual supplied with each machine will explain these features in detail. There are, however, certain basic procedures that are followed regardless of what make of equipment is used, and these are discussed below.

The important and basic points that must be attended to when setting up such a machine tool are as follows:

1. Mounting the cutter (see Art. 21-2)
2. Setting cutter head to worm lead angle
3. Mounting change gears
4. Mounting speed gears
5. Mounting workholder (see Art. 21-3)

The change gears tie in the rotation of the work spindle with the linear advance of the cutter head, as explained in the beginning of the present article. The equation used to compute the required change gears depends upon the design of the machine itself. This is always supplied by the manufacturer. After the five steps have been completed and checked out to make certain that all are correct and properly functioning, the machine is ready for cutting the worm threads.

The cutter tip is brought into contact with the outside diameter of the worm blank, to score it about 0.002″ deep. The gear trains are then engaged. A cut is taken on

FIG. 21-1. Thread-milling machine. (*Courtesy of Waltham Machine Works, Waltham, Mass.*)

the outside diameter of the blank to make certain that the equipment is working properly and is producing the proper lead. When this has been checked and found to be correct, the machined is stopped, the cutter is retracted from the work, and the cutter head is returned to the starting position. Care must be taken not to disengage the gear trains. If the gear trains are accidentally disengaged, the cutter is no longer in proper time relation with the cut taken in the worm blank.

The machine is now ready to perform its work. The cutter is now set to the full depth of thread less about 0.005″, depending on pitch, and a complete pass through the work is made. Using measuring wires or special thread gages, the worm is checked for dimensional accuracy. Several passes may be found necessary before the correct dimensions are obtained. Figure 21-1 shows a typical thread miller more suited for cutting fine-pitch threads. Figure 21-2 shows a thread miller which is adapted to coarser-pitch work.

Milling Single-thread Worms. The foregoing discussion would be the procedure to follow for the production of single-thread worms. After the finished worm has been removed from the machine, the cutter is returned to the starting position. A new blank is mounted in place and the cutting cycle is started. Throughout the operations care must be taken, when mounting a new blank, that all mounting surfaces are clean and free from chips and burrs.

Milling Multiple-thread Worms. When milling multiple-thread worms, the same procedures and precautions are followed. The added operations are those which are made for purposes of indexing. After the first thread is cut the cutter is retracted from the work so that the cutter tip is moved slightly beyond the outside diameter of

FIG. 21-2. Large thread-milling machine. (*Courtesy of Lees-Bradner Co., Cleveland, Ohio.*)

FIG. 21-3. Use of gang cutter on a worm.

the worm. With the gear trains still engaged, the cutter head is brought back to the starting position. A handwheel is provided for this purpose. After the cutter head is in starting position, the indexing device is actuated to rotate the work spindle to the next thread. The indexing mechanism is then secured and the cutter is again set to the full depth of thread. Another complete pass is made through the work. This process is continued until all threads have been cut.

Some machines are equipped with a special switch for actuating a rapid-traverse mechanism. After the cutter has been withdrawn from the work, pushing the switch will cause the cutter head to return to the starting position at a rapid rate and, at the same time, maintain the proper time relation between the work and the cutter head.

Adaptability of Thread Millers. Thread millers, as shown in Figs. 21-1 and 21-2, are well adapted to cut external and internal threads. Figure 21-3 shows a setup for cutting external threads using a gang cutter, without lead. Figure 21-4 shows a setup using a similar type of cutter for producing internal threads on a production basis. Both these processes are commonly used and, in many instances, provide the only means for producing such components easily and economically.

FIG. 21-4. Production setup using gang cutter on thread.

Figure 21-5 shows the versatility of this type of equipment for producing the unusual type of work sometimes required to meet certain design specifications. The machine may be adapted for producing parts not commonly associated with a thread miller, and its usefulness, in any shop, is dependent upon the skill and ingenuity of those handling the machine.

Milling Worms on a Gear Hobber. When worms have three or more threads, the hobbing process is the preferred method of cutting. However, single-thread worms may be milled on a gear hobber very conveniently and economically. When the required accuracy is not too great double-thread worms may also be milled using a *duplex cutter* which completes both threads in a single pass. We shall therefore confine this discussion to the procedures used to cut single-thread worms or double-thread worms where only a single pass of the cutter is required.

The hobbing machine was designed primarily for the production of spurs, helicals, splines, sprockets, wormgears, and other types of components where teeth of some kind are cut on the periphery of a blank. The use of a hobbing machine as a thread miller is incidental and merely emphasizes its adaptability to normal and special shop demands. In the production of single-thread worms, or more precisely, worms that

FIG. 21-5. Cutting high-lead-angle thread on a very long shaft. (*Courtesy of Lees-Bradner Co., Cleveland, Ohio.*)

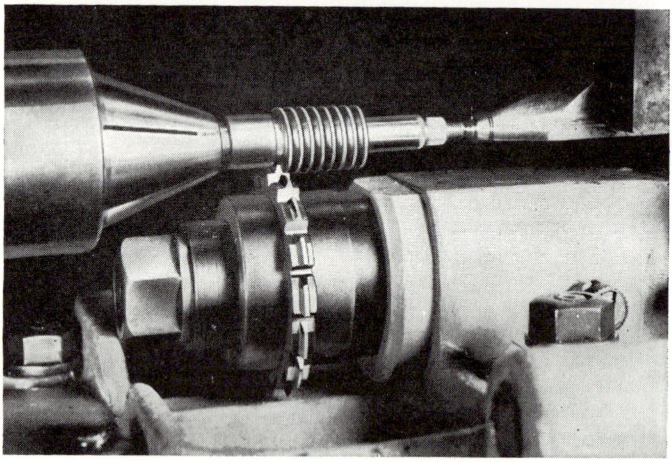

FIG. 21-6. Cutting single-thread worm on a hobbing machine. (*Courtesy of Barber-Colman Co., Rockford, Ill.*)

may be produced in a single pass of the cutter, this machine may be found to be an excellent performer.

Setting Up Hobber as Thread Miller. The procedures used to set up a thread miller, described earlier, are to be followed in setting up a hobbing machine to mill threads. However, because of the arrangement of the gear trains in the hobber, special attention should be directed to this part of the setup. On the standard hobbing machine we have the *index gear train* and the *feed gear train*. The index gears are used to provide the necessary chip load on the cutter teeth. See Art. 21-6 for details regarding chip-load selection and calculations for gears to provide proper rate of removal of material. The feed gears are selected and arranged to traverse the cutter head one lead for each revolution of the work blank.

The operator's manual supplied with each machine will give the formulas that must be used to determine each gear train for a particular job.

The cutter head, usually supplied with the machine, may not be designed for high angle settings and a 90° *hob swivel* will be required when worms are to be cut. Single- or double-thread worms have relatively low lead angles, which makes the helix angles rather high. In some instances a standard hob swivel will interfere with some part of the machine if set to these high helix angles. It is necessary, therefore, to use a 90° hob swivel to provide the clearance needed under these conditions. Figure 21-6 shows a 90° hob swivel arrangement.

The swivel will be set to the lead angle of the worm to be cut, or to the helix angle, depending upon the design and arrangement of the vernier setting scale. After the cutter swivel is secured, the milling cutter is mounted on the arbor and secured, after making certain that the precautions outlined in Art. 21-2 have been followed.

The workholder or fixture is then mounted, as explained in Art. 21-3.

When the index gears and feed gears have been selected and mounted in place, the machine is ready for cutting the worm. The following has been done up to this point:

1. Mounting cutter (see Art. 21-2)
2. Setting cutter head to worm lead angle
3. Mounting workholder (see Art. 21-3)
4. Mounting index gears
5. Mounting feed gears
6. Mounting and securing worm blank

We are now ready to cut the worm. The procedure will be exactly as outlined in Art. 21-4 except for one feature. The hobbing machine, as already noted, was not designed for cutting worms. This was, in essence, an afterthought. There is no arrangement* in the standard hobbing machine for a rapid return of the cutter to its starting point after it has taken a pass through the work. The only way that the cutter can be returned to its starting point without destroying the time relation between the cutter and the thread that has been cut is to withdraw the cutter from the work, making sure that the gear trains remain locked in, and then reverse the rotation of the motor so that the machine is driven backward. This is time-consuming, but under these circumstances one has no choice but to follow this routine.

Some machines have incorporated in their design a *differential gear train*. When such a machine is used for milling threads, the cutter head may be rapidly traversed without unlocking the gear trains.

Attention is again called to Fig. 21-6. In this illustration the relationship between the 90° hob swivel, work, and machine elements is excellently depicted. The cutter has completed its pass through the work. Without the 90° hob swivel this job could not be conveniently done on this machine.

Using Duplex-type Milling Cutter. When high accuracy is not required, such as for commercial worms, and there is a requirement for a low-lead-angle double-thread worm, a *duplex-type milling cutter* may be used. The angular setting of the cutter head is the same as that for a single-type milling cutter; the head is set to the lead angle of the worm.

Cutters of this type are usually restricted to producing the worm for which they are designed. Great care must be exercised in setting this cutter, as any deviation

* A special registering means has been developed for standard Gould and Eberhardt hobbers to permit recapturing the lead after disconnecting the gear trains to return the cutter with the regular power traverse.

from the arrangement shown in Fig. 21-7 will greatly reduce the effectiveness of the working profile. Any misalignment will cause excessive deviation in thread profile and this will show up in lack of symmetry of the thread form.

Once the cutter has been properly set up and secured the balance of the process is the same as described above.

Using Multiple Worm-milling Cutter. The use of *multiple-type worm-milling cutters* is recommended when commercial worms are to be produced in large quantities. In this type of operation the multiple (or gang) type of cutter is fed radially into the work and then, after it has gone to the correct full depth, the work makes one revolution and is completed. The cutter, unlike the cutters discussed previously, is set to zero lead angle. In other words, the axis of the work is parallel to the axis of the cutter.

Except for setting the cutter axis to zero lead angle the setup of the machine is performed as previously described. However, when a worm is to be cut the radial feed is performed simultaneously with the axial feed. Before the cutter is driven into the work the index gears and the feed gears are engaged, and then, while they are in motion, the cutter is set into the work to full depth.

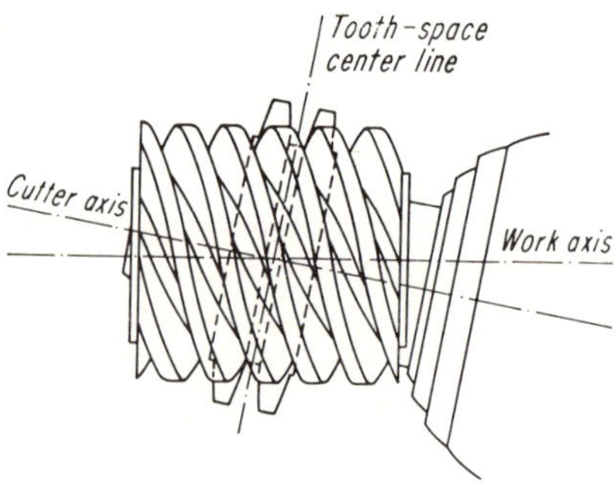

FIG. 21-7. Relationship between duplex cutter and work. (*Courtesy of Barber-Colman Co., Rockford, Ill.*)

The cutter used has several annular rows of teeth depending upon the length of the worm being cut, and the lead angle is zero (see Fig. 21-9).

Driving the cutter into the work radially may be done using a *vertical-feed attachment* to provide the radial motion, or for long production runs a cam may be used to accomplish this task (see Fig. 21-12).

21-5. Thread-hobbing Process. Hobbing worm threads is usually done when there are three or more threads in the worm. Under certain conditions double-thread worms may be cut in this manner. Hobbing worms requires the setting up of the hobbing machine the same way as will be done for a helical gear except a 90° hob swivel is used in most cases because of the low lead angle. If, as in some instances, the number of threads is such that the lead angle permits the use of a standard hob swivel, then, of course, this may be used.

The setup of the machine is the same as has been described for setting up a hobbing machine for thread milling except for setting the hob swivel and computing the index and feed gear trains.

Setting the Hob Swivel. When hobbing worms, the hob is made with cutting teeth arranged along a helix, and the hob has a lead angle. It will be recalled that the single, duplex, and multiple thread-milling cutters all had zero lead angle.

In the case of a hob used to hob worms, there is a lead angle depending upon the hob specifications. The setting of the hob swivel to the correct angle must be determined from both the lead angle of the worm and the lead angle of the hob. The lead angles of work and hob may or may not have the same hand of helix. One may be right hand; the other may be left hand. The rules to apply in determining the correct angular setting of the hob swivel are as follows:

1. When hob and work are of the same hand, add the lead angle of the hob to that of the work.
2. When the hob and work are of opposite hands, subtract the lead angle of the hob from that of the work.

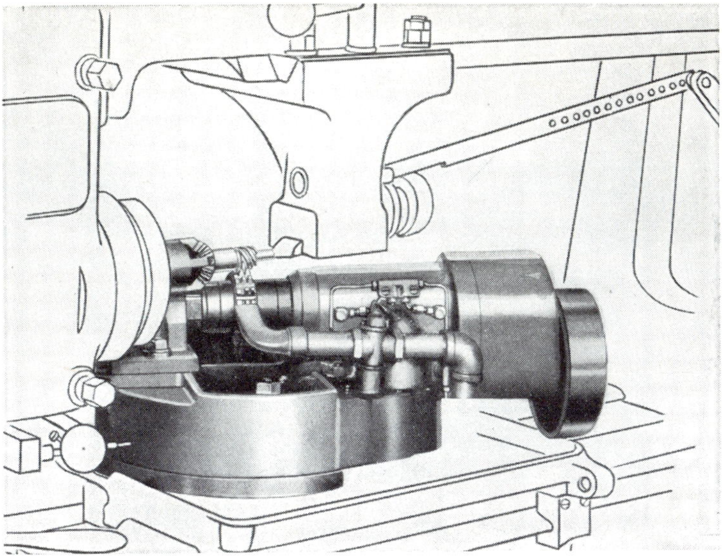

Fig. 21-8. Hob swivel capable of 90° swing setup for worm hobbing. (*Courtesy of Barber Colman Co., Rockford, Ill.*)

The Index and Feed Gear Trains. Each make of hobbing machine has designed into it the manufacturer's conception and decision as to the machine constants which are necessary for use in computing the index and feed gear trains. These constants will be found in the operator's manual. Each gear train has its own constant.

In hobbing worms the computations for the gear trains must include both the index and lead gears as these are tied together and must have the proper relationship in order to produce the correct number of teeth and lead of the work. Because of this "built-in" relationship the gear trains must be mutually engaged through a clutching mechanism and remain so engaged until the worm has been completely cut.

In hobbing worms, particularly those having low numbers of threads, the full length of the hob is used in the hobbing of the worm. The *leading* hob teeth are chamfered so that they will not pound the work blank when entering, and this chamfering allows for a more gradual entry. The teeth immediately after the chamfered, or leading, teeth are the finishing teeth. About three finishing teeth are provided as any more than that will possibly drag through the finished teeth, which is an undesirable condition.

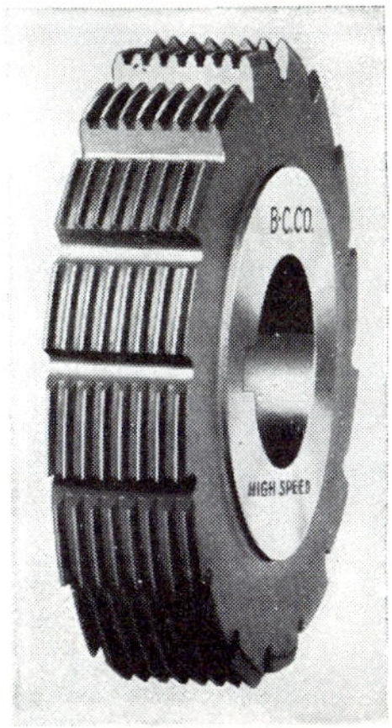

Fig. 21-9. Multiple-type worm cutter having zero lead angle. (*Courtesy of Barber-Colman Co., Rockford, Ill.*)

Reference is made to Fig. 21-8, showing a view of the 90° hob swivel and its application in hobbing worms of more than two threads.

Figure 21-10 shows a close-up of another worm being hobbed where radial infeed and axial feed are necessary because the worm is cut into a shaft with no free end, such as is provided in Fig. 21-8.

The surface finish of a hobbed worm is a function of the number of flutes or gashes in the hob. In order to provide a maximum of flutes the hobs are made as large as is practicable to enable the production of worm threads with smooth surfaces.

Using hobs for producing multiple-thread worms makes it unnecessary to index each thread as in thread milling. The hob automatically develops the indexing, and the process is continuous until the worm is completed.

FIG. 21-10. Worm-hobbing operation. (*Courtesy of Barber-Colman Co., Rockford, Ill.*)

21-6. Determination of Chip Load. Because of the many variables involved in the removal of metal when producing gears of various kinds, the amount of chip load to use for any particular job is a matter of judgment based on long experience in the field. The following definition of this term may be helpful to understand its true meaning.

Chip load is the amount of material each cutter tooth removes per revolution of the cutter.

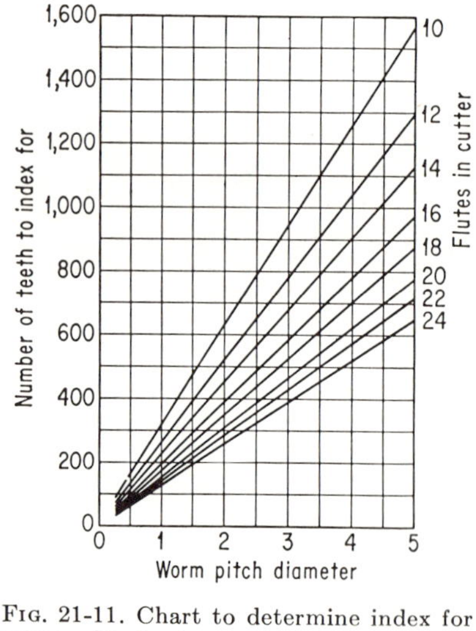

FIG. 21-11. Chart to determine index for proper chip load.

The chip load used will help to determine the surface finish obtained during cutting. It would be rather difficult to lay down rigid rules for determining the chip load required for different materials. As stated above, one's judgment and experience must be used, as there is no other way to establish these data. However, the logical approach to this problem, and one that is used in many instances, is to start with a chip load of 0.001″ and keep making adjustments in speeds and feeds until the results needed are obtained.

For milling worms the chip loads used vary from 0.0005 to 0.0030″.

For hobbing worms the chip load used may vary from 0.0005 to 0.0020″.

On thread-milling machines the chip load is varied by changing the rate at which the cutter passes through the work. The speed of rotation of the cutter is independent of the rate of cutter-head traverse. Therefore, the higher the rate of traverse, the greater the chip load. Experience will dictate what may be used under given circumstances.

When milling threads on a hobbing machine, the problem of determining and varying the chip load is somewhat different. Here we have a machine that was designed, originally, without any thought of using it to mill worm threads. In hobbing machines the rate of rotation of the worm blank is determined by the gears that make up the index gear train.

The equation that may be used to determine the make-up of the index gear train to give the required chip load is shown below.

$$N = \frac{\pi D}{TC_T} \tag{21-1}$$

where N = number of teeth to index for
D = pitch diameter of worm
T = number of flutes in cutter
C_T = chip load

A chart may be constructed to simplify the procedure. The chart shown in Fig. 21-11 may be used as a first choice. The chart itself is based on a chip load of 0.001″ per cutter tooth per revolution of cutter.

FIG. 21-12. Gear hobber showing vertical-feed attachment. (*Courtesy of Barber-Colman Co., Rockford, Ill.*)

If a finer or coarser chip load is required, the number of teeth, as represented by the ordinate, must be multiplied or divided by the required multiple of 0.001. In other words, if a chip load of 0.002″ was required the number of teeth will be divided by 2. If a finer chip load of, say, 0.0005″ was required, the number of teeth will be multiplied by 2.

21-7. Speeds and Feeds. In the manufacture of any product *time* is an important element. The time it takes to produce a worm or gear is a function of the rate of removal of material in a unit interval of time. The rate at which material is removed from a blank is a function of the *speed* of the cutter in revolutions per minute and the *feed* of the cutter in inches per minute or in inches per revolution of the work.

Many different kinds of material are used, and each shop has its own peculiarities which affect the speeds and feeds that may be used. The kind and age of the available equipment must be considered as well as the ingenuity and experience of the available working force. Whether *climb* or *conventional* milling or hobbing is used may well affect the time taken to produce a part. The type of workholding fixtures used may be another factor.

The determination of speeds and feeds used in a machine shop should be established with care and by those well grounded in the various processes and materials used for producing their products. The hob and cutter manufacturers can be valuable advisors in this regard. However, each shop should experiment with various combinations of speeds and feeds and determine what is best for their purposes. These experiments need not be formal laboratory tests. They may be effectively carried out during the initial run of the parts. The important point, however, is to keep and maintain proper records of such tests for future reference and use.

The capacity of the machine to deliver power at the point of cutting and the rigidity of the equipment are of vital importance and must be taken into account. Other factors that must be considered are as follows:

1. Depth of cut
2. Material and its cutting properties
3. Hardness of the material
4. Surface finish required
5. Accuracy required
6. Type of coolant used
7. Type of cutter used
8. Material from which cutter is made

21-8. Calculate Thread-milling Time.

Single- and Duplex-type Cutters. The following procedure may be used for computing the floor-to-floor time, in minutes, when using single- or duplex-type cutters. It may also be used for computing the floor-to-floor time for multistart worms when using a single-type cutter.

The feed rate is usually between 0.0005 and 0.0030″ chip load per cutter tooth. The cutter speeds usually run between 60 and 200 sfpm.

The following equation will be found very useful for these computations.

$$t = \frac{\pi DN}{nCG \cos \lambda}\left(\frac{B}{L} + 1.2\right) + F + I(N - 1) \tag{21-2}$$

where t = floor-to-floor time, min.
B = length of full thread, in.
D = outside diameter of worm, in.
L = lead of worm, in.
C = chip load per cutter tooth, in.
G = number of teeth in cutter
n = speed of cutter, rpm
λ = lead angle of thread, deg
N = number of threads in worm
F = loading plus unloading time, min
I = indexing time, min

It will be noticed that, when cutting single-thread worms with a single-type cutter or when cutting double-thread worms with a duplex-type cutter, the expression $I(N - 1)$ in the equation becomes equal to zero.

In such instances Eq. (21-2) will reduce to

$$t = \frac{\pi DN}{nCG \cos \lambda}\left(\frac{B}{L} + 1.2\right) + F \tag{21-3}$$

Multiple-type Thread-milling Cutters. The multiple-thread-type milling cutter, as shown in Fig. 21-9, is used for high-production commercial-class worms having small lead angles. This process is discussed in Art. 21-4. Remembering that this process involves a vertical feed motion and a circumferential feed motion to complete the cycle, the following equation may be found useful in determining the floor-to-floor production time.

$$t = \frac{C_H h_t + C_V(\pi D/\cos \lambda)}{C_H C_V G n} + F \tag{24-4}$$

where t = floor-to-floor time, min
D = outside diameter of worm, in.
G = number of flutes in cutter
n = speed of cutter, rpm
F = loading plus unloading time, min
λ = lead angle of worm, deg
C_H = chip load per cutter tooth, circumferential feed, in.
C_V = chip load per cutter tooth, vertical feed, in.
h_t = whole depth of worm thread, in.

21-9. Calculating Hobbing Time for Worms. The worm-hobbing process has been discussed in Art. 21-5. The following procedure may be used to determine the floor-to-floor hobbing time when using this process. The following equations may be used in the order given.

$$A = \frac{\sqrt{h_t(D - h_t)}}{\tan H} + \frac{W}{2 \sin H} \tag{21-5}$$

$$C = \frac{a}{\tan \phi_n} \tag{21-6}$$

$$t = \frac{N(A + B + C)}{nfT} + F \tag{21-7}$$

where t = floor-to-floor time, min
A = hob approach, in.
B = length of worm, in.
C = extra cut for full generation, in.
D = worm outside diameter, in.
a = addendum of worm, in.
n = speed of hob, rpm
N = number of threads in worm
f = feed, in. of hob advance per revolution of worm
T = number of threads in hob
F = loading plus unloading time, min
H = setting angle of hob swivel, deg
W = width of hob face, in.
h_t = whole depth of worm thread, in.
ϕ_n = normal pressure angle of worm, deg

21-10. Thread Generating. The *thread-generating process* is one that permits the cutting of worms, either single or multiple threads, using a continuous cutting cycle. It eliminates the need for indexing each thread and provides an accurate means of developing thread forms. It may be considered a combination of cutting and shaving, both processes incorporated in a single continuous cycle. A helical-pinion-type gear-shaper cutter is used, and the relationship between the cutter and the work is as shown in Fig. 21-13. The worm may be considered as a rack running with the cutter, which simulates a gear. The axis of the worm blank is at right angles to the axis of the cutter.

The worm rotates about its axis and the cutter, in addition to rotating about its own axis, is also translated in a direction parallel to the axis of the worm. The combination of rotation and translation dictates the use of a differential gear train which is incorporated in the design of the machine.

There are three sets of gears that must be determined for each job or worm to be cut. The *work change gears*, the *feed gears*, and the *differential gears*. The work-change gears control the proper relationship between the work and the cutter in terms of the

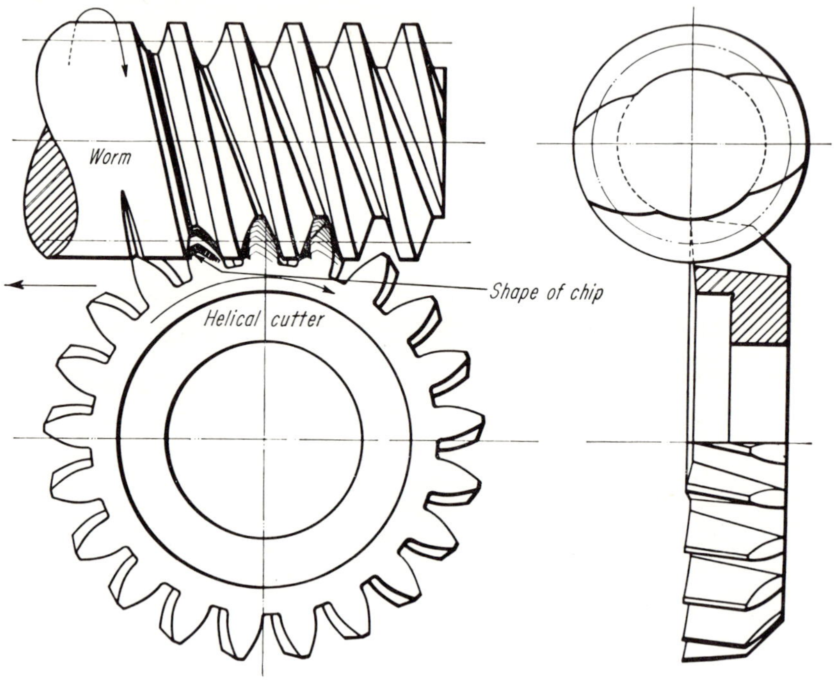

FIG. 21-13. Cutting action of helical-type cutter when applied to generating worm threads and also shape of chip taken. (*Courtesy of Fellows Gear Shaper Company, Springfield, Vt.*)

number of threads in the work and the number of teeth in the cutter. The feed gears control the advance or traverse of the cutter along the length of the worm blank. The traverse motion of the cutter disturbs the proper time relationship between the cutter and the work, and this must be compensated for by the introduction of the differential gear train to provide a correction.

The precautions and procedures to be exercised and used for mounting the cutter and workholders are the same as outlined in Arts. 21-2 and 21-3, respectively.

Any hobbing machine equipped with a differential gear train may be used to perform this operation. This is done by mounting the cutter on the work spindle and mounting the work on the cutter spindle. There are, however, limitations involved, and in order to provide the maximum of flexibility, special machines are designed and built for this work such as is shown in Fig. 21-14.

It must be noted that when worms are cut, such as is illustrated in Fig. 21-13, where the blank diameter adjacent to the worm is such as to allow the cutter to run into it, a shallow thread must be allowed for in the design. If, however, the blank diameters adjacent to the worm ends are less than the worm root diameter, then no such interference will exist. In the former case the cutter will have to be retracted in order to remove the work from the machine. In the latter case retraction of the cutter may not be necessary.

This method may also be used in cutting the hourglass or enveloping-type worm similar to that used in the Cone-Drive®.* The same procedures may be followed, with one exception. No traverse of the cutter is necessary and the time relation between the worm blank and the cutter is strictly a matter of ratio as determined by the number of teeth in the cutter and the number of threads in the worm. In both processes the cutting face of the cutter lies in the plane containing the axis of the work.

FIG. 21-14. Thread generator for generating worms and threads. (*Courtesy of Fellows Gear Shaper Co., Springfield, Vt.*)

In the case of the shell-type or cylindrical worm the size of the cutter or the number of teeth it contains leaves no particular restrictions, provided that the cutter size is within the limits of the design of the worm and the size that the machine will accommodate.

With the enveloping-type worm the cutter must be, substantially, a duplicate of the gear that the worm will run with except that it has cutting edges for generating the proper thread form.

21-11. Hobbing Wormgears. The hobbing of wormgears may be accomplished by either one of two convenient methods: the *radial* or *infeed method* or the *tangential-feed method.* Figure 21-15 shows substantially the principle of each of these methods. Wormgears produced by either of these methods will have "throated" teeth, which immediately establishes the location of the axis of the mating worm. Also established is the location of the axis of the hob used to produce these wormgears.

* Cone-Drive is a registered trade-mark of the Cone-Drive Div. of Michigan Tool Co., Detroit, Mich.

Obviously, the hob cannot travel across the face of the wormgear blank as is done in the case of spurs, spirals, and helicals. The hob slide must therefore be fixed in one position in order to obtain the results required. The hob itself may travel radially or tangentially but not parallel to the axis of the gear being cut. This introduces another precaution in setting up the hobbing machine for producing wormgears. The usual precautions, as outlined in Arts. 21-2 and 21-3, should be carefully followed for either method.

The additional precaution of locating the proper position of the axis of the hob with respect to the throat of the gear being cut is extremely important in the hobbing of wormgears. The procedures used to find the proper relationship are applicable to both the methods described, namely, infeed and tangential hobbing.

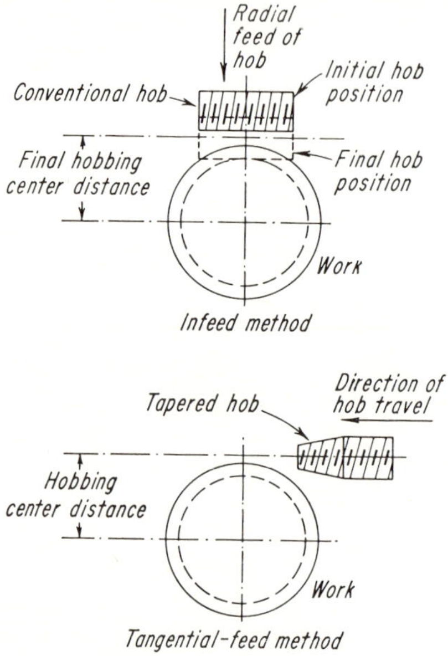

FIG. 21-15. Illustrating infeed and tangential-feed methods of cutting worm gears.

The hob setting requires special attention because of the nature of the contact between the profiles of the wormgear and its mating worm. The profiles of both these elements are warped surfaces, and great care must be exercised in their production. Because of the throated feature of the wormgear, it would be assumed that the hobs used to produce them should have the exact specifications of the mating worm, and the hob axis should be set to the same angle as the worm is to the gear when operating.

If this is done the hob would lose its usefulness after the first sharpening because it would then be smaller than the worm. Hobs made to such specifications are not practical, nor are they economical from the standpoint of unit cost per gear.

The ideal contact between the worm and its mating gear is line contact across the face of the gear. This would be obtained if the hob had the same specifications as those of the worm. As noted above, on the first sharpening the hob would be smaller than the worm, resulting in reduced radius of curvature of the throat, and contact would then be at the ends of the gear teeth with none in the center. From a practical and economical viewpoint it would be best to have a hob slightly larger than the mating worm. A wormgear produced with such a hob will provide contact between the mating elements at the center of the gear teeth, which is more desirable than at the ends.

A wormgear hob made slightly larger than the mating worm will require a hob setting, in angular terms, slightly different from the angle between the axes of the worm and its mating gear. As the hob is used and progressively sharpened it will approach the exact specifications of the worm, but it will always produce a gear that will provide contact at the center of the gear face. Furthermore, as the hob is sharpened this contact pattern will gradually spread out from the center of the gear face in a progressively wider pattern until the hob reaches the point where it is now the same size as the worm. When this point is reached, the hob has served its purpose and must be replaced with a new one.

In producing wormgears and worms it is the general practice to provide a correct worm first and use this to check the contact pattern when the gear is being cut. With any hob, provided for the purpose, the angular setting of the hob slide is a matter of trial until the proper contact pattern is obtained.

When a wormgear hob is to the same specifications as the mating worm, the angular setting of the hob slide is zero for a pair of mating components that operate together on 90° shaft angles. For the slightly larger hobs that are used when new the angle setting would be larger and the amount of increase is determined by trial as noted above.

When the hob setting has been determined, the hob slide is secured in position and kept there until the hob has been resharpened. The setting angle may be obtained by calculating the lead angle of the hob. The pitch diameter of the hob may be obtained by subtracting two gear dedendums from the measured outside diameter of the hob. Even when using this method the tooth contact pattern should be checked to be sure that proper contact will be achieved. Reference to Fig. 21-16 will show the basic relationship between hob and work.

After the hob setting has been established in terms of the setting angle and the location of the hob with respect to the gear blank the hob slide is secured.

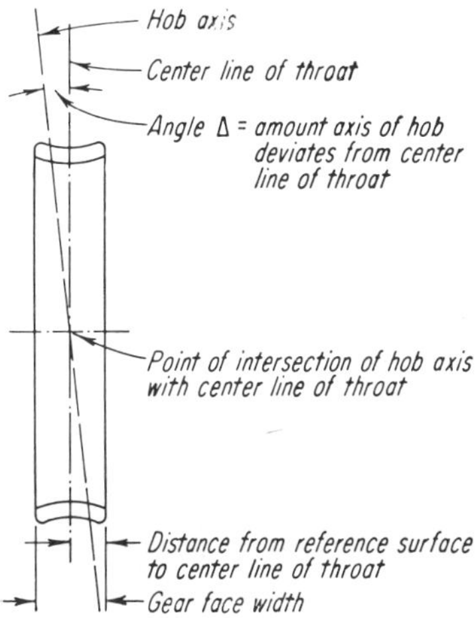

FIG. 21-16. Showing relationship between hob axis and wormgear when hob is larger than mating worm.

21-12. Infeed Hobbing of Wormgears. Hobbing wormgears by the *infeed method* may be done on a machine such as is shown in Fig. 21-12. The *vertical-feed attachment*, used to provide the *radial infeed* motion, is shown prominently in the illustration. Other hobbing machines, particularly those having a vertical work spindle, have the infeed-motion mechanism designed into the machine.

For this infeed method of hobbing, the index gears are selected from a chart which accompanies the machine, and the change gears used to provide the radial feed are also selected from a chart provided for the purpose. The change gears for the infeed motion should be selected to give the proper chip load per hob tooth, and this will be dictated by the material being cut. The hob speed will also be governed by the material being cut and the finish required.

The infeed mechanism is usually adjusted to cut the tooth slightly less than full depth in order to make certain that everything is operating properly. The actual depth may be set by calculating the proper center distance between the hob and the work for the conditions that exist. This center distance will be computed from the known specifications of the hob being used.

After the hob and work have been properly secured in their respective positions and before any cutting is done, a check must be made to make sure that both hob and work are rotating in their proper directions to suit the conditions. The diagrams in Fig. 21-17 show the proper direction of rotation for both right- and left-hand hobs and work.

When it has been established that the proper directions of rotation have been obtained, the hob is fed in until the outside diameter is very slightly scored by the tips of the hob teeth. If everything is in order, the tips of the hob teeth will develop as many scratches or score marks on the outside diameter as there are teeth in the gear.

After it has been established that everything is functioning properly, a cut is taken, as noted above, to just a little less than the full depth of tooth of the gear. This provides an opportunity to check the surface finish and to check the tooth form visually to make certain that nothing serious has developed. The hob is then fed in to full depth (or to the correct center distance between hob and work). After the hob has reached full depth the work must be allowed to make at least one full revolution to complete the generation of the gear teeth, and then the hob is retracted from the work and the machine stopped. Inspection then follows to determine correct size and tooth contact pattern.

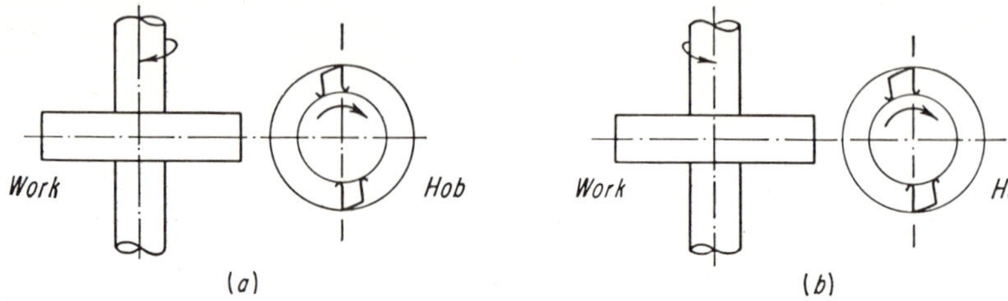

FIG. 21-17. Showing relative rotations between work and hob. (*a*) Right-hand hob; right-hand work. (*b*) Left-hand hob; left-hand work.

The infeed method of hobbing is usually used only for relatively low lead angles particularly for wormgears that may be considered to be of commercial grades. It must be remembered that wormgear hobs have relatively few cutting flutes because their sizes are restricted to the size of the worm. The gear-tooth profiles consist of a series of flats or facets, not the smooth continuous curve desired for tooth profiles. Each flute generates a flat. In order to have a relatively smooth continuous curve, many flutes must be built into a hob, and the restrictions due to the worm size do not permit a large number of flutes. It is for this reason that a low number of threads in the worm and a small lead angle allow the use of the infeed method under these circumstances. When the lead angle is high and where accuracy is required, some arrangement must be made to develop a smooth continuous tooth profile for the gear teeth. The tangential method (Art. 21-13) provides the means for achieving this end.

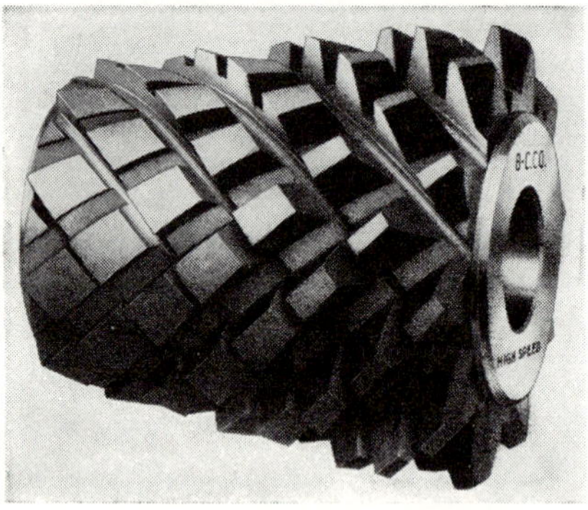

FIG. 21-18. Tapered hob for tangential hobbing of wormgears. (*Courtesy of Barber-Colman Co., Rockford, Ill.*)

21-13. Tangential Hobbing of Wormgears. When hobbing wormgears using the *infeed method,* the design of the hob is substantially cylindrical in form and similar to hobs used to produce spur gears. When using the *tangential method,* the hob is designed as shown in Fig. 21-18. This hob has a tapered section and a cylindrical section. The tapered section provides cutting teeth for roughing out the wormgear teeth, and most of the material is removed by this portion of the hob.

The cylindrical section provides cutting teeth for finishing the wormgear teeth.

When this hob is used the hob slide is fixed in position, as was discussed in Art. 21-11. However, in order to perform the complete cutting cycle, the hob travels along its axis and tangent to the wormgear blank. The combined rotary motion of the hob about its axis and its longitudinal motion along its axis creates a condition similar to that of a hob having a great many cutting flutes. This action develops wormgear teeth with a great many extremely small flats, which results in a tooth profile of high accuracy. The quality of the tooth profile is a function of the feed of the hob along its axis.

Tangential hobbing is not so common as infeed hobbing. The cost of producing wormgears by tangential hobbing is greater than by infeed hobbing, and the deciding factors will be quality required and unit cost.

In order to use this method, a hobbing machine equipped with *differential change gears* will be required. Such a machine is shown in Fig. 21-19. When setting up a machine to do this work, three sets of change gears are needed, the index gear train, the feed gear train, and the differential gear train. The latter gear train is needed to make the corrections required because of the motion of the hob along its axis.

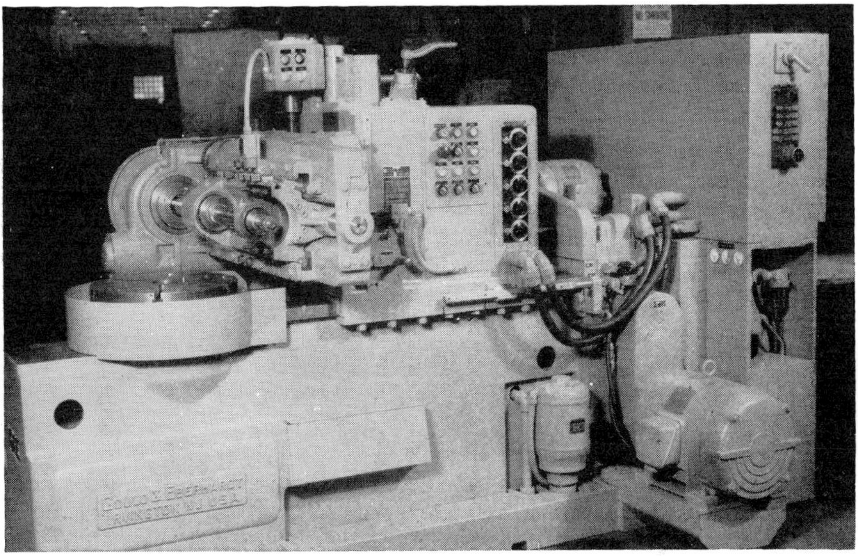

FIG. 21-19. Wormgear hobbing machine with completely automatic cycle, variable feed, and dual control. (*Courtesy of Gould & Eberhardt Div., Norton Co., Worcester, Mass.*)

The center distance between the work and the hob must be determined by calculation, and based upon the specifications of both. The final center distance and hob-setting angle will be determined when the proper contact pattern is obtained, as explained in Art. 21-11.

The specific details for any particular tangential-hobbing machine may be obtained from the operator's manual, and these manuals provide the various machine constants that are required for determining the tooth numbers or ratios for the various gear trains. In general, the gears for the index and feed gear trains are selected from a chart made up for the equipment being used. No calculations are usually necessary. However, calculations must be made to obtain the necessary gears to be used in the differential gear train. These calculations must provide a gear train to give the desired ratio to as high a degree of accuracy as can be obtained. An actual setup illustrating this method of hobbing is shown in Fig. 21-20.

This particular illustration shows a gear being cut by a combination of infeed, for the beginning of the roughing cycle, and then tangential feeding for the finishing phase of the cycle. This procedure reduces the total cutting time as compared with tangential hobbing only.

FIG. 21-20. Tangential hobbing of bronze wormgear of 77 teeth, 3 thread, 0.9586 C.P. at 97 rpm of hob and 0.016 to 0.024″ feed. (*Courtesy of Gould & Eberhardt Div., Norton Co., Worcester, Mass.*)

21-14. Adaptability of Tangential-feed Method. It will be found that a tangential hobber is a very versatile machine. Remembering that a wormgear hob must be made to the same specifications as those of the worm, or slightly larger, we can easily see that if only one or two wormgears were needed the cost of one hob to do this work could be prohibitive from the viewpoint of unit gear cost.

The tangential-feed method provides a very useful and economical answer to such situations.

Fly Cutting. A hob may be designed to do a particular job and then one finishing tooth is made in a toolroom and mounted on a suitable arbor. This one tooth is then

FIG. 21-21. Tangential hobber fly-tooling a steel-mill wormgear of 23 teeth and 3.5″ C.P. (*Courtesy of Gould & Eberhardt Div., Norton Co., Worcester, Mass.*)

FIG. 21-22. Tangential fly-tooling of camera wormgear. (*Courtesy of Gould & Eberhardt Div., Norton Co., Worcester, Mass.*)

FIG. 21-23. Multiple fly-tool cutters used to rough and finish in tandem a 16-tooth wormgear of 14 threads, 0.443″ C.P. (*Courtesy of Gould & Eberhardt Div., Norton Co., Worcester, Mass.*)

traversed through the work, tangentially, as shown in Figs. 21-21 and 21-22. This type of tool is known as a "fly tool."

A fly tool has some advantages over a complete wormgear hob in addition to cost. It has no lead errors, no spacing errors, and no errors due to profile variation (from tooth to tooth). All these errors are inherent in a wormgear hob, to some degree, and are reflected in the gear being cut. The fly tool, properly designed, fabricated, and mounted, could have decided advantages over a hob in terms of accuracy.

The fly cutter will rough and finish a wormgear in a single pass, unless the ratio between the number of teeth in the gear and the threads in the worm is an integral number. Under such circumstances the fly tool will have to be indexed for each thread in the worm until the gear has been completed.

Another advantage in using a fly tool is in the constancy of the cutting center distance. The fly-cutter specifications can be exactly the same as those of the worm so that the cutting center distance is the same as the operating center distance. When the fly tool is sharpened, it is reset in its arbor to the correct position so that no change is necessary in the cutting center distance.

Multiple Fly-tool Cutters. *Multiple fly tools* that have the appearance of gear-shaper cutters may also be used to great advantage and in some instances are preferable to a conventional wormgear hob. In Fig. 21-23 such a setup is shown using two cutters, one for roughing and one for finishing.

21-15. Feeds for Use in Cutting Wormgears.

Infeed Method. When using this method, the feeds used must, of necessity, be less than those used when hobbing spur or helical gears because fewer teeth are doing the work. Depending upon the conditions, as outlined in Arts. 21-6 and 21-7, the feeds may vary from 0.003 to 0.020″ per revolution of the work. The material used for the wormgear blank will be a deciding factor.

Tangential-feed Method. The feeds used when this method is being employed may be somewhat greater than for the infeed method. The tapered hob has substantial roughing teeth and the finishing teeth have relatively little metal to remove. Furthermore, the action inherent in the process presents considerably more teeth to the cutting path, which distributes the chip load accordingly.

Certain types of equipment have built into their controls a sequence that permits a rapid approach of the hob to the work, then a coarse feed during the roughing phase and a fine feed for the finishing cut. They may also have a rapid and coarse infeed motion, and then when this has been completed the tangential motion commences at a finer feed to complete the work.

The roughing-phase feed rates will vary from 0.003 to 0.020″ per revolution of the work to 0.010 to 0.040″ per revolution of the work for the finishing cut.

Fly-tool-feed Method. Hobbing by this method must be necessarily slower, by far, than the methods using a complete hob. The reason is that one tooth is removing all the material from the spaces. For gears that run with single-thread worms the feed rate may vary from 0.004 to 0.005″ per revolution of the work. Multiple-thread work, particularly that which has a prime or nonintegral ratio, needs finer feeds still.

A single-thread fly tool will pass through each consecutive tooth space during the cutting action. If the feed rate is 0.004″ per revolution of the work, then in one revolution of the gear blank the tool will have advanced 0.004″.

Let us assume a double-thread job and 51 teeth in the gear. The blank would have to make two revolutions before the fly tool would again pass through the first space. This means that the tool would have advanced 0.008″ as compared with the single-thread job noted above, which had 0.004″ per revolution. For the double-thread job the feed must be reduced to 0.002″ per revolution so that each time the tool entered the first space the tool advance would be 0.004″.

In other words, for prime- or nonintegral-ratio jobs the feed rate for multiple-thread work should be equal to the feed rate for single-thread jobs divided by the number of threads in the mating worm.

The feed rate for multiple-thread jobs having an integral ratio will be the same as for single-thread work. We must remember that for this kind of work the fly tool must be indexed after each complete cycle to the next thread. The action is therefore exactly like a single-thread job and must be handled in the same way.

21-16. Time Required to Cut Wormgears. The time required to hob a wormgear may be determined by the following equation. It may be used for either infeed or tangential hobbing. It may also be used for work cut with fly tools.

$$t = \frac{N_G L_t}{nfT} + F \tag{21-8}$$

where t = floor-to-floor time, min
N_G = number of teeth in wormgear being cut
L_t = total length of travel of hob, in.
n = speed of hob or fly tool, rpm
f = feed, in. of hob or fly tool advance, per revolution of work
T = number of threads in hob or fly tool
F = loading plus unloading time, min

21-17. Double-enveloping Worms and Wormgears. The discussions covered so far in this chapter, with regard to worms and wormgears, involved gearsets having one throated or enveloping member. In the present article the manufacture of *double-enveloping wormgear sets* will be our prime concern. With drives of this kind, both members are throated and therefore envelop the mate. This means that both members must be accurately located with regard to center distance, shaft angle, and in the axial position, with respect to the mate. Cutting positions, therefore, between the cutting tools and the member being cut must also be accurately located in the hobbing machine. Figure 21-24 shows the hobbing machine used to produce both components.

The hob used to cut the wormgear is as shown in Fig. 21-25, and it will be noticed that the hob is throated the same as the mating worm. The hob is made the same as the mating worm except the teeth are made somewhat thinner. The *infeed method* is used exclusively for this process.

The hob and the blank are at expanded machine center distance so that both may rotate freely without interference. The hob is then fed into the blank until the worm and wormgear center distance has been reached. It is at this final center distance that the correct helix angle is produced. If the hob was made to produce the gear with the correct tooth thickness, parts of the gear teeth will be removed during the infeed phase of the cutting cycle. This is why the hob teeth are made thinner.

After the hob reaches the position of the designed center distance of the gearset, the gear blank is rotated first in one direction and then in the other direction about its axis, which permits the hob to side-cut the gear teeth to the required thickness. This permits proper generation of the gear teeth at the correct operating center distance.

It is also important that the proper relationship between the end and side positions of the hob be maintained and the side-cutting procedure as outlined permits this relationship to be maintained throughout the cutting cycle and at the same time produce the correct tooth form.

Cutting the Throated Worm. Cutting the throated worm is an operation performed on the same machine used to produce its mating gear. The procedure is similar to that used for cutting the wormgear, and the setup is shown in Fig. 21-26. It will be noticed, however, that the worm blank is mounted on the spindle that pre-

Fig. 21-24. Hobbing machine used to cut double-enveloping wormgears and worms. (*Courtesy of Cone-Drive Div., Michigan Tool Company, Detroit, Mich.*)

Fig. 21-25. Hob used to cut double-enveloping worm. (*Courtesy of Cone-Drive Div., Michigan Tool Company, Detroit, Mich.*)

viously carried the hob. The cutter, similar in construction to a gear-shaper cutter, is mounted on the work table.

The cutter used to cut the worm also has thin teeth for the reasons previously explained. The cutting cycle is also the same, consisting of an *infeed phase* and a

FIG. 21-26. Cutting a throated double-enveloping worm. (*Courtesy of Cone-Drive Div., Michigan Tool Company, Detroit, Mich.*)

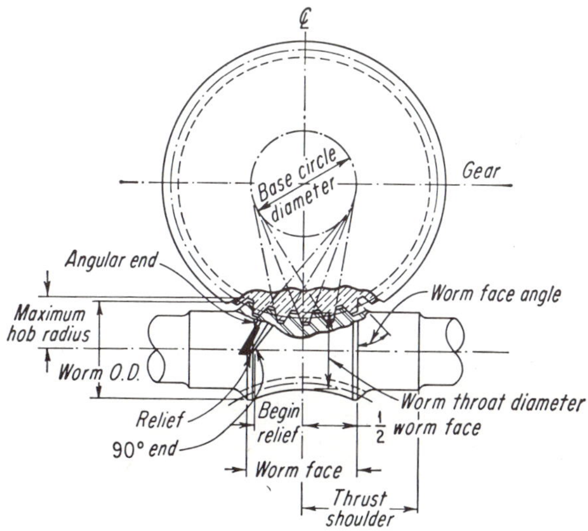

FIG. 21-27. Portion of worm thread that is removed to prevent undesirable contact.

side-cutting phase to complete the cycle. The cutter, in this instance, is made slightly larger in diameter than the mating wormgear which it represents. The radius of the throat of the worm is therefore slightly larger than the radius at the throat of the mating wormgear. This is done to minimize the possibility of interference between the worm and its mating wormgear when under load and the larger worm throat radius helps to compensate for the deflections of the gearset when subjected to such strains.

In addition to the enlarged throat radius of the worm, one other operation must be performed to ensure that no improper contact takes place at the ends of the worm. The drawing shown in Fig. 21-27 shows that portion of the worm thread which must be removed to avoid the undesirable contact at that point. Removing the portion of the worm thread to avoid undesirable contact may be accomplished by using a *relieving fixture*.

That portion of the worm thread which contacts the gear-tooth profile while entering or leaving engagement cannot extend beyond a distance equal in length to the radius of the base circle and measured in both directions from the common center line and in an axial direction with respect to the worm. As a general rule, relief should begin at a distance equal to the radius of the base circle minus 0.01 times the center distance.

Because of the special nature of this type of drive, it is suggested that the design details of the gears and their respective cutting tools be reviewed or established by the machine-tool and cutting-tools manufacturer for the final specifications and method of processing.

21-18. Cutting Time for Double-enveloping Gearsets. It would be rather difficult to quote speeds and feeds for gears of this type because of the many variables involved and because of the special nature of the components themselves. In general, the average cutting speed used for the gear blank is about 100 fpm for hobbing time and about 50 fpm when cutting the worm. These figures are for gears that operate on center distances of about 6.000″. These speeds would be increased for the smaller sizes and decreased for the larger sizes. This is due to the variation in the removal of material per revolution, which is related to the circular pitch.

How the cutting time varies can best be shown in terms of typical time required to produce two gearsets of different size. The following values are a good measure of the time, in hours, that it takes to produce such gears.

2.000″ Center Distance Gearset	
Worm:	
Blank	0.59
Generate and relieve	1.11
Gear:	
Blank	0.20
Generate, chamfer, and match	1.52
Total	3.42
6.000″ Center Distance Gearset	
Worm:	
Blank	1.72
Generate and relieve	3.20
Gear:	
Blank	0.32
Generate, chamfer, and match	4.07
Total	9.31

These figures cover complete manufacturing time from raw material to final inspection—ready for shipment.

21-19. Worm-thread Grinding. The grinding of worm threads is accomplished by using the same basic principle as is described in Art. 21-4, the difference being the use of a grinding wheel in place of a thread-milling cutter. A grinding machine, used for grinding worm threads, is shown in Fig. 21-28.

Fig. 21-28. Machine to grind a high-lead worm. (*Courtesy of Jones & Lamson Machine Company, Springfield, Vt.*)

Fig. 21-29. Grinding a worm with a multiple-rib wheel. (*Courtesy of Jones & Lamson Machine Company, Springfield, Vt.*)

The usual procedure is to rough-mill the worm threads, harden, and then finish-grind the threads to the final and desired specifications. In some instances, the worm is ground from the solid when the worm is through-hardened. Casehardened worms, however, must be rough-milled first, then ground after the casehardening operation in order to maintain the desired case on the active surfaces of the worm threads.

In many instances worms of a high precision are required, and this precision can be maintained only by grinding. Some materials are so soft and the microstructure is so constituted that when milled the surfaces are not smooth but are torn by the action of the milling cutter. In such instances the thread-grinding process is the best method to use.

Some of the methods used in thread milling may be used in thread grinding. For the manufacture of fine-pitch worms having a lead less than 0.2000″, a grinding wheel with multiple ribs may be used for high production.

When designing worms that are to be ground the blank must be made to allow clearance for the grinding wheel to run out and avoid any damage to the wheel. A commonly used wheel size is 20″ diameter.

The grinding wheels may be dressed to their proper dimensions and forms using either *diamond dressers* or by using the *crush-dressing method.* Dressing a wheel by whatever method is an important and precise operation. The work being ground cannot be more precise than the wheel which does the grinding. The dressing is performed by special devices designed for the purpose.

The accuracy obtained by using a single ribbed grinding wheel is greater than when multiribbed wheels are used. Also, crush-dressed wheels do not, in general, give such good accuracy as a diamond-dressed wheel. Of all the methods used the single-ribbed wheel provides the greatest accuracy and finest surface finish.

The coolant used while grinding is usually an insoluble thread-grinding oil and gives better surface finish, accuracy, and longer life to the wheel. A copious flow of oil should always be provided.

The speeds used in thread grinding will vary with worm pitch, material, hardness, and wheel size and type. In terms of surface feet per minute the speeds may vary from 8,000 to 12,000, and the values shown in Table 21-1 may be used as a guide in this regard.

Table 21-1. Feed Rates Normally Used in Worm-thread Grinding

Linear pitch	Feed rate, in. per min.	
	10″ wheel	20″ wheel
0.250	25	40
0.500	20	35
0.750	15	30
1.000	12	25
1.250	10	20

Production Time. The following equation may be used to estimate the actual grinding time. To this must be added the time required for loading and unloading.

$$T = \frac{n}{C_T}\left(I + \frac{L_d}{R_f}\right) N \tag{21-9}$$

where T = actual grinding time, min
n = number of threads
C_T = threads per cut
I = indexing time, min
L_d = developed length of worm, in.
R_f = feed rate, in. per min
N = number of cuts

The length of the thread L_d, as developed, may be computed as follows:

$$L_d = \frac{\pi d F}{L \cos \lambda} \tag{21-10}$$

where L_d = developed length of worm, in.
d = diameter of worm, in.
F = actual length of worm face, in.
L = lead of worm, in.
λ = lead angle of worm, deg

In mounting the grinding wheel great care must be exercised to make sure that it is properly balanced and secured. The workholder or, as is usually the case, the centers upon which the worm blank is mounted must be true and properly lined up with each other.

REFERENCES

1. Colvin, Fred H., and Frank A. Stanley: "Gear Cutting Practice," 3d ed., McGraw-Hill Book Company, Inc., New York, 1950.
2. Dudley, Darle W.: "Practical Gear Design," McGraw-Hill Book Company, Inc., New York, 1954.
3. Buckingham, Earle: "Analytical Mechanics of Gears," McGraw-Hill Book Company, Inc., New York, 1949.
4. Merritt, H. E.: "Gears," Sir Isaac Pitman and Sons, Ltd., London, 1942.
5. Barber-Colman Company: "Hob Handbook," Rockford, Ill., 1954.
6. Barber-Colman Company: "Operator's Manual," Rockford, Ill.
7. Michigan Tool Company: "Cone Drive Gears," Catalog 70, Cone Drive Gears Division, Detroit, Mich.
8. Norton Company: "Bulletins 230 and 239," Gould & Eberhardt Division, Worcester, Mass.
9. Norton Company: "New System for Automatically Hobbing Worms," Gould & Eberhardt Division, Worcester, Mass.

Chapter 22

GEAR-CUTTING TOOLS

By

RICHARD S. HILDRETH, *Chief Engineer, Michigan Tool Company, Detroit, Mich.*

This chapter is primarily intended to cover the metal-cutting tools used to produce teeth on spur and helical gears. Some of the tools used on spur or helical gears may be adapted or modified to cover the cutting of certain kinds of worms and wormgears, bevel-gears, and some other types.

Chapter 17 covers the tools used to form or cast various kinds of gear teeth. Chapter 21 has some detail information on the kinds of tools and the uses of these tools in the making of worms and wormgears. Chapter 23 gives detail information on the tools and machines used in the inspection or measurement of gears. Chapter 20 covers bevel-gear tools.

Gears that are finished by shaving or lapping require special tools. These are covered in Chap. 18.

The tools used to cut spur and helical gears are very important in the gear trade since these gears are so widely used. Usually the gear teeth are produced by cutting. Even teeth to be finally ground or shaved are cut first.

HOBS FOR SPUR AND HELICAL GEARS

The hob is one of the most versatile tools used for gear cutting. Without additional tooling, such as guides or toolholders, the simple hob will cut either spur or helical gears. The manufacture of hobs is a straightforward, well-developed art. The hob is available at a high level of accuracy, and if properly mounted and used, it will produce a high order of accuracy in the finished gear.

Much has been written about hobs and their use, and a very complete set of industry standards has been established (see Arts. 22-5 and 22-6).

22-1. General Principles of the Hob. The hob is used to cut a gear by a continuous indexing process.* It is possible to obtain a high order of accuracy in both spacing and profile on the finished gear. The sources of error are fewer and more controllable in the hobbing process than in some other methods of gear cutting.

The hob is used to cut teeth in external gears, splines, and worms. Either spur or helical teeth may be produced. As a tool, the hob simulates a worm. It has one or more threads and may therefore be called "single thread" or "multiple thread." The hob thread is gashed lengthwise to form a series of cutting teeth. This makes the thread an interrupted one rather than a continuous one.

* See Arts. 16-4 to 16-7 for detail information on the hobbing process.

The cutting edges of the hob teeth lie on a helicoid, conjugate to the form produced on the part. The hob tooth is relieved top, sides, and bottom to create clearance beyond the cutting edge.

Types of Hobs by Mounting. Arbor-type, or *shell*-type hobs have straight or tapered holes for mounting on an arbor. They usually have a keyway in the bore, or sometimes on the end face, for driving purposes.

Shank-type hobs are equipped with straight or tapered shanks to fit the hobbing-machine spindle or adapter. It is usual practice to have a pilot diameter or center to support the end opposite the shank end.

Hand of Hob. This defines the thread helix of the hob teeth. A right-hand hob has a thread helix corresponding to a right-hand screw. A left-hand hob has a thread helix corresponding to a left-hand screw. It is sometimes necessary to qualify the hand specification further by specifying "top coming" or "top going." This is used when a hob is equipped with a special feature and it is necessary to define its location.

Method of Profile Generation. The hob threads may be excluded as a factor in understanding the manner in which a hob generates. As opposed to a milling cutter, a hob has a conjugate form to produce the part form. A milling cutter has a counterpart form. In determining the conjugate form, a simple rack principle is utilized. The straight-line elements of the rack are combined to produce involute profiles, the pitch line of the rack representing the hob pitch line. For hobbing noninvolute forms on a part, the rack pitch line is the basis of determining the conjugate hob form. From the engineering standpoint, the rack pitch line is brought into tangent contact with the selected generating diameter of the noninvolute part. The rack contour is then back-generated from a layout of the part to be formed by hobbing. This back-generated form becomes the radial form of the hob cutting edge.

Manufacturing. Knowing the necessary hob form to generate the part, this form is transferred to a "form tool." The *form tool* is identical to a lathe tool and is used in almost the same manner. The tool is accurately ground with a form equivalent to the space of the hob space which was generated (see Fig. 22-1). This tool is used then on a relieving lathe to cut the hob form. The relieving lathe is equipped with a cam-controlled saddle that plunges the tool deeper into the hob during the tooth forming. The hob has been previously gashed axially for the desired number of flutes. The thread lead has usually been roughed in on a thread mill. The relieving lathe is geared to feed the cam-controlled saddle parallel to the hob axis on the desired thread lead. The saddle is also geared for the same number of relieving cycles as there are flutes in the hob. The form tool meets the proposed cutting edge of the hob and feeds in toward the hob axis a predetermined amount in time with the hob rotation. Upon clearing the heel of the hob tooth, the tool is in the gash space. It then retracts to meet the next cutting edge. This cycle is cam-controlled. The cam used is selected to produce the desired side clearance and relief at the outside diameter of the hob. The process is frequently referred to as "backing off" a hob. At this point, the hob is ready for heat-treatment, and subsequent operations following heat-treat.

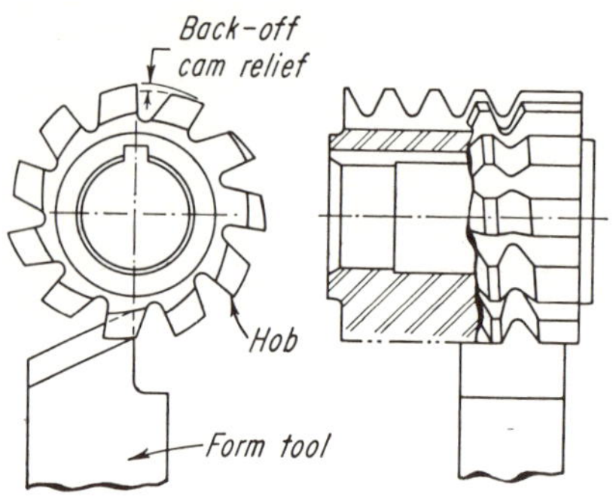

FIG. 22-1. "Backing-off" hob with form tool.

If the hob is to be unground, it will not have a grinding operation on the tooth profile. The bore, hubs, and proof diameter will be ground after heat-treat. Subsequent to sharpening, if the hob meets classification tolerances, it is ready for use.

Heat-treat growth is allowed for at "back-off time" by using a slightly shorter thread lead during forming. The hob-manufacturing process is reliant on consistency of steel and heat-treat practices.

22-2. Hob Nomenclature. Figure 22-2 shows the basic elements of a shell-type hob. Figures 22-3 and 22-4 show relative hob features and the resulting gear contour after hobbing. Features such as semitopping and protuberance should be confined to the tooth-number range designated by the hob manufacturer. For example, if a given semitopping chamfer is designed into a hob for a 16-tooth pinion, a greatly increased chamfer would be produced on a 60-tooth gear. This results from the lower contact of the gear outside diameter in the hob space.

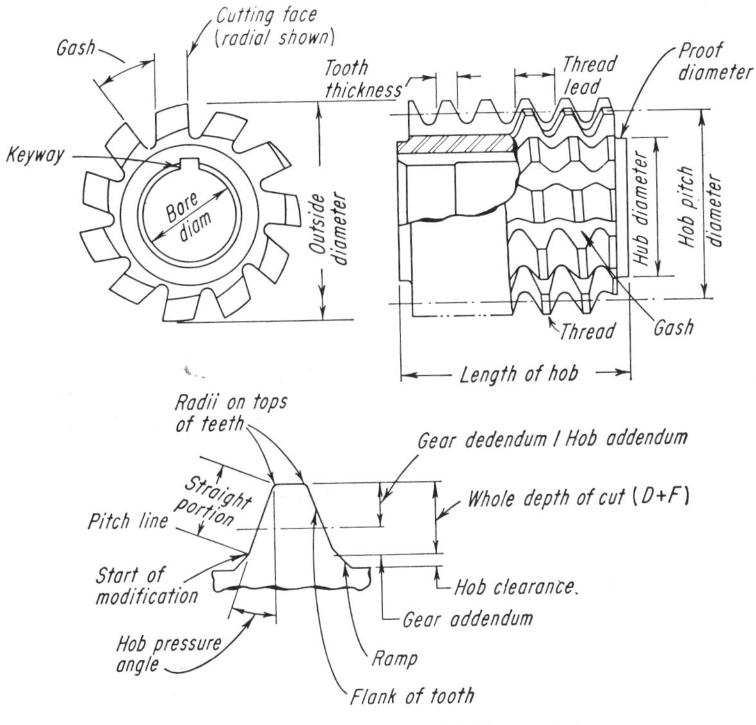

FIG. 22-2. Basic elements of shell-type hob.

22-3. Hob Classification by Threads. Hobs are classified as single- or multiple-thread hobs.

The single-thread hob, when used in conjunction with a gear, is interconnected for rotation by means of a positive mechanical system. For example, if a single-thread hob is used on a 30-tooth gear, each revolution of the hob must be the equivalent of a one-tooth index of the gear being hobbed. Therefore, 30 revolutions of the hob would be required while the 30-tooth gear is making one revolution. A double-thread hob in one revolution would be in timed relation to a two-tooth index of the gear. A double-thread hob would be making 15 revolutions while the 30-tooth gear is making one revolution. As may be noted, there is some advantage in the use of a multiple-thread hob in terms of production time. This production time is not proportionate to the increase in the number of threads and could not be so unless the number of flutes were increased in proportion to the number of threads.

The gashes or flutes can be straight and parallel with the hob axis or can be of the spiral type on a lead, such that the spiral would be normal to the thread angle of the hob. The spiral gash is usually incorporated in hob design when the thread angle is in excess of 4°, as caused by a coarse pitch or multiple threads.

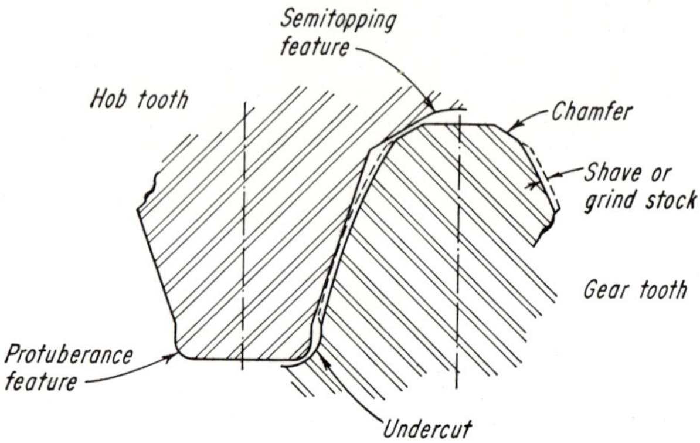

FIG. 22-3. Semitopping hob with protuberance.

The greatest accuracy is obtained by the single-thread design as opposed to the multiple-thread design. This results from the "hunting-tooth" condition of a single-thread hob. It is very desirous to avoid the use of multiple-thread hobs when the number of threads in the hob are divisible into the number of teeth in the gear being hobbed. Inherent accuracy of the hobbing process is negated if this condition exists.

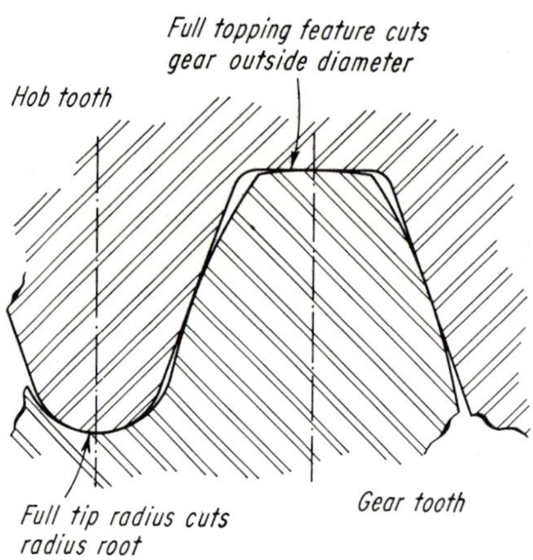

FIG. 22-4. Full-topping hob with full tip radius.

Another consideration in the use of multiple-thread hobs is the fact that the speed of the indexing worm gear increases when multiple-thread hobs are used. In the case of low numbers of teeth, it is necessary to check the speed of the index worm gear to ascertain that it does not exceed the maximum speed recommended by the hobbing-machine manufacturer.

It is noteworthy that improved methods of manufacturing have increased the tolerance quality of the class C accurate unground hobs, enough so that frequently they are compared with a class B hob in terms of tolerance and are frequently used for a finishing operation if the accuracy of the part permits this application.

The selection of the number of flutes to be introduced into a hob depends on tooth depth and hob diameter. It is pretty well standardized with the cutting-tool manufacturers in order to give maximum tool life for any particular size of hob. Increasing the number of flutes in a hob has the tendency of reducing the amount of available sharpening life on the hob.

Many features can be added into a hob, such as protuberance, semitopping, and special involute control, including modified involute, or full-fillet radii, etc.

22-4. Classification by Tolerance. Hobs are generally classified into four common areas:

1. Class A, precision
2. Class B, commercial ground
3. Class C, accurate unground
4. Class D, commercial unground

The class A hob is a ground precision-type hob used most often as a finishing hob, where no subsequent operations, such as shaving or grinding, would be involved.

The class B commercial ground hob is used for both finishing and semifinishing preparatory to grinding or shaving.

The class C accurate unground hob is the most commonly used as a semifinishing hob, preparatory to grinding or shaving as the finishing operation.

The class C hob is an unground hob, formed by "backing off" in an annealed state and then heat-treating. Except for the grinding of the bore, proof diameter, and hub faces, no further refinement of the tooth form, spacing, or lead is made.

With the class A and B ground hobs, the subsequent grinding operation is for the refinement of the limits on profile, lead, and spacing.

Class D hobs are of commercial unground tolerances, generally used as roughing-type hobs.

Ultraprecision, or class AA, hobs are available from the ground-hob manufacturers. Because of manufacturing problems, it is advisable to consider ultraprecision hobs as single-thread hobs only.

22-5. When and How to Select a Hob. Of primary importance are the physical characteristics of the part to be hobbed. Will it be arbor-mounted or shank-mounted? In either case, common-sense consideration in providing rigid support will be rewarding. The part to be hobbed must not contain interfering adjacent shoulders or diameters. In some cases it is permissible to hob into adjacent bearing diameters. In this event, the hob depth is increased as shown in Fig. 22-5.

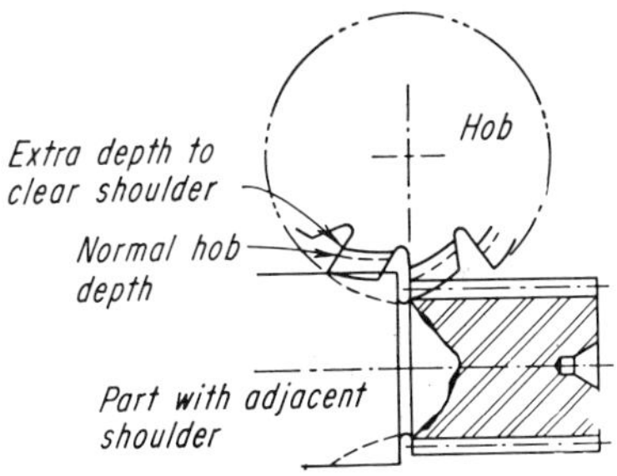

FIG. 22-5. Hob designed with extra depth to avoid complete mutilation of shoulder.

Practically all profiles that are equally spaced on a cylindrical base surface can be hobbed. The profiles need not be symmetrical, such as with involute gear teeth or straight-sided splines. Nonsymmetrical part profiles such as ratchet teeth can be readily hobbed. It is desirable that all profiles to be generated by hobbing have greater than a radial form.

When it is impossible to hob by conventional multiposition design of hobs, a single-position nongenerating hob may be used. A single-position nongenerating hob allows cutting of contours that exceed the capabilities of a generating-type hob. Small fillets and radii can be produced by single-position hobs that are impossible to produce by generating.

Hob Size. The sizes of hobs used to generate involute spur and helical gears have been standardized. Table 22-1 shows the standard sizes for hobs in the coarse-pitch series. Table 22-2 shows standard-size fine-pitch hobs. These tables are based on standards of the American Gear Manufacturer's Association.[1,2,*]

* Superscript numbers refer to references at the end of the chapter.

Table 22-1. Dimensions for Standard Single-thread Involute Spur and Helical Gear Hobs

(All dimensions are in inches)

Diam. pitch	Hole dia.	OD	Length	Keyway
1	$2\frac{1}{2}$	$10\frac{3}{4}$	15	$\frac{5}{8} \times \frac{5}{16}$
$1\frac{1}{4}$	2	$8\frac{3}{4}$	12	$\frac{1}{2} \times \frac{1}{4}$
$1\frac{1}{2}$	2	8	10	$\frac{1}{2} \times \frac{1}{4}$
$1\frac{3}{4}$	2	$7\frac{1}{4}$	9	$\frac{1}{2} \times \frac{1}{4}$
2	$1\frac{1}{2}$	$5\frac{3}{4}$	8	$\frac{3}{8} \times \frac{3}{16}$
$2\frac{1}{4}$	$1\frac{1}{2}$	$5\frac{1}{2}$	$7\frac{1}{2}$	$\frac{3}{8} \times \frac{3}{16}$
$2\frac{1}{2}$	$1\frac{1}{2}$	5	7	$\frac{3}{8} \times \frac{3}{16}$
$2\frac{3}{4}$	$1\frac{1}{2}$	5	6	$\frac{3}{8} \times \frac{3}{16}$
3	$1\frac{1}{4}$	$4\frac{1}{2}$	5	$\frac{1}{4} \times \frac{1}{8}$
$3\frac{1}{2}$	$1\frac{1}{4}$	$4\frac{1}{4}$	$4\frac{3}{4}$	$\frac{1}{4} \times \frac{1}{8}$
4	$1\frac{1}{4}$	4	4	$\frac{1}{4} \times \frac{1}{8}$
5	$1\frac{1}{4}$	$3\frac{1}{2}$	$3\frac{1}{2}$	$\frac{1}{4} \times \frac{1}{8}$
6	$1\frac{1}{4}$	$3\frac{1}{2}$	$3\frac{1}{2}$	$\frac{1}{4} \times \frac{1}{8}$
7	$1\frac{1}{4}$	$3\frac{1}{4}$	$3\frac{1}{4}$	$\frac{1}{4} \times \frac{1}{8}$
8	$1\frac{1}{4}$	3	3	$\frac{1}{4} \times \frac{1}{8}$
9	$1\frac{1}{4}$	3	3	$\frac{1}{4} \times \frac{1}{8}$
10	$1\frac{1}{4}$	3	3	$\frac{1}{4} \times \frac{1}{8}$
12	$1\frac{1}{4}$	$2\frac{3}{4}$	$2\frac{3}{4}$	$\frac{1}{4} \times \frac{1}{8}$
12	$\frac{3}{4}$	2	2	$\frac{1}{8} \times \frac{1}{16}$
14	$1\frac{1}{4}$	$2\frac{1}{2}$	$2\frac{1}{2}$	$\frac{1}{4} \times \frac{1}{8}$
14	$\frac{3}{4}$	$1\frac{7}{8}$	$1\frac{7}{8}$	$\frac{1}{8} \times \frac{1}{16}$
16	$1\frac{1}{4}$	$2\frac{1}{2}$	$2\frac{1}{2}$	$\frac{1}{4} \times \frac{1}{8}$
16	$\frac{3}{4}$	$1\frac{7}{8}$	$1\frac{7}{8}$	$\frac{1}{8} \times \frac{1}{16}$
18	$1\frac{1}{4}$	$2\frac{1}{2}$	$2\frac{1}{2}$	$\frac{1}{4} \times \frac{1}{8}$
18	$\frac{3}{4}$	$1\frac{7}{8}$	$1\frac{7}{8}$	$\frac{1}{8} \times \frac{1}{16}$

1. Hub extensions on standard hobs are $\frac{1}{8}''$ minimum, exclusive of edge radius or undercut and may be either tapered or cylindrical.
2. Standard roughing hobs are made with thin teeth to provide finishing stock equal to 0.200/diametral pitch on tooth thickness for hobs 1 to 8 diametral pitch, inclusive, unless otherwise specified by the user.

Tables 22-1 and 22-2 show hob sizes for *single*-thread hobs. Somewhat larger[3] hobs are used when the hob is multiple-threaded. It should also be pointed out that larger hobs are sometimes needed when extra large gears are cut or when there is a need for extreme accuracy. A larger hob will permit the use of a larger and stiffer arbor. Also, a larger hob will permit more flutes and therefore less chip load per cutting tooth.

Hob Features. The normal involute cutting hob may be centered in several different positions and still generate an involute tooth. As a general rule it is only necessary to center the hob so that there is at least an axial pitch and a half of cutting teeth on either side of the centering point. For some noninvolute jobs the hob has a *fixed* centering position and the hob teeth are not ground the same on either side of the centering point.

The single-position hob is critical in setup and is usually of short tool life. The single-position hob envelops the part form and is similar to a milling cutter except for the continuous-indexing feature.

Hobs for involute gears can be obtained within the classifications as follows and with features listed:

1. Finishing type (may be class A, B, or C depending upon accuracy required)
   a. Topping
   b. Semitopping
2. Semifinishing (may be class A, B, or C)
   a. Topping
   b. Semitopping
   c. Protuberance type
   d. Extra depth for shaving clearance
   e. Thin teeth, leaving grinding or shaving stock
3. Pregrinding (usually class D or C)
   a. Extra depth for grinding-wheel clearance
   b. Thin teeth, leaving grinding stock
   c. Protuberance type, if desired
4. Roughing (class D, and usually multilead hob)
   a. Thin teeth leaving stock for next operation of semifinishing or finishing

Table 22-2. Dimensions for Standard Single-thread Fine-pitch Gear Hobs

Diam. pitch	Hole dia.	OD	Length	Hole dia.	OD	Length	Hole dia.	OD	Length
20	1¼	2½	2½	¾	1⅞	1⅞			
22	1¼	2½	2	¾	1⅞	1⅞			
24	1¼	2½	2	¾	1⅞	1⅞	½	1¼	1¼
26	1¼	2½	2	¾	1⅞	1⅞	½	1¼	1¼
28	1¼	2½	2	¾	1⅞	1⅞	½	1¼	1¼
30	1¼	2½	2	¾	1⅞	1½	½	1⅛	1⅛
32				¾	1⅞	1½	½	1⅛	1⅛
34				¾	1⅞	1½	½	1⅛	1⅛
36				¾	1⅞	1½	½	1⅛	1⅛
38				¾	1⅞	1½	½	1⅛	1⅛
40				¾	1⅞	1½	½	1⅛	1⅛
42				¾	1⅞	1½	½	1⅛	1⅛
44				¾	1⅞	1½	½	1⅛	1⅛
46				¾	1⅞	1½	½	1⅛	1⅛
48				¾	1⅞	1½	½	1⅛	1⅛
50–55				¾	1⅞	1½	½	1⅛	1⅛
56–85				¾	1⅝	1⅛	½	1⅛	1⅛
86–130				¾	1½	¾	½	1⅛	⅞
131 and finer				¾	1⅜	⅝	½	1⅛	½

22-6. Accuracy of Hobs and the Importance of Accuracy. Machines for inspection of lead variation in hobs have been developed by various manufacturers of hobs. These inspection machines provide a recorded chart of each cutting edge and its relative position in respect to any other cutting edge on the same thread of the hob for its full length. Inspection occurs by coordination of rotation of the hob with advance of the inspection finger, so that the true helical path of the thread is developed and variations recorded.

The tolerances of hobs have been established by the Metal Cutting Tool Institute (MCTI) and the American Gear Manufacturers Assoc. (AGMA). This has been achieved through cooperation of both the manufacturers and the users. Most hobs are shipped with a certified inspection chart, defining the lead variation in the following categories:

1. Tooth to tooth
2. Any one axial pitch
3. Any three axial pitches

The method of graphic recorded inspection provides an excellent monitoring system of hob quality vs. end product. The method of interpretation is described in the following:

Tooth to Tooth. This is the total indicator variation on successive teeth when traversing along the true helical path. The chart in Fig. 22-6 shows the readings taken on a 12-flute hob. The scale is as indicated. Each arc line represents one complete revolution of the hob. In reading the chart for the three convolutions of the hob, three axial pitches are recorded. In the first convolution, the spacing between flutes 5 and 6 registers 0.0004″ spacing error. This is the maximum shown. Therefore, this establishes maximum tooth-to-tooth error. This could occur many times in as many convolutions available in the full length of the hob. As long as the 0.0004″ error is not exceeded, the tooth-to-tooth classification is fixed. By comparing this maximum error with the tables of hob tolerances under the correct column for range of diametral pitch, the classification can be confirmed for this element only.

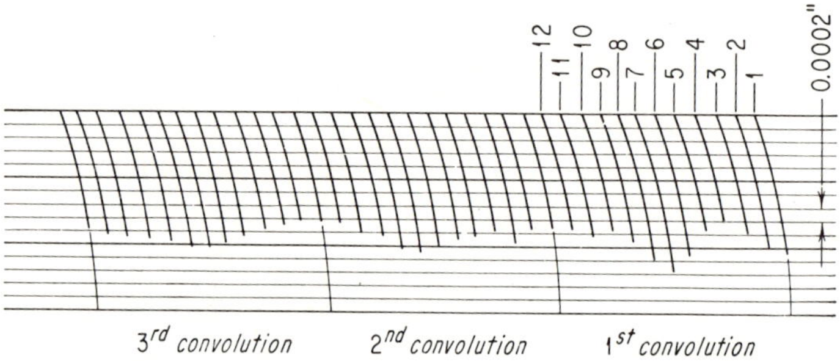

FIG. 22-6. Sample chart of hob lead check showing tooth-to-tooth readings on a 12-flute hob.

Lead Variation. In any one axial pitch, determine the total indicator variation in one complete convolution along the true helical path (360°) on a single-thread hob: one-half revolution (180°) on a two-thread hob, one-third revolution (120°) on a three-thread hob, and one-fourth revolution (90°) on a four-thread hob, etc. Referring to the sample chart in Fig. 22-6, if considering a single-thread hob, one axial pitch would be developed in 360° of rotation. All 12 flutes must be considered. The greatest total variation occurs between flutes 4 and 7 in the first convolution. The amount demonstrated is 0.0008″. The total variation must not exceed the tolerance of the accuracy classification. Use any convolution in the length-of-hob chart on either the right or left side.

In any three axial pitches, determine the total indicator variation in three convolutions along the true helical path (1,080°) on a single-thread hob, one and one-half revolutions (540°) on a two-thread hob, one revolution (360°) on a three-thread hob, and three-fourths revolution (270°) on a four-thread hob, etc.

Referring to the sample chart in Fig. 22-6, if considering a single-thread hob, three axial pitches are represented by the three groups of 12 flutes. The 36 flutes are analyzed on the total variation of the highest and lowest flute. For obvious reasons, the three axial pitches are always successive axial pitches. This specification can be applied to any three successive axial pitches in the length of the hob. This results from the use of hob shifting mechanisms that distribute hob usage across its full length. Generally, up to three teeth of the hob are in simultaneous contact with the part being hobbed: thus the specification of three axial pitches. European hobbing machines occasionally use a combination of axial and tangential hob feed. These feeds are timed continually to shift the hob axially as the hob progresses in its cut across the gear's face width (axially). For this reason, a further specification of lead variation in total length of hob is sometimes used. It is controversial whether this

Table 22-3. Inspection Tolerances for Single-thread Spur and Helical Gear Hobs

(Limits given in ten-thousandths of an inch)

Hole Dia. Size Tolerance

Hole dia.*	Class	2.500	2.000	1.500	1.250	0.750
Size tolerance (plus only)	A	8	8	5	2	2
	B	10	10	8	3	3
	C	10	10	8	3	3
	D	10	10	8	5	4

Hob Elements Affected by Pitch

Diam. pitch	Class	1–1.999	2–2.999	3–3.999	4–4.999	5–5.999	6–8.999	9–12.999	13–19.999
Hub									
Face runout, TIV†	A	8	5	2	2	2	2	2	2
	B	10	8	4	4	3	3	2	2
	C	10	8	4	4	3	3	2	2
	D	10	8	5	5	4	4	3	3
Dia. runout, TIV†	A	10	5	4	3	3	3	2	2
	B	12	8	6	5	4	4	3	2
	C	12	8	6	5	4	4	3	2
	D	15	10	8	8	6	6	5	5
Outside Diameter									
Runout, TIV†	A	30	20	15	15	10	10	10	10
	B	40	30	25	20	15	15	15	10
	C	50	45	40	25	20	17	17	12
	D	60	55	50	45	35	35	30	25
Flutes									
Adjacent flute-spacing error, TIV†	A	40	30	25	20	15	10	10	10
	B	50	45	40	30	20	15	15	10
	C	50	45	40	30	20	15	15	10
	D	60	60	50	50	30	25	25	20
Nonadjacent flute-spacing error in one turn, TIV†	A	80	60	50	40	30	30	30	25
	B	100	90	80	60	50	50	50	40
	C	100	90	80	60	50	50	50	40
	D	120	120	100	100	80	80	70	60

Flute-parallelism error, or flute-lead error, TIV†	Length	0.000–1.000	1.001–2.000	2.001–4.000	4.001–7.000	Over 7.000
	A, B, and C	10	15	25	30	50
	D	15	23	38	45	75

Table 22-3. Inspection Tolerances for Single-thread Spur and Helical Gear Hobs *(Continued)*

Diam. pitch	Class	1–1.999	2–2.999	3–3.999	4–4.999	5–5.999	6–6.999	9–12.999	13–19.999
Cutting Face									
Error (radial), TIV†	A	30	15	10	8	6	5	5	3
	B	50	25	15	10	8	7	7	5
	C	50	25	15	10	8	7	7	5
	D	100	75	50	40	30	20	20	15
Thread									
Lead variation, tooth-to-tooth, TIV†	A	7	5	4	3	2	2	2	2
	B	10	8	6	4	3	3	3	3
	C	15	12	8	6	5	4	4	4
	D	25	20	16	14	12	10	10	8
Lead variation in one turn, TIV†	A	25	18	10	8	6	5	5	4
	B	35	25	17	11	9	7	7	6
	C	45	35	22	14	11	9	9	8
	D	60	50	40	30	25	20	20	18
Accumulated error in three turns, TIV†	A	38	26	15	12	9	8	8	7
	B	53	38	22	16	12	11	10	9
	C	70	50	30	21	16	14	13	12
	D	120	100	80	60	50	40	35	25
Tooth Profile and Pressure Angle									
Depth to profile modification	A	200	180	160	140	120	100	80	60
	B	220	200	180	160	140	120	100	80
	C	220	200	180	160	140	120	100	80
	D	260	240	220	200	180	160	140	120
Involute-generating portion, TIV†	A	10	5	3	3	2	2	2	2
	B	16	8	5	5	4	3	3	3
	C	25	15	10	5	4	3	3	3
	D	80	55	30	18	12	8	8	6
Symmetry of modification (expressed as the algebraic difference of indicator readings on two sides of tooth)	A	150	130	120	100	90	80	60	50
	B	180	150	130	120	100	90	80	70
	C	180	150	130	120	100	90	80	70
	D	200	180	160	140	120	110	100	90
Tooth-thickness error (minus only)	A	30	20	15	15	10	10	10	10
	B	30	20	15	15	10	10	10	10
	C	35	25	20	20	15	15	15	15
	D	40	35	30	25	20	20	20	20

* Must be lapped straight and parallel and have 75 per cent minimum bearing area.
† TIV signifies "total indicator variation."

method is superior to the American type of hob shift, occurring after a predetermined number of pieces. The two methods have their own advantages as predicated by the hobbing application.

The tables of hob tolerance contained here and with the inspection definitions are self-explanatory. Tables 22-3 and 22-4 are taken from standards of the American Gear Manufacturers Association.[1,2]

22-7. Climb Hobbing vs. Conventional Hobbing. In conventional hobbing of spur gears, the hob rotation is in the direction of feed, as shown in the sketch. In

Table 22-4. Inspection Tolerances for Single-thread Fine-pitch Gear Hobs
(Limits given in ten-thousandths of an inch)

Hole Dia. Size Tolerance				
Hole dia.*	Class	1.250	0.750	0.500
Size tolerance (plus only)	A	2	2	2
	B	3	3	2
	C	3	3	3
Hob Elements Affected by Pitch				
Diam. pitch	Class	20–29.999	30–50.999	51 and finer
Hub				
Face runout, TIV†	A	2	2	2
	B	2	2	
	C	2	2	2
Dia. runout, TIV†	A	2	2	2
	B	2	2	
	C	2	2	2
Outside Diameter				
Runout, TIV†	A	10	7	5
	B	10	7	
	C	12	10	8
Flutes				
Adjacent flute-spacing error, TIV†	A	10	10	10
	B	10	10	
	C	10	10	10
Nonadjacent flute-spacing error in one turn, TIV†	A	25	20	20
	B	35	30	
	C	35	30	30
Flute-parallelism error, or flute-lead error, TIV†	Length	0.000–1.000	1.001–2.000	2.001–4.000
	A, B, and C	10	15	25
Cutting Faces				
Diam. pitch	Class	20–29.999	30–50.999	51 and finer
Error (radial), TIV†	A	3	3	3
	B	5	5	
	C	5	5	5

Table 22-4. Inspection Tolerances for Single-thread Fine-pitch Gear Hobs (*Continued*)

Diam. pitch	Class	20–29.999	30–50.999	51 and finer
Thread				
Lead variation, tooth-to-tooth, TIV†	A	2	2	2
	B	3	2	
	C	4	3	3
Lead variation in one turn, TIV†	A	4	3	3
	B	6	4	
	C	8	8	6
Accumulated error in three turns, TIV†	A	7	5	5
	B	9	7	
	C	12	12	8
Tooth Profile and Pressure Angle				
Straight portion	A	2	2	2
	B	3	2	
	C	3	3	3
Symmetry of approach	A	35	25	
	B	45	35	
	C	55	45	
Start of approach	A	40	30	
	B	50	40	
	C	60	50	
Tooth thickness (minus only)	A	10	5	5
	B	10	5	
	C	15	10	10

The tolerances given in this table are in substantial agreement with those of the Metal Cutting Tool Institute.

* Must be lapped straight, parallel, and have 75 per cent minimum bearing area.

† TIV means total indicator variation.

climb hobbing of spur gears, the hob rotation (and its tangential vector) oppose the direction of feed as shown in the sketch.

The selection of one method over the other is usually based on surface finish. There is no set rule to predetermine which method will give the best finish. Occasionally, machine characteristics themselves influence the choice. There is a condition of chip flow in climb hobbing that is advantageous in that the chip is directed into the tooth space already hobbed and reduces fouling of chips in the cut (see Figs. 22-7 and 22-8).

In cutting large helix angles, observation of part rotation and hob rotation and feed direction will ascertain whether climb or conventional hobbing is occurring.

Feeds and Speeds.* In this area it is difficult to find common ground for agreement. Tool engineers, manufacturing engineers, inspectors, and production personnel can always find basis for argument on this subject. Most gear-cutting departments

* See Chap. 16, section on hobbing, for more information on hobbing practices and how to estimate production time.

have their own approach to this, based on good solid experience. Feeds and speeds have been determined as a result of "cut and try," records of previous jobs, and common sense. The hob manufacturers have made suggestions, which are usually a good starting point. When the question of chip load is raised relative to hobbing, it is next to impossible to use a straightforward approach. The leading convolution of hob teeth in the cut does the majority of metal removing. Unlike a milling cutter, the hobbed chip per tooth, its area and thickness, varies with each flute engagement. It does repeat with each tooth index of the part, but each flute in the convolution removes a different-sized chip. This results from the series of generating flats combining to form an involute on the hobbed part.

Hobbing Feeds. The feed pattern of hobbing is clearly discernible in the root area of the hobbed gear. On the tooth flanks, the pattern of flats becomes more pronounced at high feed rates and with multiple-thread hobs. These flats are not detrimental if shaving or grinding is a subsequent operation. In a finish-cut gear, the flats may be objectionable to visual inspection. If of sufficient magnitude, the flats

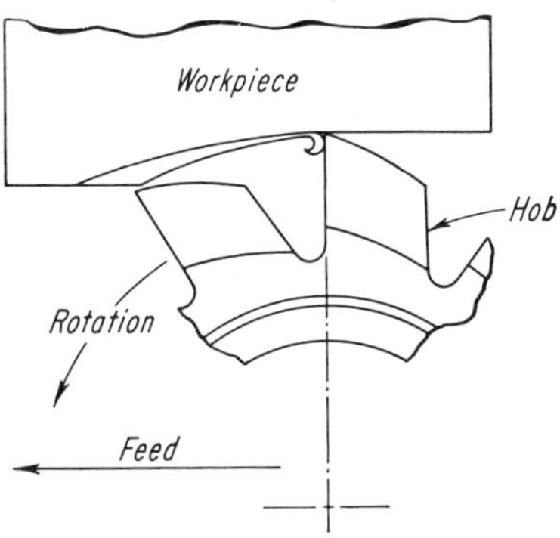

FIG. 22-7. Conventional hobbing.

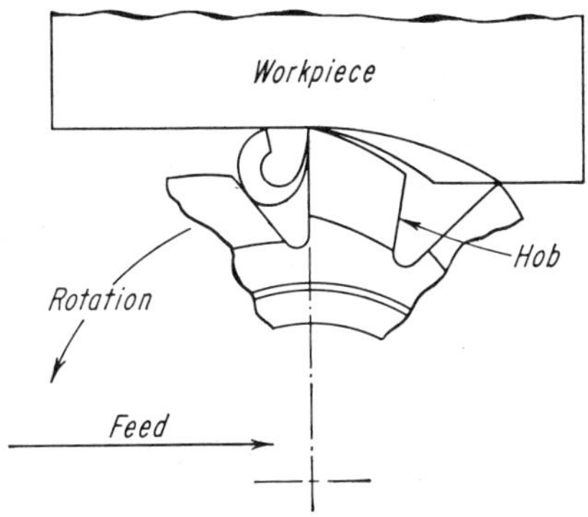

FIG. 22-8. Climb hobbing.

can influence ultimate gear accuracy. A single-thread hob with the "hunting-tooth" action will disperse the flats to a high degree.

The leading convolutions of hob teeth indicate hob dulling as a general rule and predicate the need of sharpening or the requirement of a hob shift to a sharp section of hob. The wear on these teeth should be carefully observed on a new setup. The wear of feeding too slowly is as serious as that of feeding too fast. Many hob users are misled in this respect, especially on materials of low machinability.

Indiscriminate use of multiple-thread hobs has cost hob users much in pieces per sharpening and in gear accuracy. It is generally recommended that a double-thread hob be used on not fewer than 31 teeth, or a three-thread hob be used on not fewer than 47 teeth in a gear. For commercial accuracy, there should be at least 15 teeth in the gear for each thread of hob. For best results, allow 30 teeth in the gear for each thread of the hob. Of course, there are exceptions in practice to these recommendations.

Figure 22-9 shows hob feeds plotted against tooth numbers for single- and multiple-thread hobs. These are suggested feeds within hob capacity based on a 12-flute hob for 3- to 12-pitch gears. The suggested feeds apply to steel only, using a sulfur-base coolant. The feeds do not necessarily apply to cast iron but are safe starting points. The feed rate is based on 0.0002″ advance of hob per flute. In other words, since feed is expressed in inches per revolution of part being hobbed, the total number of flutes exposed to the cut in one revolution of part is reduced to the share of feed for one flute.

Example. Hob 20-tooth pinion with single-thread hob of 12 flutes. Twenty revolutions of hob occur in one revolution of pinion. Assuming 0.050 feed per revolution of pinion, 240 flutes are exposed to cut. Share of feed for one flute is 0.050" divided by 240 flutes, or 0.0002" per flute.

There are fallacies to this approach when large numbers of teeth are considered, as can be shown by calculating the graph; nevertheless, it can be used as a starting point to determine the most satisfactory feed rate compatible to the machine, material, and hob wear. As indicated on the graph, the flats from generating will be very pronounced in the upper feed area. Also, the machine may be incapable of these feeds except as a second cut following a roughing operation.

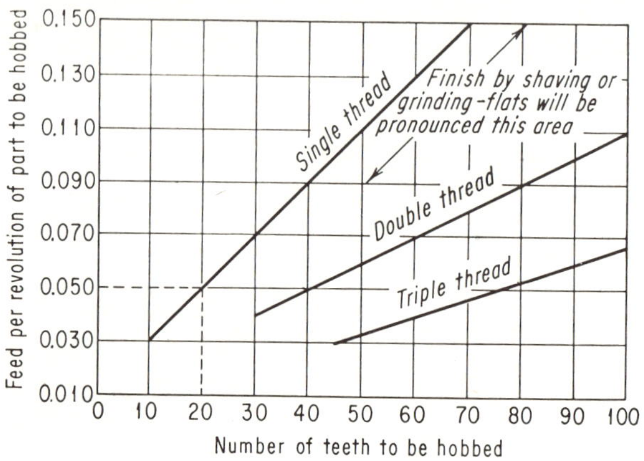

FIG. 22-9. Chart for selection of trial hob feed.

Hobbing Speeds. The speed of hobbing should be based on machinability ratings in conjunction with the following table:

Diam. pitch	*Sfm* (100% *machinable*)
1–4	160–180
5–16	180–200
17–32	200–220
32 and finer	220–240

Machinability is based on AISI B.1112 cold-drawn steel. If the machinability rating of the steel to be hobbed is known, adjust the surface feed per minute accordingly. For example, if hobbing 8620 with a 60 per cent rating, the sfm for a 10-pitch gear would be 110 to 120.

The necessary hob rpm can be determined in the formula

$$\text{Rpm} = \frac{\text{sfm} \times 12}{\pi \times \text{hob diameter}} \tag{22-1}$$

Hob Sharpening. The user of hobs should recognize limitations in his own plant for hob sharpening. Usually, for spiral gashed hobs, a hob sharpener is required in order to maintain the original accuracy of sharpening. Most hob manufacturers furnish straight gashed (in line with hob axis) hobs when the thread angle is less than 4°. After 4°, a spiral gash is introduced to maintain the cutting face normal to the thread angle. Failure to follow the original sharpening detracts from hob accuracy and causes physical changes in the form of the hobbed tooth.

Attention to rake angle of the cutting face is important, especially so in topping hobs. Most hobs are sharpened radially. Positive or negative rake angles are used as predicated by their need, depending upon the characteristics of the material being hobbed.

Assuming a hob was originally sharpened with a radial face, the error of sharpening reflects the physical changes listed:

Positive sharpening error	*Negative sharpening error*
a. Decreases hobbed pressure angle	*a*. Increases hobbed pressure angle
b. Increases whole depth of hob	*b*. Decreases whole depth of hob

These changes are the major ones occurring and result from the error of sharpening and its effect on the hob's tooth relief. Sketches in Fig. 22-10 show various effects of sharpening errors.

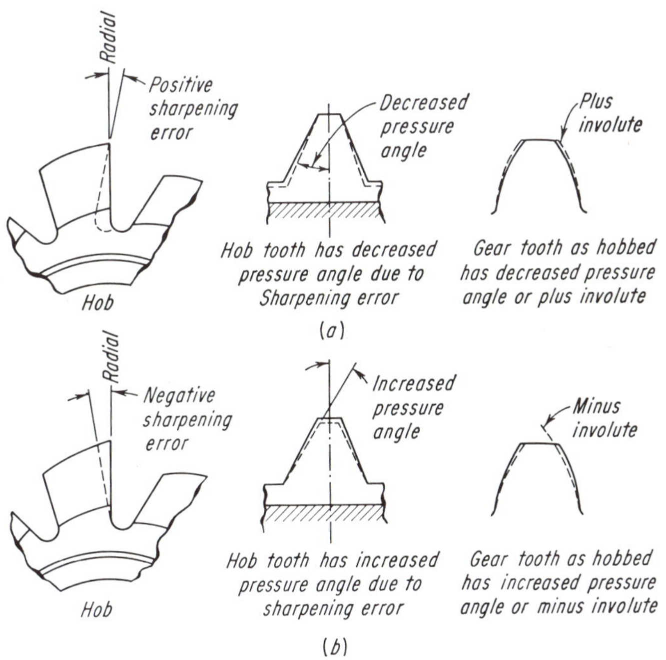

FIG. 22-10. Effect of hob-sharpening errors. (*a*) Hob sharpened with positive error. (*b*) Hob sharpened with negative error.

22-8. Hob Materials. The user of hobs is seldom concerned with the high-speed steel used in his hob; however, in the case of hob failure, it is natural to blame the hob as the source of trouble. Most often the fault lies in the setup, coolant, and variations in material being hobbed. Hob manufacturers have an extremely tight control on chemical composition of their high-speed steel and its heat-treatment. Because of the control needed in manufacturing, all steel, as furnished by suppliers, is certified with a report of chemical analysis. This is monitored by the metallurgical department of the hob manufacturer.

The high-speed steel used in hobs must have several cutting properties and combinations of properties. From the hob manufacturer's viewpoint, consistent machinability in the annealed state is important. When a hob form is produced by the "back-off" method for accurate unground hobs, the material must cut freely without tearing out on the tooth flanks. The material must separate cleanly at the heel of the tooth without flaking out. Grindability is essential to hob production. As the percentage of vanadium in the molybdenum-class steels increases, the grindability decreases.

From the hob user's viewpoint, cutting ability is paramount. Cutting ability is dependent upon properties including edge toughness, red hardness, and resistance to abrasion. Hardness and abrasion resistance are improved by increased amounts of carbon and vanadium. As mentioned before, increased vanadium goes hand in hand with increased difficulty in sharpening. The specification of the part material to be hobbed determines the most desirable hob properties for maximum hob performance.

The majority of today's hobs are made of M-2 high-speed steel. Certain applications require the use of M-3 and M-4 steels, and occasional use of the tungsten class of steels. The hob manufacturer will offer suggestions on hob steel when advised of the need for special consideration based on prior machine operations. The behavior of tools in turning and boring operations is a good guide to the need of special consideration in selection of hob material.

SHAPER CUTTERS*

The shaping process is one of the two basic methods of manufacturing gears. The shaper cutters can be designed to produce pregrind, preshave, or finished gears. This section will cover the different types of shaper cutters, their modifications and uses, along with the basic design and calculations.

While many types of parts, such as cams, sprockets, and square holes, can be produced on a gear-shaper machine, this writing will be confined to the field of involute-form spur and helical gears.

22-9. General Principles of Shaper Cutters. The method of cutting gears on the gear-shaper machine can be compared with a planing or shaping operation, in that the metal removal occurs when the tool is traveling in one direction only.

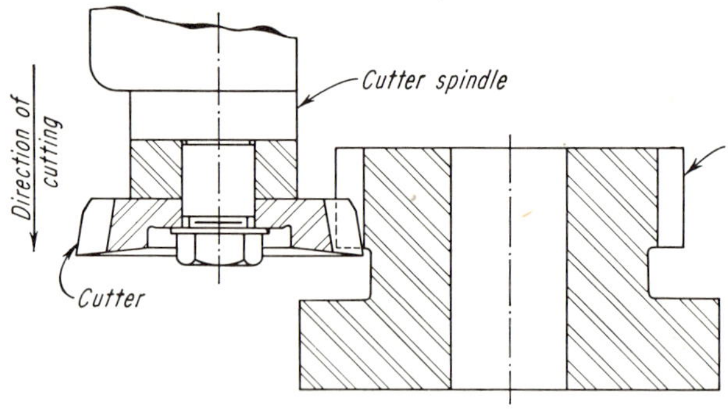

FIG. 22-11. Shaper cutter set up for push stroke.

Push-type Shaping. Figure 22-11 illustrates the shaper cutter mounted on the cutter spindle for a push stroke setup where the cutting is done on the downstroke. The tool or the work is then backed away so that the shaper cutter is relieved and does not drag on the return stroke.

* This section was written by Roy E. Landstrom, Chief Tool Designer, Michigan Tool Co., Detroit, Mich.

Pull-type Shaping. Figure 22-12 illustrates the tool and work relationship on a pull-type setup. Here the metal removal is done on the upstroke and the tool and work are relieved on the downstroke. In addition to the planing or shaping action of the tool, the work and the shaper cutter are rotated in a timed relationship, which is accomplished through the use of change gears. The generating process which takes place between the shaper cutter and the gear being cut can be compared with the rolling action of two gears in mesh. This is illustrated in Fig. 22-13.

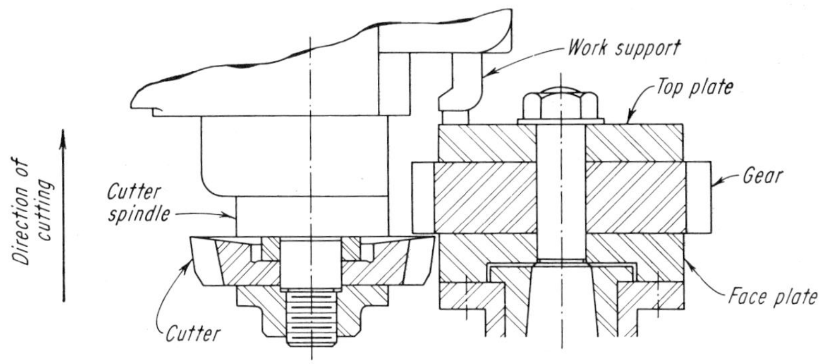

FIG. 22-12. Shaper cutter set up for pull stroke.

Rake Angles. Basically, the difference between a shaper cutter and a gear is that the shaper cutter is designed with a side-clearance angle ranging from 1°30′ to 3°30′, and a top angle to suit the side-clearance angle. Theoretically, the top angle should be a convex curve on external cutters, and a concave curve on internal cutters to ensure a consistent whole depth of the gear. Because the deviation between the required curve and a straight line is usually very little and the resultant change in whole depth very small, the straight-line design is universally used and accepted.

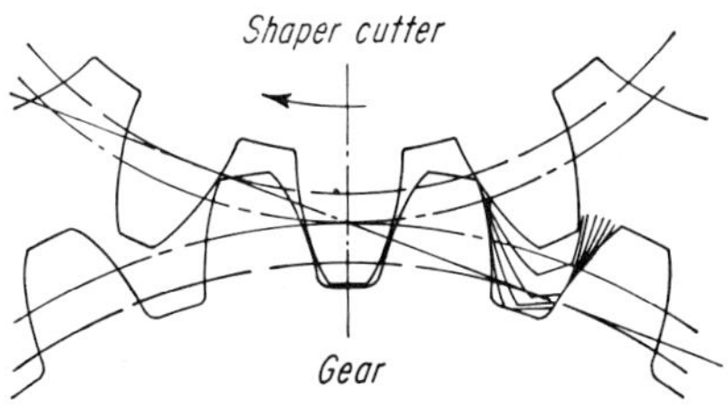

FIG. 22-13. Rotating in mesh as the cutter generates the teeth.

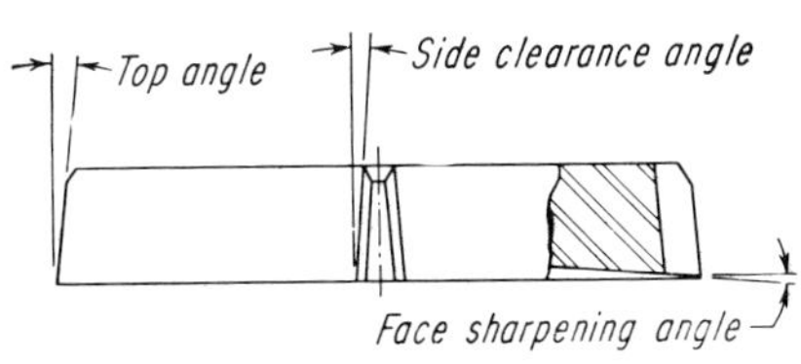

FIG. 22-14. Side clearance and top angles of a shaper cutter.

The side-clearance angle and the top angle are illustrated in Fig. 22-14 along with the face-sharpening angle.

Standard spur shaper cutters incorporating the usual 2° side-clearance angle have top angles according to Table 22-5.

Table 22-5. Shaper-cutter Top Angles

If pressure angle is	*Top angle will be*
14½°	7°56′
20°	5°49′
25°	4°43′
27½°	4°20′
30°	4°

When a shaper cutter is designed for a specific number of teeth, the top angle will vary slightly from the tables. The tables are based on an infinite number of teeth or a rack. For calculating the top angle for a specific number of teeth, see the formula in Art. 22-14.

Occasionally it is necessary to shaper-cut a part maintaining a close relationship between tooth thickness and root diameter. When this is the case, the top angle is usually provided with a lesser angle than theoretically required. With each sharpening of the cutter, the top angle can be resized giving an accurate cutter addendum for holding the proper depth in the part.

In general, a 5° positive sharpening angle is incorporated in the design of shaper cutters; however, in special cases, as little as 0° and as much as 10° is used.

FIG. 22-15. Shaping an internal gear.

It is important to remember that the sharpening angle is to be maintained throughout the life of the tool, as any deviation will result in a change in the involute form. A further explanation of this is contained in the section covering the sharpening of shaper cutters.

22-10. Field of Application. The shaper cutter is a tool of high versatility, being able to, under certain conditions, produce mating external, internal, and rack-tooth forms with the same tool.

There are certain types of gears, such as those with close adjacent shoulders, herringbone gears, and narrow-gap double-helical gears, which are particularly suitable for the shaper-cutter method. For the most part, internal gears of low- and medium-production requirements are produced by the shaper-cutter method. On high-production items, the internal gear is usually broached. The shaper cutter and internal gear relationship is illustrated in Fig. 22-15.

22-11. Types of Shaper Cutters. The four most commonly used shaper cutters are illustrated in Fig. 22-16.

Disk-type Cutters. The disk type is the most commonly used and least expensive, comparatively speaking, in the shaper-cutter group. This type of tool is used on both external and internal gears where the following conditions are satisfied:

1. The gear being cut must not have any interfering shoulder, which would interfere with the clamping nut used to secure the tool to the cutter spindle (see Fig. 22-17).
2. The workholding fixture must also be designed to clear the clamping nut.

Deep-counterbore Cutters. The purpose of the deep-counterbore cutter is to provide clearance between the clamping nut and the interfering shoulder or workholding adapter. Care must be used that the necessary clearance is provided to ensure that the clamping nut does not interfere at the end of the cutter's life or last resharpening (see Fig. 22-18).

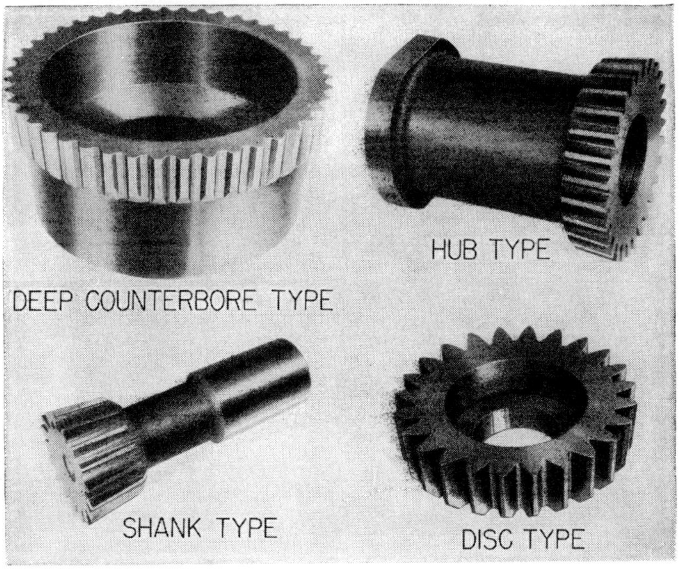

FIG. 22-16. Four common types of shaper cutters; disk type, deep counterbore, hub type, and shank type (composite required).

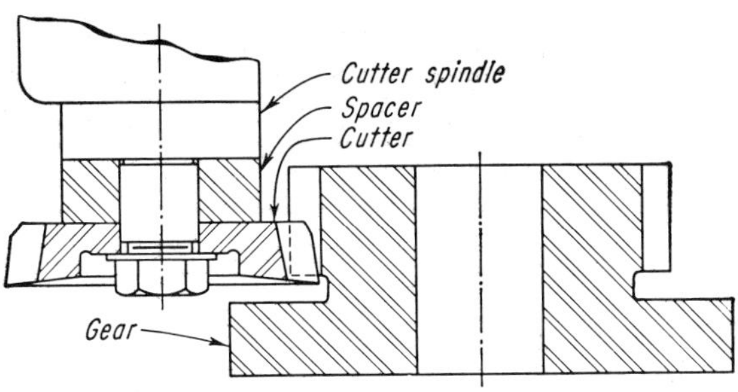

FIG. 22-17. Disk-type shaper cutter, cutting near a shoulder.

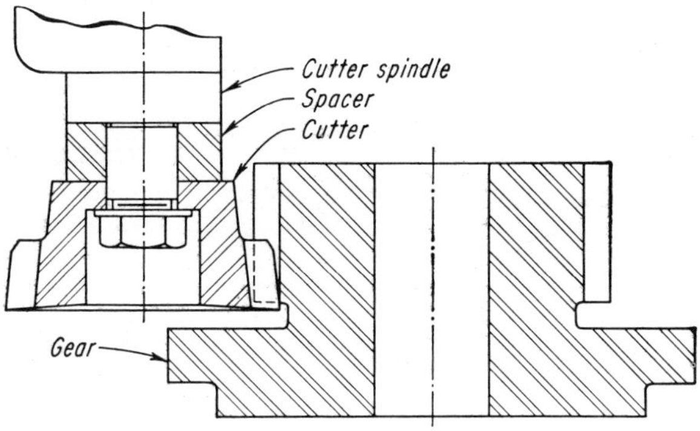

FIG. 22-18. Deep-counterbore shaper cutter designed to avoid interference of clamping nut and shoulder.

Hub-type Shaper Cutters. The hub-type shaper cutter is used primarily on internal gears where the diameter of the tool is restricted. (For a further explanation of this, see Art. 5-7.) The diameter of the tool in these cases often becomes smaller than the diameter of the cutter spindle. Figure 22-19 illustrated the use of a hub-type cutter cutting an internal gear.

The hub-type cutter offers more rigidity than the shank-type tool and should be used, when possible, where extreme accuracies are required.

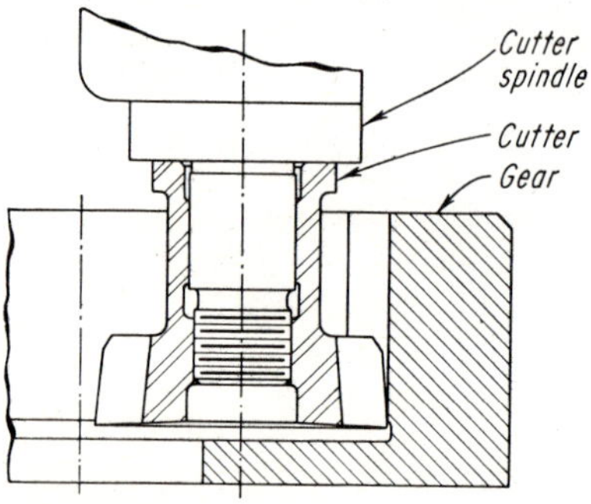

FIG. 22-19. Hub-type shaper cutter cutting internal gear.

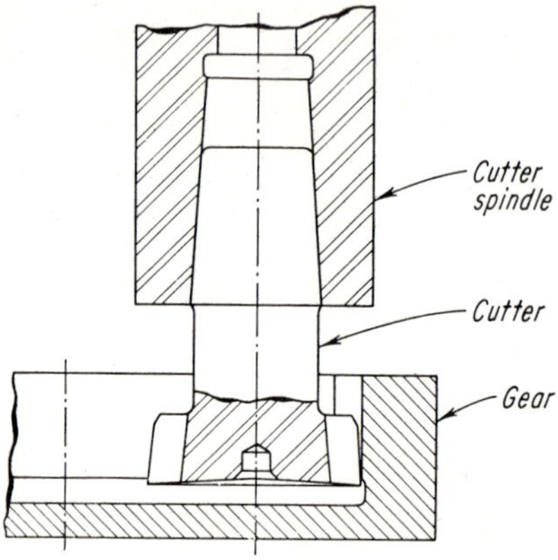

FIG. 22-20. Shank-type shaper cutter cutting internal gear.

Shank-type Shaper Cutters. The shank-type shaper cutter is used primarily on small internal gears. Here the diameter of the tool becomes so small it is not feasible to use a hole or hub for mounting purposes (see Fig. 22-20).

The shank can be made tapered or straight, depending on the model of machine on which it will be used. The over-all length of the tool is dependent on the face width of the part.

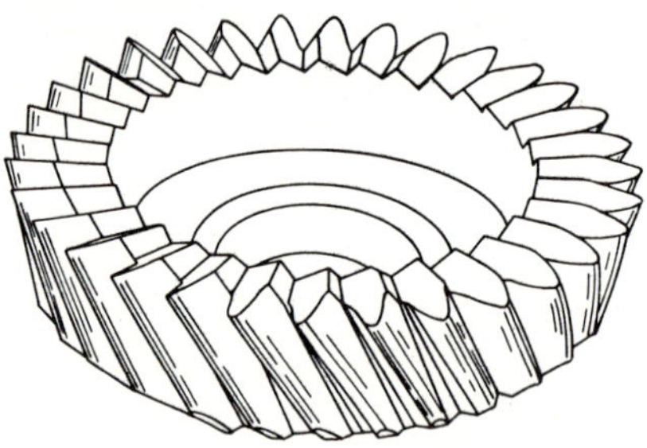

FIG. 22-21. Typical helical shaper cutter.

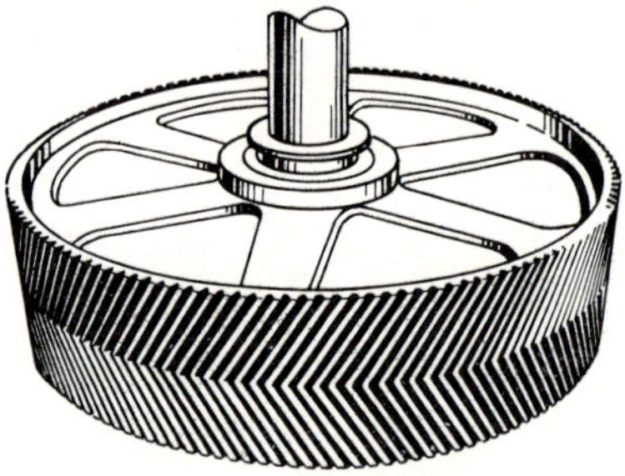

FIG. 22-22. Herringbone gear has no gap separating the two helices.

Where a drawbar is employed to hold taper-shank cutters, a flange is incorporated in the design. After the shank end is seated in the socket, the flange permits the drawbar to retract the tool only 0.003 to 0.004 into socket. This safeguards against damage to the hollow cutter spindle and offers added rigidity to the cutter.

Helical Shaper Cutters. The helical shaper cutter can be made in all the previously mentioned types, namely, disk, deep-counterbore, hub, and shank type. When cutting right-hand external gears and left-hand internal gears, a left-hand helical shaper cutter is used. If a left-hand external gear or right-hand internal gear is to be cut, a right-hand helical shaper cutter is used. The helix angle of the gear and tool must be common. A helical guide must be used in conjunction with the helical shaper cutter. This guide has the same lead and hand as the shaper cutter. It is possible then with the use of a left- and right-hand helical shaper cutter and corresponding guides to produce an entire range of tooth numbers of mating parts having the same helix angles. An exception to the foregoing is when extremely small numbers of teeth are to be cut. Here specially designed tools may be required to prevent excessive undercuts or possible flank interference. Figure 22-21 illustrates a typical helical shaper cutter.

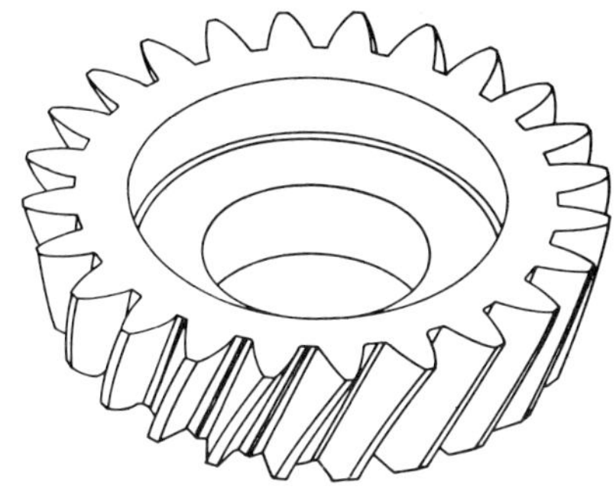

FIG. 22-23. Spur type sharpening of herringbone shaper cutter.

Herringbone-gear Shaper Cutters. The herringbone gear is differentiated from a double-helical gear in the fact that a clearance groove for the tool is not employed (see Fig. 22-22). This means that both the right- and left-hand gear are cut simultaneously with the tools working alternately. Because of the lack of clearance groove, it is necessary to design the herringbone-gear shaper cutters with a spur-type sharpening (see Fig. 22-23). The tools operate on the same spindle; so it is necessary to use matched sets of right- and left-hand cutters. The keyway is also held in relation within a set. Care must be taken when sharpening herringbone cutters so that an equal amount of stock is removed from each tool and that the outside diameter is the same after sharpening on both tools.

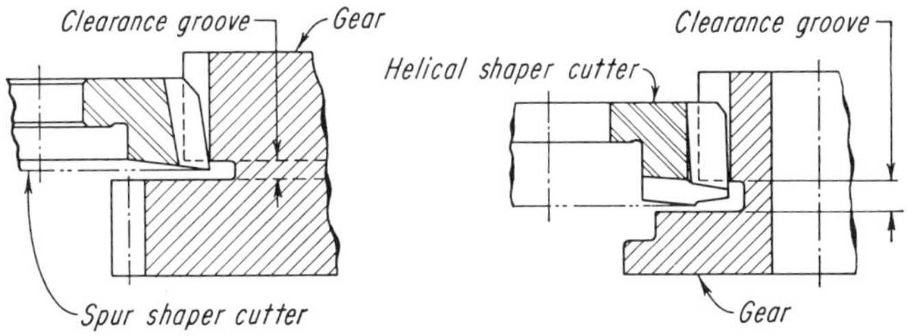

FIG. 22-24. Clearance groove for shaping.

Tool Clearance. It is important to remember in the design of gears having a close shoulder interference that a sufficient clearance groove should be incorporated. Figure 22-24 illustrates the clearance groove required for spur and helical shaper cutters.

For a spur gear, the minimum clearance groove can be calculated as follows:

$$(\text{Whole depth of gear} \times \tan \text{sharpening angle}) + \tfrac{1}{16}'' = \text{groove width} \qquad (22\text{-}2)$$

When calculating the clearance groove for a helical gear, the following formula should be used:

$$(\text{Whole depth of gear} \times \tan \text{sharpening angle}) + (\text{normal circular pitch} \times \sin \text{helix angle}) + \tfrac{1}{16}'' = \text{groove width} \qquad (22\text{-}3)$$

In cases where the clearance groove must be held to an absolute minimum on a helical gear, an off-normal sharpening can be incorporated (see Fig. 22-25). The off-normal sharpening can vary from slightly less than the helix angle to a spur-type sharpening as used on herringbone-gear shaper cutters. Any changes from a normal sharpening will result in a change in the involute form. For this reason, if an off-normal sharpening is desired, the tool manufacturer should be so advised.

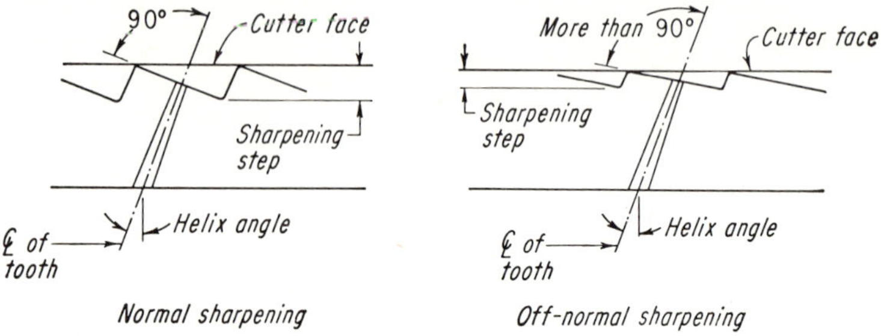

FIG. 22-25. Off-normal sharpening of helical shaper cutters.

22-12. Shaper-cutter Tooth Modifications. There are many tooth modifications which can be incorporated in the design of shaper cutters. These are ordinarily described as "special features" by the manufacturers and are priced slightly higher than tools without modification. In most cases, the modified tool is tailor-made for one specific gear and is not applicable over a wide range of gears having different tooth numbers. The most common modifications are protuberance, semitopping, approach type, topping cutters, and full tip radius. The following is a description of each of the above-mentioned features, along with the resultant change in the gear-tooth profile.

Protuberance. When a gear is to be finished by shaving or grinding, it is advantageous to have an undercut in the prefinished gear. This allows the shaving cutter, or grinding wheel, to blend smoothly with the fillet radius (see Fig. 22-26).

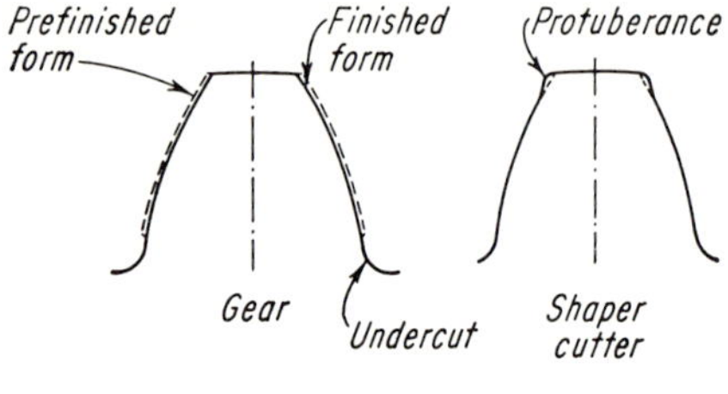

FIG. 22-26. Undercut in gear as cut by protuberance of shaper cutter.

The amount and position of the undercut are very important. Too little undercut will leave a step after the finishing operation. This step would then be in a very critical area on highly stressed gears and would possibly be the starting point for fatigue cracks. It should also be pointed out that any step presents a point of interference for the shaving cutter and can cause an erratic gear profile. As a general rule, on gears of 18 pitch and finer, the step is not large enough to interfere with the shaving operation, and the protuberance feature is not used.

If too much protuberance is designed into the tool, the undercut may extend too high on the tooth profile and result in the finishing tool not cleaning up above the start of active profile. This is detrimental in two ways: one, it reduces the involute contact ratio between mating parts, and two, it reduces the involute contact ratio between the shaving cutter and the preshave gear. In the first case, this can cause a higher noise level and/or an earlier fatigue breakdown between mating gears. In the second case, the loss of involute overlap can cause loss of involute control between the shaving cutter and the gear, and result in undesirable profile errors.

Semitopping. The semitopping feature is used to produce a chamfer at the point where the tooth profile and the outside diameter intersect.

One of the prime problems of gear manufacturers is the nicks which occur in the handling of gears after the teeth have been finished and prior to heat-treatment. This is particularly true on shaved gears where a slightly dull cutter will push up a small

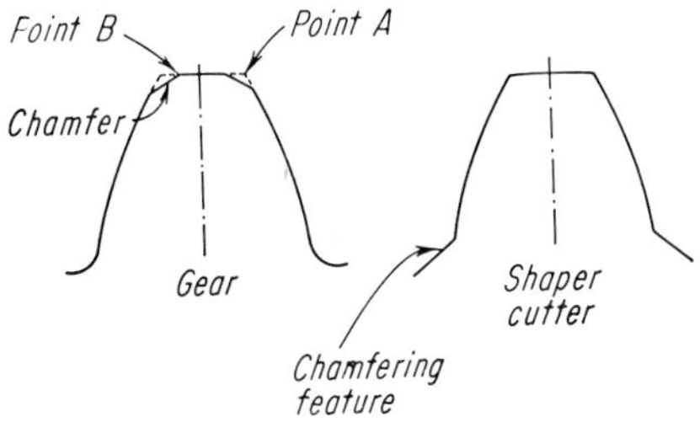

FIG. 22-27. Shaper cutter with semitopping feature.

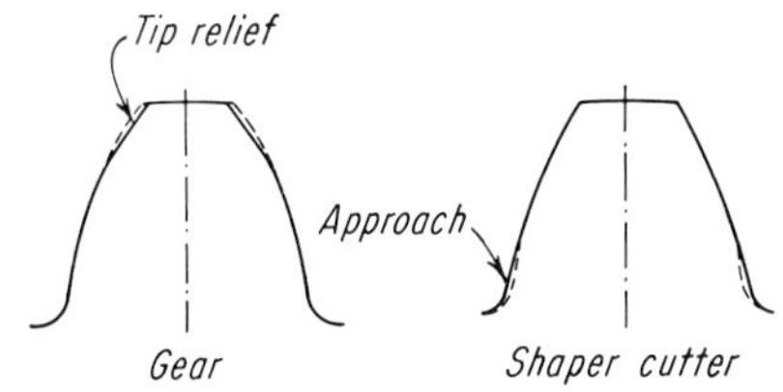

FIG. 22-28. Shaper cutter with approach feature for tip relief.

burr at the outside diameter. If the gear does not have a chamfer and comes in contact with another object, a nick will usually appear at point A in Fig. 22-27. This nick then interferes with the rolling action between the mating parts and results in a noisy gearset. If the gear has a chamfering feature and is mishandled in any way, the nick will usually appear at point B in Fig. 22-27. A nick at point B will not affect the rolling action of the gears, nor will it increase or affect the noise level.

Approach. The purpose of the approach feature is to provide a tip relief on the gear (see Fig. 22-28). The tip relief is used by gear manufacturers to provide a smoother rolling action on certain gearsets. There are two types of approach features:

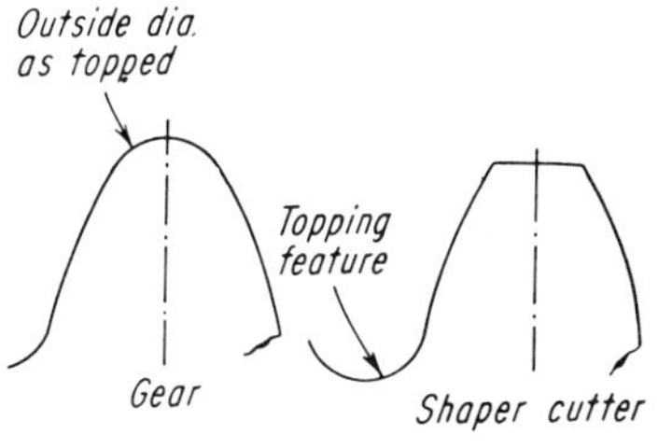

FIG. 22-29. Shaper cutter with full-topping feature.

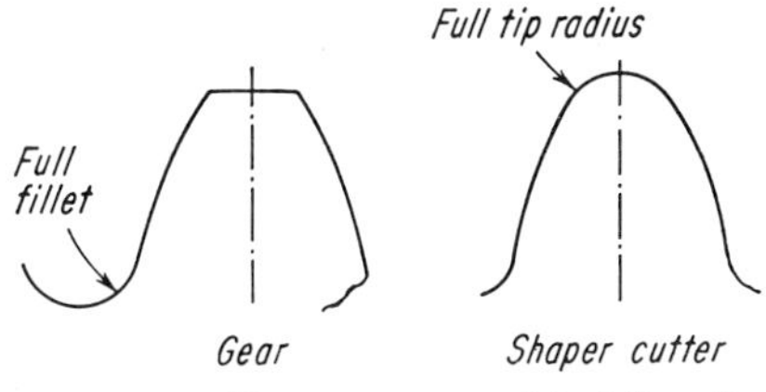

FIG. 22-30. Shaper cutter with full-radius tip.

The first is called the flank type, or the tangent approach, where the amount of tip relief will vary as the tool is sharpened back. The second type is called the H type, or the constant-approach type. Here the tool is designed in such a manner that the tip relief remains nearly constant throughout the life of the tool.

Full-topping Shaper Cutters. The full-topping shaper cutter, in addition to producing the tooth form, sizes the outside diameter, thereby establishing a relationship of pitch diameter to outside diameter (see Fig. 22-29). This feature is seldom used because most shaper-cutter applications do not require it. There is also a problem with a continuous chip which retards the chip flow, sometimes results in a poor finish, and can also decrease the accuracy of the gear.

Full Tip Radius. When a full fillet or round root is desired on the gear, a full tip radius is incorporated on the shaper cutter (see Fig. 22-30).

22-13. Sharpening Shaper Cutters. The importance of the sharpening of shaper cutters cannot be overemphasized, for an incorrectly sharpened tool can cause involute-profile errors, pitch-diameter runout, and spacing errors. In most cases, the face angle of the sharpening angle is 5°. Any deviation from this is indicated on the marking of the tool by the manufacturer.

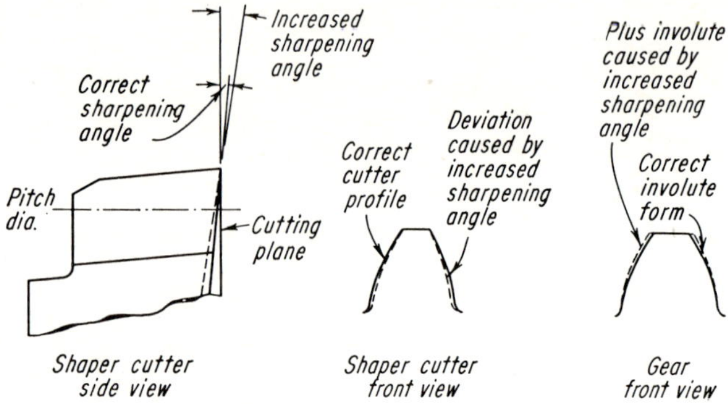

FIG. 22-31. Shaper cutter with an increased sharpening angle on face.

Figure 22-31 illustrates a tool sharpened with a larger face angle than called for and the resultant change in the involute profile on the gear.

Because the side-clearance angle causes the tooth thickness to become progressively thinner at any given diameter, a sharpening angle larger than the tool is designed for will produce a plus involute form on the gear.

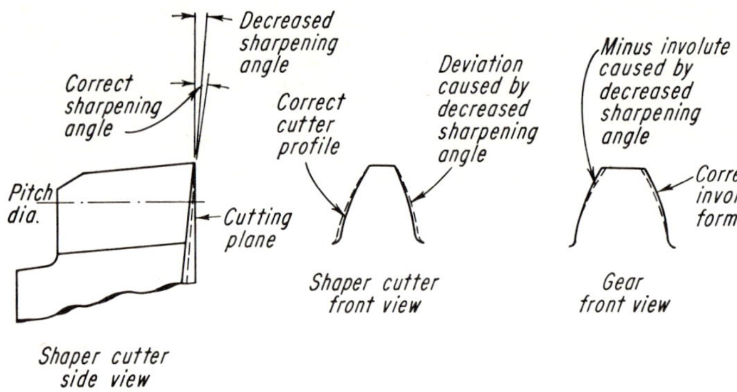

FIG. 22-32. Shaper cutter with a decreased sharpening angle on face.

Figure 22-32 illustrates that a tool sharpened with a smaller face angle than the cutter is designed for will produce a minus involute on the gear.

In the sharpening of spur shaper cutters, the tool is mounted on a rotary surface grinder (see Fig. 22-33) which is tipped to the correct sharpening angle.

It is important that a rigid machine be used, because of the varying area of contact, as the tooth and space alternately pass under the grinding wheel. It is also important that the shaper cutter be properly centralized, because any misalignment will cause the tool to produce pitch-line runout.

Rough Sharpening. It is also recommended that the tool be "rough" sharpened using a vitrified grinding wheel with a 12 to 14″ diameter of 60 grain size, grade H. The surface speed of the grinding wheel should be 5,000 to 6,000 fpm, traveling at about 45 strokes per minute, removing approximately 0.001″ per stroke. The amount of stock left for finishing should be from 0.001 to 0.004″, depending on the pitch of the tool.

Finish Sharpening. The "finish" grind should be done with a shellac or resinoid-bond silicon-carbon grinding wheel of about 400 grain size in a soft grade with free-cutting properties.

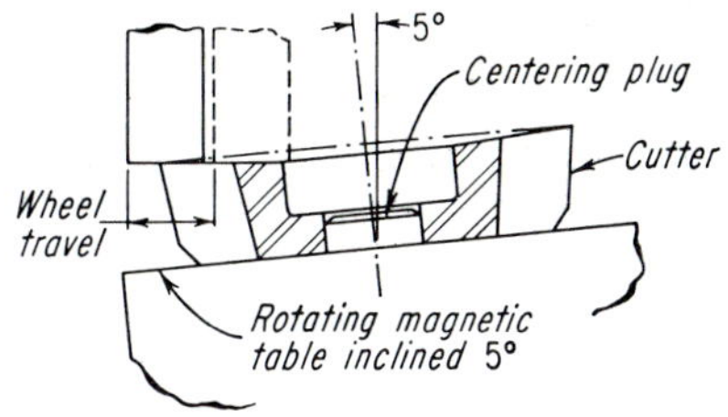

FIG. 22-33. Sharpening of spur shaper cutter.

It is possible under optimum conditions to sharpen shaper cutters in a single grinding operation. For maximum tool life, the cutting edges must be free from burning or heat checks and have a 10 microfinish or less.

In the sharpening of shank- or hub-type shaper cutters having small diameters, the diameter of the grinding wheel must be reduced. If too large a wheel is used, the sharpening angle deviates from a straight line, becomes concave-shaped, and as a result produces an involute error on the part.

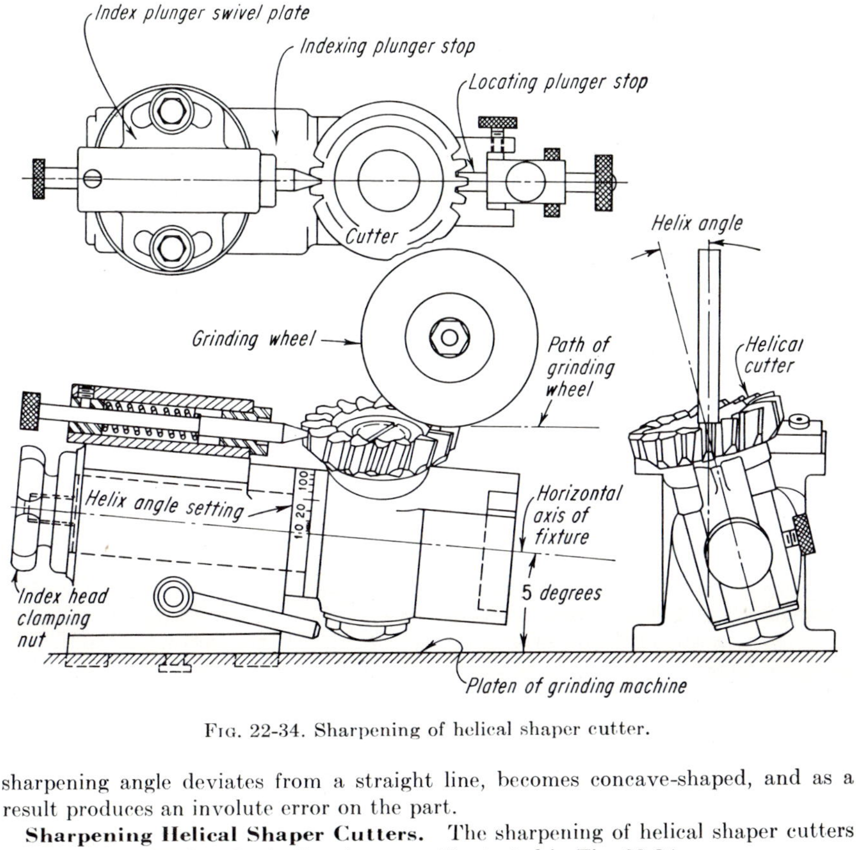

FIG. 22-34. Sharpening of helical shaper cutter.

Sharpening Helical Shaper Cutters. The sharpening of helical shaper cutters requires a hand-indexed grinding fixture as illustrated in Fig. 22-34.

Automatic fixtures are also available. Because the teeth are ground in a fixed position, one at a time, it is not possible to obtain high-quality finish. For coarse-pitch tools, free-cutting 90-grain I or J grade aluminum oxide wheels are recommended. On finer pitches, 150 grain in J or K grades is usually used. It is very important that all teeth be ground to the same height, as any deviation will result in tooth-spacing errors. In most cases, the face-sharpening angle is 5°. Any deviation from this, and also any change from a "normal" sharpening, will be marked on the tool. If the correct angles are not adhered to, an error in the involute profile will result.

22-14. Basic Shaper-cutter Calculations. A legend of the letter symbols used in gear work can be found in Chap. 4. See Art. 22-20 for specific symbols used in cutter work.

The Pitch Diameter. Each shaper cutter has a theoretical pitch diameter, which is calculated as follows: For spur-gear shaper cutters, the pitch diameter or $D = N/P$. For helical-gear shaper cutters $D = N/P_n \cos \psi$. In reality, a shaper cutter has many pitch diameters, which are called "operating" pitch diameters. Usually when the cutter is new it is *operating* at an oversized operating pitch diameter. This indicates that the operating pitch diameter is larger than the theoretical pitch diameter. As the tool is sharpened back, the operating pitch diameter decreases because of the decrease in the tooth thickness, which in turn is caused by the side-clearance angles. As this occurs, the operating pitch diameter can become smaller than the theoretical pitch diameter and is commonly referred to as operating at an undersized diameter.

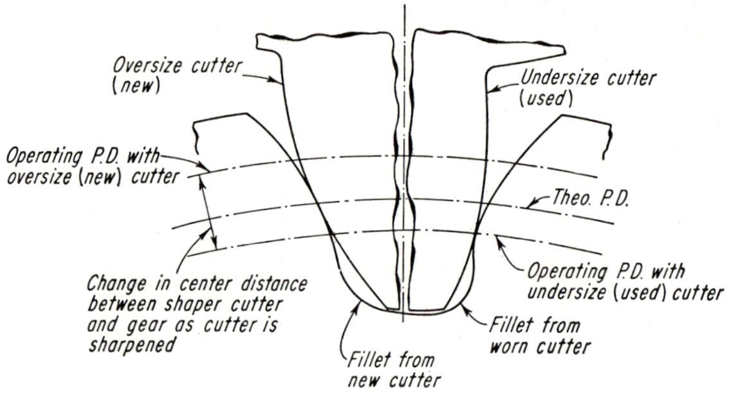

FIG. 22-35. Change of fillet on gear as cutter size changes from sharpening.

Oversize or Enlargement in Cutters. The amount of oversize which is incorporated in a new tool must be carefully considered. If too much oversize is used, the amount of flat or tooth thickness at the outside diameter of the tool becomes excessively small and will not support the cutting edge.

Another factor to be considered in determining the amount of oversize to be used is that the greater the amount of oversize, the higher the fillet will extend on the involute form.

If the fillet extends too high, it will interfere with the mating part. As the tool is sharpened back, the operating pitch diameter decreases and the fillet extends lower on the involute form (see Fig. 22-35).

Undersize or Reduction in Operating Pitch Diameters. It is sometimes necessary to design a shaper cutter which has an undersize operating pitch diameter when the tool is new. Care must be taken that when the tool is sharpened back the outside diameter of the part being cut does not contact below the base diameter of the shaper cutter. When this occurs, the portion of the involute profile which contacts below the base diameter of the tool can be trimmed off.

The Base Diameter. While the outside diameter, operating pitch diameter, and tooth thickness decrease as the shaper cutter is sharpened, the base diameter remains constant, as does the base pitch. The calculation for the base diameter differs from the normal formula used in gearing, since the face-sharpening angle and the side-clearance angles must be considered. The formula for spur shaper cutters is as follows:

$$\text{tan corrected pressure angle, or } \phi_c = \tan\phi + (\text{tan side-clearance angle} \times \text{tan face-sharpening angle}) \quad (22\text{-}4)$$

Then

$$D_{bc} \text{ or base diameter} = D_c \times \cos\phi_c \quad (22\text{-}5)$$

Example. $N_c = 30$, $P = 10$, $\phi = 20°$, face-sharpening angle $= 5°$, side-clearance angle $= 2°$.

$$\tan\phi_c = \tan 20° + (\tan 2° \times \tan 5°)$$
$$\tan\phi_c = 0.3639702 + (0.0349208 \times 0.0874887)$$
$$\tan\phi_c = 0.3670254$$
$$D_c = 30/10 = 3.000$$
$$D_{bc} = 3.000 \times 0.9387674 = 2.8163022$$

The formula for helical shaper cutters is more involved, since the base diameter on the uphill side is different from the base diameter on the downhill side because of

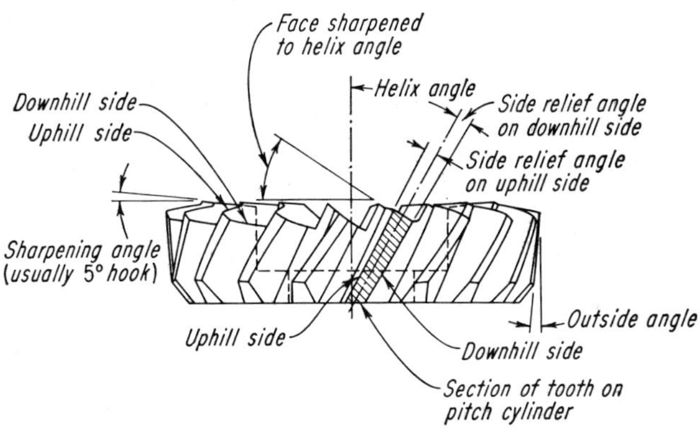

FIG. 22-36. Relief angles of helical shaper cutter.

side clearance and sharpening. As a matter of note, the uphill side is the side of the tooth which can be seen when looking at the face of the tool. The downhill side is hidden (see Fig. 22-36). To ensure freer cutting and eliminate side rubbing, it is common practice to increase the amount of side clearance on the uphill side.

Formula for both sides:

$$\text{tan corrected } \phi_n \text{ or } \phi_{n_1} = \tan\phi_n + (\text{tan side-clearance angle} \times \text{tan face-sharpening angle}) \quad (22\text{-}6)$$

For uphill side:

$$\text{tan corrected } \phi \text{ or } \phi_3 = \frac{\tan\phi_{n_1} \times \text{cos uphill side-clearance angle}}{\cos(\psi_c - \text{uphill side-clearance angle})} \quad (22\text{-}7)$$

Then

$$D_{bc} \text{ uphill side} = D \times \cos\phi_3 \quad (22\text{-}8)$$

For downhill side:

$$\tan \text{ correct } \phi \text{ or } \phi_2 = \frac{\tan \phi_{n_1} \times \cos \text{ downhill side-clearance angle}}{\cos (\psi_c + \text{downhill side-clearance angle})} \tag{22-9}$$

Then $$D_{bc} \text{ downhill side } = D \times \cos \phi_2 \tag{22-10}$$

Example. $N_c = 35$, $P_n = 10$, $\phi_n = 20°$, $\psi_c = 17°$, face-sharpening angle = 5°, side-clearance angle uphill side = 3°, downhill side = 2°.

Then $$D = \frac{N_c}{P_n \times \cos \psi_c} = \frac{35}{10 \times 0.9563048} = 3.6599209$$

For downhill side:

$$\tan \phi_{n1} = \tan 20° + (\tan 2° \times \tan 5°)$$
$$\tan \phi_{n1} = 0.3639702 + (0.0349208 \times 0.0874887)$$
$$\tan \phi_n \text{ downhill side } = 0.3670254$$

Then $$\tan \phi_2 = \frac{0.3670254 \times 0.9993908}{0.9455186}$$

$$\tan \phi_2 = 0.3879371$$
$$\phi_2 = 21.2031225 \quad 21°12'11''$$
$$\cos \phi_2 = 0.9323041$$

Then $$D_{bc} \text{ downhill side } = 3.6599209 \times 0.9323041 = 3.4121593$$

For uphill side:

$$\tan \phi_{n1} = \tan 20° + (\tan 3° \times \tan 5°)$$
$$\tan \phi_{n1} = 0.3639702 + (0.0524078 \times 0.0874887)$$
$$\tan \phi_{n1} \text{ uphill side } = 0.3685553$$

Then $$\tan \phi_3 = \frac{0.3685553 \times 0.9986295}{0.9702957}$$

$$\tan \phi_3 = 0.3793175$$
$$\phi_3 = 20.7726139 \quad 20°46'21''$$
$$\cos \phi_3 = 0.9349953$$

Then $$D_{bc} \text{ uphill side } = 3.6599209 \times 0.9349953 = 3.4220088$$

Top-angle Formula for Spur Shaper Cutters

For external gear:

C_1 = original center distance between shaper cutter and gear being cut when tool is new

ϕ = operating pressure angle when shaper cutter is new

W = usable face width of cutter

$$\text{inv } \phi_2 = \text{inv } \phi - \frac{W \times \tan \text{ side-clearance angle}}{R_b \text{ cutter} + R_b \text{ part}} \tag{22-11}$$

$$C_2 = \frac{R_b \text{ cutter} + R_b \text{ part}}{\cos \phi_2} \tag{22-12}$$

$$K = C_1 - C_2 \tag{22-13}$$

$$\tan \text{ top angle } = \frac{K}{W} \tag{22-14}$$

For internals:

$$\text{inv } \phi_2 = \text{inv } \phi + \frac{W \times \tan \text{ side-clearance angle}}{R_b \text{ part} - R_b \text{ cutter}} \tag{22-15}$$

$$C_2 = \frac{R_b \text{ part} - R_b \text{ cutter}}{\cos \phi_2} \tag{22-16}$$

$$K = C_2 - C_1 \tag{22-17}$$

$$\tan \text{ top angle } = \frac{K}{W} \tag{22-18}$$

Top-angle Formula for Helical Shaper Cutters

Symbols:

C_1 = center distance with part when cutter is new
C_2 = center distance with part when cutter is at life
UH = uphill side of cutter tooth
DH = downhill side of cutter tooth
W = usable face width of cutter
ϕ_1 = operating pressure angle when cutter is new
ϕ_2 = operating pressure angle when cutter is at life

As previously explained, the side-clearance angles will range from 1°30′ to 3°30′ depending on pitch, pressure angle, material to be cut, etc. It is a function of the tool designer to select the amount of side clearance which will best suit the particular application. The side clearance on the uphill side is usually larger by 30′ to 1° than the downhill side, as this is the side which is susceptible to rubbing.

$$\cos \psi_b = \frac{\sin \phi_n}{\sin \phi} \tag{22-19}$$

Find ψ_b.
Then

$$\psi_{b\mathrm{UH}} = \psi_b - \text{side-clearance angle} \tag{22-20}$$

$$\psi_{b\mathrm{DH}} = \psi_b + \text{side-clearance angle} \tag{22-21}$$

External:

$$\text{inv } \phi_2 = \text{inv } \phi_1 - \left(W \times 0.5 \frac{\tan \psi_{b\mathrm{DH}} - \tan \psi_{b\mathrm{UH}}}{R_b \text{ cutter} + R_b \text{ part}}\right) \tag{22-22}$$

Find $\cos \phi_2$.

$$C_2 = \frac{R_b \text{ cutter} + R_b \text{ part}}{\cos \phi_2} \tag{22-23}$$

$$\tan \text{top angle} = \frac{C_1 - C_2}{W} \tag{22-24}$$

Internal:

$$\text{inv } \phi_2 = \text{inv } \phi_1 + \left(W \times 0.5 \frac{\tan \psi_{b\mathrm{DH}} - \tan \psi_{b\mathrm{UH}}}{R_b \text{ part} - R_b \text{ cutter}}\right) \tag{22-25}$$

Find $\cos \phi_2$.

$$C_2 = \frac{R_b \text{ part} - R_b \text{ cutter}}{\cos \phi_2} \tag{22-26}$$

$$\tan \text{top angle} = \frac{C_2 - C_1}{W} \tag{22-27}$$

GEAR-MILLING CUTTERS

The use of involute gear-milling cutters has decreased appreciably in recent years. This results from the need for higher-quality gears, subsequent finishing operations, and increased production rates. The majority of present-day usages are on coarse-pitch applications. The gear-milling cutters are applied to low-piece part production or as dictated by machine availability and part capacity. A common application on coarse-pitch gears has been a roughing operation, leaving stock for finish hobbing or preshave hobbing.

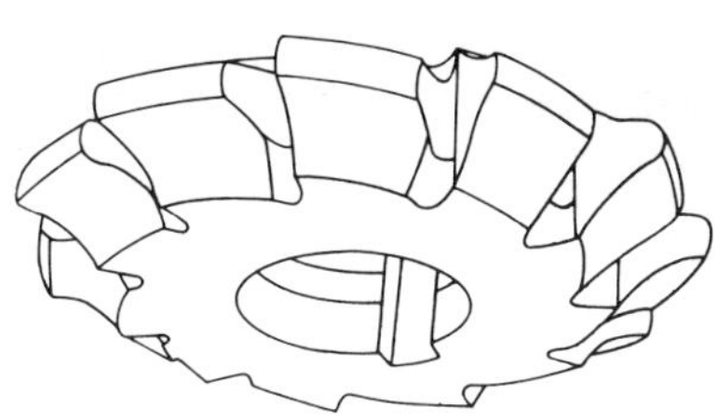

FIG. 22-37. Standard involute-gear cutter.

22-15. Types of Gear-milling Cutters in Common Use. *The standard involute gear cutter* (see Fig. 22-37) will cut a range of gear teeth. For each pitch, eight cutters are made as follows:

Cutter No.	*Will cut gears from*
1	135 teeth to rack
2	55–134 teeth
3	35–54 teeth
4	26–34 teeth
5	21–25 teeth
6	17–20 teeth
7	14–16 teeth
8	12–13 teeth

The cutters are made in either 14½ or 20° pressure angles, radial face sharpening, and a "backed-off" unground form. The cutter is designed for the lowest number of teeth in the range.

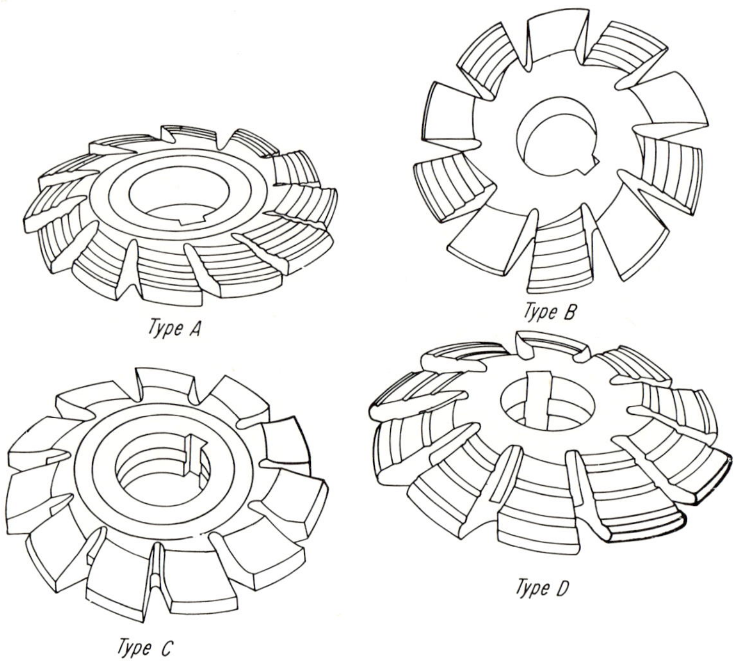

FIG. 22-38. Involute-gear stocking cutters.

As higher accuracy of tooth form is frequently required, the "half number" gear-milling cutters are used. These provide a better tooth form for the high end of each tooth range.

Cutter No.	*Will cut gears from*
1½	80–134 teeth
2½	42–54 teeth
3½	30–34 teeth
4½	23–25 teeth
5½	19–20 teeth
6½	15–16 teeth
7½	13 teeth

Involute-gear Stocking Cutters. These are roughing cutters designed for high-speed cutting. They are used singly and in duplex or triplex setups. The stocking cutters are made in four styles, as shown in Fig. 22-38.

Type A cutters have steps in both sides of all teeth and will produce steps in the gear-tooth space. These steps cut only on the diameter, giving clearance for rapid removal of stock. Type A is used for 4 pitch and coarser.

Type B cutters are stepped on one side of each tooth alternately. The stepped teeth are in plane with the smooth tooth profile, giving the advantage of chip breakers while cutting a smooth tooth surface. Type B is used for 4- to 10-pitch gears.

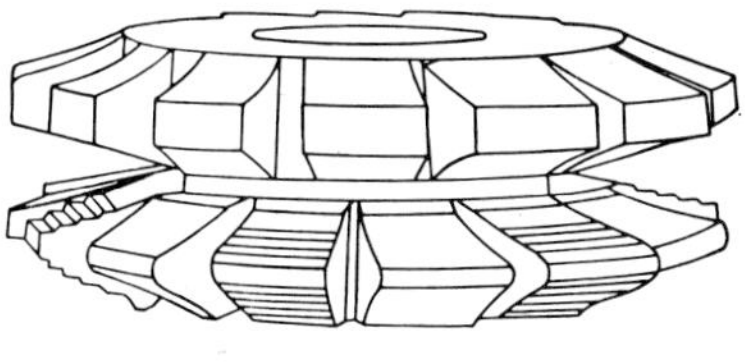

FIG. 22-39. Multiple-type involute-gear cutters.

Type C cutters are used for 10 pitch and finer. They are made with smooth teeth on both sides.

Type D cutters have three steps on one side and two on the other, alternating every other tooth. These are generally used on the coarse-pitch applications. Stocking cutters are usually furnished with a positive rake angle.

Multiple-type Involute-gear Cutters (see Fig. 22-39). These are made in sets to cut two or three teeth at one pass. These are usually single-purpose cutters used only for cutting the number of teeth for which they are designed. They are used in duplex and triplex in one of the following three combinations:

1. Roughing: with cutters centered correctly in relation to the gear.
2. Roughing and finishing: the finishing cutter is set on the gear center line.
3. Finishing: The center-line setting is the same as for roughing.

Figure 22-40 shows a duplex and triplex arrangement generally used for roughing. Finishing cutters are set up in the same arrangement. When roughing and finishing, the duplex and triplex arrangement as in Fig. 22-41 is generally used.

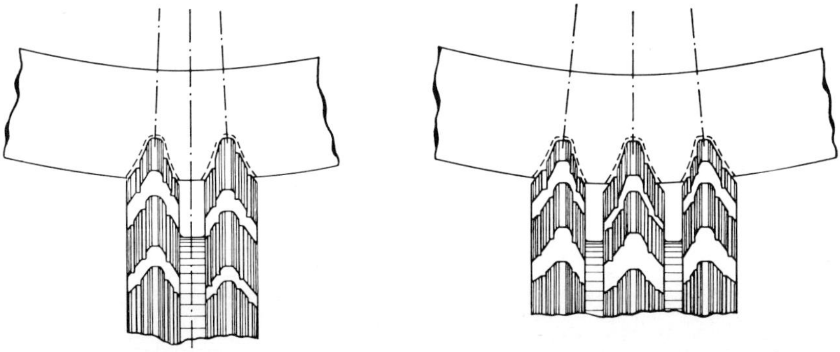

FIG. 22-40. Duplex and triplex arrangement of cutters for roughing or finishing.

22-16. Milling-cutter Design Practices. As mentioned in Art. 22-15, the cutter is designed for the lowest number of teeth in the range. The reason for this is apparent when the tooth-space configuration of a pinion is compared with the space configuration of a large gear or a rack. A No. 8 cutter designed for 12 teeth and used on 13 teeth produces a tooth curvature of predominately minus involute at the tip of the 13-tooth gear. This is generally desirable since a minus-involute curvature rolls smoothly in conjugate action. If the design practice were reversed and the cutter was designed for the highest number of teeth, then plus-involute curvature would result in the low range. Conjugate rolling action under these conditions would be rough and probably noisy.

The cutter is a formed type, as opposed to a generating cutter, such as a hob. Therefore, size or tooth thickness of the gear is an important factor. If the gear is milled oversize, it is likely that tooth curvature will be of plus-involute type. If undersize from milling too deep, the tooth curvature will be of minus-involute type.

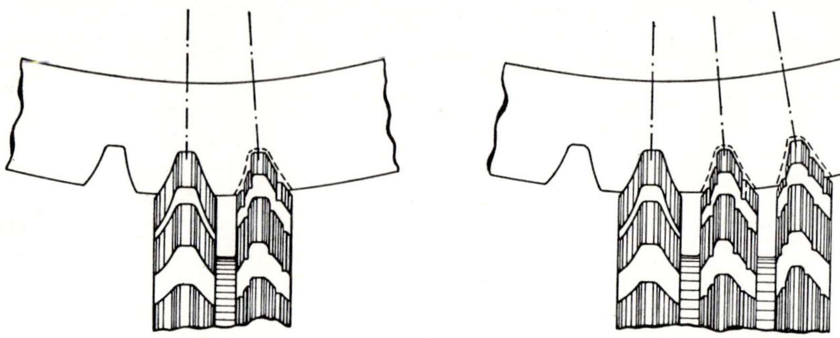

FIG. 22-41. Duplex and triplex arrangement of cutters for roughing and finishing.

When generating gears with a hob, the involute curvature is unchanged, regardless of size.

Allowances are made in the design of standard-type cutters to accommodate normal limits of tooth thickness so that, even when milling at the high side of tooth-thickness tolerance, the tooth curvature will be compatible for good conjugate rolling action.

The cutters vary in size from 8½″ diameter with a 2″ bore to 1¾″ diameter with a ⅞″ bore. Cutters are normally unground but can be obtained with ground forms.

When ordering, it is beneficial to furnish an up-to-date part print so that the supplier of the gear-milling cutter can provide the best possible tools.

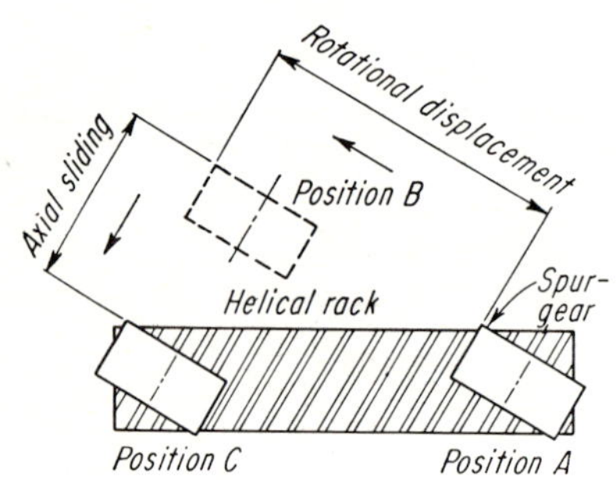

FIG. 22-42. Principle of shaving shows a spur gear and helical rack in engagement. Spur gear at position *A* is at rest. Leaving the rack stationary, gear is rolled to position *B*. Note its inherent tendency to roll in a plane parallel to its own sides and out of confines of rack, through rotational displacement *A* to *B*. To keep gear within confines of rack, it must slide axially in direction shown to position *C*. In actual shaving operation, gear is restrained from rolling to position *B*, since it is fixed on a spindle. By this action, rolling and sliding of gear are simultaneous.

GEAR-SHAVING CUTTERS

Gear shaving has developed into a highly sophisticated art. There are two basic methods and several variations in the way the shaving operation is performed. In this section we shall cover the general principles of shaving-cutter design and specific calculation procedures for a detail design. For general information on shaving machinery, see Chap. 18.

22-17. General Principles of Shaving Cutters (see Fig. 22-42). Shaving is accomplished by either of two methods. *Rack-type shaving* uses rack shaving tools comprised of ground blades assembled in a rack box and held by end blocks and a drawbar, as shown in Fig. 22-43. *Rotary shaving* uses rotary-type tools on a spindle, the tool being either the driver or driven member, and basically designed to be a mating part. The tools are precision manufactured to tolerances equal to but more often finer than master gears. The tools are serrated on both flanks of the tooth.

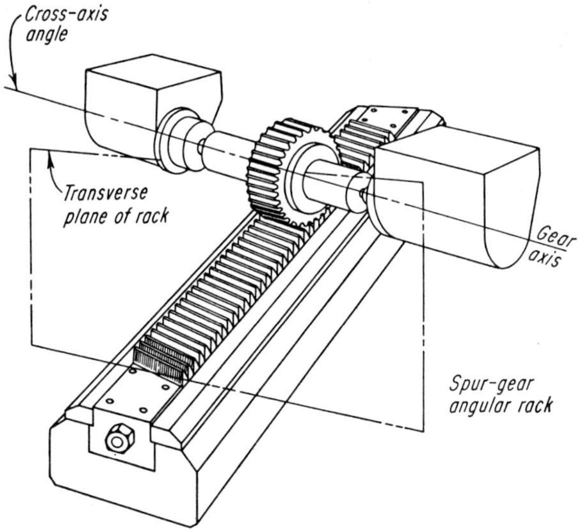

FIG. 22-43. Rack-type shaving.

The helix angle of the tool does not correspond to the helix angle of the part to be shaved. Instead, it is selected to cause meshing of tool and work on nonparallel axis, or what is referred to as *cross axis*. The condition of cross axis promotes side sliding, which, in conjunction with involute sliding, causes metal removal. The serrated flanks of the tooth act as the cutting edges (see Fig. 22-44).

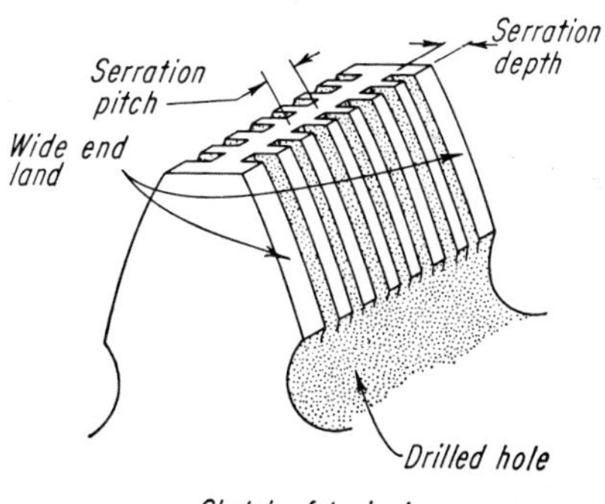

FIG. 22-44. Typical serrated tooth.

The Cross-axis Angle. The range of the cross-axis angle is generally restricted from 4 to 15°, the optimum angle being between 8 and 12°. The lower range near 4° is used to accommodate physical restrictions of adjacent shoulders. The higher range, above 12°, is used to increase the shaving-tool usage in shaving a wider range of helical gears utilizing the same shaving tool.

Spur gears would be shaved by a right- or left-hand tool of 10° helix angle, causing a 10° cross-axis angle. Usually, a right-hand tool is furnished for spur gears as a standard procedure. The same 10° right-hand shaving tool would be suitable for a gear with a 20° left-hand helix angle. The normal plane characteristics of pitch and pressure angle must be common between the spur gear, 20° left-hand gear, and 10° right-hand shaving tool. This pertains to pitch and pressure angle. The selection of the cross-axis angle is usually the tool designer's responsibility. For shoulder gears, when designing, to avoid interference, the angle may be determined by making the layouts shown in Figs. 22-45 and 22-46.

Design of Tool. The rotary-shaving cutter is usually made with a prime number of teeth. This "hunting-tooth" condition further refines spacing errors and increases accuracy of the shaved part. Average recommended size is an 8″ pitch diameter tool for most external gears. A 9″ pitch diameter tool is frequently used for low tooth numbers in the part. The choice of diameter is predicated on the usually noticeable control of the shaved involute profiles. Face widths of the shaving tools vary from 1 to 4″, depending on the requirements and type of application. Fine-pitch shaving cutters are furnished with face widths 3/8 to 5/8″ and with diameter of 3 to 4″. The shaving tools are priced according to pitch, pitch diameter, and face width. In some cases of extreme tolerances required in the tool, and special modifications, the price of tools is adjusted as needed. The tooth-flank serrations are cut serrations and are spaced according to the pitch of part and type of shaving to be done. The end lands are designed to be wider than the inner serrating lands. Depending on the type of shaving, whether transverse, modified underpass, or full underpass, the serrations

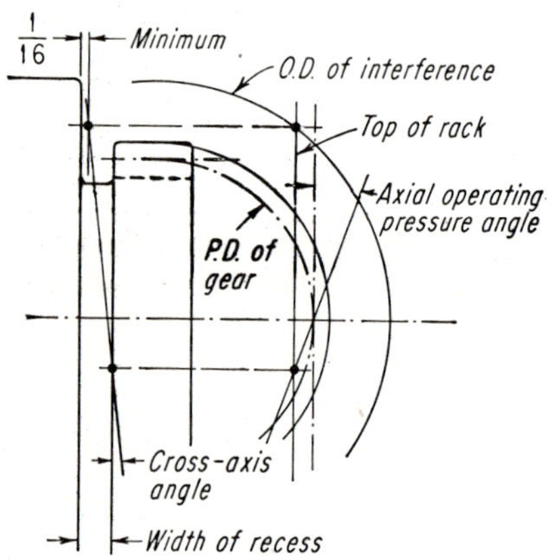

FIG. 22-45. How to determine cross-axis angle for shoulder gears for a rack finisher.

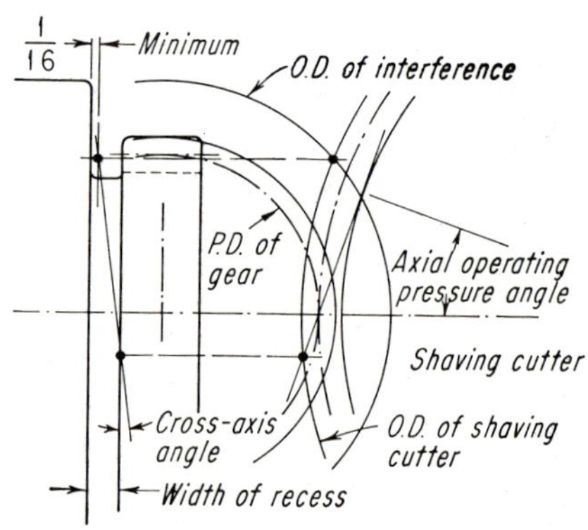

FIG. 22-46. How to determine cross-axis angle for shoulder gears for a rotary finisher.

will be annular or with a staggered pattern. The annular serration is used for transverse or modified underpass methods of shaving. The staggered type of serration for full underpass shaving compensates for the reduced axial sliding and prevents the tracking of serration lands on the tooth flank of the shaved part. The dedendum area of the shaving tool is supplemented with a drilled hole to form an undercut to allow the serrating tool to clear itself during manufacturing.

Transverse Shaving. This method is desirable for extreme face widths of gears. A standard 1″ face width of cutter may be used. With use of the shaving-machine feed slide, the tool can transfer shaving contact from one edge of the gear to the other edge. It is customary design to move the work past the cutter by means of the feed slide. On some of the larger shaving machines, for obvious reasons, the shaving cutter is mounted on a feed slide. In this discussion, the workpiece is the moving member. The table supporting the part being shaved can be rocked about a pivot point in timed relation with the feed. This method of shaving, while feeding and rocking, will produce a crowned lead on the part being shaved. Straight shaving of the lead with combinations of crown, if desired, can be done by the transverse method. Production time is the slowest by this method, since multiple feed passes are required to remove stock and properly size the gear. In transverse shaving, the pivot point of the cutter must pass both edges of the gear in order to obtain lead control as described in Fig. 22-47.

Modified-underpass Shaving (*Diagonal Shaving*). This is the most widely used method of shaving. It permits shaving gears with face widths slightly in excess of the tool width (see Fig. 22-48). Its production time in shaving is high because of the short stroke of the feed slide. One shaving cycle is usually sufficient to complete the shaving. The stock-removal method is efficient because of the higher cross-axis

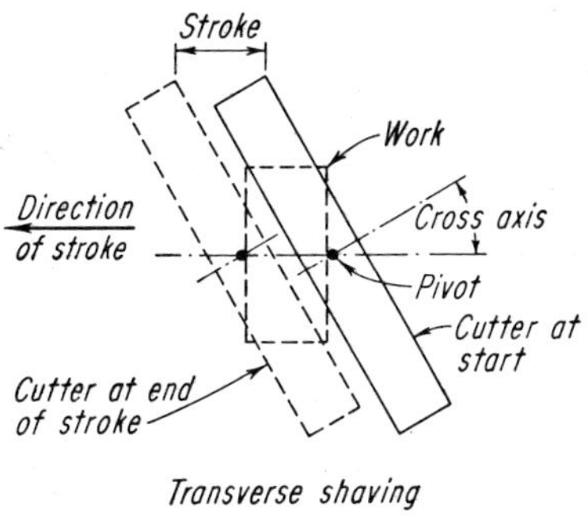

FIG. 22-47. Transverse shaving method.

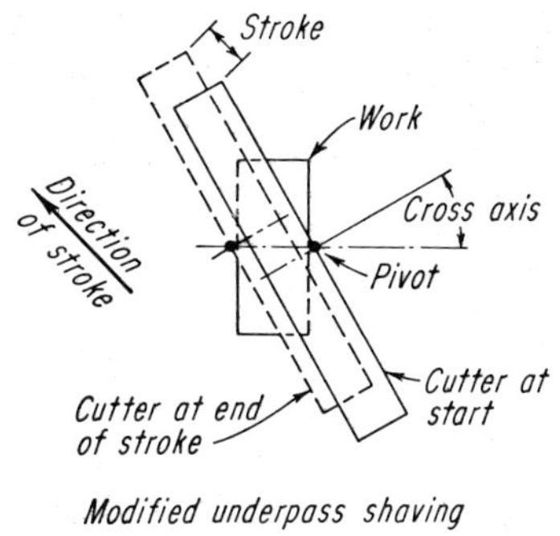

FIG. 22-48. Modified-underpass shaving method.

angle possible when compared with underpass shaving. Crowning can be accomplished by use of either a crowning attachment, as with transverse shaving, or by use of a tool ground with a concave lead. The angle of feed recommended is from 40 to 70°, measured from a normal to the work axis. When crown shaving with a tool ground with a concave lead, the amount of crown can be varied by altering the feed angle. The greatest amount of crown is obtained by feeding on an angle approaching the full-underpass condition. Crown can be decreased by feeding on an angle approaching the transverse condition. As mentioned before, the tool will have annular serrations. Excluding the ground-concave-lead feature in a shaving tool, the same tool may be used for either transverse or modified-underpass shaving, if the gear to be shaved permits.

Underpass Shaving. This method is usually dictated when shaving cluster gears or gears with close adjacent shoulders. The cross-axis angle is usually in the area of 4 to 6° in order to avoid interference of the tool with the shoulder. In all cases of underpass shaving, the tool must be wider in face width than the gear to be shaved. This is obvious by study of the diagram in Fig. 22-49 showing the direction of feed. Crowning of the gear is feasible by the underpass method, but only when the shaving tool is ground with a concave lead. The underpass method of shaving is usually the fastest method of shaving, because of the short feed stroke and the one cycle required.

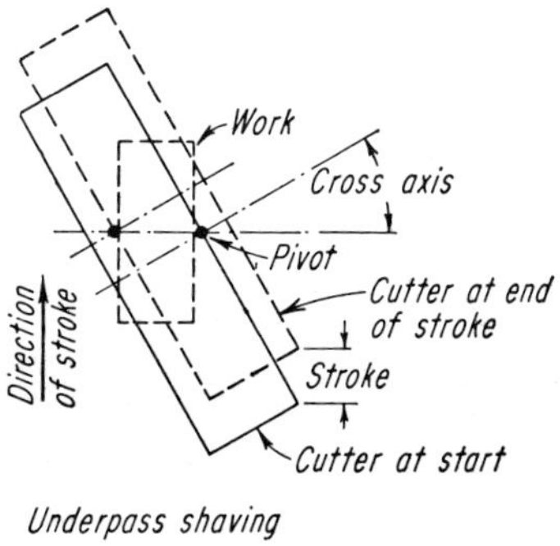

FIG. 22-49. Underpass shaving method.

The low cross-axis angle has its disadvantages in that the helical side sliding, promoting stock removal, is reduced. To compensate for this, the amount of shaving stock should be maintained at a minimum, but still allow for a full clean-up on the part tooth flanks. In this method of shaving, it may be expected that the number of pieces per sharpening of the tool will be lower than for transverse or modified-underpass shaving. This is a result of the lower cross-axis angle and the consequent lower helical side sliding. It is common to find that, in spite of good intentions in maintaining control of the preshave gear, the amount of stock left for shaving will become critical because of negligence. The amount of stock removal is critical whenever crown shaving is introduced to this method. The amount of crown per side of tooth at the gear faces must be considered as shave stock along with the normal preshave stock. The serrations must be the type with staggered pattern.

Plunge-feed Shaving. This method is not commonly used but is worth mention as to a possible method. It is an adaptation of the underpass method, applied to both internal and external gears usually of narrow face widths. The staggered pattern of serrations is used on the tool. Instead of feed-slide motion, as with the other types of shaving, the plunge-feed method uses strictly an infeed to obtain size. As would be expected, the cutting pressures are high and workholding becomes important to obtain rigidity.

Parallel-axis Shaving. In this application, the tool and work are rotated in mesh with the axes parallel. Without benefit of the cross-axis angle, there is no helical side sliding. Therefore, a reciprocating motion must be induced at a comparatively high frequency. This requires a special machine. The tool cuts only on its face, as it is rotated in one direction and then the other, as the reciprocation is taking place. It is generally applied to internal shaving. This method is not commonly used.

The Pivot Point. It is impossible to design tools or make intelligent setups without giving consideration to the pivot point. Each shaving machine is built with a swivel rotatable for cross-axis setting between the tool and work. In the construction of the shaving machine, the pivot point of this swivel is so located that its axis, if projected, would intersect both the work and shaving-tool axis. Its position of intersection with the tool axis is a known distance from the mounting shoulder of the shaving cutter spindle. Usually, the distance of pivot location is 1″ from the mounting face. If a 1″ face width shaving cutter was to be set up with the pivot point in the center of the tool, a ½″ spacer should be mounted between the cutter face and the mounting shoulder. By substituting spacers of different widths, the pivot point can be transferred to various positions within the face width of the tool. This procedure is valuable to the setup man when he sets up any of the three methods of shaving, in terms of the stroke of the feed slide. The pivot point must pass beyond both faces of the gear being shaved in order to control the lead of the tooth. Shaving is not complete unless the pivot point passes the gear faces (see Fig. 22-50). Since the pivot point defines the shortest distance between the part and tool axis, complete sizing is impossible in any other position. In trouble shooting, failure to observe the fundamental of the pivot point in setting up the machine and determining stroke has been the major cause of down time. The operator's manual for each machine will describe the location of the pivot for each machine model.

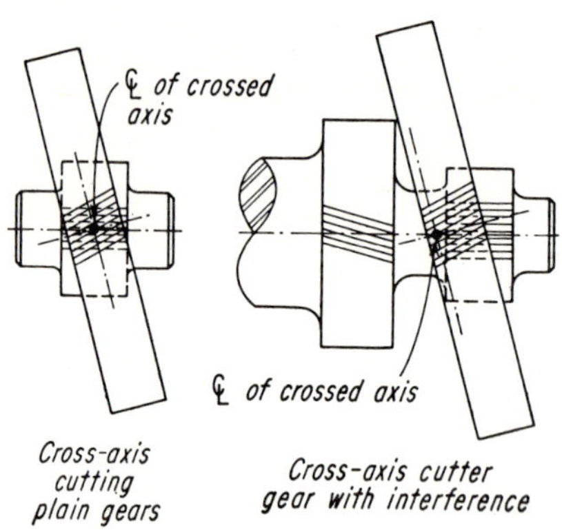

FIG. 22-50. In cutting plain or shoulder gears, the pivot point must pass beyond both faces of the gear.

Direction of Feed. In shaving either spur or helical gears by the modified-underpass shaving method, there is a choice of direction of feed and rotation. Refer to Fig. 22-51*a* for shaving spur gears, with either a right- or left-hand shaving cutter. The direction of the feed slide should be set to feed in the general direction of the cutter helix. Rotation of the cutter should be in a direction opposing the feed.

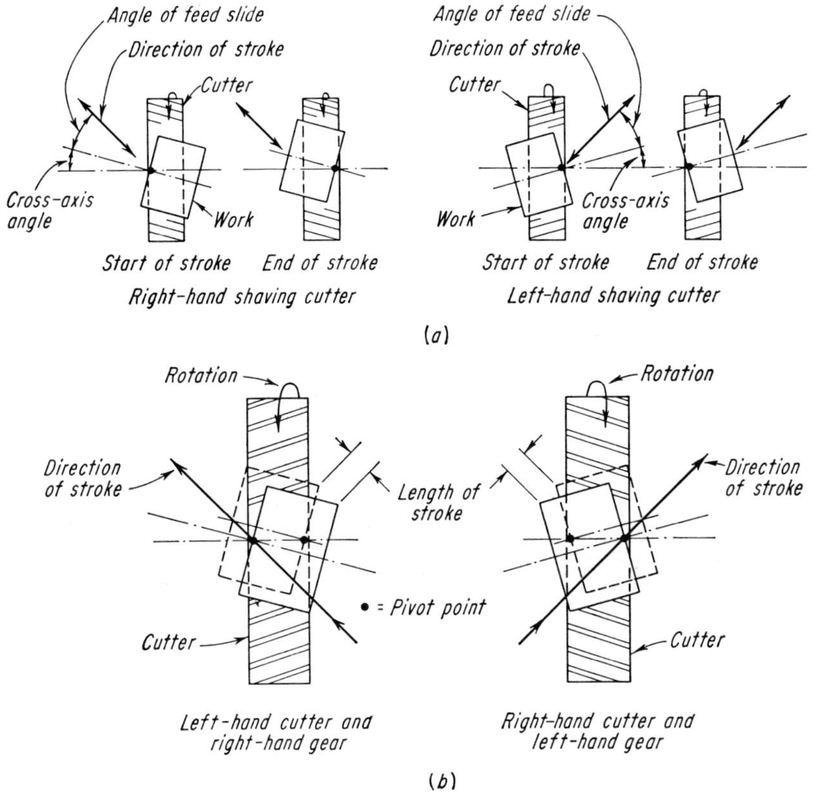

FIG. 22-51. Direction of feed and rotation for modified-underpass shaving. (*a*) Spur gears. (*b*) Helical gears.

Refer to Fig. 22-51*b* for shaving helical gears. The direction of the feed slide should be set to feed in the general direction opposing the helix of the cutter. Rotation of the cutter should be in a direction opposing the feed. Occasionally, a machine setup is made contrary to these recommendations with satisfactory results. However, these setups are in a minority.

As a rule of thumb, for spur gears:

1. Feed in direction of cutter helix.
2. Rotate cutter against direction of feed.

As a rule of thumb, for helical gears:

1. Feed against helix of cutter.
2. Rotate cutter against direction of feed.

As a matter of note, in underpass shaving, always rotate against direction of feed.

Tool Geometry. The latter part of this section will cover in detail the shaving-tool design and the calculations. The purpose of comment here is for general coverage of operating pressure angle and basic principles. Either the gear or the tool is the driving member in the shaving operation. The model of the machine predicates which is the driving member. Since the other member depends on the driver for rotation, special consideration must be given to the tool geometry, the most important consideration being the choice of the operating pressure angle to obtain the most consistent results.

The operating pressure angle is not usually the same as the theoretical pitch-line pressure angle. The tool is designed to operate and shave at a diameter of rolling other than the pitch diameter—thus the special operating pressure angle. The angle used is a calculated element, selected to contribute to optimum conditions of tooth contact. Failure to provide optimum conditions results in lack of involute control of the shaved part. One of the symptoms of lack of involute control is the hollow involute at the pitch-line area of the part. Other factors such as improper preshave can contribute to hollow irregular involutes. The tool engineer with sufficient experience can differentiate as to what factor is the cause of trouble. To eliminate tool design as a factor narrows the area of tool development. The tool throughout its life is designed to maintain the same "true form diameter" on the part. This is accomplished by retaining the correct proportions of tooth addendum and tooth thickness. The tooth thickness of the tool is decreased with each resharpening of the tool. This causes a reduction in operating center distance and the consequent change in operating pressure angle. The addendum must be decreased to contact at the original "true form diameter" or TIF diameter. This does not change in the same proportion as the change in center distance. It is a common misunderstanding to believe the base diameter of the tool changes with each sharpening. The base diameter is constant and remains so throughout the life of the tool. The operating pressure angle is the element changing as the center distance changes. The cross-axis angle decreases slightly with each resharpening. This is a function of the operating pitch diameter, decreasing with each sharpening as the lead of the tool remains constant. This is obvious from examination of the formula for lead:

$$\text{Lead} = \frac{\pi \times \text{diameter (operating pitch diameter)}}{\text{tan helix angle}} \tag{22-28}$$

Since lead and π are constants, the helix angle decreases along with the operating pitch diameter with each sharpening. Only in rare cases is the involute of the shaving cutter altered during the entire life of the shaving tool. This, when it is required, is a result of experience and tool development, based on case history and records of the tool's past performance. This practice is discouraged, since the development is costly and sometimes is a result of inadequate tool design. The entire life of the shaving tool is carefully planned by the designer. Having full knowledge of the part to be shaved, its mating part, and the preshave tool, it is possible to predict performance of the tool in terms of expected regrinds, involute characteristics, and sources of deviations that may result from preshave. The basic geometry of the tool will parallel the design of the mating gear, except for cross-axis angle and the no-backlash operating condition. With shaving tools, troubles generally stem from "action and reaction." If a fillet interference exists, as a result of an improper hob job, a crowding of contact will result, and a "reaction" one base pitch away on the part can be observed. This is the reason that the shaving-tool manufacturers press for satisfactory preshave conditions. The tool is capable of a great deal of correction, but only if the preshave part will allow it to do so. Shaving, therefore, is not a "cure-all" but a means of making a good gear better.

22-18. Preshave Gear. The preparation of the gear blank is of utmost importance in obtaining the most economical shaving operation. Concentricity of gear pitch line and bore, squareness and parallelism of the gear faces, and the bore itself are important factors. An oversize bore, for example, can completely mislead tool development for production. Profiles will be erratic, making it next to impossible to make adjustments in the grind of the shaving tool; or, if the tool has been previously developed, the erratic profile may cause a needless redevelopment.

Gear faces that are not square with the bore and parallel with each other will produce leads that are misleading. The profile and lead in the shaved gears are the important elements. The damaging factors mentioned are the most common to shaving trouble. The depth of the preshave gear should be the recommended amount of 2.35/pitch.

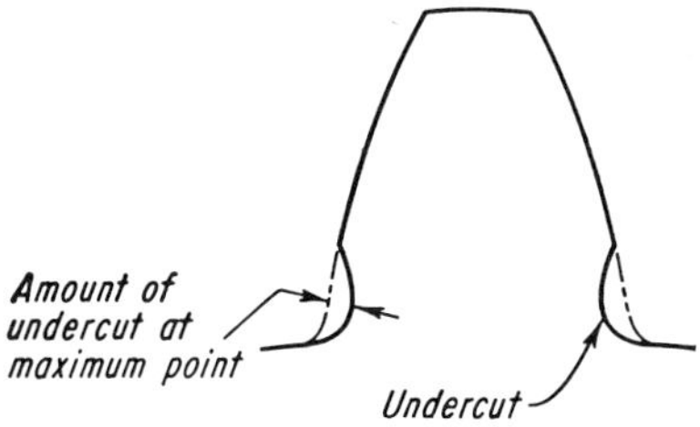

FIG. 22-52. Undercut on a gear tooth.

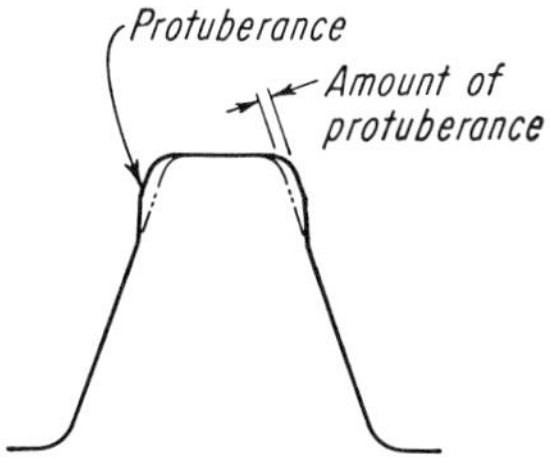

FIG. 22-53. Protuberance on a gear-hob tooth.

It is extremely important that an undercut in the fillet area or at base of the part tooth be provided for the blending of the shaved profile as shown in Fig. 22-52. A protuberance-type hob or shaper cutter is required for this purpose (see Fig. 22-53).

The stock allowance for shaving varies between users of shaving equipment. Table 22-6 is intended only as a guide. The amounts are based on circular tooth

Table 22-6. Stock Allowance for Shaving

Diam. pitch	Min.	Max.
2	0.003	0.004
6	0.002	0.004
10	0.002	0.003
16	0.001	0.003
20	0.0005	0.002
48	0.0003	0.001

thickness and may be measured by balls, rolls, or blocks, or a master gear on a rolling fixture. This table covers a pitch range from marine gears to instrument and control gears, as observed in the field. One manufacturer, on a production basis is controlling stock to 0.001″ allowance on 16 pitch. Another manufacturer, on an identical gear, finds it difficult to clean up the tooth, even with a 0.003″ stock allowance. The answer is in the preshave operation throughout. The condition of the machines, the skill of the operators, and the desire to keep tool costs down determine how much stock allowance the individual plants can tolerate. A well-known tractor manufacturer reduced transmission rejections from an average of four a day to an equal number per working week simply by requalifying his hobbing-machine spacers. The hobs, as mounted, were weaving, the shaving tool could not correct the lead variations, inspection had to compromise to maintain production, and consequently, the entire operation suffered by failure to observe a simple fundamental of good gear cutting.

22-19. Resharpening. The tool, upon dulling, will react in a manner to warn that a tool change is needed. One of the symptoms will be failure to maintain size. Upon making size adjustments, the response of the tool is inadequate to the change made. If monitoring with an involute checker, a common symptom is a plus involute near the tip of the tooth. This will grow steadily worse. A rolling fixture utilizing a master gear will signal rough rolling action. Deterioration of the surface finish is usually visible. When it has been determined that the tool is dull, the tool should be returned to the manufacturer for resharpening. Continued operation of a dull tool will cause damage that will seriously detract from the life of the tool. Serrations may flake out or excessive stock may have to be removed to recondition the cutting edges. Most users keep a record of pieces shaved per grind with each cutter in order to avoid overdulling. A frequent complaint about shaving tools is that the resharpened tool does not give so many shaved pieces as a new tool. Contrary to this, it is not uncommon to hear that the resharpened tool produced more pieces than the new tool. There are so many factors that influence tool life that at this point a careful study is necessary to determine the reason. The reason is usually found in the usage, operator technique, change in material, contaminated coolant, stock removal, or any number of things or combinations. The greatest step forward has been the clinics, or educational programs, within the user's plants.

The amount of sharpening life depends primarily on the depth of the serration. This varies with the pitch. A coarse-pitch tool can be designed with a deep serration and provide many resharpenings. A fine-pitch tool will have a shallow serration and a limited number of usages. The following table is representative of serration depth, and the amount a tooth may be reduced in tooth thickness by successive sharpenings. A shaving tool is not resharpened when its serration depth has been

Table 22-7. Amount a Shaving-cutter Tooth May Be Reduced in Sharpening

Pitch	Depth of serration	Amount tooth thickness may be reduced
6	0.045	0.066
10	0.030	0.036
20	0.020	0.020

reduced to approximately 0.012″. In the case of the 6-pitch gear-shaving tool, the 0.045″ depth of serration provides a possible reduction of tooth thickness of 0.090″. The serrating tool requires radius corners on its cutting edge and reduces the possible reduction of tooth thickness from 0.090 to 0.066″. By observing the 0.012″ minimum, the user can be assured the shaving tool will present the proper cutting edge to the part to be shaved.

The chip from a shaving tool is a thin hairlike chip. This is flushed out of the serration by the coolant and removed by a magnetic chip separator. Failure to clean the coolant by chip separating can cause early tool dulling. Evidence of this can be found by studying the shaved tooth flank, watching for chip impressions in the shaved profile.

Even though the tool is dull, take the precaution of handling as a new tool in storage and in shipping. Before shipping out for resharpening, study the tool serrations with a glass, watching for chipped serrations or impacted teeth. These are clues to whether overdulling is occurring, if gear is being deburred properly, or whether the coolant is doing its job. Improper coolant and flaked out serrations go hand in hand. Broken serrations, resulting from impacted serration slots, are indicative of lack of deburring. The impacted metal will bear no resemblance to the fine hairlike chips from shaving; instead they will be larger masses that can come only from burrs left in hobbing or shaping.

Coolants. Tests have proved that a coolant must always be used, even when shaving cast iron. Shaving dry, or with a water-soluble coolant, results in flaking of the cutting edges. When shaving cast iron, coolants become a problem. The heavier fatty oils hold the cast-iron particles in suspension and abrade the serration edges. The best coolant is kerosene, if safety regulations do not forbid its use. Mineral spirits are frequently used for cast iron, but again the flash point is low. For shaving steel, sulfurized mineral oils are recommended. When mixed production on the same shaving machine is necessary, compromises that suit both cast iron and steel must be worked out. Otherwise it is necessary to change coolants as the material changes. The manufacturer of the tools can assist in recommending the proper coolant when he has been made aware of the requirements.

22-20. Shaving-tool Calculations. Because of the continued interest in the shaving-tool design and the fact that many users of shaving tools provide their own design calculations, this section will be devoted to design calculations. The derivation of all the formulas cannot be supplied, because of limitation of space. In cases where a better understanding is needed, some derivations will be provided.

To Calculate Points of Contact. Universal involute checking and recording equipment utilizes angular measurements of part rotation to determine involute positions. This method is an easy means of expression. The angular measurement is known as rotation of the gear about its axis and is sometimes referred to as "degrees of involute" or "roll angle." In either case, it is descriptive of the amount of involute developed for a given amount of angular displacement about the gear axis. Since the origin of involute is at the base diameter, the roll angle describing the base-diameter position is zero degrees. As the involute is developed by rotation about the gear axis, any increment of rotation can be related to the involute and any point on the involute. The roll angle ϵ° is the sum of the pressure angle ϕ and the vectorial, or polar angle θ. To calculate the roll angle ϵ_x° to any radius R_x,

$$\epsilon_x^\circ = 57.29578 \left[\left(\frac{R_x}{R_b} \right)^2 - 1 \right]^{1/2} \tag{22-29}$$

when R_b = base radius

Note that R_x/R_b is the secant of the pressure angle ϕ to the radius R_x. This is an application of the identity $\tan \phi = (\sec^2 \phi - 1)^{1/2}$. The tangent ϕ multiplied by the base radius is the length of the generating line, which is tangent to the base diameter and intersects the involute at radius R_x. To convert any pressure angle ϕ to roll angle ϵ°,

$$\epsilon^\circ = \tan \phi \times 57.3 \tag{22-30}$$

The roll angle at the pitch point in a parallel-axis system is common to both members. In a cross-axis system, the roll angles at the pitch point are dissimilar. This results from the difference in helix angle ψ of the two members. This poses no problem if each pressure angle is treated separately.

Determine roll angle at the outside radius of the gear $\epsilon_{R_o}^\circ$ by Eq. (22-29). This value is $\epsilon_{R_{og}}^\circ$. The purpose is to calculate where on the pinion this outside radius will contact. The point of contact on the pinion will be defined as the "lowest point of contact," or R_{lpc}. By using roll angles as the medium of calculating, the description is better suited for the corresponding analysis of an involute chart from a universal involute checker. Calculate the common roll angle for a parallel-axis system as follows:

$$\epsilon^{\circ}_{op} = 57.29578 \left[\left(\frac{C}{R_{bp} + R_{bg}} \right)^2 - 1 \right]^{1/2} \tag{22-31}$$

Bear in mind that ϵ°_{op} is at the common pitch point and that $C/(R_{bp} + R_{bg})$ is the conventional means of determining the secant function of the operating pressure angle. The value ϵ°_{op} when substituted in the following equation will provide the roll angle at the lowest point of contact on the pinion tooth flank.

$$\epsilon^{\circ}_{lpc} = \epsilon^{\circ}_{op} - \frac{R_{bg}}{R_{bp}} (\epsilon^{\circ}_{R_{og}} - \epsilon^{\circ}_{op}) \tag{22-32}$$

The degrees of involute between the pitch point and outside radius of the gear are calculated by $\epsilon^{\circ}_{R_{og}} - \epsilon^{\circ}_{op}$. The pinion and gear are rotating at the ratio of R_{bg}/R_{bp}. When $(R_{bg}/R_{bp})(\epsilon^{\circ}_{R_{og}} - \epsilon^{\circ}_{op})$ is determined, it is known how much involute is needed below the pitch point of the pinion to be conjugate to $(\epsilon^{\circ}_{R_{og}} - \epsilon^{\circ}_{op})$. These degrees of involute, when subtracted from ϵ°_{op}, denote how far from the origin of involute of the pinion that lowest point of contact occurs. The radius of lowest point of contact R_{lpc} may be obtained from

$$R_{lpc} = R_{bp} \left[\left(\frac{\epsilon^{\circ}_{lpc}}{57.29578} \right)^2 + 1 \right]^{1/2} \tag{22-33}$$

In a cross-axis gear system, the equation is modified slightly, but the modified equation will suit a parallel-axis system.

$$\epsilon^{\circ}_{lpc} = \epsilon^{\circ}_{opp} - \frac{R_{bg} \sin \phi_g}{R_{bp} \sin \phi_p} (\epsilon^{\circ}_{R_{og}} - \epsilon^{\circ}_{opg}) \tag{22-34}$$

The expression ϵ°_{opg} is the roll angle for the gear's operating pressure angle, and ϵ°_{opp} is the roll angle for the pinion's operating pressure angle. The corrective factor of $\sin \phi_g/\sin \phi_p$ must be applied to provide the correct ratios of degrees of involute to compensate for cross-axis calculating. Equation (22-34) is compatible for parallel axis also. The value of $\sin \phi_g/\sin \phi_p$ cancels to 1, since in parallel-axis systems the pressure angles are equal and therefore roll angles will also be equal, making ϵ°_{opp} equal to ϵ°_{opg}. By inverting the ratio $R_{bg} \sin \phi_g/R_{bp} \sin \phi_p$, any point on the pinion involute can be transferred to the gear. If calculating where a point below the pitch point of the pinion contacts on the gear, Eq. (22-34) is rewritten as in the following:

$$\epsilon^{\circ}_{R_{og}} = \epsilon^{\circ}_{opg} + \frac{R_{bp} \sin \phi_p}{R_{bg} \sin \phi_g} (\epsilon^{\circ}_{opp} - \epsilon^{\circ}_{lpc}) \tag{22-35}$$

The subscripts in the foregoing indicate that the roll angle of the outside radius required to contact the lowest point of contact of the pinion is to be found. Any point on the pinion could be used that is pertinent, and the point on the gear that contacts the pinion can be determined accordingly. The use of the contacts is essential to shaving-tool design. The contact of the mating gear is calculated on the part to be shaved, to determine the lowest point of contact. The shaving-tool outside diameter is determined as needed, to contact lower than the mating gear does. When doing volume calculations and planning a design, all factors must be known and kept for a reference when the actual shaved involutes are examined and evaluated.

The cross-axis system of calculating is not difficult, as applied to shaving tools, if the basic fundamentals are observed. The normal plane elements are identical for a gear and shaving tool, regardless of the cross axis. The transverse plane elements differ, because of the difference in helix angles, but are easy to deal with if handled separately. Generally, in shaving-tool design, using the cross-axis method, a spur gear is considered as a special case of a helical gear and is treated accordingly.

Letter Symbols Used for Calculating. The following is basically in full agreement with those symbols suggested by AGMA.[9]

$N(N_g, N_p, N_c)$ = number of teeth (in gear, pinion, or cutter)
P_n = normal pitch
P = transverse pitch
p_n = normal circular pitch at pitch line
p = transverse circular pitch at pitch line
p_b = transverse circular pitch at base diameter
ϕ_n = normal plane pressure angle
$\phi(\phi_g, \phi_p, \phi_c)$ = transverse pressure angle at theoretical pitch diameter of gear, pinion, or cutter
$\phi_1(\phi_{g1}, \phi_{p1}, \phi_{c1})$ = subscripts for transverse pressure angles at operating diameters and not at theoretical pitch diameters
$\psi(\psi_g, \psi_p, \psi_c)$ = helix angle at theoretical pitch diameter of gear, pinion, or cutter
$tn(tn_g, tn_p, tn_c)$ = normal circular tooth thickness at theoretical pitch diameter of gear, pinion, or cutter
S_n = normal circular space width at theoretical pitch diameter. Assume use for gear to be shaved only
$D_o(D_{og}, D_{op}, D_{oc})$ = outside diameter of gear, pinion, or cutter
$R_o(R_{og}, R_{op}, R_{oc})$ = outside radius of gear, pinion, or cutter
$D(D_g, D_p, D_c)$ = pitch diameter of gear, pinion, or cutter
$R(R_g, R_p, R_c)$ = pitch radius of gear, pinion, or cutter
$D_b(D_{bg}, D_{bp}, D_{bc})$ = base diameter of gear, pinion, or cutter
$R_b(R_{bg}, R_{bp}, R_{bc})$ = base radius of gear, pinion, or cutter
D_r = root diameter
C = center distance
Q = cross-axis angle or shaft angle
R_{lpc} = radius to lowest point of contact on an involute

NOTE: $S_n = (p_n - t_n)$ used as symbol for space width is arbitrarily selected.

Basic Formulas for Calculating. These are commonly known except for Eq. (22-52) and can be found in many reference books and catalogues. They are repeated here for easy reference.

$$P_n = \frac{P}{\cos \psi_g} \quad \text{usually given} \tag{22-36}$$

$$P = P_n \cos \psi_g \quad \text{usually given} \tag{22-37}$$

$$\psi = \text{given} \quad \text{or} \quad \cos \psi = \frac{\tan \phi_n}{\tan \phi} \tag{23-38}$$

$$\phi = \text{given} \quad \text{or} \quad \tan \phi = \frac{\tan \phi_n}{\cos \psi} \tag{22-39}$$

$$\phi_n = \text{given} \quad \text{or} \quad \tan \phi_n = \tan \phi \cos \psi \tag{22-40}$$

t_n = given or calculate from

$$(a)\ t_n = t \cos \psi \tag{22-41}$$

$$(b)\ t_n = p_n - t_n \text{ of mating gear}$$

t = given or calculate from

$$(a)\ t = \frac{t_n}{\cos\psi} \tag{22-42}$$

(b) over-pin dimensions

$$p = \frac{\pi}{P} = \frac{p_n}{\cos\psi} \tag{22-43}$$

$$p_n = \frac{\pi}{P_n} = p\cos\psi \tag{22-44}$$

$$S = p - t \tag{22-45}$$

$$S_n = p_n - t_n \tag{22-46}$$

$$D_o = \text{given on print or } D + 2 \text{ addendums} \tag{22-47}$$

$$D = \frac{N}{P} \tag{22-48}$$

$$D_b = D\cos\phi \tag{22-49}$$

$$D_r = \text{given} \quad \text{or} \quad D_o - 2 \times \text{whole depth} \tag{22-50}$$

$$C = R_g + R_p = \frac{R_{bg} + R_{bp}}{\cos\phi} \tag{22-51}$$

$$t_{nc} = S_n - \frac{N_g(\text{inv}\,\phi_g - \text{inv}\,\phi_{g1}) + N_c(\text{inv}\,\phi_c - \text{inv}\,\phi_{c1})}{P_n} \tag{22-52}$$

Parallel-axis Calculation. This is a sample calculation demonstrating a complete calculation between a pinion with 27 teeth and a gear with 53 teeth. Following the calculation of these two spur mating members will clarify the use of calculating contact points. Immediately following will be the calculations of the same 27-tooth pinion, but the shaving tool with a 10° helix angle has been substituted for the 53-tooth spur gear. This will point up the slight variations between parallel and cross-axis calculations.

Pinion	Common elements	Gear
$N_p = 27$	$P = 6$ (22-37)	$N_g = 53$
$D_{op} = 4.833$ (22-47)	$\phi = 20°$ (22-39)	$D_{og} = 9.167$ (22-47)
$D = 4.500$ (22-48)	$\sin\phi = 0.34202$	$D = 8.833$ (22-48)
$D_{bp} = 4.2286$ (22-49)	$\tan\phi = 0.36397$	$D_{bg} = 8.3006$ (22-49)
$D_{rp} = 4.050$ (22-50)	$\cos\phi = 0.93969$	$D_r = 8.384$ (22-50)
$t = 0.2588$ (22-42)	$\text{inv}\,\phi = 0.01490$	$t = 0.2588$ (22-42)
$S = 0.2648$ (22-45)	$p = 0.5236$ (22-43)	$\epsilon°_{ro} = 26.85°$ (22-29)
$\epsilon°_{lpc} = 9.07°$ (22-32)	$C = 6.6667$ (22-51)	
$R_{lpc} = 2.1406$ (22-33)	$\epsilon°_{op} = 20.85°$ (22-31)	

Preparation of the pinion data in this manner provides data essential to the shaving-tool design. The shaving tool will have a normal pitch P_n, normal pressure angle ϕ_n, equal to the transverse values of P and ϕ of the pinion. The shaving tool must have a normal circular tooth thickness t_{nc} that will give metal-to-metal contact on tooth flanks in the circular space width S of the pinion. The outside diameter D_o of the shaving tool must contact the radius on the pinion R_{lpc}, which is the lowest point of the pinion involute contacted by the outside radius R_{og} of the 53-tooth gear. In actual practice, the shaving tool's outside diameter would contact lower on the pinion involute. This would be predicted by the preshave preparation of the pinion in respect to the undercut left by hobbing or shaping. If the shaving-cutter design was selected to cause the theoretical pitch diameters of pinion and cutter to be operating rolling diameters, then the tooth thickness of the cutter t_{nc} would be equal to the space width of the pinion S. The theoretical pressure angles at the pitch diameters would also be the operating pressure angles ϕ_1. In actual shaving-cutter design, this condition of pitch-diameter operation is more the exception than the rule. Several factors, such as physical characteristics of the shaving-tool tooth and the optimum tooth-contact condition, influence the selection of the operating pressure angle as an immediate design decision. In the example to follow, a pressure angle for operation on the pinion ϕ_{p1} has been arbitrarily selected to be less than the theoretical pressure angle ϕ. For the sake of clarity, the pinion data have been retabulated adjacent to the shaving-tool data for quick reference.

Cross-axis Calculation

Pinion to be shaved	Common elements	Shaving tool
$N_p = 27$	$P_n = 6$ (22-36)	$P = 5.9089$ (22-37)
$D_{op} = 4.833$ (22-47)	$\phi_n = 20°$ (22-40)	$\psi_c = 10°$
$D = 4.500$ (22-48)	$\sin \phi_n = 0.34202$	$\sin \psi_c = 0.17365$
$D_{bp} = 4.2286$ (22-49)	$\tan \phi_n = 0.36397$	$\tan \psi_c = 0.17633$
$D_{rp} = 4.050$ (22-50)	$\cos \phi_n = 0.93969$	$\cos \psi_c = 0.98481$
$t = 0.2588$ (22-41)	$\text{inv } \phi_n = 0.01490$	$\phi_c = 20°17'$
$S_n = 0.2648$ (22-46)	$p_n = 0.5236$ (22-44)	$\sin \phi_c = 0.34666$
$\epsilon°_{lpc} = 9.07°$ (22-32)		$\tan \phi_c = 0.36958$ (22-39)
$R_{lpc} = 2.1406$ (22-33)		$\cos \phi_c = 0.93799$
		$\text{inv } \phi_c = 0.01557$
$\phi_{p1} = 18°30'$ (selected)		$\phi_{c1} = 18°45'36''$
$\sin \phi_{p1} = 0.31730$		$\sin \phi_{c1} = 0.32160$ (22-54)
$\tan \phi_{p1} = 0.33460$		$\tan \phi_{c1} = 0.33965$
$\cos \phi_{p1} = 0.94832$		$\cos \phi_{c1} = 0.94687$
$\text{inv } \phi_{p1} = 0.01171$		$\text{inv } \phi_{c1} = 0.01223$
$\epsilon°_{opp} = 19.17$ (22-30)		$\epsilon°_{opc} = 19.46°$ (22-30)
		$N_c = 53$
		$D_{oc} = 9.1483$ (22-33)
		$D_c = 8.9695$ (22-48)
		$D_{bc} = 8.4133$ (22-49)
		$t_{nc} = 0.2209$ (22-52)
		$\epsilon \overset{\circ}{R}_{oc} = 24.47°$ (22-35)

The sine functions of pressure angle have a constant relationship regardless of operating pressure angles. From theoretical pressure angles at the pitch diameters in transverse planes of the individual members, this relationship is established.

$$\frac{\sin \phi_p}{\sin \phi_c}$$

At any time a pressure angle of one member is known or assumed, the corresponding pressure angle of the other member can be established by substitution as follows:

$$\frac{\sin \phi_p}{\sin \phi_c} = \frac{\sin \phi_{p1}}{\sin \phi_{c1}} \tag{22-53}$$

In the sample calculation, $\sin \phi c_1$ was found by

$$\sin \phi_{c1} = \frac{\sin \phi_c \times \sin \phi_{p1}}{\sin \phi_p} = \frac{0.34666 \times 0.31730}{0.34202} = 0.32160 \tag{22-54}$$

Perhaps the most important formula in usefulness is the one for tooth thickness. Its function is to furnish the normal circular tooth thickness t_{nc} at the theoretical pitch diameter. It does this without the need of transposing the tooth thickness of either the gear being shaved or the shaving tool. It has been derived from the transposition equations and satisfied every need of the designer, who otherwise would have lengthy calculations to make. Substituting in Eq. (22-52) the numerical equation will be as follows (note: because the pinion is spur, $t = t_n$ and $s = s_n$):

$$t_{nc} = 0.2648 - \frac{27(0.01490 - 0.01171) + 53(0.01557 - 0.01223)}{6} = 0.2209$$

In calculating the outside radius R_{oc} of the shaving tool, reversal of the procedure of calculating the lowest point of contact is used. See Eq. (22-35) where it was explained for a gear. By changing the subscripts, it will adapt to the shaving tool.

$$\epsilon°_{R_{oc}} = \epsilon°_{opc} + \frac{R_{bp} \times \sin \phi_p}{R_{bc} \times \sin \phi_c} \times (\epsilon°_{opp} - \epsilon°_{lpc}) \qquad (22\text{-}55)$$

$$24.47° = 19.46° + \frac{4.1143 \times 0.34202}{4.20665 \times 0.34666} \times (19.17 - 9.07)$$

The value $(\epsilon°_{opp} - \epsilon°_{lpc})$ determines that 10.10° of involute must be shaved below the operating diameter or pitch point of the pinion. This is also the amount of pinion rotation about its axis to develop this quantity of involute angle. The shaving tool must have sufficient involute above its operating diameter or pitch point to suit the pinion's 10.10° of roll angle. The exact amount required is found by the ratio shown below:

$$\frac{R_{bp} \times \sin \phi_p}{R_{bc} \times \sin \phi_c} \times (\epsilon°_{opp} - \epsilon°_{lpc})$$

This amount added to the roll angle at the pitch point of the shaving tool determines the amount of involute from base diameter to outside diameter of the shaving tool. Converting $\epsilon°_{R_{oc}}$ from roll angle to a radius by Eq. (22-33) provides the outside radius of the shaving tool.

The calculations of the shaving tool to its last sharpening are not included here. A comparison may be had as to the effect of sharpening by repeating the shaving-tool calculation, using a pinion operating pressure angle (ϕ_{p1}) of 17°30′. By comparing the outside diameter of the cutter and tooth thickness of the cutter with the original calculation, an understanding of the import of serration depth can be had.

Understanding the fundamentals of shaving, from both a practical and a theoretical standpoint adds much to the ability of the user to shave economically. When shaving troubles arise, a little logic will couple with knowledge and supply the answer in the majority of tool problems.

GEAR-TOOTH BROACHES

Broaching is a metal-cutting process, combining both roughing and finishing operations in stock removal. It is applied to gears, cylinder blocks, connecting rods, internal splines, holes, rifle barrels, wrenches, etc. The application is restricted to elements of the broached surface being within the plane of the broach axis. Limitations depend on part design and stock removal. In any case of part processing for stock removal on a production basis, the designer should always weigh the possibility of broaching and design accordingly. Many applications of production broaching are inadvertently overlooked, whereby the broach manufacturer could have shown a piece-part cost lower than cost of processing by a more conventional method. Broaching is generally thought of for high production, but there are low-production applications suitable only to broaching. These applications are economical in view of alternative methods.

22-21. Types of Broaches. Broaches are basically classified as follows:

1. By their method—push or pull broaches
2. By operation—internal or external
3. By construction—solid, built-up, progressive, circular, inserted tooth, and overlapping tooth

The consideration of broaches in this chapter will be confined strictly to internal-gear and internal-involute-spline broaches.

Pull Broaches. Pull broaches are used for most internal operations. They are capable of more stock removal than push broaches and are more suitable for longer cuts on internal work. The work-clamping pressure and broaching thrust are of importance, especially on work having thin walls. In most of these cases, a pull broach is preferable. A push broach is necessarily shorter and less expensive than the pull broach; however, based on stock removal, a pull broach can do the work which would require several push broaches.

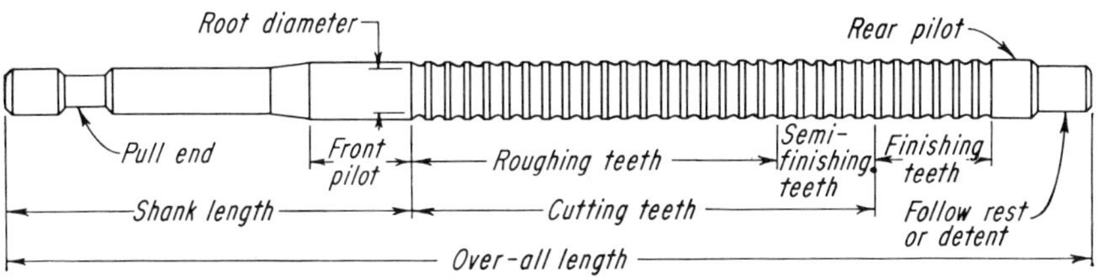

FIG. 22-54. Typical internal pull broach.

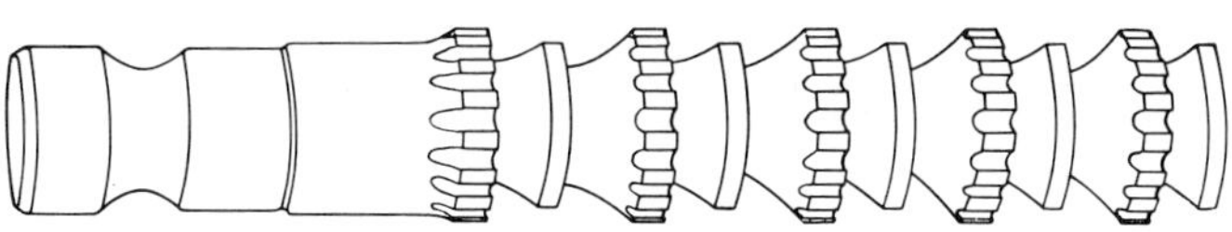

FIG. 22-55. Alternate round and spline broach, sizes minor diameter and maintains concentricity of minor, pitch, and major diameters.

Push Broaches. The push broach is used to a great extent for sizing and burnishing to correct heat-treat distortion and improve surface finish. Depending upon application, push broaches are occasionally used for blind-hole broaching. Most internal-spline broaches are of the solid type for pull broaching as shown in Fig. 22-54.

22-22. Alternate Round and Spline Broaches. Frequently it is desirable to maintain concentricity between pitch line of a splined hole and pitch line of an integral external involute gear. A broach having alternate round and splined teeth at the finishing end is used. This serves to maintain concentricity between minor diameter, pitch diameter, and major diameter of the internal spline. To facilitate processing of subsequent operations, the minor diameter may be used with a plain round arbor for locating to cut the external teeth, ensuring the concentricity desired. This eliminates the need of a splined arbor and using the major diameter to accomplish the same qualified location for subsequent external gear cutting (see Fig. 22-55).

22-23. Design Modifications. In broaching splines utilizing a major-diameter fit in final assembly, a modification of involute form of the broach is made to avoid interference with the mating part (as shown in Fig. 22-56). The profile of the broach at its major diameter is altered with a tip protuberance. This provides an undercut at the major diameter of the internal spline to eliminate corner-radius interference of the mating members. By making this modification in the broached spline and not on the mating external part, the top land of the external part is preserved for location.

As a note of warning to the designer, be certain when using a major-diameter fit that tooth proportions of male and female splines are such that a pitch-line fit cannot overcome the major-diameter fit. This can be accomplished by providing sufficient backlash in excess of a loose effective fit.

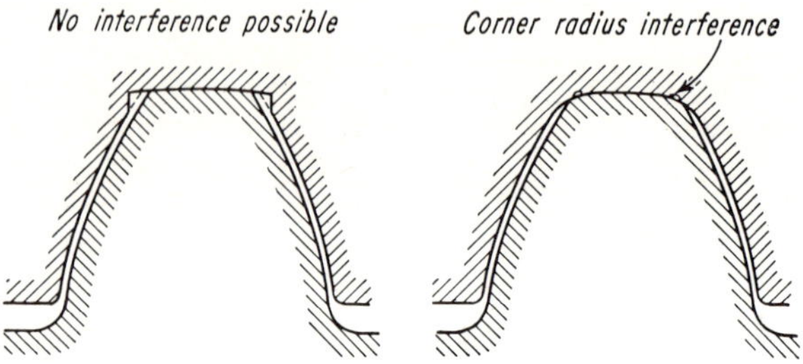

FIG. 22-56. Means of avoiding corner-radius interference.

Broaches of the pull type can be furnished with many types of pull ends. Figure 22-57 shows four of the most common: (*a*) threaded type, (*b*) key type, (*c*) automatic-sleeve type, and (*d*) pin type.

The most common type for spline production is the automatic-sleeve type for pull-up and pull-down vertical machines and on horizontal machines. Figure 22-58 shows the method of engaging and releasing the broach end. This type of puller can also be furnished with a locating flat at the pull end.

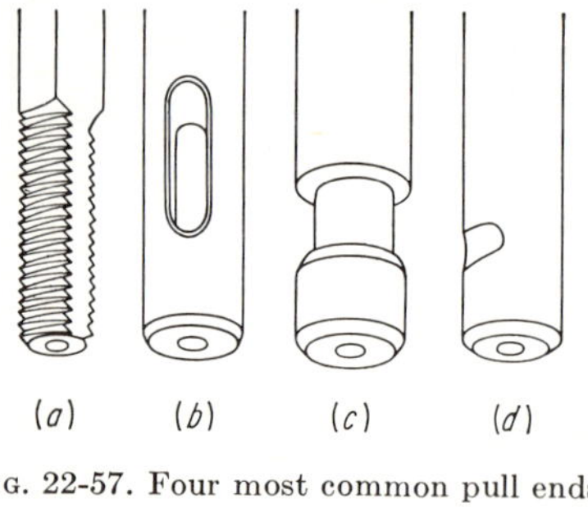

FIG. 22-57. Four most common pull ends.

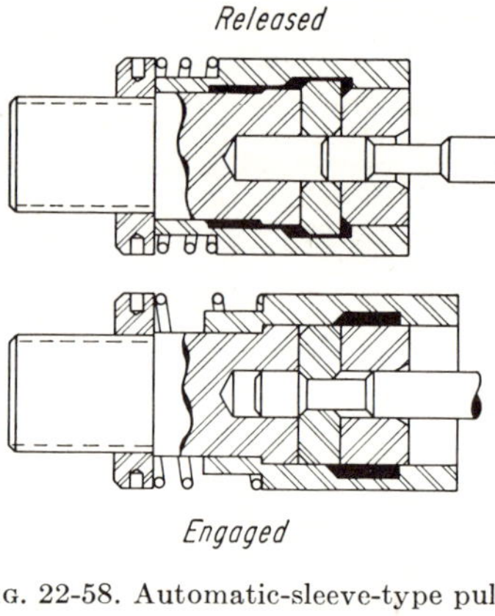

FIG. 22-58. Automatic-sleeve-type puller.

22-24. Spiral Involute Splines and Internal Spiral Gears. The broaching process has made huge gains in the area of production of internal spiral gears. Usually broach manufacturers recommend that the helix angle of the part not exceed 60° measured from the axis of the part. A lead bar is usually employed when the helix angle exceeds 15°. The part is clamped against rotation when a lead bar is used.

A particular area of interest is the planetary ring gears of automatic transmissions for the automotive industry (as shown in Fig. 22-59). The broaches are pulled through the proper helical path by guiding with a lead bar. The lead bar turns the broach as it is pulled through the clamped work. Finishing broaches and preshave broaches have been made and put into production with results difficult to duplicate by other methods of production. Excellent lead and involute control have been achieved, providing all the elements of a good finished part or a good preshave part. The parts invariably reflect the original high-accuracy components developed into the broach.

Once the broach and lead bar have been developed, allowing for heat-treat distortion, high production with good duplication can be expected. The broach user must observe good practices in sharpening, blank preparation, machine maintenance,

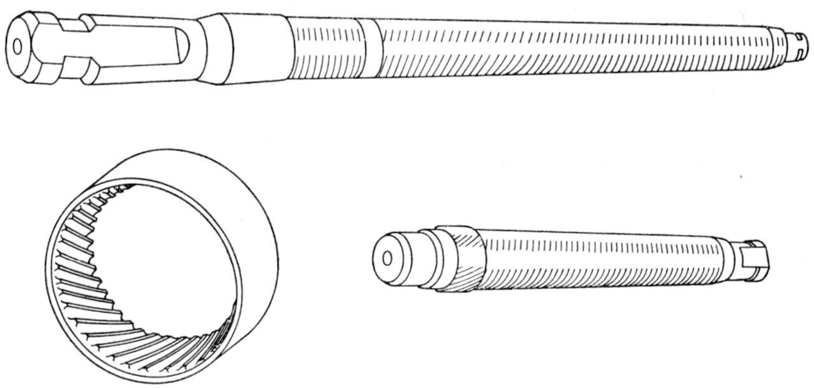

FIG. 22-59. Ring gear broached in two passes.

and consistent steel characteristics in order to retain production repeatability. This is paramount in any method of gear production to obtain tool efficiency.

22-25. Broaches for External Gears. The broaching application for external gears is feasible but limited to production quantities and size of part. The broaches have been made by series of broach rings through which the work is pushed. The broach rings are stacked, in proper sequence, held in a female adapter. The ram containing the part and piloting in bushings carries the part through the stacked broach rings.

Recently one broach manufacturer succeeded in constructing a solid broach for external gears. The results were very satisfactory, and broaches of this type have been put in production. There is at present a limitation as to diameter for the ring-type broaches. Diameter must be sufficiently large to permit introduction of a grinding wheel and grinding-wheel support internally.

22-26. Design Elements. The length of the broach, along with the diameter of the broach, governs the price. Special features, of course, have an influence on price, such as profile deviations and alternate round and splined teeth.

Broach Pitch. The pitch, or tooth spacing, determines the construction and strength of the tooth, the allowable chip space, and the number of teeth in contact at one time. At least two broach teeth, preferably more, should be in contact at one time. In cases of inadequate machine capacity, the step per tooth is reduced rather than the teeth in contact.

The formula for pitch frequently used is

$$\text{Pitch} = 0.35 \times \text{length of cut} \tag{22-56}$$

For cast iron a shorter pitch may be used.

The chip has no place to go while the tooth is in contact with the part. Therefore, the chip space must be designed to curl the chip and contain it until the tooth is clear of the work and can unload the curled chip (see Fig. 22-60).

Chip Breakers. Chip breakers are ground into round broach teeth to prevent ring formations of teeth that cannot be readily removed. Usually they are staggered in a pattern on the roughing and semifinishing teeth. Improper chip breaking can result in producing oversize splines, as well as dangerous tearing and galling in the workpiece.

Chip Load. The cut per tooth in free-cutting metal is usually 0.004 to 0.006″ measured on diameter for splines. Characteristics of the application and material being broached have considerable influence on the values of cut per tooth. Too heavy a cut will overload a broach and often result in a complete tool failure. Then, too, an insufficient cut per tooth will drag and burnish, causing heat, cutting-edge wear, and galling. A practical minimum cut is in the area of 0.0007″ on diameter.

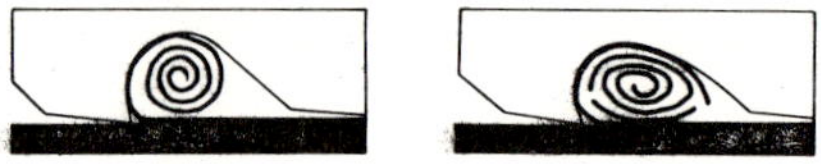

(a) (b)

FIG. 22-60. Tooth space must be designed to control chip. (*a*) Curled chip in well-designed tooth space. (*b*) Chips in improperly designed tooth space.

Material Hardness. The broach manufacturers advise that part hardness be within 10 to 38 Rockwell C. Higher-hardness material can be broached, but in these cases, the broach manufacturer should be consulted. It is likely that his case histories will provide life expectancy of the broach and recommendations on processing, as well as the possibility of a premium-steel broach.

Most broach manufacturers are using molybdenum high-speed steels of the M-2 designation for 2 per cent vanadium. The molybdenum steels have proved to be fully adequate for shock loads encountered. They respond well to surface treatment for increasing hardness of cutting edges. Special applications of broaching in ductile iron, stainless steel, and other high-hardness materials have required broaches made of M-3 steel. The M-3 steel is receptive to heat-treatment to obtain a high hardness and a tough cutting edge that is resistant to abrasive wear and early dulling by part hardness.

Face Angle of Broaches. The face angle of the broach is determined by the material being broached. Hardness of part, toughness, and ductility are the influencing factors.

A rough guide for face angles is as follows:

Material	*Face angle, deg*
Cast iron	6–8
Hard steel	8–12
Soft steel	15–20
Aluminum	10 and plus
Brittle brass	−5 to +5

These are general values and are subject to varying, as needed, to accommodate a specific material.

The Broach Pilot. The internal broach contains a front and rear pilot for the purpose implied. The front pilot is designed to extend completely through the work to guide the broach in its cutting stroke. It also functions as a protective measure to avoid overloading the first roughing tooth. If the front pilot does not pass through the unbroached hole of the part, it is an indication that the part is not properly prepared for broaching. Never grind down the front pilot to fit unbroached holes. The hole should be altered; otherwise an excessive cutting load will be placed on the first roughing tooth.

The rear pilot supports the broach as the last finish teeth pass through the work part. Without proper piloting, the rear end of the broach may whip, causing an unnatural form in the finished workpiece.

Broach-tooth Side Clearance. Depth of spline tooth being broached determines whether or not broach should be relieved on the tooth flanks. For spline depths exceeding 0.100″, side relief is generally incorporated to avoid galling or pick-up of material. Because of the manufacturing limitations on shallow-depth broaches, it is difficult to side-relieve a broach cutting less than 0.100″ depth. Occasionally, side relief is omitted to prevent broach drift when close limits are required in the broached part. The amount of side relief varies to suit the application and should therefore be left to the discretion of the broach manufacturer.

22-27. Tooling Tips. *If the broach becomes stuck in workpiece.* Before a broach completes its cutting stroke, it may become stuck in the part. *Never* try to back the broach out by reversing the machine. Free the broach and work from the machine and fixture. Try to remove the part by tapping lightly on all sides, trying to work the part toward the starting end of the broach. Failing in this, the part may be reduced by putting the broach and part in a lathe. Another means is by sawing the part in two, if practical. Several factors may have caused sticking, such as dull teeth, power failure, or insufficient machine capacity.

Broach Cutting Fluids. The fluid lubricates chip-tooth flow, reducing frictional heat and abrasion. The chips are washed out of chip spaces. Heat generated by normal broaching is dissipated.

Internal broaching requires a heavy cutting fluid, as compared with surface broaching. A sulfur-base oil is generally preferred for steel. Cast iron may be broached dry; however, finish is improved by use of a very light coolant, such as mineral spirits or kerosene.

The broach manufacturer may have specific recommendations for coolants, depending upon the application.

Broach Care. As with other tools used in gear production, the broach is a high-precision cutting tool, subject to damage and loss of efficiency by careless handling and improper sharpening.

Upon receipt of a new broach, check the last finishing tooth for size, also check the first cutting tooth to be sure the broach will not be overloaded. Measure and record, for sharpening purposes, the face angle, tooth depth, and root radius. Try the front pilot with a workpiece. Do not drive the pilot into the workpiece hole.

In storing or shipping the broaches, a little common sense will be rewarding. Store the broach in racks with partitions so that no broach can contact another. If stored for future use, or if returning for sharpening, always coat with a rustproofing lubricant. Never ship a broach with any other steel item loose in the case or carton, such as a sample part.

Sharpening. Overdulling of a broach will reduce its life at a high rate. If possible, keep a record of number of pieces per broach sharpening vs. amount of grind-back on cutting face. In a short time, an adequate history can be obtained that will profit by extending tool life. The broach manufacturer can make excellent recommendations based on his experience. The best results, however, are obtained by actual case history and records taken from the production floor.

For internal broaches, grind the front face only. When sharpening, make certain that the ground face blends with the chip-space contour (see Fig. 22-61). An interrupted surface or rough surface adds resistance to the chip curl and disposal

It is not necessary to sharpen all finishing teeth with each resharpening. Sharpen only the first finishing tooth, or the two initial finishing teeth as required. As they become undersize from subsequent sharpenings, treat the second or third or fourth tooth in sharpening as the first or second teeth were treated. The undersize finishing teeth can have chip breakers added to avoid wide chips.

Proper setup precautions should be observed in resharpening so as not to permit runout when the broach is between centers. It is advisable to contact the broach manufacturer for sharpening recommendations as to wheel grit, size of wheel, and whether to grind wet or dry. Normally, dry grinding is employed, but care must be exercised to avoid burning of the cutting edges. On long broaches, a steady rest is recommended to avoid deflection or sag of the broach.

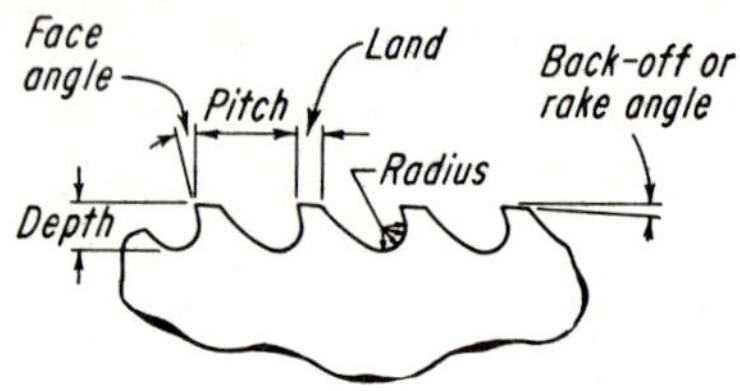

FIG. 22-61. Broach-tooth form.

Today's broach manufacturers have developed a great deal of "know-how" in broach manufacturing. This is reflected in the high precision achieved in both straight and spiral splines being cut in production. They have kept abreast of the precision of the easier-to-make external spur and helical involute splines produced by hobbing, cold forming, or grinding. It is recommended that, for new applications and chronic broaching problems in production, a broach engineer be consulted. His services invariably will be rewarding in terms of instruction and application advice.

Information Required in Ordering Broaches. The following list shows the vital information needed by the broach manufacturer.

1. Print of the part, giving dimensions and range of limits, if possible.
2. Material specifications, casting, forging, etc.
3. Prior machining operations on the part; also whether the machined surface is to be used for locating during the broaching operation.
4. Prior heat-treatment of the part. State hardness.
5. Subsequent heat-treatment of the part which may affect the accuracy of the broached spline.
6. Production rate and quantity desired.
7. Finish required.
8. Broaching machine to be used, and information as to special fixtures or guides that are to be used.
9. Specify type of puller and whether rear support is used.
10. Radius or chamfer, if required on the corner of splines at either inside or outside diameter.
11. Is the broach to size the inside diameter as well as cut the splines?
12. Concentricity requirements and relationship to reference diameters.
13. Type of fit desired.
14. Mating-part information, if available.

REFERENCES

1. American Gear Manufacturers Assoc.: "Single-thread Coarse-pitch Hobs," AGMA 121.02, February, 1956.

2. American Gear Manufacturers Assoc.: "Single-thread Fine-pitch Hobs," AGMA 122.02, February, 1956.
3. American Gear Manufacturers Assoc.: "Multiple-thread Coarse-pitch Hobs," AGMA 123.01, December, 1956.
4. American Gear Manufacturers Assoc.: "Worm Gear Hobs," AGMA 124.01, July, 1957.
5. Moncrieff, A. D.: "Pre-shave Gear Cutting Tools," AGMA 129.07, October, 1949.
6. Staub, C. R.: Be Kind to Your Hobs, *Machinery*, 1952.
7. Hildreth, Richard S.: Gear Cutting Savvy, *Am. Machinist*, 1953.
8. Dudley, Darle W.: "Practical Gear Design," McGraw-Hill Book Company, Inc., New York, 1954.
9. American Gear Manufacturers Assoc.: "Letter Symbols for Gear Engineering," AGMA 111.03, November, 1954.
10. American Gear Manufacturers Assoc.: "Terms and Definitions," AGMA 112.03, November, 1954.
11. Vogel, Werner F.: "Involutometry and Trigonometry," Michigan Tool Co., 1945.
12. Merritt, H. E.: "Gears," Pitman Publishing Corporation, New York, 1954.
13. Buckingham, Earle: "Analytical Mechanics of Gears," McGraw-Hill Book Company, Inc., New York, 1949.
14. Broaching Tool Institute: "Broaches and Broaching," BTI *Publ.* 1944.
15. Continental Tool Works: Catalog 60681, CTW *Publ.*, 1950.
16. Metal Cutting Tool Institute: "Metal Cutting Tool Handbook," MCTI *Publ.*, 1959.
17. Michigan Tool Co.: General Catalog, MTC *Publ.*, 1959.
18. Michigan Tool Co.: Gear Odontograph for Gear Milling Cutters, MTC *Publ.*, 1944.

Chapter 23

GEAR-INSPECTION DEVICES

By

ALBERT S. BEAM, *Staff Engineer, Vinco Corporation, Detroit, Mich.*

A wide variety of equipment and techniques are used to inspect the many kinds of gear teeth. This chapter presents detail information on the equipment and methods used in gear inspection. The question of how much tolerance to allow for the various gear elements was covered in Chap. 9. Chapter 13 gave information on the effect of gear tolerances on the power capacity of a given size of gearset.

The field of gear inspection can be divided into two broad categories: the analytical inspection of dimensional elements such as profile, spacing, lead, concentricity, and tooth thickness and the functional inspection by rolling with a mating master gear or production gear to determine composite variation, noise level, or variation in action. In addition to checking the teeth, the gear blank must be inspected for face width, outside diameter, bore size, hub length, etc. On very critical gears, many other details like tip radius, shape of root fillet, and surface finish are subject to close control and direct measurement.

This chapter deals primarily with the inspection devices to measure tooth elements and gear rolling action.

SPUR- AND HELICAL-GEAR INSPECTION

Spur- and helical-gear teeth are usually of *involute* tooth form in a plane transverse to the axis of the gear. Lengthwise, a *spur* tooth is parallel to the axis of the gear. A *helical* tooth, in a lengthwise direction, follows a helix. These two facts make it possible to measure any part of a spur or helical tooth by measuring its involute profile in the transverse section and by measuring its lengthwise straightness or helix.

The ease with which spur- and helical-gear teeth can be measured has led to the development of a wide variety of checking machines.

23-1. Involute-profile Measurement. The profile of gears can be inspected by using a dividing head and a height gage, by mechanical involute checkers, by projection, by measurements with pins, and by instruments which use a series of probes which contact the tooth surface.

Profile Inspection by Means of a Dividing Head and Height Gage. This method, as well as the principle of most mechanical involute checkers, is derived from a property of the involute form illustrated in Fig. 23-1.* The distance $\overline{AB}$, measured

* The writer is indebted to Thomas W. Edwards of the Vinco engineering department for his valuable assistance in preparing many of the illustrations in this chapter.

on the generatrix tangent to the base circle, is equal to the arc from the tangent point A to the origin O of the involute on the base circle. If an original tooth position is assumed in which the involute or its extension passes through point A, the angle ϵ_z which subtends the arc $\widehat{AO}$ is not only the roll angle of the generatrix from the origin O to point B but also the angle of rotation of the involute from A to O. It follows that

$$\widehat{AO} = \overline{AB} = \frac{D_b \pi \epsilon_z^\circ}{360} \qquad (23\text{-}1)$$

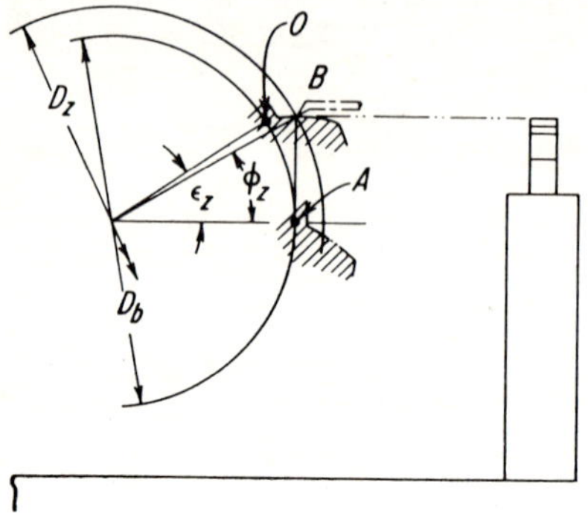

FIG. 23-1. Involute-profile inspection by increments.

Rotation of the involute profile on a dividing head through consecutive roll angles brings consecutive profile points B onto the vertical tangent line. Suitable roll angles can be computed for any selected diameter D_z from the corresponding pressure angle by the equations

$$\frac{D_z}{D_b} = \sec \phi_z \qquad (23\text{-}2)$$

$$\epsilon_z = \frac{180}{\pi} \tan \phi_z \quad \text{deg} \qquad (23\text{-}3)$$

These equations are the basis of Table 23-1, which lists involute roll angles for values of D_z/D_b from 1.00001 to 1.4500, covering practically the whole required range. The dimension $\overline{AB}$ is first determined for the form diameter D_F and a height gage set to zero at this height. The starting position of the tooth profile is obtained by rotating the gear tooth toward the indicating stylus until a zero reading is obtained. In this position the driving dog is locked to the gear or its arbor and connected with the driver of the dividing head, which has also been set to zero. Other diameters may then be assumed and new settings made for the increase in the roll angle and in the distance $\overline{AB}$. Deviations from a true involute profile are registered by the height gage which compares, for each roll-angle setting, the actual tooth height over the gear center with the computed distance $\overline{AB}$. To expedite dividing-head settings, roll-angle increments can be assumed as multiples of degrees, half degrees, or 10 minutes, as applicable, as long as the corresponding distance $\overline{AB}$ is computed for the exact roll angle which would be equal to the roll angle at the form diameter plus the assumed roll-angle increment. Figure 23-2 illustrates a profile-calibration chart which will be helpful in organizing this procedure and recording results.

Since point B of a true involute profile must be a high point, many operators determine its position by moving the height gage transversely over the involute profile and recording the maximum reading. This practice is permissible on precision involute profiles of master gears, involute templates, or highly accurate gearing, but may lead to false readings on the majority of gear teeth which have errors in the involute slope and waviness which cause the profile high point to shift from the vertical line $\overline{AB}$. These false readings can be prevented by making provisions for localizing the contact of the indicating stylus along the vertical base-circle tangent.

Profile inspection by this method is confined to involute master cams and other precision components because it requires a series of individual settings and is time-consuming. On the other hand, it is one of the most reliable approaches to the calibration of master involutes used for qualifying the action of mechanical involute checkers. It can be used for spur, helical, and internal components.

Involute-profile-measuring Machines. These machines, commonly referred to as "involute checkers," offer the preferred means of routine profile inspection, although

Table 23-1. Involute Roll Angles

$\frac{D_z}{D_b}$	ϵ_z deg	Dif	$\frac{D_z}{D_b}$	ϵ_z deg	Dif	$\frac{D_z}{D_b}$	ϵ_z deg	Dif	$\frac{D_z}{D_b}$	ϵ_z deg	Dif
1.00001	0.26	0.106	1.0006	1.99	0.159	1.0056	6.07	0.054	1.0130	9.27	0.178
1.00002	0.36	0.081	1.0007	2.14	0.148	1.0057	6.13	0.054	1.0135	9.45	0.175
1.00003	0.44	0.069	1.0008	2.29	0.139	1.0058	6.18	0.053	1.0140	9.62	0.172
1.00004	0.51	0.061	1.0009	2.43	0.132	1.0059	6.23	0.053	1.0145	9.79	0.169
1.00005	0.57	0.055	1.0010	2.56	0.125	1.0060	6.29	0.052	1.0150	9.96	0.166
1.00006	0.63	0.050	1.0011	2.69	0.120	1.0061	6.34	0.052	1.0155	10.13	0.163
1.00007	0.68	0.047	1.0012	2.81	0.115	1.0062	6.39	0.052	1.0160	10.29	0.161
1.00008	0.72	0.044	1.0013	2.92	0.110	1.0063	6.44	0.051	1.0165	10.45	0.159
1.00009	0.77	0.042	1.0014	3.03	0.107	1.0064	6.49	0.051	1.0170	10.61	0.156
1.00010	0.81	0.040	1.0015	3.14	0.103	1.0065	6.54	0.050	1.0175	10.77	0.154
1.00011	0.85	0.038	1.0016	3.24	0.100	1.0066	6.59	0.050	1.0180	10.92	0.152
1.00012	0.89	0.036	1.0017	3.34	0.097	1.0067	6.64	0.050	1.0185	11.07	0.150
1.00013	0.92	0.035	1.0018	3.44	0.094	1.0068	6.69	0.049	1.0190	11.22	0.148
1.00014	0.96	0.034	1.0019	3.53	0.092	1.0069	6.74	0.049	1.0195	11.37	0.146
1.00015	0.99	0.033	1.0020	3.63	0.090	1.0070	6.79	0.049	1.0200	11.52	0.145
1.00016	1.03	0.032	1.0021	3.72	0.088	1.0071	6.84	0.048	1.0205	11.66	0.143
1.00017	1.06	0.031	1.0022	3.80	0.086	1.0072	6.89	0.048	1.0210	11.80	0.141
1.00018	1.09	0.030	1.0023	3.89	0.084	1.0073	6.94	0.048	1.0215	11.94	0.140
1.00019	1.12	0.029	1.0024	3.97	0.082	1.0074	6.98	0.047	1.0220	12.08	0.138
1.00020	1.15	0.028	1.0025	4.05	0.080	1.0075	7.03	0.047	1.0225	12.22	0.137
1.00021	1.17	0.028	1.0026	4.13	0.079	1.0076	7.08	0.047	1.0230	12.36	0.135
1.00022	1.20	0.027	1.0027	4.21	0.077	1.0077	7.12	0.046	1.0235	12.49	0.134
1.00023	1.23	0.027	1.0028	4.29	0.076	1.0078	7.17	0.046	1.0240	12.63	0.132
1.00024	1.26	0.026	1.0029	4.37	0.075	1.0079	7.22	0.046	1.0245	12.76	0.131
1.00025	1.28	0.025	1.0030	4.44	0.074	1.0080	7.26	0.045	1.0250	12.89	0.130
1.00026	1.31	0.025	1.0031	4.52	0.072	1.0081	7.31	0.045	1.0255	13.02	0.129
1.00027	1.33	0.025	1.0032	4.59	0.071	1.0082	7.35	0.045	1.0260	13.15	0.128
1.00028	1.36	0.024	1.0033	4.66	0.070	1.0083	7.40	0.045	1.0265	13.28	0.126
1.00029	1.38	0.024	1.0034	4.73	0.069	1.0084	7.44	0.044	1.0270	13.40	0.125
1.00030	1.40	0.023	1.0035	4.80	0.068	1.0085	7.49	0.044	1.0275	13.53	0.124
1.00031	1.43	0.023	1.0036	4.87	0.067	1.0086	7.53	0.044	1.0280	13.65	0.123
1.00032	1.45	0.023	1.0037	4.93	0.066	1.0087	7.57	0.044	1.0285	13.78	0.122
1.00033	1.47	0.022	1.0038	5.00	0.066	1.0088	7.62	0.043	1.0290	13.90	0.121
1.00034	1.49	0.022	1.0039	5.07	0.065	1.0089	7.66	0.043	1.0295	14.02	0.119
1.00035	1.52	0.022	1.0040	5.13	0.064	1.0090	7.70	0.043	1.0300	14.14	0.118
1.00036	1.54	0.021	1.0041	5.19	0.063	1.0091	7.75	0.043	1.0305	14.26	0.117
1.00037	1.56	0.021	1.0042	5.26	0.062	1.0092	7.79	0.042	1.0310	14.38	0.116
1.00038	1.58	0.021	1.0043	5.32	0.062	1.0093	7.83	0.042	1.0315	14.49	0.116
1.00039	1.60	0.020	1.0044	5.38	0.061	1.0094	7.87	0.042	1.0320	14.61	0.115
1.00040	1.62	0.020	1.0045	5.44	0.060	1.0095	7.92	0.042	1.0325	14.73	0.114
1.00041	1.64	0.020	1.0046	5.50	0.060	1.0096	7.96	0.042	1.0330	14.84	0.113
1.00042	1.66	0.020	1.0047	5.56	0.059	1.0097	8.00	0.041	1.0335	14.95	0.112
1.00043	1.68	0.020	1.0048	5.62	0.058	1.0098	8.04	0.041	1.0340	15.07	0.112
1.00044	1.70	0.019	1.0049	5.68	0.058	1.0099	8.08	0.041	1.0345	15.18	0.111
1.00045	1.72	0.019	1.0050	5.74	0.057	1.0100	8.12	0.202	1.0350	15.29	0.110
1.00046	1.74	0.019	1.0051	5.79	0.057	1.0105	8.32	0.197	1.0355	15.40	0.109
1.00047	1.76	0.019	1.0052	5.85	0.056	1.0110	8.52	0.193	1.0360	15.51	0.109
1.00048	1.78	0.018	1.0053	5.91	0.056	1.0115	8.71	0.189	1.0365	15.62	0.108
1.00049	1.79	0.018	1.0054	5.96	0.055	1.0120	8.90	0.185	1.0370	15.73	0.107
1.00050	1.81	0.173	1.0055	6.02	0.055	1.0125	9.09	0.181	1.0375	15.84	0.107

Table 23-1. Involute Roll Angles (*Continued*)

$\dfrac{D_z}{D_b}$	ϵ_z deg	Dif	$\dfrac{D_z}{D_b}$	ϵ_z deg	Dif	$\dfrac{D_z}{D_b}$	ϵ_z deg	Dif	$\dfrac{D_z}{D_b}$	ϵ_z deg	Dif
1.0380	15.94	0.107	1.0810	23.52	0.150	1.1360	30.88	0.121	1.1910	37.06	0.105
1.0385	16.05	0.106	1.0820	23.67	0.150	1.1370	31.00	0.120	1.1920	37.17	0.105
1.0390	16.16	0.105	1.0830	23.82	0.149	1.1380	31.12	0.120	1.1930	37.28	0.105
1.0395	16.26	0.105	1.0840	23.97	0.148	1.1390	31.24	0.120	1.1940	37.38	0.105
1.0400	16.37	0.104	1.0850	24.12	0.147	1.1400	31.36	0.119	1.1950	37.49	0.105
1.0405	16.47	0.103	1.0860	24.27	0.147	1.1410	31.48	0.119	1.1960	37.59	0.104
1.0410	16.57	0.103	1.0870	24.41	0.146	1.1420	31.60	0.118	1.1970	37.69	0.104
1.0415	16.68	0.102	1.0880	24.56	0.145	1.1430	31.72	0.118	1.1980	37.80	0.104
1.0420	16.78	0.102	1.0890	24.71	0.144	1.1440	31.84	0.118	1.1990	37.90	0.104
1.0425	16.88	0.101	1.0900	24.85	0.144	1.1450	31.95	0.118	1.2000	38.01	0.104
1.0430	16.98	0.101	1.0910	24.99	0.143	1.1460	32.07	0.117	1.2010	38.11	0.103
1.0435	17.08	0.100	1.0920	25.14	0.142	1.1470	32.19	0.117	1.2020	38.22	0.103
1.0440	17.18	0.100	1.0930	25.28	0.142	1.1480	32.31	0.116	1.2030	38.32	0.103
1.0445	17.28	0.099	1.0940	25.42	0.141	1.1490	32.42	0.116	1.2040	38.42	0.103
1.0450	17.38	0.099	1.0950	25.56	0.140	1.1500	32.54	0.116	1.2050	38.52	0.103
1.0455	17.48	0.098	1.0960	25.70	0.140	1.1510	32.65	0.116	1.2060	38.62	0.102
1.0460	17.58	0.097	1.0970	25.84	0.139	1.1520	32.77	0.115	1.2070	38.73	0.102
1.0465	17.67	0.097	1.0980	25.98	0.138	1.1530	32.88	0.115	1.2080	38.83	0.102
1.0470	17.77	0.096	1.0990	26.12	0.138	1.1540	33.00	0.115	1.2090	38.93	0.102
1.0475	17.87	0.096	1.1000	26.26	0.137	1.1550	33.11	0.114	1.2100	39.03	0.102
1.0480	17.96	0.096	1.1010	26.39	0.137	1.1560	33.23	0.114	1.2110	39.13	0.102
1.0485	18.06	0.095	1.1020	26.53	0.136	1.1570	33.34	0.114	1.2120	39.24	0.101
1.0490	18.15	0.095	1.1030	26.67	0.136	1.1580	33.46	0.114	1.2130	39.34	0.101
1.0495	18.25	0.094	1.1040	26.80	0.135	1.1590	33.57	0.113	1.2140	39.44	0.101
1.0500	18.34	0.187	1.1050	26.94	0.134	1.1600	33.68	0.113	1.2150	39.54	0.101
1.0510	18.53	0.185	1.1060	27.07	0.134	1.1610	33.80	0.113	1.2160	39.64	0.101
1.0520	18.72	0.184	1.1070	27.20	0.133	1.1620	33.91	0.112	1.2170	39.74	0.100
1.0530	18.90	0.182	1.1080	27.34	0.133	1.1630	34.02	0.112	1.2180	39.84	0.100
1.0540	19.08	0.181	1.1090	27.47	0.132	1.1640	34.13	0.112	1.2190	39.94	0.100
1.0550	19.26	0.179	1.1100	27.60	0.132	1.1650	34.24	0.112	1.2200	40.04	0.100
1.0560	19.44	0.178	1.1110	27.74	0.131	1.1660	34.36	0.111	1.2210	40.14	0.100
1.0570	19.62	0.176	1.1120	27.87	0.131	1.1670	34.47	0.111	1.2220	40.24	0.100
1.0580	19.80	0.175	1.1130	28.00	0.130	1.1680	34.58	0.111	1.2230	40.34	0.100
1.0590	19.97	0.174	1.1140	28.13	0.130	1.1690	34.69	0.111	1.2240	40.44	0.099
1.0600	20.14	0.172	1.1150	28.26	0.129	1.1700	34.80	0.110	1.2250	40.54	0.099
1.0610	20.32	0.171	1.1160	28.39	0.129	1.1710	34.91	0.110	1.2260	40.64	0.099
1.0620	20.49	0.170	1.1170	28.52	0.128	1.1720	35.02	0.110	1.2270	40.74	0.099
1.0630	20.66	0.168	1.1180	28.64	0.128	1.1730	35.13	0.110	1.2280	40.84	0.099
1.0640	20.82	0.167	1.1190	28.77	0.128	1.1740	35.24	0.109	1.2290	40.94	0.098
1.0650	20.99	0.166	1.1200	28.90	0.127	1.1750	35.35	0.109	1.2300	41.03	0.098
1.0660	21.16	0.165	1.1210	29.03	0.127	1.1760	35.46	0.109	1.2310	41.13	0.098
1.0670	21.32	0.164	1.1220	29.15	0.126	1.1770	35.57	0.108	1.2320	41.23	0.098
1.0680	21.49	0.163	1.1230	29.28	0.126	1.1780	35.67	0.108	1.2330	41.33	0.098
1.0690	21.65	0.162	1.1240	29.40	0.125	1.1790	35.78	0.108	1.2340	41.43	0.098
1.0700	21.81	0.161	1.1250	29.53	0.125	1.1800	35.89	0.108	1.2350	41.52	0.098
1.0710	21.97	0.160	1.1260	29.65	0.125	1.1810	36.00	0.108	1.2360	41.62	0.097
1.0720	22.13	0.159	1.1270	29.78	0.124	1.1820	36.11	0.107	1.2370	41.72	0.097
1.0730	22.29	0.158	1.1280	29.90	0.124	1.1830	36.21	0.107	1.2380	41.82	0.097
1.0740	22.45	0.157	1.1290	30.03	0.123	1.1840	36.32	0.107	1.2390	41.91	0.097
1.0750	22.60	0.156	1.1300	30.15	0.123	1.1850	36.43	0.107	1.2400	42.01	0.097
1.0760	22.76	0.155	1.1310	30.27	0.123	1.1860	36.53	0.106	1.2410	42.11	0.097
1.0770	22.91	0.154	1.1320	30.40	0.122	1.1870	36.64	0.106	1.2420	42.20	0.097
1.0780	23.07	0.153	1.1330	30.52	0.122	1.1880	36.75	0.106	1.2430	42.30	0.096
1.0790	23.22	0.152	1.1340	30.64	0.121	1.1890	36.85	0.106	1.2440	42.40	0.096
1.0800	23.37	0.151	1.1350	30.76	0.121	1.1900	36.96	0.106	1.2450	42.49	0.096

Table 23-1. Involute Roll Angles (*Continued*)

D_z/D_b	ϵ_z deg	Dif	D_z/D_b	ϵ_z deg	Dif	D_z/D_b	ϵ_z deg	Dif	D_z/D_b	ϵ_z deg	Dif
1.2460	42.59	0.096	1.3010	47.68	0.090	1.3510	52.05	0.085	1.4010	56.22	0.082
1.2470	42.68	0.096	1.3020	47.77	0.089	1.3520	52.13	0.085	1.4020	56.30	0.082
1.2480	42.78	0.096	1.3030	47.86	0.089	1.3530	52.22	0.085	1.4030	56.38	0.082
1.2490	42.88	0.096	1.3040	47.95	0.089	1.3540	52.30	0.085	1.4040	56.47	0.082
1.2500	42.97	0.096	1.3050	48.04	0.089	1.3550	52.39	0.085	1.4050	56.55	0.082
1.2510	43.07	0.095	1.3060	48.13	0.089	1.3560	52.47	0.085	1.4060	56.63	0.082
1.2520	43.16	0.095	1.3070	48.22	0.089	1.3570	52.56	0.085	1.4070	56.71	0.081
1.2530	43.26	0.095	1.3080	48.31	0.089	1.3580	52.64	0.085	1.4080	56.79	0.081
1.2540	43.35	0.095	1.3090	48.40	0.089	1.3590	52.73	0.085	1.4090	56.87	0.081
1.2550	43.45	0.095	1.3100	48.49	0.089	1.3600	52.81	0.085	1.4100	56.95	0.081
1.2560	43.54	0.095	1.3110	48.57	0.089	1.3610	52.90	0.084	1.4110	57.04	0.081
1.2570	43.64	0.095	1.3120	48.66	0.089	1.3620	52.98	0.084	1.4120	57.12	0.081
1.2580	43.73	0.094	1.3130	48.75	0.088	1.3630	53.07	0.084	1.4130	57.20	0.081
1.2590	43.83	0.094	1.3140	48.84	0.088	1.3640	53.15	0.084	1.4140	57.28	0.081
1.2600	43.92	0.094	1.3150	48.93	0.088	1.3650	53.23	0.084	1.4150	57.36	0.081
1.2610	44.01	0.094	1.3160	49.02	0.088	1.3660	53.32	0.084	1.4160	57.44	0.081
1.2620	44.11	0.094	1.3170	49.10	0.088	1.3670	53.40	0.084	1.4170	57.52	0.081
1.2630	44.20	0.094	1.3180	49.19	0.088	1.3680	53.49	0.084	1.4180	57.60	0.081
1.2640	44.30	0.094	1.3190	49.28	0.088	1.3690	53.57	0.084	1.4190	57.68	0.081
1.2650	44.39	0.094	1.3200	49.37	0.088	1.3700	53.65	0.084	1.4200	57.76	0.081
1.2660	44.48	0.093	1.3210	49.46	0.088	1.3710	53.74	0.084	1.4210	57.84	0.081
1.2670	44.58	0.093	1.3220	49.54	0.088	1.3720	53.82	0.084	1.4220	57.93	0.081
1.2680	44.67	0.093	1.3230	49.63	0.087	1.3730	53.90	0.084	1.4230	58.01	0.081
1.2690	44.76	0.093	1.3240	49.72	0.087	1.3740	53.99	0.084	1.4240	58.09	0.080
1.2700	44.86	0.093	1.3250	49.81	0.087	1.3750	54.07	0.084	1.4250	58.17	0.080
1.2710	44.95	0.093	1.3260	49.89	0.087	1.3760	54.16	0.083	1.4260	58.25	0.080
1.2720	45.04	0.093	1.3270	49.98	0.087	1.3770	54.24	0.083	1.4270	58.33	0.080
1.2730	45.13	0.093	1.3280	50.07	0.087	1.3780	54.32	0.083	1.4280	58.41	0.080
1.2740	45.23	0.092	1.3290	50.15	0.087	1.3790	54.41	0.083	1.4290	58.49	0.080
1.2750	45.32	0.092	1.3300	50.24	0.087	1.3800	54.49	0.083	1.4300	58.57	0.080
1.2760	45.41	0.092	1.3310	50.33	0.087	1.3810	54.57	0.083	1.4310	58.65	0.080
1.2770	45.50	0.092	1.3320	50.41	0.087	1.3820	54.65	0.083	1.4320	58.73	0.090
1.2780	45.60	0.092	1.3330	50.50	0.087	1.3830	54.74	0.083	1.4330	58.81	0.080
1.2790	45.69	0.092	1.3340	50.59	0.087	1.3840	54.82	0.083	1.4340	58.89	0.080
1.2800	45.78	0.092	1.3350	50.67	0.086	1.3850	54.90	0.083	1.4350	58.97	0.080
1.2810	45.87	0.092	1.3360	50.76	0.086	1.3860	54.99	0.083	1.4360	59.05	0.080
1.2820	45.96	0.092	1.3370	50.85	0.086	1.3870	55.07	0.083	1.4370	59.13	0.080
1.2830	46.05	0.091	1.3380	50.93	0.086	1.3880	55.15	0.083	1.4380	59.21	0.080
1.2840	46.15	0.091	1.3390	51.02	0.086	1.3890	55.23	0.083	1.4390	59.29	0.080
1.2850	46.24	0.091	1.3400	51.11	0.086	1.3900	55.32	0.083	1.4400	59.37	0.080
1.2860	46.33	0.091	1.3410	51.19	0.086	1.3910	55.40	0.082	1.4410	59.45	0.080
1.2870	46.42	0.091	1.3420	51.28	0.086	1.3920	55.48	0.082	1.4420	59.53	0.080
1.2880	46.51	0.091	1.3530	51.36	0.086	1.3930	55.56	0.082	1.4430	59.61	0.079
1.2890	46.60	0.091	1.3440	51.45	0.086	1.3940	55.65	0.082	1.4440	59.68	0.079
1.2900	46.69	0.091	1.3450	51.54	0.086	1.3950	55.73	0.082	1.4450	59.76	0.079
1.2910	46.78	0.091	1.3460	51.62	0.086	1.3960	55.81	0.082	1.4460	59.84	0.079
1.2920	46.87	0.090	1.3470	51.71	0.085	1.3970	55.89	0.082	1.4470	59.92	0.079
1.2930	46.96	0.090	1.3480	51.79	0.085	1.3980	55.97	0.082	1.4480	60.00	0.079
1.2940	47.05	0.090	1.3490	51.88	0.085	1.3990	56.06	0.082	1.4490	60.08	0.079
1.2950	47.14	0.090	1.3500	51.96	0.085	1.4000	56.14	.082	1.4500	60.16	
1.2960	47.23	0.090									
1.2970	47.32	0.090									
1.2980	46.41	0.090									
1.2990	47.50	0.090									
1.3000	47.59	0.090									

they do not always yield the extreme degree of precision available with a dividing head and height gage. They are usually equipped with dial indicators graduated in increments of 0.0001 or with recording pickups, the sensitivity of which may be as fine as 0.00002. On the other hand, the error of the instrument may amount to 0.0001 or more, depending on the product. All these machines may be calibrated through the use of a highly accurate master involute profile which is checked on the involute checker to be tested and the results compared with the increment readings obtained with the dividing head and the height gage. Calibration by a plain cylindrical member located at a predetermined distance from the machine centers was recently proposed by the writer and is used successfully by one major automotive manufacturer.

Base diameter D_b = ...

$$\text{Constant } K = \frac{D_b \pi}{360} = 0.008727 \times D_b = \text{......................}$$

Dia.	$\sec \phi_z = \frac{D_z}{D_b}$	Roll angle ϵ_z from Table 23-1	Angular setting of dividing head $\epsilon_z - \epsilon_f$	Distance AB $K\epsilon_z$	Indicator reading
Form dia. D_f		ϵ_F =	Set to zero		Set to zero
Pitch dia. D		ϵ =			
Outside dia. D_o		ϵ_o =			

Pressure angle	Roll angle at pitch dia.
14½°	14.8177
17½°	18.0653
20°	20.8540
22½°	23.7327
25°	26.7175
27½°	29.8263
30°	33.0797

FIG. 23-2. Profile-calibration chart.

Except for some designs which employ master templates and other principles, the mechanism of these machines provides a timed relation between the rotation of the tooth profile and the straight motion of the indicating member. According to the design principle, this group of machines may be subdivided into:

1. Machines using individual base-circle disks
2. Machines using a permanent built-in master base-circle disk and a proportional compensator
3. Machines using a built-in master involute profile and a linkage
4. Instruments employing individual master templates
5. Special designs providing a pivot from the center of a substitute circle, possibly in combination with additional master cams

Involute Checkers Using Individual Base-diameter Disk. An involute-profile-measuring machine using the base-circle-disk method is shown in Fig. 23-3. The gear to be checked is mounted between the centers A and is rotated by friction on the lower center which turns with the base-circle disk B. The upper center is a live center. The distance between centers is adjustable. Base-circle disks are interchangeable to accommodate gears with different base diameters. Rotation of the base-circle disk B causes a linear movement of apron C which carries the straight edge E and is mounted on antifriction slides which are parallel to the straight edge E. Handwheel F is used to drive, through a series of crossed axes and wormgearing, the main spindle carrying the base-circle disk. To maintain contact between the base-circle disk B and straight edge E, the spindle housing, which is integral with the vertical column and mounted in slides of the main housing perpendicular to the slides of the apron C, is urged against the straight edge E through the spring-loaded anti-friction roller H actuated by handwheel G. The worm, which meshes with the worm wheel driving the main spindle, slides on its shaft to allow for the transverse motion of the spindle housing. Friction between the base-circle disk and the straight edge causes rolling motion without slippage. Hence the indicating stylus, which is mounted on the apron C and contacts the gear to be checked in the contact plane of

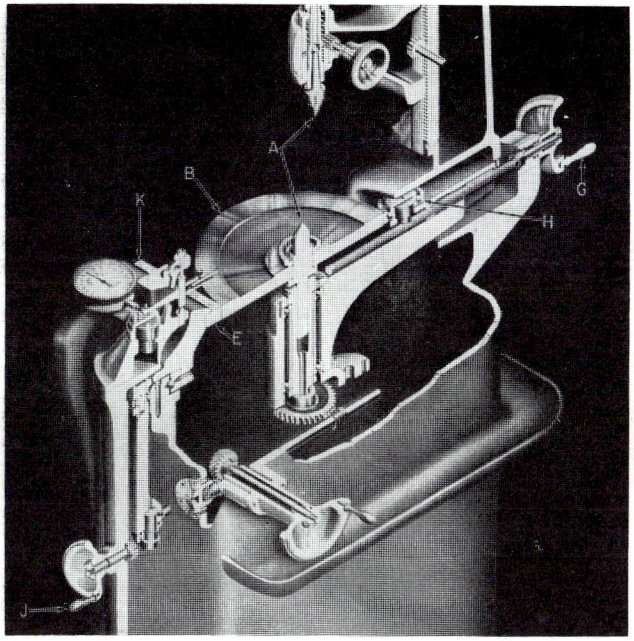

FIG. 23-3. Involute-profile-measuring machine with individual base-circle disks. (*Courtesy of Illinois Tool Works, Inc., Chicago, Ill.*)

the straight edge, describes an involute profile with respect to the rotating gear. Deviations from the involute are registered on the dial indicator D, or by an electrical recording system which is also available. Some machines are designed to drive the base-circle disk by the straight-line motion of the apron, rather than the worm drive shown in the illustration.

The upper surface of the straight edge E is graduated in increments of 0.050″, which furnishes scale readings indicative of the amount of travel at the circumference of the base circle. The relation between the scale reading S and the roll angle ϵ is given by the formula

$$S = \frac{D_b \epsilon_R}{2 \times 0.05} = 10 D_b \epsilon_R = \frac{\pi}{18} D_b \epsilon^\circ \tag{23-4}$$

where ϵ_R, ϵ° is the roll angle in radians or degrees, respectively, and D_b is the base diameter. Some machines are also equipped with a segment, graduated in degrees of roll angle, which represents the generally accepted means of expressing locations on an involute profile. The roll-angle ranges of the active profiles of mating gears are not equal, but proportional to the gear ratio. Profile charts of mating gears based on roll angle will therefore have different lengths, while profile charts based on scale readings, which are proportional to the distance traveled by the contact point along the path of contact, would have the same length. Charts based on scale readings have the advantage that they permit the evaluation of the contact between mating profiles without changing the scale of the chart. Neither scale readings nor roll angles are proportional to the corresponding diameters or to the length of the involute profile.

To enable inspection of the entire tooth profile, handwheel *J* permits raising and lowering of the indicating mechanism through a pinion-and-rack drive.

The indicating pointer which contacts the involute profile must be set longitudinally, so that the contact point lies in the plane of the straight edge which rolls with the base-circle disk. In addition, the contact point must be in an axial plane at right angles to the straight edge when the scale reading corresponds to the origin of the involute. Both settings are performed by means of the centering gage shown in Fig. 23-4. This gage is essentially a cylindrical bar placed between the centers of the machine and has a flat surface *A* that is brought into contact with the straight edge. A second vertical surface *B* of the centering gage at right angle to the flat surface *A* contacting the straight edge serves to zero the indicator at the origin of the involute to be checked. Setting surface *C* is provided to enable the corresponding zero setting for inspection of the opposite sides of the gear teeth. To bring this surface *C* into position, after 180° rotation of the indicating stylus about its axis, V plunger *D* is disengaged to allow rotation of the sleeve incorporating surfaces *B* and *C* until plunger *D* engages a second detent groove.

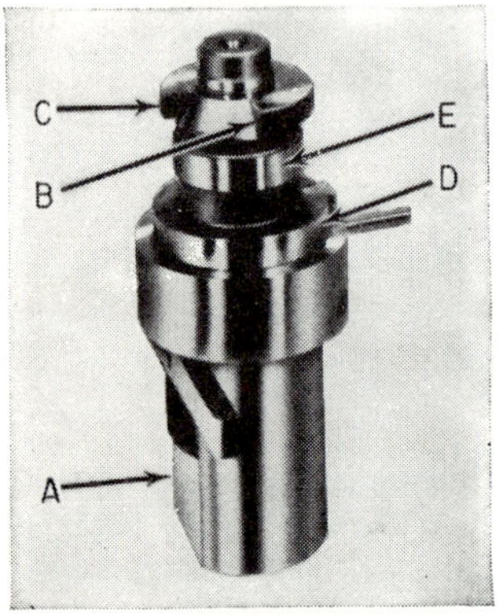

FIG. 23-4. Finger centering gage for involute machines. (*Courtesy of Illinois Tool Works, Inc. Chicago, Ill.*

If the contact area or tip of the stylus does not lie on the pointer axis, the tip will switch positions after 180° of pointer rotation. However, as long as the pointer pivot is in line with setting surface *C* or *D*, as applicable, the angle of the pointer axis with respect to the plane through the machine center and the pointer pivot will be equal though of opposite direction. To utilize this geometric condition, the indicator pickup is mounted on a swivel, the axis of which lies in a plane through the pointer pivot normal to the plane of the straight edge *E*. When the pointer is rotated 180°, the indicator pickup is also swiveled 180° about the rounded end of the pointer holder and the straight edge is adjusted until indicator readings are the same in either position. The indicator is then set to zero and index blocks *K* in Fig. 23-3 are set to the zero lines at both ends of the straight edge. The pointer is then brought down and forward to cylindrical surface *E* of the master gage, the radius of which is equal to the distance of flat surface *A* from the centers.

After these settings are made with the master gage, the machine is ready for operation with any base-circle disk. Its accuracy may be tested by means of a master involute cam which, in turn, may be calibrated on a dividing head as described in the previous section. Although both spur and helical master involutes are available, in either case with appropriate base-circle disk, the test with a nonhelical master involute may be sufficient to qualify the machine for the inspection of either spur or helical gears, since the profile is traced, in both cases, in a plane normal to the gear axis.

To check any particular gear, the proper base-circle disk is mounted on the lower center. Rotation of handwheel *G* (Fig. 23-3) brings spring pressure to bear against the base-circle disk through the antifriction roller *H* and thereby causes the spindle housing and vertical column to slide forward until rolling contact is established between the disk and the straight edge. Apron and straight edge are set by means of handwheel F to the scale reading or roll angle which corresponds to the form diameter of the gear which, possibly mounted on an accurate inspection arbor, is placed between the centers and rotated against the indicating stylus until the indicator reads zero. The spindle is then rotated by means of handwheel F and readings are taken at any desired scale reading or roll angle along the active profile of the tooth.

When checking helical gears, the indicating pointer, which can be loosened by the thumbscrew shown in Fig. 23-3, is rotated to the helix angle of the gear. Visual approximation of this setting by observation of the pointer position on the tooth surface is sufficient. A stop screw and lock nut at the end of the pointer preserve its axial setting. However, the gaging pressure is no longer at right angles to the pivot axis of the indicating lever, which causes an undesirable side-cramming effect.

Internal gears are checked with special gooseneck pointers and also require special mounting adapters.

The recording system, available with this machine, is synchronized with the spindle or by means of a "line-of-action" attachment with the apron travel. A selection of drive ratios permits adjustment of the chart scale to suit the pitch of the gear. The electronic-indicator pickup provides a choice of several magnification ratios: 200, 400, and 800 or 400, 800, 1,600, and greater.

The machine requires a base-circle disk for each gear, which is not objectionable on repetitive production operations, while undesirable where short runs require frequent change-overs. The disks are usually hardened and ground to prevent damage in storage and use. Soft steel disks may be adequate when severe usage is not expected. When checking gears with small base diameters, the base-circle disk may have to be an integral part of a special lower center. Gears which must be checked from journal diameters or similar locating surfaces are mounted in special holding fixtures which may take the place of the lower center and then include cylindrical surfaces representing the base-circle disk.

Except for the initial settings of the indicating pointer, the machine requires no gage-block settings and no calculations other than the computation of scale readings. The task of ascertaining by inspection the base diameter of a physical gear, the specifications of which are not available, would be inconvenient, though not forbidding. It could be solved by moving the spindle housing and vertical column without base-circle disk backward and forward on their slide to find the position corresponding to the profile high point and measuring the distance from the centers to the straight edge.

The minimum base diameter is limited and ranges from ⅜ to 2″ depending on the type of machine and whether a special center is used. Pointers furnished with the machine may be used on pitches as fine as 32, yet there is no valid reason for not using this or any other machine on finer pitches if a suitable special pointer is made. The maximum apron travel from either side of zero setting varies for different models from 3½ to 8¾″, which corresponds to 70 and 175 scale readings, or $401.07/D_b$ and $1{,}002.68/D_b$ degrees of roll, respectively. This roll-angle capacity should accommodate practically any gear. The inverse proportional relation to the base diameter is logical, since large gears have less roll angle than small gears. This type of machine is available in physical capacity from 12″ diameter and 18″ between centers to 36″ diameter and 36″ between centers. Vertical indicator travel ranges from 8 to 18″. Extremely long gears require special adaptations.

Involute Checkers with Permanent Master Base-circle Disk and Proportional Compensator. Machines in this group employ also a straight edge which rolls on a master base-circle disk, but this disk is part of the machine and need not be changed from one gear to the other. In order to make the travel of the indicator carrier equal to the circular travel at the base circle of the gear to be checked, the carrier is mounted on a second slide, the displacement of which is actuated by the straight-edge slide through proportional compensating devices which provide an adjustable ratio equal to the ratio D_b/D_d where D_b is the base diameter of the work gear and D_d the constant diameter of the master disk.

The machine shown in Fig. 23-5 is an example of this type of involute checker. The gear is mounted between centers *A* and driven by the lower center which is part of the main spindle. The master base-circle disk *B* mounted on the spindle is in rolling contact with straight edge *C*, which is carried by the compensator slide *D*. The motion of this slide is actuated by lead screw *E* through handwheel *F*. The lead of the screw *E* and the diameter of the disk *B* are proportioned so that one turn of the

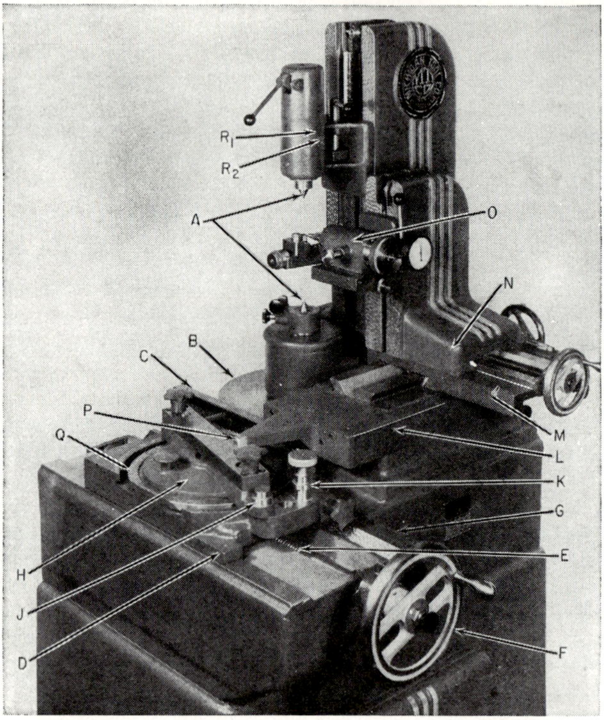

FIG. 23-5. Involute-profile-measuring machine with permanent base-circle disk. (*Courtesy of Michigan Tool Co., Detroit, Mich.*)

handwheel corresponds to 2 degrees roll angle on smaller models, 1 degree on larger models of this machine. Graduations *G* on the machine bed adjacent to the slide *D* show the amount of spindle rotation in degrees, while a collar of the handwheel *F* has 120 or 60 graduations, respectively, to permit the reading of finer increments. Slide *D* carries the pivoted compensating disk and bar *H*, which is provided with a pin *J* at a distance of 8.000″ from the pivot center. A second pin *K*, parallel to and of the same diameter as pin *J*, is mounted on slide *D*, also at 8.000″ distance from the pivot. The plane through the axis of pin *K* and the pivot axis is parallel to the direction of the slide and the straight edge *C*. The cross slide *L*, at right angles to slide *D*, with two additional slides *M* and *N*, carries the indicating assembly *O* and is provided with a follower *P*, held in contact with the compensating bar by spring pressure. Slide *N* permits adjustment of the indicating pointer so that it travels in a plane tangent to the base cylinder of the work gear. To cause the pointer to trace the required involute, the travel of cross slide *L* must be equal to the circular displacement at the gear base circle. If the base diameter of the gear were equal to the diameter D_d of disk B, the cross-slide travel would have to be equal to the travel of the slide D and the compensator bar would have to be set at 45°. For any other base diameter of the work gear, the angle setting must establish a ratio of cross-slide to compensator-slide travel equal to the ratio of the work base diameter D_b to the disk diameter D_d, or

$$\tan \alpha = \frac{D_b}{D_d} \tag{23-5}$$

The corresponding distance to be set by gage blocks between the pins J and K is

$$2 \times 8 \sin \frac{\alpha}{2} - d = 16 \sin \frac{\alpha}{2} - d \tag{23-6}$$

where d is the diameter of pins J and K. The compensating bar and disk are clamped in this position by clamp bolts in the circular T slot Q. Newer models have angular graduations on the compensating disk H, permitting settings to 0.001° through a high-powered scope.

Tailstock and indicating assembly are vertically adjustable to suit the blank dimensions of the gear and its mounting arbor. Both are counterweighted to facilitate upward sliding.

Setting surfaces R_1 and R_2 are provided on the tailstock to zero the indicator when the contact point is in a plane through the centers at right angles to the direction of cross slide L. Two opposite surfaces are available, enabling setups for checking the front sides and the rear sides of the teeth. In the first case, cross slide L travels forward for increasing roll angles and the compensating slide D to the left; in the second case, both motions are reversed. Slide M permits forward or rearward positioning of the indicator carrier so that the full range of machine motion is available in either case. This range covers 67 degrees of roll for small compensating angles, decreases proportional to cos α. This roll-angle limitation is no handicap because it reduces the available angular capacity only for large gears, which almost never have a great roll angle. Slide M may also be used to change the zero-roll-angle position of the indicating assembly with respect to the master disk, straight edge, and compensating mechanism, in order to distribute the wear of these components. Slide N permits longitudinal setting of the pointer by means of the setting master. This slide motion may also be used to determine the profile high point of a gear with unknown base diameter, as previously described.

When checking helical gears, it is possible to rotate either the pointer only, or the whole indicating assembly to the helix angle. In either case, the pointer traces the involute in the plane of rotation. In the first case, it registers deviations in the plane of rotation; in the second, readings refer to the normal plane. The second method is preferable because the gaging pressure is applied at right angles to the indicating pivot without side cramming. In addition, the profile error normal to the tooth gives a direct account of interference or clearance when the tooth surface is in contact with the mating gear. It is also a more realistic expression of shop performance in terms of the thickness of excessive or lacking material. To convert readings taken normal to the tooth to transverse readings possibly obtained on another machine, divide by the cosine of the base helix angle.

Different models offer diameter capacities from 12 to 36″, distance between centers from 14 to 84″, and face-width coverage for the full distance between centers. In selecting a model of suitable capacity, it will be advantageous to assure a relatively large ratio between the diameter D_d of the built-in master disk and the anticipated work-base diameters, in order to keep the compensator-angle setting below 45° and avoid high side pressure on the cross slide. Pointers furnished with the machine will accommodate up to 20 diametral pitch; finer pitches require a special pointer. For the inspection of internal gears, gooseneck pointers and workholding adapters are required. The machine is available with or without electrical recording system. A special attachment is also available to check tooth-to-tooth spacing; it clamps on the indicating head.

Another involute checker based on the same principle is shown in Fig. 23-6. This machine mounts the gear with horizontal axis between centers A. The master base-circle disk B of the 20″ model has the form of a counterweighted sector with 10″ radius, that of the 10″ model is a 10″ diameter disk. Rolling contact between master disk and straight edge C, carried by the cross slide D, is maintained through pre-loaded pressure roller E. Pads on the headstock enable mounting of this roller to suit

Fig. 23-6. Involute-profile-measuring machine with permanent base-circle disk. (*Courtesy of Vinco Corp., Detroit, Mich.*)

either the 20″ or the 10″ model, but conversion requires factory service. Handwheel F actuates the lead-screw drive of the cross slide which has five threads per inch so that each turn of the wheel corresponds to 0.02 radian roll angle on the 20″ model, 0.04 radian on the 10″ model. Fixed pointer G permits reading of the roll angle in degrees on the graduated disk H, which is adjustable against the master disk B, but carried along by friction. The vertical slide J next to the cross slide D carries the indicator base K and, through the follower roller L, maintains contact with the 10″ long sine bar M which is fastened to the cross slide D.

The indicating stand N is mounted on ways of the indicator base K which are parallel to the line of centers to enable profile inspection anywhere along the gear face. Axial tailstock adjustment is also available to set the required distance between centers.

The indicator base K is transversely movable through spoked knob O to bring the indicating pointer into the vertical tangent plane of the base cylinder. Once this setting is made, either by means of the removable setup bracket P with vertical reference surface 1″ behind the line of centers, or preferably with a setting master, plunger Q is pushed to an internal stop and clamped by the thumbscrew R, in order to permit convenient restoration of the transverse setting after each withdrawal of the indicator mount. Plunger Q also incorporates a fine-adjustment screw.

The sine bar is then set to the required angle and the cross slide moved to its extreme position away from the operator. Since the sine bar has to provide the ratio between the rate of rotation at the work-base circle and at the circumference of the master disk, its angular setting α can be obtained from Eq. (23-5). The corresponding sine-bar setting is 10 sin α. The indicator is then zeroed at the height of the line of centers by means of a special height block furnished with the instrument. For the rough setting, the indicator carrier S is vertically adjusted on the post of the indicating stand N by means of hand knob T, then clamped through lock knob U before final zeroing of the indicator dial. This positions the pointer at the origin of the involute at the base circle. Graduated disk H is then set to zero roll angle. When the roll angle from the base circle to the outside diameter exceeds the capacity of the machine, while the roll angle from the form diameter D_f to the outside diameter can be accommodated, the initial setting is made at a height above the line of centers equal to

$$\tfrac{1}{2}\sqrt{D_f^2 - D_b^2} \tag{23-7}$$

and the graduated disk H is set to the roll angle ϵ_f at the form diameter. The independent zeroing of the graduated disk also permits distribution of disk and sine-bar wear by starting the cross slide D short of its extreme positions.

For inspection of the opposite tooth side, either the gear or the machine operation is reversed. In the latter case, the starting position of cross slide D is toward the operator, the indicating pointer is rotated 180° about its axis, and the indicating pressure is reversed. Reversal of the gear is preferable to reversal of the machine motion because the downward motion of the slide J, to maintain contact between the roller L and sine bar M, is assured only by the weight of the slide with the indicator assembly.

When checking helical gears, this machine also provides the choice of setting either the pointer only, or the whole indicator carrier to the helix angle. Internal gears can be checked by using special gooseneck pointers and workholding adapters. The base diameter of a gear for which no specifications are given is ascertained by moving the indicator base transversely over the gear tooth until a profile high point is registered and measuring the horizontal distance of the pointer from the line of centers.

The maximum diameter capacity is 24″, the maximum distance between centers 18″. The roll-angle capacity is determined by the projected sine-bar length and the diameter D_d of the master disk B as

$$\epsilon_{\max} = \frac{10}{R_d}\frac{180}{\pi}\cos\alpha = \frac{3{,}600}{\pi\sqrt{D_d^2 + D_b^2}} \quad \text{deg} \tag{23-8}$$

Except for very large base diameters, this will be about 55° for the 20″ model, which is adequate for most gears. Since this model offers the mechanical advantage of smaller sine-bar angles, it is usually preferred. The 10″ model, with about twice the roll-angle capacity, is limited to small gears but will handle pump gears or helical gears with large roll angles. Even though fine-pitch gears are usually of small size, the 20″ model is more suitable for them because of its relatively lower rate of rotation, which yields more discriminating readings. Standard pointers can be used up to 48 diametral pitch. Pitches up to 180 have been checked with special or altered, relieved pointers. A master involute cam, held within 0.00005, is available for either model as well as an electronic recorder, which is driven from the lead screw so that the roll-angle scale is the same for gears of any size.

A third machine, employing a permanent master base-circle disk and proportional compensator, was developed by the Illinois Tool Works in 1958 for applications where a large variety of gears must be checked. The general principle of this instrument is schematically illustrated in Fig. 23-7. The master base-circle disk *A* is in rolling contact with straight edge *B* mounted on a sliding apron. Straight edge *B* contacts the ratio lever *C*, which in turn controls the motion of an indicator slide *D*. This indicator slide is parallel to the straight edge *B* and its distance from the work center can be set to a scale through the medium of an optical system. This setting distance is equal to the base radius of the gear to be checked and requires no calculations beyond the determination of gear data. In order to minimize wear on the proportional compensating linkage, the machine can also be furnished to accommodate interchangeable base-circle disks, which are mounted on the main spindle carrying the fixed disk *A* and are brought into contact with a second straight edge provided on the indicator slide *D* at the level of the interchangeable base-circle disk. When this alternate mechanism is used, the drive of the indicator slide *D* from the ratio lever *C* is disconnected. This machine comes in two models, accommodating, respectively, 12 or 24″ diameters, 18 or 30″ between centers, 9 or 12″ axial indicator travel. Minimum base diameter is $1\frac{1}{2}$″ with standard lower center, $\frac{3}{8}$″ with special center. The instrument may be equipped with an electrical recorder.

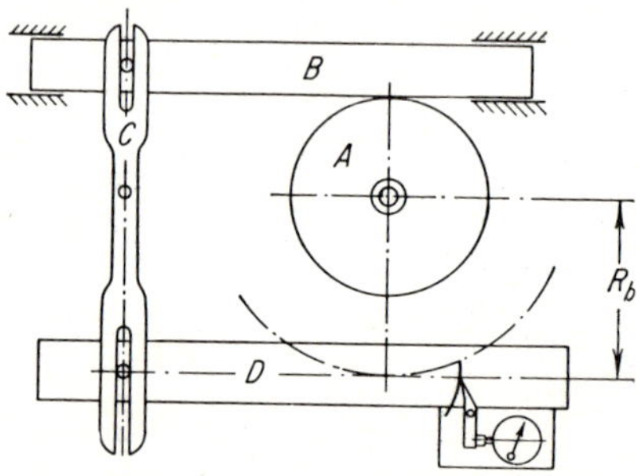

FIG. 23-7. Schematic of involute checker with either permanent or adjustable base-circle disk.

Involute Checkers Employing a Master Involute Cam and Compensating Linkage. The principle of operation of this type of machine is shown in Fig. 23-8. The upper illustration is a plan view of a typical machine; the lower cross section shows the master involute cam and drive mechanism. Shaft *B* is rotated by handwheel *A* and, through worm *C*, drives the worm wheel *E* on the work spindle. The work spindle is equipped with a center opposed by an adjustable live tailstock center. As in the case of the previously described machines, the workpiece is driven by friction with the work-spindle center. The master involute cam *G* is attached to the work spindle and drives the straight block *H*, which represents the side of a rack tooth and the angle of which is adjustable. The adjustable straight-edge block *H* is mounted on the drive slide *I* to which pin *J* is attached which bears against the ratio drive bar *K* of the ratio shaft *L*. The distance *R* of pin *J* from the center of the ratio shaft *L*, measured normal to the direction of the drive slide *I*, is equal to the base radius of the master cam divided by the cosine of the pressure-angle setting of block *H* so that its rate of travel conforms to that of a circle with radius *R* which rotates with the workpiece. The tension spring *M* urges the ratio bar *K* against pin *J*, which assures contact between the straight-edge block *H* and the master involute cam *G*. A second ratio bar *O*, parallel to drive bar *K*, is also fastened to the ratio shaft, but arranged above the ratio drive bar *K*. It is shown in the upper illustration actuating the indicator slide *P* through the pin *Q*. Contact between ratio bar *O* and pin *Q* is maintained through weights acting on the indicator slide *P*. Since the motion of the drive slide *I* is equal to the travel at the base circle of the master cam divided by the cosine of the pressure-angle setting of block *H*, while the indicator slide must travel at the rate of motion at the base circle of the workpiece, an adjustment must be available to provide for this ratio. For this purpose the indicator slide *P* is mounted on main slide *S* which is adjustable at right angles to the motion of the indicator slide *P*. The settings are

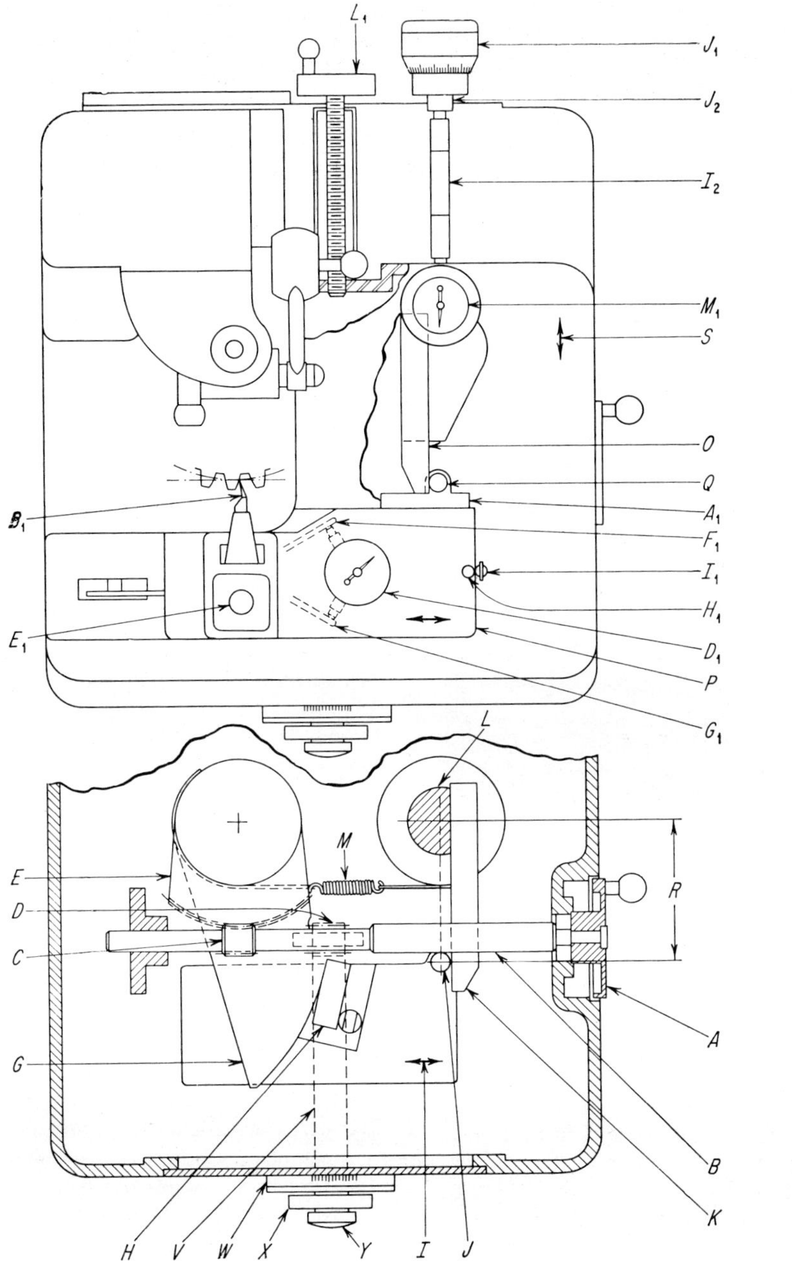

FIG. 23-8. Schematic of involute checker using involute cam. (*Courtesy of Fellows Gear Shaper Co., Springfield, Vt.*)

made by moving the main slide S through hand-operated lead screw L_1 against size blocks I_2 which bear against micrometer spindle J_2. To provide for a sensitive setting, contact between the main slide S and size blocks I_2 is made through an indicating mechanism. Main slide S is in the desired position when indicating dial M_1 reads zero. The required length of gage blocks I_2 is equal to the base radius of the work-piece, thus avoiding the need for any calculations other than the determination of such gear data as the base diameter and the roll angle. The shaft carrying the indicating pointer B_1 is pivoted at E_1 and carries arms F_1 and G_1. The holder of the indicating pointer is vertically adjustable within the range of the pivot bracket. The indicator D_1 can be swung so that its plunger contacts either arm for checking the right or left sides of the teeth, respectively. When checking the left side of a tooth, the pointer would have to be rotated 180° about its axis and the slide P would travel from right to left. The starting position of the indicator slide P would have to be the same for either direction of operation, while the motion of the involute cam and drive slide would have to start at, or near, one or the other of their extreme positions. To provide the maximum range of roll angle for either mode of operation, the bracket A_1 which carries pin Q is adjustable on the indicator slide P. This feature also permits profile inspection of gears which have a roll angle from the base circle to the outside diameter exceeding the roll-angle capacity of the machine, as long as the roll angle from the form diameter to the outside diameter can be accommodated. (The roll angle at the form diameter is termed "angular offset" in the manufacturer's literature.) It also permits distribution of wear on operating components.

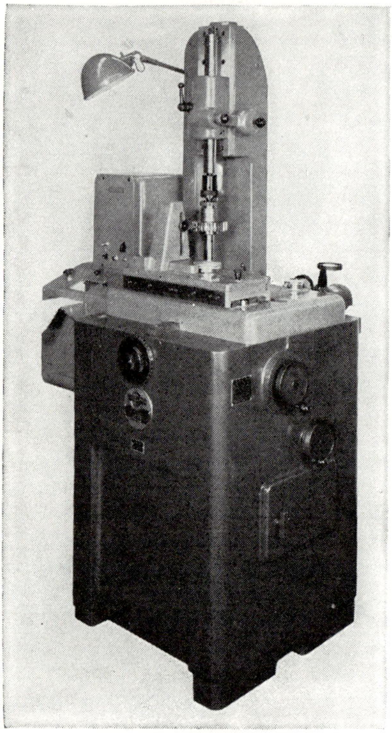

FIG. 23-9. Involute measuring instrument machine using involute cam. (*Courtesy of Fellows Gear Shaper Co., Springfield, Vt.*)

The roll angle is indicated on dial W, which is graduated in degrees and driven by shaft B through worm drive D. Dial W is adjustable to any desired position. Stop shaft V limits the amount of machine rotation from a fixed stop, which is not shown, to the adjustable stop X, which may be clamped by nut Y. This device assures duplication of roll-angle ranges when inspecting production parts.

When checking helical gears, the indicating pointer is set to the approximate helix angle of the gear. Since the pivot of the pointer holder remains parallel to the gear axis, readings are registered in the plane of rotation. In the case of high helix angles, there will be an axial component of the indicating pressure on the pivot.

Figure 23-10 shows involute setting masters. For the profile inspection of internal gears, the manufacturer furnishes not only special gooseneck pointer holders and workholding adapters but also an internal setting gage. Pin A is positioned so that the smallest distance from its cylindrical surface to the gage center is a 4.000″ base radius. It serves for a rough base-radius setting of the adjustable gooseneck stylus. The radial surfaces B_1 and B_2 are used for the "zero" setting. Surfaces C_1 and C_2 are used to establish the correct distance of the line of contact of the traveling pointer from the machine centers.

This machine has the advantage of requiring neither special disks nor setup calculations, which is attained at the price of a more sophisticated design. Once the original check-out operations are performed, conversion from one gear to another requires only one setting. This setting could even be determined by trial if the base diameter of the gear to be checked were not known. The correct function of the machine depends not only on the accuracy of the master involute drive cam but also on a number of other essential components.

The instrument is available in several models with optional electrical recorder. The paper-drive drum of the recorder is geared to the work spindle so that graphs are based on roll angle. Typical machines have 12 and 24" diameter capacity.

Instruments Employing Individual Master Templates. This relatively simple principle is employed in a portable instrument shown in Fig. 23-11, which was designed by the General Electric Company for the profile inspection of large gears, the size of which precludes mounting in conventional workholding devices, such as arbors, centers, or other fixtures; also for gears which are inaccessible except for a portion of

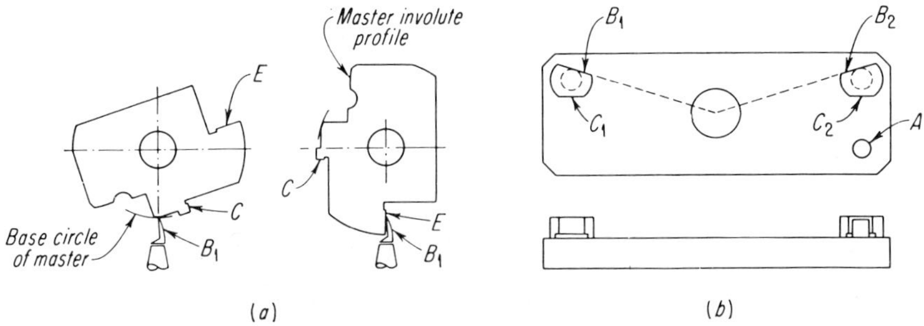

FIG. 23-10. Involute setting masters. (*a*) External gears. (*b*) Internal gears.

their circumference. It is applicable to spur and helical, external and large internal gears of relatively coarse pitch.

The instrument compares the transverse profile of the gear teeth with that of an accurate template A, which may be mounted adjustably at either side of the instrument for the inspection of either side of the gear teeth. The instrument rests on the tooth spaces of the gear with three ball-point legs B, the centers of which form a right triangle with sides of adjustable length. The length of the side which is parallel to the gear axis is optional on spur gears and equal to a suitable multiple of axial pitches on helical gears. The distance of the two front legs is adjusted so that both ball points are seated in tooth spaces at equal distance from the tooth to be checked. This adjustment varies from gear to gear with the tooth thickness and also the accuracy of the tooth spacing. The horizontal micrometer slide C is provided with opposite template mounting pads D_1 and D_2 and can be adjusted to centralize the indicating assembly E, which swivels about a vertical axis for zeroing in at either side of the central tooth. The indicating assembly E is attached to a horizontal ball-bearing slide F which also mounts the two followers G_1 and G_2 which are urged by reversible spring pressure against the profile of the template located on one or the other mounting pad. Slide H, operated by the graduated hand knob I, is mounted on the micrometer slide C and serves as a base of the horizontal slide F. It provides the vertical component of the follower motion registered on scale J and on the graduations of hand knob I. The distance of the indicating stylus from the swivel axis of the indicating assembly may be adjusted to the gear-tooth thickness by means of screw K.

The instrument requires one master template for each gear to be checked. The profile of the template must be of the highest order of accuracy, since its deviations from the required condition are reflected directly in the readings of the indicator. On the other hand, the tooth profiles of large gears have little change in curvature and can be approximated to a great degree of accuracy by simple circular arcs. It is also possible to incorporate desired profile modifications directly into the template. Care must be taken to attain the correct correlation between the starting position of the follower on the template and that of the indicator on the gear tooth. A flat surface L, which represents the theoretical outside diameter of the gear, is provided for this purpose on the template. The template mounting pad is vertically adjusted until the flat bottom of the follower G_1 or G_2 rests on surface L when the indicating stylus contacts

FIG. 23-11. Portable involute checker. *(Courtesy of General Electric Co., Lynn, Mass.)*

the tip of the gear tooth. Further adjustment may be required for the measured outside diameter of the gear unless there is little change in the angularity of the tooth profile within the range of the outside-diameter variation. These adjustments may have to be repeated for each gear because of the effect of tooth-thickness variations on the relation between the master template and the gear tooth.

Special Tooth-profile-checking Instruments. A number of single-purpose devices and special instruments are also available for the profile inspection of very large gears, of fine-pitch gears, and of gears with other than involute profiles.

For the inspection of large gears, the Vinco Corporation has developed a single-purpose device which rotates an indicating head about a suitably located pivot, utilizing the fact that the involute form of gears with high numbers of teeth can be matched very accurately by a substitute circle. The geometric foundation of this approach was originally developed by Douglas Hughson, in a paper presented in 1950 to the Gear Study Group of the Industrial Mathematics Society. If three points A, B, and C (Fig. 23-12) are selected within the active involute profile extending from F to D, a circle is defined with a radius ρ which passes through A, B, and C. Since the curvature of the involute is greater than that of the circle at point F, less at point D, the dash-dotted line representing the substitute circle must be outside the full-line involute from F to A and from B to C, inside the involute from A to B and from C to D. The center M of the circle will always be outside the base circle because the normal to the involute at B is the base-circle tangent, while the normal of the circle at the same point B is tipped counterclockwise. If $P\ (x,y)$ is a point of the involute and the center

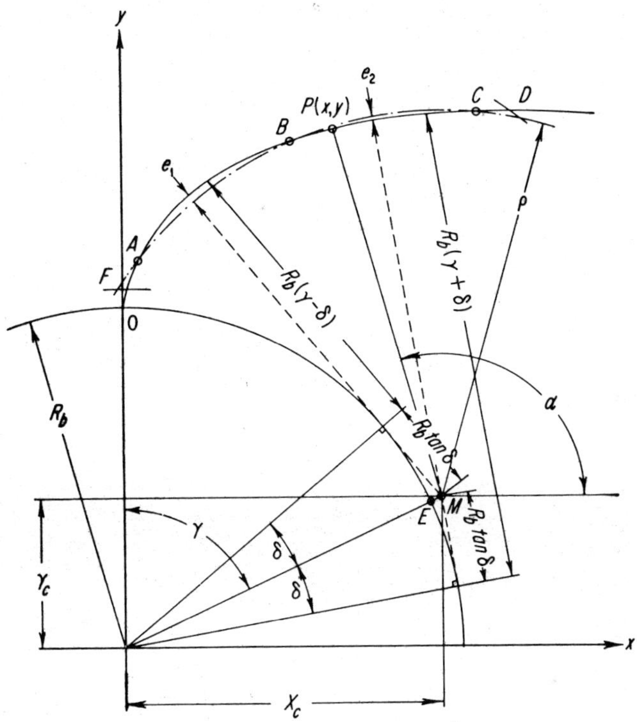

FIG. 23-12. Substitution of involute by an arc.

M of the substitute circle has the coordinates x_c, y_c, the deviation of the circle from the involute is

$$e = \rho - \sqrt{(x - x_c)^2 + (y - y_c)^2} \tag{23-9}$$

The maximum deviation of arc $\widehat{AB}$ and $\widehat{BC}$ is defined by the condition

$$\frac{de}{dx} = -\frac{(x - x_c) + (y - y_c)(dy/dx)}{\sqrt{(x - x_c)^2 + (y - y_c)^2}} = 0 \tag{23-10}$$

$$\frac{dy}{dx} = -\frac{x - x_c}{y - y_c} = -\cot\alpha$$

The last equation requires that the slope of the involute be normal to PM, or that PM is tangent to the base circle. The positions satisfying this condition are shown in dotted lines. The maximum deviations e_1 and e_2 can also be expressed in terms of angles γ snd δ as follows:

$$e_1 = R_b(\gamma - \delta) + R_b \tan\delta - \rho \tag{23-11}$$

$$e_2 = \rho - R_b(\gamma - \delta) + R_b \tan\delta \tag{23-12}$$

e_1 and e_2 are equal when

$$\rho = R_b\gamma = \widehat{OE} \tag{23-13}$$

in which case the deviation of the circle from the involute is

$$e = e_1 = e_2 = R_b \operatorname{inv} \delta \tag{23-14}$$

A tangent to the base circle at point E of length $R_b\alpha$ terminates in a point of the involute. One intersection of the involute and the substitute circle lies, therefore, approximately on the perpendicular bisectant of EM. Since the roll angle from e_1 to e_2 is 2δ, the angle δ is a measure of the range of substitution and it is seen that the deviation e increases with the range. The angle γ is roughly equal to the roll angle at the mid-point B. Deviations at F and D which may be greater or smaller than $e_1 = e_2$ can be determined from the coordinates of points F and D of the involute profile. Equations for exact equalization of all four deviations can be formulated, but are cumbersome.

In the practical application of this approach, γ is selected somewhat smaller than the roll angle at the mid-point of the profile and δ assumed approximately one-third of the total roll-angle range. Deviations are then computed at F and D and compared with the intermediate error R_b inv δ. A satisfactory solution is usually found after a few trial calculations. The method can usually also be adapted to modified profiles.

The advantage of this method lies in the high ratio between the radius ρ, which is approximately equal to the mean radius of curvature of the involute, and the depth of the gear tooth. As a consequence, relatively liberal tolerances can be allowed on the radius and also on the distance of the pivot center M from the gear axis. In the case of a 528-tooth 8-diametral-pitch, 66″ pitch diameter, 14.5° pressure angle gear, a substitute circle with $\rho = 8.3013''$ radius matches the involute within 4 microinches. A change of the radius by $\pm 0.005''$ causes only ½ millionth additional error, while a $\pm 0.001''$ change in the center distance of the pivot adds only 14 millionths inaccuracy. The distance of the pivot location from the tooth to be checked should be adjustable to the radius ρ.

The device itself is rather simple to design and build. The specific design execution for any particular gear should provide a circumferentially adjustable pivot, positioned at the computed distance from the gear center by locating on machined cylindrical surfaces such as an inside rim, outside diameter, or a large inside diameter. The deliberate provision of such surface, which may also be useful as a concentricity band during manufacturing and assembly operations, may be considered by the gear designer.

The profile-checking head developed by the Illinois Tool Works for large gears also employs a pivot located outside the base circle but compensates for the difference between the involute profile and the circle by a cam and a follower acting through an adjustable-ratio lever on an indicator slide. For a 60″ base diameter the ratio is 1:1, and it becomes more advantageous for smaller diameters where greater accuracy may be required. The instrument is clamped on a surface plate next to the gear or directly on the bed of a large hobbing or shaving machine. Its distance from the gear center must be set accurately by means of a special setting gage. As in the case of the Vinco approach, other settings are not critical. The Illinois Tool instrument is more complex than the use of a pivot to trace a substitute circle but enables inspection of any type of gear or lobe profile such as cycloids or spirals.

The profile-checking instrument F-360 of the Maag Gear Wheel Company in Zurich, Switzerland, is another portable device for large gears from 39 to 142″ outside diameter, 0.85 to 12.7″ diametral pitch. It employs a magnetized straight edge which carries the indicating mechanism and rolls on an interchangeable base-circle segment. The instrument is equipped with a mechanical recorder.

The Micrite®* involute-profile-measuring instrument, also developed by the Illinois Tool Works, is especially designed for fine-pitch gears from the very smallest up to 6″ diameter and from 20 to 120 diametral pitch. The relative motion of the indicating head and the gear to be checked is obtained by means of a system of change gears. The adjustable load on the contact stylus may be as small as 2.5 grams. The recording system produces charts based either on the roll angle or on the length of the path of contact.

Profile Inspection by Enlarged Projection. Enlarged projection of the tooth profile on a screen provided with an accurate profile layout is sometimes employed when mechanical involute-measuring machines are not available.

When this method is used, it is imperative that the centers of the base circles of the projected profile and of the layout coincide, even though the magnification ratio is usually so great that the gear center lies outside the screen layout. The importance of this condition stems from the fact that the involute curve loses its ability to transmit conjugate motion as a consequence of any radial or tangential shift with respect to the gear center. It can be accepted as true involute only when all its points have their respective correct distance from the center. It may therefore be rotated about the gear center to obtain a change in tooth thickness, but must not be translated relative to this center.

Correct correlation between projection and layout is obtained by means of staging fixtures, the design of which depends on the configuration of the gear blank. For instance, if the gear is mounted between centers, the staging fixture may include a special arbor which incorporates one radial straight edge, one at right angles to it, both located in the general area of the gear pitch diameter. The arbor is mounted between the same centers as the gear to be checked, but is oriented in the staging fixture so that the two straight edges project as vertical and horizontal lines. Since the relation of these lines to the gear center is known, they can be matched with corresponding lines on the gear layout by appropriate adjustments of the bed on which the staging fixture is mounted. The gear to be checked is then substituted for the setup arbor and rotated until the projected profile meets the profile layout at the form diameter marked on the screen.

The reliability of measurements by projection depends on the magnification ratio, the accuracy of the profile layout, the quality of the projector, and also the skill of the operator. Projection is rarely used in production operations, where measuring machines with automatic recorders are preferred. On the other hand, it offers the best practical approach to the inspection of some gears with other than involute profiles.

Profile Inspection by Measurements with Pins. Where neither profile-measuring machines nor projectors are available, the profile may be inspected by checking the width of the tooth space at various diameters through a series of measurements with three or more pins of different diameters. When using this method, allowance must be made for the fact that the angle between the profile tangent at the pin contact point and the center line of the tooth space is not the same for each pin. Also, measurements over or between sets of two opposite pins will be less conclusive than measurements from the gear center to each of three or more different pins.

A nominal dimension from the gear center to each pin is first computed for the mean tooth thickness of the gear and the difference between each actual measurement and the respective nominal dimension multiplied by the sine of the pressure angle of an involute at the pin center. This angle is always known from the original computation. The difference between the products obtained from each measurement is considered as profile error between respective contact diameters. However, this is true only if both profiles of the tooth space are symmetrical. Otherwise, a positive profile deviation at

* Micrite is a registered trade-mark of the Illinois Tool Works, Chicago, Ill.

one contact point may fully or partially prevent detection of a negative deviation at the opposite contact point of the same pin.

To prevent such false readings, the gear may be rotated 90° until one pin center is level with the gear center. Lack of profile symmetry could then cause the other pins to register different center heights. This second check would also reveal a condition sometimes encountered in form grinding where the profiles are symmetrical with respect to the center line of the space, but the center line is offset against the gear center.

Evaluation of all these conditions is not convenient for the purpose of acceptance inspection. Yet this approach is used to good advantage for process control of form-grinding setups because it permits profile inspection before the grinding wheel is fed to finish depth.

23-2. Measurement of Gear-tooth Spacing. A variety of devices are used to measure gear-tooth spacing. To understand the different kinds of equipment used to

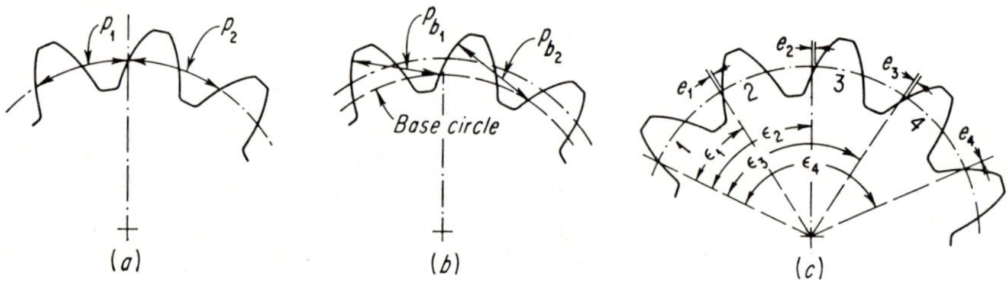

FIG. 23-13. Some kinds of spacing checks. (*a*) Pitch error of p_1 is variation between p_1 and p_2. (*b*) Base pitch error of p_{b_1} and p_{b_2}. (*c*) N = number of teeth. Angle $\epsilon_1 = 360°/N$, $\epsilon_2 = 360°/N \times 2$, $\epsilon_3 = 360°/N \times 3$. e_1 = index error of tooth 1, e_2 = index error of tooth 2. Total index error of gear is difference between largest and smallest value of e. If the smallest e is *negative*, take the *algebraic* difference.

measure spacing, it is first necessary to understand the different ways in which tooth spacing may be evaluated. These are:

1. *Pitch variation.* The difference between one circular pitch and the pitch immediately before or after the pitch in question is called "pitch variation." The maximum allowable pitch variation for a gear is the greatest variation between any two adjacent pitches. Pitch variation is widely used as a measure of gear accuracy.
2. *Pitch variation over a span of teeth.* Sometimes the accuracy of a gear is specified by a tolerance on pitch variation over a span of so many teeth, for instance, over four teeth or five teeth. The variation is determined by comparing the span length over the specified number of teeth with the span over a like number of teeth anywhere on the gear. This measurement is not used so often as the pitch measurement.
3. *Total pitch variation.* This is the difference between the *longest* and *shortest* circular pitch in the whole gear. In the case of power gears, for instance, aircraft power gears, this concept, which was originally proposed by Darle W. Dudley, seems more significant than the pitch variation since it establishes control over the worst possible pitch difference between meshing pairs of adjacent teeth in two mating gears.
4. *Base-pitch variation.* This is the variation in pitch taken along the line of action (see Fig. 23-13). Like pitch variation, base-pitch variation is the difference between a base pitch and an adjacent base pitch. Base-pitch variation is not

used much because it tends to reveal only the base-pitch error of the cutting tool and not the error in spacing produced by the generating machine.

5. *Index variation.* This is the "accumulated" spacing error over any number of teeth. Total index variation may be defined as the greatest out of position of any tooth with respect to any other tooth in the whole gear. Total index variation is important in any case where absolute trueness of angular motion is required. Radar gears, index gears for machine tools, gears of planetary systems, and high-speed power gears must be held to a high order of index accuracy.

A spacing check may be taken from any point on the surface of one tooth to a corresponding point on the surface of another tooth. It is customary, though, to make the spacing check near the center of the face width and approximately at the pitch line.

Since pitch variation and index variation are the two most important of the different kinds of spacing variation, the discussion of spacing-checking devices will be centered around these two kinds of measurements.

Measurement of Pitch Variation by Machine. Several kinds of devices or machines are on the market which will check the pitch of a gear. Figure 23-14 shows the basic elements of a pitch-checking machine. Note that the checking head has a fixed finger, a movable finger attached to an indicating device, a spring device to hold the fingers in contact with the teeth, and a stop to hold the checking head at the same position each time a reading is taken.

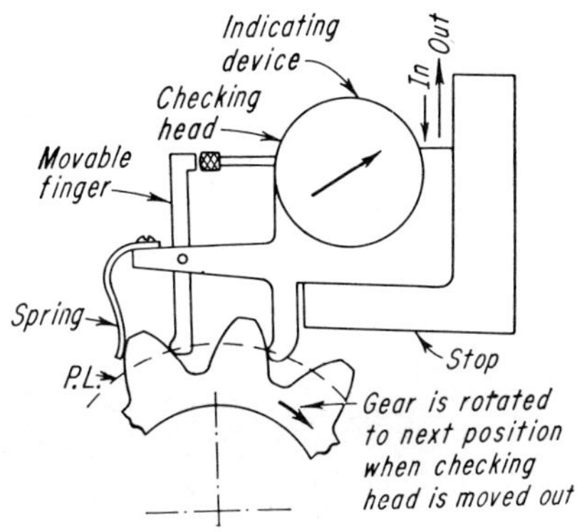

FIG. 23-14. Schematic arrangement of a pitch-checking device.

A gear to be checked by a spacing-checking machine is mounted in the machine on a rotatable axis. Preferably this axis should be the *reference* axis of the gear. If the part is not mounted on its reference axis, the runout of its reference axis with respect to the checking axis must be noted and allowed for in interpreting the spacing check. The gear is checked by moving the head in against the stop, taking a reading, withdrawing the head and rotating to the next set of teeth, moving the head in again, and taking another reading.

Figure 23-15 shows a typical pitch-checking machine. The machine has a headstock and a tailstock for mounting the gears between centers. The gaging head is mounted on a cross slide and contacts corresponding sides of adjacent teeth with one fixed contact member A and one movable contact member B. The contact is maintained by a third, spring-actuated finger C. Helical gears are checked by swiveling the barrel D, positioning the indicating mechanism to the correct helix angle. When checking the circular pitch of spur, helical, or wormgears, the gaging head is adjusted by means of handwheel E so that the gaging fingers contact the gear approximately at the pitch line. When checking base pitch, both fingers must be brought into a vertical tangent plane of the base circle. This is accomplished by gage-block settings. The indicating assembly is carried in a slide F at right angle to the barrel axis of the gaging head to permit adjustment normal to the teeth of spur or helical gears. A hand lever H permits withdrawal of the gaging head by a preset distance, which causes an index pawl I to rotate the gear through one angular pitch for the check on the next tooth.

Figure 23-16 shows a spacing-checking machine for internal gears. Besides checking spacing, this machine has equipment to permit runout checks and checks tooth

FIG. 23-15. Machine for making pitch checks or base-pitch checks on spur or helical gears. (*Courtesy of Michigan Tool Co., Detroit, Mich.*)

FIG. 23-16. Spacing-checking machine for internal gears. (*Courtesy of National Broach & Machine Co., Detroit, Mich.*)

parallelism on spur gears. Figure 23-17 shows an automatic tooth-spacing checker. This machine is a motor-driven instrument which mounts an indexing and sensing finger on a slide which performs a cyclic variable-speed motion along an oval path in a plane normal to the gear axis. Through an electrical recording system, pitch variations are plotted on a chart. The chart is read by noting the peak points for each tooth.

Automatic-type machines like the one just mentioned are coming into widespread use in high-production gear shops such as shops making automobile transmissions. The manually operated machines are used widely in jobbing gear shops doing high-precision work.

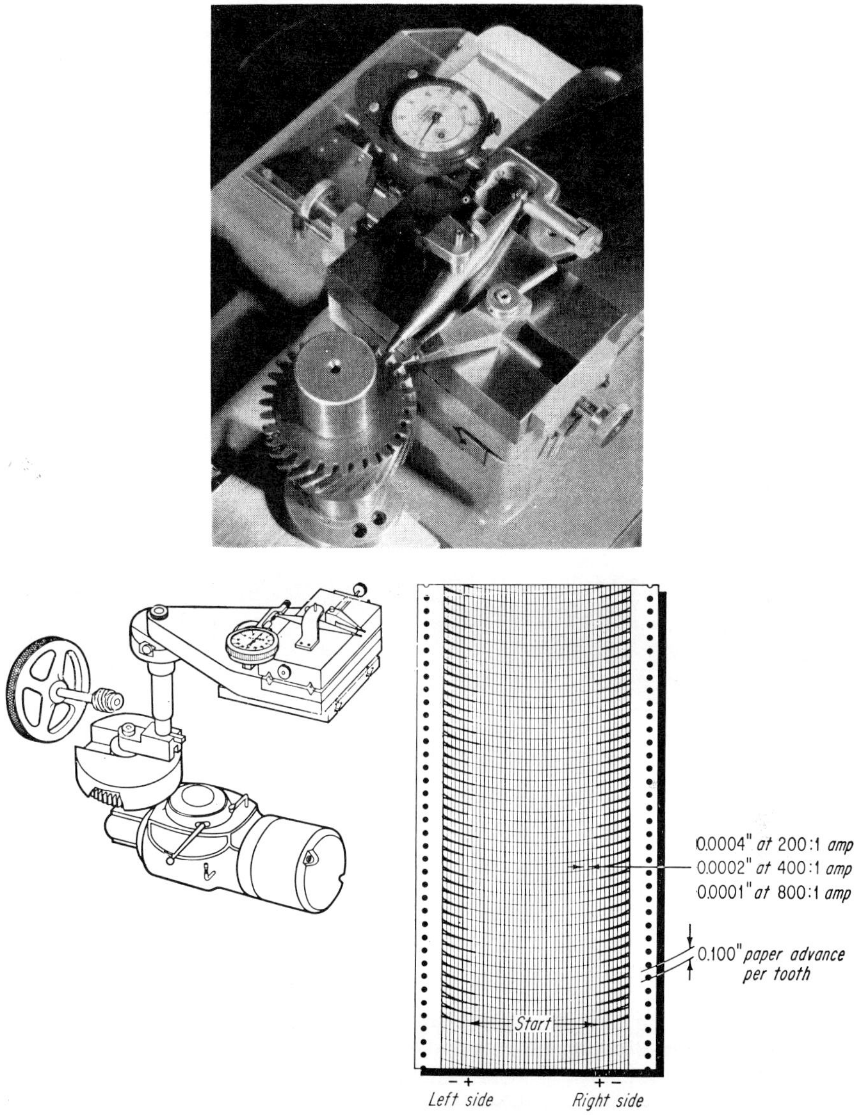

FIG. 23-17. Automatic tooth-spacing comparator. (*Courtesy of Illinois Tool Works, Inc., Chicago, Ill.*)

The most popular size of gear-checking machine on the market will handle gears up to 12″ diameter. Some kinds of checkers are built in 18-″ diameter sizes and 24″-diameter size. Above 24″ size there is very little available except that "custom" designed machines can be obtained for larger work. The larger gears pose problems in tending to be heavy and not produced in quantity like smaller gears.

Most of the manually operated machines can be obtained with either dial indicators or electrical recording equipment to measure the variation. The more popular amplification ratios for electrical recorders are 200:1, 400:1, and 800:1.

Portable Pitch-variation Checkers. A wide variety of portable devices have been used to measure pitch. In many cases gear shops have made their own devices. Some of these have been very skillfully made and used. Long before it was possible to buy any machine equipment to check spacing, a high order of spacing accuracy on gear index wheels was achieved by using portable devices to measure the spacing.

In current gear practice the portable checking device serves several purposes. It is less expensive than a machine and is used in many shops where the kind and volume of gear work does not warrant the expense of more elaborate equipment.

The portable checker is not limited to small-diameter gears. A 48″ gear or a 72″ gear can be readily checked with portable checkers. The lower-cost spacing-checking machines will handle only up to 12″ gears. Even the few larger and more expensive spacing checkers are generally built with a 24″ maximum capacity. The manufacturer of large gears of high precision is faced with either the use of portable spacing checkers or the use of other means to measure tooth spacing indirectly (see Arts. 23-4 and 23-6).

FIG. 23-18. Portable spacing checker. (*Courtesy of the Maag Gear Wheel Co., Zurich, Switzerland.*)

Figure 23-18 shows a portable spacing checker being used on a pinion. It should be noted that the portable checker needs at least two more fingers than the spacing-checking machine. These extra fingers are used to position the device with respect to the gear teeth.

The portable checker needs to be positioned radially by some cylindrical surface. If the outside diameter of the gear is machined very smooth and true and the burr is carefully removed from the gear-tooth corners, the outside diameter makes a good reference surface to mount the spacing checker. If the outside diameter is not good enough, the root-diameter land may be used. The root land is generally cut by the same tool that cuts the sides of the gear teeth. Sometimes the errors in gear-tooth cutting cause variations in the root land. If this happens, there will be some error in a spacing check made by a portable spacing checker set to mount on the root land. Generally the outside diameter is a better mounting surface than the root diameter.

Another practical difficulty in getting high-accuracy checks with a portable instrument is that the device must be hand-held. A certain amount of "human error" may creep in because the device is not held steady as it is moved from pitch to pitch.

Even when a good check is obtained by the portable device, there is a problem of interpretation. The spacing that counts is that relative to the reference axis (usually axis of rotation) of the gear. If the surface used to mount the portable device is not concentric with the reference axis, error is introduced into the check, and this should be taken out of the check by allowing for the known eccentricity of the mounting surface. In actual practice this is a rather difficult mathematical problem, and most inspectors have trouble in correcting their readings to the reference axis. This makes it highly desirable to make all checks of spacing from mounting surfaces that are true with the reference axis.

The technique of taking a spacing check is the same for either a spacing-checking machine or a portable spacing-checking device. The instrument is set on a circular pitch and the distance between checking fingers set to read "zero." The machine is then moved from pitch to pitch and readings recorded. Since the pitch used to zero the machine may not have been dimensionally correct, the readings are not of absolute value but are rather comparative. Since the definition of pitch error is based on a *variation from pitch to pitch* it is quite all right to set to zero on the first pitch.

Table 23-2 shows a sample check taken on a 12-tooth pinion. The maximum pitch variation was 0.0003″. The pitch from tooth 5 to tooth 6 was this much longer than the adjacent pitch from tooth 4 to tooth 5. Even though the other pitch variations were less, the gear is rated by its *worst* pitch variation.

Table 23-2. Sample Pitch-check Readings
(Readings in ten-thousandths of an inch)

Tooth span	Reading	Variation in reading	Corrected (absolute) values*
1–2	0		$-1\frac{1}{2}$
2–3	+2	+2	$+\frac{1}{2}$
3–4	+3	+1	$+1\frac{1}{2}$
4–5	+3	0	$+1\frac{1}{2}$
5–6	+6	+3	$+4\frac{1}{2}$
6–7	+4	−2	$+2\frac{1}{2}$
7–8	+3	−1	$+1\frac{1}{2}$
8–9	+1	−2	$-\frac{1}{2}$
9–10	0	−1	$-1\frac{1}{2}$
10–11	−2	−2	$-3\frac{1}{2}$
11–12	−1	+1	$-2\frac{1}{2}$
12–1	−1	0	$-2\frac{1}{2}$
1–2	0	+1	

Maximum pitch variation = 0.0003″. Total pitch variation = 0.0008.
* The sum of the readings is +18. By subtracting $^{18}/_{12}$ from each reading the "corrected" readings are obtained. These can be compared with Tables 23-2 and 23-3.

Table 23-2 shows the longest pitch reading to be +6. The shortest pitch is −2. This makes the total pitch variation from longest to shortest 0.0008″.

Measurement of Index Variation. There are a variety of ways to measure the index variation of gear teeth. Some of the more popular are:

1. Use dividing-head equipment.
2. Use precision index plate having same number of notches as gear to be checked.
3. Use autocollimator and precision polygon.
4. Make pitch check and then accumulate readings to find index variation.

Angular dividing heads measure the out of position of a spot on the side of a tooth with respect to a true angular setting from the starting point. Figure 23-13 shows how index variation is measured by angular settings.

Table 23-3 shows a sample index check made with a dividing head. The same 12-tooth pinion was used for this example as was used for the example of pitch-check readings shown in Table 23-2. Note that this part has an index variation of 0.0012″ as compared to 0.0003″ pitch variation. The index variation is obtained by taking the algebraic difference between the highest *plus* reading and the lowest *minus* reading.

Table 23-3. Sample Index Check with Dividing Head
(All readings are in ten-thousandths of an inch)

Tooth	Angle, deg	Index reading
1	0	0
2	30	−1½
3	60	−1
4	90	+½
5	120	+2
6	150	+6½
7	180	+9
8	210	+10½
9	240	+10
10	270	+8½
11	300	+5
12	330	+2½
1	360	0

Maximum index variation is 0.0012″.

Figure 23-19 shows a tyipcal dividing head used in precision gear-checking work. This head has a precision spindle mounted on patented ball bearings which are guaranteed not to run out more than 25 millionths of an inch (total indicator reading). The spindle carries a master disk with diamond-scribed marks for each degree. These marks are 0.0002″ wide and 0.0002″ deep.

FIG. 23-19. Optical dividing head. (*Courtesy of Vinco Corp., Detroit, Mich.*)

Adjacent to the master disk is a stationary reference disk of equal diameter. This disk has seven lines spaced in increments of 10 minutes. The area of the reference disk and the main disk is visible through a 40-power microscope A. The field of view is shown in the upper left-hand corner of Fig. 23-19.

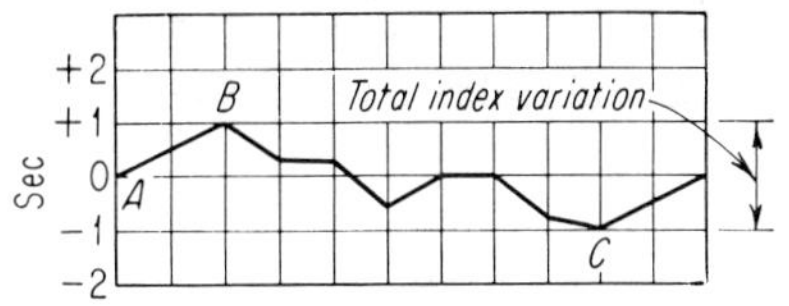

FIG. 23-20. Typical dividing head index-variation graph.

Certain numbers of gear teeth will require angular settings that do not come out to even degrees—or even to the last 10 minutes of arc. The dividing head is provided with a micrometer thimble E for making fine settings. The two sets of graduations at F permit settings to be adjusted

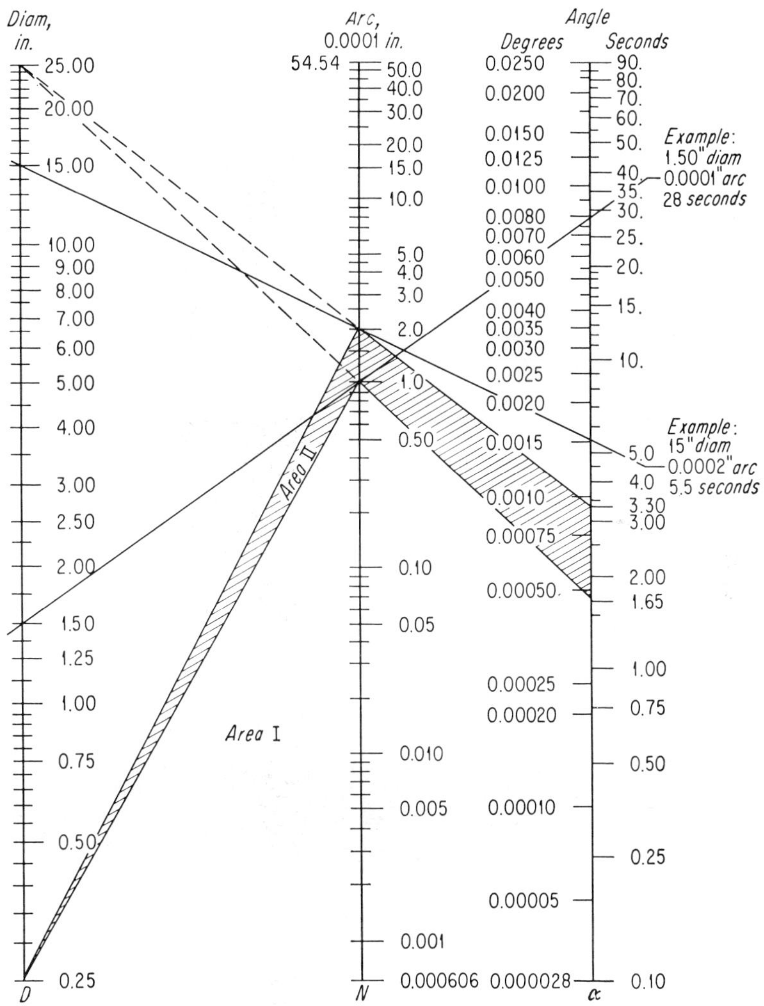

FIG. 23-21. Linear-to-angular conversion nomogram. *(Courtesy of Equitable Engineering Co., Royal Oak, Mich.)*

to the last second of arc. The mounting arbor of the gear to be checked is centered on *I* and driven by the drive assembly *H*.

The best dividing heads on the market are usually guaranteed to 2 seconds of arc. Some of the smaller and less expensive dividing heads are guaranteed to 15 seconds of

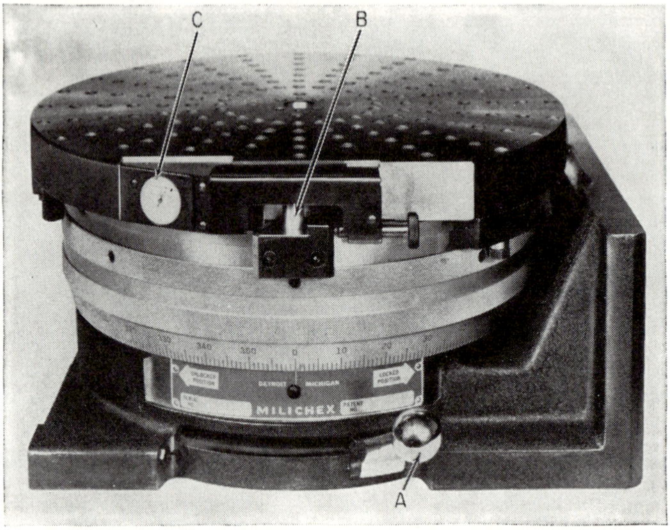

FIG. 23-22. Indexing head. (*Courtesy of Michigan Tool Co., Detroit, Mich*)

arc. There are a few very special dividing heads used in radar gear work, steam-turbine gear work, and astronomy work that are accurate to better than 1 second of arc.

The accuracy specification of a dividing head, index plate, or gear should always be expressed in terms of maximum deviation between two positions, not by a bilateral tolerance like ±2 seconds of arc or ±0.0003″. Figure 23-20 shows an error graph with 2 seconds total index variation. Note that this graph could be interpreted as ±1 second; +0, −2 seconds; or +2, −0 seconds depending on whether the original setting was made at point *A*, *B*, or *C*!

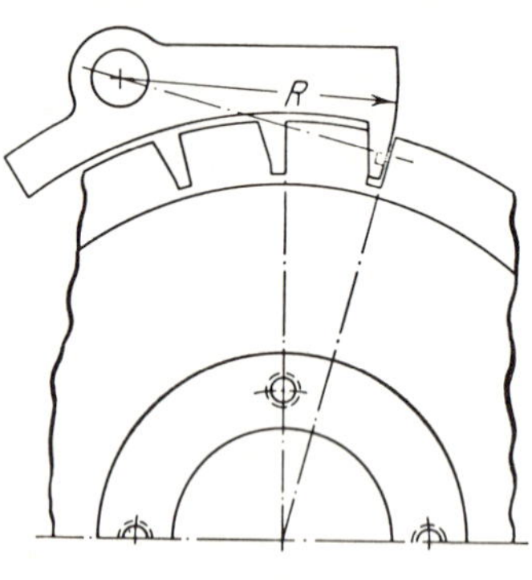

FIG. 23-23. Typical index plate. (*Courtesy of Vinco Corp., Detroit, Mich.*)

The accuracy of a dividing head may be verified by checking a large gear and then rotating the gear through some angle (like 90°) and taking another check. As the gear is shifted with respect to the dividing head, the change in the index readings of the gear reveals the error in the dividing head. Of course, the gear must not shift in its centering and the readings have to be taken with highest accuracy since the error in the dividing head is usually very small.

In judging accuracy requirements of dividing equipment, it is necessary to consider the size of the gear to be checked and the accuracy of the indicating equipment which takes the reading on the gear teeth. For instance, a 6″ gear to be checked with indicating equipment which would read to 0.00005″ would not require a 2-second dividing head. A head of this accuracy would have an error of only 29 millionths at the gear pitch line. Figure 23-21 shows a nomogram which is very helpful in establishing the relation between angular tolerances, diameters, and linear deviations.

Figure 23-22 shows an indexing table that was developed for high-precision gear work. This table has two plates with 360 radial notches at their engaging faces. Lever A actuates a cam to separate the plates for indexing. Pin B can be used with a tangent bar to index for fractions of a degree. A dial indicator C is used to help obtain uniform pressure between the tangent bar and the gage block used to set the angle. This indexing head can be used in the same manner that a dividing head is used. Indexing heads of this type can also be built to have the same number of notches as the

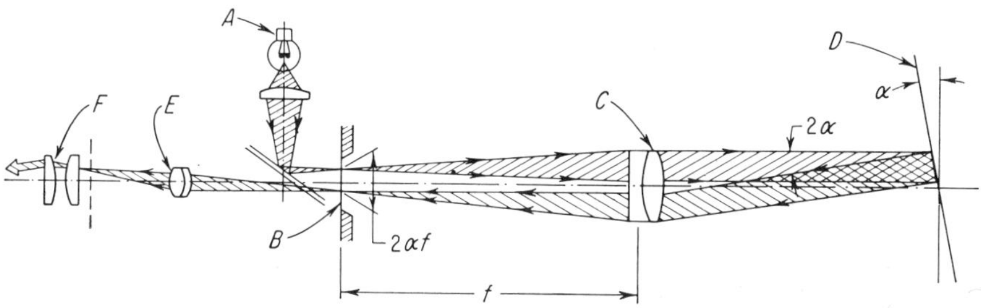

FIG. 23-24. Light-beam path of an autocollimator.

gear to be checked. This eliminates the tangent-bar settings. The indexing head can also be built with three plates for *differential* indexing.

Simple index plates may be used where the plate has the same number of notches as the gear has teeth. Figure 23-23 shows a typical style of precision index-plate notch.

One of the optical devices that is often used for high-precision index measurements is the autocollimator and precision polygon. The precision polygon may be made of glass or steel. If it has 12 sides the angle between each side would be 360/12°. The autocollimator lens arrangement sends a beam of light to the side of the polygon and returns the beam back upon itself. The polygon and autocollimator permit a rotatable table to be indexed with very high precision.

This system of dividing has the advantages that the readings are not affected by eccentric mounting of the polygon or tilt of its axis. The system has the disadvantage that polygons with many sides become very expensive.

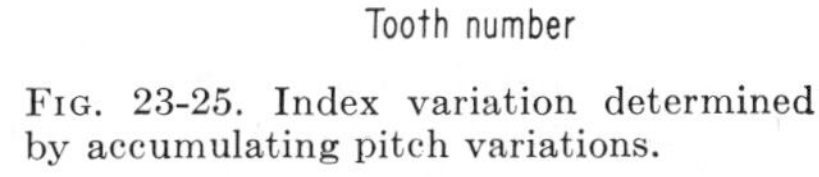

FIG. 23-25. Index variation determined by accumulating pitch variations.

Figure 23-24 shows the lens system of an autocollimator. Imagine D to be a side of a polygon. A beam of light is reflected by D through an objective lens C which forms an image of the target wires B in the focal plane of the instrument. If the polygon has been set perpendicular to the autocollimator, the image of the wires will coincide with the wires themselves. If there is an error in setting, the image of the target wires is shifted by the distance $2\alpha f$ where f is the focal length of the lens. The image is magnified by the microscope E and the error in setting is measured by the eyepiece F. The light source is A.

With the best equipment and technique of use, an autocollimator can read angles to 0.1 second. The high accuracy possibilities of this device make it very useful in calibrating dividing heads, measuring gears* of the most extreme accuracy, and checking out the accuracy of certain highly critical radar gear drives.

An old-time method of obtaining index measurements on a gear is that of making a pitch check and then "accumulating" the error. Figure 23-25 shows how this can be done with the pitch readings from the 12-tooth pinion shown in Table 23-2. The first pitch reading was 0. This is plotted as 0 in Fig. 23-25. The second reading was +2 and the third +3. In Fig. 23-25 the +2 is added to the +3 and plotted as +5. In similar manner each reading is added on to the preceding total.

The readings in Fig. 23-25 accumulate to a value of +0.0018″. A line is drawn from the start of the plot to the last plotted point. The maximum index variation is the vertical distance from this line to plotted points farthest from the line. This happens to come between tooth 7 and tooth 1.

The reason for the line is that the pitch setting of the checker was wrong by an amount equal to 0.0018/12 or 0.00015″. The sloping line corrects for the fact that the plotted points did not close to zero. It is interesting to compare the plot of Fig. 23-25 with Table 23-3 for a dividing-head check of the same gear.

The method of accumulating pitch readings has the serious disadvantage that small errors in reading each pitch tend to accumulate and throw a serious error into the index variation. If the number of teeth is not large, the method works quite well. As the number of teeth increases, the method becomes poorer and poorer. Table 23-4 gives a rough indication of how good this method is in general practice.

Table 23-4. Relative Accuracy of Method of Accumulating Pitch Readings to Determine Index Accuracy

No. of pitch readings accumulated	*Relative accuracy of index variation determined*
10	Relatively good
20	Fair
40	Poor
80	Usually no good
160	Worthless

Where dividing equipment is not available, a large gear can still be checked fairly well by accumulating pitch readings. Suppose a gear had 100 teeth. A direct accumulation of 100 pitches would be very poor. A good strategy, though, would be to set a span measurement to cover 10 pitches. The gear could be spanned with 10 readings and these could be accumulated to give the index position of 10 points around the gear. Then a spacing check of 10 pitch measurements could be made between each of the span points. Each block of 10 pitches could be accumulated to obtain the local index variations. In this way a reasonably good index measurement of the whole gear could be obtained.

23-3. Measurement of Lead. The normal definition of "lead" is that the *lead* is the axial advance of a screw thread in one turn (360°). As applied to gears, the check of "lead" has a different connotation. A lead check of a gear is a check taken to determine the alignment of the tooth across the face width. If each tooth of a mating pair is positioned properly across the face width, then the load will be distributed uniformly across the face width.

In lead-checking helical gears, the measurement of lead variation is usually taken *normal to the tooth surface*. This makes it clear that the so-called lead check of a gear is only indirectly related to the true lead of the part. In the case of a spur gear (a

* Generally it is not possible to check every tooth. For instance, an eight-sided polygon could check every *tenth* tooth on an 80-tooth gear.

helical gear with 0° helix angle), the lead is *infinite*. A lead check of a spur gear is in reality a check of whether or not the tooth is aligned *parallel with the reference axis* of the part. The lead variation of a spur gear is measured normal to the tooth surface. This direction is actually 90° away from the lead direction of a spur gear. Figure 23-26 illustrates how lead checks are made on gears.

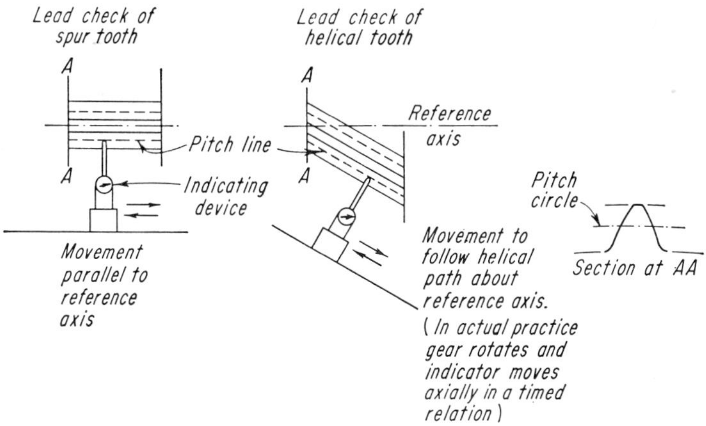

FIG. 23-26. Schematic method for making a lead check of a gear tooth.

The lead accuracy of gears may be determined by several methods. The most widely used methods are:

1. Check lead by following the helix.
2. Check lead by axial-pitch measurements.
3. Determine lead accuracy by contact checks with a mating gear or a master gear.

Helix-checking Machines. There are a variety of helix-checking machines on the market. These machines mount the gear to be checked between centers. The check is made by rotating the gear and moving a carriage in an axial direction at the right rate of lead advance per degree of rotation for the gear to be checked. The carriage carries an indicating means which is set to read deviations normal to the tooth surface. The deviation reading may be made by dial indicator, or an electrical pickup may be used and the variation recorded on chart paper just as involute checkers chart involute variation.

Figure 23-27 shows a typical helix-checking machine. Note gear *A*, the carriage *B* that moves axially, and the electronic pickup *C* that measures the lead variation.

Helix-checking machines for a gear of low helix angle (long lead) generally have a disk that rolls against a straight edge. The tangential movement of the straight edge is translated into an axial movement by sine bar or similar device. In some cases these machines are built in combination with involute checkers and the roll of the base-circle disk is used to create both an involute curve for profile checking and a helix for lead checking. Other machines employ a master lead bar and follower to obtain the helical motion.

An ordinary engine lathe has change gearing and a cross slide that can be driven axially by a lead screw in timed relation to follow a helical path. Helix-checking machines for gears of high helix angle (short lead) and for worms may be made on the change-gearing and lead-screw principle of a lathe. Figure 23-62 shows such a machine.

Like involute-checking machines, the more popular and lower-cost machines are made to handle gears up to 12″ pitch diameter. Some models are built in sizes going up to 24″ diameter capacity. A very few have been built in the size range of 36 to 48″ capacity.

Spur gears have infinite lead (no helix angle). They can be given lead checks with very simple equipment. All that is needed is a cross slide that will move axially and

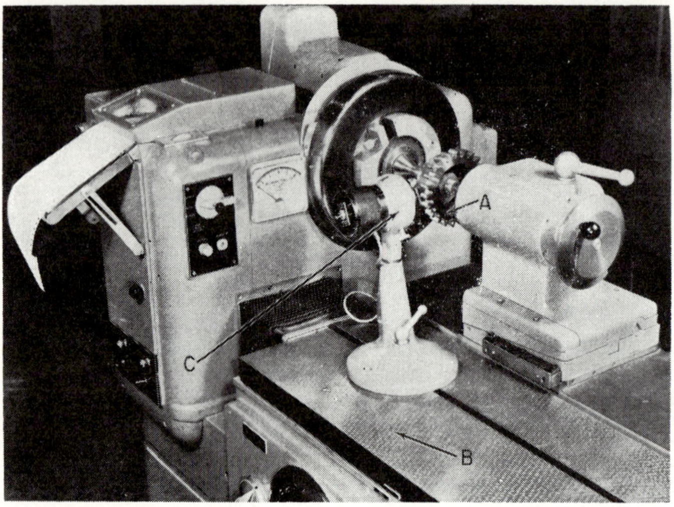

Fig. 23-27. **Lead-checking machine.** (*Courtesy of Michigan Tool Co., Detroit, Mich.*)

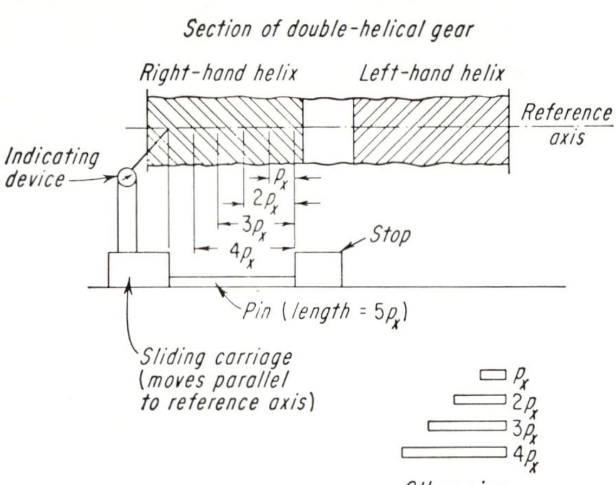

Fig. 23-28. Schematic method for making an axial-pitch check.

carry an indicating mechanism parallel to the reference axis of the gear. The spur gear can be locked against rotation while a tooth is getting a lead check.

Axial-pitch Checking. Large helical gears can have their lead accuracy determined rather well by an *axial-pitch check*. Typical herringbone gears having from 5″ to as much as 40″ face width per helix may have from 4 to as many as 15 axial pitches in the face width. A cross slide can be moved parallel to the reference axis of the gear in increments of an axial pitch. The axial-pitch distances may be set precisely by gage blocks, or a series of pins may be ground whose lengths are 1 axial pitch, 2 axial pitches, etc. Figure 23-28 shows the schematic setup of an axial-pitch check.*

The axial-pitch check has the advantage of being readily adaptable to large-diameter wide-face-width gears. Herringbone gears of even 200″ diameter and having double helices of 40″ face width per helix can readily be given an axial-pitch check.

When a gear and pinion contact, their band of contact is an *axial* band. A helix check that may spiral around the pinion as much as 20 to 60° will tend to pick up the effect of concentricity variation and will not truly measure the ability of the pinion to contact properly across a wide face width. The axial-pitch check is a true measurement in the zone where contact actually takes place. Any concentricity variation has only a negligible effect on the axial-pitch check. This means that the axial-pitch check is a more suitable check for wide-face helical gears than a check made by following the helical path.

Contact Checking. Any given pair of gears that are to run together may have their effective lead accuracy determined by a simple contact check of the two parts. This may be done in a checking stand where adjustments can be made to get the two reference axes exactly parallel to each other. The check may be made in the casing that the gears are to run in. If it is made in the casing, errors in location of casing bores will show up in the contact check; a poor contact may be due to gear-cutting error or due to error machining the casing.

Where interchangeability is required, a gear may be contacted with a master in a checking stand. Pairs of gears that properly fit their respective masters can be assumed to be capable of fitting each other (provided that the casing is good).

Contact checks are often made by putting a thin layer of "red lead" on one member and a thin layer of "prussian blue" on the other member. The parts are then rolled together and the blue makes a clear marking on the red. The thickness of the marking paste enters directly into the check. If the paste is thinned down with solvent to an extreme, a lack of contact marking may happen with a gap between the surfaces of as little as 0.0001″! On the other hand, thick paste can be used, and contact may be indicated when a separation of as much as 0.0005″ occurs. A proper application of the colors should show contact at 0.0002″ and a lack of contact at 0.0003″.

Some manufacturers will put a flash of copper plate or a flash of silver on one member. Then the gears are run under power for 15 min to ½ hr in their casing. This will show up a clear contact marking. This kind of test gives very clear-cut indications and is easier to control than red and blue colors. It has the hazard, though, that a bad pinion might have a "wobble" type of lead error. After a few minutes of running the wobbling pinion might mark up the gear all across its face width and give the illusion that the gear was fitting the pinion all right. If the gear had a little spacing error, it might in turn mark the pinion so that a wobble type of contact on the pinion could not be seen.

Contact checks on gears require the least equipment. Gears of low cost and low accuracy are frequently checked for lead accuracy by this method.

The contact check has the disadvantage that the gap between the teeth when no contact exists is not measured. Some manufacturers have developed unique and proprietary techniques to measure the tiny gap that may exist between contacting gear teeth. Measurement of gap as little as 0.0001″ can be made.

23-4. Measurement of Eccentricity. Gears that are eccentric tend to have a vibration once per revolution. If the backlash is rather small, a badly eccentric gear

* Some lead-checking machines permit disconnecting the carriage and positioning it axially by gage blocks to perform this type of check.

may jam the teeth when the "high point" goes through mesh and thereby cause an abrupt gear failure.*

The eccentricity of a gear is usually measured in terms of "runout variation." The runout is measured in a radial direction and is taken as the full indicator variation between the highest and lowest readings going all the way around the gear. If the gear is primarily eccentric, the highest and lowest readings should be 180° apart. Local spacing errors may make the highest and lowest readings come closer together. The runout of the gear is taken as the difference between the highest and lowest readings regardless of where they occur on the gear.

Runout checks are widely used in the gear trade. Even low-accuracy low-cost gears that are not given profile, spacing, or lead checks may still be given a runout check, a contact check, and a tooth-thickness check.

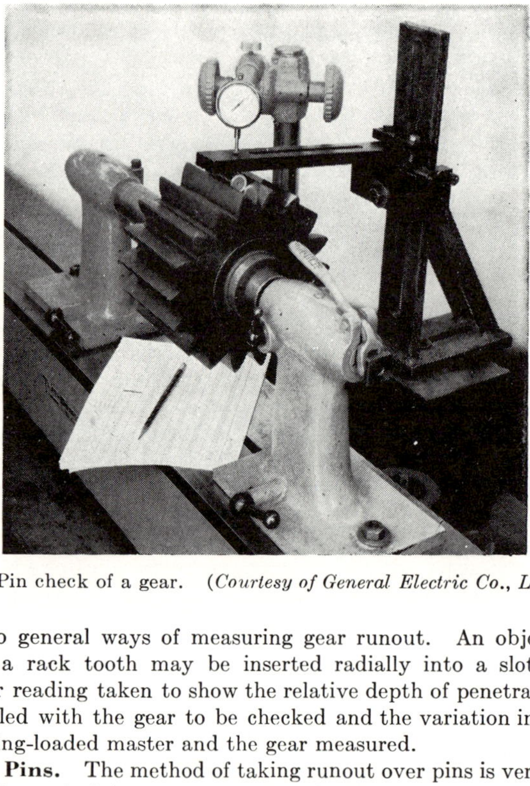

FIG. 23-29. Pin check of a gear. (*Courtesy of General Electric Co., Lynn, Mass.*)

There are two general ways of measuring gear runout. An object like a pin, a cone point, or a rack tooth may be inserted radially into a slot between teeth and an indicator reading taken to show the relative depth of penetration. A master gear may be rolled with the gear to be checked and the variation in center distance between the spring-loaded master and the gear measured.

Runout over Pins. The method of taking runout over pins is very popular in the gear trade. This method is very accurate and requires a minimum of equipment. The method can be applied to either spur or helical gears and it is not limited by gear size. Gears up to 200″ in diameter are checked by pins just as a gear 1″ in diameter is checked.

The runout readings over pins may be interpreted for both total runout and tooth-to-tooth runout. The tooth-to-tooth runout is quite similar to a pitch check (see Art. 23-2). Both give good indications of spacing accuracy of adjacent teeth. It has been found that generated gear teeth (hobbed, shaped, shaved, and generated ground) will have good pitch accuracy if the tooth-to-tooth pin check is good. The only

* See Chap. 6 for engineering calculations on this problem.

FIG. 23-30. Checking a small gear by rolling with a master. *(Courtesy of Vinco Corp., Detroit, Mich.)*

FIG. 23-31. Checking a large gear by rolling with a master. *(Courtesy of General Electric Co., Lynn, Mass.)*

exceptions to this are milled gear teeth and form-ground gear teeth. Here spacing can be bad even when the depth of the machined slot is correct.

Figure 23-29 shows a pin check being taken on a helical gear. Since the pin tends to roll slightly in a helical notch, it is necessary to roll the gear and pin under a flat plate and take the reading as the plate reaches its highest point. A simpler setup can be used to pin-check spur gears.

Runout with a Master. A spring-loaded master will indicate the runout of the gear being checked. Figure 23-35 shows the schematic setup of this check. Figure 23-30 shows a checking machine being used on a small gear. Figure 23-31 shows a large gear being checked with a portable device.

There are some problems in interpreting the runout with a master gear. Local tooth-to-tooth variations give width to the curve and make a larger difference between the high and low points of the curve (see Fig. 23-32). To be technically correct the runout should be read as the "mean" of the curve.

Another problem lies in the runout of the master gear itself. Frequently the error of the master is as much as one-third to one-fifth the variation of the gear being checked. This means that the variation of the master must be allowed for in interpreting the check.

Suppose a 30-tooth master having a runout of 0.0002″ is used to check a 150-tooth gear having a runout of 0.001″. The variation curve of the gear would have one peak

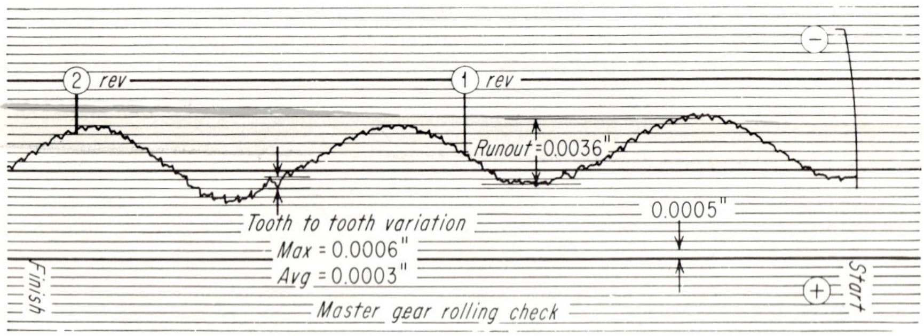

FIG. 23-32. Typical chart obtained by rolling a gear with a master.

in 150 teeth. The superimposed variation of the master would have five peaks of 0.0002″ on top of the true curve of the gear. In this case a little deduction would show how to correct the curve and get the true gear curve.

If the gear had had only 30 teeth, the problem would be more difficult. In this case, the true variation of the gear could be obtained by taking three checks and reindexing the master 10 teeth with the gear each time. This would tend to make the master-gear variation come both additive and subtractive to the gear variation. Properly taken checks of this nature and a proper analysis of the results make it possible to segregate out the true runout of the gear from the runout of the master.

23-5. Measurement of Tooth Thickness. Almost all gears require a tooth-thickness measurement. Even crude gears that receive none of the inspection measurements mentioned earlier in this chapter will have a tooth-thickness check taken. High-accuracy gears may have tooth-thickness checks taken under controlled measuring pressure and temperature with high-precision equipment.

If the teeth on a gear are all about the same thickness, and the application is not too critical, any one tooth may be measured for tooth thickness. On critical aircraft gears, it is common practice to measure four teeth 90° apart and average the readings. In some very critical work, it is necessary to measure all the teeth in the gear and have *every* tooth fall within the allowable size limits for the part. This latter case might apply to radar gears where the allowed variation in tooth thickness might be as little as 0.0002″. A measurement from center over one pin might be used.

Measurement with Tooth Calipers. A widely used instrument in gear jobbing shops is the vernier tooth calipers. Figure 23-33 shows a gear being checked by calipers. This device measures *chordal* distances. It is necessary to have a *chordal* *addendum* and a *chordal tooth thickness*. The instrument is set for the chordal addendum and then the chordal tooth thickness is measured. (Article 7-12 shows how to calculate chordal dimensions.)

With moderate care vernier calipers are accurate to ±0.0005″. With extreme care and a well-calibrated instrument somewhat closer readings can be obtained.

Measurement by Dimension over Pins. High-precision* gears made in jobbing shops are generally checked for tooth thickness by dimensions over pins. Figure 23-34 shows a measurement of this type being taken.

The dimension-over-pins reading is mathematically related to the tooth thickness. Articles 7-9 and 7-10 show how to convert specified tooth-thickness limits into dimension-over-pins limits (or dimension over balls).

A reading of dimension over pins is in effect an average of two tooth thicknesses 180° apart. A second dimension-over-pins reading taken 90° from the first reading

FIG. 23-33. Checking gear-tooth thickness with vernier calipers. (*Courtesy of General Electric Co., Lynn, Mass.*)

FIG. 23-34. Checking tooth thickness by measuring over pins. (*Courtesy of General Electric Co., Lynn, Mass.*)

can be averaged with the first reading to give the result of four averaged tooth-thickness readings 90° apart.

The dimension-over-pins method is very accurate. With moderate care the tooth thickness determined is accurate within ±0.0001″. The method is not limited by size until the micrometer required becomes too heavy too handle and too responsive to small temperature variations. Gears up to about 48″ in diameter can be handled quite well by this technique.

Measurement with a Master. High-precision gears that are mass-produced are frequently checked for tooth thickness by a master gear. The center distance established by a known master and a gear indicates the functional tooth thickness of the gear. Figure 23-35 shows how a minimum and maximum setup center distance† can be established. Any gear whose curve falls between the two limit lines has correct tooth thickness.

The check with a master gear has the advantages that all teeth in the gear are checked and the check is made very rapidly. In fact, some of the machines which automatically cut or finish gear teeth are equipped to check each gear automatically with a master and reject any that are not within limits.

With moderate care and good masters, tooth thickness can be determined by the master-gear method within ±0.0002″.

* This method is not limited to high-precision gears.

† For method of calculation see Art. 6-1 and Eqs. (6-5) and (6-9).

Span Measurement. Tooth thickness can be determined by measuring a span over several teeth. Article 7-11 shows how to calculate the span distances corresponding to given tooth thicknesses. Figure 7-14 shows the checking setup.

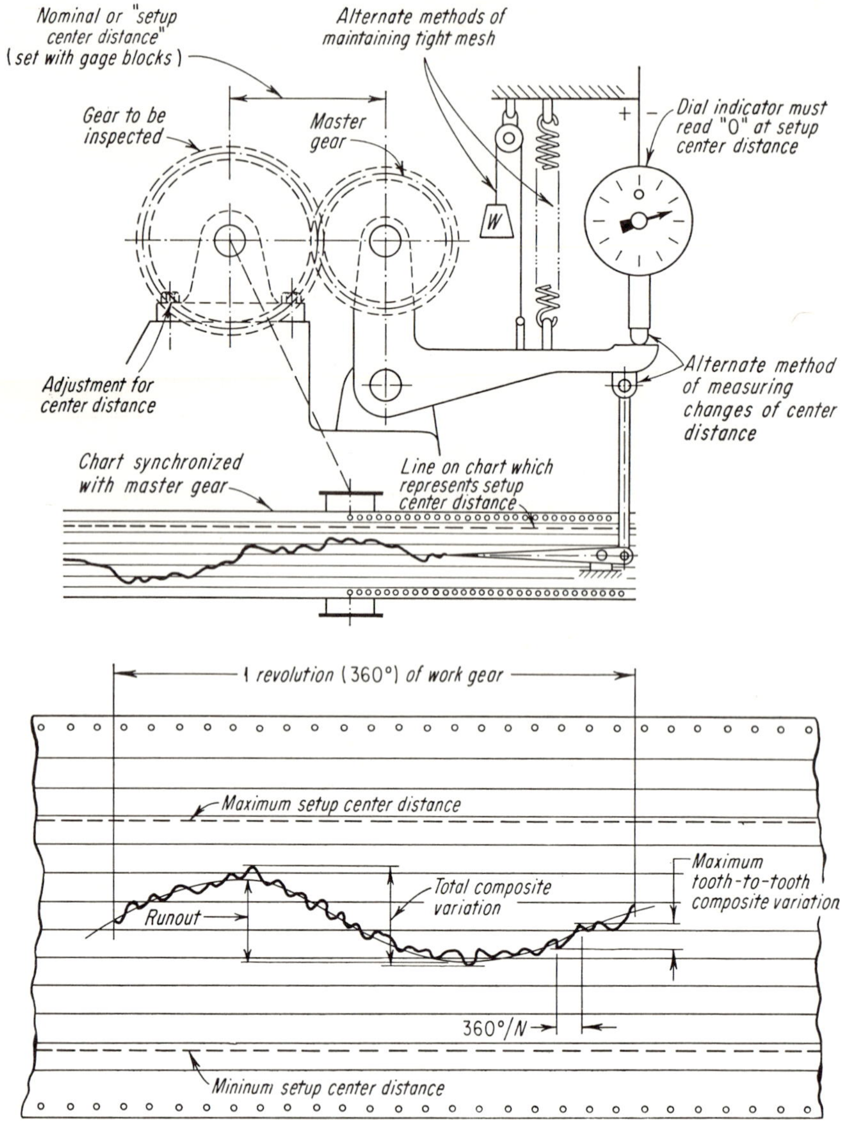

FIG. 23-35. Composite gear-checking setup and composite gear chart.

If the span measurement is taken fairly high on the tooth, the reading can be made by a micrometer instead of vernier calipers. Micrometer-read span measurements will indicate tooth thickness within about $\pm 0.0002''$.

The span measurement (with a micrometer) is more accurate than a single-tooth measurement with vernier calipers. This method can be used on very large gears

(gears over 48" diameter) where it becomes awkward to use the dimension-over-pins method. It is probable that this method will find increasing use in the future.

23-6. Composite Gear Checking. Fine-pitch gears are largely checked by the "composite" method. The composite method is used quite extensively on medium-pitch radar gears and on medium-pitch gears mass-produced for automobile transmissions. The composite check is used widely on all sizes of gears made in quantity.

Figure 23-35 shows the schematic setup for a composite check. The check itself is a check of a gear with a spring-loaded (or weight-loaded) master. The name "composite" comes from the interpretation of the check. Two determinations are made from the plotted curve. These are

Total composite variation
Maximum tooth-to-tooth composite variation

The tooth-to-tooth composite variation is a local variation that is influenced by profile variations, pitch variations, and lead variations in the length of engagement. Even surface finish may have a slight effect on the local tooth-to-tooth variation. Since the variation is affected by so many things, it was called—quite logically—a *composite* variation.

The total composite variation is made up of the runout of the part plus local variations corresponding to each tooth-engagement cycle.

Both the tooth-to-tooth composite reading and the total composite reading are affected by variations in the master gear. It is possible to correct the total composite for variation in the master just as it is possible to correct runout readings for variation in the master (see Art. 23-4). There is *no handy way* to correct tooth-to-tooth composite variation for variation in the master. The general trend in the gear trade is to blame all the tooth-to-tooth variation observed on the gear being checked. This puts the gear maker under strong pressure to get really good masters. The variation he permits in the master is going to be charged against him in the tooth-to-tooth composite readings of his product!

The tooth-to-tooth composite readings can be analyzed to give reasonably good indications of profile and spacing variations. A *uniform* variation as each tooth rolls through mesh indicates profile variation. A *sudden* jump indicates a pitch variation.

Since spacing-checking machines (see Art. 23-2) are limited in the size of gear they can check, it is common practice to check large gears by the composite method and use this as a control over tooth-to-tooth spacing. In high production of precision gears it is common practice to check *all gears* by the composite method. Any that look questionable on tooth-to-tooth variations are given additional involute-profile and pitch-spacing checks to see if they are really within limits. Since the composite check is fast, the practice just described is a relatively low cost way of ensuring high quality in all parts.

The composite type of check has some definite limitations. High-capacity gears with design profile modification cannot get the profile modification checked in a composite type of check. Gears that require very close pitch accuracy such as high-speed turbine gears, aircraft power gears, and rocket gears do not get a close enough pitch check by the composite method—particularly if the parts are right up to or slightly beyond allowed limits. The composite check will reflect lead variations only if the master gear has as much or more face width than the production gear. In this case the master can be blued in with the gear to give an indication of lead accuracy.

Composite-type checks can be made on almost any size of gear (see Figs. 23-30 and 23-31).

While the composite check is usually made with a round gear it can be made with a *straight rack*. One checker uses a rack as a master for composite checking.

23-7. Master Gears. Master gears are usually thought of as gears of extreme accuracy level, but are better defined as gages to check the meshing action of production gears. This is usually not recognized because most mechanical gages are associated with static measurements rather than having the form of machine elements used in a functional check involving machine motion.

Size of Master Gears. Most of the master gears in use have a bore size of 1.2500″. A few of the coarser-pitch masters have bore sizes of 1.7500 or 2.5000″. Frequently wide master gears are made to contact the full face width of the production gear.

Master gears will vary in pitch diameter. In most cases they range from about 2 to 5″. A few masters are made in sizes as large as 6 or 8″. It is desirable for a master to have from about 30 to 50 teeth. In the finer pitches it is possible to have

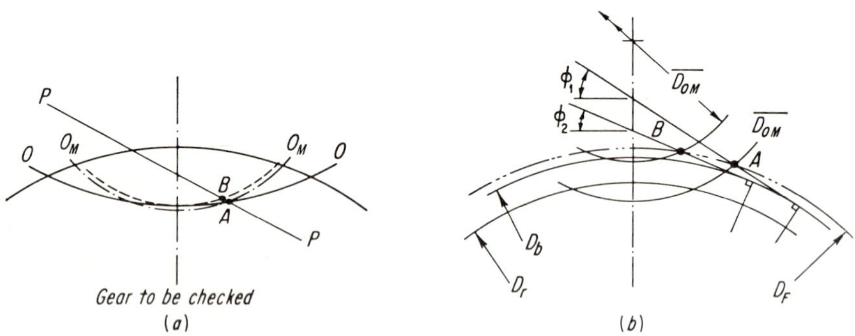

FIG. 23-36. Master-gear-meshing diagrams. (*a*) Depth of contact and checking depth. (*b*) Geometry of low-pressure-angle master gear.

plenty of teeth and therefore have a fairly straight-sided profile on the master gear. With coarse pitches this is not possible. At 2 diametral pitch, for example, an 8″ master would have only 16 teeth. A still larger master would be heavy to handle and rather costly. The coarse-pitch master tends to have a low number of teeth, tooth action near the base circle, and the high curvature of tooth form that is characteristic of low tooth numbers. The things just mentioned have to be tolerated to keep the cost and weight of the master within reasonable limits.

Master-gear-design Problems. Any gear-shop man buying master gears would like to buy one master of a given pitch and use this one master to check all gears of that pitch that are in production. There are several practical problems that make a *single* master gear quite inadequate to check a variety of gears of a given pitch. These problems are:

1. Meshing depth varies with numbers of teeth in the mating gear.
2. Meshing depth varies with the tooth thickness of the gear being checked.
3. A master with a small number of teeth may foul the root fillet of the gear being checked even though the gear may have a root fillet that will not foul the tip of the mating pinion.

About the only way to calibrate a master gear precisely is to have a set of three identical masters and roll these masters with each other in different combinations so that the variation in each one can ultimately be deduced.

Figure 23-36*a* illustrates the problem of numbers of teeth and meshing depth. In this case the gear to be checked meshes with another gear having outside circle *OO*. The lowest point of involute tooth action is at *A*. A master with a standard addendum and fewer teeth than the mating gear has involute action only down to point B. A special master can be designed with longer addendum and an outside circle O_mO_m. This master will mesh down to the desired point A.

The method of calculating a special outside diameter for a master gear is given by the following equation:

$$D_{om} = \sqrt{D_{bm}^2 + (2C \sin \phi' - \sqrt{D_f^2 - D_b^2})^2} \qquad (23\text{-}15)$$

where D_{bm} = base-circle diameter of master
C = tight-mesh center distance
ϕ' = transverse operating pressure angle [see Eq. (6-9)]
D_f = form diameter of gear to be checked (see Art. 7-4 for form-diameter calculations)
D_b = base circle of gear to be checked

The best way to make sure the meshing depth of the master is just right is to buy a special master for each gear to be checked. A compromise method is to buy a wide-face master with a "stepped *OD*" and check narrow gears by using the part of the master having the right outside diameter. A less desirable approach is to buy a relatively standard proportion master and accept the fact that the master will miss a small amount of involute profile on most of the gears being checked (or may at times check deeper than required).

Variations in tooth thickness—if large—quite obviously will affect the depth of contact in tight mesh.

Figure 23-36*b* shows how the problem of a master fouling the root fillet may be solved by a special design. In this case the form diameter of the gear to be checked is D_f. A regular master with enough outside diameter to get down to point A has an outside-diameter circle D_{om}. This circle strikes the root diameter D_r! Obviously this is an impossible situation in which to make a successful check. The solution lies in a special master having a smaller operating pressure angle ϕ_2. This master has an outside circle D_{om} and it has involute action to point B. Since point B lies on circle D_f, this master meshes just as deep as it needs to. It clears the root circle by a substantial amount (it should also be able to clear the *root fillet*). Special masters made to a low pressure angle are called "thin-tooth masters" since the low operating pressure angle is obtained by making the teeth thin at the pitch line.

When a set of three masters are bought, composite checks can be taken of master 1 with master 2, master 2 with master 3, and master 1 with master 3. Reindexing may also be used. The resulting charts make it possible to deduce the relative amount of total composite variation that is in each master. After the masters are calibrated for variation, it is customary to put one master into service checking gear production; one master is used to double-check doubtful results, and the third is held in reserve until the first master becomes too worn for further service. The first master is normally checked for wear by the third master, which is not in service and should have had no wear.

Accuracy of Master Gears. It is very desirable that a "master" gear should be perfect. Article 23-6 points out how the error in the master may be charged against the gear being checked. Master gears are made on machines similar to those being used for the highest quality of production gears. This means that the highest-accuracy production gears will always have variations which are only slightly greater than that of the best masters that can be purchased.

So far only one trade standard[4,*] on master-gear accuracy has been established. This standard specifies a level of accuracy that is satisfactory for most gear work, but the level is not good enough for the highest-quality gears being produced. Table 23-5 shows data taken from this standard.

* Superscript numbers refer to references at the end of the chapter.

Table 23-5. Arbor-type Master-gear Tolerances, Coarse-pitch Series

Grade A	Diam. pitch*		
	4 through 5.999	Over 5.999 through 11.999	Over 11.999 through 19.999
Active involute profile (TIV)	0.0003	0.0002	0.0002
Measurement over pins:			
13 through 19° pressure angle	±0.0014	±0.0011	±0.0007
Over 19° pressure angle	±0.0010	±0.0008	±0.0005
	Pitch dia.		
	Through 4.250	Over 4.250 through 6.250	Over 6.250 through 8.250
Runout (TIV):†			
13 through 19° pressure angle	0.0007	0.0008	0.0009
Over 19° pressure angle	0.0005	0.0006	0.0007
Spacing,† max. variation between adjacent teeth	0.0002	0.0002	0.0003
Lead or parallelism (TIV)†	Max. of 0.0001 deviation in central 80% of face width. 0.0001 additional allowed adjacent to each face		
Face runout (TIV)	0.0001 per in. of diameter		
Hole dia.	±0.0002–0.0000		
Outside dia.	±0.0010		
Grade B	Diam. pitch*		
	4 through 5.999	Over 5.999 through 11.999	Over 11.999 through 19.999
Active involute profile (TIV)	0.0003	0.0002	0.0002
Measurement over pins	±0.0050	±0.0050	±0.0050
	Pitch dia.		
	Through 4.250	Over 4.250 through 6.250	Over 6.250 through 8.250
Runout (TIV):†			
13 through 19° pressure angle	0.0007	0.0008	0.0009
Over 19° pressure angle	0.0005	0.0006	0.0007
Spacing,† max. variation between adjacent teeth	0.0002	0.0002	0.0003
Lead or parallelism (TIV)†	Max. of 0.0001 deviation in central 80% of face width. 0.0001 additional allowed adjacent to each face		
Face runout (TIV)	0.0001 per in. of diameter		
Hole dia.	±0.0002–0.0000		
Outside dia.	Compensated within ±0.0010 to final measurement over pins		

* For application of these tolerances to helical master gears use the normal diametral pitch and the normal pressure angle. Variations may be necessary depending upon helix angle and face width.

† Measured at the approximate pitch diameter.

Special masters can be obtained at extra expense. In medium pitches it is possible to obtain master gears from a few specialists in this work that are accurate to a tooth-to-tooth composite tolerance of 0.000025″, a total composite tolerance of 0.00005″, and a runout tolerance of 0.000025″.

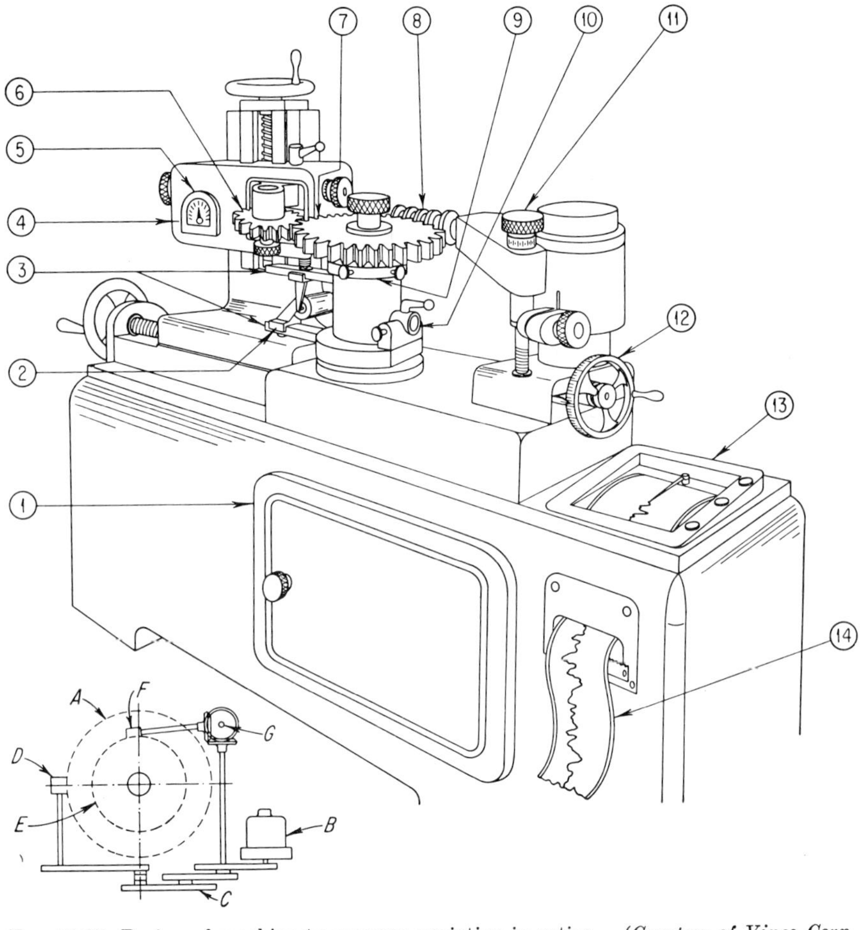

FIG. 23-37. Design of machine to measure variation in action. (*Courtesy of Vinco Corp., Detroit, Mich.*) (1) Door for access to change gears. (2) Electronic pickup. (3) Differential arms. Upper connected to work gear 7; lower connected to vertical spindle and master reference gear. (4) Gear-rolling accessory. (5) Meter for reading work-gear composite error. (6) Master rolling gear. (7) Gear being checked (work gear). (8) Master worm driving work gear. (9) Centering head for work gear. (10) Spring-pressure reversing device for checking opposite tooth flank of work gear. (11) Knob for vertical positioning of work-driving worm. (12) Control wheel for contact of driving worm with work gear. (13) Recorder. (14) Recorder tape.

23-8. Error in Action. Consideration has often been given to a kind of check that would measure the "error in action." From the standpoint of nomenclature that an *error* in a gear is a variation that is beyond the permissible tolerance, it is more correct to consider *variation in action* than *error in action.*

The variation in action can be defined as the tangential movement of a driven gear that is ahead of or behind its proper position when the driving gear moves perfectly uniformly. This kind of check would measure variation from the *driving* sides of the teeth only. One drawback to the composite check is that *both* sides of the teeth play a part in the observed variation; yet the quality of most gear drives is only a function of one side of the teeth.

Equipment to check error in action is built by some European manufacturers under the name "single-flank gear testers." These instruments consist of interchangeable friction disks which have the same ratio as the gears to be checked. One is attached to the driving gear, the other coaxial with, but independent of the driven gear. An indicating mechanism measures the variations between the driven gear and driven friction disk.

Figure 23-37 shows a more universal design proposed by an American manufacturer. A large master worm wheel *A* is driven by a motor *B* through change gears *C* and a precision worm *D*. The gear *E* to be checked is mounted freely on the same spindle as the large master worm wheel *A* and driven by the interchangeable precision master worm *F*. The center distance between the work gear *E* and master worm *F* is adjustable, since the master worm pivots at *G*. Master worms could be designed for any desired pitch and pressure angle. The change gears are selected to provide the same theoretical rate of rotation for the master worm wheel *A* and work gear *E*. An indicating device measures the variation between the rotations of the master worm wheel *A* and the work gear *E*. The machine may include means for the composite check with a conventional master gear and electronic recorders.

BEVEL- AND HYPOID-GEAR INSPECTION

By Theodore F. Klem, *Gear Engineer, Gleason Works, Rochester, N.Y.*

The curvatures of bevel teeth are more complex than those of spur or helical teeth. Checks for profile and lead are made somewhat indirectly by means of contact checks between mating gears or checks between a gear and a master. Bevel teeth may be checked for spacing by equipment somewhat similar to that used for spur gears. This section covers the inspection procedures and equipment that are peculiar to the bevel- and hypoid-gear field.

23-9. Testing Procedures and Equipment. Smooth, quiet-running gears must have uniform tooth spacing, teeth concentric with axis of rotation, and proper tooth shape. Inaccuracies in any of these can lead to rapid wear, tooth breakage, or a nonuniformity in the transmission of angular velocity.

Even well-constructed bevel-gear mountings will have some deflection. These deflections tend to concentrate the load at one end of the teeth with resultant rapid wear or breakage. Bevel-gear-testing machines make it possible to make allowances for reasonable deflections and obtain a satisfactory tooth contact under full-load conditions.

Running Test. Since the most conclusive test of bevel-gear operation is a *running* test, bevel-inspection machines (usually called "bevel-gear testers") are designed to provide a test simulating running conditions. In these testing machines, the gears are rigidly mounted in the correct operating position and run under power and a light brake load.

The smoothness and quietness of operation, the tooth bearing, the tooth size, the surface finish, and appreciable runout and tooth-to-tooth spacing errors are all checked by the running test. These tests are made both when the gears are in the soft state and after they have been hardened.

Tooth Bearing. The tooth bearing is an indication of correct tooth shape both up and down the tooth profile and lengthwise on the tooth. It is that portion of the gear-tooth surface which actually makes contact with its mate and can readily be observed by painting the teeth with a marking compound and running the gears for a few seconds.

Testing Machines. Bevel-gear-testing[17] machines are provided with axial adjustments for both the gear and pinion, and usually with a vertical offset adjustment for raising and lowering the axis of one member vertically with respect to the other for the testing of hypoid gears. In addition, the *axial adjustment of the pinion* is used to produce the same effect on the tooth bearing as a change in pressure angle. The *axial adjustment of the gear* is used to control backlash. The *vertical offset of the axes* is used to produce the same effect on the tooth bearing as a change in spiral angle.

With these adjustments, it is possible to obtain the equivalent position of the gears in the actual mounting and to measure the changes which must be applied to the cutting machines for thus locating the tooth bearing correctly.

The checking of involute tooth profiles, which is applied so commonly in the case of spur and helical gears, is not carried out on bevel gears for two principal reasons. First, the shape of a bevel-gear tooth varies at every position from end to end, instead of being constant, as in a spur gear. Hence a different reading is obtained at every checking position along the tooth. Second, even if this check were practical for bevel gears, it would not show the true tooth contact along the entire length of the tooth as shown in a bevel-gear-testing machine.

Bevel-gear-testing machines are considered to be superior to specially built test fixtures. Not only do fixtures fail to indicate how the gears will operate in service, but they lack the adjustment necessary for determining slight changes in setup of the cutting machine for tooth-bearing control. Furthermore, fixtures soon become an investment problem, as well as a space problem, every new design requiring a different fixture. However, in some special instances, a good design of a special fixture has been used advantageously.

General Testing Procedure. When a new bevel-gear ratio is produced, it is first necessary to *develop* the pair—that is, to secure the desired location and shape of the tooth bearing to ensure a smooth- and quiet-running pair of gears under final operating conditions—before the gears can be cut in production. The gear and pinion are first cut to the calculated machine setting, as recorded on a summary. Then a check is made by running the two together on the testing machine under power, to determine what changes must be made on the pinion in order to make proper allowances for deformation in heat-treatment and for mounting deflections. It is customary to make all corrections on the pinion, since changes may be made more rapidly and satisfactorily on this member. The required changes, as determined on the test machine, are then made on the pinion-generating machine, and the pinion is recut and again run with the gear. When approved tooth bearing and desired backlash are obtained, the production gears are cut.

Similarly, by means of this development procedure on the testing machine, it is possible to cut gears to match those cut at some time previously. This is particularly important where duplication is required. Once the development is completed, several sets of *control gears* should be made which duplicate exactly the newly developed pair. These control gears are used on the gear-testing machine for testing the remaining gears as they are cut, to assure uniform quality and interchangeability.

A *final test* is made on each pair of gears when they are completed. This final test serves to find the best running position of the gears and provides the final mounting distances for assembly. This in turn eliminates the need of *fitting* or *tearing down* and rebuilding of an assembly to reposition the gear.

FIG. 23-38. Hypoid tester for testing bevel gears at 90° shaft angle. (*Courtesy of Gleason Works, Rochester, N.Y.*)

FIG. 23-39. Universal bevel-gear tester for testing bevel or hypoid gears at any shaft angle. (*Courtesy of Gleason Works, Rochester, N.Y.*)

Operation of Testing Machines. There are primarily two types of bevel-gear-testing machines, namely, the *right-angle tester* for 90° shaft angle work and the *universal tester* for gears operating on both 90° and other than 90° shaft angles.

The Right-angle Testing Machine. The right-angle testers (Fig. 23-38) are equipped with adjusting screws for raising or lowering the gear head vertically, for displacing the gear head horizontally at right angles to the gear axis (that is, in the direction of the pinion axis), and for displacing the pinion head horizontally at right angles to the pinion axis (that is, in the direction of the gear axis).

The Universal Testing Machine. This machine (Fig. 23-39) has all the movements described for the right-angle tester with one additional feature, that is, the angle between the gear and pinion axes may be adjusted by swinging the gear head on a slide to the desired shaft angle.

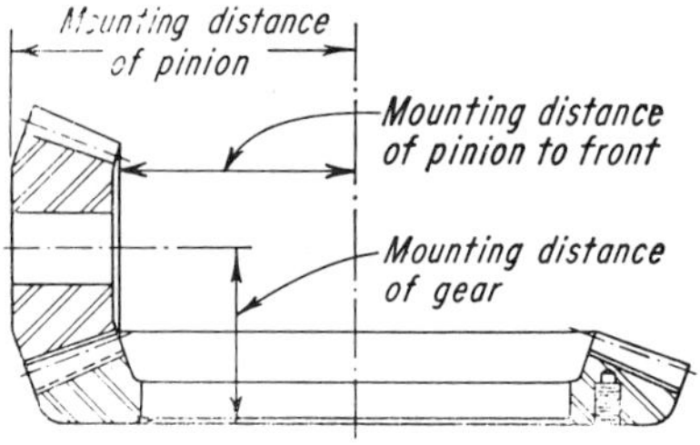

FIG. 23-40. Mounting distance of bevel gears.

Development and Testing Procedure. Throughout the development, three things should be borne in mind:

1. A smooth- and quiet-running pair of gears is required.
2. Some adjustability to prevent load concentration along the tooth edges must be provided.
3. The size, shape, and location of the tooth bearing must be consistent with the application.

To obtain smoothness and quietness, the gear teeth must have as large an area of contact as possible. This will also increase surface durability. However, with the increase in contact area, there will be a decrease in adjustability. For these reasons, the balance between smoothness and adjustability must be determined by the requirements of the specific job.

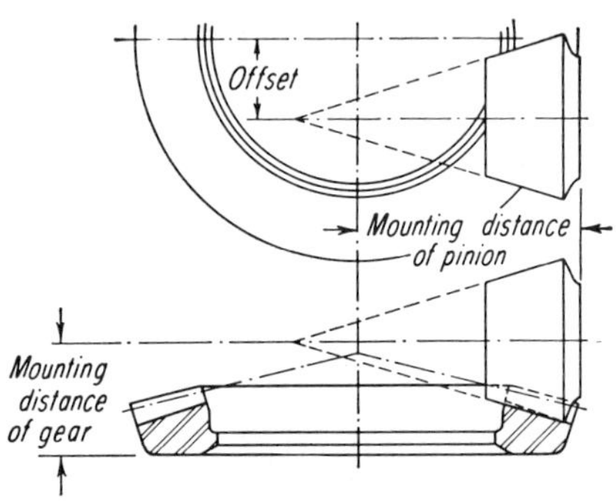

FIG. 23-41. The mounting distance of a *hypoid* gear is the linear dimension from its mounting surface to the common perpendicular from its axis to the axis of its mate.

Before the gears to be tested can be run together in the testing machine, it is necessary to locate them in the same relative position as in their final mountings. This is done by positioning the gear and pinion to their respective *mounting distances* as specified on the drawings. The workheads on the testing machines are positioned so that the gears, when mounted, will be located at their correct mounting distances (see Figs. 23-40 to 23-44).

During the development of the tooth bearing, the gears are usually set to a greater mounting distance than the final one, because the teeth are not cut to the correct size until a satisfactory development has been obtained.

Checking Tooth Thickness. On medium- and coarse-pitch gears, the measurement of tooth thickness is usually made with a gear-tooth vernier. This vernier has one setting for the (normal) chordal addendum and a second setting for the (normal) chordal thickness. When the correct settings have been made on the verniers, the

Fig. 23-42. Illustration of use of setup gages for precisely locating testing-machine workheads. These gages are specially made for each job.

Fig. 23-43. Use of gage blocks with proof bar (or mating arbor in many cases) to position testing-machine workheads correctly. Gage blocks are employed only where the quantities of gears are too small and the variety is too large to warrant the purchase of special setup gages for each job.

Fig. 23-44. The same setup as Fig. 23-43 with the gears in place.

caliper is drawn along the gear tooth from the inner and toward the outer end of the tooth. The caliper should touch both sides of the tooth and the top land at the outer end when the tooth is the correct size (see Fig. 23-43). In order to use this vernier, the outside diameter and face angle of the gear blanks must be held to recommended tolerances. It will usually be found necessary to reduce the tooth thickness slightly from the theoretical value, in order to mount the gears at their correct mounting distances with the desired backlash. Generally, the gear member is cut to the theoretical size (as measured by the tooth vernier), and the pinion is then cut to give desired backlash. This method is applied only to the first gears cut. Production checking methods are covered in the next article.

Checking Backlash. Bevel gears are cut to have a definite amount of backlash, which varies according to the pitch and operating conditions. In production testing, backlash measurement is used as a gage of tooth thickness. Backlash should be measured at the tightest point, with the gears mounted on their correct centers. In order to make this measurement on the testing machine, the pinion is held solidly against rotation and a rigidly mounted dial indicator is placed against the tooth at the extreme heel perpendicular to the surface (Fig. 23-46). The backlash is determined by moving the gear back and forth.

FIG. 23-45. Use of gear-tooth vernier to check tooth thickness of a spiral bevel pinion. Note that the vernier is held perpendicular to the top and sides of the tooth.

FIG. 23-46. Measurement of backlash in a pair of gears by means of a dial indicator.

Backlash variation is measured by locating the points of maximum and minimum backlash in the pair of gears and obtaining the difference. For precision gears, this variation in general should not exceed 0.001″.

After checking backlash by the above method on the finally approved pair of gears, the production control of backlash may be checked more easily by the following method.

The approved pair of gears (control gears) are mounted in the testing machine at their correct mounting distance. The gear is then moved axially into metal-to-metal contact with the pinion.

The amount of movement of the head from its initial position is observed. Then, when testing the production gears, the head should move past the correct mounting distance by this same amount, allowance being made for the specified backlash tolerance.

Allowances for Heat-treat Changes and Mounting Deflections. Gear teeth which are hardened usually undergo minor changes during the heat-treating process. The degree of these changes depends on the shape and size of the gear and the method of hardening. In addition, as a result of deflections in the gearbox or carrier, the gears do not remain in their theoretical mounting position when the load is applied. For these reasons, in the shop development of bevel and hypoid gears, the tooth bearing is located with an allowance for heat-treat changes and known mounting deflections.

Heat-treat Changes. In the heat-treating of curved-tooth bevel gears, the pinion teeth usually change more than the gear teeth in such a direction as to reduce the spiral angle. This change in spiral angle results in a shift of the bearing lengthwise on the tooth, toward the heel on the convex side of the pinion tooth and toward

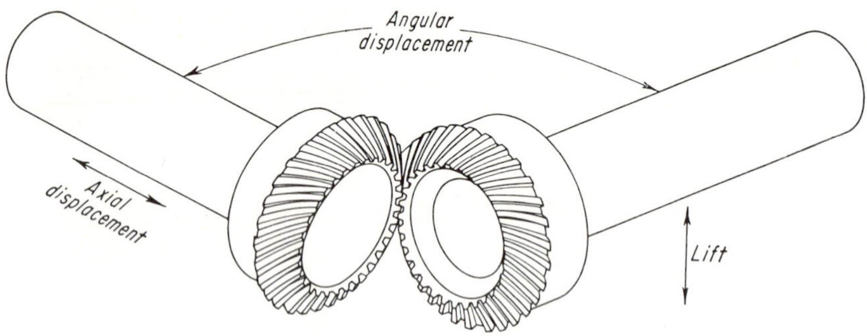

FIG. 23-47. This drawing illustrates the displacements to which bevel gears are subject in their mountings.

the toe on the concave side of the pinion tooth. Accordingly, allowance on the right-angle testing-machine settings for heat-treatment is made by raising the gear axis on a left-hand pinion and by lowering the gear axis on a right-hand pinion from the standard position. (By standard position is meant the relative position of the gear with respect to the pinion when the two members are assembled in their specified operating positions.) The amount of displacement of the gear or pinion axis from standard usually varies from 0.000 to 0.015″. The exact value for any particular job can be determined only by trial. That is, the gears are developed to have the tooth bearing in the desired location (i.e., central toe) when tested with these modified settings. Hence, during heat-treatment, the tooth bearing will move into the proper position, so that when the hardened gears are given a final test on correct centers, they will run with the correct tooth contact.

In the case of straight bevel gears, these allowances are unnecessary, since there is no spiral angle. This, however, does not mean that there is no change in straight bevel gears as a result of the heat-treatment. On the contrary, the change which does occur is not readily predictable and thus is difficult to allow for in cutting.

Mounting Deflections. Bevel gears running in their mountings are subject to axial displacements, lift, and angular displacements (see Fig. 23-47). The *axial displacement* is movement in the direction of the gear or pinion axis; the *lift* is the relative linear movement between the gear and pinion axes, and at right angles to both axes; and the *angular displacement* is a change in the shaft angle. Values of the above displacements will vary according to the size of the gear, the rigidity of the mountings, and the accuracy of machining. In general, the bearing will shift toward the heel (large end of the tooth) under load. To compensate for this, the tooth bearing under light load is placed slightly nearer the toe (small end of the tooth). The resulting bearing is called a *central toe bearing*.

Mounting deflection cannot readily be predetermined, unless experience has been gained from the effect on the same size gears of deflections in similar mountings. Two methods are open for determining the actual deflections after a preliminary set of gears has been made. The first of these is the deflection test, in which the gears are mounted in their case and the whole unit is subjected to dynamic loading (Fig. 23-48). Indicators are placed in the required positions to record relative displacement of the gear and pinion.

The second, and more common method, involves a running test of the gears under actual load in their final mountings and observation of the tooth bearing, which usually moves to the heel under load.

FIG. 23-48. Deflection test of main drive gears in an automotive rear axle.

These gears are then disassembled and mounted in the testing machine. The testing-machine settings are adjusted to duplicate the position of the tooth bearing observed on the gears in their final mountings under load. These new settings on the testing machine are the new operating positions for the gears and are equivalent to the final mounting position of the gears under load. The gears should be redeveloped to have the desired tooth bearings (i.e., central) in these new operating positions. Then, when the gears are installed in the final assembly, they will run with a full-length tooth contact under load, allowance having been made for deflection.

When the bearing has been located in the proper position on the tooth, it should be possible to move the pinion axially in or out a slight amount, approximately 0.003″, without causing a heavy bearing along the top edge or deep in the bottom of the tooth.

Vertical and Horizontal Movements. The readings for all dials on the testing machine may be considered as *zero* readings when the gears are mounted in their specified mounting positions. All horizontal and vertical movements are measured from these *zero* positions. The following rules will determine the correct signs to be used with these movements:

1. Horizontal movements of the gear on the right-angle testing machines away from the pinion head (out) = (+) plus
2. Horizontal movements of the gear head on the right-angle testing machines toward the pinion head (in) = (−) minus
3. Vertical movements of the gear head up or the pinion head down = (+) plus
4. Vertical movements of the gear head down or the pinion head up = (−) minus

As will be seen in the next article, it is occasionally very desirable to determine the vertical and horizontal settings necessary to place the tooth bearing in the center of the tooth profile at the extreme toe and extreme heel. The total movement between the toe and heel readings is obtained by subtracting the heel reading from the toe reading algebraically. To obtain the average reading between toe and the heel readings, add the values algebraically and divide by 2.

Example

	Reading at toe	Reading at heel	Total movement toe to heel	Average reading
Vertical movement..........	+0.010	−0.018	0.028	−0.004
Horizontal movement........	−0.014	+0.020	0.034	+0.003

$$\text{Total vertical movement} = (+0.010) - (-0.018) = +0.028$$
$$\text{Total horizontal movement} = (-0.014) - (+0.020) = -0.034$$

The algebraic signs of these totals are ignored, since the magnitude of these quantities is the item of interest.

$$\text{Average vertical setting} = \frac{(+0.010) + (-0.018)}{2} = -0.004$$

$$\text{Average horizontal setting} = \frac{(-0.014) + (+0.020)}{2} = +0.003$$

When obtaining the average readings, both the magnitude and the direction (sign) are important. The average reading is used to place the tooth bearing in the center of the tooth to observe its appearance or to compare it with the average reading for a master set of gears for the same job.

V and H Check. The V and H check is a method for measuring the amount and direction of the vertical and axial displacements of the pinion from its standard position, to obtain a tooth bearing in the middle of the tooth profile at the extreme toe and at the extreme heel of the tooth. In the preceding example, the first three columns entitled, "Reading at toe," "Reading at heel," and "Total movement toe to heel" constitute the V and H check.

It is customary to specify a tooth bearing approximately equal to one-half the tooth length under light load. This, in general, will allow adequate compensation for mounting deflections under load and sufficient adjustability in assembly. However, as stated above, the V and H check is a more accurate method of determining the required length of tooth bearing.

Referring to the example just given the line entitled "Vertical movement" gives the relative displacement from the central position to the toe (in example, +0.010″), from the central position to the heel (in example, −0.018″), and from toe to the heel (in example, 0.028″). This last value ("Total movement from toe to heel") is a measure of the bearing length in terms of displacement and is sometimes referred to as a ". . . -length bearing" (in example, 28-length bearing). When the total vertical movement of the V and H check is too large, it indicates that the tooth bearing is too short and that the load will therefore concentrate on too small an area of the tooth surface, causing danger of excessive wear. When the total vertical movement is too small, it indicates that the tooth bearing is too long and hence the gears will lack sufficient adjustability to compensate for mounting deflection, which may lead to load concentration at the ends of the teeth.

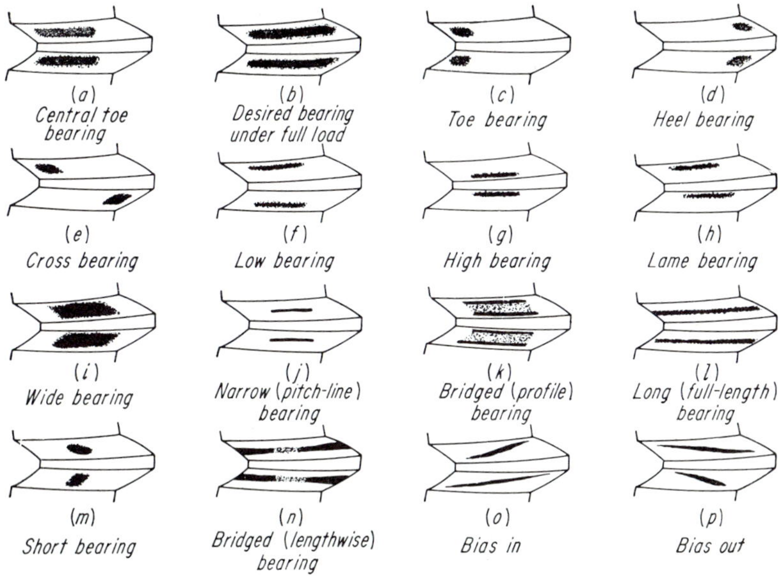

FIG. 23-49. The above sketches illustrate tooth bearing on the pinion tooth. Although a left-hand pinion is used throughout, the bearings are representative of those on a right-hand pinion or a straight bevel pinion as well. Figure *a* shows a central toe bearing. Note that the bearing extends along approximately one-half the tooth length and that it is nearer the toe of the tooth than the heel. In addition, the bearing is relieved slightly along the face and flank of the tooth. Under the light loads the tooth bearing should be in this position on the tooth. Figure *b* shows the same tooth with a bearing as it should be under full load. It should show slight relief at the ends and along the face and flank of the teeth. There should be no load concentration at the extreme edges of the teeth. Figures *c*, *d*, and *e* show differences in spiral angle between the gears tested. Figures *f*, *g*, and *h* show differences in pressure angle between the gears tested. Figures *i*, *j*, and *k* illustrate width of tooth bearing. Figures *l*, *m*, and *n* illustrate length of tooth bearing. Figures *o* and *p* illustrate bias bearings. Regardless of the hand of spiral on the pinion, "bias in" will always run from the flank at the toe to the top at the heel on the convex side and from the top at the toe to the flank at the heel on the concave side.

To determine whether the tooth bearing is correctly placed on the tooth, when a central toe bearing is desired after hardening:

1. On the concave side of the pinion tooth, the toe vertical reading should equal the allowance in vertical offset of the testing machine for hardening, *plus* approximately one-quarter of the total vertical movement from toe to heel.
2. On the convex side of the pinion tooth, the toe vertical reading should equal the allowance in the vertical offset on the testing machine for hardening *minus* approximately one-quarter of the total vertical movement from toe to heel (see Fig. 23-49).

A bias tooth bearing tends to cross the tooth on a slant (see Fig. 23-49) and is undesirable principally because of the danger of load concentration at the corners of the teeth.

However, it is sometimes necessary on spiral bevel and hypoid gears to have a slight amount of *bias out* in soft gears because, when gears are hardened, bias will be introduced.

The visual method of determining bias in a tooth bearing by observing the bearing when placed in the central position for its diagonal direction on the tooth is, of course, helpful but rather inaccurate. In determining bias by the V and H check, the

Table 23-6. Representative V and H Checks, Spiral Bevel and Hypoid Gears

Concave	Toe	Heel	Total	Convex	Toe	Heel	Total
			Passenger car: 8″ gear				
Soft:							
V	+8	−12	20	V	−3	+17	20
H	−8	+16	24	H	+4	−23	27
Hard:							
V	+3	−9	12	V	−3	+9	12
H	−3	+8	11	H	+3	−11	14
			Passenger car: 9½″ gear				
Soft:							
V	+8	−16	24	V	−3	+19	22
H	−11	+22	33	H	+2	−26	28
Hard:							
V	+4	−9	13	V	−4	+10	14
H	−4	+8	12	H	+3	−13	16
			Truck: 10″ gear				
Soft:							
V	+10	−15	25	V	−8	+22	30
H	−10	+19	29	H	+7	−34	41
Hard:							
V	+4	−12	16	V	−5	+13	18
H	−4	+12	16	H	+4	−16	20
			Truck: 13″ gear				
Soft:							
V	+7	−30	37	V	−6	+32	38
H	−8	+32	40	H	+7	−37	44
Hard:							
V	+2	−36	38	V	−3	+38	41
H	−3	+37	40	H	+4	−40	44

amount of vertical movement on the testing machine that is necessary to shift the tooth bearing from toe to heel is compared with the corresponding amount of horizontal movement necessary to keep the bearing in the middle of the tooth profile.

For spiral bevels with a 90° shaft angle and a spiral angle between 35 and 40°, the ratio of total vertical to total horizontal movement from toe bearing to heel bearing will usually be about 1:1 on the concave side of the pinion tooth, and about 3:4 on the convex side of the pinion tooth, to produce a tooth bearing with freedom from bias. If the ratio is greater than specified above, it denotes *bias in,* and if the ratio is less, it denotes *bias out.* As the spiral angle decreases, the ratio of total vertical to total horizontal movement should increase until, in the case of a Zerol®,* the horizontal movement becomes zero and the ratio becomes infinite. All the movement will be vertical. As the shaft angle decreases from 90°, the ratio of total vertical to total horizontal movement should decrease until, in the case of very low shaft angles, the vertical movement becomes zero.

Table 23-7. Representative V and H Checks, Zerol Bevel Gears

Concave	Toe	Heel	Total	Convex	Toe	Heel	Total
			Tractor: 13″ gear				
Soft:							
V	+4	−24	28	V	−3	+25	28
H	0	0	0	H	0	0	0
Hard:							
V	+4	−20	24	V	−4	+24	28
H	0	0	0	H	0	0	0
			Tractor: 19″ gear				
Soft:							
V	+15	−55	70	V	−12	+56	68
H	0	0	0	H	0	0	0
Hard:							
V	+5	−55	60	V	−6	+55	61
H	0	0	0	H	0	0	0
			Aircraft planetary 10″ gear				
Ground:							
V	+3	−10	13	V	−2	+12	14
H	0	0	0	H	0	0	0
			Aircraft 2 to 6″ gears				
Ground:							
V	+3	−10	13	V	−2	+12	14
H	0	0	0	H	0	0	0
			Low-spiral-angle propeller gears 5 to 9″ gear				
Ground:							
V	+3	−10	13	V	−3	+11	14
H	0	0	0	H	0	0	0

A few representative V and H checks for gears operating at 90° shaft angles are included to serve as a guide. The examples given show both the soft and corresponding hard (after lapping) check, with the exception of the aircraft gears, which are for gears after the teeth have been ground. References to concave and convex side of the tooth apply to the pinion. The values given are average, and it may therefore be necessary to alter them considerably, depending on the distortion from heat-treatment and on the rigidity of the gear mountings (see Tables 23-6 and 23-7).

* Registered trade-mark of Gleason Works, Rochester, N.Y.

23-10. Control Gears. A bevel or hypoid control gear is a hardened gear made to the necessary specifications of size, mounting distance, and tooth bearing to ensure proper tooth action in final assembly under load. Its purpose is to serve as a working model for the duplication of gears in regular service.

Gear teeth which are hardened usually undergo minor changes during the heat-treating process. The degree of these changes depends upon the shape and size of the gear and the method of hardening. In addition, as a result of deflections in the housing and bearings under load, the gears do not remain in their theoretical mounting position. For these reasons, in the shop development of bevel and hypoid gears, the tooth bearing is located with an allowance for heat-treat changes and known mounting deflections. These allowances are made in the manufacture of the control gears.

Types and Uses

1. *Master control gears*—used for checking *inspection control gears* only.
2. *Inspection control gears*—used for checking finished production gears.
3. *Green production control gears*—used for checking gears in the *green* or *unhardened* state. In the manufacture of these gears, compensation must be made for the effects of the final mounting and for the changes in the production gears during hardening as determined by actual trial.

   Production control gears, for checking production gears which are to have their teeth hardened and ground, are made thin by the amount of grinding stock allowed on the teeth of mating production gears. These production control gears must also allow for heat-treat changes and final mounting conditions.
4. *Pregrind production control gears.* Used for checking after the production gears have been *hardened,* but before the teeth are ground. They are also used during the grinding development. These control gears are made thin by the amount of grinding stock on the mating production gears. The tooth bearing is made the same as on the master control gears. The primary purpose of these gears is to check the uniformity of stock distribution on the sides of the gear teeth prior to grinding.

Manufacture of Control Gears. In the manufacture of these control gears, it is imperative that the effects of the final mountings be known and allowed for in the control gears.

A minimum of two control gears are made for each item, one set being suitably marked and held as a *master control gear,* and used only for checking the *inspection control gear* periodically and for reference when ordering replacement inspection control gears. The other set is used as an inspection control gear.

Production control gears are similar to the master control gears, with the exception that the size, mounting distance, and tooth bearing have been altered, as explained below, to serve as working models during various stages of the gear manufacture.

The master control gear and inspection control gear, as well as the green and pregrind production control gears for all bevel gears, should be selected gears which have been made in the same plant and under the same conditions as the production gears. If the production gears are not hardened, the control gears in general should be made either of Nitralloy N®,* nitrided to a case depth of 0.005 to 0.010″, or of a surface-hardening steel with flame-hardened teeth.

Certain special cases, such as control gears for gears which operate with little or no backlash on exact mounting distances, should be made with a minimum of backlash on exact mounting distances.

Production Testing. When testing bevel gears in production, the gears should be run with a control gear on correct operating centers established by a setup gage.

* Registered trade-mark of the Nitralloy Corp., New York, N.Y.

That is, the production gear is run with the control pinion and the production pinion is run with the control gear. The tooth bearing should be identical in shape and position with that of the masters, and the gears should run smoothly and quietly. The tooth thickness and backlash may be checked, as previously explained, by comparing the amount of axial movement of the production gear with the control pinion for metal-to-metal contact with the amount of axial movement of the control gear with the control pinion for metal-to-metal contact.

Errors in tooth spacing and runout may also be found by this test. Runout is characterized by a periodic variation in sound during each revolution—and by the tooth bearing shifting progressively around the gear from heel to toe and toe to heel. Tooth-spacing errors in which there is a cumulative error which builds up around the gear, causing a large error between the first and last tooth cut, are usually traceable to poor chucking equipment or faulty machine setup. These are indicated by a knocking sound and by either a very light or a very heavy tooth bearing on only one tooth.

FIG. 23-50. Tooth-spacing tester. (*Courtesy of Gleason Works, Rochester, N.Y.*)

On certain high-precision gear work, such as for instruments and for aircraft, extremely close tolerances of tooth spacing and concentricity must be held. Checking these values by sight and sound on a running test is not sufficient; they must be inspected by direct measurement. For this purpose, there are manual and automated inspection devices available.

Manual Tooth-spacing Tester. Error in tooth spacing can be measured by the tooth-spacing tester shown in Fig. 23-50. This device has two contact points which can be adjusted to contact corresponding sides of adjacent teeth. One contact point is used as a fixed stop and the other actuates a dial indicator reading to ten-thousandths of an inch. The two contact points are carried on a swinging bracket, which can be lifted out of position and then returned to allow successive teeth in the gear to be checked. The reading of the indicator is noted for each tooth.

Manual Concentricity Tester. Concentricity is the condition which exists when the center of the pitch circle on which the teeth are cut coincides with the center about which the gear turns. When this condition is not fulfilled, the gear is said to *run out* or to *be eccentric.* The distance between these centers is the measure of eccentricity and is measured by the concentricity tester. The runout is twice the eccentricity.

As shown in Fig. 23-51, this device has two contact points 180° apart, one a fixed stop, the other actuating a dial indicator reading to ten-thousandths of an inch. The gear to be checked is mounted on a dead center held with a clamp, and indicator readings are taken on diametrically opposite teeth with the contacts acting on corresponding sides of these teeth. The contact points are mounted on a swinging bracket which can be lifted while the gear is turned from tooth to tooth.

FIG. 23-51. Gear-concentricity tester. (*Courtesy of Gleason Works, Rochester, N.Y.*)

FIG. 23-52. Automatic spacing and runout checker. (*Courtesy of Gleason Works, Rochester, N.Y.*)

The runout is equal to one-half of the difference between the high and low readings, which makes this method of inspection quite accurate. Because of the amount of time required in inspecting gears by these last two devices, they are used primarily for periodic and not general production checking.

Automatic Spacing and Runout Checker. The automatic tooth-spacing and runout checker, as shown in Fig. 23-52, provides an extremely accurate automated and graphic check of tooth spacing and runout. The machine is equipped with separate indicator gage heads for checking gears and pinions. The gear gage head has three probes; one locates the position of the gear and checks *radial* and *axial runout;* the other two probes check spacing.

The method used to check pinions is similar to that on the manual machine. A spring-loaded positive stop is located 180° from the recording gage head, thereby making a true eccentricity check. An electronic recorder, used as an integral part of the test procedure, provides a permanent graphic record of the complete results. The recorder makes a mark on a chart for each check of runout. It is calibrated so that the error can be read directly from the chart, counting the spaces between the lines.

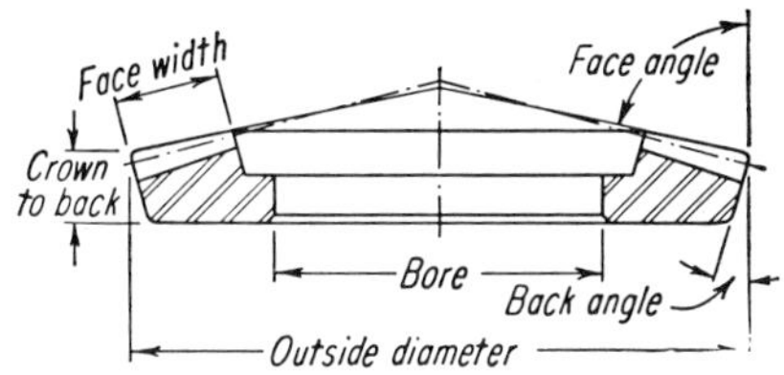

FIG. 23-53. Gear-blank dimensions.

The tooth spacing is recorded in a similar manner. In addition, the recorder establishes a reference line. The readings are averaged electronically, and after the test is complete, the operator can take the average reading shown on a computer and

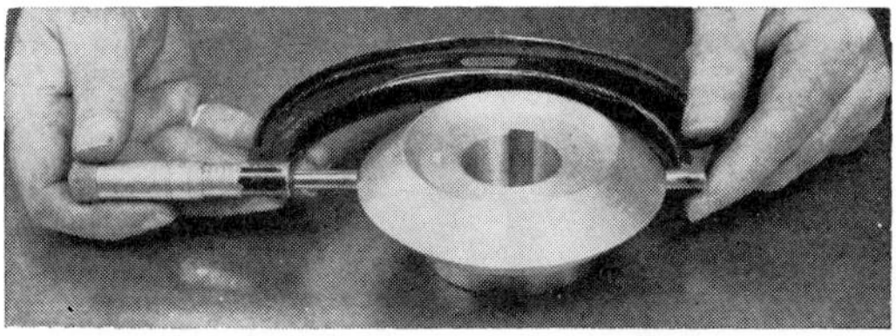

FIG. 23-54. Checking outside diameter.

FIG. 23-55. Checking face angle.

draw a norm from the reference line. From the norm line, the actual spacing error can be determined, as well as the greatest tooth-to-tooth error.

23-11. Gear-blank Inspection. It is essential that the manufacture of a gear begin with a correctly machined gear blank. The important dimensions of a gear blank are shown in Fig. 23-53. There are several methods open for the inspection of these various dimensions. The most common method is through the use of such tools as micrometers, bevel protractors, and depth gages, as shown in Fig. 23-54 to 23-55. Although this method is sufficiently accurate, it is quite slow and depends to some degree upon the skill of the individual making the check. In the use of this type of inspection, it is also necessary that the radius on the outside diameter of the blank be omitted until after the blank has been checked. This usually necessitates an extra turning operation, adding to the cost of manufacture.

Another method of blank inspection is through the use of an optical comparator,

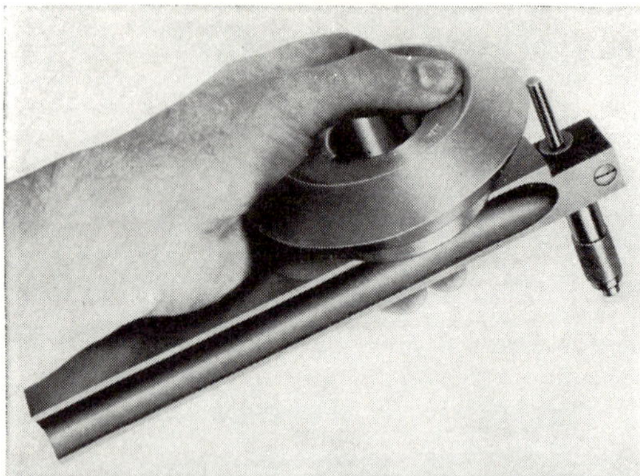

Fig. 23-56. Crown-to-back measurement with backing gage.

Fig. 23-57. Checking blanks with optical comparator.

as shown in Fig. 23-57. This method is accurate and is limited only by the size of the blank that the comparator will accommodate. Readings taken on a simple fixture, such as shown in Fig. 23-58, when compared with the *sample* or *master gear*, will also indicate whether angles, crown to back, and outside diameter are correct. A more

FIG. 23-58. Checking blanks against a master. The indicators are set to read zero on the master blank.

recent method for inspecting gear blanks is the one employing the use of a gear-blank-checking machine, as shown in Fig. 23-59. The machine consists of a work-spindle table on which the gear blank is mounted by means of a suitable workholding arbor, and a measuring head on which is mounted a dial indicator. The indicator is arranged so that the face angle and back angle of the gear blank may be trammed.

The measuring head has a vertical adjustment, with a double dial for setting the crown to back. The work-spindle table has a horizontal adjustment, with a double dial for setting the outside diameter. An angular scale, with two graduated dials, is provided for setting the indicator slide to the face-angle and back-angle values. These two dials are mounted on two separate worms, either of which may be engaged with the worm wheel on the angular scale. By shifting from one to the other, it is a simple matter to change from the face-angle to the back-angle setting. By moving the indicator head along the slide, one can tram across the face cone or along the back cone of the blank.

FIG. 23-59. Gear-blank-checking machine. (*Courtesy of Gleason Works, Rochester, N.Y.*)

Since the indicator slide moves parallel to the face cone, or back cone, as the case may be, a perfect blank will register zero on the indicator in all positions.

Face Angle and Crown to Back. When the indicator registers a *positive* value at the crown, and a uniform value along the face cone, this indicates that the face angle is correct, but the crown to back is large.

If the indicator registers a decreasing value when moving along the face cone from the crown to the inner end of the face width, this indicates that the face angle is large.

Back Angle and Outside Diameter. When the indicator registers a positive value at the crown, and a uniform value along the back cone, this indicates that the back angle is correct but the outside diameter is large.

If the indicator registers an increasing value when moving along the back cone from the crown to the back of the blank, this indicates that the back angle is large.

System of Tolerances. For medium-sized blanks, the conventional system of tolerances for crown to back, outside diameter, face angle, and back angle are as shown

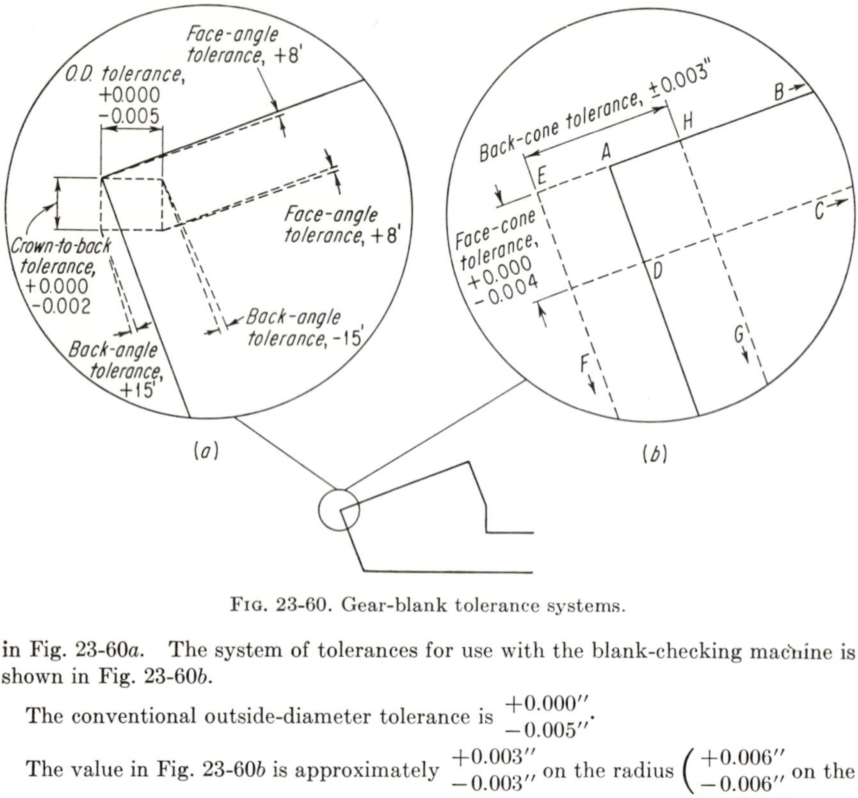

FIG. 23-60. Gear-blank tolerance systems.

in Fig. 23-60*a*. The system of tolerances for use with the blank-checking machine is shown in Fig. 23-60*b*.

The conventional outside-diameter tolerance is $\begin{matrix}+0.000''\\-0.005''\end{matrix}$.

The value in Fig. 23-60*b* is approximately $\begin{matrix}+0.003''\\-0.003''\end{matrix}$ on the radius $\left(\begin{matrix}+0.006''\\-0.006''\end{matrix}\right.$ on the outside diameter). The conventional crown-to-back tolerance is $\begin{matrix}+0.000''\\-0.002''\end{matrix}$. The value in Fig. 23-60*b* is approximately $\begin{matrix}+0.000''\\-0.004''\end{matrix}$. The conventional face-angle tolerance is $\begin{matrix}+0°8'\\-0°0'\end{matrix}$. The value in Fig. 23-60*b* allows the face cone to lie anywhere within the rectangle *ABCD*. This, in effect, allows both a plus and minus tolerance on the face angle.

The conventional back-angle tolerance is $\begin{matrix}+0°15'\\-0°15'\end{matrix}$. The value in Fig. 23-60*b* allows the back cone to lie anywhere within the rectangle *EFGH*.

The blank-checking-machine limits are, in general, wider than the conventional limits but are so set up that no sacrifice in the quality of the finished gears will result.

WORMGEAR INSPECTION

By Darle W. Dudley, *Manager, Advance Gear Engineering, General Electric Co., Lynn, Mass.*

The inspection of worms and wormgears has some similarity to both helical-gear inspection and bevel-gear inspection. Profile and lead checks can be taken on the worm member even though the profile is generally not involute. Contact checks are frequently used between mating members or between a worm part and a master.

23-12. Cylindrical Wormgear Inspection. The wormgear member has a complex curvature that is generated from a hob or other tool that duplicates the mating worm.

FIG. 23-61. Contact check between a cylindrical worm and wormgear. (*Courtesy of General Electric Co., Lynn, Mass.*)

The resulting warped surface does not lend itself to either profile or lead checks since any section through the tooth varies from a neighboring section.

Wormgears are normally checked by meshing them with their mate—or a *master worm*—in a checking fixture. The contact check with a worm reveals the following things:

1. Width, length, and location of contact pattern
2. Amount of backlash and variation in backlash
3. Height setting of worm with respect to gear to make the worm lie in the throated-out cup of the gear teeth

Figure 23-61 shows a contacting setup between a worm and wormgear. Red lead is put on the worm threads and prussian blue on the gear teeth. The worm is raised to the correct height setting. The worm is rotated to drive the gear, and this marks the contact all the way around the gear. The contact pattern is then judged for length, width, and location. Design requirements may call for contact across two-thirds of the face width. Generally it is desirable that contact does not start right where the worm enters mesh with the gear tooth. A heavy contact here serves to scrape the lubricant off the worm threads and impair lubrication. In the middle of the face width, the contact should spread out to a fairly full profile—say two-thirds of the tooth height.

Wormgears wear in more than most gear types; so it is quite likely that the design requirements will permit a more limited contact pattern on a new set than would be the case for other gear types where the wearing in of the tooth surfaces is much more limited. The designer will specify appropriate contact-pattern limits based on the material of the wormgear, the severity of the tooth loading, and the job requirements of speed, reliability, and smoothness.

The tooth thickness of wormgear teeth cannot be measured conveniently with calipers or pins (because of the warped shape). Measurement of backlash with a

FIG. 23-62. Checking a worm on a universal measuring machine. (*Courtesy of General Electric Co., Lynn, Mass.*)

worm of known thread thickness serves indirectly to measure the gear-tooth thickness. Variation in backlash indicates eccentricity or tooth-spacing variations.

If satisfactory tooth contact is not obtained at the specified height setting of the worm, a contact check may be repeated at a different height setting. If satisfactory contact is found at some other height setting, the gear has a *variation from design height* setting. It may be permissible to stamp the observed height setting on the gear and pass the gear as acceptable.

Wormgears may be checked for tooth spacing by means similar to those described in Art. 23-2. Highly precise worm wheels used for index gears may require the most precise measurement of pitch and index-spacing variations.

Cylindrical-worm Inspection. The worm may be measured by direct checks. Its lead and spacing may be measured by equipment similar to that used for helical pinions (see Arts. 23-2 and 23-3). Its profile is usually not involute; so involute* checkers are not used. The thread form is usually straight-sided—or almost straight-sided.† A checking machine that can traverse the profile at the right angle of inclination for the normal pressure angle can measure the profile by finding the deviation

* Single-thread, noninvolute worms can be checked with certain involute checkers.

† See Art. 8-8 for the method to calculate the exact normal section profile of a worm.

from a straight line. Figure 23-62 shows a general-purpose worm and hobchecking machine. This machine can measure worm profile, worm lead, and (with an index plate) thread-to-thread spacing.

The thread thickness of worms is determined by caliper checks or by diameter over rolls. The practice is similar to that for helical pinions described in Art. 23-5.

There is some limited use of a "composite" type of check for wormgears. The techniques used are similar to that used for spur or helical gears (see Art. 23-6).

23-13. Double-enveloping Wormgear Inspection.* Most double-enveloping[18] wormgears are checked for profile, lead, and tooth thickness by the indirect means of

FIG. 23-63. Machine to check double-enveloping wormgears. (*Courtesy of Cone-Drive Div., Michigan Tool Co., Detroit, Mich.*)

checking with a mating worm or a master worm. The general technique is similar to that used on cylindrical wormgears (see Art. 23-12).

A contact check of a double-enveloping wormgear has two important differences over that for a cylindrical worm. These are:

1. The worm must be set for both *height* and *axial position* before it will fit into the double throating of both the worm and the gear.
2. The contact pattern of a good set is somewhat different. The lengthwise contact of a good set tends to be somewhat shorter than that of a comparably good cylindrical wormgear. The shape of the contact pattern is somewhat different.

All double-enveloping gears are generated from the unmarked side face to an exact side-position dimension shown on the part drawing. The double-enveloping worm is generated from a fixed end-position dimension to a bearing shoulder. Double-enveloping wormgear sets should always be assembled to these closely held dimensions.

* Material for Art. 23-13 was furnished by F. E. Birtch, Manager, Cone-Drive Div., Michigan Tool Co., Detroit, Mich.

Double-enveloping wormgear sets generally have more teeth in simultaneous contact than do similar cylindrical wormgear sets. This means that a very high load-carrying capacity can be obtained even when the contact of new parts is a fairly short band. These gears have excellent wearing-in capacity; so that contact pattern will broaden rapidly in service. Figure 23-63 shows a machine used to check the spacing and pressure angle of double-enveloping wormgears. The Cone-Drive®* gear has a relatively straight-sided profile in an axial plane. The indicator can be traversed across the gear tooth in the center of the face width at an inclination corresponding to the pressure angle to check the profile. By the use of a fixed finger and a movable finger, pitch checks can also be made.

FIG. 23-64. Machine to check double-enveloping worms. (*Courtesy of Cone-Drive Div., Michigan Tool Co., Detroit, Mich.*)

Double-enveloping Worm Inspection. The worm member may be checked for lead, thread spacing, and profile by a machine like the one shown in Fig. 23-64.

By the use of the two index plates shown, the lead may be checked in any desired increments. If it is desired to check the lead for each 90° rotation of the worm, a four-tooth index plate may be used on the worm spindle and a plate having four times the number of gear teeth may be used on the gear spindle. By continuously indexing each plate one tooth at a time, the worm may be checked (by increments) for its entire length.

This system also permits a check of thread spacing. By setting the indicator slide to the axial pressure angle of the worm and traversing the indicator in and out, it is possible to check the profile of the thread. The worm thread is *straight-sided in the axial section.* Its slope is defined by a tangent line to the *base circle* of the mating gear.

Checkers of the type shown in Fig. 23-64 make it possible to give a Cone-Drive worm a complete check for lead, spacing, and profile.

* Registered trade-mark of Cone-Drive Div., Michigan Tool Co., Detroit, Mich.

SPIROID-GEAR INSPECTION*

The inspection of Spiroid®† gears makes use of equipment and techniques that have some aspects of bevel-gear inspection, some aspects of wormgear inspection, and some things that are entirely different. The Helicon®† gears can be checked almost the same as Spiroids. The inspection of the somewhat related Planoid®† gears quite closely resembles that of Spiroids.

23-14. Spiroid-gear Inspection.‡ The Spiroid gear[19] may be checked for the following things:

1. Face runout
2. Mounting distance
3. Contact pattern
4. Tooth spacing

Spiroid gears are normally finished by a "topping" hob. Face runout can be measured by rotating the gear on its axis and indicating over the tips of the teeth at

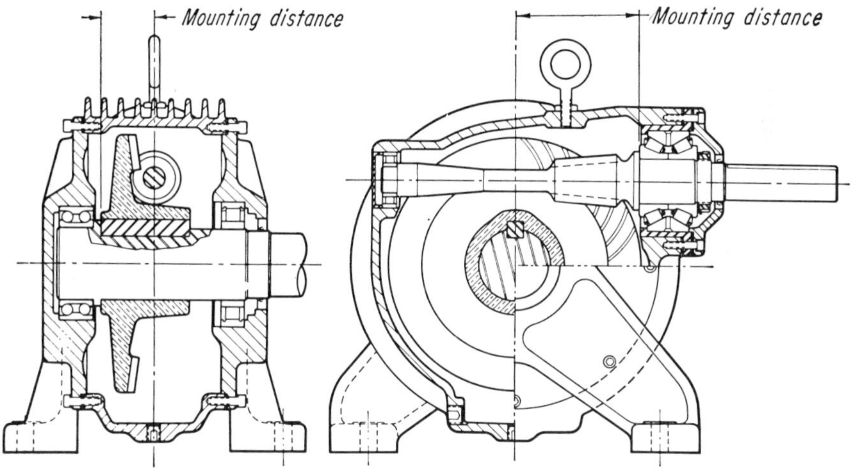

Fig. 23-65. Spiroid-gear mounting distance.

mid-face. When the tips are finished by a topping hob, the runout of the tips is the same as the runout of the working sides of the teeth. The face-runout check can be considered somewhat analogous to a concentricity check of a spur gear.

In some cases Spiroid gears are finished by a "shaving" hob which does not top off the tips of the teeth. In this case a face-runout check has to be made by plunging a cone point into the slots between teeth and reading the variation in depth of penetration as each tooth slot is probed.

The tooth size of Spiroid gears is indirectly determined by the mounting distance. Figure 23-65 shows mounting distance of a Spiroid pair in a casing. Figure 23-66 shows how mounting distance is shown on gear-part drawings. Figure 23-66 also

* By Darle W. Dudley, Manager Advance Gear Engineering, General Electric Co.

† Spiroid, Helicon, and Planoid are registered trade-marks of the Spiroid Div. of Illinois Tool Works, Inc., Chicago, Ill.

‡ The material for Arts. 23-14, 23-15, and 23-16 was furnished by Gary L. Vesey, Mechanical Engineer, Spiroid Div. of Illinois Tool Works, Inc., Chicago, Ill.

shows typical blank tolerances for Spiroid pinions and gears. These blank tolerances are inspected by conventional means.

The tooth size of a Spiroid gear is determined by mounting the gear in a checking machine and determining what mounting distance takes out all the backlash with the Spiroid pinion. A Spiroid pinion of known tooth thickness can measure the tooth thickness of a Spiroid gear. Conversely, a Spiroid gear of known thickness can be used to measure the tooth thickness of a Spiroid pinion. In each case, though, the

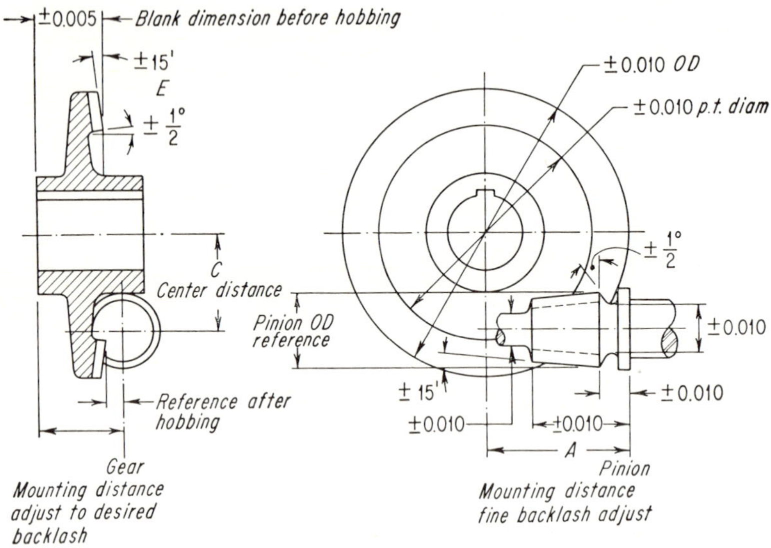

FIG. 23-66. Spiroid-gear inspection tolerances.

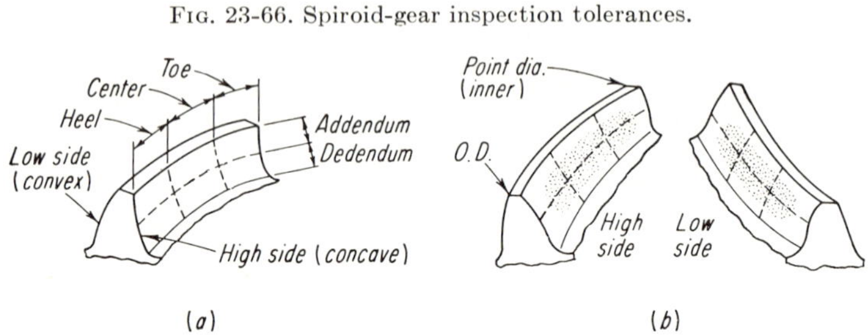

FIG. 23-67. Spiroid-gear contact pattern.

measurement of tooth thickness is handled by the indirect means of determining a tight-mesh mounting distance.

Figure 23-67 shows the nomenclature associated with a Spiroid-gear contact pattern. In general an ideal pattern would be fairly central on the tooth, favoring the toe, and extending for about 75 per cent of the tooth face. In the direction of tooth depth (profile) the pattern should have some "ease-off" at the tip and root amounting to about one-fourth of an addendum in each place. The pattern should have no hard lines or interference points and should "fade out" evenly around the periphery. Further the pattern should be the same on all gear teeth rather than tending to wander from toe to heel, which would indicate wobble or eccentricity.

Cutting scallops are readily noticeable on the gear. Ratios below 30:1 will have a slight scallop pattern. The contact pattern can be judged as to whether the scallops represent normal good workmanship or whether they represent an inferior quality of cutting and therefore make the parts subject to rejection.

The type of contact pattern will vary with the application. If loads are very heavy a somewhat biased contact will have to be specified at no load so that the full-load deflections can shift the contact into the desired position at full load. Heavy

FIG. 23-68. Machine for Spiroid-gear-contact-pattern checking. (*Courtesy of Spiroid Div. of Illinois Tool Works, Inc., Chicago, Ill.*)

FIG. 23-69. Machine for checking Spiroid-pinion lead, profile, and spacing. (*Courtesy of Spiroid Div. of Illinois Tool Works, Inc., Chicago, Ill.*)

load applications require a definite fade-out of the contact in the lengthwise direction of the tooth and an ease-off in the profile contact (crosswise direction). Light load applications may be permitted to vary from a fairly limited contact to a full-face contact that has no ease-off or fade-out.

The contact pattern on the pinion member may be judged for uniformity going around the pinion. If the pinion has a runout condition the contact may vary from high to low in 180°. The contact should also be studied to determine if serious spacing variations exist from thread to thread. A variation in contact intensity from tooth to tooth may require that the part be given direct spacing checks to determine if the spacing variation is within allowable limits. Figure 23-68 shows a typical checking machine for contact patterns and mounting distances of Spiroid gears.

Spacing checks of Spiroid gears may be taken with equipment similar to that used on other gear types. See Art. 23-2 for general information on tooth-spacing checks.

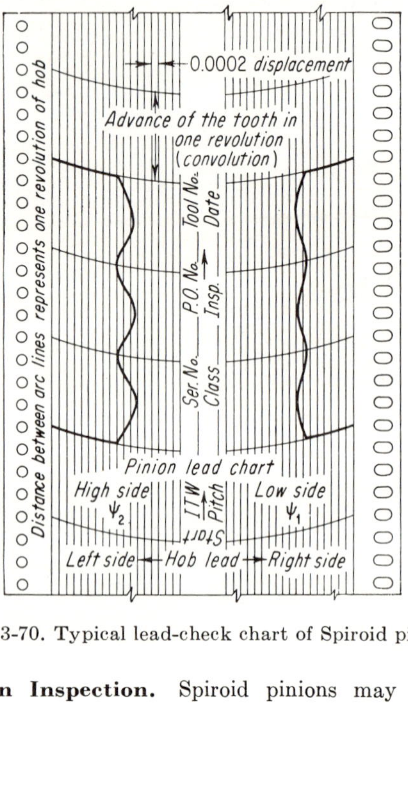

FIG. 23-70. Typical lead-check chart of Spiroid pinion.

23-15. Spiroid-pinion Inspection. Spiroid pinions may be checked for the following:

1. Lead
2. Profile
3. Thread spacing
4. Contact pattern
5. Mounting distance

Figure 23-69 shows a machine that can be used to check Spiroid-pinion lead, profile, and spacing. These checks are similar to that of a cylindrical worm (see Art. 23-12). The indicator follows the thread to measure lead variation. The machine is set to traverse the indicator across the thread to measure profile. Generally the design calls for a straight-sided profile. (A slight ease-off may be worked into the design dimensions.) The spacing can be determined by using an index plate and reindexing for each thread.

The contact check is part of the check of the mating gear and was discussed in Art. 23-14.

Figure 23-70 shows a typical lead-check chart for a Spiroid pinion. Figure 23-71 shows how a profile check may be made on a sample cut of a Spiroid-pinion blank.

As in the case of the Spiroid gear, the tooth thickness is determined indirectly by determining the mounting distance. The mounting distance may be determined on

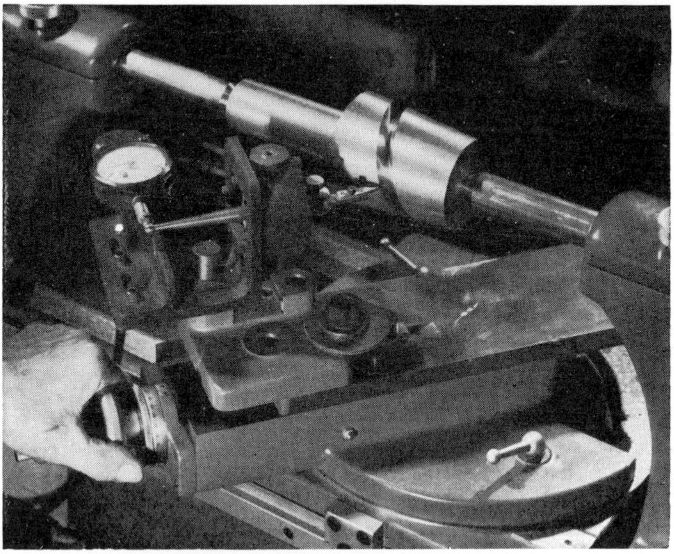

FIG. 23-71. Check of Spiroid-pinion profile on a test cut.

FIG. 23-72. Device for checking mounting distance of a Spiroid pinion. (*Courtesy of Spiroid Div. of Illinois Tool Works, Inc., Chicago, Ill.*)

the contacting machine—provided a gear of known size is used. A special tool may be used to measure the mounting distance of the pinion alone. Such a tool is shown in use in Fig. 23-72. The tool has an internal taper and will fit the outside cone of the Spiroid pinion. Such a tool is handy to calibrate Spiroid pinions when alone. The calibrated pinion can then be used in a contact-checking setup to calibrate Spiroid gears for mounting distance (or indirectly tooth size).

23-16. Spiroid Gearsets. So far there have been no formal trade standards established for Spiroid gearsets. The current practice in the gear industry, however, recognizes three general classes of accuracy. These are:

1. *Class A*. A precision-ground pinion and a precision-cut, and sometimes lapped, gear of aircraft quality, but not so precise as some instrument or missile-guidance-system gearing.

2. *Class B.* A high-class commercial-quality ground or cut set of pinion and gear.
3. *Class C.* Commercial quality. Pinion may be cut, thread-rolled, or ground from the solid. Gears may be production-cut, die-cast, sintered, or injection molded.

Table 23-7 shows the relative tolerances used for each of the three classes just mentioned.

Table 23-8. Spiroid-gear Tolerances

Center distance *C*	Class	Pinion mounting distance *A* ±	Center distance *C* ±	Gear face *R/o E*, TIR	Av. backlash TIR
0.500	A	0.006	0.0010	0.0005	0.002
	B	0.012	0.0015	0.0010	0.003
	C	0.018	0.0020	0.0020	0.004
1.00	A	0.007	0.0010	0.0010	0.003
	B	0.014	0.0015	0.0025	0.004
	C	0.021	0.0020	0.0045	0.005
2.00	A	0.009	0.0015	0.0020	0.004
	B	0.018	0.0020	0.0040	0.005
	C	0.027	0.0040	0.0080	0.007
3.00	A	0.011	0.0015	0.0030	0.005
	B	0.022	0.0020	0.0060	0.006
	C	0.033	0.0040	0.0110	0.009
4.00	A	0.014	0.0020	0.0035	0.006
	B	0.028	0.0030	0.0070	0.008
	C	0.042	0.0050	0.0140	0.012
5.00	A	0.015	0.0020	0.0040	0.007
	B	0.030	0.0030	0.0080	0.010
	C	0.045	0.0050	0.0160	0.015

It is important that the housings for Spiroid gearsets make the pinion axis square with the gear axis. The pinion axis should be square with the gear axis within 0.0003 in./in. for a class A application, within 0.001 in./in. for a class B system, and within 0.002 in./in. for a class C system.

The Spiroid gearset has the unique property that the backlash can be adjusted conveniently and without any particular effect on the quality of the contact pattern. The adjustment is made by changing mounting distance. Fine adjustment of backlash is most conveniently done by axially shimming the pinion. Since it is readily possible to set the backlash, this becomes a regular procedure in the assembly and inspection of Spiroid gearsets. Other right-angle drives can have backlash changed by axial change of mounting distance, but, except for Beveloid®* gears, they tend to be sensitive to axial position from the standpoint of contact pattern. This means that Spiroid and Beveloid gearsets are the only kind of gear where the setting of backlash can be done quite freely and without hazard to the contact pattern.

INSPECTION OF SPECIAL GEAR TYPES

Each special gear type requires some special procedure in its inspection. Generally, though, much of the inspection is similar to the inspection of the more common

* Registered trade-mark of Vinco Corp., Detroit, Mich.

types already discussed. Some brief comments will be given on the inspection of face gears and Beveloid gears.

23-17. Face-gear Inspection. Since the profile of a face gear is not an involute, it cannot be checked on an involute checker. However, tooth spacing may be inspected with a dividing head, and a runout check may be made with a cone point. In addition, a composite check may be performed in tight mesh with a master gear which is representative of the mating pinion. The tooth thickness of the master gear should exceed that of the mating pinion by the mean backlash of the assembled gearset in order to maintain the assembly center distance as closely as possible. This is necessary because conjugate action of the face gear is limited to an involute gear with the same number of teeth as the mating pinion meshing at the same center distance.

The mating pinion is a conventional spur or helical gear of involute form. It may be inspected by the methods given earlier in the chapter. (See Arts. 23-1 through 23-6.)

23-18. Beveloid-gear Inspection. The tooth sides of Beveloid gears are involute helicoids, and inspection methods discussed for helical involute gears are applicable. No special equipment is required for the analytical inspection of tooth elements, such as profile, lead, and spacing. Since Beveloid gears are conjugate to the basic rack of the involute family, they can also be roll tested for composite error or functional tooth thickness with the same master gears that are used for checking other involute gears. The only requirement is a special adapter which mounts the Beveloid gear at its Beveloid angle.

REFERENCES

1. American Gear Manufacturers Association: Gear Classification Manual for Spur, Helical and Herringbone Gears, AGMA *Publ.* 390.01, August, 1960.
2. American Gear Manufacturers Association: Inspection of Coarse-pitch Spur and Helical Gears, AGMA *Publ.* 231.02, July, 1956.
3. American Gear Manufacturers Association: Inspection of Fine-pitch Gears, AGMA *Publ.* 236.04, June, 1956.
4. American Gear Manufacturers Association: Standard Master Gears, AGMA *Publ.* 235.02, April, 1960.
5. American Gear Manufacturers Association: Data Gears Accuracy Specifications, AGMA *Publ.* 237.01, March, 1961.
6. American Gear Manufacturers Association: Inspection of Coarse-pitch Bevel and Hypoid Gears, AGMA *Publ.* 232.02, August, 1956.
7. American Gear Manufacturers Association: Inspection of Coarse-pitch Cylindrical Worms and Wormgears, AGMA *Publ.* 234.01, August, 1956.
8. Berndt, G.: "Grundlagen fuer die Messung von Stirnraedern," Springer-Verlag, Berlin, 1938.
9. Buckingham, Earle: "Spur Gears," McGraw-Hill Book Company, Inc., 1928.
10. Henriot, G.: "Traite theorique et pratique des engrenages," Dunod, Paris, 1949.
11. Houghton, P. S.: "Gears, Spur, Helical, Bevel and Worm," The Technical Press, Ltd., 1952.
12. Krumme, Walter: "Praktische Verzahnungstechnik," Carl Hanser, Munich, 1946.
13. Parkinson, A. C., and W. H. Dawney: "Gears, Gear Production and Measurement," Pitman Publishing Corporation, New York, 1948.
14. Trautschold, Reginald: "Gear Design and Production," Columbia Graphs, Connecticut, 1955.
15. Trautschold, Reginald: "Standard Gear Book," McGraw-Hill Book Company, Inc., 1935.
16. VDI-Berichte: "Messen an Zahnraedern und Getrieben Band 32," 1959.
17. Gleason Works: "How to Test Bevel Gears," Gleason Works, Rochester, N.Y.
18. Michigan Tool Co.: "Cone-Drive Gears—Catalog No. 70," Cone-Drive Div., Michigan Tool Co., Detroit, Mich.
19. Illinois Tool Works: "Spiroid Gears," Spiroid Div., Illinois Tool Works, Chicago, Ill.

Chapter 24

NUMERICAL DATA TABLES

By

ROBERT T. PARSONS, *President, The Van Keuren Company, Boston, Mass.*

There are several tables that are used very frequently in gear-design work. These tables and the directions for using them are given in this chapter.

24-1. Gear-measuring Tables. The following gear-measuring tables* are used extensively by gear manufacturers and users and provide a quick means of checking the tooth thickness of gears at the pitch diameter. A measurement over wires is made and compared with the measurement for a perfect gear which is obtainable from the

Table 24-1. Guide to Wire Selection

Series	Relation	Use
(1) $\frac{1.92''}{P_d}$		(*a*) Enlarged pinions. Wires project above OD* (*b*) Alternate for 14½ and 20° pressure angle standard-addendum external gears
(2) $\frac{1.728''}{P_d}$	$\frac{1.92 - 10\%}{P_d}$	(*a*) External standard-addendum spur and helical gears. Wires project above OD*
(3) $\frac{1.68''}{P_d}$	$\frac{1.92 - 12\frac{1}{2}\%}{P_d}$	(*a*) Suitable for 14½° internal gears above 31 teeth and 20° internal gears above 29 teeth. Wires project below ID* (*b*) Alternate for standard-addendum external gears
(4) $\frac{1.44''}{P_d}$	$\frac{1.92 - 25\%}{P_d}$	(*a*) Internal standard-addendum gears. Wires make good contact on gear teeth but do not project below ID

* Included in "Inspection of Fine Pitch Gears," ASA B6.11-1956 and "Standard Pin Measurement Tables," AGMA 231.51.

tables. The values in the tables are accurate to 0.0001″ for 1-diametral-pitch gears. After the value is selected in the proper table for the specified pressure angle, number of teeth, and wire size to be used, this value is divided by the diametral pitch of the gear being measured. If the actual measurement is greater than the theoretical value the teeth are thick, and if it is less the teeth are thin. The change factors K_m which

* Gear-measuring tables and explanations in Art. 24-1 are taken from The Van Keuren Handbook 36 with permission.

permit easy conversion of oversize or undersize readings into the approximate amount the teeth are thick or thin are tabulated alongside the measurement values.

In this chapter the measurement tables (except for Table 24-7 covering 25° gears) are presented so that for each pressure-angle measurement values are available based on two or more wire diameters. In general, it is recommended that wires from the 1.728″ series be used for measuring external gears and that wires from the 1.44″ series be used for internals. The tabulated values for the other wire diameters, however, are available for checking involute profiles and to permit greater freedom in wire selection. Table 24-1 will serve as a guide to the selection of a wire diameter for a particular application.

Table 24-2. Gear Wire Sizes

P_d	$\frac{1.92''}{P_d}$	$\frac{1.728''}{P_d}$	$\frac{1.68''}{P_d}$	$\frac{1.44''}{P_d}$
1	1.92	1.728	1.68	1.44
2	0.960	0.864	0.840	0.720
2½	0.768	0.6912	0.672	0.576
3	0.640	0.576	0.560	0.480
4	0.480	0.432	0.420	0.360
5	0.384	0.3456	0.336	0.288
6	0.320	0.288	0.280	0.240
7	0.27428	0.24686	0.240	0.20571
8	0.240	0.216	0.210	0.180
9	0.21333	0.192	0.18666	0.160
10	0.192	0.1728	0.168	0.144
11	0.17454	0.15709	0.15273	0.13091
12	0.160	0.144	0.140	0.120
14	0.13714	0.12343	0.120	0.10286
16	0.120	0.108	0.105	0.090
18	0.10667	0.096	0.09333	0.080
20	0.096	0.0864	0.084	0.072
22	0.08727	0.07855	0.07636	0.06545
24	0.080	0.072	0.070	0.060
28	0.06857	0.06171	0.060	0.05143
32	0.060	0.054	0.0525	0.045
36	0.05333	0.048	0.04667	0.040
40	0.048	0.0432	0.042	0.036
48	0.040	0.036	0.035	0.030
64	0.030	0.027	0.02625	0.0225
72	0.02667	0.024	0.02333	0.020
80	0.024	0.0216	0.021	0.018
96	0.020	0.018	0.0175	0.015
100	0.0192	0.01728	0.0168	0.0144
120	0.0160	0.0144	0.014	0.012

NOTE: For stub-tooth gears use numerator diametral pitch for selecting wire size.

Numerical Examples

1. An even-tooth external gear of 26 teeth, 10 diametral pitch, 14½° pressure angle is measured with 0.1728″ diameter wires. From Table 24-4, column 2, the measurement for 1 P_d is 28.4314″. For 10 P_d the measurement over wires is 28.4314/10 = 2.8431″. If the teeth are to be cut 0.003″ thin, the reduced

measurement over wires can be determined by using the change factor K_m. From column 3, Table 24-4, the change factor is 2.90. Thus the reduction in measurement for teeth 0.003″ thin would be 0.003″ × 2.90, or 0.0087″. The measurement over wires with teeth 0.003″ thin would be 2.8431 − 0.0087, or 2.8344″.

2. An even-tooth external gear of 20 teeth, 20 diametral pitch, 20° pressure angle with teeth to be cut 0.004″ thin is measured with three sizes of wires 0.096″, 0.0864″, and 0.084″. Using Tables 24-2 and 24-3 the following simple computations are made:

Wire size 1 P_d	1.92″	1.728″	1.68″
Wire size 20 P_d	0.096″	0.0864″	0.084″
Measurement 1 P_d	23.0227″	22.3900″	22.2249″
Measurement 20 P_d	1.1511″	1.1195″	1.1112″
Change factor K_m	2.07″	2.26″	2.33″
Teeth to be thin	0.004″	0.004″	0.004″
Reduction in measurement for thin teeth 0.004″ × K_m	0.0083″	0.0090″	0.0093″
Measurement with teeth 0.004″ thin	1.1428″	1.1105″	1.1019″

In using three sizes of wires in this case it is possible to measure the tooth thickness at the pitch diameter and also to obtain a check on the involute profile.

3. An odd-tooth internal gear of 51 teeth, 8 diametral pitch, 20° pressure angle is measured with 0.180″ diameter wires, which, when placed in the tooth spaces of the gear are not diametrically opposite by ½ tooth interval. From Table 24-5, column 2, the measurement between wires for 1 P_d is 49.6404″. For 8 P_d the measurement is 49.6404″/8, or 6.20505″. The change factor K_m from Table 24-5 is 2.71, and this when multiplied by the amount the teeth are to be thin will indicate the amount of increase in measurement between wires.

Interpretation of Results. The approximate amount the teeth are thick or thin may be determined by the change factors K_m and the amount the actual measurement differs from the computed measurement. To determine the variations in tooth thickness the following relationship is noted:

$$\Delta t = \frac{\Delta M}{K_m} \tag{24-1}$$

where Δt equals the amount the teeth are thick or thin, ΔM equals the difference between computed and actual measurement, and K_m equals the change factor taken from the appropriate table.

Odd-tooth Gears. It should be noted in Tables 24-3 to 24-7 that the measurement values for odd numbers of teeth represent dimensions over two wires that are not diametrically opposite to each other by ½ tooth interval. Thus to determine the dimension from the top of one wire to the *center of the gear* the following computation should be used:

$$\text{Dimension over wires} - \text{wire diameter} = \text{dimension between wire centers} \tag{24-2}$$

$$\frac{\text{Dimension between wire centers}}{\cos\,(90°/N)} \times 0.5 + \frac{\text{wire diameter}}{2} = \text{radius from top of one wire to center of gear} \tag{24-3}$$

Special Pressure Angles and Number of Teeth. Because of the relative simplicity of the exact formulas (see Arts. 7-9 and 7-10) it is not recommended that interpolation be made from the tables in the case of special pressure angles. Where the number of teeth is greater than is to be found in the appropriate table an interpolation may easily be made. Thus from Table 24-3 the measurement over 1.728″ diameter wires for a 1 diametral pitch 335 tooth 20° external gear would be

$$303.4538 + \left(\frac{34}{100} \times 100.0027\right) = 337.4547 \tag{24-4}$$

Measurement of Racks. For 1.728″ wires the decimal values of Tables 24-3, 24-4, and 24-7 show the amount the measurement over wires exceeds the outside diameter of a standard even-tooth gear. Therefore, if the decimal value given for infinity (∞) is divided by 2 the amount the 1.728″ wire projects above the top of the 1-pitch rack is obtained. These values are as follows:

14½°	20°	25°
0.2779″	0.2323″	0.2241″

Since the addendum of a 1-pitch gear is 1.0000″ the amount the wire projects above the pitch line is 1.0000 plus the above values.

For any other diametral pitch P_d divide the values by P_d and use wires 1.728″/P_d in diameter.

Index to Tables

Pressure angle	Type	Wire diameters	Table	Page
		Standard Addendum		
20°	External	1.728, 1.92, 1.68	24–3	24–5
14½°	External	1.728, 1.92, 1.68	24–4	24–9
20°	Internal	1.68, 1.44	24–5	24–13
14½°	Internal	1.68, 1.44	24–6	24–16
25°	External	1.728, 1.68	24–7	24–17
		Long Addendum		
20°	External	1.92	24–9	24–18
14½°	External	1.92	24–10	24–19

All measurement values in tables are in inches. For numerical examples see pages 24–2 and 24–3. For basic formulas and equations used to compute values in the tables, see Arts. 7-9 and 7-10.

Table 24-3. External Gears. 20° Pressure Angle

Measurements over wires and change factors for 1 diametral pitch external standard-addendum 20° pressure angle spur gears with 1.728, 1.92, and 1.68 diameter wires.
For any other diametral pitch divide measurements given in table by the diametral pitch.
Change factors are the same for all diametral pitches.

No. of teeth	1.728 wire dia.		Alternate 1.92 wire dia.		Alternate 1.68 wire dia.	
	M	K_m	M	K_m	M	K_m
5	7.0153	1.67	7.5271	1.48	6.8800	1.75
6	8.3032	1.83	8.8449	1.62	8.1600	1.91
7	9.1260	1.88	9.6702	1.67	8.9822	1.96
8	10.3271	1.94	10.8907	1.72	10.1783	2.01
9	11.1905	1.98	11.7573	1.77	11.0410	2.06
10	12.3445	2.01	12.9252	1.81	12.1914	2.10
11	13.2332	2.05	13.8173	1.84	13.0795	2.13
12	14.3578	2.09	14.9525	1.87	14.2013	2.16
13	15.2639	2.12	15.8618	1.91	15.1068	2.19
14	16.3683	2.14	16.9748	1.93	16.2091	2.21
15	17.2871	2.17	17.8964	1.96	17.1273	2.24
16	18.3768	2.19	18.9934	1.98	18.2154	2.26
17	19.3053	2.21	19.9244	2.01	19.1432	2.28
18	20.3840	2.23	21.0091	2.03	20.2205	2.29
19	21.3200	2.25	21.9475	2.05	21.1561	2.31
20	22.3900	2.26	23.0227	2.07	22.2249	2.33
21	23.3321	2.28	23.9670	2.08	23.1665	2.34
22	24.3952	2.29	25.0346	2.10	24.2286	2.35
23	25.3423	2.30	25.9837	2.12	25.1754	2.36
24	26.3997	2.32	27.0450	2.13	26.2318	2.38
25	27.3511	2.33	27.9982	2.15	27.1828	2.39
26	28.4036	2.34	29.0543	2.16	28.2346	2.40
27	29.3586	2.35	30.0109	2.17	29.1892	2.41
28	30.4071	2.36	31.0626	2.19	30.2371	2.42
29	31.3652	2.37	32.0222	2.20	31.1948	2.43
30	32.4102	2.38	33.0701	2.21	32.2392	2.43
31	33.3710	2.39	34.0323	2.22	33.1997	2.44
32	34.4130	2.40	35.0768	2.23	34.2412	2.45
33	35.3761	2.41	36.0413	2.24	35.2041	2.45
34	36.4155	2.41	37.0830	2.25	36.2430	2.46
35	37.3807	2.42	38.0495	2.26	37.2079	2.47
36	38.4178	2.43	39.0886	2.27	38.2445	2.48
37	39.3849	2.43	40.0569	2.28	39.2115	2.48
38	40.4198	2.44	41.0938	2.29	40.2460	2.49
39	41.3886	2.45	42.0636	2.29	41.2147	2.49
40	42.4217	2.45	43.0986	2.30	42.2473	2.50
41	43.3920	2.46	44.0699	2.31	43.2174	2.50
42	44.4234	2.46	45.1030	2.32	44.2485	2.51
43	45.3951	2.47	46.0756	2.32	45.2200	2.51
44	46.4250	2.47	47.1071	2.33	46.2496	2.52
45	47.3980	2.48	48.0809	2.34	47.2224	2.52
46	48.4265	2.48	49.1109	2.34	48.2506	2.53
47	49.4007	2.49	50.0858	2.35	49.2246	2.53
48	50.4279	2.49	51.1144	2.35	50.2516	2.53
49	51.4031	2.50	52.0903	2.36	51.2266	2.54
50	52.4292	2.50	53.1177	2.37	52.2525	2.54

Table 24-3. External Gears. 20° Pressure Angle (*Continued*)

No. of teeth	1.728 wire dia.		Alternate 1.92 wire dia.		Alternate 1.68 wire dia.	
	M	K_m	M	K_m	M	K_m
51	53.4053	2.50	54.0945	2.37	53.2284	2.54
52	54.4304	2.51	55.1208	2.38	54.2533	2.55
53	55.4074	2.51	56.0985	2.38	55.2302	2.55
54	56.4315	2.52	57.1237	2.39	56.2541	2.55
55	57.4093	2.52	58.1022	2.39	57.2318	2.56
56	58.4325	2.52	59.1265	2.40	58.2548	2.56
57	59.4111	2.53	60.1057	2.40	59.2333	2.56
58	60.4335	2.53	61.1291	2.41	60.2555	2.56
59	61.4128	2.53	62.1090	2.41	61.2347	2.57
60	62.4344	2.54	63.1315	2.41	62.2561	2.57
61	63.4144	2.54	64.1121	2.42	63.2360	2.57
62	64.4352	2.54	65.1338	2.42	64.2567	2.57
63	65.4159	2.54	66.1150	2.43	65.2372	2.58
64	66.4361	2.55	67.1360	2.43	66.2572	2.58
65	67.4173	2.55	68.1177	2.43	67.2383	2.58
66	68.4369	2.55	69.1381	2.44	68.2577	2.58
67	69.4186	2.55	70.1203	2.44	69.2394	2.59
68	70.4376	2.56	71.1401	2.44	70.2582	2.59
69	71.4198	2.56	72.1228	2.45	71.2405	2.59
70	72.4383	2.56	73.1420	2.45	72.2587	2.59
71	73.4210	2.56	74.1252	2.45	73.2414	2.59
72	74.4390	2.57	75.1438	2.46	74.2592	2.60
73	75.4221	2.57	76.1274	2.46	75.2423	2.60
74	76.4396	2.57	77.1455	2.46	76.2596	2.60
75	77.4232	2.57	78.1295	2.47	77.2432	2.60
76	78.4402	2.57	79.1471	2.47	78.2600	2.60
77	79.4242	2.58	80.1316	2.47	79.2440	2.60
78	80.4408	2.58	81.1486	2.48	80.2604	2.61
79	81.4252	2.58	82.1336	2.48	81.2448	2.61
80	82.4413	2.58	83.1501	2.48	82.2607	2.61
81	83.4262	2.58	84.1354	2.48	83.2456	2.61
82	84.4418	2.58	85.1516	2.49	84.2611	2.61
83	85.4271	2.59	86.1372	2.49	85.2463	2.61
84	86.4423	2.59	87.1529	2.49	86.2614	2.61
85	87.4279	2.59	88.1389	2.49	87.2470	2.62
86	88.4428	2.59	89.1542	2.50	88.2617	2.62
87	89.4287	2.59	90.1405	2.50	89.2476	2.62
88	90.4433	2.59	91.1555	2.50	90.2620	2.62
89	91.4295	2.60	92.1421	2.50	91.2482	2.62
90	92.4437	2.60	93.1567	2.50	92.2624	2.62
91	93.4303	2.60	94.1436	2.51	93.2489	2.62
92	94.4441	2.60	95.1579	2.51	94.2626	2.63
93	95.4310	2.60	96.1450	2.51	95.2494	2.63
94	96.4445	2.60	97.1590	2.51	96.2629	2.63
95	97.4317	2.60	98.1464	2.52	97.2500	2.63
96	98.4449	2.61	99.1601	2.52	98.2632	2.63
97	99.4323	2.61	100.1477	2.52	99.2506	2.63
98	100.4453	2.61	101.1611	2.52	100.2635	2.63
99	101.4329	2.61	102.1490	2.53	101.2511	2.63
100	102.4456	2.61	103.1621	2.53	102.2638	2.63

Table 24-3. External Gears. 20° Pressure Angle (*Continued*)

No. of teeth	1.728 wire dia.		Alternate 1.92 wire dia.		Alternate 1.68 wire dia.	
	M	K_m	M	K_m	M	K_m
101	103.4335	2.61	104.1502	2.53	103.2516	2.64
102	104.4460	2.61	105.1631	2.53	104.2640	2.64
103	105.4341	2.61	106.1514	2.53	105.2520	2.64
104	106.4463	2.62	107.1640	2.53	106.2642	2.64
105	107.4346	2.62	108.1526	2.53	107.2525	2.64
106	108.4466	2.62	109.1649	2.54	108.2644	2.64
107	109.4352	2.62	110.1537	2.54	109.2529	2.64
108	110.4469	2.62	111.1658	2.54	110.2645	2.64
109	111.4357	2.62	112.1548	2.54	111.2533	2.64
110	112.4472	2.62	113.1666	2.54	112.2647	2.64
111	113.4362	2.62	114.1558	2.54	113.2537	2.65
112	114.4475	2.62	115.1675	2.55	114.2649	2.65
113	115.4367	2.63	116.1568	2.55	115.2541	2.65
114	116.4478	2.63	117.1683	2.55	116.2651	2.65
115	117.4372	2.63	118.1578	2.55	117.2544	2.65
116	118.4481	2.63	119.1690	2.55	118.2653	2.65
117	119.4376	2.63	120.1587	2.55	119.2548	2.65
118	120.4484	2.63	121.1698	2.55	120.2655	2.65
119	121.4380	2.63	122.1597	2.56	121.2552	2.65
120	122.4486	2.63	123.1705	2.56	122.2656	2.65
121	123.4384	2.63	124.1605	2.56	123.2555	2.65
122	124.4489	2.63	125.1712	2.56	124.2658	2.65
123	125.4388	2.63	126.1614	2.56	125.2558	2.65
124	126.4491	2.64	127.1719	2.56	126.2660	2.65
125	127.4392	2.64	128.1622	2.56	127.2562	2.66
126	128.4493	2.64	129.1725	2.56	128.2661	2.66
127	129.4396	2.64	130.1630	2.57	129.2565	2.66
128	130.4496	2.64	131.1732	2.57	130.2663	2.66
129	131.4400	2.64	132.1638	2.57	131.2568	2.66
130	132.4498	2.64	133.1738	2.57	132.2664	2.66
131	133.4404	2.64	134.1646	2.57	133.2571	2.66
132	134.4500	2.64	135.1744	2.57	134.2666	2.66
133	135.4408	2.64	136.1653	2.57	135.2574	2.66
134	136.4502	2.64	137.1750	2.57	136.2667	2.66
135	137.4411	2.64	138.1661	2.58	137.2577	2.66
136	138.4504	2.64	139.1756	2.58	138.2669	2.66
137	139.4414	2.64	140.1668	2.58	139.2579	2.66
138	140.4506	2.65	141.1761	2.58	140.2670	2.66
139	141.4418	2.65	142.1674	2.58	141.2582	2.66
140	142.4508	2.65	143.1767	2.58	142.2671	2.66
141	143.4421	2.65	144.1681	2.58	143.2584	2.67
142	144.4510	2.65	145.1772	2.58	144.2672	2.67
143	145.4424	2.65	146.1688	2.58	145.2587	2.67
144	146.4512	2.65	147.1777	2.58	146.2674	2.67
145	147.4427	2.65	148.1694	2.59	147.2589	2.67
146	148.4513	2.65	149.1782	2.59	148.2675	2.67
147	149.4430	2.65	150.1700	2.59	149.2591	2.67
148	150.4515	2.65	151.1787	2.59	150.2676	2.67
149	151.4433	2.65	152.1706	2.59	151.2594	2.67
150	152.4516	2.65	153.1792	2.59	152.2677	2.67

Table 24-3. External Gears. 20° Pressure Angle (*Continued*)

No. of teeth	1.728 wire dia.		Alternate 1.92 wire dia.		Alternate 1.68 wire dia.	
	M	K_m	M	K_m	M	K_m
151	153.4435	2.65	154.1712	2.59	153.2596	2.67
152	154.4518	2.65	155.1797	2.59	154.2678	2.67
153	155.4438	2.65	156.1718	2.59	155.2598	2.67
154	156.4520	2.66	157.1801	2.59	156.2679	2.67
155	157.4440	2.66	158.1723	2.60	157.2600	2.67
156	158.4521	2.66	159.1806	2.60	158.2680	2.67
157	159.4443	2.66	160.1729	2.60	159.2602	2.67
158	160.4523	2.66	161.1810	2.60	160.2681	2.67
159	161.4445	2.66	162.1734	2.60	161.2604	2.67
160	162.4524	2.66	163.1814	2.60	162.2682	2.67
161	163.4448	2.66	164.1739	2.60	163.2606	2.68
162	164.4526	2.66	165.1818	2.60	164.2683	2.68
163	165.4450	2.66	166.1744	2.60	165.2608	2.68
164	166.4527	2.66	167.1822	2.60	166.2684	2.68
165	167.4453	2.66	168.1749	2.60	167.2610	2.68
166	168.4528	2.66	169.1826	2.60	168.2685	2.68
167	169.4455	2.66	170.1754	2.60	169.2611	2.68
168	170.4529	2.66	171.1830	2.61	170.2686	2.68
169	171.4457	2.66	172.1759	2.61	171.2613	2.68
170	172.4531	2.66	173.1834	2.61	172.2687	2.68
171	173.4459	2.66	174.1763	2.61	173.2615	2.68
172	174.4532	2.66	175.1838	2.61	174.2688	2.68
173	175.4461	2.66	176.1768	2.61	175.2617	2.68
174	176.4533	2.67	177.1842	2.61	176.2689	2.68
175	177.4463	2.67	178.1773	2.61	177.2619	2.68
176	178.4535	2.67	179.1846	2.61	178.2690	2.68
177	179.4465	2.67	180.1777	2.61	179.2621	2.68
178	180.4536	2.67	181.1849	2.61	180.2691	2.68
179	181.4467	2.67	182.1781	2.61	181.2622	2.68
180	182.4537	2.67	183.1852	2.61	182.2691	2.68
181	183.4469	2.67	184.1785	2.61	183.2623	2.68
182	184.4538	2.67	185.1856	2.62	184.2692	2.68
183	185.4471	2.67	186.1789	2.62	185.2625	2.68
184	186.4539	2.67	187.1859	2.62	186.2693	2.68
185	187.4473	2.67	188.1793	2.62	187.2627	2.68
186	188.4540	2.67	189.1862	2.62	188.2694	2.68
187	189.4474	2.67	190.1797	2.62	189.2628	2.68
188	190.4541	2.67	191.1865	2.62	190.2694	2.68
189	191.4476	2.67	192.1801	2.62	191.2629	2.69
190	192.4542	2.67	193.1868	2.62	192.2694	2.69
200	202.4548	2.68	203.1883	2.63	202.2698	2.69
201	203.4487	2.68	204.1822	2.63	203.2636	2.69
300	302.4579	2.70	303.1977	2.66	302.2719	2.71
301	303.4538	2.70	304.1937	2.66	303.2678	2.71
400	402.4596	2.71	403.2026	2.68	402.2730	2.72
401	403.4565	2.71	404.1996	2.68	403.2699	2.72
500	502.4606	2.72	503.2056	2.70	502.2736	2.72
501	503.4581	2.72	504.2032	2.70	503.2711	2.72
∞	$(N+2)$.4646	2.75	$(N+3)$.2180	2.75	$(N+2)$.2762	2.75

Table 24-4. External Gears. 14½° Pressure Angle

Measurements over wires and change factors for 1 diametral pitch external standard-addendum 14½° pressure angle spur gears with 1.728, 1.92, and 1.68 diameter wires. For any other diametral pitch divide measurements given in table by the diametral pitch. Change factors are the same for all diametral pitches.

No. of teeth	1.728 wire dia.		Alternate 1.92 wire dia.		Alternate 1.68 wire dia.	
	M	K_m	M	K_m	M	K_m
5	6.9936	1.90	7.5296	1.61	6.8485	2.03
6	8.2846	2.09	8.8551	1.77	8.1298	2.23
7	9.1116	2.17	9.6871	1.84	8.9555	2.31
8	10.3160	2.24	10.9147	1.90	10.1535	2.39
9	11.1829	2.31	11.7872	1.96	11.0189	2.45
10	12.3399	2.36	12.9617	2.01	12.1712	2.51
11	13.2317	2.42	13.8590	2.06	13.0615	2.57
12	14.3590	2.47	15.0002	2.11	14.1851	2.62
13	15.2677	2.51	15.9143	2.15	15.0925	2.66
14	16.3746	2.55	17.0327	2.19	16.1964	2.70
15	17.2957	2.59	17.9588	2.23	17.1163	2.74
16	18.3877	2.63	19.0606	2.26	18.2058	2.78
17	19.3182	2.66	19.9958	2.29	19.1351	2.81
18	20.3989	2.70	21.0850	2.33	20.2137	2.84
19	21.3368	2.73	22.0272	2.36	21.1505	2.87
20	22.4087	2.76	23.1065	2.39	22.2205	2.90
21	23.3524	2.78	24.0543	2.41	23.1634	2.93
22	24.4172	2.81	25.1257	2.44	24.2265	2.95
23	25.3658	2.83	26.0781	2.46	25.1743	2.98
24	26.4247	2.86	27.1430	2.49	26.2317	3.00
25	27.3774	2.88	28.0992	2.51	27.1836	3.02
26	28.4314	2.90	29.1586	2.53	28.2363	3.04
27	29.3876	2.92	30.1181	2.56	29.1918	3.06
28	30.4374	2.94	31.1729	2.58	30.2404	3.08
29	31.3966	2.96	32.1351	2.60	31.1990	3.09
30	32.4429	2.98	33.1859	2.62	32.2441	3.11
31	33.4047	2.99	34.1506	2.63	33.2053	3.13
32	34.4478	3.01	35.1979	2.65	34.2475	3.14
33	35.4119	3.03	36.1647	2.67	35.2110	3.16
34	36.4523	3.04	37.2090	2.69	36.2505	3.17
35	37.4185	3.06	38.1777	2.70	37.2161	3.18
36	38.4565	3.07	39.2193	2.72	38.2533	3.19
37	39.4245	3.08	40.1896	2.73	39.2208	3.21
38	40.4603	3.10	41.2288	2.75	40.2558	3.22
39	41.4299	3.11	42.2007	2.76	41.2249	3.23
40	42.4638	3.12	43.2378	2.78	42.2582	3.24
41	43.4348	3.13	44.2110	2.79	43.2287	3.25
42	44.4671	3.14	45.2461	2.80	44.2604	3.26
43	45.4394	3.15	46.2205	2.82	45.2323	3.27
44	46.4701	3.17	47.2539	2.83	46.2624	3.28
45	47.4437	3.18	48.2294	2.84	47.2355	3.29
46	48.4729	3.19	49.2613	2.85	48.2642	3.30
47	49.4477	3.20	50.2378	2.87	49.2385	3.31
48	50.4756	3.21	51.2682	2.88	50.2660	3.32
49	51.4514	3.21	52.2457	2.89	51.2413	3.33
50	52.4781	3.22	53.2748	2.90	52.2676	3.33

Table 24-4. External Gears. 14½° Pressure Angle *(Continued)*

No. of teeth	1.728 wire dia.		Alternate 1.92 wire dia.		Alternate 1.68 wire dia.	
	M	K_m	M	K_m	M	K_m
51	53.4547	3.23	54.2531	2.91	53.2439	3.34
52	54.4804	3.24	55.2810	2.92	54.2691	3.35
53	55.4579	3.24	56.2601	2.93	55.2463	3.35
54	56.4826	3.25	57.2868	2.94	56.2705	3.36
55	57.4609	3.26	58.2667	2.95	57.2485	3.37
56	58.4847	3.27	59.2924	2.96	58.2719	3.37
57	59.4637	3.28	60.2729	2.97	59.2506	3.38
58	60.4866	3.28	61.2977	2.98	60.2731	3.39
59	61.4664	3.29	62.2788	2.99	61.2526	3.39
60	62.4884	3.30	63.3027	3.00	62.2743	3.40
61	63.4689	3.31	64.2844	3.01	63.2545	3.40
62	64.4902	3.31	65.3075	3.01	64.2755	3.41
63	65.4712	3.32	66.2898	3.02	65.2562	3.42
64	66.4918	3.32	67.3121	3.03	66.2765	3.42
65	67.4734	3.33	68.2949	3.04	67.2579	3.43
66	68.4933	3.34	69.3165	3.05	68.2775	3.43
67	69.4755	3.34	70.2998	3.05	69.2594	3.44
68	70.4948	3.35	71.3206	3.06	70.2785	3.44
69	71.4775	3.35	72.3044	3.07	71.2609	3.45
70	72.4963	3.36	73.3247	3.07	72.2794	3.45
71	73.4795	3.37	74.3089	3.08	73.2623	3.46
72	74.4977	3.37	75.3285	3.09	74.2803	3.46
73	75.4813	3.38	76.3131	3.10	75.2636	3.46
74	76.4990	3.38	77.3322	3.10	76.2811	3.47
75	77.4830	3.39	78.3172	3.11	77.2649	3.47
76	78.5002	3.39	79.3357	3.12	78.2819	3.48
77	79.4847	3.40	80.3211	3.12	79.2661	3.48
78	80.5014	3.40	81.3391	3.13	80.2827	3.49
79	81.4863	3.41	82.3249	3.13	81.2673	3.49
80	82.5026	3.41	83.3423	3.14	82.2834	3.50
81	83.4877	3.42	84.3285	3.15	83.2684	3.50
82	84.5037	3.42	85.3455	3.15	84.2841	3.50
83	85.4892	3.42	86.3319	3.16	85.2694	3.51
84	86.5047	3.43	87.3485	3.16	86.2847	3.51
85	87.4906	3.43	88.3352	3.17	87.2704	3.51
86	88.5057	3.44	89.3515	3.17	88.2854	3.52
87	89.4919	3.44	90.3384	3.18	89.2714	3.52
88	90.5067	3.44	91.3543	3.18	90.2860	3.52
89	91.4932	3.45	92.3415	3.19	91.2723	3.52
90	92.5076	3.45	93.3570	3.19	92.2866	3.53
91	93.4944	3.45	94.3445	3.20	93.2732	3.53
92	94.5085	3.45	95.3596	3.20	94.2872	3.54
93	95.4956	3.46	96.3474	3.21	95.2741	3.54
94	96.5094	3.46	97.3621	3.21	96.2877	3.54
95	97.4967	3.46	98.3502	3.22	97.2749	3.54
96	98.5102	3.46	99.3646	3.22	98.2882	3.55
97	99.4978	3.47	100.3529	3.23	99.2757	3.55
98	100.5110	3.47	101.3670	3.23	100.2887	3.55
99	101.4988	3.47	102.3555	3.24	101.2764	3.55
100	102.5118	3.48	103.3693	3.24	102.2892	3.56

Table 24-4. External Gears. 14½° Pressure Angle (*Continued*)

No. of teeth	1.728 wire dia.		Alternate 1.92 wire dia.		Alternate 1.68 wire dia.	
	M	K_m	M	K_m	M	K_m
101	103.4998	3.49	104.3580	3.24	103.2771	3.56
102	104.5125	3.49	105.3715	3.25	104.2897	3.56
103	105.5008	3.49	106.3604	3.25	105.2778	3.57
104	106.5132	3.50	107.3737	3.26	106.2901	3.57
105	107.5017	3.50	108.3628	3.26	107.2785	3.57
106	108.5139	3.50	109.3758	3.27	108.2905	3.57
107	109.5026	3.50	110.3651	3.27	109.2791	3.58
108	110.5146	3.51	111.3778	3.27	110.2910	3.58
109	111.5035	3.51	112.3673	3.28	111.2798	3.58
110	112.5152	3.51	113.3798	3.28	112.2914	3.58
111	113.5044	3.52	114.3695	3.29	113.2804	3.59
112	114.5159	3.52	115.3817	3.29	114.2918	3.59
113	115.5052	3.52	116.3716	3.29	115.2809	3.59
114	116.5165	3.53	117.3835	3.30	116.2921	3.59
115	117.5060	3.53	118.3736	3.30	117.2815	3.59
116	118.5171	3.53	119.3853	3.30	118.2925	3.60
117	119.5068	3.53	120.3756	3.31	119.2821	3.60
118	120.5177	3.53	121.3871	3.31	120.2929	3.60
119	121.5075	3.53	122.3775	3.31	121.2826	3.60
120	122.5182	3.54	123.3888	3.32	122.2932	3.60
121	123.5082	3.54	124.3793	3.32	123.2831	3.61
122	124.5188	3.54	125.3905	3.32	124.2936	3.61
123	125.5089	3.54	126.3811	3.33	125.2836	3.61
124	126.5193	3.55	127.3921	3.33	126.2939	3.61
125	127.5096	3.55	128.3829	3.33	127.2841	3.61
126	128.5198	3.55	129.3937	3.34	128.2941	3.61
127	129.5103	3.55	130.3846	3.34	129.2846	3.62
128	130.5203	3.55	131.3952	3.34	130.2945	3.62
129	131.5109	3.56	132.3863	3.34	131.2851	3.62
130	132.5208	3.56	133.3967	3.35	132.2948	3.62
131	133.5115	3.56	134.3879	3.35	133.2855	3.62
132	134.5213	3.56	135.3982	3.35	134.2951	3.62
133	135.5121	3.57	136.3895	3.36	135.2859	3.63
134	136.5217	3.57	137.3996	3.36	136.2954	3.63
135	137.5127	3.57	138.3911	3.36	137.2863	3.63
136	138.5221	3.57	139.4010	3.37	138.2957	3.63
137	139.5133	3.57	140.3926	3.37	139.2867	3.63
138	140.5226	3.57	141.4023	3.37	140.2960	3.63
139	141.5139	3.58	142.3940	3.37	141.2871	3.64
140	142.5230	3.58	143.4037	3.38	142.2962	3.64
141	143.5144	3.58	144.3955	3.38	143.2875	3.64
142	144.5234	3.58	145.4050	3.38	144.2965	3.64
143	145.5149	3.58	146.3969	3.39	145.2879	3.64
144	146.5238	3.58	147.4062	3.39	146.2967	3.64
145	147.5154	3.59	148.3982	3.39	147.2883	3.64
146	148.5242	3.59	149.4074	3.39	148.2970	3.64
147	149.5159	3.59	150.3996	3.40	149.2887	3.65
148	150.5246	3.59	151.4087	3.40	150.2972	3.65
149	151.5164	3.59	152.4009	3.40	151.2890	3.65
150	152.5250	3.59	153.4098	3.40	152.2974	3.65

Table 24-4. External Gears. 14½° Pressure Angle *(Continued)*

No. of teeth	1.728 wire dia.		Alternate 1.92 wire dia.		Alternate 1.68 wire dia.	
	M	K_m	M	K_m	M	K_m
151	153.5169	3.60	154.4022	3.41	153.2893	3.65
152	154.5254	3.60	155.4110	3.41	154.2977	3.65
153	155.5174	3.60	156.4034	3.41	155.2897	3.65
154	156.5257	3.60	157.4121	3.41	156.2979	3.65
155	157.5179	3.60	158.4046	3.41	157.2900	3.66
156	158.5261	3.60	159.4132	3.42	158.2981	3.66
157	159.5183	3.60	160.4058	3.42	159.2903	3.66
158	160.5264	3.61	161.4143	3.42	160.2983	3.66
159	161.5188	3.61	162.4070	3.42	161.2906	3.66
160	162.5267	3.61	163.4153	3.43	162.2985	3.66
161	163.5192	3.61	164.4081	3.43	163.2909	3.66
162	164.5270	3.61	165.4164	3.43	164.2987	3.66
163	165.5196	3.61	166.4092	3.43	165.2912	3.67
164	166.5273	3.61	167.4174	3.43	166.2989	3.67
165	167.5200	3.62	168.4103	3.44	167.2915	3.67
166	168.5276	3.62	169.4184	3.44	168.2990	3.67
167	169.5204	3.62	170.4114	3.44	169.2917	3.67
168	170.5279	3.62	171.4194	3.44	170.2992	3.67
169	171.5208	3.62	172.4125	3.44	171.2920	3.67
170	172.5282	3.62	173.4203	3.45	172.2994	3.67
171	173.5212	3.62	174.4135	3.45	173.2922	3.67
172	174.5286	3.62	175.4212	3.45	174.2996	3.68
173	175.5215	3.63	176.4145	3.45	175.2926	3.68
174	176.5288	3.63	177.4221	3.45	176.2998	3.68
175	177.5219	3.63	178.4155	3.45	177.2928	3.68
176	178.5291	3.63	179.4230	3.46	178.3000	3.68
177	179.5223	3.63	180.4165	3.46	179.2930	3.68
178	180.5294	3.63	181.4239	3.46	180.3002	3.68
179	181.5226	3.63	182.4174	3.46	181.2933	3.68
180	182.5297	3.63	183.4248	3.46	182.3003	3.68
181	183.5230	3.63	184.4183	3.47	183.2936	3.68
182	184.5299	3.64	185.4256	3.47	184.3005	3.68
183	185.5233	3.64	186.4193	3.47	185.2938	3.69
184	186.5302	3.64	187.4265	3.47	186.3006	3.69
185	187.5236	3.64	188.4202	3.47	187.2940	3.69
186	188.5304	3.64	189.4273	3.47	188.3008	3.69
187	189.5239	3.64	190.4210	3.48	189.2942	3.69
188	190.5307	3.64	191.4281	3.48	190.3010	3.69
189	191.5243	3.64	192.4219	3.48	191.2944	3.69
190	192.5310	3.64	193.4289	3.48	192.3011	3.69
200	202.5321	3.66	203.4326	3.50	202.3018	3.70
201	203.5260	3.66	204.4267	3.50	203.2957	3.70
300	302.5395	3.72	303.4573	3.60	302.3063	3.75
301	303.5355	3.72	304.4534	3.61	303.3022	3.75
400	402.5434	3.75	403.4706	3.66	402.3087	3.78
401	403.5404	3.75	404.4676	3.66	403.3056	3.78
500	502.5458	3.78	503.4788	3.70	502.3101	3.80
501	503.5433	3.78	504.4763	3.70	503.3076	3.80
∞	$(N+2)$.5558	3.87	$(N+3)$.5145	3.87	$(N+2)$.3159	3.87

Table 24-5. Internal Gears. 20° Pressure Angle

Measurements between wires and change factors for 1 diametral pitch internal standard-addendum 20° pressure angle spur gears with 1.44 and 1.68 diameter wires.
For any other diametral pitch divide measurements given in table by the diametral pitch. Change factors are the same for all diametral pitches.

No. of teeth	1.44 wire dia.		Alternate 1.68 wire dia.		No. of teeth	1.44 wire dia.		Alternate 1.68 wire dia.	
	M	K_m	*M*	K_m		*M*	K_m	*M*	K_m
5	3.4090	2.30							
6	4.6595	2.46			56	54.6647	2.71	53.6965	3.02
7	5.4823	2.43			57	55.6431	2.71	54.6756	3.01
8	6.6608	2.52			58	56.6648	2.71	55.6975	3.01
9	7.5230	2.50	1.68 wires too large below 15 teeth		59	57.6438	2.71	56.6774	3.00
10	8.6617	2.56			60	58.6648	2.71	57.6985	3.00
11	9.5490	2.55			61	59.6445	2.71	58.6789	2.99
12	10.6623	2.59			62	60.6648	2.71	59.6994	2.99
13	11.5669	2.58			63	61.6452	2.71	60.6805	2.98
14	12.6627	2.61			64	62.6648	2.71	61.7003	2.98
15	13.5801	2.60	12.4222	8.58	65	63.6458	2.71	62.6819	2.97
16	14.6630	2.63	13.5459	5.95	66	64.6649	2.72	63.7011	2.97
17	15.5902	2.62	14.5026	5.09	67	65.6464	2.72	64.6832	2.96
18	16.6633	2.64	15.5901	4.65	68	66.6649	2.72	65.7018	2.96
19	17.5981	2.63	16.5414	4.35	69	67.6469	2.72	66.6844	2.96
20	18.6635	2.65	17.6147	4.15	70	68.6649	2.72	67.7025	2.95
21	19.6045	2.64	18.5668	3.98	71	69.6475	2.72	68.6856	2.95
22	20.6636	2.66	19.6311	3.87	72	70.6649	2.72	69.7032	2.95
23	21.6099	2.65	20.5854	3.76	73	71.6480	2.72	70.6867	2.94
24	22.6638	2.67	21.6429	3.69	74	72.6649	2.72	71.7038	2.94
25	23.6143	2.66	22.5997	3.61	75	73.6484	2.72	72.6878	2.94
26	24.6639	2.67	23.6520	3.56	76	74.6649	2.72	73.7044	2.94
27	25.6181	2.67	24.6112	3.50	77	75.6489	2.72	74.6888	2.93
28	26.6640	2.67	25.6591	3.46	78	76.6649	2.72	75.7049	2.93
29	27.6214	2.67	26.6206	3.42	79	77.6493	2.72	76.6897	2.93
30	28.6641	2.68	27.6649	3.38	80	78.6649	2.72	77.7054	2.92
31	29.6242	2.68	28.6285	3.35	81	79.6497	2.72	78.6905	2.92
32	30.6642	2.68	29.6699	3.32	82	80.6649	2.72	79.7059	2.92
33	31.6267	2.68	30.6353	3.29	83	81.6501	2.72	80.6914	2.92
34	32.6642	2.69	31.6739	3.27	84	82.6649	2.72	81.7064	2.91
35	33.6289	2.69	32.6411	3.25	85	83.6505	2.72	82.6922	2.91
36	34.6643	2.69	33.6773	3.23	86	84.6650	2.72	83.7068	2.91
37	35.6310	2.69	34.6462	3.21	87	85.6508	2.72	84.6929	2.91
38	36.6643	2.69	35.6804	3.20	88	86.6650	2.72	85.7072	2.91
39	37.6327	2.69	36.6507	3.18	89	87.6511	2.72	86.6936	2.90
40	38.6644	2.70	37.6831	3.16	90	88.6650	2.72	87.7076	2.90
41	39.6343	2.69	38.6547	3.15	91	89.6514	2.72	88.6943	2.90
42	40.6644	2.70	39.6855	3.14	92	90.6650	2.72	89.7080	2.90
43	41.6357	2.70	40.6582	3.13	93	91.6517	2.72	90.6950	2.90
44	42.6645	2.70	41.6875	3.11	94	92.6650	2.72	91.7084	2.89
45	43.6371	2.70	42.6614	3.10	95	93.6520	2.72	92.6956	2.89
46	44.6645	2.70	43.6893	3.09	96	94.6650	2.73	93.7087	2.89
47	45.6383	2.70	44.6644	3.08	97	95.6523	2.73	94.6962	2.89
48	46.6646	2.70	45.6910	3.08	98	96.6650	2.73	95.7090	2.89
49	47.6394	2.70	46.6670	3.07	99	97.6526	2.73	96.6968	2.89
50	48.6646	2.71	47.6926	3.06	100	98.6650	2.73	97.7093	2.88
51	49.6404	2.71	48.6694	3.05	101	99.6528	2.73	98.6974	2.88
52	50.6646	2.71	49.6940	3.04	102	100.6650	2.73	99.7096	2.88
53	51.6414	2.71	50.6716	3.04	103	101.6531	2.73	100.6979	2.88
54	52.6647	2.71	51.6953	3.03	104	102.6650	2.73	101.7099	2.88
55	53.6422	2.71	52.6737	3.02	105	103.6533	2.73	102.6984	2.88

Table 24-5. Internal Gears. 20° Pressure Angle (*Continued*)

No. of teeth	1.44 wire dia.		Alternate 1.68 wire dia.	
	M	K_m	M	K_m
106	104.6650	2.73	103.7102	2.88
107	105.6535	2.73	104.6989	2.88
108	106.6650	2.73	105.7105	2.87
109	107.6537	2.73	106.6994	2.87
110	108.6651	2.73	107.7107	2.87
111	109.6539	2.73	108.6998	2.87
112	110.6651	2.73	109.7110	2.87
113	111.6541	2.73	110.7002	2.87
114	112.6651	2.73	111.7112	2.87
115	113.6543	2.73	112.7006	2.87
116	114.6651	2.73	113.7114	2.86
117	115.6545	2.73	114.7010	2.86
118	116.6651	2.73	115.7117	2.86
119	117.6547	2.73	116.7014	2.86
120	118.6651	2.73	117.7119	2.86
121	119.6548	2.73	118.7018	2.86
122	120.6651	2.73	119.7121	2.86
123	121.6550	2.73	120.7022	2.86
124	122.6651	2.73	121.7123	2.86
125	123.6552	2.73	122.7026	2.86
126	124.6651	2.73	123.7125	2.85
127	125.6554	2.73	124.7029	2.85
128	126.6651	2.73	125.7127	2.85
129	127.6556	2.73	126.7032	2.85
130	128.6652	2.73	127.7129	2.85
131	129.6557	2.73	128.7036	2.85
132	130.6652	2.73	129.7130	2.85
133	131.6559	2.73	130.7039	2.85
134	132.6652	2.73	131.7132	2.85
135	133.6560	2.73	132.7042	2.85
136	134.6652	2.73	133.7134	2.85
137	135.6561	2.73	134.7045	2.84
138	136.6652	2.73	135.7135	2.84
139	137.6563	2.73	136.7047	2.84
140	138.6652	2.73	137.7137	2.84
141	139.6564	2.73	138.7050	2.84
142	140.6652	2.73	139.7139	2.84
143	141.6565	2.73	140.7053	2.84
144	142.6652	2.73	141.7140	2.84
145	143.6566	2.73	142.7055	2.84
146	144.6652	2.73	143.7141	2.84
147	145.6568	2.73	144.7058	2.84
148	146.6652	2.73	145.7143	2.84
149	147.6569	2.73	146.7061	2.84
150	148.6652	2.73	147.7144	2.84
151	149.6570	2.73	148.7063	2.84
152	150.6652	2.73	149.7145	2.83
153	151.6571	2.73	150.7065	2.83
154	152.6652	2.73	151.7146	2.83
155	153.6572	2.73	152.7068	2.83
156	154.6652	2.73	153.7148	2.83
157	155.6573	2.73	154.7070	2.83
158	156.6652	2.73	155.7149	2.83
159	157.6574	2.73	156.7072	2.83
160	158.6652	2.73	157.7150	2.83
161	159.6575	2.73	158.7074	2.83
162	160.6652	2.73	159.7151	2.83
163	161.6576	2.73	160.7076	2.83
164	162.6652	2.73	161.7152	2.83
165	163.6577	2.73	162.7078	2.83
166	164.6652	2.73	163.7153	2.83
167	165.6578	2.73	164.7080	2.83
168	166.6652	2.73	165.7154	2.83
169	167.6579	2.73	166.7082	2.83
170	168.6652	2.73	167.7156	2.82
171	169.6580	2.73	168.7084	2.82
172	170.6652	2.73	169.7157	2.82
173	171.6581	2.73	170.7086	2.82
174	172.6652	2.73	171.7158	2.82
175	173.6582	2.73	172.7087	2.82
176	174.6652	2.73	173.7158	2.82
177	175.6583	2.73	174.7089	2.82
178	176.6652	2.73	175.7159	2.82
179	177.6584	2.73	176.7090	2.82
180	178.6652	2.73	177.7160	2.82
181	179.6584	2.73	178.7092	2.82
182	180.6652	2.73	179.7161	2.82
183	181.6585	2.73	180.7094	2.82
184	182.6652	2.73	181.7162	2.82
185	183.6586	2.73	182.7095	2.82
186	184.6652	2.73	183.7162	2.82
187	185.6587	2.73	184.7097	2.82
188	186.6652	2.73	185.7163	2.82
189	187.6588	2.73	186.7098	2.82
190	188.6652	2.73	187.7164	2.82
200	198.6652	2.74	197.7168	2.81
201	199.6591	2.74	198.7107	2.81
300	298.6654	2.74	297.7192	2.79
301	299.6612	2.74	298.7151	2.79
400	398.6654	2.74	397.7203	2.78
401	399.6623	2.74	398.7172	2.78
500	498.6654	2.74	497.7210	2.77
501	499.6629	2.74	498.7185	2.77
∞	$(N-2).6655$	2.75	$(N-3).7238$	2.75

Table 24-6. Internal Gears. 14½° Pressure Angle

Measurement between wires and change factors for 1 diametral pitch internal standard-addendum 14½° pressure angle spur gears with 1.44 and 1.68 diameter wires. For any other diametral pitch divide measurements given in table by the diametral pitch. Change factors are the same for all diametral pitches.

No. of teeth	1.44 wire dia.		Alternate 1.68 wire dia.		No. of teeth	1.44 wire dia.		Alternate 1.68 wire dia.	
	M	K_m	M	K_m		M	K_m	M	K_m
5	3.5517	2.38							
6	4.8157	2.61			56	54.8701	3.58	53.6109	4.93
7	5.6393	2.63			57	55.8486	3.59	54.5913	4.90
8	6.8262	2.77			58	56.8705	3.59	55.6144	4.87
9	7.6894	2.79			59	57.8497	3.60	56.5953	4.85
10	8.8337	2.89			60	58.8709	3.60	57.6175	4.82
11	9.7219	2.91			61	59.8508	3.60	58.5990	4.80
12	10.8394	2.99			62	60.8712	3.61	59.6204	4.77
13	11.7449	3.01			63	61.8517	3.61	60.6024	4.75
14	12.8438	3.07			64	62.8715	3.61	61.6230	4.73
15	13.7620	3.09			65	63.8526	3.62	62.6055	4.71
16	14.8474	3.13	1.68 wires too large		66	64.8718	3.62	63.6254	4.69
17	15.7752	3.15			67	65.8535	3.62	64.6083	4.67
18	16.8504	3.19			68	66.8721	3.63	65.6277	4.65
19	17.7858	3.20			69	67.8543	3.63	66.6110	4.64
20	18.8529	3.24			70	68.8724	3.63	67.6297	4.62
21	19.7945	3.25			71	69.8551	3.63	68.6135	4.60
22	20.8550	3.27			72	70.8727	3.64	69.6316	4.59
23	21.8017	3.29			73	71.8558	3.64	70.6158	4.58
24	22.8569	3.31			74	72.8729	3.65	71.6334	4.56
25	23.8078	3.32			75	73.8565	3.65	72.6179	4.55
26	24.8585	3.34			76	74.8731	3.65	73.6351	4.54
27	25.8130	3.35			77	75.8572	3.65	74.6199	4.52
28	26.8599	3.37			78	76.8734	3.65	75.6366	4.51
29	27.8176	3.38			79	77.8578	3.66	76.6218	4.50
30	28.8612	3.40			80	78.8736	3.66	77.6381	4.49
31	29.8216	3.40			81	79.8584	3.66	78.6236	4.48
32	30.8623	3.42	29.4774	9.11	82	80.8738	3.66	79.6395	4.47
33	31.8251	3.43	30.4589	8.18	83	81.8590	3.67	80.6253	4.46
34	32.8633	3.44	31.5095	7.57	84	82.8740	3.67	81.6408	4.45
35	33.8282	3.45	32.4867	7.14	85	83.8595	3.67	82.6270	4.44
36	34.8642	3.46	33.5314	6.81	86	84.8742	3.67	83.6420	4.43
37	35.8311	3.46	34.5071	6.55	87	85.8600	3.67	84.6285	4.42
38	36.8650	3.47	35.5472	6.34	88	86.8743	3.68	85.6431	4.41
39	37.8336	3.48	36.5229	6.15	89	87.8605	3.68	86.6299	4.41
40	38.8658	3.49	37.5599	6.00	90	88.8745	3.68	87.6442	4.40
41	39.8359	3.50	38.5357	5.86	91	89.8610	3.68	88.6313	4.39
42	40.8665	3.51	39.5702	5.75	92	90.8747	3.68	89.6452	4.38
43	41.8380	3.51	40.5463	5.65	93	91.8614	3.69	90.6326	4.38
44	42.8672	3.52	41.5788	5.56	94	92.8749	3.69	91.6462	4.37
45	43.8399	3.52	42.5556	5.48	95	93.8619	3.69	92.6338	4.36
46	44.8678	3.53	43.5861	5.41	96	94.8750	3.69	93.6472	4.36
47	45.8416	3.54	44.5635	5.34	97	95.8623	3.69	94.6350	4.35
48	46.8683	3.54	45.5924	5.28	98	96.8752	3.69	95.6481	4.34
49	47.8432	3.55	46.5704	5.22	99	97.8627	3.70	96.6361	4.34
50	48.8688	3.56	47.5979	5.17	100	98.8753	3.70	97.6489	4.33
51	49.8447	3.56	48.5765	5.13	101	99.8631	3.70	98.6372	4.33
52	50.8692	3.57	49.6027	5.08	102	100.8754	3.70	99.6497	4.32
53	51.8461	3.57	50.5820	5.04	103	101.8635	3.70	100.6382	4.31
54	52.8697	3.58	51.6070	5.00	104	102.8756	3.70	101.6505	4.31
55	53.8474	3.58	52.5869	4.97	105	103.8638	3.70	102.6392	4.30

Table 24-6. Internal Gears. 14½° Pressure Angle (*Continued*)

No. of teeth	1.44 wire dia.		Alternate 1.68 wire dia.		No. of teeth	1.44 wire dia.		Alternate 1.68 wire dia.	
	M	K_m	M	K_m		M	K_m	M	K_m
106	104.8757	3.71	103.6512	4.30	156	154.8777	3.75	153.6627	4.14
107	105.8642	3.71	104.6401	4.29	157	155.8699	3.75	154.6551	4.14
108	106.8758	3.71	105.6519	4.29	158	156.8778	3.75	155.6630	4.13
109	107.8645	3.71	106.6410	4.28	159	157.8701	3.75	156.6555	4.13
110	108.8759	3.71	107.6526	4.28	160	158.8778	3.75	157.6633	4.13
111	109.8648	3.71	108.6419	4.27	161	159.8702	3.76	158.6559	4.13
112	110.8760	3.71	109.6532	4.27	162	160.8779	3.76	159.6636	4.13
113	111.8651	3.71	110.6427	4.27	163	161.8704	3.76	160.6563	4.12
114	112.8761	3.72	111.6538	4.26	164	162.8779	3.76	161.6639	4.12
115	113.8654	3.72	112.6434	4.26	165	163.8705	3.76	162.6567	4.12
116	114.8762	3.72	113.6544	4.25	166	164.8780	3.76	163.6642	4.12
117	115.8657	3.72	114.6442	4.25	167	165.8707	3.76	164.6570	4.12
118	116.8763	3.72	115.6550	4.24	168	166.8780	3.76	165.6645	4.12
119	117.8660	3.72	116.6450	4.24	169	167.8708	3.76	166.6573	4.11
120	118.8764	3.72	117.6556	4.24	170	168.8781	3.76	167.6647	4.11
121	119.8662	3.72	118.6457	4.23	171	169.8710	3.76	168.6576	4.11
122	120.8765	3.73	119.6561	4.23	172	170.8782	3.76	169.6649	4.11
123	121.8663	3.73	120.6464	4.23	173	171.8711	3.77	170.6579	4.11
124	122.8766	3.73	121.6566	4.22	174	172.8782	3.77	171.6651	4.11
125	123.8668	3.73	122.6470	4.22	175	173.8712	3.77	172.6582	4.10
126	124.8767	3.73	123.6571	4.22	176	174.8783	3.77	173.6653	4.10
127	125.8670	3.73	124.6477	4.21	177	175.8713	3.77	174.6585	4.10
128	126.8768	3.73	125.6575	4.21	178	176.8783	3.77	175.6655	4.10
129	127.8672	3.73	126.6483	4.21	179	177.8715	3.77	176.6588	4.10
130	128.8769	3.73	127.6579	4.20	180	178.8784	3.77	177.6658	4.10
131	129.8675	3.73	128.6489	4.20	181	179.8717	3.77	178.6591	4.10
132	130.8769	3.74	129.6583	4.20	182	180.8784	3.77	179.6659	4.10
133	131.8677	3.74	130.6495	4.19	183	181.8717	3.77	180.6593	4.09
134	132.8770	3.74	131.6588	4.19	184	182.8785	3.77	181.6662	4.09
135	133.8679	3.74	132.6500	4.19	185	183.8718	3.77	182.6596	4.09
136	134.8771	3.74	133.6592	4.19	186	184.8785	3.77	183.6665	4.09
137	135.8681	3.74	134.6505	4.18	187	185.8719	3.77	184.6599	4.09
138	136.8772	3.74	135.6596	4.18	188	186.8786	3.77	185.6667	4.09
139	137.8683	3.74	136.6510	4.18	189	187.8721	3.77	186.6601	4.09
140	138.8773	3.74	137.6600	4.17	190	188.8786	3.77	187.6668	4.09
141	139.8685	3.74	138.6515	4.17	200	198.8788	3.78	197.6678	4.07
142	140.8773	3.74	139.6604	4.17	201	199.8727	3.78	198.6618	4.07
143	141.8687	3.74	140.6520	4.17	300	298.8795	3.81	297.6725	4.00
144	142.8774	3.75	141.6608	4.16	301	299.8759	3.81	298.6694	4.00
145	143.8689	3.75	142.6525	4.16	400	398.8803	3.82	397.6752	3.96
146	144.8774	3.75	143.6612	4.16	401	399.8776	3.82	398.6731	3.96
147	145.8691	3.75	144.6530	4.16	500	498.8810	3.83	497.6767	3.94
148	146.8775	3.75	145.6615	4.16	501	499.8786	3.83	498.6753	3.94
149	147.8693	3.75	146.6534	4.15	∞	$(N-2)$.8827	3.87	$(N-3)$.6841	3.87
150	148.8775	3.75	147.6618	4.15					
151	149.8694	3.75	148.6538	4.15					
152	150.8776	3.75	149.6621	4.15					
153	151.8696	3.75	150.6543	4.14					
154	152.8776	3.75	151.6624	4.14					
155	153.8698	3.75	152.6547	4.14					

Table 24-7. External Gears. 25° Pressure Angle

Measurements over wires and change factors for 1 diametral pitch external standard-addendum 25° pressure angle spur gears with 1.728 and 1.68 diameter wires.
For any other diametral pitch divide measurements given in table by the diametral pitch.
Change factors are the same for all diametral pitches.

No. of teeth	1.728 wire dia.		Alternate 1.68 wire dia.		No. of teeth	1.728 wire dia.		Alternate 1.68 wire dia.	
	M	K_m	M	K_m		M	K_m	M	K_m
5	7.0472	1.47	6.9202	1.59					
6	8.3340	1.60	8.2003	1.64	56	58.4287	2.03	58.2726	2.05
7	9.1536	1.64	9.0199	1.68	57	59.4071	2.03	59.2509	2.05
8	10.3533	1.67	10.2155	1.72	58	60.4293	2.04	60.2730	2.05
9	11.2142	1.70	11.0762	1.75	59	61.4084	2.04	61.2521	2.05
10	12.3667	1.73	12.2260	1.77	60	62.4299	2.04	62.2735	2.05
11	13.2536	1.75	13.1126	1.79	61	63.4097	2.04	63.2532	2.06
12	14.3768	1.77	14.2338	1.81	62	64.4304	2.04	64.2739	2.06
13	15.2814	1.79	15.1381	1.83	63	65.4109	2.04	65.2543	2.06
14	16.3846	1.81	16.2397	1.85	64	66.4309	2.05	66.2742	2.06
15	17.3021	1.83	17.1570	1.86	65	67.4120	2.05	67.2553	2.06
16	18.3908	1.84	18.2445	1.88	66	68.4314	2.05	68.2746	2.06
17	19.3181	1.85	19.1716	1.89	67	69.4130	2.05	69.2562	2.06
18	20.3959	1.86	20.2483	1.90	68	70.4319	2.05	70.2749	2.06
19	21.3310	1.88	21.1832	1.91	69	71.4140	2.05	71.2571	2.06
20	22.4002	1.88	22.2515	1.92	70	72.4323	2.05	72.2752	2.07
21	23.3415	1.89	23.1926	1.92	71	73.4150	2.05	73.2579	2.07
22	24.4038	1.90	24.2542	1.93	72	74.4327	2.05	74.2755	2.07
23	25.3502	1.91	25.2005	1.94	73	75.4159	2.06	75.2586	2.07
24	26.4069	1.92	26.2566	1.95	74	76.4331	2.06	76.2758	2.07
25	27.3576	1.93	27.2071	1.95	75	77.4167	2.06	77.2594	2.07
26	28.4096	1.93	28.2586	1.96	76	78.4335	2.06	78.2761	2.07
27	29.3640	1.94	29.2128	1.96	77	79.4175	2.06	79.2601	2.07
28	30.4120	1.94	30.2603	1.97	78	80.4339	2.06	80.2763	2.07
29	31.3695	1.95	31.2177	1.97	79	81.4183	2.06	81.2607	2.07
30	32.4141	1.95	32.2619	1.98	80	82.4342	2.06	82.2766	2.08
31	33.3743	1.96	33.2220	1.98	81	83.4190	2.06	83.2614	2.08
32	34.4159	1.96	34.2632	1.99	82	84.4345	2.07	84.2768	2.08
33	35.3786	1.97	35.2258	1.99	83	85.4196	2.07	85.2620	2.08
34	36.4176	1.97	36.2664	2.00	84	86.4348	2.07	86.2771	2.08
35	37.3824	1.98	37.2292	2.00	85	87.4203	2.07	87.2625	2.08
36	38.4191	1.98	38.2655	2.00	86	88.4351	2.07	88.2773	2.08
37	39.3858	1.98	39.2323	2.01	87	89.4209	2.07	89.2631	2.08
38	40.4205	1.99	40.2666	2.01	88	90.4354	2.07	90.2775	2.08
39	41.3889	1.99	41.2349	2.01	89	91.4215	2.07	91.2636	2.08
40	42.4217	2.00	42.2675	2.02	90	92.4357	2.07	92.2777	2.08
41	43.3917	2.00	43.2374	2.02	91	93.4221	2.07	93.2641	2.08
42	44.4228	2.00	44.2683	2.02	92	94.4359	2.07	94.2779	2.08
43	45.3942	2.00	45.2396	2.02	93	95.4227	2.07	95.2646	2.08
44	46.4239	2.01	46.2690	2.02	94	96.4362	2.07	96.2780	2.08
45	47.3965	2.01	47.2417	2.02	95	97.4232	2.08	97.2650	2.09
46	48.4248	2.01	48.2697	2.03	100	102.4369	2.08	102.2785	2.09
47	49.3986	2.01	49.2435	2.03	101	103.4247	2.08	103.2663	2.09
48	50.4257	2.02	50.2704	2.03	200	202.4424	2.11	202.2825	2.11
49	51.4006	2.02	51.2452	2.03	201	203.4363	2.11	203.2764	2.11
50	52.4265	2.02	52.2710	2.04	300	302.4443	2.12	302.2839	2.12
51	53.4024	2.02	53.2468	2.04	301	303.4402	2.12	303.2798	2.12
52	54.4273	2.03	54.2716	2.04	400	402.4453	2.13	402.2845	2.13
53	55.4041	2.03	55.2483	2.04	401	403.4422	2.13	403.2815	2.13
54	56.4280	2.03	56.2721	2.05	500	502.4458	2.13	502.2850	2.13
55	57.4056	2.03	57.2497	2.05	∞	$(N + 2)$.4482	2.14	$(N + 2)$.2866	2.14

Table 24-8

Pressure angle of gear, deg	E, extra depth of cut to thin teeth (Δt)	Δt, amount teeth are cut too thin due to excess depth of cut E
$14\frac{1}{2}$	$E = 1.93 \times \Delta t$	$\Delta t = \dfrac{E}{1.93} = 0.52E$
$17\frac{1}{2}$	$E = 1.59 \times \Delta t$	$\Delta t = \dfrac{E}{1.59} = 0.63E$
20	$E = 1.37 \times \Delta t$	$\Delta t = \dfrac{E}{1.37} = 0.73E$
25	$E = 1.07 \times \Delta t$	$\Delta t = \dfrac{E}{1.07} = 0.93E$
30	$E = 0.87 \times \Delta t$	$\Delta t = \dfrac{E}{0.87} = 1.15E$

Table 24-9. 20° Pressure Angle Involute Spur Gears

Pin measurements of standard long-addendum pinions with diameters and tooth thickness according to ASA B6.7–1956. Dimensions required when using enlarged pinions. Dimensions are in inches for 1 diametral pitch. For other pitches divide table values by diametral pitch.

Enlarged pinion					Standard-center-distance system, short-addendum gear dimensions			Enlarged-center-distance system, standard mating gear, increase over standard center distance
No. of teeth n	Outside dia. D_o	Circ. tooth thickness at std. pitch dia. t	Measurement over 1.92″ dia. wires M	Change* factor K_m	Decrease in outside dia. ΔD_o	Circ. tooth thickness at std. pitch dia. T $\pi - t$	Recommended min. No. of teeth	
7	9.8286	2.0007	10.3198	1.45	1.1812	1.1409	42	0.5906
8	10.8738	1.9581	11.5123	1.51	1.0642	1.1835	39	0.5321
9	11.8380	1.9155	12.3199	1.57	0.9472	1.2261	36	0.4736
10	12.8302	1.8730	13.4408	1.62	0.8302	1.2686	33	0.4151
11	13.7132	1.8304	14.2678	1.68	0.7132	1.3112	30	0.3566
12	14.5963	1.7878	15.3428	1.73	0.5963	1.3538	27	0.2982
13	15.4793	1.7452	16.1807	1.79	0.4793	1.3964	25	0.2397
14	16.3623	1.7027	17.2233	1.84	0.3623	1.4389	23	0.1812
15	17.2453	1.6601	18.0674	1.89	0.2453	1.4815	21	0.1227
16	18.1284	1.6175	19.0851	1.95	0.1284	1.5241	19	0.0642
17	19.0114	1.5749	19.9326	2.00	0.0114	1.5667	18	0.0057

These values are suitable only for 20° full-depth involute gears. They cannot be used for other pressure angles.

* The change factors are approximately correct for all diametral pitches. Do not divide by diametral pitch.

Table 24-10. 14½° Pressure Angle Involute Spur Gears (Full Depth). Dimensions for Enlarged Pinions (31 and Fewer Teeth) and Mating Gears

Values in inches at 1 diametral pitch

Enlarged pinion (long addendum)					Short-addendum gear, mates enlarged pinion at standard center distance			Enlarged-center-distance system, standard mating gear, increase over standard center distance
No. of teeth	Outside dia.	Circ. tooth thickness at standard pitch dia.	Measurement over 1.92 dia. wires,	Change* factor	Decrease in outside dia.	Circ. tooth thickness at standard pitch dia.	Recommended min. No. of teeth	
n	D_o	t	M	K_m	ΔD_o	T		
10	13.3731	1.9259	13.6186	1.72	1.3731	1.2157	54	0.6866
11	14.3104	1.9097	14.4966	1.78	1.3104	1.2319	53	0.6552
12	15.2477	1.8935	15.6290	1.82	1.2477	1.2481	52	0.6239
13	16.1850	1.8773	16.5211	1.87	1.1850	1.2643	51	0.5925
14	17.1223	1.8611	17.6244	1.92	1.1223	1.2805	50	0.5612
15	18.0597	1.8449	18.5260	1.96	1.0597	1.2967	49	0.5299
16	18.9970	1.8286	19.6075	2.01	0.9970	1.3130	48	0.4985
17	19.9343	1.8124	20.5156	2.05	0.9343	1.3292	47	0.4672
18	20.8716	1.7962	21.5806	2.09	0.8716	1.3454	46	0.4358
19	21.8089	1.7800	22.4934	2.13	0.8089	1.3616	45	0.4045
20	22.7462	1.7638	23.5451	2.17	0.7462	1.3778	44	0.3731
21	23.6835	1.7476	24.4611	2.22	0.6835	1.3940	43	0.3418
22	24.6208	1.7314	25.5018	2.26	0.6208	1.4102	42	0.3104
23	25.5581	1.7151	26.4201	2.30	0.5581	1.4265	41	0.2791
24	26.4954	1.6989	27.4515	2.34	0.4954	1.4427	40	0.2477
25	27.4328	1.6872	28.3718	2.38	0.4328	1.4589	39	0.2164
26	28.3701	1.6665	29.3952	2.42	0.3701	1.4751	38	0.1851
27	29.3074	1.6503	30.3168	2.46	0.3074	1.4913	37	0.1537
28	30.2447	1.6341	31.3333	2.49	0.2447	1.5075	36	0.1224
29	31.1820	1.6179	32.2558	2.53	0.1820	1.5237	35	0.0910
30	32.1193	1.6017	33.2661	2.57	0.1193	1.5399	34	0.0597
31	33.0566	1.5854	34.1889	2.61	0.0566	1.5562	33	0.0283

* The change factors are approximately correct for all diametral pitches. Do not divide by diametral pitch.

Relation between Depth of Cut and Tooth Thickness. If the generating gear cutter* is fed into the blank an amount exactly equal to the whole depth of tooth, the gear is cut without backlash. To provide backlash the teeth must be made thinner by an equal amount Δt (increment of t). The formulas for finding the excess depth of cut E from Δt and vice versa are as given in Table 24-8. The relationships are exact for hobs or rack-shaped tools. They are approximate for gear shaper cutters.

Example. A total backlash of 0.016″ is desired between a 30-tooth 14½° pinion and a 100-tooth 14½° gear, and 0.008″ backlash will be taken on the pinion and 0.008 on the gear. From Table 24-4, the wire measurement for the pinion should be 2.98 × 0.008″ = 0.0238″ under the computed value, and for the gear 3.48 × 0.008″ = 0.0278″ undersize. If after a trial cut, the wire measurement on the gear is only 0.010″ under the computed value, the

* Backlash in gears milled with a form-milling cutter is usually produced by side cutting a desired amount to thin the teeth which have been cut to standard depth.

backlash provided is 0.010/3.48 = 0.0029″. This is 0.008″ − 0.0029″ = 0.0051″ too little, and from Table 24-8 the cutter must be fed in deeper by 1.93 × 00.051″ = 0.0098″.

Enlarged Pinions and Reduced Gears. It is recommended in published standards for involute gear-tooth forms that generated pinions with small numbers of teeth be designed with enlarged diameters. This is done mainly because a sufficient increase in root diameter prevents undercut; and if the outside diameter is made larger by the same amount, as is usually done, the whole depth remains standard for the hob or cutter being used. The tooth thickness in enlarged pinions can be checked by wire measurements either taken from tables or determined by computation. Tables 24-9 and 24-10 give measurement values and change factors for pinions enlarged according to the published standards.

Tables of dimensions for enlarged pinions were given in "Spur Gear Tooth Forms," ASA B6.1–1932, for both 20 and 14½° pressure angles. A newer but similar table for 20° pressure angle appears in "20 Degree Involute Fine Pitch System for Spur and Helical Gears," ASA B6.7–1956. Measurements over wires for the enlarged pinions in these standard publications are given in Tables 24-9 and 24-10.

Numerical Example. *Pinion*, 12 tooth, $3P_d$, 20°, enlargement, Table 24-9.

(1) D_o Outside diameter, from Table 24-9 $= \frac{14.5963}{3} = 4.8654''$

(2) D Pitch diameter $= \frac{12}{3} = 4.0000''$

(3) t Circular tooth thickness at D, from Table 24-9 $= \frac{1.7878}{3} = 0.5959''$

(4) T Standard circular tooth thickness from Table 24-17 $= \frac{1.3538}{3} = 0.5236''$

(5) Δt Increase in tooth thickness (3) − (4) $= 0.0723''$

(6) G Wire diameter (Table 24-2) $= \frac{1.92}{3} = 0.640''$

(7) M Measurement over 0.640 wires from Table 24-9 $= \frac{15.3428}{3} = 5.1143''$

The pin measurement of the mating reduced gear may be calculated as outlined in Art. 7-9. In addition, pin-measurement tables for gears of 20° pressure angle reduced in accordance with ASA B6.7-1956 are contained in The Van Keuren Co. *Circular* G38, 1956. In this publication tabular values are given for reduced gears mating with each of the 11 enlarged pinions. The tables start with approximately the minimum recommended number of teeth and continue to infinity.

Helical Gears. The pin measurement of helical gears may be computed as outlined in Chap. 7. A simple but approximate calculation may also be made which utilizes the spur-gear-measuring tables (see Table 24-11).

Even-tooth helical gears should be measured with two wires. Odd-tooth helical gears should not be measured with two wires but may be measured with three wires under certain conditions, or indirectly with one wire. Three wires may be used, provided the face width and helix angle of the gear will permit the arrangement of two wires in adjacent tooth spaces on one side of the gear and a third wire in a tooth space on the opposite side, so that all three wires will automatically adjust themselves to line contact with the micrometer's measuring surfaces. In cases where three wires cannot be used, the gear should be mounted on an arbor and a radial measurement made from the top of one wire to the axis of the gear.

It should be emphasized that the practice of using two wires for measuring odd-tooth helical gears and correcting the results by the application of a cos $90°/N$ factor is not valid and may result in serious errors.

Table 24-11. Wire Measurement of Standard Helical Gear

For use when normal pressure angle is $14\frac{1}{2}$, 20, or 25°

For: A-Z Gear Co. — Date: Aug. 12, 1954

Data from Drawing — Drawing No.: N90-1536

Functions of ψ (psi)

1. No. of teeth	$N = 14$	$\cos 40°$	$= 0.7660444$
2. Diametral pitch in plane of rotation	$P_d = 18.3850656$	$\frac{1}{\cos 40°}$	$= 1.305407$
3. Pitch diameter	$D = 0.7615''$		
4. Helix angle (psi) with axis ($\tan \psi = \pi D/L$)	$\psi = 40°$	$\frac{1}{(\cos 40°)^3}$	$= 2.224529$
5. Normal pressure angle (phi sub n)	$\phi_n = 14.5$		
6. Normal tooth thickness	$T_n = 0.0654''$		

Computations

7. Find normal diametral pitch (may show on drawing) $\quad P_n = P_d \times \frac{1}{\cos \psi} = 18.38 \times 1.305407 = 24$

8. Find wire diameter $\quad G = \frac{1.728''}{P_n} = \frac{1.728''}{24} = 0.072''$

9. Find number of teeth in equivalent spur gear $\quad N_e = N \times \frac{1}{(\cos \psi)^3} = 14 \times 2.224529 = 31.143406$

Data from Tables (see Table 24-4)

10. Refer to proper table and using *even-tooth values* only interpolate for values (M-D) for N_e teeth (for 1 diametral pitch gears, $D = N$):

1 Diametral Pitch Values

	N	D	M	(M-D)	Diff.
Next lower (even)	(a) 30	(c) 30″	(d) 32.4429″	(e) 2.4429″	(f) 0.0049″
Next higher (even)	(b) 32	32″	34.4478″	2.4478″	

N_e value (9) = 31.1434
N value (a) = 30
Diff. (g) = 1.1434

$$(M\text{-}D) = \overset{(e)}{2.4429''} + \frac{\overset{(g)}{1.1434}}{2} \times \overset{(f)}{0.0049''} = 2.4457''$$

11. Find (M-D) for specified normal diametral pitch $\quad (M\text{-}D)_1 = \frac{(M\text{-}D)}{P_n} = \frac{2.4457''}{24} = 0.1019''$

12. Find measurement over wires $\quad M_1 = (M\text{-}D)_1 + D = 0.1019'' + 0.7615'' = 0.8634''$

13. Measurement for odd-tooth gear $\quad \frac{M_1}{2} =$ (odd-tooth helical gears should be mounted on an arbor and a radial measurement should be made from one wire to the center of the gear)

14. Reduction in measurement over wires to thin teeth 0.002″ in normal plane (Table 24-12) $\quad \Delta M = \frac{K \times \Delta t_n}{\cos \psi} = \frac{2.33 \times 0.002''}{0.7660444} = 0.0061''$

15. Measurement over wires for teeth 0.002″ thin in normal plane $\quad M_2 = M_1 - \Delta M = 0.8634'' - 0.0061'' = 0.8573''$

Table 24-12. Change Factors K for 1.728″ Wires Used on External Helical Gears

$$K = \frac{\cos \phi}{\sin \phi_w}$$

Reduction in measurement = $K \times$ amount teeth are to be thin (in normal plane) $\div \cos \psi$
or Reduction in measurement = $K \times$ amount teeth are to be thin (in plane of rotation of the gear)

No. of teeth	Normal pressure angle = 14½°							Normal pressure angle = 20°						
	ψ = 15°	ψ = 20°	ψ = 25°	ψ = 30°	ψ = 35°	ψ = 40°	ψ = 45°	ψ = 15°	ψ = 20°	ψ = 25°	ψ = 30°	ψ = 35°	ψ = 40°	ψ = 45°
6	2.07	2.06	2.05	2.03	2.00	1.96	1.91	1.81	1.79	1.77	1.73	1.69	1.64	1.58
10	2.34	2.33	2.30	2.27	2.23	2.17	2.10	2.02	2.00	1.96	1.92	1.85	1.78	1.69
15	2.56	2.56	2.54	2.49	2.42	2.37	2.25	2.14	2.12	2.07	2.01	1.93	1.84	1.75
20	2.72	2.69	2.65	2.59	2.52	2.44	2.32	2.22	2.18	2.13	2.07	1.99	1.90	1.79
25	2.81	2.77	2.73	2.66	2.58	2.49	2.35	2.28	2.24	2.18	2.11	2.03	1.93	1.81
30	2.89	2.84	2.80	2.71	2.63	2.53	2.39	2.33	2.28	2.22	2.14	2.05	1.95	1.83
35	2.95	2.90	2.86	2.76	2.68	2.56	2.42	2.36	2.31	2.25	2.17	2.08	1.97	1.84
40	3.01	2.96	2.91	2.81	2.72	2.59	2.44	2.40	2.34	2.28	2.20	2.10	1.98	1.85
45	3.07	3.01	2.96	2.85	2.75	2.62	2.46	2.42	2.36	2.30	2.21	2.12	2.00	1.86
50	3.12	3.06	3.00	2.88	2.78	2.64	2.48	2.44	2.38	2.32	2.23	2.13	2.01	1.87
60	3.20	3.13	3.07	2.95	2.83	2.69	2.52	2.47	2.41	2.34	2.25	2.15	2.03	1.89
70	3.27	3.20	3.12	3.00	2.88	2.72	2.55	2.49	2.43	2.36	2.27	2.16	2.04	1.90
80	3.32	3.25	3.16	3.05	2.92	2.76	2.58	2.51	2.44	2.38	2.28	2.17	2.05	1.90
90	3.37	3.29	3.20	3.08	2.95	2.79	2.61	2.53	2.46	2.39	2.29	2.18	2.06	1.91
100	3.40	3.33	3.23	3.12	2.98	2.81	2.63	2.54	2.47	2.40	2.30	2.19	2.06	1.91
120	3.45	3.37	3.26	3.16	3.02	2.84	2.65	2.56	2.49	2.42	2.31	2.20	2.07	1.91
140	3.48	3.40	3.29	3.18	3.04	2.86	2.67	2.57	2.50	2.42	2.32	2.21	2.07	1.91
160	3.51	3.42	3.31	3.20	3.06	2.87	2.68	2.58	2.51	2.43	2.32	2.22	2.07	1.91
180	3.53	3.45	3.33	3.22	3.07	2.88	2.69	2.59	2.52	2.44	2.33	2.23	2.07	1.91
200	3.55	3.46	3.35	3.23	3.08	2.89	2.69	2.60	2.52	2.44	2.34	2.23	2.07	1.92
250	3.59	3.50	3.38	3.25	3.09	2.90	2.70	2.61	2.53	2.45	2.35	2.23	2.08	1.92
300	3.61	3.52	3.40	3.26	3.10	2.91	2.70	2.62	2.54	2.46	2.37	2.23	2.09	1.93
350	3.63	3.53	3.41	3.27	3.11	2.92	2.70	2.62	2.54	2.46	2.37	2.23	2.09	1.93
400	3.64	3.54	3.42	3.28	3.11	2.92	2.71	2.63	2.55	2.47	2.37	2.24	2.09	1.94
450	3.65	3.55	3.43	3.29	3.12	2.93	2.71	2.64	2.56	2.47	2.37	2.24	2.09	1.94
500	3.66	3.56	3.44	3.30	3.12	2.93	2.71	2.64	2.56	2.47	2.37	2.24	2.09	1.94

NOTE: The amount of reduction is the same for all diametral pitches for a given tooth-thickness reduction.

24-2. Trigonometric Functions. A short table of trigonometric tables is given for general reference. Those doing an extensive amount of gear calculating will find it necessary to have a whole book of seven-place trigonometric tables. (Ref. 2, 3.)

Table 24-13. Trigonometric Functions

$\phi°$	arc ϕ	sin ϕ	tan ϕ	cot ϕ	cos ϕ	sec ϕ	cosec ϕ
1°	0.01745	0.01745	0.01746	57.290	0.99985	1.0002	57.299
2	0.03491	0.03490	0.03492	28.636	0.99939	1.0006	28.654
3	0.05236	0.05234	0.05241	19.081	0.99863	1.0014	19.107
4	0.06981	0.06976	0.06993	14.301	0.99756	1.0024	14.336
5	0.08727	0.08716	0.08749	11.430	0.99619	1.0038	11.474
6°	0.10472	0.10453	0.10510	9.5144	0.99452	1.0055	9.5668
7	0.12217	0.12187	0.12278	8.1444	0.99255	1.0075	8.2055
8	0.13963	0.13917	0.14054	7.1154	0.99027	1.0098	7.1853
9	0.15708	0.15643	0.15838	6.3138	0.98769	1.0125	6.3925
10	0.17453	0.17365	0.17633	5.6713	0.98481	1.0154	5.7588
11°	0.19199	0.19081	0.19438	5.1446	0.98163	1.0187	5.2408
12	0.20944	0.20791	0.21256	4.7046	0.97815	1.0223	4.8097
13	0.22689	0.22495	0.23087	4.3315	0.97437	1.0263	4.4454
14	0.24435	0.24192	0.24933	4.0108	0.97030	1.0306	4.1336
15	0.26180	0.25882	0.26795	3.7321	0.96593	1.0353	3.8637
16°	0.27925	0.27564	0.28675	3.4874	0.96126	1.0403	3.6280
17	0.29671	0.29237	0.30573	3.2709	0.95630	1.0457	3.4203
18	0.31416	0.30902	0.32492	3.0777	0.95106	1.0515	3.2361
19	0.33161	0.32557	0.34433	2.9042	0.94552	1.0576	3.0716
20	0.34907	0.34202	0.36397	2.7475	0.93969	1.0642	2.9238
21°	0.36652	0.35837	0.38386	2.6051	0.93358	1.0711	2.7904
22	0.38397	0.37461	0.40403	2.4751	0.92718	1.0785	2.6695
23	0.40143	0.39073	0.42447	2.3559	0.92050	1.0864	2.5593
24	0.41888	0.40674	0.44523	2.2460	0.91355	1.0946	2.4586
25	0.43633	0.42262	0.46631	2.1445	0.90631	1.1034	2.3662
26°	0.45379	0.43837	0.48773	2.0503	0.89879	1.1126	2.2812
27	0.47124	0.45399	0.50953	1.9626	0.89101	1.1223	2.2027
28	0.48869	0.46947	0.53171	1.8807	0.88295	1.1326	2.1301
29	0.50615	0.48481	0.55431	1.8040	0.87462	1.1434	2.0627
30	0.52360	0.50000	0.57735	1.7321	0.86603	1.1547	2.0000
31°	0.54105	0.51504	0.60086	1.6643	0.85717	1.1666	1.9416
32	0.55851	0.52992	0.62487	1.6003	0.84805	1.1792	1.8871
33	0.57596	0.54464	0.64941	1.5399	0.83867	1.1924	1.8361
34	0.59341	0.55919	0.67451	1.4826	0.82904	1.2062	1.7883
35	0.61087	0.57358	0.70021	1.4281	0.81915	1.2208	1.7434
36°	0.62832	0.58779	0.72654	1.3764	0.80902	1.2361	1.7013
37	0.64577	0.60182	0.75355	1.3270	0.79864	1.2521	1.6616
38	0.66323	0.61566	0.78129	1.2799	0.78801	1.2690	1.6243
39	0.68068	0.62932	0.80978	1.2349	0.77715	1.2868	1.5890
40	0.69813	0.64279	0.83910	1.1918	0.76604	1.3054	1.5557
41°	0.71559	0.65606	0.86929	1.1504	0.75471	1.3250	1.5243
42	0.73304	0.66913	0.90040	1.1106	0.74314	1.3456	1.4945
43	0.75049	0.68200	0.93252	1.0724	0.73135	1.3673	1.4663
44	0.76794	0.69466	0.96569	1.0355	0.71934	1.3902	1.4396
45	0.78540	0.70711	1.00000	1.0000	0.70711	1.4142	1.4142
46°	0.80285	0.71934	1.0355	0.96569	0.69466	1.4396	1.3902
47	0.82030	0.73135	1.0724	0.93252	0.68200	1.4663	1.3673
48	0.83776	0.74314	1.1106	0.90040	0.66913	1.4945	1.3456
49	0.85521	0.75471	1.1504	0.86929	0.65606	1.5243	1.3250
50	0.87266	0.76604	1.1918	0.83910	0.64279	1.5557	1.3054

24-3. Involute Functions. The involute function is widely used in gear calculations. Chapter 1 in Art. 1-7 defines the involute function. Chapter 7 in Art. 7-8 discusses the involute and gives a table of involute functions from 1 to 90°. A short table of involute functions for general reference is given in this article. The involute function of an angle is called "inv ϕ" or simply "θ." Inv ϕ equals θ in radians.

Table 24-14. Involute Functions (inv ϕ = tan ϕ − arc ϕ)

ϕ	inv ϕ	Diff.	ϕ	inv ϕ	Diff.	ϕ	inv ϕ	Diff.	ϕ	inv ϕ	Diff.
0.0°	.0000000	0	5.0°	.0002222	136	10.0°	.0017941	548	15.0°	.0061498	1262
0.1	00000	0	5.1	02358	142	10.1	18489	559	15.1	62760	1279
0.2	00000	0	5.2	02500	147	10.2	19048	571	15.2	64039	1298
0.3	00000	1	5.3	02647	154	10.3	19619	582	15.3	65337	1315
0.4	00001	1	5.4	02801	158	10.4	20201	594	15.4	66652	1333
0.5	00002	2	5.5	02959	165	10.5	20795	605	15.5	67985	1352
0.6	00004	2	5.6	03124	171	10.6	21400	617	15.6	69337	1369
0.7	00006	3	5.7	03295	177	10.7	22017	629	15.7	70706	1389
0.8	00009	4	5.8	03472	183	10.8	22646	642	15.8	72095	1406
0.9	00013	5	5.9	03655	190	10.9	23288	653	15.9	73501	1426
1.0°	.0000018	6	6.0°	.0003845	196	11.0°	.0023941	666	16.0°	.0074927	1445
1.1	00024	7	6.1	04041	203	11.1	24607	678	16.1	76372	1463
1.2	00031	8	6.2	04244	209	11.2	25285	690	16.2	77835	1483
1.3	00039	10	6.3	04453	216	11.3	25975	703	16.3	79318	1502
1.4	00049	11	6.4	04669	223	11.4	26678	716	16.4	80820	1522
1.5	00060	13	6.5	04892	230	11.5	27394	729	16.5	82342	1541
1.6	00073	14	6.6	05122	237	11.6	28123	742	16.6	83883	1561
1.7	00087	16	6.7	05359	245	11.7	28865	755	16.7	85444	1581
1.8	00103	19	6.8	05604	252	11.8	29620	769	16.8	87025	1601
1.9	00122	20	6.9	05856	259	11.9	30389	782	16.9	88626	1621
2.0°	.0000142	22	7.0°	.0006115	267	12.0°	.0031171	795	17.0°	.0090247	1642
2.1	00164	25	7.1	06382	275	12.1	31966	809	17.1	91889	1662
2.2	00189	27	7.2	06657	282	12.2	32775	823	17.2	93551	1683
2.3	00216	29	7.3	06939	291	12.3	33598	836	17.3	95234	1703
2.4	00245	32	7.4	07230	298	12.4	34434	851	17.4	96937	1725
2.5	00277	35	7.5	07528	307	12.5	35285	865	17.5	98662	1745
2.6	00312	37	7.6	07835	315	12.6	36150	879	17.6	.0100407	1767
2.7	00349	40	7.7	08150	323	12.7	37029	894	17.7	102174	1789
2.8	00389	44	7.8	08473	332	12.8	37923	908	17.8	103963	1810
2.9	00433	46	7.9	08805	340	12.9	38831	923	17.9	105773	1831
3.0°	.0000479	50	8.0°	.0009145	349	13.0°	.0039754	938	18.0°	.0107604	1854
3.1	00529	52	8.1	09494	358	13.1	40692	952	18.1	109458	1875
3.2	00581	57	8.2	09852	367	13.2	41644	968	18.2	111333	1898
3.3	00638	60	8.3	10219	376	13.3	42612	983	18.3	113231	1920
3.4	00698	63	8.4	10595	385	13.4	43595	998	18.4	115151	1943
3.5	00761	67	8.5	10980	395	13.5	44593	1014	18.5	117094	1965
3.6	00828	71	8.6	11375	404	13.6	45607	1029	18.6	119059	1989
3.7	00899	75	8.7	11779	413	13.7	46636	1045	18.7	121048	2011
3.8	00974	79	8.8	12192	423	13.8	47681	1061	18.8	123059	2034
3.9	01053	83	8.9	12615	433	13.9	48742	1077	18.9	125093	2058
4.0°	.0001136	88	9.0°	.0013048	443	14.0°	.0049819	1093	19.0°	.0127151	2081
4.1	01224	92	9.1	13491	453	14.1	50912	1109	19.1	129232	2104
4.2	01316	96	9.2	13944	463	14.2	52021	1126	19.2	131336	2129
4.3	01412	101	9.3	14407	473	14.3	53147	1142	19.3	133465	2152
4.4	01513	106	9.4	14880	483	14.4	54289	1159	19.4	135617	2177
4.5	01619	110	9.5	15363	494	14.5	55448	1176	19.5	137794	2200
4.6	01729	116	9.6	15857	505	14.6	56624	1193	19.6	139994	2226
4.7	01845	120	9.7	16362	515	14.7	57817	1210	19.7	142220	2250
4.8	01965	126	9.8	16877	526	14.8	59027	1227	19.8	144470	2274
4.9	02091	131	9.9	17403	538	14.9	60254	1244	19.9	146744	2300
5.0°	.0002222		10.0°	.0017941		15.0°	.0061498		20.0°	.0149044	

Table 24-14. Involute Functions (inv ϕ = tan ϕ − arc ϕ) (*Continued*)

ϕ	inv ϕ	Diff.	ϕ	inv ϕ	Diff.	ϕ	inv ϕ	Diff.	ϕ	inv ϕ	Diff.
20.0°	.0149044	2325	25.0°	.0299753	3813	30.0°	.0537515	5841	35.0°	.0893423	8589
20.1	151369	2350	25.1	303566	3847	30.1	543356	5889	35.1	902012	8653
20.2	153719	2375	25.2	307413	3882	30.2	549245	5936	35.2	910665	8717
20.3	156094	2401	25.3	311295	3918	30.3	555181	5983	35.3	919382	8783
20.4	158495	2427	25.4	315213	3953	30.4	561164	6032	35.4	928165	8847
20.5	160922	2453	25.5	319166	3988	30.5	567196	6080	35.5	937012	8913
20.6	163375	2479	25.6	323154	4025	30.6	573276	6129	35.6	945925	8979
20.7	165854	2505	25.7	327179	4060	30.7	579405	6177	35.7	954904	9045
20.8	168359	2532	25.8	331239	4097	30.8	585582	6227	35.8	963949	9112
20.9	170891	2558	25.9	335336	4134	30.9	591809	6277	35.9	973061	9179
21.0°	.0173449	2585	26.0°	.0339470	4170	31.0°	.0598086	6326	36.0°	.0982240	9247
21.1	176034	2612	26.1	343640	4207	31.1	604412	6376	36.1	991487	9315
21.2	178646	2640	26.2	347847	4245	31.2	610788	6427	36.2	1000802	9383
21.3	181286	2667	26.3	352092	4282	31.3	617215	6477	36.3	1010185	9452
21.4	183953	2694	26.4	356374	4320	31.4	623692	6529	36.4	1019637	9522
21.5	186647	2722	26.5	360694	4357	31.5	630221	6580	36.5	1029159	9591
21.6	189369	2750	26.6	365051	4396	31.6	636801	6631	36.6	1038750	9662
21.7	192119	2778	26.7	369447	4434	31.7	643432	6684	36.7	1048412	9732
21.8	194897	2806	26.8	373881	4473	31.8	650116	6735	36.8	1058144	9803
21.9	197703	2835	26.9	378354	4512	31.9	656851	6789	36.9	1067947	9875
22.0°	.0200538	2863	27.0°	.0382866	4550	32.0°	.0663640	6841	37.0°	.1077822	9947
22.1	203401	2892	27.1	387416	4590	32.1	670481	6895	37.1	1087769	10019
22.2	206293	2922	27.2	392006	4630	32.2	677376	6948	37.2	1097788	10092
22.3	209215	2950	27.3	396636	4670	32.3	684324	7002	37.3	1107880	10166
22.4	212165	2980	27.4	401306	4709	32.4	691326	7057	37.4	1118046	10239
22.5	215145	3009	27.5	406015	4750	32.5	698383	7110	37.5	1128285	10314
22.6	218154	3039	27.6	410765	4790	32.6	705493	7166	37.6	1138599	10388
22.7	221193	3069	27.7	415555	4832	32.7	712659	7221	37.7	1148987	10464
22.8	224262	3099	27.8	420387	4872	32.8	719880	7277	37.8	1159451	10539
22.9	227361	3130	27.9	425259	4913	32.9	727157	7332	37.9	1169990	10615
23.0°	.0230491	3160	28.0°	.0430172	4956	33.0°	.0734489	7389	38.0	.1180605	10692
23.1	233651	3191	28.1	435128	4996	33.1	741878	7446	38.1	1191297	10769
23.2	236842	3221	28.2	440124	5039	33.2	749324	7502	38.2	1202066	10847
23.3	240063	3253	28.3	445163	5082	33.3	756826	7559	38.3	1212913	10925
23.4	243316	3284	28.4	450245	5124	33.4	764385	7618	38.4	1223838	11004
23.5	246600	3316	28.5	455369	5166	33.5	772003	7675	38.5	1234842	11082
23.6	249916	3347	28.6	460535	5210	33.6	779678	7733	38.6	1245924	11163
23.7	253264	3378	28.7	465745	5253	33.7	787411	7793	38.7	1257087	11242
23.8	256642	3411	28.8	470998	5297	33.8	795204	7851	38.8	1268329	11323
23.9	260053	3444	28.9	476295	5341	33.9	803055	7911	38.9	1279652	11404
24.0°	.0263497	3476	29.0°	.0481636	5384	34.0°	.0810966	7970	39.0°	.1291056	11486
24.1	266973	3508	29.1	487020	5430	34.1	818936	8031	39.1	1302542	11568
24.2	270481	3542	29.2	492450	5474	34.2	826967	8091	39.2	1314110	11651
24.3	274023	3575	29.3	497924	5518	34.3	835058	8152	39.3	1325761	11734
24.4	277598	3608	29.4	503442	5564	34.4	843210	8214	39.4	1337495	11818
24.5	281206	3642	29.5	509006	5610	34.5	851424	8275	39.5	1349313	11903
24.6	284848	3675	29.6	514616	5655	34.6	859699	8337	39.6	1361216	11987
24.7	288523	3709	29.7	520271	5702	34.7	868036	8399	39.7	1373203	12072
24.8	292232	3744	29.8	525973	5748	34.8	876435	8463	39.8	1385275	12159
24.9	295976	3777	29.9	531721	5794	34.9	884898	8525	39.9	1397434	12245
25.0°	.0299753		30.0°	.0537515		35.0°	.0893423		40.0°	.1409679	

NOTE: For any value of ϕ not given in table, special working tables may be easily constructed. Thus:

inv 46.6 = tan 46.6 − arc 46.6 = 1.0574704 − .8133234 = .2441470
inv 46.7 = tan 46.7 − arc 46.7 = 1.0611742 − .8150688 = .2461054
inv 46.8 = tan 46.8 − arc 46.8 = 1.0648918 − .8168141 = .2480777

24-4. Special Constants for Involute-gear Calculations. Many gear calculations can be simplified by using special constants.[4,*] Article 7-10 shows how the measurement over rolls can be simplified by m and μ functions.

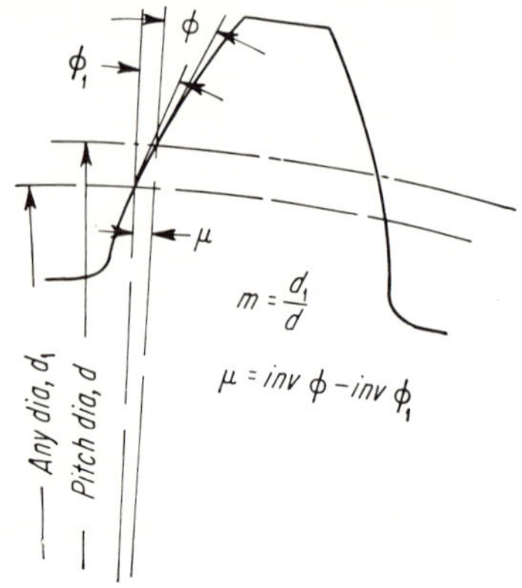

FIG. 24-1. Definition of special constants for involute calculations.

The term m is the ratio of the pitch diameter to any other diameter. The term μ is the angle in radians between the pitch point and any point on the involute curve. Figure 24-1 illustrates these relations.

Table 24-15 shows m and μ functions for the standard pressure angles of 14°30′, 20°, and 25°.

* Superscript numbers refer to references at the end of the chapter. See pp. 248 to 252 of ref. 4.

Table 24-15. Special Constants for Involute-gear Calculations

ϕ_1	$\phi = 14°30'$		$\phi = 20°$		$\phi = 25°$	
	m	μ	m	μ	m	μ
0	0.968148	0.005545	0.939693	0.014904	0.906308	0.029975
0.5	0.968184	0.005545	0.939728	0.014904	0.906342	0.029975
1.0	0.968295	0.005543	0.939835	0.014902	0.906445	0.029973
1.5	0.968480	0.005539	0.940015	0.014898	0.906618	0.029969
2.0	0.968737	0.005531	0.940265	0.014890	0.906860	0.029961
2.5	0.969070	0.005517	0.940588	0.014876	0.907171	0.029947
3.0	0.969476	0.005497	0.940982	0.014856	0.907551	0.029927
3.5	0.969956	0.005469	0.941448	0.014828	0.908001	0.029899
4.0	0.970512	0.005431	0.941987	0.014790	0.908521	0.029861
4.5	0.971142	0.005383	0.942599	0.014742	0.909110	0.029813
5.0	0.971845	0.005323	0.943282	0.014682	0.909769	0.029753
5.5	0.972625	0.005249	0.944039	0.014608	0.910499	0.029679
6.0	0.973480	0.005161	0.944868	0.014520	0.911300	0.029591
6.5	0.974411	0.005056	0.945772	0.014415	0.912171	0.029486
7.0	0.975418	0.004933	0.946750	0.014292	0.913114	0.029363
7.5	0.976501	0.004792	0.947801	0.014151	0.914128	0.029222
8.0	0.977662	0.004631	0.948927	0.013990	0.915214	0.029061
8.5	0.978900	0.004447	0.950129	0.013806	0.916373	0.028877
9.0	0.980216	0.004240	0.951406	0.013599	0.917605	0.028670
9.5	0.981609	0.004009	0.952759	0.013368	0.918909	0.028439
10.0	0.983082	0.003751	0.954189	0.013110	0.920289	0.028181
10.2	0.983694	0.003640	0.954782	0.012299	0.920861	0.028070
10.4	0.984319	0.003525	0.955389	0.012884	0.921446	0.027955
10.6	0.984956	0.003405	0.956007	0.012764	0.922042	0.027835
10.8	0.985606	0.003280	0.956637	0.012639	0.022650	0.027710
11.0	0.986268	0.003151	0.957281	0.012510	0.923271	0.027581
11.2	0.986944	0.003017	0.957936	0.012376	0.923903	0.027447
11.4	0.987633	0.002877	0.958605	0.012236	0.924548	0.027307
11.6	0.988334	0.002733	0.959286	0.012092	0.925205	0.027163
11.8	0.989049	0.002583	0.959980	0.011942	0.925874	0.027013
12.0	0.989776	0.002428	0.960685	0.011787	0.926555	0.026858
12.2	0.990517	0.002268	0.961405	0.011627	0.927248	0.026698
12.4	0.991272	0.002102	0.962137	0.011461	0.927955	0.026532
12.6	0.992039	0.001930	0.962882	0.011289	0.928673	0.026360
12.8	0.992820	0.001753	0.963640	0.011112	0.929404	0.026183
13.0	0.993614	0.001570	0.964410	0.010929	0.930147	0.026000
13.2	0.994421	0.001381	0.965194	0.010740	0.930903	0.025811
13.4	0.995242	0.001186	0.965991	0.010545	0.931671	0.025616
13.6	0.996077	0.000984	0.966801	0.010343	0.932453	0.025414
13.8	0.996925	0.000777	0.967624	0.010136	0.933247	0.025207
14.0	0.997786	0.000563	0.968460	0.009922	0.934053	0.024993
14.2	0.998662	0.000343	0.969310	0.009702	0.934873	0.024773
14.4	0.999550	0.000116	0.970172	0.009475	0.935705	0.024546
14.5	1.000	0	0.970608	0.009359	0.936125	0.024430
14.6	1.000453	−0.000117	0.971049	0.009242	0.936550	0.024313
14.8	1.001370	−0.000358	0.971939	0.009001	0.937408	0.024072
15.0	1.002300	−0.000605	0.972841	0.008754	0.938278	0.023825
15.2	1.003245	−0.000859	0.973758	0.008500	0.939163	0.023571
15.4	1.004203	−0.001120	0.974689	0.008239	0.940060	0.023310
15.6	1.005175	−0.001389	0.975632	0.007970	0.940970	0.023041

Table 24-15. Special Constants for Involute-gear Calculations (*Continued*)

ϕ_1	ϕ = 14°30′		ϕ = 20°		ϕ = 25°	
	m	μ	m	μ	m	μ
15.8	1.006162	−0.001664	0.976590	0.007695	0.941894	0.022766
16.0	1.007163	−0.001948	0.977561	0.007411	0.942831	0.022482
16.2	1.008178	−0.002239	0.978547	0.007120	0.943781	0.022191
16.4	1.009208	−0.002537	0.979546	0.006822	0.944745	0.021893
16.6	1.010252	−0.002843	0.980599	0.006516	0.945722	0.021587
16.8	1.011310	−0.003157	0.981587	0.006202	0.946713	0.021273
17.0	1.012384	−0.003480	0.982628	0.005879	0.947718	0.020950
17.2	1.013472	−0.003810	0.983685	0.005549	0.948737	0.020620
17.4	1.014574	−0.004149	0.984775	0.005210	0.949769	0.020281
17.6	1.015691	−0.004496	0.985839	0.004863	0.950814	0.019934
17.8	1.016824	−0.004851	0.986938	0.004508	0.951875	0.019579
18.0	1.017970	−0.005215	0.988051	0.004144	0.952948	0.019215
18.2	1.019133	−0.005588	0.989179	0.003771	0.954036	0.018842
18.4	1.020310	−0.005970	0.990322	0.003389	0.955138	0.018460
18.6	1.021503	−0.006361	0.991479	0.002998	0.956255	0.018069
18.8	1.022710	−0.006761	0.992651	0.002598	0.957385	0.017669
19.0	1.023932	−0.007170	0.993838	0.002189	0.958529	0.017260
19.2	1.025172	−0.007589	0.995041	0.001770	0.959689	0.016841
19.4	1.026425	−0.008017	0.996257	0.001342	0.960862	0.016413
19.6	1.027695	−0.008454	0.997490	0.000905	0.962052	0.015976
19.8	1.028980	−0.008902	0.998737	0.000457	0.963254	0.015528
20.0	1.030281	−0.009359	1.000	0	0.964473	0.015071
20.2	1.031598	−0.009827	1.001278	−0.000468	0.965705	0.014603
20.4	1.032931	−0.010305	1.002572	−0.000946	0.966953	0.014125
20.6	1.034279	−0.010792	1.003881	−0.001433	0.968215	0.013638
20.8	1.035645	−0.011291	1.005206	−0.001932	0.969493	0.013139
21.0	1.037027	−0.011800	1.006547	−0.002441	0.970787	0.012630
21.2	1.038424	−0.012320	1.007903	−0.002961	0.972095	0.012110
21.4	1.039838	−0.012850	1.009276	−0.003491	0.973419	0.011580
21.6	1.041270	−0.013392	1.010666	−0.004033	0.974759	0.011038
21.8	1.042716	−0.013945	1.012070	−0.004586	0.976113	0.010485
22.0	1.044181	−0.014509	1.013491	−0.005150	0.977484	0.009921
22.2	1.045661	−0.015084	1.014928	−0.005725	0.978870	0.009346
22.4	1.047160	−0.015672	1.016383	−0.006313	0.980273	0.008758
22.6	1.048675	−0.016270	1.017854	−0.006911	0.981692	0.008160
22.8	1.050208	−0.016881	1.019341	−0.007522	0.983126	0.007549
23.0	1.051757	−0.017504	1.020845	−0.008145	0.984576	0.006926
23.2	1.053325	−0.018139	1.022366	−0.008780	0.986044	0.006291
23.4	1.054908	−0.018787	1.023903	−0.009428	0.987527	0.005643
23.6	1.056511	−0.019447	1.025459	−0.010088	0.989027	0.004983
23.8	1.058611	−0.020119	1.027031	−0.010760	0.990543	0.004311
24.0	1.059770	−0.020805	1.028622	−0.011446	0.992078	0.003625
24.2	1.061426	−0.021503	1.030229	−0.012144	0.993628	0.002927
24.4	1.063099	−0.022215	1.031854	−0.012856	0.995194	0.002215
24.6	1.064792	−0.022940	1.033497	−0.013581	0.996779	0.001490
24.8	1.066504	−0.023678	1.035158	−0.014319	0.998381	0.000752
25.0	1.068232	−0.024430	1.036836	−0.015071	1.000	0
25.2	1.069981	−0.025196	1.038533	−0.015837	1.001636	−0.000766
25.4	1.071748	−0.025976	1.040248	−0.016617	1.003291	−0.001546
25.6	1.073533	−0.026770	1.041981	−0.017411	1.004962	−0.002340

Table 24-15. Special Constants for Involute-gear Calculations *(Continued)*

ϕ_1	ϕ = 14°30′		ϕ = 20°		ϕ = 25°	
	m	μ	m	μ	m	μ
25.8	1.075338	−0.027579	1.043733	−0.018220	1.006652	−0.003149
26.0	1.077163	−0.028402	1.045505	−0.019042	1.008360	−0.003972
26.2	1.079007	−0.029240	1.047294	−0.019881	1.010086	−0.004810
26.4	1.080869	−0.030092	1.049101	−0.027033	1.011829	−0.005662
26.6	1.082753	−0.030960	1.050929	−0.021601	1.013592	−0.006530
26.8	1.084655	−0.031843	1.052775	−0.022484	1.015373	−0.007413
27.0	1.086577	−0.032742	1.054641	−0.023383	1.017172	−0.008312
27.2	1.088521	−0.033656	1.056528	−0.024297	1.018992	−0.009226
27.4	1.090483	−0.034586	1.058433	−0.025227	1.020829	−0.010156
27.6	1.092466	−0.035531	1.060357	−0.026172	1.022685	−0.011101
27.8	1.094470	−0.036494	1.062302	−0.027135	1.024561	−0.012064
28.0	1.096494	−0.037472	1.064267	−0.028113	1.026456	−0.013042
28.2	1.098541	−0.038467	1.066254	−0.029108	1.028372	−0.014037
28.4	1.100607	−0.039479	1.068259	−0.030120	1.030306	−0.015049
28.6	1.102695	−0.040509	1.070286	−0.031150	1.032261	−0.016079
28.8	1.104804	−0.041555	1.072333	−0.032196	1.034235	−0.017125
29.0	1.106935	−0.042619	1.074401	−0.033260	1.036230	−0.018189
29.2	1.109088	−0.043700	1.076491	−0.034341	1.038246	−0.019270
29.4	1.111263	−0.044799	1.078601	−0.035440	1.040281	−0.020369
29.6	1.113460	−0.045917	1.080734	−0.036558	1.042338	−0.021487
29.8	1.115679	−0.047052	1.082888	−0.037693	1.044416	−0.022622
30.0	1.117921	−0.048206	1.085064	−0.038847	1.046514	−0.023776
30.5	1.123625	−0.051175	1.090600	−0.041816	1.051854	−0.026745
31.0	1.129474	−0.054264	1.096277	−0.044905	1.057329	−0.029834
31.5	1.135471	−0.057477	1.102098	−0.048118	1.062943	−0.033047
32.0	1.141619	−0.060819	1.108065	−0.051460	1.068698	−0.036389
32.5	1.147923	−0.064293	1.114184	−0.054934	1.074600	−0.039863
33.0	1.154383	−0.067904	1.120454	−0.058545	1.080647	−0.043474
33.5	1.161007	−0.071655	1.126884	−0.062926	1.086848	−0.047225
34.0	1.167796	−0.075552	1.133473	−0.066193	1.093204	−0.051122
34.5	1.174757	−0.079597	1.140229	−0.070238	1.099720	−0.055167
35.0	1.181890	−0.083797	1.147153	−0.074438	1.106397	−0.059367
35.5	1.189201	−0.088156	1.154249	−0.078797	1.113241	−0.063726
36.0	1.196696	−0.092679	1.161524	−0.083320	1.120258	−0.068249
36.5	1.204378	−0.097371	1.168980	−0.088012	1.127449	−0.072941
37.0	1.212251	−0.102237	1.176622	−0.092878	1.134819	−0.077807
37.5	1.220324	−0.107284	1.184457	−0.097925	1.142376	−0.082854
38.0	1.228597	−0.112516	1.192487	−0.103157	1.150121	−0.088086
38.5	1.237079	−0.117939	1.200719	−0.108580	1.158061	−0.093509
39.0	1.245773	−0.123561	1.209158	−0.114202	1.166200	−0.099131
39.5	1.254687	−0.129386	1.217810	−0.120027	1.174544	−0.104956
40.0	1.263828	−0.135423	1.226682	−0.126064	1.183101	−0.110993
41.0	1.282807	−0.148157	1.245104	−0.138798	1.200869	−0.123727
42.0	1.302771	−0.161821	1.264481	−0.152462	1.219557	−0.137391
43.0	1.323744	−0.176479	1.284867	−0.167120	1.239219	−0.152049
44.0	1.345883	−0.192199	1.306326	−0.182840	1.259916	−0.167769
45.0	1.369167	−0.209057	1.328926	−0.199698	1.281712	−0.184627
46.0	1.393704	−0.227134	1.352741	−0.217775	1.304682	−0.202704
47.0	1.419576	−0.246519	1.377853	−0.237160	1.328901	−0.222089
48.0	1.446873	−0.267310	1.404348	−0.257951	1.354455	−0.242880

Table 24-15. Special Constants for Involute-gear Calculations (*Continued*)

ϕ_1	ϕ = 14°30′		ϕ = 20°		ϕ = 25°	
	m	μ	m	μ	m	μ
49.0	1.475702	−0.289612	1.432330	−0.280253	1.381442	−0.265182
50.0	1.506169	−0.313544	1.461901	−0.304185	1.409964	−0.289114
51.0	1.538403	−0.339234	1.493187	−0.329875	1.440138	−0.314804
52.0	1.572534	−0.366825	1.526315	−0.357466	1.472089	−0.342395
53.0	1.608713	−0.396475	1.561431	−0.387116	1.505957	−0.372045
54.0	1.647112	−0.428359	1.598701	−0.419000	1.541903	−0.403929
55.0	1.687915	−0.462672	1.638305	−0.453313	1.580100	−0.438242
56.0	1.731330	−0.499632	1.680444	−0.490273	1.620742	−0.475202
57.0	1.777595	−0.539482	1.725350	−0.530123	1.664052	−0.515052
58.0	1.826973	−0.582499	1.773276	−0.573140	1.710276	−0.558069
59.0	1.879760	−0.628990	1.824511	−0.619631	1.759691	−0.604560
60.0	1.936296	−0.679308	1.879386	−0.669949	1.812616	−0.654878
61.0	1.996963	−0.733853	1.938270	−0.724494	1.869408	−0.709423
62.0	2.062206	−0.793077	2.001595	−0.783718	1.930483	−0.768647
63.0	2.132526	−0.857509	2.069849	−0.848150	1.996312	−0.833079
64.0	2.208512	−0.927749	2.143602	−0.918390	2.067445	−0.903319
65.0	2.290834	−1.004498	2.223504	−0.995139	2.144508	−0.980068
66.0	2.380285	−1.088575	2.310320	−1.079216	2.228240	−1.064145
67.0	2.477786	−1.180937	2.404961	−1.171578	2.319519	−1.156507
68.0	2.584436	−1.282719	2.508476	−1.273360	2.419356	−1.258289
69.0	2.701547	−1.395267	2.622145	−1.385908	2.528986	−1.370837
70.0	2.830676	−1.520202	2.747479	−1.510843	2.649868	−1.495772
71.0	2.973710	−1.659483	2.886318	−1.650124	2.783774	−1.635053
72.0	3.132992	−1.815502	3.040910	−1.806143	2.932874	−1.791072
73.0	3.311356	−1.991218	3.214032	−1.981859	3.099845	−1.966788
74.0	3.512402	−2.190326	3.409168	−2.180967	3.288049	−2.165896
75.0	3.740637	−2.417510	3.630695	−2.408151	3.501705	−2.393080
76.0	4.001901	−2.678786	3.884280	−2.669427	3.746281	−2.654356
77.0	4.303817	−2.982028	4.177323	−2.972669	4.028912	−2.957598
78.0	4.656527	−3.337729	4.519666	−3.328370	4.359094	−3.313299
79.0	5.073911	−3.760199	4.924783	−3.750840	4.749817	−3.735769
80.0	5.575347	−4.269474	5.411481	−4.260115	5.219225	−4.245044
81.0	6.188859	−4.894491	6.006961	−4.885132	5.793548	−4.870061
82.0	6.956435	−5.678656	6.751977	−5.669297	6.512096	−5.654226
83.0	7.944169	−6.690178	7.710681	−6.680819	7.436739	−6.665748
84.0	9.262092	−8.042744	8.989868	−8.033385	8.285100	−8.018314
85.0	11.108219	−9.940931	10.781736	−9.931572	10.398687	−9.916501
86.0	13.879064	−12.794138	13.471142	−12.784779	12.992545	−12.769708
87.0	18.498700	−17.557156	17.955002	−17.547797	17.317104	−17.532726
88.0	27.740630	−27.094819	26.925300	−27.085460	25.968710	−27.070389
89.0	55.474902	−55.731074	53.844430	−55.721715	51.922543	−55.706644
90.0	∞	−∞	∞	−∞	∞	−∞

24-5. Pitch Data and Relationships. The basic elements of gears were defined in Chap. 4. Article 4-2 gave specific information on spur gears. Article 5-4 gave recommendations for the proportions of spur-gear teeth. Tables 24-16 and 24-17 show basic formulas for some of the commonly used gear-tooth forms and the basic tooth thickness for a series of popular pitches. The tooth thickness given is the *arc tooth thickness,* and it is given without allowance for backlash. In most cases it is necessary to make an allowance for backlash and then in addition show a tolerance range.

Table 24-16. Tooth Proportions for Spur Gears

Included in this table are relationships to determine tooth dimensions for several common gear systems

	Proportion	System			
		Full-depth involute*	Stub-tooth involute*	Fine-pitch involute†	Fellows stub-tooth involute‡
a	Addendum	$\frac{1}{P_d}$	$\frac{0.8}{P_d}$	$\frac{1}{P_d}$	$\frac{1}{P_d(\text{denominator})}$
b	Dedendum	$\frac{1.250}{P_d}$ (min.)	$\frac{1}{P_d}$ (min.)	$\frac{1.200}{P_d} + 0.002$	See Ref. 4
h_k	Working depth	$\frac{2}{P_d}$	$\frac{1.6}{P_d}$	$\frac{2}{P_d}$	See Ref. 4
h_t	Whole depth	$\frac{2.250}{P_d}$ (min.)	$\frac{1.8}{P_d}$ (min.)	$\frac{2.200}{P_d} + 0.002$	See Ref. 4
D	Pitch dia.	$\frac{N}{P_d}$	$\frac{N}{P_d}$	$\frac{N}{P_d}$	$\frac{N}{P_d}$
D_o	Outside dia.§	$\frac{N+2}{P_d}$	$\frac{N+1.6}{P_d}$	$\frac{N+2}{P_d}$	$\frac{N}{P_d} + \frac{2}{P(\text{denominator})}$
T	Tooth thickness at pitch dia.	$\frac{\pi}{2P_d}$	$\frac{\pi}{2P_d}$	$\frac{\pi}{2P_d}$	$\frac{\pi}{2P_d}$
p	Circular pitch	$\frac{\pi}{P_d}$	$\frac{\pi}{P_d}$	$\frac{\pi}{P_d}$	$\frac{\pi}{P_d}$
c	Clearance	$\frac{0.250}{P_d}$ (min.)	$\frac{0.2}{P_d}$ (min.)	$\frac{0.200}{P_d} + 0.002$	See Ref. 4

* "Tooth Proportions for Coarse-pitch Involute Spur Gears," AGMA standard 201.02.

† "American Standard 20° Involute Fine Pitch System," ASA B6.7-1956.

‡ "The Involute Curve and Involute Gearing," The Fellows Gear Shaper Co., Springfield, Vt.

§ This relation must be used with caution. If a nonstandard addendum is used, the outside diameter will change.

Table 24-17. Pitch and Tooth-thickness Data

Values apply to all four systems shown in Table 24-16

P_d	p	T	P_d	p	T	P_d	p	T
½	6.2832	3.1416	7	0.4488	0.2244	32	0.0982	0.0491
¾	4.1888	2.0944	8	0.3927	0.1964	36	0.0873	0.0436
1	3.1416	1.5708	9	0.3491	0.1745	40	0.0785	0.0393
1¼	2.5133	1.2566	10	0.3142	0.1571	44	0.0714	0.0357
1½	2.0944	1.0472	11	0.2856	0.1428	48	0.0654	0.0327
1¾	1.7952	0.8976	12	0.2618	0.1309	50	0.0628	0.0314
2	1.5708	0.7854	14	0.2244	0.1122	56	0.0561	0.0280
2¼	1.3963	0.6981	16	0.1964	0.0982	60	0.0524	0.0262
2½	1.2566	0.6283	18	0.1745	0.0873	64	0.0491	0.0245
2¾	1.1424	0.5712	20	0.1571	0.0785	72	0.0436	0.0218
3	1.0472	0.5236	22	0.1428	0.0714	80	0.0393	0.0196
3½	0.8976	0.4488	24	0.1309	0.0654	96	0.0327	0.0164
4	0.7854	0.3927	26	0.1208	0.0604	100	0.0314	0.0157
5	0.6283	0.3142	28	0.1122	0.0561	120	0.0262	0.0139
6	0.5236	0.2618	30	0.1047	0.0524	128	0.0245	0.0123

24-6. Arc and Chord Data. Gear teeth are often measured by vernier calipers. The addendum setting for the calipers has to be made a slight amount larger than the gear addendum. Conversely, the chordal-tooth-thickness value is a slight amount smaller than the arc-tooth thickness of the part. Table 24-18 illustrates the "rise of arc" h_c and the "chordal thickness" t_c. Tabulated values are given for 1 diametral pitch. The chordal thicknesses given have no allowance for backlash. The values shown should be divided by the pitch and then given a proper allowance for backlash and an appropriate tolerance.

Table 24-18. Chordal Tooth Thickness and Chordal Addendum for Gears of 1 Diametral Pitch—Divide Values by Diametral Pitch for Other Pitches

(For notation, see illustration)

Notation

Chordal addendum—a_c
Chordal thickness —t_c
Rise of arc —h_c
Addendum —a
Pitch diameter —d

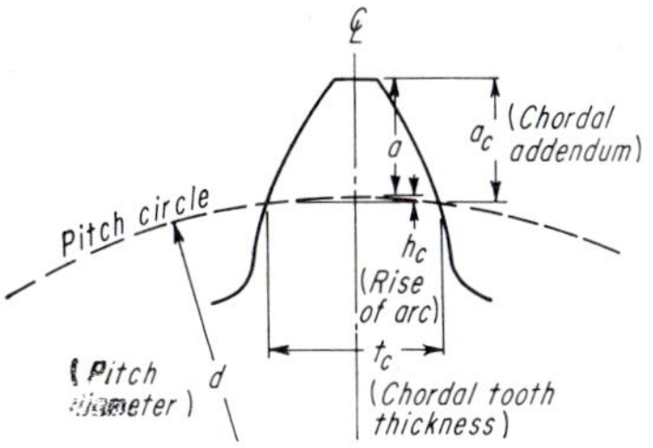

No. of teeth N	Chordal thickness, t_c	Chordal addendum, a_c	Rise of arc, h_c	No. of teeth N	Chordal thickness, t_c	Chordal addendum, a_c	Rise of, arc, h_c
8	1.56072	1.07686	0.07686	43	1.57045	1.01435	0.01435
9	1.56283	1.06836	0.06836	44	1.57046	1.01401	0.01401
10	1.56434	1.06158	0.06158	45	1.57048	1.01371	0.01371
11	1.56546	1.05597	0.05597	46	1.57049	1.01318	0.01318
12	1.56631	1.05133	0.05133	47	1.57050	1.01312	0.01312
13	1.56697	1.04733	0.04733	48	1.57051	1.01285	0.01285
14	1.56750	1.04401	0.04401	49	1.57052	1.01259	0.01259
15	1.56793	1.04109	0.04109	50	1.57053	1.01234	0.01234
16	1.56827	1.03852	0.03852	51	1.57053	1.01209	0.01209
17	1.56856	1.03623	0.03623	52	1.57055	1.01186	0.01186
18	1.56880	1.03425	0.03425	53	1.57056	1.01164	0.01164
19	1.56901	1.03245	0.03245	54	1.57057	1.01142	0.01142
20	1.56918	1.03083	0.03083	55	1.57058	1.01121	0.01121
21	1.56933	1.02936	0.02936	56	1.57059	1.01102	0.01102
22	1.56946	1.02803	0.02803	57	1.57059	1.01082	0.01082
23	1.56957	1.02681	0.02681	58	1.57060	1.01063	0.01063
24	1.56967	1.02569	0.02569	59	1.57061	1.01045	0.01045
25	1.56976	1.02466	0.02466	60	1.57062	1.01028	0.01028
26	1.56984	1.02372	0.02372	61	1.57062	1.01011	0.01011
27	1.56991	1.02284	0.02284	62	1.57062	1.00995	0.00995
28	1.56997	1.02202	0.02202	63	1.57063	1.00979	0.00979
29	1.57003	1.02126	0.02126	64	1.57064	1.00964	0.00964
30	1.57008	1.02056	0.02056	65	1.57064	1.00949	0.00949
31	1.57012	1.01989	0.01989	66	1.57064	1.00934	0.00934
32	1.57016	1.01927	0.01927	67	1.57065	1.00921	0.00921
33	1.57020	1.01869	0.01869	68	1.57065	1.00907	0.00907
34	1.57024	1.01814	0.01814	69	1.57065	1.00884	0.00884
35	1.57027	1.01762	0.01762	70	1.57066	1.00881	0.00881
36	1.57030	1.01713	0.01713	71	1.57066	1.00869	0.00869
37	1.57032	1.01667	0.01667	72	1.57066	1.00857	0.00857
38	1.57035	1.01623	0.01623	73	1.57067	1.00845	0.00845
39	1.57037	1.01581	0.01581	74	1.57067	1.00834	0.00834
40	1.57039	1.01541	0.01541	75	1.57068	1.00822	0.00822
41	1.57041	1.01504	0.01504	76	1.57068	1.00812	0.00812
42	1.57043	1.01468	0.01468	77	1.57068	1.00801	0.00801

NOTE: Table values reprinted from "The Involute Curve and Involute Gearing," with permission of the publishers, The Fellows Gear Shaper Co., Springfield, Vt.

Table 24-18. Chordal Tooth Thickness and Chordal Addendum for Gears of 1 Diametral Pitch—Divide Values by Diametral Pitch for Other Pitches
(Continued)

No. of teeth N	Chordal thickness, t_c	Chordal addendum, a_c	Rise of arc, h_c	No. of teeth N	Chordal thickness, t_c	Chordal addendum, a_c	Rise of arc, h_c
78	1.57068	1.00791	0.00791	93	1.57071	1.00664	0.00664
79	1.57068	1.00781	0.00781	94	1.57071	1.00657	0.00657
80	1.57068	1.00771	0.00771	95	1.57072	1.00649	0.00649
81	1.57068	1.00762	0.00762	96	1.57072	1.00641	0.00641
82	1.57069	1.00752	0.00752	97	1.57072	1.00634	0.00634
83	1.57069	1.00743	0.00743	98	1.57072	1.00631	0.00631
84	1.57069	1.00734	0.00734	99	1.57072	1.00623	0.00623
85	1.57069	1.00725	0.00725	100	1.57073	1.00617	0.00617
86	1.57070	1.00717	0.00717	110	1.57074	1.00561	0.00561
87	1.57070	1.00713	0.00713	120	1.57075	1.00514	0.00514
88	1.57070	1.00701	0.00701	Rack	1.57080	1.00000	0.00000
89	1.57070	1.00693	0.00693				
90	1.57070	1.00685	0.00685				
91	1.57071	1.00678	0.00678				
92	1.57071	1.00671	0.00671				

24-7. Hardness-testing Data. Almost all drawings for gear parts will need a hardness-testing specification. The method of hardness testing will depend on the material and the size and shape of piece. Table 24-19 gives general recommendations for the more commonly used hardness-test methods. Table 24-20 shows the approximate conversion between the different hardness-test scales.

Table 24-19. Hardness-testing Apparatus and Application for Gears

Instrument	Shape and type of indenter	Loading	Recommended use
Brinell	10-mm steel or tungsten carbide ball	3,000 kg	For large gears and shafts in range of hardness from 100 to 400 BHN. (If gears have a hard case, the case depth must be sufficient and the core strength adequate to support the area under test. Rockwell C tests may be used to reveal any error from such irregular conditions.) Brinell tests, when fairly representative of the general hardness, are a good measure of the ultimate tensile strength of the material
Rockwell C	Diamond Brale penetrater (120° diamond cone)	150 kg	For gears medium to large in size. Range approximately 25 to 68 Rockwell C
Rockwell A	Diamond Brale penetrater (120° diamond cone)	60 kg	For small gears and tips of large gear teeth. Range 62 to 85 Rockwell A
Rockwell 30-N	Diamond Brale penetrater (120° diamond cone, special indenter)	30 kg	For small parts and shallow casehardened parts
Rockwell 15-N	Diamond Brale penetrater (120° diamond cone, special indenter)	15 kg	Lightest Rockwell load. For testing very small parts and checking the working sides of teeth. Check for very thinly casehardened parts. Rockwell 15-N, 72-93
Vickers, pyramid	136° diamond pyramid	50 kg	All applications where piece will not be too heavy for machine. For use in testing hardness of shallow cases, etc.
Scleroscope	Does not indent surface. Diamond-tipped tup bounced on specimen	....	For applications permitting no damage or indentation to surfaces (results not always comparable with indentation-hardness tests)
Tukon	Knoop indenter	500–1,000 g for practical use	A laboratory instrument used only for finding the hardness of material on pieces of cross sections, e.g., hardness from surface inward, every 0.001″, or hardness of individual microconstituents. All test areas must have flat mirror-polished surfaces. An extremely delicate and precise test. Used for any hardness

Table 24-20. Approximate Relation between Hardness-test Scales

Brinell, 3,000 kg, 10 mm	Rockwell C	Rockwell A	Rockwell 30-N	Rockwell 15-N	Vickers pyramid	Scleroscope (Shore)	Tukon (Knoop)
	70	86.5	86.0	94.0	1076		
	65	84.0	82.0	92.0	820	90	840
	63	83.0	80.0	91.5	763	87	790
614	60	81.0	77.5	90.0	695	81	725
587	58	80.0	75.5	89.3	655	79	680
547	55	78.5	73.0	88.0	598	74	620
522	53	77.5	71.0	87.0	562	71	580
484	50	76.0	68.5	85.5	513	67	530
460	48	74.5	66.5	84.5	485	64	500
426	45	73.0	64.0	83.0	446	61	460
393	42	71.5	61.5	81.5	413	56	425
352	38	69.5	57.5	79.5	373	51	390
301	33	67.0	53.0	76.5	323	45	355
250	24	62.5	45.0	71.5	257	37.5	
230	20	60.5	41.5	69.5	236	34	
	Rockwell B		Rockwell 30-T	Rockwell 15-T			
200	93		78.0	91.0	210	30	
180	89		75.5	89.5	189	28	
150	80		70.0	86.5	158	24	
100	56		54.0	79.0	105		
80*	47		47.7	75.7			
70*	34		38.5	71.5			

NOTE: Tables 24-19 and 24-20 are reproduced from "Practical Gear Design" by Darle W. Dudley, McGraw-Hill Book Company, Inc., New York, 1954, with permission of the author and publisher.

* Based on 500-kg load and 10-mm ball.

REFERENCES

1. The Van Keuren Handbook 36, The Van Keuren Co., Watertown, Mass.
2. Vogel, Werner F.: "Involutometry and Trigonometry," Michigan Tool Co., Detroit, Mich., 1954.
3. Buckingham, Earle: "Manual of Gear Design," Sec. 1, Mathematical Tables, The Industrial Press, New York, 1935.
4. Dudley, Darle W.: "Practical Gear Design," McGraw-Hill Book Company, Inc., New York, 1954.

INDEX